IDEAS OF QUANTUM CHEMISTRY

Second edition

"Things appear, ideas persist"

Plato

IDEAS OF QUANTUM CHEMISTRY

Second edition

by

LUCJAN PIELA

Department of Chemistry, University of Warsaw, Warsaw, Poland

AMSTERDAM • BOSTON • HEIDELBERG • LONDON • NEW YORK • OXFORD
PARIS • SAN DIEGO • SAN FRANCISCO • SINGAPORE • SYDNEY • TOKYO

Elsevier
225 Wyman Street, Waltham, MA 02451, USA
525 B Street, Suite 1800, San Diego, CA 92101-4495, USA

Second edition **2014**

Library of Congress Cataloging-in-Publication Data
A catalog record for this book is available from the Library of Congress

British Library Cataloguing in Publication Data
A catalogue record for this book is available from the British Library

For information on all **Elsevier** publications
visit our web site at store.elsevier.com

Printed and bound by CPI Group (UK) Ltd, Croydon, CR0 4YY

ISBN: 978-0-444-59436-5

To all on the quest for the Truth

Contents

Online Appendices

Sources of Photographs and Figures

* The Figures in this book, except those listed below or acknowledged in situ, are manufactured by the author and reproduced thanks to the courtesy of Wydawnictwo Naukowe PWN, Poland from " Idee chemii kwantowej", © 2012 PWN, * the postal stamps of several countries have been used (Austria pp. 30, 77, 106, 615, Canada 591, 886, Denmark 7, France 979, e17, e121, 371, Gabon 354, Gambia 4, 927, Germany 302, Great Britain 1003, Greece 795, Guiné-Bissau 111, 874, Guinée 12, 74, 155, 764, 270, 764, Guyana 512, Holland 11, 796, Hungary 250, 524, Ireland 1004, Komi 328, Mali 1, 4, 26, 124, Marshall Islands 446, Micronesia 36, Nevis 340, Sweden 952, Uganda 9, Uruguay 110, USA 107, 260, 594, 612, 723, Vatican 796) * courtesy of The Nobel Foundation (John Rayleigh 5, Niels Bohr 7, Albert Einstein 107, Carl Anderson 126, James Watson 345, Francis Crick 345, Tjalling Koopmans 466, John Pople 501, Walter Kohn 676, John Van Vleck 721, Norman Ramsey 771, Edwards Purcell 771, Yuan Lee 886, Dudley Herschbach 886, Rudolph Marcus 950, Ilya Prigogine 982) * Wikipedia - the web encyclopedia, public domain (7, 27, 62, 64, 110, 111, 172, 286, 347, 363, 505, 507, 516, 585, 771, 975, 991, 745, e10) * photo of Aleksander Jabłoński p.460, courtesy of Physics Department, Nicolaus Copernicus University, Toruń, Poland * courtesy of Professor Roald Hoffmann (Cornell University, USA) 533, 536, 542, 546, 547 * courtesy of Professor Jean-Marie André (Université de Namur, Belgium) 543, 544 * courtesy of Professor Hiroshi Nakatsuji, Japan 658 * courtesy of Professor Sadlej's family, Poland p.749 * photo of Charles Galton Darwin p.124 - courtesy of Dr.R.C.McGuiness, UK * photo of Christopher Longuet-Higgins, p.261- courtesy of Professor J.D.Roberts, California Institute of Technology, USA * photo of Friedrich Hund p.461- courtesy of Mr Gerhard Hund, Germany * photo of Richard Bader 672 - courtesy of Professor Richard Bader, Canada * portrait of Hans Hellmann p.722 reproduced from a painting by Ms Tatiana Livschitz, courtesy of Professor W.H. Eugen Schwartz, Germany * photo, courtesy of Jean-Marie Lehn, France 976 * photo of Gregory Breit - courtesy of Alburtus Yale News Bureau (through Wikipedia) - 146 * Figs. 11.10-11.12 reused from S. Kais, D.R. Herschbach, N.C. Handy, C.W. Murray, G.J. Laming, J. Chem. Phys., 99 (1993) 417. * Tables 14.1-14.5, courtesy of Professor Sason Shaik, Israel * Fig. 8.33 - courtesy of Dr.Witold Mizerski, Poland 493, photographs by the author (Per-Olov Löwdin 445, Rudolph Peierls 535, Włodzimierz Kołos 591, Lutosław Wolniewicz 591, Szkocka Café in Lwów 372, Roald Hoffmann 925).

Despite reasonable attempts made, we were unable to contact the owners of the copyright of the following images: * photo of Boris Belousov p.995 from " Geroi i zladiei rossiyskoi nauki", Kronpress, Moscow, 1997, Figs on p.874 and 875 reproduced from " Biology Today", CRM Books, Del Mar, USA, © 1972 Communications Research Machines, MPn method 653 * in the website of St Andrews University, United Kingdom http://www-gap.dcs.st-and.ac.uk/~history (Sommerfeld 8, Bose 26, Bell 48, Weyl 80, Minkowski 119, Klein 123, Hartree 393, Riemann 560, Friedmann 593, Thom 671, Feigenbaum 978, 873 (Tomaglia), Shannon 876, Adleman 1002, Turing 879, Lagrange 997 * in the website of Physics Department, Moscow University http://nuclphys.sinp.msu.ru (Edward Condon 302) * in the website of Duke University (USA) www.phy.duke.edu, photo Lotte Meitner-Graf: Fritz London 611 * in the website of www.volny.cz Gilbert Lewis 7 * in the website met www.epfl.ch: Brillouin 438 * in the website http://osulibrary.orst.edu Slater 397 * in the website http://www.mathsoc.spb.ru: Fock 394 * in the website www.stetson.edu Ulam 372 ✶ in the website http://www.quantum-chemistry-history.com (Hückel 427, Roothaan 432, Hall 432).

If you are the copyright owner to any of the images we have used without your explicit permission (because we were unable to reach you during our search), please contact prof.Lucjan Piela, Chemistry Department, Warsaw University, 02093 Warsaw, Poland e-mail: piela@chem.uw.edu.pl, phone (48)-22-7226692. We will be pleased to place the appropriate information on our website at booksite.elsevier.com/978-0-444-59436-5 which supports the present book and represents its integral part.

Introduction

Quantum scimus – gutta est, ignoramus mare.
What we know is a drop, what we do not know is a sea.
(Latin sentence)

We and the Universe: A Potent Interaction

Here are a few ways that photons play a part in my life: Crocuses first appear after winter and look breathtaking on the snow, then ultramarine of the violets. Later, magnolia flowers seem like proud queens–a bright white with a subtle rosy tint. A week later, the lilacs, the ecru of acacia, and finally the rich, extraordinary kingdom of roses all make their debuts. The buds of the hydrangea (the beloved flowers of this author) are white, but when they first open, the white reacts with light quanta, and the flowers acquire vibrant, clean colors, ranging from light to dark blue. Why does all this happen? Not only do colors create a sense of wonder, but unusual shapes, textures, and hues do as well. What is in our brain that can use photons to translate our interactions with the Universe into an unimaginable variety of complex phenomena, already in our body, that can affect our decisions and actions? *Sight* represents the most powerful (highly directional and long-range) and, at the same time, the most subtle information channel to our brain.

Hearing. What could compete with the fantasy of the thrush, which sings different masterpieces every spring in my three pine trees? Why does a finch sing completely differently from the thrush, and why does it repeat its melody with amazing regularity? Why do all finches sing similar songs? What kind of internal programming compels them to do so? The program must be quite robust, being insensitive to thousands of details of the neighborhood, but not to some particular signals of danger. Birdsong is still less interesting than human verbal communication, though. A person pronounces a particular word, which may have the strength of a tornado for others. How is it possible that a local sequence of tiny air pressure amplitudes (sound) can change our world in the global scale?

Spring also provides fantastic *fragrances*: Just after winter, one can smell the heavenly aroma of violets and hyacinths, sometimes the subtle scent of bird cherry is brought by the wind from far away, then a variety of other exciting fragrances follow. What is the mechanism of recognizing and remembering smells, admiring some of them and being repelled by others?

The *taste* of fresh bread is unforgettable and is linked to a feeling of happiness, not only for me, but for many people. There must be a program imprinted in us with some chemical hardware that lets us appreciate the way things taste. What does this hardware look like?

Touching, which is based on the Pauli exclusion principle, has changed the history of the world many times (just think about kissing etc.). Such giant consequences from such a small cause?

What Do We Know?

Our senses connect us to what we think of as the Universe. Using them, we are aware of its presence, while at the same time we become a part of it. Sensory operations are the direct result of interactions between molecules, as well as between light and matter. All of these phenomena deal first with information processing, but at the same time with chemistry, physics, biology, and even psychology. In these complex events, it is impossible to discern precisely where the disciplines of chemistry, physics, biology, and psychology begin and end. Any separation of these domains is artificial. The only reason for making such separations is to focus our attention on some aspects of one indivisible phenomenon. Touch, taste, smell, sight, and hearing–are these our only links and information channels to the Universe? How little we know about it! To realize that, just look up at the sky. A myriad of stars around us point to new worlds, which will remain unknown forever. On the other hand, imagine how incredibly complicated must be the chemistry of friendship. Science cannot answer all legitimate questions that a human being may ask. Science is able to discover laws of nature, but is unable to answer a question like "*Why does our world conform to any laws at all*[1]?" This goes beyond science.

We try to understand what is around us by constructing in our minds pictures representing a "*reality*," which we call *models*. Any model relies on the perception of reality (on the appropriate scale of mass and time) emanating from our experience, and, on the other hand, on our ability to abstract by creating ideal beings. Many such models will be described in this book.

It is fascinating that humans are able to magnify the realm of the senses by using sophisticated tools (e.g., to see quarks sitting in a proton[2]), to discover an amazingly simple equation of motion[3] that describes cosmic catastrophes, with intensity beyond our imagination, and the delicate flight of a butterfly equally well. A water molecule has exactly the same properties in the Pacific Ocean as it does on Mars, or in another galaxy. The conditions in those environments

1 "*The most incomprehensible thing about the world is that it is at all comprehensible.*" (Albert Einstein).

2 A proton is 10^{15} times smaller than a human being.

3 Acceleration is directly proportional to force. Higher derivatives of the trajectory with respect to time do not enter this equation, and neither does the nature or cause of the force. The equation is also invariant with respect to any possible starting point (position, velocity, and mass). What remarkable simplicity and generality there is (within limits, see Chapter 3).

may sometimes be quite different from those in the laboratory, but we assume that if these conditions could be imposed in the lab, the molecule would behave in exactly the same way. We hold out hope that a set of physical laws apply to the entire Universe.

The model for these basic laws is not yet complete or unified. However, given the progress and important generalizations of physics, much is currently understood. For example, forces with seemingly disparate sources have been reduced to only three kinds:

- Those attributed to *strong interactions* (acting in nuclear matter)
- Those attributed to *electroweak interactions* (the domains of chemistry, biology, and β-decay)
- Those attributed to *gravitational interaction* (showing up mainly in astrophysics)

Many scientists believe that other reductions are possible, perhaps up to a single fundamental interaction, one that explains everything. This assertion is based on the conviction (which seems to be supported by developments in modern physics) that the laws of nature are not only universal, but simple.

Which of the three basic interactions is the most important? This is an ill-conceived question. The answer depends on the external conditions imposed (pressure, temperature) and the magnitude of the energy exchanged among the interacting objects. A measure of the energy exchanged (ΔE) may be taken to be the percentage of the accompanying mass deficiency (Δm) according to Einstein's relation $\Delta E = \Delta mc^2$. At a given magnitude of exchanged energies, some particles are stable. Strong interactions produce the huge pressures that accompany the gravitational collapse of a star and lead to the formation of neutron stars, where the mass deficiency Δm approaches 40%. At smaller pressures, where individual nuclei may exist and undergo nuclear reactions (strong interactions[4]), the mass deficiency is of the order of 1%. At much smaller pressures, the electroweak forces dominate, nuclei are stable and atomic, and molecular structures emerge. Life (as we know it) becomes possible. The energies exchanged are much smaller and correspond to a mass deficiency of the order of only about 10^{-7}%. The weakest of the basic forces is gravitation. Paradoxically, this force is the most important on the macro scale (galaxies, stars, planets, etc.). There are two reasons for this. Gravitational interactions share with electric interactions the longest range known (both decay as $1/r$). However, unlike electric interactions[5], those due to gravitation are not shielded. For this reason, the Earth and Moon attract each other by a huge gravitational force[6], while their electric interaction is negligible. This is how David conquers Goliath, since at any distance, electrons and protons attract each other by electrostatic forces that are about 40 orders of magnitude stronger than their gravitational attraction.

[4] With a corresponding large energy output, the energy coming from the fusion D + D→He taking place on the Sun makes our existence possible.

[5] In electrostatic interactions, charges of opposite sign attract each other, while charges of the same sign repel each other (Coulomb's law). This results in the fact that large bodies (built of a huge number of charged particles) are nearly electrically neutral and interact electrically only very weakly. This dramatically reduces the range of their electrical interactions.

[6] Huge tides and deformations of the whole Earth are witness to that.

Gravitation does not have any measurable influence on the collisions of molecules leading to chemical reactions, since reactions are due to much stronger electric interactions[7].

Narrow Temperature Range

Due to strong interactions, protons overcome mutual electrostatic repulsion and form (together with neutrons) stable nuclei, leading to the variety of chemical elements. Therefore, strong interactions are the prerequisite of any chemistry (except hydrogen chemistry). However, chemists deal with already prepared stable nuclei[8], and these strong interactions have a very small range (of about 10^{-13} cm) as compared to interatomic distances (of the order of 10^{-8} cm). This is why a chemist may treat nuclei as stable point charges that create an electrostatic field. Test-tube conditions allow for the presence of electrons and photons, thus completing the set of particles that one might expect to see (some exceptions are covered in this book). This has to do with the order of magnitude of energies exchanged, under the conditions where we carry out chemical reactions, the energies exchanged exclude practically all nuclear reactions.

On the vast scale of attainable temperatures[9], chemical structures may exist in the narrow temperature range of 0 K to thousands of degrees Kelvin. Above this range, one has plasma, which represents a soup made of electrons and nuclei. Nature, in its vibrant living form, requires a temperature range of about 200 – 320 K, a margin of only 120 K. One does not require a chemist for chemical structures to exist. However, to develop a chemical science, one has to have a chemist. This chemist can survive a temperature range of 273 K ±50 K; i.e., a range of only 100 K. The reader has to admit that a chemist may think of the job only in the narrow range of 290 – 300 K (i.e., only 10 K).

An Unusual Mission of Chemistry

Suppose our dream comes true and the grand unification of the three remaining basic forces is accomplished one day. We would then know the first principles of constructing everything. One of the consequences of such a feat is a catalogue of all the elementary particles. Perhaps the catalogue will be finite[10], it also might be simple. We might have a catalogue of the conserved

[7] This does not mean that gravitation has *no* influence on reactants' concentration. Gravitation controls the convection flow in liquids and gases (and even solids), and therefore, a chemical reaction or even crystallization may proceed in a different manner on the Earth's surface, in the stratosphere, in a centrifuge, or in space.

[8] At least, this is true in the time scale of chemical experiments. Instability of some nuclei is used by nuclear chemistry and radiation chemistry.

[9] Think of this in millions of degrees.

[10] None of this is certain. Much of elementary particle research relies on large particle accelerators. This process resembles discerning the components of a car by dropping it from increasing heights from a large building. Dropping it from the first floor yields five tires and a jack. Dropping it from the second floor reveals an engine and 11 screws of similar appearance. Eventually, though, a problem emerges: after landing from a very high floor, new components appear (which have ... exactly nothing to do with the car) and reveal that some of the collision energy has been converted to new particles!

symmetries (which seem to be more elementary than the particles). Of course, knowing such first principles would have an enormous impact on all the physical sciences. It could create an impression that everything is clear and that physics is complete. Even though such structures and processes are governed by first principles, it would still be very difficult to predict their existence by such principles alone. The resulting structures would depend not only on the principles, but also on the initial conditions, complexity, self-organization, etc.[11] Therefore, if it does happen, the Grand Unification will not change the goals of chemistry.

The author of this book is convinced that chemistry currently faces the enormous challenge of information processing, which is done in a very different way than it is performed now by computers. This unusual perspective is discussed in the last chapter of this book, which differs significantly from other chapters. It shows some exciting possibilities of chemistry, including theoretical chemistry, and it also poses some general questions about what limits are to be imposed on science.

Book Guidelines

TREE

Any book has a linear appearance; i.e., the text goes page after page, and the page numbers remind us of that. However, the *logic* of virtually any book is nonlinear, and in many cases, it can be visualized by a diagram connecting the chapters that (logically) follow one another. Such a diagram allows for multiple branches emanating from a given chapter, particularly if the branches are placed on an equal footing. Such logical connections are illustrated in this book as a TREE diagram (cover's reverse). This TREE diagram plays a very important role in this book and is intended to be a study guide. An author leads the reader in a certain direction, and the reader expects to know what this direction is, why he or she needs this direction, what will follow, and what benefits will be gained after such study. If studying were easy and did not require time, a TREE diagram might be of little importance. However, the opposite is usually true. In addition, knowledge represents much more than a registry of facts. Any understanding gained from seeing relationships among those facts and methods plays a key role[12]. The primary function of the TREE diagram is to make these relationships clear.

The use of hypertext in information science is superior to a traditional linear presentation. It relies on a tree structure. However, it has a serious drawback. Looking at a branch, we have no idea what it represents in the whole diagram, whether it is an important branch or a remote tiny one; does it lead further to important parts of the book or it is just a dead end, and so on. At the same time, a glimpse at the TREE shows us that the thick trunk is the most important structure. But what do we mean by *important?* At least two criteria may be used: it is important for the

[11] The fact that Uncle John likes to drink coffee with cream at 5 p.m. possibly follows from first principles, but it would be very difficult to trace that dependence.

[12] This advice comes from *Antiquity*: "*Knowledge is more precious than facts, understanding is more precious than knowledge, wisdom is more precious than understanding.*"

majority of readers, or important because the material is fundamental for an understanding of the laws of nature. I have chosen the first criterion[13]. Thus, the trunk of the TREE corresponds to the pragmatic way to study this book.

The trunk is the backbone of this book:

- It begins by presenting postulates, which play a vital role in formulating the foundation of quantum mechanics.
- It goes through the Schrödinger equation for stationary states, thus far the most important equation in quantum chemical applications.
- It covers the separation of nuclear and electronic motion.
- It then develops the mean-field theory of electronic structure.
- Finally, it develops and describes methods that take into account electronic correlation.

The trunk thus corresponds to a traditional course in quantum chemistry for undergraduates. This material represents the necessary basis for further extensions into other parts of the TREE (which are appropriate rather for graduate students). In particular, it makes it possible to reach the crown of the TREE, where the reader may find tasty fruit. Examples include the theory of molecule-electric field interactions, as well as the theory of intermolecular interactions (including chemical reactions), which form the very essence of chemistry. We also see that our TREE has an important branch concerned with nuclear motion, including molecular mechanics and several variants of molecular dynamics. At its base, the trunk has two thin branches: one pertains to relativity mechanics and the other to the time-dependent Schrödinger equation. The motivation for this presentation is different in each case. I do not highlight relativity theory; its role in chemistry is significant[14], but not crucial. The time-dependent Schrödinger equation is not highlighted because, for the time being, quantum chemistry accentuates stationary states. I am confident, however, that the 21st century will see significant developments in the methods designed for time-dependent phenomena.

The TREE Helps Readers Tailor Their Own Book

The TREE serves not only as a diagram of logical chapter connections, but also enables the reader to make important decisions, to wit:

- The choice of a logical path of study ("*itinerary*") leading to topics of interest
- Elimination of chapters that are irrelevant to the goal of study[15].

This means that each reader can tailor the book to his or her own needs.

[13] For example, relativity theory plays a pivotal role as a foundation of the physical sciences, but for the vast majority of chemists, its practical importance and impact are much smaller. Therefore, should relativity be represented as the base of the trunk, or as a minor branch? I have decided to make the second choice, *not* to create the impression that this topic is absolutely necessary for the student.

[14] Contemporary inorganic chemistry and metallo-organic chemistry concentrate currently on heavy elements, where relativity effects are important.

[15] It is, therefore, possible to prune some of the branches.

Of course, all readers are welcome to find their own itineraries when traversing the TREE; i.e., to create their own customized books. Some readers might wish to take into account our suggestions of how the book can be shaped.

Minimum Minimorum and Minimum

First, we can follow two basic paths:

- *Minimum minimorum* for those who want to proceed as quickly as possible to get an idea of what quantum chemistry is all about by following the chapters designated by (▲). I picture someone studying material science, biology, biochemistry, or a similar subject, who has heard that quantum chemistry explains chemistry, and want to get the flavor of it and grasp the most important information. Following ▲ signs they should read only 47 pages.
- *Minimum* for those who seek basic information about quantum chemistry; e.g., in order to use popular computer packages for the study of molecular electronic structure, they may follow the chapters designated by the symbols ▲ and △. Here, I picture a student of chemistry, specializing in, say, analytical or organic chemistry (not quantum chemistry). This path involves reading something like 300 pages + the appropriate appendices (if necessary).

Other paths proposed consist of the *minimum itinerary* (i.e., ▲ *and* △), plus special excursions, which we call "*additional itineraries.*"

Those who want to use the existing computer packages in a knowledgeable fashion or just want to know more about the chosen subject may follow the chapters designated by the following special symbols:

- Large molecules (□)
- Molecular mechanics and molecular dynamics (♠)
- Solid-state chemistry/physics (■)
- Chemical reactions (Ʊ)
- Spectroscopy (Ⓢ)
- Exact calculations on atoms or small molecules[16] (♦)
- Relativistic and quantum electrodynamics effects (▶)
- Most important computational methods of quantum chemistry (◇)

For readers interested in particular aspects of this book rather than any systematic study, the following itineraries are proposed.

- Just before an exam, read these sections of each chapter: "*Where Are We?*" "*An Example,*" "*What Is It All About?*" "*Why Is This Important?*" "*Summary,*" "*Questions,*" and "*Answers.*"
- For those interested in recent progress in quantum chemistry, we suggest reading the section "*From the Research Front*" in each chapter.

[16] Suppose that readers are interested in an accurate theoretical description of small molecules. (I picture a Ph.D. student working in quantum chemistry.) Following their itinerary, they should read, in addition to the minimum program (300 pages), an additional 230 pages, which gives about 530 pages plus the appropriate appendices, totaling about 700 pages.

- For those interested in the future of quantum chemistry, we propose the "*Ad Futurum*" sections in each chapter, and the chapters designated by (⅃)
- For people interested in the "*magical*" aspects of quantum physics (e.g., bilocation, reality of the world, teleportation, creation of matter, or tunneling) we suggest reading sections labeled with ✠

The Target Audience

I hope that the TREE structure presented above will be useful for those with varying levels of knowledge in quantum chemistry, as well as for those whose goals and interests differ from those of traditional quantum chemistry.

This book is a direct result of my lectures at the Department of Chemistry, University of Warsaw, for students specializing in theoretical rather than experimental chemistry. Is that the target audience of this book? Yes, but not exclusively. At the beginning, I assumed that the reader would have completed a basic quantum chemistry course[17] and, therefore, in the first version of the book, I omitted the basic material. However, that version became inconsistent and devoid of several fundamental problems. This is why I have decided to explain, mainly very briefly[18], these problems as well in this edition. Therefore, a student who chooses the *minimum* path along the TREE diagram (mainly along the TREE trunk) will essentially be taking an introductory course in quantum chemistry. On the other hand, the complete collection of chapters provides students with a set of advanced topics in quantum chemistry, appropriate for graduate students. For example, a number of chapters on subjects such as relativity mechanics, global molecular mechanics, solid-state physics and chemistry, electron correlation, density function theory, intermolecular interactions, and the theory of chemical reactions present material that is usually accessible in monographs or review articles.

My Goal

In writing this book, I imagined students sitting in front of me. In discussions with students, I often saw their enthusiasm, their eyes giving me a glimpse of their curiosity. First of all, this book is an acknowledgment of my young friends, my students, and an expression of the joy of being with them. Working with them formulated and influenced the way I decided to write this book. When reading textbooks, one often gets the impression that all the outstanding problems in a particular field have been solved, that everything is complete and clear, and that students are just supposed to learn and absorb the material at hand. But in science, the opposite is true. All areas can benefit from careful probing and investigation. Your insight, your different perspective or point of view may open new doors for others, even on a fundamental question.

---

[17] Such a course might be, at the level of P.W. Atkins, "*Physical Chemistry*", 6th ed. (Oxford University Press, Oxford, 1998), Chapters 11–14.

[18] This is true except where I wanted to stress some particular topics.

Fostering this kind of new insight is one of my main goals. I have tried, whenever possible, to present the reasoning behind a particular method and to avoid rote citation of discoveries. I have tried to avoid writing too much about details because I know how easy it is for a new student to miss the forest for the trees. I wanted to focus on the main ideas of quantum chemistry.

I have tried to stress this integral point of view, which is why the book sometimes deviates from what is normally considered as quantum chemistry. I sacrificed "*quantum cleanness*" in favor of exposing the interrelationships of problems. In this respect, any division between physics and chemistry, organic chemistry and quantum chemistry, quantum chemistry for chemists and quantum chemistry for biologists, or intermolecular interactions for chemists, for physicists, or for biologists is completely artificial, and sometimes even absurd[19]. I tried to cross these borders by supplying examples and comparisons from the various disciplines, as well as from everyday life, by incorporating into intermolecular interactions not only supramolecular chemistry, but also molecular computers, and particularly the latter, by writing a "*holistic*" chapter (the last chapter of this book) about the mission of chemistry.

My experience tells me that talented students who love mathematics but are new to the subject of quantum chemistry courts danger. They like complex derivations of formulas so much that it seems that the more complex the formalism, the happier the students. However, all these formulas represent no more than an approximation of reality, and sometimes it would be better to have a simple formula instead. The simple formula, even if less accurate, may tell us more and bring more understanding than a very complicated one. Behind complex formulas usually hide some very simple concepts; e.g., that two molecules do not occupy the same space, or that in a tedious iteration process, we approach the final ideal wave function in a way similar to a sculptor shaping a masterpiece. All the time, in everyday life, we unconsciously use these variational and perturbational methods–the most important tools in quantum chemistry. This book may be considered by some students as too easy. However, I prize easy explanations very highly. In later years, the student will not remember long derivations, but will know exactly why something *must* happen. Also, when deriving formulas, I try to avoid presenting the final result right away, but instead proceed with the derivation step by step[20]. The reason is psychological. Students have much stronger motivation knowing that they control everything, even by simply accepting every step of derivation. It gives them a kind of psychological integrity that is very important in any study. Some formulas may be judged to be correct just by inspection. This is especially valuable for students, and I always try to stress this.

In the course of study, students should master material that is both simple and complex. Much of this involves familiarity with the set of mathematical tools repeatedly used throughout this book. The appendices provide ample reference to such a toolbox. These include matrix algebra, determinants, vector spaces, vector orthogonalization, secular equations, matrix diagonalization,

---

[19] The abovementioned itineraries cross these borders.

[20] Sometimes this is not possible. Some formulas require painstaking effort to be derived. This was the case, for example, in the coupled cluster method on p. 636.

point group theory, delta functions, finding conditional extrema (Lagrange multipliers, penalty function methods), and Slater-Condon rules, as well as secondary quantization. I would suggest that the reader review (before reading this book) the elementary introduction to matrix algebra (Appendix A) and to vector spaces and operators (Appendix B). The material in these appendices is often used throughout this book.

The book contains numerical examples in many places. Their function is always a semi-quantitative description of a *phenomenon*, not so much the description of a particular system. This is because I prefer to get a trend of changes and an order of magnitude of the things to be illustrated, rather than highly accurate numbers. My private conviction behind this approach is quite strange and unusual: nature is so rich (think of all elements as possible substitutions, influence of neighboring atoms that could modify the properties, using pressure, etc.), that there is a good probability of finding a system exhibiting the phenomenon we found in our calculations…well, at least we hope there is.

As I have said, I imagined students sitting in a lecture hall as I wrote. The tone of this book should make you think of a lecture in interactive mode. To some, this is not the way books are supposed to be. I apologize to any readers who may not feel comfortable with this approach.

Computations Are Easy

On the webpage www.webmo.net (webMO is a free world wide web-based interface to computational chemistry packages), the reader will find a way to carry out quantum mechanical calculations (up to 60 seconds CPU time). Nowadays, this is a sufficiently long time to perform computations for molecules, even for those that have several dozens of atoms. This webpage offers several powerful professional computer programs. Using this tool is straightforward and instructive. I suggest that the reader check this as soon as possible.

Web Annex booksite.elsevier.com/978-0-444-59436-5

The role of the Web Annex is to expand the readers' knowledge after they read a given chapter. At the heart of the Web Annex are links to other people's websites. The Annex will be updated every several months. The Annex adds at least four new dimensions to my book: color, motion, an interactive mode of learning, and connection to the web (with a plethora of possibilities to go even further). When on the web, the reader may choose to come back (automatically) to the Annex at any time.

How to Begin

It is suggested that the reader starts reading this book by doing the following:

- Study the TREE diagram.
- Read the table of contents and compare it with the TREE.
- Address the question of what is your goal–i.e., why you would like to read such a book?

- Choose your own personal path on the TREE (the suggested itineraries may be of some help[21]).
- Become acquainted with the organization of a chapter before you read it.

Chapter Organization

Once an itinerary is chosen, students will cover different chapters. All the chapters have the same structure and are divided into sections as follows:

- **Where Are We?**
  In this section, readers are made aware of their current position on the TREE diagram. In this way, they know the relationship of the current chapter to other chapters, what chapters they are expected to have covered already, and the remaining chapters for which the current chapter provides a preparation. The position shows whether they should invest time and effort in studying the current chapter. In this section, a mini-TREE is also shown, indicating the current position.
- **An Example**
  Here, the reader is confronted with a practical problem that the current chapter addresses.
- **What Is It All About?**
  In this section, the essence of the chapter is presented and a detailed exposition follows. The recommended paths are also provided.
- **Why Is This Important?**
  Not all chapters are of equal importance for the reader. At this point, he or she has the opportunity to judge whether the arguments presented about the importance of a current chapter are convincing.
- **What Is Needed?**
  This section lists the prerequisites necessary for the successful completion of the current chapter. Material required for understanding the text is provided in the appendices. The reader is asked not to take this section too literally, since a tool may be needed only for a minor part of the material covered and is of secondary importance.
- **Classical Works**
  Every field of science has a founding parent or parents, who have identified the seminal problems, introduced basic ideas and selected the necessary tools. Wherever appropriate, we mention these classical investigators and their most important contributions.

[21] This choice may still be tentative and may become clear in the course of reading this book. The index at the end may serve as a significant help. For example, readers interested in drug design, which is based in particular on enzymatic receptors, should cover the chapters with ▲ (those considered most important) and then those with △ (at the very least, intermolecular interactions). They will gain the requisite familiarity with the energy that is minimized in computer programs. Readers should then proceed to those branches of the TREE diagram labeled with □. Initially, they may be interested in force fields (where the abovementioned energy is approximated), and then in molecular mechanics and molecular dynamics (♠). Students may begin this course with only the ♠ labels, but such a course would leave them without any link to quantum mechanics.

- **The Chapter's Body**
  The main body of each chapter is presented in this section.
- **Summary**
  The main body of a chapter is still a big thing to digest, and a student may be lost when reviewing the logical structure of each chapter[22]. A short summary communicates to the student the motivation for presenting the material at hand, why one should expend the effort to understand it, what the main benefits are, and why the author has attached importance to this subject. This is a useful point for reflection and consideration. What we have learned, where we are heading, and where this knowledge will be used and applied are covered here.
- **Main Concepts, New Terms**
  New terms, definitions, concepts, relationships introduced in the chapter are listed here.
- **From the Research Front**
  It is often ill advised to present state-of-the-art results to students. For example, what is the value of presenting a wave function consisting of thousands of terms for the helium atom? The logistics of such a presentation are difficult to contemplate. There is significant didactic value in presenting a wave function with one term or only a few terms where significant concepts are communicated. On the other hand, the student should be made aware of recent progress in generating new results and how well these results agree with experimental observations.
- **Ad Futurum**
  The reader deserves to have a learned assessment of the subject matter covered in a given chapter. For example, is this field stale or new? What is the prognosis for future developments in this area? These are often perplexing questions, and the reader deserves an honest answer (the present author is trying to give in this section).
- **Additional Literature**
  The present text offers only a general panorama of quantum chemistry. In most cases, there exists an extensive literature, where the reader will find more detailed information. Some of the best sources are given here.
- **Questions**
  In this section, the reader will find 10 topics, each containing four yes-or-no questions related to the current chapter.
- **Answers**
  Here, the answers to the problems in the "*Questions*" section are provided.

---

[22] This is most dangerous. A student at *any* stage of study has to be able to answer easily what the purpose of each stage is.

Acknowledgments

The second edition of this book owes much to a few unusual people. I would like to thank first of all my friend Professor Andrzej Sadlej (1941–2010) from the Nicolaus Copernicus University in Toruń, who, despite his sickness, worked to the fullest to improve the book until his last days, always smiling. I also would like to thank another friend of mine, Professor Leszek Z. Stolarczyk from the University of Warsaw, for everyday stimulating discussions that helped both of us to understand what we see around us. I am also very indebted to Professor Jacek Klinowski from the University of Cambridge (UK) and to Professor Stanisław Kucharski from Silesian University, for their kind help when I was really overwhelmed.

Many thanks to my wonderful wife, Basia, for her understanding and love.

CHAPTER 1

The Magic of Quantum Mechanics

"Imagination is more important than knowledge. Knowledge is limited. Imagination encircles the world."
Albert Einstein

Where Are We?

We are at the beginning of all the paths, at the base of the TREE.

An Example

Since 1911, we have known that atoms and molecules are built of two kinds of particles: electrons and nuclei. Experiments show the particles may be treated as pointlike objects of a certain mass and electric charge. The electronic charge is equal to $-e$, while the nuclear charge amounts to Ze, where $e = 1.6 \cdot 10^{-19}$ C and Z is a natural number. Electrons and nuclei interact according to Coulomb's law, and classical mechanics and electrodynamics predict that any atom or molecule is bound to collapse in just a femtosecond, emitting an infinite amount of energy. Hence, according to the classical laws, the complex matter we see around us should simply not exist at all.

Charles Augustin de Coulomb (1736–1806), French military engineer and one of the founders of quantitative physics. In 1777, he constructed a torsion balance for measuring very weak forces, with which he was able to demonstrate the inverse square (of the distance) law for electric and magnetic forces. He also studied charge distribution on the surfaces of dielectrics.

However, atoms and molecules do exist, and their existence may be described in detail by quantum mechanics using what is known as the *wave function*. The postulates of quantum mechanics provide the rules for finding this function and for the calculation of all the observable properties of atoms and molecules.

Ideas of Quantum Chemistry, Second Edition. http://dx.doi.org/10.1016/B978-0-444-59436-5.00001-5

What Is It All About?

Any branch of science has a list of postulates, on which the entire construction is built.[1] For quantum mechanics, six such postulates have been established in the process of reconciling theory and experiment, they may sometimes be viewed as non-intuitive. They stand behind any tool of quantum mechanics used in practical applications. They also lead to some striking conclusions concerning the reality of our world, such as the possibilities of bilocation, teleportation, etc. These unexpected conclusions have recently been experimentally confirmed.

Why Is This Important?

The postulates given in this chapter represent the *foundation* of quantum mechanics and justify all that follows in this book. In addition, our ideas of what the world is really like will acquire a new and unexpected dimension.

What Is Needed?

- Complex numbers
- Operator algebra and vector spaces, p. e7
- Angular momentum, p. e73
- Some background in experimental physics: Black body radiation, photoelectric effect (recommended, but not absolutely necessary)

Classical Works

The beginning of quantum theory was the discovery by Max Planck of the electromagnetic energy quanta emitted by a black body. His work was "*Über das Gesetz der Energieverteilung im Normalspektrum*"[2] in *Annalen der Physik, 4,* 553 (1901). ★ Four years later, Albert Einstein published a paper called "*Über die Erzeugung und Verwandlung des*

[1] These postulates are not expected to be proved.

[2] This title translates as "*On the energy distribution law in the normal spectrum.*" It was published with a note saying that the material had already been presented (in another form) at the meetings of the German Physical Society on October 19 and December 14, 1900.

On p. 556, one can find the following historical sentence on the total energy denoted as U_N which translates as: "*Therefore, it is necessary to assume that* U_N does not represent any continuous quantity that can be divided without any restriction. Instead, one has to understand that it as a discrete quantity composed of a finite number of equal parts."

Lichtes betreffenden heuristischen Gesichtspunkt" in *Annalen der Physik, 27*, 132 (1905), in which he explained the photoelectric effect by assuming that the energy is absorbed by a metal as quanta of energy. ★ In 1911, Ernest Rutherford discovered that atoms are composed of a massive nucleus and electrons: "*The Scattering of the α and β Rays and the Structure of the Atom*," in *Proceedings of the Manchester Literary and Philosophical Society*, IV, *55*, 18 (1911). ★ Two years later, Niels Bohr introduced a planetary model of the hydrogen atom in "*On the Constitution of Atoms and Molecules*" in *Philosophical Magazine*, Series 6, vol.26 (1913). ★ Louis de Broglie generalized the corpuscular and wave character of any particle in his Ph.D. thesis "*Recherches sur la théorie des quanta*," at the Sorbonne, 1924. ★ The first mathematical formulation of quantum mechanics was developed by Werner Heisenberg in "*Über quantentheoretischen Umdeutung kinematischer und mechanischer Beziehungen*," *Zeitschrift für Physik, 33*, 879 (1925). ★ Max Born and Pascual Jordan recognized matrix algebra in the formulation [in "*Zur Quantenmechanik*," Zeitschrift für Physik, *34*, 858 (1925)] and then all three [the famous "*Dreimännerarbeit*" entitled "*Zur Quantenmechanik. II.*" and published in *Zeitschrift für Physik, 35*, 557 (1925)] expounded a coherent mathematical basis for quantum mechanics. ★ Wolfgang Pauli introduced his "*two-valuedness*" for the non-classical electron coordinate in "*Über den Einfluss der Geschwindigkeitsabhängigkeit der Elektronenmasse auf den Zeemaneffekt*," published in Zeitschrift für Physik, *31*, 373 (1925), and the next year, George Uhlenbeck and Samuel Goudsmit described their concept of particle spin in "*Spinning Electrons and the Structure of Spectra*," *Nature, 117*, 264 (1926). ★ Wolfgang Pauli published his famous exclusion principle in "*Über den Zusammenhang des Abschlusses der Elektronengruppen im Atom mit der Komplexstruktur der Spektren*," which appeared in *Zeitschrift für Physik B, 31*, 765 (1925). ★ The series of papers by Erwin Schrödinger, called "*Quantisierung als Eigenwertproblem*," in *Annalen der Physik, 79*, 361 (1926). (also see other references in Chapter 2) was a major advance. He proposed a different mathematical formulation (from Heisenberg's) and introduced the notion of the wave function. ★ In the same year, Max Born, in "*Quantenmechanik der Stossvorgänge*," which appeared in *Zeitschrift für Physik, 37*, 863 (1926), gave an interpretation of the wave function. ★ The uncertainty principle was discovered by Werner Heisenberg and described in "*Über den anschaulichen Inhalt der quantentheoretischen Kinematik und Mechanik*," *Zeitschrift für Physik, 43*, 172 (1927). ★ Paul Adrien Maurice Dirac reported an attempt to reconcile quantum and relativity theories in a series of papers from 1926 to 1928 (also see the references in Chapter 3). ★ Albert Einstein, Boris Podolsky, and Natan Rosen proposed a (then a Gedanken or thinking - experiment, now a real one) test of quantum mechanics "*Can quantum-mechanical description of physical reality be considered complete?*" published in *Physical Review, 47*, 777 (1935). ★ Richard Feynman, Julian Schwinger, and Shinichiro Tomonaga developed quantum electrodynamics in the late forties. ★ John Bell, in "*On the Einstein-Podolsky-Rosen Paradox*," *Physics, 1*, 195 (1964). reported inequalities that were able to verify the very foundations of quantum mechanics. ★ Alain Aspect, Jean Dalibard, and Gérard Roger, in "*Experimental Test of Bell's Inequalities Using Time-Varying Analyzers*," *Physical Review Letters, 49*, 1804 (1982), reported measurements that violated the Bell inequality and proved the non-locality or/and (in a sense) non-reality of our world. ★ The first two-slit interference experiments proving the wave nature of electrons were performed in 1961 by Claus Jönsson from Tübingen Universität in Germany [publication " *Elektroneninterferenzen an mehreren künstlich hergestellter Feinspalten*" in Zeitschrift für Physik, *161*, 454 (1961)], while the experimental proof for interference of a single electron was presented by Pier Giorgio Merli, Gianfranco Missiroli, and Gulio Pozzi from the University of Milan in the article "*On the Statistical Aspect of electron interference phenomena*", *American Journal of Physics, 44*, 306 (1976). ★ Charles H. Bennett, Gilles Brassart, Claude Crépeau, Richard Jozsa, Asher Peres, and William K. Wootters, in "*Teleporting an unknown quantum state via dual classical and Einstein-Podolsky-Rosen channels*" in *Physical Review Letters, 70*, 1895 (1993), designed a teleportation experiment which subsequently was successfully performed by Dik Bouwmeester, Jan-Wei Pan, Klaus Mattle, Manfred Eibl, Harald Weinfurter, and Anton Zeilinger ["*Experimental Quantum Teleportation*," in *Nature, 390*, 575 (1997).]

1.1 History of a Revolution

The end of the nineteenth century was a proud period for physics, which seemed to finally achieve a state of coherence and clarity. At that time, physicists believed that the world consisted of two kingdoms: a kingdom of particles and a kingdom of electromagnetic waves. The motion of

James Clerk Maxwell (1831–1879), British physicist, professor at the University of Aberdeen, Kings College, London, and Cavendish Professor at the university of Cambridge, Cambridge. His main contributions are several famous equations for electromagnetism (1864) and the discovery of velocity distribution in gases (1860).

particles had been described by Isaac Newton's equation, with its striking simplicity, universality, and beauty. Similarly, electromagnetic waves had been described accurately by James Clerk Maxwell's simple and beautiful equations.

Young Max Planck was advised to abandon the idea of studying physics because everything had already been discovered. This beautiful idyll was only slightly incomplete, because of a few annoying details: the strange black body radiation, the photoelectric effect, and the mystery of atomic spectra. These were just a few rather exotic problems to be fixed in the near future…

As it turned out, these problems opened a whole new world. The history of quantum theory, one of most revolutionary and successful theories ever designed by man, will briefly be given below. Many of these facts are discussed further in this textbook.

1900–Max Planck

Max Karl Ernst Ludwig Planck (1858–1947), German physicist, professor at the universities in Munich, Kiel and Berlin, first director of the Institute of Theoretical Physics in Berlin. Planck was born in Kiel, where his father was a university professor of law. He was a universally talented student in grade school, and then an outstanding physics student at the University of Berlin, where he was supervised by Gustaw Kirchhoff and Hermann Helmholz. Music was his passion throughout his life, and he used to play piano duets with Einstein (who played the violin). This hard-working, middle-aged, old-fashioned professor of thermodynamics made a major breakthrough as if in an act of scientific desperation. In 1918 Planck received the Nobel Prize *"for services rendered to the advancement of Physics by his discovery of energy quanta"*. Einstein recalls jokingly Planck's reported lack of full confidence in general relativity theory: *"Planck was one of the most outstanding people I have ever known, (…) In reality, however, he did not understand physics. During the solar eclipse in 1919 he stayed awake all night, to see whether light bending in the gravitational field will be confirmed. If he understood the very essence of the general relativity theory, he would quietly go to bed, as I did."* (cited by Ernst Straus in *"Einstein: A Centenary Volume,"* p. 31).

Black Body Radiation

Planck wanted to understand black body radiation. The black body may be modeled by a box, with a small hole (shown in Fig. 1.1). We heat the box up, wait for the system to reach a stationary state (at a fixed temperature), and see what kind of electromagnetic radiation (intensity as a function of frequency) comes out of the hole. In 1900, Rayleigh and Jeans[3] tried to apply classical mechanics to this problem, and they calculated correctly that the black body would emit the electromagnetic radiation with a distribution of frequencies. However, the larger the frequency, the larger its intensity - an absurd conclusion, what is known as an *ultraviolet catastrophe*. Experiments contradicted theory (as shown in Fig. 1.1).

John William Strutt, Lord Rayleigh (1842–1919), British physicist and Cavendish Professor at the University of Cambridge, contributed greatly to physics (wave propagation, light scattering theory - Rayleigh scattering). In 1904 Rayleigh received the Nobel Prize *"for his investigations of the densities of the most important gases and for his discovery of argon in connection with these studies."*

At a given temperature T, the intensity distribution has a single maximum (at a given frequency ν, as shown in Fig. 1.1b). As the temperature increases, the maximum should shift toward higher frequencies (a piece of iron appears red at 500 °C, but bluish at 1000 °C). Just as Rayleigh and Jeans did, Planck was unable to derive this simple qualitative picture from classical theory–clearly, something had to be done. On December 14, 1900, the generally accepted date for the birth of quantum theory, Planck presented his theoretical results for the black body treated as an ensemble of harmonic oscillators. With considerable reluctance, he postulated[4] that matter cannot emit radiation except by equal portions (*"quanta"*) of energy $h\nu$, proportional to the frequency ν of vibrations of a single oscillator of the black body.

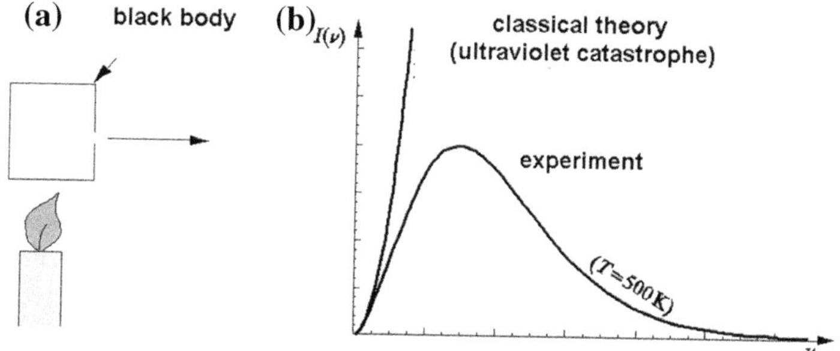

Fig. 1.1. Black body radiation. (a) As one heats a box to temperature T, the hole emits electromagnetic radiation with a wide range of frequencies. The distribution of intensity as a function of frequency ν is given in panel (b). There is a serious discrepancy between the results of classical theory and the experiment, especially for large frequencies. Only after assuming the existence of energy quanta can theory and experiment be reconciled.

[3] James Hopwood Jeans (1877–1946) was a British physicist and professor at the University of Cambridge and at the Institute for Advanced Study in Princeton. Jeans also made important discoveries in astrophysics (e.g., the theory of double stars).

[4] However, note that Planck felt uncomfortable with this idea for many years.

The famous Planck constant h followed soon after. (The actual equation for the Planck constant is $h = 6.62607 \cdot 10^{-34}$ J s, but in this book, we will use a more convenient constant[5] $\hbar = \frac{h}{2\pi}$.) This hypothesis about energy quanta led to the agreement of theory with experiment and the elimination of the ultraviolet catastrophe.

1905–Albert Einstein

The Photoelectric Effect

The second worrying problem, apart from the black body, was the photoelectric effect.[6] Light knocks electrons[7] out of metals, but only when its frequency exceeds a certain threshold. In classical theory, light energy should be stored in the metal in a continuous way and independent of the frequency used, after a sufficient period of time, the electrons should be ejected from the metal. Nothing like that was observed, however, and classical physics was unable to explain this. Einstein introduced the idea of electromagnetic radiation quanta as *particles*, which later were dubbed *photons* by Gilbert Lewis. Note that Planck's idea of a quantum concerned energy transfer from the black body to the electromagnetic field, while Einstein introduced it for the opposite direction, with the energy corresponding to Planck's quantum. Planck considered the quantum to be a portion of energy, while for Einstein, the quantum meant a particle.[8] Everything became clear: energy goes to electrons by quanta, and this is why only quanta exceeding some threshold (the binding energy of an electron in the metal) are able to eject electrons from a metal.

---

5 This constant is known as "*h bar*."

6 Experimental work on this effect had been done by Philipp Eduard Anton Lenard (1862–1947), German physicist and professor at Breslau (now Wrocław), Köln, and Heidelberg. Lenard discovered that the number of photoelectrons is proportional to the intensity of light, and that their kinetic energy does not depend at all on the intensity, depending instead on the *frequency* of light. Lenard received the Nobel Prize in 1905 "*for his work on cathode rays*." A faithful follower of Adolf Hitler, and devoted to the barbarous Nazi ideas, Lenard terrorized German science. He demonstrates that scientific achievement and decency are two separate human characteristics.

7 The electron was already known, having been predicted as early as 1868 by the Irish physicist George Johnstone Stoney (1826–1911), and finally discovered in 1897 by the British physicist Joseph John Thomson (1856–1940). Thomson also discovered a strange pattern: the number of electrons in light elements was equal to about half of their atomic mass. Free electrons were obtained much later (1906). The very existence of atoms was still a hypothesis. The atomic nucleus was discovered only later, in 1911. Physicists were also anxious about the spectra of even the simplest substances, such as hydrogen. Johann Jacob Balmer, a teacher from Basel, was able to design an astonishingly simple formula which fitted perfectly some of the observed lines in the hydrogen spectrum ("*Balmer series*"). All that seemed mysterious and intriguing.

8 It is true that Einstein wrote about "*point-like quanta*" four years later, in a careful approach identifying the quantum with the particle. Modern equipment enables us to count photons, the individual particles of light, but the human eye is also capable of detecting 6 to 8 photons striking a neuron.

In 1905, the accuracy of experimental data was too poor to confirm Einstein's theory as the only one which could account for the experimental results. Besides, the wave nature of light was supported by thousands of crystal clear experiments. Einstein's argument was so breathtaking (…particles???), that Robert Millikan decided to disprove experimentally Einstein's hypothesis. However, after ten years of investigations, Millikan acknowledged that he is forced to support Einstein's explanation, "*however absurd it may look*" (*Rev. Modern Phys., 21*, 343 (1949)). This conversion of a skeptic inclined the Nobel Committee to award Einstein the Nobel Prize in 1923 "*for his work on the elementary charge of electricity and on the photo-electric effect*".

Gilbert Newton Lewis (1875–1946) was the greatest American chemist, who advanced American chemistry internationally through his research and teaching. In a 1926 article in *Nature*, Lewis introduced the word "*photon*." He also developed an early theory of chemical bonding ("*Lewis structures*") based on counting the valence electrons and forming "*octets*" from them. The idea that atoms in molecules tend to form octets in order to complete their electron shells turned out to be surprisingly useful in predicting bond patterns in molecules. A drawback for this concept is that it was not closely connected to the ideas of theoretical physics. It is an example of an extremely clever concept rather than a coherent theory. Lewis also introduced a new definition of acids and bases, which is still in use.

1911–Ernest Rutherford

Rutherford proved experimentally that atoms have a massive nucleus, but the nucleus is very small compared to the size of the atom. The positive charge is concentrated in the nucleus, which is about 10^{-13} cm in size. The density of the nuclear matter boggles the imagination: 1 cm^3 has a mass of about 300 million tons. This is how researchers found out that an atom is composed of a massive nucleus and electrons.

1913–Niels Bohr

Niels Hendrik Bohr (1885–1962), Danish physicist and professor at Copenhagen University, played a key role in the creation and interpretation of quantum mechanics. Bohr was born in Copenhagen, the son of a professor of physiology. He graduated from Copenhagen University and in 1911, he obtained his doctorate there. Then he went to Cambridge to work under the supervision of J. J. Thomson, the discoverer of the electron. The collaboration did not work out, and in 1912, Bohr began to collaborate with Ernest Rutherford at the University of Manchester. At Manchester, Bohr made a breakthrough by introducing a planetary model of hydrogen atom. Bohr reproduced the experimental spectrum of the hydrogen atom with high accuracy. In 1922, Bohr received the Nobel Prize "*for his investigation of the structure of atoms.*" That same year, he became the father of Aage Niels Bohr, a future winner of the Nobel Prize (1975, for his studies of the structure of nuclei). In October 1943, Bohr and his family fled from Denmark to Sweden, and then to Great Britain and the United States, where he worked on the Manhattan Project.

The Model of the Hydrogen Atom

Atomic spectra were the third great mystery of early twentieth-century physics. Even interpreting the spectrum of the hydrogen atom represented a challenge. In 1913, at the age of 28, Bohr proposed a simple planetary model of this atom in which the electron, *contrary to classical mechanics*, did not fall onto the nucleus. Instead, it changed its orbit, with accompanying absorption or emission of energy quanta. Bohr assumed that angular orbital momentum is quantized and that the centrifugal force is compensated by the Coulomb attraction between the electron and the nucleus. He was able to reproduce part of the spectrum of the hydrogen atom very accurately. Bohr then began work on the helium atom (which turned out to be a disaster), but he was successful again with the helium cation[9] He^+.

Niels Bohr played an inspiring role in the development and popularization of quantum mechanics. The Copenhagen Institute for Theoretical Physics, which he founded in 1921, was the leading world center in the 1920s and 1930s, where many young theoreticians from all over the world worked on problems in quantum mechanics.[10] Bohr, with Werner Heisenberg, Max Born, and John von Neumann, contributed greatly to the elaboration of the philosophical foundations of quantum mechanics. According to this, quantum mechanics represents a coherent and complete model of reality ("*the world*"), and the discrepancies with classical mechanics have a profound and fundamental character.[11] Both theories coincide in the limit $h \rightarrow 0$ (where h is the Planck constant), and thus, the predictions of quantum mechanics reduce to those of classical mechanics (known as *Bohr's correspondence principle*).

Arnold Sommerfeld (1868–1951), German physicist and professor at the Mining Academy in Clausthal, then at the Technical University of Aachen, in the key period 1906–1938, was professor at Munich University. Sommerfeld considered not only circular (Bohr-like) orbits, but also elliptical ones, and introduced the angular quantum number. He also investigated X-rays and the theory of metals. The scientific father of many Nobel Prize winners, he did not earn this distinction himself.

1916–Arnold Sommerfeld

Old Quantum Theory

In 1916, Arnold Sommerfeld generalized the Bohr quantization rule

[9] Bohr did not want to publish without good results for all other atoms, something he would never achieve. Rutherford argued: "*Bohr, you explained hydrogen, you explained helium, people will believe you for other atoms.*"

[10] John Archibald Wheeler recalls that when he first came to the institute, he met a man working in the garden and asked him where he could find Professor Bohr. The gardener answered: "*That's me.*"

[11] The center of the controversy was that quantum mechanics is indeterministic, while classical mechanics is deterministic, although this indeterminism is not all that it seems. As will be shown later in this chapter, quantum mechanics is a fully deterministic theory in the Hilbert space (the space of all possible wave functions of the system), its indeterminism pertains to the physical space in which we live.

beyond the problem of the one-electron atom. Known as "*old quantum theory*," it did not represent any coherent theory of general applicability. As a matter of fact, this quantization was achieved by assuming that for every periodic variable (like an angle), an integral is equal to an integer times the Planck constant.[12] Sommerfeld also tried to apply the Bohr model to atoms with a single valence electron (he had to modify the Bohr formula by introducing the quantum defect; i.e., a small change in the principal quantum number, see p. 204).

1923–Louis de Broglie

Louis-Victor Pierre Raymond de Broglie (1892–1987) was studying history at the Sorbonne, carefully preparing himself for a diplomatic career, which was a very natural pursuit for someone from a princely family, as he was. His older brother Maurice, a radiographer, aroused his interest in physics. World War I (Louis did military service in a radio communications unit) and the study of history delayed his start in physics.

He was 32 when he presented his doctoral dissertation, which embarrassed his supervisor, Paul Langevin. The thesis, on the wave nature of all particles, was so revolutionary that only a positive opinion from Einstein, who was asked by Langevin to take a look of the dissertation, convinced the doctoral committee. Only five years later (in 1929), Louis de Broglie received the Nobel Prize "*for his discovery of the wave nature of electrons.*"

Waves of Matter

In his doctoral dissertation, stuffed with mathematics, Louis de Broglie introduced the concept of "*waves of matter.*" He postulated that not only photons, but also any other particle, has, besides its corpuscular characteristics, some wave properties (those corresponding to light had been known for a long, long time). According to de Broglie, the wave length corresponds to momentum p:

$$p = \frac{h}{\lambda}$$

where h is again the Planck constant! What kind of momentum can this be, in view of the fact that momentum depends on the laboratory coordinate system chosen? Well, it is the momentum measured in the same laboratory coordinate system as that used to measure the corresponding wave length.

[12] Similar periodic integrals were used earlier by Bohr.

1923–Arthur Compton[13]

Electron-Photon Scattering

It turned out that an electron-photon collision obeys the same laws of dynamics as those describing the collision of two particles: the energy conservation law and the momentum conservation law. This result confirmed the wave-corpuscular picture emerging from experiments.

1925–George E. Uhlenbeck and Samuel A. Goudsmit

Discovery of Spin

These two Dutch students explained an experiment, in which a beam of silver atoms passing through a magnetic field splits into two beams. In a short paper, they suggested that the silver atoms have (besides their orbital angular momentum) an additional internal angular momentum (spin), which was *similar* to a macroscopic body, which besides its center-of-mass motion, also has a rotational (spinning) motion.[14] Moreover, the students demonstrated that the atomic spin follows from the spin of the electrons: among the 47 electrons of the silver atom, 46 have their spin compensated (23 "*down*" and 23 "*up*"), while the last "*unpaired*" electron gives the net spin of the atom.

1925–Wolfgang Pauli[15]

Pauli Exclusion Principle

Pauli postulated that in any system, two electrons cannot be in the same state (including their spins). This "*Pauli exclusion principle*" was deduced from spectroscopic data (some states were not allowed).

---

[13] Arthur Holly Compton (1892–1962) was an American physicist and professor at the universities of Saint Louis and Chicago. He obtained the Nobel Prize in 1927 "*for the discovery of the effect named after him*"; i.e., for investigations of electron-photon scattering.

[14] Caution: Identifying the spin with the rotation of a rigid body leads to physical inconsistencies.

[15] Pauli also introduced the idea of spin when interpreting spectra of atoms with a single valence electron. He was inspired by Sommerfeld, who interpreted the spectra by introducing the quantum number $j = l \pm \frac{1}{2}$, where the quantum number l quantized the orbital angular momentum of the electron. Pauli described spin as *a bivalent non-classical characteristic of the electron* [W.Pauli, *Zeit.Phys.B*, *3*, 765 (1925)].

1925–Werner Heisenberg

Matrix Quantum Mechanics

A paper by 24-year-old Werner Heisenberg turned out to be a breakthrough in quantum theory.[16] He wrote in a letter: "*My whole effort is to destroy without a trace the idea of orbits.*" Max Born recognized matrix algebra in Heisenberg's formulation (who, himself, had not yet realized it), and in the same year, a more solid formulation of the new mechanics ("*matrix mechanics*") was proposed by Werner Heisenberg, Max Born, and Pascual Jordan.[17]

1926–Erwin Schrödinger

Schrödinger Equation

In November 1925, Erwin Schrödinger delivered a lecture at the Technical University in Zurich (ETH), in which he presented de Broglie's results. Professor Peter Debye stood up and asked the speaker:

Peter Josephus Wilhelmus Debye (1884–1966), Dutch physicist and chemist and professor in the Technical University (ETH) of Zurich (1911, 1920–1937), as well as at Göttingen, Leipzig, and Berlin, won the Nobel Prize in chemistry in 1936 "*for his contribution to our knowledge of molecular structure through his investigations on dipole moments and on the diffraction of X-rays and electrons in gases.*" Debye emigrated to the United States in 1940, where he obtained a professorship at Cornell University in Ithaca, NY (and remained in this beautiful town to the end of his life). His memory is still alive there.

[16] On June 7, 1925, Heisenberg was so tired after a bad attack of hay fever that he decided to go relax on the North Sea island of Helgoland. Here, he divided his time between climbing the mountains, learning Goethe's poems by heart, and (despite his intention to rest) hard work on the spectrum of the hydrogen atom, with which he was obsessed. It was at night on June 7 or 8 that he saw something–the beginning of the new mechanics. In later years, he wrote in his book *Der Teil and das Ganze*: " *It was about three o' clock in the morning when the final result of the calculation lay before me. At first I was deeply shaken. I was so excited that I could not think of sleep. So I left the house and awaited the sunrise on the top of a rock.*" The first man with whom Heisenberg shared his excitement a few days later was his schoolmate Wolfgang Pauli, and, after another few days, with Max Born.

[17] Jordan, despite his talents and achievements, felt underestimated and even humiliated in his native Germany. For example, he had to accept a position at Rostock University, which the German scientific elite used to call the "*Outer Mongolia of Germany.*" The best positions seemed to be reserved. When Hitler came to power, Jordan became a fervent follower.

"You are telling us about waves, but where is the wave equation in your talk?" Indeed, there wasn't any! Schrödinger began to work on this problem, and the next year formulated what is now called *wave mechanics* based on the wave equation. Both formulations, Heisenberg's and Schrödinger's,[18] turned out to be equivalent and are now known as (non-relativistic) quantum mechanics.

1926–Max Born

Statistical Interpretation of Wave Function

Max Born (1882–1970) German physicist and professor at the universities of Göttingen, Berlin, Cambridge, and Edinburgh, was born in Breslau (now Wrocław) to the family of a professor of anatomy. Born studied first in Wrocław, then at Heidelberg and Zurich. He received his Ph.D. in physics and astronomy in 1907 at Göttingen, where he began his swift academic career. Born obtained a chair at the University of Berlin in 1914 and returned to Göttingen in 1921, where he founded an outstanding school of theoretical physics, which competed with the famous institute of Niels Bohr in Copenhagen. Born supervised Werner Heisenberg, Pascual Jordan, and Wolfgang Pauli. It was Born who recognized, in 1925, that Heisenberg's quantum mechanics could be formulated in terms of matrix algebra. Together with Heisenberg and Jordan, he created the first consistent quantum theory (the famous *"drei Männer Arbeit"*). After Schrödinger's formulation of quantum mechanics, Born proposed the probabilistic interpretation of the wave function. Despite such seminal achievements, the Nobel Prizes in the 1930s were received by his colleagues, not him. Finally, when Born obtained the Nobel Prize *"for his fundamental research in quantum mechanics, especially for his statistical interpretation of the wave-function,"* in 1954 there was a great relief among his famous friends.

Born proposed interpreting the square of the complex modulus of Schrödinger's wave function as the probability density for finding the particle.

1927–Werner Heisenberg

Uncertainty Principle

Heisenberg concluded that it is not possible to measure simultaneously the position (x) and momentum of a particle (p_x) with any desired accuracy. The more exactly we measure the position (small Δx), the larger the error we make in measuring the momentum (large Δp_x), and vice versa.

[18] The formulation proposed by Paul A.M. Dirac was another important finding.

1927–Clinton Davisson, Lester H.Germer, and George Thomson[19]

Electron Diffraction

Davisson and Germer, and Thomson demonstrated in separate ingenious experiments that electrons indeed exhibit wave properties (using crystals as diffraction gratings).

1927–Walter Heitler and Fritz Wolfgang London

The Birth of Quantum Chemistry

Walter Heitler and Fritz Wolfgang London convincingly explained why two neutral atoms (like hydrogen) attract each other with a force so strong as to be comparable to the Coulomb forces between ions. Applying the Pauli exclusion principle when solving the Schrödinger equation is of key importance. Their paper was received on June 30,1927, by *Zeitschrift für Physik*, and this may be counted as the birth date of quantum chemistry.[20]

1928–Paul Dirac

Dirac Equation for the Electron and Positron

Paul Dirac's main achievements are the foundations of quantum electrodynamics and construction of the relativistic wave equation (1926–1928) that now bears his name. This equation described not only the electron, but also its antimatter counterpart, the positron (predicting antimatter). Spin was also inherently present in the equation.

1929–Werner Heisenberg and Wolfgang Pauli

Quantum Field Theory

Two classmates developed a theory of matter, and the main features still survive. In this theory, the elementary particles (the electron, photon, and so on) were viewed as excited states of the corresponding fields (the electron field, electromagnetic field, and so on).

[19] Clinton Joseph Davisson (1881–1958) was an American physicist at Bell Telephone Laboratories. He discovered the diffraction of electrons with L.H. Germer, and together they received the Nobel Prize in 1937 *"for their experimental discovery of the diffraction of electrons by crystals."* The prize was shared with G.P. Thomson, who used a different diffraction method. George Paget Thomson (1892–1975), son of the discoverer of the electron, Joseph John Thomson, and professor at Aberdeen, London, and Cambridge Universities.

[20] The term *"quantum chemistry"* was first used by Arthur Haas in his lectures to the Physicochemical Society of Vienna in 1929 (A. Haas, *"Die Grundlagen der Quantenchemie. Eine Einleitung in vier Vortragen,"* Akademische Verlagsgesellschaft, Leipzig, 1929).

1932–Carl Anderson[21]

Discovery of Antimatter (the Positron)

One of Dirac's important results was the observation that his relativistic wave equation is satisfied not only by the electron, but also by a mysterious unknown particle, the positive electron (which become known as the *positron*). This antimatter hypothesis was confirmed by Carl Anderson, who found the positron experimentally.

1948–Richard Feynman, Julian Schwinger, and Shinichiro Tomonaga[22]

Quantum Electrodynamics

The Dirac equation did not take all the physical effects into account. For example, the strong electric field of the nucleus polarizes a vacuum so much that electron-positron pairs emerge from the vacuum and screen the electron-nucleus interaction. The quantum electrodynamics (QED) developed by Feynman, Schwinger, and Tomonaga accounts for this and similar effects and brings theory and experiment to an agreement of unprecedented accuracy.

1964–John Bell

Bell Inequalities

The mathematician John Bell proved that if particles had certain properties before measurement (so that they were small but classical objects), then the measurement results would have to satisfy some inequalities that contradict the predictions of quantum mechanics (further details at the end of this chapter).

1982–Alain Aspect

Is the World Non-Local?

Experiments with photons showed that the Bell inequalities are *not* satisfied. This means that either there is instantaneous communication even between extremely distant particles ("*entangled states*"), or that the particles do not have some definite properties before the measurement is performed (more details about this are given at the end of this chapter).

---

[21] More details of this topic are given in Chapter 3.

[22] Feynman, Schwinger, and Tomonaga received the Nobel Prize in 1965 "*for their fundamental work in quantum electrodynamics, with fundamental implications for the physics of elementary particles.*"

1997–Anton Zeilinger

Teleportation of the Photon State

A research group at the University of Innsbruck used entangled quantum states to perform teleportation of a photon state[23]; that is, to prepare at a distance any state of a photon with simultaneous disappearance of this state from the teleportation site (details are given at the end of this chapter).

1.2 Postulates of Quantum Mechanics

All science is based on a number of postulates. Quantum mechanics has also elaborated a system of postulates that have been formulated to be as simple as possible and yet to be consistent with experimental results. Postulates are not supposed to be proved–their justification is efficiency. Quantum mechanics, the foundations of which date from 1925 and 1926, still represents the basic theory of phenomena within atoms and molecules. This is the domain of chemistry, biochemistry, and atomic and nuclear physics. Further progress (quantum electrodynamics, quantum field theory, and elementary particle theory) permitted deeper insights into the structure of the atomic nucleus but did not produce any fundamental revision of our understanding of atoms and molecules. Matter as described by non-relativistic[24] quantum mechanics represents a system of electrons and nuclei, treated as pointlike particles with a definite mass and electric

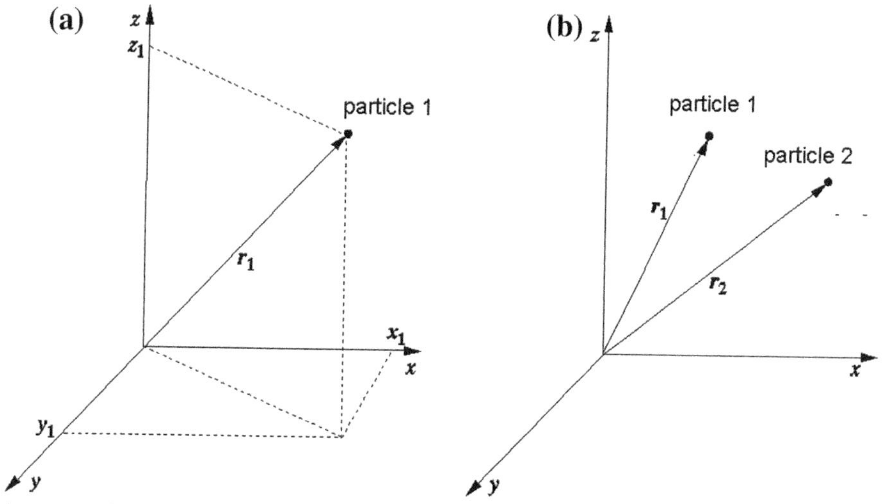

Fig. 1.2. An atom (molecule) in non-relativistic quantum mechanics. (a) A Cartesian ("*laboratory*") coordinate system is introduced in three-dimensional space. We assume (see panel b) that all the particles (electrons and nuclei) are pointlike (their instantaneous positions are shown here) and interact only by electrostatic (Coulomb) forces.

---

[23] M. Eibl, H. Weinfurter, and A. Zeilinger, *Nature, 390*, 575 (1997).
[24] This assumes that the speed of light is infinite.

charge and moving in three-dimensional space and interacting by *electrostatic* forces.[25] This model of matter (shown in Fig. 1.2) is at the core of quantum chemistry.

The assumptions on which quantum mechanics is based may be given in the form of postulates I–VI, which are described next. For simplicity, we will restrict ourselves to a single particle moving along a single coordinate axis x (the mathematical foundations of quantum mechanics are given in Appendix B available at booksite.elsevier.com/978-0-444-59436-5 on p. e7).

Postulate I (on the quantum mechanical state):

The state of the system is described by the wave function $\Psi = \Psi(x, t)$, which depends on the coordinate of particle x at time t. Wave functions in general are complex functions of real variables. The symbol $\Psi^*(x, t)$ denotes the complex conjugate of $\Psi(x, t)$. The quantity

$$p(x, t) = \Psi^*(x, t)\Psi(x, t)\, dx \tag{1.1}$$

gives the probability that at time t the x coordinate of the particle lies in the small interval $[x, x + dx]$ (see Fig. 1.3a). The probability of the particle being in the interval (a, b) on the x-axis is given by Fig. 1.3b:

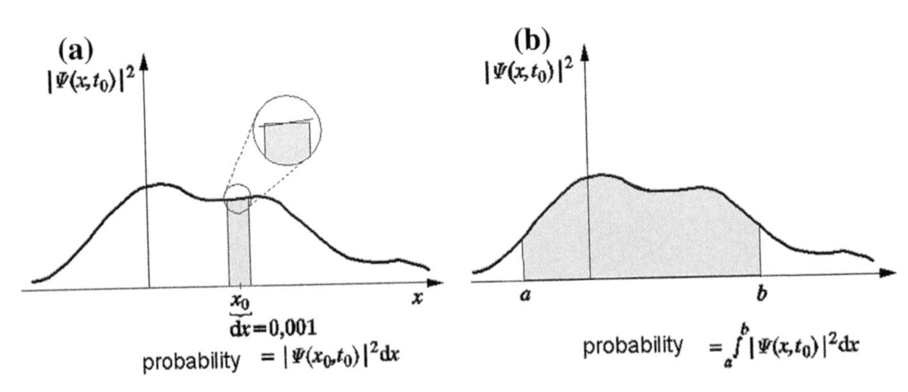

Fig. 1.3. A particle moves along the x-axis and is in the state described by the wave function $\Psi(x, t)$. (a) shows how the probability of finding particle in an infinitesimally small section of the length dx at x_0 (at time $t = t_0$) is calculated. It is not important where exactly in section $[x, x + dx]$ the number x_0 really is because the length of the section is infinitesimally small. Here, the number is positioned in the middle of the section. (b) shows how to calculate the probability of finding the particle at $t = t_0$ in a *section* (a, b).

[25] Yes, we are taking only electrostatics–that is, Coulomb interactions. It is true that a moving charged particle creates a magnetic field, which influences its own and other particles' motion. However, the Lorentz force is taken into account in the *relativistic* approach to quantum mechanics.

The probabilistic interpretation of the wave function was proposed by Max Born.[26] Analogous with the formula mass = density × volume, the quantity $\Psi^*(x,t)\Psi(x,t)$ is called the *probability density* that a particle at time t has position x.

In order to treat the quantity $p(x,t)$ as a probability, at any instant t, the wave function must satisfy the *normalization condition*:

$$\int_{-\infty}^{\infty} \Psi^*(x,t)\Psi(x,t)\, dx = 1. \tag{1.2}$$

All this may be generalized for more complex situations. For example, in three-dimensional space, the wave function of a single particle depends on position $\boldsymbol{r} = (x, y, z)$ and time $\Psi(\boldsymbol{r}, t)$, and the *normalization condition* takes the form

$$\int_{-\infty}^{\infty} dx \int_{-\infty}^{\infty} dy \int_{-\infty}^{\infty} dz \Psi^*(x,y,z,t)\Psi(x,y,z,t) \equiv \int \Psi^*(\boldsymbol{r},t)\Psi(\boldsymbol{r},t)\, dV$$
$$\equiv \int \Psi^*(\boldsymbol{r},t)\Psi(\boldsymbol{r},t) d\boldsymbol{r} = 1. \tag{1.3}$$

For simplicity, the last two integrals are given without the integration limits, but they are there *implicitly*, and this convention will be used throughout the book unless stated otherwise.

For n particles (see Fig. 1.4), shown by vectors $\boldsymbol{r}_1, \boldsymbol{r}_2, \ldots \boldsymbol{r}_n$ in three-dimensional space, the interpretation of the wave function is as follows. The probability P that at a given time $t = t_0$,

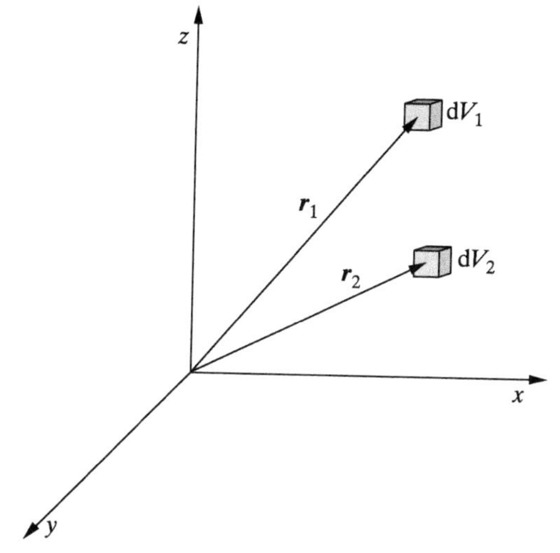

Fig. 1.4. Interpretation of a many-particle wave function, an example for two particles. The number $|\psi(\boldsymbol{r}_1, \boldsymbol{r}_2, t_0)|^2 dV_1 dV_2$ represents the probability that at $t = t_0$, particle 1 is in its box of volume dV_1 shown by vector $\boldsymbol{r}_1$ and particle 2 is in its box of volume dV_2 indicated by vector $\boldsymbol{r}_2$.

[26] M. Born, Zeitschrift fuř Physik, *37*, 863 (1926).

particle 1 is in the domain V_1, particle 2 is in the domain V_2, etc., is computed as

$$P = \int_{V_1} dV_1 \int_{V_2} dV_2 \cdots \int_{V_n} dV_n \Psi^*(\boldsymbol{r}_1, \boldsymbol{r}_2, \ldots, \boldsymbol{r}_n, t_0)\Psi(\boldsymbol{r}_1, \boldsymbol{r}_2, \ldots, \boldsymbol{r}_n, t_0)$$

$$\equiv \int_{V_1} d\boldsymbol{r}_1 \int_{V_2} d\boldsymbol{r}_2 \cdots \int_{V_n} d\boldsymbol{r}_n \Psi^*(\boldsymbol{r}_1, \boldsymbol{r}_2, \ldots, \boldsymbol{r}_n, t_0)\Psi(\boldsymbol{r}_1, \boldsymbol{r}_2, \ldots, \boldsymbol{r}_n, t_0).$$

Often in this book, we will perform what is called *normalization* of a function, which is required if a probability is to be calculated. Suppose that we have a unnormalized function[27] ψ; that is,

$$\int_{-\infty}^{\infty} \psi(x,t)^* \, \psi(x,t) \, dx = A, \tag{1.4}$$

with $0 < A \neq 1$. To compute the probability ψ, it must be normalized; i.e., multiplied by a normalization constant N, such that the new function $\Psi = N\psi$ satisfies the normalization condition: $1 = \int_{-\infty}^{\infty} \Psi^*(x,t)\Psi(x,t)\,dx = N^*N \int_{-\infty}^{\infty} \psi^*(x,t)\psi(x,t)\,dx = A|N|^2$. Hence, $|N| = \frac{1}{\sqrt{A}}$. How is N computed? One person may choose it as equal to $N = \frac{1}{\sqrt{A}}$, another might select $N = -\frac{1}{\sqrt{A}}$, and yet a third might select $N = e^{1989i}\frac{1}{\sqrt{A}}$. There are, therefore, an infinite number of legitimate choices of the phase ϕ of the wave function $\Psi(x,t) = e^{i\phi}\frac{1}{\sqrt{A}}\psi$. Yet, when $\Psi^*(x,t)\Psi(x,t)$, is calculated, everyone will obtain the same result, $\frac{1}{A}\psi^*\psi$, because the phase disappears. In most applications, this is what will happen, so the computed physical properties will not depend on the choice of phase. There are cases, however, where the phase will be important.

Postulate II (on operator representation of mechanical quantities)

The mechanical quantities that describe the particle (energy, the components of vectors of position, momentum, angular momentum, etc.) are represented by linear operators acting in the Hilbert space (see Appendix B available at booksite.elsevier.com/978-0-444-59436-5). There are two important examples of the operators: the operator of the particle's position $\hat{x} = x$ (i.e., multiplication by x, or $\hat{x} = x\,\cdot$; see Fig. 1.5), as well as the operator of the (x-component) momentum $\hat{p}_x = -i\hbar\frac{d}{dx}$, where i stands for the imaginary unit.

Note that the mathematical form of the operators is always defined with respect to a Cartesian coordinate system.[28] From the given operators (Fig. 1.5), the operators of some other quantities may be constructed. The potential energy operator $\hat{V} = V(x)$, where $V(x)$ [the multiplication operator by the function $\hat{V}f = V(x)f$] represents a function of x called a *potential*. The kinetic

[27] In this example, Eq. (1.3) has not been satisfied.

[28] Nevertheless, they may then be transformed to other coordinate systems.

Mechanical quantity	Classical formula	Operator acting on f
coordinate	x	$\hat{x}f \stackrel{\text{def}}{=} xf$
momentum component	p_x	$\hat{p}_x f \stackrel{\text{def}}{=} -i\hbar\frac{\partial f}{\partial x}$
kinetic energy	$T = \frac{mv^2}{2} = \frac{p^2}{2m}$	$\hat{T}f = -\frac{\hbar^2}{2m}\Delta f$

Fig. 1.5. Mechanical quantities and the corresponding operators.

energy operator of a single particle (in one dimension) is $\hat{T} = \frac{\hat{p}_x{}^2}{2m} = -\frac{\hbar^2}{2m}\frac{d^2}{dx^2}$, and in three dimensions, it is as follows:

$$\hat{T} = \frac{\hat{p}^2}{2m} = \frac{\hat{p}_x{}^2 + \hat{p}_y{}^2 + \hat{p}_z{}^2}{2m} = -\frac{\hbar^2}{2m}\Delta, \tag{1.5}$$

where the Laplacian Δ is

$$\Delta \equiv \frac{\partial^2}{\partial x^2} + \frac{\partial^2}{\partial y^2} + \frac{\partial^2}{\partial z^2} \tag{1.6}$$

and m denotes the particle's mass. The total energy operator, or *Hamiltonian* is the most frequently used:

$$\hat{H} = \hat{T} + \hat{V}. \tag{1.7}$$

An important feature of operators is that they may not commute[29]; i.e., for two particular operators $\hat{A}$ and $\hat{B}$, one may have $\hat{A}\hat{B} - \hat{B}\hat{A} \neq 0$. This property has important physical consequences (see the upcoming discussion of Postulate IV and the Heisenberg uncertainty principle). Because of the possible non-commutation of the operators, transformation of the classical formula (in which the commutation or non-commutation did not matter) may not be unique. In such a case, from all the possibilities, one has to choose an operator, which is Hermitian. The operator $\hat{A}$ is Hermitian if for any functions ψ and ϕ from its domain, one has

$$\int_{-\infty}^{\infty} \psi^*(x)\hat{A}\phi(x)\,dx = \int_{-\infty}^{\infty} [\hat{A}\psi(x)]^*\phi(x)\,dx. \tag{1.8}$$

[29] Commutation means $\hat{A}\hat{B} = \hat{B}\hat{A}$.

$\int \psi^* \phi \, d\tau \equiv \langle\psi\|\phi\rangle$	Scalar product of two functions
$\int \psi^* \hat{A}\phi \, d\tau \equiv \langle\psi\|\hat{A}\phi\rangle$ or $\langle\psi\|\hat{A}\|\phi\rangle$	Scalar product of ψ and $\hat{A}\phi$ or a matrix element of the operator $\hat{A}$
$\hat{Q} = \|\psi\rangle\langle\psi\|$	Projection operator on the direction of the vector ψ
$1 = \sum_k \|\psi_k\rangle\langle\psi_k\|$	Spectral resolution of identity. Its sense is best seen when acting on χ: $\chi = \sum_k \|\psi_k\rangle\langle\psi_k\|\chi\rangle = \sum_k \|\psi_k\rangle c_k$.

Fig. 1.6. Dirac notation.

Using what is known as *Dirac notation* (see Fig. 1.6), the above equality may be written in a concise form:

$$\langle\psi \mid \hat{A}\phi\rangle = \langle\hat{A}\psi \mid \phi\rangle. \tag{1.9}$$

In Dirac notation,[30] the key role is played by vectors *bra*: $\langle \,|$ and *ket*: $|\, \rangle$ denoting respectively $\psi^* \equiv \langle\psi|$ and $\phi \equiv |\phi\rangle$. Writing the bra and ket as $\langle\psi|\,|\phi\rangle$ denotes $\langle\psi|\phi\rangle$, or the scalar product of ψ and ϕ in a unitary space (see Appendix B available at booksite.elsevier.com/978-0-444-59436-5), while writing it as $|\psi\rangle\,\langle\phi|$ means the operator $\hat{Q} = |\psi\rangle\,\langle\phi|$, because of its action on function $\xi = |\xi\rangle$ shown as $\hat{Q}\xi = |\psi\rangle\,\langle\phi|\,\xi = |\psi\rangle\,\langle\phi|\xi\rangle = c\psi$, where $c = \langle\phi|\xi\rangle$:

- $\langle\psi \mid \phi\rangle$ denotes a scalar product of two functions (i.e., vectors of the Hilbert space) ψ and ϕ, also known as the overlap integral of ψ and ϕ.
- $\left\langle\psi \mid \hat{A}\phi\right\rangle$ or $\left\langle\psi \mid \hat{A} \mid \phi\right\rangle$ stands for the scalar product of two functions: ψ and $\hat{A}\phi$, or the matrix element of operator $\hat{A}$.
- $\hat{Q} = |\psi\rangle\,\langle\psi|$ means the projection operator on the vector ψ (in the Hilbert space).
- The last formula (with $\{\psi_k\}$ representing the complete set of functions) represents what is known as "*spectral resolution of identity*," best demonstrated when acting on an arbitrary function χ:

$$\chi = \sum_k |\psi_k\rangle\,\langle\psi_k \mid \chi\rangle = \sum_k |\psi_k\rangle\, c_k.$$

We have obtained the decomposition of the function (i.e., a vector of the Hilbert space) χ on its components $|\psi_k\rangle\, c_k$ along the basis vectors $|\psi_k\rangle$ of the Hilbert space. The coefficient $c_k = \langle\psi_k \mid \chi\rangle$ is the corresponding scalar product, and the basis vectors ψ_k are normalized. This formula says something trivial: any vector can be retrieved when adding *all* its components together.

[30] The deeper meaning of this notation is discussed in many textbooks about quantum mechanics; e.g., A. Messiah, *Quantum Mechanics*, vol. I, Amsterdam (1961), p. 245. Here, we treat the Dirac notation as a convenient tool.

Postulate III (on time evolution of the state)

Time-Dependent Schrödinger Equation
The time-evolution of the wave function Ψ is given by the equation

$$i\hbar\frac{\partial\Psi(x,t)}{\partial t} = \hat{H}\Psi(x,t), \tag{1.10}$$

where $\hat{H}$ is the system Hamiltonian; see Eq. (1.7).

$\hat{H}$ may depend on time (energy changes in time, interacting system) or may be time-independent (energy conserved, isolated system). Equation (1.10) is called the time-dependent Schrödinger equation (Fig. 1.7).

When $\hat{H}$ is time independent, the general solution to Eq. (1.10) can be written as

$$\Psi(x,t) = \sum_{n=1}^{\infty} c_n \Psi_n(x,t), \tag{1.11}$$

where $\Psi_n(x,t)$ represent special solutions to Eq. (1.10), that have the form

$$\Psi_n(x,t) = \psi_n(x)\, e^{-i\frac{E_n}{\hbar}t}, \tag{1.12}$$

and c_n stands for some constants. Substituting the special solution to Eq. (1.10) leads to[31] what is known as the *time-independent Schrödinger equation*:

$$\Psi(x,t_0)$$
$$\downarrow$$
$$\hat{H}\Psi(x,t_0)$$
$$\downarrow$$
$$i\hbar\left(\frac{\partial\Psi}{\partial t}\right)_{t=t_0}$$
$$\downarrow$$
$$\Psi(x,t_0+\mathrm{d}t) = \Psi(x,t_0) - \frac{i}{\hbar}\hat{H}\Psi\,\mathrm{d}t$$

Fig. 1.7. Time evolution of a wave function. Knowing $\Psi(x,t)$ at a certain $t = t_0$ makes it possible to compute $\hat{H}\Psi(x,t_0)$, and from this [using Eq. (1.10)] one can calculate the time derivative $\frac{\partial\Psi(x,t_0)}{\partial t} = -\frac{i\hat{H}\Psi(x,t_0)}{\hbar}$. Knowledge of the wave function at time $t = t_0$ and of its time derivative is sufficient to calculate the function a little later ($t = t_0 + \mathrm{d}t$).

31 $i\hbar\frac{\partial\Psi_n(x,t)}{\partial t} = i\hbar\frac{\partial\psi_n(x)\,e^{-i\frac{E_n}{\hbar}t}}{\partial t} = i\hbar\psi_n(x)\,\frac{\partial e^{-i\frac{E_n}{\hbar}t}}{\partial t} = i\hbar\psi_n(x)\left(-i\frac{E_n}{\hbar}\right)\frac{\partial e^{-i\frac{E_n}{\hbar}t}}{\partial t} = E_n\psi_n e^{-i\frac{E_n}{\hbar}t}$. However, $\hat{H}\Psi_n(x,t) = \hat{H}\psi_n(x)\,e^{-i\frac{E_n}{\hbar}t} = e^{-i\frac{E_n}{\hbar}t}\hat{H}\psi_n(x)$, because the Hamiltonian does not depend on t. Hence, after dividing both sides of the equation by $e^{-i\frac{E_n}{\hbar}t}$, one obtains the time-independent Schrödinger equation. Therefore,

Time-independent Schrödinger Equation for Stationary States

$$\hat{H}\psi_n = E_n\psi_n \quad n = 1, 2, \ldots, \infty. \tag{1.13}$$

The equation represents an example of an eigenvalue equation of the operator; the functions ψ_n are called the *eigenfunctions*, and E_n are the *eigenvalues* of the operator $\hat{H}$ (we have assumed here that their number is equal to ∞). It can be shown that E_n is real (see Appendix B available at booksite.elsevier.com/978-0-444-59436-5, p. e7). The eigenvalues are the permitted energies of the system, and the corresponding eigenfunctions Ψ_n are defined in Eqs. (1.12) and (1.13). These states have a special character because the probability given by Eq. (1.1) does not change in time[32] (Fig. 1.8):

$$p_n(x, t) = \Psi_n^*(x, t)\Psi_n(x, t)\, dx = \psi_n^*(x)\psi_n(x)\, dx = p_n(x). \tag{1.14}$$

Therefore, in determining these states, known as stationary states, one can apply the time–independent formalism based on the Schrödinger equation (1.13).

Postulate IV (on interpretation of experimental measurements)

This postulate pertains to ideal measurements, such that no error is introduced through imperfections in the measurement apparatus. We assume the measurement of the physical quantity A,

the stationary state $\psi_n(x)\exp\left(-i\frac{E_n}{\hbar}t\right)$ is time-dependent, but this time dependence comes only from the factor $\exp\left(-i\frac{E_n}{\hbar}t\right)$. When, in the future, we calculate the probability density $\Psi_n^*\Psi_n$, we would not need the factor $\exp\left(-i\frac{E_n}{\hbar}t\right)$, only its modulus whereas $\left|\exp\left(-i\frac{E_n}{\hbar}t\right)\right| = 1$ for any value of t. The time dependence through $\exp\left(-i\frac{E_n}{\hbar}t\right)$ means that as time goes by, function ψ_n is multiplied by an oscillating complex number, which never attains zero. The number oscillates on a circle of radius 1 within the complex plane. For example, limiting ourselves to the angles $m\cdot 90°$, $m = 0, 1, 2, 3, \ldots$ in the time evolution of a stationary state, function $\psi_n(x)$ is multiplied by: $1, i, -1, -i, \ldots$, respectively. It is a bit frustrating though, that the frequency of this rotation $\omega = \frac{E_n}{\hbar}$ depends on E_n because this number depends on you (you may add an arbitrary constant to the potential energy and the world will be functioning exactly as before). This is true, but whatever you compare with experimental results, you calculate $\Psi_m^*\ \Psi_n$, which annihilates the arbitrary constant and we get as time dependence: $\exp(-i\omega t)$ with $\omega = \frac{E_n - E_m}{\hbar}$. Already at this point, one can see that such oscillations might be damped by something, preferably of frequency just equal to ω. We will show this in detail in Chapter 2, and indeed it will turn out that such a damping by oscillating external electric field represents the condition to change the state Ψ_n to Ψ_m (and vice versa).

[32] There is a problem though. Experiments show that this is true only for the ground state, not for the excited states, which turn out to be quasi-stationary only. These experiments prove that in excited states, the system emits photons until it achieves the ground state. This excited state instability (which goes beyond the non-relativistic approximation that this book is focused on) comes from coupling with the electromagnetic field of the vacuum, the phenomenon ignored in presenting the postulates of quantum mechanics. The coupling is a real thing because there are convincing experiments showing that the vacuum is not just nothing. It is true that the vacuum's mean electric field is zero, but the electromagnetic field fluctuates even in the absence of photons (the mean square of the electric field does not equal zero).

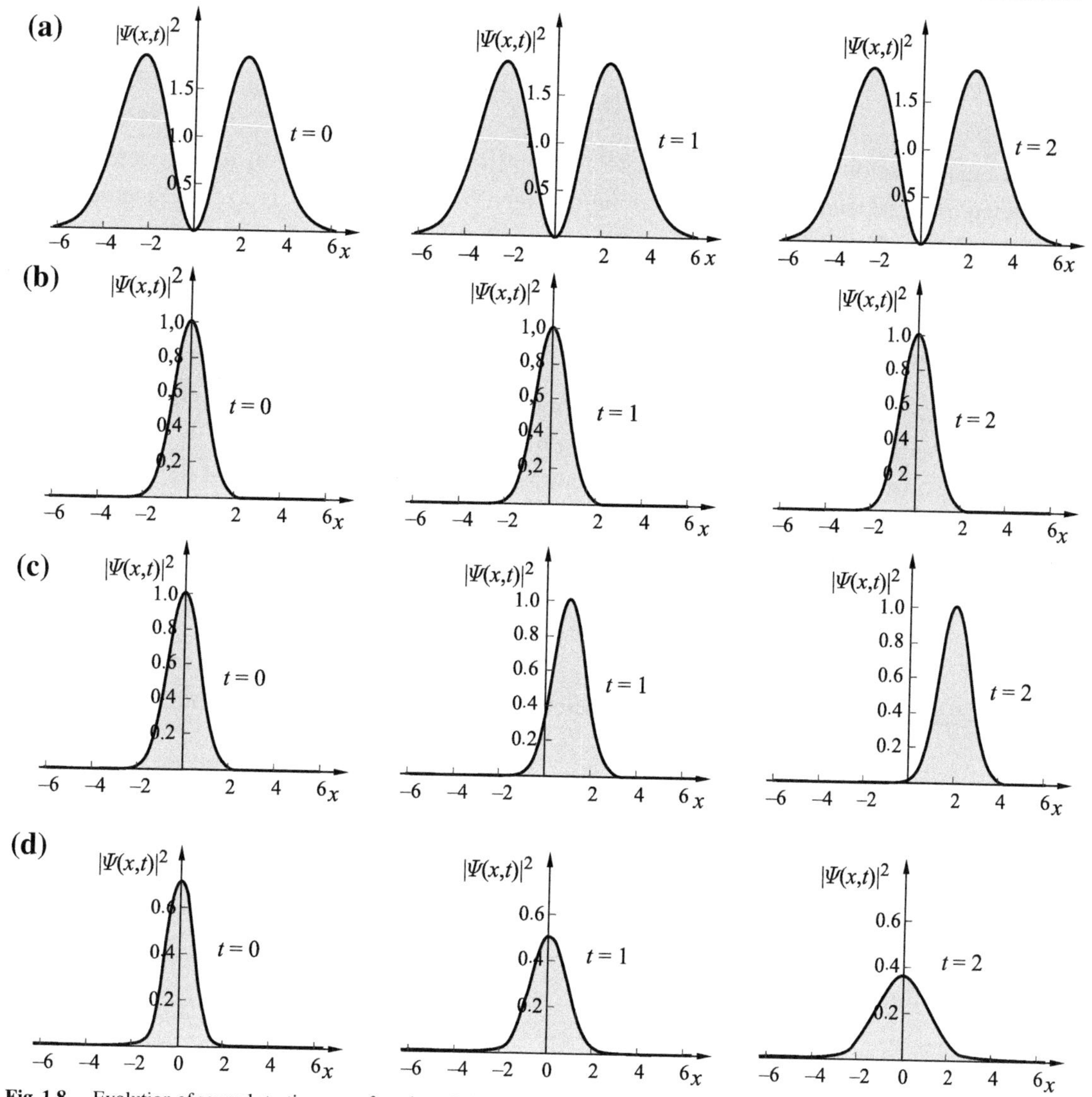

Fig. 1.8. Evolution of several starting wave functions Ψ (rows a-d) for a system shown as $|\Psi(x,t)|^2$ for three snapshots $t = 0, 1, 2$ (columns). In all cases, the area under $|\Psi(x,t)|^2$ equals 1 (normalization of $\Psi(x,t)$). Cases (a) and (b) show $|\Psi(x,t)|^2$ as time-independent; these are stationary states satisfying the time-independent Schrödinger equation. Contrary to this, in cases (c) and (d), function $|\Psi(x,t)|^2$ changes very much as time passes: in a translational motion with shape preserving in case (c) and irregularly in case (d). These are non-stationary states. The non-stationary states always represent linear combinations of stationary ones.

represented by its time-independent operator $\hat{A}$ and, for the sake of simplicity, that the system is composed of a single particle (with one variable only).

> The result of a single measurement of a mechanical quantity A can be *only* an eigenvalue a_k of the operator $\hat{A}$.

The eigenvalue equation for operator $\hat{A}$ reads

$$\hat{A}\phi_k = a_k\phi_k, \quad k = 1, 2, \ldots, M. \tag{1.15}$$

The eigenfunctions ϕ_k are orthogonal[33]. When the eigenvalues do not form a continuum, they are *quantized*, and then the corresponding eigenfunctions $\phi_k, k = 1, 2, \ldots, \infty$ satisfy the orthonormality relations[34]:

$$\int_{-\infty}^{\infty} \phi_k^*(x)\phi_l(x)\,dx \equiv \langle \phi_k \mid \phi_l \rangle \equiv \langle k \mid l \rangle = \delta_{kl} \equiv \begin{cases} 1 \text{ when } k = l, \\ 0 \text{ when } k \neq l, \end{cases} \tag{1.16}$$

where we have given several equivalent notations of the scalar product, which will be used in the present book, δ_{kl} is known as the *Kronecker delta.*

Since eigenfunctions $\{\phi_k\}$form the complete set, then the wave function of the system may be expanded as (M is quite often equal to ∞)

$$\psi = \sum_{k=1}^{M} c_k\phi_k, \tag{1.17}$$

where the c_k are in general complex coefficients. From the normalization condition for ψ we have[35]

$$\sum_{k=1}^{M} c_k^* c_k = 1. \tag{1.18}$$

According to the axiom, the probability that the result of the measurement is a_k is equal to $c_k^* c_k$.

If the wave function that describes the state of the system has the form given by Eq. (1.17) and does not reduce to a single term $\psi = \phi_k$, then the result of the measurement of the quantity A *cannot* be foreseen. We will measure *some* eigenvalue of the operator $\hat{A}$, but cannot predict which one. After the measurement is completed, the wave function of the system represents the eigenstate that corresponds to the measured eigenvalue (known as the *collapse of the wave function*). According to the postulate, the only thing to say about the measurements is that the

[33] If two eigenfunctions correspond to the same eigenvalue, they are not necessarily orthogonal, but they *can* still be orthogonalized (if they are linearly independent, as discussed in the Appendix J available at booksite.elsevier.com/978-0-444-59436-5, p. e99). Such orthogonal functions still remain the eigenfunctions of $\hat{A}$. Therefore, one can always construct the orthonormal set of the eigenfunctions of a Hermitian operator.

[34] If ϕ_k belongs to a continuum, they cannot be normalized, but they still can be made mutually orthogonal.

[35] $\langle \psi \mid \psi \rangle = 1 = \sum_{k=1}^{M} \sum_{l=1}^{M} c_k^* c_l \langle \phi_k \mid \phi_l \rangle = \sum_{k,l=1} c_k^* c_l \delta_{kl} = \sum_{k=1}^{M} c_k^* c_k$.

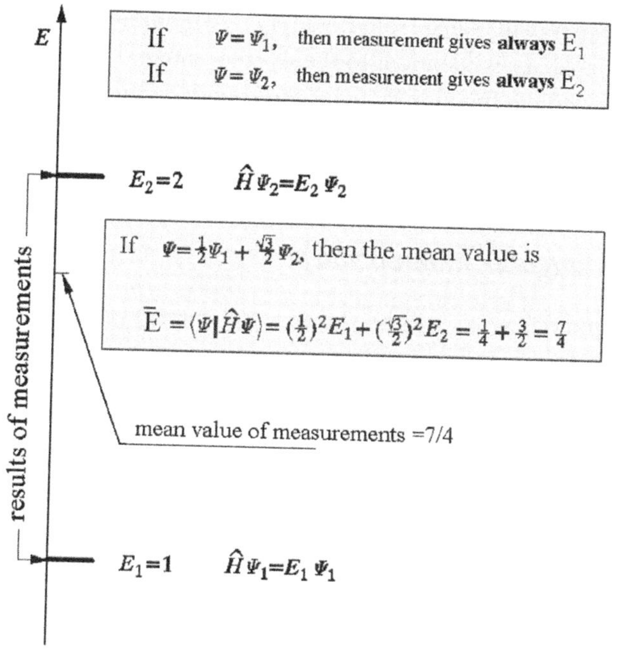

Fig. 1.9. The results of measurements of a quantity A are the eigenvalues of the operator $\hat{A}$.

mean value $\overline{a}$ of the quantity A (from many measurements) is to be compared with the following theoretical result[36] (Fig. 1.9).

$$\overline{a} = \sum_{k=1}^{M} c_k^* c_k a_k = \frac{\left\langle \psi \mid \hat{A}\psi \right\rangle}{\langle \psi \mid \psi \rangle}. \tag{1.19}$$

If we have a special case, $\psi = \phi_k$ (all coefficients $c_l = 0$, except $c_k = 1$), the measured quantity is exactly equal to a_k. From this, it follows that, if the wave function is an eigenfunction of operators of several quantities (this happens when the operators commute; see Appendix B available at booksite.elsevier.com/978-0-444-59436-5), then all these quantities when measured produce with certainty the eigenvalues corresponding to the eigenfunction.

[36] $\left\langle \psi \mid \hat{A}\psi \right\rangle = \left\langle \sum_{l=1}^{M} c_l\phi_l \mid \hat{A}\sum_{k=1}^{M} c_k\phi_k \right\rangle = \sum_{k=1}^{M}\sum_{l=1}^{M} c_k^* c_l \left\langle \phi_l \mid \hat{A}\phi_k \right\rangle = \sum_{k=1}^{M}\sum_{l=1}^{M} c_k^* c_l a_k \langle \phi_l \mid \phi_k \rangle = \sum_{k=1}^{M}\sum_{l=1}^{M} c_k^* c_l a_k \delta_{kl} = \sum_{k=1}^{M} c_k^* c_k a_k$ (here we assume that $\langle \psi \mid \psi \rangle = l$). In case of degeneracy ($a_k = a_l = \cdots$), the probability is $c_k^* c_k + c_l^* c_l + \cdots$ This is how one computes the mean value of anything. Just take all possible distinct results of measurements, multiply each by its probability, and sum up all resulting numbers.

The coefficients c can be calculated from Eq. (1.17). After multiplying by ϕ_l^* and integration, one has $c_l = \langle \phi_l \mid \psi \rangle$; i.e., c_l is identical to the overlap integral of the function ψ describing the state of the system and the function ϕ_l that corresponds to the eigenvalue a_l of the operator $\hat{A}$. In other words, the more the eigenfunction corresponding to a_l resembles the wave function ψ, the more frequently a_l will be measured.

Postulate V (spin angular momentum)

Spin of elementary particles. As will be shown in Chapter 3 (about relativistic effects), spin angular momentum will appear in a natural way. However, in nonrelativistic theory the existence of spin is postulated.[37]

Besides its orbital angular momentum $\boldsymbol{r} \times \boldsymbol{p}$, an elementary particle has an internal angular momentum (analogous to that associated with the rotation of a body about its own axis) called spin $\boldsymbol{S} = (S_x, S_y, S_z)$. Two quantities are measurable: the square of the spin length: $(|\boldsymbol{S}|^2 = S_x^2 + S_y^2 + S_z^2)$ and one of its components, by convention, S_z. These quantities only take some particular values: $|\boldsymbol{S}|^2 = s(s+1)\hbar^2$, $S_z = m_s \hbar$, where the spin magnetic quantum number $m_s = -s, -s+1, \ldots, s$. The spin quantum number s, characteristic of the type of particle (often called simply its spin), can be written as: $s = n/2$, where n may be zero or a natural number ("*an integer or half-integer*" number).

Enrico Fermi (1901–1954), Italian physicist and professor at universities in Florence, Rome, New York, and Chicago. Fermi introduced the notion of statistics for the particles with a half-integer spin number (called *fermions*) during the Florence period. Dirac made the same discovery independently, so this property is called the *Fermi-Dirac statistics*.

Young Fermi was notorious for being able to derive a formula from any domain of physics faster than someone using textbooks. His main topic was nuclear physics. He played an important role in the A bomb construction at Los Alamos, and in 1942 he built the world's first nuclear reactor on a tennis court at the University of Chicago. Fermi was awarded the Nobel Prize in 1938 "*for his demonstration of the existence of new radioactive elements and for results obtained with them, especially with regard to artificial radioactive elements.*"

[37] This has been forced by experimental facts; e.g., the energy level splitting in a magnetic field suggested two possible electron states connected to the internal angular momentum.

The particles with a half-integer[38] s (e.g., $s = \frac{1}{2}$ for electrons, protons, neutrons, and neutrinos) are called *fermions*, and the particles with an integer s (e.g., $s = 1$ for deuteron, photon[39]; $s = 0$ for meson π and meson K) are called *bosons*.

The magnetic[40] spin quantum number m_s quantizes the z component of the spin angular momentum.

Satyendra Nath Bose (1894–1974), Indian physicist and professor at Dakka and Calcutta, a polyglot and a self-taught expert in many fields; first recognized that particles with integer spin number have different statistical properties. Einstein contributed to a more detailed description of this statistic.

Thus,

a particle with spin quantum number s has an additional (spin) degree of freedom, or an additional coordinate–*spin coordinate* σ. The spin coordinate differs widely from a spatial coordinate because it takes only $2s + 1$ *discrete values* (Fig. 1.10) associated to $-s, -s+1, \ldots, 0, \ldots + s$.

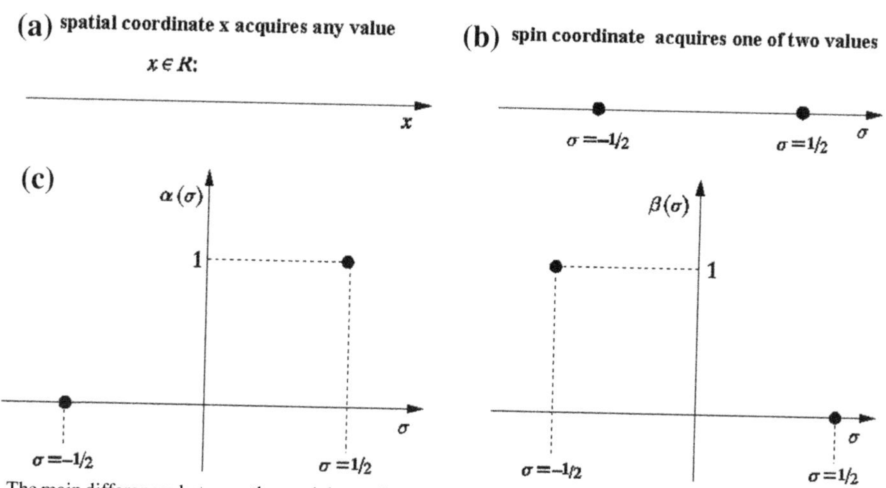

Fig. 1.10. The main differences between the spatial coordinate (x) and spin coordinate (σ) of an electron. (a) The spatial coordinate is *continuous:* it may take any value of a real number. (b) The spin coordinate σ has a granular character (*discrete* values): for $s = \frac{1}{2}$, it can take only one of two values. One of the values is represented by $\sigma = -\frac{1}{2}$, the other one by $\sigma = \frac{1}{2}$. Panel (c) shows two widely used basis functions in the spin space: $\alpha(\sigma)$ and $\beta(\sigma)$, respectively.

[38] Note that the length of the spin vector for an elementary particle is given by nature once and for all. Thus, if there is any relation between the spin and the rotation of the particle about its own axis, it has to be a special relation. One cannot change the angular momentum of such a rotation.

[39] The photon represents a particle of zero mass. One can show that instead of three possible m_s there are only two: $m_s = 1, -1$. We call these two possibilities *polarizations* ("*parallel*" and "*perpendicular*").

[40] This name is related to the energy level splitting in a magnetic field, from which the number is deduced. A nonzero s value is associated with the magnetic dipole, which in a magnetic field acquires $2s + 1$ energetically non-equivalent positions.

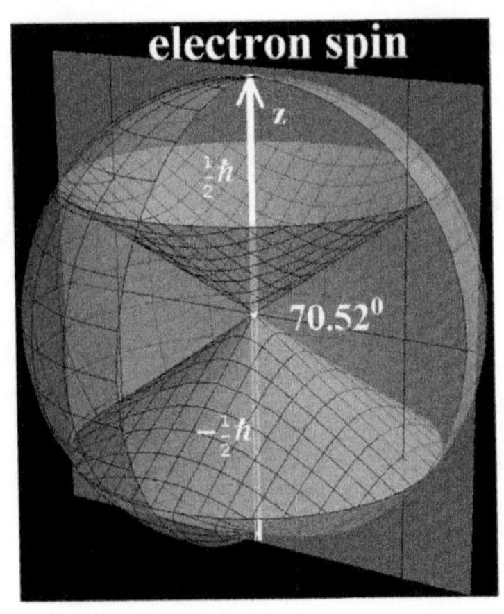

Fig. 1.11. Diagram of the spin angular momentum vector for a particle with spin quantum number $s = \frac{1}{2}$. The only measurable quantities are the spin length $\sqrt{s(s+1)}\hbar = \frac{\sqrt{3}}{2}\hbar$ and the projection of the spin on the quantization axis (chosen as coincident with the perpendicular axis z), which takes only the values $-s, -s+1, \ldots +s$ in units $\hbar$ i.e., $S_z = -\frac{1}{2}\hbar, \frac{1}{2}\hbar$. Since the x and y components of the spin remain indefinite, one may visualize the same by locating the spin vector (of constant length $\sqrt{s(s+1)}\hbar$) *anywhere* on a cone surface that assures a given z component. Thus, one has $2s + 1 = 2$ such cones.

Most often, one will have to deal with electrons. For electrons, the spin coordinate σ takes two values, often called "*up*" and "*down*." We will (arbitrarily) choose $\sigma = -\frac{1}{2}$ and $\sigma = +\frac{1}{2}$ Figs. (1.11).

According to the postulate (p. 26), the square of the spin length is always the same and equal to $s(s+1)\hbar^2 = \frac{3}{4}\hbar^2$. The maximum projection of a vector on a chosen axis is equal to $\frac{1}{2}\hbar$, while the length of the vector is larger, equal to $\sqrt{s(s+1)}\hbar = \frac{\sqrt{3}}{2}\hbar$. We conclude that the vector of the spin angular momentum makes an angle θ with the quantization axis, with $\cos\theta = \frac{1}{2}/\frac{\sqrt{3}}{2} = \frac{1}{\sqrt{3}}$. From this, one obtains[41] $\theta = \arccos\frac{1}{\sqrt{3}} \approx 54°44'$. Fig. 1.11 shows that the spin angular momentum has indefinite x and y components, while always preserving its length and projection on the z-axis.

Spin basis functions for $s = \frac{1}{2}$

One may define (see Fig. 1.10c) the complete set of orthonormal basis functions of the spin space of an electron:

$$\alpha(\sigma) = \begin{cases} 1 & \text{dla} \quad \sigma = \frac{1}{2} \\ 0 & \text{dla} \quad \sigma = -\frac{1}{2} \end{cases} \quad \text{and} \quad \beta(\sigma) = \begin{cases} 0 & \text{dla} \quad \sigma = \frac{1}{2} \\ 1 & \text{dla} \quad \sigma = -\frac{1}{2} \end{cases}$$

[41] In the general case, the spin of a particle may take the following angles with the quantization axis: $\arccos\frac{m_s}{\sqrt{s(s+1)}}$ for $m_s = -s, -s+1, \ldots +s$,.

or, in a slightly different notation, as orthogonal unit vectors[42]:

$$|\alpha\rangle = \begin{pmatrix} 1 \\ 0 \end{pmatrix}; \quad |\beta\rangle = \begin{pmatrix} 0 \\ 1 \end{pmatrix}.$$

Orthogonality follows from $\langle \alpha \mid \beta \rangle \equiv \sum_\sigma \alpha(\sigma)^* \beta(\sigma) = 0{\cdot}1 + 1{\cdot}0 = 0$. Similarly, normalization means that $\langle \alpha \mid \alpha \rangle \equiv \sum_\sigma \alpha(\sigma)^* \alpha(\sigma) = \alpha\left(-\frac{1}{2}\right)^* \alpha\left(-\frac{1}{2}\right) + \alpha\left(\frac{1}{2}\right)^* \alpha\left(\frac{1}{2}\right) = 0{\cdot}0 + 1{\cdot}1 = 1$, etc.

We shall now construct operators of the spin angular momentum.

The following definition of spin operators is consistent with the postulate about spin:

$$\hat{S}_x = \frac{1}{2}\hbar\sigma_x$$
$$\hat{S}_y = \frac{1}{2}\hbar\sigma_y$$
$$\hat{S}_z = \frac{1}{2}\hbar\sigma_z,$$

where the Pauli matrices of rank 2 are defined as

$$\sigma_x = \begin{pmatrix} 0 & 1 \\ 1 & 0 \end{pmatrix} \quad \sigma_y = \begin{pmatrix} 0 & -i \\ i & 0 \end{pmatrix} \quad \sigma_z = \begin{pmatrix} 1 & 0 \\ 0 & -1 \end{pmatrix}.$$

Indeed, after applying $\hat{S}_z$ to the spin basis functions, one obtains:

$$\hat{S}_z |\alpha\rangle \equiv \hat{S}_z \begin{pmatrix} 1 \\ 0 \end{pmatrix} = \frac{1}{2}\hbar \begin{pmatrix} 1 & 0 \\ 0 & -1 \end{pmatrix} \begin{pmatrix} 1 \\ 0 \end{pmatrix} = \frac{1}{2}\hbar \begin{pmatrix} 1 \\ 0 \end{pmatrix} = \frac{1}{2}\hbar |\alpha\rangle,$$
$$\hat{S}_z |\beta\rangle \equiv \hat{S}_z \begin{pmatrix} 0 \\ 1 \end{pmatrix} = \frac{1}{2}\hbar \begin{pmatrix} 1 & 0 \\ 0 & -1 \end{pmatrix} \begin{pmatrix} 0 \\ 1 \end{pmatrix} = \frac{1}{2}\hbar \begin{pmatrix} 0 \\ -1 \end{pmatrix} = -\frac{1}{2}\hbar |\beta\rangle.$$

Therefore, functions α and β represent the eigenfunctions of the $\hat{S}_z$ operator with corresponding eigenvalues $\frac{1}{2}\hbar$ and $-\frac{1}{2}\hbar$. How to construct the operator $\hat{S}^2$? From the Pythagorean theorem, after applying Pauli matrices,

[42] This is in the same spirit as wave functions represent vectors: vector components are values of the function for various values of the variable.

Wolfgang Pauli (1900–1958), German physicist and professor in Hamburg, Technical University of Zurich, Institute for Advanced Studies in Princeton, New Jersey, the son of a physical chemistry professor at Vienna and a classmate of Werner Heisenberg. At the age of 20, he wrote a famous article on relativity theory for *Mathematical Encyclopedia*, which was published later as a book. A year later, Pauli defended his doctoral dissertation under the supervision of Arnold Sommerfeld in Munich. The renowned Pauli exclusion principle was proposed in 1924. Pauli received the Nobel Prize in 1945 "*for the discovery of the Exclusion Principle, also called the Pauli Principle.*" Pauli was famous for his harsh opinions exclaimed during seminars ("*it is not even wrong*", "*I did not understand a single word*", etc.).

one obtains:

$$\begin{aligned}
\hat{S}^2 |\alpha\rangle &= \hat{S}^2 \begin{pmatrix} 1 \\ 0 \end{pmatrix} = \left(\hat{S}_x^2 + \hat{S}_y^2 + \hat{S}_z^2\right) \begin{pmatrix} 1 \\ 0 \end{pmatrix} \\
&= \frac{1}{4}\hbar^2 \left\{ \begin{matrix} \begin{pmatrix} 0 & 1 \\ 1 & 0 \end{pmatrix}\begin{pmatrix} 0 & 1 \\ 1 & 0 \end{pmatrix} + \begin{pmatrix} 0 & -i \\ i & 0 \end{pmatrix}\begin{pmatrix} 0 & -i \\ i & 0 \end{pmatrix} \\ + \begin{pmatrix} 1 & 0 \\ 0 & -1 \end{pmatrix}\begin{pmatrix} 1 & 0 \\ 0 & -1 \end{pmatrix} \end{matrix} \right\} \begin{pmatrix} 1 \\ 0 \end{pmatrix} \\
&= \frac{1}{4}\hbar^2 \begin{pmatrix} 1+1+1 & 0+0+0 \\ 0+0+0 & 1+1+1 \end{pmatrix} \begin{pmatrix} 1 \\ 0 \end{pmatrix} = \frac{3}{4}\hbar^2 \begin{pmatrix} 1 & 0 \\ 0 & 1 \end{pmatrix} \begin{pmatrix} 1 \\ 0 \end{pmatrix} \\
&= \frac{3}{4}\hbar^2 \begin{pmatrix} 1 \\ 0 \end{pmatrix} = \left[\frac{1}{2}\left(\frac{1}{2} + 1\right)\hbar^2\right] |\alpha\rangle .
\end{aligned}$$

The function $|\beta\rangle$ gives an identical result. These are, therefore, the pure states. However, one should remember that a particle also can be prepared in a mixed spin state, which is a common procedure in the modern nuclear magnetic resonance technique.

Therefore, both basis functions α and β represent the eigenfunctions of $\hat{S}^2$ and correspond to the same eigenvalue. Thus, the definition of spin operators through Pauli matrices gives results identical to those postulated for S^2 and S_z, and the two formulations are equivalent. From Pauli matrices, it follows that the functions α and β are not eigenfunctions of $\hat{S}_x$ and $\hat{S}_y$ and that the following relations are satisfied[43]:

$$\begin{aligned}
[\hat{S}^2, \hat{S}_z] &= 0, \\
[\hat{S}_x, \hat{S}_y] &= i\hbar\hat{S}_z, \\
[\hat{S}_y, \hat{S}_z] &= i\hbar\hat{S}_x,
\end{aligned}$$

43 The last three formulae are easy to memorize, since the sequence of the indices is always "*rotational*"; i.e., $x, y, z, x, y, z, \ldots$

$$[\hat{S}_z, \hat{S}_x] = i\hbar\hat{S}_y,$$

which is in agreement with the general properties of angular momenta[44] (Appendix on p. e73).

Spin of non-elementary particles. The postulate on spin pertains to an elementary particle. What about a system composed of such particles? Do such systems have spin? Spin represents angular momentum (a vector), and therefore, the angular momentum vectors of the elementary particles have to be added. A system composed of a number of elementary particles (each with its spin $\boldsymbol{s}_i$) has as an observable (a measurable quantity) the square

$$|\boldsymbol{S}|^2 = S(S+1)\hbar^2$$

of the total spin vector

$$\boldsymbol{S} = \boldsymbol{s}_1 + \boldsymbol{s}_2 + \cdots \boldsymbol{s}_N$$

and one of the components of $\boldsymbol{S}$ (denoted by S_z):

$$S_z = M_S\hbar, \text{ for } M_S = -S, -S+1, \ldots, S,$$

whereas the number S stands, just as for a single particle, for an integer or half-integer nonnegative number. Particular values of S (often called simply *spin*) and of the spin magnetic number M_S depend on the directions of vectors $\boldsymbol{s}_i$. It follows that no excitation of a nonelementary boson (that causes another summing of the individual spin vectors) can change the particle to a fermion and vice versa. Systems with an even number of fermions are always bosons, while those with an odd number of fermions are always fermions.

Nuclei. The ground states of the important nuclei ^{12}C and ^{16}O correspond to $S = 0$, while those of ^{13}C, ^{15}N, and ^{19}F have $S = \frac{1}{2}$.

Atoms and molecules. Does an atom as a whole represent a fermion or a boson? This depends on which atom and which molecule are involved. Consider the hydrogen atom, which is composed of two fermions (proton and electron, both with spin number $\frac{1}{2}$). This is sufficient to deduce that one is dealing with a boson. For similar reasons, the sodium atom, with 23 nucleons (each of spin $\frac{1}{2}$) in the nucleus and 11 electrons moving around it, also represents a boson.

When one adds together two electron spin vectors $\boldsymbol{s}_1 + \boldsymbol{s}_2$, then the maximum z component of the spin angular momentum will be (in $\hbar$ units): $|M_S| = m_{s1} + m_{s2} = \frac{1}{2} + \frac{1}{2} = 1$. This corresponds to the vectors $\boldsymbol{s}_1, \boldsymbol{s}_2$, called "*parallel*" to each other, while the minimum $|M_S| = m_{s1} + m_{s2} = \frac{1}{2} - \frac{1}{2} = 0$ is an "*antiparallel*" configuration of $\boldsymbol{s}_1$ and $\boldsymbol{s}_2$ (Fig. 1.12).

[44] Also, note that the mean values of S_x and S_y are both equal to zero in the α and β states; e.g., for the α state, one has $\left\langle \alpha \mid \hat{S}_x\alpha \right\rangle = \left\langle \begin{pmatrix} 1 \\ 0 \end{pmatrix} \mid \hat{S}_x \begin{pmatrix} 1 \\ 0 \end{pmatrix} \right\rangle = \frac{1}{2}\hbar \left\langle \begin{pmatrix} 1 \\ 0 \end{pmatrix} \mid \begin{pmatrix} 0 & 1 \\ 1 & 0 \end{pmatrix} \begin{pmatrix} 1 \\ 0 \end{pmatrix} \right\rangle = \frac{1}{2}\hbar \left\langle \begin{pmatrix} 1 \\ 0 \end{pmatrix} \mid \begin{pmatrix} 0 \\ 1 \end{pmatrix} \right\rangle = 0.$

This means that in an external vector field (of direction z), when the space is no longer isotropic, only the projection of the total angular momentum on the field direction is conserved. A way to satisfy this is to recall the behavior of a top in a gravitational field. The top rotates about its own axis, but the axis precesses about the field axis. This means that the total electron spin momentum moves on the cone surface, making an angle of 54°44′

(a)

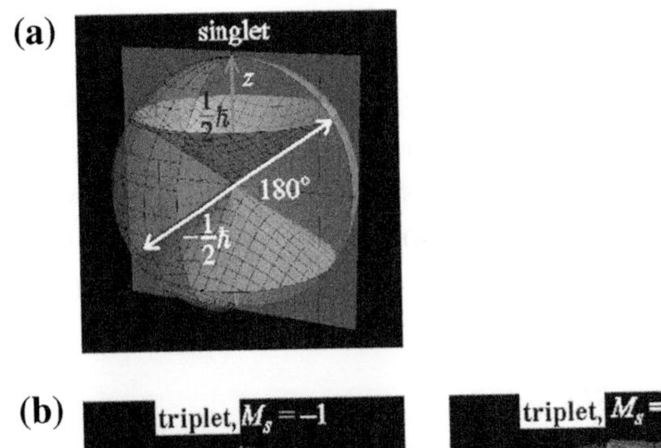

(b)

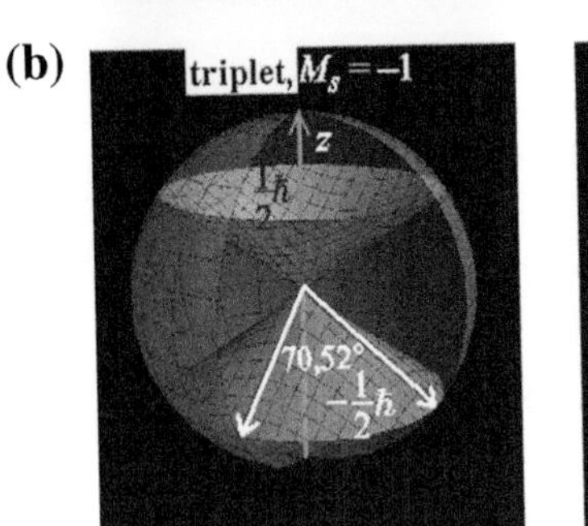

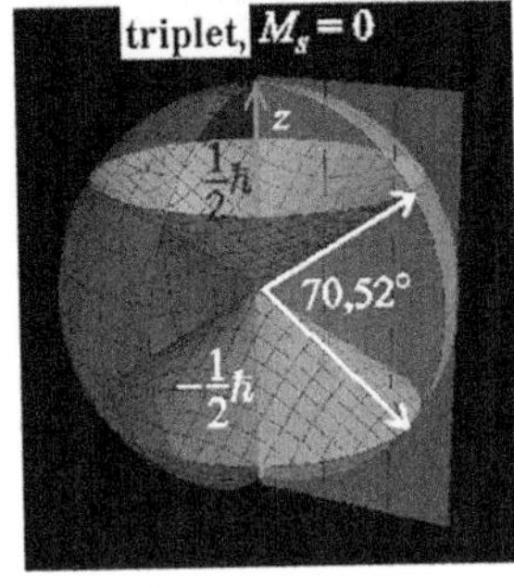

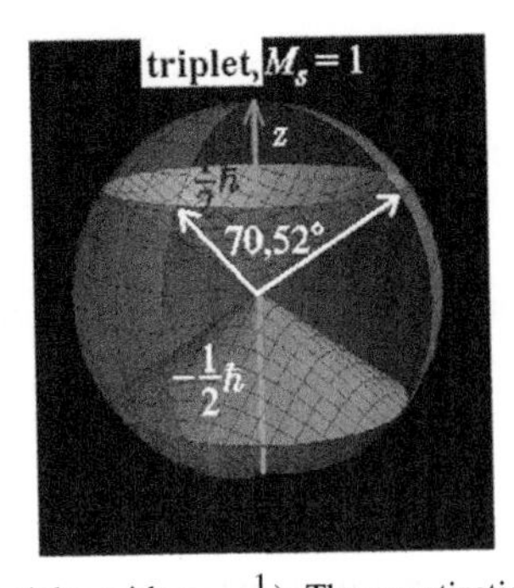

Fig. 1.12. Spin angular momentum for a system with two electrons (in general, particles with $s = \frac{1}{2}$). The quantization axis is arbitrarily chosen as the vertical axis z. Then, the spin vectors of individual electrons (see Fig. 1.11) may be thought to reside somewhere on the upper cone that corresponds to $m_{s1} = \frac{1}{2}$, or on the lower cone corresponding to $m_{s1} = -\frac{1}{2}$. For two electrons, there are two spin eigenstates of $\hat{S}^2$. One has total spin quantum number $S = 0$ (singlet state); the other is triply degenerate (triplet state), and the three components of the state have $S = 1$ and $S_z = 1, 0, -1$ in $\hbar$ units. In the singlet state, (a) the vectors s_1 and s_2 remain on the cones of different orientation and have the opposite (antiparallel) orientations, so that $s_1 + s_2 = \mathbf{0}$. Although their exact positions on the cones are undetermined (and moreover, the cones themselves follow from the arbitrary choice of the quantization axis in space), they are always pointing in opposite directions. The three triplet components (b) differ by the direction of the total spin angular momentum (of constant length $\sqrt{S(S+1)}\hbar = \sqrt{2}\hbar$). The three directions correspond to three projections $M_S\hbar$ of spin momentum. In each of the three cases, the angle between the two spins equals $\omega = 70.52^0$ (although in many textbooks–including this one–they are said to be parallel, but in fact, they are not, as is discussed in the text).

The first situation indicates that, for the state with parallel spins $S = 1$, and for this S, the possible $M_S = 1, 0, -1$. This means there are three states: $(S, M_S) = (1, 1), (1, 0), (1, -1)$. If no direction in space is privileged, then all the three states correspond to the same energy (triple degeneracy). This is why such a set of three states is called a *triplet state*. The second situation witnesses the existence of a state with $S = 0$, which obviously corresponds to $M_S = 0$. This state is called a *singlet state*.

---

with the external field axis in α state and an angle $180° - 54°44'$ in the β state. Whatever the motion, it must satisfy $\left\langle \alpha \mid \hat{S}_x\alpha \right\rangle = \left\langle \alpha \mid \hat{S}_y\alpha \right\rangle = 0$ and $\left\langle \beta \mid \hat{S}_x\beta \right\rangle = \left\langle \beta \mid \hat{S}_y\beta \right\rangle = 0$. No more information is available, but one may imagine the motion as a precession just like that of the top.

Let us calculate the angle ω between the individual electronic spins:

$$\begin{aligned}|\mathbf{S}|^2 &= \left(\mathbf{s}_1+\mathbf{s}_2\right)^2 = s_1^2+s_2^2+2\mathbf{s}_1\cdot\mathbf{s}_2 = s_1^2+s_2^2+2s_1\cdot s_2\cos\omega\\ &= \frac{1}{2}\left(\frac{1}{2}+1\right)\hbar^2\cdot 2+2\sqrt{\frac{1}{2}\left(\frac{1}{2}+1\right)}\sqrt{\frac{1}{2}\left(\frac{1}{2}+1\right)}\hbar^2\cos\omega\\ &= \left(\frac{3}{2}+\frac{3}{2}\cos\omega\right)\hbar^2 = \frac{3}{2}\left(1+\cos\omega\right)\hbar^2.\end{aligned}$$

Singlet and Triplet States:
For the singlet state, $|\mathbf{S}|^2 = S\left(S+1\right)\hbar^2 = 0$, so $1+\cos\omega = 0$ and $\omega = 180°$. This means that the two electronic spins in the singlet state are antiparallel. For the triplet state, $|\mathbf{S}|^2 = S\left(S+1\right)\hbar^2 = 2\hbar^2$, and so $\frac{3}{2}\left(1+\cos\omega\right)\hbar^2 = 2\hbar^2$ i.e., $\cos\omega = \frac{1}{3}$, or $\omega = 70.52°$, see Fig. 1.12. Despite forming the angle $\omega = 70.52°$ the two spins in the triplet state are said to be parallel.

The two electrons that we have considered may, for example, be part of a hydrogen molecule. Therefore, when considering electronic states, we may have to deal with singlets or triplets. However, in the same hydrogen molecule, we have two protons, whose spins may also be parallel (orthohydrogen) or antiparallel (parahydrogen). In parahydrogen the nuclear spin is $S=0$, while in orthohydrogen $S=1$. In consequence, there is only one state for parahydrogen ($M_S=0$), and three states for orthohydrogen ($M_S=1,0,-1$).[45]

Postulate VI (on the permutational symmetry)

Unlike classical mechanics, quantum mechanics is radical: it requires that two particles of the same kind (two electrons, two protons, etc.) should play the same role in the system, and therefore also in their description enshrined in the wave function.[46] Quantum mechanics *guarantees* that the roles played in the Hamiltonian by two identical particles are identical. By this philosophy, exchange of the labels of two identical particles [i.e., the exchange of their coordinates $x_1, y_1, z_1, \sigma_1 \leftrightarrow x_2, y_2, z_2, \sigma_2$, in short, $1 \leftrightarrow 2$] leads, at most, to a change of the phase ϕ of the wave function: $\psi\left(2,1\right) \rightarrow e^{i\phi}\psi\left(1,2\right)$, because in such a case, $|\psi(2,1)| = |\psi(1,2)|$ (and this guarantees equal probabilities of both situations). However, when we exchange the two labels once more, we have to return to the initial situation: $\psi\left(1,2\right) =$

[45] Since all the states have very similar energy (and therefore at high temperatures, the Boltzmann factors are practically the same), there are three times as many molecules of orthohydrogen as of parahydrogen. Both states (ortho and para) differ slightly in their physicochemical characteristics.

[46] Everyday experience tells us the opposite; e.g., a car accident involving a Mercedes does not cause all editions of that particular model to have identical crash traces.

$e^{i\phi}\psi(2,1) = e^{i\phi}e^{i\phi}\psi(1,2) = \left(e^{i\phi}\right)^2\psi(1,2)$. Hence, $\left(e^{i\phi}\right)^2 = 1$, i.e., $e^{i\phi} = \pm 1$. Postulate VI says that $e^{i\phi} = +1$ refers to bosons, while $e^{i\phi} = -1$ refers to fermions.

> The wave function ψ which describes identical *bosons* i and j has to be *symmetric* with respect to the exchange of coordinates x_i, y_i, z_i, σ_i and x_j, y_j, z_j, σ_j; i.e., if $x_i \leftrightarrow x_j$, $y_i \leftrightarrow y_j$, $z_i \leftrightarrow z_j$, $\sigma_i \leftrightarrow \sigma_j$, then $\psi(1,2,\ldots,i,\ldots j,\ldots,N) = \psi(1,2,\ldots,j,\ldots i,\ldots,N)$. If particles i and j denote identical *fermions*, the wave function must be *antisymmetric*; i.e., $\psi(1,2,\ldots,i,\ldots j,\ldots,N) = -\psi(1,2,\ldots,j,\ldots i,\ldots,N)$.

Let us see the probability density that two fermions (particles 1 and 2) occupy the same position in space, and additionally, that they have the same spin coordinate $(x_1, y_1, z_1, \sigma_1) = (x_2, y_2, z_2, \sigma_2)$. We have $\psi(1,1,3,4,\ldots,N) = -\psi(1,1,3,4,\ldots,N)$; hence $\psi(1,1,3,4,\ldots,N) = 0$ and, of course, $|\psi(1,1,3,4,\ldots,N)|^2 = 0$. Conclusion: two electrons of the same spin coordinate (we will sometimes say: "*of the same spin*") avoid each other. This is called the *exchange* or *Fermi hole* around each electron.[47] The reason for the hole is the antisymmetry of the electronic wave function, or in other words, the Pauli exclusion principle.[48]

---

[47] Electrons represent fermions $\left(s = \frac{1}{2}, \sigma = +\frac{1}{2}, -\frac{1}{2}\right)$ and therefore, there are two forms of repulsion among them: a Coulomb one because of their electric charge, and that one between same spin electrons only, which follows from the Pauli exclusion principle. As it will be shown in Chapter 11, the Coulomb repulsion is much less important than the effect of the Pauli exclusion principle.

Positions of the nuclei determine those parts of the space, which are preferred by electrons (just because of the Coulomb attraction electron-nucleus): the vicinity of the nuclei. However, considering one such part (a particular nucleus), the probability of finding electrons of the same spin coordinate is negligible (from the Pauli exclusion principle together with the continuity condition for the wave function). However, the Pauli exclusion principle does not pertain to the electrons of opposite spin coordinates, therefore two such electrons can occupy the same small volume even despite the energy increase because of their Coulomb repulsion. Hence, we have a compromise of two opposite effects: the electronic pair is *attracted* by a nucleus (proportionally to the nuclear charge), but the interelectronic distance should not be too small, because of the electron-electron Coulomb repulsion. A third electron seeking its low energy in the vicinity of this nucleus has no chance: because its spin coordinate is necessarily the same as that of one of the electrons. The third electron has therefore to quit this area even at the expense of a large increase in energy, and find another low-energy (Coulomb) region, maybe together with his colleague of the opposite spin (thus forming another electron pair). This picture does not mean that one is able to discern the electrons–they play the same role in the system, this is guaranteed by the antisymmetry of the wave function. Therefore, we will not know *which* electrons form a particular electronic pair, but only that there are two of them in it and they have the *opposite* spin coordinates.

Now let us turn to molecules. For electrons, which are very weakly bound (valence electrons) a space with Coulomb attraction of *two* nuclei might be a good choice (we will see this in Chapter 8). This space may also be shared by two electrons of the opposite spins, since this is still better than to send the partner elsewhere in space.

Thus, already at this stage, we foresee a fundamental role of electronic pairs (the opposite spin coordinates within the pair) leading to the electronic shell structure in atoms and molecules and to chemical bonds in molecules.

[48] The Pauli exclusion principle is sometimes formulated in another way: two electrons cannot be in the same state (including spin). The connection of this strange phrasing (what does "*electron state*" mean?) with the above will become clear in Chapter 8.

Thus, the probability density of finding two identical fermions in the same position and with the same spin coordinate is equal to zero. There is no such restriction for two identical bosons or two identical fermions with *different* spin coordinates. They can be at the same point in space.

This is related to what is known as *Bose condensation*.[49]

***

Among the above postulates, the strongest controversy has always been associated with postulate IV, which says that, except for some special cases, one cannot predict the result of a particular single measurement, but only its probability. More advanced considerations devoted to postulate IV lead to the conclusion that there is no way (neither experimental protocol nor theoretical reasoning), to predict when and in which direction an excited atom will emit a photon. This means that quantum mechanics is not a deterministic theory.

The indeterminism appears, however, only in the physical space, while in the space of all states (Hilbert space), everything is perfectly deterministic. The wave function evolves in a deterministic way according to the time-dependent Schrödinger Eq. (1.10).

The puzzling way in which indeterminism operates will be shown below.

1.3 The Heisenberg Uncertainty Principle

Consider two mechanical quantities A and B, for which the corresponding Hermitian operators (constructed according to Postulate II), $\hat{A}$ and $\hat{B}$, give the commutator $[\hat{A}, \hat{B}] = \hat{A}\hat{B} - \hat{B}\hat{A} = i\hat{C}$, where $\hat{C}$ is a Hermitian operator.[50] This is what happens, for example, for $A = x$ and $B = p_x$. Indeed, for any differentiable function ϕ, one has $[\hat{x}, \hat{p_x}]\phi = -xi\hbar\phi' + i\hbar(x\phi)' = i\hbar\phi$, and therefore, the operator $\hat{C}$ in this case means simply multiplication by $\hbar$.

[49] This was carried out by Eric A. Cornell, Carl E. Wieman, and Wolfgang Ketterle (who received the Nobel Prize in 2001 *"for discovering a new state of matter"*). In the Bose condensate, the bosons (alkali metal atoms) are in the same place in a peculiar sense. The total wave function for the bosons was, to a first approximation, a product of *identical* nodeless wave functions for the particular bosons (this ensures proper symmetry). Each of the wave functions extends considerably in space (the Bose condensate is as large as a fraction of a millimetre), but all have been centered in the same point in space.

[50] This is guaranteed. Indeed, $\hat{C} = -i\left[\hat{A}, \hat{B}\right]$ and then the Hermitian character of $\hat{C}$ is shown by the following chain of transformations: $\left\langle f \mid \hat{C}g\right\rangle = -i\left\langle f \mid \left[\hat{A}, \hat{B}\right]g\right\rangle = -i\left\langle f \mid (\hat{A}\hat{B} - \hat{B}\hat{A})g\right\rangle = -i\left\langle(\hat{B}\hat{A} - \hat{A}\hat{B})f \mid g\right\rangle = \left\langle -i(\hat{A}\hat{B} - \hat{B}\hat{A})f \mid g\right\rangle = \left\langle \hat{C}f \mid g\right\rangle$.

From axioms of quantum mechanics, one can prove that a product of errors (in the sense of standard deviation) of measurements of two mechanical quantities is greater or equal to $\frac{1}{2}\left\langle[\hat{A}, \hat{B}]\right\rangle$, where $\left\langle[\hat{A}, \hat{B}]\right\rangle$ is the mean value of the commutatator $[\hat{A}, \hat{B}]$.

This is known as the *Heisenberg uncertainty principle.*

Werner Karl Heisenberg (1901–1976) was born in Würzburg (Germany), attended high school in Munich, and then (with his friend Wolfgang Pauli) studied physics at the University of Munich under Arnold Sommerfeld's supervision. In 1923, he defended his doctoral thesis on turbulence in liquids. Reportedly, during the doctoral examination, he had problems writing down the chemical reaction in lead batteries. He joined the laboratory of Max Born at Göttingen (following his friend Pauli) and in 1924, he joined the Institute of Theoretical Physics in Copenhagen, working under the supervision of Niels Bohr. A lecture delivered by Niels Bohr decided the future direction of his work. Heisenberg later wrote: "*I was taught optimism by Sommerfeld, mathematics in Göttingen, physics by Bohr.*" In 1925 (only a year after being convinced by Bohr), Heisenberg developed a formalism that became the first successful quantum theory. Then, in 1926, Heisenberg, Born, and Jordan elaborated the formalism, which resulted in a coherent theory ("*matrix mechanics*"). In 1927, Heisenberg obtained a chair at Leipzig University, which he held until 1941 (when he became director of the Kaiser Wilhelm Physics Institute in Berlin). Heisenberg received the Nobel Prize in 1932 "*for the creation of quantum mechanics, the application of which has, inter alia, led to the discovery of the allotropic forms of hydrogen.*"

In 1937, Werner Heisenberg was at the height of his powers. He became a professor and got married. However, just after returning from his honeymoon, the president of the university called him, saying that there was a problem. In Der Stürmer, an article by Professor Johannes Stark (a Nobel Prize winner and faithful Nazi) was about to appear, claiming that Professor Heisenberg is not such a good patriot as he pretended because he socialized in the past with Jewish physicists. Soon Professor Heisenberg was invited to SS headquarters at Prinz Albert Strasse in Berlin. The interrogation took place in the basement. On the raw concrete wall, there was the scoffing slogan "*Breathe deeply and quietly.*" One of the questioners was a Ph.D. student from Leipzig, who had once been examined by Heisenberg. The terrified Heisenberg told his mother about the problem. She recalled that in her youth, she had made the acquaintance of Heinrich Himmler's mother. Frau Heisenberg paid a visit to Frau Himmler and asked her to pass a letter from her son to Himmler. At the beginning, Himmler's mother tried to separate her maternal feelings for her beloved son from politics. She was finally convinced after Frau Heisenberg said "*We mothers should care about our boys.*" After a certain time, Heisenberg received a letter from Himmler saying that his letter "*coming through unusual channels*" has been examined especially carefully. He promised to stop the attack. In the *post scriptum* there was a precisely tailored phrase: "*I think it best for your future, if for the benefit of your students, you would carefully separate scientific achievements from the personal and political beliefs of those who carried them out. Your faithfully, Heinrich Himmler*" (after D. Bodanis, "*$E = mc^2$*", Fakty, Warsaw, 2001, p. 130).

Werner Heisenberg did not carry out any formal proof. Instead, he analyzed a *Gedankenexperiment* (an imaginary ideal experiment) with an electron interacting with an electromagnetic wave ("*Heisenberg's microscope*").

The formal proof goes as follows.

Recall the definition of the variance, or the square of the standard deviation $(\Delta A)^2$, of measurements of the quantity A:

$$(\Delta A)^2 = \langle \hat{A}^2 \rangle - \langle \hat{A} \rangle^2, \tag{1.20}$$

where $\langle \hat{X} \rangle$ means the mean value of many measurements of the quantity X. The standard deviation ΔA represents the width of the distribution of A; i.e., the error made. Eq. (1.20) is equivalent to[51]

$$(\Delta A)^2 = \langle (\hat{A} - \langle \hat{A} \rangle)^2 \rangle. \tag{1.21}$$

Consider the product of the standard deviations for the operators $\hat{A}$ and $\hat{B}$, taking into account that $\langle \hat{u} \rangle$ denotes (Postulate IV) the integral $\langle \Psi \mid \hat{u} \mid \Psi \rangle$ according to Eq. (1.19). One obtains (denoting $\hat{\mathcal{A}} = \hat{A} - \langle \hat{A} \rangle$ and $\hat{\mathcal{B}} = \hat{B} - \langle \hat{B} \rangle$; and $[\hat{\mathcal{A}}, \hat{\mathcal{B}}] = [\hat{A}, \hat{B}]$):

$$(\Delta A)^2 \cdot (\Delta B)^2 = \langle \Psi \mid \hat{\mathcal{A}}^2 \Psi \rangle \langle \Psi \mid \hat{\mathcal{B}}^2 \Psi \rangle = \langle \hat{\mathcal{A}} \Psi \mid \hat{\mathcal{A}} \Psi \rangle \langle \hat{\mathcal{B}} \Psi \mid \hat{\mathcal{B}} \Psi \rangle,$$

where the Hermitian character of the operators $\hat{\mathcal{A}}$ and $\hat{\mathcal{B}}$ is used. Now, let us use the Schwartz inequality (see Appendix B available at booksite.elsevier.com/978-0-444-59436-5) $\langle f_1 \mid f_1 \rangle \langle f_2 \mid f_2 \rangle \geq |\langle f_1 \mid f_2 \rangle|^2$:

$$(\Delta A)^2 \cdot (\Delta B)^2 = \langle \hat{\mathcal{A}} \Psi \mid \hat{\mathcal{A}} \Psi \rangle \langle \hat{\mathcal{B}} \Psi \mid \hat{\mathcal{B}} \Psi \rangle \geq |\langle \hat{\mathcal{A}} \Psi \mid \hat{\mathcal{B}} \Psi \rangle|^2.$$

Next,

$$\begin{aligned} \langle \hat{\mathcal{A}} \Psi \mid \hat{\mathcal{B}} \Psi \rangle &= \langle \Psi \mid \hat{\mathcal{A}} \hat{\mathcal{B}} \Psi \rangle = \langle \Psi | \{ [\hat{\mathcal{A}}, \hat{\mathcal{B}}] + \hat{\mathcal{B}} \hat{\mathcal{A}} \} \Psi \rangle = i \langle \Psi \mid \hat{C} \Psi \rangle + \langle \Psi \mid \hat{\mathcal{B}} \hat{\mathcal{A}} \Psi \rangle \\ &= i \langle \Psi \mid \hat{C} \Psi \rangle + \langle \hat{\mathcal{B}} \Psi \mid \hat{\mathcal{A}} \Psi \rangle = i \langle \Psi \mid \hat{C} \Psi \rangle + \langle \hat{\mathcal{A}} \Psi \mid \hat{\mathcal{B}} \Psi \rangle^*. \\ \text{Hence,} \quad & i \langle \Psi \mid \hat{C} \Psi \rangle = 2i \, \mathrm{Im}\{ \langle \hat{\mathcal{A}} \Psi \mid \hat{\mathcal{B}} \Psi \rangle \}. \end{aligned}$$

This means that $\mathrm{Im}\{\langle \hat{\mathcal{A}} \Psi \mid \hat{\mathcal{B}} \Psi \rangle\} = \frac{\langle \Psi | \hat{C} \Psi \rangle}{2}$, which gives $|\langle \hat{\mathcal{A}} \Psi \mid \hat{\mathcal{B}} \Psi \rangle| \geq \frac{|\langle \Psi | \hat{C} \Psi \rangle|}{2}$. Hence,

$$(\Delta A)^2 \cdot (\Delta B)^2 \geq |\langle \hat{\mathcal{A}} \Psi \mid \hat{\mathcal{B}} \Psi \rangle|^2 \geq \frac{|\langle \Psi \mid \hat{C} \Psi \rangle|^2}{4}, \tag{1.22}$$

or, taking into account that $|\langle \Psi \mid \hat{C} \Psi \rangle| = |\langle \Psi | [\hat{A}, \hat{B}] \Psi \rangle|$, we have

$$\Delta A \cdot \Delta B \geq \frac{1}{2} |\langle \Psi \mid [\hat{A}, \hat{B}] \Psi \rangle|. \tag{1.23}$$

[51] This is because $\langle (\hat{A} - \langle \hat{A} \rangle)^2 \rangle = \langle \hat{A}^2 - 2\hat{A} \langle \hat{A} \rangle + \langle \hat{A} \rangle^2 \rangle = \langle \hat{A}^2 \rangle - 2 \langle \hat{A} \rangle^2 + \langle \hat{A} \rangle^2 = \langle \hat{A}^2 \rangle - \langle \hat{A} \rangle^2$.

There are two important special cases:

(a) $\hat{C} = 0$; i.e., the operators $\hat{A}$ and $\hat{B}$ commute. We have $\Delta A \cdot \Delta B \geq 0$; i.e., the errors can be arbitrarily small. Both quantities therefore can be measured simultaneously without error.

(b) $\hat{C} = \hbar$, as in the case of $\hat{x}$ and $\hat{p}_x$.

Then, $(\Delta A) \cdot (\Delta B) \geq \frac{\hbar}{2}$.

In particular, for $\hat{A} = \hat{x}$ and $\hat{B} = \hat{p}_x$, if quantum mechanics is valid, one cannot measure the exact position and the exact momentum of a particle. When precision with which x is measured increases, the particle's momentum has so wide a distribution that the error of determining p_x is huge (see Fig. 1.13).[52]

The power of the Heisenberg uncertainty principle is seen, when it is used for estimation, just from scratch, of the size of some systems.[53]

Example: Size of the Hydrogen Atom

How on earth can this be estimated from virtually no information? Let us see.

We assume that the electron moves, while the nucleus does not.[54] Whatever the electron does in the hydrogen atom, it has to conform to the uncertainty principle:

$\Delta x \cdot \Delta p_x \geq \frac{\hbar}{2}$; i.e., in the most compact state, we may expect the equality $\Delta x \cdot \Delta p_x = \frac{\hbar}{2}$. We may estimate Δx as the radius r of the atom, while $\Delta p_x = \sqrt{\langle p_x^2 \rangle - \langle p_x \rangle^2} = \sqrt{\langle p_x^2 \rangle - 0} = \sqrt{\langle p_x^2 \rangle}$. Therefore, we have an estimation

$r \cdot \sqrt{\langle p_x^2 \rangle} = \frac{\hbar}{2}$, or $\sqrt{\langle p_x^2 \rangle} = \frac{\hbar}{2r}$. The total energy may be estimated as the sum of the kinetic and potential energies: $E = \frac{\langle p^2 \rangle}{2m} - \frac{e^2}{r} = \frac{\langle p_x^2 + p_y^2 + p_z^2 \rangle}{2m} - \frac{e^2}{r} = \frac{3\hbar^2}{8mr^2} - \frac{e^2}{r}$. Now, let us find the minimum of $E(r)$ as its probable value: $\frac{\mathrm{d}E}{\mathrm{d}r} = 0 = -2\frac{3\hbar^2}{8mr^3} + \frac{e^2}{r^2}$, or $e^2 = \frac{3}{4}\frac{\hbar^2}{mr}$. Hence, we have an estimation: $r = \frac{3}{4}\frac{\hbar^2}{me^2} = 0.75 \cdot a_0$, where $a_0 = 0.529$ Å (as will be shown in Chapter 4, Eq. (4.41), p. 202) is known as the "*Bohr first orbit radius of the hydrogen atom.*" Thus, just from the Heisenberg uncertainty principle, we got a value, which has the correct order of magnitude!

Example: Size of Nucleus

How large is an atomic nucleus? Well, again, we may estimate its size from the Heisenberg uncertainty principle knowing only one thing: that the binding energy per nucleon is of the

---

[52] There is an apocryphal story about a Polizei patrol stopping Professor Heisenberg for speeding. The very serious man asks: "*Do you know how fast you were going when I stopped you?*" Professor Heisenberg answered: "*I have absolutely no idea, Herr Oberleutnant, but I can tell you precisely where you stopped me.*"

[53] Energy and mass are also included here.

[54] Note that from the momentum conservation law, we have that the nucleus moves 1840 times slower than the electron. This practically means that the electron moves in the electric field of the immobilized nucleus.

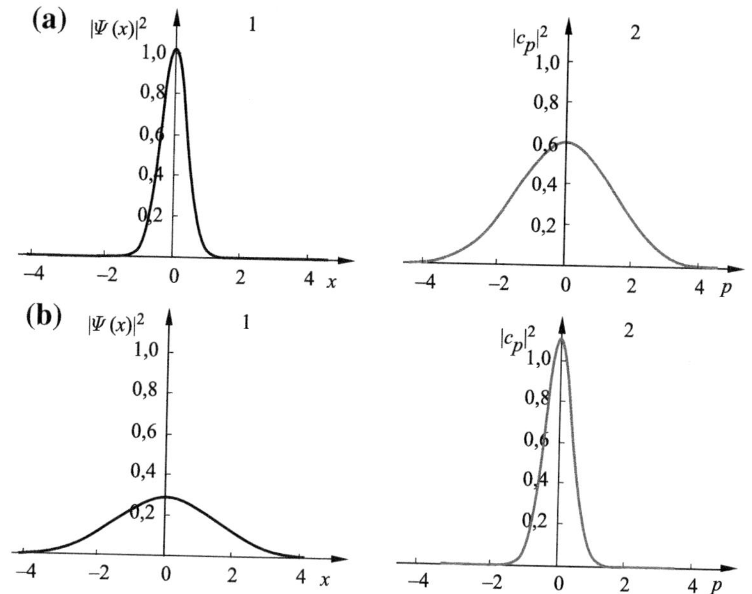

Fig. 1.13. Illustration of the Heisenberg uncertainty principle in the case of a single particle moving on the x axis. (a1) $|\Psi(x)|^2$ (i.e., the probability density distribution of finding the particle) as a function of coordinate x. The width of this distribution is related to the expected error of determining x. The curve is narrow, which means that the error in determining the particle's position is expected to be small. Wave function $\Psi(x)$ can be expanded in the infinite series $\Psi(x) = \sum_p c_p \exp(ipx)$, where p denotes the momentum. Note that each individual function $\exp(ipx)$ is an eigenfunction of momentum, and therefore, if $\Psi(x) = \exp(ipx)$, a measurement of momentum would give exactly p. If, however, $\Psi(x) = \sum_p c_p \exp(ipx)$, then such a measurement yields a given p with the probability $|c_p|^2$. Panel (a2) shows c_p as function of p. As one can see, a broad range of p (large uncertainty of momentum) ensures a sharp $|\Psi(x)|^2$ distribution (small uncertainty of position). Simply, the waves $\exp(ipx)$ to obtain a sharp peak of $\Psi(x)$ should exhibit a perfect constructive interference in a small region and a perfect destructive interference elsewhere. This requires a lot of different p's i.e., a broad momentum distribution. Panels (b1-b2) show the same, but this time, a narrow p distribution gives a broad x distribution. This means that a small error in determining the particle's position is necessarily associated with a large error in determining the particle's momentum (Heisenberg uncertainty principle). The principle has nothing mysterious in it and does not represent a unique feature of quantum mechanics. If one wanted to construct from the ocean waves (of various directions and wavelengths) a tall water pole in the middle of the ocean, one would be forced to use many wavelengths to get the constructive interference in one spot and destructive in all the others.

order of 8 MeV. We are now in the realm of nuclear forces acting among nucleons. It seems, therefore, that our task is extremely difficult... We just ignore all these extremely complex forces, considering instead a single nucleon moving in a mean potential, and bound with the energy $E = 8$ MeV, and therefore has to be of the same order that its kinetic energy. We focus on this kinetic energy[55] of the nucleon, assuming simply that $E = \frac{\langle p^2 \rangle}{2m_N} = \frac{\langle (\Delta p)^2 \rangle}{2m_N}$. From this, we calculate $\sqrt{\langle (\Delta p)^2 \rangle} = \Delta p = \sqrt{2m_N E}$ and, using the Heisenberg uncertainty principle, we have the uncertainty of the nucleon's position as $a = \frac{\hbar}{2\Delta p} = \frac{\hbar}{2\sqrt{2m_N E}}$. Calculating this in atomic units, we get ($\hbar = 1, e = 1, m = 1$, where m is the electron mass: $a = \frac{\hbar}{2\sqrt{2m_N E}} = \frac{1}{2\sqrt{2 \cdot 1840 \cdot 8 \cdot 10^6 \cdot 3.67516 \cdot 10^{-2}}} = 0.0000152$ a.u.$\approx 10^{-13}$ cm, 100000 times smaller than

[55] We are interested in the order of its magnitude only.

the hydrogen atom. Experiments confirm our number: this is indeed the order of magnitude of the sizes of atomic nuclei.

1.4 The Copenhagen Interpretation of the World[56]

The picture of the world that emerged from quantum mechanics was "*diffuse*" (Heisenberg uncertainty principle) with respect to classical mechanics. In classical mechanics one could measure a particle's position and momentum with a desired accuracy,[57] whereas the Heisenberg uncertainty principle states that this is simply impossible.

Bohr presented a philosophical interpretation of the world, which at its foundation had in a sense a non-reality of the world.

> According to Bohr, before a measurement on a particle is made, *nothing* can be said about the value of a given mechanical quantity, unless the wave function represents an eigenfunction of the operator of this mechanical quantity. Moreover, except this case, the particle does not have any fixed value of mechanical quantity at all.

A measurement gives a value of the mechanical property (A). Then, according to Bohr, after the measurement is completed, the state of the system changes (the called wave function collapse or decoherence) to the state described by an eigenfunction of the corresponding operator $\hat{A}$, and as the measured value one obtains the eigenvalue corresponding to the wave function. According to Bohr, there is no way to foresee which eigenvalue one will get as the result of the measurement. However, one can calculate the *probability*of getting a particular eigenvalue. This probability may be computed as the square of the overlap integral (cf., p. 26) of the initial wave function and the eigenfunction of $\hat{A}$.

1.5 Disproving the Heisenberg Principle–Einstein-Podolsky-Rosen's Recipe

The Heisenberg uncertainty principle came as a shock, many scientists felt a strong impulse to prove that it was false. One of them was Albert Einstein, who used to play with ideas by performing some imaginary ideal experiments (in German, *Gedankenexperiment*) in order to demonstrate internal contradictions in theories. Einstein believed in the reality of our world. With his colleagues Podolsky and Rosen (the "*EPR team*"), he designed a special Gedanken experiment.[58] It represented an attempt to disprove the Heisenberg uncertainty principle and

[56] Schrödinger did not like the Copenhagen interpretation. Once Bohr and Heisenberg invited him for a Baltic Sea cruise, they indoctrinated him so strongly that he became ill and stopped participating in their discussions.

[57] This is an exaggeration. Classical mechanics also has its own problems with uncertainty. For example, obtaining the same results for a game of dice would require a perfect reproduction of the initial conditions, which is never feasible.

[58] A. Einstein, B. Podolsky, and N. Rosen, *Phys.Rev., 47*, 777 (1935).

to show that one *can* measure the position and momentum of a particle without any error. To achieve this, they invoked the help of a second particle.

The key statement of the whole reasoning, given in the EPR paper, was the following: "*If, without in any way disturbing a system, we can predict with certainty (i.e., with probability equal to unity) the value of a physical quantity, then there exists an element of physical reality corresponding to this physical quantity.*" EPR considered a coordinate system fixed in space and two particles: 1 with coordinate x_1 and momentum p_{x1} and 2 with coordinate x_2 and momentum p_{x2}, the total system being in a state with well-defined total momentum: $P = p_{x1} + p_{x2}$ and well-defined relative position $x = x_1 - x_2$. The meaning of the words *well-defined* is that, according to quantum mechanics, there is a possibility of the exact measurement of the two quantities (x and P), because the two operators $\hat{x}$ and $\hat{P}$ do commute.[59] At this point, Einstein and his colleagues and the great interpreters of quantum theory agreed.

We now come to the crux of the real controversy.

The particles interact, then separate and fly far away (at any time, we are able to measure exactly both x and P). When they are extremely far from each other (e.g., one close to us, the other one millions of light years away), we begin to suspect that each of the particles may be treated as free. Then, we decide to measure p_{x1}. However, after we do it, we know *with absolute certainty* the momentum of the second particle $p_{x2} = P - p_{x1}$, and this knowledge has been acquired without any perturbation of particle 2. According to the above cited statement, one has to admit that p_{x2} represents an element of physical reality. So far, so good. However, *we might have decided* with respect to particle 1 to measure its *coordinate* x_1. If this happened, then we would know with absolute certainty the position of the second particle, $x_2 = x - x_1$, without disturbing particle 2 at all. Therefore, x_2, as p_{x2}, is an element of physical reality. The Heisenberg uncertainty principle says that it is *impossible* for x_2 and p_{x2} to be exactly measurable quantities. The conclusion was that the Heisenberg uncertainty principle is wrong, and quantum mechanics is at least incomplete.

A way to support the Heisenberg principle was to treat the two particles as an indivisible total system and reject the supposition that the particles are independent, even if they are millions of light years apart. This is how Niels Bohr defended himself against Einstein (and his two colleagues). He said that the state of the total system in fact never fell apart into particles 1 and 2, and still is in what is known as an *entangled quantum state*[60] of the system of particles 1 and 2, and

any measurement influences the state of the system as a whole, independently of the distance of particles A and B.

[59] Indeed, $\hat{x}\hat{P} - \hat{P}\hat{x} = (\hat{x}_1 - \hat{x}_2)(\hat{p}_{x1} + \hat{p}_{x2}) - (\hat{p}_{x1} + \hat{p}_{x2})(\hat{x}_1 - \hat{x}_2) = [\hat{x}_1, \hat{p}_{x1}] - [\hat{x}_2, \hat{p}_{x2}] + [\hat{x}_1, \hat{p}_{x2}] - [\hat{x}_2, \hat{p}_{x1}] = -i\hbar + i\hbar + 0 - 0 = 0$.

[60] In honour of Einstein, Podolsky, and Rosen, the entanglement of states is sometimes called the *EPR effect*.

This reduces to the statement that measurement manipulations on particle 1 influence the results of measurements on particle 2. This correlation between measurements on particles 1 and 2 has to take place immediately, regardless of the space that separates them.[61] This is a shocking and non-intuitive feature of quantum mechanics. This is why it is often said, also by specialists, that quantum mechanics cannot be understood. One can apply it successfully and obtain an excellent agreement with experiment, but there is something strange in its foundations. This represents a challenge: an excellent theory, but based on some unclear foundations.

In the following, some precise experiments will be described, in which it is shown that quantum mechanics is right, however absurd it looks.

1.6 Schrödinger's Cat

The above mentioned paper by Einstein, Podolsky, and Rosen represented a severe critique of quantum mechanics in the form that has been presented by its fathers. After its publication, Erwin Schrödinger published a series of works[62] showing some other problematic issues in quantum mechanics. In particular, he described a *Gedanken experiment*, later known as the Schrödinger's cat paradox. According to Schrödinger, this paradox shows some absurd consequences of quantum mechanics.

Here is the paradox. There is a cat closed in an isolated steel box (filled with air). Together with the cat (but protected from it), there is a Geiger counter in the box. We put in the counter a bit of a radioactive substance, carefully prepared in such a way that every hour, two events happen with the same probability: either a radioactive nucleus decays or no radioactive nucleus decays. If a nucleus decays, it causes ionization and electric discharge in the counter tube, which in turn results in a hammer hitting a glass capsule with the hydrocyanic acid (HCN) gas, which kills the cat. If one puts the cat in for an hour, then *before* opening the box, one cannot say the cat is dead or alive. According to the Copenhagen interpretation, this is reflected by a proper wave function Ψ (for the box with everything in it), which is a superposition of two states with equal probability amplitudes: one state corresponds to the cat being alive, the other to it being dead. Therefore, function Ψ describes a cat in an intermediate state, neither alive nor dead, just in the middle between life and death, which, according to Schrödinger, would represent a totally absurd description. Einstein joined Schrödinger enthusiastically in this mockery from the Copenhagen interpretation, adding in his style an "*even better*" idea about how to kill the cat (using gunpowder instead of the cyanide). All in all, in this way an atomic scale phenomenon

---

[61] Nevertheless, the correlation is not quite clear. One may pose some questions. The statement about the instantaneous correlation between particles 1 and 2 in the EPR effect cannot be correct, because the measurements are separated in the space-time manifold and the simultaneity is problematic (see Chapter 13). What is the laboratory fixed coordinate system? How is information about particle 2 transferred to where we carry out the measurement on particle 1? This takes time. After that time, particle 1 is elsewhere. Is there anything to say about the separation time? In which coordinate system is the separation time measured?

[62] E. Schrödinger, *Naturwissenschaften 23*, 807, 823, 844 (1935).

(decay of a nucleus) can result in some drastic events in the macroscopic world. If, after an hour, an observer opens the box ("*measurement*" of the state of the content of the box), then according to the Copenhagen interpretation, this will result in the collapse of the corresponding wave function, and the observer will find the cat either alive or dead. This is the crux of the paradox.

Since the early days of quantum mechanics, many scholars were trying to rationalize the paradox, always relying on some particular interpretation of quantum mechanics. In one of the interpretations, quantum mechanics does not describe a single system, but rather an infinite set of systems. We have therefore plenty of cats and the same number of boxes, each of them with the same macabre gear inside. Then, the paradox disappears, because, after the boxes are open, in 50% of cases, the cats will be alive, and in 50%, they will be dead. In another interpretation, it is criticized that Schrödinger treats the box as a quantum system, while the observer is treated classically. In this interpretation, not only Schrödinger plays the role of the observer, but also the cat, and even the box itself (since it may contain a camera). What happened may be described differently by each of the observers, depending on what information they have about the whole system. For example, in the cat (alive or dead), there is information about what has happened even *before* the box is open. The human observer does not have this information. Therefore, the collapse of the wave function happened earlier for the cat and later for the observer! Only after the box is open, it will turn out for both the cat and the observer that the collapse happened to the same state.

Hugh Evereth stepped out with another, truly courageous, interpretation. According to Evereth, we do not have a single universe, as most of us might think naively, but a plethora of coexisting universes, and in each of them, some other things happen! When the box is opened, the state of the box is entangled with the state of the observer, but then a collapse of the wave function happens, and in one of the universes the cat is alive, but in the other one, it is dead (bifurcation). These two universes evolve independently ("*parallel universes*"), and they do not know about each other. Similar bifurcations happen massively in other events; hence, according to Evereth, the number of parallel universes is astronomic.

The quantity of possible interpretations, with large differences among them and some of them with a desperate character, indicate that the problem of understanding quantum mechanics is still unsolved.

1.7 Bilocation

Assume that the world and everything in it (stars, Earth, Moon, ... me, table, proton, electron, etc.) exist objectively. One may suspect from everyday observations that this is the case. For example, the Moon is seen by many people, who describe it in a similar way.[63] Instead of the

[63] This *may* indicate that the Moon exists independently of our observations and overcome importunate suspicions that the Moon ceases to exist when we do not look at it. Besides, there are people who claim to have seen the Moon from very close and even touched it (admittedly through a glove) and this slightly strengthens our belief in the Moon's existence. First of all, one has to be cautious. For example, some chemical substances, hypnosis, or an ingenious set of mirrors may cause some people to be convinced about the reality of some phenomena, while

Moon, let us begin with something simpler: how does this idea apply to electrons, protons, or other elementary particles? This is an important question because the world as we know it–including the Moon–is mainly composed of protons.[64] Here, one encounters a mysterious problem. I will try to describe it by reporting the results of several experiments.

Following Richard Feynman,[65] imagine two slits in a wall. Every second (the time interval has to be large enough to be sure that we deal with properties of a *single* particle), we send an electron towards the slits. There is a screen behind the two slits, and when an electron hits the screen, there is a flash (fluorescence) at the point of collision. Nothing special happens. Some electrons will not reach the screen at all, but traces of others form a pattern, which seems quite chaotic. The experiment looks monotonous and boring–just a flash here, and another there. One cannot predict where a particular electron will hit the screen. But suddenly we begin to suspect that there is some regularity in the traces (see Fig. 1.14).

A strange pattern appears on the screen: a number of high concentrations of traces are separated by regions of low concentration. This resembles the interference of waves; e.g., a stone thrown into water causes interference behind two slits: an alternation of high and low amplitudes of water level. Well, but what does an electron have in common with a wave on the water surface? The interference on water was possible because there were two sources of waves (the Huygens principle) — that is, two slits.

Common sense tells us that nothing like this could happen with the electron, because first, the electron could not pass through *both* slits, and, second, unlike the waves, the electron has hit a tiny spot on the screen (transferring its energy). Let us repeat the experiment with a *single* slit. The electrons go through the slit and make flashes on the screen here and there, but there is only a single major concentration region (just facing the slit) fading away from the center (with some minor minima).

This result should make you feel faint. Why? You would like the Moon, a proton, or an electron to be solid objects, wouldn't you? All investigations made so far indicate that the electron is a pointlike elementary particle. If, in the experiments we have considered, the electrons were to be divided into two classes (those that went through slit 1 and those that passed through slit 2), then the electron patterns would be different. The pattern with the two slits *had to be* the sum of the patterns corresponding to only one open slit (facing slit 1 and slit 2). *We do not have that picture.*

---

others do not see them. Yet, would it help if even everybody saw something? We should not verify serious things by voting. The example of the Moon also intrigued others; cf., D. Mermin, "*Is the Moon there, when nobody looks?*" *Phys. Today, 38*, 38 (1985).

[64] In the darkest Communist times, a colleague of mine came to my office. Conspiratorially, very excitedly, he whispered: "*The proton decays!!!*" He just read in a government newspaper that the lifetime of protons turned out to be finite. When asked about the lifetime, he gave an astronomical number, something like 10^{30} years or so. I said: "*Why do you look so excited then and why all this conspiracy?*" He answered: "*The Soviet Union is built of protons, and therefore is bound to decay as well!*"

[65] This is after Richard Feynman, *The Character of Physical Law,* MIT Press, Cambridge, MA, 1967.

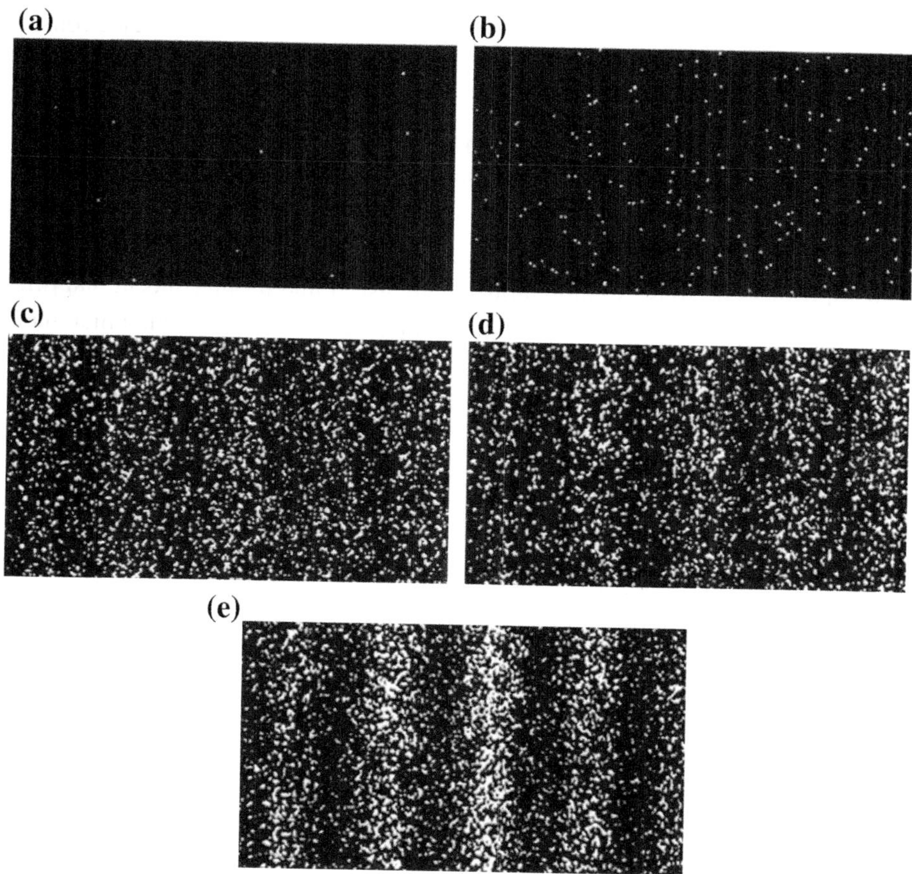

Fig. 1.14. Two-slit electron interference pattern registered by Akiro Tonomura. (a) 10 electrons; (b) 100 electrons; (c) 3000 electrons (one begins to suspect something); (d) 20000 electrons (no doubt, we will have a surprise); (e) 70000 electrons (here it is!). The conclusion is that there is only one possibility–each electron went through the two slits (according to J. Gribbin, *Q is for Quantum: An Encyclopedia of Particle Physics*, Weidenfeld and Nicolson, 1998).

The *only* explanation for this interference of the electron with itself is that with the two slits open, it went through *both*.

Clearly, the two parts of the electron united somehow and caused the flash at a single point on the screen. The quantum world is really puzzling. Despite the fact that the wave function is delocalized, the measurement gives its single point position (decoherence). How could an electron pass simultaneously through two slits? We do not understand why, but this is what has happened.

Maybe it is possible to pinpoint the electron passing through two slits. Indeed, one may think of the Compton effect: a photon collides with an electron and changes its direction, and this can be detected ("*a flash on the electron*"). When one prepares two such ambushes at the two open slits, it turns out that the flash is always on a single slit, not on both. This cannot be true!

If it were true, then the pattern would be of a non-interference character (and had to be the sum of the two one-slit patterns), but we have the interference: No, there is no interference. *Now*, the pattern does not show the interference. The interference was when the electrons were not observed. When we observe them, there is no interference. Somehow we have disturbed the electron's momentum (the Heisenberg principle), and the interference disappears.

We have to accept that the electron passes through two slits. Maybe it only pertains to the electron, or maybe the Moon is something completely different. But this is a weak hope. The same thing happens to protons. Sodium atoms were also found to interfere.[66] A sodium atom, of a few Å in diameter, looks like an ocean liner compared to a child's toy boat of a tiny electron (which is 42000 times less massive). And this ocean liner passed through two slits separated by thousands of Å. At the end of 1999, similar interference was observed for the fullerene,[67] a giant C_{60} molecule that is about a million times more massive than the electron. It is worth noting that after this adventure, the fullerene molecule remained intact. Somehow all its atoms, with the details of their chemical bonds, preserved their nature. There is something intriguing in this.

1.8 The Magic of Erasing the Past

John Archibald Wheeler was not completely satisfied with the description of the hypothetical two-slit experiment, if it were performed in an astronomically large distance scale. Indeed, there was something puzzling about it. Suppose that the two slits are far away from us. A photon goes through the slit region, and then flies toward us for a long time, and finally, it arrives to our screen. With a large number of such photons, we obtain an interference picture on the screen that witnesses each photon went through two slits. However, when a photon passed the slit region, leaving the slits behind and heading toward us, we had plenty of time to think. In particular, we might have an idea to replace the screen by two telescopes, each directed on one slit. In similar situations, we never observed half of the photon in one telescope and half in the other one: a photon was always seen in *one* telescope only. The telescopes allow us to identify unambiguously the slit that the particular photon went through. Therefore, we may divide all such photons into two classes (depending on the slit they went through) and their distribution cannot be interference-like. Rather, it has to be bullet-like (it is a sum of distributions of both classes). And now we have a paradox:

Quantum Eraser
Our decision (taken *after* the electrons passed the slits) to replace the screen by the telescopes visibly changed the way the electrons have been passing through the slits. This means our action changed the past!

[66] To observe such phenomena, the slit distance has to be of the order of the de Broglie wavelength, $\lambda = h/p$, where h is the Planck constant and p is the momentum. Cohen–Tannoudji lowered the temperature to such an extent that the momentum was close to 0, and λ could be of the order of thousands of Å.

[67] M. Arndt, O. Nairz, J. Voss-Andreae, C. Keller, G. van der Zouw, and A. Zeilinger, *Nature, 401,* 680 (1999).

This strange behavior was confirmed experimentally in 2007.[68] Note that we have changed the past only in a very special sense, since we are speaking here about two separate experiments: one with the telescopes and then the other one with the screen. This is not like undoing an airplane catastrophe that has already taken place, but rather like changing something before the catastrophe takes place in order not to have it happen.

1.9 A Test for a Common Sense: The Bell Inequality

John Bell proved a theorem in 1964 that pertains to the results of measurements carried out on particles and some of the inequalities they have to fulfill. The theorem pertains to the basic logic of the measurements and is valid independently of the kind of particles and of the nature of their interaction. The theorem soon became very famous because it turned out to be a useful tool allowing us to verify some fundamental features of our knowledge about the world.[69]

Imagine a launching gun (Fig. 1.15), which ejects a series of pairs of identical rectangular bars flying along a straight line (no gravitation) in opposite directions (opposite velocities). The axes of the bars are always parallel to each other and always perpendicular to a straight line.

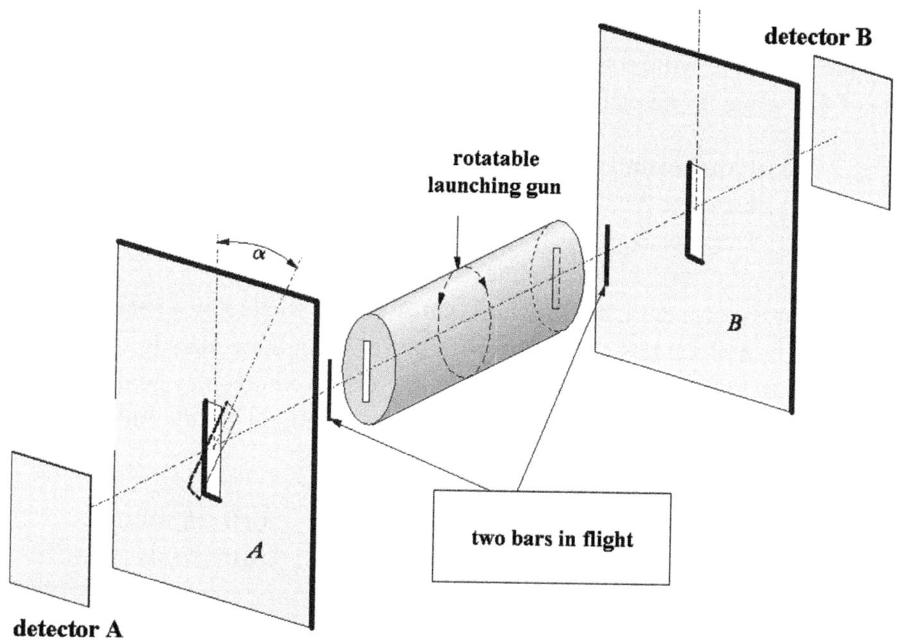

Fig. 1.15. Bell inequalities. A bar launching gun adopts stochastic positions (of equal probability) when rotating about the axis. Each time, the full magazine of bars is loaded. The slits also may be rotated about the axis. The bars arrive at slits A and B. Some will go through and be detected behind the slits.

[68] V. Jacques, E. Wu, F. Grosshans, F. Treussart, Ph. Grangier, A. Aspect, and J.-F. Roch, *Science, 315*, 966 (2007).
[69] W. Kołos, Proceedings of the IV Castel Gandolfo Symposium of John Paul II, 1986.

John Stuart Bell (1928–1990), Irish mathematician at Centre Européen de la Recherche Nucleaire (CERN) in Geneva. In the 1960s, Bell attacked an old controversion of locality versus non-locality, hidden variables, etc., apparently disappointed after an exchange of ideas between Einstein and Bohr.

The launching machine is constructed in such a way that the probabilities of all orientations of the bars are equal, and that any two launching series are absolutely identical. At a certain distance from the launching machine, there are two rectangular slits A and B (which are the same on both sides). If the bar's longer axis coincides with the longer dimension of the slit, then the bar will go through for sure and will be registered as "1"; i.e., "*it has arrived*" by the detector. If the bar's longer axis coincides with the shorter axis of the slit, then the bar will not go through for sure, and will be detected as "0." For other angles between the bar and slit axes, the bar will sometimes go through (when it fits the slit), sometimes not (when it does not).[70]

Having prepared the launching gun (our magazine contains 16 pairs of bars), we begin our experiments. Four experiments will be performed. Each experiment will need the full magazine of bars. In the first experiment, the two slits will be parallel. This means that the fate of both bars in any pair will be exactly the same: if they go through, then both will; and if they are stopped by the slits, both will be stopped. Our detectors have registered (we group the 16 pairs in clusters of 4 to make the sequence more transparent):

Experiment I (angle 0)				
Detector A:	1001	0111	0010	1001
Detector B:	1001	0111	0010	1001

Now, we repeat Experiment I, but this time, slit A will be rotated by a small angle α (Experiment II). At slit B, nothing has changed, so we must obtain there exactly the same sequence of zeros and ones as in Experiment I. At slit A, however, the results may be different. Since the rotation angle is small, the difference list will be short. We might get the following result:

Experiment II (angle α)				
Detector A:	10**1**1	0111	0010	**0**001
Detector B:	1001	0111	0010	1001

There are two differences (highlighted in bold) between the lists for the two detectors.

[70] Simple reasoning shows that for a bar of length L, two possibilities: "*to go through*" and "*not to go through*" are equally probable (for a bar of zero width) if the slit width is equal to $\frac{L}{\sqrt{2}}$.

Now, we move on to Experiment III. This time, slit A comes back to its initial position, but slit B is rotated by $-\alpha$. Because of our perfect gun, we must obtain at detector A the same result as in Experiment I. However, at slit B, we find some difference with respect to Experiments I and II:

Experiment III (angle $-\alpha$)				
Detector A:	1001	0111	0010	1001
Detector B:	1001	0**0**11	0**1**10	1001

There are two differences (shown in bold) between the two detectors.

We now carry out Experiment IV. We rotate slit A by angle α, and slit B by angle $-\alpha$. Therefore, at detector A, we obtain the same results as in Experiment II, while at detector B, it is the same as in Experiment III. Therefore, we detect the following:

Experiment IV (angle 2α)				
Detector A:	10**1**1	0111	0010	**0**001
Detector B:	1001	0**0**11	0**1**10	1001

Now there are four differences between Detector A and Detector B (shown in bold). In Experiment IV, the number of differences could not be larger (Bell inequality). In our case, it could be four or fewer. When would it be fewer? That would happen when accidentally the bold figures (i.e., the differences of Experiments II and III with respect to those of Experiment I) coincide. In this case, this would be counted as a difference in Experiments II and III, while in Experiment IV, it would not be counted as a difference.

Thus, we have demonstrated

Bell Inequality:

$$N(2\alpha) \leq 2N(\alpha), \tag{1.24}$$

where N stands for the number of measurement differences. The Bell inequality was derived under the assumption that whatever happens at slit A, it does not influence what happens at slit B (this is how we constructed the counting tables) and that the two flying bars have (perhaps unknown by the observer) only a real (definite) direction in space, the same for both bars.

It would be interesting to perform a real experiment similar to Bell's to confirm the Bell inequality, as discussed in the next section. This opens the door for deciding in a physical experiment whether:

- Elementary particles are classical (though extremely small) objects that have some well-defined attributes regardless of whether we observe them (Einstein's view)
- Elementary particles do not have such attributes, and only measurements themselves make them acquire measured values (Bohr's view).

1.10 Photons Violate the Bell Inequality

French scientists from the Institute of Theoretical and Applied Optics in Orsay published the results of their experiments with photons.[71] The excited calcium atom emitted pairs of photons (analogs of our bars), which moved in opposite directions and had the same polarization. After flying about 6 m, they both met the polarizers — analogs of slits A and B in the Bell procedure. A polarizer allows a photon with polarization state $|0\rangle$, or parallel (to the polarizer axis), always pass through, and always rejects any photon in the polarization state $|1\rangle$, or perpendicular (indeed perpendicular to the above parallel setting). When the polarizer is rotated about the optical axis by an angle, it will pass through a percentage of the photons in state $|0\rangle$ and a percentage of the photons in state $|1\rangle$. When both polarizers are in the parallel setting, there is perfect correlation between the two photons of each pair; i.e., exactly as in Bell's Experiment I. In the photon experiment, this correlation was checked for 50 million photons every second for about 12000 seconds.

Bell's experiments II–IV have been carried out. Common sense indicates that, even if the two photons in a pair have random polarizations (perfectly correlated though always the same, like the bars), they still have *some* polarizations; i.e., it may be unknown but definite (as in the case of the bars; i.e., what E,P, and R believed happens). *Hence, the results of the photon experiments would have to fulfil the Bell inequality.* However, the photon experiments have shown that the Bell inequality is violated, but the results remain in accordance with the prediction of quantum mechanics.

There are therefore only two possibilities (compare the two points at the end of the previous section):

(a) either the measurement on a photon carried out at polarizer A(B) results in some instantaneous interaction with the photon at polarizer B(A)

(b) or/and, the polarization of any of these photons is completely indefinite (even if the polarizations of the two photons are fully correlated i.e., the same) and only the measurement on one of the photons at A (B) determines its polarization, which results in the automatic determination of the polarization of the second photon at B(A), even if they are separated by millions of light years.

Both possibilities are sensational. The first assumes a strange form of communication between the photons or the polarizers. This communication must be propagated with a velocity *exceeding* the speed of light because an experiment was performed in which the polarizers were switched (this took something like 10 nanoseconds) *after* the photons started (their flight took about 40 nanoseconds). Despite this, communication between the photons did exist.[72] Possibility b) as a matter of fact represents Bohr's interpretation of quantum mechanics: elementary particles do not have definite attributes (e.g., polarization).

---

[71] A. Aspect, J. Dalibard, and G. Roger, *Phys.Rev.Lett. 49*, 1804 (1982).

[72] This again is the problem of delayed choice. It seems that when starting, the photons have a knowledge of the *future* setting of the apparatus (the two polarizers).

As a result, there is a dilemma: either the world is "*non-real*" (in the sense that the properties of particles are not determined before measurement) or/and there is instantaneous (i.e., faster than light) communication between particles which operates independently of how far apart they are ("*non-locality*").

This dilemma may make everybody's metaphysical shiver.

1.11 Teleportation

The idea of teleportation comes from science fiction. In general, it refers to the following:

- Acquisition of full information about an object located at A,
- Its transmission to B,
- Creation (materialization) of an identical object at B,
- At the same time, the disappearance of the object at A.

At first sight, it *seems* that this contradicts quantum mechanics. The Heisenberg uncertainty principle says that it is not possible to prepare a perfect copy of the object, because in the case of mechanical quantities with non-commuting operators (like positions and momenta), there is no way to have them measured exactly, in order to rebuild the system elsewhere with the same values of the quantities.

The trick is, however, that the quantum teleportation we are going to describe, will not violate the Heisenberg principle because the mechanical quantities needed will not be measured and the copy made based on their values.

The teleportation protocol was proposed by Bennett and coworkers,[73] and applied by Anton Zeilinger's group.[74] The latter used the entangled states (EPR effect) of two photons described above.[75]

Assume that photon A (number 1) from the entangled state belongs to Alice, and photon B (number 2) to Bob. Alice and Bob introduce a common fixed coordinate system. Both photons have *identical* polarizations in this coordinate system, *but neither Alice nor Bob knows which.* Alice may measure the polarization of her photon and send this information to Bob, who may prepare his photon in that state. This, however, does not amount to teleportation because the original state could be a linear combination of the $|0\rangle$ (parallel) and $|1\rangle$ (perpendicular) states.

---

[73] C.H. Benneth, G. Brassard, C. Crépeau, R. Josza, A. Peres, and W.K. Wootters, *Phys.Rev.Letters, 70*, 1895 (1993).

[74] At the time, this group was working at the University of Innsbruck (Austria).

[75] An ultraviolet (UV) laser beam hits a barium borate crystal (known for its birefringence). Photons with parallel polarization move along the surface of a cone (with the origin at the beam-surface collision point), and the photons with perpendicular polarization move on another cone, with the two cones intersecting. From time to time, a single UV photon splits into two photons of equal energies and different polarizations. Two such photons when running along the intersection lines of the two cones, and therefore not having a definite polarization (i.e., being in a superposition state composed of both polarizations), represent the two entangled photons.

In such a case, Alice's measurement would "*falsify*" the state due to wave function collapse (it would give either $|0\rangle$ or $|1\rangle$; cf. p. 24).

Since Alice and Bob have two entangled photons of the same polarization, let us assume that the state of the two photons is the following superposition[76]: $|00> + |11>$, where the first position in every ket pertains to Alice's photon, and the second one pertains to Bob's.

Now, Alice wants to carry out teleportation of her additional photon (number 3) in an unknown quantum state $\phi = a|0\rangle + b|1\rangle$ (known as *qubit*), where a and b stand for unknown coefficients[77] satisfying the normalization condition $a^2 + b^2 = 1$. Therefore, the state of three photons (Alice's: the first and the third position in the three-photon ket; and, Bob's: the second position) will be $\left[|00\rangle + |11\rangle\right]\left[a|0\rangle + b|1\rangle\right] = a|000\rangle + b|001\rangle + a|110\rangle + b|111\rangle$.

Alice prepares herself for teleportation of the qubit ϕ corresponding to her second photon. She first prepares a device called the *XOR gate*.[78]

What is the XOR gate? The device manipulates two photons. One is treated as the steering photon, the second as the steered photon (see Table 1.1). The device operates thus: if the steering photon is in state $|0\rangle$, then no change is introduced for the state of the steered photon. If, however, the steering photon is in the state $|1\rangle$, the steered photon will be switched over; i.e., it will be changed from 0 to 1 or from 1 to 0. Alice chooses the photon in the state ϕ as her steering photon 3, and photon 1 as her steered photon.

After the XOR gate is applied, the state of the three photons will be as follows: $a|000\rangle + b|101\rangle + a|110\rangle + b|011\rangle$.

Alice continues her preparation by using another device called the *Hadamard gate*, which operates on a single photon and does the following:

$$|0\rangle \rightarrow \frac{1}{\sqrt{2}}\left(|0\rangle + |1\rangle\right),$$

$$|1\rangle \rightarrow \frac{1}{\sqrt{2}}\left(|0\rangle - |1\rangle\right).$$

Table 1.1. The XOR gate changes the state of the steered photon, only when the steering photon is on.

Steering	Steered before XOR	Steered after XOR
$\|0\rangle$	$\|0\rangle$	$\|0\rangle$
$\|0\rangle$	$\|1\rangle$	$\|1\rangle$
$\|1\rangle$	$\|0\rangle$	$\|1\rangle$
$\|1\rangle$	$\|1\rangle$	$\|0\rangle$

[76] The teleportation result does not depend on the state.

[77] Neither Alice nor Bob will know these coefficients up to the end of the teleportation procedure, but Alice still will be able to send her qubit to Bob!

[78] In the expression, *XOR* is the abbreviation of "*eXclusive OR.*"

Alice applies this operation to her photon 3, and after that, the three-photon state is changed to the following:

$$\frac{1}{\sqrt{2}}[a|000\rangle + a|001\rangle + b|100\rangle - b|101\rangle + a|110\rangle + a|111\rangle + b|010\rangle - b|011\rangle]$$
$$= \frac{1}{\sqrt{2}}[|0\left(a|0\rangle + b|1\rangle\right)0\rangle + |0\left(a|0\rangle - b|1\rangle\right)1\rangle - |1\left(a|1\rangle + b|0\rangle\right)0\rangle + |1\left(a|1\rangle - b|0\rangle\right)1\rangle]. \tag{1.25}$$

There is a superposition of four three-photon states in the last row. Each state shows the state of Bob's photon (number 2 in the ket), at any given state of Alice's two photons. Finally, Alice carries out the measurement of the polarization states of her photons (1 and 3). This inevitably causes her to get (for each of the photons) either $|0\rangle$ or $|1\rangle$. This causes her to know the state of Bob's photon from the three-photon superposition [Eq. (1.25)]:

- Alice's photons 00; i.e., Bob has his photon in state $\left(a|0\rangle + b|1\rangle\right) = \phi$.
- Alice's photons 01; i.e., Bob has his photon in state $\left(a|0\rangle - b|1\rangle\right)$.
- Alice's photons 10; i.e., Bob has his photon in state $\left(a|1\rangle + b|0\rangle\right)$.
- Alice's photons 11; i.e., Bob has his photon in state $\left(a|1\rangle - b|0\rangle\right)$.

Then Alice calls Bob and tells him the result of her measurements of the polarization of her two photons.

Bob knows therefore, that if Alice sends him $|00\rangle$, this means that the teleportation is over: he already has his photon in state ϕ! If Alice sends him one of the remaining possibilities, he would know exactly what to do with his photon to prepare it in state ϕ, and he does this with his equipment. The teleportation is over: Bob has the teleported state ϕ, Alice has lost her photon state ϕ when performing her measurement (wave function collapse).

Note that to carry out the successful teleportation of a photon state, Alice had to communicate something to Bob by a classical channel (like a telephone).

1.12 Quantum Computing

Richard Feynman pointed out that contemporary computers are based on the "*all-or-nothing*" philosophy (two bits: $|0\rangle$ or $|1\rangle$), while in quantum mechanics, one may also use a linear combination (superposition) of these two states with arbitrary coefficients a and b: $a|0\rangle + b|1\rangle$. Would a quantum computer based on such superpositions be better than the traditional one? The hope associated with quantum computers relies on a multitude of the quantum states (those obtained using variable coefficients $a, b, c, \ldots$) and possibility of working with many of them using a single processor. It was (theoretically) proved in 1994 that quantum computers could factorize natural numbers much faster than traditional computers. This sparked intensive research on the concept of quantum computation, which uses the idea of entangled states. According to many researchers, any entangled state (a superposition) is extremely sensitive to the slightest interaction with the environment, and as a result, decoherence takes place very easily, which is devastating for quantum computing.[79] The first attempts at constructing quantum computers were based on protecting the entangled states, but, after a few simple operations, decoherence took place.

In 1997, Neil Gershenfeld and Isaac Chuang discovered that any routine nuclear magnetic resonance (NMR) measurement represents nothing but a simple quantum computation. The breakthrough was recognizing that a qubit may be also represented by the huge number of molecules in a liquid.[80] The nuclear spin angular momentum (say, corresponding to $s = \frac{1}{2}$) is associated with a magnetic dipole moment (Chapter 12), and those magnetic dipole moments interact with an external magnetic field and with themselves. An isolated magnetic dipole moment has two states in a magnetic field: a lower energy state corresponding to the antiparallel configuration (state $|0\rangle$) and a higher energy state related to the parallel configuration (state $|1\rangle$). By exposing a sample to a carefully tailored nanosecond radiowave impulse, one obtains a rotation of the nuclear magnetic dipoles, which corresponds to their state being a superposition $a|0\rangle + b|1\rangle$.

Here is a prototype of the XOR gate. Take chloroform[81] $[^{13}CHCl_3]$ as our example. Due to the interaction of the magnetic dipoles of the proton and of the carbon nucleus (both either in parallel or antiparallel configurations with respect to the external magnetic field), a radiowave impulse of a certain frequency causes the carbon nuclear spin magnetic dipole to rotate by 180 degrees, provided that the proton spin dipole moment is parallel to that of the carbon.[82] Similarly, one may conceive other logical gates. The spins change their orientations according to a sequence of impulses, which play the role of a computer program. There are many technical problems to overcome in the quantum computers "*in liquid*" : the magnetic interaction of distant nuclei is very weak, decoherence remains a worry, and for the time being, the number of operations is limited to several hundred. However, this is only the beginning of a new computer technology. It is most important that chemists know the future computers well–they are simply molecules.

---

[79] It pertains to an entangled state of (already) distant particles. When the particles interact strongly, the state is more stable. The wave function for H_2 also represents an entangled state of two electrons with opposite spins, yet the decoherence does not take place even at short internuclear distances. As we will see, entangled states can also be obtained in liquids.

[80] Interaction of the molecules with the environment does not necessarily result in decoherence.

[81] The NMR operations on spins pertain in practice to a tiny fraction of the nuclei of the sample (of the order of 1 : 1000000)

[82] This is indeed consistent with the XOR gate logical table because for the parallel spins (entries: 00 and 11), the output is 0 (meaning transition), while for the opposite spins (entries: 01 and 10) the output is 1 (meaning no transition).

Summary

Classical mechanics was unable to explain certain phenomena: black body radiation, the photoelectric effect, and the stability of atoms and molecules, as well as their spectra. Quantum mechanics, created mainly by Werner Heisenberg and Erwin Schrödinger, explained these effects. The new mechanics was based on six postulates:

- Postulate I says that all information about the system follows from the wave function ψ. The quantity $|\psi|^2$ represents the probability density of finding particular values of the coordinates of the particles that the system is composed of.
- Postulate II allows mechanical quantities (e.g., energy) to be ascribed to operators. One obtains the operators by writing down the classical expression for the corresponding quantity and replacing momenta (e.g., p_x) by momenta operators $\left(\text{here, } \hat{p}_x = -i\hbar\frac{\partial}{\partial x}\right)$.
- Postulate III gives the time evolution equation for the wave function ψ (time-dependent Schrödinger equation $\hat{H}\psi = i\hbar\frac{\partial\psi}{\partial t}$), using the energy operator (*Hamiltonian* $\hat{H}$). For time-independent $\hat{H}$ one obtains the time-independent Schrödinger equation $\hat{H}\psi = E\psi$ for the stationary states.
- Postulate IV pertains to ideal measurements. When making a measurement of a quantity A, one can obtain only an eigenvalue of the corresponding operator $\hat{A}$. If the wave function ψ represents an eigenfunction of $\hat{A}$ [i.e., $(\hat{A}\psi = a\psi)$], then one obtains always the eigenvalue corresponding to ψ (i.e., a) as a result of the measurement. If, however, the system is described by a wave function, which does not represent any eigenfunction of $\hat{A}$, then one obtains also an eigenvalue of $\hat{A}$, but there is no way to predict which eigenvalue. The only thing one can predict is the mean value of many measurements, which may be computed as $\bar{a} = \frac{\langle\psi|\hat{A}\psi\rangle}{\langle\psi|\psi\rangle}$.
- Postulate V says that an elementary particle has an internal angular momentum (spin). One can measure only two quantities: the square of the spin length $s(s+1)\hbar^2$ and one of its components $m_s\hbar$, where $m_s = -s, -s+1, \ldots, +s$, with spin quantum number $s \geq 0$ that is characteristic of the type of particle (integers for bosons, half-integers for fermions). The spin magnetic quantum number m_s takes $2s+1$ values, related to the $2s+1$ values of the (granular) spin coordinate σ.
- Postulate VI has to do with the symmetry of the wave function with respect to different labeling identical particles. If one exchanges the labels of two identical particles (the exchange of all the coordinates of the two particles), then for two identical fermions, the wave function has to change its sign (antisymmetric), while for two identical bosons, the function does not change (symmetry). As a consequence, two identical fermions with the same spin coordinate cannot occupy the same point in space.

Quantum mechanics is one of the most peculiar theories. It gives numerical results that agree extremely well with experiments, but on the other hand, the relation of these results to our everyday experience sometimes seems shocking. For example, it turned out that a particle or even a molecule may somehow exist in two locations (it passes through two slits simultaneously), but when one verifies this, it is always in one place. It also turned out that

- Either a particle has no definite properties ("*the world is unreal*"), and the measurement fixes them somehow
- Or/and, there is instantaneous communication between particles, however distant they are from each other ("*non-locality of interactions*").

It turned out that in the Bohr-Einstein controversy, Bohr was right. The Einstein-Podolsky-Rosen paradox resulted (in agreement with Bohr's view) in the concept of entangled states. These states have been used experimentally teleport a photon state without violating the Heisenberg uncertainty principle. Also, the entangled states stand behind the idea of quantum computing: with a superposition (qubit) of two states $a|0\rangle + b|1\rangle$ instead of $|0\rangle$ and $|1\rangle$ as information states.

Main Concepts, New Terms

antisymmetric function (p. 34)
axis of quantization (p. 24)
Bell inequality (p. 47)
bilocation (p. 43)
decoherence (p. 40)
delayed choice (p. 50)
Dirac notation (p. 20)
eigenfunction (p. 22)
eigenvalue problem (p. 22)
entangled states (p. 41)
EPR effect (p. 40)
experiment of Aspect (p. 50)
Gedankenexperiment (p. 40)
Heisenberg uncertainty principle (p. 12)
Hilbert space (p. 8)
interference of particles (p. 44)
locality of the world (p. 49)
logical gate (p. 54)
mean value of an operator (p. 25)
measurement (p. 22)
normalization (p. 18)
operator of a quantity (p. 18)
Pauli exclusion principle (p. 34)
Pauli matrices (p. 29)
quantum eraser (p. 46)
qubit (p. 52)
"*reality of the world*" (p. 40)
Schrödinger's cat (p. 42)
singlet state (p. 32)
spin angular momentum (p. 26)
spin coordinate (p. 27)
stationary state (p. 22)
symmetric function (p. 34)
symmetry of wave function (p. 34)
teleportation (p. 51)
time evolution equation (p. 21)
triplet state (p. 32)
wave function (p. 16)
wave function collapse (p. 24)

From the Research Front

Until recently, the puzzling foundations of quantum mechanics could not be verified directly by experimentation. As a result of enormous technological advances in quantum electronics and quantum optics, it became possible to carry out experiments on single atoms, molecules, photons, etc. In 2004 a group of researchers has teleported for the first time an atomic state[83], while another group[84] has successfully performed teleportation of a photon state across the Danube river (at a distance of 600 m). Even molecules such as fullerene were subjected to successful interference experiments. Quantum computer science is just beginning to prove that its principles are correct.

Ad Futurum

Quantum mechanics has been shown in the past to give excellent results, but its foundations are still unclear.[85] There is no successful theory of decoherence that would explain why and how a delocalized state becomes localized after the measurement. It is possible to make fullerene interfere, and it may be that in the near future, we will be able to do this with a virus.[86] It is interesting that fullerene passes instantaneously through two slits with its whole complex electronic structure, as well as a nuclear framework, although the de Broglie wavelength is quite different for the electrons and for the nuclei. Visibly, the "*overweighted*" electrons interfere differently than free ones. After fullerene passes the slits, one sees it in a single spot on the screen (decoherence). It seems that there are cases when even strong

[83] M. Riebe, H. Häffner, C. F. Roos, W. Hänsel, M. Ruth, J. Benhelm, G. P. T. Lancaster, T. W. Körber, C. Becher, F. Schmidt-Kaler, D. F. V. James, and R. Blatt, *Nature 429*, 734 (2004).

[84] R. Ursin, T. Jennewein, M. Aspelmeyer, R. Kaltenbaek, M. Lindenthal, P. Walther and A. Zeilinger, *Nature 430*, 849 (2004).

[85] A pragmatic viewpoint is shared by the vast majority: "*Do not wiseacre, just compute!*"

[86] A similar statement has been made by Anton Zeilinger.

interaction does not make decoherence necessary. Sławomir Szymański presented his theoretical and experimental results[87] and showed that the functional group $-CD_3$ exhibits a delocalized state (which corresponds to its rotation instantaneously in both directions; i.e., coherence) and, which makes the thing more peculiar, interaction with the environment *not only does not destroy the coherence, but makes it more robust*. This type of phenomenon might fuel investigations towards future quantum computer architecture.

Additional Literature

The Ghost in the Atom: A Discussion of the Mysteries of Quantum Physics, P.C.W. Davies and J.R. Brown, eds, Cambridge University Press, Cambridge, UK, (1986).

Two BBC journalists interviewed eight outstanding physicists: Alain Aspect (photon experiments), John Bell (Bell inequalities), John Wheeler (Feynman's Ph.D. supervisor), Rudolf Peierls ("*Peierls metal-semiconductor transition*"), John Taylor ("*black holes*"), David Bohm ("*hidden parameters*"), and Basil Hiley ("*mathematical foundations of quantum physics*"). It is most striking that all these physicists give *very* different theoretical interpretations of quantum mechanics (summarized in Chapter 1).

R. Feynman, *QED–The Strange Theory of Light and Matter,* Princeton University Press, Princeton, NJ (1985).

Excellent popular presentation of quantum electrodynamics written by one of the outstanding physicists of the 20th century.

A. Zeilinger, "*Quantum teleportation*," *Scientific American*, 282, 50 (2000).

The leader in teleportation describes this new field of study.

N. Gershenfeld, and I.L. Chuang "*Quantum computing with molecules*," *Scientific American, 278*, 66 (1998).

First-hand information about NMR computing.

Ch.H. Bennett "*Quantum information and computation*," *Physics Today, 48*, 24 (1995).

Another first-hand description of NMR computing.

Questions

1. When we insert into ψ some particular values of the variables, we obtain a number c. Which of the following statements about c is correct?
   a. $c = 1$ for a normalized ψ
   b. c may be a complex number and c^*c means the probability density for the system having the variables at these particular values
   c. $|c|^2$ means the probability for having the system with variables equal to these particular values
   d. c is either positive or zero
2. For a Hermitian operator $\hat{A}$ and all functions ψ_i from its domain, one has:
   a. $\hat{A}\psi_i = \hat{A}\psi_j$
   b. $\left\langle \hat{A}\psi_i \mid \psi_j \right\rangle = \left\langle \psi_j \mid \hat{A}\psi_i \right\rangle$
   c. $\left\langle \hat{A}\psi_i \mid \psi_j \right\rangle = \left\langle \psi_i \mid \hat{A}\psi_j \right\rangle$
   d. $\left\langle \hat{A}\psi_i \mid \hat{A}\psi_j \right\rangle = \langle \psi_i \mid \psi_j \rangle$
3. If operators $\hat{A}$ and $\hat{B}$ commute:
   a. any eigenfunction of $\hat{A}$ represents also an eigenfunction of $\hat{B}$

[87] S. Szymański, *J. Chem. Phys. 111*, 288 (1999).

b. one can choose such a set of functions, that any eigenfunction of $\hat{A}$ is also an eigenfunction of $\hat{B}$
c. the quantities A and B are both exactly measurable (without making any error)
d. $\hat{A}\hat{B} - \hat{B}\hat{A} = 0$

4. For a Hermitian operator $\hat{A} : \hat{A}\psi_1 = a_1\psi_1$ and $\hat{A}\psi_2 = a_2\psi_2$.
   a. then $\psi_1 = \psi_2$ and $a_1 = a_2$
   b. function $\psi = c_1\psi_1 + c_2\psi_2$ (c_1 and c_2 are complex numbers) also represents an eigenfunction of $\hat{A}$
   c. If $a_1 = a_2$, function $\psi = c_1\psi_1 + c_2\psi_2$ (c_1 and c_2 are complex numbers) represents an eigenfunction of $\hat{A}$
   d. $a_1^* = a_1$ and $a_2^* = a_2$
5. For functions ψ_1 and ψ_2 satisfying $\hat{A}\psi_1 = a_1\psi_1$ and $\hat{A}\psi_2 = a_2\psi_2$ ($\hat{A}$ is a Hermitian operator), one has:
   a. $\langle\psi_1 \mid \psi_1\rangle = 1$, $\langle\psi_2 \mid \psi_2\rangle = 1$ and $\langle\psi_1 \mid \psi_2\rangle = 0$
   b. $\langle\psi_1 \mid \psi_2\rangle = 0$
   c. $\langle\psi_1 \mid \psi_1\rangle > 0$ and $\langle\psi_2 \mid \psi_2\rangle > 0$
   d. $\langle\psi_1 \mid \psi_2\rangle = 0$, if $a_1 \neq a_2$.
6. Hamiltonian $\hat{H}$ for a particle of mass m and with the potential energy $-\frac{1}{r^6}$, where r stands for the distance of the particle from the center of the coordinate system, has which of the following forms:
   a. $\frac{\hbar^2}{2m}\Delta - \frac{1}{r^6}$
   b. $-\frac{\hbar^2}{2m}\left(\frac{\partial^2}{\partial x^2} + \frac{\partial^2}{\partial y^2} + \frac{\partial^2}{\partial z^2}\right) - \frac{1}{r^6}$
   c. $-\frac{\hbar^2}{2m}\Delta\frac{1}{r^6}$
   d. $-\frac{\hbar^2}{2m}\Delta - \frac{1}{r^6}$
7. The angle between spins of two electrons in the singlet state is equal γ, while in their triplet state ω. These angles are:
   a. $\gamma = 180°$ $\omega = 0°$
   b. $\gamma = 180°$ ω is different in any of the three triplet states
   c. γ and ω are undetermined
   d. $\gamma = 0°$ $\omega = \arccos\frac{1}{3}$
8. The mean value of quantity A from measurements should be compared to $\bar{a}$ (wave function ψ normalized) expressed in which of the following ways?
   a. $\bar{a} = \frac{\langle\psi \mid \hat{A}\psi\rangle}{\langle\psi \mid \psi\rangle}$
   b. $\bar{a} = \langle\psi \mid \hat{A}\psi\rangle$
   c. $\bar{a} = \frac{1}{N}\sum_{k=1}^{k=N} a_k$, where N stands for the number of the eigenvalues a_k
   d. $\bar{a} = \int \psi^* \hat{A}\psi d\tau$, where the integration is over the whole range of the variables
9. The error ΔA of measurement of the physical quantity A, which the Heisenberg uncertainty principle is talking about, means (wave function ψ is normalized):
   a. $\Delta A = \sqrt{\langle\psi \mid \hat{A}^2\psi\rangle - \langle\psi \mid \hat{A}\psi\rangle^2}$
   b. $\Delta A = \sqrt{\langle\psi \mid \Delta\hat{A}^2\psi\rangle}$

c. $\Delta A = \sqrt{\left(\left\langle \psi \mid \hat{A} - \left\langle \psi \mid \hat{A}\psi \right\rangle \mid \psi \right\rangle\right)^2}$

d. $\Delta A = \sqrt{\left\langle \psi \mid (\hat{A} - \left\langle \psi \mid \hat{A}\psi \right\rangle)^2 \mid \psi \right\rangle}$

10. The teleportation experiments carried out so far:
    a. pertain to sending a photon from place A to B without violation of the Heisenberg uncertainty principle
    b. the teleported state of the photon disappears from spot A and appears in spot B
    c. require a classical communication channel between the sender and the receiver
    d. pertain to exchange of the entangled particles between the sender and the receiver

Answers

1b, 2c, 3b,c,d, 4c,d, 5c,d, 6b,d, 7d, 8a,b,d, 9a,d, 10b,c

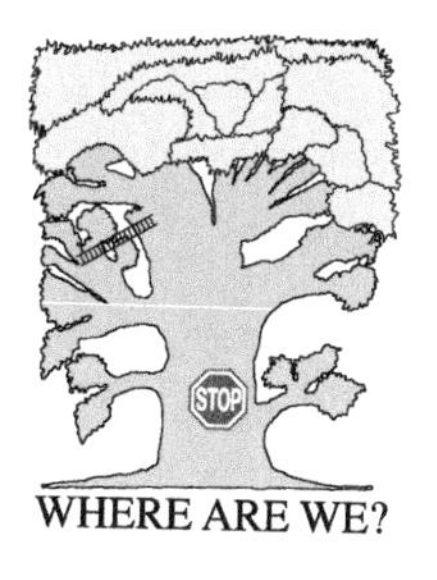

WHERE ARE WE?

CHAPTER 2

The Schrödinger Equation

Litterarum radices amarae sunt, fructus iucundiores.
The roots of science are bitter, while sweet are the fruits.
(Latin maxim)

Where Are We?

The postulates constitute the foundation of quantum mechanics (the base of the TREE trunk). One of their consequences is the Schrödinger equation for stationary states. Thus, we begin our itinerary with the TREE. The second part of this chapter is devoted to the time-dependent Schrödinger equation, which, from the pragmatic point of view, is outside the scope of this book (which is why it is a side branch on the left side of the TREE).

An Example

A friend asked us to predict what the UV spectrum of anthracene[1] (see Fig. 2.1) looks like. One can predict any UV spectrum if one knows the electronic stationary states of the molecule. The only way to obtain such states and their energies is to solve the time-independent Schrödinger equation. Thus, one has to solve the equation for the Hamiltonian for anthracene, then find the ground (the lowest) and the excited stationary states. The energy differences of these states will tell us where (in the energy scale) to expect light absorption, then the wave functions will enable us to compute the intensity of this absorption.

What Is It All About?

[1] Anthracene consists of three condensed benzene rings (⌬⌬⌬).

Ideas of Quantum Chemistry, Second Edition. http://dx.doi.org/10.1016/B978-0-444-59436-5.00002-7

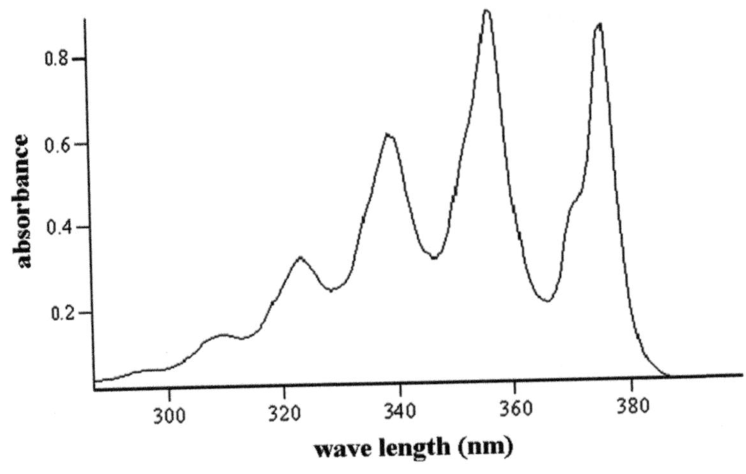

Fig. 2.1. The UV-VIS spectrum of anthracene.

The *time-independent* Schrödinger equation is the one place where stationary states can be produced as solutions of the equation. The *time-dependent* Schrödinger equation is the equation of motion, describing the evolution of a given wave function as time passes. As always with an equation of motion, one has to provide an initial state (starting point); i.e., the wave function for $t = 0$. Both the stationary states and the evolution of the non-stationary states depend on the energy operator (the Hamiltonian). If one finds some symmetry with the Hamiltonian, this will influence the symmetry of the wave functions. At the end of this chapter, we will be interested in the evolution of a wave function after applying a perturbation.

Why Is This Important?

The wave function is a central notion in quantum mechanics and is obtained as a solution of the Schrödinger equation. Hence, this chapter is necessary for understanding quantum chemistry.

What Is Needed?

- Postulates of quantum mechanics; Chapter 1 (necessary)
- Matrix algebra; see Appendix A available at booksite.elsevier.com/978-0-444-59436-5, on p. e1 (advised)
- Center-of-mass separation; see Appendix I available at booksite.elsevier.com/978-0-444-59436-5, on p. e93 (necessary)
- Translation vs momentum and rotation vs angular momentum; see Appendix F available at booksite.elsevier.com/978-0-444-59436-5, on p. e73 (necessary)
- Dirac notation; p. 20 (necessary)
- Two-state model; see Appendix D available at booksite.elsevier.com/978-0-444-59436-5, on p. e65 (necessary)
- Dirac delta; see Appendix E available at booksite.elsevier.com/978-0-444-59436-5, on p. e69 (necessary)

Classical Works

A paper by the mathematician Emmy Noether "*Invariante Variationsprobleme,*" published in *Nachrichten von der Gesellschaft der Wissenschaften zu Göttingen*, (1918), pp. 235–257, was the first to follow the conservation laws of certain physical quantities with the symmetry of theoretical descriptions of the system. ★ Four papers by Erwin Schrödinger, which turned out to cause an "*earthquake*" in science: *Annalen der Physik*, 79(1926)361; *ibid.*, 79(1926)489; *ibid.* 80(1926)437; *ibid.* 81(1926)109, all under the title "*Quantisierung als Eigenwertproblem,*" presented quantum mechanics as an eigenvalue problem (known from the developed differential equation theory), instead of an abstract Heisenberg algebra. Schrödinger proved the equivalence of both theories, gave the solution for the hydrogen atom, and introduced the variational principle. ★ The time-dependent perturbation theory described in this chapter was developed by Paul Dirac in 1926. Twenty years later, Enrico Fermi, lecturing at the University of Chicago, coined the term "*the Golden Rule*" for these results. From then on, they became known as the "*Fermi Golden Rule.*"

2.1 Symmetry of the Hamiltonian and Its Consequences

2.1.1 The Non-Relativistic Hamiltonian and Conservation Laws

From classical mechanics, it follows that for an isolated system (and assuming the forces to be central and obeying the action-reaction principle), its *energy*, *momentum*, and *angular momentum*, are conserved.

Emmy Noether (1882–1935), German mathematician, informally professor, formally only the assistant of David Hilbert at the University of Göttingen (in the first quarter of the 20th century, women were not allowed to be professors in Germany). Her outstanding achievements in mathematics meant nothing to the Nazis, because Noether was Jewish, and in 1933, Noether was forced to emigrate to the United States and join the Institute for Advanced Study at Princeton University.

Imagine a well-isolated spaceship observed in a space-fixed coordinate system. Its energy is preserved, its center of mass moves along a straight line with constant velocity (the total, or center-of-mass, momentum vector is preserved), and it preserves the total angular momentum[2]. The same is true for a molecule or atom, but the conservation laws have to be formulated in the language of quantum mechanics.

Where did the conservation laws come from? Emmy Noether proved that they follow from the symmetry operations, with respect to which the equation of motion is invariant.[3]

Thus, it turns out that invariance of the equation of motion with respect to an arbitrary translation in time (time homogeneity) results in the energy conservation principle; with respect to translation in space (space homogeneity) gives the total momentum conservation principle; and with respect to rotation in space (space isotropy) implies the total angular momentum conservation principle.

---

[2] That is, its length and direction are key. Think of a skater performing a spin: extending the arms sideways slows down the rotation, while stretching them along the axis of rotation results in faster rotation. But all the time, the total angular momentum vector is the same. Well, what happens to the angular momentum when the dancer finally stops rotating due to friction? The angular momentum attains zero? No–that is simply impossible. When the dancer increases her angular velocity, the Earth's axis changes its direction a bit to preserve the previous angular momentum of the total system (the earth + the dancer). When the dancer stops, the Earth's axis comes back to its previous position, but not completely, because a part of the angular momentum is hidden in the rotation of molecules caused by the friction. Whatever happens, the total angular momentum must be preserved! If the spaceship captain wanted to stop the rotation of the ship, which is making the crew sick, he could either throw something (e.g., gas from a steering jet) away from the ship, or spin a well-oriented body fast inside the ship. But even the captain is unable to change the *total* angular momentum.

[3] In the case of a one-parameter family of operations, $\hat{S}_\alpha \hat{S}_\beta = \hat{S}_{\alpha+\beta}$; e.g., translation ($\alpha$, β stand for the translation vectors), rotation (α, β are rotational angles), etc. Some other operations may not form such families, and then the Noether theorem is no longer valid. This was an important discovery. Symmetry of a theory is much more fundamental than the symmetry of an object. The symmetry of a theory means that phenomena are described by the same equations no matter what laboratory coordinate system is chosen.

These principles may be regarded as the foundations of science.[4] The homogeneity of time allows the expectation that repeating experiments will give the same results. The homogeneity of space makes it possible to compare the results of the same experiments carried out in two different laboratories. Finally, the isotropy of space allows one to reject any suspicion that a different orientation of our laboratory bench changes the result.

Conservation laws represent most precious information about our system. It is not important what happens to the *isolated* system, what it is composed of, how complex the processes taking place in it are, whether they are slow or violent, whether there are people in the system or not, or whether they think how to cheat the conservation laws or not. Nothing can happen if it violates the conservation of energy, momentum, or angular momentum.

Now, let us incorporate this into quantum mechanics.

All symmetry operations (e.g., translation, rotation, reflection in a plane) are isometric; i.e., $\hat{U}^\dagger = \hat{U}^{-1}$ and $\hat{U}$ does not change the distance between points of the transformed object (Figs. 2.2 and 2.3).

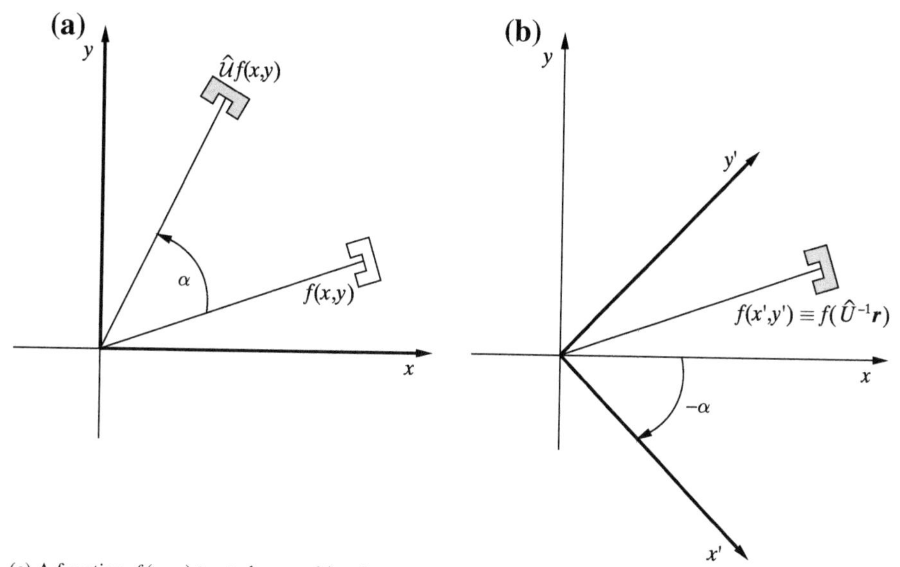

Fig. 2.2. (a) A function $f(x, y)$ treated as an object is rotated by angle α. (b) the coordinate system is rotated by angle $-\alpha$. The new position of the object in the old coordinate system (a) is the same as the initial position of the object in the new coordinate system (b).

[4] Well, this is true to some extent. For example, the Universe does not show an exact isotropy because the matter there does not show spherical symmetry. Moreover, even if only one object were in the Universe, this very object would itself destroy the anisotropy of the Universe. We should rather think of this as a kind of idealization (approximation of reality).

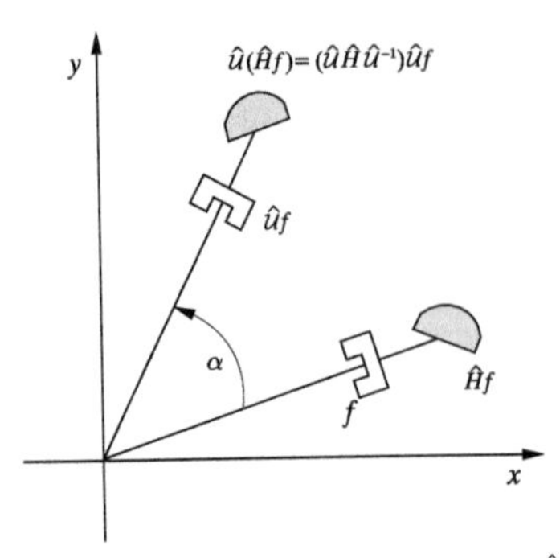

Fig. 2.3. A schematic view from the Hilbert space of functions $f(x, y)$. The f and $\hat{H}f$ represent different functions. Rotation (by α) of function $\hat{H}f$ gives function $\hat{\mathcal{U}}(\hat{H}f)$ and, in consequence, is bound to denote the *rotation of* f (i.e., $\hat{U}f$) and the transformation $\hat{U}\hat{H}\hat{U}^{-1}$ of the operator $\hat{H}$. Indeed, only then $\hat{\mathcal{U}}\hat{H}\hat{\mathcal{U}}^{-1}$ acting on the rotated function (i.e., $\hat{\mathcal{U}}f$) gives $\hat{\mathcal{U}}\hat{H}\hat{\mathcal{U}}^{-1}(\hat{\mathcal{U}}f) = \hat{\mathcal{U}}(\hat{H}f)$; i.e., the rotation of the result. Because of $\hat{\mathcal{U}}(\hat{H}f) = (\hat{\mathcal{U}}\hat{H})(\hat{\mathcal{U}}f)$, when verifying the invariance of $\hat{H}$ with respect to transformation $\hat{\mathcal{U}}$, it is sufficient to check whether $\hat{\mathcal{U}}\hat{H}$ has the same formula as $\hat{H}$, but expressed in the new coordinates. Only this $\hat{\mathcal{U}}\hat{H}$ will fit f expressed in the new coordinates; i.e., to $\hat{\mathcal{U}}f$. This is how we will proceed shortly.

The operator $\hat{U}$ acting in 3D Cartesian space corresponds to the operator $\hat{\mathcal{U}}$ acting in the Hilbert space, cf., Eq. (C.2), p. e20. Thus, the function $f(\mathbf{r})$ transforms to $f' = \hat{\mathcal{U}}f = f(\hat{U}^{-1}\mathbf{r})$, while the operator $\hat{A}$ transforms to $\hat{A}' = \hat{\mathcal{U}}\hat{A}\hat{\mathcal{U}}^{-1}$ (Fig. 2.3). The formula for $\hat{A}'$ differs in general from $\hat{A}$, but when it does not (i.e., $\hat{A}' = \hat{A}$), then $\hat{\mathcal{U}}$ commutes with $\hat{A}$.

Indeed, then $\hat{A} = \hat{\mathcal{U}}\hat{A}\hat{\mathcal{U}}^{-1}$ (i.e., one has the commutation relation $\hat{A}\hat{\mathcal{U}} = \hat{\mathcal{U}}\hat{A}$), which means that $\hat{\mathcal{U}}$ and $\hat{A}$ share their eigenfunctions (see Appendix B available at booksite.elsevier.com/978-0-444-59436-5, on p. e7).

Let us take the Hamiltonian $\hat{H}$ as the operator $\hat{A}$. Before writing it down, let us introduce *atomic units*. Their justification comes from something similar to laziness. The quantities one calculates in quantum mechanics are stuffed up by some constants: $\hbar = \frac{h}{2\pi}$, where h is the Planck constant; electron charge $-e$; its (rest) mass m_0; etc. These constants appear in clumsy formulas with various powers, in the numerator and denominator (see Table of Units, end of this book). One always knows, however, that the quantity one computes is energy, length, time, etc. and knows how the unit energy, the unit length, etc. are expressed by $\hbar$, e, m_0.

Atomic Units

If one inserts $\hbar = 1$, $e = 1$, $m_0 = 1$, this gives a dramatic simplification of the formulas. One has to remember, though, that these units have been introduced and, whenever needed, one can evaluate the result in other units (see Table of conversion coefficients, end of this book).

The Hamiltonian for a system of M nuclei (with charges Z_I and mass m_I, $I = 1, \ldots, M$) and N electrons, in the non-relativistic approximation and assuming point-like particles without any internal structure[5], takes in atomic units (a.u.) the following form (see p. 18)

$$\hat{H} = \hat{T}_n + \hat{T}_e + \hat{V}, \tag{2.1}$$

where the kinetic energy operators for the nuclei and electrons (in a.u.) read as

$$\hat{T}_n = -\frac{1}{2}\sum_{I=1}^{M} \frac{1}{m_I}\Delta_I \tag{2.2}$$

$$\hat{T}_e = -\frac{1}{2}\sum_{i=1}^{N} \Delta_i, \tag{2.3}$$

where the Laplacians are

$$\Delta_I = \frac{\partial^2}{\partial X_I^2} + \frac{\partial^2}{\partial Y_I^2} + \frac{\partial^2}{\partial Z_I^2},$$

$$\Delta_i = \frac{\partial^2}{\partial x_i^2} + \frac{\partial^2}{\partial y_i^2} + \frac{\partial^2}{\partial z_i^2},$$

and the Cartesian coordinates of the nuclei and electrons are indicated by vectors $\boldsymbol{R}_I = (X_I, Y_I, Z_I)$ and $\boldsymbol{r}_i = (x_i, y_i, z_i)$, respectively.

---

[5] No internal structure of the electron has yet been discovered. The electron is treated as a point-like particle. Contrary to this, nuclei have a rich internal structure and nonzero dimensions.

A clear multi-level-like structure appears (which has to a large extent forced a similar structure on the corresponding scientific methodologies):

- Level I. A nucleon (neutron, proton) consists of three (valence) quarks, clearly seen on the scattering image obtained for the proton. Nobody has yet observed a free quark.
- Level II. The strong forces acting among nucleons have a range of about $1-2$ fm($1\text{ fm} = 10^{-15}$ m). Above $0.4-0.5$ fm, they are attractive, while at shorter distances they correspond to repulsion. One need not consider their quark structure when computing the forces among nucleons, but they may be treated as particles *without internal structure*. The attractive forces between nucleons practically do not depend on the nucleon's charge and are so strong that they may overcome the Coulomb repulsion of protons. Thus, the nuclei composed of many nucleons (various chemical elements) may be formed, which in mean field theory exhibit a shell structure (analogous to electronic structure, cf., Chapter 8) related to the packing of the nucleons. The motion of the nucleons is strongly correlated. A nucleus may have various energy states (ground and excited), may be distorted, may undergo splitting, etc. About 2000 nuclei are known, of which only 270 are stable. The smallest nucleus is the proton, and the largest known so far is ^{209}Bi (209 nucleons). The largest observed number of protons in a nucleus is 118. Even the largest nuclei have diameters about 100000 times smaller than the electronic shells of the atom. Even for an atom with atomic number 118, the first Bohr radius is equal to $\frac{1}{118}$ a.u. or $5 \cdot 10^{-13}$ m, which is still about 100 times larger than the nucleus.
- Level III. Chemists *can* neglect the internal structure of nuclei. A nucleus *can* be treated as a structureless point-like particle, and using the theory described in this book, one is able to predict extremely precisely virtually all the chemical properties of atoms and molecules. Some interesting exceptions will be given at the end of Chapter 6.

The operator $\hat{V}$ corresponds to the electrostatic interaction of all the particles (nucleus-nucleus, nucleus-electron, electron-electron)[6]:

$$\hat{V} = \sum_{I=1}^{M}\sum_{J>I}^{M}\frac{Z_I Z_J}{|\boldsymbol{R}_I - \boldsymbol{R}_J|} - \sum_{I=1}^{M}\sum_{i=1}^{N}\frac{Z_I}{|\boldsymbol{r}_i - \boldsymbol{R}_I|} + \sum_{i=1}^{N}\sum_{j>i}^{N}\frac{1}{|\boldsymbol{r}_i - \boldsymbol{r}_j|}, \tag{2.4}$$

or, in a simplified form,

$$\hat{V} = \sum_{I=1}^{M}\sum_{J>I}^{M}\frac{Z_I Z_J}{R_{IJ}} - \sum_{I=1}^{M}\sum_{i=1}^{N}\frac{Z_I}{r_{iI}} + \sum_{i=1}^{N}\sum_{j>i}^{N}\frac{1}{r_{ij}}. \tag{2.5}$$

If the Hamiltonian turned out to be invariant with respect to a symmetry operation $\hat{U}$ (translation, rotation, etc.), this would imply the commutation of $\hat{U}$ and $\hat{H}$. We will check this in more detail below.

Note that the distances R_{IJ}, r_{iI} and r_{ij} in the Coulombic potential energy (2.5) witness about assumption of instantaneous interactions in non-relativistic theory (infinite speed of traveling the interaction through space).

2.1.2 Invariance with Respect to Translation

Translation by vector $\boldsymbol{T}$ of function $f(\boldsymbol{r})$ in space means that the function $\hat{\mathcal{U}} f(\boldsymbol{r}) = f(\hat{U}^{-1}\boldsymbol{r}) = f(\boldsymbol{r} - \boldsymbol{T})$ i.e., an opposite (by vector $-\boldsymbol{T}$) translation of the coordinate system (Fig. 2.4).

Transformation $\boldsymbol{r}' = \boldsymbol{r} + \boldsymbol{T}$ does not change the Hamiltonian. This is evident for the potential energy $\hat{V}$, because the translations $\boldsymbol{T}$ cancel out, leaving the interparticle distances unchanged. For the kinetic energy, one obtains $\frac{\partial}{\partial x'} = \sum_{\sigma=x,y,z}\frac{\partial \sigma}{\partial x'}\frac{\partial}{\partial \sigma} = \frac{\partial x}{\partial x'}\frac{\partial}{\partial x} = \frac{\partial}{\partial x}$, and all the kinetic energy operators (Eqs. (2.2) and (2.3)) are composed of the operators having this form.

> The Hamiltonian is therefore invariant with respect to any translation of the coordinate system.

There are two main consequences of translational symmetry:

- Whether the coordinate system used is fixed in Trafalgar Square or in the center of mass of the system, one has to solve the same mathematical problem.
- The solution to the Schrödinger equation corresponding to the space-fixed coordinate system (SFS) located in Trafalgar Square is Ψ_{pN}, whereas Ψ_{0N} is calculated in the body-fixed coordinate system (see Appendix I available at booksite.elsevier.com/978-0-444-59436-5)

[6] We do not include in this Hamiltonian tiny magnetic interactions of electrons and nuclei coming from their spin and orbital angular momenta, because they are of relativistic nature (see Chapter 3). In Chapter 12, we will be interested just in such small magnetic effects and the Hamiltonian will have to be generalized to include these interactions.

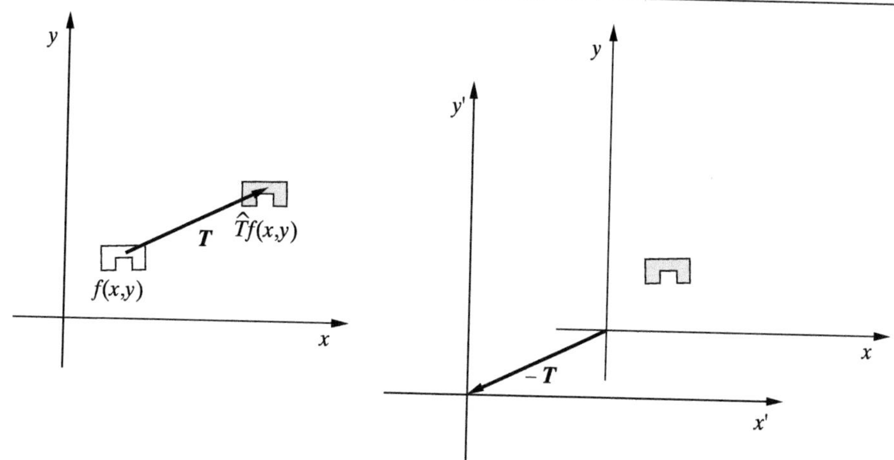

Fig. 2.4. A function f shifted by vector $\boldsymbol{T}$ (symmetry operation $\hat{T}$); i.e., $\hat{T} f(x, y)$ in the coordinate system (x, y) is the same as function $f(x', y')$, in the coordinate system (x', y') shifted by $-\boldsymbol{T}$.

located in the center of mass at $\boldsymbol{R}_{CM}$, which moves in the SFS with the total momentum $\boldsymbol{p}_{CM}$. These two solutions are related by[7] $\Psi_{pN} = \Psi_{0N} \exp(i\boldsymbol{p}_{CM} \cdot \boldsymbol{R}_{CM})$. The number $N = 0, 1, 2, \ldots$ counts the energy states after the center-of-mass motion is separated.

This means that the energy spectrum represents a continuum, because the center of mass may have any non-negative kinetic energy $p_{CM}^2/(2m)$. If, however, one assumes that $p_{CM} = 0$, then the energy spectrum is discrete for low-energy eigenvalues (see Eq. 1.13).

This spectrum corresponds to the bound states; i.e., those states that do not correspond to any kind of dissociation (including ionization). Higher-energy states lead to dissociation of the molecule, and the fragments may have any kinetic energy. Therefore, above the discrete spectrum, one has a continuum of states. The states Ψ_{0N} will be called *spectroscopic states.* The bound states Ψ_{0N} are square-integrable, as opposed to Ψ_{pN}, which are not because of function $\exp(i\boldsymbol{p}_{CM} \cdot \boldsymbol{R}_{CM})$, which describes the free motion of the center of mass.

2.1.3 Invariance with Respect to Rotation

The Hamiltonian is also invariant with respect to any rotation in space $\hat{U}$ of the coordinate system about a fixed axis. The rotation is carried out by applying an orthogonal matrix transformation[8] $\boldsymbol{U}$ of vector $\boldsymbol{r} = (x, y, z)^T$ that describes any particle of coordinates x, y, z. Therefore, all the

[7] This follows from the separation of the center-of-mass motion (see Appendix I available at booksite.elsevier.com/978-0-444-59436-5) and noting that $\exp(i\boldsymbol{p}_{CM} \cdot \boldsymbol{R}_{CM})$ represents a solution for the motion of a free particle (Chapter 4).

[8] $\boldsymbol{U}^T = \boldsymbol{U}^{-1}$.

particles undergo the same rotation and the new coordinates are $\boldsymbol{r}' = \hat{U}\boldsymbol{r} = \boldsymbol{U}\boldsymbol{r}$. Again, there is no problem with the potential energy because a rotation does not change the interparticle distances. What about the Laplacians in the kinetic energy operators? Let us see:

$$
\begin{aligned}
\Delta &= \sum_{k=1}^{3} \frac{\partial^2}{\partial x_k^2} = \sum_{k=1}^{3} \frac{\partial}{\partial x_k}\frac{\partial}{\partial x_k} = \sum_{k=1}^{3}\left(\sum_{i=1}^{3}\frac{\partial}{\partial x_i'}\frac{\partial x_i'}{\partial x_k}\right)\left(\sum_{i=1}^{3}\frac{\partial}{\partial x_i'}\frac{\partial x_i'}{\partial x_k}\right) \\
&= \sum_{i=1}^{3}\sum_{j=1}^{3}\sum_{k=1}^{3}\left(\frac{\partial}{\partial x_i'}\frac{\partial x_i'}{\partial x_k}\right)\left(\frac{\partial}{\partial x_j'}\frac{\partial x_j'}{\partial x_k}\right) \\
&= \sum_{i=1}^{3}\sum_{j=1}^{3}\sum_{k=1}^{3}\left(\frac{\partial}{\partial x_i'}U_{ik}\right)\left(\frac{\partial}{\partial x_j'}U_{jk}\right) = \sum_{i=1}^{3}\sum_{j=1}^{3}\sum_{k=1}^{3}\left(\frac{\partial}{\partial x_i'}U_{ik}\right)\left(\frac{\partial}{\partial x_j'}U_{kj}^{\dagger}\right) \\
&= \sum_{i=1}^{3}\sum_{j=1}^{3}\left(\frac{\partial}{\partial x_i'}\right)\left(\frac{\partial}{\partial x_j'}\right)\sum_{k=1}^{3} U_{ik}U_{kj}^{\dagger} \\
&= \sum_{i=1}^{3}\sum_{j=1}^{3}\left(\frac{\partial}{\partial x_i'}\right)\left(\frac{\partial}{\partial x_j'}\right)\delta_{ij} = \sum_{k=1}^{3}\frac{\partial^2}{\partial\left(x_k'\right)^2} = \Delta'.
\end{aligned}
$$

Thus, one has invariance of the Hamiltonian with respect to any rotation about the origin of the coordinate system. This means (see p. e73) that the Hamiltonian and the operator of the square of the total angular momentum $\hat{J}^2$ (as well as of one of its components, denoted by $\hat{J}_z$) commute. One is able, therefore, to measure simultaneously the energy, the square of total angular momentum, and one of the components of total angular momentum, and (as will be shown in Chapter 4) one has ($\boldsymbol{r}$ and $\boldsymbol{R}$ denote the electronic and the nuclear coordinates, respectively)

$$\hat{J}^2\Psi_{ON}\left(\boldsymbol{r},\boldsymbol{R}\right) = J(J+1)\hbar^2\Psi_{ON}\left(\boldsymbol{r},\boldsymbol{R}\right) \tag{2.6}$$

$$\hat{J}_z\Psi_{ON}\left(\boldsymbol{r},\boldsymbol{R}\right) = M_J\hbar\Psi_{ON}\left(\boldsymbol{r},\boldsymbol{R}\right), \tag{2.7}$$

where $J = 0, 1, 2, \ldots$ and $M_J = -J, -J+1, \ldots + J$.

Any rotation matrix may be shown as a product of "*elementary*" rotations, each about axes x, y, or z. For instance, rotation about the y axis by angle θ corresponds to the matrix $\begin{pmatrix} \cos\theta & 0 & -\sin\theta \\ 0 & 1 & 0 \\ \sin\theta & 0 & \cos\theta \end{pmatrix}$. The pattern of such matrices is simple: one has to put in some places sines, cosines, zeros, and 1s with the proper signs.[9] This matrix is orthogonal[10]; i.e., $U^T = U^{-1}$, which the reader may easily check. The product of two orthogonal matrices represents an orthogonal matrix; therefore, any rotation corresponds to an orthogonal matrix.

2.1.4 Invariance with Respect to Permutation of Identical Particles (Fermions and Bosons)

The Hamiltonian also has permutational symmetry. This means that if one exchanged labels numbering the identical particles, independently of how it was done, one would always obtain the identical mathematical expression for the Hamiltonian. This implies that any wave function has to be either symmetric (for bosons) or antisymmetric (for fermions) with respect to the exchange of labels between two identical particles (cf., p. 34).

2.1.5 Invariance of the Total Charge

In addition to energy, momentum, and angular momentum, strict conservation laws are obeyed exclusively for the total electric charge and the baryon and lepton numbers (a given particle contributes $+1$, the corresponding antiparticle -1).[11] The charge conservation follows from the gauge symmetry. Total electric charge conservation follows from the fact that description of the system has to be invariant with respect to the mixing of the particle and antiparticle states, which is analogous to rotation.

2.1.6 Fundamental and Less Fundamental Invariances

The conservation laws described are of a fundamental character because they are related to the homogeneity of space and time, the isotropy of space, and the non-distinguishability of identical particles.

Besides these strict conservation laws (energy, momentum, angular momentum, permutation of identical particles, charge, and baryon and lepton numbers), there are also some approximate laws. Two of these: parity and charge conjugation, will be discussed below. They are rooted in these strict laws, but are valid only in some conditions. For example, in most experiments, not only the baryon number, but also the number of nuclei of each kind, are conserved. Despite the importance of this law in chemical reaction equations, this does not represent any strict conservation law, as shown by radioactive transmutations of elements.

---

[9] Clockwise and anticlockwise rotations and two possible signs at sines make memorizing the right combination difficult. In order to choose the correct one, one may use the following trick. First, decide that what moves is an *object* (e.g., a function, not the coordinate system). Then, you have to have in pocket my book. With Fig. 2.2a, one sees that the rotation of the point with coordinates $(1, 0)$ by angle $\theta = 90°$ should give the point $(0, 1)$, and this is assured only by the rotation matrix: $\begin{pmatrix} \cos\theta & -\sin\theta \\ \sin\theta & \cos\theta \end{pmatrix}$.

[10] Therefore, it is also unitary (see Appendix A available at booksite.elsevier.com/978-0-444-59436-5, on p. e1).

[11] For instance, in the Hamiltonian 2.1, it is assumed that whatever might happen to our system, the numbers of the nucleons and electrons will remain constant.

Some other approximate conservation laws will soon be discussed.

2.1.7 Invariance with Respect to Inversion–Parity

There are orthogonal transformations that are not equivalent to any rotation; e.g., the matrix of inversion

$$\begin{pmatrix} -1 & 0 & 0 \\ 0 & -1 & 0 \\ 0 & 0 & -1 \end{pmatrix},$$

which corresponds to changing $\boldsymbol{r}$ to $-\boldsymbol{r}$ for all the particles, does not represent any rotation. If one performs such a symmetry operation, the Hamiltonian remains invariant and $|\Psi_{0N}(-\boldsymbol{r}, -\boldsymbol{R})|^2 = |\Psi_{0N}(\boldsymbol{r}, \boldsymbol{R})|^2$. This is evident both for $\hat{V}$ (the interparticle distances do not change), and for the Laplacian (single differentiation changes sign, while double does not). Two consecutive inversions indicate an identity operation: $\Psi_{0N}(\boldsymbol{r}, \boldsymbol{R}) = \exp(i\alpha)\Psi_{0N}(-\boldsymbol{r}, -\boldsymbol{R}) = [\exp(i\alpha)]^2 \Psi_{0N}(\boldsymbol{r}, \boldsymbol{R})$. Hence, $[\exp(i\alpha)]^2 = 1$, $\exp(i\alpha) = \pm 1$, and one has

$$\Psi_{0N}\left(-\boldsymbol{r}, -\boldsymbol{R}\right) = \Pi \Psi_{0N}\left(\boldsymbol{r}, \boldsymbol{R}\right), \quad \text{where } \Pi \in \{1, -1\}.$$

Therefore,

> the wave function of a stationary state represents an eigenfunction of the inversion operator, and the eigenvalue can be either $\Pi = 1$ or $\Pi = -1$. This property is known as parity (P).

Now the reader will be taken by surprise. From what we have said, it follows that no molecule has a nonzero dipole moment. Indeed, the dipole moment is calculated as the mean value of the dipole moment operator [i.e., $\boldsymbol{\mu} = \langle \Psi_{0N} | \hat{\boldsymbol{\mu}} \Psi_{0N} \rangle = \langle \Psi_{0N} | \left(\sum_i q_i \boldsymbol{r}_i\right) \Psi_{0N} \rangle$]. This integral will be calculated very easily: the integrand is antisymmetric with respect to inversion,[12] and therefore, $\boldsymbol{\mu} = \boldsymbol{0}$.

So, is the very meaning of the dipole moment, a quantity often used in chemistry and physics, a fantasy? If HCl has no dipole moment, then it is more understandable that H_2 does not either. All this seems absurd, though. What about this dipole moment?

Let us stress that our conclusion pertains to the *total* wave function, which has to reflect the space isotropy leading to the zero dipole moment, because all orientations in space are equally probable. If one applied the transformation $\boldsymbol{r} \rightarrow -\boldsymbol{r}$ only to *some* particles in the molecule (e.g., electrons), and not to others (e.g., the nuclei), the wave function will show no parity (it would be neither symmetric nor antisymmetric). We will introduce the Hamiltonian in Chapter 6, which corresponds to immobilizing the nuclei (the adiabatic or Born-Oppenheimer approximation) in certain positions in space, and in such a case, the wave function depends on the electronic

[12] Ψ_{0N} may be symmetric or antisymmetric, but $|\Psi_{0N}|^2$ is bound to be symmetric. Therefore, since $\sum_i q_i \boldsymbol{r}_i$ is antisymmetric, then indeed, the integrand is antisymmetric (while the integration limits are symmetric).

coordinates only. This wave function may be neither symmetric nor antisymmetric with respect to the *partial* inversion transformation $\boldsymbol{r} \rightarrow -\boldsymbol{r}$ (for the electrons only). To give an example, let us imagine the HF molecule in a coordinate system, with its origin in the middle between the H and F nuclei. Consider a particular configuration of the 10 electrons of the molecule, all close to the fluorine nucleus in some well-defined points. One may compute the value of the wave function for this configuration of electrons. Its square gives us the probability density of finding this particular configuration of electrons. Now, imagine the (*partial*) inversion $\boldsymbol{r} \rightarrow -\boldsymbol{r}$ applied to all the electrons. Now they will all be close to the proton. If one computes the probability density for the new situation, one would obtain a completely different value (which is much, much smaller because the electrons prefer fluorine, not hydrogen). There is neither symmetry nor antisymmetry. No wonder, therefore, that if one computed $\boldsymbol{\mu} = \langle \Psi_{0N} | \hat{\boldsymbol{\mu}} \Psi_{0N} \rangle$ with such a function (integration is over the electronic coordinates only), the result would differ from zero. This is why chemists believe the HF molecule has a nonzero dipole moment.[13] On the other hand, if the molecule taken as the example were B_2 (also 10 electrons), then the two values have had to be equal because they describe the same physical situation. This corresponds, therefore, to a wave function with definite parity (symmetric or antisymmetric), and therefore, in this case, $\boldsymbol{\mu} = \boldsymbol{0}$. This is why chemists believe such molecules as H_2,B_2,O_2 have no dipole moment.

Product of Inversion and Rotation

The Hamiltonian is also invariant with respect to some other symmetry operations, like changing the sign of the x coordinates of all particles, or similar operations that are products of inversion and rotation. If one changed the sign of all the x coordinates, it would correspond to a mirror reflection. Since rotational symmetry stems from space isotropy (which we will treat as trivial), the mirror reflection may be identified with parity P.

Enantiomers

A consequence of inversion symmetry is that the wave functions have to be eigenfunctions of the inversion operator with eigenvalues $\Pi = 1$ (i.e., the wave function is symmetric), or $\Pi = -1$ (i.e., the wave function is antisymmetric). Any asymmetric wave function corresponding to a stationary state is therefore excluded (illegal). However, two optical isomers (enantiomers), corresponding to an object and its mirror image, do exist (Fig. 2.5).[14]

[13] Therefore, what they do measure? The answer will be given in Chapter 12.

[14] The property that distinguishes them is known as *chirality* (human hands are an example of chiral objects). The chiral molecules (enantiomers) exhibit optical activity; i.e., polarized light passing through a solution of one of the enantiomers undergoes a rotation of the polarization plane always in the same direction (which may be seen easily by reversing the direction of the light beam). The enantiomeric molecules have the same properties, provided one is checking this by using non-chiral objects. If the probe were chiral, one of the enantiomers would interact with it differently (for purely sterical reasons). Enantiomers (e.g., molecular associates) may be formed from chiral or non-chiral subunits.

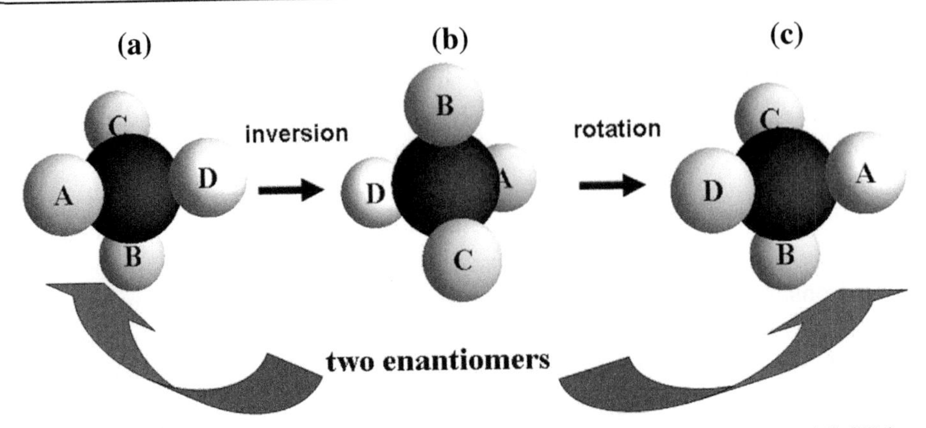

Fig. 2.5. Example of chirality. (a) A molecule central atom-(ABCD) with four different substituents: A,B,C,D in a nonplanar configuration; (b) the same molecule after applying the inversion operation (with respect to the central atom); (c) an attempt of superposing the initial and the transformed molecules by rotation, to get matching, fails; (a) and (c) represent an example of a pair of enantiomers. Each of these isomers, after reflection in a mirror (e.g., the central atom-BC plane) becomes identical to its partner.

We ask in a pharmacy for D-glucose (strangely enough the pharmacist is fully cooperative and does not make trouble). We pay a small sum, and he gives us something that should not exist[15] – a substance with a single enantiomer. We should obtain a substance composed of molecules in their stationary states, which therefore have to have definite parity, either as a sum of the wave functions for the two enantiomers D and L ($\Pi = 1$, cf., Appendix D available at booksite.elsevier.com/978-0-444-59436-5, on p. e65, Example I) : $\psi_+ = \psi_D + \psi_L$ or as the difference ($\Pi = -1$): $\psi_- = \psi_D - \psi_L$. The energies corresponding to ψ_+ and ψ_- differ, but the difference is extremely small (quasi-degeneracy). The brave shopkeeper has given us something with the wave function $\psi = N\left(\psi_+ + \psi_-\right) = \psi_D$ (the result of decoherence), which therefore describes a non-stationary state.[16] As we will see later in this chapter (p. 94), the approximate lifetime of the state is proportional to the inverse of the integral $\left\langle \psi_D \right| \hat{H}\psi_L \Big\rangle$. If one computed this integral, one would obtain an extremely small

Chen Ning Yang (b. 1922) and Tsung Dao Lee (b. 1926) American physicists and professors at the Advanced Study Institute in Princeton, predicted in 1956 that parity would break in the weak interactions, which a few months later was confirmed experimentally by Madame Chien-Shung Wu. In 1957, Yang and Lee received the Nobel Prize *"for their penetrating investigation of parity laws, which led to important discoveries regarding elementary particles."*

[15] Stated more exactly, it should be unstable.

[16] Only ψ_+ and ψ_- are stationary states.

number.[17] It would turn out that the pharmacy could safely keep the stock of glucose for millions of years. Maybe the reason for decoherence is interaction with the rest of the Universe, maybe even interaction with a vacuum. The very existence of enantiomers, or even if one of them prevails on Earth, does not mean breaking parity symmetry. This would happen if one of the enantiomers corresponded to a lower energy than the other.[18]

---

[17] This is seen even after attempting to overlap two molecular models physically (see Fig. 2.5). The overlap of the wave functions will be small for the same reasons (the wave functions decay exponentially with distance).

[18] This is what happens in reality, although the energy difference is extremely small. Experiments with β-decay have shown that nature breaks parity in weak interactions. Parity conservation law, therefore, has an approximate character.

With no preference for any of the enantiomers, one of them may spontaneously increase its concentration until it reaches 100%; i.e., with a complete elimination of the other one (0%). Is something like that possible at all? It seems to contradict common sense, since one of the enantiomers won, while the other lost, whereas their chances were exactly equal.

This phenomenon occurs in reality, if autocatalysis is involved. The key information is the following: in such a system, a large, random, and self-augmenting fluctuation is possible. Indeed, let us imagine just for simplicity 50 molecules of D and 50 molecules of L, together with a certain number of molecules N (let us call them "*neutral*"), with the following possible reactions (giving equal chances to D and L):

$$D + L \rightarrow 2N$$
$$D + N \rightarrow 2D$$
$$L + N \rightarrow 2L.$$

The last two reactions (of autocatalytic character) represent the induction of an enantiomer ("*forcing chirality*") through the interaction of N with a molecule of a given enantiomer. No preference of D or L is assumed in this induction. Despite of this, one of the enantiomers will defeat the other. To explain this, let us assume that the elementary reactions 1,2, or 3 of the individual molecules form a random chain in time (for example: 1, 2, 2, 3, 1, 3, ...). If the chain starts by the first reaction, we have a situation similar to the starting one (49 : 49 instead of 50 : 50, in the number of D:L molecules). If, however, the chain starts from the second reaction, there will be a 51 : 50 preference of D, while when the third reaction takes place, one will have 50 : 51 preference of L. In the last two cases, we have a deviation from the equal chances of two enantiomers (fluctuation). Suppose that we have the second case; i.e., 51 : 50 preference of D. Note, that now, when continuing the reaction chain at random, the chance the reaction 2 to happen, is greater than the chance of reaction 3. It is seen, therefore, that any fluctuation from the "*equal-chances situation*" has a tendency to self-augment, although there is always a chance that the fluctuation disappears. The racemic mixture, therefore, represents an instable system. Sooner or later, there will be a transition (first mild, but then more and more violent) from racemate to the absolute prevailing of one of the enantiomers. Their chances were and are equal, but just by chance, one of them won.

Sometimes random processes may lead to large fluctuations. For instance, Louis Pasteur was able to crystallize the crystals of one enantiomer and the crystals of the other enantiomer and then to separate them. This is possible not only without Pasteur, but also without any human, just spontaneously. Suppose that two such pieces of crystals were created where the primordial life was to be started, and that afterward, a volcano explosion took place, in which the volcanic lava covered (or even destroyed) one of the crystals, eliminating it from anything what happened later (e.g., the chiral induction during biological evolution). This simple argument, proposed by Leszek Stolarczyk, is worth considering in light of the fact that in nature, there is a strong preference for one chirality.

2.1.8 Invariance with Respect to Charge Conjugation

If one changed the signs of the charges of all particles, the Hamiltonian would not change.

This therefore corresponds to exchanging particles and antiparticles. Such a symmetry operation is called the *charge conjugation* and denoted as C symmetry. This symmetry will be not marked in the wave function symbol (because as a rule, we are dealing with matter, not antimatter), but we will want to remember it later. Sometimes it may turn out unexpectedly to be useful (see Chapter 13, p. 820). After Wu's experiment, physicists tried to save the hypothesis that what is conserved is the CP symmetry; i.e., the product of charge conjugation and inversion. However, analysis of experiments with the meson K decay has shown that even this symmetry is approximate (although the deviation is extremely small).

2.1.9 Invariance with Respect to the Symmetry of the Nuclear Framework

In many applications, the positions of the nuclei are fixed (clamped nuclei approximation; see Chapter 6), often in a high-symmetry configuration (see Appendix C available at booksite.elsevier.com/978-0-444-59436-5, on p. e17). For example, the benzene molecule in its ground state (after minimizing the energy with respect to the positions of the nuclei) has the symmetry of a regular hexagon. In such cases, the electronic Hamiltonian additionally exhibits invariance with respect to some symmetry operations, and therefore, the wave functions are the eigenstates of these molecular symmetry operations. Therefore, any wave function may have an additional label: namely, the symbol of the irreducible representation[19] it belongs to.

2.1.10 Conservation of Total Spin

In an isolated system, the *total* angular momentum $\boldsymbol{J}$ is conserved. However, $\boldsymbol{J} = \boldsymbol{L} + \boldsymbol{S}$, where $\boldsymbol{L}$ and $\boldsymbol{S}$ stand for the orbital and spin angular momenta (sum over all particles), respectively. The spin angular momentum $\boldsymbol{S}$, being a sum over all particles, is not conserved.

However, the (non-relativistic) Hamiltonian does not contain any spin variables. This means that it commutes with the operator of the square of the total spin, as well as with the operator of one of the spin components (by convention, the z component). Therefore, in the non-relativistic approximation, one can simultaneously measure the energy E, the square of the spin S^2, and one of its components: S_z.

[19] This is the representation of the symmetry group composed of the symmetry operations mentioned above.

2.1.11 Indices of Spectroscopic States

In summary, assumptions about the homogeneity of space and time, isotropy of space, and parity conservation lead to the following quantum numbers (indices) for the spectroscopic states:

- N quantizes energy.
- J quantizes the length of total angular momentum.
- M quantizes the z component of total angular momentum.
- Π determines parity:

$$\Psi_{N,J,M,\Pi}\left(\boldsymbol{r}, \boldsymbol{R}\right).$$

Besides these indices that follow from the fundamental laws (in the case of parity, it is a little too exaggerated), there may be also some indices related to less fundamental conservation laws, as follows:

- The irreducible representation index of the symmetry group of the clamped nuclei Hamiltonian (see Appendix C available at booksite.elsevier.com/978-0-444-59436-5)
- The values of S^2 (traditionally, one gives the *multiplicity* $2S+1$) and S_z

2.2 Schrödinger Equation for Stationary States

Erwin Schrödinger (1887–1961), Austrian physicist and professor at Jena, Stuttgart, Graz, Breslau, Zurich, Berlin and Vienna.

In later years, Schrödinger recalled the Zurich period most warmly, especially discussions with mathematician Hermann Weyl and physicist Peter Debye. In 1927, Schrödinger succeeded Max Planck in the Chair of Theoretical Physics at the University of Berlin, and in 1933 received the Nobel Prize "*for the discovery of new productive forms of atomic theory.*" Hating the Nazi regime, he left Germany in 1933 and moved to the University of Oxford. However, homesick for his native Austria, he went back in 1936 and took a professorship at the University of Graz. Meanwhile, Hitler carried out his Anschluss with Austria in 1938, and Schrödinger, even though not a Jew, could have been an easy target as one who fled Germany because of the Nazis. He emigrated to the United States (Princeton), and then to Ireland (Institute for Advanced Studies in Dublin), worked there till 1956, then returned to Austria and taught at the University of Vienna until his death.

In his scientific work, as well as in his personal life, Schrödinger did not strive for big goals; rather, he worked by himself. Maybe what characterizes him best is that he always was ready to leave having belongings ready in his rucksack. Among the goals of this textbook listed in the Introduction there is no demoralization of youth. This is why I will stop here, limit myself to the carefully selected information given above and refrain from describing the circumstances, in which quantum mechanics was born. For those students who read the recommended in the Additional Literature, I provide some useful references: W. Moore, "*Schrödinger: Life and Thought*," Cambridge University Press, 1989, and the comments on the book given by P.W. Atkins, *Nature*, 341(1989), also http://www-history.mcs.st-andrews.ac.uk/history/Mathematicians/Schrodinger.html

Schrödinger's curriculum vitae found in Breslau (now Wrocław):

"Erwin Schrödinger, born on Aug., 12, 1887 in Vienna, the son of the merchant Rudolf Schrödinger and his wife née Lauer. The family of my father comes from the Upper Palatinate and Wirtemberg region, and the family of my mother from German Hungary and (from the maternal side) from England. I attended a so called "*academic*" high school (once part of the university) in my native town. Then during 1906–1910 I studied physics at Vienna University, where I graduated in 1910 as a doctor of physics. I owe my main inspiration to my respectable teacher Fritz Hasenöhrl, who by an unlucky fate was torn from his diligent students - he fell gloriously as an attack commander on the battlefield of Vielgereuth. As well as Hasenöhrl, I owe my mathematical education to Professors Franz Mertens and Wilhelm Wirtinger, and a certain knowledge of experimental physics to my principal of many years (1911–1920) Professor Franz Exner and my intimate friend R.M.F. Rohrmuth. A lack of experimental and some mathematical skills oriented me basically towards theory. Presumably the spirit of Ludwig Boltzmann (deceased in 1906), operating especially intensively in Vienna, directed me first towards the probability theory in physics. Then, (...) a closer contact with experimental works of Exner and Rohrmuth oriented me to the physiological theory of colors, in which I tried to confirm and develop the achievements of Helmholtz. In 1911–1920 I was a laboratory assistant under Franz Exner in Vienna, of course, with $4\frac{1}{2}$ years long pause caused by war. I have obtained my habilitation in 1914 at the University of Vienna, while in 1920 I accepted an offer from Max Wien and become his assistant professor at the new theoretical physics department in Jena. This lasted, unfortunately, only one semester, because I could not refuse a professorship at the Technical University in Stuttgart. I was there also only one semester, because in April 1921 I came to the University of Hessen in succession to Klemens Schrafer. I am almost ashamed to confess, that at the moment I sign the present curriculum vitae I am no longer a professor at the University of Breslau, because on Oct.15. I received my nomination to the University of Zurich. My instability may be recognized exclusively as a sign of my ingratitude!

Breslau, Oct., 5, 1921.
Dr Erwin Schrödinger

(found in the archives of the University of Wrocław (Breslau) by Professor Zdzisław Latajka and Professor Andrzej Sokalski, translated by Andrzej Kaim and the author. Since the manuscript was hardly legible due to Schrödinger's handwriting, some names may have been misspelled.)

It may be instructive to see how Erwin Schrödinger invented his famous equation 1.13 for stationary states ψ of energy E ($\hat{H}$ denotes the Hamiltonian of the system):

$$\hat{H}\psi = E\psi. \tag{2.8}$$

Schrödinger surprised the contemporary quantum elite (associated mainly with Copenhagen and Göttingen) with his clear formulation of quantum mechanics as wave mechanics. Many scientists regard January 27, 1926, when Schrödinger submitted a paper entitled "*Quantisierung als Eigenwertproblem*"[20] to *Annalen der Physik,* as the birthday of wave mechanics.

Most probably Schrödinger's reasoning was as follows. De Broglie discovered that what people called a particle also had a wave nature (as described in Chapter 1). That is really puzzling. If a wave is involved, then according to Debye's suggestion at the November seminar in Zurich, it might be possible to write the standing wave equation with $\psi(x)$ as its amplitude at position x:

$$v^2 \frac{d^2\psi}{dx^2} + \omega^2 \psi = 0, \tag{2.9}$$

where v stands for the (phase) velocity of the wave and ω represents its angular frequency ($\omega = 2\pi\nu$, where ν is the usual frequency), which is related to the wavelength λ by the well-known formula[21]

$$\omega/v = \frac{2\pi}{\lambda}. \tag{2.10}$$

Besides, Schrödinger knew from de Broglie, who had lectured in Zurich on this subject, that the wavelength, λ, is related to a particle's momentum p through $\lambda = h/p$, where $h = 2\pi\hbar$ is the Planck constant. This equation is the most famous achievement of de Broglie, and it relates the corpuscular (p) character and the wave (λ) character of any particle.

On the other hand, the momentum p is related to the total energy (E) and the potential energy (V) of the particle through $p^2 = 2m(E - V)$, which follows from the expression for the kinetic energy $T = \frac{mv^2}{2} = p^2/(2m)$ and $E = T + V$. Therefore, Eq. (2.9) can be rewritten as

$$\frac{d^2\psi}{dx^2} + \frac{1}{\hbar^2}[2m(E - V)]\psi = 0, \tag{2.11}$$

The most important step toward the great discovery was transferring the term with E to the right side. Let us see what Schrödinger obtained:

$$\left[-\frac{\hbar^2}{2m}\frac{d^2}{dx^2} + V\right]\psi = E\psi. \tag{2.12}$$

This was certainly a good moment for a discovery. Schrödinger obtained a kind of eigenvalue equation 1.13, recalling his experience with eigenvalue equations in the theory of liquids.[22] What

[20] This title translates as "*Quantization as an Eigenproblem.*" Well, once upon a time, quantum mechanics was discussed in German. Some traces of that period remain in the nomenclature. One is the "*eigenvalue*" problem, or "*eigenproblem,*" which is a German-English hybrid.

[21] In other words, $\nu = \frac{v}{\lambda}$ or $\lambda = vT$ (i.e., wavelength equals the velocity times the period). Equation (2.9) represents an oscillating function $\psi(x)$. Indeed, it means that $\frac{d^2\psi}{dx^2}$ and ψ differ by sign; i.e., if ψ is above the x axis, then it curves down, while if it is below the x axis, then it curves up.

[22] This is a very interesting coincidence: Werner Heisenberg was also involved in fluid dynamics. At the beginning, Schrödinger did not use operators. They appeared after he established closer contacts with the University of Göttingen.

is striking in Eq. (2.12) is the odd fact that an operator $-\frac{\hbar^2}{2m}\frac{d^2}{dx^2}$ amazingly plays the role of kinetic energy. Indeed, we see the following: *something* plus potential energy, all that multiplied by ψ, equals total energy times ψ. Therefore, this *something* clearly must be the kinetic energy! But, wait a minute. The kinetic energy is equal to $\frac{p^2}{2m}$. From this, it follows that, in the equation obtained instead of p, there is a certain *operator* $i\hbar\frac{d}{dx}$ or $-i\hbar\frac{d}{dx}$, because only then does the squaring give the right answer.

Hermann Weyl (1885–1955), German mathematician, professor at ETH Zurich, the University of Göttingen, and the Institute for Advanced Studies at Princeton (in the United States), expert in the theory of orthogonal series, eigenvalue problems, geometric foundations of physics, group theory and differential equations. Weyl adored Schrödinger's wife, was a friend of the family, and provided an ideal partner for Schrödinger in conversations about the eigenvalue problem.

Would the key to the puzzle be simply taking the classical expression for total energy and inserting the above operators instead of the momenta? What was Schrödinger supposed to do? The best choice is always to begin with the simplest toys, such as the free particle, the particle in a box, the harmonic oscillator, the rigid rotator, or the hydrogen atom. Nothing is known about whether Schrödinger himself had a sufficiently deep knowledge of mathematics to be able to solve the (sometimes non-trivial) equations related to these problems, or whether he had advice from a friend versed in mathematics, such as Hermann Weyl.

It turned out that instead of p, $-i\hbar\frac{d}{dx}$ had to be inserted rather than $i\hbar\frac{d}{dx}$ (Postulate II, Chapter 1).

2.2.1 Wave Functions of Class Q

The postulates of quantum mechanics, especially the probabilistic interpretation of the wave function given by Max Born, limit the class of functions allowed ("*class Q*," or "*quantum*").

Any wave function conforms to the following:

- It cannot be zero everywhere (Fig. 2.6a), because the system is somewhere in space.
- It has to be continuous (Fig. 2.6b). This also means that it cannot take infinite values at any point in space[23] (Fig. 2.6c,d).

[23] If this happened in any nonzero volume of space (Fig. 2.6d), the probability would tend to infinity (which is prohibited). However, the requirement is stronger than that: a wave function cannot take an infinite value even at a single point (Fig. 2.6c). Sometimes such functions appear among the solutions of the Schrödinger equation, and those have to be rejected. The formal argument is that, if not excluded from the domain of the Hamiltonian, the latter would be non-Hermitian when such a function was involved in $\left\langle f|\hat{H}g\right\rangle = \left\langle \hat{H}f|g\right\rangle$. A non-Hermitian Hamiltonian might lead to complex energy eigenvalues, which is prohibited.

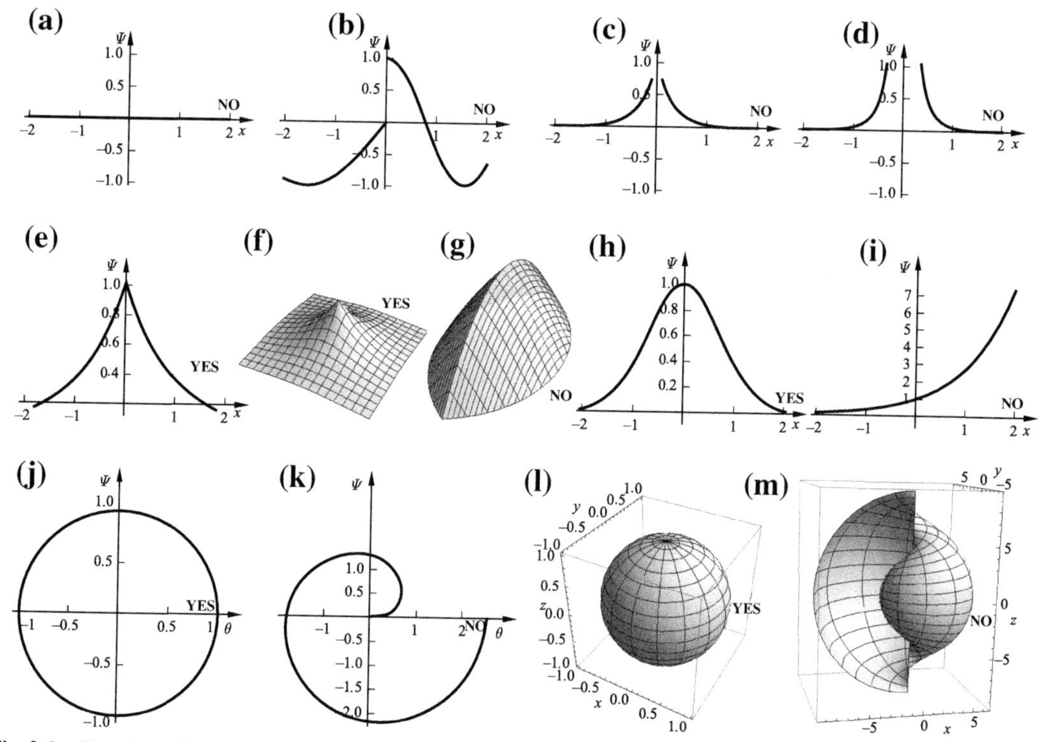

Fig. 2.6. Functions of class Q (i.e., wave functions allowed in quantum mechanics) — examples and counterexamples. A wave function (a) must not be zero everywhere in space; (b) has to be continuous; (c) cannot tend to infinity even at a single point; (d) cannot tend to infinity; (e,f,g) its first derivative cannot be discontinuous for infinite number of points; (h,i) must be square integrable (j,k,l,m) has to be defined uniquely in space (for angular variable θ).

- It has to have a continuous first derivative as well (everywhere in space except isolated points (Fig. 2.6e,f,g), where the potential energy tends to $-\infty$), because the Schrödinger equation is a second-order differential equation and the second derivative must be defined.
- For bound states, it has to tend to zero at infinite values of any of the coordinates (Fig. 2.6h,i), because such a system is compact and does not disintegrate in space. In consequence (from the probabilistic interpretation), the wave function is square integrable; i.e., $\langle\Psi|\Psi\rangle < \infty$.
- It has to have a uniquely defined value in space[24] (Fig. 2.6j,k,l,m).

2.2.2 Boundary Conditions

The Schrödinger equation is a differential equation. In order to obtain a special solution to such equations, one has to insert the particular boundary conditions to be fulfilled. Such conditions follow from the physics of the problem; i.e., with which kind of experiment we are going to compare the theoretical results. For example:

[24] At any point in space, the function has to have a single value. This plays a role only if we have an angular variable, such as ϕ. Then, ϕ and $\phi + 2\pi$ have to give the same value of the wave function. We will encounter this problem in the solution for the rigid rotator in Chapter 4.

- For the bound states (i.e., square-integrable states), we put the condition that the wave function has to vanish at infinity; i.e., if any of the coordinates tends to infinity: $\psi(x=\infty)=\psi(x=-\infty)=0$.
- For cyclic systems of circumference L, the natural conditions will be $\psi(x)=\psi(x+L)$ and $\psi'(x)=\psi'(x+L)$, because they ensure a smooth matching at $x=0$ of the wave function for $x<0$ and of the wave function for $x>0$.
- For scattering states (not discussed here), the boundary conditions are more complex.[25]

There is a countable number of bound states. Each state corresponds to eigenvalue E.

An energy level may be degenerate; that is, more than one wave function may correspond to it, all the wave functions being linearly independent (their number is the degree of degeneracy). The eigenvalue spectrum is usually represented by putting a single horizontal section (in the energy scale) for each wave function:

E_3

E_2

E_1

E_0

An Analogy

Let us imagine all the stable positions of a chair on the floor (Fig. 2.7).

Consider a simple chair, which is very uncomfortable for sitting, but very convenient for a mathematical exercise. Each of the four "*legs*" represents a rod of length a, the "*seat*" is simply a square built of identical rods, the "*back*" consists of three such rods making a C shape. The

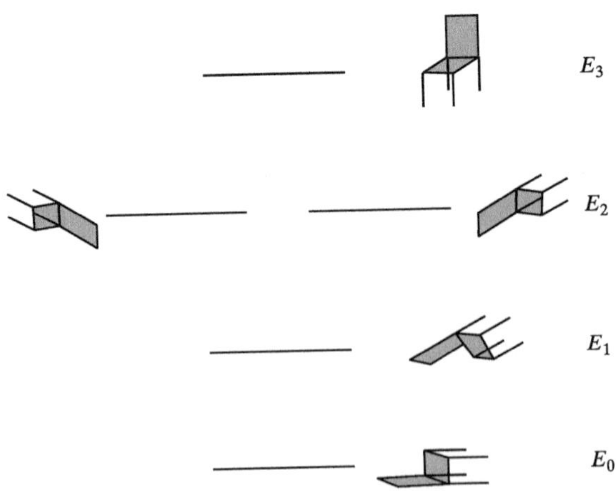

Fig. 2.7. The stable positions of a chair on the floor. In everyday life, we most often use the third excited state.

[25] J.R. Taylor, *Scattering Theory*, Wiley, New York (1972) is an excellent reference.

potential energy of the chair (in position i) in gravitational field equals mgh_i, where m stands for the mass of the chair, g is the standard gravity, and h_i the height of the center of mass with respect to the floor. We obtain the following energies, E_i, of the stationary states (in mga):

— The chair is lying on the support: $E_0 = \frac{4}{11}$
— The chair is lying inclined: the support and the seat touch the floor $E_1 = \frac{7\sqrt{2}}{22}$.
— The chair is lying on the side: $E_2 = \frac{1}{2}$.
Note, however, that we have *two* sides. The energy is the same for the chair lying on the first and second side (because the chair is symmetric), but these are *two* states of the chair, not one. The degree of degeneracy equals 2, and therefore on the energy diagram, we have two horizontal sections. Note how the problem of degeneracy naturally has appeared. The degeneracy of the energy eigenstates of molecules results from their symmetry, exactly as in the case of the chair. In some cases, one may obtain an *accidental degeneracy* (cf. p. 203), which does not follow from the symmetry of an object like a chair, but from the properties of the potential field. This is called *dynamic symmetry*.[26]
— The chair is in the normal position: $E_3 = 1$.

There are no more stable states of the chair, and there are only four energy levels (Fig. 2.7). The stable states of the chair are analogs of the stationary quantum states of Fig. 1.8a,b, on p. 23, while unstable states of the chair on the floor are analogs of the non-stationary states of Fig. 1.8c,d. Of course, there are plenty of unstable positions of the chair with respect to the floor. The stationary states of the chair have more in common with chemistry that one might think. A chairlike molecule (organic chemists have already synthesized much more complex molecules) interacting with a crystal surface would very probably have similar stationary states, Fig. 2.8.

Note that a chair with one leg removed made by a very impractical cabinetmaker, a fan of surrealism (in our analogy, a strange chairlike molecule: just another pattern of chemical bonds; i.e., another electronic state), will result in a different set of energy levels for such a weird chair. Thus, we see that the vibrational levels depend in general on the electronic state.

2.2.2.1 Mathematical and Physical Solutions

> It is worth noting that not all solutions of the Schrödinger equation are physically acceptable.

For example, for bound states, all other solutions than those of class Q (see p. 80) must be rejected. Also, the solution ψ, which does not exhibit the proper symmetry, must be rejected.

[26] Cf. C. Runge, *Vektoranalysis* vol. I, p. 70, S. Hirzel, Leipzig (1919); W. Lenz, *Zeit. Physik*, 24, 197(1924); as well as L.I. Schiff, *Quantum Mechanics*, McGraw-Hill, New York (1968).

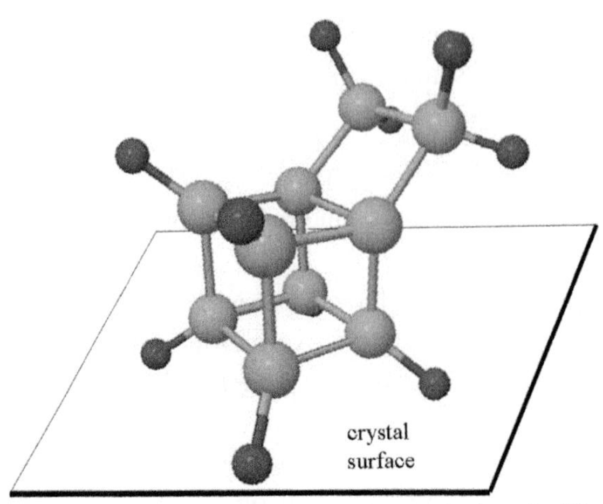

Fig. 2.8. A quantum-mechanical analogy of the stable positions of a chair on the floor. A stiff molecule $C_{10}H_{10}$ with the shape shown here, when interacting with a crystal surface, would acquire several stable positions similar to those of the chair on the floor. They would correspond to some vibrational states (the molecule would vibrate about these positions) of a given electronic state ("*the same bond pattern*"), which in this analogy would correspond to the fixed structure of the chair.

In particular, such illegal, unacceptable functions are asymmetric with respect to the label exchange for electrons (e.g., symmetric for some pairs and antisymmetric for others). Also, a fully symmetric function would also be such a non-physical (purely mathematical) solution. They are called mathematical, but non-physical, solutions to the Schrödinger equation. Sometimes such mathematical solutions correspond to a lower energy than any physically acceptable energy (in such a case, they are called the underground states).

2.3 The Time-Dependent Schrödinger Equation

What would happen if one prepared the system in a given state ψ, which does not represent a stationary state? For instance, one may deform a molecule by using an electric field and then switch the field off.[27] The molecule will turn out to be suddenly in state ψ; i.e., not in its stationary state. Then, according to quantum mechanics, the state of the molecule will start to change according to the time evolution equation (time-dependent Schrödinger equation):

$$\hat{H}\psi = i\hbar\frac{\partial\psi}{\partial t}. \tag{2.13}$$

The equation plays a role analogous to Newton's equation of motion in classical mechanics. In Newton's equation, the position and momentum of a particle evolve. In the time-dependent Schrödinger equation, the evolution proceeds in a completely different space–the space of states or the Hilbert space (cf., Appendix B available at booksite.elsevier.com/978-0-444-59436-5, on p. e7).

[27] Here, we disregard the influence of the magnetic field that accompanies any change of electric field.

Therefore, in quantum mechanics, one has absolute determinism, but in the state space. Indeterminism begins only in our space, when one asks about the coordinates of a particle.

2.3.1 Evolution in Time

As it is seen from Eq. (2.13), knowledge of the Hamiltonian and of the wave function at a given time (left side) contains sufficient information to determine the time derivative of the wave function (right side). This means that we may compute the wave function after an infinitesimal time dt:

$$\psi + \frac{\partial \psi}{\partial t}\mathrm{d}t = \psi - \frac{i}{\hbar}\hat{H}\psi \mathrm{d}t = \left[1 + \left(-i\frac{t}{N\hbar}\right)\hat{H}\right]\psi,$$

where $\mathrm{d}t = t/N$ with N (natural number) very large. Thus, the new wave function results from action of the operator $[1+(-i\frac{t}{N\hbar})\hat{H}]$ on the old wave function. Now, we may apply the operator again and again. We assume that $\hat{H}$ is time-independent. The total operation is nothing but the action of the operator:

$$\lim_{N\to\infty}\left[1 + \left(-i\frac{t}{N\hbar}\right)\hat{H}\right]^N.$$

Please recall that $e^x = \lim_{N\to\infty}\left[1+\frac{x}{N}\right]^N$.

Hence, the time evolution corresponds to action on the initial ψ of the operator $\exp\left(-\frac{it}{\hbar}\hat{H}\right)$:

$$\psi' = \exp\left(-\frac{it}{\hbar}\hat{H}\right)\psi. \tag{2.14}$$

Quantity $exp\hat{A}$ is defined through the Taylor expansion: $e^{\hat{A}} = 1 + \hat{A} + \hat{A}^2/2 + \cdots$

Our result satisfies the time-dependent Schrödinger equation,[28] if $\hat{H}$ does not depend on time (as we assumed when constructing ψ').

[28] One may verify inserting ψ' into the Schrödinger equation. Differentiating ψ' with respect to t, the leftside is obtained.

Inserting the spectral resolution of the identity[29] (cf., postulate II in Chapter 1), one obtains[30]

$$\psi' = \exp\left(-i\frac{t}{\hbar}\hat{H}\right) 1\psi = \exp\left(-i\frac{t}{\hbar}\hat{H}\right)\sum_n |\psi_n\rangle\langle\psi_n|\psi\rangle$$
$$= \sum_n \langle\psi_n|\psi\rangle \exp\left(-i\frac{t}{\hbar}\hat{H}\right)|\psi_n\rangle = \sum_n \langle\psi_n|\psi\rangle \exp\left(-i\frac{t}{\hbar}E_n\right)|\psi_n\rangle.$$

This is how the state ψ will evolve. It will be similar to one or another stationary state ψ_n, more often to those ψ_n, which overlap significantly with the starting function (ψ) and/or correspond to low energy (low frequency). If the overlap $\langle\psi_n|\psi\rangle$ of the starting function ψ with a stationary state ψ_n is zero, then during the evolution, no admixture of the ψ_n state will be seen; i.e., only those stationary states that constitute the starting wave function ψ contribute to the evolution.

2.3.2 Time Dependence of Mechanical Quantities

Let us take a mechanical quantity A and the corresponding (Hermitian) operator $\hat{A}$, and check whether the computed mean value (the normalization of the wave function ψ is assumed) $\overline{A} \equiv \left\langle\hat{A}\right\rangle \equiv \left\langle\psi(t)|\hat{A}(t)\psi(t)\right\rangle$ depends on time. The time derivative of $\left\langle\hat{A}\right\rangle$ reads as (we use the time-dependent Schrödinger equation: $\hat{H}\psi = i\hbar\frac{\partial\psi}{\partial t}$):

$$\begin{aligned}
\frac{\mathrm{d}\left\langle\hat{A}\right\rangle}{\mathrm{d}t} &= \left\langle\frac{\partial}{\partial t}\psi|\hat{A}\psi\right\rangle + \left\langle\psi|\left[\frac{\partial}{\partial t}\hat{A}\right]\psi\right\rangle + \left\langle\psi|\hat{A}\frac{\partial}{\partial t}\psi\right\rangle \\
&= \left\langle-\frac{i}{\hbar}\hat{H}\psi|\hat{A}\psi\right\rangle + \left\langle\psi|\left[\frac{\partial}{\partial t}\hat{A}\right]\psi\right\rangle + \left\langle\psi|\hat{A}\left(-\frac{i}{\hbar}\right)\hat{H}\psi\right\rangle \\
&= \frac{i}{\hbar}\left\langle\hat{H}\psi|\hat{A}\psi\right\rangle + \left\langle\psi|\left[\frac{\partial}{\partial t}\hat{A}\right]\psi\right\rangle - \frac{i}{\hbar}\left\langle\psi|\hat{A}\hat{H}\psi\right\rangle \\
&= \frac{i}{\hbar}\left\langle\psi|\hat{H}\hat{A}\psi\right\rangle + \left\langle\psi|\left[\frac{\partial}{\partial t}\hat{A}\right]\psi\right\rangle - \frac{i}{\hbar}\left\langle\psi|\hat{A}\hat{H}\psi\right\rangle \\
&= \frac{i}{\hbar}\left\langle\psi|\left[\hat{H},\hat{A}\right]\psi\right\rangle + \left\langle\psi|\left[\frac{\partial}{\partial t}\hat{A}\right]\psi\right\rangle = \frac{i}{\hbar}\left\langle\left[\hat{H},\hat{A}\right]\right\rangle + \left\langle\frac{\partial}{\partial t}\hat{A}\right\rangle.
\end{aligned}$$

[29] The use of the spectral resolution of the identity in this form is not fully justified. A sudden cut in the electric field may leave the molecule with a nonzero translational energy. However, in the above spectral resolution, one has the time-independent stationary states computed in the center-of-mass coordinate system, and therefore translation is not taken into account.

[30] Here, we used the property of an analytical function f, that for any eigenfunction ψ_n of the operator $\hat{H}$, one has $f(\hat{H})\psi_n = f(E_n)\psi_n$. This follows from the Taylor expansion of $f(\hat{H})$ acting on eigenfunction ψ_n.

It is seen, that the mean value of a mechanical quantity in general depends on time through two components: the first contains the mean value of the commutator $\left[\hat{H}, \hat{A}\right] = \hat{H}\hat{A} - \hat{A}\hat{H}$, and the second one represents the mean value of the time derivative of the operator:

$$\frac{\mathrm{d}\left\langle\hat{A}\right\rangle}{\mathrm{d}t} = \frac{i}{\hbar}\left\langle\left[\hat{H}, \hat{A}\right]\right\rangle + \left\langle\frac{\partial}{\partial t}\hat{A}\right\rangle. \tag{2.15}$$

Thus, even if $\hat{A}$ does not depend on time explicitly $\left(\frac{\partial}{\partial t}\hat{A} = 0\right)$, but does not commute with $\hat{H}$, the expected value of A (i.e., $\left\langle\hat{A}\right\rangle$ is time-dependent[31]).

2.3.3 Energy Is Conserved

For any isolated system, $\hat{H} \neq f(t)$, and when we take $\hat{A} = \hat{H}$, both terms equal zero and we get $\frac{\mathrm{d}\left\langle\hat{H}\right\rangle}{\mathrm{d}t} = 0$.

The mean value of the Hamiltonian is conserved during evolution.

2.3.4 Symmetry Is Conserved

The time-dependent Schrödinger equation says the following: we have the wave function at time $t = 0$ [i.e., $\psi(x, 0)$]. If you want to see what it will look like at time t, you just have to apply to function $\psi(x, 0)$ an evolution operator $\exp(-i\frac{\hat{H}}{\hbar}t) \equiv \hat{U}(t)$ and you get the answer: $\hat{U}(t)\psi(x, 0) = \psi(x, t)$.

There remains, however, a small problem: how function $\psi(x, t)$ will be related to $\psi(x, 0)$? What kind of question is this? This will be in general just another function (however, it will preserve the normalization condition and the mean value of the energy). It is as if somebody asked about the evolution of an Arabian horse. It will preserve its weight (analog of the normalization), and it will move with the same kinetic energy (analog of conserving total energy). Let us, however, consider some more subtle features; e.g., is it possible that at $t = 0$, the horse is in a

[31] Eq. (2.15) looks a bit suspicious. The quantity $\left\langle\frac{\partial}{\partial t}\hat{A}\right\rangle$ is certainly a real function as the mean value of a Hermitian operator, but what about $\frac{i}{\hbar}\left\langle\left[\hat{H}, \hat{A}\right]\right\rangle$ with this imaginary unit i? Well, everything is all right because the operator $\left[\hat{H}, \hat{A}\right]$ is antihermitian; i.e., $\left\langle\psi|\left[\hat{H}, \hat{A}\right]\psi\right\rangle = -\left\langle\left[\hat{H}, \hat{A}\right]\psi|\psi\right\rangle$. This, however, means that $\left\langle\psi|\left[\hat{H}, \hat{A}\right]\psi\right\rangle = -\left\langle\psi|\left[\hat{H}, \hat{A}\right]\psi\right\rangle^*$ and, therefore, for the complex number $z = \left\langle\psi|\left[\hat{H}, \hat{A}\right]\psi\right\rangle = \left\langle\left[\hat{H}, \hat{A}\right]\right\rangle$, we have $z + z^* = 0$. Therefore, $\left\langle\left[\hat{H}, \hat{A}\right]\right\rangle$ is necessarily an imaginary number of the type ib, with real b, and $\frac{i}{\hbar}\left\langle\left[\hat{H}, \hat{A}\right]\right\rangle = \frac{i}{\hbar}ib = -\frac{b}{\hbar}$ is a real number.

symmetric state,[32] while after a while [action of $\hat{U}(t)$], the horse is in an *asymmetric* state, such as one that has a tendency to bend its head to the right?

Let us consider a symmetry operator $\hat{R}$ that commutes with $\hat{H}$: $\hat{R}\hat{H} = \hat{H}\hat{R}$. Assume that the initial state $\psi(x, 0)$ *exhibits* a symmetry. This means satisfying the following equation:

$$\hat{R}\psi(x, 0) = \exp(i\alpha)\psi(x, 0), \tag{2.16}$$

where α represents a certain real number.[33] The symmetry is guaranteed in such a case because the complex modulus of the transformed function (which decides about probability) does not change: $|\hat{R}\psi(x, 0)| = |\exp(i\alpha)\psi(x, 0)| = |\exp(i\alpha)| \cdot |\psi(x, 0)| = 1 \cdot |\psi(x, 0)| = |\psi(x, 0)|$. This symmetry is characterized by the value of α. What can we say about the symmetry of the final state? Let us see[34]:

$$\hat{R}\psi(x, t) = \hat{R}\hat{U}(t)\psi(x, 0) = \hat{R}\exp\left(-i\frac{\hat{H}}{\hbar}t\right)\psi(x, 0) = \exp\left(-i\frac{\hat{H}}{\hbar}t\right)\hat{R}\psi(x, 0)$$

$$= \exp\left(-i\frac{\hat{H}}{\hbar}t\right)\exp(i\alpha)\psi(x, 0) = \exp(i\alpha)\exp\left(-i\frac{\hat{H}}{\hbar}t\right)\psi(x, 0) = \exp(i\alpha)\psi(x, t).$$

Thus, the final state exhibits at any time t the same symmetry (due to the same value of α):

$$\hat{R}\psi(x, t) = \exp(i\alpha)\psi(x, t), \tag{2.17}$$

Evolution in time causes neither the appearance nor the disappearance of the symmetry of the wave function. The symmetry is conserved.

---

[32] This means that the horse in its movements does not prefer its right (left) side over its left (right) side. This, does not mean that taking a picture of such a horse results in a perfect symmetry in the picture. Such a state means only, that after taking many such pictures and after superposing all of them to get one picture, we see a perfectly symmetric creature.

[33] If this symmetry operation means

- an arbitrary time-independent translation of the coordinate system, say along the x axis, $\alpha \sim p_x$ (p_x means the x component of the momentum, see Appendix F available at booksite.elsevier.com/978-0-444-59436-5, on p. e73) and p_x represents a constant of motion (i.e., does not change).
- an arbitrary translation on the time axis gives $\alpha \sim E$(E stands for the total energy) and E is also a constant of motion.
- for an arbitrary rotation of the coordinate system, say, about z axis, one obtains $\alpha \sim J_z$ (J_z denotes the z component of the angular momentum, see Appendix F available at booksite.elsevier.com/978-0-444-59436-5, on p. e73) with J_z as a constant of motion.

[34] Note that from $\hat{R}\hat{H} = \hat{H}\hat{R}$ also follows $\hat{R}\exp\left(-i\frac{\hat{H}}{\hbar}t\right) = \exp\left(-i\frac{\hat{H}}{\hbar}t\right)\hat{R}$ (recall the definition of $\exp\left(-i\frac{\hat{H}}{\hbar}t\right)$ through the Taylor expansion).

2.3.5 Meditations at a Spring

In chemistry, one assumes (tacitly) that two molecules, say, a water molecule created in a chemical reaction a millisecond ago and a water molecule from the Oligocene well (e.g., created more than 23 million years ago), represent identical objects. How could we know this?

In liquid water, the molecules are subject to intermolecular interactions, which complicate things. Let us consider the same molecules, but isolated in outer space. It is generally believed that even a molecule created a millisecond ago (not speaking about one from the Oligocene well) had enough time to achieve the ground state via emission of photons. If this is true, we can consider them to be described by identical ground-state wave functions.

What about 1 femtosecond (10^{-15} s) instead of 1 millisecond? Well, very probably, the first molecule would be in a non-stationary state[35] and it would have no time to emit photons. These two molecules would be different (i.e., distinguishable).

Let us consider larger molecules–e.g., two molecules of hemoglobin in interstellar space, created by two different methods a femtosecond ago.[36] With a probability very close to 1, these two molecules would be created in two different conformational states. Now, both states evolve in time. Even if they would lower their energies and get the same ground state by emitting photons, this would take virtually an infinite amount of time due to the plethora of kinetic traps (metastable conformations) on their trajectories within the configurational space. These metastable conformations are separated by important energy barriers, difficult to overcome. A hemoglobin molecule, unlike a water molecule, will in general have a long memory of their initial states. Thus, we see that all systems evolve, but the evolution time spans an incredibly large time scale.[37]

Well, let us consider water aggregates (treated as large molecules) formed by a net of hydrogen bonds, like those in liquid water. As shown by Margarita Rodnikova, the hydrogen bond network in liquid water exhibits a kind of elastic properties[38] (i.e., it behaves like molecular aggregates with some stability). What about the lifetime of such aggregates?[39] The contemporary approach to this problem is just ignoring it or saying arbitrarily that liquid water has no memory. It would be certainly more appropriate to leave the answer to experiment.[40]

---

[35] A single vibration in a molecule is a matter of femtoseconds.

[36] This example looks surrealistic, but science relies on questions of the "*what if?*" type.

[37] Think of a shell visible in a rock, or the rock structure itself. These structures were created many millions of years ago, but evolve so slowly that we see them today.

[38] M.N. Rodnikova, *J. Phys. Chem. (Russ.)*, **67**, 275(1993).

[39] One has to define somehow the lifetime ("*memory*"). For example, it could be the relaxation time τ, after which the root mean square deviation from the starting structure (in atomic resolution) exceeds 1 Å. If the structure is stable, both τ and the memory of the molecular aggregate are large.

[40] Some water aggregates bound by the hydrogen bonds have very large lifetimes. How can I know this? Simply, I have seen my footprints in the snow. They certainly represented nothing else but structures that exist only because the hydrogen bonds. These structures did not disappear in a femtosecond, but were there for many hours. Interestingly, when the temperature was raised by a few degrees (above 0°C) my footprints disappeared in an hour. Did they disappear instantaneously upon melting?

2.3.6 Linearity

The most mysterious feature of the Schrödinger equation is its linear character. The world is nonlinear because the effect is never strictly proportional to its cause. However, if $\psi_1(x,t)$ and $\psi_2(x,t)$ satisfy the time-dependent Schrödinger equation, then their arbitrary linear combination also represents a solution.[41]

2.4 Evolution After Switching a Perturbation

Let us suppose that we have a system with the Hamiltonian $\hat{H}^{(0)}$ and its stationary states $\psi_k^{(0)}$:

$$\hat{H}^{(0)}\psi_k^{(0)} = E_k^{(0)}\psi_k^{(0)}, \tag{2.18}$$

which form the orthonormal complete set[42]

$$\psi_k^{(0)}(x,t) = \phi_k^{(0)}(x)\exp\left(-i\frac{E_k^{(0)}}{\hbar}t\right), \tag{2.19}$$

where x represents the coordinates, and t denotes time.

Let us assume that at time $t=0$, the system is in the stationary state $\psi_m^{(0)}$.

> At t = 0, a drama begins: One switches on the perturbation $V(x,t)$, that depends in general on all the coordinates (x) and time (t), and after time τ, the perturbation is switched off. Now we are asking about the probability of finding the system in stationary state $\psi_k^{(0)}$.

After the perturbation is switched on, wave function $\psi_m^{(0)}$ is no longer stationary and begins to evolve in time according to the time-dependent Schrödinger equation $\left(\hat{H}^{(0)}+\hat{V}\right)\psi = i\hbar\frac{\partial\psi}{\partial t}$. This is a differential equation with partial derivatives with the boundary condition $\psi(x,t=0) = \phi_m^{(0)}(x)$. The functions $\left\{\psi_n^{(0)}\right\}$ form a complete set, and therefore the wave function that fulfills the Schrödinger equation $\psi(x,t)$ at any time can be represented as a linear combination with time-dependent coefficients c:

$$\psi(x,t) = \sum_n^{\infty} c_n(t)\,\psi_n^{(0)}(x,t). \tag{2.20}$$

---

[41] Indeed, $\hat{H}(c_1\psi_1 + c_2\psi_2) = c_1\hat{H}\psi_1 + c_2\hat{H}\psi_2 = c_1 i\hbar\frac{\partial\psi_1}{\partial t} + c_2 i\hbar\frac{\partial\psi_2}{\partial t} = i\hbar\frac{\partial(c_1\psi_1 + c_2\psi_2)}{\partial t}$.

[42] This can always be assured (by suitable orthogonalization and normalization) and follows from the Hermitian character of the operator $\hat{H}^{(0)}$.

Inserting this on the left side of the time-dependent Schrödinger equation, one obtains

$$\left(\hat{H}^{(0)} + \hat{V}\right)\psi = \sum_n c_n \left(\hat{H}^{(0)} + \hat{V}\right)\psi_n^{(0)} = \sum_n c_n \left(E_n^{(0)} + V\right)\psi_n^{(0)},$$

whereas its right side gives

$$i\hbar\frac{\partial\psi}{\partial t} = i\hbar\sum_n \left[\psi_n^{(0)}\frac{\partial c_n}{\partial t} + c_n\frac{\partial\psi_n^{(0)}}{\partial t}\right]$$
$$= i\hbar\sum_n \left[\psi_n^{(0)}\frac{\partial c_n}{\partial t} + c_n\left(-\frac{i}{\hbar}E_n^{(0)}\right)\psi_n^{(0)}\right] = \sum_n \left[i\hbar\psi_n^{(0)}\frac{\partial c_n}{\partial t} + c_n E_n^{(0)}\psi_n^{(0)}\right].$$

Both sides give

$$\sum_n c_n \hat{V}\psi_n^{(0)} = \sum_n \left(i\hbar\frac{\partial c_n}{\partial t}\right)\psi_n^{(0)}.$$

Let us multiply the left side by $\psi_k^{(0)*}$ and integrate the result in

$$\sum_n^{\infty} c_n V_{kn} = i\hbar\frac{\partial c_k}{\partial t}, \tag{2.21}$$

for $k = 1, 2, \ldots$, where

$$V_{kn} = \left\langle \psi_k^{(0)} | \hat{V}\psi_n^{(0)} \right\rangle. \tag{2.22}$$

The formulas obtained are equivalent to the Schrödinger equation. These are differential equations, which we would generally wish to see, provided that the summation is not infinite.[43] In practice, however, one has to keep the summation finite.[44] If the assumed number of terms in the summation is not too large, then solving the problem using computers is feasible.

2.4.1 The Two-State Model–Time-Independent Perturbation

For the sake of simplicity, let us take the *two-state model* (cf., Appendix D available at booksite. elsevier.com/978-0-444-59436-5, on p. e65) with two orthonormal eigenfunctions $\left|\phi_1^{(0)}\right\rangle = |1\rangle$and $\left|\phi_2^{(0)}\right\rangle = |2\rangle$of the Hamiltonian $\hat{H}^{(0)}$

$$\hat{H}^{(0)} = E_1^{(0)} |1\rangle\langle 1| + E_2^{(0)} |2\rangle\langle 2|$$

[43] In fact, only then is the equivalence to the Schrödinger equation ensured.

[44] This is typical for expansions into the complete set of functions (the so-called algebraic approximation).

with the following perturbation $\left(\text{to assure } \hat{V} \text{ to be Hermitian}\right)$:

$$\hat{V} = v\,|1\rangle\,\langle 2| + v^*\,|2\rangle\,\langle 1|,$$

with the corresponding matrix $\mathbf{V} = \begin{pmatrix} 0 & v \\ v^* & 0 \end{pmatrix}$.

This model has an exact solution (even for a large perturbation $\hat{V}$). One may introduce various time dependencies of $\hat{V}$, including various regimes for switching on the perturbation.

The differential equations (2.21) for the coefficients $c_1(t)$ and $c_2(t)$ are (in a.u., $\omega_{21} = E_2^{(0)} - E_1^{(0)}$ and $v = v^*$):

$$\begin{aligned} c_2 v \exp(-i\omega_{21}t) &= i\frac{\partial c_1}{\partial t} \\ c_1 v \exp(i\omega_{21}t) &= i\frac{\partial c_2}{\partial t}. \end{aligned} \tag{2.23}$$

Now we assume that v is time-independent and the initial wave function is $|1\rangle$; i.e., $c_1(0) = 1$, $c_2(0) = 0$. In such a case, one obtains[45]

$$\begin{aligned} c_1(t) &= \exp\left(-i\frac{1}{2}\omega_{21}t\right)\left[\cos(avt) + i\frac{\omega_{21}}{2av}\sin(avt)\right], \\ c_2(t) &= -\frac{1}{a}\exp\left(i\frac{1}{2}\omega_{21}t\right)\sin(avt), \end{aligned} \tag{2.24}$$

where $a = \sqrt{1 + \left(\frac{\omega_{21}}{2v}\right)^2}$.

2.4.2 Two States–Degeneracy

One of the most important cases corresponds to the degeneracy $\omega_{21} = E_2^{(0)} - E_1^{(0)} = 0$. One obtains $a = 1$ and

$$\begin{aligned} c_1(t) &= \cos(vt) \\ c_2(t) &= -\sin(vt). \end{aligned}$$

This is a very interesting result. After applying symmetric orthogonalization (see Appendix J available at booksite.elsevier.com/978-0-444-59436-5), the functions $|1\rangle$ and $|2\rangle$ may be identified with the ψ_D and ψ_L for the D and L enantiomers (cf., p. 74) or, with the wave

[45] For example, use Mathematica software to do this. Let us check the conservation of normalization (i.e., its time independence): $|c_1(t)|^2 + |c_2(t)|^2 = \cos^2(avt) + \frac{\omega^2}{(2av)^2}\sin^2(avt) + \frac{1}{a^2}\sin^2(avt) = \cos^2(avt) + \frac{\omega^2+4v^2}{(2av)^2}\sin^2(avt) = \cos^2(avt) + \sin^2(avt) = 1$.

functions 1s, centered on the two nuclei in the H_2^+ molecule. As one can see from the last two equations, the two wave functions oscillate, transforming one to the other, with an oscillation period $T = \frac{2\pi}{v}$. If v were very small (as in the case of D- and L-glucose), then the oscillation period would be very large. This happens to D- and L-enantiomers of glucose, where changing the nuclear configuration from one to the other enantiomer means breaking a chemical bond (a high and wide energy barrier to overcome). This is why the drugstore owner can safely stock a single enantiomer for a very long time.[46] But this may not be true for other enantiomers. For instance, imagine a pair of enantiomers that represent some intermolecular complexes, and a small change of the nuclear framework may cause one of them to transform into the other. In such a case, the oscillation period may be much smaller than the lifetime of the Universe–e.g., it may be comparable to the time of an experiment. In such a case, one could observe the oscillation between the two enantiomers. This is what happens in reality. One observes a spontaneous racemization, which is of dynamic character (i.e., a single molecule oscillates between D and L forms).

2.4.3 The Two-State Model - An Oscillating Perturbation

It is interesting what happens to the wave function as time passes, when a perturbation varying with time as $\exp(i\omega t)$ is switched on. Formula (for $\hat{V}$) has to be modified to (with the second term having an $\exp(-i\omega t)$ time-dependence to ensure that the operator (2.25) is Hermitian)

$$\hat{V} = v[\exp(i\omega t)\,|1\rangle\,\langle 2| + \exp(-i\omega t)\,|2\rangle\,\langle 1|], \tag{2.25}$$

with $|1\rangle \equiv \psi_1^{(0)}$ and $|2\rangle \equiv \psi_2^{(0)}$ and real v.

Inserting such a perturbation into Eqs. (2.21) results in Eqs. (2.23), but replacing $\omega_{21} \to \omega_{21} - \omega$. Assuming that function $|1\rangle$ is the starting one, that means $c_1(0) = 1$, $c_2(0) = 0$, and therefore, Eq. (2.24) takes the form

$$c_1(t) = \exp\left[-i\frac{1}{2}(\omega_{21} - \omega)\,t\right]\left[\cos(avt) + i\frac{(\omega_{21} - \omega)}{2av}\sin(avt)\right],$$

$$c_2(t) = -\frac{1}{a}\exp\left[i\frac{1}{2}(\omega_{21} - \omega)\,t\right]\sin(avt),$$

where this time $a = \sqrt{1 + \left(\frac{\omega_{21}-\omega}{2v}\right)^2}$.

2.4.4 Two States–Resonance Case

For $\omega = \omega_{21}$ (the energy of the photon matches the energy level difference), we obtain

$$c_1(t) = \cos(vt),$$
$$c_2(t) = -\sin(vt); \tag{2.26}$$

[46] It cannot, however, be kept longer than the sell-by date.

i.e., the system oscillates between state $\psi_1^{(0)}$ and state $\psi_2^{(0)}$, with period $\frac{2\pi}{v}$, and no one of these states is privileged.[47]

It is intriguing to see that for oscillating perturbation in the case of two levels of different energies, we got exactly the same behavior as in the case of degenerate levels. It looks as if the two levels have equal energy (degeneracy) after the resonance photon energy of one of them is counted, depending on absorption or emission.

2.4.5 Short-Time Perturbation–The First-Order Approach

If one is to apply first-order perturbation theory, two things have to be assured: the perturbation $\hat{V}$ has to be small, and the time of interest has to be small (switching the perturbation in corresponds to $t = 0$). This is what we are going to assume from now on. At $t = 0$, one starts from the mth state and therefore $c_m = 1$, while other coefficients $c_n = 0$. Let us assume that to the first approximation, the domination of the mth state continues even *after* switching the perturbation on, and we will be interested in detecting the most important tendencies in time evolution of c_n for $n \neq m$. These assumptions (they give first-order perturbation theory[48]) lead to a considerable simplification of Eqs. (2.21):

$$V_{km} = i\hbar \frac{\partial c_k}{\partial t} \quad \text{for } k = 1, 2, \ldots N.$$

In this, and the further equations of this chapter, the coefficients c_k will depend implicitly on the initial state m. The change of $c_k(t)$ is therefore proportional to V_{km}. A strong coupling for the expansion function $\psi_k^{(0)}$ (i.e., the system becomes a bit similar to that described by $\psi_k^{(0)}$) corresponds to large values of the coupling coefficient V_{km}, which happens when function $\psi_k^{(0)}$

47 Such oscillations necessarily mean that the energy of the system changes periodically: after time $\tau = \frac{\pi}{2v}$ the system absorbs a photon of energy $\hbar\omega_{12}$ from the electromagnetic field. Then, again, after τ, emits the same photon, and then absorption, emission, etc., this scenario repeats periodically (excitations and deexcitations). Such behavior is possible only because of continuous supplying of the photons from the field given by Eq. (2.25). If the system interacted with a single photon instead of the field (2.25), the excited state would change to the ground state and the photon in the form of a spherical wave expanding to infinity (equal probability of detecting the photon in any spot on a sphere). If we had only our system in the Universe and the Universe were limited by a mirror, the photon would finally come back to the system, causing its excitation, then emission, etc., similar to our solution. However, our Universe is not limited by a mirror and the photon spherical wave would expand to infinity without finding any obstacles. There would be no chance for the photon to come back and the atom would not be excited.

48 For the sake of simplicity, we will not introduce a new notation for the coefficients c corresponding to the first-order procedure. If the above simplified equation were introduced to the left side of Eq. (2.21), then its solution would give c accurate up to the second order, etc.

resembles function $\hat{V}\psi_m^{(0)}$. This represents a strong constraint both for $\psi_k^{(0)}$ and $\hat{V}$, only some special perturbations $\hat{V}$ are able to couple effectively two states,[49] such as $\psi_k^{(0)}$ and $\psi_m^{(0)}$.

The quantity V_{km} depends on time for two or even three reasons. First and second, the stationary states $\psi_m^{(0)}$ and $\psi_k^{(0)}$ do have a time dependence, and third, the perturbation $\hat{V}$ may also depend on time. Let us highlight the time dependence of the wave functions by introducing the frequency

$$\omega_{km} = \frac{E_k^{(0)} - E_m^{(0)}}{\hbar}$$

and the definition

$$v_{km} \equiv \left\langle \phi_k^{(0)} | \hat{V} \phi_m^{(0)} \right\rangle.$$

One obtains

$$-\frac{i}{\hbar} v_{km}\, e^{i\omega_{km} t} = \frac{\partial c_k}{\partial t}.$$

Subsequent integration with the boundary condition $c_k(\tau = 0) = 0$ for $k \neq m$ gives

$$c_k(\tau) = -\frac{i}{\hbar} \int_0^\tau dt\; v_{km}(t)\, e^{i\omega_{km} t}. \tag{2.27}$$

The square of $c_k(\tau)$ represents (to the accuracy of first-order perturbation theory) the probability that at time τ, the system will be found in state $\psi_k^{(0)}$. Let us calculate this probability for a few important cases of perturbation $\hat{V}$.

2.4.6 Time-Independent Perturbation and the Fermi Golden Rule

From Eq. (2.27), one has

$$c_k(\tau) = -\frac{i}{\hbar} v_{km} \int_0^\tau dt\; e^{i\omega_{km} t} = -\frac{i}{\hbar} v_{km} \frac{e^{i\omega_{km}\tau} - 1}{i\omega_{km}} = -v_{km} \frac{e^{i\omega_{km}\tau} - 1}{\hbar\omega_{km}}. \tag{2.28}$$

[49] First of all, there must be something in $\hat{V}$, which influences the particles of the system, like electric field interaction with electrons of an atom. If the atom stays spherically symmetric, there is no coupling with the field. Only by making a shift of electrons do we get some interaction, proportional to this shift. Therefore, roughly speaking, $\hat{V}$ is in this case proportional to the shift x of the most weakly bound electron. Let us assume that we start from $\psi_m^{(0)}$ (describing this electron, we neglect the other electrons), which is a spherically symmetric function. Therefore, $\hat{V}\psi_m^{(0)} \sim x \cdot$ (spherically symmetric) and the most promising function $\psi_k^{(0)}$ would be of the p_x type, since only then the integral V_{km} would have a chance to be nonzero. But still, if $\psi_m^{(0)}$ and $\psi_k^{(0)}$ correspond to different energies, they have their phase factors multiplied in the integrand, which results in their product oscillating in time with the frequency $\omega_{km} = (E_k^{(0)} - E_m^{(0)})/\hbar$. The only way to damp these oscillations (making V_{km} having a substantial value for any t) is to use $\hat{V}$, and therefore x, oscillating itself, preferably with the same frequency to damp effectively. To damp oscillations $\exp(i\omega_{km} t)$, we have to have $\hat{V}$ oscillating as $\exp(-i\omega_{km} t)$, because these factors give no oscillations when multiplied. The last conclusion is independent of the nature of $\hat{V}$. Therefore, even from such a simple reasoning, we have to change quantum state using light frequency, which matches the difference of the corresponding energy levels.

Now let us calculate the probability density $P_m^k = |c_k|^2$, that at time τ, the system will be in state k (the initial state is m):

$$P_m^k(\tau) = |v_{km}|^2 \frac{\left(-1+\cos\omega_{km}\tau\right)^2 + \sin^2\omega_{km}\tau}{\left(\hbar\omega_{km}\right)^2} = |v_{km}|^2 \frac{\left(2-2\cos\omega_{km}\tau\right)}{\left(\hbar\omega_{km}\right)^2}$$

$$= |v_{km}|^2 \frac{\left(4\sin^2\frac{\omega_{km}\tau}{2}\right)}{\left(\hbar\omega_{km}\right)^2} = |v_{km}|^2 \frac{1}{\hbar^2} \frac{\left(\sin^2\frac{\omega_{km}\tau}{2}\right)}{\left(\frac{\omega_{km}}{2}\right)^2}.$$

In order to undergo the transition from state m to state k, one has to have a large v_{km} (i.e., a large coupling of the two states through perturbation $\hat{V}$). Note that probability P_m^k strongly depends on the time τ chosen; the probability oscillates as the square of the sine when τ increases. For some τ it is large, while for others it is zero. From example 4 in Appendix E available at booksite.elsevier.com/978-0-444-59436-5, on p. e69 one can see that for large values of τ, one may write the following approximation[50] to P_m^k:

$$P_m^k(\tau) \cong |v_{km}|^2 \pi \frac{\tau}{\hbar^2} \delta\left(\frac{\omega_{km}}{2}\right) = \frac{2\pi\tau}{\hbar^2} |v_{km}|^2 \delta\left(\omega_{km}\right) = \frac{2\pi\tau}{\hbar} |v_{km}|^2 \delta\left(E_k^{(0)} - E_m^{(0)}\right),$$

where we have used twice the Dirac delta function property that $\delta(ax) = \frac{\delta(x)}{|a|}$.

As one can see, P_m^k is proportional to time, which makes sense only because time τ has to be relatively small (i.e., first-order perturbation theory has to be valid). Note that the Dirac delta function forces the energies of both states (the initial and the final) to be equal because of the time independence of $\hat{V}$.

A time-independent perturbation is unable to change the state of the system when it corresponds to a change of its energy.

A very similar formula is systematically derived in several important cases. Probably this is why the probability per unit time is called the *Fermi golden rule*[51]:

Fermi Golden Rule

$$w_m^k \equiv \frac{P_m^k(\tau)}{\tau} = |v_{km}|^2 \frac{2\pi}{\hbar} \delta\left(E_k^{(0)} - E_m^{(0)}\right). \tag{2.29}$$

50 We assume that τ is large when compared to $2\pi/\omega_{km}$, but it is not too large to keep the first-order perturbation theory valid.

51 E. Fermi, *Nuclear Physics*, University of Chicago Press, Chicago (1950) p. 142.

2.4.7 The Most Important Case: Periodic Perturbation

Let us assume a time-dependent periodic perturbation:

$$\hat{V}\left(x,t\right)=\hat{v}\left(x\right)e^{\pm i\omega t}.$$

Such a perturbation corresponds to an oscillating electric field[52] of angular frequency ω.

Let us take a look at successive equations, which were obtained at the time-independent $\hat{V}$. The only change will be that V_{km} will take the form

$$V_{km}\equiv\left\langle\psi_k^{(0)}|\hat{V}\psi_m^{(0)}\right\rangle=v_{km}e^{i(\omega_{km}\pm\omega)t}\text{ instead of }V_{km}\equiv\left\langle\psi_k^{(0)}|\hat{V}\psi_m^{(0)}\right\rangle=v_{km}e^{i\omega_{km}t}.$$

The whole derivation will be therefore identical, except that the constant ω_{km} will be replaced by $\omega_{km}\pm\omega$. Hence, we have a new form of the Fermi golden rule for the probability per unit time of transition from the mth to the kth state:

Fermi Golden Rule

$$w_m^k\equiv\frac{P_m^k\left(\tau\right)}{\tau}=|v_{km}|^2\frac{2\pi}{\hbar}\delta\left(E_k^{(0)}-E_m^{(0)}\pm\hbar\omega\right).\qquad(2.30)$$

Note that $\hat{V}$ with $\exp(+i\omega t)$ needs the equality $E_k^{(0)}+\hbar\omega=E_m^{(0)}$, which means that $E_k^{(0)}\leq E_m^{(0)}$; and therefore, one has emission from the mth to the kth states. On the other hand, $\hat{V}$ with $\exp(-i\omega t)$ forces the equation $E_k^{(0)}-\hbar\omega=E_m^{(0)}$, which corresponds to absorption from the mth to the kth state.

Therefore, a periodic perturbation is able to make a transition between states of different energy.

Summary

The Hamiltonian of any isolated system is invariant with respect to the following transformations (operations):

- Any translation in time (homogeneity of time)
- Any translation of the coordinate system (space homogeneity)
- Any rotation of the coordinate system (space isotropy)
- Inversion ($\boldsymbol{r}\rightarrow-\boldsymbol{r}$)

[52] In the homogeneous field approximation, the field interacts with the dipole moment of the molecule (cf. Chapter 12)

$$V\left(x,t\right)=V\left(x\right)e^{\pm i\omega t}=-\hat{\boldsymbol{\mu}}\cdot\boldsymbol{\mathcal{E}}e^{\pm i\omega t},$$

where $\boldsymbol{\mathcal{E}}$ denotes the electric field intensity of the light wave and $\hat{\boldsymbol{\mu}}$ is the dipole moment operator.

- Reversing all charges (charge conjugation)
- Exchanging labels of identical particles

This means that the wave function corresponding to a stationary state (the eigenfunction of the Hamiltonian) also has to be an eigenfunction of the following:

- Total momentum operator (due to the translational symmetry)
- Total angular momentum operator and one of its components (due to the rotational symmetry)
- Inversion operator
- Any permutation (of identical particles) operator (due to the non-distinguishability of identical particles)
- $\hat{S}^2$ and $\hat{S}_z$ operators (for the non-relativistic Hamiltonian, p. 66, due to the absence of spin variables in it)

Such a wave function corresponds to the energy belonging to the energy continuum.[53] Only after separation of the center-of-mass motion does one obtain the spectroscopic states (belonging to a discrete spectrum) $\Psi_{N,J,M,\Pi}(\boldsymbol{r},\boldsymbol{R})$, where $N = 0, 1, 2, \ldots$ denotes the quantum number of the electronic state, $J = 0, 1, 2, \ldots$ quantizes the total angular momentum, $M_J, -J \le M_J \le J$ quantizes its component along the z axis, and $\Pi = \pm 1$ represents the parity with respect to the inversion. As to the invariance with respect to permutations of identical particles, an acceptable wave function has to be antisymmetric with respect to the exchange of identical fermions, whereas it has to be symmetric when exchanging identical bosons.

The time-independent Schrödinger equation $\hat{H}\psi = E\psi$ has been derived from the wave equation and the de Broglie formula. Solving this equation results in the stationary states and their energies. This is the basic equation of quantum chemistry. The prevailing weight of research in this domain is concentrated on solving this equation for various systems.

The time-dependent Schrödinger equation $\hat{H}\psi = i\hbar\frac{\partial\psi}{\partial t}$ represents the time evolution of an arbitrary initial wave function. The assumption that translation in time is a unitary operator leads to preserving the normalization of the wave function and of the mean value of the Hamiltonian. If the Hamiltonian is time-independent, then one obtains the formal solution to the Schrödinger equation by applying the operator $\exp(-\frac{it}{\hbar}\hat{H})$ to the initial wave function. The time evolution of the stationary state $\phi_m^{(0)}$ is most interesting in the case of suddenly switching the perturbation $\hat{V}$. The state is no longer stationary, and the wave function begins to change as time passes. Two cases have been considered:

- Time-independent perturbation.
- Periodic perturbation.

Only in the case of a time-dependent perturbation may the system change the energy state.

Main Concepts, New Terms

atomic units (p. 66)
algebraic approximation (p. 91)
baryon number (p. 71)
bound state (p. 82)
charge conjugation (p. 76)
dipole moment (p. 72)
dynamic symmetry (p. 83)
enantiomers (p. 73)
Fermi golden rule (p. 96)
first-order perturbation theory (p. 94)
functions of class Q (p. 80)
gauge symmetry (p. 71)
invariance of theory (p. 64)
inversion (p. 72)
lepton number (p. 71)
mathematical solution (p. 84)
mirror reflection (p. 73)
molecular symmetry (p. 76)
periodic perturbation (p. 97)
physical solutions (p. 83)

[53] This is because the molecule as a whole (i.e., its center of mass) may have an arbitrary kinetic energy. Sometimes it is rewarding to introduce the notion of the *quasicontinuum* of states, which arises if the system is enclosed in a large box instead of considering it in infinite space. This simplifies the underlying mathematics.

From the Research Front

The overwhelming majority of research in the domain of quantum chemistry is based on the solution of the time-independent Schrödinger equation. Without computers, it was possible to solve (in an approximate way) the equation for H_2^+ by conducting a hall full of secretaries with primitive calculators for many hours (what a determination). Thanks to computers, solving such problems became easy as early as the 1960s. Despite the enormous progress in computer science, until the end of the 1980s, the molecules studied were rather small when compared to researchers' expectations. They could be treated only as models because they were usually devoid of substituents that theoreticians were forced to consider irrelevant. The last years of the twentieth century were marked by the unprecedented delivery by theoreticians of powerful high-tech efficient tools of quantum chemistry to other specialists: chemists, physicists, etc., as well as laypeople. The software computes millions of integrals, uses sophisticated mathematics (literally the whole arsenal of quantum chemistry), but users need not know about it. All they have to do is click a mouse on a quantum chemistry method icon.[54] Despite such progress, the time-dependent Schrödinger equation is solved extremely rarely. For the time being, researchers are interested mainly in stationary states. The quality of results depends on the size of the molecules investigated. Very accurate computations (accuracy ~ 0.01 kcal/mol) are feasible for the smallest molecules containing a dozen of electrons, less accurate ones use first principles (*ab initio methods*) and are feasible for hundreds of atoms (accuracy to a few kcals/mol). Semi-empirical quantum calculations[55] of even poorer accuracy are applicable to thousands of atoms.

Ad Futurum

The numerical results routinely obtained so far indicate that, for the vast majority of chemical problems (yet not all, cf., Chapter 3) there is no better tool than the Schrödinger equation. Future progress will be based on more and more accurate solutions for larger and larger molecules. The appetite for progress is unlimited here, but the numerical difficulties increase much faster than the size of the system. However, progress in computer science has systematically opened new possibilities, which always are many times larger than previous ones. Some simplified alternatives to the Schrödinger equation (e.g., such as described in Chapter 11) will also be more important.

Undoubtedly methods based on the time-dependent Schrödinger equation will also be developed. A typical possible target might be to plan a sequence of laser pulses[56] such that the system undergoes a change of state from ψ_1 to ψ_2 (a state-to-state reaction). The way we carry out chemical reactions, usually by rather primitive heating, may change to a precise transformation of the system from state to state.

[54] I hope all students understand that a quantum chemist has to be equipped with something more than a strong forefinger for clicking.

[55] In such calculations, many integrals are approximated by simple formulas (sometimes involving experimental data), the main goal of which is efficiency.

[56] That is a sinusoidal impulse for each of the sequences: switching on time, duration, intensity, and phase. In terms of contemporary laser techniques, it is an easy task. Now chemists should consider transforming the reagents into products. The beginnings of such an approach are already present in the literature (e.g., J. Manz, G.K. Paramonov, M. Polášek, and C. Schütte, *Isr. J. Chem.*, **34**, 115(1994).

It seems that at the essence of science is the fundamental question of why, and a clear answer to this question follows from a deep understanding the Nature's machinery. We *cannot* tell a student, "*Well, this is what the computer says,*" because it is more important for you and me to understand than that the computer cranks out an answer. Hence, interpretation of the results will be of crucial importance (a sort of Bader analysis; cf., Chapter 11). Progress in this realm seems to be rather modest for the time being.

Additional Literature

R. Feynman, *The Character of Physical Law,* Cox and Wyman, Ltd, London, (1965).

The best recommendation is that the Feynman's books need no recommendation.

J. Cioslowski, in "Pauling's Legacy: Modern Modelling of the Chemical Bond," eds. Z.B. Maksić, W.J. Orville-Thomas, Elsevier, Amsterdam, (1999) p.1.

Questions

1. The momentum conservation law is a consequence of
   a. space isotropy
   b. homogeneity of time
   c. homogeneity of space
   d. invariance of the time-dependent Schrödinger equation with respect to change of the sign of time
2. Invariance of the time-independent Hamiltonian with respect to any rotation of the coordinate system results in
   a. momentum conservation
   b. the square of the total angular momentum and only one of the components of the total angular momentum being measurable
   c. the total energy and the square of the total angular momentum being measurable
   d. the total energy and only one of the components of the total angular momentum being measurable
3. The non-relativistic Hamiltonian contains
   a. Coulombic interaction of all charged particles
   b. interaction of magnetic fields created by moving charged particles
   c. interaction of point-like particles
   d. the interaction of the spin magnetic moments of the individual particles
4. The wave function
   a. has to be continuous
   b. has to have the same value for an angular variable α and for $\alpha + 2\pi$
   c. must not tend to infinity
   d. must have continuous derivative (in the whole domain)
5. Wave function for a bound state
   a. tends to zero as any Cartesian coordinate of particles tends to infinity
   b. must have a continuous first derivative
   c. being related to probability, takes only non-negative values
   d. has to be defined for all values of Cartesian coordinates of all the particles of the system
6. Time-independent Schrödinger equation
   a. is satisfied by a wave function that is an eigenfunction of the Hamiltonian of the system
   b. any solution of this equation corresponds to a physically acceptable state of the system

c. an eigenvalue may correspond to several linearly independent eigenfunctions
d. energy levels represent eigenvalues of this equation

7. Time-dependent Schrödinger equation [$\hat{H}$ stands for the Hamiltonian, $\psi(x, t)$ is the wave function, x symbolizes the spatial coordinates, t denotes time]

   a. describes the evolution in time of any quantum state
   b. takes the form: $\hat{H}\psi = \hbar\frac{\partial\psi}{\partial t}$
   c. allows to calculate the time derivative of a wave function from its spatial-coordinate dependence at a fixed time
   d. if $\hat{H}$ does not depend on time, its solution reads as $\psi(x, t) = \exp(-i\frac{\hat{H}}{\hbar}t)\psi(x, t = 0)$

8. In evolution of any wave function,

   a. for some values of time the wave function is equal to one of the eigenfunctions of the Hamiltonian
   b. the mean value of the Hamiltonian decreases
   c. normalization of the wave function is satisfied at any time
   d. when starting from an excited state we end up in the ground state

9. In the two-state model (two states with energies $E_1^{(0)}$ and $E_2^{(0)}$) with a time-independent perturbation,

   a. the system always oscillates between these two states: for some values of time the wave function represents either of these two states
   b. the system oscillates between these two states only in the case of degeneracy: for some values of time the wave function represents either of these two states
   c. in the case of degeneracy the larger the coupling energy of the states the larger the oscillation period is
   d. the coupling energy for D and L isomers of glucose is small, because the transition from one isomer to the other requires breaking a chemical bond (large barrier)

10. The Fermi golden rule

    a. was derived by Dirac
    b. pertains to the probability of changing the state due to perturbation applied
    c. a time-independent perturbation enables a transition from a state of higher energy to a state of lower energy
    d. a time-dependent periodic perturbation may change a state to another state of different energy

Answers

1c, 2b,c,d, 3a,c, 4a,b,c, 5a,d, 6a,c,d, 7a,c,d, 8c, 9b,d, 10a,b,d

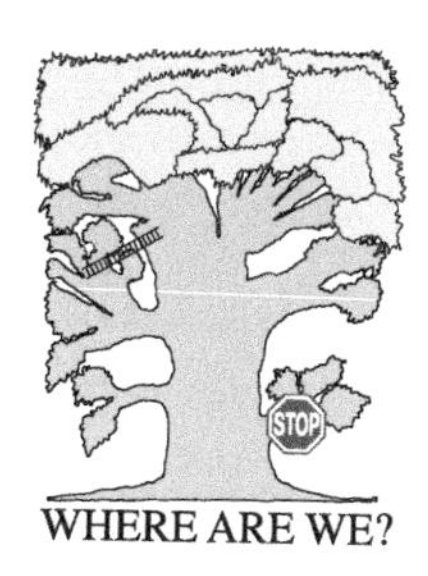

Beyond the Schrödinger Equation

"Newton, forgive me..."
Albert Einstein

Where Are We?

The problems considered in the present chapter are shown as a small side-branch at the base of the tree[1].

An Example

Imagine that you are with some colleagues in your brand-new luxury car made to your very particular specifications, with an extravagant feature: the unlimited speed (of light). The manual says proudly that its exclusive quantum-mechanical construction is based purely on the famous Schrödinger equation, in which $c = \infty$. Your colleagues are sure that your car is much better than the corresponding cheap relativistic model for everybody. Well, it turned out recently,[2] your non-relativistic wonder car would not start at all: when pushed, the starter would be able to make only a unpleasant sound indicating a dead battery. The reason is a large relativistic effect in the electric potential difference between the lead electrode and the lead dioxide electrode. Your non-relativistic battery would attain only something like 20% of the voltage that the relativistic battery usually produces. Well, I am sure you will enthusiastically agree with me (together with millions of car drivers and passengers all over the world), that there is a need to abandon the non-relativistic theory and quest for a more accurate one.

Now, here is still another argument to ponder. Many people would want to know everything about the precious metal gold. The yellow shine of this metal has hypnotized humanity for centuries. Few people know that the color of gold, as calculated assuming the infinite velocity of light, would actually be silver-like[3].

---

[1] This chapter owes much to the presentation given by L. Pisani, J.-M. André, M.-C. André, and E. Clementi, *J. Chem. Educ.*, **70**, 894–901 (1993), as well as to the work of my friends J.-M. André, D.H. Mosley, M.-C. André, B. Champagne, E. Clementi, J.G. Fripiat, L. Leherte, L. Pisani, D. Vercauteren, and M. Vracko, *Exploring Aspects of Computational Chemistry: Vol. I, Concepts*, Presses Universitaires de Namur, pp. 150–166 (1997); *Vol. II, Exercises,* Presses Universitaires de Namur, pp. 249–272(1997).

[2] R. Ahuja, A. Blomqvist, P. Larsson, P. Pyykkö, and P. Zaleski-Ejgierd, *Phys. Rev. Lett.*, 106, 18301 (2011).

[3] P. Pyykkö, *Chem. Rev.*, **88**, 563 (1988); also P. Pyykkö, *ibid.*, **97**, 597 (1997).

Ideas of Quantum Chemistry, Second Edition. http://dx.doi.org/10.1016/B978-0-444-59436-5.00003-9

The Schrödinger equation fails especially for heavy elements. Here is an example of three diatomics: Cu_2, Ag_2, Au_2 ($Z_{Cu} = 29$, $Z_{Ag} = 47$, $Z_{Au} = 79$)[4].

Bond Length (Å)	Cu	Ag	Au
Non-relativistic calculations	2.26	2.67	**2.90**
Relativistic calculations	2.24	2.52	2.44
Experimental results	2.22	2.48	**2.47**

The heavier the element, the larger is the error of the non-relativistic approach. This is a huge discrepancy (see boldface) for such a quantity as bond length.

What Is It All About?

The greater the velocity of an object, the greater the errors in Newton dynamics. Electrons have greater velocity when close to the nuclei of a large electric charge[5]. This is why relativistic corrections may turn out to be important for heavy elements.

---

4 J.-M. André and M.-C. André, "*Une introduction à la théorie de la relativité classique et quantique à l'usage des chimistes*", Universitè de Namur, Namur, 1999, p. 2.

5 This is easy to estimate. From Appendix H available at booksite.elsevier.com/978-0-444-59436-5 on p. e91 it follows that the mean value of the kinetic energy of an electron described by the 1s orbital in an atom of atomic

The Schrödinger equation is incompatible with special relativity theory. This flaw has to be corrected somehow. The problem is far from being solved, but progress so far shows the Schrödinger equation, the spin of a particle, etc. in a new light.

Why Is This Important?

This chapter addresses the very foundations of physics, and in principle, this subject has to be treated on an equal footing with the postulates of quantum mechanics. The Schrödinger equation of Chapter 2 does not fulfill (as will be shown in this chapter) the requirements of relativity theory, and therefore is in principle "*illegal.*" In this chapter, Dirac's attempt to generalize the Schrödinger equation to adapt it to relativity theory will be described. If one assumes that particle velocities are small compared to that of light, then from this more general theory, one obtains the Schrödinger equation. Also, the notion of spin, which was introduced as a postulate in Chapter 1, follows as a natural consequence of the relativistic theory. One may draw the conclusion that this chapter addresses "*the foundations of foundations*," and therefore should occupy a prominent position in the TREE instead of representing a small side branch (as it does now). However, the relativistic effects, even if visible in chemistry, do not play an important role in the case of the light elements (almost the whole of organic chemistry, as well as almost the whole of biology). This is why I have chosen a rather pragmatic ("*non-fundamental*") way of presentation. This chapter is mainly for those readers who are interested in

- "*The foundations of foundations*"
- Very accurate calculations for small atoms and molecules
- Calculations for the systems containing heavy elements.

What Is Needed?

- The postulates of quantum mechanics (Chapter 1, necessary)
- Operator algebra (see Appendix A available at booksite.elsevier.com/978-0-444-59436-5, p. e1 necessary)
- Vector and scalar potentials (see Appendix G available at booksite.elsevier.com/978-0-444-59436-5, p. e81, necessary)

Classical Works

The American physicist Albert Michelson (by himself in 1881 and in 1887 with Edward Morley) carried out some experiments showing that the speed of light is the same in the directions perpendicular and parallel to the Earth's orbit; i.e., the Earth's orbital velocity did not change the speed of light with respect to the Earth. The results were published in the *American Journal of Science*, **22**, 120 (1881) with the title "*The relative motion of the Earth and the luminiferous aether*," and *ibid.*, **34**, 333 (1887) (with a similar title). ★ In 1889, the Irish physicist George Francis FitzGerald made the conjecture that if all moving objects were foreshortened in the direction of their motion, this would account for the strange results of the Michelson-Morley experiment. This was published in *Science*, **13**, 390 (1889), with the title "*The ether and the Earth's atmosphere.*" ★ The revolutionary special relativity theory (which explained this in detail) was developed by Albert Einstein in an article entitled "*Zur Elektrodynamik bewegter*

number Z is equal to $\bar{T} = \frac{1}{2}Z^2$ (in a.u.). On the other hand, for a rough estimation of the electron velocity v, one may write $\bar{T} = \frac{mv^2}{2}$. This results in the expression $v = Z$ valid in a.u., while the velocity of light $c = 137.036$ a.u. The largest Z known hardly exceeds 100. It is seen, therefore, that if an atom with $Z > 137$ existed, then the 1s electrons would attain velocities exceeding the velocity of light. Even if this calculation is nothing but a rule of thumb, there is no doubt that when Z increases, a certain critical Z value is approached (the so-called *relativistic mass effect*).

Körper," published in *Annalen der Physik (Leipzig)*, **17**, 891 (1905). ★ Einstein's article is based largely on the ideas of the Dutchman Hendrik Antoon Lorentz, who independently of FitzGerald proposed the Lorentz transformation (of space and time) in 1904. The transformation accounted for the contraction of moving objects, as predicted by FitzGerald. The paper "*Electromagnetic Phenomena in a System Moving with any Velocity smaller than that of Light*" was published in *Proceedings of the Academy of Sciences of Amsterdam*, **6**, 809 (1904). ★ The German mathematician Hermann Minkowski realized that the work of Lorentz and Einstein could best be understood using a non-Euclidean space of the space and time variables. His first paper on this subject was "*Die Grundgleichungen für die elektromagnetischen Vorgänge in bewegten Körper,*" published in *Nachrichten der königlichen Gesellschaft der Wissenschaften zu Göttingen* (1908). ★ The Soviet physicist Vladimir A.Fock derived the first relativistic wave equation for a particle [published in *Zeitschrift für Physik*, **39**, 226 (1926)], then the German scientist Walter Gordon did the same and also published in *Zeitschrift für Physik*, **40**, 117 (1926). Finally, a similar theory was proposed independently by the Swede Oskar Klein in *Zeitschrift für Physik*, **41**, 407 (1927). The Austrian scientist Erwin Schrödinger also derived the same equation, and this is why it is sometimes called "*the equation with many fathers*." ★ A more advanced quantum mechanical theory (for a single particle) adapted to the principles of relativity was given by the British physicist Paul Adrien Maurice Dirac in several articles in *Proceedings of the Royal Society (London)* entitled "*The fundamental equations of quantum mechanics,*" **A109**, 642 (1926); "*Quantum mechanics and a preliminary investigation of the hydrogen atom, ibid.* **A110**, 561 (1926); "*The quantum theory of radiation,*" *ibid.*" **A114**, 243 (1927); and "*The quantum theory of the electron,*" *ibid.* **A117** (1928) 610 and, "*The Quantum Theory of the Electron. Part II*" *ibid.* **A118**, 351 (1928). ★ An extension of relativistic quantum theory to many-electron problems (still approximate) was published by the American physicist Gregory Breit in *Physical Review* with the title "*The effect of retardation on the interaction of two electrons,*" **34**, 553 (1929), and then in two other papers entitled "*Fine structure of He as a test of the spin interaction of two electrons,*" *ibid.* **36**, 383 (1930), and "*Dirac's equation and the spin-spin interactions of two electrons,*" *ibid.* **39**, 616 (1932). ★ In 1948, the American physicists Richard Feynman and Julian Schwinger, as well as the Japanese physicist Shinichiro Tomonaga independently invented the quantum electrodynamics (QED), which successfully combined quantum theory with the special theory of relativity and produced extremely accurate results. ★ The relativistic approach has been introduced to quantum chemistry by Pekka Pyykkö and Jean-Paul Desclaux in a paper "*Relativity and Periodic System of Elements*" published in Accounts of Chemical Research, **12**, 276, 1979.

3.1 A Glimpse of Classical Relativity Theory

3.1.1 The Vanishing of Apparent Forces

Ernest Mach (1838–1916) was an Austrian physicist and philosopher, professor at the Universities of Graz, Prague, and Vienna, and godfather of Wolfgang Pauli. Mach investigated supersonic flows. In recognition of his achievements, the velocity of sound in air (1224 km/hour) is called Mach 1.

The three principles of Newtonian[6] dynamics were taught to us in school. The first principle, that a free body (with no acting force) moves uniformly along a straight line, seems to be particularly simple. It was not so simple for Ernest Mach, though.

[6] For Newton's biography, see Chapter 7.

Mach wondered how one recognizes that no force is acting on a body. The contemporary meaning of the first principle of Newton dynamics is the following. First, we introduce a Cartesian coordinate system x, y, z to the Universe, then remove from the Universe all objects except one, to avoid any interactions. Then, we measure equal time intervals using a spring clock and put the corresponding positions of the body to the coordinate system (thus, we are there with our clock and our ruler). The first principle says that the positions of the body are along a straight line and equidistant. What a crazy procedure! The doubts and dilemmas of Mach were implanted in the mind of Albert Einstein.

Albert Einstein (1879–1955), born in Ulm, Germany, studied at the ETH, Zurich. He is considered by many as a genius for all times. Einstein had the ability to look at difficult problems from an unusual perspective, seeing the simplicity behind the complex reality. The beginnings were however hard. He failed the entrance exam to the ETH Zurich. He wrote: *"If I have the good fortune to pass my examinations, I would study mathematics and physics. I imagine myself becoming a teacher."* Three of Einstein's fellow students were appointed assistants at ETH, but not him. As a teenager and student, Einstein rejected many social conventions. This is why he was forced to begin his scientific career at a secondary position in the federal patent office. Being afraid of his supervisor, he used to read books that he kept hidden in a drawer (he called the drawer the "*Department of Physics*").

The year of his 26th birthday, 1905, was particularly fruitful. He published three fundamental papers: about relativity theory, about Brownian motion, and about the photoelectric effect. For the latter work, Einstein received the Nobel Prize in physics in 1921. After these publications, he was appointed professor at the University of Zurich and then at the University of Prague. Starting in 1914, Einstein headed the Physics Institute in Berlin, which was founded especially for him. He emigrated to the United States in 1933 because he was relentlessly persecuted by the Nazis for his Jewish origin. Einstein worked at the Institute for Advanced Study in Princeton, New Jersey, and died there in 1955. According to his will, his ashes were dispersed over the United States from the air.

This Bern Patent Office employee also knew about the dramatic dilemmas of Hendrik Lorentz, which will be discussed shortly. Einstein recalls that there was a clock at a tram stop in Bern. Whenever his tram moved away from the stop, the modest patent office clerk asked himself what the clock would show if the tram had the velocity of light. Other passengers probably read their newspapers, but Einstein had questions that led humanity on new pathways.

Let us imagine two coordinate systems (each in 1-D): O "*at rest*" (we assume it to be inertial[7]), while the coordinate system O' moves with respect to the first in a certain way (possibly very complicated). The position of the moving point may be measured in O, giving the number x as the result, while in O', one gets the result x'. These numbers are related to one another as follows (t is time, f is a function of time):

$$x' = x + f(t). \tag{3.1}$$

If a scientist working in a lab associated with the coordinate system O would like to calculate the force acting on the abovementioned point body, he would get a result proportional to the acceleration (i.e., to $\frac{d^2x}{dt^2}$). If the same were done by another scientist working in a lab in O', then he would obtain another force, this time proportional to the acceleration computed as $\frac{d^2x'}{dt^2} = \frac{d^2x}{dt^2} + \frac{d^2f}{dt^2}$. The second term in this force is the *apparent* force. One encounters such apparent forces in elevators, on a carousel, etc., where a body moves as if for no reason.

Let us note an important consequence: if one postulates the same forces (and therefore the same dynamics) in two-coordinate systems, $f(t)$ has to be a *linear* function (because its second derivative is equal to zero). This means that a family of all coordinate systems that moved uniformly with respect to one another would be characterized by the same description of phenomena because the forces computed would be the same (*inertial systems*).

Physics textbooks written in the two laboratories associated with O and O' would describe all the phenomena in the same way.

The linearity condition gives $x' = x + vt$. Let us take a fresh look of this equation: x' represents a linear combination of x and t, which means that time and the linear coordinate mix together. One has two coordinates: one in the O coordinate system and the other in the O' coordinate system. Wait a minute! Since the time and the coordinate are on an equal footing (they mix together), maybe one may also have the time (t) appropriate for (i.e., running in) the

[7] That is, the Newton equation is satisfied. A coordinate system associated with an accelerating train is not inertial because there is a nonzero force acting on everybody on the train, while people sit.

O and the time (t') running in the O' coordinate system?

Now, a crucial step in the reasoning. Let us write in a most general way a linear transformation of coordinates and time (the forces computed in both coordinate systems are the same):

$$x' = Ax + Bt$$
$$t' = Cx + Dt.$$

First, the corresponding transformation matrix has to be invertible (i.e., non-singular), because inversion simply means exchanging the roles of the two coordinate systems and of the observers flying with them. Thus, one has:

$$x = \bar{A}x' + \bar{B}t'$$
$$t = \bar{C}x' + \bar{D}t'.$$

Next, A has to be equal to $\bar{A}$ because the measurements of length in O and O' (i.e., x and x'), cannot depend on whether one looks at the O coordinate system from O', or at the O' system from O. If the opposite were true, then one of the coordinate systems would be privileged (treated in a special way). This, however, is impossible because the two coordinate systems differ *only* in that O' flies from O with velocity v, while O flies from O' with velocity $-v$, but the space is isotropic. The same has to happen with the time measurements: on board O (i.e., t), and on board O' (i.e., t'), therefore $D = \bar{D}$. Since (from the inverse transformation matrix) $\bar{A} = \frac{D}{AD-BC}$ and $\bar{D} = \frac{A}{AD-BC}$, we have

$$\frac{D}{AD - BC} = A$$
$$\frac{A}{AD - BC} = D.$$

From this, $\frac{D}{A} = \frac{A}{D}$ follows, or

$$A^2 = D^2. \tag{3.2}$$

From the two solutions, $A = D$ and $A = -D$, one has to choose only $A = D$, because the second solution would mean that the times t and t' have opposite signs; i.e., when time runs forward in O it would run backward in O'. Thus, we have

$$A = D. \tag{3.3}$$

3.1.2 The Galilean Transformation

Galileo Galilei (1564–1642) was an Italian scientist and professor of mathematics at Pisa. Only those who have visited Pisa are able to appreciate the inspiration (for studying the free fall of bodies of different materials) from the incredibly leaning tower. Galileo's opus magnum (right-hand side) was published by Elsevier in 1638.

DISCORSI
E
DIMOSTRAZIONI
MATEMATICHE,
intorno à due nuoue ſcienze
Attenenti alla
MECANICA & I MOVIMENTI LOCALI,
del Signor
GALILEO GALILEI LINCEO,
Filoſofo e Matematico primario del Sereniſſimo
Grand Duca di Toſcana.
Con vna Appendice del centro di grauità d'alcuni Solidi.

IN LEIDA,
Appreſſo gli Elſevirii. M. D. C. XXXVIII.

The equality condition $A = D$ is satisfied by the Galilean transformation, in which the two coefficients equal 1 :

$$x' = x - vt$$
$$t' = t,$$

where position x and time t (say, of a passenger on a train), is measured in a fixed platform coordinate system, while x' and t' are measured in a train-fixed coordinate system. There are no apparent forces in the two coordinate systems related by the Galilean transformation. Also, the Newtonian equation is consistent with our intuition that time flows at the same pace in any coordinate system.

3.1.3 The Michelson-Morley Experiment

Hendrik Lorentz (1853–1928) was a Dutch scientist and a professor at Leiden. Lorentz was very close to formulating the special theory of relativity. It was pointed out to Lorentz in 1894 that FitzGerald had published something similar. He wrote to FitzGerald, but the latter replied that indeed he has sent a half-page article to *Science*, but he did not know "*whether they ever published it*." After this, Lorentz took every opportunity to stress that FitzGerald was the first to present the idea.

Hendrik Lorentz indicated that the Galilean transformation represents only *one possibility* of making the apparent forces vanish (i.e., assuring that $A = D$). Both constants need not equal 1. As it happens, such a generalization was found by an intriguing experiment performed in 1887.

Albert Michelson (1852–1931) was an American physicist and professor at the University of Chicago, USA. He specialized in the precise measurements of the speed of light.

Edward Williams Morley (1838–1923) was an American physicist and chemist and professor of chemistry at Western Reserve University in Cleveland, Ohio.

Michelson and Morley were interested in whether the speed of light differs when measured in two laboratories moving with respect to one another. According to the Galilean transformation, the two velocities of light should be different, in the same way as the speed of train passengers (measured with respect to the platform) differs depending on whether they walk in the same or the opposite direction with respect to the train motion. Michelson and Morley replaced the train by the Earth, which moves along its orbit around the Sun with a speed of about 40 km/s. Fig. 3.1 shows the Michelson-Morley experimental framework schematically. Let us imagine two identical right-angle V-shaped objects with all the arm lengths equal to L.

Each of the objects has a semitransparent mirror at its vertex[8], and ordinary mirrors at the ends. We will be interested in how much time it takes the light to travel along the arms of our objects (back and forth). One of the two arms of any object must be oriented along the x axis, while the other one must be orthogonal to it. The mirror system enables us to overlap the light beam from the horizontal arm (x axis) with the light beam from the perpendicular arm. If there were any difference in phase between them, we would immediately see the interference

[8] Such a mirror is made by covering glass with a silver coating.

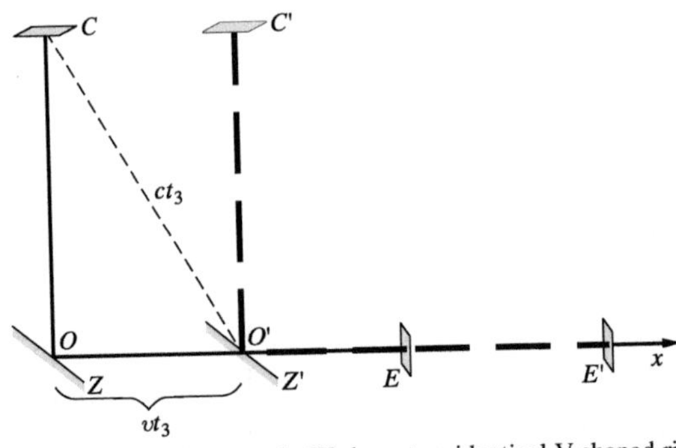

Fig. 3.1. The Michelson-Morley experimental framework. We have two identical V-shaped right-angle objects, each associated with a Cartesian coordinate system (with origins O and O'). The first is at rest (solid line), while the second (dashed) moves with velocity v with respect to the first (along coordinate x). We are going to measure the velocity of light in two laboratories rigidly bound to the two coordinate systems. The mirrors are at the ends of the objects: C, E in O and C', E' in O', while at the origins, two semitransparent mirrors Z and Z' are installed. Time $2t_3 \equiv t_\downarrow$ is the time it takes for light to go down and up the vertical arm.

pattern[9]. The second object moves along x with velocity $\boldsymbol{v}$ (and is associated with coordinate system O') with respect to the first ("*at rest*," associated with coordinate system O).

3.1.4 The Galilean Transformation Crashes

In the following section, we will suppose that the Galilean transformation is true. In coordinate system O, the time required for light to travel (in a round trip) the length of the arm along the x axis ($T_\rightarrow$) and that required to go perpendicularly to the axis ($T_\downarrow$) are the same:

$$T_\rightarrow = \frac{2L}{c}$$
$$T_\downarrow = \frac{2L}{c}.$$

Thus, in the O coordinate system, there will be no phase difference between the two beams (one coming from the parallel, the other from the perpendicular arm) and therefore, no interference will be observed. Let us consider now a similar measurement in O'. In the arm collinear with x , when light goes in the direction of $\boldsymbol{v}$, it has to take more time (t_1) to get to the end of the arm:

$$ct_1 = L + vt_1, \tag{3.4}$$

than the time required to come back (t_2) along the arm:

$$ct_2 = L - vt_2. \tag{3.5}$$

[9] From my own experience, I know that interference measurement is very sensitive. A laser installation was fixed to a steel table 10 cm thick set into the foundations of the Chemistry Department building, and the interference pattern was seen on the wall. My son Peter (then five years old) just touched the table with his finger. Everybody could see immediately a large change in the pattern, because the table . . . bent.

Thus, the total horizontal round-trip time $t_{\rightarrow}$ is[10]

$$t_{\rightarrow} = t_1 + t_2 = \frac{L}{c-v} + \frac{L}{c+v} = \frac{L(c+v)+L(c-v)}{(c-v)(c-v)} = \frac{2Lc}{c^2-v^2} = \frac{\frac{2L}{c}}{1-\frac{v^2}{c^2}}. \tag{3.6}$$

What about the perpendicular arm in the coordinate system O'? In this case, the time for light to go down (t_3) and up will be the same (let us denote total flight time by $t_{\downarrow} = 2t_3$; see Fig. 3.1). Light going down requires time $t_3 = \frac{t_{\downarrow}}{2}$ to travel along the arm. Light goes along the hypotenuse of the rectangular triangle with sides L and $\frac{vt_{\downarrow}}{2}$ (because it goes down, but not only down, since after $\frac{t_{\downarrow}}{2}$, it is found at $x = \frac{vt_{\downarrow}}{2}$). We will find, therefore, the time $\frac{t_{\downarrow}}{2}$ from the Pythagorean theorem:

$$\left(c\frac{t_{\downarrow}}{2}\right)^2 = L^2 + \left(v\frac{t_{\downarrow}}{2}\right)^2, \tag{3.7}$$

or

$$t_{\downarrow} = \sqrt{\frac{4L^2}{c^2-v^2}} = \frac{2L}{\sqrt{c^2-v^2}} = \frac{\frac{2L}{c}}{\sqrt{1-\frac{v^2}{c^2}}}. \tag{3.8}$$

The times $t_{\downarrow}$ and $t_{\rightarrow}$ do not equal each other for the moving system and there will be the interference that has previously been discussed in this chapter.

However, there is actually no interference!
Thanks to this result, Lorentz was forced to doubt the Galileian transformation (apparently the foundation of the whole science).

3.1.5 The Lorentz Transformation

The interference predicted by the Galilean transformation is impossible because physical phenomena would experience the two systems in a different way, while they differ only by their relative motions ($\boldsymbol{v}$ has to be replaced by $-\boldsymbol{v}$).

To have everything back in order, Lorentz assumed that when a body moves, its length $L_{\rightarrow}$ (measured by using the unit length at rest in the coordinate system O) along the direction of the motion *contracts* according to the following equation:

$$L_{\rightarrow} = L\sqrt{1-\frac{v^2}{c^2}}. \tag{3.9}$$

[10] Those who have some experience with relativity theory will certainly recognize the characteristic term $1-\frac{v^2}{c^2}$.

If we insert such a length, instead of L, in the expression for $t_{\rightarrow}$, then we obtain

$$t_{\rightarrow} = \frac{\frac{2L_{\rightarrow}}{c}}{1-\frac{v^2}{c^2}} = \frac{\frac{2L\sqrt{1-\frac{v^2}{c^2}}}{c}}{1-\frac{v^2}{c^2}} = \frac{\frac{2L}{c}}{\sqrt{1-\frac{v^2}{c^2}}}, \tag{3.10}$$

and everything is perfect again: $t_{\downarrow} = t_{\rightarrow}$. There is no interference, which means that x' (i.e., the position of a point belonging to a rigid body as measured in O') and x (the position of the same point measured in O) have to be related by the following formula. The coordinate x measured by an observer in his O is composed of the intersystem distance OO' (i.e., vt plus the distance $O'-$ point), but measured using the length unit of the observer in O [i.e., the unit that resides in O (thus, *non-contracted* by the motion)]. Because of the contraction $1 : \sqrt{1-\frac{v^2}{c^2}}$ of the rigid body, the latter result will be *smaller* than x' (recall that x' is what the observer measuring the position in his O' obtains); hence:

$$x = x'\sqrt{1-\frac{v^2}{c^2}} + vt \tag{3.11}$$

or

$$x' = \frac{x}{\sqrt{1-\frac{v^2}{c^2}}} - \frac{vt}{\sqrt{1-\frac{v^2}{c^2}}}, \tag{3.12}$$

which means that in the linear transformation, assuring no apparent forces,

$$A = \frac{1}{\sqrt{1-\frac{v^2}{c^2}}}, \tag{3.13}$$

$$B = -\frac{v}{\sqrt{1-\frac{v^2}{c^2}}}. \tag{3.14}$$

Of course, Professor O'Connor in his laboratory O' would not believe Professor Oconnor (sitting in his O lab) that he (O'Connor) has a contraction of the rigid body. And indeed, if Professor O'Connor measured the rigid body using his standard length unit (he would not know that his unit is contracted), then the length measured would be exactly the same as that measured just before separation of the two systems, when both systems were at rest. In a kind of retaliation, Professor O'Connor could say that it is certainly not him who has the contraction, but his colleague, Oconnor. He would be right, because for him, his system is at rest, and his colleague Oconnor flies away from him with velocity $-v$. Indeed, the formula (3.11) makes that very clear: an exchange $x \leftrightarrow x'$, $t \leftrightarrow t'$ and $v \leftrightarrow -v$, leads to the point of view of Professor O'Connor

$$x' = x\sqrt{1-\frac{v^2}{c^2}} - vt', \tag{3.15}$$

and one can indeed see an evident contraction of the rigid body of Professor Oconnor. This way, neither of these two coordinate systems is privileged. That is very, very good.

As we have already shown, in linear transformation $(x', t') \rightarrow (x, t)$, the diagonal coefficients have to equal $(A = D)$; therefore,

$$t' = Cx + Dt \tag{3.16}$$

$$\text{and } D = \frac{1}{\sqrt{1 - \frac{v^2}{c^2}}}. \tag{3.17}$$

To complete the determination of the linear transformation, we have to calculate the constant C (p. 109). Albert Einstein assumed that if Professors Oconnor and O'Connor began (in their own coordinate systems O and O') to measure the velocity of light, then despite the different distances (x and x') and different flight times[11] (t and t'), both scientists would get the same velocity of light (denoted by c).

In other words, $x = ct$ and $x' = ct'$.

Using this assumption and Eqs. (3.12) and (3.17), we obtain

$$ct' = Dct - vDt, \tag{3.18}$$

while multiplying Eq. (3.16) for t' by c, we get

$$ct' = cCx + Dct. \tag{3.19}$$

Subtracting both equations, we have

$$0 = -vDt - cCx \tag{3.20}$$

or

$$C = -\frac{vtD}{cx} = -\frac{vtD}{cct} = -\frac{vD}{c^2}. \tag{3.21}$$

Thus, we obtain the full Lorentz transformation, which on one hand assures no apparent forces and on the other, assures the same speed of light in *both* systems:

$$x' = \frac{1}{\sqrt{1 - \frac{v^2}{c^2}}}x - \frac{v}{\sqrt{1 - \frac{v^2}{c^2}}}t$$

$$t' = -\frac{v}{c^2}\frac{1}{\sqrt{1 - \frac{v^2}{c^2}}}x + \frac{1}{\sqrt{1 - \frac{v^2}{c^2}}}t.$$

[11] At the moment of separation, $t = t' = 0$.

Let us check first whether if $v = 0$, then everything is OK. Yes, it is. Indeed, the denominator equals 1, and we have $t' = t$ and $x' = x$. Let us see what would happen if the velocity of light were equal to infinity. Then, the Lorentz transformation becomes identical to the Galilean one. In general, after expanding t' and x' in a power series of v^2/c^2, we obtain

$$x' = -vt + x + \frac{1}{2}(-vt + x)\frac{v^2}{c^2} + \cdots$$

$$t' = t + \left(-\frac{x}{v} + \frac{t}{2}\right)\frac{v^2}{c^2} + \cdots$$

This means that only at very high velocity v may we expect differences between both transformations.

3.1.6 New Law of Adding Velocities

Our intuition has worked out for velocities that are much smaller than the velocity of light. The Lorentz transformation teaches us something that seems to contradict our intuition, though What does it mean that the velocity of light is constant? Suppose that we are flying with the velocity of light and send the light in the direction of our motion. Our intuition tells us that the light will have the velocity equal to 2c–but that has to be wrong. How will it happen?

We would like to have the velocity in the coordinate system O, but first, let us find the velocity in the coordinate system O' (i.e., $\frac{dx'}{dt'}$). From the Lorentz transformation, one obtains, step by step:

$$\frac{dx'}{dt'} = \frac{\frac{1}{\sqrt{1-\frac{v^2}{c^2}}}dx - \frac{v}{\sqrt{1-\frac{v^2}{c^2}}}dt}{-\frac{v}{c^2}\frac{1}{\sqrt{1-\frac{v^2}{c^2}}}dx + \frac{1}{\sqrt{1-\frac{v^2}{c^2}}}dt} = \frac{\frac{dx}{dt} - v}{1 - \frac{v}{c^2}\frac{dx}{dt}}. \quad (3.22)$$

By extracting $\frac{dx}{dt}$ or using the symmetry relation (when $O' \to O$, then $v \to -v$), we obtain:

$$\frac{dx}{dt} = \frac{\frac{dx'}{dt'} + v}{1 + \frac{v}{c^2}\frac{dx'}{dt'}} \quad (3.23)$$

or

Velocity Addition Law

$$V = \frac{v' + v}{1 + \frac{vv'}{c^2}}. \quad (3.24)$$

In this way, we have obtained a new rule of adding the velocities of the train and its passenger. Everybody naively thought that if the train velocity is v and the passenger velocity with respect to the train corridor is v', then the velocity of the passenger with respect to the platform is

$V = v + v'$. It turned out that this is not true. On the other hand, when both velocities are negligibly small with respect to c, then indeed one restores the old rule

$$V = v' + v. \tag{3.25}$$

Now, let us try to fool Mother Nature. Suppose that our train is running with the velocity of light (i.e., $v = c$), and we take out a flashlight and shine the light forward (i.e., $\frac{dx'}{dt'} = v' = c$). What will happen? What will the velocity V of the light be with respect to the platform? $2c$? From Eq. (3.24), we have $V = \frac{2c}{2} = c$. This is precisely what is called the universality of the speed of light. Now, let us make a bargain with Nature. We are dashing through the train at the speed of light ($v = c$) and walking along the corridor with velocity $v' = 5$ km/h. What will our velocity be with respect to the platform? Let us calculate again:

$$\frac{dx}{dt} = \frac{5 + c}{1 + \frac{c}{c^2}5} = \frac{5 + c}{1 + \frac{5}{c}} = c\frac{5 + c}{5 + c} = c. \tag{3.26}$$

Once more we have been unable to exceed the speed of light c. So let's make one last attempt. Let us take the train velocity as $v = 0.95c$, and fire along the corridor a powerful missile with speed $v' = 0.10c$. Will the missile exceed the speed of light or not? We have

$$\frac{dx}{dt} = \frac{0.10c + 0.95c}{1 + \frac{0.95c}{c^2}0.10c} = \frac{1.05c}{1 + 0.095} = \frac{1.05}{1.095}c = 0.9589c, \tag{3.27}$$

and c is not exceeded. What a wonderful formula!

3.1.7 The Minkowski Space-Time Continuum

The Lorentz transformation may also be written as $\begin{bmatrix} x' \\ ct' \end{bmatrix} = \frac{1}{\sqrt{1-\frac{v^2}{c^2}}}\begin{bmatrix} 1 & -\frac{v}{c} \\ -\frac{v}{c} & 1 \end{bmatrix}\begin{bmatrix} x \\ ct \end{bmatrix}$.

What would happen if the roles of the two systems were interchanged? To this end, let us express x, t as x', t'. By inversion of the transformation matrix, we obtain[12]

$$\begin{bmatrix} x \\ ct \end{bmatrix} = \frac{1}{\sqrt{1 - \frac{v^2}{c^2}}}\begin{bmatrix} 1 & \frac{v}{c} \\ \frac{v}{c} & 1 \end{bmatrix}\begin{bmatrix} x' \\ ct' \end{bmatrix}. \tag{3.28}$$

We have perfect symmetry because it is clear that the sign of the velocity has to change. Therefore,

none of the systems is privileged.

[12] You may check this by multiplying the matrices of both transformations; doing this, we obtain the unit matrix.

Now let us come back to Einstein's morning tram meditation[13] about what he would see on the clock at the tram stop if the tram had the velocity of light. Now we have the tools to solve the problem. It concerns two events–two ticks of the clock observed in the coordinate system associated with the tram stop (i.e., $x_1 = x_2 \equiv x$), but happening at two different times t_1 and t_2 [differing by, say, 1 second (i.e., $t_2 - t_1 = 1$), this is associated with the corresponding movement of the clock hand]. What will Einstein see when his tram leaves the stop with velocity v with respect to the stop, or in other words, when the tram stop moves with respect to him with velocity $-v$? He will see the same two events, but in his coordinate system, they will happen at $t_1' = \frac{t_1}{\sqrt{1-\frac{v^2}{c^2}}} - \frac{\frac{v}{c^2}x}{\sqrt{1-\frac{v^2}{c^2}}}$ and $t_2' = \frac{t_2}{\sqrt{1-\frac{v^2}{c^2}}} - \frac{\frac{v}{c^2}x}{\sqrt{1-\frac{v^2}{c^2}}}$. That is, according to the tram passenger, the two ticks at the tram stop will be separated by the time interval

$$t_2' - t_1' = \frac{t_2 - t_1}{\sqrt{1 - \frac{v^2}{c^2}}} = \frac{1}{\sqrt{1 - \frac{v^2}{c^2}}}.$$

Thus, when the tram ran through the streets of Bern with velocity $v = c$, the hands on the tram-stop clock (when seen from the tram) would not move at all, and this second would be equivalent to eternity.

This effect is known as *time dilation*. Of course, for the passengers waiting at the tram stop and watching the clock, its two ticks would be separated by exactly 1 second. If Einstein took his watch out of his waistcoat pocket and showed it to them through the window, *they* would be amazed. The seconds will pass at the tram stop, while Einstein's watch would seem to have stopped. This effect has been double-checked experimentally many times. For example, the meson lives such a short time (in the coordinate system associated with it), that when created by cosmic rays in the stratosphere, it would have no chance of reaching a surface laboratory before decaying. Nevertheless, as seen from the laboratory coordinate system, the meson's clock ticks very slowly and mesons are observable.

---

[13] Even today, Bern looks quite provincial. In the center, Albert Einstein lived at Kramgasse 49, in a small house, squeezed by others, next to a small café, with Einstein's achievements on the walls. Einstein's small apartment is on the second floor, containing a room facing the backyard, in the middle a child's room, and a large living room facing the street. A museum employee with oriental features says the apartment looks as it did in the "*miraculous year 1905*," everything is the same (except the wallpaper), and then, she concludes by saying "*Maybe this is the most important place in the history of science.*"

Hermann Minkowski introduced the seminal concept of the four-dimensional space-time continuum (x, y, z, ct)[14]. In our one-dimensional space, the elements of the Minkowski space-time continuum are events [i.e., vectors (x, ct)], something happens at space coordinate x at time t, when the event is observed from coordinate system O. When the same event is observed in two coordinate systems, then the corresponding x, t and x', t' satisfy the Lorentz transformation. It turns out that in both coordinate systems *the distance of the event from the origin of the coordinate system is preserved.* The square of the distance is calculated in a strange way:

Hermann Minkowski (1864–1909) was a German mathematician and physicist, a professor in Bonn, Königsberg, and Technische Hochschule Zurich, and from 1902, a professor at the University of Göttingen. This teacher of Einstein concluded: "*Space of itself and time of itself will sink into mere shadows, and only a kind of union between them shall survive.*"

$$(ct)^2 - x^2 \tag{3.29}$$

for the event (x, ct). Indeed, let us check carefully:

$$\begin{aligned}(ct')^2 - (x')^2 &= \frac{1}{1-\frac{v^2}{c^2}}\left(-\frac{v}{c}x + ct\right)^2 - \frac{1}{1-\frac{v^2}{c^2}}\left(x - \frac{v}{c}ct\right)^2 \\ &= \frac{1}{1-\frac{v^2}{c^2}}\left[\frac{v^2}{c^2}x^2 + c^2t^2 - 2vxt - x^2 - \frac{v^2}{c^2}c^2t^2 + 2vxt\right] \\ &= \frac{1}{1-\frac{v^2}{c^2}}\left[\frac{v^2}{c^2}x^2 + c^2t^2 - x^2 - \frac{v^2}{c^2}c^2t^2\right] = (ct)^2 - (x)^2. \end{aligned} \tag{3.30}$$

This equation and Eq. (3.28) enabled Minkowski to interpret the Lorentz transformation [Eq. (3.28)] as a *rotation* of the event (x, ct) in the Minkowski space about the origin of the coordinate system (since any rotation preserves the distance from the rotation axis).

[14] Let me report a telephone conversation between the Ph.D. student Richard Feynman and his supervisor, Professor Archibald Wheeler from the Princeton Advanced Study Institute (according to Feynman's Nobel lecture, 1965): Wheeler: "*Feynman, I know why all electrons have the same charge and the same mass!*" Feynman: "*Why?*" W: "*Because they are all the same electron!*" Then, Wheeler explained: "*Suppose that the world lines which we were ordinarily considering before in time and space–instead of only going up in time were a tremendous knot, and then, when we cut through the knot by the plane corresponding to a fixed time, we would see many, many world lines and that would represent many electrons (...)*"

3.1.8 How Do We Get $E = mc^2$?

The Schrödinger equation is invariant with respect to the Galilean transformation. Indeed, the Hamiltonian contains the potential energy, which depends on interparticle distances (i.e., on the differences of the coordinates), whereas the kinetic energy operator contains the second derivative operators that are invariant with respect to the Galilean transformation. Also, since $t = t'$, the time derivative in the time-dependent Schrödinger equation does not change.

Unfortunately, both Schrödinger equations (time-independent and time-dependent) are not invariant with respect to the Lorentz transformation, and therefore, they are illegal. As a result, one cannot expect the Schrödinger equation to describe accurately objects that move with velocities comparable to the speed of light.

Let us consider a particle moving in the potential V. The Schrödinger equation has been derived (see p. 79) from the total energy expression

$$E = \frac{p^2}{2m} + V, \tag{3.31}$$

where $\boldsymbol{p}$ is the momentum vector and m is the mass.

Einstein was convinced that nothing could be faster than light[15]. Therefore, what would happen if a particle were subject to a constant force? It would eventually attain the velocity of light, and what would happen afterward? There was a problem, and Einstein assumed that in the laboratory coordinate system in which the particle is accelerated, the particle will increase its mass. In the coordinate system fixed on the particle no mass increase will be observed, but in the laboratory system, it will. We have to admire Einstein's courage. For millions of people, the mass of a body represented an invariant characteristic of the body. How was the mass supposed to increase? Well, Einstein reasoned that the increase law should be such that the particle was able to absorb *any* amount of the kinetic energy. This means that when $v \to c$, then we have to have $m(v) \to \infty$. One of the possible formulas for $m(v)$ may contain a factor typical of relativity theory [cf. Eq. (3.17)]:

$$m(v) = \frac{m_0}{\sqrt{1 - \frac{v^2}{c^2}}}, \tag{3.32}$$

where m_0 is the so-called rest mass of the particle (i.e., its mass measured in the coordinate system residing on the particle). It is seen that if v/c were zero (as it is in the non-relativistic world), then m would be equal to m_0 (i.e., to a constant, as it is in non-relativistic physics)[16].

---

[15] Maybe this is true, but nothing in the special theory of relativity compels such a statement.

[16] Therefore, no corrections to the Schrödinger equation are needed. At the beginning of this chapter, we arrived at the conclusion that the electron velocity in an atom is close to its atomic number Z (in a.u.). Hence, for the hydrogen atom ($Z_H = 1$), one may estimate $v/c \simeq 0.7\%$; i.e., v of the electron in the $1s$ state represents a velocity of the order of 2100 km/s, which is probably very impressive for a car driver, but not for an electron. However, for

For the time being, the legitimacy of Eq. (3.32) is questionable as being just one of the possible *ad hoc* suppositions. However, Einstein has shown that this particular formula fits the existing equation of motion. First, after expanding the mass into the Taylor series, one obtains something interesting:

$$m(v) = m_0 \left\{ 1 + \frac{1}{2}\frac{v^2}{c^2} + \frac{3}{8}\frac{v^4}{c^4} + \cdots \right\}, \tag{3.33}$$

especially after multiplying the result by c^2:

$$mc^2 - m_0c^2 = \frac{m_0 v^2}{2} + \text{smaller terms.} \tag{3.34}$$

It looks as if the kinetic energy indeed was stored directly in the mass m. Einstein deduced that the total kinetic energy of the body may be equal to

$$E = mc^2.$$

He convinced himself about this after calculating its time derivative. After assuming that Eq. (3.32) is correct, one obtains:

$$\frac{dE}{dt} = c^2\frac{dm}{dt} = c^2\frac{d}{dt}\frac{m_0}{\sqrt{1-\frac{v^2}{c^2}}} = m_0c^2\frac{d}{dt}\frac{1}{\sqrt{1-\frac{v^2}{c^2}}} = m_0c^2\left(-\frac{1}{2}\right)\left(1-\frac{v^2}{c^2}\right)^{-\frac{3}{2}}\frac{-2v}{c^2}\frac{dv}{dt}$$

$$= m_0\left(1-\frac{v^2}{c^2}\right)^{-\frac{3}{2}} v\frac{dv}{dt} = \frac{m_0}{\sqrt{\left(1-\frac{v^2}{c^2}\right)}}\frac{1}{1-\frac{v^2}{c^2}}v\frac{dv}{dt} = \frac{m_0}{\sqrt{\left(1-\frac{v^2}{c^2}\right)}}\left(1+\frac{\frac{v^2}{c^2}}{1-\frac{v^2}{c^2}}\right)v\frac{dv}{dt}$$

$$= \frac{m_0}{\sqrt{\left(1-\frac{v^2}{c^2}\right)}}v\frac{dv}{dt} + \frac{v^2}{c^2}m_0\left(1-\frac{v^2}{c^2}\right)^{-\frac{3}{2}}v\frac{dv}{dt} = mv\frac{dv}{dt} + v^2\frac{dm}{dt} = v\frac{d(mv)}{dt}.$$

gold ($Z_{Au} = 79$), we obtain $v/c \simeq 51\%$. This means that in an atom of gold, the electron mass is larger by about 15% with respect to its rest mass; and therefore, the relativistic effects are non-negligible. For such important elements as C, O, N (biology), the relativistic corrections may be safely neglected. A young graduate student, Grzegorz Łach, posed an interesting, purely academic question (such questions and the freedom to discuss them represent the cornerstone and the beauty of university life): Will the human body survive if relativistic effects are switched off? Most of the biomolecules would function practically without significant changes, but the heavy metal atoms in enzyme-active sites might direct differently the chemical reactions in which they are involved. But would they? Would the new direction be destructive for the body? Nobody knows. On the other hand, we have forgotten about the spin concept, which follows consequently only from relativistic quantum theory. Without spin, no world similar to ours is conceivable.

Precisely the same equation is satisfied in non-relativistic mechanics if E denotes the kinetic energy:

$$\frac{dE}{dt} = \frac{d(\frac{mv^2}{2})}{dt} = \frac{1}{2}m2v\frac{dv}{dt} = v\frac{d(mv)}{dt}. \tag{3.35}$$

Therefore, in relativity theory,

$$E_{kin} = mc^2. \tag{3.36}$$

This formula has been verified in laboratories many times. For example, it is possible nowadays to speed electrons in cyclotrons up to a velocity that differs from c by $\frac{1}{8000000}c$. That corresponds to $1 - \frac{v^2}{c^2} = \frac{1}{4000000}$, and the electron's mass m becomes 2000 times larger than its m_0. This means that the electron is pumped up with energy to such an extent that its mass is similar to that of the proton. The energy stored in mass is huge. If, from the mass of a 20000 TNT atomic bomb, one subtracted the mass of its ashes after an explosion[17], then one would obtain only about 1 g! The energy freed from this 1 g gives an effect similar to the apocalypse.

3.2 Toward Relativistic Quantum Mechanics

The Equation of Many Fathers

We would like to express the kinetic energy E_{kin} through the particle's momentum p, because we would then know how to obtain the corresponding quantum mechanical operators (Chapter 1, p. 18). To this end, let us consider the expression

$$\begin{aligned} E_{kin}^2 - (m_0c^2)^2 &= m^2c^4 - m_0^2c^4 = m_0^2c^4\left(\frac{1}{1-v^2/c^2} - 1\right) = \frac{m_0^2c^4}{1-v^2/c^2}\frac{v^2}{c^2} \\ &= m^2v^2c^2 = p^2c^2. \end{aligned} \tag{3.37}$$

Therefore,

$$E_{kin} = c\sqrt{p^2 + m_0^2c^2} \tag{3.38}$$

and the total energy E in the external potential V is as follows:

$$E = c\sqrt{p^2 + m_0^2c^2} + V. \tag{3.39}$$

What if the particle is subject to an electromagnetic field, given by the electric field $\boldsymbol{\mathcal{E}}$ and the magnetic field $\boldsymbol{H}$ (or the magnetic induction $\boldsymbol{B}$) in every point of the space? Instead of $\boldsymbol{\mathcal{E}}$ and $\boldsymbol{H}$ (or $\boldsymbol{B}$), we may introduce two other quantities: the vector field $\boldsymbol{A}$ and the scalar field ϕ (see Appendix G available at booksite.elsevier.com/978-0-444-59436-5). As we can show

[17] R. Feynman, R.B. Leighton, and M. Sands, "*Feynman Lectures on Physics,*" Addison-Wesley Publishing Company, Boston (1964).

in classical electrodynamics[18], the kinetic energy of the particle subject to an electromagnetic field is very similar to the same expression without the field [Eq. (3.38)]; namely, for a particle of charge q, the momentum $\mathbf{p}$ is to be replaced by $\mathbf{p} - \frac{q}{c}\mathbf{A}$ and the potential V by $q\phi$. Therefore, we obtain the following expression for the total energy of the particle in an electromagnetic field:

$$E = c\sqrt{\left(\mathbf{p} - \frac{q}{c}\mathbf{A}\right)^2 + m_0^2c^2} + q\phi, \tag{3.40}$$

where $\mathbf{A}$ and ϕ represent functions of the particle's position.

If we wanted to use the last expression to construct the Hamiltonian, then we would find serious difficulty; namely, the momentum operator $\hat{\mathbf{p}} = -i\hbar\nabla$ (replacing $\mathbf{p}$ according to postulate II, as described in Chapter 1) is under the square root, leading to nonlinear operators. A few brave scientists noted, however, that if someone made a square, then the danger would disappear. We would obtain

$$(E - q\phi)^2 = c^2\left[\left(\mathbf{p} - \frac{q}{c}\mathbf{A}\right)^2 + m_0^2c^2\right]. \tag{3.41}$$

All that has been and still is a sort of guessing from some traces or indications.

The equations corresponding to physical quantities will be transformed to the corresponding operator equations, and it will be assumed that both sides of them will act on a wave function.

What, then, should be inserted as the operator $\hat{H}$ of the energy E? This was done by Schrödinger (before by Fock, Klein, and Gordon, which is why it is also known as the "*equation of many fathers*"). Schrödinger inserted what he had on the right side of his time-dependent equation $\hat{H}\Psi = i\hbar\frac{\partial}{\partial t}\Psi$; i.e., $\hat{H} = i\hbar\frac{\partial}{\partial t}$.

Oskar Klein (1894–1977) was the youngest son of the chief rabbi of Sweden and a professor of mathematics and physics at Stockholm Högskola. Walter Gordon (1893–1940) until 1933 was a professor at the University of Hamburg, and after that, he resided in Sweden.

This way,

$$\left(i\hbar\frac{\partial}{\partial t} - q\phi\right)^2 = c^2\left[\left(-i\hbar\nabla - \frac{q}{c}\mathbf{A}\right)^2 + m_0^2c^2\right], \tag{3.42}$$

or after acting on the wave function, we obtain the Fock-Klein-Gordon equation:

$$\left(i\hbar\frac{\partial}{\partial t} - q\phi\right)^2 \Psi = c^2\left[\left(-i\hbar\nabla - \frac{q}{c}\mathbf{A}\right)^2 + m_0^2c^2\right]\Psi. \tag{3.43}$$

[18] For example, H.F. Hameka, *Advanced Quantum Chemistry*, Addison-Wesley, Reading, p. 40 (1965).

This equation has at least one advantage over the Schrödinger equation: ct and x, y, z enter the equation on equal footing, which is required by special relativity. Moreover, the Fock-Klein-Gordon equation is invariant with respect to the Lorentz transformation, whereas the Schrödinger equation is not. This is a prerequisite of any relativity-consistent theory, and it is remarkable that such a simple derivation made the theory invariant. The invariance, however, does not mean that the equation is accurate. The Fock-Klein-Gordon equation describes a boson particle because Ψ is a usual scalar-type function, in contrast to what we will see shortly in the Dirac equation.

3.3 The Dirac Equation

3.3.1 The Dirac Electronic Sea and the Day of Glory

Paul Adrien Maurice Dirac (1902–1984) was a British physicist, theoretician, and professor at universities in Cambridge, and then in Oxford. Dirac was very interested in on hiking and climbing. He used to practice before expeditions by climbing trees near Cambridge, in the black outfit in which always gave his lectures.

He spent his last years at the University of Tallahassee in Florida. On being guided through New York City, Dirac remembered old times. The guide remarked that there were visible changes, among others the buses had been painted pink. Dirac quietly agreed, adding that indeed they had, at least from one side...

Charles Galton Darwin (1887–1962) was a British physicist and mathematician, professor at the University of Edinburgh, Scotland, and the grandson of the famed evolutionist Sir Charles Robert Darwin. Darwin investigated the scattering of α particles on atoms. Courtesy of Dr. R.C. McGuiness, National Physical Laboratory, UK.

Paul Dirac used the Fock-Klein-Gordon equation to derive a Lorentz transformation invariant equation[19] for a single fermion particle. The Dirac equation is solvable only for several very simple cases. One of them is the free particle (Dirac), and the other is an electron in the electrostatic field of a nucleus (Charles Darwin–but not *the* one you are thinking of). One may add here a few other systems (e.g., the harmonic oscillator), and that's it.

From Eq. (3.38), in the case of a free particle $V = 0$, one obtains two sets of energy eigenvalues, one corresponding to the negative energies,

$$E = -\sqrt{p^2c^2 + m_0^2c^4}, \tag{3.44}$$

[19] See J.D. Bjorken and S.D. Drell, *Relativistic Quantum Mechanics*, McGraw-Hill, New York, (1964).

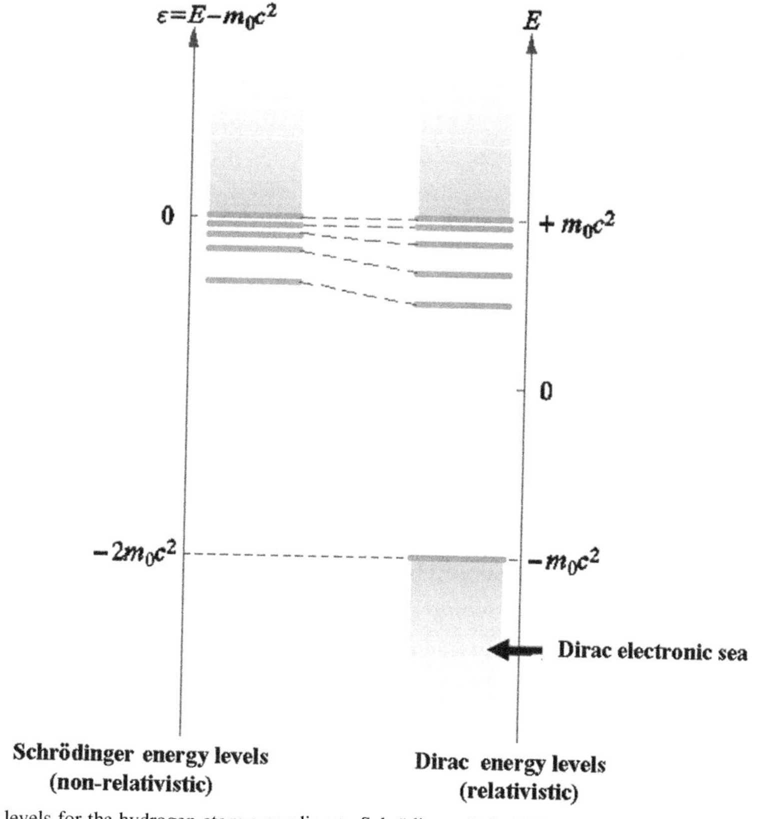

Fig. 3.2. Energy levels for the hydrogen atom according to Schrödinger (left side) and Dirac (right side). The shadowed areas correspond to the positive and negative energy continua.

and the other corresponding to the positive energies,

$$E = +\sqrt{p^2c^2 + m_0^2c^4}. \tag{3.45}$$

Dirac was not worried by the fact that both roots appear after an *ad hoc* decision to square the expression for the energy [Eqs. (3.40) and (3.41)]. As we can see, since the momentum may change from 0 to ∞ ($p = mv$, and for $v \to c$, we have $m \to \infty$), we therefore have the *negative energy continuum* and symmetrically located *positive energy continuum*, both of which are separated by the energy gap $2m_0c^2$ (Fig. 3.2).

When he was 26 years old, Dirac made the absurd assumption that what people call a vacuum is in reality a sea of electrons occupying the negative energy continuum (known as the *Dirac electronic sea*). The sea was supposed to consist of an infinite number of electrons, which had to imply catastrophic consequences concerning, for example, the infinite mass of the Universe, but Dirac did not feel any doubt about his notion. "*We see only those electrons that have positive energy*," said Dirac. Why he was so determined? Well, Dirac's concept of the sea was proposed to convince us that due to the Pauli exclusion principle, the doubly occupied sea electronic states

Carl David Anderson (1905–1991) was an American physicist and a professor at the Pasadena Institute of Technology. In 1932, Anderson discovered the positron when studying cosmic rays (using the cloud chamber). He received for this the Nobel Prize in 1936. He was also a co-discoverer of the muon.

are simply inaccessible for an extra electron, which is therefore forced to have the positive energies. All this looks like no more than an *ad hoc* speculation. Dirac could remove these difficulties by resigning from the negative spectrum, but he did not.[20]

Then, consequently continuing his reasoning, Dirac asked whether he could somehow see those electrons that occupy the sea. His answer was that yes, it is possible. According to Dirac, it is sufficient to excite such an electron by providing the energy of the order of $2m_0c^2$ to cover the energy gap (the energy $2m_0c^2$ is very large, of the order of 1 MeV). Then the sea electron would have the positive energy and could be observed as other electrons with positive energy. However, besides the electron, there would be a hole in the Dirac sea. Dirac has been severely molested about what this strange hole would correspond to in experimental physics. Once, when pushed too strongly, he said desperately that this was a proton. Some seemed to be satisfied, but others began to attack him furiously. However, Dirac has demonstrated that the hole would have the dynamic and electric properties of an electron, except that its sign would be opposite[21]. This has been nothing but a hypothesis for the existence of antimatter, a state of matter unknown at that time. Imagine the shock of the scientific community, when three years later, Carl Anderson reported the creation of electron-positron pairs from a vacuum after providing energy $2m_0c^2$. That was a glorious day for Dirac and his quantum theory.

In a moment, we will see the determination with which Dirac attacked the Fock-Klein-Gordon operator equation, which we will write a little differently here:

$$\left[\frac{i\hbar\frac{\partial}{\partial t} - q\phi}{c}\right]^2 - \left[\left(-i\hbar\nabla - \frac{q}{c}\mathbf{A}\right)^2 + m_0^2c^2\right] = 0. \tag{3.46}$$

[20] None of the many fathers of the Fock-Klein-Gordon equation dared to take into account another possibility, the one with the negative square root in Eq. (3.40), a step made by Paul Dirac. In this case the Dirac's argument about the electron sea and the Pauli exclusion principle would not work, since we have to do with the bosons! We would have an abyss of negative energies, a disaster for the theory.

[21] Paul Dirac, when a pupil in primary school, made his reputation after solving a riddle that goes very well with the person who thought out the positively charged electron (positron):

Three fishermen went fishing and camping overnight at a lake. After fishing all day, when evening came, they put the fish in a bucket and, tired, fell asleep in the tent. At midnight, one of the fishermen woke up and, tired of the whole escapade, decided to take one-third of all the fish, leave the tent quietly, and go home. When he counted the fish in the bucket, it turned out that the number of fish was indivisible by 3. However, when he threw one fish back in the lake, the number was divisible by 3, he took his one-third and went away. After a while, a second fisherman woke up and did the same, and then the third. The question was, how many fish were in the bucket? Dirac's answer was... −2.

Let us first introduce the following abbreviations:

$$\pi_0 = \frac{i\hbar\frac{\partial}{\partial t} - q\phi}{c}, \quad \pi_\mu = -i\hbar\frac{\partial}{\partial \mu} - \frac{q}{c}A_\mu \tag{3.47}$$

for $\mu = x, y, z$ or 1, 2, 3.

Dirac persisted in treating Eq. (3.46) as $a^2 - b^2$; therefore, he rewrote it in the form $(a+b)(a-b)$; i.e.,

$$\left(\pi_0 + \sum_{\mu=x,y,z} \alpha_\mu \pi_\mu + \alpha_0 m_0 c\right)\left(\pi_0 - \sum_{\mu=x,y,z} \alpha_\mu \pi_\mu - \alpha_0 m_0 c\right) = 0. \tag{3.48}$$

He was so confident that he said Eq. (3.48) had to be satisfied at any price by finding suitable unknowns α_i (independent of coordinates and time). The αs had to satisfy the following relations (anticommutation relations):

$$\alpha_\mu^2 = 1, \tag{3.49}$$

$$\alpha_\mu \alpha_\nu + \alpha_\nu \alpha_\mu = 0 \quad for \quad \mu \neq \nu. \tag{3.50}$$

Indeed, using the anticommutation relations, one recovers the Fock-Klein-Gordon equation:

$$\left(\pi_0 + \sum_{\mu=x,y,z}^{3} \alpha_\mu \pi_\mu + \alpha_0 m_0 c\right)\left(\pi_0 - \sum_{\mu=x,y,z}^{3} \alpha_\mu \pi_\mu - \alpha_0 m_0 c\right)$$

$$= \pi_0^2 - \left[\sum_{\mu=x,y,z}^{3} \alpha_\mu \pi_\mu + \alpha_0 m_0 c\right]^2 = \pi_0^2 - \sum_{\mu,\nu=x,y,z}^{3} \alpha_\mu \alpha_\nu \pi_\mu \pi_\nu$$

$$- \sum_{\mu=x,y,z}^{3} (\alpha_\mu \alpha_0 + \alpha_0 \alpha_\mu)\, \pi_\mu m_0 c - \alpha_0^2 m_0^2 c^2 = \pi_0^2 - \sum_{\mu,\nu=x,y,z}^{3} (\alpha_\mu \alpha_\nu + \alpha_\mu \alpha_\nu)\pi_\mu \pi_\nu$$

$$- m_0^2 c^2 = \pi_0^2 - \sum_{\mu=x,y,z}^{3} \pi_\mu^2 - m_0^2 c^2.$$

Note that the αs cannot be just numbers, because no numbers can satisfy the anticommutation relation. They have to be matrices. Since we have four variables x, y, z, and t, then we may expect matrices of order 4, but they could be larger. Here is one of the consistent choices of matrices ($\mathbf{0}, \mathbf{1}$ are the zero and unit $2x2$ matrices, respectively, while $\sigma_x, \sigma_y, \sigma_z$ are the Pauli matrices, defined on p. 29, that determine electron spin):

$$\alpha_x = \begin{pmatrix} \mathbf{0} & \sigma_x \\ \sigma_x & \mathbf{0} \end{pmatrix} \quad \alpha_y = \begin{pmatrix} \mathbf{0} & \sigma_y \\ \sigma_y & \mathbf{0} \end{pmatrix}$$

$$\alpha_z = \begin{pmatrix} \mathbf{0} & \sigma_z \\ \sigma_z & \mathbf{0} \end{pmatrix} \quad \alpha_0 \equiv \beta = \begin{pmatrix} \mathbf{1} & \mathbf{0} \\ \mathbf{0} & -\mathbf{1} \end{pmatrix}.$$

The Pauli matrices represent the first sign of what will happen later on: the Dirac equation will automatically describe the spin angular momentum.

3.3.2 The Dirac Equations for Electrons and Positrons

After the factorization described above, Dirac obtained two operator equations. The Dirac equations (for the positron and electron) correspond to these operators acting on the wave function Ψ. Thus, we obtain the equation for the negative electron energies i.e. for the *positron* (we use the abbreviation, a kind of the "*dot product*" of the α matrices and the operator π : $\sum_{\mu=x,y,z} \alpha_\mu \pi_\mu \equiv \boldsymbol{\alpha} \cdot \boldsymbol{\pi}$)

$$\left(\pi_0 + \boldsymbol{\alpha} \cdot \boldsymbol{\pi} + \alpha_0 m_0 c\right) \Psi = 0 \tag{3.51}$$

and for the positive electron energies (*electron*):

$$\left(\pi_0 - \boldsymbol{\alpha} \cdot \boldsymbol{\pi} - \alpha_0 m_0 c\right) \Psi = 0. \tag{3.52}$$

These two equations are coupled together through the same function Ψ, which has to satisfy both of them. This coupling caused a lot of trouble in the past. First, people assumed that the equation with the negative electron energies (positron equation) may be ignored because the energy gap is so large that the Dirac sea is occupied whatever a chemist does with a molecule. This assumption turned out to cause some really vicious or odd performances of numerical procedures (discussed later in this chapter). The electron equation alone reads as

$$i\hbar \frac{\partial \Psi}{\partial t} = (q\phi + c\boldsymbol{\alpha} \cdot \boldsymbol{\pi} + \alpha_0 m_0 c^2)\Psi. \tag{3.53}$$

If one is interested in *stationary states* (cf., p. 22), the wave function has the form $\Psi(x, y, z, t) = \Psi(x, y, z)e^{-i\frac{E}{\hbar}t}$, where we have kept the same symbol for the time independent factor $\Psi(x, y, z)$. After dividing by $e^{-i\frac{E}{\hbar}t}$, we obtain

The Dirac Equation for Stationary Electronic States

$$(E - q\phi - \alpha_0 m_0 c^2 - c\boldsymbol{\alpha} \cdot \boldsymbol{\pi})\Psi = 0. \tag{3.54}$$

The quantity $q\phi = V$ in future applications will denote the Coulomb interaction of the particle under consideration with the external potential.

3.3.3 Spinors and Bispinors

The last equation needs to be commented on. Because the matrices α have dimension 4, then Ψ has to be a four-component vector (known as *bispinor*, its connection to the spin concept will be shown later in this chapter):

$$\Psi = \begin{pmatrix} \psi_1 \\ \psi_2 \\ \phi_1 \\ \phi_2 \end{pmatrix} = \begin{pmatrix} \boldsymbol{\psi} \\ \boldsymbol{\phi} \end{pmatrix},$$

where the first two components (ψ_1 and ψ_2, functions of class Q), which for reasons that will become clear in a moment are called *large components,* are hidden in vector $\boldsymbol{\psi}$, while the two *small components* (ϕ_1 and ϕ_2, functions of class Q)[22] are labeled as vector $\boldsymbol{\phi}$. Vectors $\boldsymbol{\psi}$ and $\boldsymbol{\phi}$ are called the *spinors.*

So how should you operate the $N-$component spinor (note that for $N = 4$, we have called them *bispinors*)? Let us construct the proper Hilbert space for the $N-$component spinors. As usual (p. e7), first we will define the sum of two spinors in the following way:

$$\begin{pmatrix} \Psi_1 \\ \Psi_2 \\ \dots \\ \Psi_N \end{pmatrix} + \begin{pmatrix} \Phi_1 \\ \Phi_2 \\ \dots \\ \Phi_N \end{pmatrix} = \begin{pmatrix} \Psi_1 + \Phi_1 \\ \Psi_2 + \Phi_2 \\ \dots \\ \Psi_N + \Phi_N \end{pmatrix},$$

and then, the product of the spinor by a number γ is given as follows:

$$\gamma \begin{pmatrix} \Psi_1 \\ \Psi_2 \\ \dots \\ \Psi_N \end{pmatrix} = \begin{pmatrix} \gamma\Psi_1 \\ \gamma\Psi_2 \\ \dots \\ \gamma\Psi_N \end{pmatrix}.$$

Next, we check that the spinors form an Abelian group with respect to the above-defined addition (cf., Appendix C available at booksite.elsevier.com/978-0-444-59436-5, p. e17) and that the conditions for the vector space are fulfilled (see Appendix B available atbooksite.elsevier.com/978-0-444-59436-5). Then, we define the scalar product of two spinors:

$$\langle \Phi \mid \Psi \rangle = \sum_{i=1}^{N} \langle \Phi_i \mid \Psi_i \rangle ,$$

where the scalar products $\langle \Phi_i \mid \Psi_i \rangle$ are defined as usual in the Hilbert space of class Q functions. Then, using the scalar product $\langle \Phi \mid \Psi \rangle$, we define the distance between two spinors: $\|\Phi - \Psi\| \equiv \sqrt{\langle \Phi - \Psi \mid \Phi - \Psi \rangle}$ and afterwards, the concept of the Cauchy series (the distances between

[22] It will be shown that in the non-relativistic approximation, the large components reduce to the wave function known from the Schrödinger equation, and the small components vanish. The constant E, as well as the function V, individually multiply each component of the bispinor Ψ, while $\boldsymbol{\sigma} \cdot \boldsymbol{\pi} \equiv \alpha_x \pi_x + \alpha_y \pi_y + \alpha_z \pi_z$ denotes the "*dot product*" of the matrices α_μ , $\mu = x, y, z$ by the operators π_μ (in the absence of the electromagnetic field, it is simply the momentum operator component; see p. e81). The matrix β is multiplied by the constant m_0c^2, and then by the bispinor Ψ.

the consecutive terms tend to zero). The Hilbert space of spinors will contain all the linear combinations of the spinors together with the limits of all the Cauchy series.

An operator acting on a spinor means that a spinor with each component resulting from action on the corresponding component is as follows:

$$\hat{A}\begin{pmatrix}\Psi_1\\ \Psi_2\\ \cdots\\ \Psi_N\end{pmatrix}=\begin{pmatrix}\hat{A}\Psi_1\\ \hat{A}\Psi_2\\ \cdots\\ \hat{A}\Psi_N\end{pmatrix}.$$

Sometimes we will use a notation showing that a matrix of operators acts on a spinor. In this case, the result corresponds to multiplication of the matrix (of operators) and the vector (spinor):

$$\begin{pmatrix}\hat{A}_{11} & \hat{A}_{12} & \cdots & \hat{A}_{1N}\\ \hat{A}_{21} & \hat{A}_{22} & \cdots & \hat{A}_{2N}\\ \cdots & \cdots & \cdots & \cdots\\ \hat{A}_{N1} & \hat{A}_{N2} & \cdots & \hat{A}_{NN}\end{pmatrix}\begin{pmatrix}\Psi_1\\ \Psi_2\\ \cdots\\ \Psi_N\end{pmatrix}=\begin{pmatrix}\sum_j \hat{A}_{1j}\Psi_j\\ \sum_j \hat{A}_{2j}\Psi_j\\ \cdots\\ \sum_j \hat{A}_{Nj}\Psi_j\end{pmatrix}.$$

3.3.4 What Next?

In the following section, we will show

a. that the first two components of the bispinor are much larger than the last two.
b. that in the limit $c \to \infty$, the Dirac equation gives the Schrödinger equation.
c. that the Dirac equation accounts for the spin angular momentum of the electron.
d. how to obtain, in a simple way, an approximate solution of the Dirac equation to the electron in the field of a nucleus (hydrogen-like atom).

3.3.5 Large and Small Components of the Bispinor

Using matrix multiplication rules, the Dirac equation [Eq. (3.54)] with bispinors can be rewritten in the form of two equations with spinors $\boldsymbol{\psi}$ and $\boldsymbol{\phi}$:

$$(E - V - m_0c^2)\boldsymbol{\psi} - c(\boldsymbol{\sigma}\cdot\boldsymbol{\pi})\boldsymbol{\phi} = \mathbf{0} \tag{3.55}$$

$$(E - V + m_0c^2)\boldsymbol{\phi} - c(\boldsymbol{\sigma}\cdot\boldsymbol{\pi})\boldsymbol{\psi} = \mathbf{0}. \tag{3.56}$$

The quantity m_0c^2 represents the energy. Let us use this energy to shift the energy scale (we are always free to choose 0 on this scale): $\varepsilon = E - m_0c^2$, in order to make ε comparable in the future to the eigenvalue of the Schrödinger equation (p. 77). We obtain

$$(\epsilon - V)\boldsymbol{\psi} - c(\boldsymbol{\sigma}\cdot\boldsymbol{\pi})\boldsymbol{\phi} = \mathbf{0} \tag{3.57}$$

$$(\epsilon - V + 2m_0c^2)\boldsymbol{\phi} - c(\boldsymbol{\sigma}\cdot\boldsymbol{\pi})\boldsymbol{\psi} = \mathbf{0}. \tag{3.58}$$

This set of equations corresponds to a single matrix equation:

$$\begin{pmatrix} V & c(\boldsymbol{\sigma} \cdot \boldsymbol{\pi}) \\ c(\boldsymbol{\sigma} \cdot \boldsymbol{\pi}) & V - 2m_0c^2 \end{pmatrix} \begin{pmatrix} \boldsymbol{\psi} \\ \boldsymbol{\phi} \end{pmatrix} = \begin{pmatrix} \epsilon & 0 \\ 0 & \epsilon \end{pmatrix} \begin{pmatrix} \boldsymbol{\psi} \\ \boldsymbol{\phi} \end{pmatrix}. \quad (3.59)$$

3.3.6 How to avoid Drowning in the Dirac Sea

When, in the past, Eq. (3.59) was solved and the energy ϵ minimized with respect to the variational parameters in the trial spinors $\boldsymbol{\psi}$ and $\boldsymbol{\phi}$ (a routine practice in the non-relativistic case; see Chapter 5), then some serious numerical problems occurred. Either the numerical procedures diverged, or the solutions obtained were physically unacceptable. The reason for this was that the existence of the Dirac sea had been totally ignored by neglecting Eq. (3.51) for the positron and taking Eq. (3.52) solely for electron motion. The variational trial functions, however, felt the presence of Dirac sea electronic states (there was nothing in the theory that would prevent the electron from attempting to occupy negative energies), and the corresponding variational energies dived down the energy scale toward the abyss of the sea without bottom[23]. The presence of the Dirac sea makes the Dirac theory, in fact, a theory of an infinite number of particles, whereas formally it is only a theory of a single particle in an external field. This kind of discomfort made people think of the possibility of describing the electron from the Dirac electronic sea by replacing the bispinors by the exact spinor (two-component) theory[24]. Such exact separation has been reported by Barysz and Sadlej[25].

An approximate (and simple) prescription was also invented to avoid the catastrophic drowning described above. Indeed, Eq. (3.58) can be transformed without any problem to

$$\boldsymbol{\phi} = \left(1 + \frac{(\epsilon - V)}{2m_0c^2}\right)^{-1} \frac{1}{2m_0c}(\boldsymbol{\sigma} \cdot \boldsymbol{\pi})\boldsymbol{\psi}.$$

[23] The possible severity of the problem has been shown by M. Stanke and J. Karwowski, "*Variational principles in the Dirac theory: Spurious solutions, unexpected extrema and other traps*," in *New Trends in Quantum Systems in Chemistry and Physics*, vol. I, pp. 175–190, eds. J. Maruani et al., Kluwer Academic Publishers, Dordrecht (2001). Sometimes an eigenfunction corresponds to a quite different eigenvalue. Nothing of that sort appears in non-relativistic calculations.

[24] This is exact within the Dirac model.

[25] M. Barysz, A.J. Sadlej, and J.G. Snijders, *Int. J. Quantum Chem.*, **65**, 225 (1997); M. Barysz, *J. Chem. Phys.*, **114**, 9315 (2001); M. Barysz and A.J. Sadlej, *J. Mol. Struct. (Theochem)*, **573**, 181 (2001); M. Barysz and A.J. Sadlej, *J. Chem. Phys.*, **116**, 2696 (2002). In the latter paper, an exact solution to the problem was given. The two-component theory, although more appealing, both from the point of view of physics as well as computationally, implies a change in definition of the operators; e.g., the position operator is replaced by a quite complex expression. This fact, ignored in computations using two-component theories, has been analyzed in the following articles: V. Kellő, A.J. Sadlej, and B.A. Hess, *J. Chem. Phys.*, **105**, 1995 (1996); M. Barysz and A.J. Sadlej, *Theor. Chem. Acc.*, **97**, 260 (1997); V. Kellő and A.J. Sadlej, *Int. J. Quantum Chem.*, **68**, 159 (1998); V. Kellő and A.J. Sadlej, *J. Mol. Struct. (Theochem)*, **547**, 35 (2001).

Since $2m_0c^2$ represents a huge energy when compared to the kinetic energy $\varepsilon - V$, then the first parenthesis on the right side is approximately equal to 1. This means, however, that

$$\boldsymbol{\phi} \approx \frac{1}{2m_0c}(\boldsymbol{\sigma} \cdot \boldsymbol{\pi})\boldsymbol{\psi}, \tag{3.60}$$

which is known as *kinetic balancing*. It was shown that the *kinetically balanced* trial function achieves the miracle[26] of the energy not tending to $-\infty$. The kinetic balancing indicates some fixed relation between $\boldsymbol{\phi}$ and $\boldsymbol{\psi}$.

Let us focus now on $\boldsymbol{\sigma} \cdot \boldsymbol{\pi}$. This is a 2×2 matrix and in the absence of an electromagnetic field ($\boldsymbol{\pi} = \boldsymbol{p}$), one has:

$$\begin{aligned}\boldsymbol{\sigma} \cdot \boldsymbol{\pi} &= \sigma_x \hat{p}_x + \sigma_y \hat{p}_y + \sigma_z \hat{p}_z = \begin{pmatrix} 0 & \hat{p}_x \\ \hat{p}_x & 0 \end{pmatrix} + \begin{pmatrix} 0 & -i\hat{p}_y \\ i\hat{p}_y & 0 \end{pmatrix} + \begin{pmatrix} \hat{p}_z & 0 \\ 0 & -\hat{p}_z \end{pmatrix} \\ &= \begin{pmatrix} \hat{p}_z & \hat{p}_x - i\hat{p}_y \\ \hat{p}_x + i\hat{p}_y & -\hat{p}_z \end{pmatrix}.\end{aligned}$$

It is seen that $\boldsymbol{\sigma} \cdot \boldsymbol{\pi}$ is on the order of momentum mv, and for the small velocities, on the order of m_0v.

Hence, one obtains $\boldsymbol{\phi} \approx \frac{1}{2m_0c}(\boldsymbol{\sigma} \cdot \boldsymbol{\pi})\boldsymbol{\psi} \approx \frac{v}{2c}\boldsymbol{\psi}$, and so the component $\boldsymbol{\phi}$ is for small v much smaller than the component $\boldsymbol{\psi}$,

which justifies the terms *small* and *large* components[27].

3.3.7 From Dirac to Schrödinger–How Is the Non-Relativistic Hamiltonian Derived?

The approximate relation ("*kinetic balance*") between the large and small components of the bispinor (that holds for small v/c) may be used to eliminate the small components[28] from Eqs. (3.57) and (3.58). We obtain

$$(\epsilon - V)\boldsymbol{\psi} - c(\boldsymbol{\sigma} \cdot \boldsymbol{\pi})\frac{1}{2m_0c}(\boldsymbol{\sigma} \cdot \boldsymbol{\pi})\boldsymbol{\psi} = \tag{3.61}$$

$$(\epsilon - V)\boldsymbol{\psi} - \frac{1}{2m_0}(\boldsymbol{\sigma} \cdot \boldsymbol{\pi})(\boldsymbol{\sigma} \cdot \boldsymbol{\pi})\boldsymbol{\psi} = \mathbf{0}. \tag{3.62}$$

Let us take a closer look at the meaning of the expression

$$(\boldsymbol{\sigma} \cdot \boldsymbol{\pi})(\boldsymbol{\sigma} \cdot \boldsymbol{\pi}) = \begin{pmatrix} \hat{p}_z & \hat{p}_x - i\hat{p}_y \\ \hat{p}_x + i\hat{p}_y & -\hat{p}_z \end{pmatrix}\begin{pmatrix} \hat{p}_z & \hat{p}_x - i\hat{p}_y \\ \hat{p}_x + i\hat{p}_y & -\hat{p}_z \end{pmatrix} = \begin{pmatrix} \hat{p}^2 & 0 \\ 0 & \hat{p}^2 \end{pmatrix} = \hat{p}^2\mathbf{1}$$

26 This remedy has not only an *ad hoc* character, but also it does not work for the heaviest atoms, which are otherwise the most important target of relativistic computations.

27 These terms refer to the positive part of the energy spectrum. For the negative continuum (the Dirac sea), the proportion of the components is reversed.

28 A more elegant solution was reported by Andrzej W. Rutkowski, *J. Phys. B*, **19**, 3431, 3443 (1986). For the one-electron case, this approach was later popularized by Werner Kutzelnigg as the direct perturbation theory (DPT).

Let us insert this expression into Eq. (3.62). We obtain what is sometimes called "*the Schrödinger equation with spin*" (because it is satisfied by a two-component spinor):

$$\left(\frac{\hat{p}^2}{2m_0} + V\right)\boldsymbol{\psi} = \epsilon\boldsymbol{\psi}.$$

Recalling that $\hat{\boldsymbol{p}}$ represents the momentum operator, we observe that each of the large components satisfies the familiar Schrödinger equation:

$$\left(-\frac{\hbar^2}{2m_0}\Delta + V\right)\psi = \epsilon\psi.$$

Therefore, the non-relativistic equation has been obtained from the relativistic one, assuming that the velocity of particle v is negligibly small with respect to the speed of light c. The Dirac equation remains valid even for larger particle velocities.

3.3.8 How Does the Spin Appear?

It will be shown that the Dirac equation for the free electron in an external electromagnetic field is leading to the spin concept. Thus, in relativistic theory, the spin angular momentum appears in a natural way, whereas in the non-relativistic formalism, it was the subject of a postulate of quantum mechanics (p. 26).

First let us introduce the following identity:

$$\left(\boldsymbol{\sigma}\cdot\boldsymbol{a}\right)\left(\boldsymbol{\sigma}\cdot\boldsymbol{b}\right) = \left(\boldsymbol{a}\cdot\boldsymbol{b}\right)\mathbf{1} + i\boldsymbol{\sigma}\cdot\left(\boldsymbol{a}\times\boldsymbol{b}\right),$$

where we have three times the product of two matrices, each formed by a "*scalar product*" of matrix $\boldsymbol{\sigma}$ and a vector[29], consecutively: $\boldsymbol{\sigma}\cdot\boldsymbol{a}$, $\boldsymbol{\sigma}\cdot\boldsymbol{b}$ and $\boldsymbol{\sigma}\cdot\left(\boldsymbol{a}\times\boldsymbol{b}\right)$. Now, taking $\boldsymbol{a}=\boldsymbol{b}=\boldsymbol{\pi}$, one obtains the relation

$$(\boldsymbol{\sigma}\cdot\boldsymbol{\pi})(\boldsymbol{\sigma}\cdot\boldsymbol{\pi}) = \left(\boldsymbol{\pi}\cdot\boldsymbol{\pi}\right)\mathbf{1} + i\boldsymbol{\sigma}\left(\boldsymbol{\pi}\times\boldsymbol{\pi}\right).$$

[29] Such a "*scalar product*" is, $\boldsymbol{\sigma}\cdot\boldsymbol{a} = \sigma_x a_x + \sigma_y a_y + \sigma_z a_z$. Let us look at the example of $\boldsymbol{\sigma}\cdot\boldsymbol{a}$, where we have:

$$\boldsymbol{\sigma}\cdot\boldsymbol{a} = \begin{pmatrix} 0 & a_x \\ a_x & 0 \end{pmatrix} + \begin{pmatrix} 0 & -ia_y \\ ia_y & 0 \end{pmatrix} + \begin{pmatrix} a_z & 0 \\ 0 & -a_z \end{pmatrix} = \begin{pmatrix} a_z & a_x - ia_y \\ a_x + ia_y & -a_z \end{pmatrix}.$$

Multiplying the matrices $(\boldsymbol{\sigma}\cdot\boldsymbol{a})$ and $(\boldsymbol{\sigma}\cdot\boldsymbol{b})$, therefore, we get

$$(\boldsymbol{\sigma}\cdot\boldsymbol{a})(\boldsymbol{\sigma}\cdot\boldsymbol{b}) = \begin{pmatrix} \boldsymbol{a}\cdot\boldsymbol{b} + i(\boldsymbol{a}\times\boldsymbol{b})_z & (\boldsymbol{a}\times\boldsymbol{b})_y + i(\boldsymbol{a}\times\boldsymbol{b})_x \\ -(\boldsymbol{a}\times\boldsymbol{b})_y + i(\boldsymbol{a}\times\boldsymbol{b})_x & \boldsymbol{a}\cdot\boldsymbol{b} - i(\boldsymbol{a}\times\boldsymbol{b})_z \end{pmatrix}$$

$$= (\boldsymbol{a}\cdot\boldsymbol{b})\mathbf{1} + i\begin{pmatrix} (\boldsymbol{a}\times\boldsymbol{b})_z & (\boldsymbol{a}\times\boldsymbol{b})_x - i(\boldsymbol{a}\times\boldsymbol{b})_y \\ (\boldsymbol{a}\times\boldsymbol{b})_x + i(\boldsymbol{a}\times\boldsymbol{b})_y + & -(\boldsymbol{a}\times\boldsymbol{b})_z \end{pmatrix} = (\boldsymbol{a}\cdot\boldsymbol{b})\mathbf{1} + i\boldsymbol{\sigma}\cdot(\boldsymbol{a}\times\boldsymbol{b}),$$

which is the right side of the identity.

If the vector $\boldsymbol{\pi}$ had numbers as its components, the last term would have had to be zero because the vector product of two parallel vectors would be zero. This, however, need not be true when the vector components are operators (as they are in this case). Since $\boldsymbol{\pi} = \boldsymbol{p} - \frac{q}{c}\boldsymbol{A}$, then $(\boldsymbol{\pi} \cdot \boldsymbol{\pi}) = \boldsymbol{\pi}^2$ and $(\boldsymbol{\pi} \times \boldsymbol{\pi}) = iq\frac{\hbar}{c} curl\boldsymbol{A}$. To check this, we will obtain the last equality for the x components of both sides (the proof for the other two components looks the same). Let the operator $(\boldsymbol{\pi} \times \boldsymbol{\pi})$ act on an arbitrary function $f(x, y, z)$. As a result, we expect the product of f and the vector $iq\frac{\hbar}{c} curl\boldsymbol{A}$ to be as follows:

$$\begin{aligned}
(\boldsymbol{\pi} \times \boldsymbol{\pi})_x f &= (\hat{p}_y - q/cA_y)(\hat{p}_z - q/cA_z)f - (\hat{p}_z - q/cA_z)(\hat{p}_y - q/cA_y)f \\
&= \left[\hat{p}_y\hat{p}_z - q/c\hat{p}_yA_z - q/cA_y\hat{p}_z + (q/c)^2 A_yA_z - \hat{p}_z\hat{p}_y + q/c\hat{p}_zA_y \right. \\
&\quad \left. + q/cA_z\hat{p}_y - (q/c)^2 A_zA_y\right] f \\
&= -q/c(-i\hbar)\left\{\frac{\partial}{\partial y}(A_z f) - A_z\frac{\partial f}{\partial y} + A_y\frac{\partial f}{\partial z} - \frac{\partial}{\partial z}(A_y f)\right\} \\
&= i\hbar q/c\left\{\frac{\partial A_z}{\partial y} - \frac{\partial A_y}{\partial z}\right\} f = \frac{iq\hbar}{c}\left(curl\boldsymbol{A}\right)_x f.
\end{aligned}$$

From the Maxwell equations (p. e81), we have $curl\boldsymbol{A} = \boldsymbol{H}$, where $\boldsymbol{H}$ represents the magnetic field intensity. Let us insert this into the Dirac equation [which is valid for kinetic energy much smaller than $2m_0c^2$; see Eq. (3.61)]:

$$\begin{aligned}
(\epsilon - V)\psi &= \frac{1}{2m_0}(\boldsymbol{\sigma} \cdot \boldsymbol{\pi})(\boldsymbol{\sigma} \cdot \boldsymbol{\pi})\psi = \frac{1}{2m_0}(\boldsymbol{\pi} \cdot \boldsymbol{\pi})\psi \\
&+ \frac{i}{2m_0}\boldsymbol{\sigma} \cdot (\boldsymbol{\pi} \times \boldsymbol{\pi})\psi = \frac{1}{2m_0}(\boldsymbol{\pi} \cdot \boldsymbol{\pi})\psi + \frac{i}{2m_0}\frac{iq\hbar}{c}(\boldsymbol{\sigma} \cdot \boldsymbol{H})\psi \\
&= \left[\frac{\pi^2}{2m_0} - \frac{q\hbar}{2m_0c}\boldsymbol{\sigma} \cdot \boldsymbol{H}\right]\psi = \left[\frac{\pi^2}{2m_0} + \frac{e\hbar}{2m_0c}\boldsymbol{\sigma} \cdot \boldsymbol{H}\right]\psi.
\end{aligned}$$

In the last parenthesis, beside the kinetic energy operator (first term), there is a strange second term. The term has the appearance of the interaction energy $-\boldsymbol{M} \cdot \boldsymbol{H}$ of a mysterious magnetic dipole moment $\boldsymbol{M}$ with magnetic field $\boldsymbol{H}$ (cf. interaction with magnetic field, p. 764). The operator of this electronic dipole moment $\boldsymbol{M} = -\frac{e\hbar}{2m_0c}\boldsymbol{\sigma} = -\mu_B\boldsymbol{\sigma}$, where μ_B stands for the Bohr magneton equal to $\frac{e\hbar}{2m_0c}$. The spin angular momentum operator of the electron is denoted by (cf. p. 29) $\boldsymbol{s}$. Therefore, one has $\boldsymbol{s} = \frac{1}{2}\hbar\boldsymbol{\sigma}$. Inserting $\boldsymbol{\sigma}$ to the equation for $\boldsymbol{M}$, we obtain

$$\boldsymbol{M} = -2\frac{\mu_B}{\hbar}\boldsymbol{s} = -\frac{e}{m_0c}\boldsymbol{s}. \tag{3.63}$$

It is exactly twice as much as we get for the orbital angular momentum and the corresponding orbital magnetic dipole.

When two values differ by an integer factor (as in this case), this should stimulate our mind because it may mean something fundamental that might depend on the number of dimensions of our space or something similar. However, one of the most precise experiments ever made by humans gave[30] $2.0023193043737 \pm 0.0000000000082$ instead of 2 (hence, the anomalous magnetic spin dipole moment of the electron). Therefore, our excitement must diminish. A more accurate theory (quantum electrodynamics, some of the effects of which will be described later in this chapter) gave a result that agreed with the experiment within an experimental error of ± 0.0000000008. The extreme accuracy achieved witnessed the exceptional status of quantum electrodynamics, because no other theory of mankind has achieved such a level of accuracy.

3.3.9 Simple Questions

How should we interpret a bispinor wave function? Does the Dirac equation describe a single fermion, an electron, a positron, an electron and a Dirac sea of other electrons (infinite number of particles), or an effective electron or effective positron (interacting with the Dirac sea)? After eighty years, these questions do not have a clear answer.

Despite the glorious invariance with respect to the Lorentz transformation and despite spectacular successes, the Dirac equation has some serious drawbacks, including a lack of clear physical interpretation. These drawbacks are removed by a more advanced theory–quantum electrodynamics.

3.4 The Hydrogen-like Atom in Dirac Theory

After this short break, we have returned to the Dirac theory. The hydrogen-like atom may be simplified by immobilizing the nucleus and considering a single particle–the electron[31] moving in the electrostatic field of the nucleus[32] $-Z/r$. This problem has an exact solution first obtained by Charles Galton Darwin (cf. p. 124). The electron state is described by four quantum numbers n, l, m, m_s, where $n = 1, 2, \ldots$ stands for the principal, $0 \le l \le n-1$ for the angular, $|m| \le l$ for the magnetic (both are integers), and $m_s = \frac{1}{2}, -\frac{1}{2}$ for the spin quantum number. Darwin

[30] R.S. Van Dyck Jr., P.B. Schwinberg, and H.G. Dehmelt, *Phys. Rev. Letters*, **59**, 26 (1990).

[31] In the Dirac equation, $\boldsymbol{A} = \boldsymbol{0}$ and $-e\phi = V = -\frac{Ze^2}{r}$ were set.

[32] The center of mass motion can be easily separated from the Schrödinger equation. Nothing like this has been done for the Dirac equation. The atomic mass depends on its velocity with respect to the laboratory coordinate system, the electron and proton mass also depend on their speeds, and there is also a mass deficit as a result of binding between both particles. All this seems to indicate that separation of the center of mass is not possible. Nevertheless, for an energy expression accurate to a certain power of c^{-1}, such a separation is, at least in some cases, possible.

obtained the following formula for the relativistic energy of the hydrogen-like atom (in a.u.):

$$E_{n,j} = -\frac{1}{2n^2}\left[1 + \frac{1}{nc^2}\left(\frac{1}{j+\frac{1}{2}} - \frac{3}{4n}\right)\right],$$

where $j = l + m_s$, and c is the speed of light (in a.u.). For the ground state (1s, $n = 1$, $l = 0, m = 0, m_s = \frac{1}{2}$), we have

$$E_{1,\frac{1}{2}} = -\frac{1}{2}\left[1 + \left(\frac{1}{2c}\right)^2\right].$$

Thus, instead of the non-relativistic energy equal to $-\frac{1}{2}$, from the Dirac theory, we obtain -0.5000067 a.u., which means a very small stabilizing correction to the non-relativistic energy. The electron energy levels for the non-relativistic and relativistic cases are shown schematically in Fig. 3.2.

3.4.1 *Step by Step: Calculation of the Hydrogen-Like Atom Ground State Within Dirac Theory*

3.4.1.1 *Matrix Form of the Dirac Equation*

We will use the Dirac equation [Eq. (3.59)]. First, the basis set composed of two bispinors will be created: $\boldsymbol{\Psi}_1 = \begin{pmatrix} \boldsymbol{\psi} \\ \mathbf{0} \end{pmatrix}$ and $\boldsymbol{\Psi}_2 = \begin{pmatrix} \mathbf{0} \\ \boldsymbol{\phi} \end{pmatrix}$, and the wave function $\boldsymbol{\Psi}$ will be sought as a linear combination $\boldsymbol{\Psi} = c_1\boldsymbol{\Psi}_1 + c_2\boldsymbol{\Psi}_2$, which represents an approximation. Within this approximation, the Dirac equation looks like this:

$$\begin{pmatrix} V-\varepsilon & c(\boldsymbol{\sigma}\cdot\boldsymbol{\pi}) \\ c(\boldsymbol{\sigma}\cdot\boldsymbol{\pi}) & V-2m_0c^2-\varepsilon \end{pmatrix}(c_1\boldsymbol{\Psi}_1 + c_2\boldsymbol{\Psi}_2) = \mathbf{0},$$

which gives

$$c_1\begin{pmatrix} V-\varepsilon & c(\boldsymbol{\sigma}\cdot\boldsymbol{\pi}) \\ c(\boldsymbol{\sigma}\cdot\boldsymbol{\pi}) & V-2m_0c^2-\varepsilon \end{pmatrix}\boldsymbol{\Psi}_1 + c_2\begin{pmatrix} V-\varepsilon & c(\boldsymbol{\sigma}\cdot\boldsymbol{\pi}) \\ c(\boldsymbol{\sigma}\cdot\boldsymbol{\pi}) & V-2m_0c^2-\varepsilon \end{pmatrix}\boldsymbol{\Psi}_2 = \mathbf{0}.$$

By making a scalar product first with $\boldsymbol{\Psi}_1$ and then with $\boldsymbol{\Psi}_2$, we obtain two equations:

$$c_1\left\langle\boldsymbol{\Psi}_1\left|\begin{pmatrix} V-\varepsilon & c(\boldsymbol{\sigma}\cdot\boldsymbol{\pi}) \\ c(\boldsymbol{\sigma}\cdot\boldsymbol{\pi}) & V-2m_0c^2-\varepsilon \end{pmatrix}\right.\boldsymbol{\Psi}_1\right\rangle + c_2\left\langle\boldsymbol{\Psi}_1\left|\begin{pmatrix} V-\varepsilon & c(\boldsymbol{\sigma}\cdot\boldsymbol{\pi}) \\ c(\boldsymbol{\sigma}\cdot\boldsymbol{\pi}) & V-2m_0c^2-\varepsilon \end{pmatrix}\right.\boldsymbol{\Psi}_2\right\rangle = 0,$$

$$c_1\left\langle\boldsymbol{\Psi}_2\left|\begin{pmatrix} V-\varepsilon & c(\boldsymbol{\sigma}\cdot\boldsymbol{\pi}) \\ c(\boldsymbol{\sigma}\cdot\boldsymbol{\pi}) & V-2m_0c^2-\varepsilon \end{pmatrix}\right.\boldsymbol{\Psi}_1\right\rangle + c_2\left\langle\boldsymbol{\Psi}_2\left|\begin{pmatrix} V-\varepsilon & c(\boldsymbol{\sigma}\cdot\boldsymbol{\pi}) \\ c(\boldsymbol{\sigma}\cdot\boldsymbol{\pi}) & V-2m_0c^2-\varepsilon \end{pmatrix}\right.\boldsymbol{\Psi}_2\right\rangle = 0.$$

Taking into account the particular structure of the bispinors $\mathbf{\Psi}_1$ and $\mathbf{\Psi}_2$, we obtain the same equations expressed in (two component) spinors:

$$c_1 \langle \boldsymbol{\psi} \left|(V - \varepsilon)\, \boldsymbol{\psi}\right\rangle + c_2 \langle \boldsymbol{\psi} \left| c(\boldsymbol{\sigma} \cdot \boldsymbol{\pi})\boldsymbol{\phi}\right\rangle = 0$$
$$c_1 \langle \boldsymbol{\phi} \left| c(\boldsymbol{\sigma} \cdot \boldsymbol{\pi})\boldsymbol{\psi}\right\rangle + c_2 \langle \boldsymbol{\phi} \left|(V - 2m_0c^2 - \varepsilon)\boldsymbol{\phi}\right\rangle = 0.$$

This is a set of homogeneous linear equations. To obtain a nontrivial solution[33], the determinant of the coefficients multiplying the unknowns c_1 and c_2 has to be zero (the secular determinant, cf., the variational method in Chapter 5):

$$\begin{vmatrix} \langle \boldsymbol{\psi} \left|(V - \varepsilon)\, \boldsymbol{\psi}\right\rangle & \langle \boldsymbol{\psi} \left| c(\boldsymbol{\sigma} \cdot \boldsymbol{\pi})\boldsymbol{\phi}\right\rangle \\ \langle \boldsymbol{\phi} \left| c(\boldsymbol{\sigma} \cdot \boldsymbol{\pi})\boldsymbol{\psi}\right\rangle & \langle \boldsymbol{\phi} \left|(V - 2m_0c^2 - \varepsilon)\boldsymbol{\phi}\right\rangle \end{vmatrix} = 0. \tag{3.64}$$

The potential V in Eq. (3.64) will be taken as $-Z/r$, where r is the electron-nucleus distance.

3.4.1.2 The Large Component Spinor

It is true that we have used an extremely poor basis in this discussion; however, we will try to compensate for it by allowing a certain flexibility within the large component spinor: $\boldsymbol{\psi} = \begin{pmatrix} 1s \\ 0 \end{pmatrix}$, where the hydrogen-like function $1s = \sqrt{\frac{\zeta^3}{\pi}} exp(-\zeta r)$. The parameter ζ will be optimized in such a way as to minimize the energy ε of the electron. This idea is similar to the variational method in the non-relativistic theory (Chapter 5 and Appendix H available at booksite.elsevier.com/978-0-444-59436-5, p. e91), however, it is hardly justified in the relativistic case. Indeed, as proved by numerical experience, the variational procedure often fails. As a remedy, we will use the kinetic balancing already described of the large and small components of the bispinor (p. 132). The spinor of the small components is therefore obtained automatically from the large components (approximation):

$$\boldsymbol{\phi} = N_\phi(\boldsymbol{\sigma} \cdot \boldsymbol{\pi}) \begin{pmatrix} 1s \\ 0 \end{pmatrix} = N_\phi \begin{pmatrix} \hat{p}_z & \hat{p}_x + i\hat{p}_y \\ \hat{p}_x - i\hat{p}_y & \hat{p}_z \end{pmatrix} \begin{pmatrix} 1s \\ 0 \end{pmatrix}$$
$$= N_\phi \begin{pmatrix} \hat{p}_z(1s) \\ (\hat{p}_x + i\hat{p}_y)(1s) \end{pmatrix},$$

where N_ϕ is a normalization constant. In the above formula, $\hat{\boldsymbol{p}}$ represents the momentum operator. The normalization constant N_ϕ will be found from

$$\langle \boldsymbol{\phi} \mid \boldsymbol{\phi} \rangle = 1 = \left|N_\phi\right|^2 \left\{\langle \hat{p}_z(1s) \mid \hat{p}_z(1s)\rangle + \langle (\hat{p}_x + i\hat{p}_y)(1s) \mid (\hat{p}_x + i\hat{p}_y)(1s)\rangle\right\} = \left|N_\phi\right|^2 \cdot$$
$$\left\{ \begin{array}{c} \langle \hat{p}_z(1s) \mid \hat{p}_z(1s)\rangle + \langle \hat{p}_x(1s) \mid \hat{p}_x(1s)\rangle + i\langle \hat{p}_x(1s) \mid \hat{p}_y(1s)\rangle - i\langle \hat{p}_y(1s) \mid \hat{p}_x(1s)\rangle \\ + \langle \hat{p}_y(1s) \mid \hat{p}_y(1s)\rangle \end{array} \right\}.$$

[33] It is easy to give a trivial solution, but it is not acceptable (the wave function cannot equal zero everywhere): $c_1 = c_2 = 0$.

In the above formula, integrals with the imaginary unit i equal zero because the integrand is an odd function. After using the Hermitian character of the momentum operator we obtain $1 = |N_\phi|^2 \langle 1s | \hat{p}^2 1s \rangle = |N_\phi|^2 \zeta^2$. The last equality follows from Appendix H available at booksite.elsevier.com/978-0-444-59436-5, p. e91. Thus, one may choose $N_\phi = 1/\zeta$.

3.4.1.3 Calculating Integrals in the Dirac Matrix Equation

Now we will calculate one by one all the integrals that appear in the Dirac matrix equation. The integral $\langle \boldsymbol{\psi} | -\frac{Z}{r} \boldsymbol{\psi} \rangle = -Z\zeta$, because the scalar product leads to the nuclear attraction integral with a hydrogen-like atomic orbital, and this gives the result above (see Appendix H available at booksite.elsevier.com/978-0-444-59436-5, p. e91). The next integral can be computed as follows:

$$\begin{aligned}
\left\langle \boldsymbol{\phi} \left| \frac{1}{r} \boldsymbol{\phi} \right.\right\rangle &= |N_\phi|^2 \left\langle \begin{pmatrix} \hat{p}_z(1s) \\ (\hat{p}_x + i\hat{p}_y)(1s) \end{pmatrix} \left| \frac{1}{r} \begin{pmatrix} \hat{p}_z(1s) \\ (\hat{p}_x + i\hat{p}_y)(1s) \end{pmatrix} \right.\right\rangle \\
&= |N_\phi|^2 \left\langle \hat{p}_z(1s) \left| \frac{1}{r} \hat{p}_z(1s) \right.\right\rangle + \left\langle (\hat{p}_x + i\hat{p}_y)(1s) \left| \frac{1}{r} (\hat{p}_x + i\hat{p}_y)(1s) \right.\right\rangle \\
&= |N_\phi|^2 \left\langle (1s) \left| \hat{p}_z \frac{1}{r} \hat{p}_z(1s) \right.\right\rangle + \left\langle (1s) \left| (\hat{p}_x - i\hat{p}_y) \frac{1}{r} (\hat{p}_x + i\hat{p}_y)(1s) \right.\right\rangle \\
&= |N_\phi|^2 \left\langle (1s) \left| \left(\hat{p}_z \frac{1}{r}\right) \hat{p}_z(1s) \right.\right\rangle + \left\langle (1s) \left| \left[(\hat{p}_x - i\hat{p}_y) \frac{1}{r} \right] (\hat{p}_x + i\hat{p}_y)(1s) \right.\right\rangle \\
&\quad + \left\langle (1s) \left| \frac{1}{r} \hat{p}_z \hat{p}_z(1s) \right.\right\rangle + \left\langle (1s) \left| \frac{1}{r} (\hat{p}_x - i\hat{p}_y)(\hat{p}_x + i\hat{p}_y)(1s) \right.\right\rangle .
\end{aligned} \tag{3.65}$$

In the second row, the scalar product of spinors is used, while in the third row, the Hermitian character of the operator $\hat{p}$. Further,

$$\begin{aligned}
\left\langle \boldsymbol{\phi} \left| \frac{1}{r} \boldsymbol{\phi} \right.\right\rangle &= |N_\phi|^2 \left[\left\langle (1s) \left| \left(\hat{p}_z \frac{1}{r}\right) \hat{p}_z(1s) \right.\right\rangle + \left\langle (1s) \left| \frac{1}{r} \left(\hat{p}_x^2 + \hat{p}_y^2 + \hat{p}_z^2\right)(1s) \right.\right\rangle \right. \\
&\quad \left. + \left\langle (1s) \left| \left[(\hat{p}_x - i\hat{p}_y) \frac{1}{r} \right] (\hat{p}_x + i\hat{p}_y)(1s) \right.\right\rangle \right] \\
&= |N_\phi|^2 \left[\left\langle (1s) \left| \left(\hat{p}_z \frac{1}{r}\right) \hat{p}_z(1s) \right.\right\rangle - \left\langle (1s) \left| \frac{1}{r} \Delta(1s) \right.\right\rangle + \left\langle (1s) \left| \left(\hat{p}_x \frac{1}{r}\right) \hat{p}_x(1s) \right.\right\rangle \right. \\
&\quad + \left\langle (1s) \left| \left(\hat{p}_y \frac{1}{r}\right) \hat{p}_y(1s) \right.\right\rangle - i \left\langle (1s) \left| \left(\hat{p}_y \frac{1}{r}\right) \hat{p}_x(1s) \right.\right\rangle \\
&\quad \left. + i \left\langle (1s) \left| \left(\hat{p}_x \frac{1}{r}\right) \hat{p}_y(1s) \right.\right\rangle \right].
\end{aligned} \tag{3.66}$$

We used the atomic units (and therefore $\hat{p}^2 = -\Delta$), and the momentum operator is equal to $-i\nabla$. The two integrals at the end cancel each other out, because each of the integrals does not change when the variables are interchanged: $x \leftrightarrow y$.

Finally, we obtain the following formula: $\langle \boldsymbol{\phi} \left| \frac{1}{r} \boldsymbol{\phi} \right\rangle = -N_\phi^2 \left\{ \langle 1s \left| \frac{1}{r} \Delta(1s) \right\rangle + \langle 1s \left| \left(\nabla \frac{1}{r}\right) \nabla(1s) \right\rangle \right\} = -\zeta^{-2} \left\{ \left(-3\zeta^3 + 2\zeta^3\right) \right\} = \zeta$, where the equality follows from a direct calculation of the two integrals[34].

The next matrix element to calculate is equal to $\langle \boldsymbol{\phi} \left| c \left(\boldsymbol{\sigma} \cdot \boldsymbol{\pi}\right) \boldsymbol{\psi} \right\rangle$. We proceed as follows (recall kinetic balancing and we also use Appendix H available at booksite.elsevier.com/978-0-444-59436-5, p. e91):

$$\begin{aligned}
\langle \boldsymbol{\phi} \left| c \left(\boldsymbol{\sigma} \cdot \boldsymbol{\pi}\right) \boldsymbol{\psi} \right\rangle &= N_\phi c \left\langle (\boldsymbol{\sigma} \cdot \boldsymbol{\pi}) \begin{pmatrix} 1s \\ 0 \end{pmatrix} \middle| (\boldsymbol{\sigma} \cdot \boldsymbol{\pi}) \begin{pmatrix} 1s \\ 0 \end{pmatrix} \right\rangle \\
&= N_\phi c \left\langle \begin{pmatrix} \hat{p}_z(1s) \\ (\hat{p}_x + i\hat{p}_y)(1s) \end{pmatrix} \middle| \begin{pmatrix} \hat{p}_z(1s) \\ (\hat{p}_x + i\hat{p}_y)(1s) \end{pmatrix} \right\rangle \\
&= N_\phi c \left[\langle \hat{p}_z(1s) \left| \hat{p}_z(1s) \right\rangle + \langle (\hat{p}_x + i\hat{p}_y)(1s) \left| (\hat{p}_x + i\hat{p}_y)(1s) \right\rangle \right] \\
&= N_\phi c \, \langle 1s \left| \hat{p}^2(1s) \right\rangle = \frac{1}{\zeta} c\zeta^2 = c\zeta.
\end{aligned}$$

The last matrix element reads as

$$\begin{aligned}
\langle \boldsymbol{\psi} \left| c \left(\boldsymbol{\sigma} \cdot \boldsymbol{\pi}\right) \boldsymbol{\phi} \right\rangle &= N_\phi c \left\langle \begin{pmatrix} 1s \\ 0 \end{pmatrix} \middle| (\boldsymbol{\sigma} \cdot \boldsymbol{\pi})^2 \begin{pmatrix} 1s \\ 0 \end{pmatrix} \right\rangle \\
&= N_\phi c \left\langle \begin{pmatrix} 1s \\ 0 \end{pmatrix} \middle| \begin{pmatrix} \hat{p}^2 & 0 \\ 0 & \hat{p}^2 \end{pmatrix} \begin{pmatrix} 1s \\ 0 \end{pmatrix} \right\rangle = N_\phi c \, \langle 1s \left| \hat{p}^2 1s \right\rangle = c\frac{1}{\zeta}\zeta^2 = c\zeta.
\end{aligned}$$

3.4.1.4 Dirac's Secular Determinant

At this point, we have all the integrals needed and may write the secular determinant corresponding to the matrix form [Eq. (3.64)] of the Dirac equation. After inserting the calculated integrals, we get

$$\begin{vmatrix} -Z\zeta - \varepsilon & c\zeta \\ c\zeta & -Z\zeta - 2c^2 - \varepsilon \end{vmatrix} = 0.$$

Expanding the determinant gives the equation for the energy ε:

$$\varepsilon^2 + \varepsilon(2Z\zeta + 2c^2) + [Z\zeta(Z\zeta + 2c^2) - c^2\zeta^2] = 0.$$

Hence, we get two solutions:

$$\varepsilon_\pm = -(c^2 + Z\zeta) \pm \sqrt{(c^4 + \zeta^2 c^2)}.$$

[34] In the first integral, we have the same situation as earlier in this chapter. In the second integral, we write the nabla operator in Cartesian coordinates, obtain a scalar product of two gradients, and then get three integrals equal to one another (they contain x, y, z), and it is sufficient to calculate one of them by spherical coordinates using the formula H.2 in Appendix H available at booksite.elsevier.com/978-0-444-59436-5, p. e91.

Note that the square root is of the order of c^2 (in a.u.), and with the (unit) mass of the electron m_0, it is of the order of m_0c^2. Therefore, the minus sign before the square root corresponds to a solution with energy of the order of $-2m_0c^2$, while the plus sign corresponds to energy on the order of zero. Recall that we have shifted the energy scale in the Dirac equation and this is the last solution, ε_+ (hereafter denoted by ε), which is to be compared to the energy of the non-relativistic hydrogen-like atom:

$$\varepsilon = -\left(c^2 + Z\zeta\right) + \sqrt{c^4 + \zeta^2 c^2} = -\left(c^2 + Z\zeta\right) + c^2\sqrt{1 + \frac{\zeta^2}{c^2}}$$
$$= -\left(c^2 + Z\zeta\right) + c^2\left(1 + \frac{\zeta^2}{2c^2} - \frac{\zeta^4}{8c^4} + \cdots\right) = -Z\zeta + \frac{\zeta^2}{2} + \left(-\frac{\zeta^4}{8c^2} + \cdots\right). \quad (3.67)$$

3.4.1.5 Non-relativistic Solution

If $c \to \infty$ (i.e., we approach the non-relativistic limit), then $\varepsilon = -Z\zeta + \frac{\zeta^2}{2}$. Minimization of this energy with respect to ζ gives its optimum value $\zeta_{opt}^{nonrel} = Z$. In this way, one recovers the result known from non-relativistic quantum mechanics (see Appendix H available at booksite.elsevier.com/978-0-444-59436-5) obtained in the variational approach to the hydrogen atom with the $1s$ orbital as a trial function.

3.4.1.6 Relativistic Contraction of Orbitals

Minimizing the relativistic energy given by Eq. (3.67) leads to an equation for optimum $\zeta \equiv \zeta_{opt}^{rel}$:

$$\frac{d\varepsilon}{d\zeta} = 0 = -Z + \frac{1}{2}\left(c^4 + \zeta^2 c^2\right)^{-\frac{1}{2}} 2\zeta c^2 = -Z + \left(c^4 + \zeta^2 c^2\right)^{-\frac{1}{2}} \zeta c^2,$$

giving

$$\zeta_{opt}^{rel} = \frac{Z}{\sqrt{1 - \frac{Z^2}{c^2}}}.$$

The result differs remarkably from the non-relativistic value $\zeta_{opt}^{nonrel} = Z$, but approaches the non-relativistic value when $c \to \infty$. Note that the difference between the two values increases with atomic number Z, and that the relativistic exponent is always larger than its non-relativistic counterpart. This means that the relativistic orbital decays faster with the electron-nucleus distance, and therefore,

> the relativistic orbital $1s$ is smaller (contraction) than the corresponding non-relativistic one.

Let us see how this situation is for the hydrogen atom. In that case, $\zeta_{opt}^{rel} = 1.0000266$ as compared to $\zeta_{opt}^{nonrel} = Z_H = 1$. And what about $1s$ orbital of gold? For gold, $\zeta_{opt}^{rel} = 96.68$, while $\zeta_{opt}^{nonrel} = Z_{Au} = 79$. This is a large relativistic contraction of the atomic orbitals. Since for a heavy atom, the effective exponent of the atomic orbitals decreases when moving from the low-energy compact $1s$ orbital to higher-energy outer orbitals, this means that the most important relativistic orbital contraction occurs for the inner shells. The chemical properties of an atom depend on what happens to its outer shells (valence shell). Therefore, we may conclude that the relativistic corrections are expected to play a secondary role in chemistry[35].

If we insert ζ_{opt}^{rel} into Eq. (3.67), we obtain the minimum value of ε:

$$\varepsilon_{min} = -\left(c^2 + Z\zeta\right) + \sqrt{c^4 + \zeta^2}. \tag{3.68}$$

Since Z^2/c^2 is small with respect to 1, we may expand the square root in the Taylor series, $\sqrt{1-x} = 1 - \frac{1}{2}x - \frac{1}{8}x^2 - \cdots$ We obtain

$$\begin{aligned}\varepsilon_{min} &= -c^2 + c^2\left\{1 - \left(\frac{1}{2}\right)\left(\frac{Z^2}{c^2}\right) - \frac{1}{8}\left(\frac{Z^2}{c^2}\right)^2 - \cdots\right\} \\ &= -\frac{Z^2}{2}\left(1 + \left(\frac{Z}{2c}\right)^2 + \cdots\right).\end{aligned} \tag{3.69}$$

In the case of the hydrogen atom ($Z = 1$), we have

$$\varepsilon_{min} = -\frac{1}{2}\left(1 + \left(\frac{1}{2c}\right)^2 + \cdots\right), \tag{3.70}$$

where the first two terms shown give Darwin's exact result[36] (discussed earlier in this chapter). Inserting $c = 137.036$ a.u., we obtain the hydrogen atom ground-state energy $\varepsilon = -0.5000067$ a.u., which agrees with Darwin's result.

3.5 Toward Larger Systems

The Dirac equation is rigorously invariant with respect to the Lorentz transformation, which is certainly the most important requirement for a relativistic theory. Therefore, it would seem to be a logically sound approximation for a relativistic description of a single quantum particle. Unfortunately, this is not true. Recall that the Dirac Hamiltonian spectrum contains a

[35] We have to remember, however, that the relativistic effects also propagate from the inner shells to the valence shell through the orthogonalization condition, which has to be fulfilled after the relativistic contraction. This is why the gold valence orbital $6s$ shrinks, which has an immediate consequence in the relativistic shortening of the bond length in Au_2, cited at the beginning of this chapter.

[36] That is, the exact solution to the Dirac equation for the electron in the external electric field produced by the proton.

continuum of states with negative energies, which is not bound from below (a kind of "*energy abyss*" with threatening consequences). Dirac desperately postulated that all these negative continuum energy levels are inaccessible for an electron occupying a "*normal*" energy level with positive energy, such as, for example, the levels of the hydrogen atom[37]. This inaccessibility was supposed to result from postulated occupation of these levels by the electrons from the Dirac sea and operation of the Pauli exclusion principle.

The first contradiction within the Dirac theory is the following: the Dirac model ceases to be of a one-particle type (by occupying the negative energy continuum by an infinite number of electrons), but despite of this it is unable to describe even two electrons when they are in the positive energy states. A second contradiction is visible from the Fock-Klein-Gordon equation. There also is a negative energy continuum [one also may have a negative solution when making the square root of the square of Eq. (3.39), p. 122], but since the Fock-Klein-Gordon equation describes bosons[38], for which the Pauli exclusion principle is not operating (see Chapter 1), the concept of Dirac about inaccessibility is useless in this case.[39]

The internal inconsistencies of the Dirac model lead to some absurd results. For example, the Dirac electron velocity, defined as the eigenvalue of the operator being the time derivative of the position operator, is everywhere equal to the velocity of light. These and other inconsistencies are removed only after going to a more advanced theory, which is quantum electrodynamics (QED). Unfortunately, QED represents a complicated machinery, which nowadays cannot be applied to large molecules of chemical interest. Therefore, in practice, we are forced to use the Dirac equation, and to be effective for larger systems, we should somehow invent its many-particle generalizations. During the more than eighty years after the Dirac theory was derived, the dangers and traps connected to the Dirac equation have been quite well recognized. Currently, the Dirac theory is able to produce results with excellent agreement with experimental data. This however is often associated with a special care during computations.

3.5.1 Non-Interacting Dirac Electrons

Just to see what kind of problems are associated with application of the Dirac equation, let us consider two non-interacting Dirac electrons. Let us assume that a two-electron Hamiltonian is simply a sum of the two one-electron Hamiltonians. Such two-electron problem can be

---

[37] Such transitions would cause not only the destruction of atoms, but also a catastrophe on a scale of the universe, since an infinite amount of energy has to be emitted.

[38] An analysis of the transformation properties of the Fock-Klein-Gordon equation and of the Dirac equation leads to the conclusion that satisfaction of the first of these equations requires the usual (i.e., scalar) wave function, whereas the second equation requires a bispinor character of the wave function. Scalar functions describe spinless particles (because they cannot be associated with the Pauli matrices), while bispinors in the Dirac equation are associated with the Pauli matrices, and describe a particle of 1/2 spin.

[39] This means that the Dirac sea is nothing more than a fairy tale. But sometimes fairy tales are useful and calming (in the short term), for children as well as for scientists.

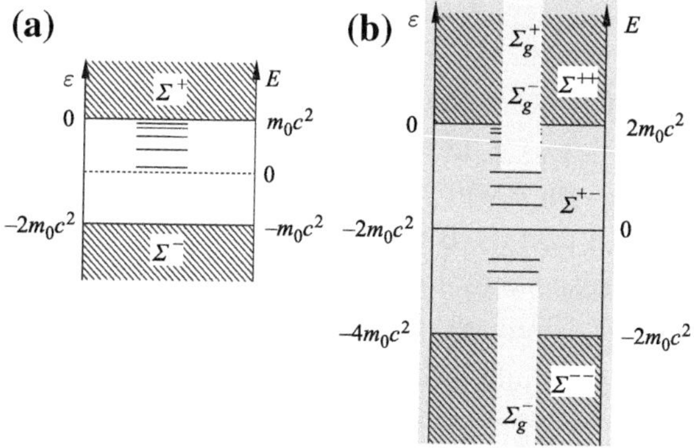

Fig. 3.3. The spectrum of the Dirac Hamiltonian for one electron (a) and two non-interacting electrons (b). On the right side, the original scale of energy (E) was used, while on the left side, the shifted scale (ε) was applied. In the one-electron case, one obtains two continua: the first one, Σ^+, describes the states with positive energies, while the second one, Σ^-, pertains to the states with negative energies. In the case of two electrons, the continua Σ^{++} and Σ^{--} appear when both electrons belong to the same continuum Σ^+ or Σ^-. However, the total scale of energy is covered by continuum Σ^{+-}, which corresponds to one electron in Σ^+ and the other in Σ^-. Continua Σ_d^+ and Σ_d^- appear when one of the electrons occupies a discrete level, while the second one is in a state belonging to a continuum (Σ^+ or Σ^-).

separated into two one-electron ones. If we are in a situation with the Schrödinger equation, nothing special happens–we construct the solution for the two-electron problem from the one-electron solutions. If we solve such a problem in the Dirac model, the negative continuum becomes a source of serious problems. Note that the sum of energies of two electrons, one with the positive and one with the negative energy, may give us any energy (see Fig. 3.3). We see also that the energy eigenvalues of the two-electron Hamiltonian that correspond to bound states are immersed in the continuum extending from $-\infty$ to $+\infty$. This is strange, but for the time being, it does not cause any numerical catastrophe, provided that we first separate the problem into the two one-electron ones (for the non-interacting electrons this is possible) and then solve the two problems separately. If we do this way we get the correct values of the discrete states.

3.5.2 Dirac-Coulomb (DC) Model

Things become very complex when we try to switch on the electronic interaction. First, what should we use as the interaction energy operator? Nobody knows, but one idea might be to add to the sum of the Dirac one-electron operators the Coulombic interaction operators of all the particles. This is what is known as the *Dirac-Coulomb (DC) model.*[40]

[40] Some researchers take into account the relativistic effects as a kind of perturbation of the Schrödinger equation. They solve the latter one with the corresponding potential energy modified in such a way as to mimic the relativistic effects.

The DC approximation is rightly seen by many scholars as methodologically inconsistent. Indeed, as we will show shortly, the Coulomb law needs the infinite velocity of light, which involves its non-relativistic character. Such a non-relativistic ingredient should not be mixed with any relativistic theory (e.g., the Dirac model), because we may get an odd theory that could lead us into scientific swamps with traps and dancing elves, which in reality do not exist. Already in the fifties of 20th century Brown and Ravenhall[41] remarked that the DC theory predicts some states that correspond to mixtures of one-particle continuum-like state and one-particle bound-like state. This is referred to in the literature as the Brown-Ravenhall sickness. This means that in the DC model, the ground state and all the excited states do not represent any bound states, but are in fact resonance states. Resonance states are unstable by definition [i.e., they have a finite lifetime and describe inevitable spontaneous disintegration of the system into pieces (although this may happen after the system has enjoyed a long life)].[42] This is in an obvious contradiction with experimental facts, because, for example, the helium atom exists, and there is nothing to indicate its internal instability.

Therefore, we have a serious problem because the modern relativistic computations for atoms and molecules are almost exclusively based on the DC approximation[43]. The problem can be at least partially removed if one applies some methods based on the one-electron model (e.g., a mean field model)[44]. A mean field means that we study the motion of a single particle in the field of other particles, but their motion is averaged (their motion and, in a sense, the other particles themselves disappear from the theory through the averaging). In such a model, the wave function for all electrons is built from some one-electron functions, which in this case represent four-component bispinors. One obtains these bispinors by solving the corresponding Dirac equation in a kinetically-balanced bispinor basis, [Eq. (3.60)], p. 132. The kinetic balance protects the model against the variational collapse, but not against the Brown-Ravenhall sickness. One avoids the symptoms of this sickness by projecting the DC equation onto the space of those states that correspond to positive energy[45]. Such a procedure assures stability of solutions and allows for using the well-known methods developed for bound states. However, we pay a price: the variational space used is not complete. Moreover, often construction of a basis set that does not contain admixtures from Σ^- is impossible[46].

---

[41] G.E. Brown and D.G. Ravenhall, *Proc. Royal Soc.* **A208**, 552 (1951).

[42] The resonance states will be discussed in more detail in Chapter 6.

[43] For small systems, like hydrogen-, helium-, and lithium-like atoms, there is a possibility that more accurate methods of QED could be used.

[44] See, Chapter 8.

[45] In practice, this looks as follows. The many-electron wave function (let us focus our attention on a two-electron system only) is constructed from those bispinors, which correspond to positive energy solutions of the Dirac equation. For example, among two-electron functions built of such bispinors, no function corresponds to Σ^{--}, Σ_g^- and, most importantly, to Σ^{+-}. This means that carrying out computations with such a basis set, we do not use the full DC Hamiltonian, but instead, its projection on the space of states with positive energies.

[46] One example of this is when we are beyond the mean-field model.

It has been realized only recently that to avoid the abovementioned traps, as well as to see how the solutions of the DC equation look like, one has to apply the methods specific to resonance states[47] because they are such states and cannot be legitimately treated otherwise. The energies obtained as solutions of the full DC equation (a correct approach) differ from the solutions of the equation projected onto the space of states with positive energies (approximation) by the terms proportional to $(Z/c)^3$, with Z meaning the nuclear charge[48]. It turned out also that the lifetimes of the resonance states, which correspond to the truly bound states, are proportional to $(Z/c)^{-3}$, and therefore are relatively long (especially for light atoms)[49]. It is worth noting that the DC equation, also with the Breit corrections described in the next section, is exact up to the terms proportional to $(Z/c)^2$. Therefore, the effects connected to the instability of the DC solution, although they have been very annoying at the numerical as well as the interpretation level, are smaller than the limits of validity of the DC model itself[50].

3.6 Beyond the Dirac Equation...

How reliable is the relativistic quantum theory presented? The Dirac or Fock-Klein-Gordon equations, as is usually the case in physics, describe only some aspects of reality. The fact that both equations are invariant with respect to the Lorentz transformation indicates *only* that the space-time *symmetry* properties are described correctly. The physical machinery represented by these equations is not so bad, since several predictions have been successfully made (antimatter, electron spin, energy levels of the hydrogen atom). Yet, in the latter case, an assumption of the external field $V = -\frac{Ze^2}{r}$ is a positively desperate step, which in fact is unacceptable in a fair relativistic theory for the proton and the electron (and not only of the electron in the external field of the nucleus). Indeed, the proton and the electron move. At a given time, their distance is equal to r, but such a distance might be inserted into the Coulombic law if the speed of light were infinite because the two particles would feel their positions instantaneously. However, since any perturbation by a position change of a particle needs time to travel to the other particle, we have to use another distance somehow, taking this into account (Fig. 3.4). The same pertains, of course, to any pair of particles in a many-body system (the so-called *retarded potential*).

There is certainly a need for a more accurate theory.

---

[47] G. Pestka, M. Bylicki, and J. Karwowski, *J. Phys. B: At. Mol. Opt. Phys.*, **39**, 2979 (2006); *ibid.*, **40**, 2249 (2007).

[48] Here, we consider an atom.

[49] In principle, one should write $Ze^2/\hbar c$ instead of Z/c. The dimensionless constant $e^2/\hbar c \approx 1/137$ is known as a *fine structure constant*, in atomic units ($e = 1$, $\hbar = 1$; see Chapter 4), it equals $\frac{1}{c}$.

[50] The corrections proportional to $(Z/c)^n$ for $n \geq 3$, which come from the QED, are nowadays being treated perturbationally (Chapter 5), also for many-electron systems, like all atoms. These corrections may become important in interpretation of some particular phenomena, like the X-ray spectra.

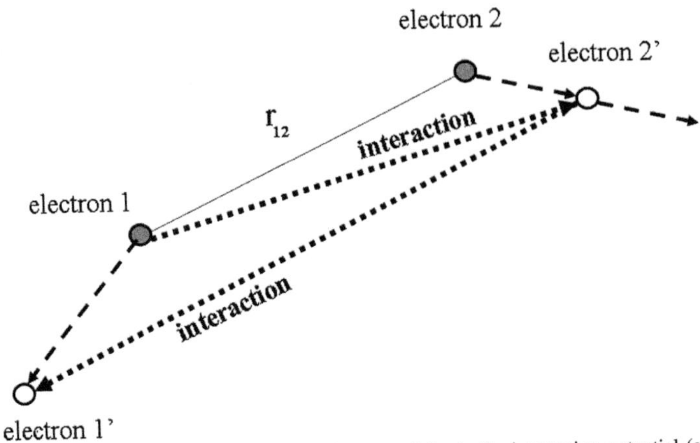

Fig. 3.4. Retardation of the interaction. The distance r_{12} of two particles in the interaction potential (as in Coulomb's law) is bound to represent an approximation because we assume an instantaneous interaction. However, when the two particles find each other (which takes time), they are already somewhere else. In Richard Feynman's wording, "*The Sun atom shakes, my eye electron shakes eight minutes later.*"

3.6.1 The Breit Equation

Gregory Breit (1899–1981) was an American physicist and professor at the New York, Wisconsin, Yale, Buffalo universities. Breit and Eugene Wigner introduced the resonance states of particles, and, with Edward Condon, created the proton-proton scattering theory. Breit measured also the height of the ionosphere.

Breit constructed a many-electron relativistic theory that takes into account such a retarded potential in an approximate way[51]. Breit explicitly considered only the electrons of an atom; its nucleus (similar to the Dirac theory) created only an external field for the electrons. This ambitious project was only partly successful because the resulting theory turned out to be approximate not only from the point of view of quantum theory (with some interactions not taken into account), but also from the point of view of relativity theory (an approximate Lorentz transformation invariance).

For two electrons, the Breit equation has the following form (r_{12} stands for the distance between electron 1 and electron 2):

$$\left\{\hat{H}(1) + \hat{H}(2) + \frac{1}{r_{12}} - \frac{1}{2r_{12}}\left[\boldsymbol{\alpha}(1)\boldsymbol{\alpha}(2) + \frac{\left[\boldsymbol{\alpha}(1)\cdot\boldsymbol{r}_{12}\right]\left[\boldsymbol{\alpha}(2)\cdot\boldsymbol{r}_{12}\right]}{r_{12}^2}\right]\right\}\Psi = E\Psi, \quad (3.71)$$

[51] The Breit equation is invariant with respect to the Lorentz transformation, but only within an accuracy up to some small terms.

where [cf., Eq. (3.54) with E replaced by the Hamiltonian]

$$\hat{H}\left(i\right) = q_i\phi\left(\boldsymbol{r}_i\right) + c\boldsymbol{\alpha}\left(i\right)\boldsymbol{\pi}\left(i\right) + \alpha_0\left(i\right)m_0c^2 = -e\phi\left(\boldsymbol{r}_i\right) + c\boldsymbol{\alpha}\left(i\right)\boldsymbol{\pi}\left(i\right) + \alpha_0\left(i\right)m_0c^2$$

is the Dirac Hamiltonian for electron i pointed by vector $\boldsymbol{r}_i$, whereas the Dirac matrices for electron i $\left[(\boldsymbol{\alpha}(i) = \left[\alpha_x\left(i\right)\right), \alpha_y\left(i\right), \alpha_z\left(i\right)\right]\right]$ and the corresponding operators $\pi_\mu\left(i\right)$ have been defined on p. 127, $\phi\left(\boldsymbol{r}_i\right)$ represent the scalar potential calculated at $\boldsymbol{r}_i$. The wave function Ψ represents a 16-component spinor (here represented by a square matrix of rank 4), because for each electron, we would have the usual Dirac bispinor (with four components) and the two-electron wave function depends on the Cartesian product of the components[52].

The Breit Hamiltonian (in our example, for two electrons in an electromagnetic field) can be approximated by the following useful formula[53], known as the *Breit-Pauli Hamiltonian*:

$$\hat{H}\left(1,2\right) = \hat{H}_0 + \hat{H}_1 + \cdots \hat{H}_6, \tag{3.72}$$

where

- $\hat{H}_0 = \frac{\hat{p}_1^2}{2m_0} + \frac{\hat{p}_2^2}{2m_0} + V$ represents the familiar non-relativistic Hamiltonian.
- $\hat{H}_1 = -\frac{1}{8m_0^3c^2}\left(\hat{p}_1^4 + \hat{p}_2^4\right)$ comes from the velocity dependence of mass, more precisely from the Taylor expansion of Eq. (3.38) for small velocities:
- $\hat{H}_2 = -\frac{e^2}{2\left(m_0c\right)^2}\frac{1}{r_{12}}\left[\hat{\boldsymbol{p}}_1 \cdot \hat{\boldsymbol{p}}_2 + \frac{\boldsymbol{r}_{12}\cdot(\boldsymbol{r}_{12}\cdot\hat{\boldsymbol{p}}_1)\hat{\boldsymbol{p}}_2}{r_{12}^2}\right]$ stands for the correction[54] that accounts in part for the abovementioned retardation. Alternatively, the term may be viewed as the interaction energy of two magnetic dipoles, each resulting from the orbital motion of an electron (*orbit-orbit term*)
- $\hat{H}_3 = \frac{\mu_B}{m_0c}\left\{\left[\boldsymbol{\mathcal{E}}\left(\boldsymbol{r}_1\right) \times \hat{\boldsymbol{p}}_1 + \frac{2e}{r_{12}^3}\boldsymbol{r}_{12} \times \hat{\boldsymbol{p}}_2\right] \cdot \boldsymbol{s}_1 + \left[\boldsymbol{\mathcal{E}}\left(\boldsymbol{r}_2\right) \times \hat{\boldsymbol{p}}_2 + \frac{2e}{r_{12}^3}\boldsymbol{r}_{21} \times \hat{\boldsymbol{p}}_1\right] \cdot \boldsymbol{s}_2\right\}$ is the interaction energy of the electronic magnetic moments (resulting from the abovementioned orbital motion) with the spin magnetic dipole moments (*spin-orbit coupling*), μ_B stands for the Bohr magneton, and $\boldsymbol{\mathcal{E}}$ denotes the electric field vector. Since we have two orbital magnetic dipole moments and two spin orbital dipole moments, there are four spin-orbit interactions. The first term in square brackets stands for the spin-orbit coupling of the same electron, while the second term represents the coupling of the spin of one particle with the orbit of the second.
- $\hat{H}_4 = \frac{ie\hbar}{(2m_0c)^2}\left[\hat{\boldsymbol{p}}_1 \cdot \boldsymbol{\mathcal{E}}\left(\boldsymbol{r}_1\right) + \hat{\boldsymbol{p}}_2 \cdot \boldsymbol{\mathcal{E}}\left(\boldsymbol{r}_2\right)\right]$ is a non-classical term peculiar to the Dirac theory (also present in the one-electron Dirac Hamiltonian) called the *Darwin term*.

---

[52] In the Breit equation [Eq. (3.71)], the operators in {} act either by multiplying the 4×4 matrix Ψ by a function (i.e., each element of the matrix) or by a 4×4 matrix resulting from α matrices.

[53] H.A. Bethe and E.E. Salpeter, *Quantum Mechanics of One- and Two-Electron Atoms*, Plenum Publishing Corporation, New York, 1977, p. 181.

[54] For the non-commuting operators: $\hat{\boldsymbol{a}}\left(\hat{\boldsymbol{a}} \cdot \hat{\boldsymbol{b}}\right)\hat{\boldsymbol{c}} = \sum_{i,j=1}^{3}\hat{a}_i\hat{a}_j\hat{b}_j\hat{c}_i$.

- $\hat{H}_5 = 4\mu_B^2 \left\{ -\frac{8\pi}{3} \left(s_1 \cdot s_2 \right) \delta \left(\boldsymbol{r}_{12} \right) + \frac{1}{r_{12}^3} \left[s_1 \cdot s_2 - \frac{(s_1 \cdot \boldsymbol{r}_{12})(s_2 \cdot \boldsymbol{r}_{12})}{r_{12}^2} \right] \right\}$ corresponds to the spin dipole moment interactions of the two electrons (*the spin-spin term*). The first term is called the *Fermi contact term*, since it is nonzero only when the two electrons touch one another (see Appendix E available at booksite.elsevier.com/978-0-444-59436-5, p. e69), whereas the second term represents the classical *dipole-dipole interaction* of the two electronic spins (cf., the multipole expansion in Appendix X available at booksite.elsevier.com/978-0-444-59436-5, p. e169 and Chapter 13); i.e., the interaction of the two spin magnetic moments of the electrons (with the factor 2, according to Eq. (3.63), p. 134).
- $\hat{H}_6 = 2\mu_B \left[\boldsymbol{H} \left(\boldsymbol{r}_1 \right) \cdot s_1 + \boldsymbol{H} \left(\boldsymbol{r}_2 \right) \cdot s_2 \right] + \frac{e}{m_0 c} \left[\boldsymbol{A} \left(\boldsymbol{r}_1 \right) \cdot \hat{\boldsymbol{p}}_1 + \boldsymbol{A} \left(\boldsymbol{r}_2 \right) \cdot \hat{\boldsymbol{p}}_2 \right]$ is known as the *Zeeman interaction* [i.e., the interaction of the spin (the first two terms) and the orbital electronic magnetic dipole moments (the second two terms) with the external magnetic field $\boldsymbol{H}$]; cf., Eq. (3.63).

The terms listed above are of prime importance in the theory of the interaction of matter with the electromagnetic field (e.g., in nuclear magnetic resonance).

3.6.2 About QED

The Dirac and Breit equations do not account for several subtle effects. They are predicted by QED, a many-particle theory.

The QED energy may be conveniently developed in a power series of $\frac{1}{c}$:

- In zero order, we have the non-relativistic approximation (solution to the Schrödinger equation).
- There are no first-order terms.
- The second order contains the Breit corrections.
- The third and further orders are called the *radiative corrections*.

Radiative Corrections

The radiative corrections include:

- *Interaction with the vacuum* (Fig. 3.5a). In contemporary physics theory, the perfect vacuum does not just represent nothing. The electric field of the vacuum itself fluctuates about zero and these instantaneous fluctuations influence the motion of any charged particle. When a strong electric field operates in a vacuum, the latter undergoes a polarization (called *vacuum polarization*), which means a spontaneous creation of matter, and more specifically, of particle-antiparticle pairs.
- *Interaction with virtual photons*. The electric field influences the motion of the electron. What about its own electric field? Does that influence its motion as well? The latter effect is

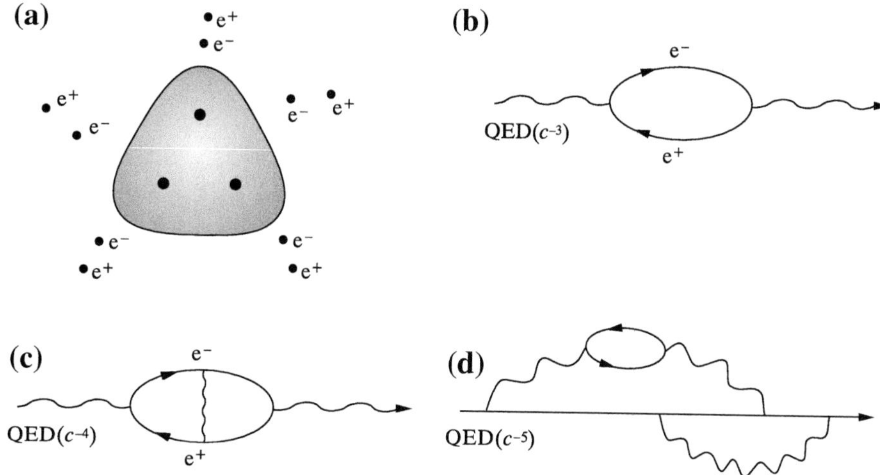

Fig. 3.5. (a) The electric field close to the proton (composed of three quarks) is so strong that it creates matter and antimatter (shown as electron-positron pairs). The three quarks visible in scattering experiments represent the valence quarks. (b) One of the radiative effects in the QED correction of the c^{-3} order (see Table 3.1). The pictures show the sequence of the events from left to right. A photon (wavy line on the left) polarizes the vacuum, an electron-positron pair (solid lines) is created, and the photon vanishes. Then the created particles annihilate each other and a photon is created. (c) A similar event (of the c^{-4} order in QED), but during the existence of the electron-positron pair, the two particles interact by exchange of a photon. (d) An electron (horizontal solid line) emits a photon, which creates an electron-positron pair that annihilates producing another photon. Meanwhile, the first electron emits a photon, then first absorbs the photon from the annihilation, and afterward the photon emitted by itself earlier. This effect is of the order c^{-5} in QED.

usually modeled by allowing the electron to emit photons and then to absorb them ("*virtual photons*")[55] (Fig. 3.5b,c,d).

The QED calculations performed to date have been focused on the energy. The first calculations of atomic susceptibilities (helium) within an accuracy including the c^{-2} terms were carried out independently by Pachucki and Sapirstein[56] and by Cencek and coworkers[57], and with accuracy up to c^{-3} (with estimation of the c^{-4} term) by Łach and coworkers. To get a sense of what subtle effects may be computed nowadays, Table 3.1 shows the components of the first ionization energy and of the dipole polarizability (see Chapter 12) of the helium atom.

Notes on Table 3.1:

- $\hat{H}_0$ denotes the result obtained from the accurate solution of the Schrödinger equation (i.e., the non-relativistic and finite nuclear mass theory). Today, the solution of the equation could

[55] For unknown reasons, physics is based on the interaction of objects of spin $\frac{1}{2}$ (like electrons or quarks) mediated by objects of spin 1 (like photons, gluons, or W particles). This principle is described by Richard Feynman (see the "*Additional Literature*" section later in this chapter).

[56] K. Pachucki and J. Sapirstein, *Phys. Rev. A*, **63**, 12,504 (2001).

[57] W. Cencek, K. Szalewicz, and B. Jeziorski, *Phys. Rev. Lett.*, **86** 5675 (2001).

Table 3.1. Contributions of various physical effects (non-relativistic, Breit, QED, and beyond QED, distinct physical contributions shown in bold) to the ionization energy and the dipole polarizability α of the helium atom, as well as comparison with the experimental values (all quantities are expressed in atomic units; i.e., $e = 1, \hbar = 1, m_0 = 1$, where m_0 denotes the rest mass of the electron). The first column gives the symbol of the term in the Breit–Pauli Hamiltonian [Eq. (3.72)] as well as of the QED corrections given order by order (first corresponding to the electron-positron vacuum polarization (QED), then, beyond quantum electrodynamics, to other particle-antiparticle pairs (non-QED): $\mu, \pi, \ldots$) split into several separate effects. The second column contains a short description of the effect. The estimated error (third and fourth columns) is given in parentheses in the units of the last figure reported.

Term	Physical Interpretation	Ionization Energy [MHz]	α[a.u.$\times 10^{-6}$][a]
$\hat{H}_0$	**Schrödinger equation**	**5 945 262 288.62(4)**	**1 383 809.986(1)**
δ	**Nonzero size of the nucleus**	**−29.55(4)**	**0.022(1)**
$\hat{H}_1$	**p^4 term**	**1 233 305.45(1)**	**−987.88(1)**
$\hat{H}_2$(el-el)	Electron-electron retardation (Breit interaction)	48 684.88(1)	−23.219(1)
$\hat{H}_2$(el-n)	Electron-nucleus retardation (Breit interaction)	319.16(1)	−0.257(3)
$\hat{H}_2$	**Breit (total)**	**49 004.04(1)**	**−23.476(3)**
$\hat{H}_3$	**Spin-orbit**	0	0
$\hat{H}_4$(el-el)	Electron-electron Darwin term	117 008.83(1)	−66.083(1)
$\hat{H}_4$(el-n)	Electron-nucleus Darwin term	−1 182 100.99(1)	864.85(2)
$\hat{H}_4$	**Darwin (total)**	**−1 065 092.16(1)**	**798.77(2)**
$\hat{H}_5$	**Spin-spin (total)**	**−234 017.66(1)**	**132.166(1)**
$\hat{H}_6$	**Spin-field**	0	0
QED(c^{-3})	Vacuum polarization correction to electron-electron interaction	−72.48(1)	0.41(1)
QED(c^{-3})	Vacuum polarization correction to electron-nucleus interaction	1 463.00(1)	−1.071(1)
QED(c^{-3})	Total vacuum polarization in c^{-3} order	1 390.52(1)	−1.030(1)
QED(c^{-3})	**Vac.pol.+ other c^{-3} QED correction**	**−40 483.98(5)**	**30.66(1)**
QED(c^{-4})	Vacuum polarization	12.26(1)	0.009(1)
QED(c^{-4})	**Total c^{-4} QED correction**	**−834.9(2)**	**0.56(22)**
QED-h.o.	**Estimation of higher order QED correction**	**84(42)**	**−0.06(6)**
non-QED	**Contribution of virtual muons, pions, etc.**	**0.05(1)**	**−0.004(1)**
$\sum$	**Theory (total)**	**5 945 204 223(42)**[b]	**1 383 760.79(23)**
	Experiment	**5 945 204 238(45)**[c]	**1 383 791(67)**[d]

[a] G. Łach, B. Jeziorski, and K. Szalewicz, *Phys. Rev. Lett.*, **92**, 233,001 (2004).

[b] G.W.F. Drake and P.C. Martin, *Can. J. Phys.*, 76, 679 (1998); V. Korobov and A. Yelkhovsky, *Phys. Rev. Lett.*, **87**, 1930 (2001).

[c] K.S.E. Eikema, W. Ubachs, W. Vassen, and W. Hogervorst, *Phys. Rev. A*, **55**, 1866 (1997).

[d] F. Weinhold, *J. Phys. Chem.*, **86**, 1111 (1982).

be obtained with greater accuracy than reported here. Imagine that here, the theory is limited by the error we know the helium atom mass, which is only 12 significant figures.

- The effect of the nonzero size of the nucleus is small, so it is almost never taken into account in computations. If we enlarged the nucleus to the size of an apple, the first Bohr orbit would be 10 km from the nucleus. And still (sticking to this analogy), the electron is able to distinguish a point from an apple? Not quite, it sees the (tiny) difference because the electron knows the region close to the nucleus: it is there that it resides most often. Anyway, the theory is able to compute such a tiny effect.
- *The term p^4 and the total Darwin effect nearly cancel each other out for unclear reasons.* This cancellation is being persistently confirmed in other systems as well. Strangely enough, this pertains not only to the ionization energy, but also to the polarizability.
- After the abovementioned cancellation (of p^4 and Darwin terms), the retardation becomes one of the most important relativistic effects. As seen from Table 3.1, the effect is about 100 times larger (both for the ionization energy and the polarizability) for the electron-electron retardation than for that of the nucleus-electron. This is quite understandable because the nucleus represents a "*massive rock*" (it is about 7000 times heavier in comparison to an electron), it moves slowly, and in the nucleus-electron interaction, only the electron contributes to the retardation effect. Two electrons make the retardation much more serious.
- Term $\hat{H}_3$ (spin-orbit coupling) is equal to zero for symmetry reasons (for the ground state).
- In the Darwin term, the nucleus-electron vs electron-electron contribution have reversed magnitudes: about 1 : 10, as compared to 100 : 1 in retardation. Again, this time it seems intuitively correct. We have the sum of the particle-particle terms in the Hamiltonian $\hat{H}_4 = \frac{ie\hbar}{(2m_0c)^2}\left[\hat{\boldsymbol{p}}_1 \cdot \boldsymbol{\mathcal{E}}(\boldsymbol{r}_1) + \hat{\boldsymbol{p}}_2 \cdot \boldsymbol{\mathcal{E}}(\boldsymbol{r}_2)\right]$, where $\boldsymbol{\mathcal{E}}$ means an electric field created by two other particles on the particle under consideration. Each of the terms is proportional to $\nabla_i \nabla_i V = \Delta_i V = 4\pi q_i \delta(\boldsymbol{r}_i)$, where δ is the Dirac δ delta function (see Appendix E available at booksite.elsevier.com/978-0-444-59436-5, p. e69), and q_i denotes the charge of the particle "i." The absolute value of the nuclear charge is twice the electron charge.
- In term $\hat{H}_5$, the spin-spin relates to the electron-electron interaction because the helium nucleus has spin angular momentum 0.
- The Coulombic interactions are modified by polarization of vacuum (just as two charges in a dielectric medium interact more weakly). Table 3.1 reports such corrections[58] to the electron-electron and the electron-nucleus interactions [QED(c^{-3})], taking into account that electron-positron pairs jump out from the vacuum. One of these effects is shown in Fig. 3.5a. As seen from Table 3.1, the nucleus polarizes the vacuum much more easily (about 10 times more than the polarization by electrons). Once again, the larger charge of the nucleus makes the electric field larger and qualitatively explains the effect. Note that the QED corrections (corresponding to e-p creation) decrease quickly with their order. One of these higher-order corrections is shown in Fig. 3.5d.

[58] However, these effects represent a minor fraction of the total QED(c^{-3}) correction.

- What about the creation of other (than e-p) particle-antiparticle pairs from the vacuum? the larger the rest mass is, the more difficult it is to squeeze out the corresponding particle-antiparticle pair. And yet we have some tiny effect (see non-QED entry) corresponding to the creation of such pairs as muon-antimuon (μ), pion-antipion[59] (π), etc. This means that the helium atom is composed of the nucleus and the two electrons only, when we look at it within a certain approximation. To tell the truth, the atom contains also photons, electrons, positrons, muons, pions, and whatever you wish, but with a smaller and smaller probability of appearance. All that has only a minor effect of the order of something like the seventh significant figure (both for the ionization potential and for the polarizability).

Summary

The beginning of the twentieth century has seen the birth and development of two revolutionary theories: relativity and quantum mechanics. These two theories turned out to be incompatible, and attempts were made to make them consistent. This chapter consists of two interrelated parts:

- An introduction of the elements of relativity theory, and
- Attempts to make quantum theory consistent with relativity (relativistic quantum mechanics).

Special Theory of Relativity

- If experiments are to be described in the same way in two laboratories that move with respect to the partner laboratory with constant velocities v and $-v$, respectively, then the apparent forces have to vanish. The same event is described in the two laboratories (by two observers) in the corresponding coordinate system (in one, the event happens at coordinate x and time t, and in the second, at x' and t'). A sufficient condition that makes the forces vanish is based on linear dependence: $x' = Ax + Bt$ and $t' = Cx + Dt$, where A, B, C, D denote some constants.
- In order to put both observers on the same footing, we have to have $A = D$.
- The Michelson-Morley experiment has shown that each of the observers will note that in the partner's laboratory, there is a *contraction* of the dimension pointing to the partner. As a consequence, there is a time *dilation*; i.e., each of the observers will note that time flows slower in the partner's laboratory.
- Einstein assumed that in spite of this, any of the observers will measure the same speed of light (c) in his coordinate system.
- This leads to the *Lorentz transformation*, which says where and when the two observers see the same event. The Lorentz transformation is especially simple after introducing the Minkowski space (x, ct): $\begin{bmatrix} x' \\ ct' \end{bmatrix} = \frac{1}{\sqrt{1-\frac{v^2}{c^2}}}\begin{Bmatrix} 1 & -\frac{v}{c} \\ -\frac{v}{c} & 1 \end{Bmatrix}\begin{bmatrix} x \\ ct \end{bmatrix}$. None of the two coordinate systems is privileged (relativity principle).
- Finally, we derived Einstein's formula $E_{kin} = mc^2$ for the kinetic energy of a body with mass m (this depends on its speed with respect to the coordinate system where the mass is measured).

---

[59] Pions are π mesons, subnuclear particles with mass comparable to that of the muon, a particle about 200 times more massive than an electron. Pions were discovered in 1947 by C.G. Lattes, G.S.P. Occhialini, and C.F. Powell.

Relativistic Quantum Mechanics

- Fock, Klein, and Gordon found the total energy for a particle using the Einstein formula for kinetic energy $E_{kin} = mc^2$, adding the potential energy and introducing the momentum[60] $p = mv$. After introducing an external electromagnetic field (characterized by the vector potential $\boldsymbol{A}$ and the scalar potential ϕ), they obtained the following relation among operators $\left[\frac{i\hbar\frac{\partial}{\partial t}-q\phi}{c}\right]^2 - \left[\left(-i\hbar\nabla - \frac{q}{c}\boldsymbol{A}\right)^2 + m_0^2c^2\right] = 0$, where m_0 denotes the rest mass of the particle.
- Paul Dirac factorized the left side of this equation by treating it as the difference of squares. This gave two continua of energy separated by a gap of width $2m_0c^2$. Dirac assumed that the lower (negative energy) continuum is fully occupied by electrons (i.e., a vacuum), while the upper continuum is occupied by the single electron (our particle). If we managed to excite an electron from the lower continuum to the upper one, then in the upper continuum, we would see an electron, while the hole in the lower continuum would have the properties of a positive electron (positron). This corresponds to the creation of the electron-positron pair from the vacuum.
- The Dirac equation for the electron has the following form: $\left(i\hbar\frac{\partial}{\partial t}\right)\Psi = (q\phi + c\sum_{\mu=x,y,z}\alpha_\mu \pi_\mu + \alpha_0 m_0c^2)\Psi$, where π_μ in the absence of magnetic field is equal to the momentum operator $\hat{p}_\mu$, $\mu = x, y, z$, while α_μ stand for the square *matrices* of the rank 4, which are related to the Pauli matrices (see the discussion of spin in Chapter 1). As a consequence, the wave function Ψ has to be a four-component vector composed of square integrable functions (*bispinor*).
- The Dirac equation demonstrated pathological behavior when a numerical solution was sought. The reason for this was the decoupling of the electron and positron equations. The exact separation of the negative and positive energy continua has been demonstrated by Barysz and Sadlej, but it leads to a more complex theory. Numerical troubles are often removed by an *ad hoc* assumption called *kinetic balancing* (i.e., fixing a certain relation among the bispinor components). By using this relation, we prove that there are two *large* and two *small* (smaller by a factor of about $2\frac{c}{v}$) components of the bispinor[61].
- The kinetic balance can be used to eliminate the small components from the Dirac equation. Then, the assumption $c = \infty$ (non-relativistic approximation) leads to the *Schrödinger equation* for a single particle.
- The Dirac equation for a particle in the electromagnetic field contains the interaction of the spin magnetic moment with the magnetic field. In this way, spin angular momentum appears in the Dirac theory in a natural way (as opposed to the non-relativistic case, where it has had to be postulated).
- The problem of an electron in the external electric field produced by the nucleus (i.e., the hydrogen-like atom) has been solved exactly within the Dirac model. It turned out that the relativistic corrections are important only for systems with heavy atoms.
- It has been demonstrated in a step-by-step calculation how to obtain an approximate solution of the Dirac equation for the hydrogen-like atom. One of the results is that the relativistic orbitals are contracted compared to the non-relativistic ones.
- Finally, the Breit equation has been given. The equation goes beyond the Dirac model by taking into account the *retardation* effects. The Breit-Pauli expression for the Breit Hamiltonian contains several easily interpretable physical effects.
- Quantum electrodynamics (QED) provides an even better description of the system by adding radiative effects that take into account the interaction of the particles with the vacuum.

[60] They wanted to involve the momentum in the formula to be able to change the energy expression to an operator ($\boldsymbol{p} \rightarrow \hat{\boldsymbol{p}}$) according to the postulates of quantum mechanics.

[61] For solutions with negative energies, this relation is reversed.

Main Concepts, New Terms

From the Research Front

Dirac-Coulomb theory within the mean field approximation (see Chapter 8) is routinely applied to molecules and allows us to estimate the relativistic effects even for large molecules. In the computer era, this means that there are computer programs available that allow anybody to perform relativistic calculations.

Much worse is the situation with the calculations going beyond the Dirac approach. The first estimation for molecules of relativistic effects beyond the Dirac approximation was carried out by Ladik,[62] and then by Jeziorski and Kołos[63]. Besides the computation of the Lamb shift for the water molecule[64], not much has been computed in this area for years.

Then it turned out that a promising approach (the test was for the hydrogen molecule) is to start with an accurate solution to the Schrödinger equation[65] and go directly toward the expectation value of the Breit-Pauli Hamiltonian with this wave function (i.e., to abandon the Dirac equation), and then to the QED corrections.[66] This Breit-Pauli contribution represents the complete first nonzero relativistic effect proportional to[67] $\left(\frac{1}{c}\right)^2$. Then the complete QED contribution proportional to $\left(\frac{1}{c}\right)^3$ is computed directly, as well as a major part of the $\left(\frac{1}{c}\right)^4$ QED term.

---

[62] J. Ladik, *Acta. Phys. Hung.*, **10**, 271 (1959).

[63] The calculations were performed for the hydrogen molecular ion H_2^+ [B. Jeziorski and W. Kołos, *Chem. Phys. Lett.*, **3**, 677 (1969).]

[64] P. Pyykkö, K.G. Dyall, A.G. Császár, G. Tarczay, O.L. Polyansky, and J. Tennyson, *Phys. Rev. A*, **63**, 24,502 (2001).

[65] This solution was done with the center of mass separated out.

[66] K. Piszczatowski, G. Łach, M. Przybytek, J. Komasa, K. Pachucki, and B. Jeziorski, *J. Chem. Theory Comput.*, **5**, 3039 (2009).

[67] That is, the $\frac{1}{c}$ contribution vanishes.

Calculated in this way, dissociation energy for the hydrogen molecule amounts to 36118.0695 ± 0.0010 cm^{-1}, as compared to the experimental result[68] 36118.0696 ± 0.0004 cm^{-1}. The estimated size of the neglected QED effects is ± 0.0004 cm^{-1}. The even more stringent and equally successful test of this approach from my colleagues will be reported in Chapter 6, in the section "*From the Research Front.*" It seems, therefore, that for now, and taking the best-known (and the simplest) molecule, we may say that the present theory describes Nature with extreme accuracy.

Ad Futurum

Compared to typical chemical phenomena, the relativistic effects remain of marginal significance in almost all instances for the biomolecules or molecules typical in traditional organic chemistry. In inorganic chemistry, however, these effects could be much more important. Probably the Dirac-Coulomb theory combined with the mean field approach will remain a satisfactory standard for the vast majority of researchers, at least for the next few decades. At the same time, there will be theoretical and computational progress for small molecules (and for atoms), where the Dirac theory will be progressively replaced by quantum electrodynamics.

Hans Albrecht Bethe (1906–2005) was an American physicist, a professor at Cornell University, and a student of Arnold Sommerfeld. Bethe contributed to many branches of physics, such as crystal field theory, interaction of matter with radiation, quantum electrodynamics, and the structure and nuclear reactions of stars (for the latter achievement, he received the Nobel Prize in physics in 1967).

In most applications, we do not need an accuracy like that reported in the section "*From the Research Front,*" earlier in this chapter, but the reason why such results are important is general, pertaining to all those, who perform any kind of quantum-mechanical calculations. In this way science tests the reliability of its most reliable tools, both experimental and theoretical. The agreement achieved makes all of us confident that we know what we are doing and that the world works much in the way that we think it should.

Sooner or later, however, as has happened so many times in the past, we will meet an irreducible discrepancy. If this happens, a better theory will have to be constructed–this is how science operates.

Additional Literature

H. Bethe and E. Salpeter, *Quantum Mechanics of One- and Two-Electron Atoms*, Plenum Publishing Corporation, New York (1957).

This book is absolutely exceptional. It is written by excellent specialists in such a competent way and with such care, that despite the passage of many decades since it was written, it remains the fundamental and best source.

I.M. Grant and H.M. Quiney, "*Foundations of the relativistic theory of atomic and molecular structure,*" *Adv. At. Mol. Phys.*, **23**, 37 (1988).

L. Pisani, J.M. André, M.C. André, and E. Clementi, *J. Chem. Educ.*, **70**, 894–901(1993); also J.M. André, D.H. Mosley, M.C. André, B. Champagne, E. Clementi, J.G. Fripiat, L. Leherte, L. Pisani, D. Vercauteren, and M. Vracko, *Exploring Aspects of Computational Chemistry, Vol. I, Concepts*, Presses Universitaires de Namur, pp. 150–166(1997), Vol. II, Exercises, Presses Universitaires de Namur, pp. 249–272(1997).

[68] J. Liu, E J. Salumbides, U. Hollenstein, J.C.J. Koelemeji, K.S.E. Eikema, W. Ubachs, and F. Merkt, *J. Chem. Phys.*, **130**, 174,306 (2009).

This is a clear article and a fine book. Their strength lies in the very simple examples of the application of the theory.

R.P. Feynman, *QED–The Strange Theory of Light and Matter*, Princeton University Press, Princeton, NJ, 1988.
Excellent book written by one of the celebrities of our times in the style "*quantum electrodynamics not only for poets*".

Questions

1. In two inertial systems, the same forces operate if the two coordinate systems
   a. are both at rest
   b. move with the same velocity
   c. are related by Galilean transformation
   d. x' and t' depend linearly on x and t
2. The Michelson-Morley experiment has shown that when an observer in the coordinate system O measures a length in O' (and both coordinate systems fly apart; $v' = -v$), then he obtains
   a. a contraction of the length unit of the observer in O'
   b. a contraction of lengths along the direction of the motion
   c. the same result that is obtained by an observer in O'
   d. a contraction of length perpendicular to the motion
3. An observer in O measures the times that a phenomenon takes in O and O' (and both coordinate systems fly apart; $v' = -v$)
   a. time goes more slowly in O' only if $|v| < \frac{c}{2}$
   b. time goes with the same speed in O'
   c. time goes more slowly in O' only if $|v| > \frac{c}{2}$
   d. the time of the phenomenon going on in O will be shorter
4. In the Minkowski space, the distance of any event from the origin (and both coordinate systems fly apart; $v' = -v$) is
   a. the same for observers in O and in O'
   b. equal to $ct - x$
   c. equal to 0
   d. is calculated according to a non-Euclidean metric
5. A bispinor represents
   a. a two-component vector with each component being a spinor
   b. a two-component vector with complex numbers as components
   c. a four-component vector with functions as components
   d. a square integrable vector function
6. Non-physical results of numerical solutions to the Dirac equation appear because
   a. the Dirac sea is neglected
   b. the electron states have components belonging to the negative energy continuum
   c. the electron has kinetic energy equal to its potential energy
   d. there is no lower limit of the energy spectrum.
7. The Schrödinger equation can be deduced from the Dirac equation under the assumption that
   a. v/c is small
   b. the speed of the electron is close to c

c. all components of the bispinor have equal length
d. the magnetic field is zero

8. In the Breit equation, there is an approximate cancellation of
   a. the retardation effect with the nonzero size of the nucleus effect
   b. the retardation effect electron-electron with that of the electron-nucleus
   c. the spin-spin effect with the Darwin term
   d. the Darwin term with the p^4 term
9. Dirac's hydrogen atom orbitals, when compared to Schrödinger's are
   a. more concentrated close to the nucleus, but have a larger mean value of r
   b. have a smaller mean value of r
   c. more concentrated close to the nucleus
   d. of the same size, because the nuclear charge has not changed
10. The Breit equation
    a. is invariant with respect to the Lorentz transformation
    b. takes into account the interaction of the magnetic moments of electrons resulting from their orbital motion
    c. takes into account the interaction of the spin magnetic moments
    d. describes only a single particle

Answers

1a,b,c,d, 2a,b, 3d, 4a,d, 5a,c,d, 6a,b,d, 7a, 8d, 9b,c, 10b,c

CHAPTER 4

Exact Solutions—Our Beacons

"Longum iter per praecepta, breve et efficax per exempla."
Teaching by precept is a long road, but short and efficient is the way by example.

Where Are We?

We are in the middle of the TREE trunk.

An Example

Two chlorine atoms stay together; they form the molecule Cl_2. If we want to know its main mechanical properties, it would very quickly be seen that the two atoms have a certain equilibrium distance and any attempt to change this (in either direction) would be accompanied by work to be done. It looks like the two atoms are coupled together by a sort of spring. If one assumes that the spring satisfies Hooke's law,[1] the system is equivalent to a harmonic oscillator. If we require that no rotation in space of such a system is allowed, the corresponding Schrödinger equation has an exact[2] analytical solution.

What Is It All About?

---

[1] Another requirement is that we limit ourselves to small displacements.
[2] Exact means ideal; i.e., without any approximation.

Ideas of Quantum Chemistry, Second Edition. http://dx.doi.org/10.1016/B978-0-444-59436-5.00004-0

Why Is This Important?

The Schrödinger equation is quite easy to solve nowadays with a desired accuracy for many systems. There are only a few ones for which the exact solutions are possible. These problems and solutions play an extremely important role in physics, since they represent a kind of beacon for our navigation in science, where as a rule we deal with complex systems. These may most often be approximated by those for which exact solutions exist. For example, a real diatomic molecule is difficult to describe in detail, and it certainly does not represent a harmonic oscillator. Nevertheless, the main properties of diatomics follow from the simple harmonic oscillator model. When chemists or physicists must describe reality they always first try to simplify the problem,[3] to make it similar to one of the simple problems described in the present chapter. Thus, from the beginning, we know the (idealized) solution. This is of prime importance when discussing the (usually complex) solution to a higher level of accuracy. If this higher-level description differs dramatically from that of the idealized one, most often this indicates that there is an error in our calculations, and no task is more urgent than to find and correct it.

What Is Needed?

- The postulates of quantum mechanics (Chapter 1)
- Separation of the center of mass motion (see Appendix I available at booksite.elsevier.com/978-0-444-59436-5 on p. e93)

[3] One of the cardinal strategies of science, when we have to explain a strange phenomenon, is first to simplify the system and create a model or series of models (more and more simplified descriptions) that still exhibit the phenomenon. The first model to study should be as simple as possible because it will shed light on the main machinery.

- Operator algebra (see Appendix B available at booksite.elsevier.com/978-0-444-59436-5 on p. e7)

In the present textbook, we assume that the reader knows most of the problems described in this chapter from a basic course in quantum chemistry. This is why the problems are given briefly–only the most important results, without derivation, are reported. On the other hand, such a presentation will be sufficient for our goals in most cases.

Classical Works

The hydrogen atom problem was solved by Werner Heisenberg in *"Über quantentheoretischen Umdeutung kinematischer und mechanischer Beziehungen,"* published in *Zeitschrift für Physik, 33*, 879 (1925). ★ Erwin Schrödinger arrived at an equivalent picture within his wave mechanics in *"Quantisierung als Eigenwertproblem.I,"* published in *Annalen der Physik, 79*, 361 (1926). Schrödinger also gave the solution for the harmonic oscillator in a paper (Quantisierung as Eigenwertproblem.II) which appeared in *Annalen der Physik, 79*, 489 (1926). ★ The Morse oscillator problem was solved by Philip McCord Morse in *"Diatomic molecules according to the wave mechanics. II. Vibrational levels,"* in *Physical Review, 34*, 57 (1929).[4] ★ The tunneling effect was first considered by Friedrich Hund in *"Zur Deutung der Molekelspektren,"* published in *Zeitschrift für Physik, 40*, 742 (1927). ★ The idea of supersymmetry was introduced to quantum mechanics by Candadi V. Sukumar in an article called *"Supersymmetry factorisation of the Schrödinger equation and a Hamiltonian hierarchy,"* published in the *Journal of Physics A, 18*, L57 (1985). ★ The Schrödinger equation for the harmonium[5] was first solved by Sabre Kais, Dudley R.Herschbach, and Raphael David Levine in *"Dimensional scaling as a symmetry operation,"* which appeared in the *Journal of Chemical Physics, 91*, 7791 (1989).

4.1 Free Particle

The potential energy for a free particle is a constant (taken arbitrarily as zero): $V = 0$; therefore, energy E represents only the kinetic energy. The Schrödinger equation takes the form

$$-\frac{\hbar^2}{2m}\frac{d^2\Psi}{dx^2} = E\Psi,$$

or in other words,

$$\frac{d^2\Psi}{dx^2} + \kappa^2\Psi = 0,$$

with $\kappa^2 = \frac{2mE}{\hbar^2}$. The constant κ in this situation[6] is a real number.

The special solutions to this equation are $\exp(i\kappa x)$ and $\exp(-i\kappa x)$, $\kappa \geq 0$. Their linear combination with arbitrary complex coefficients A' and B' represents the general solution:

$$\Psi = A'\exp(i\kappa x) + B'\exp(-i\kappa x). \tag{4.1}$$

This is a wave of wavelength $\lambda = \frac{2\pi}{\kappa}$. Function $\exp(i\kappa x)$ represents the eigenfunction of the momentum operator: $\hat{p}_x \exp(i\kappa x) = -i\hbar\frac{d}{dx}\exp(i\kappa x) = -i\hbar i\kappa\exp(i\kappa x) = \kappa\hbar\exp(i\kappa x)$. For eigenvalue $\hbar\kappa > 0$, therefore, the eigenfunction $\exp(i\kappa x)$ describes a particle moving toward $+\infty$. Similarly, $\exp(-i\kappa x)$ corresponds to a particle of the same energy, but moving in

[4] Note the spectacular speed at which the scholars worked (Heisenberg's first paper appeared in 1925).
[5] In this case, we are talking about a harmonic model of the helium atom.
[6] The kinetic energy is always positive.

the opposite direction. The function $\Psi = A' \exp(i\kappa x) + B' \exp(-i\kappa x)$ is a superposition of these two states. A measurement of the momentum can give only two values: $\kappa\hbar$ with probability proportional to $|A'|^2$ or $-\kappa\hbar$ with probability proportional to $|B'|^2$.

4.2 Box with Ends

The problem pertains to a single particle in a potential (Fig. 4.1a):

$$V(x) = 0 \text{ for } 0 \leq x \leq L$$
$$V(x) = \infty \text{ for other } x.$$

Just because the particle will never go outside the section $0 \leq x \leq L$ (where it would find $V(x) = \infty$), the value of the wave function outside the section is equal to 0. It remains to find the function in $0 \leq x \leq L$.

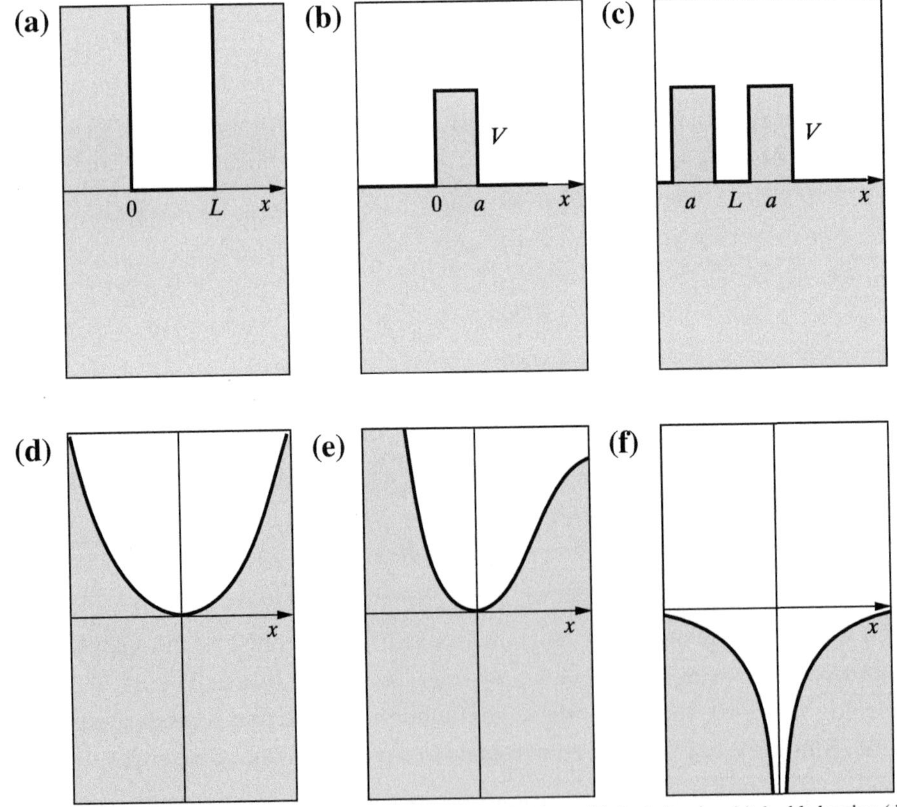

Fig. 4.1. The potential energy functions for the following: (a) particle in a box; (b) single barrier; (c) double barrier; (d) harmonic oscillator; (e) Morse oscillator; (f) hydrogen atom.

Let us write the Schrödinger equation for $0 \le x \le L$ with the Hamiltonian containing kinetic energy only (since $V = 0$, one has $E \ge 0$):

$$-\frac{\hbar^2}{2m}\frac{d^2\Psi}{dx^2} = E\Psi. \tag{4.2}$$

The solution is given by Eq. (4.1), which also may be written as (with A and B as complex numbers)

$$\Psi = A \sin \kappa x + B \cos \kappa x, \tag{4.3}$$

with

$$\kappa^2 = \frac{2mE}{\hbar^2}. \tag{4.4}$$

Now, the key is to recall (as discussed in Chapter 2), that the wave function has to be continuous, and therefore, two conditions have to be fulfilled: (1) $\Psi = 0$ for $x = 0$ and (2) $\Psi = 0$ for $x = L$. The first condition immediately gives $B = 0$, the second in this situation[7] is equivalent to $\kappa L = n\pi$, for $n = 0, 1, \ldots$. From this follows energy quantization because κ contains energy E. One obtains, therefore, the following solution (a standing wave[8]):

$$E_n = \frac{n^2 h^2}{8mL^2}, \tag{4.5}$$

$$\Psi_n = \sqrt{\frac{2}{L}} \sin \frac{n\pi}{L} x, \tag{4.6}$$

$n = 1, 2, 3, \ldots$

because $n = 0$ has to be excluded, as it leads to the wave function equal to zero everywhere, while $n < 0$ may be safely excluded, as it leads to the same wave functions as[9] $n > 0$. We have *chosen* the normalization constant as a positive real number. Fig. 4.2 shows the wave functions for $n = 1, 2,$ and 3.

Note that the wave functions (plotted as functions of $x, 0 \le x \le L$) are similar to what one may expect for a vibrating string of length L, with *immobilized* ends. If someone had a generator of vibrations with variable frequency and tried to transfer the vibrational energy to the string, he would succeed only for frequencies close to some particular resonance frequencies. The lowest-resonance frequency corresponds to such motion of the string that at any given time, all its parts have the same sign of the amplitude (this corresponds to Ψ_1). Increasing frequency will not result in energy transfer until the next resonance is reached, with the amplitude of the string such as that shown by Ψ_2 (single node, where $\Psi_2 = 0$). This is similarly true for $\Psi_n, n = 3, 4, \ldots,$ with the number of nodes equal to $n - 1$.

[7] A has to be nonzero; otherwise, $\Psi = 0$, which is forbidden.

[8] Recall that any stationary state has a trivial time-dependence through the factor $\exp(-i\frac{E}{\hbar}t)$. A standing wave at any time t has a frozen pattern of the nodes; i.e., the points x with $\Psi = 0$.

[9] It has the opposite sign, but that does not matter.

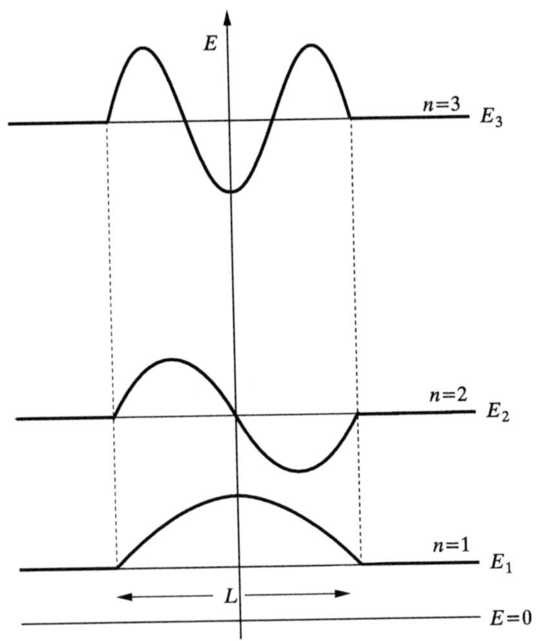

Fig. 4.2. The wave functions for the particle in a box corresponding to $n = 1, 2$, and 3. Note the increasing number of nodes, when the energy E_i of the stationary state increases.

Example 1. ***Butadiene Naively***

The particle-in-box problem has more to do with chemistry than would appear at first glance.

In organic chemistry, we consider some molecules with conjugated double and single carbon-carbon bonds. One of the simplest of these is butadiene:

```
        H\       /H
          C=C
H\      /    \H
  C=C
H/    \H
```

What does this molecule have to do with the particle in a box? Nothing, it seems. First, we have not a single particle, but 40 particles (10 nuclei and 30 electrons). Second, where is this constant potential for the motion of the particle? It doesn't exist. Third, a molecule does not represent a 1-D but a 3-D object, and in addition, a curved one instead of a section of the x-axis. It would seem that any attempt to apply such a primitive theory to this molecule is ridiculous, and yet in such a difficult situation, we will see the power of the exact solutions reported in this chapter. All the above objections are perfectly justified, but let us try to simplify our system a little.

In the molecule being studied here, the CC bonds are "*averaged*," which facilitates the motion of the π electrons along the system (this notion will become clear in Chapter 8). The four π electrons are loosely bound to the molecule, and we may assume that other electrons are always rigidly bound and therefore will be ignored.

Suppose the following:

- We removed the π electrons from the molecule (and put them temporarily into a safe).
- We "*ground up*" the remaining (positively charged) molecular core and distributed the ground mass uniformly along the x-axis within a section of length L equal to the length of the molecule (averaging the potential energy for a charged particle) to construct a kind of highway for the π electrons.
- We added the first π electron from the safe.

In such a case, this single electron would represent something similar to a particle in a box.[10] Assuming this simplified model, we know all the details of the electron distribution, including the ground state ψ_1 and excited wave functions: $\psi_2, \psi_3, \ldots$ (in the one-particle case, they are called the *orbitals*). If we now took the π electrons from the safe and added them one by one to the system, assuming that they would not see one another[11] and then taking into account the Pauli exclusion principle (the double occupancy by electrons of the individual orbitals is described in more detail in Chapter 8), we would obtain information about the electron density distribution in the molecule.[12]

In our example, the orbitals are real and the total electron density distribution (normalized to four π electrons; i.e., giving 4 after integration over x) is given as[13]

$$\rho(x) = 2\psi_1^2 + 2\psi_2^2 = 2\frac{2}{L}\sin^2\frac{\pi}{L}x + 2\frac{2}{L}\sin^2\frac{2\pi}{L}x = \frac{4}{L}\left(\sin^2\frac{\pi}{L}x + \sin^2\frac{2\pi}{L}x\right).$$

The function $\rho(x)$ is shown in Fig. 4.3a.

It is seen that

1. $\rho(x)$ is the largest on the outermost bonds in the molecule, exactly where chemists put two short lines to symbolize a double bond
2. π-electron density [i.e., $\rho(x)$] is nonzero in the center. This means that the bond there is not strictly a single bond.

This key information about the butadiene molecule has been obtained at practically no cost from the primitive FEMO model.

---

[10] This is not quite so, though, because the potential is not quite constant. Also, one might remove the particle from the box at the expense of a large but finite energy (ionization), which is not feasible for the particle in a box.

[11] As we will see in Chapter 8, this approximation is more realistic than it sounds.

[12] The idea we are describing is called the Free Electron Molecular Orbitals (FEMO) method.

[13] The student "i" is characterized by a probability density distribution $\rho_i(x)$ of finding him at coordinate x (we limit ourselves to a single variable, measuring the student's position, say, on his way from the dormitory to the university). If all students moved independently, the sum of their individual probability densities at point x_0 [i.e., $\rho(x_0) = \sum_i \rho_i(x_0)$] would be proportional to the probability density of finding any student at x_0. The same pertains to electrons, when assumed to be independent.

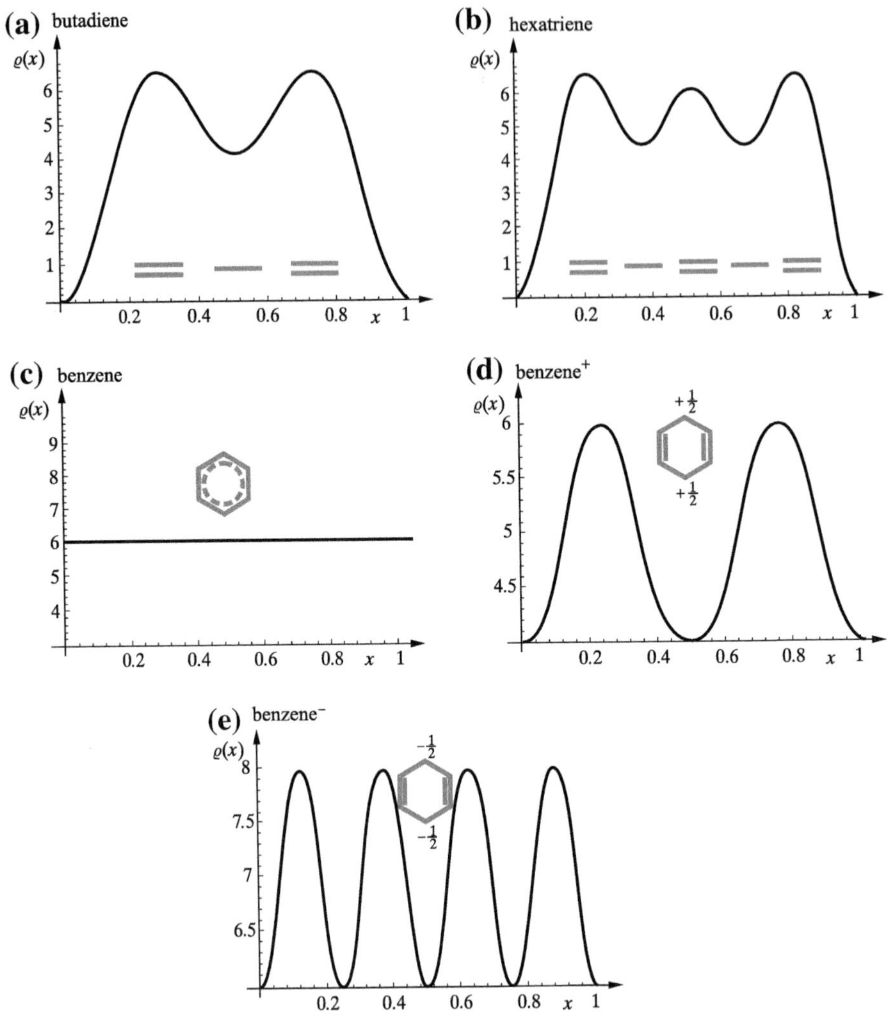

Fig. 4.3. $\pi-$ electron density charge distributions for several molecules computed by the Free Electron Molecular Orbitals (FEMO) method. The length L of each molecule has been assumed to equal 1. For other lengths, the charge distributions are similar. The electron density for four electrons in butadiene (a) and of six electrons in hexatriene (b). The electron density maxima coincide with the positions chemists write as double bonds. The six electron density distribution in the benzene molecule is peculiar because it is constant along the perimeter of the molecule (c). If we subtract an electron from benzene (d) or add an electron to it (e), then maxima and minima of the π electron density appear. If an electron is subtracted (d), there are two maxima (double bonds) and two π electron deficient regions denoted as having charge $+\frac{1}{2}$. After one electron π is added (e), then we obtain four maxima (two double bonds and two electron-rich regions denoted by charge $-\frac{1}{2}$).

Of course, we cannot expect the description to reflect all the details of the charge distribution in the butadiene molecule, but one may expect this approach to be able to reflect at least some rough features of the π electron distribution. If the results of more advanced calculations contradicted the rough particle-in-box results, then we should take a closer look at them and search for an error. This is the strength of the simple exact model systems. They play the role of the beacons–points of reference.

4.3 Cyclic Box

The 1-D box described above is similar to a stick in which the particle can move. The butadiene molecule is rather similar to such a stick and, therefore, the 1-D box models it quite well.

And what can model the benzene molecule? In a crude approximation, we may think of benzene as a stick with the two ends *joined* in such a way as to be unable to recognize where the union has taken place. Limiting ourselves to this effect,[14] we may use the solution given by Eq. (4.3) and impose appropriate boundary conditions. What could these boundary conditions be? The wave functions at the two ends of the box have to be stitched together without leaving any trace of the seam. This is achieved by two boundary conditions: $\Psi(0) = \Psi(L)$ forcing the two wave function values to match and $\Psi'(0) = \Psi'(L)$ making the seam "*invisible*."[15] The two conditions mean

$$A \sin \kappa 0 + B \cos \kappa 0 = A \sin \kappa L + B \cos \kappa L$$
$$A\kappa \cos \kappa 0 - B\kappa \sin \kappa 0 = A\kappa \cos \kappa L - B\kappa \sin \kappa L$$

or

$$B = A \sin \kappa L + B \cos \kappa L$$
$$A = A \cos \kappa L - B \sin \kappa L.$$

To find a non-trivial solution, the determinant of the coefficients at the unknown quantities A and B has to vanish[16]: $\begin{vmatrix} \sin \kappa L & \cos \kappa L - 1 \\ \cos \kappa L - 1 & -\sin \kappa L \end{vmatrix} = 0$, which is equivalent to

$$\cos \kappa L = 1.$$

The last condition gives $\kappa L = 2\pi n, n = 0, \pm 1, \pm 2, \ldots$. This immediately produces a formula for the energy very similar to that for the box with ends, Eq. (4.5), but with the replacement $n \rightarrow 2n$:

[14] This also neglects such effects as the particular shape of the benzene (curvature, etc.).

[15] There is no such a thing in nature as infinitely steep potential energy walls or infinite values of the potential energy (as in the particle-in-a-box problem). This means we should treat such idealized cases as limit cases of possible continuous potential energy functions. From the Schrödinger equation $-\frac{\hbar^2}{2m}\frac{\partial^2 \psi}{\partial x^2} = E\psi - V\psi$, we see that in such a case, the continuity of V implies (ψ must be continuous) the continuity of $\frac{\partial^2 \psi}{\partial x^2}$. The continuity of the second derivative results in the continuity of the first derivative as well. What about a non-physical case of a discontinuous V, as in the rectangular barrier case? Well, then we lose the continuity of the second derivative by definition, but still we may have the continuity of the first derivative (this we force successfully in the case being described). In a more drastic (and non-physical) case of the discontinuity, as in the particle in a box (infinitely steep V and, on top of that $V = \infty$), we lose continuity of the first derivative (at $x = 0, L$).

[16] This is a set of homogeneous linear equations.

$$E_n = \frac{(2n)^2 h^2}{8mL^2}, \tag{4.7}$$

where this time $n = 0, \pm 1, \pm 2, \ldots$

The corresponding wave functions are

$$\psi_0 = \sqrt{\frac{1}{L}} \text{ for } n = 0,$$
$$\psi_{n>0} = A \sin \frac{2\pi n}{L} x + B \cos \frac{2\pi n}{L} x,$$
$$\psi_{n<0} = -A \sin \frac{2\pi |n|}{L} x + B \cos \frac{2\pi |n|}{L} x.$$

Since $\psi_{n>0}$ and $\psi_{n<0}$ correspond to the same energy, any combination of them also represents an eigenfunction of the Schrödinger equation corresponding to the same energy (see Appendix B available at booksite.elsevier.com/978-0-444-59436-5 on p. e7). Therefore, taking as the new wave functions (for $n \neq 0$) the normalized sum and difference of the above wave functions, we finally obtain the solutions to the Schrödinger equation in the simplest form:

$$\Psi_0 \equiv \psi_0 = \sqrt{\frac{1}{L}} \text{ for } n = 0$$
$$\Psi_{n>0} = \sqrt{\frac{2}{L}} \sin \frac{2\pi n}{L} x \text{ for } n > 0$$
$$\Psi_{n<0} = \sqrt{\frac{2}{L}} \cos \frac{2\pi n}{L} x \text{ for } n < 0$$

4.3.1 Comparison of Two Boxes: Hexatriene and Benzene

Let us now take an example of two molecules: hexatriene and benzene (i.e., the cyclohexatriene). Let us assume for simplicity that the length of the hexatriene L is equal to the perimeter of the benzene.[17] Both molecules have six π-electrons (any of them). The electrons doubly occupy (the Pauli exclusion principle) three one-electron wave functions corresponding to the lowest energies. Let us compute the sum of the electron energies as the "*total electron energy*"[18] (in the units $\frac{h^2}{8mL^2}$, to have the formulas as compact as possible):

[17] This is to some extent an arbitrary assumption, which simplifies the final formulas nicely. In such cases, we have to be careful that the conclusions are valid.

[18] As will be shown in Chapter 8, this method represents an approximation.

- Hexatriene: $E_{heks} = 2 \times 1 + 2 \times 2^2 + 2 \times 3^2 = 28$
- Benzene: $E_{benz} = 2 \times 0 + 2 \times 4 + 2 \times 4 = 16$.

We conclude that six π electrons in the benzene molecule correspond to lower energy (i.e., they are more stable) than the six π electrons in the hexatriene molecule. Chemists find this experimentally: the benzene ring with its π electrons survives in many chemical reactions, whereas this rarely happens to the π- electron system of hexatriene.

Our simple theory predicts that the benzene π- electron system is more stable than that of the hexatriene molecule.

And what about the electronic density in both cases? We obtain (Fig. 4.3b-c)

- Hexatriene: $\rho(x) = 2 \times \frac{2}{L}\left[\sin^2 \frac{\pi}{L}x + \sin^2 \frac{2\pi}{L}x + \sin^2 \frac{3\pi}{L}x\right]$
- Benzene: $\rho(x) = 2 \times \frac{1}{L} + 2 \times \frac{2}{L}[\sin^2 \frac{2\pi}{L}x + \cos^2 \frac{2\pi}{L}x] = \frac{6}{L}$.

This is an extremely interesting result.

The π-electron density is constant along the perimeter of the benzene molecule.

No single and double bonds - all CC bonds are equivalent (Fig. 4.3c). Previous experience had already led chemists to the conclusion that all the $C-C$ bonds in benzene are equivalent. This is why they decided to write down the benzene formula in the form of a regular hexagon with a circle in the middle (i.e., not to show the single and double bonds, ⏣). The FEMO method reflected that feature in a naive way. Don't the π electrons see where the carbon nuclei are? Of course they do. We will meet some more exact methods in further chapters of this textbook, which give a more detailed picture, but it will turn out that all CC bonds would have the same density distribution, similar to the solution given by the primitive FEMO method. From the wave functions (p. 168), it follows that this will happen not only for benzene, but also for all the systems with $(4n+2)$–electrons, $n = 1, 2, ...$ because of a very simple (and, therefore, very beautiful) reason that $\sin^2 x + \cos^2 x = 1$ for any x.

The addition or subtraction of an electron makes the distribution non-uniform (Fig. 4.3d-e). Also in six π-electron hexatriene molecules, uniform electron density is out of the question (Fig. 4.3b). Note that the maxima of the density coincide with the double bonds chemists like to write down. However, even in this molecule, there is still a certain equalization of bonds, since the π electrons are also where chemists write a single bond (although the π electron density is smaller over there[19]).

Again, important information has been obtained at almost no cost.

[19] This is a location where, in the classical picture, no π electron should be.

2-D Rectangular Box

Let us consider a rectangular box (Fig. 4.4) with sides L_1 and L_2 and $V = 0$ inside and $V = \infty$ outside. We first separate the variables x and y, which leads to the two 1-D Schrödinger equations (solved as shown above).

The energy eigenvalue is, therefore, equal to the sum of the energies for the 1-D problems:

$$E_n = \frac{h^2}{8m}\left(\frac{n_1^2}{L_1^2} + \frac{n_2^2}{L_2^2}\right), \tag{4.8}$$

while the wave function has the form of the product of both 1-D solutions:

$$\Psi_{n_1 n_2} = 2\sqrt{\frac{1}{L_1 L_2}} \sin\frac{n_1\pi}{L_1}x \cdot \sin\frac{n_2\pi}{L_2}y, \tag{4.9}$$

where $n_1, n_2 = 1, 2, \ldots$

If someone cut out a plywood square, immobilized the square sides, and tried to transfer vibrations using a vibration generator, he would be most effective for some resonance frequencies. The distribution of amplitudes of vibrations of the plywood (as functions of x and y) would be very similar to what we see in Fig. 4.4. Note that for the plywood *square*, any frequency except the lowest one corresponds to *two* vibrational modes, exactly as in the 2-D square box ($L_1 = L_2 = L$) one has two states $\Psi_{n_1 n_2}$ and $\Psi_{n_2 n_1}$ (double degeneracy). If one made the sides different (of the plywood or of the 2-D box) the degeneracy would be lifted. Thus, the degeneracy is a consequence of symmetry.

4.4 Carbon Nanotubes

Graphite is an allotropic form of carbon with an extraordinary structure: a stack of honeycomb-looking, identical sheets of carbon atoms (graphenes); see Fig. 4.5a. Inside any graphene sheet, the carbon atoms are bound by chemical bonds (see Chapter 8) forming a hexagonal lattice, the graphene sheets are attracted to each other by relatively weak intermolecular forces (described in Chapter 13). These systems look as being difficult subjects for solving the Schrödinger equation. We will try, however, to simplify them, in order to make this solution manageable. A single graphene sheet, similar to the benzene ring, has the most mobile π-electrons, one per carbon atom. These electrons may be treated in the spirit of the FEMO method as a set of independent electrons.

The fullerenes (the most important of them is C_{60} of a football shape) are related to the graphene sheets, and they correspond to some bending of a single graphene sheet (built of carbon regular hexagons), which is possible most effectively by replacing in the structure some

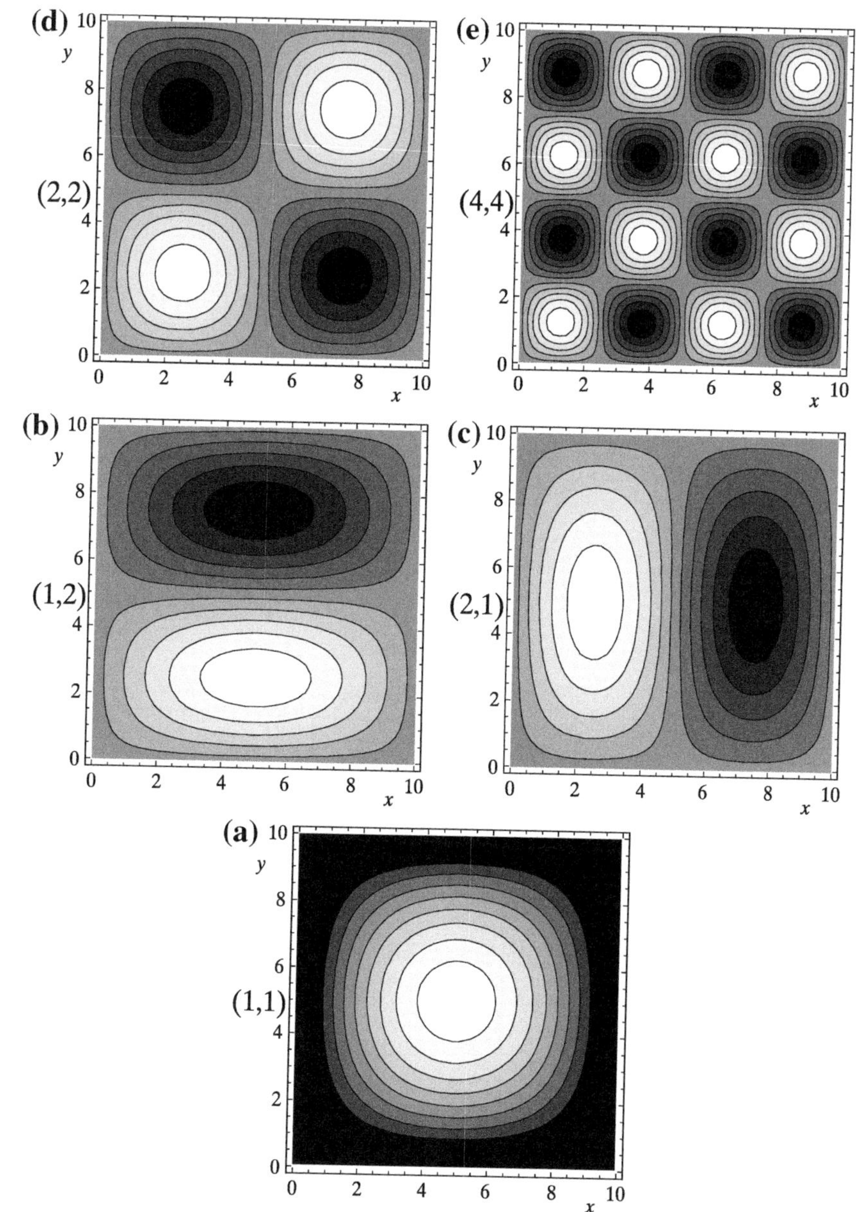

Fig. 4.4. Examples of the wave functions for a particle in a square box, the quantum numbers (n_1, n_2) correspond to (a) $(1, 1)$; (b) $(1, 2)$; (c) $(2, 1)$; (d) $(2, 2)$; (e) $(4, 4)$. In the case shown, the higher the energy, the more nodes there are in the wave function. This rule is not generally true. For example, in a rectangular box with $L_1 \gg L_2$, even a large increase of n_1 does not raise the energy too much, while introducing a lot of nodes. On the other hand, increasing n_2 by 1 raises the energy much more, while introducing only one extra node. A reader acquainted with hydrogen atom orbitals will easily recognize the resemblance of these images to some of them (cf., pp. 204–208), because of the rule mentioned in the text.

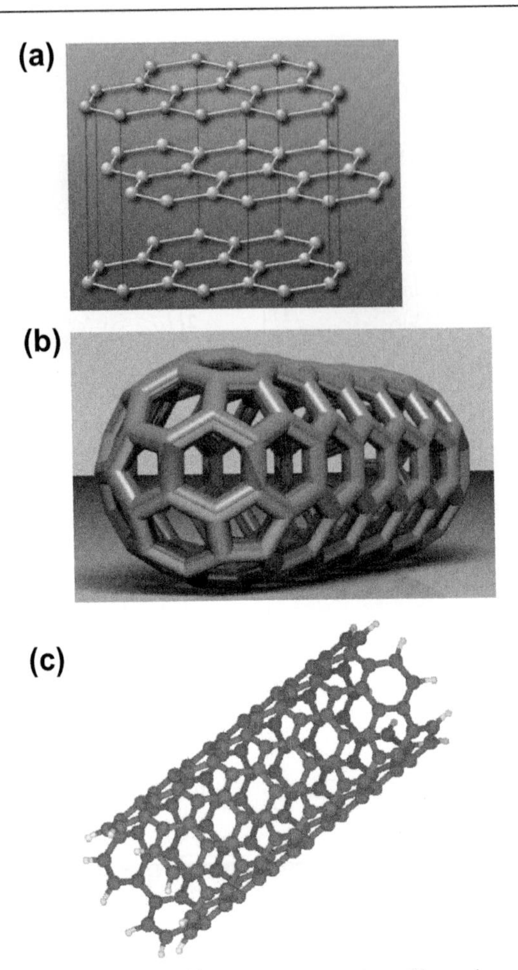

Fig. 4.5. Graphene, graphite, and nanotubes. (a) Graphite represents a structure of layered graphene sheets; (b) by replacing some hexagons by pentagons in the graphene, one obtains a curvature of the monoatomic carbon surface; (c) nanotube.

hexagons by pentagons (see Fig. 4.5b). The bending together with the possibility of seaming the sides of the graphene may lead theoreticians to a plethora of possible structures, including closed ones such as fullerenes. There are, however, some other interesting possibilities. Keeping the hexagonal structure intact, one may just roll the graphene sheet and seam the sides by the carbon-carbon bonds to make a cylinder[20] (see Fig. 4.5c).

[20] The rolling up may be accomplished in several different ways: making the seam directly "*head-to-head*"; i.e., without any shift when seaming, with one carbon shift, two carbons shift, etc. The resulting nanotubes preserve the locally hexagonal structure, but their physical and chemical properties strongly depend not only on the radius of the nanotube, but primarily on the abovementioned shift.

Using the FEMO method, Massimo Fusaro[21] treated the nanotube as a surface for free motion of independent electrons. The calculated energy levels were then occupied by a corresponding number of electrons (one from each carbon atom). The radius $R = \frac{3ja}{2\pi}$ of the cylinder is defined through the number $(2j)$ of the carbon atoms terminating the nanotube, where $a = 1.42$ Å is the CC nearest-neighbor distance in typical systems with the conjugated single and double CC bonds. The length of the cylinder can be calculated as $L = (\frac{N_C}{2j} + 1)\frac{\sqrt{3}a}{2}$, where N_C stands for the total number of the carbon atoms in the nanotube. The total number of electrons (also N_C) considered in the FEMO model can be calculated from the radius and the length of the nanotube as $N_C = \frac{4}{3}\left(\frac{2\pi R \cdot L}{a^2\sqrt{3}} - \frac{\pi R}{a}\right)$.

A nanotube is a complicated, many-electron molecule and, it seems, its electronic structure can be revealed only after long computations that give an approximation to the solution of the Schrödinger equation. And again, we may admire the power of simple models. After assuming the continuous cylinder model, one may expect for $R = 0$ that we have to recover the solution for the box with ends, while for $L = 0$ and $R \neq 0$, one should get the solution for the cyclic box. It turns out that in principle, we do get something like that, except that the shortest cylinder built of carbon atoms cannot have zero length.

The Schrödinger equation for a particle on a side surface of a cylinder (assuming the cylinder axis as x, the position of the particle is given by x, ϕ, where ϕ is an angle measuring rotation about x) can be solved exactly. After applying the boundary conditions (the wave function has to vanish at the ends of the cylinder), one gets the following solution (A_{nl} is a normalization constant):

$$\psi_{n,l>0}(x,\phi) = A_{nl} \sin(l\phi) \sin\left(\frac{n\pi}{L}x\right) \quad \text{dla} \quad l > 0,$$
$$\psi_{n,l<0}(x,\phi) = A_{nl} \cos(l\phi) \sin\left(\frac{n\pi}{L}x\right) \quad \text{dla} \quad l < 0,$$
$$\psi_{n,l=0}(x,\phi) = A_{n0} \sin\left(\frac{n\pi}{L}x\right) \quad \text{dla} \quad l = 0,$$

where $n = 1, 2, \ldots, \frac{N_C}{2j}$ (due to the atomic resolution of our model, a larger n would make the repeating of the already-obtained solutions), while $l = 0, \pm1, \pm2, \ldots$ The wave functions given above represent nothing else but products of the wave function for the box with ends and for the cyclic box.

The calculated energy eigenvalues are

$$E_{nl} = \frac{n^2h^2}{8mL^2} + \frac{(2l)^2 h^2}{8m\,(2\pi R)^2},$$

[21] M. Fusaro, *J.Comput.Theor.Nanoscience, 6*, 1175 (2009). A general approach to infinite nanotubes using their translational symmetry (see Chapter 9) is described in D.J. Klein, W.A. Seitz, and T.G. Schmalz, *J. Phys. Chem., 97*, 1231 (1993).

which means that we have a sum of energy for the box with ends and of the cyclic box (for the length of the box equal to $2\pi R$). For $l \neq 0$, we get double degeneracy; for $l = 0$, the levels are non-degenerate.

Thus, a particle confined on the side surface of the cylinder belongs also to those problems of quantum mechanics that can be solved exactly.

Nanotube in More Detail

Does the FEMO orbital model describe a real nanotube? Maybe it is just a nice exercise unrelated to the reality. In contemporary quantum chemistry, we have a toolbox with many methods that are much more accurate than FEMO. One of these methods is known as *density functional theory* (DFT; see Chapter 11) and proved to be a reliable tool. Fusaro carried out DFT computations of what are called *reactivity indices*,[22] which measure which positions on the nanotube surface are more reactive than other ones. These particular reactivity indices have been associated with what is known as the *Fukui function*,[23] which describes the rate at which the electric charge *at a particular point* changes when the total electric charge of the nanotube varies. Therefore, this is an indication where an attacking ion would be directed when it approaches the nanotube (it will choose the place with the largest charge that is opposite to its own charge). It turned out that the Fukui function for the nanotube as a function of position changes quasi-periodically (not counting the end effects) along the nanotube, as well as around the nanotube axis. The values of the Fukui function calculated within the FEMO model and those computed by the much more advanced DFT approach agreed semi-quantitatively. Both these results reveal the abovementioned quasi-periodicity.

Such results are encouraging. We see that at least in this case, we can understand the machinery of the world by seeing that its main features appear from something very simple–a kind of primitive theory in which we control and understand every detail. A more sophisticated theory may add some new (and sometimes useful and important) features, but in many cases, such a theory does not have the power of showing convincingly why something happens.

4.5 Single Barrier

Is it possible to pass through a barrier with less energy than the barrier height? Yes, as we will soon see.

4.5.1 Tunneling Effect Below the Barrier Height

Let us imagine a rectangular potential energy barrier (as shown previously in Fig. 4.1): $V(x) = V_0$ for $0 \leq x \leq a$, with $V(x) = 0$ for other values of x ($V_0 > 0$ is a number). Let us assume a

[22] M. Fusaro, *J. Comp. Theor. Nanoscience.*, *11*, 2393 (2010).

[23] K. Fukui, T. Yonezawa, and H. Shingu, *J. Chem. Phys.*, *20*, 722 (1952); K. Fukui, T. Yonezawa, and C. Nagata, *J. Chem. Phys.*, *21*, 174 (1953); T. Yonezawa and C. Nagata, *Bull. Chem. Soc. Japan*, *27*, 423 (1954).

particle of mass m going from the negative values of x (i.e., from the left side), with its kinetic energy equal to E.

The x axis will be divided into three regions:

1. $-\infty < x < 0$, then $V(x) = 0$.
2. $0 \le x \le a$, then $V(x) = V_0$.
3. $a < x < \infty$, then $V(x) = 0$.

So, the barrier region has length a. In each of these regions, the Schrödinger equation will be solved, and then the solutions will be stitched together in such a way as to make it smooth at any boundary between the regions. The solutions for each region separately can be written very easily as[24]

$$\Psi_1(x) = A_1 e^{i\kappa_0 x} + B_1 e^{-i\kappa_0 x}, \tag{4.10}$$
$$\Psi_2(x) = A_2 e^{i\kappa x} + B_2 e^{-i\kappa x}, \tag{4.11}$$
$$\Psi_3(x) = A_3 e^{i\kappa_0 x} + B_3 e^{-i\kappa_0 x}, \tag{4.12}$$

where A and B represent amplitudes of the de Broglie waves running right and left, respectively, where in regions 1 and 3 $\kappa_0^2 = \frac{2mE}{\hbar^2}$, and in region[25] 2: $\kappa^2 = \frac{2m(E-V_0)}{\hbar^2}$. The quantities $|A_i|^2$ and $|B_i|^2$ are proportional to the probability of finding the particle going right ($|A_i|^2$) and left ($|B_i|^2$) in the region i.

In regions 1 and 2, one may have the particle going right, but also left because of reflections from the boundaries, hence in 1 and 2 A and B will have nonzero values. However, in region 3, we will have $B_3 = 0$ because there will be no wave going back (it has nothing to reflect on). We have to work on choosing such A and B, as to wave functions of connected regions match at the boundaries (the functions should "*meet*") and, at each boundary, have the same value of the first derivative (they should meet smoothly). Satisfaction of these requirements is sufficient to determine the ratios of the coefficients, and this is what we are looking for when aiming at describing what will happen more often than something else.

Therefore, such a perfect stitching means that for $x = 0$ and $x = a$, one has

$$\Psi_1(x=0) = \Psi_2(x=0),$$
$$\Psi_2(x=a) = \Psi_3(x=a)$$
$$\Psi_1'(x=0) = \Psi_2'(x=0)$$
$$\Psi_2'(x=a) = \Psi_3'(x=a).$$

[24] This is the wave function for a free particle. The particle has the possibility (and therefore also the probability) to move left or right. The formulas are natural because in a particular region, the potential energy represents a constant [in regions 1 and 3, $V(x) = 0 = const$, while in region 2, $V(x) = V_0 = const'$].

[25] We will obtain these formulas after solving in each region the Schrödinger equation $\frac{\partial^2 \Psi}{\partial x^2} + \kappa^2 \Psi = 0$.

Using Eqs. (4.10) through (4.12), we may rewrite this as

$$A_1 + B_1 = A_2 + B_2$$
$$A_2 e^{i\kappa a} + B_2 e^{-i\kappa a} = A_3 e^{i\kappa_0 a}$$
$$\kappa_0 A_1 - \kappa_0 B_1 = \kappa A_2 - \kappa B_2$$
$$A_2 \kappa e^{i\kappa a} - B_2 \kappa e^{-i\kappa a} = A_3 \kappa_0 e^{i\kappa_0 a}.$$

We have five unknowns: A_1, B_1, A_2, B_2, and A_3, but only four equations to determine them. Note that we are interested only in $\frac{|A_3|^2}{|A_1|^2} \equiv D$, because this is the probability of passing the barrier assuming that the particle was sent from the left. Therefore, after dividing all equations by A_1, we have only four unknowns: $\frac{B_1}{A_1} \equiv b_1$, $\frac{A_2}{A_1} \equiv a_2$, $\frac{B_2}{A_1} \equiv b_2$, $\frac{A_3}{A_1} \equiv a_3$, and four equations to determine them:

$$1 + b_1 = a_2 + b_2,$$
$$a_2 e^{i\kappa a} + b_2 e^{-i\kappa a} = a_3 e^{i\kappa_0 a},$$
$$\kappa_0 - \kappa_0 b_1 = \kappa a_2 - \kappa b_2,$$
$$a_2 \kappa e^{i\kappa a} - b_2 \kappa e^{-i\kappa a} = a_3 \kappa_0 e^{i\kappa_0 a}.$$

The solution of this set of equations (for a given $E > 0$) gives the transmission coefficient $D(E) = |a_3|^2$ as a function of energy:

$$D = \frac{1}{1 + \frac{1}{4\beta(\beta-1)} \sin^2 \kappa a}, \tag{4.13}$$

where $\beta = \frac{E}{V_0}$ is a ratio of the impact energy and the barrier height. The quantity $\kappa = \sqrt{\frac{2m(E-V_0)}{\hbar^2}}$ is positive for $E > V_0$, and we use directly Eq. (4.13). For $E < V_0$, there is a problem because κ is imaginary [$\kappa = \sqrt{\frac{2m(E-V_0)}{\hbar^2}} = ik$, where $k = \sqrt{\frac{2m(V_0-E)}{\hbar^2}} > 0$]. However, calculation of D is simple again; for $E < V_0$, we obtain[26]

$$D = \frac{1}{1 + \frac{1}{4\beta(1-\beta)} sh^2 ka}. \tag{4.14}$$

[26] This is because $\sin \kappa a = \frac{\exp(i\kappa a) - \exp(-i\kappa a)}{2i} = \frac{\exp(-ka) - \exp(+ka)}{2i} = -\frac{1}{i} sh(ka) = i\ sh(ka)$, so $\sin^2 \kappa a = -sh^2(ka)$.

When $E = V_0$, there is a problem because $\beta = 1$ and $\frac{1}{(\beta-1)}\sin^2\kappa a$ is an expression of the type $\frac{0}{0}$ (i.e., division by zero). Using, however, the de l'Hospital rule, we get[27]

$$D = \frac{1}{1 + \frac{ma^2}{2\hbar^2}V_0}.$$

Here, common sense tells us that if $E > V_0$, the particle certainly will pass the barrier: we say that it will go "*over the barrier.*" For the time being, let us just consider the cases for which the energy is smaller than the barrier: $E < V_0$. If our pointlike particle behaved according to classical mechanics, it would not pass the barrier. It is like a car having velocity[28] v and mass m (therefore, its kinetic energy $E = \frac{mv^2}{2}$) cannot pass a hill of height h, if $E < V_0$ with $V_0 = mgh$ (g is the gravitational acceleration). Let us see how it will look in quantum mechanics. From Eq. (4.14), it can be seen that for any $E < V_0$, we have $D > 0$ except if[29] $E = 0$. This means that the particle can pass through a wall like the car in our analogy passed through a tunnel made in the hill. This intriguing phenomenon is known as *tunneling*.[30]

Let us see what the effectiveness of the tunneling (D) depends on. Fig. 4.6 shows that

for tunneling,

- The larger the impact energy of the particle, the larger D is.
- A higher barrier (V_0) or a wider barrier (larger a) decreases D.
- Tunneling of a lighter particle is easier.

[27] The de l'Hospital rule gives a series of transformations

$$\frac{1}{(\beta-1)}\sin^2\kappa a = \lim_{E\to V_0}\frac{\sin^2\left(\sqrt{\frac{2m(E-V_0)}{\hbar^2}}a\right)}{\frac{E}{V_0}-1}$$
$$= \lim_{E\to V_0}\frac{2\sin\left(\sqrt{\frac{2m(E-V_0)}{\hbar^2}}a\right)\frac{\frac{2m}{\hbar^2}}{2\sqrt{\frac{2m(E-V_0)}{\hbar^2}}a}a^2}{\frac{1}{V_0}} = \frac{2ma^2}{\hbar^2}V_0.$$

[28] Here, we assume that the engine is switched off.

[29] In such a case, in Eq. (4.14), $\beta = 0$, with the consequence that $D = 0$.

[30] The wonder I am trying to convince you of will fade a bit, however. What kinetic energy is really necessary for an object to be transported over the hill? Should it be always equal to or greater than mgh? Well imagine that our car is divided into a series of pieces (say 1000 of them), tied one to the next by thin threads of a certain length. What about transporting the car now? Well, we will need to expend energy to transport the first piece to the top of the barrier (say $\frac{mgh}{1000}$), but the second piece would be then pulled by the first one descending behind the hill! The same happens for other pieces. We see therefore, that when the object to transport is not pointlike, but instead has some dimension in space, the energy cost of passing the barrier may be smaller! A quantum particle is described by a wave function extended in space; therefore, its passing the barrier is possible even if the particle itself has insufficient kinetic energy. This is what we observe in the tunneling effect.

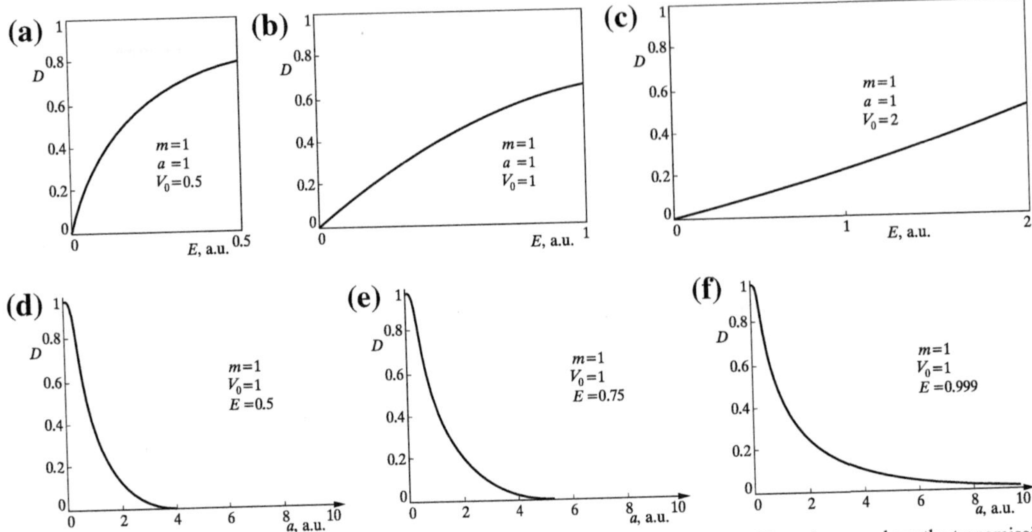

Fig. 4.6. A classical particle cannot tunnel through a barrier, while a quantum particle *can*. These images show the transmission coefficient (tunneling) of the electron having various energies (always lower than the barrier) and passing through a barrier of various heights and widths. Panels (a)–(c) show that the greater the energy, the easier it is to tunnel, and the higher the barrier, the harder it is to pass the barrier (at the same energy of the particle). Panels (d)–(f) show the dependence of the transmission coefficient on the barrier width: the wider the barrier, the harder it is to go through.

All these changes are monotonic. This means there are no magic values of the impact energy, of the barrier height or width, for which the tunneling would be easier.

What about the wave function for a tunnelling particle ($E < V_0$)? The answer is in Fig. 4.7. As one can see:

- The real as well as the imaginary parts of the wave function do not equal zero along the barrier (i.e., the particle will appear there).
- The real as well as the imaginary parts of the wave function vanish exponentially along the barrier.
- The latter means that since the barrier has a finite width, the wave function will not vanish at the exit from the barrier region. Then, after going out of the barrier region, the wave function again begin to oscillate having the same wavelength, *but a smaller amplitude*! This means that the particle passes the barrier with a certain probability p and reflects from the barrier with probability $1 - p$.

4.5.2 Surprises for Energies Larger than the Barrier

Now let us see what will happen if the impact energy E is larger than the barrier height V_0? The transmission coefficient is given by the general Eq. (4.13).

Fig. 4.8 shows the probability of passing the barrier (D) as a function of E. A common sense suggests that, if a particle has kinetic energy larger than the barrier height (V_0), it will always pass the barrier. We have, however, a surprise: it passes the barrier with probability

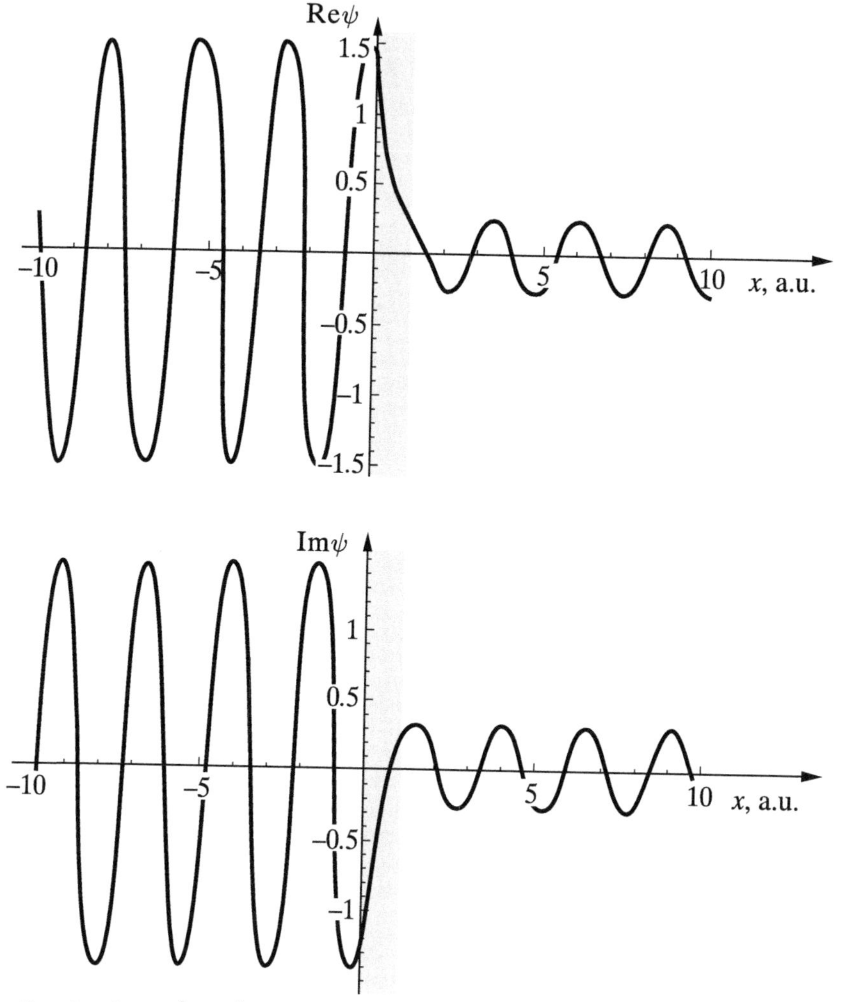

Fig. 4.7. Tunneling of an electron ($m = 1$) with energy $E = 2.979$ a.u. through a single barrier of height 5 a.u., and width 1 a.u. The wave function plot (real and imaginary parts) corresponds to the following values of coefficients $A_1 = 1$ (as a reference); $B_1 = 0.179 - 0.949i$; $A_2 = 1.166 - 0.973i$; $B_2 = 0.013 + 0.024i$; $A_3 = -0.163 - 0.200i$ and represents a wave.

equal to 100% only for certain particular impact energies E, when κ satisfies $\kappa a = n\pi$, where $n = 1, 2, \ldots$ For other values of κ, the particle has only a certain chance to pass (see Fig. 4.8). This result is a bit puzzling. What is special in these magic values of κ, for which the particle does not see the barrier? For the time being we do not know, but for now, here is something interesting. For these magic impact energies, we have $\frac{2m(E-V_0)}{\hbar^2}a^2 = n^2\pi^2$ or

$$E \equiv E_n = V_0 + \frac{n^2h^2}{8ma^2}, \tag{4.15}$$

but this means that energy levels of the particle are in a box of a length that is equal to the barrier width and placed exactly on top of the barrier! This seems unbelievable. The barrier

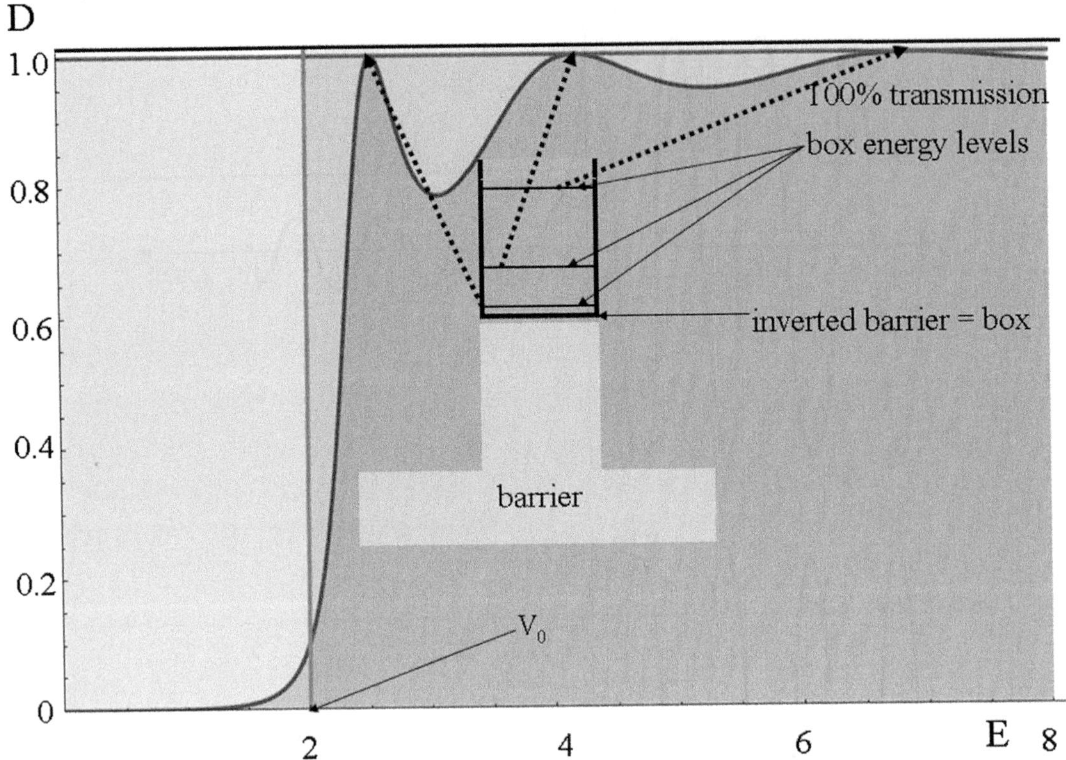

Fig. 4.8. Probability (D) of passing the barrier by a particle (its mass is 1 a.u. and is equal to the mass of an electron, the barrier: height $V_0 = 2$ a.u., width $a = 3$ a.u.) as a function of the impact energy (E). For $E < V_0$, one has the tunneling probability D increasing with E. The vertical line corresponds to $E = V_0$; for a particle with this energy, the probability of passing the barrier is only about 0.1. Then, for $E > V_0$ (i.e., the impact energy being larger than the barrier height), the probability of reflecting is generally large. However, if the impact energy matches a particle-in-a-box eigenvalue of energy, the transmission coefficient increases to 100%.

and the box of the same length represent extremes in their character. But despite of that, here is the miracle coming true! It looks as if the de Broglie wave describing the particle should fit a half-integer number of its wavelengths within the region of perturbation (either the energy barrier or the energy well). Only then does the wave not lose its rhythm, a fact important for constructive interference, and only then the particle passes over the barrier as if without seeing it. It is interesting that to get this effect, one has to have $E > V_0$. There was no such effect for tunneling (for $E < V_0$); the interference was impossible due to the exponential decay of the wave function instead of its oscillatory behavior.

Here is one more intriguing problem. Eq. (4.15) tells us that if the impact energy increases, the magiclike 100% passages will be rarer and rarer. We might expect a reverse behavior, because for very large E, the particle should not pay attention to the barrier and pass through it without seeing it. Fig. 4.8 makes the situation clear: the 100% passages indeed will be rarer and rarer with increasing E, but it will be easier and easier to pass through for impact energies, which do *not* satisfy $\kappa a = n\pi$. For large values of the impact energies, the particle practically will not see the barrier (in agreement with common sense).

4.6 The Magic of Two Barriers

If we take *two* rectangular barriers of height V_0 with a well between them (Fig. 4.1c), then we will also have some magic. This time, we allow for any energy of the particle ($E > 0$).

How Will the Problem Be Solved?

We have five non-overlapping sections of the x-axis. In each section, the wave function will be assumed to be in the form $\Psi(x) = Ae^{i\kappa x} + Be^{-i\kappa x}$ with some A and B coefficients, and with the corresponding values of $\kappa^2 = \frac{2m(E-V)}{\hbar^2}$. In section 5, however, the particle goes right and never left; hence $B_5 = 0$. Now, the other coefficients A and B will be determined by stitching the wave function nicely at each of the four boundaries in order to have it move smoothly through the boundary (the wave function values and the first derivative values have to be equal for the left and right sections to meet at this boundary). In this way, we obtain a set of eight linear equations with eight unknown ratios: $\frac{A_i}{A_1}, i = 2, 3, 4, 5$ and $\frac{B_i}{A_1}, i = 1, 2, 3, 4$. The most interesting ratio is A_5/A_1 because this coefficient determines the transmission coefficient through the two barriers. Using the program Mathematica,[31] we obtain an amazing result.

Transmission Coefficient

Let us check how the transmission coefficient (which in our case is identical to the transmission probability) changes through two identical barriers of height $V_0 = 5$ (all quantities given in a.u.), each of width $a = 1$, when increasing the impact energy E from 0 to $V_0 = 5$. In general, the transmission coefficient is very small. For example, for $E = 2$, the transmission coefficient through the *single* barrier (D_{single}) amounts to 0.028; that is, the chance of transmission is about 3%, while the transmission coefficient through the double barrier (D_{double}) is equal to 0.00021 (i.e., about 100 times smaller). It stands to reason, it is encouraging. It is fine that it is harder to cross two barriers than a single barrier.[32] And the story will certainly be repeated for other values of E. To be sure, let us scan the whole range $0 \leq E < V_0$. The result is shown in Fig. 4.9 and represents a surprise.

4.6.1 Magic Energetic Gates (Resonance States)

There is something really exciting going on. In our case, we have three energies $E \leq V_0$, at which the transmission coefficient D_{double} increases dramatically in a narrow energy range. These energies are 0.34, 1.364, and 2.979. Thus, there are three secret energetic gates for going

[31] See the Web Annex at booksite.elsevier.com/978-0-444-59436-5, and the file Mathematica\Dwiebar.ma.

[32] This is even more encouraging for a prison governor; of course, a double wall is better than a single one.

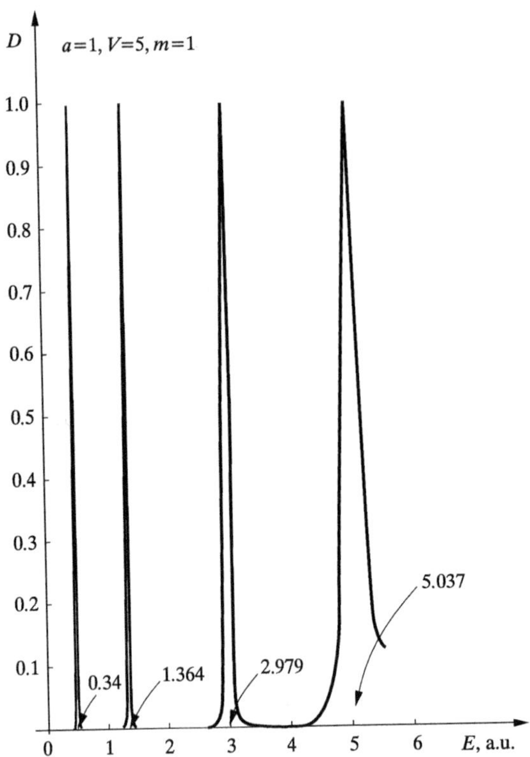

Fig. 4.9. The transmission coefficient (D) for a particle going through a potential double barrier (of height 5 a.u.) as a function of the particle impact energy E. We see some sudden increases of the transmission coefficient (resonance states).

through the double barrier. It is sufficient just to hit the right energy (resonance energy). Is the chance of passing two barriers large? Let us take a look. For all three resonances, the transmission coefficient is equal to $D_{\text{double}} = 1$, but it drops down differently when going off resonance. Thus, there are three particle energies, for which the particle goes through the two barriers like a knife through butter, as if the barriers did not exist.[33] Moreover, as we can see for the third resonance, the transmission coefficient through the single barrier amounts to $D_{single} = 0.0669$ (i.e., only 7%), but through two barriers, it is 100%. It looks as if it would be hard for a prisoner to pass through a single armored prison door, but when the anxious prison governor made a second armored door behind the first, the prisoner (trained in quantum mechanics) disappeared through the two doors like a ghost.[34]

So what happens over there? Let us stress once more that the phenomenon is 100 % of a quantum nature because a classical particle would tunnel neither through the double nor through

[33] This news should be strictly confidential in penitentiary departments. This phenomenon was described first by David Bohm, *Quantum Theory*, Prentice-Hall, New York (1951).

[34] There is experimental evidence for such resonance tunneling through two energy barriers in semiconductors. One of the first reports on this topic was a paper by T.C.L.G. Soliner, W.D. Goodhue, P.E. Tannenwald, C.D.Parker, and D.D.Peck, *Appl. Phys. Lett., 43,* 588 (1983).

the single barrier. Why do we observe such dramatic changes in the transmission coefficient for the two barriers? We may have some suspicions. From the time the second barrier is created, a new situation appears: a well *between* the two barriers, something similar to the box discussed earlier.[35] A particle-in-a-box has some peculiar energy values: the energies of the stationary states (cf., p. 163). In our situation, all these states correspond to a continuum, but something magic might happen if the particle had just one of these energies. Let us calculate the stationary-state energies assuming that $V_0 = \infty$. Using the atomic units in the energy formula, we have $E_n = \frac{h^2}{8m}\frac{n^2}{L^2} = \frac{\pi^2}{L^2}\frac{n^2}{2}$. To simplify the formula even more, let us take $L = \pi$. Finally, we have simply $E_n = \frac{n^2}{2}$. Hence, we might expect something strange for the energy E equal to $E_1 = \frac{1}{2}$, $E_2 = 2$, $E_3 = \frac{9}{2}$, $E_4 = 8$ a.u., etc. (the last energy level, $E_4 = 8$, is already higher than the barrier height). Note, however, that the resonance states obtained appear at quite different energies: 0.34, 1.364, and 2.979.

But perhaps this intuition nevertheless contains a grain of truth. Let us concentrate on E_1, E_2, E_3, and E_4. One may expect that the wave functions corresponding to these energies are similar to the ground-state (nodeless), first (single node), and second (two nodes) excited states of the particle-in-a-box. What then happens to the nodes of the wave function for the particle going through two barriers? Here are the plots for the off-resonance (Fig. 4.10) and resonance (of the highest energy; see Fig. 4.11) cases.

These figures and similar figures for lower-energy resonances support the hypothesis: if an integer number of the half-waves of the wave function fit the region of the "*box*" between the barriers ("*barrier-box-barrier*"[36]), in this case, we may expect resonance–a secret gate to go through the barriers.[37] As we can see, indeed we have been quite close to guessing the reason for the resonances. On the other hand, it turned out that the box length should include not only the box itself, but also the barrier widths. Perhaps to obtain the right resonance energies, we simply have to adjust the box length. Since, instead of resonance at $E_1 = \frac{1}{2}$, we have resonance at energy 0.34, then we may guess that it is sufficient to change the box width L to $L' = \sqrt{\frac{0.5}{0.34}}L = 1.21L$, to make the first resonance energies match. Then, instead of $E_1 = \frac{1}{2}$, we have exactly the first resonance energy equal to $E_1' = 0.34$. This agreement was forced by us, but later, instead of $E_2 = 2$ we obtain $E_2' = 1.36$, which agrees very well with the second resonance energy 1.364. Then, instead of $E_3 = 4.5$, we obtain $E_3' = 3.06$, a good approximation

[35] Note, however, that the box has finite well depth and finite width of the walls.

[36] We see one more time that a quantum particle is a kind of magic object. Even its passing over a single barrier revealed already some unexpected difficulties. Now we see some strange things happen for two barriers. It looks as if the particle's passing (through a single or a double barrier) goes smoothly given the condition that the particle does not "*lose its rhythm*" in the barrier region. This "*not losing the rhythm*" reflects its wave nature and means fitting an integer number of its de Broglie half-waves within the obstacle region.

[37] As one can see in this case, contrary to what happened with a single barrier, the wave function does not vanish exponentially within the barriers.

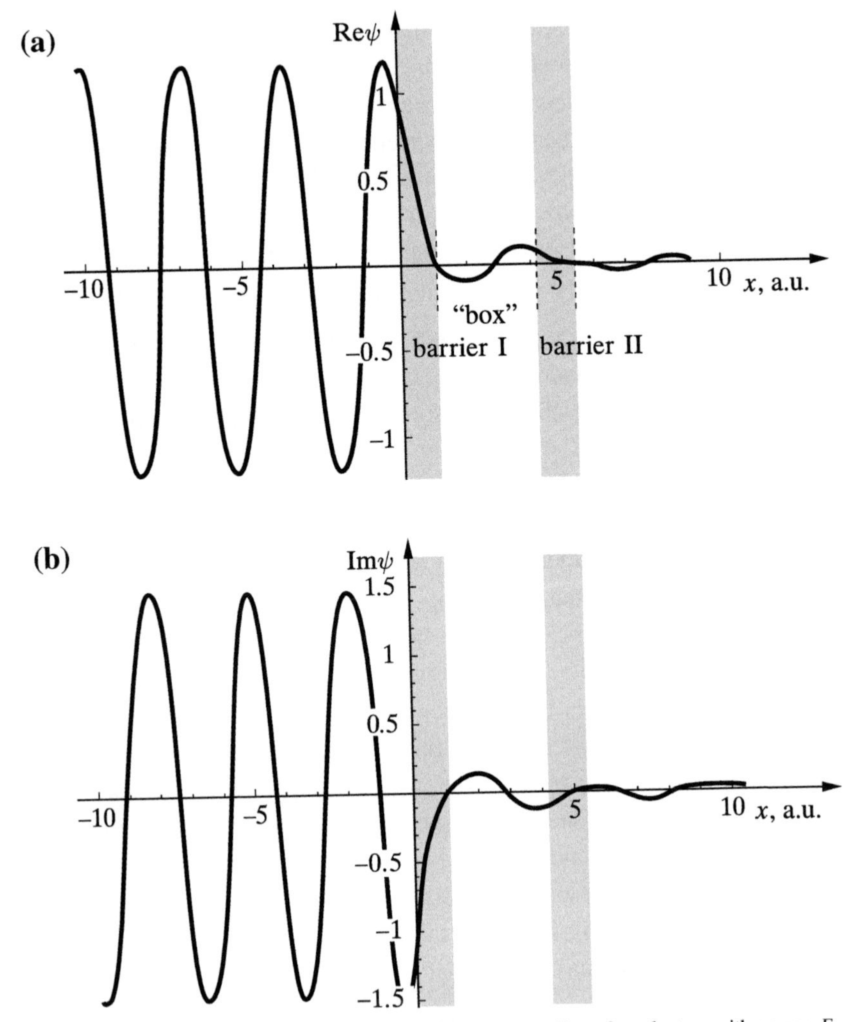

Fig. 4.10. The off-resonance case below the barriers' height. With the tunneling of an electron with energy $E = 2$ a.u. through two barriers of height $V_0 = 5$ and width $a = 1$, the barrier separation is $L = \pi$ (all quantities in a.u.). The real part of the wave function (a) oscillates before the first barrier and is reduced by an order of magnitude in the first barrier. Between the barriers, the function oscillates for about one period, decays in the second barrier, and goes out of the barrier region with an amplitude representing about 5% of the starting amplitude. A similar picture follows from the imaginary part of the wave function (b).

to 2.979, but evidently the closer the barrier energy, the harder it is to obtain agreement.[38] The next resonance state is expected to occur at $E'_4 = 8 \times 0.68 = 5.44$, but we have forgotten that this energy already exceeds the barrier height ($V_0 = 5$ a.u.). We will come back to this state in a moment.

[38] Note, please, that resonance width is different for each resonance. The most narrow resonance corresponds to the lowest energy, the widest to the highest energy. The width of resonances is related to the notion of the resonance lifetime τ (τ is proportional to the inverse of the resonance width).

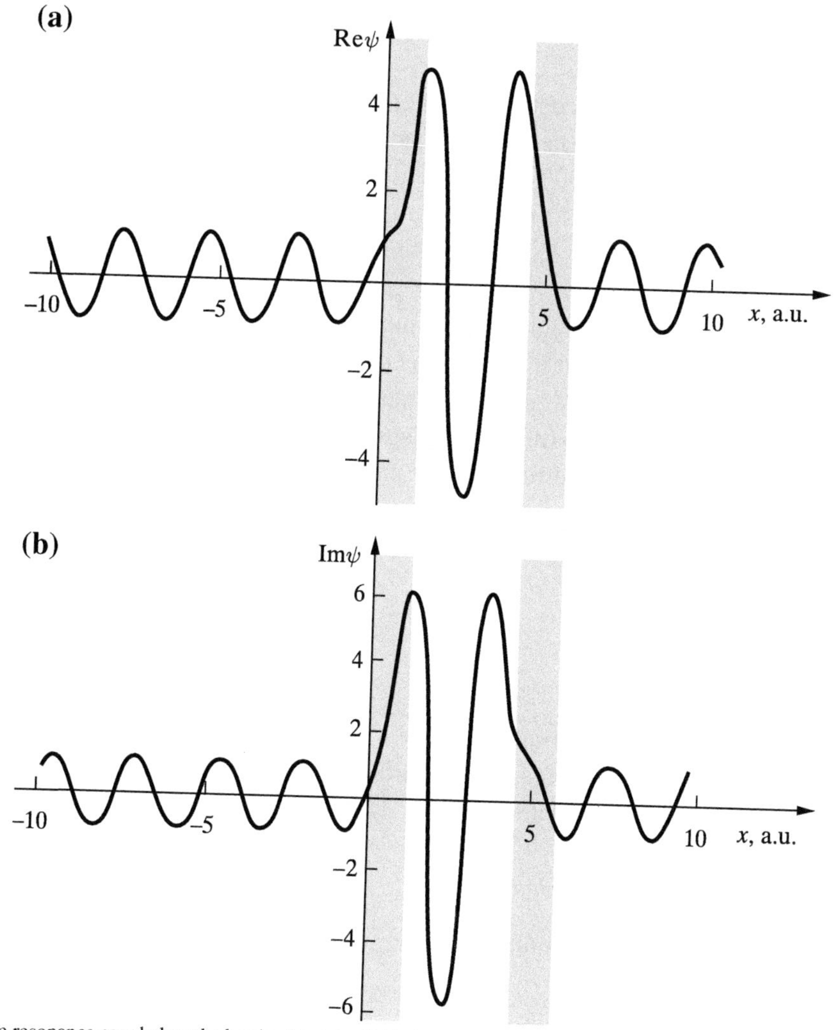

Fig. 4.11. The resonance case below the barriers' height. With the tunneling of an electron with energy $E = 2.979$ a.u. through two barriers of height $V_0 = 5$ and width $a = 1$, the barrier separation is $L = \pi$ (all quantities in a.u.). The real part of the wave function (a) oscillates before the first barrier with amplitude 1 and increases by a factor of about 3.5 within the first barrier. Between the barriers, the function makes slightly more than about one period, decays in the second barrier, and goes out of the barrier region with an amplitude representing about 100% of the starting amplitude. A similar picture follows from the imaginary part of the wave function (b).

4.6.2 Strange Flight Over the Barriers

Let us consider the two barriers and an electron with higher energy than the barrier height V_0. What will happen? Well, we may say that this means the particle energy is sufficient to pass the barrier. Let us see.

Let us assume the barrier height $V_0 = 5$ and the particle energy is equal to 5.5 a.u. We solve our equations and we obtain a transmission coefficient equal to 0.138; hence the electron will bounce back with a probability of about 86%. How it did bounce off? That's hard to say.

Fig. 4.9 shows the transmission coefficient for energies higher than the barrier height. It turns out that at energy $E = 5.037$ a.u. (i.e., higher than the barrier height), another resonance state is hidden, which assures almost 100% certainty of transmission (whereas the particle energies in the energetic neighborhood lead to a considerable reflection rate, as described above). We expected such behavior for all $E > V_0$, but it turned out to be true for the resonance state. Let us recall that we have already predicted "*by mistake*" a box stationary state with energy $E_4' = 5.44$, which is higher than the barrier height V_0. This, and the number of the nodes within the barrier range seen in Fig. 4.12, tells us that indeed this is the state.[39]

What makes the difference between the resonance and off-resonance states for $E > V_0$? The corresponding wave functions (real and imaginary parts) are given in Figs. 4.12 and 4.13.

Thus, resonance states may also hide in that part of the continuum that has energy higher than the barriers (with a short lifetime because such resonances are wide; cf., Fig. 4.9). They are also a reminder of the stationary states of the particle in a box longer than the separation of the barriers and infinite well depth.

4.7 Harmonic Oscillator

A 1-D harmonic oscillator is a particle of mass m, subject to thc force $-kx$, where the force constant $k > 0$, and x is the deviation of the particle from its equilibrium position[40] ($x = 0$). The potential energy is given as a parabola $V = \frac{1}{2}kx^2$.

New Variable and the Transformed Hamiltonian

First, let us write down the Hamiltonian:

$$\hat{H} = -\frac{\hbar^2}{2m}\frac{d^2}{dx^2} + \frac{1}{2}kx^2.$$

[39] It corresponds to a lower energy than we predicted (similar to the case of E_3). No wonder that due to finite well depth, the states corresponding to the upper part of the well "*feel*" the box is longer.

[40] A harmonic oscillator represents a single particle. Sometimes, however, we mean by it *two* pointlike particles bound together by a spring, having a certain equilibrium length, and when the length deviates from the equilibrium, the potential energy is proportional to the square of the deviation. This two-particle model looks much more attractive to us since diatomic molecules, like H_2, HCl, etc., resemble it. As a consequence, one might apply to diatomics conclusions from the solution of the Schrödinger equation for the harmonic oscillator. The diatomics will be discussed in detail in Chapter 6. It will turn out there that a relative motion of two atoms can indeed be reduced (in a certain approximation) to a harmonic motion of a single particle; however, its deviation from the equilibrium position cannot be lower than $-R_0$, where R_0 stands for the equilibrium length of the diatomic. The reason is very simple: even if the spring connecting the two atoms were an ideal one, the harmonicity can pertain to extending the spring, but certainly not to squeezing it. When squeezing by $-R_0$, the two atoms seat one on the other, and this is the end of squeezing at all. This means that one cannot have a harmonic molecule for some basic reasons.

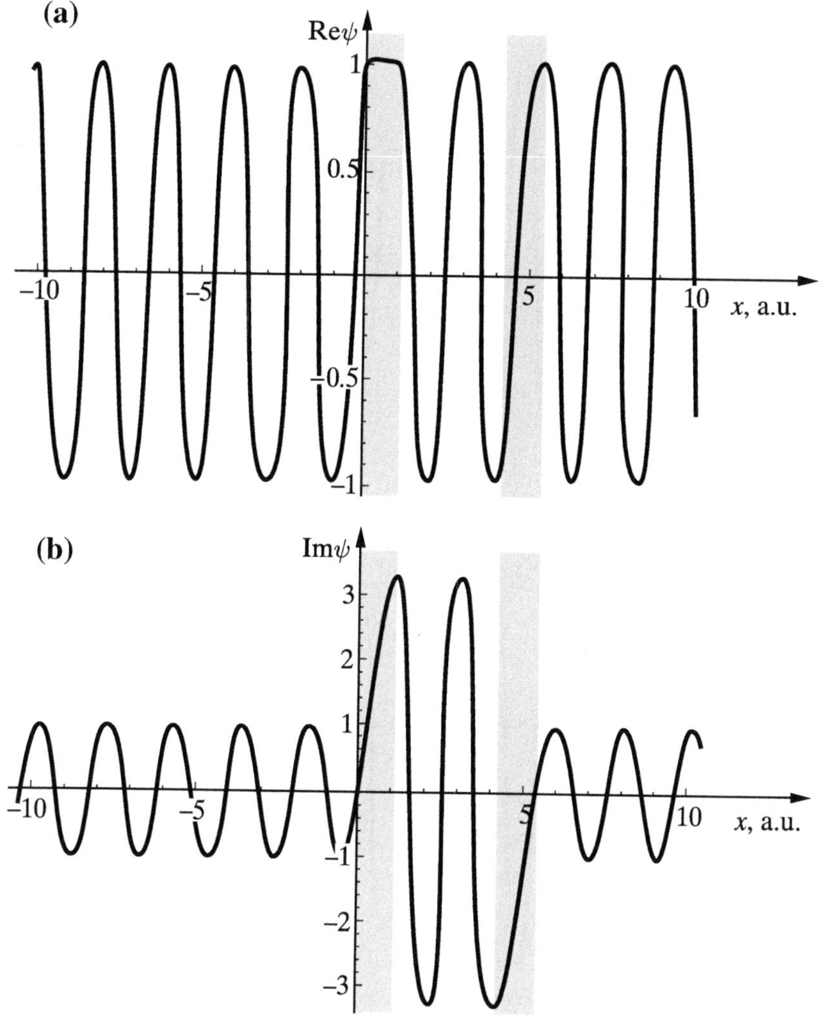

Fig. 4.12. The resonance case over the barriers. The wave function for an electron with energy $E = 5.037$ a.u.; i.e., over the barrier $V_0 = 5$. As we can see, the amplitude is nearly the same for the wave function before and after the barriers (this means the transmission coefficient of the order of 100%). The real part, and especially the imaginary part, wobble within the range of the barrier's range; i.e., within section (0, 5.14). (Note: the imaginary part has a large amplitude.) We may guess that the state is related to the three-node stationary state.

Now, let us introduce a very useful scaled coordinate: $q = \sqrt[4]{\frac{km}{\hbar^2}}x$, with $-\infty < q < \infty$. The Hamiltonian written with using the new variable reads very interestingly as $\left(\omega = \sqrt{\frac{k}{m}}\right)$

$$\hat{H} = -\frac{\hbar^2}{2m}\left(\frac{dq}{dx}\frac{d}{dq}\right)^2 + \frac{1}{2}k\frac{1}{\sqrt{\frac{km}{\hbar^2}}}q^2 = -\frac{1}{2}\hbar\sqrt{\frac{k}{m}}\frac{d^2}{dq^2} + \frac{1}{2}\hbar\sqrt{\frac{k}{m}}q^2 = \hbar\omega\left[-\frac{1}{2}\frac{d^2}{dq^2} + \frac{1}{2}q^2\right]. \tag{4.16}$$

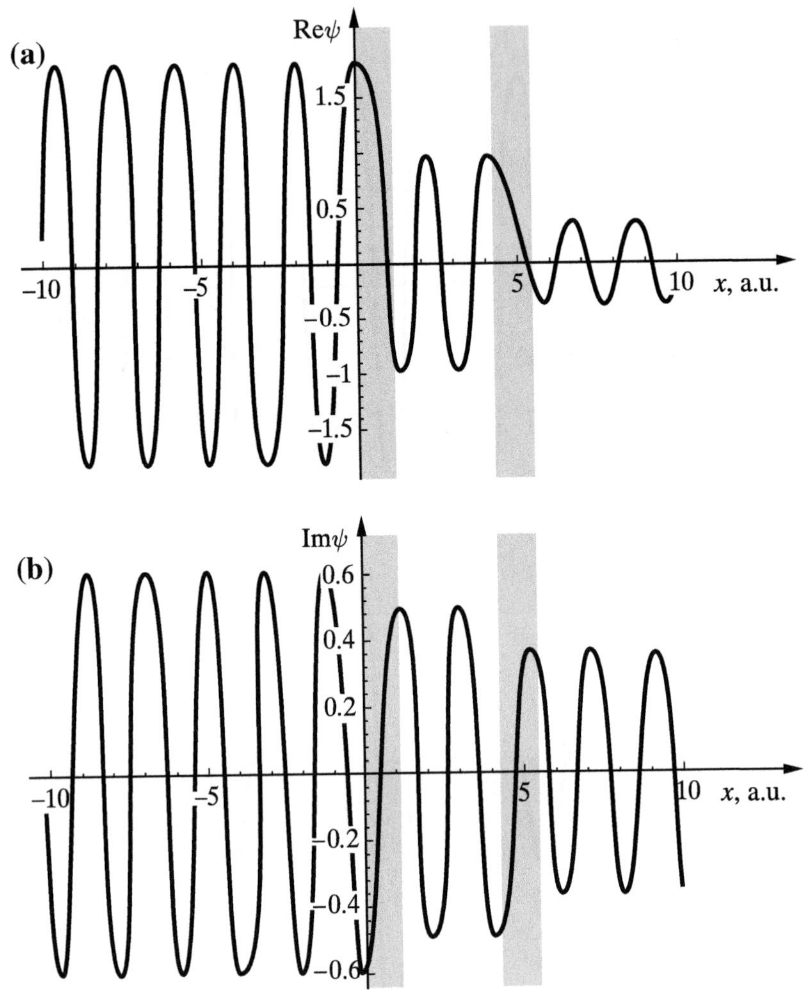

Fig. 4.13. The off-resonance case over the barriers; the wave function for an electron ($E = 5.5$ a.u.; i.e., *over* the barrier height $V_0 = 5$). Despite the fact that $E > V_0$, the amplitude of the outgoing wave is considerably reduced after passing the range of the barriers (0, 5.14). This means that the particle flying over the barriers may reflect from them.

Creation and Annihilation Operators

Now let us prepare two tools, which are the operators: $\hat{B} = \frac{1}{\sqrt{2}}\left(\frac{d}{dq} + q\right)$, known as a creation operator, and $\hat{b} = \frac{1}{\sqrt{2}}\left(-\frac{d}{dq} + q\right)$, being an annihilation operator (their names will become clear in a minute). It turns out[41] that $\hat{b} = \hat{B}^{\dagger}$. As we will see in a moment, these operators

[41] Indeed, for f and g being arbitrary functions of class Q (please recall that $i\frac{d}{dq}$ represents a Hermitian operator): $\left\langle \hat{b} f | g \right\rangle = \frac{1}{\sqrt{2}} \left\langle (-\frac{d}{dq} + q) f | g \right\rangle = \frac{1}{i}\frac{1}{\sqrt{2}} \left\langle i\frac{d}{dq} f | g \right\rangle + \frac{1}{\sqrt{2}} \langle q f | g \rangle = \frac{1}{i}\frac{1}{\sqrt{2}} \left\langle f | i\frac{d}{dq} g \right\rangle + \frac{1}{\sqrt{2}} \langle q f | g \rangle = \frac{1}{\sqrt{2}} \left\langle f | \frac{d}{dq} g \right\rangle + \frac{1}{\sqrt{2}} \langle q f | g \rangle = \frac{1}{\sqrt{2}} \left\langle f | (\frac{d}{dq} + q) g \right\rangle = \left\langle f | \hat{B} g \right\rangle$, which means $\hat{b} = \hat{B}^{\dagger}$.

will create or annihilate the energy quanta of our system, which will be called *phonons*, each of energy $h\nu = \hbar\omega$.

Let us see what the operator $\hat{B}^\dagger \hat{B}$ does. For an arbitrary function[42] f, we have

$$\hat{B}^\dagger \hat{B} f = \left(\frac{1}{\sqrt{2}}(-\frac{d}{dq} + q)\right)\left(\frac{1}{\sqrt{2}}(\frac{d}{dq} + q)\right) f = -\frac{1}{2}\frac{d^2 f}{dq^2} - \frac{1}{2}\frac{d}{dq}(qf) + \frac{1}{2}q\frac{df}{dq} + \frac{1}{2}q^2 f$$
$$= -\frac{1}{2}\frac{d^2 f}{dq^2} - \frac{1}{2}(f + q\frac{df}{dq}) + \frac{1}{2}q\frac{df}{dq} + \frac{1}{2}q^2 f = \left(-\frac{1}{2}\frac{d^2}{dq^2} + \frac{1}{2}q^2 - \frac{1}{2}\right) f.$$

Hence, we recognize immediately that the Hamiltonian has the form

$$\hat{H} = h\nu(\hat{B}^\dagger \hat{B} + \frac{1}{2}). \tag{4.17}$$

Eigenfunctions and Eigenvalues

Now we will find the eigenfunctions [Eq. (4.22)] and eigenvalues [Eq. (4.21)] of the Hamiltonian $\hat{H}$. To this end, we note first that[43]

$$[\hat{B}, \hat{B}^\dagger] = 1 \text{ or } \hat{B}\hat{B}^\dagger = \hat{B}^\dagger \hat{B} + 1. \tag{4.18}$$

Let us consider the following function:

$$\psi_v = (v!)^{-\frac{1}{2}} (\hat{B}^\dagger)^v \psi_0 \tag{4.19}$$

where ψ_0 stands for the normalized Gaussian function[44]: $\psi_0(q) = \frac{1}{\sqrt[4]{\pi}} \exp(-\frac{1}{2}q^2)$. Let us note first that $\hat{B}\psi_0 = \left(\frac{1}{\sqrt{2}}(\frac{d}{dq} + q)\right) \frac{1}{\sqrt[4]{\pi}} \exp(-\frac{1}{2}q^2) = \frac{1}{\sqrt[4]{\pi}} \left(\frac{1}{\sqrt{2}}(-q + q)\right) \exp(-\frac{1}{2}q^2) = 0$. This

[42] When transforming operators, one has to remember that they act on a function (this is why we consider such a function explicitly); otherwise, it is easy to make a mistake.

[43] We have

$$\hat{B}\hat{B}^\dagger f = \left(\frac{1}{\sqrt{2}}(\frac{d}{dq} + q)\right)\left(\frac{1}{\sqrt{2}}(-\frac{d}{dq} + q)\right) f$$
$$= -\frac{1}{2}\frac{d^2 f}{dq^2} + \frac{1}{2}\frac{d}{dq}(qf) - \frac{1}{2}q\frac{df}{dq} + \frac{1}{2}q^2 f$$
$$= -\frac{1}{2}\frac{d^2 f}{dq^2} + \frac{1}{2}(f + q\frac{df}{dq}) - \frac{1}{2}q\frac{df}{dq} + \frac{1}{2}q^2 f = \left(-\frac{1}{2}\frac{d^2}{dq^2} + \frac{1}{2}q^2 + \frac{1}{2}\right) f,$$

and therefore, we get the commutation relation needed.

[44] We briefly check the normalization: $\int_{-\infty}^{\infty} |\psi_0(q)|^2 dq = \frac{1}{\sqrt{\pi}} \int_{-\infty}^{\infty} \exp(-q^2) dq = \frac{1}{\sqrt{\pi}} \int_{-\infty}^{\infty} \exp(-q^2) dq = \frac{1}{\sqrt{\pi}}\sqrt{\pi} = 1$.

result will be used in a minute. We have also[45]

$$\hat{B}^\dagger \hat{B} \psi_v = v \psi_v. \tag{4.20}$$

Now, recall that what we have in the Hamiltonian [Eq. (4.17)] is just $\hat{B}^\dagger \hat{B}$. Therefore, function ψ_v given by Eq. (4.19) for $v = 0, 1, 2, \ldots$ represents an eigenfunction of the Hamiltonian: $\hat{H}\psi_v = h\nu(\hat{B}^\dagger \hat{B} + \frac{1}{2})\psi_v = E_v \psi_v$, with the energy

$$E_v = h\nu(v + 1/2), \tag{4.21}$$

with $v = 0, 1, 2, \ldots$ Note, that the oscillator energy is never equal to zero and is additive; i.e., the phonons, each of energy $\hbar\omega$, do not interact. The smaller the oscillating mass or the larger the force constant, the larger the energy of the phonon; see Fig. 4.14.

The harmonic oscillator has an infinite number of the energy levels, all of them non-degenerate, their separation is constant and equals to $h\nu$.

The state ψ_0 describes the absence of phonons (i.e., a phonon vacuum); therefore, an attempt of annihilation of the vacuum gives zero: $\hat{B}\psi_0 = 0$. The function $\psi_v = (v!)^{-\frac{1}{2}} (\hat{B}^\dagger)^v \psi_0$, which is a general solution of the Schrödinger equation, represents therefore a state with v non-interacting phonons, each of energy $h\nu$, created from the phonon vacuum (by using v creation operators $\hat{B}^\dagger$).

---

[45] Let us see what we will get when applying the operator $\hat{B}^\dagger \hat{B}$ [and using the commutation relation (4.18)] to the wave function ψ_v:

$$\hat{B}^\dagger \hat{B}\psi_v = (v!)^{-\frac{1}{2}} \hat{B}^\dagger \hat{B}(\hat{B}^\dagger)^v \psi_0 = (v!)^{-\frac{1}{2}} \hat{B}^\dagger \hat{B}\hat{B}^\dagger (\hat{B}^\dagger)^{v-1}\psi_0 = (v!)^{-\frac{1}{2}} \hat{B}^\dagger \left(\hat{B}^\dagger \hat{B} + 1\right)(\hat{B}^\dagger)^{v-1}\psi_0$$
$$= (v!)^{-\frac{1}{2}} (\hat{B}^\dagger)^2 \hat{B}(\hat{B}^\dagger)^{v-1}\psi_0 + (v!)^{-\frac{1}{2}} \hat{B}^\dagger (\hat{B}^\dagger)^{v-1}\psi_0 = (v!)^{-\frac{1}{2}} (\hat{B}^\dagger)^2 \hat{B}(\hat{B}^\dagger)^{v-1}\psi_0 + \psi_v.$$

The commutation relation replaces $\hat{B}\hat{B}^\dagger$ by $\hat{B}^\dagger \hat{B} + 1$. The $\hat{B}^\dagger \hat{B}$ part made that in the sequence of v operators $\hat{B}^\dagger$ in ψ_v the operator $\hat{B}$ has been shifted right by one position (see the first term) and on its right side, it has only $v - 1$ of $\hat{B}^\dagger$ operators. On the other hand, the presence of the unit operator made that we got the second term in the form $1 \cdot \psi_v = \psi_v$. Now we will use once more the commutation relation. Also this time it will cause a one-position right shift of $\hat{B}$ resulting in $(v!)^{-\frac{1}{2}} (\hat{B}^\dagger)^3 \hat{B}(\hat{B}^\dagger)^{v-2}\psi_0$ together with a new additive term ψ_v (altogether we have already $2\psi_v$). This procedure is repeated v times up to the point, when we get (note the first term turns out to be zero): $(v!)^{-\frac{1}{2}} (\hat{B}^\dagger)^{v+1}\hat{B}\psi_0 + v\psi_v = 0 + v\psi_v = v\psi_v$, where we have used the relation $\hat{B}\psi_0 = 0$. Hence, $\hat{B}^\dagger \hat{B}\psi_v = v\psi_v$.

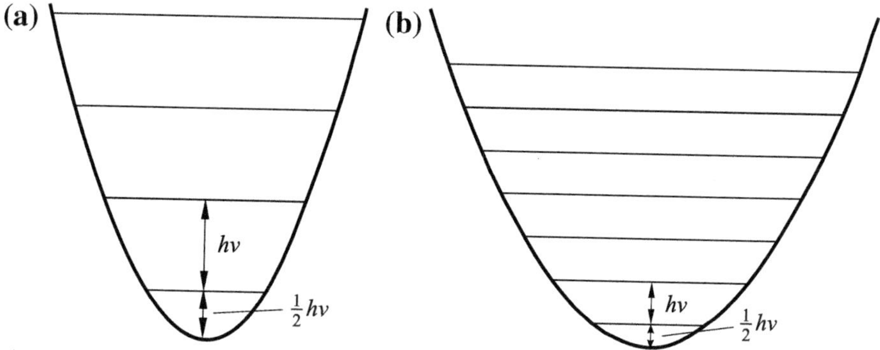

Fig. 4.14. A "*fingerprint*" of the harmonic oscillator–an infinite number of equidistant energy levels. These images represent in fact several independent figures one on top of the other: the parabola means the potential energy as a function of deviation of the oscillating particle from its equilibrium position, and the energy levels denote in principle some points on the axis of the total energy (traditionally and arbitrarily shown as horizontal sections). (a) Energy levels of an oscillator with a large force constant k; (b) energy levels of an oscillator with a small force constant k.

The eigenfunctions of the Schrödinger equation can be also shown explicitly:

$$\Psi_v(q) = N_v H_v(q) \exp\left(-\frac{q^2}{2}\right) \tag{4.22}$$

where $v = 0, 1, 2, \ldots.$ is the oscillation quantum number, H_v represent the Hermite polynomials[46] (of degree v) defined as

$$H_v(q) = (-1)^v \exp(q^2) \frac{d^v \exp(-q^2)}{dq^v},$$

and N_v is the normalization constant $N_v = \sqrt{\left(\frac{\alpha}{\pi}\right)^{\frac{1}{2}} \frac{1}{2^v v!}}$.

The first Hermite polynomials are

$$H_0(q) = (-1)^0 \exp(q^2) \frac{\mathrm{d}^0(\exp(-q^2))}{\mathrm{d}^0 q} = \exp(q^2)\exp(-q^2) = 1,$$

$$H_1(q) = (-1)^1 \exp(q^2) \frac{\mathrm{d}(\exp(-q^2))}{\mathrm{d}q} = -\exp(q^2)\left(-2q\right)\exp(-q^2) = 2q,$$

$$H_2(q) = (-1)^2 \exp(q^2) \frac{\mathrm{d}^2(\exp(-q^2))}{\mathrm{d}q^2} = 4q^2 - 2, \ldots$$

Figs. 4.15 shows how the wave functions for the 1-D harmonic oscillator look like, and the plots for a 2-D harmonic oscillator (one obtains the solution by a simple separation of variables,

[46] Charles Hermite was a French mathematician (1822–1901), professor of Sorbonne. The Hermite polynomials were defined half a century earlier by Pierre Laplace.

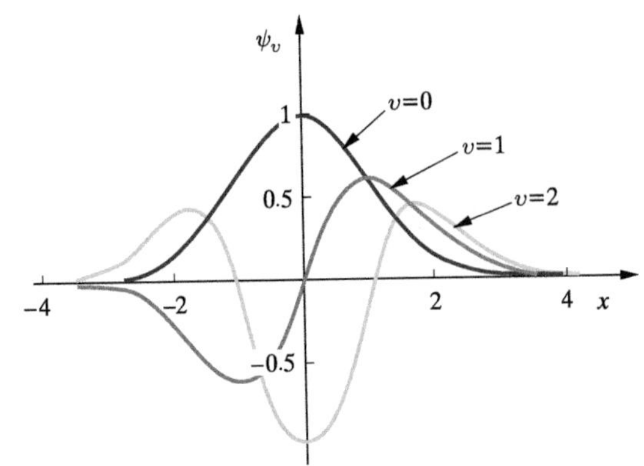

Fig. 4.15. Some of the wave functions Ψ_v for a 1-D oscillator. The number of nodes increases with the oscillation quantum number v.

the wave function is a product of the two wave functions for the harmonic oscillators with x and y variables, respectively) (see Fig. 4.16).

The harmonic oscillator is one of the most important and beautiful models in physics. When almost nothing is known, except that the particles are held by forces, then the first model considered is the harmonic oscillator. This happened for the black body problem (discussed in Chapter 1), and now it is the case with the quantum dots,[47] string theory,[48] solvated electron,[49] etc.

4.8 Morse Oscillator

4.8.1 Morse Potential

Diatomic molecules differ from harmonic oscillators mainly in that they may dissociate. If we pull a diatomic molecule with internuclear distance R equal to the equilibrium distance R_e, then at the beginning, displacement $x = R - R_e$ is indeed proportional to the force applied. However, afterward the pulling becomes easier and easier. Finally, the molecule dissociates; i.e., we separate the two parts without any effort at all. This fundamental difference with respect to the harmonic oscillator is qualitatively captured by the potential proposed by Morse (parameter $\alpha > 0$)[50]:

$$V(x) = De^{-\alpha x}\left(e^{-\alpha x} - 2\right). \tag{4.23}$$

[47] Quantum dots are part of the "*nanotechnology*": on a solid surface, some atomic clusters are placed (quantum dots), lines of such atoms (nanowires), etc. Such systems may exhibit unusual properties.

[48] Quarks interact through exchanging gluons. An attempt to separate two quarks leads to such a distortion of the gluon bond (string) that the string breaks down and separates into two strings with new quarks at their ends created from the distortion energy.

[49] Many polar molecules may lower their energy in a liquid by facing an extra electron with their positive poles of the dipole. This is the *solvated electron.*

[50] Philip McCord Morse (1903 – 1985) was an American theoretical physicist.

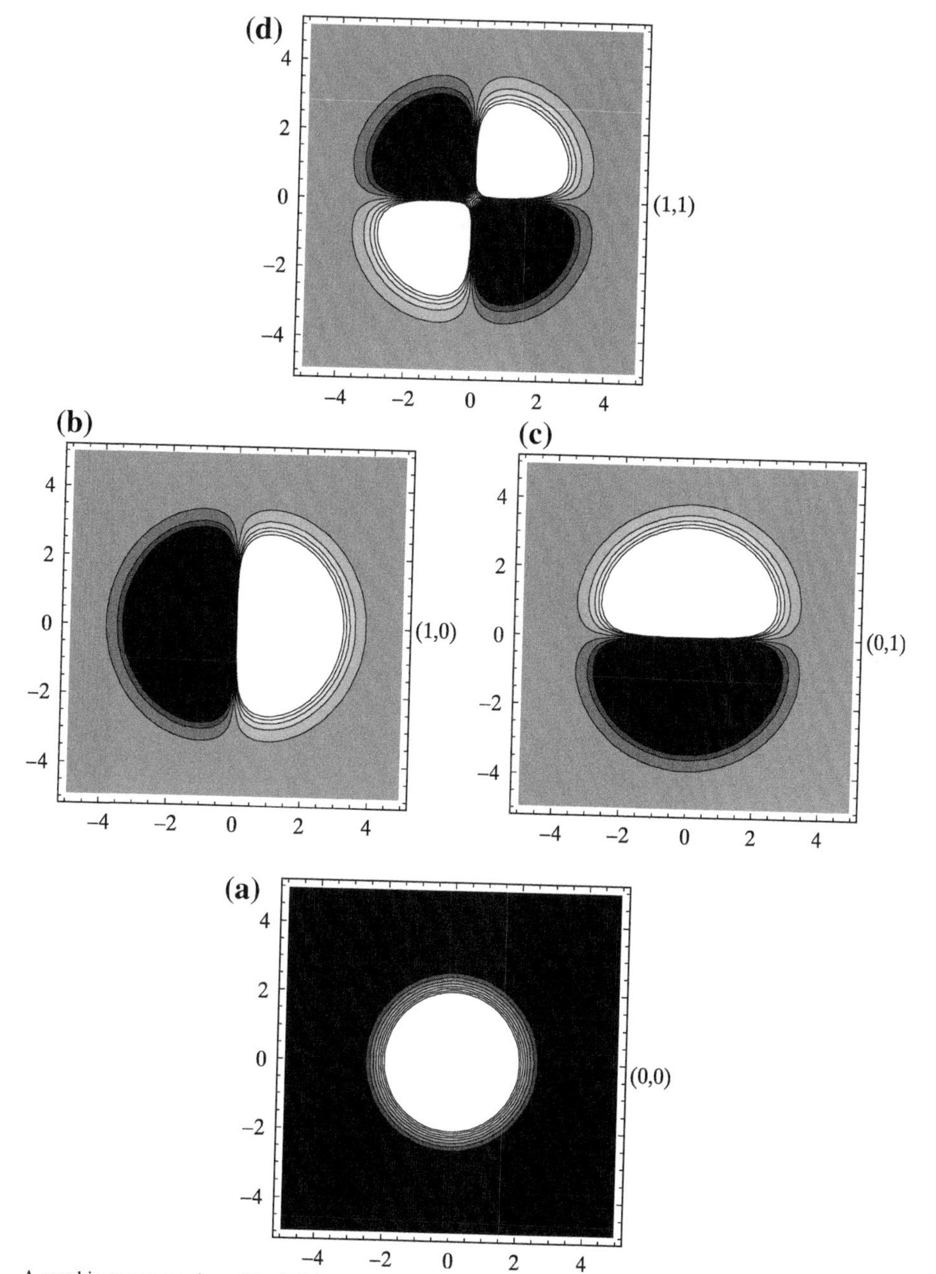

Fig. 4.16. A graphic representation of the 2-D harmonic oscillator wave function (isolines). Panels (a) through (i) show the wave functions labeled by a pair of oscillation quantum numbers (v_1, v_2). The higher the energy, the larger the number of node planes. A reader acquainted with the wave functions of the hydrogen atom will easily recognize a striking resemblance between these figures and the orbitals.

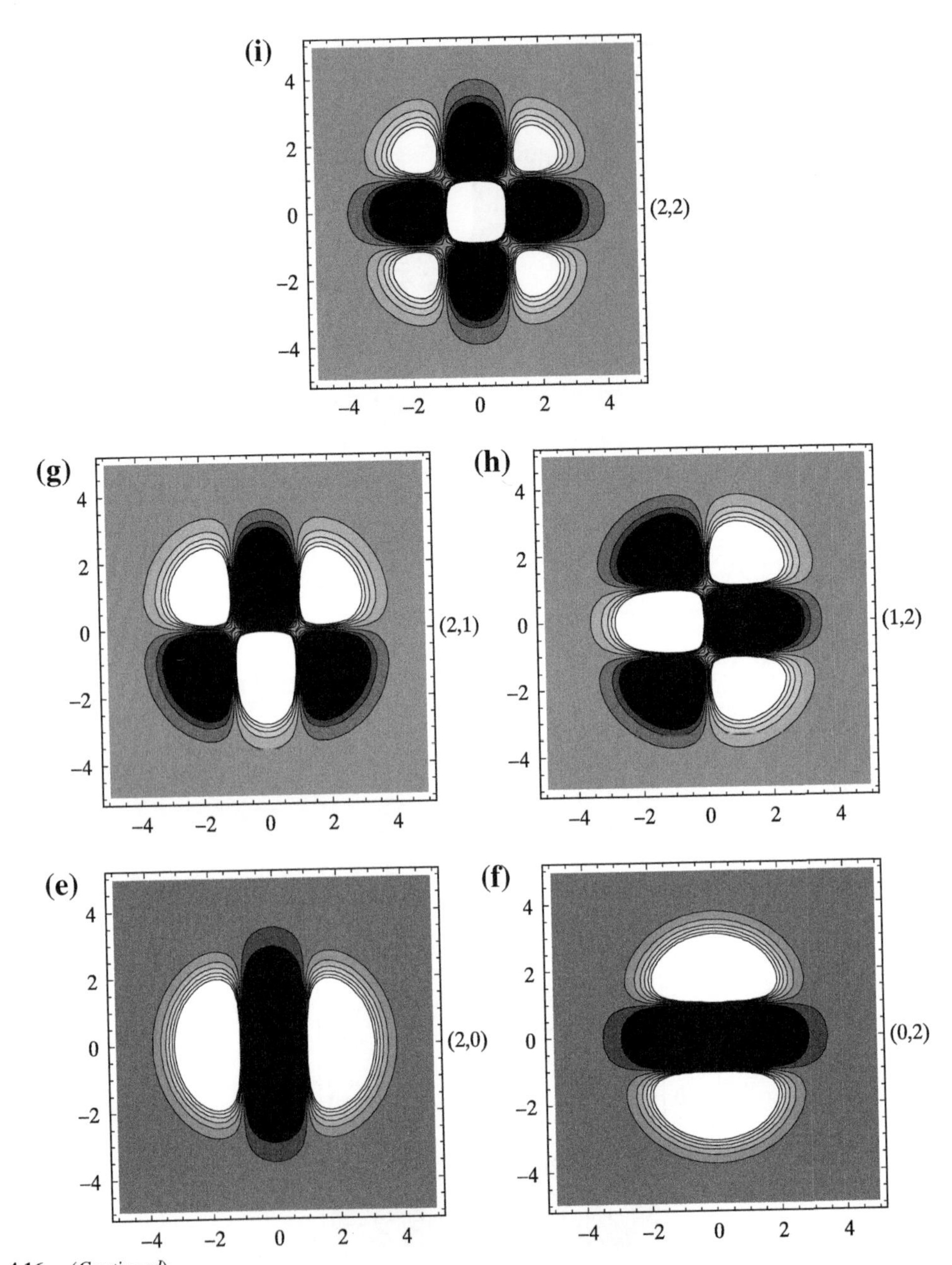

Fig. 4.16. *(Continued)*

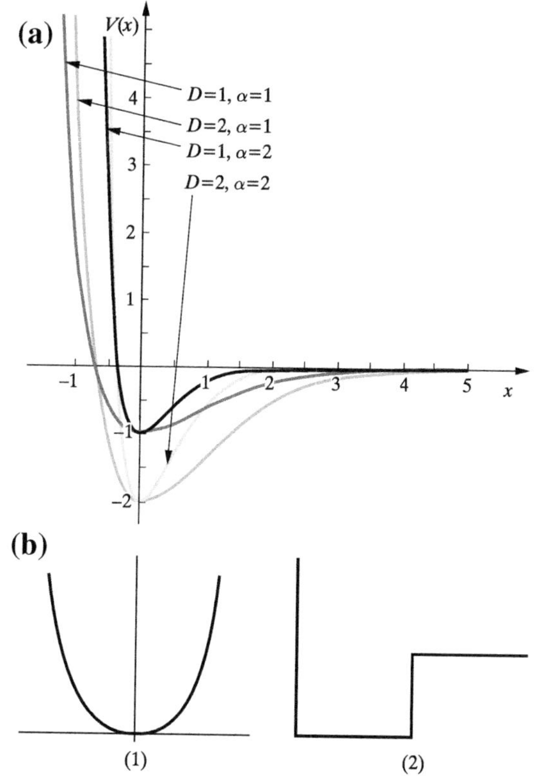

Fig. 4.17. (a) The Morse potential energy curves have the shape of a hook. How does the shape depend on the Morse parameters? The figures show the curves for $D = 1, 2$ and $\alpha = 1, 2$. As we can see, D controls the well depth and α its width. (b) The Morse oscillator is a kind of compromise between the harmonic oscillator (b1) and a rectangular well (b2). Both potentials correspond to exact solutions of the Schrödinger equation. Model b2 gives the discrete spectrum, as well as the continuum and the resonance states. The latter ones are only very rarely considered for Morse oscillators, but they play an important role in scattering phenomena (primarily in reactive collisions).

As we can see, D represents the well depth and the parameter α decides its width. When the displacement $x = 0$, the function attains the minimum $V = -D$, when $x \to \infty$, then $V \to 0$ (see Fig. 4.17).

Besides the abovementioned similarity, the Morse oscillator differs from real diatomics mainly by two qualitative features. First, for $R = 0$, we obtain a *finite* potential energy for the Morse oscillator. Second, the asymptotic behavior of the Morse oscillator for $x \to \infty$ means exponential asymptotics, while the atomic and molecular systems at large distances interact as $\frac{1}{R^n}$.

The second derivative of $V(x)$ calculated at the minimum of the well represents the force constant k of the Morse oscillator:

$$k = 2\alpha^2 D. \tag{4.24}$$

The parabola $-D + \frac{1}{2}kx^2$ best approximates $V(x)$ close to $x = 0$ and represents the harmonic oscillator potential energy (with the force constant k). The Morse oscillator is hard to squeeze–the potential energy goes up faster than that of the harmonic oscillator with the same force constant k.

Solution

One had to have courage to presume that analytical solution with such a potential energy exists. Morse found the solution, which represents a rare example of an exact solution to a nonlinear problem. Exact solutions exist not only for the ground (oscillation quantum number $v = 0$), but also for all the excited states ($v = 1, 2, \ldots v_{max}$) belonging to the discrete spectrum. The energy levels are non-degenerate and are given by the following formula:

$$E_v = -D + h\nu\left(v + \frac{1}{2}\right) - h\nu\left(v + \frac{1}{2}\right)^2 \beta, \tag{4.25}$$
$$v = 0, 1, 2, \ldots, v_{max},$$

where, using atomic units, we obtain

$$h\nu = 2\alpha\left(\frac{D}{2\mu}\right)^{\frac{1}{2}}. \tag{4.26}$$

This formula follows from the parabolic approximation of the Morse potential (which is valid for small displacements x),[51] while

$$\beta = \frac{\hbar\omega}{4D}, \tag{4.27}$$

where μ is the mass of the oscillating particle. When the Morse oscillator serves as a model of a diatomic molecule, μ stands for the reduced mass of both nuclei $\mu = (1/m_1 + 1/m_2)^{-1}$ (see Appendix I available at booksite.elsevier.com/978-0-444-59436-5 on p. e93). As we can see, the energy of the oscillator never equals zero (similar to the harmonic oscillator) and

the separation between consecutive energy levels decreases.

The wave functions are slightly more complicated than those for the harmonic oscillator and are given by the following formula:

$$\psi_v(z) = N_v e^{-\frac{z}{2}} z^{b_v} L_v^{2b_v}(z), \tag{4.28}$$

[51] Let us recall that, for the harmonic oscillator $2\pi\nu = \sqrt{\frac{k}{\mu}}$; therefore, from Eq. (4.24), $h\nu = \hbar\alpha\sqrt{\frac{2D}{\mu}}$, while $\hbar = 1$ a.u.

where $v = 0, 1, \ldots v_{max}$, variable z is a real number related to displacement x by the formula $z = 2ae^{-\alpha x}$, and the normalization coefficient $N_v = \sqrt{\frac{2b_v v!}{\Gamma(2b_v+v+1)}}$ with $\Gamma(z) = \int_0^\infty e^{-t}t^{z-1}dt$, while

$$a = \frac{2D}{\hbar\omega} \tag{4.29}$$

$$b_v = a - \frac{1}{2} - v > 0. \tag{4.30}$$

The above condition gives maximum $v = v_{max}$; therefore, $v_{max} + 1$ is the number of eigenfunctions. Thus, we always have a finite number of energy levels.

L stands for the polynomial given by the formula

$$L_n^c(z) = \frac{1}{n!}e^z z^{-c} \frac{d^n}{dz^n}\left(e^{-z}z^{n+c}\right), \tag{4.31}$$

where $n = 0, 1, 2, \ldots$ is the polynomial degree.[52] A short exercise gives

$$L_0^c(z) = 1$$
$$L_1^c(z) = (c+1) - z$$
$$L_2^c(z) = \frac{1}{2}z^2 - (c+2)\,z + \frac{1}{2}(c+1)(c+2).$$
$$\ldots$$

This means the number of nodes in a wave function is equal to v (as in the harmonic oscillator). The wave functions resemble those of the harmonic oscillator but are slightly deformed (have a bit larger values for $x > 0$).

For very large well depths (D), the parameter β of Eq. (4.27) becomes very small. This results in E_v approaching the corresponding formula for the harmonic oscillator $-D + h\nu(v + 1/2)$, and the energy levels become equidistant from the nearest neighbor separation equal to $h\nu$. The potential is highly anharmonic (of the "*hook-type*"), but the energy levels would be equidistant, as in the harmonic oscillator. Is it possible? Yes, it is. The key is that, for small values of v, the term $-h\nu(v + 1/2)^2\beta$ does not yet enter into play, and low energy levels correspond to small amplitudes (x) of vibrations. For small x, the potential is close to parabolic,[53] as for the harmonic oscillator with force constant k.

[52] Indeed, n-time derivation gives $e^{-z}z^{n+c}$ as a term with the highest power of z. Multiplication by $e^z z^{-c}$ gives z^n

[53] This is as witnessed by a Taylor expansion of $V(x)$ for $x = 0$.

Example 1. ***Hydrogen Molecule***

The hydrogen molecule has been investigated in detail. As will be seen in Chapters 6 and 10, the theory challenges there some very subtle experiments. Let us approximate the most accurate theoretical potential energy curve[54] (as a function of the internuclear distance R) by a Morse curve.

Is such an approximation reasonable? Let us see. From Wolniewicz's calculations, we may take the parameter $D = 109.52$ kcal/mol $= 38293$ cm^{-1}, while the parameter α is chosen in such a way as to reproduce the theoretical binding energy for $R = R_e + 0.4$ a.u.,[55] where $R_e = 1.4$ a.u. is the position of the minimum binding energy. It turns out that, say, "by chance" this corresponds to $\alpha = 1$. From Eqs. (4.29) and (4.30), we obtain $a = 17.917$, and the allowed v are those satisfying the inequality $b_v = 17.417 - v > 0$. We expect, therefore, 18 energy levels with $v = 0, 1, \ldots, 17$ for H_2 and 25 energy levels for T_2 (in the last case, $b_v = 24.838 - v > 0$.). Accurate calculations of Wolniewicz give 14 vibrational levels for H_2, and 25 levels for T_2. Thus, decreasing the reduced mass makes the vibrational levels less dense and some vibrational levels even disappear. This means that isotope substitution by a heavier isotope leads to stabilization. Moreover, from Eq. (4.26) we obtain for H_2 : $h\nu = 0.019476$ a.u.$= 4274$ cm^{-1}, while from Eq. (4.27), we have $\beta = 0.0279$. From these data, one may calculate the energetic gap between the ground ($v = 0$) and the first excited state ($v = 1$) for H_2, $\Delta E_{0\to 1}$, as well as between the first and the second excited states, $\Delta E_{1\to 2}$. We get:

$$\Delta E_{0\to 1} = h\nu - h\nu[(1 + 1/2)^2 - (0 + 1/2)^2]\beta = h\nu(1 - 2\beta)$$
$$\Delta E_{1\to 2} = h\nu - h\nu[(2 + 1/2)^2 - (1 + 1/2)^2]\beta = h\nu(1 - 4\beta).$$

Inserting the calculated $h\nu$ and β gives $\Delta E_{0\to 1} = 4155$ cm^{-1} and $\Delta E_{1\to 2} = 3797$ cm^{-1}. The first value agrees very well with the experimental value[56] of 4161 cm^{-1}. However, comparison of the second value with the measured 3926 cm^{-1} is a little bit worse, although it is still not bad for our simple theory. The quantity D represents the *binding energy*; i.e., the energy difference between the well bottom and the energy of the dissociated atoms. In order to obtain the dissociation energy, we have to consider that the system does not start from the energy corresponding to the bottom of the curve, but from the level with $v = 0$ and energy $\frac{1}{2}h\nu$. Hence, our estimation of the *dissociation energy* is $E_{\text{diss}} = D - \frac{1}{2}h\nu = 36156$ cm^{-1}, while the experimental value amounts to 36118 cm^{-1}.

Example 2. ***Two Water Molecules***

Our first example pertains to a chemical bond. Now let us take in the same way a quite different situation, where we have relatively weak intermolecular interactions; namely, the hydrogen bond between two water molecules. The binding energy in such a case is of the order of $D = 6$ kcal mol^{-1} $= 0.00956$ a.u.$= 2097$ cm^{-1}; i.e., about twenty times smaller

[54] L.Wolniewicz, *J. Chem. Phys., 103,* 1792 (1995).

[55] Of course, this choice is arbitrary.

[56] I.Dabrowski, *Can. J. Phys., 62,* 1639 (1984).

than before. To stay within a single oscillator model, let us treat each water molecule as a pointlike mass. Then, $\mu = 16560$ a.u. Let us stay with the same value of $\alpha = 1$. We obtain (p. 197) $a = 17.794$, and hence $b_0 = 17.294, b_1 = 16.294, \ldots, b_{17} = 0.294, b_{n>17} < 0$. Thus (accidentally), we also have 18 vibrational levels.

This time, from Eq. (4.26), we have $h\nu = 0.001074$ a:u: $= 235$ cm^{-1}, and $\beta = 0.02810$ a.u.; therefore, $\Delta E_{0\to1} = 222$ cm^{-1} and $\Delta E_{1\to2} = 209$ cm^{-1}. These numbers have the same order of magnitude as those appearing in the experiments (cf., p. 362).

4.9 Rigid Rotator

A rigid rotator is a system of two pointlike masses, m_1 and m_2, with a constant distance R between them. The Schrödinger equation may be easily separated into two equations, one for the center of mass motion and the other for the relative motion of the two masses (see Appendix I available at booksite.elsevier.com/978-0-444-59436-5 on p. e93). We are interested only in the second equation, which describes the motion of a particle of mass μ equal to the reduced mass of the two particles, and the position in space given by the spherical coordinates R, θ, ϕ, where $0 \le R < \infty, 0 \le \theta \le \pi, 0 \le \phi \le 2\pi$. The kinetic energy operator is equal to $-\frac{\hbar^2}{2\mu}\Delta$, where the Laplacian Δ represented in the spherical coordinates is given in Appendix H available at booksite.elsevier.com/978-0-444-59436-5 on p. e91. Since R is a constant, the part of the Laplacian that depends on the differentiation with respect to R is absent.[57] In this way, we obtain the equation (equivalent to the Schrödinger equation) for the motion of a particle on a sphere:

$$-\frac{\hbar^2}{2\mu R^2}\left\{\frac{1}{\sin\theta}\frac{\partial}{\partial\theta}\left(\sin\theta\frac{\partial}{\partial\theta}\right)+\frac{1}{(\sin\theta)^2}\frac{\partial^2}{\partial\phi^2}\right\}Y = EY, \tag{4.32}$$

where $Y(\theta, \phi)$ is the wave function to be found and E represents the energy. This equation may be rewritten as

$$\hat{J}^2 Y = 2\mu R^2 EY, \tag{4.33}$$

where the square of the angular momentum operator equals

$$\hat{J}^2 = -\hbar^2\left[\frac{1}{\sin\theta}\frac{\partial}{\partial\theta}\left(\sin\theta\frac{\partial}{\partial\theta}\right)+\frac{1}{\sin^2\theta}\frac{\partial^2}{\partial\phi^2}\right].$$

The equation may be rewritten as

$$\frac{1}{Y}\left\{\frac{1}{\sin\theta}\frac{\partial}{\partial\theta}\left(\sin\theta\frac{\partial Y}{\partial\theta}\right)+\frac{1}{(\sin\theta)^2}\frac{\partial^2 Y}{\partial\phi^2}\right\} = \lambda,$$

[57] This reasoning has an heuristic character, but the conclusions are correct. Removing an operator is a subtle matter. In the correct solution to this problem, we have to consider the two masses with a variable distance R with the full kinetic energy operator and potential energy in the form of the Dirac delta function (see Appendix E available at booksite.elsevier.com/978-0-444-59436-5 on p. e69) $-\delta(R - R_0)$.

where $\lambda = -\frac{2\mu R^2}{\hbar^2}E$. The solution of the equation is known in mathematics as a *spherical harmonic*.[58] It exists if $\lambda = -J(J+1)$, $J = 0, 1, 2, \ldots$:

$$Y_J^M(\theta, \phi) = N_{JM} \cdot P_J^{|M|}(\cos\theta) \cdot \frac{1}{\sqrt{2\pi}} \exp(iM\phi), \tag{4.34}$$

where $N_{JM} = \sqrt{\frac{2l+1}{2}\frac{(l-|m|)!}{(l+|m|)!}}$ is the normalization coefficient, and P is the *associated Legendre polynomial*,[59] defined as

$$P_J^{|M|}(x) = (1-x^2)^{\frac{|M|}{2}} \frac{d^{|M|}}{dx^{|M|}} P_J(x), \tag{4.35}$$

with the *Legendre polynomial*

$$P_J(x) = \frac{1}{2^J J!} \frac{d^J}{dx^J} \left(x^2 - 1\right)^J. \tag{4.36}$$

From the uniqueness of the solution (Fig. 2.6j), it follows that M has to be an integer.[60] The solution exists if $J = 0, 1, 2, 3, \ldots$, and from the analysis of the associate Legendre polynomials, it follows that M cannot exceed[61] J because otherwise $Y = 0$. The energy levels are given by

$$E_J = J(J+1)\frac{\hbar^2}{2\mu R^2} \tag{4.37}$$

for $J = 0, 1, 2, \ldots$

It is seen that the lowest energy level $(J = 0)$ corresponds to $Y_0^0 = \text{const}$ (the function is, of course, nodeless, as shown in Fig. 4.18a). This means that all orientations of the rotator have equal probability. The first excited state corresponds to $J = 1$ and is triply degenerate, since $M = 0, \pm 1$. The corresponding wave functions are: $Y_1^0 = \sqrt{\frac{3}{4\pi}}\cos\theta$,

---

[58] There are a few definitions of the spherical harmonics in the literature [see E.O.Steinborn and K.Ruedenberg, *Advan.Quantum Chem.*, *7*, 1 (1973)]. The Condon-Shortley convention often is used, and is related to the definition given above in the following way: $Y_J^M = \varepsilon_M \left[Y_J^M\right]_{CS}$, $Y_J^J = (-1)^J \left[Y_J^J\right]_{CS}$, where $\varepsilon_M = i^{|M|+M}$.

[59] Adrien Legendre (1752–1833) was a French mathematician and professor at the Ecole Normale Superieure, an elite school of France founded by Napoleon Bonaparte.

[60] Indeed, since ϕ is an angle, we have $\exp(iM\phi) = \exp[iM(\phi + 2\pi)]$. Hence, $\exp(iM2\pi) = 1$, and therefore, $\cos(2\pi M) = 1$ and $\sin(2\pi M) = 0$. This is fulfilled only if M is an integer.

[61] $P_J(x)$ is a polynomial of the Jth degree, while $\frac{d^{|M|}}{dx^{|M|}}$ in $P_J^{|M|}(x)$ decreases the degree by M. If M exceeds J, then $P_J^{|M|}(x)$ automatically becomes equal to zero.

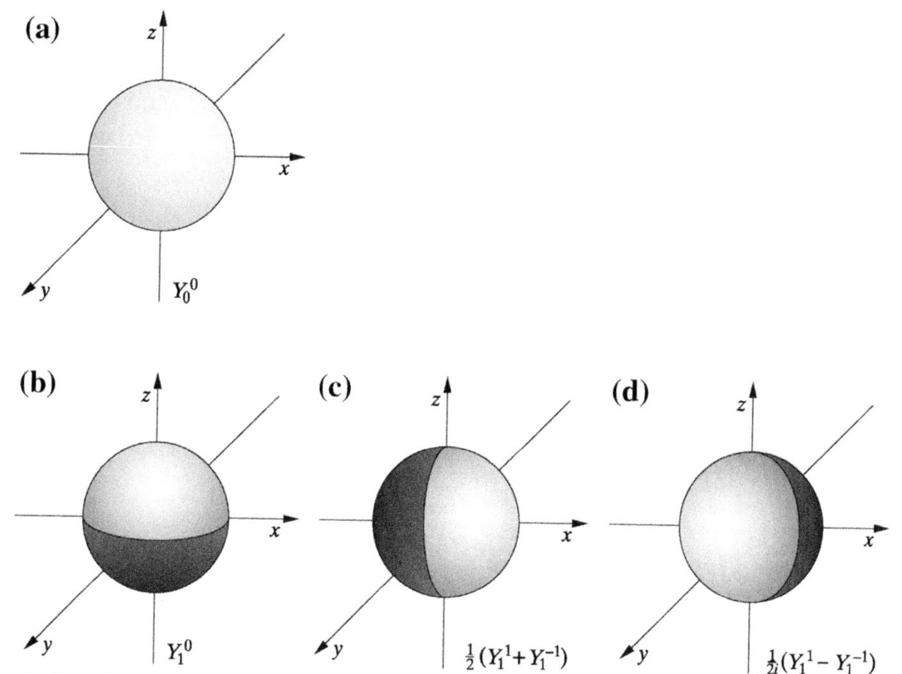

Fig. 4.18. A schematic representation of the nodes for rigid rotator wave functions: (a) ground state (nodeless); (b) triply degenerate first excited state (single node).

$Y_1^1 = \sqrt{\frac{3}{8\pi}} \sin\theta \exp(i\phi)$, $Y_1^{-1} = \sqrt{\frac{3}{8\pi}} \sin\theta \exp(-i\phi)$. The first function, being real, may be easily plotted (Fig. 4.18b), while the second and the third are not (they are complex). Since they both correspond to the same eigenvalue of the Hamiltonian, their arbitrary linear combination is an equally good eigenfunction of this operator. Therefore, we may take Y_1^1 and Y_1^{-1} as $\psi_1 = \frac{1}{2}\left(Y_1^1 + Y_1^{-1}\right) = \sqrt{\frac{3}{8\pi}} \sin\theta \cos\phi$ and $\psi_1 = \frac{1}{2i}\left(Y_1^1 - Y_1^{-1}\right) = \sqrt{\frac{3}{8\pi}} \sin\theta \sin\phi$. Both functions are real, and they are shown in Figs. 4.18c-d. Note that again, we have the usual situation: the ground state is nodeless, the first excited state has a single node, etc.

Y_J^M is not only the eigenfunction of the Hamiltonian $\hat{H}$ and of the square of the angular momentum $\hat{J}^2$, but also of the z component of the angular momentum operator, $\hat{J}_z = -i\hbar\frac{\partial}{\partial\phi}$:

$$\hat{J}_z Y_J^M = M\hbar Y_J^M. \tag{4.38}$$

4.10 Hydrogen-Like Atom

We have two particles: an electron of mass m and charge $-e$ and a nucleus of mass M and charge $+Ze$. The Hamiltonian contains two kinetic energy operators and the Coulombic interaction $-Ze^2/r$, where r is the electron-nucleus separation. We have, therefore, six coordinates. In Appendix I available at booksite.elsevier.com/978-0-444-59436-5 on p. e93, it is shown how the center-of-mass motion can be separated (we are not interested in this motion). There remain three coordinates, x, y, and z, showing where the electron is with respect to the nucleus. The resulting Schrödinger equation contains a single kinetic-energy operator of a particle of reduced mass μ (almost equal to the electron mass) with coordinates x, y, and z, and Coulombic interaction of the electron and the nucleus (as before). Now, instead of x, y, and z, we introduce the spherical coordinates r, θ, and ϕ. Then, as the class Q solution, we obtain

$$\psi_{nlm}(r,\theta,\phi) = N_{nl} R_{nl}(r) Y_l^m(\theta,\phi), \tag{4.39}$$

where Y_l^m is identical to the solution [Eq. (4.34)] of a rigid rotator of length r, and the function R_{nl} has the following form in a.u.:

$$R_{nl}(r) = r^l L_{n+l}^{2l+1}\left(\frac{2Zr}{na_0}\right) \exp\left(-\frac{Zr}{na_0}\right), \tag{4.40}$$

where the *Bohr first orbit radius* is

$$a_0 = \frac{1}{\mu} \simeq 1\text{a.u.}, \tag{4.41}$$

where

principal quantum number $n = 1, 2, 3 \ldots$
azimuthal quantum number $l = 0, 1, 2, \ldots, n-1$
magnetic quantum number $m = -l, -l+1, \ldots, 0, \ldots +l.$

and the *associated Laguerre polynomial* $L_\alpha^\beta(x)$ is defined as

$$L_\alpha^\beta(x) = \frac{d^\beta}{dx^\beta} L_\alpha(x), \tag{4.42}$$

while the *Laguerre polynomial* is given by[62]

$$L_\alpha(x) = \exp(x) \frac{d^\alpha}{dx^\alpha}[x^\alpha \exp(-x)]. \tag{4.43}$$

Since the Hamiltonian commutes with the square of the total angular momentum operator $\hat{J}^2$ and with the operator of $\hat{J}_z$ (cf., Chapter 2 and Appendix F available at booksite.elsevier.com/978-0-444-59436-5 on p. e73), then the functions ψ_{nlm} are the eigenfunctions of the following operators:

$$\hat{H}\psi_{nlm} = E_n \psi_{nlm} \tag{4.44}$$

[62] $L_\alpha^\beta(x)$ are indeed polynomials of the $\alpha - \beta$ degree. If $\beta > \alpha$, from Eq. (4.42), it follows that $L_\alpha^\beta = 0$.

$$\hat{J}^2\psi_{nlm} = l(l+1)\hbar^2\psi_{nlm}, \tag{4.45}$$

$$\hat{J}_z\psi_{nlm} = m\hbar\psi_{nlm}, \tag{4.46}$$

where, expressed in a.u.,

$$E_n = -\frac{Z^2}{2n^2}\left(\frac{1}{1+\frac{1}{M_p}}\right), \tag{4.47}$$

with M_p representing the proton mass (in a.u.; i.e., about 1840). The content of the parentheses in the last formula works out as 0.999457 (i.e., almost 1), which would be obtained for an infinite mass of the nucleus.

Each of the energy levels is n^2–fold degenerate. Note that the hydrogen atom energy depends solely on the principal quantum number n. The fact that the energy does not depend on the projection of the angular momentum $m\hbar$ is natural because the space is isotropic and no direction is privileged. However, the fact that it does not depend on the length of the angular momentum $\sqrt{l(l+1)}\hbar$ is at first sight strange. The secret is in the Coulombic potential $\frac{1}{r}$ produced by the pointlike nucleus and connected with the notion of dynamic symmetry mentioned on p. 83. If we considered a non-pointlike nucleus or were interested in the orbital 2s of such a quasi-hydrogen atom as lithium,[63] then the energy *would* depend on the quantum number l.

The one-electron wave functions (orbitals) of the hydrogen atom with $l = 0$ are traditionally denoted as ns: $1s$, $2s$,...,with $l = 1$, as np: $2p, 3p, 4p, \ldots$, with $l = 2, 3, \ldots$, as nd: $3d, 4d$,, nf: $4f, 5f, \ldots$.

The wave functions ψ_{nlm} can be plotted in several ways. For example, the function $(nlm) =$ (100) or 1s, given by the formula

$$1s \equiv \psi_{100}(r,\theta,\phi) = \sqrt{\frac{Z^3}{\pi}}\exp(-Zr), \tag{4.48}$$

and can be visualized in several alternative forms, as shown in Fig. 4.19.

We see that what the electron likes most is to sit on the nucleus. Indeed, if we chopped the space into tiny cubes and then computed the value of $(1s)^2$ in each cube (the function is real, so the complex modulus sign is irrelevant), and multiplied the number obtained by the volume of the cube, the resulting number in each cube would mean the probability of finding the electron in a particular cube. Evidently, this number will be largest for the cube that contains the

[63] In this case, the nucleus is screened by a cloud of two 1s electrons. The 2s electron thinks that it is in a hydrogen atom with a spatious nucleus of the size of the 1s orbital and an effective charge +1.

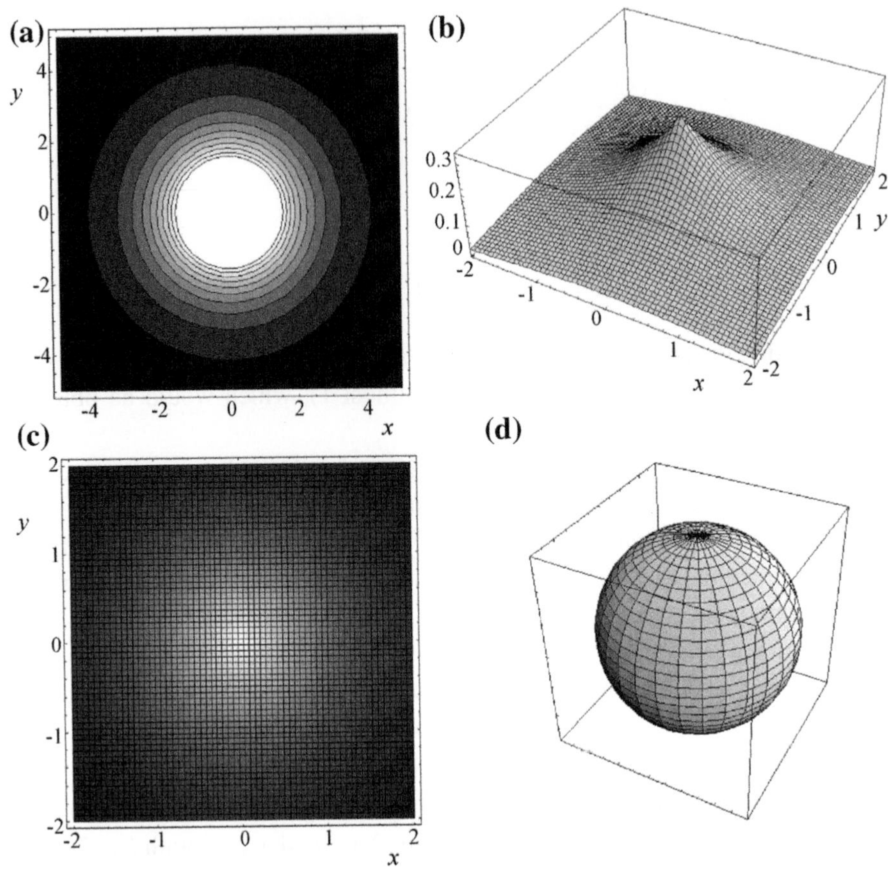

Fig. 4.19. Various ways of visualization the $1s$ hydrogen orbital, which is a function of electron position in 3-D space (coordinates in panels a-c are in a.u.). (a) Isolines of the $z = 0$ section of the wave function (going through the nucleus). Black means the value zero, white color means a high value. This is a map of a mountain. The center of (a) shows a large white plateau that represents an artefact. In fact, in panel (b), the section of the $1s$ orbital as a function of r represents a mountain with a sharp summit (a discontinuity of the first derivative). Panel (c) is similar to (a), but instead of isolines, we have a white mist with the highest concentration in the center, disappearing exponentially with increasing distance r. Panel (d) shows a spherically symmetric isosurface of the wave function.

nucleus (the origin). In school, we were told about the Bohr model[64] the orbits, and the first Bohr orbit (corresponding to the atom ground state). Do we relegate all this to mythology? Not completely. If we changed the question to "*What is the distance at which the electron is most likely to be found?*" then the answer should indeed be as we were taught in school: the first Bohr orbit. This is easy to show by computing the *radial probability density* of finding the electron (i.e., integrating over all orientations, leaving the dependence on the distance):

---

[64] Nobody is perfect–not even geniuses. Here is a story by John Slater: "*Brillouin delivered an interesting lecture concerning his relations. When he finished, Bohr stood up and attacked him with an inhuman fury. I have never heard any adult scold another person in public with such an emotional engagement without any reason whatsoever. After this show, I have decided that my antipathy with respect to Bohr dating since* 1924 *continues to be justified.*"

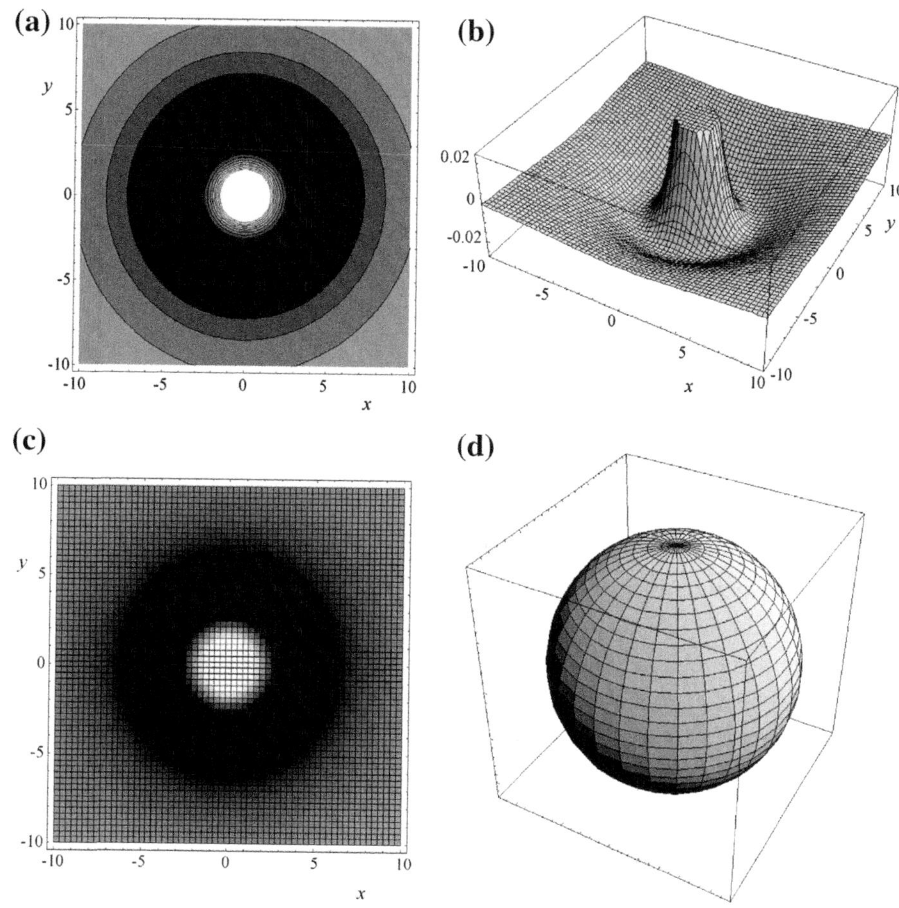

Fig. 4.20. Various graphical representations of the hydrogen $2s$ orbital (coordinates in panels a-c are expressed in a.u.). (a) Isolines of the $z = 0$ section of the orbital. Gray means zero, white a high positive value, and black a negative value. Note that gray is not only at the peripheries, but also around the center. This means that $2s$ orbital exhibits a nodal sphere of radius 2 a.u. [see Eq. (4.49)], that would contain a little more than 5% of the electronic density (whereas for the $1s$ orbital, the same sphere contains about 75% of electron cloud). The center of the figure (a) shows a large white plateau, which represents an artefact. In fact, panel (b), showing the section of $2s$ orbital represents a mountain with a sharp peak (a discontinuity of the first derivative is shown incorrectly in the figure; instead of a sharp summit, one has an artefact plateau) with a depression at its base. Panel (c) is similar to (a), but instead of isolines, one has a white mist with the largest concentration in the center, then taking the negative values (black mist) and finally disappearing exponentially with increasing distance r. Panel (d) shows a spherically symmetric isosurface of the wave function (the sphere was shown as larger than the $1s$ orbital because the $2s$ orbital decays more slowly than $1s$).

$\rho(r) = \int d\theta d\phi r^2 \sin\theta\, |\psi_{100}|^2 = 4Z^3 r^2 \exp(-2Zr)$. The maximum of $\rho(r)$ corresponds exactly to $r = 1$ a.u. or the first Bohr orbit radius.[65]

The $2s$ orbital ($n = 2, l = 0, m = 0$) reads as (see Fig. 4.20):

$$2s \equiv \psi_{200}(r, \theta, \phi) = N_{2s}\left(Zr - 2\right)\exp(-Zr/2), \tag{4.49}$$

[65] The computed maximum position does not coincide with the mean value of r (see Appendix H available at booksite.elsevier.com/978-0-444-59436-5 on p. e91) $\langle r\rangle = \langle\psi_{100}|r\psi_{100}\rangle = \int_0^\infty dr r\rho(r) = \frac{3}{2}$ a.u.

with the normalization constant $N_{2s} = \frac{Z^{\frac{3}{2}}}{4\sqrt{2\pi}}$. A sphere of radius $2/Z$ (representing the nodal sphere) contains[66] only a little more than 5% of the total electronic density (independently of Z).[67]

The wave functions (orbitals) with $m \neq 0$ are difficult to draw because they are complex. However, we may plot the real part of ψ_{nlm} (i.e. $Re\psi_{nlm}$) by taking the sum of ψ_{nlm} and ψ_{nl-m} [i.e., $2\text{Re}\psi_{nlm}$ and the imaginary part of ψ_{nlm} (i.e. $\text{Im}\psi_{nlm}$) from the difference of ψ_{nlm} and ψ_{nl-m} equal to $2i\text{Im}\psi_{nlm}$]. These functions are real and can be easily plotted. In this way, we obtain the orbitals $2p_x$ and $2p_y$ from the functions ψ_{211} and ψ_{21-1}. The orbital ψ_{210} is identical to $2p_z$:

$$2p_x = N_{2p}x \exp(-Zr/2)$$
$$2p_y = N_{2p}y \exp(-Zr/2)$$
$$2p_z = N_{2p}z \exp(-Zr/2),$$

where an easy calculation (just five lines long) gives the normalization constant $N_{2p} = ZN_{2s}$. The $2p$ orbitals are shown in Fig. 4.21.

Note that a linear combination of eigenfunctions is an eigenfunction, if the functions mixed correspond to the same eigenvalue. This is why $2p_x$ and $2p_y$ are the eigenfunctions of the Hamiltonian and of the square of the angular momentum operator, but they are not eigenfunctions of $\hat{J}_z$.

Similarly, we obtain the five real $3d$ orbitals. They can be easily obtained from Eq. (4.39) and subsequently making them real by choosing $\text{Re}\psi_{nlm}$ and $\text{Im}\psi_{nlm}$. As a result, we have the following normalized $3d$ orbitals ($N_{3d} = \frac{Z^{\frac{7}{2}}}{81}\sqrt{\frac{2}{\pi}}$):

$$3d_{xy} = N_{3d}xy \exp(-Zr/3),$$
$$3d_{xz} = N_{3d}xz \exp(-Zr/3),$$
$$3d_{yz} = N_{3d}yz \exp(-Zr/3),$$
$$3d_{x^2-y^2} = \frac{1}{2}N_{3d}(x^2 - y^2) \exp(-Zr/3)$$
$$3d_{3z^2-r^2} = \frac{1}{2\sqrt{3}}N_{3d}(3z^2 - r^2) \exp(-Zr/3).$$

The $2p$ and $3d$ orbitals are shown[68] in Figs. 4.21, 4.22, and 4.23. A summary of the hydrogen atomic orbitals is shown in Fig. 4.24.[69]

66 See the Mathematica files for Chapter 4 in the Web Annex.

67 A sphere of the same radius encloses about 75% of the electron density for the $1s$ orbital.

68 It pays to memorize the abbreviations $2p_x, 2p_y, 2p_z$ and the five $3d$ orbitals. Indeed, we may then easily write down their mathematical formulas (even neglecting the normalization constants). Having the formulas, we may draw any section of them; i.e., we can predict the form of Figs. 4.22 and 4.23.

69 A night bus ride might give us some unexpected impressions. Of all atomic orbitals, you may most easily see orbital $1s$. Just look through the condensation on a bus window at a single street lamp. You will see a gleam that

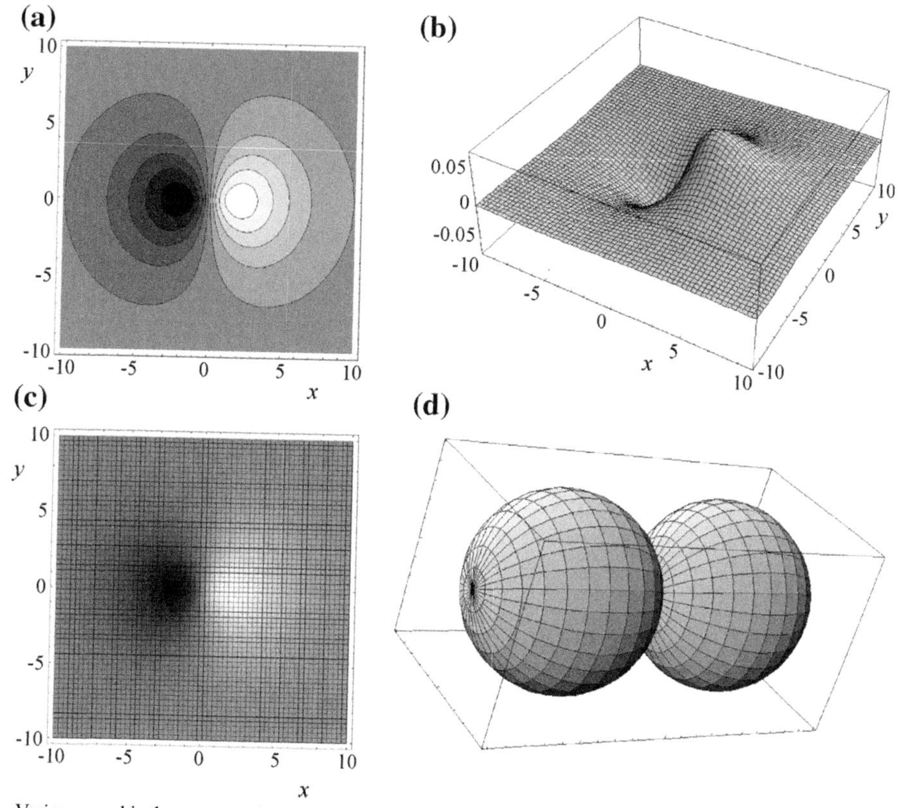

Fig. 4.21. Various graphical representations of the hydrogen $2p_x$ orbital (coordinates in panels a-c are expressed in a.u.). The two other $2p$ orbitals: $2p_y$ and $2p_z$ look the same as $2p_x$, but are oriented along axes y and z, respectively. Note that for the hydrogen atom, all four orbitals $2s$, $2p_x$, $2p_y$, and $2p_z$ correspond to the same energy, and all have a single nodal surface. For $2s$, the surface (Fig. 4.20) is a sphere of radius 2, for the $2p_x$, $2p_y$, and $2p_z$ orbitals, the nodal surfaces are the planes $x, y, z = 0$. (a) isolines of the $z = 0$ section of the orbital. Gray means zero, white means a high value, and black means a negative value. (b) The values of the section $z = 0$. Note in panels (a) and (b), that the right side of the orbital is positive, while the left side is negative. The maximum (minimum) value of the orbital is at $x = 2$ ($x = -2$) a.u. Panel (c) is similar to (a), but instead of isolines, we have a mist with the largest value (white) on the right and the smallest (and negative, black) value on the left. The orbital finally disappears exponentially with increasing distance r from the nucleus. Panel (d) shows an isosurface of the absolute value of the angular part of the wave function $|Y_1^0|$. As for Y_1^0 itself, one of the lobes takes negative values and the other positive values, and they touch each other at the position of the nucleus. To obtain the orbital, we have to multiply this angular function by a spherically symmetric function of r. This means that an isosurface of the absolute value of the wave function will also have two lobes (for the wave function itself, one will be positive and the other negative), but they will not touch each other in accordance with Panel a.

---

decays to black night. You may also quite easily find a double lamp that will offer you a $2p$ orbital and sometimes has the chance to see some of the $3d$ orbitals. Once I have even found the $2s$ orbital, but I do not understand how it was possible. I was looking at a single lamp, which made an intense gleam in the center, which gradually decayed and then again an annular gleam appeared that finally vanished. This is what the square of the $2s$ orbital looks like.

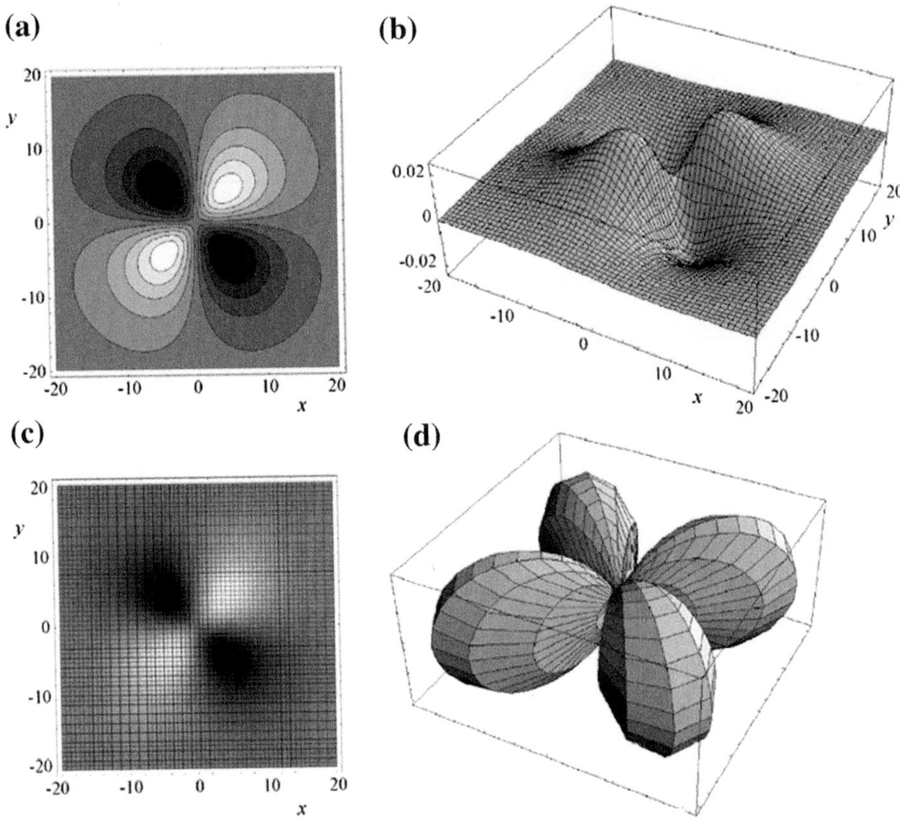

Fig. 4.22. Various graphical representations of the hydrogen $3d_{xy}$ orbital (coordinates in panels a-c are expressed in a.u.). The three other $3d$ orbitals: $3d_{yz}$, $3d_{xz}$ and $3d_{x^2-y^2}$, look the same as $3d_{xy}$, but they are oriented in space according to their indices, see Fig. 4.24. (a) Isolines of the $z = 0$ section of the orbital. Gray means zero, white means a positive value, and black means a negative value. Note in panels a and b that $3d$ orbitals are symmetric with respect to inversion. One may imagine the $z = 0$ section of $3d_{xy}$ as two hills and two valleys [as shown in panel (b)]. Panel (c) is similar to (a), but instead of isolines, one has a white mist with the highest value on the northeast line and the smallest (and negative) value on the northwest line (black mist). The orbital finally disappears exponentially with increasing distance r from nucleus. Panel (d) shows an isosurface of the absolute value of the angular part of the wave function: $|Y_2^2 - Y_2^{-2}|$. As for $Y_2^2 - Y_2^{-2}$ itself, two of the lobes take negative values, the other two take positive values, and they touch each other at the nucleus. To obtain the orbital, one has to multiply this angular function by a spherically symmetric function of r. This means that an isosurface of the absolute value of the wave function will have also four lobes (for the wave function itself, two will be positive and the other two negative), but they will not touch in accordance with panel a.

4.10.1 Positronium and Its Short Life...in Molecules

A bound pair of an electron and a positron (the "*positive electron*" predicted by Dirac and discovered by Anderson; see Chapter 3) is known as *positronium*. The positronium can be detected in the radiation of some radioactive elements, as well as a product of some collisions of atomic nuclei. Its lifetime is of the order of a millisecond. After such a time, a mutual annihilation of the two objects takes place with the emission of the γ photons.

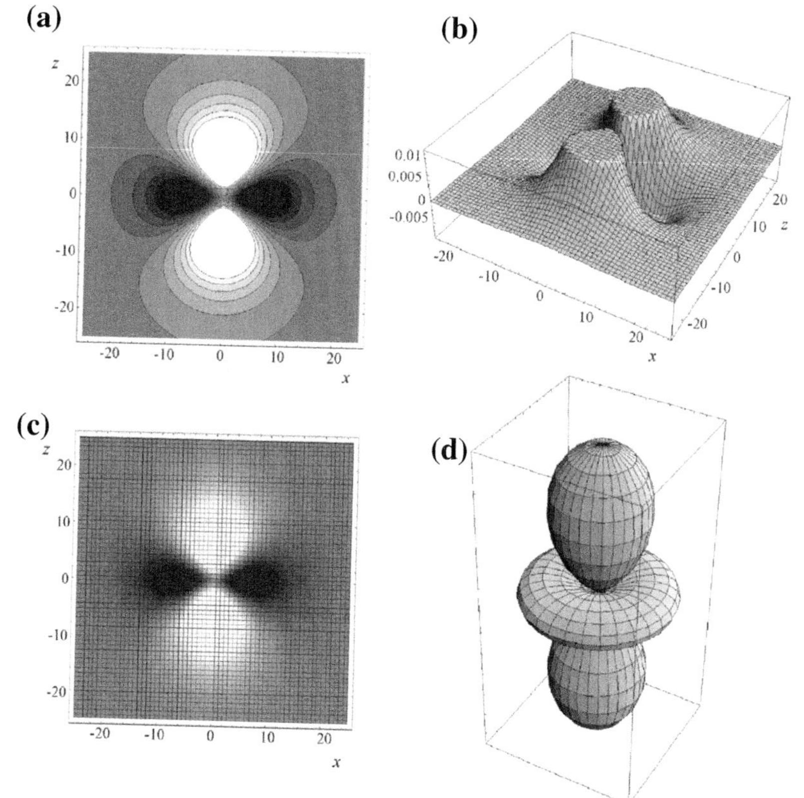

Fig. 4.23. Various graphical representations of the hydrogen $3d_{3z^2-r^2}$ orbital (coordinates in panels a-c are expressed in a.u.). The z-axis is vertical, and the x-axis is horizontal. (a) Isolines of the xz section of the orbital. Gray means zero, white means a high positive value, and black means a negative value. Note that in panels a-b, $3d_{3z^2-r^2}$ orbitals are symmetric with respect to inversion. We may imagine the xz section of the $3d_{3z^2-r^2}$ as two hills and two valleys [panel (b)], and the hills are higher than the depth of the valleys (the plateaus in panel b are artificial). Panel (c) is similar to (a), but instead of isolines, one has a mist with the highest value (white) on the north-south line and the smallest (and negative, black mist) value on the east-west line. The orbital finally disappears exponentially with increasing distance r from the nucleus. Panel (d) shows an isosurface of the absolute value of the angular part of the wave function ($|Y_2^0|$). As for Y_2^0 itself, there are two positive lobes and a negative ring, and they touch each other at the nucleus. To obtain the orbital, we have to multiply this angular function by a spherically symmetric function of r. This means that an isosurface of the absolute value of the wave function will have two lobes along the z axis as well as the ring, but they will not touch in accordance with panel a. The lobes along the z-axis are positive, and the ring is negative. A peculiar form of $3d_{3z^2-r^2}$ becomes more familiar when one realizes that it simply represents a sum of two "*usual*" $3d$ orbitals. Indeed, $3d_{3z^2-r^2} = \frac{1}{2\sqrt{3}}N_1[2z^2 - \left(x^2+y^2\right)]\exp(-Zr/3) = \frac{1}{2\sqrt{3}}N_1[\left(z^2-x^2\right)+\left(z^2-y^2\right)]\exp(-Zr/3) = \frac{1}{2\sqrt{3}}N_1\frac{2}{N_1}\left(3d_{z^2-x^2}+3d_{z^2-y^2}\right) = \frac{1}{\sqrt{3}}\left(3d_{z^2-x^2}+3d_{z^2-y^2}\right)$.

The Schrödinger equation gives only the stationary states (the ground state and the excited ones), which live forever. It cannot describe, therefore, the annihilation process, but it is able to describe the positronium before the annihilation takes place. This equation for the positronium is identical to that for the hydrogen atom, except that the reduced mass changes from $\mu \approx 1$ a.u. to $\mu = \frac{1}{2}$ a.u. ($\frac{1}{\mu} = \frac{1}{1} + \frac{1}{1} = 2$). As a result (see the definition of a_0 on p. 202), one may say that "the size of the orbitals of the positronium doubles" when compared to that of the

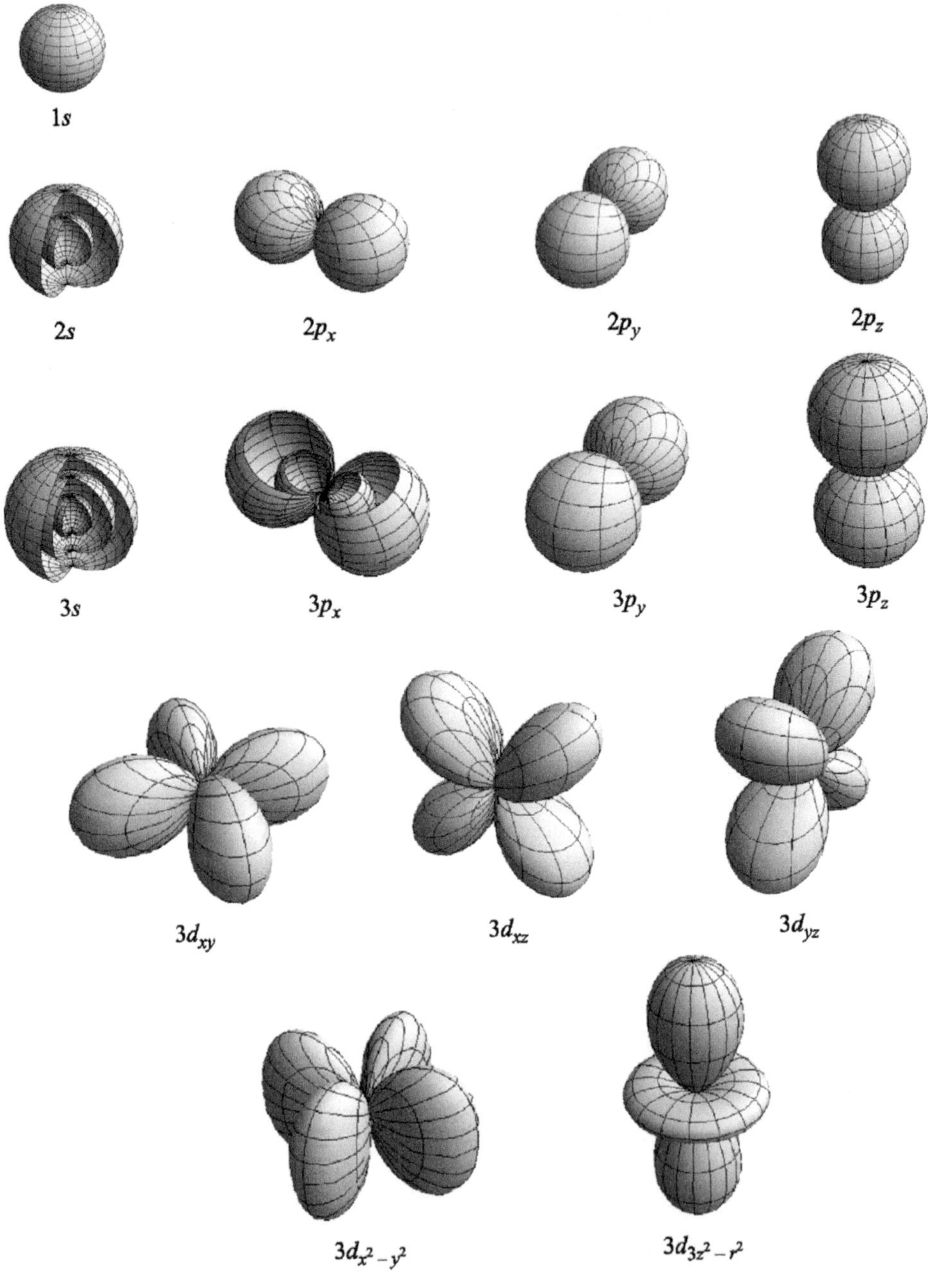

Fig. 4.24. A review of all orbitals with the quantum number $n = 1, 2, 3$. The x-, y-, and z-axes (not shown) are oriented in the same way (as the directions of the $2p_x$, $2p_y$, and $2p_z$ orbitals, respectively). The figures 2s and 3s are schematic; their cross sections are intended to underline that these orbitals are spherically symmetric, but they possess a certain "*internal structure.*" The 1s orbital decays monotonically with r (this is shown by a limiting sphere), but 2s and 3s change sign one and two times, respectively. The internal spheres displayed symbolize the corresponding nodal spheres. The orbital $3p_x$ (representing also $3p_y$ and $3p_z$) is shown in a similar convention: there is an extra nodal surface inside (besides the plane $x = 0$), resembling a smaller orbital p.

hydrogen atom, since their exponential decay coefficient is now multiplied by $\frac{1}{2}$. It does not matter whether one puts the positron or the electron in the center of the positronium (where the positronium orbitals are centered)–the view of the positronium from such a center is the same.[70] In both cases, what moves is a "*quasi-particle*" of the opposite charge to that kept in the center, its mass being $\frac{1}{2}$.

A millisecond is a short time for humans, but a very long time in the molecular world. Let us take a look of a system consisting of proton+two electrons+positron, where the positron might be seen to play a role of an extremely light second nucleus in such a "hydrogen molecule." If the corresponding "atoms" (the hydrogen atom and the positronium) did not interact, one would have the total energy $-\frac{1}{2} - \frac{1}{4} = -0.75$ a.u., while an accurate result for the total system under consideration[71] is equal to -0.789 a.u. Therefore, the binding energy of this unusual "hydrogen molecule" is of the order of 25 kcals/mol; i.e., about $\frac{1}{4}$ of the binding energy of the hydrogen molecule.

4.11 What Do All These Solutions Have in Common?

- In all the systems considered (except the tunneling effect, where the wave function is non-normalizable), the stationary states are similar: the number of their nodes increases with their energy (the nodeless function corresponds to the lowest energy).
- If the potential energy is a *constant* (particle-in-a-box, rigid rotator), then the energy level (nearest-neighbor) separation *increases* with the energy.[72] The energy levels get closer for larger boxes, longer rotators, etc.
- *A parabolic* potential energy well (harmonic oscillator) reduces this tendency, and the energy levels are *equidistant*. The distance decreases if the parabola gets wider (less restrictive).
- The Morse potential energy curve may be seen as a function that may be approximated (as the energy increases) by wider and wider parabolic sections. No wonder, therefore, that the level separation *decreases*. The number of energy levels is finite.[73]
- The Coulomb potential, such as that for the hydrogen atom, resembles vaguely the Morse potential (dissociation limit, but infinite depth). We expect, therefore, that the energy levels for the hydrogen-like atom will become closer and closer when the energy increases, and we are right. Is the number of these energy levels finite, as for the Morse potential? This is a more subtle question. Whether the number is finite or not decides the asymptotics (the behavior at infinity). The Coulomb potential makes the number infinite.

[70] The situation is similar to what we have seen in the case of the hydrogen atom and the observer sitting on the electron or on the proton, as described in Appendix I available at booksite.elsevier.com/978-0-444-59436-5, p. e93.

[71] K. Strasburger, H. Chojnacki, *J. Chem. Phys., 108,* 3218 (1998).

[72] In both cases, the distance goes as the square of the quantum number.

[73] This type of reasoning prepares us for confronting real situations. Practically, we will never deal with the abstract cases described in the present chapter, and yet in later years, we may say something like this: "*Look, this potential energy function is similar to case X in Chapter 4 of that thick boring book we have been forced to study. So the distribution of energy levels and wave functions has to be similar to those given there.*"

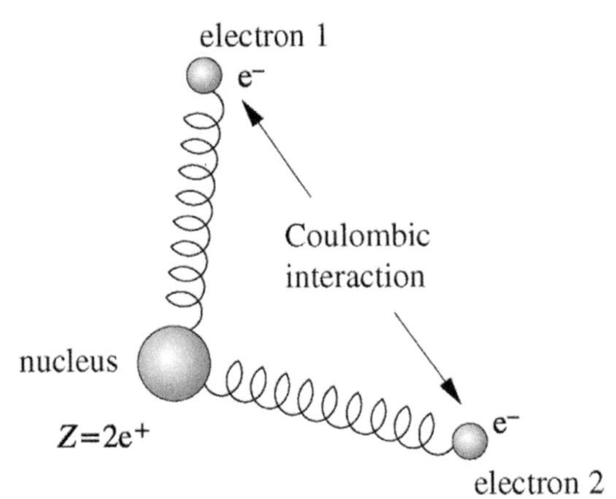

Fig. 4.25. The Hooke helium atom. The electrons repel by Coulombic forces and are attracted by the nucleus by harmonic (non-Coulombic) forces.

4.12 Hooke Helium Atom (Harmonium)

Two-electron systems already represent a serious problem for quantum chemistry because the mutual correlation of electronic motions must be carefully taken into account. As we will see in further chapters, such calculations are feasible, but the wave functions are very complicated; e.g., they may represent linear combinations of thousands of terms and still only be *approximations* of the exact solution to the Schrödinger equation. This is why people were surprised when Kais et al. showed that a two-electron system has an exact analytical solution.[74]

Unfortunately, this wonderful two-electron system is (at least partially) non-physical. It represents a strange helium atom, in which the two electrons (with their distance denoted by r_{12}) interact through the Coulombic potential, but each is attracted to the nucleus by a harmonic spring [i.e., satisfying the Hooke law (of equilibrium length 0 and force constant k, with electron-nucleus distances denoted by r_1 and r_2); see Fig. 4.25].

The Hamiltonian of this problem (atomic units are used[75]) has the form:

$$\hat{H} = -\frac{1}{2}\Delta_1 - \frac{1}{2}\Delta_2 + \frac{1}{2}k(r_1^2 + r_2^2) + \frac{1}{r_{12}}.$$

It is amazing in itself that the Schrödinger equation for this system has an analytical solution (for $k = \frac{1}{4}$). It could be an extremely complicated analytical formula. It is a sensation that the solution is dazzlingly beautiful and simple

$$\psi\left(\boldsymbol{r}_1, \boldsymbol{r}_2\right) = N\left(1 + \frac{1}{2}r_{12}\right)\exp\left[-\frac{1}{4}\left(r_1^2 + r_2^2\right)\right],$$

[74] S. Kais, D.R. Herschbach, N.C. Handy, C.W. Murray, and G.J. Laming, *J. Chem. Phys., 99*, 417 (1993).

[75] Here, the nucleus is assumed to have an infinite mass.

where

$$|N|^2 = \frac{\pi^{\frac{3}{2}}}{8 + 5\sqrt{\pi}}.$$

The wave function represents the product of the two harmonic oscillator wave functions (Gaussian functions), but also an additional extremely simple correlation factor $\left(1 + \frac{1}{2}r_{12}\right)$. As we will see in Chapter 10, exactly such a term is required for the ideal solution. In this exact function, there is nothing else–just what is absolutely necessary.[76]

4.13 Hooke Molecules

The readers together with the author admired an exact solution of the Schrödinger equation for the harmonium. It is truly a wonder: two electrons, but still we get an exact (and simple) solution. Is it the end of such wonders?

Well, let us consider two particles for a while longer. The Hamiltonian for a system of two point masses m_1 and m_2 with positions shown by vectors $\boldsymbol{r}_1$ and $\boldsymbol{r}_2$ and interacting according to the potential energy term $V_{12}\left(|\boldsymbol{r}_1 - \boldsymbol{r}_2|\right)$ (therefore, being a function of their distance only) has the form $\hat{H}(\boldsymbol{r}_1, \boldsymbol{r}_2)$:

$$\hat{H}(\boldsymbol{r}_1, \boldsymbol{r}_2) \equiv \hat{H}_{12}(\boldsymbol{r}_1, \boldsymbol{r}_2) = -\frac{\hbar^2}{2m_1}\Delta_1 - \frac{\hbar^2}{2m_2}\Delta_2 + V_{12}\left(|\boldsymbol{r}_1 - \boldsymbol{r}_2|\right). \tag{4.50}$$

By introducing the center-of-mass coordinates

$$\mathbf{R}_{CM} = \frac{m_1\boldsymbol{r}_1 + m_2\boldsymbol{r}_2}{m_1 + m_2}, \tag{4.51}$$

and the coordinates describing the position of particle 1 with respect to the position of particle 2 as follows:

$$\boldsymbol{\rho} = \boldsymbol{r}_1 - \boldsymbol{r}_2, \tag{4.52}$$

we get[77] the Hamiltonian in the new coordinates (see Appendix I available at booksite.elsevier.com/978-0-444-59436-5, Example 1):

$$\hat{H}(\mathbf{R}_{CM}, \boldsymbol{\rho}) = -\frac{\hbar^2}{2M}\Delta_{CM} - \frac{\hbar^2}{2\mu}\Delta_\rho + V_{12}\left(\rho\right), \tag{4.53}$$

where $M = m_1 + m_2$ and the reduced mass $\mu = \frac{m_1 m_2}{M}$.

The resulting Schrödinger equation can be separated into two equations: one describing the motion of the center of mass (with the Hamiltonian $-\frac{\hbar^2}{2M}\Delta_{CM}$) and the second one describing the relative motion of the particles $\left(\text{with the Hamiltonian} -\frac{\hbar^2}{2\mu}\Delta_\rho + V_{12}\left(\rho\right)\right)$. This looks as if

---

[76] However, we might have millions of complicated terms.

[77] We replace a set of six coordinates (the components of $\mathbf{r}_1$ and $\mathbf{r}_2$) by a new set of six coordinates.

someone described the whole motion as the independent motion of two "*particles*": one of them has the position of the center of mass and the mass equal to M, while the other "*particle*" has the position of particle 1, seen from particle 2, and has the same mass as the reduced mass μ.

This result has been already used by us (e.g., when describing the rigid rotator). We just want now to recall it and to introduce a useful notation before continuing the discussion in the same vein.

Three Particles

Is a similar separation of the total Schrödinger equation into three one-particle independent equations still possible? It turns out that in principle it is, but under certain specific conditions. It is possible if the Hamiltonian has the form

$$\hat{H}(\boldsymbol{r}_1, \boldsymbol{r}_2, \boldsymbol{r}_3) = \hat{H}_{12}(\boldsymbol{r}_1, \boldsymbol{r}_2) - \frac{\hbar^2}{2m_3}\Delta_3 + a_{13}r_{13}^2 + a_{23}r_{23}^2, \tag{4.54}$$

where $r_{ab} = |\boldsymbol{r}_a - \boldsymbol{r}_b|$ for $a, b = 1, 2, 3$, and $a_{13} = km_1m_3, a_{23} = km_2m_3, k \geq 0$.

Therefore, the separation is possible at any potential[78] V_{12} describing the interaction of 1 and 2, but particle 3 has to interact pairwise with the other particles by harmonic forces and, in addition, the harmonic force constants a_{13} and a_{23} must be proportional to the corresponding masses.

Now the separation should be demonstrated, but we will not do that because all steps resemble very much what was shown in Appendix I available at booksite.elsevier.com/978-0-444- 59436-5, p. e93. The separation is achieved by introducing the following new coordinates (each of the vectors given below has three components treated as new coordinates):

$$\mathbf{R}_0 = \frac{m_1\boldsymbol{r}_1 + m_2\boldsymbol{r}_2 + m_3\boldsymbol{r}_3}{m_1 + m_2 + m_3},$$

$$\boldsymbol{\rho} = \boldsymbol{r}_1 - \boldsymbol{r}_2,$$

$$\mathbf{R}_1 = \frac{m_1\boldsymbol{r}_1 + m_2\boldsymbol{r}_2}{m_1 + m_2} - \boldsymbol{r}_3.$$

The procedure of transforming the Laplacians from the ones expressed in the old coordinates to those given in the new coordinates (the same one shown in Appendix I available at booksite.elsevier.com/978-0-444-59436-5), leads to three mutually independent Hamiltonians:

$$\hat{H} = \hat{h}_0(\boldsymbol{R}_0) + \hat{h}_1(\boldsymbol{\rho}) + \hat{h}_2(\boldsymbol{R}_1), \tag{4.55}$$

where

$$\hat{h}_0(\boldsymbol{R}_0) = -\frac{\hbar^2}{2M}\Delta_{R_0} \tag{4.56}$$

[78] The potential has to depend on the interparticle distance *only*. However, such a situation is common in physics.

corresponds to the center-of-mass motion,

$$\hat{h}_1(\boldsymbol{\rho}) = -\frac{\hbar^2}{2\mu_1}\Delta_\rho + V(\rho) + k\mu_1 m_3 \rho^2 \tag{4.57}$$

describes the relative motion of particles 1 and 2, and

$$\hat{h}_2(\boldsymbol{R}_1) = -\frac{\hbar^2}{2\mu_2}\Delta_{R_1} + k(m_1+m_2)m_3\, R_1^2, \tag{4.58}$$

represents the spherical harmonic oscillator Hamiltonian, which in this case describes the motion of particle 3 with respect to the center of mass of particles 1 and 2. In the above equations, the following abbreviations have been used: $M = m_1 + m_2 + m_3$, $\mu_1 = \frac{m_1 m_2}{m_1+m_2}$, $\mu_2 = \frac{(m_1+m_2)m_3}{M}$.

So here is something really interesting. Indeed, we can manipulate the masses arbitrarily and therefore create links among some apparently unrelated systems. Note that if V is chosen as the Coulombic repulsion, Eq. (4.50) describes either two-electron Hooke atom (i.e., harmonium[79]), and then we put $m_1 = m_2 \ll m_3$ or a one-electron Hooke diatomic ($m_1 = m_2 \gg m_3$).

Four Particles

In the case of four particles, a separable Hamiltonian has the form

$$\hat{H}(\boldsymbol{r}_1,\boldsymbol{r}_2,\boldsymbol{r}_3,\boldsymbol{r}_4) = \hat{H}_{12}(\boldsymbol{r}_1,\boldsymbol{r}_2) + \hat{H}_{34}(\boldsymbol{r}_3,\boldsymbol{r}_4) + k\left(m_1 m_3\, r_{13}^2 + m_1 m_4\, r_{14}^2 + m_2 m_3\, r_{23}^2 + m_2 m_4\, r_{24}^2\right).$$

New coordinates (on the left side) are introduced:

$$\begin{aligned}
\boldsymbol{\rho}_1 &= \boldsymbol{r}_1 - \boldsymbol{r}_2,\\
\boldsymbol{\rho}_2 &= \boldsymbol{r}_3 - \boldsymbol{r}_4,\\
\boldsymbol{\rho}_{34} &= \boldsymbol{\rho}_3 - \boldsymbol{\rho}_4,\\
\boldsymbol{R}_0 &= \frac{M_1\boldsymbol{\rho}_3 + M_2\boldsymbol{\rho}_4}{M_1+M_2},
\end{aligned}$$

with

$$\begin{aligned}
\boldsymbol{\rho}_3 &= \frac{m_1\boldsymbol{r}_1 + m_2\boldsymbol{r}_2}{M_1},\\
\boldsymbol{\rho}_4 &= \frac{m_3\boldsymbol{r}_3 + m_4\boldsymbol{r}_4}{M_2},
\end{aligned}$$

where $M_1 = m_1 + m_2$, $M_2 = m_3 + m_4$. After the corresponding changes in the Laplacians, we get in the new coordinates

$$\hat{H}(\boldsymbol{r}_1,\boldsymbol{r}_2,\boldsymbol{r}_3,\boldsymbol{r}_4) = \hat{h}_0(\boldsymbol{R}_0) + \hat{h}_1(\boldsymbol{\rho}_1) + \hat{h}_2(\boldsymbol{\rho}_2) + \hat{h}_{34}(\boldsymbol{\rho}_{34}), \tag{4.59}$$

[79] Harmonium represents the two-electron Hooke atom. A Hooke diatomic molecule means two heavy particles (nuclei) interacting by Coulomb forces. The same is true with electrons, but the heavy particle-light particle interactions are harmonic.

where

$$\hat{h}_i(\boldsymbol{\rho}_i) = -\frac{\hbar^2}{2\mu_i}\Delta_{\rho_i} + V(\rho_i) + k\mu_i M_{3-i}\,\rho_i^2, \qquad i = 1, 2, \tag{4.60}$$

$$\hat{h}_{34}(\boldsymbol{\rho}_{34}) = -\frac{\hbar^2}{2\mathfrak{M}}\Delta_{\rho_{34}} + kM_1M_2\rho_{34}^2, \tag{4.61}$$

with $\mu_i = \frac{m_{2i-1}m_{2i}}{m_{2i-1}+m_{2i}}$, $\mathfrak{M} = \frac{M_1M_2}{M_1+M_2}$, $\hat{h}_0(\boldsymbol{R}_0)$ describes the center-of-mass motion. If $m_1 = m_2 \gg m_3 = m_4$, we obtain a model of a homonuclear Hooke's diatomic.

We are tempted now to go further and think about an even larger number of particles. Maybe we will conquer more and more, step by step, until a number of particles for which no separation is possible. Then we might say we did our best, but this is the limit... For the reader who shares this concern, we have a good news: the separation is sometimes possible even for an infinite number of particles!

N Particles

Let us suppose we have K pairs of particles in a system of N particles, their interaction within each pair given by potential energy expressions[80] $V_i(\rho_i), i = 1, 2, \ldots, K$ (where $\boldsymbol{\rho}_i = \boldsymbol{r}_{2i-1} - \boldsymbol{r}_{2i}$). The other interactions are described by quadratic forms of coordinates[81] (satisfying some conditions that ensure separability). The separation of variables gives K one-particle Hamiltonians:

$$\hat{h}_i(\boldsymbol{\rho}_i) = -\frac{\hbar^2}{2\mu_i}\Delta_{\rho_i} + V_i(\rho_i) + A_i\,\rho_i^2, \qquad i = 1, 2, \ldots, K, \tag{4.62}$$

where A_i is a constant that depends on particles' masses, and $N - K - 1$ Hamiltonians describing spherical harmonic oscillators with properly defined coordinates (known as normal coordinates; see Chapter 7). We also get, of course, the one-particle Hamiltonian for the center-of-mass motion. In such a case of separability, we obtain N Schrödinger equations, each one for one "*particle*."

Solution of the problem of N interacting particles and a systematic procedure leading to the separation of variables has been given by my friend, Jacek Karwowski, a professor at the

[80] V_i with index i means that each of the two-particle potentials may have different mathematical forms.

[81] The assumption of the quadratic form may lead (provided that some conditions are satisfied) to harmonic motions. We see once more a peculiar role of harmonic approximation in physics. They often lead to unexpectedly simple (and therefore beautiful) expressions.

Nicolaus Copernicus University (Torun, Poland).[82] The solutions already known in the literature turn out to be the special cases of this general one: the already-described harmonium ($N = 3$), H_2^+ Hooke's molecule,[83] and H_2 Hooke's molecule.[84]

4.14 Charming SUSY and New Solutions

How can we find exact solutions for new systems? Our exciting story about SUSY is reserved for real fans of her quantum beauty, and certainly may be omitted by readers with a more pragmatic than romantic attitude toward quantum mechanics.

We start from some banal observations (tension in our story will be built up gradually): two identical molecules have the same energy levels. But what about two *different* systems? Can the sets of energy levels of these systems be identical? Well, it is like saying that a zither and a jar would have the same set of vibrational normal modes. If this strange thing happened, we would guess that this might come from some invisible common feature of these two objects. This feature might be called *supersymmetry (SUSY)* just to stress that the two systems are related one to the other by a hidden symmetry operation.[85]

Theoretical physics already has considered such a problem. There was the possibility of reducing the number of different kinds of interactions by introducing a relation between the theoretical description of a fermion of spin s and a boson of spin $s \pm \frac{1}{2}$. This relation has been named *supersymmetry*. Such an idea seems very courageous because the particles have very

[82] J. Karwowski, *Int. J. Quantum Chem., 108,* 2253 (2008). Jacek Karwowski was the first to notice that the separability can be performed sequentially (by pairs of particles interacting through V_i) in this way including any number of particles pairs. He proved then that the remaining interactions being quadratic forms may in some cases be reduced to a canonical form (i.e., to normal modes; see Chapter 7) in such a way that does not pertain to the variables of already separated pairs. The first step is always feasible, while the second one is only if the quadratic form satisfies some conditions.

[83] X. Lopez, J.M. Ugalde, L. Echevarria, and E.V. Ludeña, *Phys. Rev., A74,* 042504 (2006).

[84] E.V. Ludeña, X. Lopez, and J.M. Ugalde, *J. Chem. Phys., 123,* 024102 (2005).

[85] Quantum chemistry includes an intriguing notion of isospectral molecules, which have identical energy levels (we mean the Hueckel method; i.e., a simplified molecular orbital model; see Chapter 8). From this, we have a long way to go to reach the same systems of realistic energy levels, and even further to the same spectra (e.g., in UV-VIS) of both molecules, which need the same transition probabilities. Nevertheless, we would not be astonished if, for example, a piece of tiger skin and the marigold flower had the same color (coming from the same spectra), although probably the molecules responsible for this would be different.

My friend Leszek Stolarczyk remarked on still another kind of symmetry in chemistry (in an unpublished paper). Namely, the alternant hydrocarbons, defined in the Hueckel theory, despite the fact that they often do not have any spatial symmetry at all, have a symmetric energy level pattern with respect to a reference energy. We meet the same feature in the Dirac theory, if the electronic and positronic levels are considered. This suggests a underlying, not yet known, internal reason common to the alternant hydrocarbons as viewed in the Hueckel theory and the Dirac model, which seems to be related somehow to the notion of supersymmetry.

different properties (due to the antisymmetry of wave functions for fermions and symmetry for bosons; see Chapter 1). Despite of that, such a relation has been introduced.[86]

4.14.1 SUSY Partners

The supersymmetry in quantum mechanics[87] will be shown in the simplest case possible: a single particle of mass m moving along the x-axis. The solution of the Schrödinger equation for the harmonic oscillator, p.188, by using the creation and annihilation operators will shine out, exposing new and unexpected beauty.

Similar to what we have done for the quantum oscillator, let us introduce the operator

$$\hat{A} = \frac{\hbar}{\sqrt{2m}} \frac{d}{dx} + W(x), \tag{4.63}$$

together with its Hermitian conjugate[88]:

$$\hat{A}^{\dagger} = -\frac{\hbar}{\sqrt{2m}} \frac{d}{dx} + W(x). \tag{4.64}$$

Function $W(x)$ is called a *SUSY superpotential.*

From these two operators, two Hamiltonians may be constructed:

$$\hat{H}_1 = \hat{A}^{\dagger} \hat{A} \tag{4.65}$$

and

$$\hat{H}_2 = \hat{A} \hat{A}^{\dagger}. \tag{4.66}$$

After inserting Eqs. (4.63) and (4.64) into Eq. (4.65), we obtain[89]

$$\begin{aligned}
\hat{H}_1 f = \hat{A}^{\dagger} \hat{A} f &= \left(-\frac{\hbar}{\sqrt{2m}} \frac{d}{dx} + W(x)\right) \left(\frac{\hbar}{\sqrt{2m}} \frac{d}{dx} + W(x)\right) f \\
&= -\frac{\hbar}{\sqrt{2m}} \frac{d}{dx} \frac{\hbar}{\sqrt{2m}} \frac{df}{dx} - \frac{\hbar}{\sqrt{2m}} \frac{d(Wf)}{dx} + W \frac{\hbar}{\sqrt{2m}} \frac{df}{dx} + W^2 f \\
&= -\frac{\hbar^2}{2m} \frac{d^2 f}{dx^2} + W^2 f - f \frac{\hbar}{\sqrt{2m}} \frac{dW}{dx} = \left(-\frac{\hbar^2}{2m} \frac{d^2}{dx^2} + V_1\right) f.
\end{aligned}$$

[86] It is not yet certain whether SUSY represents indeed a symmetry seen in nature. Since in what is known as the standard model, a rigorous SUSY does not appear, it is believed that SUSY (if it exists at all) "*is broken.*" If it would appear that SUSY exists for very high energies, this would be very important for unification of the electro-weak and strong interactions in physics.

[87] The idea of supersymmetry has been introduced to quantum mechanics by C. V. Sukumar, *J. Phys. A, 18,* L57 (1985). My inspiration for writing this section came from a beautiful 2004 lecture by Avinash Khare, who also belongs to the pioneers of SUSY in quantum mechanics [F.Cooper, A.Khare, and U.Sukhatme, "*Supersymmetry and quantum mechanics,*" World Scientific, Singapore (2001)].

[88] The proof is almost identical to that shown for the harmonic oscillator.

[89] In order to avoid a mistake, we will show all this when acting on an arbitrary function f of class Q.

Hence, we have

$$\hat{H}_1 = -\frac{\hbar^2}{2m}\frac{d^2}{dx^2} + V_1, \tag{4.67}$$

where

$$V_1 = W^2 - \frac{\hbar}{\sqrt{2m}} W'. \tag{4.68}$$

Similarly, we get

$$\hat{H}_2 = -\frac{\hbar^2}{2m}\frac{d^2}{dx^2} + V_2, \tag{4.69}$$

$$V_2 = W^2 + \frac{\hbar}{\sqrt{2m}} W'. \tag{4.70}$$

The potentials V_1 and V_2 are known as *SUSY partner potentials.*

4.14.2 Relation Between the SUSY Partners

Let us write down the Schrödinger equation for Hamiltonian $\hat{H}_1$ (the SUSY partners will be indicated by superscript numbers):

$$\hat{H}_1\psi_n^{(1)} = \hat{A}^\dagger\hat{A}\psi_n^{(1)} = E_n^{(1)}\psi_n^{(1)}.$$

After multiplying this from the left by $\hat{A}$, we get something interesting: $\hat{A}\hat{A}^\dagger(\hat{A}\psi_n^{(1)}) = E_n^{(1)}(\hat{A}\psi_n^{(1)})$, which means $\hat{H}_2(\hat{A}\psi_n^{(1)}) = E_n^{(1)}(\hat{A}\psi_n^{(1)})$. After introducing a symbol for the wave functions of Hamiltonian $\hat{H}_2$, we have $\hat{H}_2\psi_n^{(2)} = E_n^{(1)}\psi_n^{(2)}$, where $\psi_n^{(2)} \equiv \hat{A}\psi_n^{(1)}$.

It turns out, therefore, that each energy eigenvalue of $\hat{H}_1$ also represents an energy eigenvalue of $\hat{H}_2$, while the wave function corresponding to $\hat{H}_1$, after transforming it by applying operator $\hat{A}$, becomes a wave function of $\hat{H}_2$. We transform similarly the Schrödinger equation for $\hat{H}_2$ as follows:

$$\hat{H}_2\psi_n^{(2)} = \hat{A}\hat{A}^\dagger\psi_n^{(2)} = E_n^{(2)}\psi_n^{(2)}$$

through multiplying by $\hat{A}^\dagger : \hat{A}^\dagger\hat{A}(\hat{A}^\dagger\psi_n^{(2)}) = E_n^{(2)}(\hat{A}^\dagger\psi_n^{(2)})$. We obtain $\hat{H}_1(\hat{A}^\dagger\psi_n^{(2)}) = E_n^{(2)}(\hat{A}^\dagger\psi_n^{(2)})$. Therefore, we have the splendid beauty of SUSY right before our eyes:

$$\psi_n^{(2)} \equiv \hat{A}\psi_n^{(1)} \quad \text{and} \quad \psi_n^{(1)} \equiv \hat{A}^\dagger\psi_n^{(2)}. \tag{4.71}$$

An eigenvalue of $\hat{H}_1$ represents also an eigenvalue of Hamiltonian $\hat{H}_2$. Also, if $\psi_n^{(1)}$ is an eigenfunction of $\hat{H}_1$, then function $\hat{A}\psi_n^{(1)}$ is also an eigenfunction of $\hat{H}_2$. On the other hand, if $\psi_n^{(2)}$ is an eigenfunction of $\hat{H}_2$, then $\hat{A}^\dagger\psi_n^{(2)}$ also represents an eigenfunction of $\hat{H}_1$.

There is one small point that we forgot to tell you: the SUSY may have a defect. Namely, the proof given above fails if for some reason, $\hat{A}\psi_n^{(1)} = 0$ (case 1) or $\hat{A}^\dagger\psi_n^{(2)} = 0$ (case 2). Indeed, if this happened, this would mean that an eigenfunction of $\hat{H}_2$ or $\hat{H}_1$, respectively, would be 0 everywhere, which is unacceptable in quantum mechanics (wave functions must be of class Q). Will such a case have a chance to happen? First of all, note[90] that $E_n^{(i)} \geq 0$.

- *Case 1*. From $\hat{A}\psi_n^{(1)} = 0$, it follows that $\hat{A}^\dagger\hat{A}\psi_n^{(1)} = 0$, which means $\hat{H}_1\psi_n^{(1)} = 0 = E_n^{(1)}\psi_n^{(1)}$. Since $E_n^{(i)} \geq 0$, this may happen for $n = 0$ only; and on top of that, we have to have $E_0^{(1)} = 0$. The SUSY partner of this state is $\psi_0^{(2)} = \hat{A}\psi_0^{(1)} = 0$ (because in the case considered, $\hat{A}\psi_0^{(1)} \equiv 0$). This, however means that the SUSY partner $\psi_0^{(2)}$ simply does not exist in this case.
- *Case 2*. We have $\hat{A}^\dagger\psi_n^{(2)} = 0$, which leads to $\hat{A}\hat{A}^\dagger\psi_n^{(2)} = 0$, and this means $\hat{H}_2\psi_n^{(2)} = 0 = E_n^{(2)}\psi_n^{(2)}$. Since $E_n^{(i)} \geq 0$, this may happen only if $n = 0$; and in addition, we should have $E_0^{(2)} = 0$. The SUSY partner of this state $\psi_0^{(1)} = \hat{A}^\dagger\psi_0^{(2)} = 0$ (because by definition, $\hat{A}^\dagger\psi_0^{(2)} \equiv 0$). This means the SUSY partner $\psi_0^{(1)}$ does not exist.

Therefore,

> at any complication (either $\hat{A}\psi_0^{(1)} = 0$ or $\hat{A}^\dagger\psi_0^{(2)} = 0$), if it happens, the ground-state Schrödinger equation has a *single* SUSY partner only. The conditions $\hat{A}\psi_0^{(1)} = 0$ and $\hat{A}^\dagger\psi_0^{(2)} = 0$ cannot be satisfied simultaneously.

The equation $\hat{A}\psi_0^{(1)} = 0$ may be used to calculate $\psi_0^{(1)}$, if $W(x)$ is known or, to find $W(x)$, if $\psi_0^{(1)}$ is known.[91] In the first case, we have (with N standing for the normalization constant)[92]

$$\psi_0^{(1)} = N \exp\left[-\frac{\sqrt{2m}}{\hbar}\int_{-\infty}^{x} W(y)\right] dy. \tag{4.72}$$

---

[90] This follows from the equation for the eigenvalues: $\hat{H}_1\psi_n^{(1)} = \hat{A}^\dagger\hat{A}\psi_n^{(1)} = E_n^{(1)}\psi_n^{(1)}$. Indeed, let us make the scalar product of both sides with function $\psi_n^{(1)}$: $E_n^{(1)} = \left\langle\psi_n^{(1)}|\hat{H}_1\psi_n^{(1)}\right\rangle = \left\langle\psi_n^{(1)}|\hat{A}^\dagger\hat{A}\psi_n^{(1)}\right\rangle = \left\langle\hat{A}\psi_n^{(1)}|\hat{A}\psi_n^{(1)}\right\rangle \geq 0$, because the square of the length of vector $\hat{A}\psi_n^{(1)}$ is non-negative.

[91] In such a case, we will force one of the SUSY partners to have one level more than the other.

[92] Indeed, let us check this:

$$\begin{aligned}\hat{A}\psi_0^{(1)} &= \left(\frac{\hbar}{\sqrt{2m}}\frac{d}{dx} + W(x)\right)\psi_0^{(1)} \\ &= N\frac{\hbar}{\sqrt{2m}}\exp\left[-\frac{\sqrt{2m}}{\hbar}\int_{-\infty}^{x} W(y)\right] dy\left(-\frac{\sqrt{2m}}{\hbar}\right)W(x) + W(x)\psi_0^{(1)} \\ &= -W(x)\psi_0^{(1)} + W(x)\psi_0^{(1)} = 0.\end{aligned}$$

As to the second case, we have[93]

$$W(x) = -\frac{\hbar}{\sqrt{2m}} \frac{\frac{d}{dx}\psi_0^{(1)}}{\psi_0^{(1)}}. \tag{4.73}$$

Example 1. ***Harmonic oscillator*** $\longleftrightarrow$ ***harmonic oscillator***

Let us see what kind of SUSY partner has a harmonic oscillator. First, let us shift the energy scale[94] in such a way that the ground state has energy 0. This means that the shifted potential has the form

$$V_1(x) = \frac{1}{2}kx^2 - \frac{1}{2}h\nu = \frac{1}{2}kx^2 - \frac{1}{2}\hbar\omega. \tag{4.74}$$

Now, from Eq. (4.73) and the expression for the ground state wave function for the harmonic oscillator, we find the superpotential from Eq. (4.73):

$$W(x) = -\frac{\hbar}{\sqrt{2m}} \frac{\frac{d}{dx}\psi_0^{(1)}}{\psi_0^{(1)}}$$
$$= -\frac{\hbar}{\sqrt{2m}} \frac{-\sqrt{\frac{km}{\hbar^2}}x \exp\left(-\frac{1}{2}\sqrt{\frac{km}{\hbar^2}}x^2\right)}{\exp\left(-\frac{1}{2}\sqrt{\frac{km}{\hbar^2}}x^2\right)} = \sqrt{\frac{k}{2}}x.$$

Then, from Eqs. (4.68) and (4.70), we find the SUSY partners: $V_1 = W^2 - \frac{\hbar}{\sqrt{2m}}W' = \frac{1}{2}kx^2 - \frac{\hbar}{\sqrt{2m}}\sqrt{\frac{k}{2}} = \frac{1}{2}kx^2 - \frac{1}{2}\hbar\omega$, which agrees with Eq. (4.74), and a new potential partner:

$$V_2 = W^2 + \frac{\hbar}{\sqrt{2m}}W' = \frac{1}{2}kx^2 + \frac{\hbar}{\sqrt{2m}}\sqrt{\frac{k}{2}} = \frac{1}{2}kx^2 + \frac{1}{2}\hbar\omega.$$

As we see, the potential V_2 also represents a harmonic oscillator potential. The only difference, a shift on the energy scale, is not important.[95] We see once more that a harmonic oscillator plays an outstanding role in physics.

[93] When we check this: the right side is equal to

$$-\frac{\hbar}{\sqrt{2m}} \frac{\frac{d}{dx}\psi_0^{(1)}}{\psi_0^{(1)}} = -\frac{\hbar}{\sqrt{2m}} \frac{-\frac{\sqrt{2m}}{\hbar}W(x)\psi_0^{(1)}}{\psi_0^{(1)}} = W(x),$$

which is equal to the left side. This fascinating formula says: "*Show me the logarithmic derivative of the ground state wave function and I will tell you what system we are talking about.*"

[94] The world looks the same after someone decides to measure energy with respect to another point on the energy axis.

[95] From a very formal point of view, the second SUSY partner is devoid of the ground state of the first partner, as discussed earlier. However, since a harmonic oscillator has an infinite number of equidistant levels, a new SUSY partner means only a shift of all energy levels up by $\hbar\omega$, so the two SUSY partners represent the same system.

Example 2. ***Particle in a box $\longleftrightarrow$ cotangent potential***

At the beginning of this chapter (p. 162), the problem of the particle in a box was solved. Now we will treat this as one of the SUSY partners–say, the first one. Our goal will be to find the second SUSY partner. Similar to the harmonic oscillator a while ago, we will shift the energy (this time by $-\frac{h^2}{8mL^2}$) in order to have the ground state of the particle in a box at energy 0. The formula for the energy levels in the new energy scale, using the SUSY-type notation for the eigenvalues and the eigenfunctions (as well as counting the ground state as corresponding to $n = 0$), has the form

$$E_n^{(1)} = \frac{(n+1)^2 h^2}{8mL^2} - \frac{h^2}{8mL^2} = \frac{n\,(n+2)\,h^2}{8mL^2}, n = 0, 1, 2, \ldots,$$

while the normalized wave function is

$$\psi_n^{(1)} = \sqrt{\frac{2}{L}} \sin \frac{(n+1)\pi}{L} x.$$

From Eq. (4.73), we find the superpotential $W(x)$ as

$$W(x) = -\frac{\hbar}{\sqrt{2m}} \frac{\frac{\mathrm{d}}{\mathrm{d}x}\psi_0^{(1)}}{\psi_0^{(1)}} = -\frac{\hbar\pi}{\sqrt{2m}L} \mathrm{ctg}\left(\frac{\pi}{L}x\right),$$

while from Eqs. (4.68) and (4.70), we obtain the potential energies for the two SUSY partners:

$$V_1(x) = W^2 - \frac{\hbar}{\sqrt{2m}} W' = \frac{\hbar^2\pi^2}{2mL^2}\left[\mathrm{ctg}\left(\frac{\pi}{L}x\right)\right]^2 - \frac{\hbar}{\sqrt{2m}}\left[-\frac{\hbar}{\sqrt{2m}}\frac{\pi}{L}\right]\frac{\mathrm{dctg}\left(\frac{\pi}{L}x\right)}{\mathrm{d}x}$$

$$= \frac{\hbar^2\pi^2}{2mL^2}\left[\mathrm{ctg}\left(\frac{\pi}{L}x\right)\right]^2 + \frac{\hbar^2\pi^2}{2mL^2}\frac{-1}{\sin^2\frac{\pi}{L}x} = \frac{\hbar^2\pi^2}{2mL^2}\left[\mathrm{ctg}^2\left(\frac{\pi}{L}x\right) - \frac{1}{\sin^2\frac{\pi}{L}x}\right]$$

$$= \frac{\hbar^2\pi^2}{2mL^2}\left[\frac{\cos^2\frac{\pi}{L}x - 1}{\sin^2\frac{\pi}{L}x}\right] = -\frac{\hbar^2\pi^2}{2mL^2} = -\frac{h^2}{8mL^2} = \mathrm{const}$$

$$V_2(x) = W^2 + \frac{\hbar}{\sqrt{2m}} W' = \frac{\hbar^2\pi^2}{2mL^2}\left[\mathrm{ctg}\left(\frac{\pi}{L}x\right)\right]^2 + \frac{\hbar}{\sqrt{2m}}\left[-\frac{\hbar}{\sqrt{2m}}\frac{\pi}{L}\right]\frac{\mathrm{dctg}\left(\frac{\pi}{L}x\right)}{\mathrm{d}x}$$

$$= \frac{\hbar^2\pi^2}{2mL^2}\left[\mathrm{ctg}\left(\frac{\pi}{L}x\right)\right]^2 - \frac{\hbar^2\pi^2}{2mL^2}\frac{-1}{\sin^2\frac{\pi}{L}x} = \frac{\hbar^2\pi^2}{2mL^2}\frac{\cos^2\frac{\pi}{L}x + 1}{\sin^2\frac{\pi}{L}x}$$

$$= \frac{\hbar^2\pi^2}{2mL^2}\left[2\mathrm{ctg}^2\frac{\pi}{L}x + 1\right].$$

The first formula says that $V_1(x)$ has a constant value in $[0, L]$ (this corresponds to the particle-in-a-box problem). The bottom of this box is at energy $-\frac{h^2}{8mL^2}$, because only then the ground state energy will equal 0 (as it was assumed). Its SUSY partner corresponds to $V_2(x) = \frac{\hbar^2\pi^2}{2mL^2}\left[2\mathrm{ctg}^2\frac{\pi}{L}x + 1\right]$. As it follows from Eq. (4.71), the wave functions corresponding

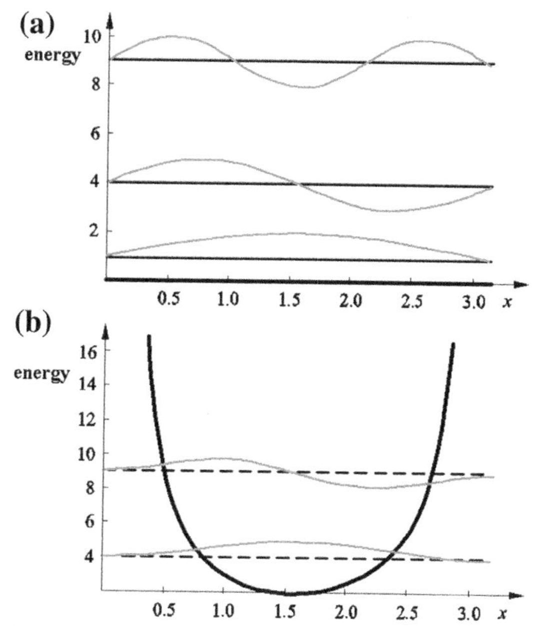

Fig. 4.26. The SUSY partners lead to the same set of energy levels, except possibly for one (the ground state of one of the partners). One of the SUSY partners corresponds to the particle in a box (see p. 162), as shown in panel (a), together with the three lowest energy levels 1, 4, and 9. In order to obtain the simplest formulas (and therefore get rid of the unnecessary luggage), $\frac{\hbar^2\pi^2}{2mL^2}$ has been used as the energy unit and, additionally, the box length has been chosen equal π. The original energy scale has been used in the figure (in which the potential energy for the particle in the box equals 0). The corresponding wave functions have been plotted for the three lowest levels (each plot put artificially at the height of the corresponding energy level). The functions have zero value beyond $(0, L)$. Panel (b) shows $V_2(x)$, calculated as the SUSY partner, together with the two lowest energy levels 4 and 9 and the corresponding wave functions (exposed at the height of their levels).

to $V_2(x)$ can be calculated from the particle-in-a-box wave functions by applying operator $\hat{A}$. Therefore,

$$\begin{aligned}
\psi_1^{(2)} = \hat{A}\psi_1^{(1)} &= \sqrt{\frac{2}{L}}\left(\frac{\hbar}{\sqrt{2m}}\frac{d}{dx} + W(x)\right)\sin\frac{2\pi}{L}x \\
&= \sqrt{\frac{2}{L}}\left(\frac{\hbar}{\sqrt{2m}}\frac{2\pi}{L}\cos\frac{2\pi}{L}x - \frac{\hbar\pi}{\sqrt{2m}L}\mathrm{ctg}\left(\frac{\pi}{L}x\right)\sin\frac{2\pi}{L}x\right) \\
&= \sqrt{\frac{2}{L}}\frac{\hbar}{\sqrt{2m}}\frac{2\pi}{L}\left(\cos\frac{2\pi}{L}x - \mathrm{ctg}\left(\frac{\pi}{L}x\right)\sin\frac{\pi}{L}x\cos\frac{\pi}{L}x\right) \\
&= \sqrt{\frac{2}{L}}\frac{\hbar}{\sqrt{2m}}\frac{2\pi}{L}\left(\cos\frac{2\pi}{L}x - \cos^2\frac{\pi}{L}x\right) = N'\sin^2\frac{\pi}{L}x,
\end{aligned}$$

where N' is a normalization constant. Similarly, one gets $\psi_2^{(2)} = N''\sin x\sin 2x$. Fig. 4.26 shows $V_1(x)$ and $V_2(x)$, as well as the energy levels corresponding to both SUSY partners. For each of the partners, the consecutive energy levels correspond to the wave functions with increasing number of nodes. As one can see, the SUSY partner potentials differ widely: $V_1(x)$

is sharp-edged, while $V_2(x)$ resembles this a bit, but is cleverly rounded in order to repel the eigenstates corresponding to $V_1(x)$ (by one level) up, and precisely to the positions of the partner's excited states.

The two SUSY partners differ by one (ground-state) level.

Our story about beautiful SUSY has a happy ending: it was possible to find about a dozen of SUSY partners (for the solutions of the Schrödinger equation known earlier).

4.15 Beacons and Pearls of Physics

Sometimes students, spoiled by the handy computers available nowadays, tend to treat the simple systems described in this chapter as primitive and out of date. A professor has taken them from the attic and after dusting them off, shows them in a class, while out in the "real world," computers with high-level science, splendid programs, and colorful graphs await. This is wrong. The simple systems considered in this chapter correspond to extremely rare, *exact* solutions of Schrödinger equations and are, therefore, precious pearls of physics by themselves. Nobody will give a better solution, and the conclusions are 100 percent certain. It is true that except the hydrogen atom, they all correspond to some idealized systems.[96] There is no such a thing as an unbreakable spring (e.g., a harmonic oscillator) or a rotator, that does not change its length, etc. And yet these problems represent our firm ground or the beacons of our native land. After reading the present chapter, we will be preparing for a long voyage. When confronted with the surprises of new lands and trying to understand them,

> the *only* points of reference or the beacons that tell us about *terra firma* will be the problems for which analytical solutions have been found.

Summary

Exact analytical solutions[97] to the Schrödinger equation play an important role as an organizer of our quantum mechanical experiences. Such solutions have only been possible to obtain for some idealized objects. This is of great importance for the interpretation of approximate solutions for real systems. Another great feature of exact solutions is that they have an extremely wide range of applications: they are useful independent of whether we concentrate on an electron in an atom, a molecule, a nucleon in a nucleus, a molecule as an entity, etc.

The main features of the solutions are:

- **Free particle.** The particle may be described as the superposition of the state $\exp(i\kappa x)$, corresponding to the particle moving right (positive values of x), and the state $\exp(-i\kappa x)$, that corresponds to the particle moving left. Both states correspond to the same energy (and opposite momenta $\pm\hbar\kappa$).
- **Particle in a box.** We consider first a particle in a 1-D box; i.e., the particle is confined to section $[0, L]$ with potential energy (for a particle of mass m and coordinate x) equal to zero and ∞ outside the section. Such a potential forces the wave function to be nonzero only within the section $[0, L]$. We solve the elementary

[96] Like Platonic ideal solids.

[97] This is to distinguish from accurate solutions (i.e., received with a desired accuracy).

Schrödinger equation and obtain $\Psi = A \sin \kappa x + B \cos \kappa x$, where $\kappa^2 = \frac{2mE}{\hbar^2}$. Quantization appears in a natural way from the condition of continuity for the wave function at the boundaries: $\Psi(0) = 0$ and $\Psi(L) = 0$. These two conditions give the expression for the energy levels $E_n = \frac{n^2 h^2}{8mL^2}$ and for the wave functions $\Psi_n = \sqrt{\frac{2}{L}} \sin \frac{n\pi}{L} x$ with quantum number $n = 1, 2, \ldots$ *Conclusion:* the successive energy levels are more and more distant and the wave function is simply a section of the sine function (with 0 value at the ends).

- **Tunneling effect**. We have a particle of mass m and a rectangular barrier (section $[0, a]$, width a and height V_0). Beyond this section, the potential energy is zero. The particle comes from the negative x values and has energy $E < V_0$. A classical particle would be reflected from the barrier. However, for the quantum particle:
  - The transmission coefficient is nonzero.
  - The passage of a large energy particle is easier.
  - A narrower barrier means a larger transmission coefficient.
  - The higher the barrier, the smaller the transmission coefficient is.

  The first feature is the most sensational, the others are intuitively quite acceptable.
- **Resonances over the barrier.** A particle with the energy $E > V_0$ either bounces off the barrier or passes over it. For some particular energies, there are transmissions of the particle with 100% certainty. These resonance states are at the energies, which correspond to the eigenvalues of a potential well with the shape of the barrier (i.e., as in the particle in a box) and placed on top of the barrier.
- **Resonance states at the double barrier.** It turns out that (for a given interbarrier distance and at energies lower than the barrier) there are some "*magic*" energies of the particle (resonance energies), at which the transmission coefficient is equal 100%. The magic energies correspond to the stationary states that would be for a particle in a box a little longer than the interbarrier distance. The resonance states are also over the barrier and nearly assure a transmission coefficient equal to 100%, whereas other energies may lead to reflection of the particle, even if they are larger than the barrier height.
- **Harmonic oscillator**. A single particle of mass m coupled by a harmonic spring (with force constant k) corresponds to potential energy $V = \frac{kx^2}{2}$. We obtain quantization of the energy: $E_\upsilon = h\nu\left(\upsilon + \frac{1}{2}\right)$, where the vibrational quantum number $\upsilon = 0, 1, 2, \ldots,$ and the angular frequency $\omega = 2\pi\nu = \sqrt{\frac{k}{m}}$. We see that the non-degenerate energy levels are equidistant, and their distance is larger for a larger force constant and smaller mass. The wave function has the form of a product of a Gaussian factor and a polynomial of degree υ. The polynomial assures the proper number of nodes, while the Gaussian factor damps the plot to zero for large displacements from the particle equilibrium position. The harmonic oscillator may be viewed (Chapter 6) as a model (for small displacements) for two masses bound by a harmonic spring (like a diatomic molecule).
- **Morse oscillator**. The harmonic oscillator does not allow for the breaking of the spring connecting two particles, while the Morse oscillator admits dissociation. This is extremely important because real diatomic molecules resemble the Morse rather than the harmonic oscillator. The solution for the Morse oscillator has the following features:
  - Energy levels are non-degenerate.
  - Their number is finite.
  - For large well depths, the low-energy levels tend to be the energy levels of the harmonic oscillator (the levels are nearly equidistant).
  - The higher the energy level, the larger the displacement from the equidistant situation is (the energy levels get closer).
  - The wave functions, especially those corresponding to deep-lying levels, are very similar to the corresponding ones of the harmonic oscillator,[98] but they do not exhibit the symmetry.[99]

[98] This is despite the fact that the formula itself is very different.

[99] The wave functions for the harmonic oscillator are either even or odd with respect to the inversion operation ($x \rightarrow -x$).

- **Rigid rotator.** This is a system of two masses m_1 and m_2 that keep their distance R fixed. After separating the center-of-mass motion (see Appendix I available at booksite.elsevier.com/978-0-444-59436-5 on p. e93), we obtain an equation of motion for a single particle of mass equal to the reduced mass μ moving on a sphere of radius R (position given by angles θ and ϕ). The energy is determined by the quantum number $J = 0, 1, 2, \ldots$ and is equal to $E_J = J(J+1)\frac{\hbar^2}{2\mu R^2}$. As we can see:
  - There is an infinite number of energy levels.
  - The separation of the energy levels increases with the energy (similar to the particle-in-a-box problem).
  - The separation is smaller for larger masses.
  - The separation is smaller for longer rotators.

  The wave functions are the spherical harmonics $Y_J^M(\theta, \phi)$, which for low J are very simple and for large J are complicated trigonometric functions. The integer quantum number M satisfies the relation $|M| \le J$. The energy levels are, therefore, $(2J+1)$-tuply degenerate.
- **Hydrogen-like atom.** We have an electron and a nucleus of charges $-e$ and $+Ze$, respectively, or -1 and $+Z$ in atomic units. The wave function is labeled by three quantum numbers: principal $n = 1, 2, \ldots,$ azimuthal $l = 0, 1, \ldots (n-1)$ and magnetic $m = -l, (-l+1), \ldots, 0, \ldots l$. The energy in atomic units is given by the formula[100] $E_n = -Z^2/(2n^2)$. The wave function represents the product of a polynomial (of r), an exponential function decreasing with r and a spherical harmonic $Y_l^m(\theta, \phi)$, where r, θ, ϕ are the spherical coordinates of the electron, and the nucleus is at the origin. The wave functions are denoted by the symbols nl_m (with s for $l = 0$, p for $l = 1$, etc.): $1s, 2s, 2p_0, 2p_1, 2p_{-1}, \ldots$ The degeneracy of the nth level is equal to n^2.
- **Hooke helium atom.** In this peculiar helium atom, the electrons are attracted to the nucleus by harmonic springs (of equal strength) of equilibrium length equal to zero. For $k = \frac{1}{4}$, an exact analytical solution exists. The exact wave function is a product of two Gaussian functions and a simple factor: $\left(1 + \frac{1}{2}r_{12}\right)$, which correlates the motions of the two electrons.
- **Hooke molecules.** If one has K pairs of particles (with the intrapair interaction of any kind) in a system of N particles and the other interactions are described by quadratic forms of coordinates satisfying some conditions, the separation of variables is possible and gives K one-particle Hamiltonians and $N - K - 1$ Hamiltonians describing spherical harmonic oscillators with properly defined coordinates (known as *normal coordinates*; see Chapter 7). Together with the one-particle Hamiltonian for the center-of-mass motion, we obtain N Schrödinger equations, each one for one "*particle*" only.
- **Supersymmetry (SUSY).** For a known exact solution of the Schrödinger equation for a system, one can define another system with the same spectrum of the eigenvalues (possibly except the ground-state eigenvalue). Such a pair of the systems is known as the *supersymmetry (SUSY) partners*.

Main Concepts, New Terms

[100] The mass of the nucleus is set to infinity.

phonon (p. 190)
phonon vacuum (p. 190)
positronium (p. 208)
radial density (p. 204)
resonance state (p. 181)
rigid rotator (p. 199)
spherical harmonics (p. 200)
SUSY partner (p. 219)
SUSY superpotential (p. 218)
transmission coefficient (p. 176)
tunnelling effect (p. 174)

From the Research Front

A field like the one discussed in the present chapter seems to be definitely closed. We think that we have been lucky enough to solve some simple problems, but others are just too complicated. But this is not true. For several decades, it has been possible to solve a series of nonlinear problems, which were thought in the past to be hopeless. What decides success is choice of the problem, quality of researchers, courage, etc.[101] It is worth noting that a systematic search for promising systems to solve is currently in progress.

Ad Futurum

It seems that the number of exactly solvable problems will continue to increase, although the pace of such research will be low. If exactly solvable problems were closer and closer to the practice of physics, it would be of great importance.

Additional Literature

J. Dvořák and L. Skála, "*Analytical solutions of the Schrödinger equation. Ground state energies and wave functions,*" *Collect. Czech. Chem. Commun., 63,* 1161 (1998).

Here is a very interesting article with the character of a synthesis. Many potentials,[102] leading to exactly solvable problems, are presented in a uniform theoretical approach. The authors give also their own generalizations.

F. Cooper, A. Khare, and U. Sukhatme, "*Supersymmetry and quantum mechanics,*" World Scientific, Singapore (2001).

Questions

1. Free particle. Wave function $\Psi(x) = A\exp(i\kappa x) + B\exp(-i\kappa x)$, where $A, B, \kappa > 0$ represent constants
   a. is the de Broglie's wave of a free particle on the x-axis
   b. describes a particle that moves right with the probability A and left with probability B
   c. and the constants A and B ensure the normalization of wave function Ψ
   d. describes a particle that moves right with the probability proportional to $|A|^2$ and left with probability proportional to $|B|^2$.
2. Particle in a box. Energy quantization follows
   a. from the normalization of the wave function
   b. from a single value of the wave function at any point of space

[101] Already the Morse potential looks very difficult to manage, to say nothing of the harmonic helium atom.
[102] This includes six that are not discussed in this textbook.

c. from the continuity of the wave function
d. from the fact that the wave function represents an eigenfunction of the Hamiltonian.

3. The energy eigenvalues and eigenfunctions for the particle in a box
   a. the energy eigenvalues are doubly degenerate because the box is symmetric
   b. the eigenfunctions are orthogonal
   c. increasing box length results in lowering energy levels and smaller distances between them
   d. the energy levels of a heavier particle are lower than those corresponding to a lighter particle.
4. The tunneling effect (E stands for the energy, V_0 means the height of the energy barrier).
   a. for $E < V_0$ the particle will bounce off the barrier with the probability smaller than 1
   b. the larger the particle's energy $E < V_0$ or smaller the barrier's width the particle will pass easier through the barrier
   c. for $E < V_0$ a heavier particle will pass the barrier more easily
   d. for $E > V_0$, the particle always passes the barrier.
5. Harmonic oscillator.
   a. when the force constant is multiplied by 4, the energy level separation doubles
   b. has an infinite number of energy levels
   c. its energy levels are non-degenerate and equidistant
   d. if a deuteron oscillates in a parabolic energy well instead of a proton, the separation of the energy levels doubles.
6. Harmonic oscillator.
   a. corresponds to a parabolic potential energy well
   b. it is only in the ground state that the mean value of the amplitude is equal zero
   c. its wave functions are all symmetric with respect to the transformation $x \to -x$.
   d. the square of the complex modulus of the wave function vanishes for $x \to \pm\infty$.
7. Rigid rotator (J is the rotational quantum number).
   a. its energy levels are equidistant and degenerate
   b. the degeneracy of the energy level corresponding to J is equal to $2J + 1$
   c. the rotational energy is proportional to $J(J + 1)$
   d. if the two rotating masses double their mass, the separation between the energy levels decreases by the factor of 2.
8. Hydrogen atom.
   a. the largest probability density of finding the electron described by the wave function $1s$ is on the nucleus
   b. for the $1s$ state the largest value of the probability of finding a given nucleus-electron separation is for the value of such separation equal to 1 a.u.
   c. the mean value of the nucleus-electron separation is equal to the radius of the first Bohr orbit
   d. a negative value of an atomic orbital means a lower value, while the positive means a higher value of the probability of finding the electron.
9. Hydrogen atom.
   a. its ground-state wave function is spherically symmetric
   b. the wave functions $1s, 2s, 3s$ are spherically symmetric, while $2p_x$ represents an eigenfunction of the Hamiltonian, of the square of the angular momentum and of the z component of the angular momentum
   c. the number of the node surfaces for the wave functions $1s, 2s, 3s, 2p_x, 3d_{xy}$ is equal to 0, 1, 2, 2, 3
   d. the wave function $2p_y$ has a cylindrical symmetry.

10. Hydrogen atom (n, l, m stand for the quantum numbers, ψ_{nlm} represent orbitals).
    a. the wave function $2\psi_{42-1} - \frac{1}{2}\psi_{32-1}$ is an eigenfunction of the Hamiltonian
    b. the wave function $\frac{1}{2}\psi_{320} + \frac{1}{2}\psi_{32-1}$ is an eigenfunction of the square of the angular momentum
    c. the wave function $i\frac{1}{3}\psi_{310} - \frac{1}{2}i\psi_{320}$ is an eigenfunction of the Hamiltonian and of the z component of the angular momentum
    d. the wave function $\psi_{310} + i\,\frac{1}{2}\psi_{32-1} - i\psi_{300}$ represents an eigenfunction of the Hamiltonian

Answers

1a,d,2c,d,3b,c,d,4a,b,5a,b,c,6a,d,7b,c,d,8a,b,9a,d,10b,c,d

Two Fundamental Approximate Methods

Even the upper end of the river believes in the ocean.
William Stafford

Where Are We?

We are moving upward into the central parts of the TREE trunk.

An Example

We are interested in properties of the ammonia molecule in its ground and excited states; e.g., we would like to know the mean value of the nitrogen-hydrogen distance. Only quantum mechanics gives a method for calculation this value (p. 26): we have to calculate the mean value of an operator with the ground-state wave function. But where could this function be taken from? Could it be a solution of the Schrödinger equation? Impossible; unfortunately, this equation is too difficult to solve (14 particles; cf. problems with exact solutions in Chapter 4).

The only possibility is somehow to obtain an approximate solution to this equation.

What Is It All About?

We need mathematical methods that will allow us to obtain approximate solutions of the Schrödinger equation. These methods are the variational method and the perturbational approach.

Ideas of Quantum Chemistry, Second Edition. http://dx.doi.org/10.1016/B978-0-444-59436-5.00005-2

- Hylleraas Equation (♦)
- Degeneracy (♦)
- Convergence of the Perturbational Series (♦)

Why Is This Important?

We have to know how to compute wave functions. The exact wave function is definitely out of our reach, so in this chapter, we will talk about how to calculate the approximations.

What Is Needed?

- Postulates of quantum mechanics (Chapter 1; needed)
- Hilbert space (see Appendix B available at booksite.elsevier.com/978-0-444-59436-5 p. e7; necessary)
- Matrix algebra (see Appendix A available at booksite.elsevier.com/978-0-444-59436-5 p. e1; needed)
- Lagrange multipliers (see Appendix N available atbooksite.elsevier.com/978-0-444-59436-5 on p. e121; needed)
- Orthogonalization (see Appendix J available at booksite.elsevier.com/978-0-444-59436-5 p. e99; occasionally used)
- Matrix diagonalization (see Appendix K available at booksite.elsevier.com/978-0-444-59436-5 p. e105; needed)
- Group theory (see Appendix C available at booksite.elsevier.com/978-0-444-59436-5 p. e17; occasionally used)

Classical Works

The variational method of linear combinations of functions was formulated by Walther Ritz in a paper "*Über eine neue Methode zur Lösung gewisser Variationsprobleme der mathematischen Physik*," published in *Zeitschrift für Reine und Angewandte Mathematik*, **135,** 1 (1909). ★ The method was applied by Erwin Schrödinger in his first works, "*Quantisierung als Eigenwertproblem*" in *Annalen der Physik*, **79,** 361 (1926); *ibid.* **79,** 489 (1926); *ibid.* 80, 437 (1926); *ibid.* 81, 109 (1926). Schrödinger also used the perturbational approach when developing the theoretical results of Lord Rayleigh for vibrating systems (hence the often-used term *Rayleigh-Schrödinger perturbation theory*). ★ Egil Andersen Hylleraas, in the work "*Über den Grundterm der Zweielektronenprobleme von* H^-, He, Li^+, Be^{++} *usw.*," published in *Zeitschrift der Physik*, **65,** 209 (1930), showed for the first time that the variational principle may be used for separate terms of the perturbational series.

5.1 Variational Method

5.1.1 Variational Principle

Let us write the Hamiltonian $\hat{H}$ of the system under consideration[1] and take an *arbitrary* function Φ, which satisfies the following conditions:

- It depends on the same coordinates as the solution to the Schrödinger equation.
- It is of class Q (which enables it to be normalized).

[1] We focus here on the non-relativistic case (Eq. (2.1)), where the lowest eigenvalue of $\hat{H}$ is bound from below ($> -\infty$). As we remember from Chapter 3, this is not fulfilled in the relativistic case (Dirac's electronic sea), and may lead to serious difficulties in applying the variational method.

We calculate the number ε that depends on Φ (i.e., ε is the mean value of the Hamiltonian and a functional of Φ):

$$\varepsilon[\Phi] = \frac{\left\langle \Phi \middle| \hat{H} \middle| \Phi \right\rangle}{\langle \Phi | \Phi \rangle}.$$

The variational principle states the following:

- $\varepsilon \geq E_0$, where E_0 is the *ground-state* energy of the system (the lowest eigenvalue of $\hat{H}$)
- in the above inequality, $\varepsilon = E_0$ happens if and only if Φ equals the exact ground-state wave function Ψ_0 of the system.

Proof (expansion into eigenfunctions)

The eigenfunctions $\{\Psi_i\}$ of the Hamiltonian $\hat{H}$ represent a complete orthonormal set (see Appendix B available at booksite.elsevier.com/978-0-444-59436-5 on p. e7) in the Hilbert space of our system.[2] This means that any function belonging to this space can be represented as a linear combination of the functions of this set

$$\Phi = \sum_{i=0}^{\infty} c_i \Psi_i, \tag{5.1}$$

where c_i assures the normalization of Φ (i.e., $\sum_{i=0}^{\infty} |c_i|^2 = 1$, because $\langle \Phi | \Phi \rangle = \sum_{i,j} c_j^* c_i \left\langle \Psi_j | \Psi_i \right\rangle = \sum_{i,j} c_j^* c_i \delta_{ij} = \sum_i c_i^* c_i = 1$). Let us insert this into the expression for the mean value of the energy $\varepsilon = \left\langle \Phi | \hat{H} \Phi \right\rangle$. Then,

$$\begin{aligned}
\varepsilon - E_0 &= \left\langle \Phi | \hat{H} \Phi \right\rangle - E_0 = \left\langle \sum_{j=0}^{\infty} c_j \Psi_j | \hat{H} \sum_{i=0}^{\infty} c_i \Psi_i \right\rangle - E_0 \\
&= \sum_{i,j=0}^{\infty} c_j^* c_i E_i \left\langle \Psi_j | \Psi_i \right\rangle - E_0 = \sum_{i,j=0}^{\infty} c_j^* c_i E_i \delta_{ij} - E_0 \\
&= \sum_{i=0}^{\infty} |c_i|^2 E_i - E_0 \cdot 1 = \sum_{i=0}^{\infty} |c_i|^2 E_i - E_0 \sum_{i=0}^{\infty} |c_i|^2 = \sum_{i=0}^{\infty} |c_i|^2 \left(E_i - E_0 \right) \geq 0.
\end{aligned}$$

Note that the *equality* (in the last step) is satisfied only if $\Phi = \Psi_0$. This therefore proves the variational principle $\varepsilon \geq E_0$.

[2] The functions are and will remain unknown; we use only the property of forming a complete set here.

Proof using Lagrange multipliers

In several places in this book, we need to use similar proofs with Lagrange multipliers. This is why we will demonstrate how to prove the same theorem using this technique (see Appendix N available at booksite.elsevier.com/978-0-444-59436-5 on p. e121).

Take the functional

$$\varepsilon[\Phi] = \left\langle \Phi | \hat{H} \Phi \right\rangle. \tag{5.2}$$

We want to find a function that assures a minimum of the functional and satisfies the normalization condition

$$\langle \Phi | \Phi \rangle - 1 = 0. \tag{5.3}$$

We will change the function Φ a little (the change will be called *variation*) and see how this will change the value of the functional $\varepsilon[\Phi]$. But for this functional, we have Φ and Φ^*. Therefore, it seems that for Φ^*, we have to take into account the variation made in Φ. In reality, however, there is no need to do that: it is sufficient to make the variation either in Φ or in Φ^* (the result does not depend on the choice[3]). This makes the formulas simpler. We decide to choose the variation of Φ^*; i.e., $\delta\Phi^*$.

Now we apply the machinery of the Lagrange multipliers (see Appendix N available at booksite.elsevier.com/978-0-444-59436-5 on p. e121). Let us multiply Eq. (5.3) by the (for the time being) unknown Lagrange multiplier E and subtract afterward from the functional ε,

---

[3] Let us demonstrate this principle, because we will use it several times in this book. In all our cases, the functional (which depends here on a single function $\phi(x)$, but later we will also deal with several functions in a similar procedure) might be rewritten as

$$\varepsilon[\phi] = \left\langle \phi | \hat{A} \phi \right\rangle, \tag{5.4}$$

where $\hat{A}$ is a Hermitian operator. Let us write $\phi(x) = a(x) + ib(x)$, where $a(x)$ and $b(x)$ are real functions. The change of ε is equal to

$$\begin{aligned}\varepsilon[\phi + \delta\phi] - \varepsilon[\phi] &= \left\langle a + \delta a + ib + i\delta b | \hat{A}(a + \delta a + ib + i\delta b) \right\rangle - \left\langle a + ib | \hat{A}(a + ib) \right\rangle \\ &= \left\langle \delta a + i\delta b | \hat{A}\phi \right\rangle + \left\langle \phi | \hat{A}(\delta a + i\delta b) \right\rangle + \text{quadratic terms} \\ &= \left\langle \delta a | \hat{A}\phi + (\hat{A}\phi)^* \right\rangle + i \left\langle \delta b | (\hat{A}\phi)^* - \hat{A}\phi \right\rangle + \text{quadratic terms.}\end{aligned}$$

The variation of a functional only represents a linear part of the change, and therefore $\delta\varepsilon = \left\langle \delta a | \hat{A}\phi + (\hat{A}\phi)^* \right\rangle + i \left\langle \delta b | (\hat{A}\phi)^* - \hat{A}\phi \right\rangle$. At the extremum, the variation has to equal zero at *any* variations of δa and δb. This may happen only if $\hat{A}\phi + (\hat{A}\phi)^* = 0$ and $(\hat{A}\phi)^* - \hat{A}\phi = 0$. This means $\hat{A}\phi = 0$ or, equivalently, $(\hat{A}\phi)^* = 0$. The first of the conditions would be obtained if in ε we made the variation in ϕ^* *only* (the variation in the extremum would then be $\delta\varepsilon = \left\langle \delta\phi | \hat{A}\phi \right\rangle = 0$); hence, from the arbitrariness of $\delta\phi^*$, we would get $\hat{A}\phi = 0$). The second, if we made the variation in ϕ *only* [then, $\delta\varepsilon = \left\langle \phi | \hat{A}\delta\phi \right\rangle = \left\langle \hat{A}\phi | \delta\phi \right\rangle = 0$ and $(\hat{A}\phi)^* = 0$] and the result is exactly the same. This is what we wanted to show: *we may vary either ϕ or ϕ^* and the result is the same.*

resulting in an auxiliary functional $G[\Phi]$:

$$G[\Phi] = \varepsilon[\Phi] - E(\langle\Phi|\Phi\rangle - 1).$$

The variation of G (which is analogous to the differential of a function) represents a linear term in $\delta\Phi^*$. For an extremum, the variation has to equal zero:

$$\delta G = \left\langle \delta\Phi|\hat{H}\Phi\right\rangle - E\left\langle\delta\Phi|\Phi\right\rangle = \left\langle\delta\Phi|(\hat{H} - E)\Phi\right\rangle = 0.$$

Since this has to be satisfied for *any* variation $\delta\Phi^*$, then it can follow only if

$$(\hat{H} - E)\Phi_{opt} = 0, \tag{5.5}$$

which means that the optimal $\Phi \equiv \Phi_{opt}$ is a solution of the Schrödinger equation[4] with E as the energy of the stationary state; i.e., Φ_{opt} together with the normalization condition.

Now let us multiply Eq. (5.5) by Φ^*_{opt} and integrate. We obtain

$$\left\langle\Phi_{opt}|\hat{H}\Phi_{opt}\right\rangle - E\left\langle\Phi_{opt}|\Phi_{opt}\right\rangle = 0, \tag{5.6}$$

or

$$E = \varepsilon\left[\frac{1}{\sqrt{\left\langle\Phi_{opt}|\Phi_{opt}\right\rangle}}\Phi_{opt}\right], \tag{5.7}$$

which means that the conditional minimum of $\varepsilon[\Phi]$ is $E = min(E_0, E_1, E_2, \ldots) = E_0$ (the ground state). Hence, for any other Φ, we obtain $\varepsilon \geq E_0$.

The same result was obtained when we expanded Φ into the eigenfunction series.

Variational Principle for Excited States

The variational principle has been proved for an approximation to the ground-state wave function. But what about excited states? If the variational function Φ is orthogonal to exact solutions to the Schrödinger equation that correspond to all the states of lower energy than the state we are interested in, the variational principle is still valid[5]. If the wave function k being sought represents the *lowest* state among those belonging to a given irreducible representation of the symmetry group of the Hamiltonian, then the orthogonality mentioned above is automatically guaranteed (see Appendix C available at booksite.elsevier.com/978-0-444-59436-5 on p. e17). For other excited states, the variational principle cannot be satisfied, except that function Φ does not contain lower-energy wave functions (i.e., is orthogonal to them, e.g., because the wave functions have been cut out of it earlier).

[4] In the variational calculus, the equation for the optimum Φ or the conditional minimum of a functional ε is called the Euler equation. One can see that in this case, the Euler equation is identical to the Schrödinger equation.

[5] The corresponding proof will be only slightly modified. Simply stated, in the expansion (5.1) of the variational function Φ, the wave functions Ψ_i that correspond to lower energy states (than the state in which we are interested) will be absent. We will therefore obtain $\sum_{i=1} |c_i|^2 \left(E_i - E_k\right) \geq 0$, because state k is the lowest of all the states i.

Watch for Mathematical States

We mentioned in Chapter 1 that not all solutions of the Schrödinger equation are acceptable. Only those that satisfy the symmetry requirements with respect to the exchange of labels corresponding to identical particles (Postulate V) are acceptable. The other solutions are called *mathematical*. If, therefore, a careless scientist takes a variational function Φ with a non-physical symmetry, the variational principle following our derivation exactly (p. 234) will still be valid, but with respect to the *mathematical ground state*. The mathematical states may correspond to energy eigenvalues *lower* than the physical ground state (they are called the *underground states*, cf., p. 84). All this would end up as a catastrophe, because the mean value of the Hamiltonian would tend toward the non-physical underground mathematical state.

5.1.2 Variational Parameters

The variational principle may seem a little puzzling. We insert an *arbitrary* function Φ into the integral and obtain a result related to the ground state of the system under consideration. And yet the arbitrary function Φ may have absolutely nothing to do with the molecule that we are considering. The problem is that the integral still contains the most important information about our system. The information resides in $\hat{H}$. Indeed, if someone wrote down the expression for $\hat{H}$, we would know right away that the system contains N electrons and M nuclei. We would also know the charges of the nuclei; i.e., the chemical elements of which the system is composed.[6] This is important information.

The variational method represents an application of the variational principle. The trial wave function Φ is taken in an analytical form (with the variables denoted by the vector $\boldsymbol{x}$ and automatically satisfying Postulate V). In the key positions in the formula for Φ, we introduce the parameters $\boldsymbol{c} \equiv (c_0, c_1, c_2, \ldots, c_P)$, which we may change smoothly. The parameters play the role of tuning, and their particular values listed in vector $\boldsymbol{c}$ result in a certain shape of $\Phi\left(\boldsymbol{x}; \boldsymbol{c}\right)$. The integration in the formula for ε pertains to the variables $\boldsymbol{x}$; therefore, the result depends uniquely on $\boldsymbol{c}$. Our function $\varepsilon(\boldsymbol{c})$ has the form

$$\varepsilon\left(c_0, c_1, c_2, \ldots c_P\right) \equiv \varepsilon(\boldsymbol{c}) = \frac{\left\langle \Phi(\boldsymbol{x}; \boldsymbol{c}) | \hat{H} \Phi(\boldsymbol{x}; \boldsymbol{c}) \right\rangle}{\langle \Phi(\boldsymbol{x}; \boldsymbol{c}) | \Phi(\boldsymbol{x}; \boldsymbol{c}) \rangle}.$$

Now the problem is to find the minimum of the function $\varepsilon\left(c_0, c_1, c_2, \ldots c_P\right)$.

---

6 And yet we would be unable to decide whether we are dealing with matter or antimatter, or whether we have to perform calculations for the benzene molecule or for six CH radicals (cf. Chapter 2).

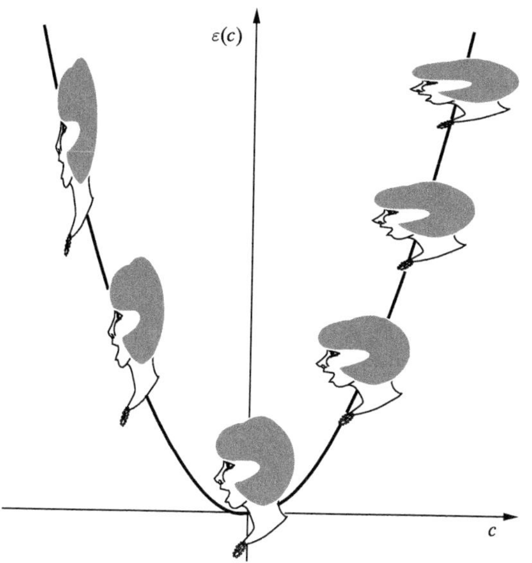

Fig. 5.1. The philosophy behind the variational method. A parameter c is changed in order to obtain the best solution possible.

In the general case, the task is not simple because what we are searching for is the *global* minimum. The relation

$$\frac{\partial \varepsilon\left(c_0, c_1, c_2, \ldots c_P\right)}{\partial c_i} = 0 \text{ for } i = 0, 1, 2, \ldots P,$$

therefore, represents only a *necessary* condition for the global minimum.[7] This problem may be disregarded in the following circumstances:

- The number of minima is small.
- In particular, when we use Φ with the linear parameters $\boldsymbol{c}$ (in this case, we have a single minimum; see below).

The above equations enable us to find the optimum set of parameters $\boldsymbol{c} = \boldsymbol{c}_{opt}$. Then,

> in a given class of the trial functions Φ the best possible approximation of Ψ_0 is $\Phi\left(\boldsymbol{x}; \boldsymbol{c}_{opt}\right)$, and the best approximation of E_0 is $\varepsilon\left(\boldsymbol{c}_{opt}\right)$.

Fig. 5.1 shows the essence of the variational method.[8]

[7] More information about global minimization may be found in Chapter 6.

[8] The variational method is used in everyday life. Usually we determine the target (say, cleaning the car), and then by trial and error, we approach the ideal, but we never fully achieve it.

Example

Let us assume that someone does not know that the hydrogen-like atom problem (where the nucleus has a charge Z) has an exact solution.[9] Let us apply the simplest version of the variational method to see what kind of problem we will be confronted with.

An important first step will be to decide which class of trial functions to choose. We take the class[10] (for $c > 0$) $\exp(-cr)$ and after normalization of the function: $\Phi(r, \theta, \phi; c) = \sqrt{\frac{c^3}{\pi}} \exp(-cr)$. The calculation $\varepsilon[\Phi] = \left\langle \Phi \middle| \hat{H} \middle| \Phi \right\rangle$ is shown in Appendix H available at booksite.elsevier.com/978-0-444-59436-5 on p. e91. We obtain $\varepsilon(c) = \frac{1}{2}c^2 - Zc$. We very easily find the minimum of $\varepsilon(c)$ and the optimum c is equal to $c_{opt} = Z$, which, as we know from Chapter 4, represents the exact result. This is exceptional: in practice (for atoms or molecules), we would never know the exact result. The optimal ε might then be obtained after many days of number crunching[11].

5.1.3 Linear Variational Parameters or the Ritz Method[12]

The Ritz method represents a special kind of variational method. The trial function Φ is represented as a *linear* combination of the *known* basis functions $\{\Psi_i\}$ with the (for the moment) *unknown* variational coefficients c_i.

$$\Phi = \sum_{i=0}^{P} c_i \Psi_i.$$

---

[9] For a larger system, we will not know the exact solution either.

[10] A particular choice is usually made by scientific discussion. The discussion might proceed as follows.
The electron and the nucleus attract themselves; therefore, they will be close in space. This is assured by many classes of trial functions [e.g., $\exp(-cr)$, $\exp(-cr^2)$, $\exp(-cr^3)$, etc., where $c > 0$ is a single variational parameter]. In the present example, we pretend that we do not know which class of functions is most promising (i.e., which will give lower ε). Let us begin with class $\exp(-cr)$, and other classes will be investigated in the years to come. With that decision made, we do the calculations.

[11] For example, for $Z = 1$, we had to decide a starting value of c [say, $c = 2$; $\varepsilon(2) = 0$]. If we try $c = 1.5$, we obtain a lower (i.e., better) value $\varepsilon(1.5) = -0.375$ a.u., so the energy goes down. This is a positive result, so let us try $c = 1.2$; $\varepsilon(1.2) = -0.48$ j.at. This is another excellent result. However, when we continue and take $c = 0.7$, we obtain $\varepsilon = -0.455$ (i.e., a higher energy). We would continue in this way and finally obtain something like $c_{opt} = 1.0000000$. We might be satisfied by eight significant figures and decide to stop the calculations. We would never be sure, however, whether other classes of trial functions would provide still better (i.e., lower) energies. In our particular case, this, of course, would never happen, because we have accidentally taken a class that contains the exact wave function.

[12] Walther Ritz was a Swiss physicist and a former student of Poincaré. His contributions, beside the variational approach, include perturbation theory and the theory of vibrations. Ritz is also known for his controversial disagreement with Einstein on the time flow problem ("*time flash*"), which was concluded by their joint article "*An agreement to disagree*" [W. Ritz and A. Einstein, *Phys. Zeit.*, **10**, 323 (1909)].

Then

$$\varepsilon = \frac{\left\langle \sum_{i=0}^{P} c_i \Psi_i | \hat{H} \sum_{i=0}^{P} c_i \Psi_i \right\rangle}{\left\langle \sum_{i=0}^{P} c_i \Psi_i | \sum_{i=0}^{P} c_i \Psi_i \right\rangle} = \frac{\sum_{i=0}^{P} \sum_{j=0}^{P} c_i^* c_j H_{ij}}{\sum_{i=0}^{P} \sum_{j=0}^{P} c_i^* c_j S_{ij}} = \frac{A}{B}. \tag{5.8}$$

In the formula above, $\{\Psi_i\}$ represents the chosen basis set.[13] The basis set functions are usually non-orthogonal, and therefore

$$\left\langle \Psi_i | \Psi_j \right\rangle = S_{ij} , \tag{5.9}$$

where $\boldsymbol{S}$ stands for the *overlap matrix*, and the integrals

$$H_{ij} = \left\langle \Psi_i | \hat{H} \Psi_j \right\rangle \tag{5.10}$$

are the matrix elements of the Hamiltonian. Both matrices ($\boldsymbol{S}$ and $\boldsymbol{H}$) are calculated once and for all. The energy ε becomes a function of the linear coefficients $\{c_i\}$. The coefficients $\{c_i\}$ and the coefficients $\left\{c_i^*\right\}$ are not independent (c_i can be obtained from c_i^*). Therefore, as the linearly independent coefficients, we may treat either $\{c_i\}$ or[14] $\left\{c_i^*\right\}$. When used for the minimization of ε, both choices would give the same result. We decide to treat $\left\{c_i^*\right\}$ as variables. For each $k = 0, 1, \ldots P$ we have to have at a minimum,

$$\begin{aligned} 0 = \frac{\partial \varepsilon}{\partial c_k^*} &= \frac{\left(\sum_{j=0}^{P} c_j H_{kj}\right) B - A \left(\sum_{j=0}^{P} c_j S_{kj}\right)}{B^2} \\ &= \frac{\left(\sum_{j=0}^{P} c_j H_{kj}\right)}{B} - \frac{A}{B} \frac{\left(\sum_{j=0}^{P} c_j S_{kj}\right)}{B} = \frac{\left(\sum_{j=0}^{P} c_j \left(H_{kj} - \varepsilon S_{kj}\right)\right)}{B}, \end{aligned}$$

which leads to the *secular equations*

$$\left(\sum_{j=0}^{P} c_j \left(H_{kj} - \varepsilon S_{kj}\right) \right) = 0 \text{ for } k = 0, 1, \ldots P. \tag{5.11}$$

The unknowns in the above equation are the coefficients c_j and the energy ε. With respect to the coefficients $\boldsymbol{c}$, Eq. (5.11) represents a homogeneous set of linear equations. Such a set has

[13] Such basis sets are available in the literature. A practical problem arises as to how many such functions should be used. In principle, we should have used $P = \infty$. This, however, is unfeasible. We are restricted to a finite, preferably small number. And this is the moment when it turns out that some basis sets are more effective than others, that this depends on the problem considered, etc.

[14] See footnote on p. 234.

a non-trivial solution if the *secular determinant* is equal to zero (see Appendix A available at booksite.elsevier.com/978-0-444-59436-5)

$$\det\left(H_{kj} - \varepsilon S_{kj}\right) = 0. \tag{5.12}$$

This happens, however, only for some particular values of ε satisfying the above equation. Since the rank of the determinant equals $P + 1$, we therefore obtain $P + 1$ solutions ε_i, $i = 0, 1, 2, \ldots P$. Due to the Hermitian character of operator $\hat{H}$, the matrix $\boldsymbol{H}$ will be Hermitian as well. In Appendices J on p. e99 and L on p. e107, we show that the problem reduces to the diagonalization of some transformed $\boldsymbol{H}$ matrix (also Hermitian). This guarantees that all ε_i will be real numbers.[15] Let us denote the lowest ε_i as ε_0, to represent an approximation[16] to the ground-state energy E_0. The other ε_i, $i = 1, 2, \ldots P$ will approximate the excited states of the system with energies E_1, E_2, E_3, ... We obtain an approximation to the ith wave function by inserting the calculated ε_i into Eq. (5.11), and then, after including the normalization condition, we find the corresponding set of c_i. The problem is solved.

5.2 Perturbational Method

5.2.1 Rayleigh–Schrödinger Approach

The idea of the perturbational approach is very simple. We know everything about a certain non-perturbed problem. Then we slightly perturb the system, and everything changes. If the perturbation is small, it seems there is a good chance that there will be no drama: the wave function and the corresponding energy will change only a little (if the changes were large, the perturbational approach would be inapplicable). The whole perturbational procedure aims at finding these tiny changes with satisfactory precision.

Perturbational theory is notorious for quite clumsy equations. Unfortunately, there is no way around this if we want to explain how to calculate things. However, in practice, only a few of these equations will be used and those equations will be highlighted in frames in this chapter.

Let us begin our story. We would like to solve the Schrödinger equation

$$\hat{H}\psi_k = E\psi_k, \tag{5.13}$$

and as a rule, we will be interested in a single state k, which most often is the ground state ($k = 0$). This particular state will play an exceptional role in the formulas of the perturbational theory.

[15] This is a very good development, because the energy of the photon that is required for excitation from one state to the other will be a real number.

[16] This approximation assumes that we used the basis functions that satisfy Postulate V (on symmetry).

We apply a perturbational approach when[17]

$$\hat{H} = \hat{H}^{(0)} + \hat{H}^{(1)},$$

where the so-called unperturbed operator $\hat{H}^{(0)}$ is large, while the perturbation operator $\hat{H}^{(1)}$ is small[18]. We assume that there is no problem whatsoever with solving the unperturbed Schrödinger equation (also for $n = k$):

$$\hat{H}^{(0)}\psi_n^{(0)} = E_n^{(0)}\psi_n^{(0)}. \tag{5.14}$$

We assume that functions $\psi_n^{(0)}$ form an orthonormal set, which is natural.[19] We are interested in the fate of the wave function $\psi_k^{(0)}$ after the perturbation is switched on (when it changes to ψ_k). We choose the intermediate normalization; i.e.,

$$\left\langle\psi_k^{(0)}|\psi_k\right\rangle = 1. \tag{5.15}$$

The intermediate normalization means that ψ_k, as a vector of the Hilbert space (see Appendix B available at booksite.elsevier.com/978-0-444-59436-5 on p. e7), has the normalized $\psi_k^{(0)}$ as the component along the unit basis vector $\psi_k^{(0)}$. In other words, $\psi_k = \psi_k^{(0)}+$ terms orthogonal to $\psi_k^{(0)}$. The intermediate normalization is convenient, but not necessary. Although convenient for derivation of perturbational equations, it leads to some troubles when the mean values of operators are to be calculated.

We are all set to proceed. First, we introduce the *perturbational parameter* $0 \leq \lambda \leq 1$ in Hamiltonian $\hat{H}$, making it, therefore, λ–dependent[20]:

$$\hat{H}(\lambda) = \hat{H}^{(0)} + \lambda\hat{H}^{(1)}.$$

When $\lambda = 0$, $\hat{H}(\lambda) = \hat{H}^{(0)}$, while $\lambda = 1$ gives $\hat{H}(\lambda) = \hat{H}^{(0)} + \hat{H}^{(1)}$. In other words, we tune the perturbation at will from 0 to $\hat{H}^{(1)}$. However, it is worth noting that $\hat{H}(\lambda)$ for $\lambda \neq 0, 1$ may not correspond to any physical system. However, it does not need to. We are interested here in a mathematical trick; we will come back to reality by putting $\lambda = 1$ in the final formulas.

We are interested in the Schrödinger equation satisfied for all values $\lambda \in [0, 1]$:

$$\hat{H}(\lambda)\psi_k(\lambda) = E_k(\lambda)\psi_k(\lambda).$$

[17] We assume that all operators are Hermitian.

[18] This is in the sense that the energy spectrum of $\hat{H}^{(0)}$ is only slightly changed after the perturbation $\hat{H}^{(1)}$ is switched on.

[19] We can always do that because $\hat{H}^{(0)}$ is Hermitian (see Appendix B available at booksite.elsevier.com/978-0-444-59436-5).

[20] Its role is technical. It will enable us to transform the Schrödinger equation into a sequence of perturbational equations, which must be solved one by one. Then the parameter λ disappears from the theory, because we put $\lambda = 1$.

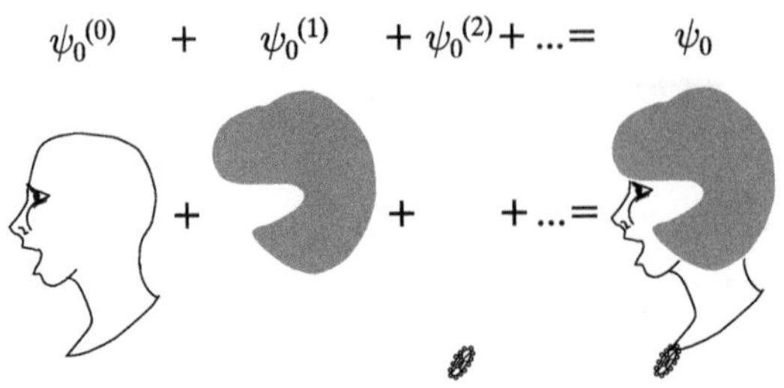

Fig. 5.2. Philosophy of the perturbational approach (the optimistic version). The ideal ground-state wave function ψ_0 is constructed as a sum of a good zero-order approximation ($\psi_0^{(0)}$) and consecutive small corrections ($\psi_0^{(n)}$). The first-order correction ($\psi_0^{(1)}$) is still quite substantial, but fortunately the next corrections amount to only small cosmetic changes. I did not dare to continue the pictures in case of possible divergence of the perturbational series.

Now this is a key step in the derivation. We expect that both the energy and the wave function can be expanded in a power series[21] of λ:

$$E_k(\lambda) = E_k^{(0)} + \lambda E_k^{(1)} + \lambda^2 E_k^{(2)} + \cdots \tag{5.16}$$

$$\psi_k(\lambda) = \psi_k^{(0)} + \lambda \psi_k^{(1)} + \lambda^2 \psi_k^{(2)} + \cdots, \tag{5.17}$$

where $E_k^{(i)}$ stands for some (unknown for the moment) coefficients, and $\psi_k^{(i)}$ represents the functions to be found. We expect the two series to converge (Fig. 5.2).

---

[21] It is in fact a Taylor series with respect to λ. The physical meaning of these expansions is the following: $E_k^{(0)}$ and $\psi_k^{(0)}$ are good approximations of $E_k(\lambda)$ and $\psi_k(\lambda)$. The rest will be calculated as a sum of small correction terms. Both series are based on a very simple idea. Imagine a rose on a long stalk. We are interested in the distance of the flower center from the ground. Suppose the corresponding measurement gives $h^{(0)}$ (an analog of $E_k^{(0)}$ or $\psi_k^{(0)}$). Now we are going to conceive a theory that predicts the distance of the flower center from the ground (h), when a fly of mass λ sits on the flower center. The first idea is that the bending should be proportional to λ. Therefore, we have our first guess that $h(\lambda) - h^{(0)} \approx \lambda h^{(1)}$, where $h^{(1)}$ stands for some suitable constant of proportionality. We find this constant with a high precision either experimentally, by weighing some small weights (less ambitious), or theoretically (more ambitious) by applying a model of the flower. Then we proudly see that our theory is working, when some small flies are sitting on the rose. After some time, we are bored by weighing small flies, and begin to test the formula $h(\lambda) - h^{(0)} \approx \lambda h^{(1)}$ for bees. Unfortunately, our theory ceases to work well (the effect of the nonlinearity). We watch our formula as a function of λ fail and, all of a sudden, we get an idea. The idea is to add a new term $\lambda^2 h^{(2)}$ with a suitably chosen $h^{(2)}$ constant: $h(\lambda) - h^{(0)} \approx \lambda h^{(1)} + \lambda^2 h^{(2)}$. And indeed, we again find success ...until we consider even heavier insects. Then we introduce $h^{(3)}$, etc. This is a good time to stop improving our theory and say that we have a theory that works only for small insects. This is reflected by the formulas for $E_k(\lambda)$ and $\psi_k(\lambda)$ given above.

An attempt to use the theory for heavy weights is bound to fail. These formulas will not work (even if we take into account an infinite number of terms) for large weights [e.g., ones that are so large that the stalk breaks (an abrupt change of its structure)]. In such cases, the series for $E_k(\lambda)$ and $\psi_k(\lambda)$ will diverge, and we will witness a catastrophic failure of the perturbation theory.

In practice, we calculate only $E_k^{(1)}$, $E_k^{(2)}$ and (quite rarely) $E_k^{(3)}$, and for the wave function, we usually limit to $\psi_k^{(1)}$.

How are these corrections calculated?

We insert the two perturbational series for $E_k(\lambda)$ and $\psi_k(\lambda)$ into the Schrödinger equation:

$$\left(\hat{H}^{(0)} + \lambda \hat{H}^{(1)}\right)\left(\psi_k^{(0)} + \lambda \psi_k^{(1)} + \lambda^2 \psi_k^{(2)} + \cdots\right)$$
$$= \left(E_k^{(0)} + \lambda E_k^{(1)} + \lambda^2 E_k^{(2)} + \cdots\right)\left(\psi_k^{(0)} + \lambda \psi_k^{(1)} + \lambda^2 \psi_k^{(2)} + \cdots\right),$$

and, since the equation has to be satisfied for *any* λ belonging to $0 \le \lambda \le 1$, then this may happen only if

the coefficients at the same powers of λ on the left and right sides must be equal.

This gives a sequence of an infinite number of perturbational equations to be satisfied by the unknown $E_k^{(n)}$ and $\psi_k^{(n)}$. These equations may be solved consecutively, allowing us to calculate $E_k^{(n)}$ and $\psi_k^{(n)}$ with larger and larger n. We have, for example,

$$\begin{aligned}
\text{for } \lambda^0 &: \hat{H}^{(0)} \psi_k^{(0)} = E_k^{(0)} \psi_k^{(0)}, \\
\text{for } \lambda^1 &: \hat{H}^{(0)} \psi_k^{(1)} + \hat{H}^{(1)} \psi_k^{(0)} = E_k^{(0)} \psi_k^{(1)} + E_k^{(1)} \psi_k^{(0)}, \\
\text{for } \lambda^2 &: \hat{H}^{(0)} \psi_k^{(2)} + \hat{H}^{(1)} \psi_k^{(1)} = E_k^{(0)} \psi_k^{(2)} + E_k^{(1)} \psi_k^{(1)} + E_k^{(2)} \psi_k^{(0)}, \\
&\ldots \\
&\text{etc.}^{22}
\end{aligned} \tag{5.18}$$

Doing the same procedure with the intermediate normalization [Eq. (5.15)], we obtain

$$\left\langle \psi_k^{(0)} | \psi_k^{(n)} \right\rangle = \delta_{0n}. \tag{5.19}$$

The first of Eqs. (5.18) is evident (the unperturbed Schrödinger equation does not contain any unknowns). The second equation involves two unknowns, $\psi_k^{(1)}$ and $E_k^{(1)}$. There is a way to eliminate $\psi_k^{(1)}$ by using the Hermitian character of the operators. Indeed, by making the scalar product of the equation with $\psi_k^{(0)}$, we obtain

$$\left\langle \psi_k^{(0)} | \left(\hat{H}^{(0)} - E_k^{(0)}\right) \psi_k^{(1)} + \left(\hat{H}^{(1)} - E_k^{(1)}\right) \psi_k^{(0)} \right\rangle$$
$$= \left\langle \psi_k^{(0)} | \left(\hat{H}^{(0)} - E_k^{(0)}\right) \psi_k^{(1)} \right\rangle + \left\langle \psi_k^{(0)} | \left(\hat{H}^{(1)} - E_k^{(1)}\right) \psi_k^{(0)} \right\rangle$$
$$= 0 + \left\langle \psi_k^{(0)} | \left(\hat{H}^{(1)} - E_k^{(1)}\right) \psi_k^{(0)} \right\rangle = 0,$$

[22] Here, we see the construction principle of these equations: we write down all the terms that give a given value of the sum of the upper indices.

i.e.,

the formula for the first-order correction to the energy

$$E_k^{(1)} = H_{kk}^{(1)}, \tag{5.20}$$

where we defined

$$H_{kn}^{(1)} = \langle \psi_k^{(0)} | \hat{H}^{(1)} | \psi_n^{(0)} \rangle. \tag{5.21}$$

Conclusion: the first-order correction to the energy, $E_k^{(1)}$, represents the mean value of the perturbation with the unperturbed wave function of the state in which we are interested (usually the ground state).[23]

Now, from the perturbation [Eq. (5.18)] corresponding to $n = 2$, we get the following through a scalar product with $\psi_k^{(0)}$:

$$\left\langle \psi_k^{(0)} | \left(\hat{H}^{(0)} - E_k^{(0)} \right) \psi_k^{(2)} \right\rangle + \left\langle \psi_k^{(0)} | \left(\hat{H}^{(1)} - E_k^{(1)} \right) \psi_k^{(1)} \right\rangle - E_k^{(2)}$$
$$= \left\langle \psi_k^{(0)} | \hat{H}^{(1)} \psi_k^{(1)} \right\rangle - E_k^{(2)} = 0,$$

and hence

$$E_k^{(2)} = \left\langle \psi_k^{(0)} | \hat{H}^{(1)} \psi_k^{(1)} \right\rangle. \tag{5.22}$$

For the time being, we cannot compute $E_k^{(2)}$ because we do not know $\psi_k^{(1)}$, but soon we will. Let us expand $\psi_k^{(1)}$ into the complete set of the basis functions $\left\{ \psi_n^{(0)} \right\}$ with as-yet-unknown coefficients c_n:

$$\psi_k^{(1)} = \sum_{n \neq k} c_n \psi_n^{(0)}. \tag{5.23}$$

Note that because of the intermediate normalization [Eqs. (5.15) and (5.19)], we did not take the term with $n = k$. We get

$$\left(\hat{H}^{(0)} - E_k^{(0)} \right) \sum_{n \neq k} c_n \psi_n^{(0)} + \hat{H}^{(1)} \psi_k^{(0)} = E_k^{(1)} \psi_k^{(0)},$$

[23] This is natural, and we use such a perturbative estimation all the time. What it really says is that we do not know what the perturbation exactly does, but let us estimate the result by assuming that all things are going on as they were before the perturbation was applied. In the first-order approach, insurance estimates your loss by averaging over similar losses of others. Every morning, you assume that the traffic on the highway will be the same as it is everyday, etc.

and then transform into

$$\sum_{n \neq k} c_n \left(E_n^{(0)} - E_k^{(0)}\right) \psi_n^{(0)} + \hat{H}^{(1)} \psi_k^{(0)} = E_k^{(1)} \psi_k^{(0)}.$$

We find c_m by making the scalar product of both sides of the last equation with $\psi_m^{(0)}$. Due to the orthonormality of functions $\left\{\psi_n^{(0)}\right\}$, we obtain

$$c_m = \frac{H_{mk}^{(1)}}{E_k^{(0)} - E_m^{(0)}},$$

which, from Eq. (5.22), gives the following formula for the first-order correction to the wave function[24]

$$\psi_k^{(1)} = \sum_{n \neq k} \frac{H_{nk}^{(1)}}{E_k^{(0)} - E_n^{(0)}} \psi_n^{(0)}, \tag{5.24}$$

and then the formula for the second-order correction to the energy

$$E_k^{(2)} = \sum_{n \neq k} \frac{|H_{kn}^{(1)}|^2}{E_k^{(0)} - E_n^{(0)}}. \tag{5.25}$$

All terms on the right side of these formulas are known (or can be calculated) from the unperturbed Eq. (5.14). For $k = 0$ (ground state), we have

$$E_0^{(2)} \leq 0. \tag{5.26}$$

[24] This formula may be seen as the foundation for almost everything we do in our everyday life, consciously or unconsciously. We start from an initial situation (say, state) $\psi_k^{(0)}$. We want to change this state, while monitoring whether the results are OK for us the entire time. The cautious thing to do is to apply a very small amount of perturbation (in this situation, the change is practically equal to $\psi_k^{(1)}$) and watch carefully how the system reacts upon it. The effects are seen mainly on $\psi_k^{(1)}$ and $E_k^{(2)}$. In practical life, we apparently forget about higher-order perturbational formulas. Instead, we need to apply the first-order perturbation theory all the time (in this way, the higher-order corrections reenter implicitly). We decide whether in the new situation $\psi_k^{(0)} + \psi_k^{(1)}$ (which now is again treated as unperturbed) the applied perturbation should be kept the same, decreased, or increased. This is how we enter a curve when driving a car: we adapt the steering-wheel position to what we see after having its previous position, and we keep doing this again and again.

From Eq. (5.24), we see that the contribution of function $\psi_n^{(0)}$ to the wave function deformation is large if the coupling between states k and n (i.e., $H_{nk}^{(1)}$) is large, and the closer in the energy scale these two states are.

We will limit ourselves to the most important corrections in the hope that the perturbational method converges quickly (we will see in a moment how surprising the perturbational series behavior can be) and further corrections are much less important.[25] The formulas for higher corrections become more and more complicated.

5.2.2 Hylleraas Variational Principle[26]

The derived formulas are rarely employed in practice because we only very rarely have at our disposal all the necessary solutions of Eq. (5.14). The eigenfunctions of the $\hat{H}^{(0)}$ operator appeared as a consequence of using them as the complete set of functions (e.g., in expanding $\psi_k^{(1)}$). There are, however, some numerical methods that enable us to compute $\psi_k^{(1)}$ using the complete set of functions $\{\phi_i\}$, which are not the eigenfunctions of $\hat{H}^{(0)}$.

Hylleraas noted[27] that the functional

$$\mathcal{E}[\tilde{\chi}] = \left\langle \tilde{\chi} \middle| \left(\hat{H}^{(0)} - E_0^{(0)}\right) \tilde{\chi} \right\rangle \tag{5.27}$$

$$+ \left\langle \tilde{\chi} \middle| \left(\hat{H}^{(1)} - E_0^{(1)}\right) \psi_0^{(0)} \right\rangle + \left\langle \psi_0^{(0)} \middle| \left(\hat{H}^{(1)} - E_0^{(1)}\right) \tilde{\chi} \right\rangle \tag{5.28}$$

exhibits its minimum at $\tilde{\chi} = \psi_0^{(1)}$, and for this function, the value of the functional is equal to $E_0^{(2)}$. Indeed, inserting $\tilde{\chi} = \psi_0^{(1)} + \delta\chi$ into Eq. (5.28) and using the Hermitian character of the operators, we have

$$\begin{aligned}
\left[\psi_0^{(1)} + \delta\chi\right] - \left[\psi_0^{(1)}\right] &= \left\langle \psi_0^{(1)} + \delta\chi \middle| \left(\hat{H}^{(0)} - E_0^{(0)}\right) (\psi_0^{(1)} + \delta\chi) \right\rangle \\
&+ \left\langle \psi_0^{(1)} + \delta\chi \middle| \left(\hat{H}^{(1)} - E_0^{(1)}\right) \psi_0^{(0)} \right\rangle \\
&+ \left\langle \psi_0^{(0)} \middle| \left(\hat{H}^{(1)} - E_0^{(1)}\right) (\psi_0^{(1)} + \delta\chi) \right\rangle \\
&= \left\langle \delta\chi \middle| \left(\hat{H}^{(0)} - E_0^{(0)}\right) \psi_0^{(1)} + \left(\hat{H}^{(1)} - E_0^{(1)}\right) \psi_0^{(0)} \right\rangle \\
&+ \left\langle \left(\hat{H}^{(0)} - E_0^{(0)}\right) \psi_0^{(1)} + \left(\hat{H}^{(1)} - E_0^{(1)}\right) \psi_0^{(0)} \middle| \delta\chi \right\rangle \\
&+ \left\langle \delta\chi \middle| \left(\hat{H}^{(0)} - E_0^{(0)}\right) \delta\chi \right\rangle = \left\langle \delta\chi \middle| \left(\hat{H}^{(0)} - E_0^{(0)}\right) \delta\chi \right\rangle \geq 0.
\end{aligned}$$

[25] Some scientists have been bitterly disappointed by this assumption.

[26] See Hylleraas biographic note in Chapter 10.

[27] E.A. Hylleraas, *Zeit. Phys.*, **65**, 209 (1930).

This proves the Hylleraas variational principle. The last equality follows from the first-order perturbational equation, and the last inequality from the fact that $E_0^{(0)}$ is assumed to be the lowest eigenvalue of $\hat{H}^{(0)}$ (see the variational principle).

What is the minimal value of the functional under consideration? Let us insert $\tilde{\chi} = \psi_0^{(1)}$. We obtain

$$\begin{aligned}
\mathcal{E}\left[\psi_0^{(1)}\right] &= \left\langle \psi_0^{(1)} | \left(\hat{H}^{(0)} - E_0^{(0)}\right) \psi_0^{(1)} \right\rangle \\
&\quad + \left\langle \psi_0^{(1)} | \left(\hat{H}^{(1)} - E_0^{(1)}\right) \psi_0^{(0)} \right\rangle + \left\langle \psi_0^{(0)} | \left(\hat{H}^{(1)} - E_0^{(1)}\right) \psi_0^{(1)} \right\rangle \\
&= \left\langle \psi_0^{(1)} | \left(\hat{H}^{(0)} - E_0^{(0)}\right) \psi_0^{(1)} + \left(\hat{H}^{(1)} - E_0^{(1)}\right) \psi_0^{(0)} \right\rangle + \left\langle \psi_0^{(0)} | \hat{H}^{(1)} \psi_0^{(1)} \right\rangle \\
&= \left\langle \psi_0^{(1)} | 0 \right\rangle + \left\langle \psi_0^{(0)} | \hat{H}^{(1)} \psi_0^{(1)} \right\rangle = \left\langle \psi_0^{(0)} | \hat{H}^{(1)} \psi_0^{(1)} \right\rangle = E_0^{(2)}.
\end{aligned}$$

5.2.3 Hylleraas Equation

The first-order perturbation equation (p. 243) after inserting

$$\psi_0^{(1)} = \sum_j^N d_j \phi_j \tag{5.29}$$

takes the form

$$\sum_j^N d_j (\hat{H}^{(0)} - E_0^{(0)}) \phi_j + (\hat{H}^{(1)} - E_0^{(1)}) \psi_0^{(0)} = 0.$$

Making the scalar products of the left and right sides of the equation with functions $\phi_i, i = 1, 2, \ldots$, we obtain

$$\sum_j^N d_j (\hat{H}_{ij}^{(0)} - E_0^{(0)} S_{ij}) = -(\hat{H}_{i0}^{(1)} - E_0^{(1)} S_{i0}) \quad \text{for} \quad i = 1, 2, \ldots, N,$$

where $\hat{H}_{ij}^{(0)} \equiv \left\langle \phi_i | \hat{H}^{(0)} \phi_j \right\rangle$, and the overlap integrals $S_{ij} \equiv \langle \phi_i | \phi_j \rangle$. Using the matrix notation, we may write the Hylleraas equation as

$$(\boldsymbol{H}^{(0)} - E_k^{(0)} \boldsymbol{S}) \boldsymbol{d} = -\boldsymbol{v}, \tag{5.30}$$

where the components of the vector $\boldsymbol{v}$ are $v_i = \hat{H}_{i0}^{(1)} - E_0^{(1)} S_{i0}$. All the quantities can be calculated and the set of N linear equations with unknown coefficients d_i remains to be solved.[28]

[28] We obtain the same equation if in the Hylleraas functional [Eq. (5.28)], the variational function χ is expanded as a linear combination [Eq. (5.29)], and then vary d_i in a similar way to that of the Ritz variational method described on p. 238.

5.2.4 Degeneracy

There is a trap in the perturbational formulas, which may lead to catastrophe. Let us imagine that the unperturbed state k (its change represents the target of the perturbation theory) is g-tuply degenerate. Then Eq. (5.14) is satisfied by g wave functions $\psi_{k,m}^{(0)}$, $m = 1, 2, \ldots, g$, which according to Appendix B available at booksite.elsevier.com/978-0-444-59436-5, p. e7 can be chosen as orthonormal; i.e., $\left\langle \psi_{k,m}^{(0)} | \psi_{k,n}^{(0)} \right\rangle = \delta_{mn}$:

$$\hat{H}^{(0)} \psi_{k,m}^{(0)} = E_k^{(0)} \psi_{k,m}^{(0)}. \tag{5.31}$$

If one decided to choose a given $\psi_{k,m}^{(0)}$ as the unperturbed state, we could not calculate either $E_k^{(2)}$ or $\psi_k^{(1)}$, because the denominators in the corresponding formulas would equal 0 and the results would "explode to infinity."

We should fix the problem in some way. Let us focus on a particular unperturbed state $\psi_{k,1}^{(0)}$. The slightest perturbation applied would force us to consider the Ritz variational method to determine the resulting state. We can choose all $\psi_{k,m}^{(0)}$ functions as possible expansion functions; however, this means also including $\psi_{k,m}^{(0)}$, $m = 2, 3, \ldots, g$, but they may enter with potentially very large weights.[29] This means the possibility of a giant change of the wave function resulting from even a very small perturbation.[30] Therefore, we cannot say the perturbation is small, while this is the most important requirement for perturbation theory to operate (Fig. 5.3).

The first thing to do is to adapt the wave function of the unperturbed system to the perturbation. To this end, we will use the Ritz variational method with the expansion functions[31] $\psi_{k,m}^{(0)}$, $m = 1, 2, \ldots, g$:

$$\phi_{k,m}^{(0)} = \sum_{i=1}^{g} c_{km} \psi_{k,m}^{(0)}. \tag{5.32}$$

[29] This is the case because they satisfy the Schrödinger equation for the unperturbed system with the same eigenvalue of energy. As a consequence, any linear combination of them satisfies this equation, as well as $\psi_{k,1}^{(0)}$ itself.

[30] Thinking about the future, this is the reason for the richness of local spatial configurations around atoms in chemical compounds (which is called the *valency*). Most often, such atoms offer a degenerate or quasi-degenerate valence orbitals. They can, therefore, mix (hybridization) easily even under small perturbation coming from the neighborhood. The product of such mixing (hybridized orbitals) is oriented toward the perturbing partner atoms. On top of that, these hybrids may offer 2, 1, 0 electrons (depending on what number of electrons the partner's orbitals carry), which leads to the coordination bond with donation of the electron pair, covalent bond, or coordination bond with acceptance of the electron pair, respectively.

[31] The same solution is obtained when inserting Eq. (5.32) into the first-order perturbational equation, carrying out the scalar products of this equation consecutively with $\psi_{k,m}^{(0)}$, $m = 1, 2, \ldots, g$ and solving the resulting system of equations.

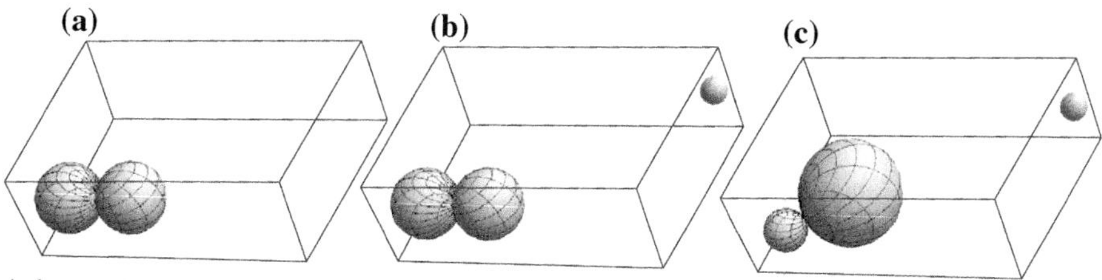

Fig. 5.3. A degenerate state may mean an instable system ("*superpolarizable*"), which adapts itself very easily to an external (even very small) perturbation. The figure shows how the hydrogen atom in a degenerate state changes when a proton approaches it. (a) the atom in a non-perturbed stationary state $2p_x$. (b) The proton interacts with the hydrogen atom, but the $2p_x$ orbital does not reflect this. It is true that the orbital $2p_x$ describes the total energy very well (because the interaction is weak), but it describes the electron charge distribution in the hydrogen atom interacting with the proton very badly. (c) Contrary to this, the function $\phi_k^{(0)} = (2s) + \frac{1}{\sqrt{2}}(2p_x + 2p_y)$, satisfies the Schrödinger equation for the unperturbed hydrogen atom as well as the function $2p_x$, but it is much more reasonable after the perturbation is switched on. The function $\phi_k^{(0)}$ shows that the electron recognizes the direction of the proton and tries to approach it. The function $\phi_k^{(0)}$ represents a giant change with respect to $2p_x$. The most reasonable unperturbed function will be determined by the variational method as a linear combination of (mostly) $2s, 2p_x, 2p_y$, and $2p_z$. After this is done, one may keep improving the function by adding perturbational corrections.

The secular equation (5.11), after inserting $\hat{H} = \hat{H}^{(0)} + \hat{H}^{(1)}$ and using Eq. (5.31), transforms to[32]

$$\det\left\{E_k^{(0)}\delta_{mn} + \hat{H}_{mn}^{(1)} - \varepsilon\delta_{mn}\right\} = 0. \tag{5.33}$$

This may be rewritten as

$$\det\left\{\hat{H}_{mn}^{(1)} - E_k^{(1)}\delta_{mn}\right\} = 0, \tag{5.34}$$

with[33] $E_k^{(1)} \equiv \varepsilon - E_k^{(0)}$. Solving this equation, we obtain the unknown energies $E_{k,i}^{(1)}, i = 1, 2, \ldots g$. Solving the secular equation (Chapter 5 and Appendix L available at booksite.elsevier.com/978-0-444-59436-5, p. e107) consecutively for $E_{k,i}^{(1)}, i = 1, 2, \ldots g$, we get the sets (numbered by $i = 1, 2, \ldots g$) of coefficients c_{km}. This means we have the quantity g of the unperturbed functions. We choose one of them, which corresponds to the target state, and the perturbational approach is applied to it in the same way as for the non-degenerate case (except we no longer include $\psi_{k,m}^{(0)}$ to the expansion functions).

5.2.5 Convergence of the Perturbational Series

The perturbational approach is applicable when the perturbation only slightly changes the energy levels, therefore not changing their order. This means that the unperturbed energy

[32] $\hat{H}_{mn} = \left\langle\psi_{k,m}^{(0)}|\hat{H}^{(0)}\psi_{k,n}^{(0)}\right\rangle + \left\langle\psi_{k,m}^{(0)}|\hat{H}^{(1)}\psi_{k,n}^{(0)}\right\rangle = E_k^{(0)}\delta_{mn} + \hat{H}_{mn}^{(1)}; S_{mn} = \delta_{mn}$.

[33] The energies $E_{k,i}^{(1)}$ have an additional superscript (1) just to stress that they are proportional to perturbation. They all stem from the unperturbed state k, but in general, they differ for different i; i.e., we usually get a splitting of the energy level k.

level separations have to be much larger than a measure of perturbation such as $\hat{H}_{kk}^{(1)} = \left\langle \psi_k^{(0)} | \hat{H}^{(1)} \psi_k^{(0)} \right\rangle$. However, even in this case, we may expect complications.

The subsequent perturbational corrections need not be monotonically decreasing. However, if the perturbational series [Eq. (5.17)] converges, for any $\varepsilon > 0$, we may choose such N_0 that for $N > N_0$, we have $\left\langle \psi_k^{(N)} | \psi_k^{(N)} \right\rangle < \varepsilon$; i.e., the vectors $\psi_k^{(N)}$ have smaller and smaller lengths in the Hilbert space.

Unfortunately, perturbational series are often divergent in a sense known as *asymptotic convergence*. A divergent series $\sum_{n=0}^{\infty} \frac{A_n}{z^n}$ is called an *asymptotic series* of a function $f(z)$, if the function $R_n(z) = z^n [f(z) - S_n(z)]$, where $S_n(z) = \sum_{k=0}^{n} \frac{A_k}{z^k}$ satisfies the following condition: $\lim_{z \to \infty} R_n(z) = 0$ for any fixed n. In other words, the error of the summation {i.e., $[f(z) - S_n(z)]$} tends to 0 as $z^{-(n+1)}$ or faster.

Despite the fact that the series used in physics and chemistry are often asymptotic (therefore, divergent), we are able to obtain results of high accuracy with them, provided that we limit ourselves to an appropriate number of terms. The asymptotic character of such series manifests itself in practice in such a way that the partial sums stabilize and we obtain numerically a situation typical for convergence. For instance, we sum up the consecutive perturbational corrections and obtain the partial sums changing on the eighth, then ninth, then tenth significant figures. This is a very good time to stop the calculations, publish the results, finish the scientific career and move on to other business. The addition of further perturbational corrections ends up in catastrophe (cf. Appendix X available at booksite.elsevier.com/978-0-444-59436-5 on p. e169). It begins by an innocent, very small, increase in the partial sums, they just begin to change the ninth, then the eighth, then the seventh significant figure. Then, it only gets worse and worse and ends with an explosion of the partial sums to ∞ and a very bad state of mind for the researcher (I did not dare to depict it in Fig. 5.2).

In perturbation theory, we assume that $E_k(\lambda)$ and $\psi_k(\lambda)$ are analytical functions of λ (p. 242). In this *mathematical* aspect of the physical problem, we may treat λ as a complex number. Then the radius of convergence ρ of the perturbational series on the complex plane is equal to the smallest $|\lambda|$, for which one has a pole of $E_k(\lambda)$ or $\psi_k(\lambda)$. If the limit exists, the convergence radius ρ_k for the energy perturbational series may be computed as[34]

$$\rho_k = \lim_{N \to \infty} \frac{\left| E_k^{(N)} \right|}{\left| E_k^{(N+1)} \right|}.$$

For physical reasons, $\lambda = 1$ is most important. It is, therefore, desirable to have $\rho_k \geq 1$. Note from Fig. 5.4, that if $\rho_k \geq 1$, then the series with $\lambda = 1$ is convergent together with the series with $\lambda = -1$.

Let us take as the unperturbed system the harmonic oscillator (with the potential energy equal to $\frac{1}{2}x^2$) in its ground state, and the operator $\hat{H}^{(1)} = -0.000001 \cdot x^4$ as its perturbation.

[34] If the limit does not exist, nothing can be said about ρ_k.

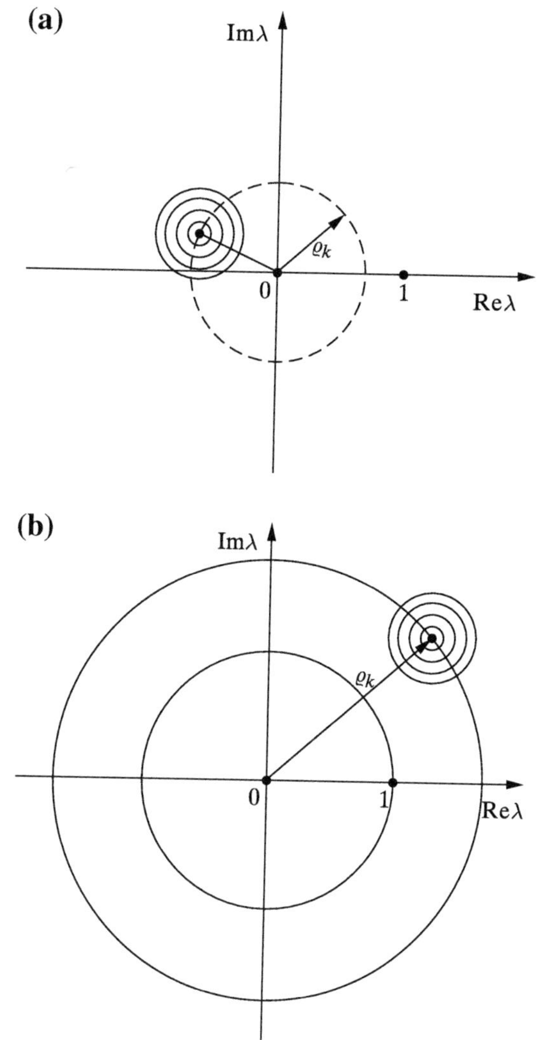

Fig. 5.4. The complex plane of the λ parameter. The physically interesting points are at $\lambda = 0, 1$. In perturbation theory, we finally put $\lambda = 1$. Because of this, the convergence radius ρ_k of the perturbational series has to be $\rho_k \geq 1$. However, if any complex λ with $|\lambda| < 1$ corresponds to a pole of the energy, the perturbational series will diverge in the physical situation ($\lambda = 1$). The figure shows the position of a pole by concentric circles. (a) the pole is too close ($\rho_k < 1$), and the perturbational series diverges; (b) the perturbational series converges because $\rho_k > 1$.

In such a case, the perturbation seems to be small[35] in comparison with the separation of the eigenvalues of $\hat{H}^{(0)}$. And yet the perturbational series sows the seeds of catastrophe. It is quite easy to see why a catastrophe has to happen. After the perturbation is added, the potential

[35] As a measure of the perturbation, we may use $\left\langle \psi_0^{(0)} | \hat{H}^{(1)} \psi_0^{(0)} \right\rangle$, which means an integral of x^4 multiplied by a Gaussian function (cf. Chapter 4). Such an integral is easy to calculate and, in view of the fact that it will be multiplied by the (small) factor 0.000001, the perturbation will turn out to be small.

becomes *qualitatively* different from $\frac{1}{2}x^2$. For large x, instead of going to ∞, it will tend to $-\infty$. The perturbation is not small at all–it is a monster. This will cause the perturbational series to diverge. How will this happen in practice? Well, in higher orders, we have to calculate the integrals $\left\langle \psi_n^{(0)} | \hat{H}^{(1)} \psi_m^{(0)} \right\rangle$, where n, m stand for the vibrational quantum numbers. As we recall from Chapter 4, high-energy wave functions have large values for large x, where the perturbation gets larger and larger as x increases. This is why the integrals will be large. Therefore, the better we do our jobs (higher orders, higher-energy states), the faster we approach catastrophe.

Let us consider the opposite perturbation $\hat{H}^{(1)} = +0.000001 \cdot x^4$. Despite the fact that everything looks good (i.e., the perturbation does not qualitatively change the potential), the series will diverge sooner or later. It is bound to happen because the convergence radius does not depend on the sign of the perturbation. A researcher might be astonished when the corrections begin to "explode."

Quantum chemistry experiences with perturbational theories look quite consistent, to wit:

- Low orders may give excellent results.
- Higher orders often make the results worse.[36]

Summary

There are basically two numerical approaches to obtain approximate solutions to the Schrödinger equation: variational and perturbational. In calculations, we usually apply variational methods, while perturbational methods are often applied to estimate some small physical effects. The result is that most concepts (practically all the ones we know of) characterizing the reaction of a molecule to an external field come from the perturbational approach. This leads to such quantities (see Chapter 12) as dipole moment, polarizability, and hyperpolarizability. The computational role of perturbational theories may, in this context, be seen as being of the second order.

- **Variational method**
  - The method is based on the variational principle, which says that, if for a system with Hamiltonian $\hat{H}$ we calculate the number $\varepsilon = \frac{\langle \Phi | \hat{H} \Phi \rangle}{\langle \Phi | \Phi \rangle}$, where Φ stands for an arbitrary function, then the number $\varepsilon \geq E_0$, with E_0 being the ground-state eigenvalue of $\hat{H}$. If it happens that $\varepsilon[\Phi] = E_0$, then there is only one possibility: Φ represents the exact ground-state wave function ψ_0.
  - The variational principle means that to find an approximate ground-state wave function, we can use the variational method: minimize $\varepsilon[\Phi]$ by changing (varying) Φ. The minimum value of $\varepsilon[\Phi]$ equals $\varepsilon[\Phi_{opt}]$; which approximates the ground-state energy E_0 and corresponds to Φ_{opt}; i.e., an approximation to the ground-state wave function ψ_0.
  - In practice, the variational method consists of the following steps:
    * Make a decision as to the trial function class, among which the $\Phi_{opt}(\boldsymbol{x})$ will be sought[37].
    * Introduce into the function the variational parameters $\boldsymbol{c} \equiv (c_0, c_1, \ldots c_P) : \Phi(\boldsymbol{x}; \boldsymbol{c})$. In this way, ε becomes a function of these parameters: $\varepsilon(\boldsymbol{c})$.
    * Minimize $\varepsilon(\boldsymbol{c})$ with respect to $\boldsymbol{c} \equiv (c_0, c_1, \ldots c_P)$ and find the optimal set of parameters $\boldsymbol{c} = \boldsymbol{c}_{opt}$.
    * The value $\varepsilon(\boldsymbol{c}_{opt})$ represents an approximation to E_0.
    * The function $\Phi(\boldsymbol{x}; \boldsymbol{c}_{opt})$ is an approximation to the ground-state wave function $\psi_0(\boldsymbol{x})$.

[36] Even orders as high as 2000 have been investigated in the hope that the series will improve the results.
[37] $\boldsymbol{x}$ symbolizes the set of coordinates (space and spin, cf. Chapter 1).

- The Ritz procedure is a special case of the variational method, in which the parameters $\boldsymbol{c}$ enter Φ linearly: $\Phi(\boldsymbol{x};\boldsymbol{c}) = \sum_{i=0}^{P} c_i \Psi_i$, where $\{\Psi_i\}$ are some known basis functions that form (or more exactly, in principle form) the complete set of functions in the Hilbert space. This formalism leads to a set of homogeneous linear equations to solve (secular equations), from which we find approximations to the ground- and excited-state energies and wave functions.

- **Perturbational method**

We assume that the solution to the Schrödinger equation for the unperturbed system is known ($E_k^{(0)}$ for the energy and $\psi_k^{(0)}$ for the wave function, usually $k = 0$; i.e., the ground state), but when a small perturbation $\hat{H}^{(1)}$ is added to the Hamiltonian, then the solution changes (to E_k and ψ_k, respectively) and is to be sought using the perturbational approach. Then the key assumption is $E_k(\lambda) = E_k^{(0)} + \lambda E_k^{(1)} + \lambda^2 E_k^{(2)} + \ldots$ and $\psi_k(\lambda) = \psi_k^{(0)} + \lambda\psi_k^{(1)} + \lambda^2\psi_k^{(2)} + \ldots$, where λ is a parameter that tunes the perturbation. The goal of the perturbational approach is to compute corrections to the energy $E_k^{(1)}, E_k^{(2)}, \ldots$ and to the wave function $\psi_k^{(1)}, \psi_k^{(2)}, \ldots$. We assume that because the perturbation is small, only a few such corrections are to be computed–in particular, $E_k^{(1)} = \left\langle \psi_k^{(0)} | \hat{H}^{(1)} \psi_k^{(0)} \right\rangle$, $\psi_k^{(1)} = \sum_{n \neq k} \frac{H_{nk}^{(1)}}{E_k^{(0)} - E_n^{(0)}} \psi_n^{(0)}$, $E_k^{(2)} = \sum_{n \neq k} \frac{|H_{kn}^{(1)}|^2}{E_k^{(0)} - E_n^{(0)}}$, where $H_{kn}^{(1)} = \left\langle \psi_k^{(0)} | \hat{H}^{(1)} \psi_n^{(0)} \right\rangle$.

Main Concepts, New Terms

asymptotic convergence (p. 250)
complete set of functions (p. 246)
convergence radius (p. 250)
corrections to energy (p. 242)
corrections to wave function (p. 242)
Hylleraas equation (p. 247)
Hylleraas functional (p. 247)
Hylleraas variational principle (p. 246)
perturbation (p. 240)
perturbational method (p. 240)
perturbed system (p. 240)
Ritz method (p. 238)
secular equation (p. 239)
secular determinant (p. 240)
trial function (p. 236)
underground states (p. 236)
unperturbed system (p. 240)
variational function (p. 232)
variational method (p. 232)
variational parameters (p. 236)
variational principle (p. 232)
variational principle for excited states (p. 235)

From the Research Front

In practice, the Ritz variational method is used most often. One of the technical problems to be solved is the size of the basis set. Enormous progress in computation and software development now facilitates investigations that 20 years ago were absolutely beyond the imagination. The world record in quantum chemistry means a few billion expansion functions. To accomplish this, quantum chemists have had to invent some powerful methods of applied mathematics.

Ad Futurum

The computational technique impetus that we witness nowadays will continue in the future (though perhaps in a modified form). It will be no problem to find some reliable approximations to the ground-state energy and wave function for a molecule composed of thousands of atoms. We will do this effectively; however, we may ask whether such effectiveness is at the heart of science. Would it not be interesting to know what these 10 billion terms in our wave function are telling us, and what we could learn from this?

Additional Literature

E Steiner, *The Chemistry Maths Book*, Oxford: Oxford University Press (1996).

A very good textbook that contains much useful information about the secular equation.

W. H. Press et al., *Numerical Recipes, The Art of Scientific Computing*, Third Edition, Cambridge University Press, New York (2007).

Probably the best textbook in computational mathematics; some chapters are very closely related to the topics of this chapter (diagonalization, linear equations).

H. Margenau and G. M. Murphy, The Mathematics of Physics and Chemistry, D. van Nostrand Co., New York; 1956.

An excellent book dealing with most mathematical problems which we may encounter in chemistry and physics, including the variational and perturbational methods.

J. O. Hirschfelder, W. Byers Brown, and S. T. Epstein, "*Recent developments in perturbation theory*," Adv. Quantum Chem., **1**, 255 (1964).

An article on perturbation theory that has been obligatory for those working in the domain for many years.

Questions

1. Variational principle
   a. says that the mean value of the Hamiltonian calculated with any trial wave function is greater than the experimental ground-state energy of the system.
   b. says that the number $\varepsilon = \frac{\langle\phi|\hat{H}\phi\rangle}{\langle\phi|\phi\rangle}$ computed with any trial wave function ϕ is greater than or equal to the lowest eigenvalue of the Hamiltonian $\hat{H}$.
   c. if $\frac{\langle\phi|\hat{H}\phi\rangle}{\langle\phi|\phi\rangle}$ is equal to the lowest eigenvalue of the Hamiltonian $\hat{H}$, this means ϕ is an exact eigenfunction of $\hat{H}$ corresponding to this eigenvalue.
   d. may lead to non-physical states of the system.
2. Variational method ($\hat{H}$ stands for the Hamiltonian, ϕ is a trial function, ε denotes the mean value of energy)
   a. means minimization of functional $\varepsilon[\phi] = \frac{\langle\phi|\hat{H}\phi\rangle}{\langle\phi|\phi\rangle}$.
   b. variational parameters are introduced in the trial function ϕ and ε becomes a function of these parameters.
   c. we minimize ϕ as a function of variational parameters.
   d. we search for the global minimum of ε as a function of variational parameters. The set of the optimum parameters gives best approximation to the ground-state wave function and to the energy of the ground state.
3. Variational method
   a. if a variational trial function is orthogonal to the exact eigenfunctions of all the states with energy $E < A$, the variational principle pertains to the state with the lowest eigenvalue of energy, which satisfies $E \geq A$.
   b. cannot be used if perturbation is large.
   c. if the variational wave function transforms according to the irreducible representation Γ of the symmetry group of the Hamiltonian, the variational method will find an approximation to the lowest-energy state with the wave function transforming according to Γ.
   d. is applicable for the ground state only.

4. Ritz method ($\hat{H}$ stands for the Hamiltonian, ϕ is a variational trial function, Ψ_i denote the basis functions)
   a. is a variational method, in which the trial function has the form $\phi = \sum_{i=0}^{P} c_i \Psi_i$, where Ψ_i are the eigenfunctions of the Hamiltonian, while c_i are unknown coefficients.
   b. is a variational method, in which the trial function has the form $\phi = \sum_{i=0}^{P} c_i \Psi_i$, where Ψ_i are known basis functions, while c_i are unknown coefficients.
   c. the optimal variational coefficients are found by solving the secular equations.
   d. approximate eigenvalues of the Hamiltonian are obtained as solutions of the equation $\det\left(H_{ij} - \varepsilon S_{ij}\right) = 0$, where $H_{ij} = \left\langle \Psi_i | \hat{H} \Psi_j \right\rangle$ and $S_{ij} = \left\langle \Psi_i | \Psi_j \right\rangle$.

5. Ritz method ($\hat{H}$ stands for the Hamiltonian, ϕ is a variational trial function, Ψ_i denote the real basis functions, P means the number of variational parameters)
   a. the number of integrals $H_{ij} = \left\langle \Psi_i | \hat{H} \Psi_j \right\rangle$ needed in computation is equal to $P(P+1)$.
   b. functions Ψ_i have to be orthonormal.
   c. the approximate eigenvalues of the Hamiltonian are computed by diagonalization of the Hamiltonian matrix: $H_{ij} = \left\langle \Psi_i | \hat{H} \Psi_j \right\rangle$.
   d. the matrix of elements H_{ij} is Hermitian and, its eigenvalues are always real numbers.

6. Perturbational approach[38]
   a. ψ_k is an eigenfunction of $\hat{H}$, while $\psi_k^{(0)}$ is an eigenfunction of $\hat{H}^{(0)}$.
   b. intermediate normalization means that in the equality $\psi_k = c_k^{(0)} \psi_k^{(0)} + \chi$, the integral $\langle \chi | \psi_k \rangle = 0$, while $c_k^{(0)} = 1$.
   c. intermediate normalization means that in equality $\psi_k = c_k^{(0)} \psi_k^{(0)} + \chi$, the integral $\left\langle \chi | \psi_k^{(0)} \right\rangle = 0$, while $c_k^{(0)} = 1$.
   d. $E_k = E_k^{(0)} + E_k^{(1)}$.

7. Perturbational approach (notation of question 6)
   a. $E_k^{(0)} + E_k^{(1)} = \left\langle \psi_k^{(0)} | \hat{H} \psi_k^{(0)} \right\rangle$.
   b. $E_0^{(2)} \leq 0$.
   c. $E_k^{(2)} \leq 0$.
   d. $E_k^{(2)} = \sum_{n(\neq k)}^{\infty} \frac{\left|\hat{H}_{kn}^{(1)}\right|^2}{E_k^{(0)} - E_n^{(0)}}$.

8. Perturbational approach (notation of question 6)
   a. from the intermediate normalization, it follows that $\left\langle \psi_k^{(n)} | \psi_k^{(0)} \right\rangle = 0$ for $n > 0$.
   b. $E_0^{(0)} + E_0^{(1)} \geq E_0$.

---

[38] $\hat{H}$ means the Hamiltonian of the system, $\hat{H}^{(0)}$ stands for the unperturbed Hamiltonian; $\hat{H}^{(1)}$ denotes the perturbation; ψ_k means the wave function of the perturbed state k; $\psi_k^{(0)}$ denotes the wave function of the unperturbed state k, with both functions normalized; E_k and $E_k^{(0)}$ mean the corresponding energies; $\psi_k^{(n)}$ and $E_k^{(n)}$ stand for the nth correction to the wave function and energy, respectively.

c. $|E_k^{(n-1)}| \geq |E_k^{(n)}|$.

d. to calculate $E_0^{(2)}$, one has to know $\psi_0^{(2)}$.

9. Perturbational approach (notation of question 6)

a. to compute $E_0^{(2)}$, it is sufficient to know $\psi_0^{(n)}$, $n = 0, 1$.

b. $\hat{H}_{kn}^{(1)}$ is a real number.

c. from the Hermitian character of $\hat{H}^{(1)}$, it follows that $\hat{H}_{kn}^{(1)*} = \left\langle \psi_n^{(0)} | \hat{H}^{(1)} \psi_k^{(0)} \right\rangle$.

d. $\left\langle \psi_k^{(0)} | \hat{H}^{(1)} \psi_n^{(0)} \right\rangle \left\langle \psi_n^{(0)} | \hat{H}^{(1)} \psi_k^{(0)} \right\rangle \geq 0$.

10. Perturbational approach (notation of question 6, λ stands for perturbational parameter in expansion of E_k and ψ_k)

a. Hylleraas variational principle pertains to the inequality $\varepsilon = \frac{\langle \phi | \hat{H} \phi \rangle}{\langle \phi | \phi \rangle} \geq E_0$ with the variational trial function ϕ proposed by Hylleraas.

b. the minimum of the Hylleraas functional is attained for the trial function equal to $\psi_k^{(1)}$, while the value of this minimum is equal to the second order correction to energy.

c. if a series representing a physical quantity tends to ∞, it cannot be applied in computation of this quantity.

d. if, for $\lambda = 0,7 + 0,7i$, the perturbational series $E_k = E_k^{(0)} + \lambda E_k^{(1)} + \lambda^2 E_k^{(2)} \ldots$ diverges, this means that it is also divergent for $\lambda = 1$.

Answers

1b,c,d, 2a,b,d, 3a,c, 4b,c,d, 5d, 6a,c, 7a,b,d, 8a,b, 9a,c,d, 10b,d

Separation of Electronic and Nuclear Motions

"Any separation is a link."
Simone Weil

Where Are We?

We are on the most important branching out of the TREE.

An Example

A colleague shows us the gas phase absorption spectra separately: of the hydrogen atom, of the chlorine atom, and of the hydrogen chloride recorded in the ultraviolet and visible (UV-VIS), infrared (IR) and microwave ranges. In the IR range, neither the hydrogen atom nor the chlorine atom have any electromagnetic wave absorption. However, on the other hand, the hydrogen chloride diatomic molecule that is formed by these two atoms has a very rich absorption spectrum with a quasi-regular and mysterious structure shown in Fig. 6.6 on p. 286. If the theory given in the previous chapters is correct, then it should explain every detail of such a strange spectrum. We also hope we will understand why such a spectrum may appear.

What Is It All About?

Ideas of Quantum Chemistry, Second Edition. http://dx.doi.org/10.1016/B978-0-444-59436-5.00006-4

Nuclei are thousands of times heavier than electrons. As an example, let us take the hydrogen atom. From the conservation of momentum law, it follows that the proton moves 1840 times slower than the electron. In a polyatomic system, while a nucleus moves a little, an electron travels many times through the molecule. It seems that much can be simplified when assuming electronic motion in a field created by immobile nuclei. This concept is behind what is called *adiabatic approximation,* in which the motions of the electrons and the nuclei are separated.[1] Only after this approximation is introduced can we obtain the fundamental concept of chemistry: the molecular structure in 3-D space.

The separation of the electronic and nuclear motions will be demonstrated in detail by taking the example of a diatomic molecule.

Why Is This Important?

The separation of the electronic and nuclear motions represents a fundamental approximation of quantum chemistry. Without this, chemists would lose their *basic model of the molecule*: the 3-D structure with the nuclei occupying some positions in 3-D space, with chemical bonds, etc. This is why this chapter occupies the *central* position on the TREE.

What Is Needed?

- Postulates of quantum mechanics (Chapter 1)
- Separation of the center-of-mass motion (see Appendix I available at booksite.elsevier.com/978-0-444-59436-5 on p. e93)
- Rigid rotator (Chapter 4)
- Harmonic and Morse oscillators (Chapter 4)
- Conclusions from group theory (see Appendix C available at booksite.elsevier.com/978-0-444-59436-5 p. e17, advised)
- Dipole moment (see Appendix X available at booksite.elsevier.com/978-0-444-59436-5 p. e169, occasionally used)

Classical Works

A fundamental approximation (called the Born-Oppenheimer approximation) was introduced in a paper called "*Zur Quantentheorie der Molekeln*" by Max Born and Julius Robert Oppenheimer in *Annalen der Physik, 84,* 457 (1927). The approximation follows from the fact that nuclei are much heavier than electrons. ★ The conical intersection problem was first recognized by three young and congenial Hungarians: Janos (later John) von Neumann and Jenó Pál (later Eugene) Wigner in the papers "*Über merkwürdige diskrete Eigenwerte*" in *Physikalische Zeitschrift, 30,* 465 (1929), and "*Über das Verhalten von Eigenwerten bei adiabatischen Prozessen*" also published in *Physikalische Zeitschrift, 30,* 467 (1929), and later in a paper "*Crossing of Potential Surfaces*" by Edward Teller published in the *Journal of Physical Chemistry 41,* 109 (1937). ★ Gerhard Herzberg was the greatest spectroscopist of the 20th century, author of the fundamental three-volume work: *Spectra of Diatomic Molecules* (1939), *Infrared and Raman Spectra of Polyatomic Molecules* (1949) and *Electronic Spectra of Polyatomic Molecules* (1966). ★ The world's first computational papers using a rigorous approach that went beyond the Born-Oppenheimer approximation for molecules were two articles by Włodzimierz Kołos and Lutosław Wolniewicz. The first was "*The coupling between electronic and nuclear motion and the relativistic effects in the ground state of the H_2 molecule,*" published in *Acta Physica Polonica, 20,* 129 (1961). The second was "*A complete non-relativistic treatment of the H_2 molecule,*" published in *Physics Letters, 2,* 222 (1962). ★ The discovery of the conical intersection and the funnel effect in photochemistry is attributed to Howard E. Zimmerman: *Molecular Orbital Correlation Diagrams, Möbius Systems, and Factors Controlling Ground- and Excited-State Reactions* [*Journal of the American Chemical Society, 88,* 1566

[1] This does not mean that electrons and nuclei move independently. We obtain two coupled equations: one for the motion of the electrons in the field of the fixed nuclei, and the other for the motion of the nuclei in the potential averaged over the electronic positions.

(1966)] and to Josef Michl [*Journal of Molecular Photochemistry, 4*, 243 (1972)]. Important contributions in this domain were also made by Lionel Salem and Christopher Longuet-Higgins.

John von Neumann (1903–1957), known as Jancsi (then Johnny), was the wunderkind of a top Hungarian banker. (Jancsi showed off at receptions by reciting from memory all the phone numbers after reading a page of the phone book.) He attended the famous Lutheran High School in Budapest, the same one as Jenó Pál Wigner (who later used the name Eugene). In 1926, von Neumann received his chemistry engineering diploma; and in the same year, he completed his Ph.D. in mathematics at the University of Budapest. He finally emigrated to the United States and founded the Princeton Advanced Study Institute. John von Neumann was a mathematical genius who contributed to the mathematical foundations of quantum theory, computers, and game theory.

von Neumann made a strange offer of a professorship at the Advanced Study Institute to Stefan Banach from the John Casimir University in Lwów. He handed him a check with "1" handwritten on it and asked Banach to add as many zeros as he wanted. "*This is not enough money to persuade me to leave Poland,*" answered mathematician Banach.

Edward Teller (1908–2004), American physicist of Hungarian origin and professor at the George Washington University, the University of Chicago, and the University of California. Teller left Hungary in 1926, received his Ph.D. in 1930 at the University of Leipzig, and fled Nazi Germany in 1935. Teller was the project leader of the U.S. hydrogen bomb project in Los Alamos, believing that this was the way to overthrow communism ("I am passionately opposed to killing, but I am even more passionately fond of freedom"). The hydrogen bomb patent is owned by Teller and Stanisław Ulam.

Eugene Paul Wigner (1902–1995), American chemist, physicist and mathematician of Hungarian origin and professor at Princeton University. At the age of 11, Wigner, a schoolboy from Budapest, was in a sanatorium in Austria with suspected tuberculosis. Lying for hours on a deck chair reading books, he was seduced by the beauty of mathematics (fortunately, it turned out that he did not have tuberculosis).

In 1915, Wigner entered the Lutheran High School in Budapest. Fulfilling the wishes of his father, who dreamed of having a successor in managing the familial tannery, Wigner graduated from the Technical University in Budapest as a chemist. In 1925, at the Technical University in Berlin, he defended his Ph.D. thesis on chemical kinetics "*Bildung und Zerfall von Molekülen,*" under the supervision of Michael Polanyi, a pioneer in the study of chemical reactions. In 1926 Wigner left the tannery.

By chance he was advised by his colleague von Neumann to focus on group theory (where he obtained the most spectacular successes). Wigner was the first to understand the main features of the nuclear forces. In 1963 he won the Nobel Prize *"for his contributions to the theory of the atomic nucleus and elementary particles, particularly through the discovery and application of fundamental symmetry principles."*

Christopher Longuet-Higgins (1923–2004), professor at the University of Sussex, Great Britain, began his scientific career as a theoretical chemist. His main achievements are connected with conical intersection, as well as with the introduction of permutational groups in the theoretical explanation of the spectra of flexible molecules.

Longuet-Higgins was elected the member of the Royal Society of London for these contributions. He turned to artificial intelligence at the age of 40, and in 1967, he founded the Department of Machine Intelligence and Perception at the University of Edinburgh. Longuet-Higgins investigated machine perception of speech and music. His contribution to this field was recognized by the award of an honorary doctorate in Music by Sheffield University. *Courtesy of Professor John D.Roberts.*

6.1 Separation of the Center-of-Mass Motion

Space-Fixed Coordinate System (SFCS)

Let us consider first a diatomic molecule with the nuclei labeled a, b, and n electrons. Let us choose a Cartesian coordinate system in our laboratory (called the *space-fixed coordinate system, or SFCS*) with the origin located at an arbitrarily chosen point and with arbitrary orientation of the axes[2]. The nuclei have the following positions: $\boldsymbol{R}_a = (X_a, Y_a, Z_a)$ and $\boldsymbol{R}_b = (X_b, Y_b, Z_b)$, while electron i has the coordinates x_i', y_i', z_i'.

We write the Hamiltonian for the system (as discussed in Chapter 1):

$$\hat{\mathcal{H}} = -\frac{\hbar^2}{2M_a}\Delta_a - \frac{\hbar^2}{2M_b}\Delta_b - \sum_{i=1}^{n}\frac{\hbar^2}{2m}\Delta_i' + V, \tag{6.1}$$

where the first two terms stand for the kinetic energy operators of the nuclei, the third term corresponds to the kinetic energy of the electrons (m is the electron mass, and all Laplacians are in the SFCS), and V denotes the Coulombic potential energy operator (interaction of all the

[2] For example, right in the center of the Norwich market square.

particles, nucleus-nucleus, nuclei-electrons, and electrons-electrons)[3]:

$$V = \frac{\mathcal{Z}_a \mathcal{Z}_b e^2}{R} - \mathcal{Z}_a \sum_i \frac{e^2}{r_{ai}} - \mathcal{Z}_b \sum_i \frac{e^2}{r_{bi}} + \sum_{i<j} \frac{e^2}{r_{ij}}. \tag{6.2}$$

While we are not interested in collisions of our molecule with a wall or similar obstruction, we may consider a separation of the motion of the center of mass, and then forget about the motion and focus on the rest (i.e., on the *relative* motion of the particles).

New Coordinates

The total mass of the molecule is $M = M_a + M_b + mn$. The components of the center-of-mass position vector are

$$X = \frac{1}{M}\left(M_a X_a + M_b X_b + \sum_i m x_i'\right)$$

$$Y = \frac{1}{M}\left(M_a Y_a + M_b Y_b + \sum_i m y_i'\right)$$

$$Z = \frac{1}{M}\left(M_a Z_a + M_b Z_b + \sum_i m z_i'\right).$$

Now, we decide to abandon this coordinate system (SFCS). Instead of the old coordinates, we will choose a new set of $3n+6$ coordinates (see Appendix I available at booksite.elsevier.com/978-0-444-59436-5 on p. e93, choice II):

- Three center-of-mass coordinates: X, Y, Z
- Three components of the vector $\boldsymbol{R} = \boldsymbol{R}_a - \boldsymbol{R}_b$ that point to nucleus a from nucleus b
- $3n$ electronic coordinates $x_i = x_i' - \frac{1}{2}(X_a + X_b)$, and similarly, for y_i and z_i, for $i = 1, 2, \ldots n$, which show the electron's position with respect to the geometric center[4] of the molecule.

Hamiltonian in the New Coordinates

The new coordinates have to be introduced into the Hamiltonian. To this end, we need the second derivative operators in the old coordinates to be expressed by the new ones. To begin

[3] Do not confuse coordinate Z with nuclear charge $\mathcal{Z}$.

[4] If the origin were chosen in the center of mass instead of the geometric center, V becomes mass-dependent (J. Hinze, A. Alijah and L. Wolniewicz, *Pol. J. Chem.*, *72*, 1293 (1998); cf. also see Appendix I available at booksite.elsevier.com/978-0-444-59436-5, example 2. We want to avoid this.

with (similarly as in Appendix I available at booksite.elsevier.com/978-0-444-59436-5), let us construct the *first* derivative operators:

$$\frac{\partial}{\partial X_a} = \frac{\partial X}{\partial X_a}\frac{\partial}{\partial X} + \frac{\partial Y}{\partial X_a}\frac{\partial}{\partial Y} + \frac{\partial Z}{\partial X_a}\frac{\partial}{\partial Z} + \frac{\partial R_x}{\partial X_a}\frac{\partial}{\partial R_x} + \frac{\partial R_y}{\partial X_a}\frac{\partial}{\partial R_y} + \frac{\partial R_z}{\partial X_a}\frac{\partial}{\partial R_z}$$
$$+\sum_i \frac{\partial x_i}{\partial X_a}\frac{\partial}{\partial x_i} + \sum_i \frac{\partial y_i}{\partial X_a}\frac{\partial}{\partial y_i} + \sum_i \frac{\partial z_i}{\partial X_a}\frac{\partial}{\partial z_i}$$
$$= \frac{\partial X}{\partial X_a}\frac{\partial}{\partial X} + \frac{\partial R_x}{\partial X_a}\frac{\partial}{\partial R_x} + \sum_i \frac{\partial x_i}{\partial X_a}\frac{\partial}{\partial x_i} = \frac{M_a}{M}\frac{\partial}{\partial X} + \frac{\partial}{\partial R_x} - \frac{1}{2}\sum_i \frac{\partial}{\partial x_i}$$

and the same goes for the coordinates Y_a and Z_a. For the nucleus b, the expression is a little bit different: $\frac{\partial}{\partial X_b} = \frac{M_b}{M}\frac{\partial}{\partial X} - \frac{\partial}{\partial R_x} - \frac{1}{2}\sum_i \frac{\partial}{\partial x_i}$.

For the first derivative operator with respect to the coordinates of the electron i, we obtain:

$$\frac{\partial}{\partial x_i'} = \frac{\partial X}{\partial x_i'}\frac{\partial}{\partial X} + \frac{\partial Y}{\partial x_i'}\frac{\partial}{\partial Y} + \frac{\partial Z}{\partial x_i'}\frac{\partial}{\partial Z} + \frac{\partial R_x}{\partial x_i'}\frac{\partial}{\partial R_x} + \frac{\partial R_y}{\partial x_i'}\frac{\partial}{\partial R_y} + \frac{\partial R_z}{\partial x_i'}\frac{\partial}{\partial R_z}$$
$$+\sum_j \frac{\partial x_j}{\partial x_i'}\frac{\partial}{\partial x_j} + \sum_j \frac{\partial y_j}{\partial x_i'}\frac{\partial}{\partial y_j} + \sum_j \frac{\partial z_j}{\partial x_i'}\frac{\partial}{\partial z_j}$$
$$= \frac{\partial X}{\partial x_i'}\frac{\partial}{\partial X} + \frac{\partial x_i}{\partial x_i'}\frac{\partial}{\partial x_i} = \frac{m}{M}\frac{\partial}{\partial X} + \frac{\partial}{\partial x_i}$$

and the same goes for y_i' and z_i'.

Now, let us create the second derivative operators:

$$\frac{\partial^2}{\partial X_a^2} = \left(\frac{M_a}{M}\frac{\partial}{\partial X} + \frac{\partial}{\partial R_x} - \frac{1}{2}\sum_i \frac{\partial}{\partial x_i}\right)^2 = \left(\frac{M_a}{M}\right)^2 \frac{\partial^2}{\partial X^2} + \frac{\partial^2}{\partial R_x^2} + \frac{1}{4}\left(\sum_i \frac{\partial}{\partial x_i}\right)^2$$
$$+2\frac{M_a}{M}\frac{\partial}{\partial X}\frac{\partial}{\partial R_x} - \frac{\partial}{\partial R_x}\sum_i \frac{\partial}{\partial x_i} - \frac{M_a}{M}\frac{\partial}{\partial X}\sum_i \frac{\partial}{\partial x_i},$$

$$\frac{\partial^2}{\partial X_b^2} = \left(\frac{M_b}{M}\frac{\partial}{\partial X} - \frac{\partial}{\partial R_x} - \frac{1}{2}\sum_i \frac{\partial}{\partial x_i}\right)^2 = \left(\frac{M_b}{M}\right)^2 \frac{\partial^2}{\partial X^2} + \frac{\partial^2}{\partial R_x^2} + \frac{1}{4}\left(\sum_i \frac{\partial}{\partial x_i}\right)^2$$
$$-2\frac{M_b}{M}\frac{\partial}{\partial X}\frac{\partial}{\partial R_x} + \frac{\partial}{\partial R_x}\sum_i \frac{\partial}{\partial x_i} - \frac{M_b}{M}\frac{\partial}{\partial X}\sum_i \frac{\partial}{\partial x_i},$$

$$\frac{\partial^2}{\partial (x_i')^2} = \left(\frac{m}{M}\frac{\partial}{\partial X} + \frac{\partial}{\partial x_i}\right)^2 = \left(\frac{m}{M}\right)^2 \frac{\partial^2}{\partial X^2} + \frac{\partial^2}{\partial x_i^2} + 2\frac{m}{M}\frac{\partial}{\partial X}\frac{\partial}{\partial x_i}.$$

After inserting all this into the Hamiltonian [Eq. (6.1)] we obtain the Hamiltonian expressed in the new coordinates[5]:

[5] The potential energy also has to be expressed using the new coordinates.

$$\hat{\mathcal{H}} = \left[-\frac{\hbar^2}{2M} \Delta_{XYZ} \right] + \hat{H}_0 + \hat{H}', \tag{6.3}$$

where the first term means the center-of-mass kinetic energy operator and $\hat{H}_0$ is the *electronic Hamiltonian* (*clamped nuclei Hamiltonian*):

$$\hat{H}_0 = -\sum_i \frac{\hbar^2}{2m} \Delta_i + V, \tag{6.4}$$

while $\Delta_i \equiv \frac{\partial^2}{\partial x_i^2} + \frac{\partial^2}{\partial y_i^2} + \frac{\partial^2}{\partial z_i^2}$ and

$$\hat{H}' = -\frac{\hbar^2}{2\mu} \Delta_{\boldsymbol{R}} + \hat{H}'', \tag{6.5}$$

with $\Delta_{\boldsymbol{R}} \equiv \frac{\partial^2}{\partial R_x^2} + \frac{\partial^2}{\partial R_y^2} + \frac{\partial^2}{\partial R_z^2}$, where

$$\hat{H}'' = \left[-\frac{\hbar^2}{8\mu} \left(\sum_i \nabla_i \right)^2 + \frac{\hbar^2}{2} \left(\frac{1}{M_a} - \frac{1}{M_b} \right) \nabla_{\boldsymbol{R}} \sum_i \nabla_i \right],$$

and μ denotes the reduced mass of the two nuclei ($\mu^{-1} = M_a^{-1} + M_b^{-1}$).

The $\hat{H}_0$ does not contain the kinetic energy operator of the nuclei, but it does contain all the other terms (this is why it is called the *electronic* or *clamped nuclei Hamiltonian*): the first term stands for the kinetic energy operator of the electrons, and V means the potential energy corresponding to the Coulombic interaction of all particles. The first term in the operator $\hat{H}'$ (i.e., $-\frac{\hbar^2}{2\mu}\Delta_{\boldsymbol{R}}$), denotes the kinetic energy operator of the nuclei[6], while the operator $\hat{H}''$ couples the motions of the nuclei and electrons[7].

---

[6] What moves is a particle of reduced mass μ and coordinates R_x, R_y, and R_z. This means that the particle has the position of nucleus a, whereas nucleus b is at the origin. Therefore, this term accounts for the vibrations of the molecule (changes in length of $\boldsymbol{R}$), as well as its rotations (changes in orientation of $\boldsymbol{R}$).

[7] The first of these two terms contains the reduced mass of the two nuclei, where ∇_i denotes the nabla operator for electron i, $\nabla_i \equiv \boldsymbol{i}\frac{\partial}{\partial x_i} + \boldsymbol{j}\frac{\partial}{\partial y_i} + \boldsymbol{k}\frac{\partial}{\partial z_i}$ with $\boldsymbol{i}, \boldsymbol{j}, \boldsymbol{k}$–the unit vectors along the x-,y-, and z-axes. The second term is nonzero only for the heteronuclear case and contains the mixed product of nablas: $\nabla_R \nabla_i$ with $\nabla_{\boldsymbol{R}} = \boldsymbol{i}\frac{\partial}{\partial R_x} + \boldsymbol{j}\frac{\partial}{\partial R_y} + \boldsymbol{k}\frac{\partial}{\partial R_z}$ and R_x, R_y, R_z as the components of the vector $\boldsymbol{R}$.

After Separation of the Center-of-Mass Motion

After separation of the center-of-mass motion [the first term in Eq. (6.3) is gone; see Appendix I available at booksite.elsevier.com/978-0-444-59436-5 on p. e93], we obtain the eigenvalue problem of the Hamiltonian:

$$\hat{H} = \hat{H}_0 + \hat{H}'. \tag{6.6}$$

This is an exact result, fully equivalent to the Schrödinger equation.

6.2 Exact (Non-Adiabatic) Theory

The total wave function that describes both electrons and nuclei can be proposed in the following form[8] ($\mathcal{N} = \infty$):

$$\Psi(\boldsymbol{r}, \boldsymbol{R}) = \sum_{k}^{\mathcal{N}} \psi_k(\boldsymbol{r}; R) f_k(\boldsymbol{R}), \tag{6.7}$$

where $\psi_k(\boldsymbol{r}; R)$ are the eigenfunctions of $\hat{H}_0$:

[8] Where would such a form of the wave function come from?

If the problem were solved exactly, then the solution of the Schrödinger equation could be sought; e.g., by using the Ritz method (p. 238). Then we have to decide what kind of basis set to use. We could use two auxiliary complete basis sets: one that depended on the electronic coordinates $\{\bar{\psi}_k(\boldsymbol{r})\}$, and another that depended on the nuclear coordinates $\{\bar{\phi}_l(\boldsymbol{R})\}$. The complete basis set for the Hilbert space of our system could be constructed as a Cartesian product $\{\bar{\psi}_k(\boldsymbol{r})\} \times \{\bar{\phi}_l(\boldsymbol{R})\}$; i.e., all possible product-like functions $\bar{\psi}_k(\boldsymbol{r})\bar{\phi}_l(\boldsymbol{R})$. Thus, the wave function could be expanded in a series, as follows:

$$\begin{aligned}\Psi(\boldsymbol{r}, \boldsymbol{R}) &= \sum_{kl} c_{kl}\bar{\psi}_k(\boldsymbol{r})\bar{\phi}_l(\boldsymbol{R}) = \sum_{k}^{\mathcal{N}} \bar{\psi}_k(\boldsymbol{r}) \left[\sum_{l} c_{kl}\bar{\phi}_l(\boldsymbol{R})\right] \\ &= \sum_{k}^{\mathcal{N}} \bar{\psi}_k(\boldsymbol{r}) f_k(\boldsymbol{R}),\end{aligned}$$

where $f_k(\boldsymbol{R}) = \sum_l c_{kl}\bar{\phi}_l(\boldsymbol{R})$ stands for a to-be-sought coefficient depending on $\boldsymbol{R}$ (rovibrational function). If we were dealing with complete sets, then both $\bar{\psi}_k$ and f_k should not depend on anything else, since a sufficiently long expansion of the terms $\bar{\psi}_k(\boldsymbol{r})\bar{\phi}_l(\boldsymbol{R})$ would be suitable to describe all possible distributions of the electrons and the nuclei.

However, we are unable to manage the complete sets. Instead, we are able to take only a few terms in this expansion. We would like them to describe the molecule reasonably well, and at the same time to have only one such term. If so, it would be reasonable to introduce a parametric dependence of the function $\bar{\psi}_k(\boldsymbol{r})$ on the position of the nuclei, which in our case of a diatomic molecule means the internuclear distance. This is equivalent to telling someone how the electrons behave when the internuclear distances have some specific values, and how they behave when the distances change.

$$\hat{H}_0(R)\psi_k(\mathbf{r};R) = E_k^0(R)\psi_k(\mathbf{r};R) \tag{6.8}$$

and depend parametrically[9] on the internuclear distance R, and $f_k(\mathbf{R})$ are yet unknown rovibrational functions (describing the rotations and vibrations of the molecule).

Averaging Over Electronic Coordinates

First, let us write down the Schrödinger equation with the Hamiltonian [Eq. (6.6)] and the wave function, as in Eq. (6.7):

$$(\hat{H}_0 + \hat{H}')\sum_l^{\mathcal{N}} \psi_l(\mathbf{r};R)\, f_l(\mathbf{R}) = E\sum_l^{\mathcal{N}} \psi_l(\mathbf{r};R)\, f_l(\mathbf{R}). \tag{6.9}$$

Let us multiply both sides by $\psi_k^*(\mathbf{r};R)$ and then integrate over the *electronic* coordinates $\mathbf{r}$ (which will be stressed by the subscript "e"):

$$\sum_l^{\mathcal{N}} \left\langle \psi_k | (\hat{H}_0 + \hat{H}')[\psi_l f_l] \right\rangle_e = E\sum_l^{\mathcal{N}} \langle \psi_k|\psi_l\rangle_e\, f_l. \tag{6.10}$$

On the right side of Eq. (6.10), we profit from the orthonormalization condition $\langle \psi_k|\psi_l\rangle_e = \delta_{kl}$, and on the left side, we recall that ψ_k is an eigenfunction of $\hat{H}_0$:

$$E_k^0 f_k + \sum_l^{\mathcal{N}} \left\langle \psi_k | \hat{H}'[\psi_l f_l] \right\rangle_e = E f_k. \tag{6.11}$$

Now, let us focus on the expression $\hat{H}'(\psi_l f_l) = -\frac{\hbar^2}{2\mu}\Delta_{\mathbf{R}}(\psi_l f_l) + \hat{H}''(\psi_l f_l)$, which we have in the integrand in Eq. (6.11). Let us concentrate on the first of these terms:

$$\begin{aligned}
-\frac{\hbar^2}{2\mu}\Delta_{\mathbf{R}}(\psi_l f_l) &= -\frac{\hbar^2}{2\mu}\nabla_{\mathbf{R}}\nabla_{\mathbf{R}}(\psi_l f_l) = -\frac{\hbar^2}{2\mu}\nabla_{\mathbf{R}}[\psi_l\nabla_{\mathbf{R}} f_l + (\nabla_{\mathbf{R}}\psi_l) f_l] \\
&= -\frac{\hbar^2}{2\mu}[\nabla_{\mathbf{R}}\psi_l\nabla_{\mathbf{R}} f_l + \psi_l\Delta_{\mathbf{R}} f_l + (\Delta_{\mathbf{R}}\psi_l) f_l + \nabla_{\mathbf{R}}\psi_l\nabla_{\mathbf{R}} f_l] \\
&= -\frac{\hbar^2}{2\mu}[2\left(\nabla_{\mathbf{R}}\psi_l\right)\left(\nabla_{\mathbf{R}} f_l\right) + \psi_l\Delta_{\mathbf{R}} f_l + (\Delta_{\mathbf{R}}\psi_l) f_l].
\end{aligned} \tag{6.12}$$

[9] For each value of R, we have a different formula for ψ_k.

After inserting the result into $\left\langle \psi_k | \hat{H}'(\psi_l f_l) \right\rangle_e$ and recalling Eq. (6.5), we have

$$\begin{aligned}
\left\langle \psi_k | \hat{H}'[\psi_l f_l] \right\rangle_e &= 2\left(-\frac{\hbar^2}{2\mu}\right) \langle \psi_k | \nabla_{\boldsymbol{R}} \psi_l \rangle_e \nabla_{\boldsymbol{R}} f_l + \langle \psi_k | \psi_l \rangle_e \left(-\frac{\hbar^2}{2\mu}\right) \Delta_{\boldsymbol{R}} f_l \\
&+ \left\langle \psi_k | \left(-\frac{\hbar^2}{2\mu}\right) \Delta_{\boldsymbol{R}} \psi_l \right\rangle_e f_l + \left\langle \psi_k | \hat{H}'' \psi_l \right\rangle_e f_l \\
&= (1 - \delta_{kl}) \left(-\frac{\hbar^2}{\mu}\right) \langle \psi_k | \nabla_{\boldsymbol{R}} \psi_l \rangle_e \nabla_{\boldsymbol{R}} f_l - \delta_{kl} \frac{\hbar^2}{2\mu} \Delta_{\boldsymbol{R}} f_l + H'_{kl} f_l, \quad (6.13)
\end{aligned}$$

with

$$H'_{kl} \equiv \left\langle \psi_k | \hat{H}' \psi_l \right\rangle_e .$$

At that point, we obtain the following form of Eq. (6.11):

$$E_k^0 f_k + \sum_l^{\mathcal{N}} \left[(1 - \delta_{kl}) \left(-\frac{\hbar^2}{\mu}\right) \langle \psi_k | \nabla_{\boldsymbol{R}} \psi_l \rangle_e \nabla_{\boldsymbol{R}} f_l - \delta_{kl} \frac{\hbar^2}{2\mu} \Delta_{\boldsymbol{R}} f_l + H'_{kl} f_l \right] = E f_k .$$

Here, we have profited from the equality $\langle \psi_l | \nabla_{\boldsymbol{R}} \psi_l \rangle_e = 0$, which follows from the differentiation of the normalization condition[10] for the function ψ_l.

Non-Adiabatic Nuclear Motion

Grouping all the terms with f_l on the left side, we obtain a set of $\mathcal{N}$ equations:

$$\left[-\frac{\hbar^2}{2\mu} \Delta_{\boldsymbol{R}} + E_k^0(R) + H'_{kk}(R) - E \right] f_k = - \sum_{l(\neq k)}^{\mathcal{N}} \Theta_{kl} \, f_l, \quad (6.14)$$

for $k = 1, 2, \ldots \mathcal{N}$ with the non-adiabatic coupling operators

$$\Theta_{kl} = -\frac{\hbar^2}{\mu} \langle \psi_k | \nabla_{\boldsymbol{R}} \psi_l \rangle_e \nabla_{\boldsymbol{R}} + H'_{kl} . \quad (6.15)$$

Note that the operator H'_{kl} depends on the length of the vector $\boldsymbol{R}$, but not on its direction.[11]

[10] We assume that the phase of the wave function $\psi_k(\boldsymbol{r}; R)$ does not depend on R; i.e., $\psi_k(\boldsymbol{r}; R) = \bar{\psi}_k(\boldsymbol{r}; R) \exp(i\phi)$, where $\bar{\psi}_k$ is a real function and $\phi \neq \phi(R)$. This immediately gives $\langle \psi_k | \nabla_{\boldsymbol{R}} \psi_k \rangle_e = \left\langle \bar{\psi}_k | \nabla_{\boldsymbol{R}} \bar{\psi}_k \right\rangle_e$, which is zero, from differentiating the normalization condition. Indeed, the normalization condition: $\int \psi_k^2 \mathrm{d}\tau_e = 1$. Hence, $\nabla_{\boldsymbol{R}} \int \psi_k^2 \mathrm{d}\tau_e = 0$, or $2 \int \psi_k \nabla_{\boldsymbol{R}} \psi_k \mathrm{d}\tau_e = 0$. Without this approximation, we will surely have trouble.

[11] This follows from the fact that we have in $\hat{H}'$ [see Eq. (6.5)] the products of nablas (i.e., scalar products). The scalar products do not change upon rotation because both vectors involved rotate in the same way and the angle between them does not change.

Equation (6.14) is equivalent to the Schrödinger equation.

Equations (6.14) and (6.15) have been derived under the assumption that ψ_k of Eq. (6.7) satisfies Eq. (6.8). If instead of $\psi_k(\mathbf{r}; R)$, we use a (generally non-orthogonal) complete set $\{\bar{\psi}_k(\mathbf{r}; R)\}$ in Eq. (6.7), Eqs. (6.14) and (6.15) would change to

$$\left[-\frac{\hbar^2}{2\mu}\Delta_{\mathbf{R}} + \bar{E}_k(R) + H'_{kk}(R) - E\right] f_k = -\sum_{l(\neq k)}^{\mathcal{N}} \Theta_{kl}\, f_l, \tag{6.16}$$

for $k = 1, 2, \ldots \mathcal{N}$ with the non-adiabatic coupling operators

$$\Theta_{kl} = -\frac{\hbar^2}{\mu}\left\langle\bar{\psi}_k|\nabla_{\mathbf{R}}\bar{\psi}_l\right\rangle_e \nabla_{\mathbf{R}} + H'_{kl} + \left\langle\bar{\psi}_k|\bar{\psi}_l\right\rangle_e\left(-\frac{\hbar^2}{2\mu}\Delta_{\mathbf{R}}\right) \tag{6.17}$$

and $\bar{E}_k(R) \equiv \left\langle\bar{\psi}_k|\hat{H}_0\bar{\psi}_k\right\rangle_e$. Functions $\bar{\psi}_k(\mathbf{r}; R)$ may be chosen as the wave functions with some chemical significance.

6.3 Adiabatic Approximation

If the curves $E_k^0(R)$ for different k are well separated in the energy scale, we may expect that the coupling between them is small, and therefore all Θ_{lk} for $k \neq l$ may be set equal to zero. This is called the *adiabatic approximation*. In this approximation, we obtain from Eq. (6.14):

$$\left[-\frac{\hbar^2}{2\mu}\Delta_{\mathbf{R}} + E_k^0(R) + H'_{kk}(R)\right] f_k(\mathbf{R}) = E f_k(\mathbf{R}), \tag{6.18}$$

where the diagonal correction $H'_{kk}(R)$ is usually very small compared to $E_k^0(R)$.

In the adiabatic approximation, the wave function is approximated by a product

$$\Psi \approx \psi_k(\mathbf{r}; R) f_k(\mathbf{R}) \tag{6.19}$$

The function $f_k(\mathbf{R})$ depends explicitly not only on R, but also on the direction of vector $\mathbf{R}$, and therefore it will describe future vibrations of the molecule (changes of R), as well as its rotations (changes of the direction of $\mathbf{R}$).

A Simple Analogy

Let us pause a moment to get a sense of the adiabatic approximation.

To some extent, the situation resembles an attempt to describe a tourist (an electron) and the Alps (nuclei). Not only the tourist moves, but also the Alps, as has been quite convincingly proved by geologists.[12] The probability of encountering the tourist may be described by a "*wave function*" computed for a fixed position of the mountains (shown by a map bought in a shop). This is a very good approximation because when the tourist wanders over hundreds of miles, the beloved Alps move a tiny distance, so the map seems to be perfect all the time. On the other hand, the probability of having the Alps in a given configuration is described by the geologists' "wave function" f, saying e.g. what is the probability that the distance between the Matterhorn and the Jungfrau is equal to R. When the tourist revisits the Alps after a period of time (say, a few million years), the mountains will have changed (the new map bought in the shop will reflect this fact). The probability of finding the tourist may again be computed from the new wave function, which is valid for the new configuration of the mountains (a parametric dependence). Therefore, the probability of finding the tourist in the spot indicated by the vector $\boldsymbol{r}$ at a given configuration of the mountains $\boldsymbol{R}$ can be approximated by a product[13] of the probability of finding the mountains at this configuration $|f_k(\boldsymbol{R})|^2 d\boldsymbol{R}$ and the probability $|\psi_k(\boldsymbol{r};\boldsymbol{R})|^2 d\boldsymbol{r}$ of finding the tourist in the position shown by the vector $\boldsymbol{r}$, when the mountains have this particular configuration $\boldsymbol{R}$. In the case of our molecule, this means the adiabatic approximation (a product-like form), Eq. (6.19).

This parallel fails in one important way: the Alps do not move in the potential created by tourists, the dominant geological processes are tourist-independent. Contrary to this, as we will soon see, nuclear motion is dictated by the electrons through the potential averaged over the electronic motion.

6.4 Born-Oppenheimer Approximation

In the adiabatic approximation, $H'_{kk} = \int \psi_k^* H' \psi_k d\tau_e$ represents a small correction to $E_k^0(R)$. Neglecting this correction results in the Born-Oppenheimer approximation:

$$H'_{kk} = 0.$$

[12] The continental plates collide like billiard balls in a kind of quasi-periodic oscillation. During the current oscillation, the India plate, which moved at a record speed of about 20 cm a year, hit the Euroasiatic plate. This is why the Himalayan mountains are so beautiful. The collision continues, and the Himalayas will be even more beautiful someday. Europe was hit from the south by a few plates moving at only about 4 cm a year, and this is why Alps are lower than Himalayas. While visiting the Atlantic coast of Maine, I wondered that the color of the rocks was very similar to those I remembered from Brittany, in France. That was it! Once upon a time, the two coasts made a common continent. Later, we had to rediscover America. The Wegener theory of continental plate tectonics, when created in 1911, was viewed as absurd, although the mountain *ranges* suggested that some plates were colliding.

[13] This is an approximation because in the non-adiabatic (i.e., fully correct) approach, the total wave function is a superposition of many such products, corresponding to various electronic and rovibrational wave functions.

Note that in the Born-Oppenheimer approximation, the potential energy for the motion of the nuclei $E_k^0(R)$ is independent of the mass of the nuclei, whereas in the adiabatic approximation, the potential energy $E_k^0(R) + H'_{kk}(R)$ depends on the mass.

Julius Robert Oppenheimer (1904–1967), American physicist and professor at the University of California in Berkeley, the California Institute of Technology in Pasadena, and the Institute for Advanced Study in Princeton. From 1943 to 1945, Oppenheimer headed the Manhattan Project (atomic bomb).

From John Slater's autobiography:

"Robert Oppenheimer was a very brilliant physics undergraduate at Harvard during the 1920s, the period when I was there on the faculty, and we all recognized that he was a person of very unusual attainments. Rather than going on for his graduate work at Harvard, he went to Germany, and worked with Born, developing what has been known as the Born-Oppenheimer approximation."

... And a Certain Superiority of Theory Over Experiment

In experiments, every chemist finds his molecule confined close to a minimum of the electronic energy hypersurface (most often of the ground state). A powerful theory might be able to predict the results of experiments even for the nuclear configurations that are far from those that are accessible for current experiments. This is the case with quantum mechanics, which is able to describe in detail what would happen to the electronic structure[14], if the nuclear configuration were very strange; e.g., the internuclear distances were close to zero, if not exactly zero. Within the Born-Oppenheimer approximation, the theoretician is free to put the nuclei wherever he wishes. This means that we are able to discuss and then just to test "*what would be if*," even if this "*if*" were crazy. For example, some small internuclear distances are achievable at extremely large pressures. At such pressures, some additional difficult experiments have to be performed to tell us about the structure and processes. A theoretician just sets the small distances and makes a computer run. This is really exceptional: we may set some conditions that are out of reach of experiments (even very expensive ones), and we are able to tell with confidence and at low cost what will be.

---

[14] We just do not have any reason to doubt it.

6.5 Vibrations of a Rotating Molecule

Our next step will be an attempt to separate rotations and oscillations within the adiabatic approximation. To this end, the function $f_k(\boldsymbol{R}) = f_k(R, \theta, \phi)$ will be proposed as a *product* of a function Y which will account for rotations (depending on θ, ϕ), and a certain function $\frac{\chi_k(R)}{R}$ describing the oscillations i.e., dependent on R

$$f_k(\boldsymbol{R}) = Y(\theta, \phi)\frac{\chi_k(R)}{R}. \tag{6.20}$$

No additional approximation is introduced in this way. We say only that the isolated molecule vibrates independently of whether it is oriented toward the Capricorn or Taurus constellations ("space is isotropic").[15] The function $\chi_k(R)$ is yet unknown, and we are going to search for it; therefore, dividing by R in (6.20) is meaningless.[16]

Now, we will try to separate the variables θ, ϕ from the variable R in Eq. (6.18); i.e., to obtain two separate equations for them. First, let us define the quantity

$$U_k(R) = E_k^0(R) + H'_{kk}(R). \tag{6.21}$$

After inserting the Laplacian (in spherical coordinates; see Appendix H available at booksite.elsevier.com/978-0-444-59436-5 on p. e91) and the product [Eq. (6.20)] into Eq. (6.18), we obtain the following series of transformations:

$$\left[-\frac{\hbar^2}{2\mu}\left(\frac{1}{R^2}\frac{\partial}{\partial R}R^2\frac{\partial}{\partial R} + \frac{1}{R^2\sin\theta}\frac{\partial}{\partial\theta}\sin\theta\frac{\partial}{\partial\theta} + \frac{1}{R^2\sin^2\theta}\frac{\partial^2}{\partial\phi^2}\right) + U_k(R)\right]Y\frac{\chi_k}{R} = EY\frac{\chi_k}{R},$$

$$-\frac{\hbar^2}{2\mu}\left(\frac{Y}{R}\frac{\partial^2\chi_k}{\partial R^2} + \frac{\chi_k}{R}\frac{1}{R^2\sin\theta}\frac{\partial}{\partial\theta}\sin\theta\frac{\partial Y}{\partial\theta} + \frac{\chi_k}{R}\frac{1}{R^2\sin^2\theta}\frac{\partial^2 Y}{\partial\phi^2}\right) + YU_k(R)\frac{\chi_k}{R} = EY\frac{\chi_k}{R},$$

$$-\frac{\hbar^2}{2\mu}\left(\frac{1}{\chi_k}\frac{\partial^2\chi_k}{\partial R^2} + \frac{1}{Y}\left(\frac{1}{R^2\sin\theta}\frac{\partial}{\partial\theta}\sin\theta\frac{\partial Y}{\partial\theta} + \frac{1}{R^2\sin^2\theta}\frac{\partial^2 Y}{\partial\phi^2}\right)\right) + U_k(R) = E,$$

$$-\left(\frac{R^2}{\chi_k}\frac{\partial^2\chi_k}{\partial R^2}\right) + \frac{2\mu}{\hbar^2}U_k(R)R^2 - \frac{2\mu}{\hbar^2}ER^2 = \frac{1}{Y}\left(\frac{1}{\sin\theta}\frac{\partial}{\partial\theta}\sin\theta\frac{\partial Y}{\partial\theta} + \frac{1}{\sin^2\theta}\frac{\partial^2 Y}{\partial\phi^2}\right).$$

The result is fascinating. The left side depends on R only, and the right side only on θ and ϕ. Both sides equal each other independently of the values of the variables. This can only happen if each side is equal to a constant (λ), the same for each. Therefore, we have

$$-\left(\frac{R^2}{\chi_k}\frac{\partial^2\chi_k}{\partial R^2}\right) + \frac{2\mu}{\hbar^2}U_k(R)R^2 - \frac{2\mu}{\hbar^2}ER^2 = \lambda \tag{6.22}$$

[15] It is an assumption about "the space", which is assumed not to be changed by the presence of the Capricorn, Taurus, or other constellation.

[16] In the case of polyatomics, the function $f_k(\boldsymbol{R})$ may be more complicated because some vibrations (e.g., a rotation of the CH_3 group) may contribute to the total angular momentum, which has to be conserved (this is related to space isotropy; cf., p. 69).

$$\frac{1}{Y}\left(\frac{1}{\sin\theta}\frac{\partial}{\partial\theta}\sin\theta\frac{\partial Y}{\partial\theta}+\frac{1}{\sin^2\theta}\frac{\partial^2 Y}{\partial\phi^2}\right)=\lambda. \tag{6.23}$$

Now, we are amazed to see that Eq. (6.23) is identical (cf., p. 199) to that which appeared as a result of the transformation of the Schrödinger equation for a rigid rotator, Y denoting the corresponding wave function. As we know from p. 200, this equation has a solution only if $\lambda = -J(J+1)$, where $J = 0, 1, 2, ...$ Since Y stands for the rigid rotator wave function, we now concentrate exclusively on the function χ_k, which describes vibrations (changes in the length of $\boldsymbol{R}$).

After inserting the permitted values of λ into Eq. (6.22), we get

$$-\frac{\hbar^2}{2\mu}\left(\frac{\partial^2\chi_k}{\partial R^2}\right)+U_k(R)\chi_k - E\chi_k = -\frac{\hbar^2}{2\mu R^2}J(J+1)\chi_k.$$

Let us write this equation in the form of the eigenvalue problem for the unidimensional motion of a particle (we change the partial into the regular derivative) of mass μ:

$$\left(-\frac{\hbar^2}{2\mu}\frac{\mathrm{d}^2}{\mathrm{d}R^2}+V_{kJ}\right)\chi_{kvJ}(R)=E_{kvJ}\chi_{kvJ}(R) \tag{6.24}$$

with potential energy (let us stress that $R > 0$)

$$V_{kJ}(R)=U_k(R)+J(J+1)\frac{\hbar^2}{2\mu R^2}, \tag{6.25}$$

which takes the effect of centrifugal force on the vibrational motion into account. The solution χ_k, as well as the total energy E_k, have been labeled by two additional indices: the rotational quantum number J (because the potential depends on it) and the numbering of the solutions $v = 0, 1, 2, \ldots$

The solutions of Eq. (6.24) describe the vibrations of the nuclei. The function $V_{kJ} = E_k^0(R) + H'_{kk}(R) + J(J+1)\hbar^2/(2\mu R^2)$ plays the role of the potential energy curve for the motion of the nuclei.

The above equation, and therefore also

the very notion of the potential energy curve for the motion of the nuclei, appears only after the adiabatic (the product-like wave function, and H'_{kk} preserved) or the Born-Oppenheimer (the product-like wave function, but H'_{kk} removed) approximation is applied. Only in the Born-Oppenheimer approximation is the potential energy $U_k(R)$ mass-independent; e.g., the same for isotopomers H_2, HD, and D_2.

If $V_{kJ}(R)$ were a parabola (as it is for the harmonic oscillator), the system would never acquire the energy corresponding to the bottom of the parabola because the harmonic oscillator energy levels (cf., p. 190) correspond to *higher* energy. The same pertains to V_{kJ} of a more complex shape.

6.5.1 One More Analogy

The fact that the electronic energy $E_k^0(R)$ plays the role of the potential energy for vibrations not only represents the result of rather complex derivations, but is also natural and understandable. The nuclei keep together thanks to the electronic "*glue*" (we will come back to this in Chapter 8). Let us imagine two metallic balls (nuclei) in a block of transparent gum (electronic cloud), as shown in Fig. 6.1.

If we were interested in the motion of the *balls*, we would have to take the potential energy as well as the kinetic energy into account. The potential energy would depend on the distance R between the balls, in the same way as the gum's elastic energy depends on the stretching or squeezing the gum to produce a distance between the balls equal to R. Thus, the potential energy for the motion of the balls (nuclei) has to be the potential energy of the gum (electronic energy).[17]

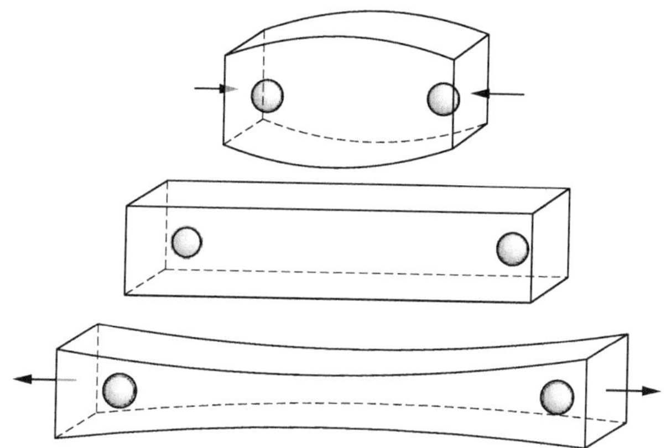

Fig. 6.1. Two metallic balls in a block of gum. How will they vibrate? This will be dictated by the elastic properties of the gum.

[17] The adiabatic approximation is of more general importance than the separation of the electronic and nuclear motions. Its essence pertains to the problem of two coexisting time scales in some phenomena: fast and slow scales. The examples below indicate that we have to do with an important and general philosophical approach:

- In Chapter 14 on chemical reactions, we will consider slow motion along a single coordinate, and fast motions along other coordinates (in the configurational space of the nuclei). "*Vibrationally adiabatic*" approximation will also be introduced, and the slow motion will proceed in the potential energy averaged over fast motions and calculated at each fixed value of the slow coordinate.
- Similar reasoning was behind vibrational analysis in systems with hydrogen bonds [Y. Marechal and A. Witkowski, *Theor. Chim. Acta, 9*, 116 (1967).] The authors selected a slow intermolecular motion proceeding in the potential energy averaged over fast intramolecular motions.

This situation corresponds to a non-rotating system. If we admit rotation, we would have to take the effect of centrifugal force on the potential energy (or elastic properties) of the gum into account. This effect is analogous to the second term in Eq. (6.25) for $V_{kJ}(R)$.

6.5.2 What Vibrates, What Rotates?

One may say that, as a result of averaging over electron coordinates, the electrons disappeared from the theory. The only effect of their presence are numbers: the potential energy term $U_k(R)$ of Eq. (6.21). Equation (6.24) says that the vibrating and rotating objects are bare nuclei, which seems strange because they certainly move somehow with electrons. Our intuition says that what should vibrate and rotate are atoms, not nuclei. In our example with the gum, it is evident that the iron balls should be a bit heavier since they pull the gum with them.

Where is this effect hidden? It has to be a part of the non-adiabatic effect, and can be taken into account within the non-adiabatic procedure described on p. 265. It looks quite strange. Such an obvious effect[18] is hidden in a theory that is hardly used in computational practice, because of its complexity?

If the excited electronic states are well separated from the ground electronic state $k = 0$, it turned out that one may catch a good part of this effect for the ground state by using the perturbation theory (see Chapter 5). It is possible to construct[19] a set of more and more advanced approximations for calculating the rovibrational levels. All of them stem from the following equation for the vibrational motion of the nuclei, a generalization of Eq. (6.24):

$$\left[-\frac{1}{R^2}\frac{\mathrm{d}}{\mathrm{d}R}\frac{R^2}{2\mu_{\|}(R)}\frac{\mathrm{d}}{\mathrm{d}R} + W_{0J}(R)\right]\chi_{0vJ}(R) = E_{0vJ}\,\chi_{0vJ}(R), \tag{6.26}$$

where the operator on the left side corresponds to the kinetic energy of vibration given in Eq. (6.24), but this time, instead of the constant reduced mass μ of the nuclei, we have the mass denoted as $\mu_{\|}(R)$ that is R-dependent. The potential energy

$$W_{0J}(R) = E_0^0(R) + H'_{00}(R) + \frac{J(J+1)}{2\mu_{\perp}R^2} + \delta\mathcal{E}_{na}(R) \tag{6.27}$$

also resembles the potential energy of Eq. (6.25), but the reduced mass of the nuclei μ in the centrifugal energy, Eq. (6.25), is replaced now by a function of R denoted by $\mu_{\perp}(R)$. Visibly the nuclei are "*dressed*" by electrons, and this dressing not only is R-dependent, which is understandable, but also depends on what the nuclei are doing (vibration[20] or rotation). There is also a non-adiabatic increment $\delta\mathcal{E}_{na}(R)$, which effectively takes into account the presence

[18] The effect is certainly small, because the mass of the electrons that make a difference (move with the nucleus) are about 1836 times smaller than the mass of the nucleus alone.

[19] K. Pachucki and J. Komasa, *J. Chem. Phys., 129*, 34102 (2008).

[20] The R-dependent μ was introduced by R.M. Herman and A. Asgharian, *J. Chem. Phys., 45*, 2433 (1966).

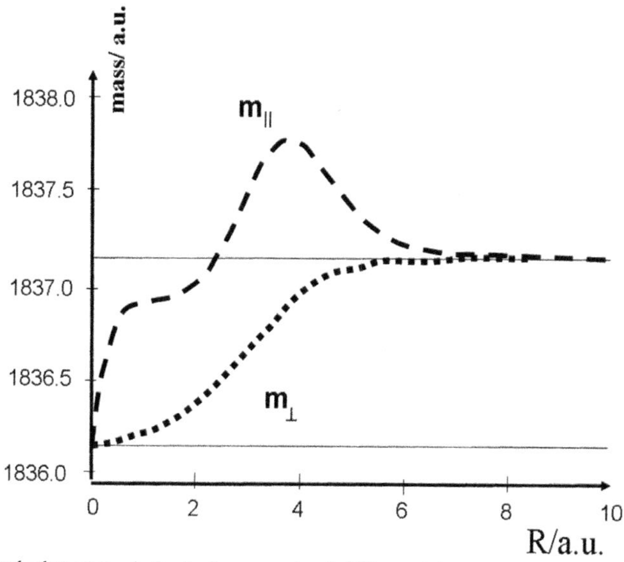

Fig. 6.2. What vibrates and what rotates in the hydrogen molecule? The variable reduced masses $\mu_{||}$ and $\mu_{\perp}$ correspond to the masses of the objects in the hydrogen molecule that vibrate ($m_{||}$) and rotate ($m_{\perp}$). Conclusion: at large distances move atoms, at very short move the bare nuclei.

of higher states. We do not give here the formulas for $\delta\mathcal{E}_{na}(R)$, $\mu_{||}$ and $\mu_{\perp}$ (they all have been derived by Pachucki and Komasa[21]).

The following sequence of approximations can be designed:

- The Born-Oppenheimer approximation: $\mu_{||} = \mu_{\perp} = \mu$; $W_{0J}(R) = E_0^0(R) + \frac{J(J+1)}{2\mu R^2}$
- The adiabatic approximation: $\mu_{||} = \mu_{\perp} = \mu$; $W_{0J}(R) = E_0^0(R) + H'_{00}(R) + \frac{J(J+1)}{2\mu R^2}$
- The effective non-adiabatic approximation: $\mu_{||}, \mu_{\perp}$ taken as R-dependent; $W_{0J}(R) = E_0^0(R) + H'_{00}(R) + \frac{J(J+1)}{2\mu_{\perp} R^2} + \delta\mathcal{E}_{na}(R)$.

It is interesting to see what kind of object vibrates and rotates in the hydrogen molecule. As one can see from Fig. 6.2, $\mu_{||}(R = \infty) = \mu_{\perp}(R = \infty) = M_p + m$, while $\mu_{||}(R = 0) = \mu_{\perp}(R = 0) = M_p$, where M_p stands for the mass of proton and m is the electron mass. Thus, for large R, the hydrogen *atoms* vibrate and rotate, while for very small R - only bare *protons* do. For finite nonzero values of R, the rotation-related effective atomic mass $m_{\perp}(R)$ changes monotonically, while the vibration-related effective atomic mass $m_{||}(R)$ undergoes peculiar changes exhibiting a maximum mass at about 4 a.u. (a bit larger than $M_p + m$) and additionally, an impressive plateau of about $M_p + \frac{3}{4}m$ just before going to $m_{||} = M_p$ at $R = 0$. This is what equations give; however, we have problems with rationalizing such things.

[21] Equation (6.27) may be treated as the most general definition of the potential energy curve for the motion of the nuclei.

The approach reported is able to produce the non-adiabatic corrections to all rovibrational levels corresponding to the ground electronic state.[22]

6.5.3 The Fundamental Character of the Adiabatic Approximation–PES

In the case of a polyatomic molecule with N atoms ($N > 2$), V_{kJ} depends on $3N - 6$ variables determining the configuration of the nuclei. The function $V_{kJ}(\boldsymbol{R})$ therefore represents a surface in $(3N - 5)$-dimensional space (a *hypersurface*). This potential energy (hyper)surface $V_{kJ}(\boldsymbol{R})$, or PES, for the motion of the nuclei represents one of the most important ideas in chemistry.

This concept makes possible contact with what chemists call the spatial "*structure*" of the molecule, identified with its nuclear configuration corresponding to the minimum of the PES for the electronic ground state. It is only because of the adiabatic approximation, that we may imagine the 3-D shape of a molecule as a configuration of its nuclei bound by an electronic cloud (see Fig. 6.3). This object moves and rotates in space, and in addition, the nuclei vibrate about their equilibrium positions with respect to other nuclei (which may be visualized as a rotation-like motion close to the minimum of an energy valley).

Without the adiabatic approximation, questions about the molecular 3-D structure of the benzene molecule could only be answered in a very enigmatic way. For example:

- The molecule does not have any particular 3-D shape.
- The motion of the electrons and nuclei is very complicated.
- Correlations of motion of all the particles exist (electron-electron, nucleus-nucleus, electron-nucleus).
- These correlations are in general very difficult to elucidate.

Identical answers would be given if we were to ask about the structure of the DNA molecule. Obviously, something is going wrong, and perhaps we should expect more help from theory.

For the benzene molecule, we could answer questions like: What is the mean value of the carbon-carbon, carbon-proton, proton-proton, electron-electron, electron-proton, and electron-carbon distances in the benzene molecule in its ground and excited states? Note that because all identical particles are indistinguishable, the carbon-proton distance pertains to any carbon and any proton, and so on. To discover that the benzene molecule is essentially a planar hexagonal object would be very difficult. What could we say about a protein? A pile of paper with such numbers would give us *the* true (though non-relativistic) picture of the benzene molecule, but it would be useless, just as a map of the world with 1:1 scale would be useless for a tourist. It is just too exact. If we relied on this, progress in the investigation of the molecular world

[22] It is worth noting that for H_2 and its lowest rovibrational level (to cite one example), making μ R-dependent [i.e., using $\mu_{||}$ and $\mu_{\perp}$ and neglecting $\delta\mathcal{E}_{na}(R)$] gives 84% of the total non-adiabatic effect, while neglecting this R-dependence [i.e., putting $\mu_{||} = \mu_{\perp} = \mu$ and taking $\delta\mathcal{E}_{na}(R)$ into account gives 15%. These two effects seem to be quite independent.

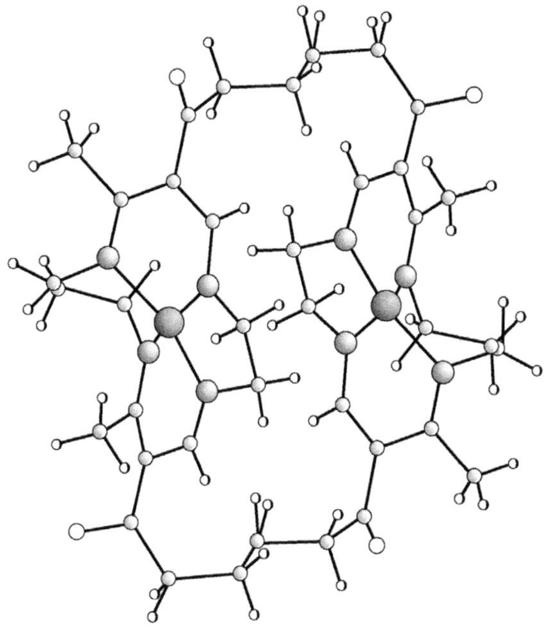

Fig. 6.3. A 3-D *model* (called the "*structure*") of a molecule allows us to focus attention on spatial and temporal relations that are similar to those we know from the macroscopic world. Although the concept of "*spatial structure*" may occasionally fail, in virtually all cases in chemistry and physics, we use a 3-D molecular model that resembles what is shown here for a particular molecule (using a 2-D projection of the 3-D model). There are "balls" and "connecting sticks." The balls represent atoms (of various sizes, and the size characterizes the corresponding element), the sticks of different length are supposed to represent what are called "chemical bonds." What should be taken seriously, and what shouldn't be? First, the scale. The real molecule is about 100000000 times smaller than the picture here. Second, the motion. This static model shows a kind of averaging over all the snapshots of the real vibrating atoms. In Chapters 8 and 11, we will see that indeed the atoms of which the molecule is composed keep together because of a pattern of interatomic chemical bonds (which characterizes the electronic state of the molecule) that to some extent resemble sticks. An atom in a molecule is never spherically symmetric (cf., Chapter 11), but can be approximated by its spherical core ("ball"). The particular molecule shown here has two tetraazaanulene macrocycles that coordinate two Ni^{2+} ions (the largest spheres). The macrocycles are held together by two $-(CH_2)_4-$ molecular links. Note that any atom of a given type binds a certain number of its neighbors. The most important message is: if such structural information offered by the 3-D molecular model were not available, it would not be possible to design and carry out the complex synthesis of the molecule. *Courtesy of Professor B. Korybut-Daszkiewicz.*

would more or less stop. A radical approach in science, even if more rigorous, is very often less fruitful or fertile. Science needs models, simpler than reality but capturing the essence of it, which direct human thought toward much more fertile regions.

> The adiabatic approximation offers a *simple 3-D model* of a molecule–an extremely useful concept with great interpretative potential.

In later chapters of this book, this model will gradually be enriched by introducing the notion of chemical bonds between *some* atoms, angles between consecutive chemical bonds, electronic lone pairs, electronic pairs that form the chemical bonds, etc. Such a model inspires our imagination (... sometimes too much).[23] This is the foundation of all chemistry, all organic syntheses, conformational analysis, most of spectroscopy, etc. Without this beautiful model, progress in chemistry would be extremely difficult.

6.6 Basic Principles of Electronic, Vibrational, and Rotational Spectroscopy

6.6.1 Vibrational Structure

Equation (6.24) represents the basis of molecular spectroscopy and involves changing the molecular electronic, vibrational, or rotational state of a diatomic molecule. Fig. 6.4 shows an example how the curves $U_k(R)$ [also $E_k^0(R)$] may appear for three electronic states $k = 0, 1, 2$ of a diatomic molecule. Two of these curves ($k = 0, 2$) have a typical for bonding states "hook-like" shape. The third ($k = 1$) is also typical, but for repulsive electronic states.

It was assumed in Fig. 6.4 that $J = 0$ and therefore $V_{kJ}(R) = U_k(R)$. Next, Eq. (6.24) was solved for $U_0(R)$ and a series of solutions $\chi_{kvJ} = \chi_{0v0}$ was found: $\chi_{000}, \chi_{010}, \chi_{020}, \ldots$ with energies $E_{000}, E_{010}, E_{020}, \ldots$, respectively. Then, in a similar way, for $k = 2$, one has obtained

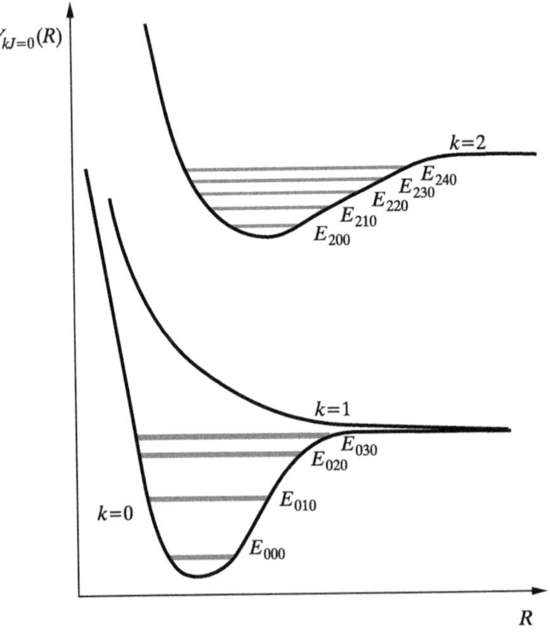

Fig. 6.4. The curves $V_{kJ}(R)$ for $J = 0$ [$V_{k0}(R) = U_k(R)$] for the electronic states $k = 0, 1, 2$ of a diatomic molecule (scheme). The energy levels E_{kvJ} for $J = 0$ corresponding to these curves are also shown. The electronic state $k = 0$ has four, $k = 1$ has zero, and $k = 2$ has five vibrational energy levels.

[23] We always have to remember that the useful model represents nothing more than a kind of pictorial representation of a more complex and unknown reality.

the series of solutions: $\chi_{200}, \chi_{210}, \chi_{220}, \ldots$ with the corresponding energies $E_{200}, E_{210}, E_{220}, \ldots$ This means that these two electronic levels ($k = 0, 2$) have a *vibrational structure* ($v = 0, 1, 2, \ldots$), the corresponding vibrational levels are shown in Fig. 6.4. Any attempt to find the vibrational levels for the electronic state $k = 1$ would fail.

The pattern of the vibrational levels looks similar to those for the Morse oscillator (p. 192). The low levels are nearly equidistant, reminding us of the results for the harmonic oscillator. The corresponding wave functions also resemble those for the harmonic oscillator. Higher-energy vibrational levels are getting closer and closer, as for the Morse potential. This is a consequence of the anharmonicity of the potential–we are just approaching the dissociation limit where the $U_k(R)$ curves differ qualitatively from the harmonic potential.

6.6.2 Rotational Structure

What would happen if we took $J = 1$ instead of $J = 0$? This corresponds to the potential energy curves $V_{kJ}(R) = U_k(R) + J(J+1)\hbar^2/(2\mu R^2)$, which in our case is $V_{k1}(R) = U_k(R) + 1(1+1)\hbar^2/(2\mu R^2) = U_k(R) + \hbar^2/(\mu R^2)$ for $k = 0, 1, 2$. The new curves therefore represent the old curves plus the term $\hbar^2/(\mu R^2)$, which is the same for all the curves. This corresponds to a *small* modification of the curves for large R and a *larger* modification for small R (see Fig. 6.5). The potential energy curves just go up a little bit on the left.[24] Of course, this is why the solution of Eq. (6.24) for these new curves will be similar to that which we had before; but this tiny shift upward will result in a tiny shift upward of all the computed vibrational levels. Therefore, the levels E_{kv1} for $v = 0, 1, 2, \ldots$ will be a little higher than the corresponding E_{kv0} for $v = 0, 1, 2, \ldots$ (this pertains to $k = 0, 2$, there will be no vibrational states for $k = 1$). This means that each vibrational level v will have its own *rotational structure* corresponding to $J = 0, 1, 2, \ldots$.

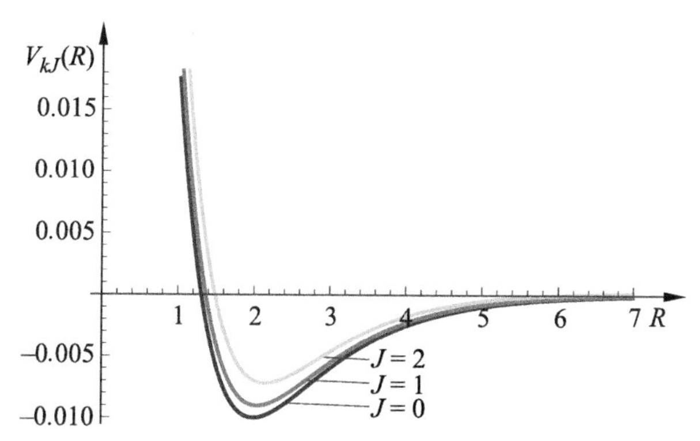

Fig. 6.5. Potential energy curves in arbitrary units corresponding to a diatomic [$V_{kJ}(R)$, k is the electronic state quantum number] for the rotational quantum numbers $J = 0, 1, 2$. One can see the bond weakening under rotational excitation.

---

[24] With an accompanying small shift to the right the position of the minimum.

Increasing J means that the potential energy curve becomes shallower[25] At some lower Js the molecule may accommodate all or part of the vibrational levels that exist for $J = 0$. It may happen that after a high-energy rotational excitation (to a large J), the potential energy curve will be so shallow that no vibrational energy level will be possible. This means that the molecule will undergo dissociation due to the excessive centrifugal force.

Separation Between Energy Levels

For molecules other than hydrides, the separation between rotational levels $(E_{kvJ+1} - E_{kvJ})$ is smaller by two to three orders of magnitude than the separation between vibrational levels $(E_{k,v+1,J} - E_{kvJ})$, and the latter is smaller by one or two orders of magnitude when compared to the separation of the electronic levels $(E_{k+1,v,J} - E_{kvJ})$.

> This is why electronic excitation corresponds to the absorption of UV or visible light, vibrational excitation to the absorption of infrared radiation, and rotational excitation to the absorption of microwave radiation.

This is what is used in a microwave oven. Food (such as chicken) on a ceramic plate is irradiated by microwaves. This causes rotational excitation of the water molecules[26] that are always present in food. The "rotating" water molecules cause a transfer of kinetic energy to protein, similar to what would happen in traditional cooking. After removing the food from the microwave, the chicken is hot, but the plate is cool (as there is nothing to rotate in the material that makes it up).

In practice, we always have to do with the absorption or emission spectra of a specimen from which we are trying to deduce the relative positions of the energy levels of the molecules involved. We may conclude that in theoretical spectra computed in the center-of-mass system, there will be allowed and forbidden energy intervals.[27] There is no energy levels corresponding to bound states in the forbidden intervals.[28] In the allowed intervals, any region corresponds to an electronic state, whose levels exhibit a pattern (i.e., clustering into vibrational series: one cluster corresponding to $v = 0$, the second to $v = 1$, etc.). Within any cluster, we have rotational levels corresponding to $J = 0, 1, 2, \ldots$ This follows from the fact that the distances between the levels with different k are large, with different v are smaller, and with different J are even smaller.

---

[25] The curve $V_{kJ}(R)$ becomes shallower and the system gets less stable, but for small J, the force constant paradoxically (the second derivative at minimum, if any) *increases*; i.e., the system becomes stiffer due to rotation. Indeed, the second derivative of the rotational energy is equal to $J(J+1)\frac{3\hbar^2}{\mu R^4} > 0$ and, if the position of the minimum of the new curve shifted only a bit (J not too large) with respect to the position of the minimum of $U_k(R)$, the force constant would increase due to the rotational excitation.

[26] Such rotation is somewhat hindered in the solid phase.

[27] In a space-fixed coordinate system (see p. e93), we always are dealing with a continuum of states (due to translations, see p. 69).

[28] The non-bound states densely fill the total energy scale above the dissociation limit of the ground state.

6.7 Approximate Separation of Rotations and Vibrations

Vibrations cannot be exactly separated from rotations for a very simple reason. During vibrations, the length R of the molecule changes, which makes the moment of inertia $I = \mu R^2$ change and influences the rotation of the molecule[29] according to Eq. (6.25).

The separation is feasible only when making an approximation (e.g., when assuming the mean value of the moment of inertia instead of the moment itself). Such a mean value is close to $I = \mu R_e^2$, where R_e stands for the position of the minimum of the potential energy V_{k0}. So we may decide to accept the potential [Eq. (6.25)] for the vibrations in the approximate form[30]:

$$V_{kJ}(R) \approx U_k(R) + J(J+1)\frac{\hbar^2}{2\mu R_e^2}.$$

Since the last term is a constant, this immediately gives the separation of the rotations from the vibrational Eq. (6.24):

$$\left(-\frac{\hbar^2}{2\mu}\frac{d^2}{dR^2} + U_k(R)\right)\chi_{kvJ}(R) = E'\chi_{kvJ}(R), \tag{6.28}$$

where the constant

$$\begin{aligned} E' &= E_{kvJ} - E_{rot}(J), \\ E_{rot}(J) &= J(J+1)\frac{\hbar^2}{2\mu R_e^2}. \end{aligned} \tag{6.29}$$

Now, we may always write the potential $U_k(R)$ as a number $U_k(R_e)$ plus the rest labeled by $V_{vibr}(R)$:

$$U_k(R) = U_k\left(R_e\right) + V_{vibr}(R). \tag{6.30}$$

Then, it is appropriate to call $U_k(R_e)$ the electronic energy $E_{el}(k)$ (corresponding to the equilibrium internuclear distance in electronic state k), while the function $V_{vibr}(R)$ stands for the vibrational potential satisfying $V_{vibr}(R_e) = 0$. After introducing this into Eq. (6.28), we obtain the equation for vibrations (in general, anharmonic):

$$\left(-\frac{\hbar^2}{2\mu}\frac{d^2}{dR^2} + V_{vibr}(R)\right)\chi_{kvJ}(R) = E_{vibr}(v)\chi_{kvJ}(R),$$

where the constant $E_{vibr}(v) = E' - E_{el}$, hence (after adding the translational energy–recalling that we have separated the center-of-mass motion), we have the final approximation:

[29] Let us recall the energetic pirouette of a dancer. Her graceful movements, stretching her arms out or aligning them along her body, immediately translate into slow or fast rotational motion.

[30] This looks reasonable for *small* amplitude vibrations only. However, this amplitude becomes larger under rotational excitations. Thus, in principle, R_e should increase if J increases and therefore the rotational energy is lower than shown by the formula.

$$E_{kvJ} \approx E_{trans} + E_{el}\left(k\right) + E_{vibr}\left(v\right) + E_{rot}\left(J\right), \tag{6.31}$$

where the corresponding quantum numbers are given in parentheses: the electronic (k), the vibrational (v), and the rotational (J).

6.8 Understanding the IR Spectrum: HCl

Assume that we have a diluted gas[31] of HCl and we are testing its optical absorption in the microwave region. It is worth noting that the H atom or the Cl atom by itself has zero absorption in this range of spectrum. The spectrum of the HCl molecules represents a strange sequence of double peaks in a peculiar quasi-periodic order. This means the absorption is a direct result of making a molecule from these atoms. We will have to deal with some relative motion of the two interacting atoms, which will be described by molecular vibrational and rotational states and optical transitions between them (with the electronic state staying the same).

6.8.1 Selection Rules

Not all transitions are allowed. All selection rules stem ultimately from conservation laws.

The conservation of energy law says that only a photon of energy $\hbar\omega$ that fits the difference of energy levels can be absorbed.

This fitting is not enough, however. There also must be a coupling (oscillating with frequency ω) between the electromagnetic field and the system. From the theory of interaction of matter and electromagnetic field, we know that the most important coupling term is equal to $-\hat{\boldsymbol{\mu}} \cdot \boldsymbol{\mathcal{E}} = -(\hat{\mu}_x\mathcal{E}_x + \hat{\mu}_y\mathcal{E}_y + \hat{\mu}_z\mathcal{E}_z)$; cf. p. 97, where $\boldsymbol{\mathcal{E}}$ is the oscillating electric field vector of the electromagnetic field and $\hat{\boldsymbol{\mu}}$ is the dipole moment operator. We will assume that the electromagnetic wave propagates along the z-axis; therefore, $\mathcal{E}_z = 0$ and only $\hat{\mu}_x$ and $\hat{\mu}_y$ will count. The quantity $\boldsymbol{\mathcal{E}}$ provides the necessary oscillations in time, while the absorption is measured by $|c|^2$ with $c = \langle \Psi_k | \hat{\boldsymbol{\mu}}(\boldsymbol{r}, \mathbf{R}) \Psi_{k'} \rangle_{e,n}$, the coupling between the initial electronic rovibrational state $\Psi_k(\boldsymbol{r}, R) = \psi_k(\boldsymbol{r}; \mathbf{R}) f_k(\boldsymbol{R}) = \psi_{k=0}(\boldsymbol{r}; \mathbf{R}) \chi_{v=0}(R) Y_J^M(\theta, \phi)$, and the final electronic rovibrational state $\Psi_{k'}(\boldsymbol{r}, R) = \psi_{k'=0}(\boldsymbol{r}; \mathbf{R}) \chi_{v'}(R) Y_{J'}^{M'}(\theta, \phi)$. Where we decided to be within the ground electronic state $(k = k' = 0)$ and start from the ground vibrational state $(v = 0)$, the symbol $\langle | \rangle_{e,n}$ denotes the integration over the coordinates

[31] No intermolecular interaction will be assumed.

of all the electrons ("e") and nuclei ("n").[32] We will integrate first within $c = \langle \Psi_k | \hat{\boldsymbol{\mu}} \Psi_{k'} \rangle_{e,n}$ over electronic coordinates: $\langle \psi_0(\boldsymbol{r};\mathbf{R}) | \hat{\boldsymbol{\mu}}(\boldsymbol{r},\mathbf{R}) \psi_0(\boldsymbol{r};\mathbf{R}) \rangle_e \equiv \boldsymbol{\mu}_{00}(\mathbf{R})$ and get $c = \left\langle \chi_{v=0}(R) Y_J^M(\theta,\phi) | \boldsymbol{\mu}_{00}(\mathbf{R}) \chi_{v'}(R) Y_{J'}^{M'}(\theta,\phi) \right\rangle_n$. The quantity $\boldsymbol{\mu}_{00}(\mathbf{R})$ is the dipole moment of the molecule in the ground electronic state and oriented in space along $\mathbf{R}$.

In case $\boldsymbol{\mu}_{00}(\mathbf{R}) = \mathbf{0}$ (also in case $\boldsymbol{\mu}_{00}(\mathbf{R}) \cdot \boldsymbol{\mathcal{E}} = 0$ for any $\mathbf{R}$), there will be no absorption. Thus, to get a nonzero absorption in rotational and vibrational (microwave or IR) spectra, one has to do with polar molecules, at least for certain $\mathbf{R}$. Therefore, all homonuclear diatomics, although they have a rich structure of rovibrational levels, are unable to absorb electromagnetic radiation in the microwave as well as in the IR range.

A vector in 3-D space may be defined in a Cartesian coordinate system by giving the x, y, and z components, but also in the spherical coordinate system by giving R, θ, ϕ polar coordinates. Now, let us write the dipole moment $\boldsymbol{\mu}_{00}(\mathbf{R})$ in spherical coordinates[33]: $\boldsymbol{\mu}_{00}(\mathbf{R}) = \sqrt{\frac{8\pi}{3}} \mu_{00}(R) Y_1^m(\theta,\phi)$ (see p. e169) with $m = \pm 1$, $m = 0$ is excluded because it represents $\boldsymbol{\mu}_{00,z} = 0$, which is irrelevant in view of $\mathcal{E}_z = 0$ (there is no coupling in such a case). Thus,

$$
\begin{aligned}
c &= \left\langle \chi_{v=0}(R) Y_J^M(\theta,\phi) | \boldsymbol{\mu}_{00}(\boldsymbol{R}) \chi_{v'}(R) Y_{J'}^{M'}(\theta,\phi) \right\rangle_n \\
&= \sqrt{\frac{8\pi}{3}} \langle \chi_0(R) | \mu_{00}(R) \chi_{v'}(R) \rangle_R \left\langle Y_J^M | Y_1^m Y_{J'}^{M'} \right\rangle_{\theta,\phi},
\end{aligned}
$$

where at each integral we have indicated the coordinates to integrate over. Now, introducing the equilibrium internuclear distance R_e (the position of the minimum of the potential energy curve) and the displacement $Q = R - R_e$, as well as expanding $\mu_{00}(R)$ in the Taylor series: $\mu_{00}(R) = \mu_{00}(R_e) + \left(\frac{\partial \mu_{00}}{\partial R}\right)_{R=R_e} Q + \ldots$ and neglecting the higher terms denoted as $+\ldots$

[32] When describing the electronic function, we have put explicitly the position in space of the nuclei ($\mathbf{R}$) instead of the usual notation with R (which does not tell us how the nuclear axis is oriented in space).

[33]

$$
\begin{aligned}
\mu^{(10)} &= qR\sqrt{\frac{4\pi}{3}} Y_1^0 = \mu(R)\sqrt{\frac{4\pi}{3}} Y_1^0, \\
\mu^{(1,\pm 1)} &= \mu(R)\sqrt{\frac{8\pi}{3}} Y_1^{\pm 1}.
\end{aligned}
$$

one obtains:

$$c = \sqrt{\frac{8\pi}{3}} \left[\mu_{00}(R_e) \langle \chi_0(R) | \chi_{v'}(R) \rangle_R + \left(\frac{\partial \mu_{00}}{\partial R} \right)_{R=R_e} \langle \chi_0(R) | Q \chi_{v'}(R) \rangle_R \right]$$
$$\times \left\langle Y_J^M | Y_1^m Y_{J'}^{M'} \right\rangle_{\theta,\phi} = 0$$
$$+ \sqrt{\frac{8\pi}{3}} \left(\frac{\partial \mu_{00}}{\partial R} \right)_{R=R_e} \langle \chi_0(R) | Q \chi_{v'}(R) \rangle_R \left\langle Y_J^M | Y_1^{\pm 1} Y_{J'}^{M'} \right\rangle_{\theta,\phi} .$$

There is only one such wave function $\chi_{v'}$ of the harmonic oscillator, for which $\langle \chi_0(R) | Q \chi_{v'}(R) \rangle_R \neq 0$: it happens only for[34] $v' = 1$.

We obtain the selection rule for the IR spectroscopy: it is necessary that during the vibration, the dipole moment changes. The main effect of the IR absorption from the $v = 0$ state is that the vibrational quantum number has to change from 0 to 1.

The integral $\left\langle Y_J^M | Y_1^{\pm 1} Y_{J'}^{M'} \right\rangle_{\theta,\phi}$ is nonzero only if[35] $M' = M - m,\ J' = J \pm 1$. This integral has to do with conservation of the total angular momentum and with the conservation of the parity of the system. Any photon has the spin quantum number[36] $s = 1$ (cf. p. 26), which means that besides its energy, it carries the angular momentum: $\hbar$ or $-\hbar$ (right or left circular polarizations of the photon, the electric field $\mathcal{E}$ rotating within the xy plane). After absorption the photon disappears, but it does not matter: the total angular momentum has to be conserved whatever happens. Therefore, the total system: molecule+photon, before as well as after absorption, has to have the total angular momentum with the quantum number equal to[37] $|J - s|,\ J,\ J + s$ i.e., $J - 1,\ J,\ J + 1$. The second possibility (with J) would mean that in the IR spectroscopy, the violation of parity occurs.[38] Indeed, the parity of Y_J^M is equal[39] to $(-1)^J$. Therefore, the case $J' = J$ in view of parity of $Y_1^{\pm 1}$ would mean that this is an odd function to integrate, which would make the integral equal to zero.[40] Thus,

---

34 Simply, $Q\chi_0(R)$ is proportional to the Hermite polynomial H_1; i.e., is proportional to $\chi_1(R)$. Due to the orthonormal character of all χ_v, this gives $v = 1$ as the only possibility.

35 The rule $M' = M - m$ follows from $\int_0^{2\pi} \exp[i(-M + m + M')\phi] d\phi = 2\pi \delta_{M', M-m}$.

36 With two polarizations: $m_s = 1$ or $m_s = -1$, the polarization $m_s = 0$ is excluded due to the zero mass.

37 We will describe this problem of quantum-mechanical adding of two angular momenta in a more general way on p. 343.

38 The conservation of parity is violated in nature, but this effect is much too small to be seen in the analyzed spectrum.

39 Recall the s,p,d,... orbitals of the hydrogen atom. They correspond to Y_l^m, $l = 0, 1, 2, \ldots$, respectively, and they are of even ($l = 0, 2$) or odd parity ($l = 1$).

40 This is why we do not have the peak ("missing"): $v = 0,\ J = 0 \rightarrow v = 1,\ J = 0$.

the selection rule for the IR and for the microwave spectroscopy reads as:
No photon absorption can happen unless $\Delta J = \pm 1$.

6.8.2 Microwave Spectrum Gives the Internuclear Distance

The lowest energy needed to excite the system would be achieved by changing J only; the related frequencies (for transitions that are allowed by selection rules[41]: $kvJ = 00J \rightarrow 00(J+1)$, $J = 0, 1, 2, \ldots$) are in the range of microwaves. From Eq. (6.29), we get the theoretical estimation of the transition energy $h\nu = hc\bar{\nu} = (J+1)(J+2)\frac{\hbar^2}{2\mu R_e^2} - J(J+1)\frac{\hbar^2}{2\mu R_e^2} = 2(J+1)\frac{\hbar^2}{2\mu R_e^2} = (J+1)2B$. Using the recorded microwave spectrum, we may estimate from this formula the equilibrium interatomic distance for HCl. For the consecutive J, we get 1.29 Å, independently of J (not too large though)[42]. Thus, from the microwave spectrum of HCl, we can read the "interatomic distance." We may compare this distance with, say, the position of minimum of the computed potential energy curve $U_0(R_e)$ of Eq. (6.30).

6.8.3 IR Spectrum and Isotopic Effect

What about the IR region? Fig. 6.6a gives the recorded absorption.

First, why are there these strange doublets in the IR spectrum for HCl? Well, the reason is quite trivial: two natural chlorine isotopes: ^{35}Cl and ^{37}Cl, which are always present in the natural specimen (with proportion 3:1). The H^{35}Cl molecule rotates (as well as oscillates) differently than the H^{37}Cl because of the reduced mass difference [see Eq. (6.29)]. This difference of μ is very small, since what decides in μ is the small mass of the proton[43]. Thus, these two molecules will correspond to two spectra that are similar, but shifted a bit with respect to one another on the frequency axis, the heavier isotope spectrum corresponding to a bit lower frequency.

Fig. 6.6a can be understood with the help of Eq. (6.28), which shows us a model of the phenomena taking place. At room temperature, most of the molecules (Boltzmann law) are in their ground electronic and vibrational states ($k = 0$, $v = 0$). IR quanta are unable to change quantum number k, but they have sufficient energy to change v and J quantum numbers. Fig. 6.6a shows what in fact has been recorded. From the transition selection rules (see above), we have $\Delta v = 1 - 0 = 1$ and either the transitions of the kind $\Delta J = (J+1) - J = +1$ (what is known as the *R branch*, right side of the spectrum) or of the kind $\Delta J = J - (J+1) = -1$ (the *P branch*, left side).

---

[41] For more about this, see Appendix C available at booksite.elsevier.com/978-0-444-59436-5.

[42] This is because the minimum of the potential energy curve V_{0J} shifts for large J.

[43] The presence of a deuterium-substituted molecule would have much more serious consequences (a larger shift of the spectrum), because what first of all counts for the reduced mass is the light atom. And the reduced mass controls rotation and vibration.

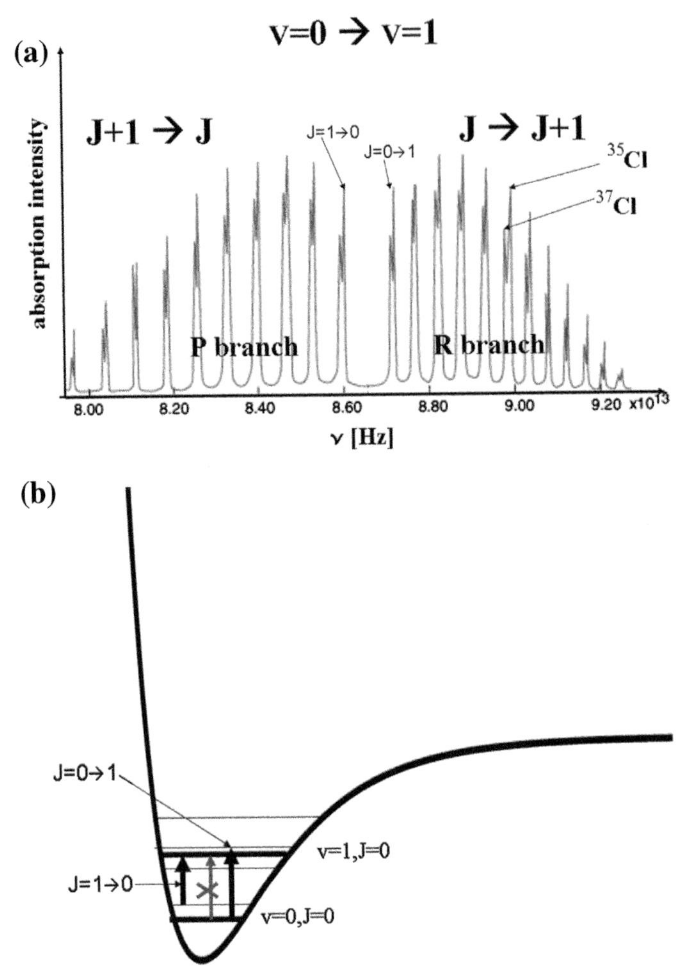

Fig. 6.6. What can we learn about the HCl molecule from its IR spectrum? (a) The IR spectrum (each doublet results from two chlorine isotopes: ^{35}Cl and ^{37}Cl present in the specimen). (b) The central position in the spectrum (between R and P branches) seems to be missing because the transition $\upsilon = 0, J = 0 \rightarrow \upsilon = 1, J = 0$ is forbidden by the selection rules (as described in the text), and its hypothetical position can be determined with high precision as the mean value of the two transitions shown: $J = 0 \rightarrow J = 1$ and $J = 1 \rightarrow J = 0$. This allows us to compute the force constant of the HCl bond. The energy difference of the same two quanta allows us to estimate the moment of inertia, and therefore the H...Cl distance. Note that the rotational levels corresponding to the vibrational state $\upsilon = 1$ are closer to each other than those for $\upsilon = 0$. This is due to the wider and wider well and longer and longer equilibrium distance corresponding to the rotationally corrected potential for the motion of the nuclei.

6.8.4 IR Spectrum Gives the Internuclear Distance

The remarkable regularity of the spectrum comes from the fact that the transition energy difference (of the nearest-neighbor peak positions) in a given branch is:

- For the R branch: $E_{excit,J} = h\nu = h\nu_0 + (J+1)(J+2)\frac{\hbar^2}{2\mu R_e^2} - J(J+1)\frac{\hbar^2}{2\mu R_e^2} = h\nu_0 + (J+1)2B$ and $E_{excit,J+1} - E_{excit,J} = (J+2)2B - (J+1)2B = 2B$.

- For the P branch: $E_{excit,J} = h\nu = h\nu_0 + J(J+1)\frac{\hbar^2}{2\mu R_e^2} - (J-1)J\frac{\hbar^2}{2\mu R_e^2} = h\nu_0 + J2B$ and $E_{excit,J} - E_{excit,J-1} = J2B - (J-1)2B = 2B$.

From the known distance $2B$, we can compute the estimation for the equilibrium distance R_e. We see that they are indeed quite equidistant in the spectrum for the R branch and for the P branch separately, but there is a small difference in the Bs for these branches. Is the theory described wrong? No, only our oversimplified theory fails a little. The B for the P branch is a bit larger because the mean interatomic distance gets larger for larger J (due to the centrifugal force).

6.8.5 Why We Have a Spectrum "Envelope"

What about the overall shape of the peaks' intensity ("*the envelope*") of the R and P branches? It looks quite strange: as if the transition from the levels with $v = 0$, $J = 2$ and $v = 0$, $J = 3$ had the largest intensity. Why? The rotational levels are so close that they are significantly populated at a given temperature. In a thermal equilibrium, the population of the levels by HCl molecules is proportional to the degeneracy of the level number J times the Boltzmann factor [i.e., to $p(J;T) = (2J+1)\exp[-\frac{J(J+1)B}{k_BT}]$]. Let us find for which J the probability[44] $p(J;T)$ attains a maximum: $\frac{\mathrm{d}p}{\mathrm{d}J} = 0 = 2\exp[-\frac{J(J+1)B}{k_BT}] - (2J+1)\frac{(2J+1)B}{k_BT}\exp[-\frac{J(J+1)B}{k_BT}]$, which gives for J_{opt} the equation $2 - (2J_{opt}+1)^2\frac{B}{k_BT} = 0$, or $(2J_{opt}+1)^2 = \frac{2k_BT}{B}$. For $T = 300$ K, this gives $J_{opt} = 2.7$; i.e., between $J = 2$ and $J = 3$. It looks as this is just what we see. We may say, therefore, that the spectrum shown has been recorded close to room temperature.

6.8.6 Intensity of Isotopomers' Peaks

One problem still remains. Since the isotopes ^{35}Cl and ^{37}Cl occur with the ratio 3:1, we might expect a similar intensity ratio of the two spectra. Why, therefore, do we have the ratio (Fig. 6.6a) looking as something like 4:3 (for low J)? There are two possible explanations: heavier rotator and heavier oscillator have lower energies and their levels are more populated at nonzero temperatures (however the effect is opposite), and/or this spectrum has too low a resolution, and we are comparing the maxima, while we should be comparing the integral intensity of the peaks (this means the area under the signal recorded). It turns out that in a higher-resolution spectra, for the integral intensities, we indeed see the ratio 3:1.

Thus, we may say that we understand the spectrum of HCl given in Fig. 6.6a.

6.9 A Quasi-Harmonic Approximation

The detailed form of $V_{vibr}(R)$ is obtained from $U_k(R)$ of Eq. (6.30) and therefore from the solution of the Schrödinger Eq. (6.24) with the clamped nuclei Hamiltonian. In principle, there

[44] It is not normalized to unity, but that does not matter here.

is no other solution but to solve Eq. (6.28) numerically. It is tempting, however, to get an idea of what would happen if a harmonic approximation were applied; i.e., when a harmonic spring was installed between both vibrating atoms. Such a model is very popular when discussing molecular vibrations. There is a unexpected complication though: such a spring cannot exist even in principle. Indeed, even if we constructed a spring that elongates according to Hooke's law, one cannot ensure the same will occur for shrinking. It is true that at the beginning, the spring may fulfill the harmonic law for shrinking as well, but when $R \to 0_+$, the two nuclei just bump into each other and the energy goes to infinity instead of being parabolic. For the spring to be strictly harmonic, we have to admit $R < 0$, which is forbidden because R means a distance. Fig. 6.7 shows the difference between the harmonic potential and the quasi-harmonic approximation for Eq. (6.28).

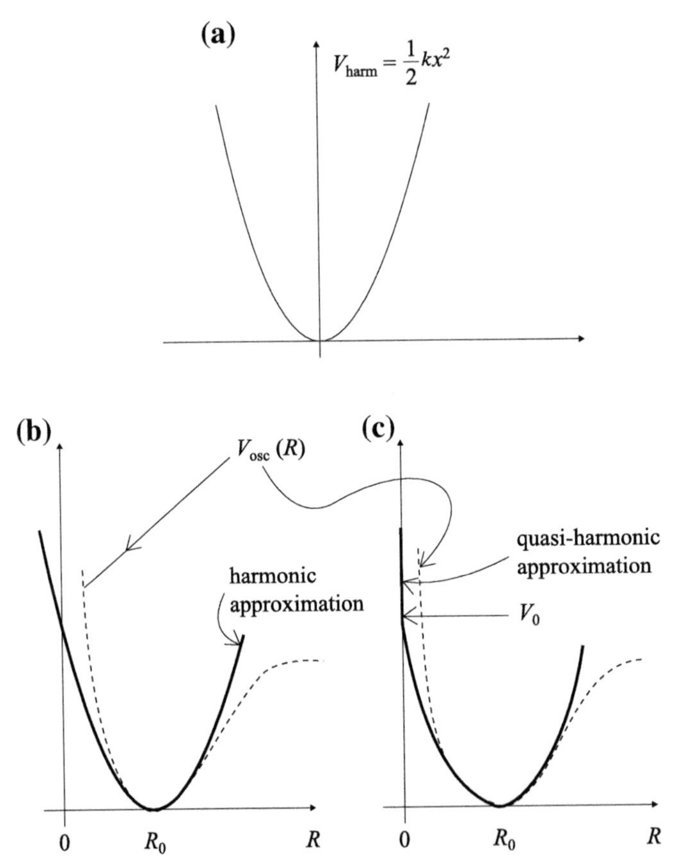

Fig. 6.7. The difference between harmonic and quasi-harmonic approximations for a diatomic molecule. (a) The potential energy for the harmonic oscillator. (b) The harmonic approximation to the oscillator potential $V_{vibr}(R)$ for a diatomic molecule is not realistic since at $R = 0$ (and at $R < 0$), the energy is finite, whereas it should go asymptotically to infinity when R tends to 0. (c) A more realistic (quasi-harmonic) approximation is as follows: the potential is harmonic up to $R = 0$, and for negative R, it goes to infinity. The difference between the harmonic and quasi-harmonic approximations pertains to such high energies (high oscillation amplitudes), that it is practically of negligible importance. In cases b and c, there is a range of small amplitudes where the harmonic approximation is applicable.

What do we do? Well, sticking to principles is always the best choice.[45] Yet, even in the case of the potential wall shown in Fig. 6.7c, we have an analytical solution.[46] The solution is quite complex, but it gets much simpler assuming $\frac{V_0}{h\nu} \equiv \alpha \gg v$, where $v = 0, 1, 2, \ldots$ stands for the vibrational quantum number that we are going to consider, and $V_0 \equiv V_{vibr}(0)$. This means that we limit ourselves to those vibrational states that are well below V_0. This is quite satisfactory because the hypothetical bump of the two nuclei would occur at vast (even unrealistic) V_0. In such a case, the vibrational energy is equal to $E_v = h\nu\left(v' + \frac{1}{2}\right)$, where the modified "quantum number" $v' = v + \varepsilon_v$ with a tiny modification:

$$\varepsilon_v = \frac{1}{\sqrt{2\pi}}\frac{1}{v!}\left(4\alpha\right)^{v+\frac{1}{2}}\exp\left(-2\alpha\right).$$

The corresponding wave functions very much resemble those of the harmonic oscillator, except that for $R \leq 0$, they are equal to zero. The strictly harmonic approximation results in $\varepsilon_v = 0$, and therefore, $E_v = h\nu\left(v + \frac{1}{2}\right)$; see Chapter 4.

Conclusion: The quasi-harmonic approximation has almost the same result as the (less realistic) harmonic one.

6.10 Polyatomic Molecule

6.10.1 Kinetic Energy Expression

A similar procedure can be carried out for a polyatomic molecule.

Let us consider an SFCS (see Appendix I available at booksite.elsevier.com/978-0-444-59436-5 on p. e93), and vector $\boldsymbol{R}_{CM}$ indicating the center of mass of a molecule composed of M atoms; see Fig. 6.8. Let us construct a Cartesian coordinate system (a body-fixed coordinate system, or BFCS) with the origin in the center of mass and the axes parallel to those of the SFCS (the third possibility in see Appendix I available at booksite.elsevier.com/978-0-444-59436-5).

In the BFCS, an atom α of mass[47] M_α is indicated by the vector $\boldsymbol{r}_\alpha$, its equilibrium position[48] by $\boldsymbol{a}_\alpha$, and the vector of displacement is $\boldsymbol{\xi}_\alpha = \boldsymbol{r}_\alpha - \boldsymbol{a}_\alpha$. If the molecule were rigid and did not rotate in the SFCS, then the velocity of the atom α would be equal to $\boldsymbol{V}_\alpha = \frac{d}{dt}(\boldsymbol{R}_{CM} + \boldsymbol{r}_\alpha) = \dot{\boldsymbol{R}}_{CM}$ (dots mean time derivatives), because the vector $\boldsymbol{r}_\alpha$, indicating the atom from the BFCS, would not change at all. If, in addition, the molecule, still preserving its rigidity, rotated about its center of mass with angular velocity $\boldsymbol{\omega}$ (the vector having the direction of the rotation axis,

[45] Let me stress once more that the problem appears when making the quasi-harmonic approximation, not in the real system we have.

[46] E. Merzbacher, *Quantum mechanics*, Wiley, New York, 2d edition (1970). The solution we are talking about has to be extracted from a more general problem in this reference. The potential energy used in the reference also has its symmetric counterpart for $R < 0$. Hence, the solution needed here corresponds to the antisymmetric solutions in the more general case (only for such solutions where the wave function is equal to zero for $R = 0$).

[47] What this mass means was discussed earlier in this chapter.

[48] We assume that such a position exists. If there are several equilibrium positions, we just choose one of them.

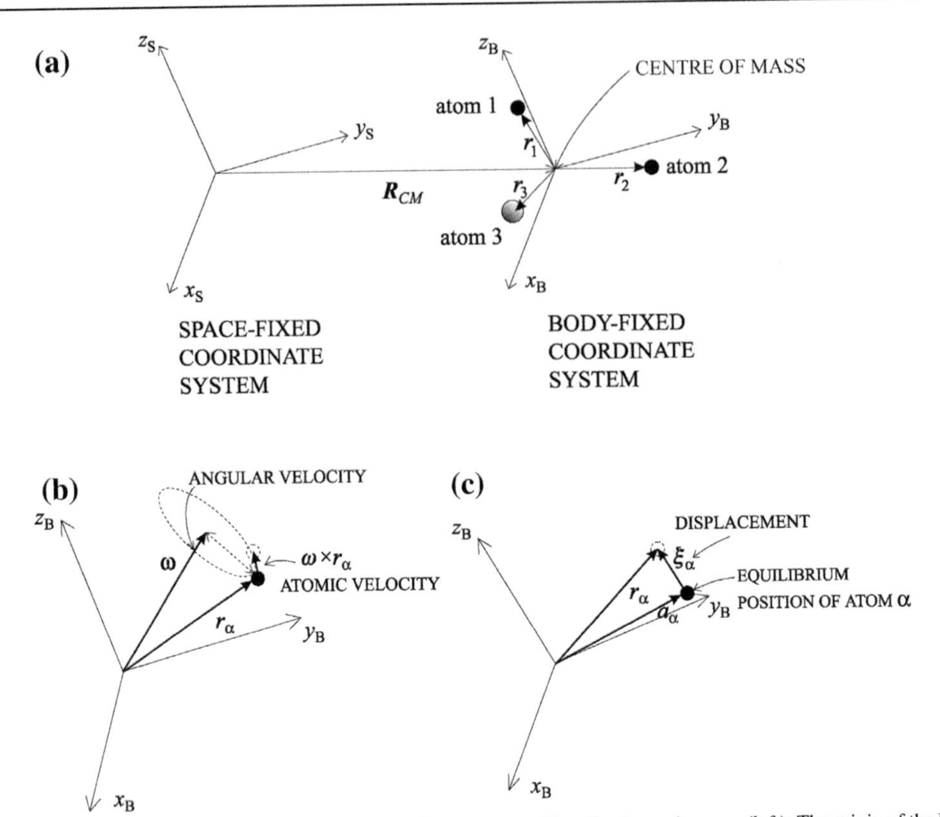

Fig. 6.8. SFCS and BFCS. (a) SFCS is a Cartesian coordinate system arbitrarily chosen in space (left). The origin of the BFCS is located in the center of mass of the molecule (right). The center of mass is shown by the vector $\boldsymbol{R}_{CM}$ from the SFCS. The nuclei of the atoms are indicated by vectors $\boldsymbol{r}_1, \boldsymbol{r}_2, \boldsymbol{r}_3 \ldots$ from the BFCS. Panel (b) shows what happens to the velocity of atom α, when the system is rotating with the angular velocity given as vector $\boldsymbol{\omega}$. In such a case, the atom acquires additional velocity $\boldsymbol{\omega} \times \boldsymbol{r}_\alpha$. Panel (c) shows that if the molecule vibrates, then atomic positions $\boldsymbol{r}_\alpha$ differ from the equilibrium positions $\boldsymbol{a}_\alpha$ by the displacements $\boldsymbol{\xi}_\alpha$.

right-handed screw orientation, and length equal to the angular velocity in radians per second), then the velocity of the atom α would equal[49] $\boldsymbol{V}_\alpha = \dot{\boldsymbol{R}}_{CM} + (\boldsymbol{\omega} \times \boldsymbol{r}_\alpha)$. However, our molecule is not rigid; everything moves inside it (let us call these motions "vibrations"[50]). Note that no restriction was made yet with respect to the displacements $\boldsymbol{\xi}_\alpha$ - they could be some giant internal motions. Then, the velocity of the atom α with respect to the SFCS is

$$\boldsymbol{V}_\alpha = \dot{\boldsymbol{R}}_{CM} + (\boldsymbol{\omega} \times \boldsymbol{r}_\alpha) + \dot{\boldsymbol{\xi}}_\alpha. \tag{6.32}$$

[49] $|\boldsymbol{\omega} \times \boldsymbol{r}_\alpha| = \omega r_\alpha \sin\theta$, where θ stands for the angle axis/vector $\boldsymbol{r}_\alpha$. If the atom α is on the rotation axis, this term vanishes ($\theta = 0$ or π). In other cases, the rotation radius is equal to $r_\alpha \sin\theta$.

[50] Such a "*vibration*" may mean an vibration of the OH bond, but also a rotation of the $-CH_3$ group or a large displacement of a molecular fragment.

When these velocities $\boldsymbol{V}_\alpha$ are inserted into the kinetic energy T of the molecule calculated in the SFCS, then we get

$$T = \frac{1}{2}\sum_\alpha M_\alpha \left(\boldsymbol{V}_\alpha\right)^2 = \frac{1}{2}\left(\dot{\boldsymbol{R}}_{CM}\right)^2 \sum_\alpha M_\alpha + \frac{1}{2}\sum_\alpha M_\alpha \left(\boldsymbol{\omega} \times \boldsymbol{r}_\alpha\right)^2 + \frac{1}{2}\sum_\alpha M_\alpha \left(\dot{\boldsymbol{\xi}}_\alpha\right)^2$$
$$+\dot{\boldsymbol{R}}_{CM}\cdot\left[\boldsymbol{\omega} \times \left(\sum_\alpha M_\alpha \boldsymbol{r}_\alpha\right)\right] + \dot{\boldsymbol{R}}_{CM}\cdot\sum_\alpha M_\alpha \dot{\boldsymbol{\xi}}_\alpha + \sum_\alpha M_\alpha \left(\boldsymbol{\omega} \times \boldsymbol{r}_\alpha\right)\cdot\dot{\boldsymbol{\xi}}_\alpha.$$

The first three ("diagonal") terms have a clear interpretation. These are the kinetic energy of the center of mass, the kinetic energy of rotation, and the kinetic energy of vibrations. The last three terms ("non-diagonal") denote the *roto-translational, vibro-translational*, and *vibro-rotational couplings*, respectively.

6.10.2 Quasi-Rigid Model–Simplifying by Eckart Conditions

There is a little problem with the expression for the kinetic energy: we have a redundancy in the coordinates. Indeed, we have three coordinates for defining translation ($\boldsymbol{R}_{CM}$), three that determine rotation ($\boldsymbol{\omega}$), and on top of that M vectors $\boldsymbol{r}_\alpha$. This is too many: six are redundant. Using such coordinates would be very annoying because we would not be sure whether they are consistent.

We may impose six relations among the coordinates and in this way (if they are correct) get rid of the redundancy. The first three relations are evident because the origin of the BFCS is simply the center of mass. Therefore,

$$\sum_\alpha M_\alpha \boldsymbol{r}_\alpha = \boldsymbol{0}, \tag{6.33}$$

which as we assume is also true when the atoms occupy equilibrium positions

$$\sum_\alpha M_\alpha \boldsymbol{a}_\alpha = \boldsymbol{0}.$$

Hence, we obtain a useful relation

$$\sum_\alpha M_\alpha \left(\boldsymbol{r}_\alpha - \boldsymbol{a}_\alpha\right) = \boldsymbol{0},$$
$$\sum_\alpha M_\alpha \boldsymbol{\xi}_\alpha = \boldsymbol{0},$$

which, after differentiation with respect to time, becomes

$$\sum_\alpha M_\alpha \dot{\boldsymbol{\xi}}_\alpha = \boldsymbol{0}. \tag{6.34}$$

If there were several sets of $\boldsymbol{a}_\alpha$'s (i.e., several minima of the potential energy), we would have a problem. This is one of the reasons we need the assumption of the quasi-rigid molecule.

Inserting Eqs. (6.33) and (6.34) into the kinetic energy expression makes the roto-translational and vibro-translational couplings vanish. Thus, we have

$$T = \frac{1}{2}\left(\dot{\boldsymbol{R}}_{CM}\right)^2 \sum_\alpha M_\alpha + \frac{1}{2}\sum_\alpha M_\alpha \left(\boldsymbol{\omega} \times \boldsymbol{r}_\alpha\right)^2 + \frac{1}{2}\sum_\alpha M_\alpha \left(\dot{\boldsymbol{\xi}}_\alpha\right)^2 + \sum_\alpha M_\alpha \left(\boldsymbol{\omega} \times \boldsymbol{r}_\alpha\right) \cdot \dot{\boldsymbol{\xi}}_\alpha.$$

Noting that $\boldsymbol{r}_\alpha = \boldsymbol{a}_\alpha + \boldsymbol{\xi}_\alpha$ and using the relation[51] $(\boldsymbol{A} \times \boldsymbol{B}) \cdot \boldsymbol{C} = \boldsymbol{A} \cdot \left(\boldsymbol{B} \times \boldsymbol{C}\right)$, we obtain immediately

$$T = \frac{1}{2}\left(\dot{\boldsymbol{R}}_{CM}\right)^2 \sum_\alpha M_\alpha + \frac{1}{2}\sum_\alpha M_\alpha \left(\boldsymbol{\omega} \times \boldsymbol{r}_\alpha\right)^2 + \frac{1}{2}\sum_\alpha M_\alpha \left(\dot{\boldsymbol{\xi}}_\alpha\right)^2 + \boldsymbol{\omega} \cdot \sum_\alpha M_\alpha \left(\boldsymbol{a}_\alpha \times \dot{\boldsymbol{\xi}}_\alpha\right) + \boldsymbol{\omega} \cdot \sum_\alpha M_\alpha \left(\boldsymbol{\xi}_\alpha \times \dot{\boldsymbol{\xi}}_\alpha\right).$$

We completely get rid of the redundancy if we agree the second *Eckart condition*[52] is introduced (equivalent to three conditions for the coordinates):

$$\sum_\alpha M_\alpha \left(\boldsymbol{a}_\alpha \times \dot{\boldsymbol{\xi}}_\alpha\right) = \boldsymbol{0}. \tag{6.35}$$

The condition means that we do not expect the internal motion to generate any angular momentum.[53] This completes our final expression for the kinetic energy T of a polyatomic quasi-rigid molecule

$$T = T_{trans} + T_{rot} + T_{vibr} + T_{Coriolis}. \tag{6.36}$$

The kinetic energy in an SFCS is composed of:

- The kinetic energy of the center of mass (translational energy), $T_{trans} = \frac{1}{2}\left(\dot{\boldsymbol{R}}_{CM}\right)^2 \sum_\alpha M_\alpha$.
- The rotational energy of the whole molecule, $T_{rot} = \frac{1}{2}\sum_\alpha M_\alpha \left(\boldsymbol{\omega} \times \boldsymbol{r}_\alpha\right)^2$.
- The kinetic energy of the internal motions ("*vibrations*"), $T_{vibr} = \frac{1}{2}\sum_\alpha M_\alpha \left(\dot{\boldsymbol{\xi}}_\alpha\right)^2$.

---

[51] These are two ways of calculating the volume of the parallelepiped according to the formula: surface of the base times the height.

[52] Carl Eckart, professor at California Institute of Technology, contributed to the birth of quantum mechanics [e.g., C. Eckart, *Phys. Rev., 28,* 711 (1926)].

[53] The problem is whether indeed we do not generate any momentum by displacing the nuclei from their equilibrium positions. A flexible molecule may have quite a number of different equilibrium positions (see Chapter 7). We cannot expect all of them to satisfy Eq. (6.35), where one of these equilibrium positions is treated as a reference. Eq. (6.35) means that we restrict the molecular vibrations to have only small amplitudes about a single equilibrium position (quasi-rigid model).

- The last term, usually very small, is known as Coriolis energy[54], $T_{Coriolis} = \boldsymbol{\omega} \cdot \sum_\alpha M_\alpha \left(\boldsymbol{\xi}_\alpha \times \dot{\boldsymbol{\xi}}_\alpha\right)$. It couples the internal motions ("*vibrations*") within the molecule with its rotation.

After the Eckart conditions are introduced, all the coordinates (i.e., the components of the vectors $\boldsymbol{R}_{CM}$, $\boldsymbol{\omega}$ and all $\boldsymbol{\xi}_\alpha$), can be treated as independent.

6.10.3 Approximation: Decoupling of Rotation and Vibration

Since the Coriolis term is small, in the first approximation we may decide to neglect it. Also, when assuming small vibrational amplitudes $\boldsymbol{\xi}_\alpha$, which is a reasonable approximation in most cases, we may replace $\boldsymbol{r}_\alpha$ by the corresponding equilibrium positions $\boldsymbol{a}_\alpha$ in the rotational term of Eq. (6.35): $\sum_\alpha M_\alpha(\boldsymbol{\omega} \times \boldsymbol{r}_\alpha)^2 \approx \sum_\alpha M_\alpha(\boldsymbol{\omega} \times \boldsymbol{a}_\alpha)^2$, in full analogy with Eq. (6.29). After these two approximations have been made, the kinetic energy represents the sum of the three independent terms (i.e., each depending on different variables)

$$T \approx T_{trans} + T_{rot} + T_{vibr} \tag{6.37}$$

with $T_{rot} \approx \frac{1}{2} \sum_\alpha M_\alpha(\boldsymbol{\omega} \times \boldsymbol{a}_\alpha)^2$.

6.10.4 Spherical, Symmetric, and Asymmetric Tops

Equation (6.37) may serve to construct the corresponding kinetic energy operator for a polyatomic molecule. There is no problem (see Chapter 1) with the translational term: $-\frac{\hbar^2}{2\sum_\alpha M_\alpha}\Delta_{R_{CM}}$; the vibrational term will be treated in Chapter 7, p. 355.

There is a problem with the rotational term. A rigid body (the equilibrium atomic positions $\boldsymbol{a}_\alpha$ are used), such as the benzene molecule, rotates, but due to symmetry, it may have some special axes characterizing the *moments of inertia*. The moment of inertia represents a tensor of rank 3 with the following components:

$$\left\{\begin{array}{lll} \sum_\alpha M_\alpha(a_{y,\alpha}^2 + a_{z,\alpha}^2) & \sum_\alpha M_\alpha a_{x,\alpha} a_{y,\alpha} & \sum_\alpha M_\alpha a_{x,\alpha} a_{z,\alpha} \\ \sum_\alpha M_\alpha a_{x,\alpha} a_{y,\alpha} & \sum_\alpha M_\alpha(a_{x,\alpha}^2 + a_{z,\alpha}^2) & \sum_\alpha M_\alpha a_{y,\alpha} a_{z,\alpha} \\ \sum_\alpha M_\alpha a_{x,\alpha} a_{z,\alpha} & \sum_\alpha M_\alpha a_{y,\alpha} a_{z,\alpha} & \sum_\alpha M_\alpha(a_{x,\alpha}^2 + a_{y,\alpha}^2) \end{array}\right\},$$

to be computed in the BFCS (see Appendix I available at booksite.elsevier.com/978-0-444-59436-5 on p. e93). The diagonalization of the matrix (see Appendix K available at booksite.elsevier.com/978-0-444-59436-5 on p. e105) corresponds to a certain rotation of the BFCS to a coordinate system rotating with the molecule (RMCS), and gives as the eigenvalues I_{xx}, I_{yy}, I_{zz}.

[54] Gaspard Gustav de Coriolis (1792 – 1843), was a French engineer and mathematician and director of the Ecole Polytechnique in Paris. In 1835, Coriolis introduced the notion of work, the equivalence of work and energy, and also a coupling of rotation and vibration.

When $I_{xx} = I_{yy} = I_{zz}$, the rotating body is called a *spherical rotator* or a *spherical top* (example: methane molecule); when $I_{xx} = I_{yy} \neq I_{zz}$, it is called a *symmetric top* (examples: benzene, ammonia molecules); when $I_{xx} \neq I_{yy} \neq I_{zz}$, then the top is asymmetric (example: water molecule).

Fig. 6.9 gives four classes of the rotators (tops).
Then, the classical expression for the kinetic energy of rotation takes the form[55]

$$\frac{1}{2}\sum_{\alpha} M_\alpha \left(\boldsymbol{\omega} \times \boldsymbol{a}_\alpha\right)^2 = \frac{1}{2}\left(I_{xx}\omega_x^2 + I_{yy}\omega_y^2 + I_{zz}\omega_z^2\right) = \frac{J_x^2}{2I_{xx}} + \frac{J_y^2}{2I_{yy}} + \frac{J_z^2}{2I_{zz}}, \tag{6.38}$$

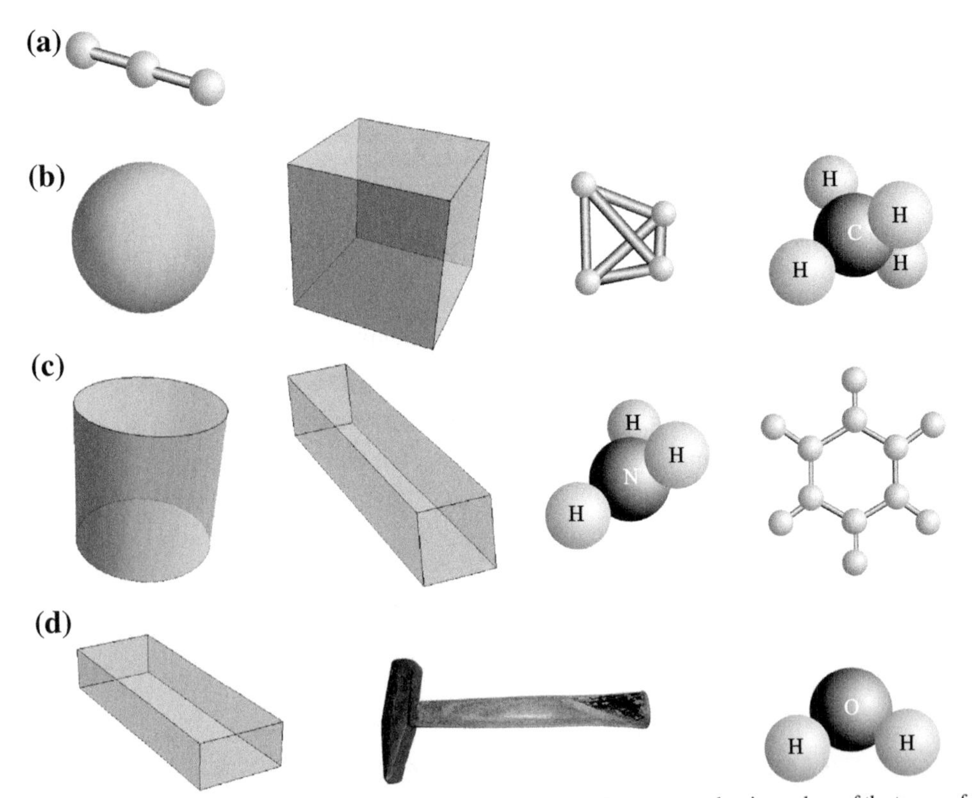

Fig. 6.9. Examples of four classes of tops (rotators). The numbers I_{xx}, I_{yy}, I_{zz} represent the eigenvalues of the tensor of inertia computed in a BFCS. There are four possibilities: (a) a linear rotator ($I_{xx} = I_{yy} = 0, I_{zz} \neq 0$); e.g., a diatomic or CO_2 molecule; (b) a spherical rotator ($I_{xx} = I_{yy} = I_{zz}$); e.g., a sphere, a cube, a regular tetrahedron, or a methane molecule; (c) a symmetric rotator ($I_{xx} = I_{yy} \neq I_{zz}$); e.g., a cylinder, a rectangular parallelepiped with square base, ammonia or benzene molecule; (d) an asymmetric rotator ($I_{xx} \neq I_{yy} \neq I_{zz}$); e.g., a general rectangular parallelepiped, a hammer, or water molecule.

[55] H. Goldstein, *Classical Mechanics*, 2d edition, Addison-Wesley (1980).

where ω_x, ω_y, and ω_z stand for the components of $\boldsymbol{\omega}$ in the RMCS, and J_x, J_y, and J_z represent the components of angular momentum also computed in the RMCS.[56]

It is not straightforward to write down the corresponding kinetic energy operator. The reason is that in the above expression, we have curvilinear coordinates (because of the rotation from BFCS to RMCS[57]), whereas the quantum mechanical operators were introduced only for the Cartesian coordinates (Chapter 1, p. 18). How do we write an operator expressed in some curvilinear coordinates q_i and the corresponding momenta p_i? Boris Podolsky solved this problem,[58] and the result is

$$\hat{T} = \frac{1}{2} g^{-\frac{1}{2}} \hat{\boldsymbol{p}}^T g^{\frac{1}{2}} \boldsymbol{G}^{-1} \hat{\boldsymbol{p}}, \tag{6.39}$$

where $\hat{p}_i = -i\hbar\frac{\partial}{\partial q_i}$, $\boldsymbol{G}$ represents a symmetric matrix (*metric tensor*) of the elements g_{rs}, defined by the square of the length element $ds^2 \equiv \sum_r \sum_s g_{rs} dq_r dq_s$, with $g = \det \boldsymbol{G}$ and g_{rs}(g and all g_{rs} being in general some functions of q_r).

6.10.5 Separation of Translational, Rotational, and Vibrational Motions

Equation (6.39) represents the kinetic energy operator. To obtain the corresponding Hamiltonian, we have to add to this energy the potential energy for the motion of the nuclei, U_k, where k labels the electronic state. The last energy depends uniquely on the variables $\boldsymbol{\xi}_\alpha$ that describe atomic vibrations and corresponds to the electronic energy $U_k(R)$ of Eq. (6.30), except that instead of the variable R, which pertains to the oscillation, we have the components of the vectors $\boldsymbol{\xi}_\alpha$. Then, in full analogy with Eq. (6.30), we may write

$$U_k(\boldsymbol{\xi}_1, \boldsymbol{\xi}_2, \ldots \boldsymbol{\xi}_M) = U_k(\mathbf{0}, \mathbf{0}, \ldots \mathbf{0}) + V_{k,vibr}(\boldsymbol{\xi}_1, \boldsymbol{\xi}_2, \ldots \boldsymbol{\xi}_M),$$

where the number $U_k(\mathbf{0}, \mathbf{0}, \ldots \mathbf{0}) = E_{el}(k)$ may be called the electronic energy in state k, and $V_{k,vibr}(\mathbf{0}, \mathbf{0}, \ldots \mathbf{0}) = 0$.

Since (after the approximations are made) the translational, rotational, and vibrational (internal motion) operators depend on their own variables, after separation the total wave function represents a product of three eigenfunctions (translational, rotational, and vibrational) and the total energy is the sum of the translational, rotational, and vibrational energies [fully analogous with Eq. (6.31)]:

[56] We recall from classical mechanics that an expression for rotational motion results from the corresponding one for translational motion by replacing mass by moment of inertia, momentum by angular momentum, and velocity by angular velocity. Therefore, the middle part of the above formula for kinetic energy represents an analog of $\frac{mv^2}{2}$ and the last part is an analog of $\frac{p^2}{2m}$.

[57] The rotation is carried out by performing three successive rotations by what are known as Euler angles. For details, see Fig. 14.5, as well as R.N. Zare, *Angular Momentum*, Wiley, New York (1988), p. 78.

[58] B. Podolsky, *Phys. Rev.*, *32*, 812 (1928).

$$E \approx E_{trans} + E_{el}(k) + E_{rot}(J) + E_{vibr}(v_1, v_2, \ldots v_{3M-6}). \tag{6.40}$$

where k denotes the electronic state, J the rotational quantum number, and v_i are the vibrational quantum numbers that describe the vibrational excitations (in Chapter 7, we will see a harmonic approximation for these oscillations).

6.11 Types of States

6.11.1 Repulsive Potential

If we try to solve Eq. (6.28) for vibrations with a repulsive potential, we would not find any solution of class Q. Among continuous, but non-square-integrable, functions, we would find an infinite number of eigenfunctions, and the corresponding eigenvalues would form a continuum, Fig. 6.10a. These eigenvalues reflect the fact that the system has dissociated and its dissociation products may have any kinetic energy larger than the dissociation limit (i.e., when having dissociated fragments with no kinetic energy). Any collision of two fragments (that correspond to the repulsive electronic state) will finally result in the fragments flying off. Imagine that the two fragments are located at a distance R_0, with corresponding total energy E, and that the system is allowed to relax according to the potential energy shown in Fig. 6.10a. The system slides down the potential energy curve (i.e., the potential energy lowers) and, since the total energy is conserved, its kinetic energy increases accordingly. Finally, the potential energy curve flattens, attaining $E_A + E_B$, where E_A denotes the internal energy of the fragment A (a similar thing happens for B). The final kinetic energy is equal to $E - (E_A + E_B)$ in SFCS.

6.11.2 "Hook-like" Curves

Another typical potential energy curve is shown in Fig. 6.10b and has the shape of a hook. Solving Eq. (6.28) for such a curve usually[59] gives a series of bound states; i.e., with their wave functions (Fig. 6.11) concentrated in a finite region of space and exponentially vanishing on leaving it. Fig. 6.10 shows the three discrete energy levels found and the continuum of states above the dissociation limit, similar to the curve in Fig. 6.10a. The continuum has, in principle, the same origin as before (any kinetic energy of the fragments).

Thus, the overall picture is that a system may have some bound states, but above the dissociation limit, it can also acquire any energy and the corresponding wave functions are non-normalizable (non-square-integrable).

[59] This applies to a sufficiently deep and wide potential energy well.

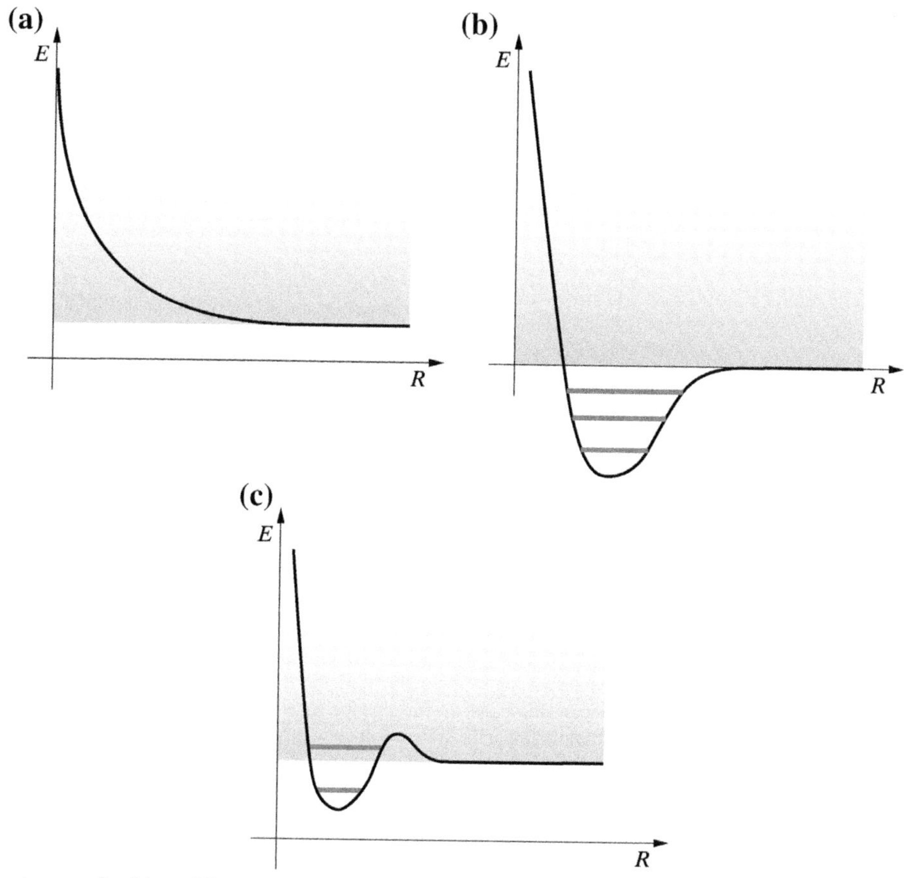

Fig. 6.10. An example of three different electronic states. (a) Repulsive state (no vibrational states, a ball representing the nuclear configuration will slide down resulting in dissociation); (b) three bound vibrational states (the ball will oscillate within the well); (c) one bound vibrational state (the ball oscillates) and one metastable vibrational state (the ball oscillates for some time and then goes to infinity, which means dissociation). A continuum of allowed states (shadowed area) with nonzero kinetic energy of the dissociation products is above the dissociation limit.

6.11.3 Continuum

The continuum may have a quite complex structure. First of all, the number of states per energy unit depends, in general, on the position on the energy scale where this energy unit is located. Thus, the continuum may be characterized by the *density of states* (the number of states per unit energy) as a function of energy. This may cause some confusion because the number of continuum states in any energy section is infinite. The problem is, however, that the infinities differ: some are "more infinite than others." The continuum does not mean a banality of the states involved (Fig. 6.10c). The continuum extends over the dissociation limit, irrespective what kind of potential energy curve one has for finite values of R. In cases similar to that of Fig. 6.10c, the continuum will exist independently of how wide and high the barrier is. But, the barrier

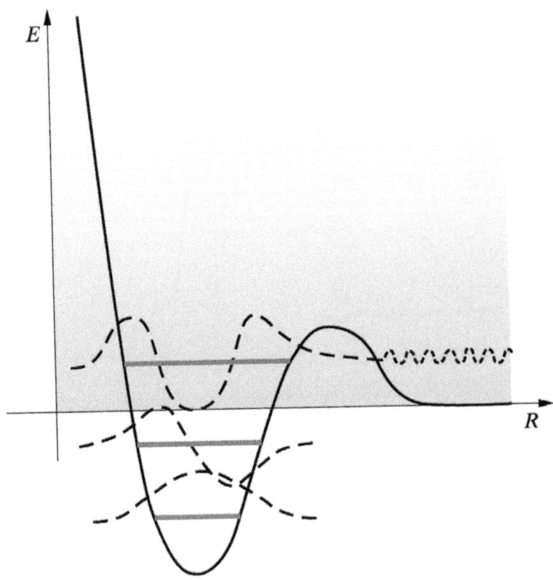

Fig. 6.11. The bound, continuum, and resonance (metastable) states of an anharmonic oscillator. Two discrete bound states are shown (energy levels and wave functions) in the lower part of the image. The continuum (shaded area) extends above the dissociation limit; i.e., the system may have any of the energies above the limit. There is one resonance state in the continuum, which corresponds to the third level in the potential energy well of the oscillator. Within the well, the wave function is very similar to the third state of the harmonic oscillator, but there are differences. One is that the function has some low-amplitude oscillations on the right side. They indicate that the function is non-normalizable and that the system will dissociate sooner or later.

may be so wide that the system will have no idea about any "extra-barrier life," and therefore it will have its "quasi-discrete" states with the energy higher than the dissociation limit. Yet, these states despite its similarity to bound states belong to the continuum (are non-normalizable). Such states are metastable and are called *resonances* (cf. p. 182), or *encounter complexes*. The system in a metastable state will sooner or later dissociate, but before this happens it may have a quite successful long life. Fig. 6.11. shows how the metastable and stationary states differ: the metastable ones do not vanish in infinity.

Fig. 6.12 shows what happens to the $V_{kJ}(R)$ curves, if J increases. A simple model potential $U_k(R)$ has been chosen for this illustration. As shown in Fig. 6.12, rotational excitations may lead to a qualitative change of the potential energy curve for the motion of the nuclei. Rotational excitations destabilize the system, but in a specific way. First, they always introduce a barrier for dissociation (*centrifugal barrier*), but despite of that, the dissociation becomes easier due to a large "pushing up" of the well region. Second, by increasing the energy for small distances, the rotational excitations either make some vibrational levels disappear or may change the character of the levels from stationary ones to metastable vibrational states (vibrational resonances in the continuum). Third, as one can see from Fig. 6.12, the equilibrium distance increases upon rotational excitations.

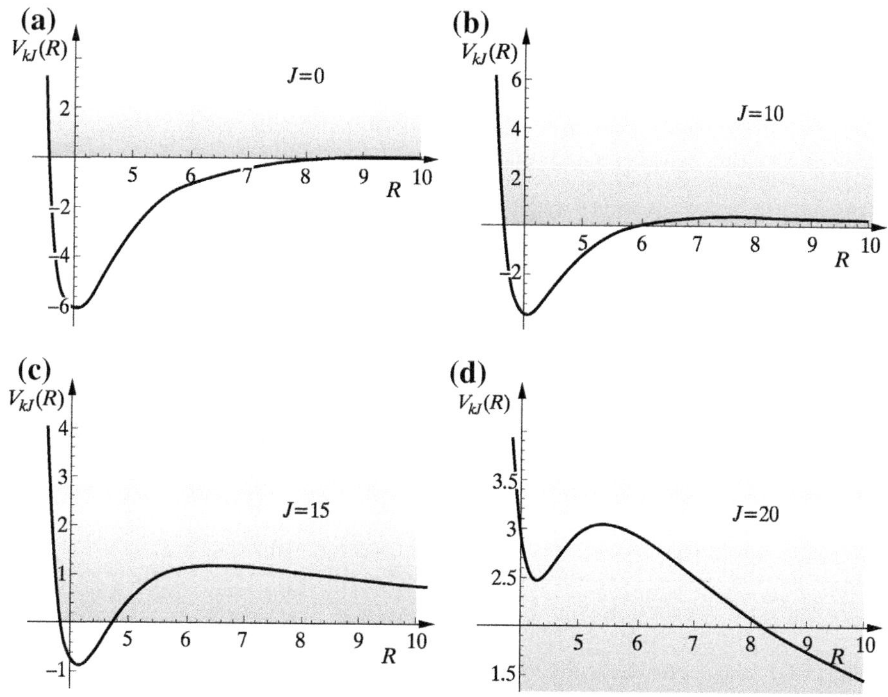

Fig. 6.12. A rotational excitation may lead to creating resonance states. As an illustration, a potential energy curve $V_{kJ}(R)$ has been chosen that resembles what we would see for two water molecules bound by the hydrogen bond. Its first component $U_k(R)$ is taken in the form of the so-called Lennard-Jones potential (cf., p. 347) $U_k(R) = \varepsilon_k \left[\left(\frac{R_{0k}}{R}\right)^{12} - 2\left(\frac{R_{0k}}{R}\right)^{6} \right]$, with the parameters for the electronic ground state ($k = 0$): $\varepsilon_0 = 6$ kcal/mol and $R_{00} = 4$ a.u. and the corresponding reduced mass $\mu = 16560$ a.u., the parameter ε_0 stands for the well depth, and the R_{00} denotes the position of the well minimum. Panels (a), (b), (c), and (d) correspond to $V_{kJ}(R) = U_k(R) + J(J+1)\hbar^2/(2\mu R^2)$ with $J = 0, 10, 15, 20$, respectively. The larger J is, the shallower the well: the rotation *weakens* the bond, but in a peculiar way. Due to the centrifugal force, the metastable resonance states appear. These are the "normal" vibrational states pushed up by the centrifugal energy beyond the energy of the dissociation limit. For $J = 20$, already *all* states (including the potential resonances) belong to the continuum.

Besides the typical continuum states, which result from the fact that the dissociation products fly slower or faster, one may have also the continuum metastable or resonance states, which resemble the bound states.

The human mind wants to translate such situations into simple pictures, which help us understand what happens. Fig. 6.13 shows an analogy associated with astronomy: the Earth and the Moon are in a bound state and the Earth and an asteroid are in a "primitive," continuum-like state, but if it happens that an asteroid went around the Earth several times and then flew away into space, then one has to do with an analog of a metastable or resonance state (characterized by a finite and nonzero lifetime).

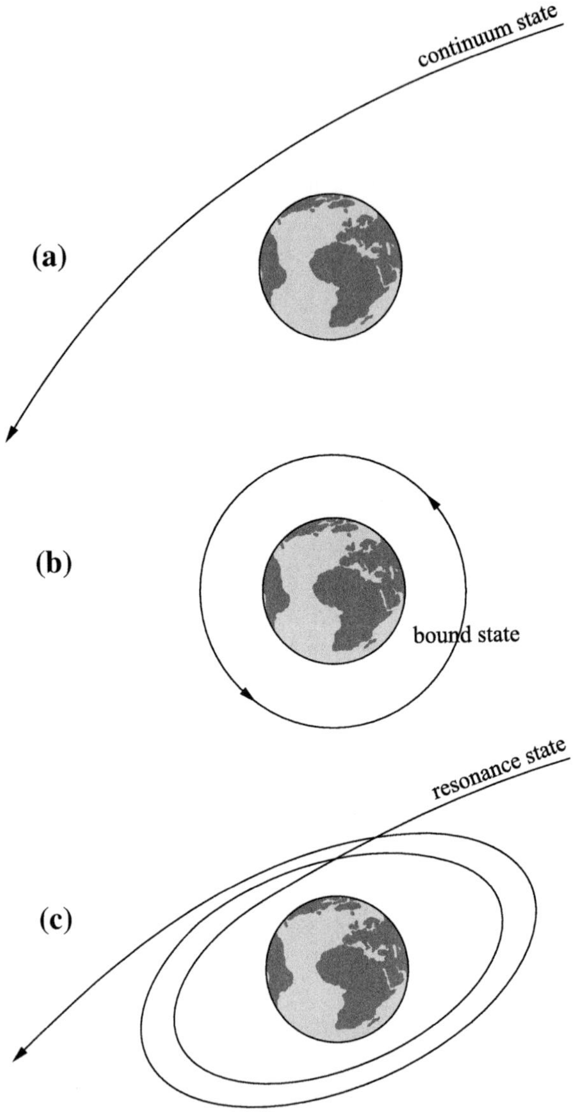

Fig. 6.13. Continuum, bound, and resonance states–an analogy involving the "states" of the Earth and an interacting body. (a) A "primitive" continuum state: an asteroid flies by the Earth and changes trajectory; (b) a bound state: the Moon is orbiting around the Earth; (c) a resonance state: the asteroid was orbiting several times about the Earth and then flew away.

The Schrödinger equation $H\psi = E\psi$ is time-independent; therefore, its solutions do not inform us about the sequence of events, but only about all the possible events with their probability amplitudes.[60] This is why the wave function for the metastable state of Fig. 6.11 exhibits oscillations at large x: they inform us about a *possibility* of dissociation.

[60] As Einstein said: "*The only reason for time is so that everything does not happen at once.*" The time-independent Schrödinger equation behaves as if "everything would happen at once."

6.11.4 Wave Function "Measurement"

Could we know the vibrational wave function in a given electronic and rotational state? It seemed that such a question could only be answered by quantum mechanical calculations. It turned out,[61] however, that the answer can also come from experimentation. In this experiment, three states are involved: the electronic ground state (G), an electronic excited state M, in particular its vibrational state. This state will be measured, and the third electronic state of a repulsive character (REP) that helps as a detector; see Fig. 6.14.

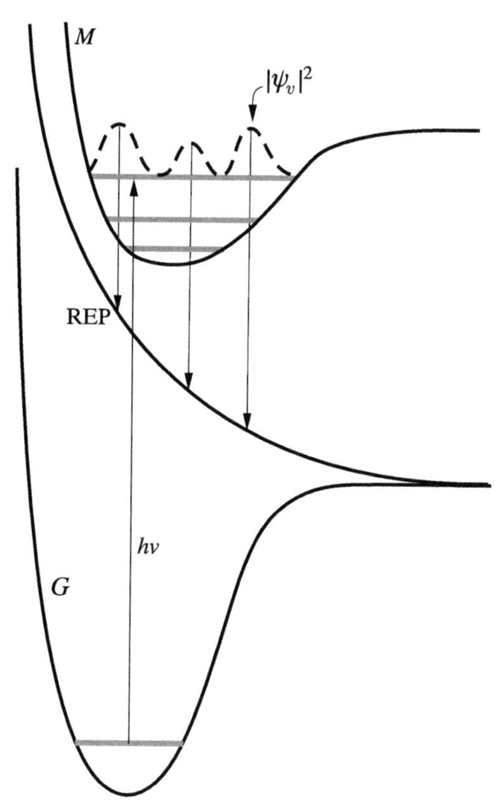

Fig. 6.14. A "measurement" of the wave function ψ_v, or more exactly of the corresponding probability density $|\psi_v|^2$. A molecule is excited from its electronic ground state G to a certain vibrational state ψ_v in the electronic excited state M. From M, the molecule undergoes a fluorescence transition to the state REP. Since the REP state is of repulsive character, the potential energy transforms into kinetic energy (the total energy being preserved). By measuring the kinetic energy of the dissociation products, one is able to calculate their starting potential energy (i.e., how high they were on the REP curve). This enables us to calculate $|\psi_v|^2$.

[61] W. Koot, P.H.P. Post, W.J. van der Zande, and J. Los, *Zeit. Physik* D, *10,* 233 (1988). The experimental data pertain to the hydrogen molecule.

James Franck (1882–1964), German physicist and professor at the Kaiser Wilhelm Institut für Physikalische Chemie in Berlin, then at the University of Göttingen. Then at John Hopkins University in Baltimore, Maryland, and from 1938 to 1949 at the University of Chicago. Frank also participated in the Manhattan Project. As a freshman at the Department of Law at the University of Heidelberg, he made the acquaintance of the student Max Born. Born persuaded him to resign from his planned career as a lawyer and pursue studies in chemistry, geology, and then physics. In 1914, Franck and his colleague Gustav Hertz used electrons to bombard mercury atoms. The young researchers noted that electrons lose 4.9 eV of their kinetic energy after colliding with mercury atoms. This excess energy is then released by emitting a UV photon. This was the first experimental demonstration that atoms have the electronic energy levels foreseen by Niels Bohr. Both scientists earned the Nobel Prize in 1925 for their work. The fact that, during World War I, Franck was twice decorated with the Iron Cross was the reason that he was one of the few Jews whom the Germans tolerated in academia. Franck, a citizen of the Third Reich, illegally deposited his Nobel Prize medal (with his engraved name) in the Niels Bohr Institute in Copenhagen, Denmark. When in April 1940, the attacking German troops marched through the streets of the Danish capital, George de Hevesy (a future Nobel laureate, 1943) was hiding the golden medal in a strange and very chemical way–he dissolved it in aqua regia. The bottle safely stayed on the shelf the whole occupation period under the nose of the Germans. After the war, the Nobel Committee exchanged the bottle for a new medal for Franck.

Edward Condon (1902–1974), American physicist and one of the pioneers of quantum theory in the United States. In 1928, Condon and Gurney discovered the tunneling. More widely known is his second great achievement–the Franck-Condon rule (discussed later in this chapter). During the WW2 he participated in the Manhattan project.

We excite the molecule from the ground vibrational state of G to a certain vibrational state ψ_v of M using a laser. Then the molecule undergoes a spontaneous *fluorescence transition* to REP. The electronic state changes so fast that the nuclei have no time to move (the *Franck-Condon rule*). Whatever falls (vertically, because of the Franck-Condon rule) on the REP state as a result of fluorescence dissociates because this state is repulsive. The kinetic energy of the dissociation products depends on the internuclear distance R when the fluorescence took place (i.e., on the length of the slide the system had down the REP). How often such an R occurs depends on $|\psi_v(R)|^2$. Therefore, investigating the kinetic energy of the dissociation products gives $|\psi_v|^2$.

6.12 Adiabatic, Diabatic, and Non-Adiabatic Approaches

Let us summarize the diabatic, adiabatic, and non-adiabatic concepts, as shown in Fig. 6.15.

Adiabatic case. Suppose that we have a Hamiltonian $\hat{\mathcal{H}}(\boldsymbol{r}; \boldsymbol{R})$ that depends on the electronic coordinates $\boldsymbol{r}$ and parametrically depends on the configuration of the nuclei $\boldsymbol{R}$. In practical

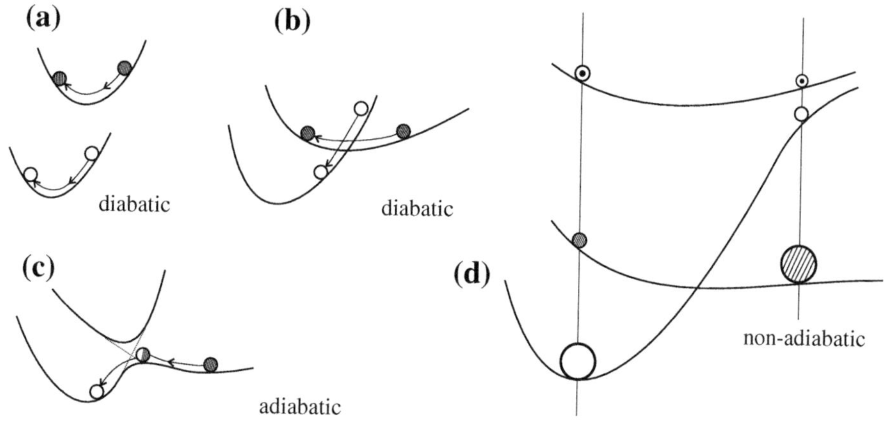

Fig. 6.15. The diabatic, adiabatic, and non-adiabatic approaches to the motion of nuclei (a schematic view). (a) A state that preserves the chemical structure for any molecular geometry is called *diabatic* (e.g., is always ionic, or always covalent). The energies of these states are calculated as the mean values of the clamped nuclei Hamiltonian. In the lower-energy state, the system is represented by a white ball (say, in the ionic state); in the second, the system is represented by the black ball (say, a covalent structure). These balls vibrate all the time in the corresponding wells, preserving the chemical structure. (b) It may happen that two diabatic states cross. If the nuclear motion is fast, the electrons are unable to adjust and the nuclear motion may take place on the diabatic curves (i.e., the bond pattern does not change during this motion). (c) The adiabatic approach, where the diabatic states mix (mainly at a crossing region). Each of the adiabatic states is an eigenfunction of the clamped nuclei Hamiltonian. If the nuclear motion is slow, the electrons are able to adjust to it instantaneously and the system follows the lower adiabatic curve. The bond pattern changes qualitatively during this motion (black ball changes to white ball; e.g., the system undergoes a transition from covalent to ionic). The total wave function is a product of the adiabatic electronic state and a rovibrational wave function. (d) The non-adiabatic approach. In this particular case, three diabatic curves come into play. The total wave function is the sum of three functions (their contributions are geometry-dependent, a larger ball means a larger contribution), each function is a product of a diabatic electronic state times a rovibrational wave function. The system is shown at two geometries. Changing the nuclear geometry, it is as if the system has moved on three diabatic surfaces at the same time. This motion is accompanied by changing the proportions (visualized by the size of the balls) of the electronic diabatic states composing it.

applications, most often $\hat{\mathcal{H}}(\boldsymbol{r};\boldsymbol{R}) \equiv \hat{H}_0(\boldsymbol{r};\boldsymbol{R})$, the electronic clamped nuclei Hamiltonian corresponding to Eq. (6.8) and generalized to polyatomic molecules. The eigenfunctions $\psi(\boldsymbol{r};\boldsymbol{R})$ and the eigenvalues $E_i(\boldsymbol{R})$ of the Hamiltonian $\hat{\mathcal{H}}(\boldsymbol{r};\boldsymbol{R})$ are called *adiabatic* (see Fig. 6.15). If we take $\hat{\mathcal{H}} = \hat{H}_0(\boldsymbol{r};\boldsymbol{R})$, then in the adiabatic approximation (p. 268), the total wave function is represented by a product

$$\Psi(\boldsymbol{r},\boldsymbol{R}) = \psi(\boldsymbol{r};\boldsymbol{R})f(\boldsymbol{R}), \tag{6.41}$$

where $f(\boldsymbol{R})$ is a rovibrational wave function that describes the rotations and vibrations of the system.

Diabatic case. Imagine now a basis set $\bar{\psi}_i(\boldsymbol{r};\boldsymbol{R}),\ i = 1, 2, 3, \ldots M$ of some particular electronic wave functions (we will call them *diabatic*) that also depend parametrically on $\boldsymbol{R}$. There are two reasons for considering such a basis set. The first is that we are going to solve the Schrödinger equation $\hat{\mathcal{H}}\Psi_i = E_i\Psi_i$ by using the Ritz method (Chapter 5) and we need a basis set of the expansion functions:

$$\psi(\mathbf{r};\mathbf{R}) \approx \sum_i^M c_i \bar{\psi}_i(\mathbf{r};\mathbf{R}). \tag{6.42}$$

The second reason pertains to chemical interpretation: usually any of the diabatic wave functions are chosen as corresponding to a particular electronic distribution (chemical bond pattern) in the system,[62] and from Eq. (6.42), we may recognize what kind of chemical structure dominates Ψ. Thus, using the diabatic basis, there is a chance of gaining insight into the chemistry going on in the system.[63]

The wave functions $\bar{\psi}_i$ are in general non-orthogonal (we assume them to be normalized). For each of them, we may compute the mean value of the energy (the integration is over the electronic coordinates) as follows:

$$\bar{E}_i(\mathbf{R}) = \langle \bar{\psi}_i | \hat{\mathcal{H}}(\mathbf{R}) \bar{\psi}_i \rangle, \tag{6.43}$$

and we will call it the *diabatic energy*.

The key point is that we may compare the eigenvalues and eigenfunctions of $\hat{\mathcal{H}}(\mathbf{R})$; i.e., the adiabatic states with $\bar{E}_i$ and $\bar{\psi}_i$, respectively. If the diabatic states are chosen in a realistic way, they are supposed to be close to the adiabatic states for most configurations $\mathbf{R}$ (see Figs. 6.15a–c). These relations will be discussed shortly.

Non-adiabatic case. The diabatic states or the adiabatic states may be used to construct the basis set for the motion of the electrons *and* nuclei in the non-adiabatic approach. Such a basis function is taken as a product of the electronic (diabatic or adiabatic) wave function and of a rovibrational wave function that depends on $\mathbf{R}$. In a non-adiabatic approach, the total wave function is a superposition of these product-like contributions:

$$\Psi(\mathbf{r};\mathbf{R}) = \sum_k \bar{\psi}_k(\mathbf{r};\mathbf{R}) f_k(\mathbf{R}). \tag{6.44}$$

---

[62] Let us take the example of the NaCl molecule: $\bar{\psi}_1$ may describe the ionic Na^+Cl^- distribution, while $\bar{\psi}_2$ may correspond to the covalent bond Na–Cl. The adiabatic wave function $\psi(\mathbf{r};\mathbf{R})$ of the NaCl molecule may be taken as a superposition of $\bar{\psi}_1$ and $\bar{\psi}_2$. The valence bond (VB) wave functions (VB structures) described in Chapter 10 may be viewed as diabatic states.

[63] This is very important for chemical reactions, in which a chemical structure undergoes an abrupt change. In chemical reactions, large changes of nuclear configuration are accompanied by motions of electrons; i.e., large changes in the chemical bond pattern [a qualitative change of c_i of Eq. (6.42)]. Such a definition leaves us liberty in the choice of diabatic states. This liberty can be substantially reduced by the following. Let us take two adiabatic states that dissociate to different products, well separated on the energy scale. However, for some reason, the two adiabatic energies are getting closer for some finite values of R. For each value of R, we define a space spanned by the two adiabatic functions for that R. Let us find in this space two normalized functions that maximize the absolute value of the overlap integral with the two dissociation states. These two (usually non-orthogonal) states may be called *diabatic*.

This sum means that in the non-adiabatic approach, the motion of the system involves many potential energy surfaces at the same time (see Fig. 6.15d).

The diabatic and the adiabatic electronic states are simply two choices of the basis set in non-adiabatic calculations. If the sets were complete, the results would be identical. The first choice underlines the importance of the chemical bond pattern and the interplay among such patterns. The second basis set highlights the order of the eigenvalues of $\hat{\mathcal{H}}(\boldsymbol{R})$ (the lower/higher-energy adiabatic state).[64]

6.13 Crossing of Potential Energy Curves for Diatomics

6.13.1 The Non-Crossing Rule

Can the adiabatic curves $E_k^0(R)$ cross when R changes?

To solve this problem in detail, let us limit ourselves to the simplest situation: the two-state model (see Appendix D available at booksite.elsevier.com/978-0-444-59436-5). Let us consider a diatomic molecule and such an internuclear distance R_0 that the two electronic adiabatic states[65] $\psi_1(\boldsymbol{r}; R_0)$ and $\psi_2(\boldsymbol{r}; R_0)$ correspond to the non-degenerate (but close in the energy scale) eigenvalues of the clamped nuclei Hamiltonian $\hat{H}_0(R_0)$:

$$\hat{H}_0(R_0)\psi_i(\boldsymbol{r}; R_0) = E_i(R_0)\psi_i(\boldsymbol{r}; R_0), \; i = 1, 2.$$

Since $\hat{H}_0$ is Hermitian and $E_1 \neq E_2$, we have the orthogonality of $\psi_1(\boldsymbol{r}; R_0)$ and $\psi_2(\boldsymbol{r}; R_0)$: $\langle\psi_1|\psi_2\rangle = 0$.

Now, we are interested in solving

$$\hat{H}_0(R)\psi(\boldsymbol{r}; R) = E\psi(\boldsymbol{r}; R)$$

for R in the vicinity of R_0 and ask whether it is possible for the energy eigenvalues to cross. The eigenfunctions of $\hat{H}_0$ will be sought as linear combinations of ψ_1 and ψ_2:

$$\psi(\boldsymbol{r}; R) = c_1(R)\psi_1(\boldsymbol{r}; R_0) + c_2(R)\psi_2(\boldsymbol{r}; R_0). \tag{6.45}$$

Note that for this distance R

$$\hat{H}_0(R) = \hat{H}_0(R_0) + V(R), \tag{6.46}$$

and $V(R)$ is certainly small because R is close to R_0 and $V(R_0) = 0$. Using the Ritz method (Chapter 5, see Appendix D, case III), we arrive at two adiabatic solutions, and the corresponding

[64] In polyatomic systems, there is a serious problem with the adiabatic basis (this is why the diabatic functions are preferred). As we will see later, the adiabatic electronic wave function is multivalued, and the corresponding rovibrational wave function, having to compensate for this (because the total wave function must be single-valued), also has to be multivalued.

[65] These states are adiabatic only for $R = R_0$, but when considering $R \neq R_0$, they may be viewed as diabatic (because they are not the eigenfunctions for that R).

energies read as

$$E_{\pm}(R) = \frac{\bar{E}_1 + \bar{E}_2}{2} \pm \sqrt{\left(\frac{\bar{E}_1 - \bar{E}_2}{2}\right)^2 + |V_{12}|^2}, \tag{6.47}$$

where $V_{ij}(R) \equiv \left\langle \psi_i | \hat{V}(R) \psi_j \right\rangle$ and

$$\bar{E}_i(R) = \left\langle \psi_i(\mathbf{r}; R_0) | \hat{H}_0(R) \psi_i(\mathbf{r}; R_0) \right\rangle = E_i(R) + V_{ii}(R). \tag{6.48}$$

The crossing of the energy curves at a given R means that $E_+ = E_-$, and from this, it follows that the expression under the square root symbol has to equal zero. Since, however, the expression is the sum of two squares, the crossing needs *two* conditions to be satisfied simultaneously:

$$\bar{E}_1 - \bar{E}_2 = 0, \tag{6.49}$$

$$|V_{12}| = 0. \tag{6.50}$$

Two conditions, and a *single* parameter R to change. If you adjust the parameter to fulfill the first condition, the second one is violated, and vice versa. The crossing $E_+ = E_-$ may occur only when, for some reason; e.g., because of the symmetry, the *coupling constant* is automatically equal to zero, $|V_{12}| = 0$, for all R. Then, we have only a *single* condition to be fulfilled, and it can be satisfied by changing the parameter R; i.e., crossing can occur. The condition $|V_{12}| = 0$ is equivalent to $|H_{12}| \equiv \left\langle \psi_1 | \hat{H}_0(R) \psi_2 \right\rangle = 0$, because $\hat{H}_0(R) = \hat{H}_0(R_0) + \hat{V}$, and $\left\langle \psi_1 | \hat{H}_0(R_0) \psi_2 \right\rangle = 0$ due to the orthogonality of both eigenfunctions of $\hat{H}_0(R_0)$.

Now we will refer to group theory (see Appendix C available at booksite.elsevier.com/978-0-444-59436-5, p. e17). The Hamiltonian represents a fully symmetric object, whereas the wave functions ψ_1 and ψ_2 are not necessarily fully symmetric because they may belong to other irreducible representations of the symmetry group. Therefore, in order to make the integral $|H_{12}| = |V_{12}| = 0$, it is sufficient that ψ_1 and ψ_2 transform according to *different* irreducible representations (have different symmetries).[66] Thus,

the adiabatic curves cannot cross if the corresponding wave functions have the same symmetry.

What will happen if such curves are heading for something that looks like an inevitable crossing? Such cases are quite characteristic and look like an avoided crossing. The two curves look as if they repel each other and avoid the crossing.

[66] H_{12} transforms according to the representation being the direct product of three irreducible representations: that of ψ_1, that of ψ_2, and that of $\hat{H}$ (the last is, however, fully symmetric, and therefore, does not count in this direct product). In order to have $H_{12} \neq 0$, this direct product, after decomposition into irreducible representations, has to contain a fully symmetric irreducible representation. This, however, is possible only when ψ_1 and ψ_2 transform according to the same irreducible representation.

If two states of a diatomic molecule correspond to different symmetries, then the corresponding potential energy curves can cross.

6.13.2 Simulating the Harpooning Effect in the NaCl Molecule

Our goal now is to show, in an example, what happens to adiabatic states (eigenstates of $\hat{\mathcal{H}}(R)$), if two diabatic energy curves (mean values of the Hamiltonian with the diabatic functions) do cross. Although we are not aiming at an accurate description of the NaCl molecule (we prefer simplicity and generality), we will try to construct a toy (a model) that mimics this particular system.

The sodium atom has 11 electrons (the electronic configuration[67]: $1s^2 2s^2 2p^6 3s^1$), and the chlorine atom contains 17 electrons ($1s^2 2s^2 2p^6 3s^2 3p^5$). The solution of the Schrödinger equation for 28 electrons is difficult. But we are not looking for trouble. Note that with NaCl, the real star is a single electron that goes from the sodium to the chlorine atom, making Na^+ and Cl^- ions. The ions attract each other by the Coulombic force and form the familiar ionic bond. But there is a problem. What is of lower energy: the two non-interacting atoms Na and Cl or the two non-interacting ions Na^+ and Cl^-? The ionization energy of sodium is $I = 495.8$ kJ/mol $= 0.1888$ a.u., whereas the electron affinity of chlorine is only $A = 349$ kJ/mol $= 0.1329$ a.u. This means that the NaCl molecule in its ground state dissociates into atoms, not ions.

To keep the story simple, let us limit ourselves to the single electron mentioned above.[68] First, let us define the two diabatic states (the basis set) of the system: only the 3s orbital of Na (when the electron resides on Na; we have atoms) denoted by $|3s\rangle$ and the 3p orbital of Cl (when the electron is on Cl; we have ions) $|3p\rangle$. Now, what about the Hamiltonian $\hat{\mathcal{H}}$? Well, a reasonable model Hamiltonian may be taken as[69]

$$\hat{\mathcal{H}}(\boldsymbol{r}; R) = -I\ |3s\rangle\langle 3s| - A\,|3p\rangle\langle 3p| - \frac{1}{R}\,|3p\rangle\langle 3p| + \exp(-R).$$

Indeed, the mean values of $\hat{\mathcal{H}}$ in the $|3s\rangle$ and $|3p\rangle$ states are equal to

$$\bar{E}_1(R) \equiv \mathcal{H}_{11} = \left\langle 3s|\hat{\mathcal{H}}(3s)\right\rangle = -I - AS^2 - \frac{1}{R}S^2 + \exp(-R),$$

$$\bar{E}_2(R) \equiv \mathcal{H}_{22} = \left\langle 3p|\hat{\mathcal{H}}(3p)\right\rangle = -IS^2 - A - \frac{1}{R} + \exp(-R),$$

where (assuming the diabatic functions to be real) the overlap integral $S \equiv \langle 3s|3p\rangle = \langle 3p|3s\rangle$. First of all, this Hamiltonian gives the correct energy limits $\bar{E}_1(R) = -I$ and $\bar{E}_2(R) = -A$, when $R \to \infty$ (the electron binding energy by the sodium and by the chlorine for dissociation into atoms and ions, respectively), which is already very important. The term $\exp(-R)$

[67] What these configurations really mean is explained in Chapter 8.

[68] The other electrons in our approach will only influence the numerical values of the interaction parameters.

[69] $\boldsymbol{r}$ stands for the coordinates of the electron, and for the diatomic molecule, R replaces $\boldsymbol{R}$.

mimics the repulsion of the inner shells of both atoms[70] and guarantees that the energies go up (which they should) at $R \to 0$. Note also that the $\bar{E}_1(R)$ and $\bar{E}_2(R)$ curves indeed mimic the approaching Na and Cl, and Na^+ and Cl^-, respectively, because in $\bar{E}_2(R)$, there is a Coulomb term $-\frac{1}{R}$, while in $\bar{E}_1(R)$, such an interaction practically disappears for large R. All this gives us a certain confidence that our Hamiltonian $\hat{\mathcal{H}}$ grasps the most important physical effects for the NaCl molecule. The resulting non-diagonal element of the Hamiltonian reads as:

$$\left\langle 3s|\hat{\mathcal{H}}(3p)\right\rangle \equiv \mathcal{H}_{12} = S\left[-I - A - \frac{1}{R} + \exp\left(-R\right)\right].$$

As to S, we could in principle calculate it by taking some approximate atomic orbitals, but our goal is less ambitious than that. Let us simply set $S = R\exp\left(-R/2\right)$. Why? Since $S = \langle 3s|3p\rangle = 0$, if $R \to \infty$ or if $R \to 0$, and $S > 0$ for other values of R, then at least our formula takes care of this. In addition, Figs. 6.16a–b show that such a formula for S also gives a quite reasonable set of diabatic curves $\bar{E}_1(R)$ and $\bar{E}_2(R)$: both curves have a single minimum, the minimum for the ionic curve is at about 5.23 a.u., close to the experimental value of 5.33 a.u., and the binding energy is 0.11 a.u. (0.13 for the adiabatic case, see below), and it is also close to the experimental value of 0.15 a.u. Thus, our model to a reasonable extent resembles the real NaCl molecule.

Our goal is the adiabatic energies computed using the diabatic basis chosen, Eq. (6.42). see Appendix D available at booksite.elsevier.com/978-0-444-59436-5 (general case) gives the eigenvalues [$E_+(R)$ and $E_-(R)$] and the eigenfunctions (ψ_+ and ψ_-). Figs. 6.16c–d, show the adiabatic compared to the diabatic curves. The avoided crossing at about 17.9 a.u. is the most important. If the two atoms begin to approach (shown in light gray in Fig. 6.16a), the energy does not change too much (flat energy curve), but if the ions do the same, the energy goes down because of the long-range Coulombic attraction (dark gray). Thus, the two adiabatic curves (that nearly coincide with the two diabatic curves, especially for large R) are going to cross each other Figs. 6.16a–b but the two states have the same symmetry with respect to the molecular axis (as witnessed by $S \neq 0$) and, therefore, the crossing cannot occur, as shown in Fig. 6.16d. As a result, the two curves *avoid the crossing* and, as shown in Figs. 6.16c–f, the "atomic" curve switches to the "ionic" curve and vice versa. This switching means an electron jumping from Na to Cl and, therefore, formation of the ions Na^+ and Cl^- (then the ions approach fast - this is the *harpooning effect,* introduced to chemistry by Michael Polanyi). This jump occurs at long distances, of the order of 9 Å.

Is this jump inevitable?

[70] It prevents the two cores collapsing; cf. Chapter 13.

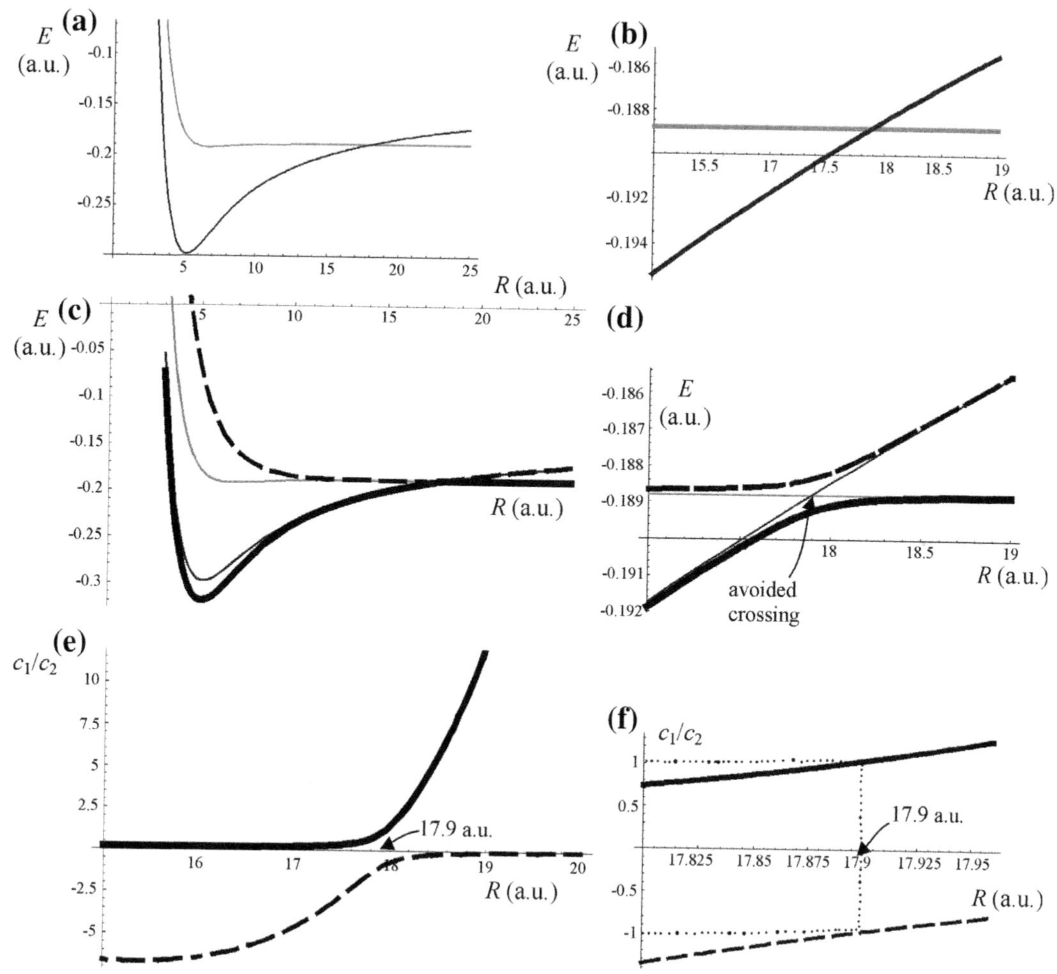

Fig. 6.16. A simple one-electron model of electron transfer in the NaCl molecule. (a) The mean values of the Hamiltonian with two *diabatic* states: one (light gray) being the $3s$ atomic orbital of the sodium atom (atomic curve), the second (dark gray) the $2p$ atomic orbital of the chlorine atom (ionic curve). The two diabatic curves intersect. (b) A closer view of the intersection. (c) The two diabatic curves [gray, as in (a,b)] and the two *adiabatic* curves (black), the lower-energy (solid), the higher-energy (dashed). Although the drawing looks like intersection, in fact the adiabatic curves "*repel*" each other, as shown in (d). (e) Each of the adiabatic states is a linear combination of two diabatic states (atomic and ionic). The ratio c_1/c_2 of the coefficients for the lower-energy (solid line) and higher-energy states (dashed line), c_1 is the contribution of the atomic function, c_2 – of the ionic function. As we can see, the lower-energy (higher-energy) adiabatic state is definitely atomic (ionic) for $R > 17.9$ a.u. and definitely ionic (atomic) for smaller R in the vicinity of the avoided crossing. (f) The ratio c_1/c_2 very close to the avoided crossing point. As we can see, at this point, one of the adiabatic states is the sum, and the other the difference of the two diabatic states.

If the electron is able to adapt instantaneously to the position of the nuclei (slow nuclear motion), the system follows the adiabatic curve. If the nuclear motion is very fast, the system follows the diabatic curve and no electron transfer takes place. The electron transfer is more probable if the gap $2|\mathcal{H}_{12}|$ between $E_+(R)$ and $E_-(R)$ is large.

In our model, for large distances, the adiabatic are practically identical with the diabatic states, except in the avoided crossing region (see Figs. 6.16c–d).

6.14 Polyatomic Molecules and Conical Intersection

Crossing for Polyatomics

The non-crossing rule for a diatomic molecule was based on Eq. (6.47). To achieve the crossing, we had to make vanish two independent terms with only one parameter (the internuclear distance R) to vary. It is important to note that in the case of a polyatomic molecule, the formula would be the same, but the number of parameters would be larger: $3M - 6$ in a molecule with M nuclei. For $M = 3$, therefore, one has already three such parameters. No doubt even for a three-atomic molecule, we would be able to make the two terms equal to zero and, therefore, achieve $E_+ = E_-$; i.e., the crossing of the two diabatic hypersurfaces would occur.

Let us investigate this possibility, which, for reasons that will become clear later, is called *conical intersection*. We will approach this concept by a few steps.

Cartesian System of 3M Coordinates ($\mathbf{O}_{3M}$)

All the quantities in Eq. (6.47) depend on $n = 3M - 6$ coordinates of the nuclei. These coordinates may be chosen in many different ways; the only thing we should bother about is that they have to determine the positions of M point objects. Just to begin, let us construct a Cartesian system of $3M$ coordinates ($\mathbf{O}_{3M}$). Let us locate (Fig. 6.17) nucleus 1 at the origin (in this way, we eliminate three degrees of freedom connected with the translation of the system), and nucleus 2 will occupy the point x_2 on the x-axis; i.e., $y_2 = z_2 = 0$. In this way, we have eliminated two rotations of the system. The total system may still be rotated about the x-axis.

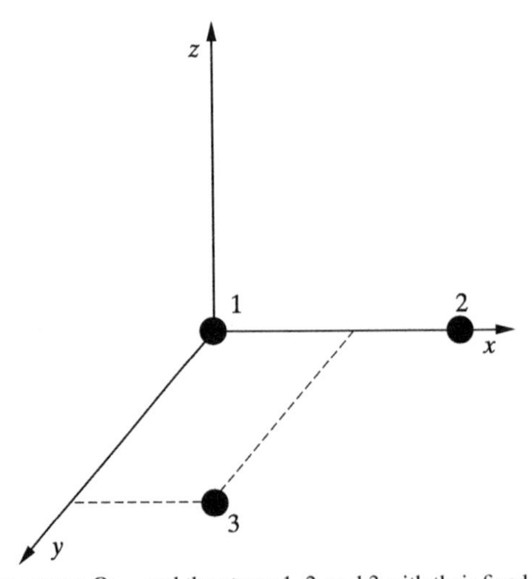

Fig. 6.17. The Cartesian coordinate system $\mathbf{O}_{3M}$ and the atoms 1, 2, and 3 with their fixed positions.

This last possibility can be eliminated when we decide to locate the nucleus 3 in the plane x, y (i.e., the coordinate $z_3 = 0$).

Thus, six degrees of freedom have been eliminated from the $3M$ coordinates. The other nuclei may be indicated by vectors (x_i, y_i, z_i) for $i = 4, 5, \ldots M$. As we can see, there has been a lot of arbitrariness in these choices.[71]

Cartesian System of 3M − 6 Coordinates ($\mathbf{O}_{3M-6}$)

This choice of coordinate system may be viewed a little differently. We may construct a Cartesian coordinate system with the origin at atom 1 and the axes x_2, x_3, and y_3 (see Fig. 6.17), and x_i, y_i, and z_i for $i = 4, 5, \ldots M$. Thus, we have a Cartesian coordinate system ($\mathbf{O}_{3M-6}$) with $3 + 3(M - 3) = 3M - 6 = n$ axes, which may be labeled (in the sequence given above) in a uniform way: $\bar{x}_i, i = 1, 2, \ldots n$. A single point $\boldsymbol{R} = (\bar{x}_1, \bar{x}_2, \ldots \bar{x}_{3M-6})$ in this n-dimensional space determines the positions of all M nuclei of the system. If necessary, all these coordinates may be expressed by the old ones, but it will not be because our goal is different.

Two Special Vectors in the $\mathbf{O}_{3M-6}$ Space

Let us consider two functions $\bar{E}_1 - \bar{E}_2$ and V_{12} of the configuration of the nuclei $\boldsymbol{R} = (\bar{x}_1, \bar{x}_2, \ldots \bar{x}_{3M-6})$; i.e., with domain being the $\mathbf{O}_{3M-6}$ space. Now, let us construct two vectors in $\mathbf{O}_{3M-6}$:

$$\nabla(\bar{E}_1 - \bar{E}_2) = \sum_{i=1}^{3M-6} \boldsymbol{i}_i \left(\frac{\partial(\bar{E}_1 - \bar{E}_2)}{\partial \bar{x}_i} \right)_0,$$

$$\nabla V_{12} = \sum_{i=1}^{3M-6} \boldsymbol{i}_i \left(\frac{\partial V_{12}}{\partial \bar{x}_i} \right)_0,$$

where $\boldsymbol{i}_i$ stands for the unit vector along axis $\bar{x}_i$, while the derivatives are calculated in a point of the configurational space for which

$$\sqrt{\left(\frac{\bar{E}_1 - \bar{E}_2}{2} \right)^2 + |V_{12}|^2} = 0;$$

i.e., where according to Eq. (6.47), one has the intersection of the adiabatic hypersurfaces.

6.14.1 Branching Space and Seam Space

We may introduce any coordinate system. We are free to do this because our object (molecule) stays immobile, but our way of determining the nuclear coordinates changes. We will change

[71] By the way, if the molecule were diatomic, the third rotation need not be determined and the number of variables would be equal to $n = 3 \times 2 - 5 = 1$.

the coordinate system in n-dimensional space once more. This new coordinate system is formed from the old one ($\mathbf{O}_{3M-6}$) by rotation.

The rotation will be done in such a way as to make the plane determined by the two first axes ($\bar{x}_1$ i $\bar{x}_2$) of the old coordinate system coincide with the plane determined by the two vectors: $\nabla(\bar{E}_1 - \bar{E}_2)$ oraz $\nabla(V_{12})$,

Let us denote the coordinates in the rotated coordinate system by $\xi_i, i = 1, 2, \ldots, n$. The new coordinates can, of course, be expressed as some linear combinations of the old ones, but these details need not concern us. The most important thing is that we have the axes of the coordinates ξ_1 and ξ_2, which determine the same plane as the vectors $\nabla\left(\bar{E}_1 - \bar{E}_2\right)$ and ∇V_{12}. This plane is known as the *branching space (plane)*. The space of all vectors $\left(0.0, \xi_3 \ldots \xi_{3M-6}\right)$ is called the *seam space*. The directions $\nabla\left(\bar{E}_1 - \bar{E}_2\right)$ and ∇V_{12} need not be orthogonal, although they look this way in illustrations shown in the literature.[72]

Now we are all set to define the conical intersection.

6.14.2 Conical Intersection

Why has this slightly weird coordinate system been chosen? We see from the formula [Eq. (6.47)] for E_+ and E_- that ξ_1 and ξ_2 correspond to the fastest change of the first term and the second term under the square-root sign, respectively.[73]

Any change of other coordinates (along the axes orthogonal to the plane $\xi_1\xi_2$) does not influence the value of the square root; i.e., it does not change the difference between E_+ and E_- (although the values of E_+ and E_- change).

Therefore, the hypersurface E_+ intersects with the hypersurface E_-, and their common part (i.e., the seam space) are all those vectors of the n-dimensional space that fulfill the condition: $\xi_1 = 0$ and $\xi_2 = 0$. The intersection represents a $(n - 2)$–dimensional subspace of the n-dimensional space of the nuclear configurations.[74] When we withdraw from the point

---

[72] See F. Bernardi, M. Olivucci, and M.A. Robb, *Chem. Soc. Rev.*, 25, 321 (1996). The authors confirmed to me that the angle between these vectors is often quite small.

[73] Let us take a scalar field V and calculate its value at the point $\boldsymbol{r}_0 + \boldsymbol{r}$, where we assume $|\boldsymbol{r}| \ll 1$. From the Taylor expansion, we have with good accuracy, $V\left(\boldsymbol{r}_0 + \boldsymbol{r}\right) = V(\boldsymbol{r}_0) + \left(\nabla V\right)_{\boldsymbol{r}=\boldsymbol{r}_0} \cdot \boldsymbol{r} = V(\boldsymbol{r}_0) + \left|\left(\nabla V\right)_{\boldsymbol{r}=\boldsymbol{r}_0}\right| \cdot r \cos\theta$. We obtain the largest absolute value of the increment of V for $\theta = 0$ and $\theta = 180°$ i.e., along the vector $\left(\nabla V\right)_{\boldsymbol{r}=\boldsymbol{r}_0}$.

[74] If the axes ξ_1 and ξ_2 were chosen in another way on the plane determined by the vectors $\nabla\left(\bar{E}_1 - \bar{E}_2\right)$ and ∇V_{12}, the conical intersection would be described in a similar simple way. If, however, the axes were chosen outside the plane, it may happen that moving along more than just two axes, they would split into E_+ and E_-. Our choice stresses that the intersection of E_+ and E_- represents a $(n - 2)$ – dimensional subspace (seam space).

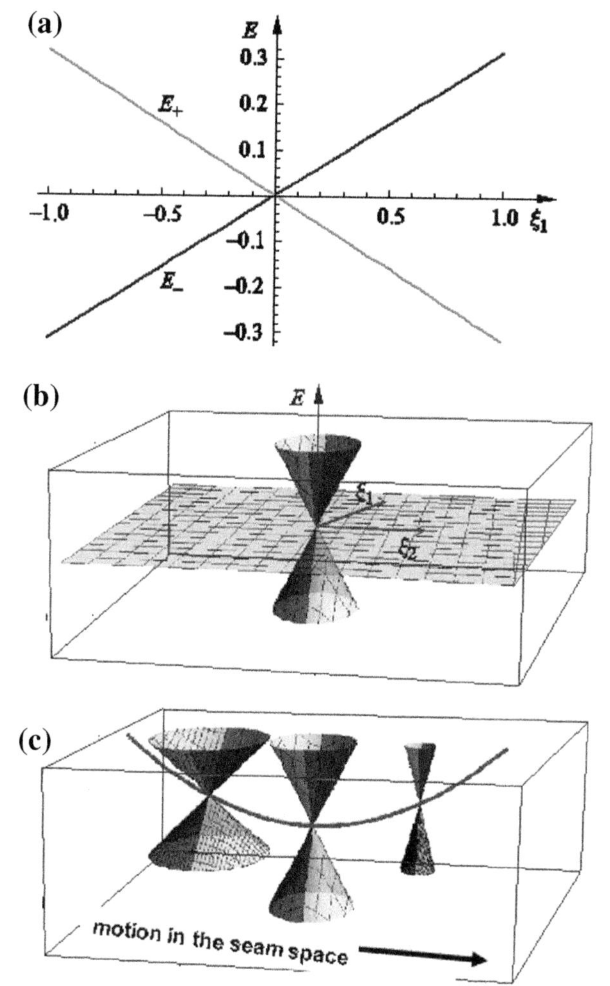

Fig. 6.18. Conical intersection (scheme). E represents the electronic energy as a function of coordinates of the nuclei: $\xi_1, \xi_2, \xi_3, \xi_4, \ldots \xi_{3M-6}$. This shows only the coordinates ξ_1 and ξ_2, which define what is known as the *branching space* $\xi_1\xi_2$, while the space of all vectors $(0, 0, \xi_3, \xi_4, \ldots \xi_{3M-6})$ known as the seam space is not shown in panel (a) or (b). (a) Section of the cones along ξ_1 at a given point of the seam space; the equality $E_+ = E_-$ holds for the conical intersection point. (b) The vectors $\nabla(\bar{E}_1 - \bar{E}_2)$ and ∇V_{12} span the branching plane (the horizontal plane; both vectors are calculated at the conical intersection). The upper cone E_+ and the lower cone E_- correspond to Eq. (6.47), and each consists of two diabatic surfaces (gray and white). (c) Staying at the branching point (0, 0), but moving in the seam space, one remains all the time in the conical intersection, but the cones look different (different cone openings) and the energy $E_+ = E_-$ changes (solid line).

$(0, 0, \xi_3, \xi_4, \ldots \xi_{3M-6})$ by changing the coordinates ξ_1 and/or ξ_2, a difference between E_+ and E_- appears. For small increments $d\xi_1$, the changes in the energies E_+ and E_- are proportional to $d\xi_1$ and for E_+ and E_- differ in sign. This means that the hypersurfaces E_+ and E_- as functions of ξ_1 (at $\xi_2 = 0$ and fixed other coordinates) have the shapes shown in Fig. 6.18a. For ξ_2, the situation is similar, but the cone may differ by its angle. From this, it follows that the ground and excited state hypersurfaces intersect with each other (the intersection set represents the subspace of all vectors $(0, 0, \xi_3, \xi_4, \ldots, \xi_n)$) and split when we go out of the intersection point according to the *cone rule*; i.e., E_+ and E_- change *linearly* when moving in the plane ξ_1, ξ_2 from the point $(0, 0)$.

This is called the *conical intersection* (see Fig. 6.18b). The cone opening angle is in general different for different points of the seam space (see Fig. 6.18c).

The conical intersection plays a fundamental role in the theory of chemical reactions (Chapter 14). The lower (ground-state) as well as the higher (excited-state) hypersurfaces are composed of two diabatic parts, which in polyatomics correspond to different patterns of chemical bonds. This means that the system (represented by a point) when moving on the ground-state adiabatic hypersurface toward the join of the two parts, passes near the conical intersection point, over the energy barrier, and goes to the products. This is the essence of a chemical reaction.

6.14.3 Berry Phase

We will focus on the adiabatic wave functions close to the conical intersection. Our goal will be to show something strange, that

when going around the conical intersection point in the configurational space, the electronic wave function changes its phase; and after coming back to the starting point, this change results in the opposite sign of the function.

First, let us prepare an itinerary in the configuration space around the conical intersection. We need a parameter, which will be an angle α and will define our position during our trip around the point. Let us introduce some abbreviations in Eq. (6.47): $\Delta \equiv \frac{\bar{E}_1-\bar{E}_2}{2}$, $h \equiv V_{12}$, and define α in the following way:

$$\sin\alpha = \Delta/\rho,$$

$$\cos\alpha = h/\rho,$$

$$\text{where } \rho = \sqrt{\Delta^2 + h^2}.$$

We will move around the conical intersection within the plane given by the vectors $\nabla\Delta$ and ∇h (branching plane). The conical intersection point is defined by $|\nabla\Delta| = |\nabla h| = 0$. Changing α from 0 to 2π, we have to go, at a distance $\rho(\alpha)$, once through a maximum of h (say, in the direction of the maximum gradient ∇h), and once through its minimum $-h$ (the opposite direction). This is ensured by $\cos\alpha = h/\rho$. Similarly, we have a single maximum and a single minimum of $\nabla\Delta$ (as must happen when going around), when assuming that $\sin\alpha = \Delta/\rho$. We do not need more information about our itinerary because we are interested in how the wave function changes after making a complete trip (i.e., 360° around the conical intersection and returning to the starting point).

The adiabatic energies are given in Eq. (6.47) and the corresponding coefficients of the diabatic states are reported in Appendix D available at booksite.elsevier.com/978-0-444-59436-5 (the first, most general case):

$$\left(\frac{c_1}{c_2}\right)_{\pm} = \frac{1}{h}\left[\Delta \pm \sqrt{\Delta^2 + h^2}\right] = \tan\alpha \pm \frac{1}{\cos\alpha}.$$

Thus,

$$\frac{c_{1,+}}{c_{2,+}} = \frac{\sin\alpha + 1}{\cos\alpha} = \frac{\left(\sin\frac{\alpha}{2} + \cos\frac{\alpha}{2}\right)^2}{\cos^2\frac{\alpha}{2} - \sin^2\frac{\alpha}{2}} = \frac{\left(\sin\frac{\alpha}{2} + \cos\frac{\alpha}{2}\right)}{\left(\cos\frac{\alpha}{2} - \sin\frac{\alpha}{2}\right)},$$

$$\frac{c_{1,-}}{c_{2,-}} = \frac{\sin\alpha - 1}{\cos\alpha} = \frac{-\left(\cos\frac{\alpha}{2} - \sin\frac{\alpha}{2}\right)^2}{\cos^2\frac{\alpha}{2} - \sin^2\frac{\alpha}{2}} = -\frac{\left(\cos\frac{\alpha}{2} - \sin\frac{\alpha}{2}\right)}{\left(\cos\frac{\alpha}{2} + \sin\frac{\alpha}{2}\right)}.$$

To specify the coefficients in $\psi_+ = c_{1,+}\psi_1 + c_{2,+}\psi_2$ and $\psi_- = c_{1,-}\psi_1 + c_{2,-}\psi_2$, with ψ_1 and ψ_2 denoting the diabatic states, we have to take the two normalization conditions into account: $c_{1,+}^2 + c_{2,+}^2 = 1$, $c_{1,-}^2 + c_{2,-}^2 = 1$ and the orthogonality of ψ_+ and ψ_- : $c_{1,+}c_{1,-} + c_{2,+}c_{2,-} = 0$. After a little algebra, we get

$$c_{1,+} = \frac{1}{\sqrt{2}}\left(\cos\frac{\alpha}{2} + \sin\frac{\alpha}{2}\right),$$

$$c_{2,+} = \frac{1}{\sqrt{2}}\left(\cos\frac{\alpha}{2} - \sin\frac{\alpha}{2}\right).$$

$$c_{1,-} = -\frac{1}{\sqrt{2}}\left(\cos\frac{\alpha}{2} - \sin\frac{\alpha}{2}\right),$$

$$c_{2,-} = \frac{1}{\sqrt{2}}\left(\cos\frac{\alpha}{2} + \sin\frac{\alpha}{2}\right).$$

Now, let us consider the wave functions ψ_+ and ψ_- at the angle α and at the angle $\alpha + 2\pi$. Note that $\cos\frac{\alpha+2\pi}{2} = \cos\left(\frac{\alpha}{2} + \pi\right) = -\cos\frac{\alpha}{2}$ and $\sin\frac{\alpha+2\pi}{2} = \sin\left(\frac{\alpha}{2} + \pi\right) = -\sin\frac{\alpha}{2}$. Therefore, both the electronic functions ψ_+ and ψ_- have to change their signs after the journey (i.e., the "*geometric*" phase or Berry phase); that is,

$$\psi_+(\alpha + 2\pi) = -\psi_+(\alpha)$$

and

$$\psi_-(\alpha + 2\pi) = -\psi_-(\alpha).$$

This is how the conical intersection is usually detected.

Since the total wave function has to be single-valued, this means the function that describes the motion of the nuclei (and multiplies the electronic function) has to compensate for that change and also undergo a change of sign.

The Berry phase has some interesting analogy to gymnastics; (see p. 902).

The Role of the Conical Intersection–Non-radiative Transitions and Photochemical Reactions

The conical intersection was underestimated for a long time. However, photochemistry demonstrated that it happens much more frequently than expected. Laser light may excite a molecule from its ground state to an excited electronic state (Fig. 6.19).

Let us assume that the nuclei in the electronic ground state have their optimal positions characterized by point 1 in the configurational space (they vibrate in its neighborhood but let us ignore the quantum nature of these vibrations[75]).

The change of electronic state takes place so fast that the nuclei do not have enough time to move. Thus the positions of the nuclei in the excited state are identical to those in the ground state (Franck-Condon rule).

Point 2 (FC) in Fig 6.19 shows the very essence of the Franck-Condon rule–a vertical transition. The corresponding nuclear configuration may differ quite significantly from the nearest potential energy minimum (point 3) in the excited-state PES. In a few femtoseconds, the system slides down from FC to the neighborhood of point 3, transforming its potential energy into kinetic energy. Usually point 3 is separated from the conical intersection configuration 5 by a barrier with the corresponding potential energy saddle point 4 ("*transition state*"). Behind the saddle point, there is usually an energy valley[76] with a deep funnel ending in the conical intersection configuration (point 5). As soon as the system overcomes the barrier at the transition state (4), by going over it or by tunneling, it will be sucked in by the conical intersection attractor with almost 100% probability.

The system goes through the "*funnel*" to the electronic ground-state hypersurface with probability 1.

[75] Electronic energy hypersurfaces represent the PES for the motion of the nuclei. In the quantum mechanical picture, only some energies will be allowed: we will have the vibrational and rotational energy levels, as for diatomics. The same energy levels corresponding to E_+ may be close in the energy scale to those of E_-. Moreover, it may happen that the vibrational wave functions of two such levels may overlap significantly in space, which means that there is a significant probability that the system will undergo a transition from one to the other vibrational state. In short, in the quantum mechanical picture, the motion of the system is not necessarily bound to a single PES, but the two PESs are quite penetrable.

[76] This is on the excited-state PES.

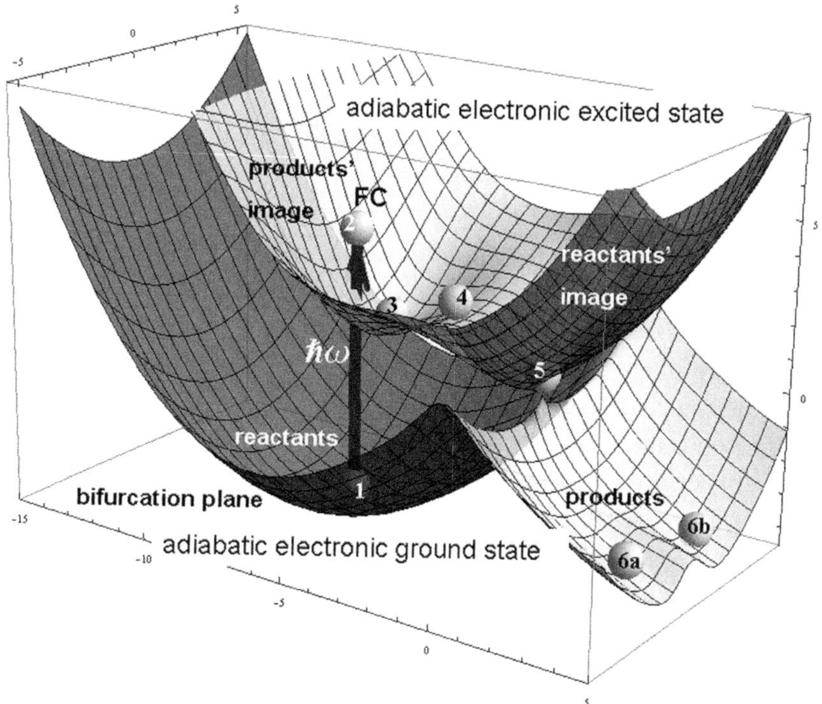

Fig. 6.19. Non-radiative transitions explained by the photochemical funnel effect (related to the conical intersection). This shows the electronic energy as a function of the coordinates ξ_1 and ξ_2 within the branching space. a) There are two adiabatic surfaces: the lower one corresponds to the ground electronic state (E_-), and the upper one pertains the excited electronic state (E_+). Each of the surfaces is composed of two parts corresponding originally to the diabatic states: the darker one corresponds to the electronic structure of the reactants, the lighter one corresponds to the electronic structure of the products. The spheres indicate some particular configuration of the nuclei. Sphere 1 indicates the reactants, and the arrow symbolizes a photoexcitation by absorption of a photon with the appropriate energy $h\nu = \hbar\omega$. The excitation takes place instantaneously at a fixed reactants' configuration (Franck-Condon rule), but the electronic excited state corresponds already to the products, and the forces acting on the nuclei correspond to the excited surface slope at the point labeled FC (sphere 2). The forces make the system move towards the minimum (sphere 3). If the kinetic energy acquired is large enough to overcome the barrier (sphere 4), the system enters the funnel, inevitably reaches the conical intersection point (sphere 5), and in a radiationless process, begins moving on the ground-state adiabatic hypersurface. The system may end up at different products (spheres 6a and 6b), or it may go back to the configuration of the reactants (sphere 1).

Then the system will continue its path in the ground-state PES, E_-, going either toward products $6a$ or $6b$, or going back to point 1 (non-reactive path).

Of course, the total energy has to be conserved. The non-radiative process described will take place if the system finds a way to dissipate its energy; i.e., to transfer an excess of electronic energy into the vibrational, rotational, and translational degrees of freedom of its own or neighboring molecules (e.g., of the solvent).[77]

[77] The energy is usually distributed among the degrees of freedom in an unequal way.

We may ask whether we will find some other conical intersections in the ground-state PES. In general, the answer is positive. There are at least two reasons for this.

In the simplest case, the conical intersection represents the dilemma of an atom C (approaching molecule AB): attach to A or attach to B?

Thus, any encounter of three atoms causes a conical intersection (we will come back to this in Chapter 14). In *each* case, the important thing is a configuration of nuclei, where a small variation may lead to distinct sets of chemical bonds. Similar "pivot points" may happen for four, five, six, or more atoms. Thus, we will encounter not only the minima, maxima, and saddle points, but also the conical intersection points when traveling in the ground-state PES.

The second reason is the permutational symmetry. Very often, the system contains the same kinds of nuclei. Any exchange of the positions of such nuclei moves the point representing the system in configuration space to some distant regions, whereas the energy does not change at all. Therefore, any PES has to exhibit the corresponding permutational symmetry. All the details of PES will repeat $M!$ times for a system with M identical nuclei. This will multiply the number of conical intersections.

More information about conical intersection will be given in Chapter 14, when we will be equipped with the theoretical tools to describe how the electronic structure changes during chemical reactions.

6.15 Beyond the Adiabatic Approximation

6.15.1 Vibronic Coupling

In polyatomic molecules, a diabatic state represents a product of an electronic wave function $\psi_i^{(\Gamma_1)}$ and a rovibrational function[78] $f_v^{(\Gamma_2)}$; i.e., a *rovibronic state*:

$$\psi_i^{(\Gamma_1)}(\boldsymbol{r};\boldsymbol{R})\, f_v^{(\Gamma_2)}(\boldsymbol{R}), \tag{6.51}$$

where the upper indices are related to the irreducible representations of the symmetry group of the clamped-nuclei Hamiltonian that the functions belong to (i.e., according to which the corresponding functions transform; see Appendix C available at booksite.elsevier.com/978-0-444-59436-5, p. e17). If one considers the electronic and the vibrational states only[79] Eq. (6.51) denotes a *vibronic state*. The product function transforms according to the direct product representation $\Gamma_1 \times \Gamma_2$.

[78] This function describes rotations and vibrations of the molecule.

[79] For the sake of simplicity, we are skipping the rotational wave function.

If one is interested in those solutions of the Schrödinger equation, which belong to the irreducible representation Γ, the function $\psi_i^{(\Gamma_1)}(\boldsymbol{r};\boldsymbol{R})f_v^{(\Gamma_2)}(\boldsymbol{R})$ is useful as a basis function only if $\Gamma_1 \times \Gamma_2$ contains Γ. For the same reason, another basis function may be useful $\psi_{i'}^{(\Gamma_3)}(\boldsymbol{r};\boldsymbol{R})f_{v'}^{(\Gamma_4)}(\boldsymbol{R})$, as well as other similar functions:

$$\psi^{(\Gamma)} = c_{i1}\psi_i^{(\Gamma_1)}\left(\boldsymbol{r};\boldsymbol{R}\right)f_1^{(\Gamma_2)}(\boldsymbol{R}) + c_{i'2}\psi_{i'}^{(\Gamma_3)}\left(\boldsymbol{r};\boldsymbol{R}\right)f_2^{(\Gamma_4)}(\boldsymbol{R}) + \cdots \tag{6.52}$$

If, say, coefficients c_{i1} and $c_{i'2}$ are large, an effective superposition of the two vibronic states is taking place, which is known as *vibronic coupling*.

We are, therefore, beyond the adiabatic approximation (which requires a single vibronic state, a product function) and the very notion of the single potential energy hypersurface for the motion of the nuclei becomes irrelevant. In the adiabatic approximation, the electronic wave function is computed from Eq. (6.8) with the clamped nuclei Hamiltonian; i.e., the electronic wave function does not depend on what the nuclei are *doing*, but only where they *are*. In other words, the electronic structure is determined [by finding a suitable $\psi_i^{(\Gamma_1)}(\boldsymbol{r};\boldsymbol{R})$ through solution of the Schrödinger equation] at fixed position $\boldsymbol{R}$ of the nuclei. This implies that in this approximation, the electrons always have enough time to adjust themselves to any instantaneous position of the nuclei. One may say that in a sense, the electrons and the nuclei are perfectly correlated in their motion: electrons follow the nuclei. Therefore,

a non-adiabatic behavior (or vibronic coupling) means a weakening of this perfect correlation, which is equivalent to saying that it may happen that the electrons do not have enough time to follow a (too-fast) motion of the nuclei.

This weakening is usually allowed by taking a linear combination of Eq. (6.52), which may be thought as a kind of frustration for electrons which vibration ("type of motion") of the nuclei to follow. If $\Gamma_1 \neq \Gamma_3$ and $\Gamma_2 \neq \Gamma_4$, one may say that we have to do with such an electronic state, which resembles $\psi_i^{(\Gamma_1)}$, when the molecule participates in a vibration of symmetry Γ_2 and resembles $\psi_{i'}^{(\Gamma_3)}$, when the molecule vibrates according to Γ_4.

This idea may be illustrated by the following examples.

Example 1: Dipole-Bound Electron

Imagine a molecular dipole. One may think of it as having a + and a − pole. We are interested in its + pole, because now we consider an extra electron, which will be bound with the dipole by the + pole-electron attraction. Obviously, such an attraction should depend on the dipole

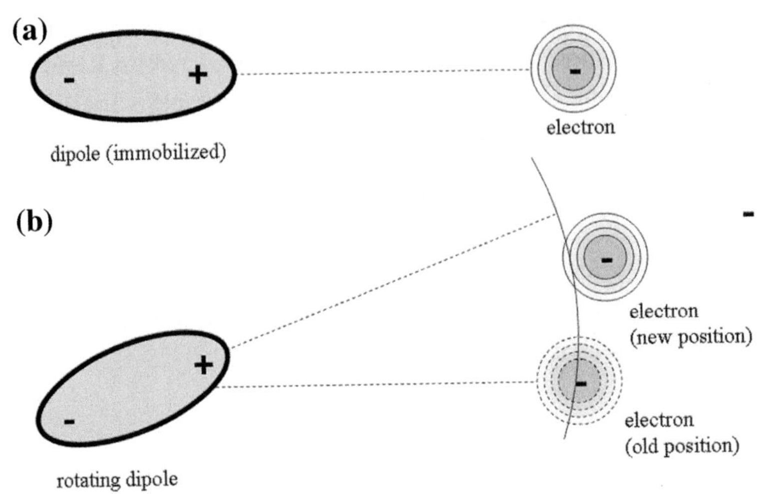

Fig. 6.20. A strange situation: An electron is unable to follow the motion of the nuclei (we are beyond the adiabatic approximation, a non-adiabatic case). (a) Some molecular dipoles with a sufficiently large dipole moment may bind an extra electron (a cloud on the right), which in such a case is far from the dipole and is attracted by its pole. The positive pole plays a role of a pseudonucleus for the extra electron. (b) When the dipole starts to rotate (a state with a nonzero angular momentum), the electron follows the motion of the pole. This is, however, difficult for high angular momenta (the electron has not enough time to adapt its position right toward the pole), and it is even harder because the centrifugal force pushes the extra electron farther away.

moment of the dipolar molecule. How strong must a point dipole be to be able to bind an electron? This question has been already asked, and the answer[80] is that this happens for the pointlike dipole moment larger than[81] 1.625 D. If the dipole itself represents an electronic closed shell molecule, the extra electron is usually very far (see Fig. 6.20a), even at distances of the order of 50 Å.

Now imagine the dipole starts to rotate (see Fig. 6.20b). At small angular momentum, the electron supposedly does not have any problem with following the motion of the positive pole. For larger angular momenta, the electron speeds up, its distance to the dipole increases due to the centrifugal force, and when this happens, it gets harder and harder to follow the motion of the positive pole. The electron does not have enough time. This means a larger and larger non-adiabatic correction.

Example 2: Hydrogen Molecule

Let us form two diabatic states: $\psi_i^{(\Gamma_1)}(\boldsymbol{r}; R)$ corresponding to the double occupation of the bonding orbital $1s_a + 1s_b$ and the other, $\psi_{i'}^{(\Gamma_3)}(\boldsymbol{r}; R)$, corresponding to the double occupation of the bonding excited orbital $2s_a + 2s_b$. In this case, we will take $\Gamma_3 = \Gamma_1$ (it does not mean

[80] E. Fermi and E. Teller, *Phys. Rev.*, *47*, 399 (1947).
[81] For non-pointlike dipoles, one may expect this limiting value to be less important, since the essence of the problem is binding an electron by a positive charge. This, however, happens even for marginally small positive charges (see the hydrogen-like atom).

the f functions are the same). The rovibrational function will be taken as (we assume the vibrational and rotational ground state) $f_1^{(\Gamma_2)}(\boldsymbol{R}) = \chi_1(R)Y_0^0(\theta,\phi) = \chi_1(R)$, but $f_2^{(\Gamma_4)}(\boldsymbol{R}) = \chi_2(R)Y_0^0(\theta,\phi) = \chi_2(R)$. The vibrational ground state $\chi_1(R)$ has its maximum at $R = R_{1s}$, where the minimum of the potential energy curve $E_{1s}(R)$ is, while $\chi_2(R)$ has its maximum at $R = R_{2s} > R_{1s}$, where the potential energy curve $E_{2s}(R)$ exhibits the minimum. The mixing coefficients c_{i1} and c_{i2} will obviously depend on R. For $R = R_{1s}$, we will have $c_{i1} \gg c_{i2}$, because the ground-state bonding orbital will describe well the electronic charge distribution, and for this R, the $2s_a+2s_b$ orbital will have a very high energy (the size of the $2s$ orbitals does not fit the distance). However, when R increases, the energy corresponding to $1s_a+1s_b$ will increase, while the energy corresponding $2s_a + 2s_b$ will decrease (because of better fitting). This will result in a more important value of $|c_{i2}|$ and a bit smaller value of $|c_{i1}|$ than it was for $R = R_{1s}$. There is, therefore, a coupling of vibration with the electronic state–a vibronic coupling.

Example 3: Harpooning Effect

The harpooning effect from p. 308 represents also an example of a vibronic coupling, if the two diabatic states: the ionic one $\psi_i^{(\Gamma_1)}(\boldsymbol{r}; R)$ and the neutral one $\psi_{i'}^{(\Gamma_3)}(\boldsymbol{r}; R)$ are considered with their corresponding vibrational states.

Example 4: Benzene

Let us take a benzene molecule. Chemists have realized for a long time that all CC bonds in this molecule are equivalent (some quantum chemical arguments for this view were presented on p. 167). The benzene molecule does not represent a static hexagonal object. The molecule undergoes $3M - 6 = 30$ vibrations (normal modes, which will be discussed in Chapter 7). One of these modes, say, described by the vibrational wave function $f_v^{(\Gamma_2)}(\boldsymbol{R})$, resembles a kind of ring pulsing ("*breathing*"), and during these vibrations, the electronic wave function $\psi_i^{(\Gamma_1)}(\boldsymbol{r}; \boldsymbol{R})$ describes the six equivalent CC bonds. There is also another vibrational mode corresponding the vibrational function $f_{v'}^{(\Gamma_4)}(\boldsymbol{R})$, that, in its certain phase, corresponds to shortening of the two opposite CC bonds and lengthening of the four other CC bonds. During such a motion, the electronic structure changes and will correspond to what is known as the *Dewar structure*:

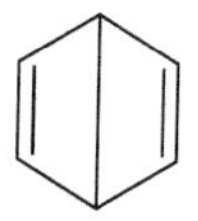

(the shortened bonds will resemble double bonds, and the others will resemble single bonds), corresponding to the electronic wave function $\psi_{i'}^{(\Gamma_3)}(\boldsymbol{r}; \boldsymbol{R})$. There will be much more such possibilities what is symbolized in Eq. (6.52) by "...".

The rovibrational functions $f_v^{(\Gamma_2)}(\boldsymbol{R})$ and $f_{v'}^{(\Gamma_4)}(\boldsymbol{R})$ must exhibit a strong asymmetry with respect to the equilibrium point (*vibronic anharmonicity*). Indeed, it is natural that the abovementioned Dewar structure is energetically favored for the vibrational deviations that shorten the to-be-double CC bond and becomes unfavorable for the opposite deviations.

6.15.2 Consequences for the Quest of Superconductors

Superconductivity, discovered by a Dutch scholar Heike Kammerlingh Onnes in 1911, is certainly a fascinating phenomenon. In some substances (like originally in mercury, tin, and lead), measurement of the electric conductivity as a function of lowering the temperature ended up by an abrupt decrease (below a critical temperature) of the electric resistance to zero value. Such a property would be great for operating technical devices or sending electric energy at large distances. The problem is that the critical temperature turned out to be extremely low–until 1987, it was always lower than about 23 K. The situation changed after discovery of what is now known as high-temperature superconductors (HTS) by J. George Bednorz and K. Alex Müller in 1987. Nowadays, after discovering hundreds of new HTSs, the highest critical temperature found is equal to about 164 K. In virtually all cases, it turned out that the HTSs have a characteristic atomic layer structure with alternating copper and oxygen atoms.

The "Magic" Cu–O Distance

No current theory explains properly the phenomena exhibited by HTSs. There are several theoretical concepts, but their striking weakness is that they provide no indication as to the class of promising materials that one should look for the HTS. After decades of research, an intriguing conclusion has been however found, that the closer the Cu–O distance to a "*magic value*" $R_{Cu-O} = 1.922$ Å is, the higher the corresponding critical temperature is.[82] This remarkable correlation went virtually unnoticed by the solid-state physics community for a long time.

How could such a precise criterion work? Well, this strongly suggests that something important happens at distance $R_{Cu-O} = 1.922$ Å, but for some reason, it does not when it is away from this value.

Primum non Nocere...[83]

Why does a bulb emit light? It happens because the motion of the electrons in a thin wire inside the bulb meets a resistance of chaotic vibrations of the nuclei. The kinetic energy of the electrons (resulting from the electric power plant operation for our money) goes partially for making collisions with the nuclei. These collisions lead to high-energy electronic states, which emit light when relaxing. In principle, this is why we pay our electric bill.

---

[82] C.N.R. Rao and A.K. Ganguli, *Chem. Soc. Review*, *24*, 1 (1995).

[83] "*First, do no harm*"–a phrase attributed to Greek physician Hippocrates (460–370 B.C.) as a suggested minimum standard for medical doctors.

And what if the nuclear motion, instead of interfering, helped electrons to move? Well, then the resistance would drop, just similarly as it does in superconductivity. Maybe there is something in it.

In 1993, Jeremy K. Burdett postulated some possible reason for superconductivity.[84] His hypothesis is related to a crossing of potential energy curves, precisely the subject of our earlier interest. According to Burdett, the "*magic Cu–O distance*" possibly corresponds to a crossing of two close-in-energy electronic diabatic states: a diabatic state characterized by the electron holes mainly on the copper atoms and another one with the electron holes mainly on the oxygen atoms. Thus, these two states differ by the electronic charge distribution[85], similarly as it was in NaCl (see p. 308). However, unlike as it was for NaCl, the minima, according to Burdett, do not differ much (if at all) in energy. Another important difference is that for NaCl, the crossing takes place for the Na-Cl distances that are several times larger than the nearest neighbor Na-Cl distance in the crystal of the rock salt, while for the HTSs, the Cu-O distance in crystals is close to the corresponding crossing point. This means that atomic vibrations may cause oscillating about the crossing point. As usually, from crossing of the diabatic curves, two adiabatic states appear: the ground state with the double minimum and an excited state (see Fig. 6.21).

Burdett's main point is the coupling of the ground vibrational state with the two diabatic electronic states. It is during such vibrations that a dramatic change of the electronic charge distribution is supposed to take place (strong *vibronic coupling*). The position of the vibrational level on the energy scale is said to be critical for superconductivity. If the position is substantially lower than the energy of the top of the barrier (Fig. 6.21b), one has to do with either of the two states localized in a given well. This corresponds to no communication between the wells, and we have to deal with either of the two different charge distributions (a "*mixed-valence*" compound). If, on the other hand, the vibrational level has large energy (Fig. 6.21c), high above the barrier top energy, one receives an averaged charge distribution, which does not change much during vibrations. According to Burdett, the superconductivity appears, when the vibrational level is close to the same energy as that of the top of the barrier (Fig. 6.21d), this causes a strong coupling of the two diabatic electronic states through the vibrational state.

Relevant Vibrations

It is natural to imagine that the electron transfer between two atoms (of type A) may be accomplished by a mediator–a third object, say atom B. This is why research began from studies of the effectivity of the transfer of an extra electron in the ABA^- system, like $Na^+\ F^-\ Na^0$ (i.e., $A \equiv Na^+$, $A^- \equiv Na^0$, $B \equiv F^-$), when B oscillates between atoms A. Therefore, the

[84] J.K. Burdett, *Inorg. Chem.*, *32,* 3915 (1993).

[85] Most probably, the key phenomena take place in the copper-oxygen layers and may be described as a reversible reaction $Cu^{3+} + O^{2-} \rightleftharpoons Cu^{2+\cdot} + O^{-\cdot}$ or $Cu^{2+\cdot} + O^{2-} \rightleftharpoons Cu^{+} + O^{-\cdot}$ (a dot means an unpaired electron). The presence of the unpaired electrons implies some ferro- or/and antiferromagnetic properties of these materials, which indeed have been discovered in the HTS phenomenon.

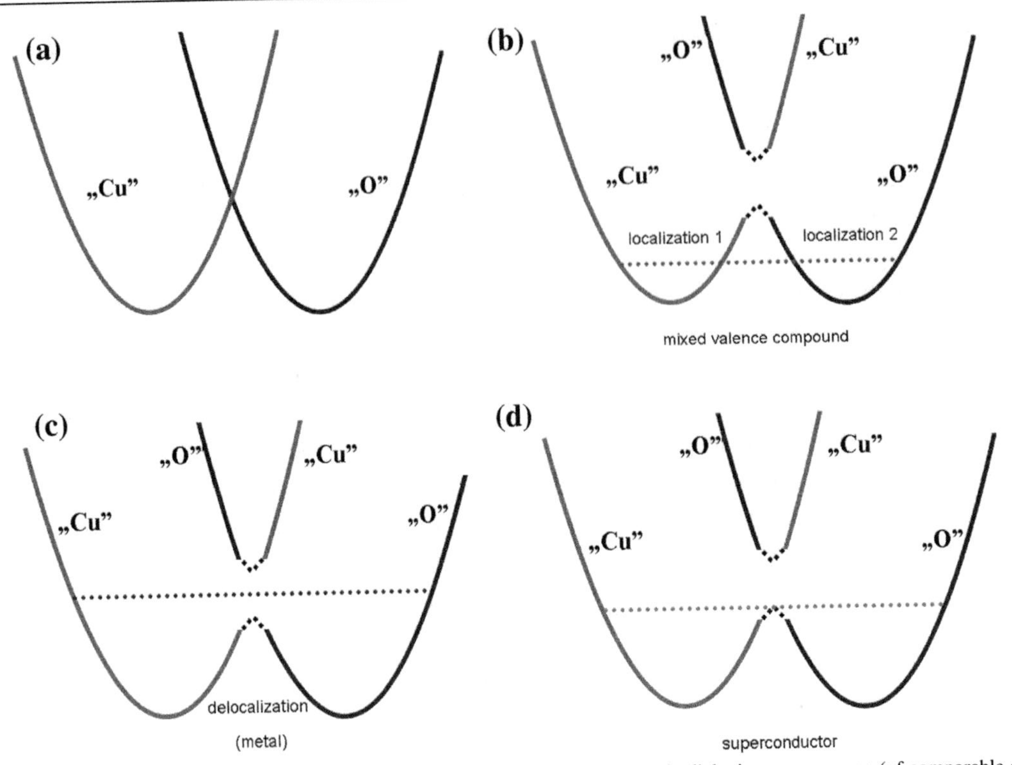

Fig. 6.21. Burdett's concept of superconductivity (scheme). (a) Two electronic diabatic energy curves (of comparable energies corresponding to their equilibrium positions) cross, resulting in two adiabatic energy curves (the ground and excited states, b,c,d). The diabatic states differ widely by the electronic charge distribution: one of them, denoted by the symbol Cu, corresponds to the electron holes on the copper atoms, while the second one, denoted by O, has such holes on the oxygen atoms. According to Burdett, the superconductivity has to do with the position of the lowest vibrational level of the ground electronic state. (b) The level is too low in energy, the vibrations are localized (either in the left- or in the right side potential energy well; this is equivalent to a quasi-degeneracy of the sum and difference of the delocalized vibrational states). The tunneling is marginal because of the exponential decay of the localized vibrational wave functions in the separating barrier. One has to do with an insulator in either of two coexisting states differing by the electronic charge distribution ("*oxidation states*")–what is known as a mixed valence compound. (c) The level is too high in energy, and the vibrations are fully delocalized and proceed in the global potential energy well. The well details do not count for much; one has to do with a state similar to averaging of the two states ("*a metallic state*"); i.e., both Cu and O have some averaged oxidation states when vibration occurs. (d) A "*magic*" position of the vibrational state, right at the height of the barrier. One may see this as two localized vibrational states that can tunnel easily through the barrier. The vibrations change the oxidation states of *Cu* and *O*; i.e., cause the electron transfer.

interesting vibration should be similar to an antisymmetric stretching vibration. In such a case, B transports an electron between the A centers. We may consider this vibration at various AA distances. If one assumes Burdett's concept, the following questions, related to the possible materials involved[86], appear:

- What would we get as the electronic charge distribution if we assume optimization of the AA distance (still keeping the constraint of linearity of ABA)? Would we get a symmetrization of

[86] Which type of chemical compounds are most promising HTSs?

the charge distribution, as in $Na^{+\frac{1}{2}}$ F^- $Na^{+\frac{1}{2}}$ ("*averaged oxidation state*"), or we would rather obtain an asymmetric distribution like Na^+ F^- Na^0 or Na^0 F^- Na^+ ("*mixed valence compound*").

- How do the above possibilities depend on the chemical character of A and B?

Where Can We Expect Superconductivity?

Well, we do not know the answer, but there are some indications. It turned out that in the vibronic coupling[87],

- Chemical identity of A and B is very important; the strongest vibronic coupling corresponds to halogens and hydrogen (A, B $\equiv$ F, Cl, Br, I, H).
- The strongest vibronic coupling corresponds to A=B (with the maximum for A, B $\equiv$ F), although this condition is not the most important one.
- To exhibit the electronic instability under oscillation of B in the ABA^- radical,
  - A and B must be strongly electronegative (this may explain why oxygen is present in all the HTSs).
  - A and B must form a strong covalent bond, whereas a large overlap of the corresponding orbitals[88] is more important than the equality of their energies (cf. p. 430, Chapter 8).

6.15.3 Photostability of Proteins and DNA

How does it happen that life flourishes under protection of the Sun, whereas it is well known the star emits some deadly radiation like charged particles and UV photons? We have two main protecting targets: one is Earth's magnetic field, and the second is Earth's atmosphere. Despite the atmospheric protection, some important part of the UV radiation attains the surface of the Earth. Substances usually are not transparent for the UV, whereas absorption of a UV photon is often harmful for chemical bonds, making their dissociation or/and creating other bonds. This is desirable for producing the vitamin D_3 in our body, but in many cases, it ends badly. For example, some substances, like DNA or some important proteins, have to be completely protected because their destruction would destroy the basis of life itself. Therefore, how do these substances function so efficiently in the vibrant life processes? What represents an additional target that protects them so well?

It turns out that this wonder target is the ubiquitous hydrogen bond, an important factor determining the 3-D shape of both DNA and proteins (see p. 870). The hydrogen bond $X - H \ldots Y$ (see p. 863) has some special features that also concern its UV properties, and this

[87] W. Grochala, R. Konecny, and R. Hoffmann, *Chem. Phys., 265,* 153 (2001); W. Grochala, R. Hoffmann, *New J. Chem., 25,* 108 (2001); W. Grochala and R. Hoffmann, *J. Phys. Chem. A, 104,* 9740 (2000); W. Grochala and R. Hoffmann, *Pol. J. Chem., 75,* 1603 (2001); W. Grochala, R. Hoffmann, and P.P. Edwards, *Chem. Eur. J., 9,* 575 (2003); W. Grochala, *J. Mater. Chem., 19,* 6949 (2009).

[88] For HTSs, these orbitals are $2p$ of oxygen and $3d$ of copper.

holds independently which electronegative atoms XY are involved. The explanation of the UV protection mechanism of the hydrogen bond given below comes from Sobolewski and Domcke.[89]

Fig. 6.22 shows three electronic energy hypersurfaces for the hydrogen bond $X - H \ldots Y$, visualized as sections along the proton position coordinate that describes the position of the proton in the hydrogen bridge. The ground state S_0 (light gray) represents a diabatic state corresponding to the resonance structure $X - H \ldots Y$, in which two electrons are at X and two at Y. The energy of this singlet state has a minimum for the proton position close to X. An absorption of the UV photon makes the transition of the system to the lowest-energy singlet excited state (^1LE, dark gray) with its electronic structure denoted as $(X - H \ldots Y)^*$. Most important, its energy curve intersects another singlet excited state, which corresponds to the electron transfer from X to Y (^1CT, black). The resulting conical intersection of the states ^1LE and ^1CT is

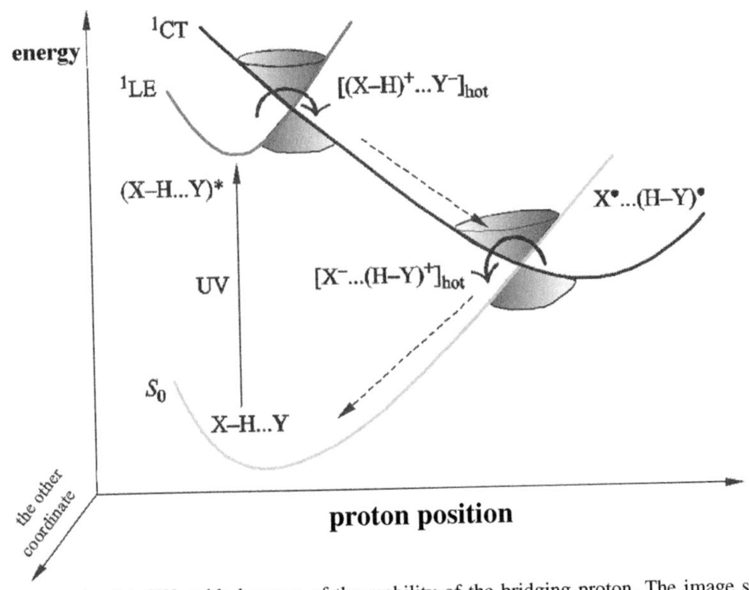

Fig. 6.22. The hydrogen bond is UV stable because of the mobility of the bridging proton. The image shows the electronic energy as a function of the proton position in the hydrogen bond $X - H \ldots Y$ (another coordinate that measures deviation of the proton from the XY axis is also marked). The electronic ground state S_0 energy curve (light gray) corresponds to the "*four-electron*" diabatic wave function corresponding to the bond pattern $X - H \ldots Y$. A UV transition to the lowest excited singlet state $(X - H \ldots Y)^*$ (^1LE, dark gray) is shown by a vertical arrow. The electronic energy curve for this state intersects (the conical intersection is shown as two cones) a singlet diabatic state (^1CT, black) that corresponds to a transfer of an electron from X to Y. The excitation energy is sufficiently large to allow the system to attain the black curve corresponding to the structure $(X - H)^+ \ldots Y^-$. After passing the conical intersection, one deals with a vibrationally excited state, which is symbolized by $\left[(X - H)^+ \ldots Y^-\right]_{hot}$. The proton continues its motion towards Y and the structure begins to be of the radical-ionic type: $X^\bullet \ldots (H - Y)^\bullet$. The system meets the second conical intersection, which allows it to attain the ground state (light gray). This time, the proton moves towards X, while its electronic energy changes to the vibrational energy of the molecule and the surrounding water. Thus, the UV photon does not harm chemical bonds, its energy goes instead to heating the surrounding water.

[89] A. Sobolewski and W. Domcke, *Chem. Phys. Chem.*, *7*, 561 (2006).

shown as a double cone. The photon energy is large enough that the system reaches the conical intersection point and ends up on the black energy curve, which means a single electron transfer symbolized by $(X - H)^+ \ldots Y^-$. Since the minimum of the black curve is shifted far to the right, after going out from the conical intersection, one has to do with a vibrational excited state denoted as an ionic "*hot*" structure: $\left[(X - H)^+ \ldots Y^-\right]_{hot}$. The system slides down the black curve changing high potential energy to the vibrational energy and kinetic energy of the surrounding water molecules. This sliding down means that as the proton moves to the right, the system remains all the time in the ionic state. This, however, means there is an ion-radical structure of the type: $X^{\bullet} \ldots (H - Y)^{\bullet}$. When sliding down, the system meets the second conical intersection, which makes it possible to continue the motion on the ground-state curve (light gray). At this value of the proton position, one has the "*hot*" structure. $[X^- \ldots H - Y^+]_{hot}$. Now the sliding down means going left (the proton comes back) and transferring the vibrational energy to the water.

Therefore, the net result is the following: the absorption of the UV photon, after some bouncing of the proton in the hydrogen bridge, results in heating the surrounding water, while the hydrogen bond stays safe in its ground state.

6.15.4 Muon-Catalyzed Nuclear Fusion

Some molecules look really peculiar. They may contain a muon instead of an electron. *A muon* is an unstable particle with the charge of an electron and mass equal to 207 electronic masses.[90] For such a mass, assuming that nuclei are infinitely heavier than muon looks like a very bad approximation. Therefore, the calculations need to be non-adiabatic. The first computations for muonic molecules were performed by Kołos, Roothaan, and Sack[91] in 1960. The idea behind the project was muon-catalyzed fusion of deuterium (d) and tritium (t); the abbreviations here pertain to the nuclei only. This fascinating problem was proposed by Andrei Sakharov. Its essence is as follows.

[90] The muon was discovered in 1937 by C.D. Anderson and S.H. Neddermeyer. Its lifetime is about $2.2 \cdot 10^{-6}$ s. The muons belong to the lepton family (with the electron and τ particle, the latter with a mass equal to about 3640 electronic masses). Nature created, for some unknown reasons, "more massive electrons". When the nuclear physicist Isidor Rabi was told about the incredible mass of the τ particle, he dramatically shouted: "*Who ordered that*?!".

[91] W. Kołos, C.C.J. Roothaan, R.A. Sack, *Rev. Mod. Phys.*, *32*, 205 (1960).

Andrei Dimitriy Sakharov (1921–1989) Russian physicist and father of the Soviet hydrogen bomb. During the final celebration of the H bomb project, Sakharov expressed his hope that the bombs would never be used. A Soviet general answered coldly that it was not scientists' business to decide such things. This was a turning point for Sakharov, and he began his fight against the totalitarian system.

The idea of muon-induced fusion was conceived by Sakharov in 1945, in his first scientific paper, under the supervision of Tamm. In 1957, David Jackson realized that muons may serve as catalysts.

If the electron in the molecule d*te* is replaced by a muon, immediately the dimension of the molecule decreases by a factor of about 200. How is this possible?

Well, the radius of the first Bohr orbit in the hydrogen atom (see, p. 202) is equal to $a_0 = \frac{\hbar^2}{\mu e^2}$. After introducing atomic units, this formula becomes $a_0 = \frac{1}{\mu}$, and when we take into account that the reduced mass $\mu \approx m$ (m stands for the electron mass), we get $a_0 \approx 1$. This approximation works for the electron because in reality, $\mu = 0.9995m$. If, in the hydrogen atom, we have a muon instead of an electron, then μ would equal about 250 m. This, however, means that such a "*muon Bohr radius*" would be about 250 times smaller. Nuclear forces begin to operate at such a small internuclear separation (strong interactions; see Fig. 6.23a), and are able to overcome the Coulombic barrier and stick the nuclei together by *nuclear fusion*. The muon, however, is released, and may serve as a catalyst in the next nuclear reaction.

Deuteron and tritium bound together represent a helium nucleus. One muon may participate in about 200–300 such muon-catalyzed fusion processes.[92] Everybody knows how much effort and money has been spent for decades (for the moment in vain) to ignite the nuclear synthesis $\mathrm{d} + \mathrm{t} \rightarrow \mathrm{He}$. Muon-catalyzed fusion might be an alternative solution. If the muon project were successful, humanity would have access to a practically unlimited source of energy. Unfortunately, theoretical investigations suggest that the experimental yield already achieved is about the maximum theoretical value.[93]

---

[92] The commercial viability of this process will not be an option unless we can demonstrate 900 fusion events for each muon. About 10 g of deuterium and 15 g of tritium fusion would then be sufficient to supply the average person with electricity for life.

[93] This has been the subject of a joint Polish-American project. More about this may be found in K. Szalewicz, S. Alexander, P. Froelich, S. Haywood, B. Jeziorski, W. Kołos, H.J. Monkhorst, A. Scrinzi, C. Stodden, A. Velenik, and X. Zhao, in *Muon Catalyzed Fusion*, eds. S.E. Jones, J. Rafelski, H.J. Monkhorst, AIP Conference Proceedings, 181, 254 (1989).

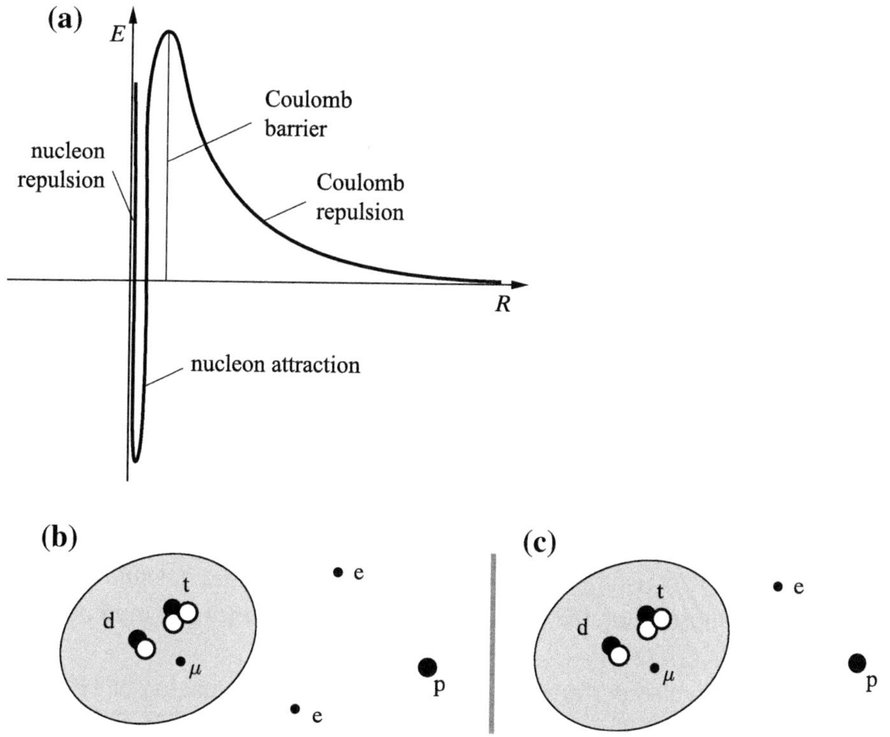

Fig. 6.23. (a) The interaction energy potential of d and t as a function of the interparticle distance (R), taking the nuclear forces into account (an outline). At large R, of the order of nanometers, we have Coulombic repulsion, at distances of the order of femtometers the internuclear attractive forces (called the *strong interaction*) are switched on and overcome the Coulombic repulsion. At a distance of a fraction of a femtometer, again we have a repulsion. (b) "*Russian dolls*" (outline): the analogs of H_2 and H_2^+.

6.15.5 "Russian Dolls," or a Molecule Within Molecule

Scrinzi and Szalewicz[94] carried out non-adiabatic calculations (p. 265) for a system of six particles: proton (p), deuterium (d), tritium (t), muon (μ), and two electrons (e) interacting by Coulombic forces (i.e., no nuclear forces are assumed). It is not easy to predict the structure of the system. It turned out that the resulting structure is a kind of "Russian doll"[95] (see Fig. 6.23b): the muon has acted according to its mass (see above) and created tdμ with a dimension of about 0.02 Å. This system may be viewed as a partly split nucleus of charge +1 or, alternatively, as a mini-model of the hydrogen molecular ion (scaled at 1:200). The "nucleus" serves as a partner to the proton, and both create a system similar to the hydrogen molecule, in which the two electrons play their usual binding role and the internuclear distance is about 0.7 Å. It turns out that the nonzero dimension of the "nucleus" makes a difference, and the energies computed

[94] A. Scrinzi and K. Szalewicz, *Phys. Rev.* A, *39*, 4983 (1989).
[95] (((woman @ woman)@ woman)@)

with and without an approximation of the pointlike nucleus differ. The difference is tiny (about 0.20 meV), but it is there.

It is quite remarkable that such small effects are responsible for the fate of the total system. The authors report that the relaxation of the "nucleus" $dt\mu$ (from the excited state to the ground state[96]) causes the ionization of the system: one of the electrons flies off. Such an effect, however, may excite those who study this phenomenon. How is it possible? The "nucleus" is terribly small when seen by an electron orbiting far away. How could the electron detect that the nucleus has changed its state and that it has no future in the molecule? Here, however, our intuition fails. For the electron, the most frequently visited regions of the molecule are nuclei. We will see this in Chapter 8 (p. 444), but even the 1s state of the hydrogen atom (the maximum of the orbital is at the nucleus; see p. 201) suggests the same. Therefore, no wonder the electron *could* recognize that something has abruptly changed on one of the nuclei and (being already excited) it received much more freedom–so much, in fact, that it could leave the molecule.

We may pose an interesting question: Does the "Russian doll" represent the global minimum of the particle system? We may imagine that the proton changes its position with the deuterium or tritium; i.e., new isomers (isotopomers[97]) appear. The authors did not study this question[98], but they investigated substituting the proton with deuterium and tritium (and obtained similar results).

Scrinzi and Szalewicz also performed some calculations for an analog of H_2^+: proton, deuterium, tritium, muon, and electron. Here, the "Russian doll" looks wonderful (Fig. 6.23c); it is a four-level object:

- The molecular ion (the analog of H_2^+) is composed of *three* objects: the proton, the "split nucleus" of charge +1 and the electron.
- The "split nucleus" is also composed of *three* objects: d,t,μ (a mini-model of H_2^+).
- The tritium is composed of *three* nucleons: the proton and the two neutrons.
- Each of the nucleons is composed of *three* quarks (called the valence quarks).

---

96 A. Scrinzi and K. Szalewicz, *Phys. Rev. A.*, *39*, 2855 (1989). The $dt\mu$ ion is created in the rovibrational state $J = 1, v = 1$, and then the system spontaneously goes to the lower energy 01 or 00 state. The energy excess causes one electron to leave the system (ionization). This is an analog of the Auger effect in spectroscopy.

97 The situation is quite typical, although we very rarely think this way. Some people say that they observe *two different systems*, whereas others say that they see *two states of the same system*. This begins with the hydrogen atom–it looks different in its $1s$ and $3p_z$ states. We can easily distinguish two different conformations of cyclohexane, two isomers of butane, and some chemists would say these are different substances. Going much further, N_2 and CO represent two different molecules, or is one of them nothing but an excited state of the other? However strange it may sound for a chemist, N_2 represents an excited state of CO because we may imagine a nuclear reaction of the displacement of a proton from one nitrogen to the other (and the energy curve per nucleon as a function of the atomic mass is convex). Such a point of view is better for viewing each object as a "new animal": it enables us to see and use some relations among these animals.

98 They focused their attention on $td\mu$.

Summary

- In the adiabatic and the Born-Oppenheimer approximations, the total wave function is taken as a product $\Psi = \psi_k(\mathbf{r}; R) f_k(\mathbf{R})$ of the function $f_k(\mathbf{R})$, which describes the motion of the nuclei (vibrations and rotations) and the function $\psi_k(\mathbf{r}; R)$ that pertains to the motion of electrons (and depends parametrically on the configuration of the nuclei; here, we give the formulas for a diatomic molecule). This approximation relies on the fact that the nuclei are thousands of times heavier than the electrons.
- The function $\psi_k(\mathbf{r}; R)$ represents an eigenfunction of the electronic Hamiltonian $\hat{H}_0(R)$; i.e., the Hamiltonian $\hat{H}$, in which the kinetic energy operator for the nuclei is assumed to be zero (the *clamped nuclei Hamiltonian*)
- The eigenvalue of the clamped nuclei Hamiltonian depends on positions of the nuclei and in the Born-Oppenheimer approximation, it is mass-independent. This energy as a function of the configuration of the nuclei represents the potential energy for the motion of the nuclei (Potential Energy Surface, or PES).
- The function $f_k(\mathbf{R})$ is a product of a spherical harmonic[99] Y_J^M that describes the rotations of the molecule (J and M stand for the corresponding quantum numbers) and a function that describes the vibrations of the nuclei.
- The diagram of the energy levels shown in Fig. 6.4 represents the basis of molecular spectroscopy. The diagram may be summarized in the following way:
  - The energy levels form some series separated by energy gaps, with no discrete levels. Each series corresponds to a single electronic state k, and the individual levels pertain to various vibrational and rotational states of the molecule in electronic state k.
  - Within the series for a given electronic state, there are groups of energy levels, each group characterized by a distinct vibrational quantum number ($v = 0, 1, 2, \ldots$), and within the group, the states of higher and higher energies correspond to the increasing rotational quantum number J.
  - The energy levels fulfill some general relations:
    * Increasing k corresponds to an electronic excitation of the molecule (UV-VIS, ultraviolet and visible spectrum).
    * Increasing v pertains to a vibrational excitation of the molecule, and requires the energy to be smaller by one or two orders of magnitude than an electronic excitation (IR, infrared spectrum).
    * Increasing J is associated with energy smaller by one or two orders of magnitude than a vibrational excitation (microwaves).
- Above the dissociation limit, one is dealing with a continuum of states of the dissociation products with kinetic energy. In such a continuum, one may have also the resonance states, which may have wave functions that resemble those of stationary states but differ from them by having finite lifetimes.
- The electronic wave functions $\psi_k(\mathbf{r}; R)$ correspond to the energy eigenstates $E_k^0(R)$, which are functions of R. The energy curves[100] $E_k^0(R)$ for different electronic states k may cross each other, unless the molecule is diatomic and the two electronic states have the same symmetry.[101] In such a case, we have what is known as an avoided crossing (see Figs. 6.15 and 6.16).
- The adiabatic states represent the eigenfunctions of $\hat{H}_0(R)$. If electrons have enough time to follow the nuclei, we may apply the adiabatic function (which may change its chemical character when varying R). The diabatic states are not the eigenfunctions of $\hat{H}_0(R)$ and preserve their chemical character when changing R. If electrons are too slow to follow the nuclei, changing R may result in keeping the same chemical character of the solution (diabatic state). In the adiabatic and diabatic approaches, the motion of the nuclei is described using a single PES.
- The non-adiabatic approach requires using several or many PESs when describing motion of the nuclei. The total wave function is a linear combination of the rovibronic functions with different and R-dependent amplitudes.

[99] This refers to the eigenfunction for the rigid rotator.

[100] These curves are expressed as functions of R.

[101] That is, they transform according to the same irreducible representation.

- For polyatomic molecules, the energy hypersurfaces $E_k^0(\boldsymbol{R})$ can cross. The most important is the *conical intersection* (Fig. 6.19) of the two (I and II) diabatic hypersurfaces; i.e., those that (each individually) preserve a given pattern of chemical bonds. This intersection results in two adiabatic hypersurfaces ("*lower PES and upper PES*"). Each of the adiabatic hypersurfaces consists of two parts: one belonging to I and the second to II. Using a suitable coordinate system in the configurational space, we obtain the adiabatic hypersurface splitting (the difference of E_- and E_+) when changing two coordinates (ξ_1 *and* ξ_2) only (the branching plane). The splitting begins by a linear dependence on ξ_1 and ξ_2, which gives a sort of cone (hence the name *conical intersection*). The other coordinates (the seam space) alone are unable to cause the splitting, although they may influence the opening angle of the cone.
- Conical intersection plays a prominent role in the photochemical reactions because the excited molecule slides down the upper adiabatic hypersurface to the funnel (just the conical intersection point) and then, with a yield close to 100%, lands on the lower adiabatic hypersurface (assuming that there is a mechanism for dissipation of the excess energy).
- The vibronic effects are the basis of many important phenomena.

Main Concepts, New Terms

From the Research Front

For the hydrogen molecule, one may currently get a very high accuracy in predicting rovibrational levels. For example, exact analytic formulas have been derived[102] that allow one to compute the Born-Oppenheimer potential with the uncertainty smaller than 10^{-9} cm^{-1} and add the correction for the nonzero size of each nucleus (the latter correction shifts the rovibrational energy levels by less than 10^{-4} cm^{-1} in all cases). The approach presented on p. 275 is able to produce the adiabatic diagonal correction, and the non-adiabatic corrections for all rovibrational states of the ground electronic state with the accuracy better than 10^{-4} cm^{-1}. One is able to test the accuracy of not only the theory given in this chapter, but also of its most sophisticated extensions, including quantum electrodynamics (QED). One may say that virtually for the first time, QED can be confronted with the most accurate experiments beyond the traditional territory of the free electron and simple atoms (hydrogen, helium, lithium); i.e., for systems with more than one nucleus. For the hydrogen molecule, one starts with an accurate solution to the Schrödinger equation[103] and then, circumventing the Dirac equation, one includes all the relativistic Breit-Pauli terms [all terms of the order of $\left(\frac{1}{c}\right)^2$, the terms of the order of $\frac{1}{c}$ vanish, where 137.0359991 a.u.] and later, the complete QED corrections of the order of $\left(\frac{1}{c}\right)^3$ and the leading terms of $\left(\frac{1}{c}\right)^4$. Just to show the accuracy achieved for the hydrogen molecule, for the $J = 0 \rightarrow 1$ rotational excitation, the theory gives[104] 118.486812(9) cm^{-1}, while the most accurate experiment to date[105] gives 118.48684(10) cm^{-1}. Some theories trying to explain the presence of black matter need the nuclear forces operating at larger distances than they are traditionally believed to do. If these theories were true, there would be no such agreement between the theory and experiment, and we would see a larger difference.

Ad Futurum

The computational effort needed to calculate the PES for an M atomic molecule is proportional to 10^{3M-6}. This strong dependence suggests that, for the next 20 years, it would be unrealistic to expect high-quality PES computations for $M > 10$. However, experimental chemistry offers high-precision results for molecules with hundreds of atoms. It seems inevitable that it will be possible to freeze the coordinates of many atoms. There are good reasons for such an approach: indeed, most atoms play the role of spectators in chemical processes. It may be that limiting ourselves to, say, 10 atoms will make the computation of rovibrational spectra feasible.

Additional Literature

J. Hinze, A. Alijah, and L. Wolniewicz, "*Understanding the adiabatic approximation; the accurate data of H_2 transferred to H_3*", *Pol. J. Chem.*, *72,* 1293 (1998).

The paper reports the derivation of the equation of motion for a polyatomic molecule. As the origin of the BFCS, unlike in this chapter, the center of mass was chosen[106].

W. Kołos, "*Adiabatic approximation and its accuracy,*" *Advan. Quantum Chem.*, *5,* 99 (1970).

---

[102] K. Pachucki, *Phys. Rev. A*, *82,* 032509 (2010).

[103] The center of mass rests at the origin. The solution of the Schrödinger equation is achieved numerically; i.e., the non-adiabatic treatment is applied with very high and controlled accuracy.

[104] J. Komasa, K. Piszczatowski, G. Łach, M. Przybytek, B. Jeziorski, and K. Pachucki, *J. Chem. Theor. Comput.*, *7,* 3105 (2011).

[105] D.E. Jennings, S.L. Bragg, and J.W. Brault, *Astrophys. J.*, *282,* L85 (1984). The uncertainty in parentheses is given in the units of the last digit reported.

[106] We have chosen the center of the ab bond.

F. Bernardi, M. Olivucci, and M. A. Robb, "*Potential energy surface crossings in organic photochemistry*," *Chem. Soc. Rev.* 321–328 (1996).

W. Domcke, D. R.Yarkony, and H. Köppel (eds.), "*Conical intersections: Electronic structure, dynamics, and spectroscopy*," *Advanced Series in Physical Chemistry*, Vol. 15, World Scientific Publishing, Singapore (2004).

Questions

1. The non-adiabatic theory for a diatomic ($\mathbf{r}$ denotes the electronic coordinates, $\mathbf{R}$ stands for the vector connecting nucleus b with nucleus a, $R \equiv |\mathbf{R}|$, N means the number of electrons, m is the electron mass, V represents the Coulombic interaction of all particles, μ is the reduced mass of the two nuclei of masses M_a and M_b).
   a. the total wave function can be represented as $\Psi(\mathbf{r}, \mathbf{R}) = \sum_k \psi_k(\mathbf{r}; R) f_k(\mathbf{R})$, where the functions ψ_k form a complete set in the Hilbert space for electrons (at a given R), and f_k are the coefficients depending on $\mathbf{R}$
   b. in the expression $\Psi(\mathbf{r}, \mathbf{R}) = \sum_k \psi_k(\mathbf{r}; R) f_k(\mathbf{R})$ the functions $f_k(\mathbf{R})$ describe rotations and vibrations of the molecule
   c. as functions $\psi_k(\mathbf{r}; R)$ one may assume the eigenfunctions of the electronic Hamiltonian
   d. may provide only some approximation of the solution to the Schrödinger equation
2. Adiabatic approximation (notation as in question 1).
   a. is also known as the Born-Oppenheimer approximation
   b. the electronic Hamiltonian can be obtained from the total Hamiltonian by neglecting the kinetic energy operator for the nuclei
   c. in the adiabatic approximation the total wave function represents a product $\psi_k(\mathbf{r}; \mathbf{R}) f_k(\mathbf{R})$, where $\psi_k(\mathbf{r}; \mathbf{R})$ stands for the eigenfunction of the clamped nuclei Hamiltonian for the configuration of the nuclei given by $\mathbf{R}$, while $f_k(\mathbf{R})$ denotes the wave function for the motion of the nuclei
   d. E_k^0 as a function of $\mathbf{R}$ represents the eigenvalue of the clamped nuclei Hamiltonian that corresponds to the wave function $\psi_k(\mathbf{r}; \mathbf{R})$.
3. A diatomic in the adiabatic approximation, the origin of the coordinate system is in the geometric center of the molecule (at $\mathbf{R}/2$). The nuclei vibrate in the potential:
   a. $E_k^0(R) + J(J+1)\frac{\hbar^2}{2\mu R^2}$
   b. $\left\langle \psi_k | \hat{H} \psi_k \right\rangle + (2J+1)\frac{\hbar^2}{2\mu R^2}$
   c. $E_k^0(R) + H'_{kk} + J(J+1)\frac{\hbar^2}{2\mu R^2}$
   d. $\left\langle \psi_k | \hat{H} \psi_k \right\rangle + J(J+1)\frac{\hbar^2}{2\mu R^2}$
4. The potential energy curves for the motion of the nuclei for electronic states computed at the Born-Oppenheimer approximation for diatomics
   a. may not intersect
   b. have to intersect at an internuclear distance
   c. cannot intersect, if the corresponding eigenfunctions belong to the same irreducible representation of the symmetry group of the Hamiltonian
   d. may intersect, if the corresponding wave functions are of different symmetry.
5. The potential energy for the motion of the nuclei in the Born-Oppenheimer approximation:
   a. contains the eigenvalue of the clamped nuclei Hamiltonian
   b. as a function of the configuration of the nuclei may exhibit many minima
   c. contains the electronic energy
   d. does not change after rotational excitations

6. Due to the rotational excitation $J \rightarrow (J+1)$ of a diatomic of bond length R
   a. one has to add to the potential energy a term proportional to $(2J+1)R^2$
   b. the potential energy for vibrations changes
   c. the molecule may dissociate due to the centrifugal force
   d. the momentum of the molecule increases
7. The adiabatic approximation
   a. takes into account the finite mass of the nuclei
   b. means the total wave function being a product of the electronic wave function and a wave function that describes the motion of the nuclei
   c. as a consequence leads to the concept of a spatial shape of a molecule
   d. is better satisfied by a molecule with muons instead of electrons.
8. Basics of spectroscopy within the Born-Oppenheimer approximation.
   a. the electronic structure changes after absorbing microwaves
   b. to excite vibrational levels (preserving the electronic state) one needs the IR radiation
   c. a red sweater witnesses about a dye that absorbs red light
   d. microwaves can excite rotations of polar molecules.
9. At the conical intersection, the following directions in the space of the nuclear configurations make splitting of E_+ and E_-
   a. $\nabla(\bar{E}_1 - \bar{E}_2)$ and $\nabla(V_{12})$
   b. $\nabla(\frac{\bar{E}_1+\bar{E}_2}{2})$ and $\nabla(V_{12})$
   c. $\nabla(\frac{\bar{E}_1-\bar{E}_2}{2})$ and $\nabla(V_{12})$
   d. any direction in the branching space.
10. At the conical intersection the opening angle of the cone
    a. equals zero
    b. in general differs along the directions of $\nabla(\bar{E}_1 - \bar{E}_2)$ and $\nabla(V_{12})$
    c. depends on the point of the seam space
    d. in the Born-Oppenheimer approximation is the same for different isotopomers.

Answers

1a,b,c, 2b,c,d, 3c,d, 4a,c,d, 5a,b,c 6b,c, 7b,c, 8b,d, 9a,c,d 10b,c,d

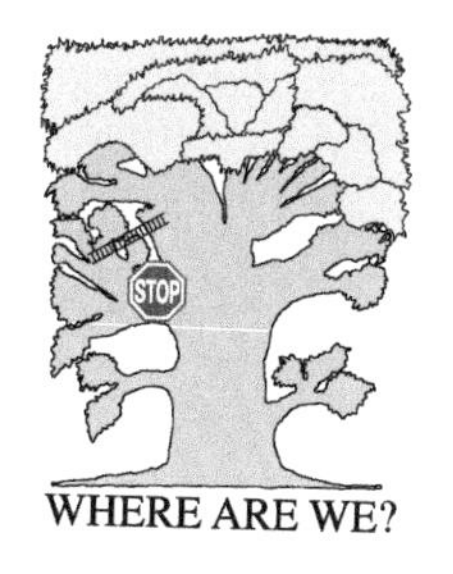

CHAPTER 7

Motion of Nuclei

"If you are out to describe the truth, leave elegance to the tailor."
Albert Einstein

Where Are We?

We are on the most important side branch of the TREE.

An Example

Which of the conformations shown in Fig. 7.1 is more stable: the "*boat*" or "*chair*" of cyclohexane C_6H_{12}? How do particular conformations look in detail (symmetry, interatomic distances, bond angles), when the electronic energy as a function of the positions of the nuclei attains a minimum value? How will the boat and chair conformations change if one of the hydrogen atoms is replaced by the benzyl substituent C_6H_5–? What is the most stable conformation of the trimer $C_6H_{11}-(CH_2)_3-C_6H_{10}-(CH_2)_3-C_6H_{11}$?

What Is It All About?

Ideas of Quantum Chemistry, Second Edition. http://dx.doi.org/10.1016/B978-0-444-59436-5.00007-6

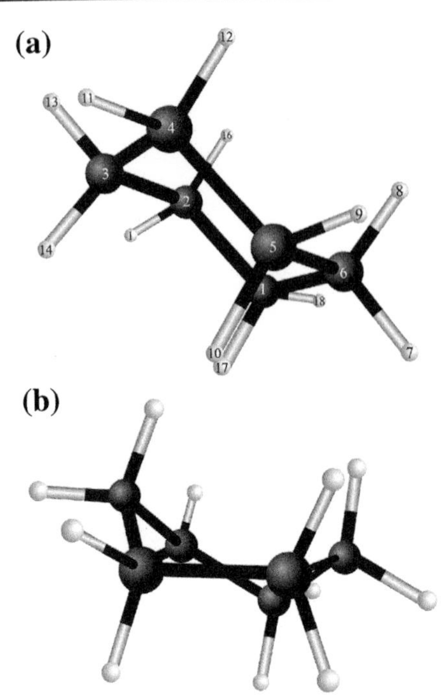

Fig. 7.1. The chair (a) and boat (b) conformations of cyclohexane. These geometries (obtained from arbitrary starting conformations) are optimized in the force field, which we will define in this chapter. The force field indicates, in accordance with experimental results, that the chair conformation is the more stable by about 5.9 kcal/mol. Thus, we obtain all the details of the atomic positions (bond lengths, bond angles, etc.). Note that the chair conformation obtained exhibits D_{3d} symmetry, while the boat conformation corresponds to D_2 (the boat has somewhat warped planks because of repulsion of the two upper hydrogen atoms).

As shown in Chapter 6, the solution of the Schrödinger equation in the Born-Oppenheimer approximation can be divided into two tasks: the problem of electronic motion in the field of the clamped nuclei (this will be the subject of the next chapters) and the problem of *nuclear motion in potential energy determined by electronic energy.*

> The ground-state electronic energy $E_k^0(\boldsymbol{R})$ of Eq. (6.8) (where $k = 0$ means the ground state) will be denoted in short as $V(\boldsymbol{R})$, where $\boldsymbol{R}$ represents the vector of the nuclear positions.

The function $V(\boldsymbol{R})$ has quite a complex structure and exhibits many basins of stable conformations (as well as many maxima and saddle points).

The problem of the shape of $V(\boldsymbol{R})$, as well as of the nuclear motion on the $V(\boldsymbol{R})$ potential energy hypersurface, will be the subject of this chapter. It will be seen that electronic energy can be computed with high accuracy as a function of $\boldsymbol{R}$ only for very simple systems (such as an atom plus a diatomic molecule system), for which quite a lot of detailed information can be obtained.

In practice, for large molecules, we are limited to only some approximations to $V(\boldsymbol{R})$ called *force fields*. After accepting such an approximation, we encounter the problem of optimization of the positions of the nuclei; i.e., of obtaining the most stable molecular conformation (or configuration[1]). The geometry of such a conformation is usually identified with a minimum on the electronic energy hypersurface, playing the role of the potential energy for the nuclei. Finding the stable conformation from a starting geometry of the nuclear framework is the subject of *local molecular mechanics*. In practice, we have the problem of having a huge number of such minima. The real challenge in such a case is finding the *most* stable structure, usually corresponding to the global minimum (*global molecular mechanics*) of $V(\boldsymbol{R})$.

Molecular mechanics does not deal with nuclear motion as a function of time, as well as with the kinetic energy of the system (related to its temperature). This is the subject of molecular dynamics, which means solving the Newton equation of motion for all the nuclei of the system interacting through potential energy $V(\boldsymbol{R})$. Various approaches to this question (of general importance) will be presented at the end of the chapter.

Why Is This Important?

The spatial structure of molecules in atomic resolution represents the most important information that decides about the chemical and physical properties of substances. Such key information is offered by a few experimental methods only: X-ray diffraction analysis, the neutron diffraction analysis and the nuclear magnetic resonance technique. Both types of diffraction analysis require crystals of good quality. *Theory not only offers a much less expensive alternative (by minimizing $E_0^0(\boldsymbol{R}) \equiv V(\boldsymbol{R})$), but in addition, it often reveals many additional structural details that explain the experimental results.*

What Is Needed?

- Laplacian in spherical coordinates (Appendix H available at booksite.elsevier.com/978-0-444-59436-5, p. e91, recommended)
- Angular momentum operator and spherical harmonics (Chapter 4, recommended)
- Harmonic oscillator (p. 186)
- Ritz method (Appendix L available at booksite.elsevier.com/978-0-444-59436-5, p. e107)
- Matrix diagonalization (Appendix K available at booksite.elsevier.com/978-0-444-59436-5, p. e105)
- Newton equation of motion
- Chapter 8 (an exception: the Car-Parrinello method needs some results which will be given in Chapter 8, marginally important)
- Entropy, free energy, sum of states

Classical Works

There is no more classical work on dynamics than Sir Issac Newton's monumental *Philosophiae Naturalis Principia Mathematica*, published in 1687 by Cambridge University Press. ★ The idea of the force field was first presented by

[1] Two *conformations* correspond to the same pattern of chemical bonds and differ by rotations of the fragments about some of these bonds. On the contrary, going from one *configuration* to another may mean changing the chemical bonds.

Mordechai Bixon and Shneior Lifson in an article called "*Potential functions and conformations in cycloalkanes*," *Tetrahedron 23*, 769 (1967). ★ The paper by Berni Julian Alder and Thomas Everett Wainwright, "*Phase transition for a hard sphere system*," *Journal of Chemical Physics*, *27*, 1208 (1957), is treated as the beginning of the molecular dynamics. ★ The work by Aneesur Rahman, "*Correlations in the motion of atoms in liquid argon*," published in *Physical Review*, *A136*, 405 (1964), used for the first time a realistic interatomic potential (for 864 atoms). ★ The molecular dynamics of a small protein was first described in a paper by Andy McCammon, Bruce Gelin, and Martin Karplus called "*Dynamics of folded proteins*," *Nature*, 267, 585 (1977). ★ The simulated annealing method is believed to have been used first by Scott Kirkpatrick, Charles D. Gellat, and Mario P. Vecchi in "*Optimization by simulated annealing*," *Science*, *220*, 671 (1983). ★ The Metropolis criterion for the choice of the current configuration in the Monte Carlo method was given by Nicolas Constantine Metropolis, Arianna W. Rosenbluth, Marshal N. Rosenbluth, Augusta H. Teller, and Edward Teller in "*Equations of state calculations by fast-computing machines*," in *Journal of Chemical Physics*, *21*, 1087 (1953). ★ The Monte Carlo method was used first by Enrico Fermi, John R. Pasta, and Stanisław Marcin Ulam during their time at Los Alamos (*Studies of Nonlinear Problems*, vol. 1, *Los Alamos Reports*, LA-1940). Ulam is also the discoverer of cellular automata.

Isaac Newton (1643–1727), English physicist, astronomer and mathematician, and professor at Cambridge University. In 1672, he became a member of the Royal Society of London, and in 1699, he became the director of the Royal Mint, who was said to be merciless to the forgers. In 1705, Newton became a Lord. In the *opus magnum* mentioned above, he developed the notions of space, time, mass, and force, gave three principles of dynamics and the law of gravity and showed that the latter pertains to problems that differ enormously in their scale (e.g., the famous apple and the planets). Newton is also a founder of differential and integral calculus (independently from G.W. Leibnitz).

In addition, Newton made some fundamental discoveries in optics. Among other things, he was the first to think that light is composed of particles. The first portrait of Newton (at the age of 46) by Godfrey Kneller is shown here. Readers are encouraged to consult the excellent book *Isaac Newton*, by G.E. Christianson (Oxford University Press, New York, 1996).

7.1 Rovibrational Spectra–An Example of Accurate Calculations: Atom–Diatomic Molecule

One of the consequences of adiabatic approximation is the potential energy hypersurface $V(\boldsymbol{R})$ for the motion of nuclei. To obtain the wave function for the motion of nuclei (and then to construct the total product-like wave function for the motion of electrons and nuclei), we have to solve the Schrödinger equation with $V(\boldsymbol{R})$ as the potential energy. This is what this hypersurface is for. We will find rovibrational energy levels and the corresponding wave functions, which will allow us to obtain rovibrational spectra (frequencies and intensities) to compare with experimental results.

7.1.1 Coordinate System and Hamiltonian

Let us consider a diatomic molecule AB plus a weakly interacting atom C (e.g., H–H $\cdots$ Ar or CO $\cdots$ He), the total system in its electronic ground state. Let us center the origin of the body-fixed coordinate system[2] (with the axes oriented as in the space-fixed coordinate system; see Appendix I available at booksite.elsevier.com/978-0-444-59436-5, p. e93) in the center of mass of AB. The problem, therefore, involves $3 \times 3 - 3 = 6$ dimensions.

However strange it may sound, six is too much for contemporary (otherwise impressive) computer techniques. Let us subtract one dimension by assuming that no vibrations of AB occur (rigid rotator). The five-dimensional problem becomes manageable. The assumption about the stiffness of AB now also pays off because we immediately exclude two possible chemical reactions C + AB $\rightarrow$ CA + B and C+AB $\rightarrow$ CB+A, and therefore admit a limited set of nuclear configurations–only those that correspond to a weakly bound complex C + AB. This approximation is expected to work better when the AB molecule is stiffer; i.e., it has a larger force constant (and therefore a larger vibration frequency).[3]

Carl Gustav Jacob Jacobi (1804–1851), German mathematical genius and the son of a banker, graduated from school at the age of 12, professor at Koenigsberg University. Jacobi made important contributions to number theory, elliptic functions, partial differential equations, and analytical mechanics. The crater Jacobi on the Moon is named after him.

We will introduce the *Jacobi coordinates* (Fig 7.2; cf. p. 897): three components of vector $\boldsymbol{R}$ pointing to C from the origin of the coordinate system (the length R and angles Θ and Φ, and both angles denoted by $\hat{\boldsymbol{R}}$) and the angles θ, ϕ show the orientation $\hat{\boldsymbol{r}}$ of vector $\boldsymbol{r} = \overrightarrow{AB}$. All together there are five coordinates–as there should be.

Now let us write down the Hamiltonian for the motion of the nuclei in the Jacobi coordinate system (with the stiff AB molecule with AB equilibrium distance equal to r_{eq})[4]:

$$\hat{H} = -\frac{\hbar^2}{2\mu R^2}\frac{d}{dR}R^2\frac{d}{dR} + \frac{\hat{l}^2}{2\mu R^2} + \frac{\hat{j}^2}{2\mu_{AB} r_{eq}^2} + V,$$

[2] Any coordinate system is equally good from the point of view of mathematics, but its particular choice may make the solution easy or difficult. In the case of a weak C $\cdots$ AB interaction (the current case), the proposed choice of the origin is one of the natural ones.

[3] A certain measure of this might be the ratio of the dissociation energy of AB to the dissociation energy of C $\cdots$ AB. The higher the ratio, the better our model will be.

[4] This is as proposed in S. Bratož and M.L. Martin, *J. Chem. Phys.*, *42*, 1051 (1965).

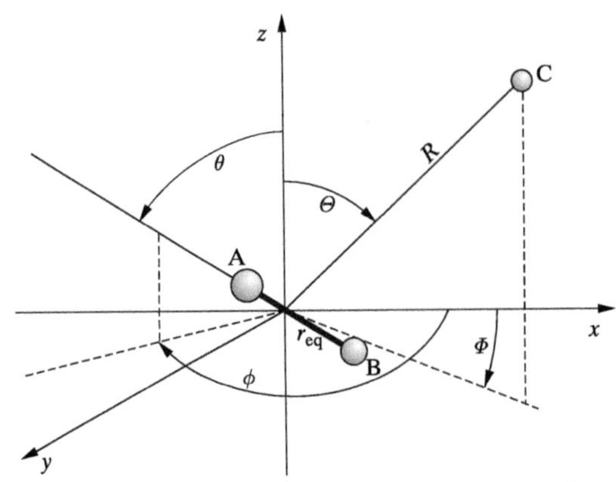

Fig. 7.2. The Jacobi coordinates for the C$\cdots$AB system. The origin is in the center of mass of AB (and the distance AB is constant and equal to r_{eq}). The positions of atoms A and B are fixed by giving the angles θ, ϕ. The position of atom C is determined by three coordinates: R, Θ and Φ. There are five coordinates: R, Θ, Φ, θ, ϕ or R, $\hat{\boldsymbol{R}}$, and $\hat{\boldsymbol{r}}$.

where $\hat{l}^2$ denotes the operator of the square of the angular momentum of the atom C, $\hat{j}^2$ stands for the square of the angular momentum (cf., p. 199) of the molecule AB:

$$\hat{l}^2 = -\hbar^2 \left[\frac{1}{\sin\Theta} \frac{\partial}{\partial\Theta} \sin\Theta \frac{\partial}{\partial\Theta} + \frac{1}{\sin^2\Theta} \frac{\partial^2}{\partial\Phi^2} \right]$$

$$\hat{j}^2 = -\hbar^2 \left[\frac{1}{\sin\theta} \frac{\partial}{\partial\theta} \sin\theta \frac{\partial}{\partial\theta} + \frac{1}{\sin^2\theta} \frac{\partial^2}{\partial\phi^2} \right],$$

μ is the reduced mass of C and the mass of (A+B), μ_{AB} denotes the reduced mass of A and, B, and V stands for the potential energy of the nuclear motion.

The expression for $\hat{H}$ is quite understandable. First of all, we have in $\hat{H}$ five coordinates, as there should be: R, two angular coordinates hidden in the symbol $\hat{\boldsymbol{R}}$; and two angular coordinates symbolized by $\hat{\boldsymbol{r}}$. The four angular coordinates enter the operators of the squares of the two angular momenta. The first three terms in $\hat{H}$ describe the kinetic energy, and V is the potential energy (the electronic ground state energy depends on the nuclear coordinates). The kinetic energy operator describes the radial motion of C with respect to the origin (first term), the rotation of C about the origin (second term), and the rotation of AB about the origin (third term).

7.1.2 Anisotropy of the Potential V

How can we figure out the shape of V? Let us first make a section of V. If we freeze the motion of AB,[5] the atom C would have (concerning the interaction energy) a sort of energetic well around AB wrapping the AB molecule, caused by the C$\cdots$AB van der Waals interaction. The bottom

[5] That is, the angles θ i ϕ are fixed.

of the well would be very distant from the molecule (van der Waals equilibrium distance), while the shape determined by the bottom points would resemble the shape of AB; i.e., it would be a little elongated. The depth of the well would vary depending on its orientation with respect to the origin.

If V were isotropic (i.e., if atom C would have C $\cdots$ AB interaction energy independent[6] of $\hat{\boldsymbol{r}}$), then of course we might say that there is no coupling between the rotation of C and the rotation of AB. We would have then a separate conservation law for the first and the second angular momentum and the corresponding commutation rules (cf. Chapter 2 and Appendix F available at booksite.elsevier.com/978-0-444-59436-5):

$$\left[\hat{H}, \hat{l}^2\right] = \left[\hat{H}, \hat{j}^2\right] = 0,$$
$$\left[\hat{H}, \hat{l}_z\right] = \left[\hat{H}, \hat{j}_z\right] = 0.$$

Therefore, the wave function of the total system would be the eigenfunction of $\hat{l}^2$ and $\hat{l}_z$ as well as of $\hat{j}^2$ and $\hat{j}_z$. The corresponding quantum numbers $l = 0, 1, 2, \ldots$ and $j = 0, 1, 2, \ldots$ which determine the eigenvalues of the squares of the angular momenta $\hat{l}^2$ and $\hat{j}^2$, as well as the corresponding quantum numbers $m_l = -l, -l+1, \ldots, l$ and $m_j = -j, -j+1, \ldots, j$, which determine the projections of the corresponding angular momenta on the z-axis, would be legal[7] quantum numbers (the full analogy with the rigid rotator is discussed in Chapter 4). The rovibrational levels could be labeled using pairs of quantum numbers: (l, j). In the absence of an external field (no privileged orientation in space), any such level would be $(2l+1)(2j+1)$-tuply degenerate, since this is the number of different projections of both angular momenta on the z axis.

7.1.3 Adding the Angular Momenta in Quantum Mechanics

However, V *is not isotropic* (although the anisotropy is small). What then? Then, of all angular momenta, only the *total* angular momentum $\boldsymbol{J} = \boldsymbol{l} + \boldsymbol{j}$ is conserved (the conservation law results from the very foundations of physics; cf. Chapter 2).[8] Therefore, the vectors $\boldsymbol{l}$ and $\boldsymbol{j}$, when added to $\boldsymbol{J}$, would make all allowed angles: from minimum angle (the quantum number $J = l + j$), through larger angles[9] and the corresponding quantum numbers $J = l + j - 1, l + j - 2$, etc., up to the maximum angle, corresponding to $J = |l - j|$). Therefore,

[6] That is, the bottom of the well would be a sphere centered in the center of mass of AB, and the well depth would be independent of the orientation.

[7] We used to say "*good*" instead of "*legal*."

[8] Of course, the momentum has also been conserved in the isotropic case, but in this case, the energy was identical independently of the quantum number J (resulting from different angles between $\boldsymbol{l}$ and $\boldsymbol{j}$).

[9] The projections of the angular momenta are quantized.

the number of all possible values of J (each corresponding to a different energy) is equal to the number of projections of the *shorter*[a] of the vectors $\boldsymbol{l}$ and $\boldsymbol{j}$ on the longer one, i.e.,

$$J = (l + j), (l + j - 1), \ldots, |l - j|. \tag{7.1}$$

For a given J, there are $2J + 1$ projections of $\boldsymbol{J}$ on the z-axis (because $|M_J| \leq J$); without any external field, all these projections correspond to identical energies.

[a] In the case of two vectors of the same length, the role of the shorter vector may be taken by either of them.

Check that the number of all possible eigenstates is equal to $(2l + 1)(2j + 1)$; i.e., exactly what we had in the isotropic case. For example, for $l = 1$ and $j = 1$, the degeneracy in the isotropic case is equal to $(2l+1)(2j+1) = 9$, while for anisotropic V, we would deal with five states for $J = 2$ (all of the same energy), three states corresponding to $J = 1$ (the same energy, but different from $J = 2$), and a single state with $J = 0$ (still another value of energy)–nine states altogether. This means that switching anisotropy partially removed the degeneracy of the isotropic level (l, j) and gave the levels characterized by quantum number J.

7.1.4 Application of the Ritz Method

We will use the Ritz variational method (see Chapter 5, p. 238) to solve the Schrödinger equation. What should we propose as the expansion functions? It is usually recommended that we proceed systematically and choose first a complete set of functions depending on R, then a complete set depending on $\hat{\boldsymbol{R}}$, and finally a complete set that depends on the $\hat{\boldsymbol{r}}$ variables. Next, one may create the complete set depending on all five variables (these functions will be used in the Ritz variational procedure) by taking all possible products of the three functions depending on R, $\hat{\boldsymbol{R}}$, and $\hat{\boldsymbol{r}}$. There is no problem with the complete sets that have to depend on $\hat{\boldsymbol{R}}$ and $\hat{\boldsymbol{r}}$, as these may serve the spherical harmonics (the wave functions for the rigid rotator) $\{Y_l^m(\Theta, \Phi)\}$ and $\{Y_{l'}^{m'}(\theta, \phi)\}$, while for the variable R, we may propose the set of harmonic oscillator wave functions $\{\chi_v(R)\}$.[10] Therefore, we may use as the variational function[11]:

$$\Psi(R, \Theta, \Phi, \theta, \phi) = \sum c_{vlml'm'} \chi_v(R) Y_l^m(\Theta, \Phi) Y_{l'}^{m'}(\theta, \phi),$$

[10] See Chapter 4. Of course, our system does not represent any harmonic oscillator, but what counts is that the harmonic oscillator wave functions form a complete set (as the eigenfunctions of a Hermitian operator).

[11] The products $Y_l^m(\Theta, \Phi) Y_{l'}^{m'}(\theta, \phi)$ may be used to produce linear combinations that are automatically the eigenfunctions of $\hat{J}^2$ and $\hat{J}_z$, and have the proper parity (see Chapter 2). This may be achieved by using the Clebsch-Gordan coefficients (D.M. Brink and G.R. Satchler, *Angular Momentum*, Clarendon, Oxford, 1975). The good news is that this way, we can obtain a smaller matrix for diagonalization in the Ritz procedure. The bad news is that the matrix elements will contain more terms to be computed. The method described here will give the same result as using the Clebsch-Gordan coefficients because the eigenfunctions of the Hamiltonian obtained within the Ritz method will automatically be the eigenfunctions of $\hat{J}^2$ i $\hat{J}_z$, as well as having the proper parity.

where c are the variational coefficients and the summation goes over the v, l, m, l', m' indices. The summation limits have to be finite in practical applications; therefore, the summations go to some maximum values of v, l, and l' (m and m' vary from $-l$ to l and from $-l'$ to $+l'$). We hope (as always in quantum chemistry) that the numerical results of a demanded accuracy will not depend on these limits. Then, as usual, the Hamiltonian matrix is computed and diagonalized (see p. e105), and the eigenvalues E_J, as well as the eigenfunctions ψ_{J,M_J} of the ground and excited states, are found. Each of the eigenfunctions will correspond to some J, M_J and to a certain parity. The problem is solved.

7.2 Force Fields (FF)

The middle of the twentieth century marked the end of a long period of determining the building blocks of chemistry: chemical elements, chemical bonds, and bond angles. The lists of these are not definitely closed, but future changes will be more cosmetic than fundamental. This made it possible to go one step further and begin to rationalize the structure of molecular systems, as well as to foresee the structural features of the compounds to be synthesized. The crucial concept is based on the Born-Oppenheimer approximation and on the theory of chemical bonds and resulted in the spatial structure of molecules. The great power of such an approach was first proved by the construction of the DNA double helix model by Watson and Crick. The first DNA model was built from iron spheres, wires, and tubes.

James Dewey Watson, born 1928, American biologist and professor at Harvard University. Francis Harry Compton Crick (1916–2004), British physicist and professor at the Salk Institute in San Diego. Both scholars won the 1962 Nobel Prize for *"their discoveries concerning the molecular structure of nucleic acids and its significance for information transfer in living material."* At the end of their historic paper, J.D. Watson, F.H.C. Crick,

Nature, 737 (1953) (which was only about 800 words long) the famous enigmatic but crucial sentence appears: *"It has not escaped our notice that the specific pairing we have postulated immediately suggests a possible copying mechanism for the genetic material."* The story behind the discovery is described in a colorful and non-conventional way by Watson in his book *Double Helix: A Personal Account of the Discovery of the Structure of DNA.* Touchstone Books, 2001, Kent.

This approach created problems, though: one of the founders of force fields, Michael Levitt, recalls[12] that a model of a Transfer Ribonucleic Acid (tRNA) fragment constructed by him with 2000 atoms weighted more than 50 kg.

The experience accumulated paid off by proposing some approximate expressions for electronic energy, which is, as we know from Chapter 6, the potential energy of the motion of the nuclei. This is what we are going to talk about next.

Suppose that we have a molecule (a set of molecules can be treated in a similar way). We will introduce the *force field,* which will be a *scalar* field–a function $V(\boldsymbol{R})$ of the nuclear coordinates $\boldsymbol{R}$. The function $V(\boldsymbol{R})$ represents a generalization (from one dimension to $3N-6$ dimensions) of the function $E_0^0(R)$ of Eq. (6.8) on p. 266. The force acting on atom j occupying position x_j, y_j, z_j is computed as the components of the vector $\boldsymbol{F}_j = -\nabla_j V$, where

$$\nabla_j = \boldsymbol{i} \cdot \frac{\partial}{\partial x_j} + \boldsymbol{j} \cdot \frac{\partial}{\partial y_j} + \boldsymbol{k} \cdot \frac{\partial}{\partial z_j}, \tag{7.2}$$

with $\boldsymbol{i}, \boldsymbol{j}, \boldsymbol{k}$ denoting the unit vectors along x, y, z, respectively.

Force Field
A force field represents a mathematical expression $V(\boldsymbol{R})$ for electronic energy as a function of the nuclear configuration $\boldsymbol{R}$. Its gradient gives the forces acting on the atoms.

Of course, if we had to write down this scalar field accurately, we would have to solve (with an accuracy of about 1 kcal/mol) the electronic Schrödinger equation 6.8 for every configuration $\boldsymbol{R}$ of the nuclei and take the lowest eigenvalue [i.e., an analog of $E_0^0(R)$] as $V(\boldsymbol{R})$. This would take so much time, even for small systems composed of a few atoms, that we would abandon this method with great relief. Even if such a calculation required huge computation time, it would give results that would have been quite simple in their overall features (assuming that the molecule has a pattern of chemical bonds). It just would turn out that V could be approximated by the following model function:

- **Chemical bonds.** $V(\boldsymbol{R})$ would be close to minimal if any chemical bond between atoms X and Y had *a certain characteristic reference length* r_0 that would depend on the chemical character of the atoms X and Y. If the bond length were changed (shortened or elongated) to a certain value r, then the energy would increase according to the harmonic law (with force constant k_{XY}) and then some deviations from the harmonic approximation would appear.[13] A harmonic term of the kind $\frac{1}{2}k_{XY}(r-r_0)^2$ incorporated additively into V *replaces the true anharmonic dependence by a harmonic approximation* (assumption of small amplitudes) as if the two atoms had been bound by a harmonic spring (in the formula, the atomic

[12] M. Levitt, *Nature Struct. Biol.*, 8, 392 (2001).

[13] These deviations from harmonicity (i.e., from the proportionality of force and displacement) are related to the smaller and smaller force needed to elongate the bond and the larger and larger force needed to shorten the bond.

indices at symbols of distances have been omitted). The most important feature is that the same formula $\frac{1}{2}k_{XY}\left(r-r_0\right)^2$ is used for all chemical bonds $X-Y$, *independently of some particular chemical neighborhood of a given* $X-Y$ *bond*. For example, one assumes that a non-distorted single C–C bond[14] has a characteristic reference length $r_0 = 1.523$ Å and a characteristic force constant $k_{XY} = 317 \frac{\text{kcal}}{\text{mol Å}^2}$; similarly, some distinct parameters pertain to the C=C bond: $r_0 = 1.337$ Å, $k_{XY} = 690 \frac{\text{kcal}}{\text{mol Å}^2}$, etc.[15]

- **Bond angles**. While preserving the distances r in the A–B and B–C bonds, we may change the bond angle α = A–B–C, and thus change the A···C distance. A corresponding change of V has to be associated with such a change. The energy has to increase when the angle α deviates from a characteristic reference value α_0. The harmonic part of such a change may be modeled by $\frac{1}{2}k_{XYZ}\left(\alpha-\alpha_0\right)^2$ (the indices for angles are omitted), which is equivalent to setting a corresponding harmonic spring for the bond angle and requires small amplitudes $|\alpha-\alpha_0|$. The k_{XYZ} are assumed to be "*universal*"; i.e., they depend on the chemical character of the X, Y, Z atoms, and do not depend on other details, such as the neighborhood of these atoms. For example, for the angle C–C–C $\alpha_0 = 109.47^0$ and $k_{XYZ} = 0.0099 \frac{\text{kcal}}{\text{mol degree}^2}$ (for all C–C–C fragments of the molecule), which means that changing the C···C distance by varying angle is about an order of magnitude easier than changing a CC bond length.
- **The van der Waals interaction**. Two atoms X and Y, that do not form a chemical bond X–Y, as well as not participating in any sequence of bonds X–A–Y, still interact. There is nothing in the formulas introduced above that would prevent X and Y *collapsing* without any change of V. However, when two such atoms approach at a distance smaller than the sum of their radii (the van der Waals radii; see p. 859), then V had to increase greatly.[16] On the other hand, at large interatomic distances, the two atoms have to attract each other by the dispersion interaction vanishing as r^{-6} (cf. Chapter 13, p. 822). Hence, there is an equilibrium distance r_0, at which the interaction energy attains a minimum equal to $-\varepsilon$. These features of the interaction are captured by the widely used Lennard-Jones potential $V_{LJ}\left(X,Y\right) = \varepsilon\left[\left(\frac{r_e}{r}\right)^{12} - 2\left(\frac{r_e}{r}\right)^6\right]$, where we skip for brevity the indices X, Y on the right side. The Lennard-Jones potential given above is called LJ 12–6

John E. Lennard-Jones (1894–1954), professor of theoretical chemistry of the University of Cambridge, UK. The reader may find a historic picture of the theoretical chemistry team in *Intern. J. Quantum Chemistry, S23*, XXXII (1989).

[14] That is, when all other terms in the force field equal zero.

[15] A CC bond involved in a conjugated single and double bonds (e.g., in benzene) also has its own parameters.

[16] A similar thing happens with cars: the repair cost increases greatly when the distance between two cars decreases below two thicknesses of the paint job.

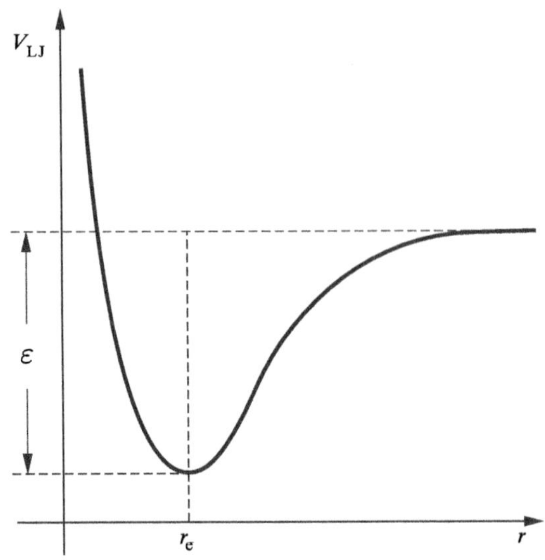

Fig. 7.3. The Lennard-Jones (LJ 12 − 6) potential. The parameter ε represents the depth of the potential well, while the parameter r_e denotes the minimum position. This r_e, corresponding to the *non-bonding* interaction of atoms X and Y, has no direct relation to the r_0 value pertaining to the *chemical* bond X–Y. The first is larger than the second by about an angstrom.

(reflecting the powers involved). Sometimes other powers are used leading to other "*LJ m-n*" potentials.[17] Due to their simplicity, the LJ potentials are widely used (see Fig. 7.3).

- **Electrostatic interaction.** All the terms that we have introduced to V so far do not take into account the fact that atoms carry net charges q_X and q_Y that have to interact electrostatically by Coulombic forces. To take this effect into account the electrostatic energy terms $q_X q_Y / r$ are added to V, where we assume that the net charges q_X and q_Y are *fixed* (i.e., independent of the molecular conformation).[18]

[17] The power 12 has been chosen for two reasons. First, the power is sufficiently large to produce a strong repulsion when the two atoms approach each other. Second, ... $12 = 6 \times 2$. The last reason makes the first derivative formula (i.e., the force) look more elegant than other powers do. A more elegant formula is usually faster to compute, and this is of practical importance.

[18] In some force fields, the electrostatic forces depend on the dielectric constant of the neighborhood (e.g., solvent) despite the fact that this quantity has a macroscopic character and does not pertain to the nearest neighborhood of the interacting atoms. If all the calculations had been carried out taking the molecular structure of the solvent into account, as well as the polarization effects, no dielectric constant would have been needed. If this is not possible, then the dielectric constant effectively takes into account the polarization of the medium (including reorientation of the solvent molecules). The next problem is *how* to introduce the dependence of the electrostatic interaction of two atomic charges on the dielectric constant. In some of the force fields, we introduce a brute force kind of damping; namely, the dielectric constant is introduced into the denominator of the Coulombic interaction as equal to the interatomic distance. This is equivalent to saying that the electrostatic interaction is practically damped down for larger distances.

In second-generation force fields, we explicitly take into account the induction interaction; e.g., the dependence of the atomic electric charges on molecular conformations. Such force fields, when explicitly taking the solvent molecules into account, should not introduce the dielectric constant.

- **Torsional interactions**. In addition to all the terms described above, we often introduce to the force field a *torsional term* $A_{X-Y-Z-W}\left(1-\cos n\omega\right)$ for each torsional angle ω showing how V changes when a rotation ω about the chemical bond YZ, in the sequence X–Y–Z–W of chemical bonds, takes place (n is the multiplicity of the energy barriers per single turn[19]). Some rotational barriers already result from the van der Waals interaction of the X and W atoms, but in practice, the barrier heights have to be corrected by the torsional potentials to reproduce experimental values.

- **Mixed terms**. Besides the abovementioned terms, one often introduces some *coupling (mixed) terms*; e.g., bond-bond angle, etc. The reasoning behind this is simple. The X-Y-Z bond angle force constant has to depend on the bond-lengths X-Y and Y-Z, etc.

Summing up a simple force field might be expressed as shown in Fig. 7.4, where for the sake of simplicity, the indices X, Y at r, r_0, as well as X, Y, Z at α, α_0 and X, Y, Z, W at ω, have been omitted:

$$V=\sum_{X-Y}\frac{1}{2}k_{XY}\left(r-r_0\right)^2+\sum_{X-Y-Z}\frac{1}{2}k_{XYZ}\left(\alpha-\alpha_0\right)^2+\sum_{X\cdots Y}V_{LJ}\left(X,Y\right)$$
$$+\sum_{X,Y}\frac{q_X q_Y}{r}+\sum_{tors}A_{X-Y-Z-W}\left(1-\cos n\omega\right)+\text{ coupling terms (if any).}$$

Hence, some simple formulas help us to figure out how the electronic energy $E_0^0(\boldsymbol{R})=V(\boldsymbol{R})$ looks as a function of the configuration $\boldsymbol{R}$ of the nuclei. Our motivation is as follows:

- Economy of computation: *ab initio* calculations of the electronic energy for larger molecules would have been many orders of magnitude more expensive.
- In addition, a force field gives $V(\boldsymbol{R})$ in the form of a simple formula for *any* positions $\boldsymbol{R}$ of the nuclei, while the calculation of the electronic energy would give us $V(\boldsymbol{R})$ numerically; i.e., for some *selected* nuclear configurations.

7.3 Local Molecular Mechanics (MM)

7.3.1 Bonds That Cannot Break

It is worth noting that the force fields correspond to a fixed (and unchangeable during computation) system of chemical bonds. The chemical bonds are treated as springs, most often satisfying Hooke's[20] law (harmonic); therefore, they are unbreakable. Similarly, the bond angles are forced to satisfy Hooke's law. Such a force field is known as *flexible molecular mechanics* (MM)[21].

[19] For example, $n=3$ for ethane.

[20] Robert Hooke, British physicist and biologist (1635–1703).

[21] There are a few such force fields in the literature. They give similar results, as far as their main features are considered. The force field concept was able to clarify many observed phenomena and even fine effects. It may also fail, as with anything in the real world.

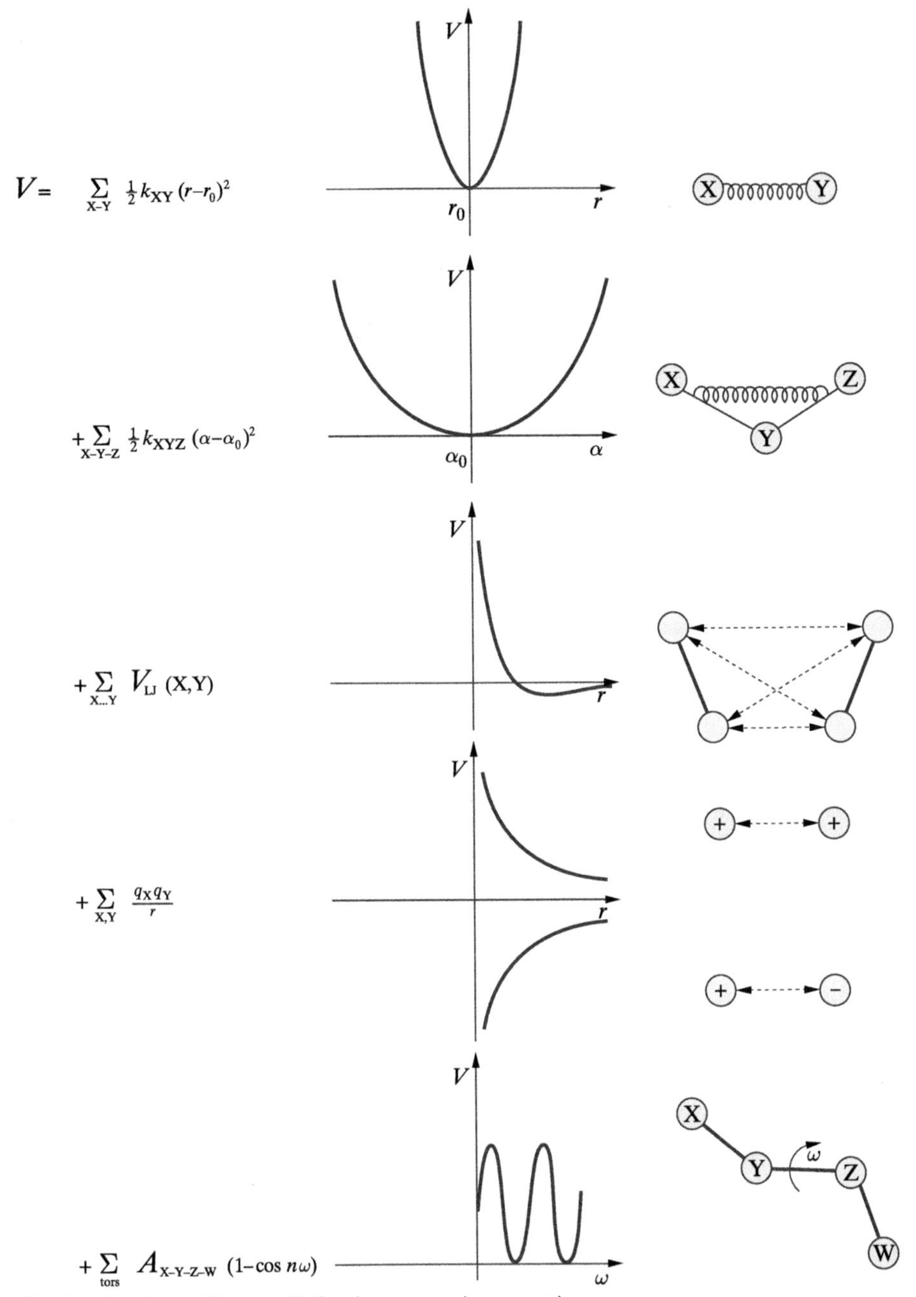

Fig. 7.4. The first force field of Bixon and Lifson in a mnemonic presentation.

To decrease the number of variables, we sometimes use *rigid molecular mechanics* (MM),[22] in

[22] The rigid molecular mechanics was a very useful tool for Paul John Flory (1910–1985), American chemist and professor at Cornell and Stanford. Using such mechanics, Flory developed a theory of polymers that explained

which the bond lengths and the bond angles are fixed at values close to experimental ones, but the torsional angles are free to change. The argument behind such a choice is that a quantity of energy that is able to make only tiny changes in the bond lengths, larger but still small changes in the bond angles, can make large changes in the torsional angles; i.e., the torsional variables determine the overall changes of the molecular geometry. Of course, the second argument is that a smaller number of variables means lower computational costs.

Molecular mechanics represents a method of finding a stable configuration of the nuclei by using a minimization of $V(\boldsymbol{R})$ with respect to the nuclear coordinates (for a molecule or a system of molecules). The essence of molecular mechanics is that we roll the potential energy hypersurface slowly downhill from a starting point chosen (corresponding to a certain starting geometry of the molecule) to the "*nearest*" energy minimum corresponding to the final geometry of the molecule.

The "*sliding down*" is carried out by a minimization procedure that traces, point by point, the trajectory in the configurational space; e.g., in the direction of the negative gradient vector calculated at any consecutive point. The minimization procedure represents a mechanism showing how to obtain the next geometry from the previous one. The procedure ends, when the geometry ceases to change (e.g., the gradient vector has zero length[23]). The geometry attained is called the *equilibrium* or *stable geometry*. The rolling described above is more like a crawling down with large friction, since in molecular mechanics, the kinetic energy is always zero and the *system is unable to go uphill* of V.

A lot of commercial software[24] offers force field packages. Unfortunately, the results depend to quite a significant degree on the force field chosen. Even using the same starting geometry, we may obtain final (equilibrium) results that differ very much one from another. Usually the equilibrium geometries obtained in one force field do not differ much from those in another one, but the corresponding energies may be very different. Therefore, the most stable geometry (corresponding to the lowest energy) obtained in a force field may turn out to be less stable in another one, thus leading to different predictions of the molecular structure.

A big problem in molecular mechanics is that the final geometry is very close to the starting one. We start from a boat (chair) conformation of cyclohexane and obtain a boat (chair) equilibrium geometry. The essence of molecular mechanics however, is that when started from a distorted boat (chair) conformation, we obtain the perfect, beautiful equilibrium boat (chair)

their physical properties. In 1974, he obtained the Nobel Prize "*for his fundamental achievements, both theoretical and experimental, in the physical chemistry of macromolecules.*"

[23] The gradient also equals zero at energy maxima and energy saddle points. To be sure that a minimum really has been finally attained, we have to calculate (at the particular point suspected to be a minimum) a Hessian [i.e., the matrix of the second derivatives of V, then diagonalize it (cf. p. e105)] and check whether the eigenvalues obtained are all positive.

[24] See the Web Annex.

conformation, which may be compared to experimental results. Molecular mechanics is extremely useful in conformational studies of systems with a small number of stable conformations, either because the molecule is small, rigid, or its overall geometry is fixed. In such cases, all or all "*reasonable*,"[25] conformations can be investigated and those with the lowest energy can be compared to experimental results.

7.3.2 Bonds That Can Break

Harmonic bonds cannot be broken, and therefore, molecular mechanics with harmonic approximation is unable to describe chemical reactions. When instead of harmonic oscillators, we use the Morse model (p. 192), then the bonds can be broken.

And yet we most often use the harmonic oscillator approximation. Why? There are a few reasons:

- The Morse model requires many computations of the exponential function, which is expensive[26] when compared to the harmonic potential.
- The Morse potential requires three parameters, while the harmonic model needs only two parameters.
- In most applications, the bonds do not break, and it would be very inconvenient to obtain breaking due to a particular starting point, for instance.
- A description of chemical reactions requires not only the possibility if breaking bonds, but also a realistic (i.e., quantum chemical) computation of the charge distributions involved (cf. p. 368). The Morse potential would be too simplistic for such purposes.

7.4 Global Molecular Mechanics

7.4.1 Multiple Minima Catastrophe

If the number of local minima is very large (and this may happen even for medium-size molecules) or even "*astronomic*", then exploring the whole conformational space (all possible geometries) by finding all possible minima using a minimization procedure becomes impossible. Hence, we may postulate another procedure which may be called *global molecular mechanics* and could find the global minimum (the most stable conformation) starting from any point in the configurational space.

If the number of local minima is small, there is in principle no problem with using theory. Usually it turns out that the quantum mechanical calculations are feasible, often even at the *ab initio* level. A closer insight leads, however, to the conclusion that only some extremely accurate and expensive calculations would give the correct energy sequence of the conformers, and that only for quite small molecules with a dozen atoms. This means that for larger molecules, we

[25] A very dangerous word!

[26] Each time requires a Taylor expansion calculation.

are forced to use molecular mechanics. For molecules with a few atoms, we might investigate the whole conformational space by sampling it by a stochastic or systematic procedure, but this approach soon becomes prohibitive for larger molecules.

For such larger molecules, we encounter difficulties that may only be appreciated by individuals who have made such computations themselves. We may say, in short, that virtually nothing helps us with the huge number of conformations to investigate. According to Schepens,[27] the number of conformations found is proportional to the time spent conducting the search. It is worth noting that this means catastrophe, because for a twenty amino acid oligopeptide, the number of conformations is of the order[28] of 10^{20}, and for a hundred amino acids, it is 10^{100}. Also, methods based on molecular dynamics (cf. p. 364) do not solve the problem, since they could cover only a tiny fraction of the total conformational space.

7.4.2 Does the Global Minimum Count?

The goal of conformational analysis is to find those conformations of the molecule that are observed under experimental conditions. At temperatures close to 300 K, the lowest-energy conformations prevail in the sample; i.e., first of all, those corresponding to the global minimum of the potential energy V.

We may ask whether indeed the global minimum of the potential energy decides the observed experimental geometry. Let us disregard the influence of the solvent (neighborhood). A better criterion would be the global minimum of the *free energy*, $E - TS$, where the entropic factor would also enter.[29] A wide potential well means a higher density of vibrational states, and a narrow well means a lower density of states (cf. Eq. (4.26), p. 196; a narrow well corresponds to a large α). If the global minimum corresponds to a wide well, the system in such a well is additionally stabilized by entropy.

For large molecules, there is a possibility that, due to the synthesis conditions, the molecule is trapped in a local minimum (*kinetic minimum*), which is different from the global minimum of the free energy (*thermodynamic minimum*); see Fig. 7.5.

For the same reason that the diamonds (kinetic minimum) in your safe do not change spontaneously into graphite (thermodynamic minimum), a molecule imprisoned in the kinetic minimum may rest there for a very long time (compared to the length of the experiment).

---

[27] Wijnand Schepens, Ph.D. thesis, University of Gand, 2000.

[28] The difficulty of finding a particular conformation among 10^{20} conformations is pretty bad. Maybe the example here will show the severity of the problem being encountered. A single grain of sand has a diameter of about 1 mm. Let us try to align 10^{20} of such sand grains side by side. What will the length of such a chain of grains be? Let us compute: 10^{20} mm $= 10^{17}$ m $= 10^{14}$ km. One light year is 300000 km/s $\times$ 3600 s $\times$ 24 $\times$ 365 $\simeq 10^{13}$ km. Hence, the length is about 10 light years–i.e., longer than the round trip from the Sun to the nearest star, Alpha Centauri.

[29] According to the famous formula of Ludwig Boltzmann, entropy $S = k_B \ln \Omega(E)$, where Ω is the number of the states available for the system at energy E. The more states there are, the larger the entropy is.

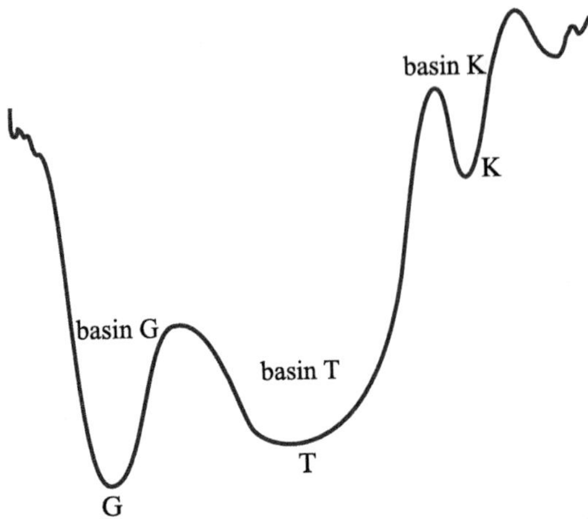

Fig. 7.5. The basins of the *thermodynamic minimum* (T), of the *kinetic minimum* (K) and of the *global* minimum (G). The deepest basin (G) should not correspond to the thermodynamically most stable conformation (T). Additionally, the system may be caught in a kinetic minimum (K), from which it may be difficult to tunnel to the thermodynamic minimum basin. Explosives and fullerenes may serve as examples of K.

Christian Anfinsen (1916 –1995), American biochemist, obtained, the Nobel Prize in 1972 "*for his work on ribonuclease, especially concerning the connection between the amino acid sequence and the biologically active conformation.*" He made an important contribution showing that after denaturation (a large change of conformation) some proteins fold back spontaneously to their native conformation.

Despite these complications, we generally assume in conformational analysis that the global minimum and other low-energy conformations play the most important role. In living matter, taking a definite (native) conformation is sometimes crucial. It has been shown[30] that the native conformation of natural enzymes has much lower energy than those of other conformations (energy gap). Artificial enzymes with stochastic amino acid sequences do not usually have this property, resulting in no well-defined conformation.

In my opinion, the global molecular mechanics is one of the most important challenges not only in chemistry, but in natural sciences in general. This is because optimization is fundamental in virtually all domains and, if it is treated mathematically, usually transforms into a problem of finding the global minimum.[31]

[30] E.I. Shakanovich and A.M. Gutin, *Proc. Natl. Acad. Sci. USA*, *90*, 7195 (1993); A. Šali, E.I. Shakanovich, and M. Karplus, *Nature*, *369*, 248 (1994).

[31] The reader may find the description of the author's adventure with this problem in L. Piela in *Handbook of Global Optimization*, vol.2, P.M. Pardalos, H.E. Romeijn (eds.), Kluwer Academic Publishers, Dordrecht (2002), p. 461.

7.5 Small Amplitude Harmonic Motion–Normal Modes

The hypersurface $V(\boldsymbol{R})$ has, in general (especially for large molecules), an extremely complex shape with many minima, each corresponding to a stable conformation. Let us choose *one* of those minima and ask *what kind of motion the molecule undergoes when only small displacements from the equilibrium geometry are allowed.* In addition, we assume that the potential energy for this motion is a *harmonic* approximation of the $V(\boldsymbol{R})$ in the neighborhood of the minimum.[32] Then we obtain the *normal vibrations* or *normal modes.*

Normal Modes

A normal mode represents a harmonic oscillation (of a certain frequency) of all the atoms of the molecule about their equilibrium positions with the same phase for all the atoms (i.e., all the atoms attain their equilibrium position at the same time).

7.5.1 Theory of Normal Modes

Suppose that we have at our disposal an analytical expression for $V(\boldsymbol{R})$ (e.g., the force field), where $\boldsymbol{R}$ denotes the vector of the Cartesian coordinates of the N atoms of the system (it has $3N$ components). Let us assume (see Fig. 7.6) that the function $V(\boldsymbol{R})$ has been minimized in the configurational space, starting from an initial position $\boldsymbol{R}_i$ and going downhill until a minimum

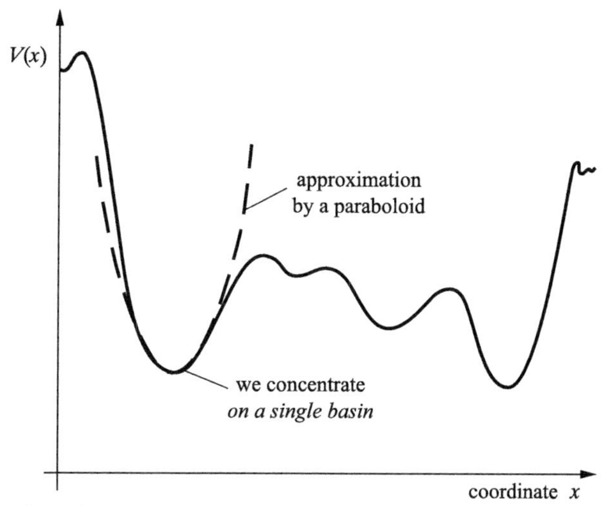

Fig. 7.6. A schematic (one-dimensional) view of the hypersurface $V(x)$ that illustrates the choice of a particular basin of V related to the normal modes to be computed. The basin chosen is then approximated by a paraboloid in $3N$ variables. This gives the $3N-6$ modes with nonzero frequencies and 6 "modes" with zero frequencies.

[32] We may note *en passant* that a similar philosophy prevailed in science until quite recent times: take only the linear approximation and forget about nonlinearities. It turned out, however, that the nonlinear phenomena (cf., Chapter 15) are really fascinating.

position $\boldsymbol{R}_0$ has been reached, the $\boldsymbol{R}_0$ corresponding to one of many minima the V function may possess[33] (we will call the minimum the "closest" to the $\boldsymbol{R}_i$ point in the configurational space). All the points $\boldsymbol{R}_i$ of the configurational space that lead to $\boldsymbol{R}_0$ represent the *basin of the attractor*[34] $\boldsymbol{R}_0$.

From this time on, all other basins of the function $V(\boldsymbol{R})$ have "disappeared from the theory"– only motion in the neighborhood of $\boldsymbol{R}_0$ is to be considered.[35] If someone is aiming to apply harmonic approximation and to consider small displacements from $\boldsymbol{R}_0$ (as we do), then it is a good idea to write down the Taylor expansion of V about $\boldsymbol{R}_0$ [hereafter, instead of the symbols $X_1, Y_1, Z_1, X_2, Y_2, Z_2, \ldots$ for the atomic Cartesian coordinates, we will use a slightly more uniform notation: $\boldsymbol{R} = (X_1, X_2, X_3, X_4, X_5, X_6, ..X_{3N})^T$]:

$$V(\boldsymbol{R}_0 + \boldsymbol{x}) = V(\boldsymbol{R}_0) + \sum_i \left(\frac{\partial V}{\partial x_i}\right)_0 x_i + \frac{1}{2}\sum_{ij}\left(\frac{\partial^2 V}{\partial x_i \partial x_j}\right)_0 x_i x_j + \cdots, \tag{7.3}$$

where $\boldsymbol{x} = \boldsymbol{R} - \boldsymbol{R}_0$ is the vector with the *displacements* of the atomic positions from their equilibria ($x_i = X_i - X_{i,0}$ for $i = 1, \ldots, 3N$), while the derivatives are computed at $\boldsymbol{R} = \boldsymbol{R}_0$.

In $\boldsymbol{R}_0$, all the first derivatives vanish. According to the harmonic approximation, the higher-order terms denoted as "$+\cdots$" are neglected. In effect, we have

$$V(\boldsymbol{R}_0 + \boldsymbol{x}) \cong V(\boldsymbol{R}_0) + \frac{1}{2}\sum_{ij}\left(\frac{\partial^2 V}{\partial x_i \partial x_j}\right)_0 x_i x_j. \tag{7.4}$$

In matrix notation, we have $V(\boldsymbol{R}_0 + \boldsymbol{x}) = V(\boldsymbol{R}_0) + \frac{1}{2}\boldsymbol{x}^T \boldsymbol{V}''\boldsymbol{x}$, where $\boldsymbol{V}''$ is a square matrix of the Cartesian *force constants*, $(\boldsymbol{V}'')_{ij} = \left(\frac{\partial^2 V}{\partial x_i \partial x_j}\right)_0$.

The Newton equations of motion for all the atoms of the system can be written in matrix form as ($\ddot{\boldsymbol{x}}$ means the second derivative with respect to time t)

$$\boldsymbol{M}\ddot{\boldsymbol{x}} = -\boldsymbol{V}''\boldsymbol{x}, \tag{7.5}$$

where $\boldsymbol{M}$ is the diagonal matrix of the atomic masses (the numbers on the diagonal are: $M_1, M_1, M_1, M_2, M_2, M_2, \ldots$), because we calculate the force component along the axis k as $-\frac{\partial V}{\partial x_k} = -\frac{1}{2}\sum_j \left(\frac{\partial^2 V}{\partial x_k \partial x_j}\right)_0 x_j - \frac{1}{2}\sum_i \left(\frac{\partial^2 V}{\partial x_i \partial x_k}\right)_0 x_i = -\sum_j \left(\frac{\partial^2 V}{\partial x_k \partial x_j}\right)_0 x_j = -(\boldsymbol{V}''\boldsymbol{x})_k$.

We may use the relation $\boldsymbol{M}^{\frac{1}{2}}\boldsymbol{M}^{\frac{1}{2}} = \boldsymbol{M}$ in order to write Eq. (7.5) in a slightly different way:

$$\boldsymbol{M}^{\frac{1}{2}}\boldsymbol{M}^{\frac{1}{2}}\ddot{\boldsymbol{x}} = -\boldsymbol{M}^{\frac{1}{2}}\boldsymbol{M}^{-\frac{1}{2}}\boldsymbol{V}''\boldsymbol{M}^{-\frac{1}{2}}\boldsymbol{M}^{\frac{1}{2}}\boldsymbol{x}, \tag{7.6}$$

[33] These are improper minima because a translation or rotation of the system does not change V.

[34] The total configurational space consists of a certain number of such non-overlapping basins.

[35] For another starting conformation $\boldsymbol{R}_i$, we might obtain another minimum of $V(\boldsymbol{R})$. This is why the choice of $\boldsymbol{R}_i$ has to have a definite relation to that which is observed experimentally.

where $M^{\frac{1}{2}}$ is a matrix similar to M, but its elements are the square roots of the atom masses instead of the masses, while the matrix $M^{-\frac{1}{2}}$ contains the inverse square roots of the masses. The last equation, after multiplying from the left by $M^{-\frac{1}{2}}$, gives

$$\ddot{y} = -Ay, \tag{7.7}$$

where $y = M^{\frac{1}{2}}x$ and $A = M^{-\frac{1}{2}}V''M^{-\frac{1}{2}}$.

Let us try to find the solution in the form[36]

$$y = c_1 \exp(+i\omega t) + c_2 \exp(-i\omega t),$$

where the vectors c_i (of the dimension $3N$) of the complex coefficients are time independent. The coefficients c_i depend on the initial conditions, as well as on the A matrix. If we decide to find a solution, in which at time $t = 0$, all the atoms are at equilibrium [i.e., $y(t = 0) = \mathbf{0}$, the same phase], then we obtain the relation $c_1 = -c_2$, leading to the formula

$$y = L \sin(\omega t), \tag{7.8}$$

where the vector[37] L and ω depend on the matrix A. Vector L is determined only to the accuracy of a multiplication constant because multiplication of L by any number does not interfere with satisfying Eq. (7.7).

When we insert the proposed solution [Eq. (7.8)] in Eq. (7.7), we immediately find that ω and L have to satisfy the following equation:

$$(A - \omega^2 \mathbf{1})L = \mathbf{0}. \tag{7.9}$$

The values of ω^2 represent the eigenvalues,[38] while the L are the eigenvectors of the A matrix. There are $3N$ eigenvalues, and each of them corresponds to its eigenvector L. This means that we have $3N$ normal modes, with each mode characterized by its angular frequency $\omega = 2\pi\nu$(ν is the frequency) and its vibration amplitudes L. Hence, it would be natural to assign a normal mode index $k = 1, \ldots, 3N$ for ω and L. Therefore, we have

$$(A - \omega_k^2 \mathbf{1})L_k = 0 \tag{7.10}$$

[36] This form (with $\omega = a + ib$) allows for a constant solution ($a = b = 0$), an exponential growth or vanishing ($a = 0, b \neq 0$), oscillations ($a \neq 0, b = 0$), oscillatory growing or oscillatory vanishing ($a \neq 0, b \neq 0$). For R_0 denoting a minimum $\det A > 0$, this ensures a solution with $a \neq 0, b = 0$.

[37] The vector is equal to $2ic_1$, but since c_1 is unknown, as for the time being is L, therefore, we can say goodbye to c_1 without feeling any discomfort whatsoever.

[38] A is a symmetric matrix, hence its eigenvalues ω^2 and therefore $\omega = a + ib$ are real ($b = 0$). Whether ω is positive, negative, or zero depends on the hypersurface V at R_0; see Fig. 7.7.

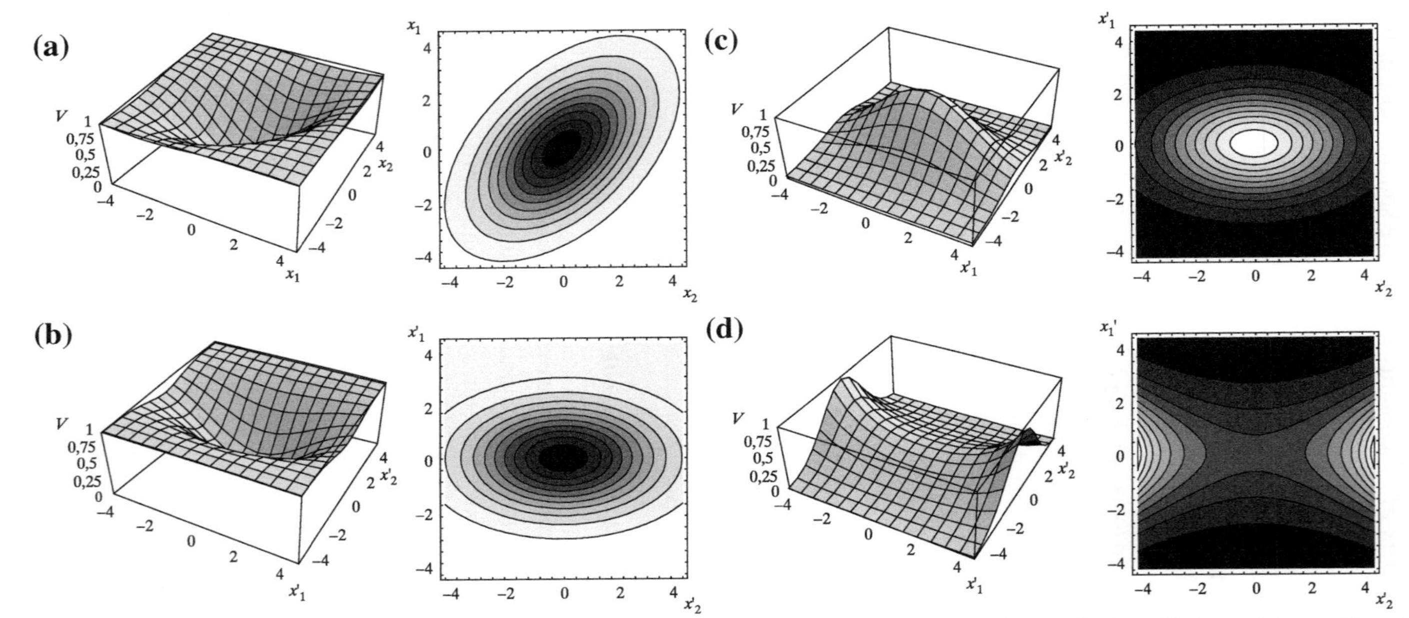

Fig. 7.7. A ball oscillating in a potential energy well (scheme). (a) and (b) show the normal vibrations (normal modes) about a point $\boldsymbol{R}_0 = \boldsymbol{0}$ being *a minimum* of the potential energy function $V(\boldsymbol{R_0}+\boldsymbol{x})$ of two variables $\boldsymbol{x} = (x_1, x_2)$. This function is first approximated by a quadratic function; i.e., a paraboloid $\tilde{V}(x_1, x_2)$. Computing the normal modes is equivalent to such a *rotation of the Cartesian coordinate system* (a), that the new axes (b) x_1' and x_2' become the principal axes of any section of $\tilde{V}$ by a plane $\tilde{V} = const$ (i.e., ellipses). Then, we have $\tilde{V}(x_1, x_2) = V(R_0 = \boldsymbol{0}) + \frac{1}{2}k_1 (x_1')^2 + \frac{1}{2}k_2 (x_2')^2$. The problem then becomes equivalent to the two-dimensional harmonic oscillator (cf., Chapter 4) and separates into two independent one-dimensional oscillators (normal modes): one of angular frequency $\omega_1 = 2\pi\nu_1 = \sqrt{\frac{k_1}{m}}$ and the other with angular frequency $\omega_2 = 2\pi\nu_2 = \sqrt{\frac{k_2}{m}}$, where m is the mass of the oscillating particle. Panels (c) and (d) show what would happen if $\boldsymbol{R}_0$ corresponded not to a minimum, but to a maximum (c) or the saddle point (d). For a maximum (c), k_1 and k_2 in $\tilde{V}(x_1', x_2') = V(\boldsymbol{0}) + \frac{1}{2}k_1 (x_1')^2 + \frac{1}{2}k_2 (x_2')^2$ would both be *negative*, and therefore the corresponding normal "*vibrations*" would have had both imaginary frequencies, while for the saddle point (d), only one of the frequencies would be imaginary.

The diagonalization of $\boldsymbol{A}$ (p. e105) is an efficient technique for solving the eigenvalue problem using commercial computer programs (diagonalization is equivalent to a rotation of the coordinate system; see Fig. 7.7).

This is equivalent to replacing V by a $3N$–dimensional paraboloid with origin $\boldsymbol{R}_0$. The normal mode analysis finds a rotation of the coordinate system such that the new axes coincide with the principal axes of the paraboloid.

There will be six frequencies (five for a linear molecule) equal to zero. They are connected to the translation and rotation of the molecule in space: three translations along x, y, and z and three rotations about x, y, and z (two in the case of a linear molecule). Such free translations/rotations do not change the energy and therefore may be thought to correspond to zero force constants.

If we are interested in what the particular atoms are doing when a single mode l is active, then the displacements from the equilibrium position as a function of time are expressed as

$$\boldsymbol{x}_k = \boldsymbol{M}^{-\frac{1}{2}}\boldsymbol{y}_k = \boldsymbol{M}^{-\frac{1}{2}}\boldsymbol{L}_k Q_k \sin(\omega_k t), \tag{7.11}$$

where $Q_k \in (-\infty, \infty)$, because the displacements $\mathbf{x}_k$ are not bound by the length of vector $\mathbf{L}_k$, which has been arbitrarily set to 1. Therefore, if one shifts all atoms according how they behave in the normal mode k, one should insert the variable Q_k, which tunes the displacement amplitude.

A given atom participates in all vibrational modes. Even if any vibrational mode makes all atoms move, some atoms move more than others. It may happen that a particular mode changes mainly the length of *one of the chemical bonds (stretching mode)*, another mode moves another bond, and yet another changes a particular *bond angle (bending mode)*, etc.

This means that some chemical bonds or some functional groups may have *characteristic* vibration frequencies, which is of great importance for the identification of these bonds or groups in spectral analysis.

In Table 7.1 these frequencies ν characteristic for some particular chemical bonds are given as the wave numbers $\bar{\nu}$ defined by the relation

$$\omega = 2\pi\nu = 2\pi\bar{\nu}c, \tag{7.12}$$

with c being the velocity of light and ν the frequency. The wave number is the number of the wave lengths covering a distance of 1 cm.

Example 1. ***Water Molecule***

A single water molecule has $3 \times 3 = 9$ normal modes. Six of them would have the angular frequencies ω equal zero (they correspond to three free translations and three free rotations of the molecule in space). Three normal modes remain, and the vectors $\boldsymbol{x}$ of Eq. (7.11) for these

Table 7.1. Characteristic frequencies (wave numbers, in cm^{-1}) typical for some chemical bonds (stretching vibrations) and bond angles (bending vibrations).

Bond	Vibration	Wave Number
C–H	Stretching	2850–3400
H–C–H	Bending	1350–1460
C–C	Stretching	700–1250
C=C	Stretching	1600–1700
C≡C	Stretching	2100–2250
C=O	Stretching	1600–1750
N–H	Stretching	3100–3500
O–H	Stretching	3200–4000

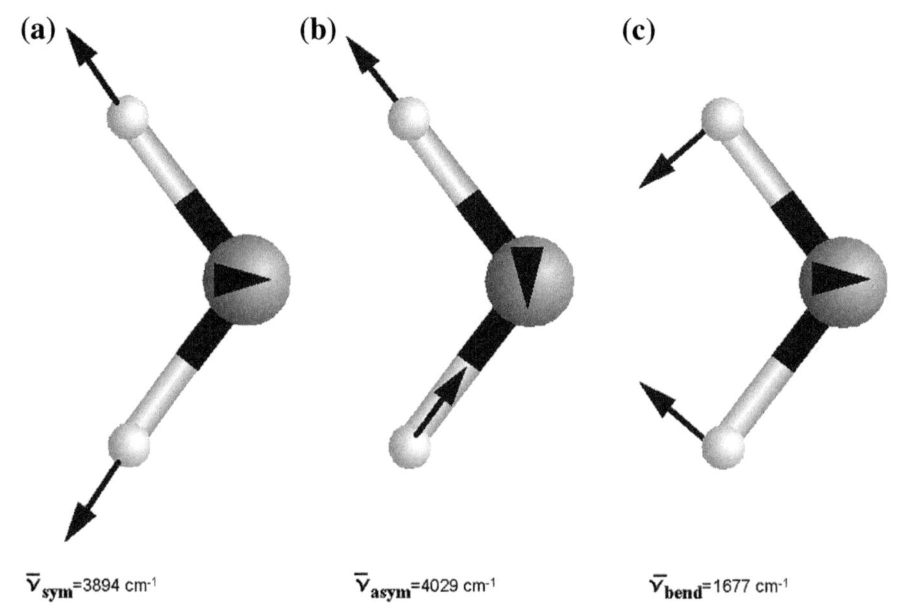

Fig. 7.8. The normal modes of the water molecule: (a) symmetric, (b) antisymmetric, (c) bending. The arrows indicate the directions and proportions of the atomic displacements.

modes can be described as follows (see Fig. 7.8; the corresponding wave numbers have been given in parentheses[39]):

- One of the modes means a *symmetric* stretching of the two OH bonds ($\bar{\nu}_{sym} = 3894$ cm^{-1}).
- The second mode corresponds to a similar, but *antisymmetric* motion; i.e., when one of the bonds shortens, the other one elongates, and vice versa[40] ($\bar{\nu}_{asym} = 4029$ cm^{-1}).

[39] J. Kim, J.Y. Lee, S. Lee, B.J. Mhin, and K.S. Kim, *J. Chem. Phys.*, *102*, 310 (1995). This paper reports normal mode analysis for potential energy hypersurfaces computed by various methods of quantum chemistry. I have chosen the coupled cluster method [CCSD(T); see Chapter 10] as an illustration.

[40] The shortening has the same value as the lengthening. This is a result of the harmonic approximation, in which both shortening and lengthening require the same energy.

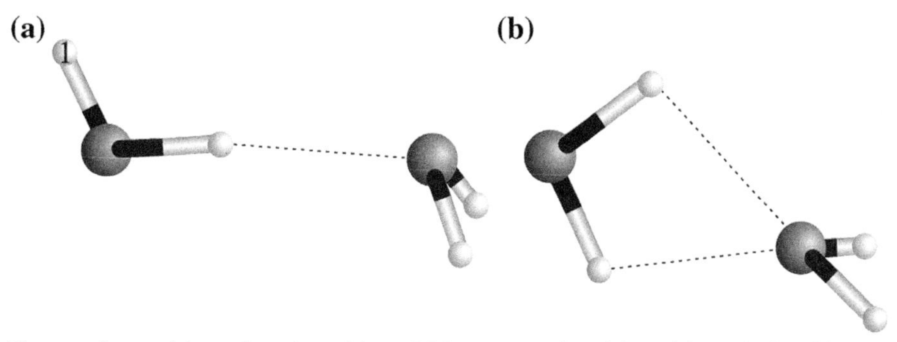

Fig. 7.9. The water dimer and the configurations of the nuclei that correspond to minima of the two basins of the potential energy V. The global minimum (a) corresponds to a single hydrogen bond O–H. . .O; the local minimum (b) corresponds to the bifurcated hydrogen bond.

- The third mode is called the *bending mode* and corresponds to an oscillation of the HOH angle about the equilibrium value ($\bar{\nu}_{bend} = 1677$ cm^{-1}).

Example 2. ***Water Dimer***

Now let us take *two interacting water* molecules. First, let us ask how many minima we can find on the electronic ground-state energy hypersurface. Detailed calculations have shown that there are two such minima (see Fig. 7.9). The global minimum corresponds to the configuration characteristic for the *hydrogen bond* (cf. p. 863). One of the molecules is a donor, and the other is an acceptor of a proton (Fig. 7.9a). A local minimum of smaller stability appears when one of the water molecules serves as a donor of two protons, while the other serves as an acceptor of them called the *bifurcated* hydrogen bond,[41] (Fig. 7.9b).

Now, *we decide* to focus on the global minimum potential well. We argue that for thermodynamic reasons, this particular well will be most often represented among water dimers. This potential energy well has to be approximated by a paraboloid. The number of degrees of freedom is equal to $6 \times 3 = 18$, and this is also the number of normal modes to be obtained. As in Example 1, six of them will have zero frequency and the number of true vibrations is 12. This is the number of normal modes, each with its frequency ω_k and the vector $\boldsymbol{x}_k = \boldsymbol{M}^{-\frac{1}{2}} \boldsymbol{L}_k Q_k \sin\left(\omega_k t\right)$ that describes the atomic motion. The two water molecules, after forming the hydrogen bond, have not lost their individual features (in other words, the OH vibration is *characteristic*).

In dimer vibrations, we will find the vibration frequencies of individual molecules changed a little by the water-water interaction. These modes should appear in pairs, but the two frequencies should differ (the role of the two water molecules in the dimer is different). The proton acceptor has something attached to its heavy atom, and the proton donor has something attached to the light hydrogen atom. Let us recall that in the harmonic oscillator, the reduced mass is relevant, which therefore is almost equal to the mass of the *light* proton. If something attaches to this atom,

[41] See, for example, a theoretical analysis given by R.Z. Khaliullin, A.T. Bell, and M. Head-Gordon, *Chem. Eur. J.*, *15*, 851 (2009), at the DFT level (see Chapter 11 for more).

it means a considerable lowering of the frequency. This is why lower frequencies correspond to the proton donor. The computed frequencies[42] are the following:

- Two stretching vibrations with frequencies of 3924 cm^{-1} and 3904 cm^{-1} (both antisymmetric; the higher frequency corresponds to the proton acceptor, the lower to the proton donor)
- Two stretching vibrations with frequencies of 3796 cm^{-1} and 3704 cm^{-1} (both symmetric; again the higher frequency corresponds to the proton acceptor, the lower to the proton donor)
- Two bending vibrations with frequencies of 1624 cm^{-1} (donor bending) and 1642 cm^{-1} (acceptor bending)

Thus, among 12 modes of the dimer, we have discovered 6 modes that are related to the individual molecules: 4 OH stretching and 2 HOH bending modes. Now, we have to identify the remaining 6 modes. These are the intermolecular vibrations (see Fig. 7.9a):

- Stretching of the hydrogen bond O–H···O (the vibration of two water molecules treated as entities): 183 cm^{-1}
- Bending of the hydrogen bond O–H···O in the plane of the figure: 345 cm^{-1}
- Bending of the hydrogen bond O–H···O in the plane perpendicular to the figure: 645 cm^{-1}
- Rocking of the hydrogen atom H_1 perpendicular to the figure plane: 115 cm^{-1}
- Rocking of the second water molecule (the right side of the Fig. a) in the figure plane: 131 cm^{-1}
- Rocking of the second water molecule (the right side of the Fig. a) about its symmetry axis: 148 cm^{-1}

As we can see, the intermolecular interactions have made the "intramolecular" vibration frequencies decrease,[43] while the "intermolecular" frequencies have very low frequencies. The last effect is nothing strange, of course, because a change of intermolecular distances does require a small expenditure of energy (which means small force constants). Note that the simple Morse oscillator model considered in Chapter 4 (p. 198) gave the correct order of magnitude of the intermolecular frequency of two water molecules (235 cm^{-1}, compared to the above, much more accurate, result of 183 cm^{-1}).

Example 3. The Formaldehyde Molecule $H_2C{=}O$

Each molecule of this useful substance is planar and consists of only four atoms, so the number of the normal vibrational modes should be equal to $3 \cdot 4 - 6 = 6$. The quantum-mechanical calculations based on the ground-state electronic energy as a function of the nuclear

[42] R.J. Reimers and R.O. Watts, *Chem. Phys. 85*, 83 (1984).

[43] This is how the hydrogen bonds behave. This seemingly natural expectation after attaching an additional mass to a vibrating system is legitimate when assuming that the force constants have not increased. An interesting example of the opposite effect for a wide class of compounds has been reported by Pavel Hobza and Zdenek Havlas [*Chem. Rev.*, *100*, 4253 (2000).].

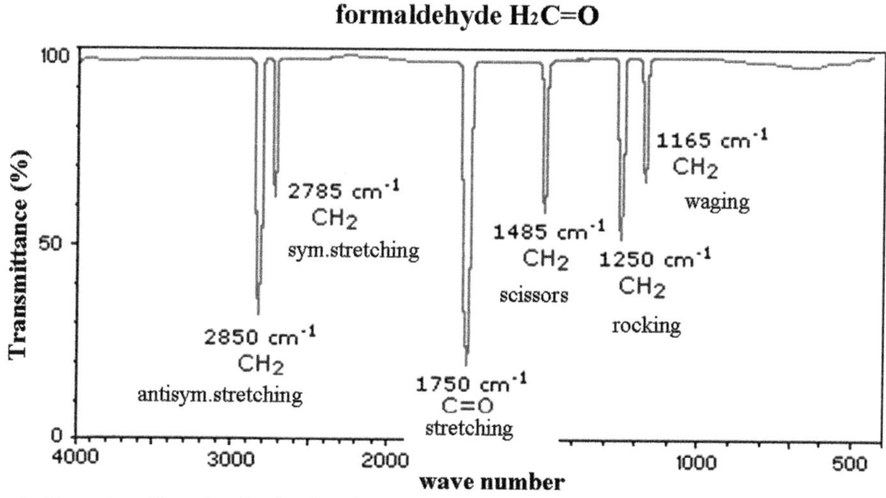

Fig. 7.10. An illustration of how the vibrational modes concept is related to the experimental IR spectrum, including an example of the formaldehyde molecule (a planar $H_2C{=}O$ system). The number of the vibrational modes $3N - 6 = 6$ is identical to the number of the absorption lines (each line corresponding to the vibrational excitation $v = 0 \rightarrow 1$). If you are interested in the intensity of these absorptions, you need to compute some integrals involving the corresponding vibrational wave functions.

coordinates[44] gave six normal modes. Fig. 7.10 shows the experimental infrared (IR) absorption spectrum. One sees six light frequencies ("*absorption lines*") for which a significant absorption occurs. Each of these lines is interpreted as an excitation of one of the six normal oscillators (modes),[45] each corresponding to a transition from the ground vibrational state with $v = 0$ to the first excited state with $v = 1$ of the corresponding harmonic oscillator. We have already some experience and may predict that the highest frequencies will correspond to the CH stretching vibrations in CH_2, because of the low mass of hydrogen atoms. One of the modes is symmetric (calculation: 2934 cm^{-1}, experiment: 2785 cm^{-1}), while the other is antisymmetric (calculation: 2999 cm^{-1}, experiment: 2850 cm^{-1}), similar to the water molecule. We expect a stretching vibration of the C=O bond to have a much lower frequency because much heavier atoms are involved (calculated: 1769 cm^{-1}, experiment: 1750 cm^{-1}). Bending vibrations are of lower energy than the stretching ones (for atoms of comparable masses). Indeed, for the scissor-like mode of CH_2, the computation gives 1546 cm^{-1}, while the experimental result is 1485 cm^{-1}. Next, the predicted by theory rocking vibration of CH_2 (that moves the CH_2 group within the molecular plane) has frequency 1269 cm^{-1}, whereas measurement gives 1250 cm^{-1}. Finally, a wagging motion (that moves the entire CH_2 group out of the molecular plane) corresponds to the calculated 1183 cm^{-1}, and measured 1165 cm^{-1} frequencies.

[44] I have used the Gaussian package with the coupled cluster method [CCSD(T), a procedure known as a rather reliable one, as described in Chapter 10; also, a reasonable quality of the atomic basis set 6-311G(d,p) was used; see Chapter 8].

[45] As one can see, these spectral lines exhibit certain widths (related to accompanying various rotational excitations).

7.5.2 Zero-Vibration Energy

The computed minimum of V (using any either the quantum-mechanical or force field method) does not represent the energy of the system for exactly the same reason why the bottom of the parabola (the potential energy) does not represent the energy of the harmonic oscillator (cf. p. 186). The reason is the kinetic energy contribution.

If all the normal oscillators are in their ground states ($v_j = 0$, called *zero-vibrations*), then the energy of the system is the energy of the bottom of the parabola $V_{\min}$ plus the zero-vibration energy (we assume no rotational contribution):

$$E = V_{\min} + \frac{1}{2}\sum_j (h\nu_j). \tag{7.13}$$

In the above formula, it has been assumed that the vibrations are harmonic. This assumption usually makes the frequencies higher by several percentage points (cf. p. 198).

Taking anharmonicity into account is a much more difficult task than normal mode analysis. Note (see Fig. 7.11) that in the anharmonic case the wave function becomes asymmetric with respect to $x = 0$ as compared to the harmonic wave function.

7.6 Molecular Dynamics (MD)

In all the methods described above, there is no such thing as temperature. It is difficult to tolerate such a situation.

This is what molecular dynamics is for. The idea is very simple.

If we knew the potential energy V as a function of the position ($\boldsymbol{R}$) of all the atoms (we may think here about a force field as its approximation[46]), then all the forces that the atoms undergo

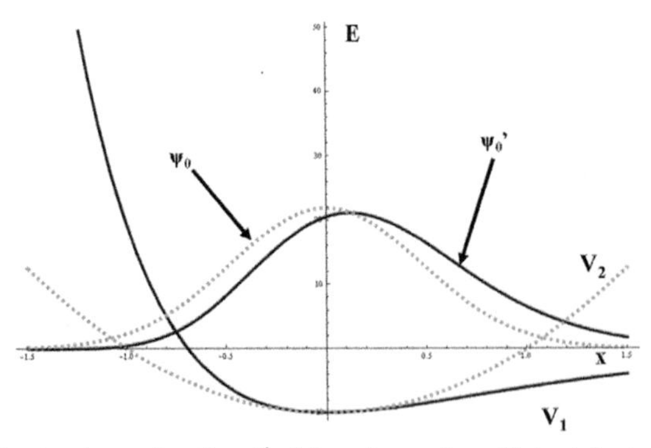

Fig. 7.11. The ground-state vibrational wave function ψ_0' of the anharmonic oscillator (of potential energy V_1, taken here as the Morse oscillator potential energy ——) is *asymmetric and shifted* toward positive values of the displacement when compared to the wave function ψ_0 for the harmonic oscillator with the same force constant (the potential energy V_2, •••).

[46] For more, refer back to p. 346.

could be easily computed. If $\boldsymbol{R} = (X_1, X_2, \ldots X_{3N})^T$ denotes the coordinates of all the N atoms (X_1, X_2, X_3 are the x, y, and z coordinates of atom 1; X_4, X_5, X_6 are the x, y, and z of atom 2; etc.), then $-\frac{\partial V}{\partial X_1}$ is the x-component of the force that atom 1 undergoes; $-\frac{\partial V}{\partial X_2}$ is the y-component of the same force; etc. When a force field is used, all this can be easily computed even analytically.[47] We had the identical situation in molecular mechanics, but there we were interested just in making these forces equal to zero (through obtaining the equilibrium geometry). In molecular dynamics, we are interested in time t, the velocity of the atoms (in this way temperature will come into play) and the acceleration of the atoms.

> Our immediate goal is collecting the atomic positions as functions of time; i.e., of the system trajectory.

The Newton equation tells us that, knowing the force acting on a body (e.g., an atom), we may compute the acceleration that the body undergoes. We have to know the mass, but there is no problem with that.[48] Hence the ith component of the acceleration vector is equal to

$$a_i = -\frac{\partial V}{\partial X_i} \cdot \frac{1}{M_i} \tag{7.14}$$

for $i = 1, 2, \ldots, 3N$ ($M_i = M_1$ for $i = 1, 2, 3$, $M_i = M_2$ for $i = 4, 5, 6$, etc.).

Now, let us assume that at $t = 0$, all the atoms have the initial coordinates $\boldsymbol{R}_0$ and the initial velocities[49] $\boldsymbol{v}_0$. Now we assume that the forces calculated act *unchanged* during a short period Δt (often 1 femtosecond, or 10^{-15} sec). We know what should happen to a body (atom) if under influence of a constant force during time Δt. Each atom undergoes a uniformly variable motion, and the new position may be found in the vector

$$\boldsymbol{R} = \boldsymbol{R}_0 + \boldsymbol{v}_0 \Delta t + \boldsymbol{a}\frac{\Delta t^2}{2}, \tag{7.15}$$

[47] That is, an analytical formula can be derived.

[48] We assume that what moves is the nucleus (see the discussion on p. 274). In MD, we do not worry about the fact that the nucleus moves together with its electrons, because the mass of all electrons of an atom is more than 1840 times smaller than that of the nucleus. Thus, the mass is negligible in view of other approximation, which have been made.

[49] Where could these coordinates be taken from? To tell the truth, it can almost be considered to happen randomly. We say "almost" because some essential things will be assured. First, we may quite reasonably conceive the geometry of a molecule, because we know which atoms form the chemical bonds, their reasonable lengths, the reasonable values of the bond angles, etc. That is, however, not all we would need for larger molecules. What do we take as dihedral angles? This is a difficult case. Usually we take a conformation, which we could call "reasonable." Shortly we will take a critical look at this problem. The next question is the velocities. Having nothing better at our disposal, we may use a random number generator, assuring, however, that the velocities are picked out according to the Maxwell-Boltzmann distribution suitable for a given temperature T of the laboratory; e.g., 300 K. In addition, we will make sure that the system does not rotate or flies off. In this way, we have our starting position and velocity vectors $\boldsymbol{R}_0$ and $\boldsymbol{v}_0$.

and its new velocity in the vector is

$$\boldsymbol{v} = \boldsymbol{v}_0 + \boldsymbol{a}\Delta t, \tag{7.16}$$

where the acceleration $\boldsymbol{a}$ is a vector composed of the acceleration vectors of all the N atoms

$$\begin{aligned} \boldsymbol{a} &= (\boldsymbol{a}_1, \boldsymbol{a}_2, \ldots \boldsymbol{a}_N)^T, \\ \boldsymbol{a}_1 &= \left(-\frac{\partial V}{\partial X_1}, -\frac{\partial V}{\partial X_2}, -\frac{\partial V}{\partial X_3}\right) \cdot \frac{1}{M_1}, \\ \boldsymbol{a}_2 &= \left(-\frac{\partial V}{\partial X_4}, -\frac{\partial V}{\partial X_5}, -\frac{\partial V}{\partial X_6}\right) \cdot \frac{1}{M_2} \\ &\text{etc.} \end{aligned} \tag{7.17}$$

Everything on the right side of Eqs. (7.15) and (7.16) is known. Therefore, the new positions and the new velocities are easy to calculate.[50] Now, we may use the new positions and velocities as starting ones and repeat the whole procedure over and over. This makes it possible to go along the time axis in a steplike way in practice, reaching even nanosecond times (10^{-9} sec), which means millions of such steps. The procedure described above simply represents the numerical integration of $3N$ differential equations. If $N = 2000$, then the task is impressive. It is so straightforward because we are dealing with a numerical solution, not an analytical one.[51]

7.6.1 What Does MD Offer Us?

The computer simulation makes the system evolve from the initial state to the final one. The position $\boldsymbol{R}$ in $3N$–dimensional space becomes a function of time, so $\boldsymbol{R}(t)$ represents the trajectory of the system in the configurational space. A similar statement pertains to $\boldsymbol{v}(t)$. Knowing the trajectory means that we know the smallest details of the motion of all the atoms. Within the approximations used, we can therefore answer any question about this motion. For example, we may ask about some mean values, like the mean value of the total energy, potential energy, kinetic energy, the distance between atom 4 and atom 258, etc. All these quantities may be computed at any step of the procedure, then added up and divided by the number of steps, giving the mean values we require. In this way, we may obtain the theoretical prediction of the mean value of the interatomic distance and then compare it to, say, the NMR result.

In this way, we may search for some correlation of motion of some atoms or groups of atoms; i.e., the *space correlation* ("*when this group of atoms is shifted to the left, then another group is most often shifted to the right*") or the *time correlation* ("*when this thing happens to the functional group* G_1, *then after a time* τ *that most often takes place with another functional group* G_2") or *time autocorrelation* ("*when this happens to a certain group of atoms, then after time* τ *the same most often happens to the same group of atoms*"). For example, is the x-coordinate of atom 1 (X_1) correlated to the coordinate y of atom 41 (X_{122}), or are these two quantities absolutely independent? The answer to this question is given by the correlation

[50] In practice, we use a more accurate computational scheme called the *leap frog algorithm*.

[51] By the way, if somebody gave us the force field for galaxies (this is simpler than for molecules), we could solve the problem as easily as in our case. This is what astronomers often do.

coefficient $c_{1,122}$ calculated for M simulation steps in the following way:

$$c_{1,122} = \frac{\frac{1}{M}\sum_{i=1}^{M}\left(X_{1,i} - \langle X_1\rangle\right)\left(X_{122,i} - \langle X_{122}\rangle\right)}{\sqrt{\left(\frac{1}{M}\sum_{i=1}^{M}\left(X_{1,i} - \langle X_1\rangle\right)^2\right)\left(\frac{1}{M}\sum_{i=1}^{M}\left(X_{122,i} - \langle X_{122}\rangle\right)^2\right)}},$$

where $\langle X_1\rangle$ and $\langle X_{122}\rangle$ denote the mean values of the coordinates indicated, and the summation goes over the simulation steps. It is seen that any deviation from independence means a nonzero value of $c_{1,122}$. What could be more correlated to the coordinate X_1 than the same X_1 (or $-X_1$)? Of course, absolutely nothing. In such a case (in the formula, we replace $X_{122,i} \to X_{1,i}$ and $\langle X_{122}\rangle \to \langle X_1\rangle$), we obtain $c_{1,1} = 1$ or -1. Hence, c always belongs to $[-1, 1]$, $c = 0$ means independence, $c \pm 1$ means maximum dependence.

Has molecular dynamics anything to do with reality?

If the described procedure were applied without any modification, then most probably we would have bad luck and our $\boldsymbol{R}_0$ would be located on a slope of the hypersurface V. Then, the solution of the Newton equations would reflect what happens to a point (representing the system) when placed on the slope–that is it would slide downhill. The point would go faster and faster, and soon the vector $\boldsymbol{v}$ would not correspond to the room temperature, but, say, to 500 K. Of course, despite such a high temperature, the molecule would not disintegrate because this is not a real molecule, but one operating with a force field that usually corresponds to unbreakable chemical bonds.[52] Although the molecule will not fall apart, such a large T has nothing to do with the temperature of the laboratory. This suggests that after some number of steps, we should check whether the atomic velocities still correspond to the proper temperature. If not, it is recommended to scale all the velocities by multiplying them by a factor in order to make them correspond again to the desired temperature. For this reason, the only goal of the first part of a molecular dynamics simulation is known as *thermalization*, in which the error connected to the nonzero Δt is averaged and the system is forced stepwise (by scaling) to behave as what is called the *canonical ensemble*. The canonical ensemble preserves the number of molecules, the volume, and the temperature (simulating contact with a thermostat at temperature T). In such a thermalized system, total energy fluctuations are allowed.

Now that the thermalization has completed, the next (main) stage of molecular dynamics begins i.e., the harvesting of data (trajectory).

7.6.2 What Should We Worry About?

- During simulation, the system has to have enough time to wander through all parts of the phase space[53] that are accessible in the experimental conditions (with which the simulation is to be compared). We are never sure that this happens. We have to check whether the computed mean values depend upon the simulation time. If they do not, then probably everything is all right–we have a description of the equilibrium state.

[52] This pertains to a single molecule bound by chemical bonds; a system of several molecules could fall apart.

[53] This is Cartesian space of all atomic positions and momenta.

- The results of the MD (the mean values) should not depend on the starting point because it has been chosen arbitrarily. This is usually satisfied for small molecules and their collections. For large and flexible molecules, we usually start from the vector $\boldsymbol{R}_0$ found from X-ray determined atomic positions. Why? Because *after* the MD, we still will stay close to this all in all experimental conformation. If the simulation started from another conformation, it would result in a conformation close to this new starting point. This is because even with the most powerful computers, simulation times are too short. In such a way, we have a simulation of one conformation evolution rather than a description of the thermodynamic equilibrium.
- The simulation time in the MD is limited on one side by the power of computers and on the other side by the time step Δt, which is not infinitesimally small, and it creates an error that cumulates during the simulation (as a result, the total energy may vary too much and the system may be heading into non-physical regions of the phase space).

7.6.3 MD of Non-Equilibrium Processes

We do not always want thermalization. We may be interested what happens, when a DNA molecule being anchored to a solid surface by one of its end functional groups is distorted by pulling the other end of the molecule. Such MD results may nowadays be compared to the corresponding experiment.

Here is yet another example. A projectile hits a wall ("armor"), Fig. 7.12. The projectile is composed of Lennard-Jones atoms (with some ε_p and $r_{e,p}$; p. 347), and we assume the same for the wall (for other values of the parameters, let us make the wall less resistant than the projectile: $\varepsilon_w < \varepsilon_p$ and $r_{e,w} > r_{e,p}$). All together, we may have hundreds of thousands (or even millions) of atoms (i.e., there are millions of differential equations to solve). Now, we prepare the input $\boldsymbol{R}_0$ and $\boldsymbol{v}_0$ data. The wall atoms are assumed to have stochastic velocities drawn from the Maxwell-Boltzmann distribution for room temperature. The same for the projectile atoms, but additionally, they have a constant velocity component along the direction pointing to the wall. At first, nothing particularly interesting happens–the projectile flies toward the wall with a constant velocity (while all the atoms of the system vibrate). Of course, it is most interesting when the projectile hits the wall. Once the front part of the projectile touches the wall, the wall atoms burst into space in a kind of eruption, the projectile's tip loses some atoms, and the spot on the wall hit by the projectile vibrates and sends a shock wave.

Among more civil applications of this theory, we may use this method to plan better drills and better steel plates, as well as about other micro-tools which have a bright future.

7.6.4 Quantum-Classical MD

A typical MD does not allow for breaking chemical bonds, and the force fields that allow this give an inadequate, classical picture, so a quantum description is sometimes a must. The systems treated by MD are usually quite large, which excludes a full quantum-mechanical description.

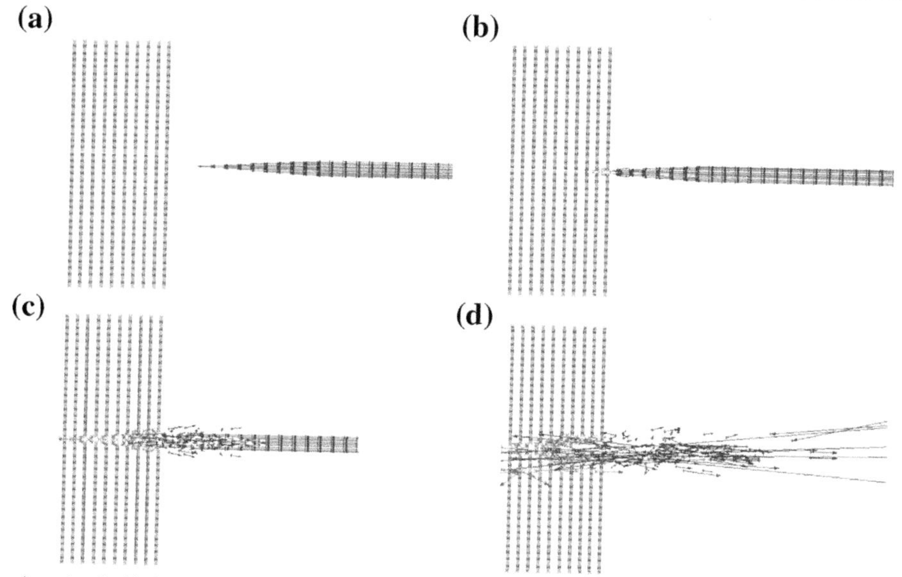

Fig. 7.12. A projectile hitting an armor–an example of simulating non-equilibrium processes in molecular dynamics. (a) On the right is a projectile flying left, on the left the armor, both objects built of particles ("*atoms*") that interact according to the Lennard-Jones interatomic potential. The Lennard-Jones parameters for the projectile atoms are as follows: $\varepsilon = 0.2$ a.u., $r_e = 2.3$ a.u., for those of the armor: $\varepsilon = 0.1$ a.u., $r_e = 2.6$ a.u. (the projectile-armor interaction parameters correspond to their mean values). The projectile atoms are denoted by $+$, the armor atoms by $\times$. At the starting point shown, the armor atoms all have the velocity equal to zero, while all the projectile atoms have only the x-component of velocity (x is the left-right axis), which was equal to $v_x = -55000\ \frac{\text{km}}{\text{h}}$ (i.e., the missile is heading right into the armor). (b) An early stage of the hit. The top of the missile undergoes a destruction, the armor atoms in the epicenter accelerate (arrows) and change their position. It is remarkable that this motion is limited for the time being to some small region, and that one can see something that is similar to an intact missile ("*a virtual missile*"), which, however, is built from the atoms of both colliding objects. (c) Half of the missile is already destroyed, and one can see quite a lot of particles that bounce off the armor and go back very fast, while the missile was able to crush the armor width. Despite of this, the abovementioned "*virtual missile*" is still visible. (d) The armor is destroyed, a lot of debris is flying off, and one does not see any virtual missile anymore. These results were obtained by Marcin Gronowski supervised by the author, the undergraduate course at the University of Warsaw, Department of Chemistry.

For enzymes (natural catalysts), researchers proposed[54] joining the quantum and the classical descriptions by making the precision of the description dependent on how far the region of focus is from the enzyme active center (where the reaction the enzyme facilitates takes place). They proposed dividing the system (enzyme + solvent) into three regions:

- Region I represents the active center atoms.
- Region II is composed of the other atoms of the enzyme molecule.
- Region III is the solvent.

[54] P. Bała, B. Lesyng, and J.A. McCammon, in *Molecular Aspects of Biotechnology: Computational Methods and Theories*, Kluwer Academic Publishers, (1992), p. 299. A similar philosophy stands behind the Morokuma's ONIOM procedure, as described in M. Svensson, S. Humbel, R.D.J. Froese, T. Matsubara, S. Sieber, and K. Morokuma, *J. Phys. Chem. 100*, 19357 (1996).

Region I is treated as a quantum mechanical object and described by the proper time-dependent Schrödinger equation, while region II is treated classically by the force-field description and the corresponding Newton equations of motion and region III is simulated by a continuous medium (no atomic representation) with a certain dielectric permittivity.

The regions are coupled by their interactions: quantum mechanical Region I is subject to the external electric field produced by Region II evolving according to its MD as well as that of Region III; and Region II feels the charge distribution changes that Region I undergoes through electrostatic interaction.

7.7 Simulated Annealing

The goal of MD may differ from simply computing some mean values; e.g., we may try to use MD to find regions of the configurational space for which the potential energy V is particularly low.[55] From a chemist's point of view, this means trying to find a particularly stable structure (conformation of a single molecule or an aggregate of molecules). To this end, MD is sometimes coupled with an idea of Kirkpatrick et al.[56], taken from an ancient method of producing metal alloys of exceptional quality (the famous steel of Damascus), and serving to find the minima of arbitrary functions.[57] The idea behind simulated annealing is extremely simple.

This goal is achieved by a series of heating and cooling procedures (called the *annealing protocol*). First, a starting configuration is chosen that, to the best of our knowledge, is of low energy, and the MD simulation is performed at a high temperature T_1. As a result, the system (represented by a point $\boldsymbol{R}$ in the configuration space) rushes through a large manifold of configurations $\boldsymbol{R}$; i.e., wanders over a large portion of the hypersurface $V(\boldsymbol{R})$. Then, a lower temperature T_2 is chosen and the motion slows down, the visited portion of the hypersurface shrinks and hopefully corresponds to some regions of low values of V–the system is confined in a large superbasin (composed of individual minima basins). Now the temperature is raised to a certain value $T_3 < T_1$, thus allowing the system eventually to leave the superbasin and to choose another one, maybe of lower energy. While the system explores the superbasin, the system is cooled again, this time to temperature $T_4 < T_2$, and so forth. Such a procedure does not give any guarantee of finding the global minimum of V, but there is a reasonable chance of getting a configuration with lower energy than the start. The method, being straightforward to implement, is very popular. Its successes are spectacular, although sometimes the results are disappointing. The highly prized swords made in ancient Damascus using annealing, prove that the metal atoms settle down in quasi-optimal positions forming a solid state of low energy, which is very difficult to break or deform.

---

[55] This is similar to global molecular mechanics.

[56] S. Kirkpatrick, C.D. Gelatt Jr., and M.P. Vecchi, *Science, 220*, 671 (1983).

[57] I recommend a very good book: W.H. Press, B.P. Flannery, S.A. Teukolsky, and W.T. Vetterling, *Numerical Recipes: The Art of Scientific Computing*, Cambridge University Press, New York, 2007.

7.8 Langevin Dynamics

In the MD, we solve Newton equations of motion for all atoms of the system. Imagine we have a large molecule in an aqueous solution (biology offers us important examples). We have no chance to solve Newton equations because there are too many of them (a lot of water molecules). What do we do then? Let us recall that we are interested in the macromolecule; the water molecules are interesting only as a medium that changes the conformation of the macromolecule. The changes may occur for many reasons, but the simplest is the most probable–just the fact that the water molecules in their thermal motion hit the atoms of the macromolecule. If so, their role is reduced to a source of chaotic strikes. The main idea behind Langevin dynamics is to ensure that the atoms of the macromolecule indeed feel some random hits from the surrounding medium *without taking this medium into consideration explicitly. This is the main advantage of the method.*

Paul Langevin (1872–1946), French physicist and professor at the College de France. His main achievements are in the theory of magnetism and in relativity theory. His Ph.D. student Louis de Broglie made a breakthrough by attributing wave properties to particles in quantum theory.

A reasonable part of this problem may be incorporated into the Langevin equation of motion:

$$M_i \ddot{X}_i = -\frac{\partial V}{\partial X_i} + F_i\left(t\right) - \gamma_i M_i \dot{X}_i, \tag{7.18}$$

for $i = 1, 2, \ldots, 3N$, where besides the force $-\nabla V$ resulting from the potential energy V for the macromolecule alone, we also have an additional stochastic force $\boldsymbol{F}\left(t\right)$, whose magnitude and direction are drawn keeping the force related to the temperature and assuring its isotropic character. The coefficient γ_i is a friction coefficient and the role of friction is proportional to atomic velocity.

The Langevin equations are solved in the same way as those of MD, with the additional stochastic force drawn using a random number generator.

7.9 Monte Carlo Dynamics

Las Vegas, Atlantic City, and Monte Carlo are notorious for their day and night use of such random number generators as billiards, roulette, and cards. Because of this, the idea and even the name of Monte Carlo (MC) have been accepted in mathematics, physics, chemistry, and biology. The key concept is that a random number, when drawn successively many times, may serve to create a sequence of system snapshots.

All this began from an idea of a mathematician from Lwów, in Poland (now Lviv in the Ukraine) named Stanisław Marcin Ulam.

Stanisław Ulam (1909–1984), first associated with the University of Lwów, then professor at Harvard University, the University of Wisconsin, the University of Colorado, and Los Alamos National Laboratory. In Los Alamos, Ulam solved the most important bottleneck in hydrogen bomb construction by suggesting that pressure is the most important factor and that sufficient pressure could be achieved by using the *atomic* bomb as a detonator. Using this idea and an idea of Edward Teller about further amplification of the ignition effect by implosion of radiation, both scholars designed the hydrogen bomb. They both owned the U.S. patent for H-bomb production.

According to the *Ulam Quarterly Journal* (http://www.ulam.usm.edu/editor.html), Ulam's contribution to science includes logic, set theory, measure theory, probability theory, computer science, topology, dynamic systems, number theory, algebra, algebraic and arithmetic geometry, mathematical biology, control theory, mathematical economy, and mathematical physics. He developed and coined the name of the Monte Carlo method, and also invented the cellular automata method (described at the end of this chapter). Ulam wrote a very interesting autobiography *Adventures of a Mathematician.* During the A-bomb project, Ulam lamented that he always worked with symbols, not numbers, he is driven so low as to use them now, even worse: these are numbers with decimal points!

The picture shows one of the "*magic places*" of the world science, the *Szkocka Café*, on Akademicka Street, Lwów. At this café, now a bank at Prospekt Szewczenki 27, before World War II, young Polish mathematicians (including Stanisław Ulam and the mathematical genius Stefan Banach) made a breakthrough thereafter called the "*Polish school of mathematics*."

Perhaps an example will best explain the MC method. I have chosen the methodology introduced to the protein folding problem by Andrzej Koliński and Jeffrey Skolnick.[58] In a version of this method, we use a simplified model of the real protein molecule, a polymer composed of monomeric peptide units ···NH–CO–CHR–···, as a chain of pointlike entities NH–CO–CH from which protrude points representing various side chains R. The polymer chain goes through the vertices of a crystallographic lattice (the side chain points can also occupy only the lattice vertices), which excludes a lot of unstable conformations and enables us to focus on those chemically/physically relevant. The lattice representation speeds computation by several orders of magnitude.

The reasoning goes as follows. The nonzero temperature of the water that the protein is immersed in makes the molecule acquire random conformations all the time. The authors

[58] J. Skolnick, and A. Koliński, *Science, 250*, 1121 (1990).

assumed that a given starting conformation is modified by a series of random micro-modifications. A micro-modification allowed has to be chosen so as to obey three rules:

- It must be chemically/physically acceptable.
- It must always be local; i.e., they have to pertain to a small fragment of the protein, because in the future we would like to simulate the kinetics of the protein chain (how a conformational change evolves).
- When repeated, they should be able to transform any conformation into any other conformation of the protein.

In this way, we can modify the molecular conformation, but we want the protein to move; i.e., to have the *dynamics* of the system (a sequence of molecular conformations, each one derived from the previous one in a physically acceptable way).

To this end, we have to be able to write down the energy of any given conformation. This is achieved by assigning an energy award (i.e., energy lowering) if the configuration corresponds to intramolecular energy gain (e.g., trans conformation, the possibility of forming a hydrogen bond or a hydrophobic effect; see Chapter 13), and an energy penalty for intramolecular repulsion (e.g., cis conformation, or when two fragments of the molecule are to occupy the same space). It is generally better if the energy awards and penalties are accompanied by an ingredient that reflects the complex reality of the Protein Data Bank, the most extended database. This database may serve to add what is known as the *statistical interaction potential*. The potential is a correction computed from the frequency of finding two kinds of amino acids close in space (e.g., glutamic acid and lysine; there are 20 natural amino acids) in the Protein Data Bank. If the frequency is larger than just the random one, we deduce that an attraction has to occur between them and assign an energy gain that takes into account the value of this higher frequency.[59]

Now we have to let the molecule move. We start from an arbitrarily chosen conformation and calculate its energy E_1. Then, a micro-modification, or even a series of micro-modifications (this way the calculations go faster), is drawn from the micro-modifications list and applied to the molecule. Thus, a new conformation is obtained with energy E_2. Now the most important step takes place. We decide to *accept* or *reject* the new conformation according to the *Metropolis criterion*[60], which gives the probability of the acceptance as

$$P_{1\to 2} = \begin{cases} 1 & \text{if } E_2 \leq E_1 \\ a = \exp\left(-\frac{(E_2 - E_1)}{k_B T}\right) & \text{if } E_2 > E_1. \end{cases}$$

Well, we have a *probability*, but what we need is a clear decision: *to be or not to be in state* 2? This is where the MC spirit comes in; see Fig. 7.13. By using a random number generator, we

[59] Such an approach may be certainly considered as a bit too pragmatic. People believe that before coming to a true force field that will give us correct answers (in the distant future), we have a lot of time to explore what kind of corrections the current force fields need to get reasonable answers.

[60] N. Metropolis, A.W. Rosenbluth, M.N. Rosenbluth, A.H. Teller, and E. Teller, *J. Chem. Phys.*, *21*, 1087 (1953).

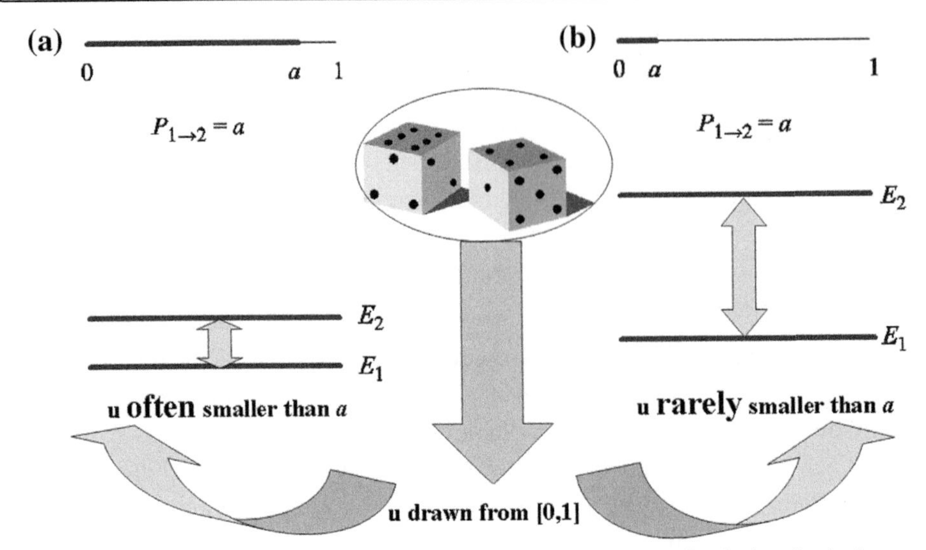

Fig. 7.13. Metropolis algorithm. (a) If E_2 is only a little higher than E_1, then the Metropolis criterion often leads to accepting the new conformation (of energy E_2). (b) On the other hand, if the energy difference is large, then the new conformation is accepted only rarely.

draw a random number[61] u from section [0, 1] and compare it with the number a. If $u \leq a$, then we accept the new conformation; otherwise, conformation 2 is rejected (and we forget about it). The whole procedure is repeated over and over again: drawing micro-modifications → a new conformation → comparison with the old one by the Metropolis criterion → accepting (i.e., the new conformation becomes the current one) or rejecting (i.e., the old conformation remains the current one), etc.

The Metropolis criterion is one of those mathematical tricks that a chemist has to know about. Note that the algorithm always accepts the conformation 2 if $E_2 \leq E_1$; therefore, it will have a tendency to lower the energy of the current conformation. On the other hand, when $E_2 > E_1$, the algorithm may decide to increase the energy by accepting the higher energy conformation 2. If $\frac{(E_2 - E_1)}{k_B T} > 0$ is small, the algorithm accepts the new conformation very easily (Fig. 7.13a). On the other hand, an attempt of a very high jump (Fig. 7.13b) in energy may be successful in practice only at very high temperatures. The algorithm prefers higher energy conformations to the same extent as the Boltzmann distribution. Thus, grinding the mill of the algorithm on and on (sometimes it takes months on the fastest computers of the world) and making statistics of the number of accepted conformations as a function of energy, we arrive at the Boltzmann distribution as it should be in thermodynamic equilibrium.

Thus, as the mill grinds, we can make a movie which would reveal how the protein molecule behaves at high temperatures: the protein acquires practically any new conformation generated by the random micro-modifications, and it looks as if the molecule is participating in a

[61] The current situation is like this: there are deterministic computer programs, which are claimed to generate random numbers. According to John von Neumann those who use such programs "*live in the state of sin.*"

kind of rodeo. However, we decide the temperature. Thus, let us decide to lower the temperature. Until a certain temperature, we will not see any qualitative change in the rodeo, but at a sufficiently low temperature, we can recognize that something has happened to the molecule. From time to time, some local structures typical of the secondary structures of proteins (the α–helices and the zigzag type β–strands, the latter like to bind together laterally by hydrogen bonds) emerge and vanish, emerge again, etc.

When the temperature decreases, at a certain critical value, T_{crit}, a stable structure suddenly emerges (an analog of the so-called native structure; i.e., the one ensuring the molecule can perform its function in nature).

The structure vibrates a little, especially at the ends of the protein, but further cooling does not introduce anything new. The native structure exhibits a unique secondary structure pattern along the polymeric chain (i.e., definite sections of the α and β structures) which packs together into a unique *tertiary structure*. In this way, a highly probable scenario for the *coil-globule* was demonstrated for the first time by Koliński and Skolnick.

One of the most successful variations of the abovementioned basic MC algorithm is what is known as Monte Carlo with Replica Exchange (MCRE). The idea is to speed up the exploration of the conformational space by making parallel computations for the same system ("*replica*"), but each computation differing by the assumed and fixed temperature, from very low to very high. Then, during the simulation, one stochastically exchanges the replicas in a way that is analogous to the Metropolis criterion (instead of E_1 and E_2, one considers $\frac{E_1}{k_B T_1}$ and $\frac{E_2}{k_B T_2}$). This simple idea makes easier to overcome even large energy barriers.

Predicting the 3-D structure of globular proteins from the sequence of amino acids[62] and using a sophisticated algorithm[63] based on all relevant physical interactions and adjusted using the accumulated structural knowledge of proteins (nearly 100000 structures in the Protein Data Bank, no bias towards the target) is feasible nowadays with a remarkable accuracy. An example is shown in the section Ad Futurum.

Example: Conformational Autocatalysis as a Model of Prion Disease Propagation

In biology, there is a concept of "*contagious misfolded proteins*" (prions), which are supposed to increase their quantity spontaneously at the expense of the native fold of the same protein.

[62] This problem is sometimes called "*the second genetic code*" in the literature. This name reflects the final task of obtaining information about protein function from the "*first genetic code*" (i.e., DNA) information that encodes protein production.

[63] J. Skolnick, Y. Zhang, A.K. Arakaki, A. Koliński, M. Boniecki, A. Szilagyi, and D. Kihara, *Proteins*, 53 (2003), Suppl. 6, 469–479.

This phenomenon is suspected to be the cause of dangerous diseases such as Creutzfeldt-Jakob disease in humans or "mad cow disease" in cattle.

The elaborated force field and the above described MCRE technique proved to predict the 3-D structure of many globular proteins to a very good accuracy (e.g., of the order of 2 Å of the root mean square deviation, rms). Having in hand such a tool, one may consider questions that are related to the prion disease propagation.

Imagine a protein with an amino-acid sequence designed in such a way that the protein folds to a certain conformation (A) being its lowest-energy conformation ("*native*"). The sequence, however, also has another state (conformation B), that widely differs structurally from A and is metastable. Thus, there is a frustration introduced on purpose in the molecule.[64] The protein will usually fold to A, but if some external factors are present, it may prefer to fold to B. These external factors might be simply the intermolecular interaction of several such molecules. For example, if the lowest energy of the dimer corresponded to the BB conformation, a molecule in conformation A, when in contact with another molecule in conformation B, might change from AB to BB. This would mean eliminating of the most stable conformation of a protein molecule just by contact with a contagious, metastable ("misfolded") form, which resembles spreading of a kind of "conformational disease."

If such a scenario were possible, then is the theory strong enough to predict the necessary amino acid sequence that would behave like that? Well, let us consider an oligopeptide composed of 32 amino acids only, which is small enough to make many series of computations feasible. There is already some computational experience accumulated, which serves as a guideline in planning the amino-acid sequence.

First of all, to make the effect seen clearly, we plan to have A and B structurally distinct as much as possible. Let one of them be of α-helical (A), the other (B) of β-sheet character, both quite compact ones ("*globular*"). To induce this compactness, let us locate two glycines in the middle of the sequence because we want the chain to bend easily there, changing its direction in space (a "*hairpin-like*" structure). Glycine is unique among the 20 amino acids given in nature; it is known for its flexibility, since it has as the side chain a very small object– the hydrogen atom (no sterical hindrance). Two other glycines have been put at positions 8 and 25, exactly where we plan to have some increased flexibility necessary for the planned β-sheet (we plan four β-strands bound together by the hydrogen bonds). We hope these latter glycines will not break the two α-helices to be formed and packed in the hairpin-like structure. Well, we would like the native structure (A) to have an additional stability. To this end, we will introduce the interactions within the four pairs of glutamic acid and lysine (they interact through the electrostatic attraction $-NH_3^+ \cdots {}^-OOC-$) in the positions that ensure that they are close in space, but only in a pattern typical for an α-helix. However, to increase a chance of forming β structures (to make the B structure stable), the sequence has been enriched by a pattern of hydrophobic valines and isoleucines (they are known for forming what is called the *valine-leucine zipper*) in the amino-acid sequence, their positions assuring a strong interaction

[64] E. Małolepsza, M. Boniecki, A. Koliński, and L. Piela, *Proc. Natl Acad. Sciences (USA)*, *102*, 7835 (2005).

within a β structure. Although the water structure (necessary for the hydrophobic effect) is not taken into account explicitly in the force field used, it is present implicitly through the force field; e.g., the statistical amino-acid interaction potential from the Protein Data Bank (which does take it into account).

To complete such a design, the amino-acid sequence has been tuned, increasing the propensity toward α-helices or β-sheets. The calculations have been repeated independently for many sequences, each time checking the behavior of single molecules, as well as dimers and trimers. Among the final 14 sequences, only the following:

GVEIAVKGAEVAAKVGGVEIAVKAGEVAAKVG

(G = glycine, V = valine, E = glutamic acid, I = isoleucine, A = alanine, K = lysine) exhibited the desired conformational autocatalysis.

It turned out in the MCRE procedure that the oligopeptide molecule always attains the global minimum conformation in the form of a two-helix bundle (see Fig. 7.14a-b), if the temperature is in a certain range. Below this range, not only the two-helix bundle, but also a (very different from the latter one) four-member β-barrel (see Fig. 7.14c), are stable. If two protein molecules interact, one of them frozen (for whatever reasons, chemical of physical) in its metastable β–barrel conformation, while the second molecule is free to move (Fig. 7.14d), the second molecule practically always folds to the β–barrel that interacts very strongly with the frozen β–barrel (Fig. 7.14e). This happens even when the second molecule starts from its native (i.e., α-helical) form: the α-helical form unfolds and then folds to the β–barrel (Fig. 7.14e). It has been also demonstrated that a third protein molecule, when in the presence of the two frozen β-barrels, folds to the β-barrel that fits very well to the two forming a stack of three β-barrel molecules. Formation of such stacks is notorious for prion diseases.

This model may be seen as a prototype of a prion disease propagation, in which a metastable "incorrect conformation" spontaneously spreads out in the system.

7.10 Car-Parrinello Dynamics

Despite the fact that this textbook is based on the assumption that the reader has completed a basic quantum chemistry course, the author (as already stated in the Introduction) does not profit from this too extensively. Car-Parrinello dynamics is an exception. It positively belongs in this chapter, while borrowing heavily from the results of Chapter 8. If the reader feels uncomfortable with this, this section may just be skipped.

We have already listed a few problems associated with the otherwise powerful MD. We have also mentioned that the force field parameters (e.g., the net atomic charges) do not vary when the conformation changes or when two molecules approach, whereas everything has to change. Car and Parrinello[65] thought of a remedy in order to make the parameters change "*in flight*."

[65] R. Car and M. Parrinello, *Phys. Rev. Letters*, *55*, 2471 (1985).

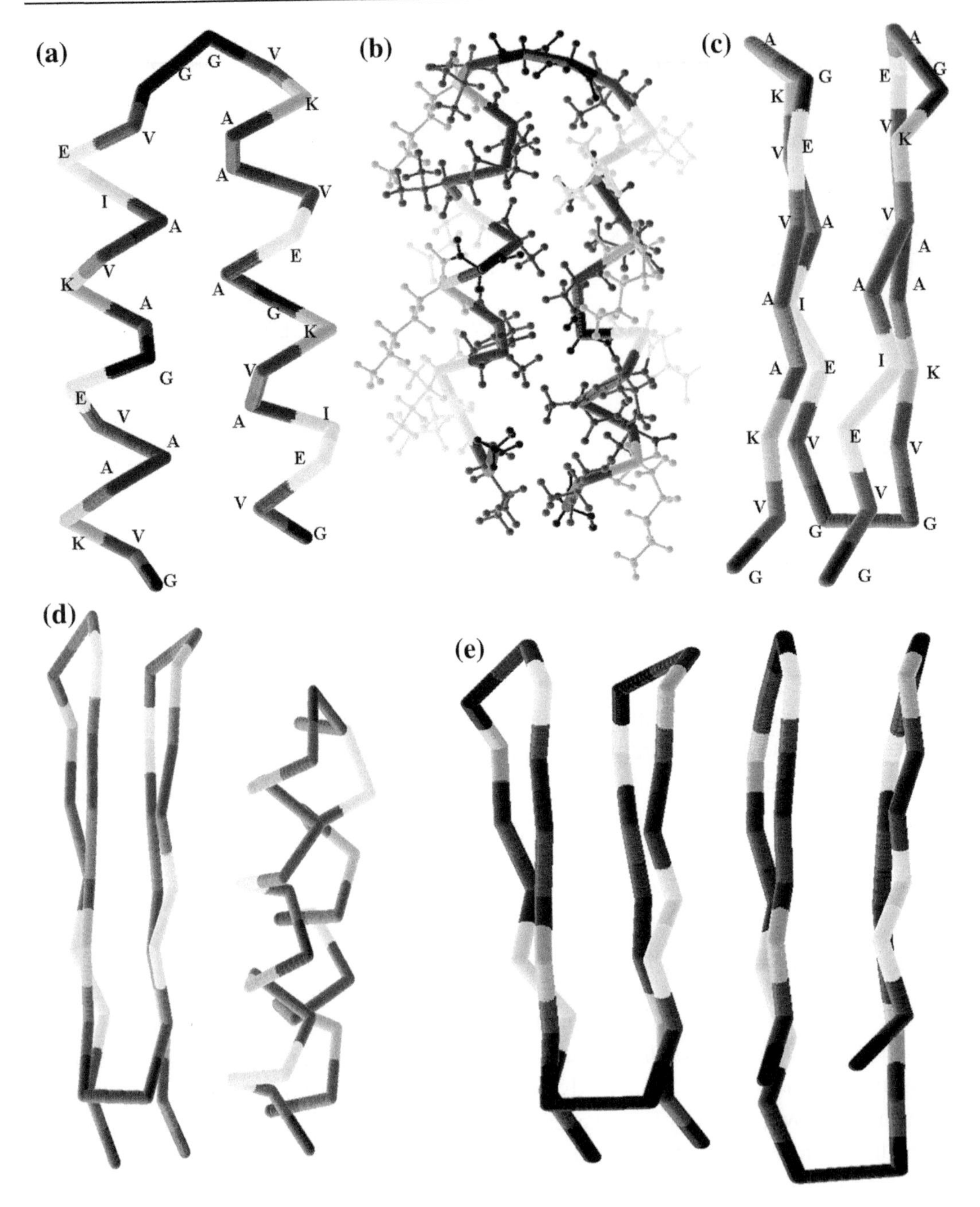

Fig. 7.14. (Continued)

Let us assume the one-electron approximation.[66] The total electronic energy $E_0^0(\boldsymbol{R})$ is (in the Born-Oppenheimer approximation) not only a function of the positions of the nuclei, but also a functional of the spinorbitals $\{\psi_i\} : V = V\left(\boldsymbol{R}, \{\psi_i\}\right) \equiv E_0^0(\boldsymbol{R})$.

The function $V = V\left(\boldsymbol{R}, \{\psi_i\}\right)$ will be minimized with respect to the positions $\boldsymbol{R}$ of the nuclei and the spinorbitals $\{\psi_i\}$, depending on the electronic coordinates.

If we are going to change the spinorbitals, we have to take care of their orthonormality at all stages of the change.[67] For this reason, Lagrange multipliers (see Appendix N available at booksite.elsevier.com/978-0-444-59436-5) appear in the equations of motion. We obtain the following set of Newton equations for the motion of N nuclei:

$$M_I \ddot{X}_I = -\frac{\partial V}{\partial X_I} \quad \text{for} \quad I = 1, \ldots, 3N$$

and an equation of motion for each spinorbital (each corresponding to the evolution of one electron probability density in time):

$$\mu \ddot{\psi}_i = -\hat{F}\psi_i + \sum_{j=1} \Lambda_{ij}\psi_j, \tag{7.19}$$

where μ is a *fictitious parameter*[68] for the electron, $\hat{F}$ is a Fock operator (see Chapter 8, p. 407), and Λ_{ij} are the Lagrange multipliers to ensure the orthonormality of the spinorbitals ψ_j.

Both equations are quite natural. The first (the Newton equation) says that a nucleus has to move in the direction of the force acting on it ($-\frac{\partial V}{\partial X_I}$) and the larger the force and the smaller the mass, the larger the acceleration achieved. This sounds correct. The left side of the second equation and the first term on the right side say the following: Let the spinorbital ψ_i change in such a way that the orbital energy has a tendency to go down (in the sense of the mean value).

Fig. 7.14. Conformational autocatalysis. (a) a two α-helix bundle conformation of a protein with a particular amino acid sequence: GVEIAVKGAEVAAKVGGVEIAVKAGEVAAKVG (G = glycine, V = valine, E = glutamic acid, I = isoleucine, A = alanine, K = lysine). This 3D structure is shown in a simplified way by representing a single amino acid by a point connected to its neighbors; (b) The all heavy atom representation of such structures can be recovered, an example for the two α-helix bundle. The structure corresponds to the global minimum of the protein molecule. Note the stabilizing interaction of the hydrophobic amino acids (AA,VA,VV) between the helices, as well as the electrostatic KE stabilization within them. (c) A β-barrel conformation (metastable one) of the same protein. The stabilization of this structure also comes from similar interactions, this time acting between the β-strands. (d) The starting configuration for the MCRE procedure: the frozen β-barrel (left) interacts with the two α-helix bundle (right), which is subject to the MCRE dynamics. (e) The final result after a long MCRE run: the α-helix bundle has unfolded and then refolded, but to a β-barrel conformation, thus producing two interacting β-barrels (autocatalysis).

[66] The approximation will be described in Chapter 8 and is based on assuming the wave function in the form of a single Slater determinant built of orthonormal spinorbitals. Car and Parrinello give their procedure for the density functional theory (DFT), where a single Slater determinant also plays an important role.

[67] This is because the formulas they satisfy are valid under this condition.

[68] We may call it "*mass*." In practical applications, μ is large, usually taken as a few hundreds of the electron mass, because this ensures the agreement of theory and experiment.

How does this follow from the equations? From a basic course in quantum chemistry (this will be recalled in Chapter 8), we know that the orbital energy may be computed as the mean value of the operator $\hat{F}$ with the spinorbital ψ_i; i.e., $\left\langle \psi_i \mid \hat{F}\psi_i \right\rangle$. To focus our attention, let us assume that ψ_i is localized in a small region of space (see Fig. 7.15).

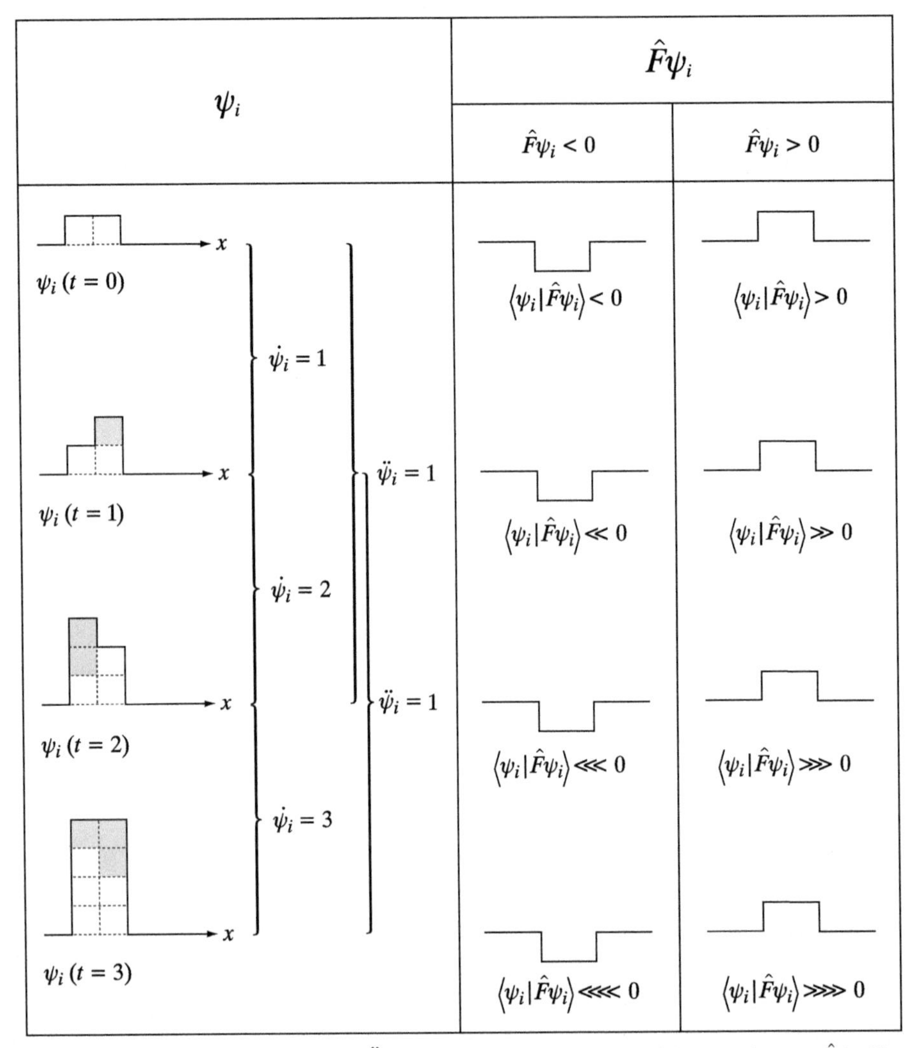

Fig. 7.15. A scheme showing why the acceleration $\ddot{\psi}_i$ of the spinorbital ψ_i has to be of the same sign as $-\hat{F}\psi_i$. Time (arbitrary units) goes from up ($t = 0$) downward ($t = 3$), where the time step is $\Delta t = 1$. On the left side, the changes (localized in 1-D space, the x-axis) of ψ_i are shown in a schematic way (▫). It is seen that the velocity of the change is not constant and the corresponding acceleration is equal to 1. Now let us imagine for simplicity that function $\hat{F}\psi_i$ has its nonzero values precisely where $\psi_i \neq 0$ and let us consider two cases: a) $\hat{F}\psi_i < 0$ and b) $\hat{F}\psi_i > 0$. In such a situation, we may easily foresee the sign of the mean value of the energy $\left\langle \psi_i \mid \hat{F}\psi_i \right\rangle$ of an electron occupying spinorbital ψ_i. In situation a), the conclusion for changes of ψ_i is maintain or, in other words, even *increase* the acceleration $\ddot{\psi}_i$ making it proportional to $-\hat{F}\psi_i$. In b), the corresponding conclusion is *suppress* these changes or *decrease the acceleration*; e.g., making it negative as $-\hat{F}\psi_i$. Thus, in *both* cases, we have $\mu\ddot{\psi}_i = -\hat{F}\psi_i$, which agrees with Eq. (7.19). In both cases, there is a trend to lower orbital energy $\varepsilon_i = \left\langle \psi_i \mid \hat{F}\psi_i \right\rangle$.

From Fig. 7.15, it is seen that it would be desirable to have the acceleration $\ddot{\psi}_i$ with the same sign as $-\hat{F}\psi_i$. This is equivalent to increasing the changes that lower the corresponding orbital energy, and to suppressing the changes that make it higher. The ψ_i spinorbitals obtained in the numerical integration have to be corrected for orthonormality, as is assured by the second term in Eq. (7.19).

The prize for the elegance of the Car-Parrinello method is the computation time, which allows to treat systems currently up to a few hundreds of atoms (while MD may even deal with a million atoms). The integration interval has to be decreased by a factor of 10 (i.e., 0.1 fs instead of 1 fs), which allows us to reach simulation times of the order of 10 to100 picoseconds only instead of nanoseconds, as in classical MD.

7.11 Cellular Automata

Another powerful tool is the cellular automata method invented by John (or Janos) von Neumann and Stanisław Marcin Ulam (under the name of "*cellular spaces*"). The cellular automata are mathematical models in which space and time both have a granular structure (similar to Monte Carlo simulations on lattices, in MD only time has such a structure). A cellular automaton consists of a periodic lattice of cells (nodes in space). In order to describe the system locally, we assume that every cell has its "*state*" representing a vector of N components. Each component is a Boolean variable; i.e., a variable having a logical value; (e.g., "0" for "*false*" and "1" for "*true*").

A cellular automaton evolves using some *propagation and collision (or actualization) rules* that are always of a *local character*. The local character means that (at a certain time step t and a certain cell), the variables change their values depending only on what happened at *the* cell and at its neighbors at time step $t-1$. The propagation rules dictate what would happen next with variables on the cell for each cell independently. But this may provoke a collision of the rules because a Boolean variable on a cell may be forced to change by the propagation rules related to two or more cells. We need a unique decision, and this comes from the collision, or actualization, rules.

For physically relevant states, the propagation and collision rules for the behavior of such a set of cells as time goes on may mirror what would happen with a physical system. This is why cellular automata are appealing. Another advantage is that due to the locality mentioned above, the relevant computer programs may be effectively parallelized, which usually significantly speeds up computations. The most interesting cellular automata are those for which the rules are of a nonlinear character (cf. Chapter 15).

Example 4. *Gas Lattice Model*

One of the simplest examples pertains to a lattice model of a gas. Let the lattice be regular two-dimensional (Fig. 7.16).

Propagation Rules:

There are a certain number of pointlike particles of equal mass that may occupy the nodes (cells) only and have only unit velocities pointing either in north-south or east-west directions,

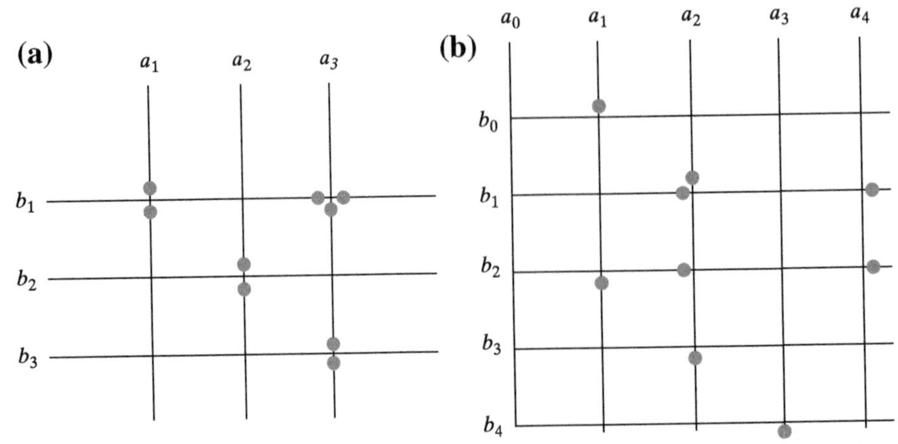

Fig. 7.16. Operation of a cellular automaton–a model of gas. The particles occupy the lattice nodes (cells). Their displacement from the node symbolizes which direction they are heading on with the velocity equal to 1 length unit per 1 time step. In the left scheme (a), the initial situation is shown. In the right scheme the result of the one step propagation and one step collision is shown. Collision only took place in one case (at a_3b_2), and the collision rule has been applied (of the lateral outgoing). The game would become more dramatic if the number of particles were larger, and if the walls of the box as well as the appropriate propagation rules (with walls) were introduced.

thus reaching the next row or column after a unit of time. We assign each cell a state which is a four-dimensional vector of Boolean variables. The first component tells us whether there is a particle moving north on the node (Boolean variables take 0 or 1), the second moving east, the third south, and the fourth west. There should be no more than one particle going in one direction at any node; therefore, a cell may correspond to 0,1,2,3,4 particles. Any particle is shifted by one unit in the direction of the velocity vector.

Collision Rules:

If two particles are going to occupy the same state component at the same cell, the two particles are annihilated and a new pair of particles is created with drawn positions and velocities. Any two particles which meet at a node with opposite velocities acquire the velocities that are opposite to each other and perpendicular to the old ones (the "*lateral outgoing*"; see Fig. 7.16).

This primitive model has nevertheless an interesting property. It turns out that such a system attains an equilibrium state. No wonder that this approach with more complex lattices and rules became popular. Using the cellular automata, we may study an extremely wide range of phenomena, such as turbulent flow of air along a wing surface, electrochemical reactions, etc. It is a simple and powerful tool of general importance.

Summary

- A *detailed* information about a molecule (in this case a three-atom complex C...AB) may be obtained making use of the potential energy hypersurface for the nuclear motion computed as the ground-state electronic energy (however, even in this simplified case, the AB distance has been frozen). After constructing the basis functions appropriate for the five variables and applying the Ritz method for solving the Schrödinger equation, we obtain the rovibrational levels and corresponding wave functions for the system. This allows us to compute the microwave spectrum.

- We may construct an approximation to the potential energy hypersurface for the motion of the nuclei by designing what is known as a *force field*, or a simple expression for the electronic energy as a function of the position of the nuclei. Most often in proposed force fields, we assume harmonicity of the chemical bonds and bond angles ("*springs*"). The hypersurface obtained often has a complex shape with many local minima.
- *Molecular Mechanics* (should have the adjective "*local*") represents
  - The choice of the starting configuration of the nuclei (a point in the configuration space)
  - Sliding downhill from the point (configuration) to the "nearest" local minimum, which corresponds to a stable conformation with respect to small displacements in the configurational space
- *Global molecular mechanics* means
  - The choice of the starting configuration of the nuclei
  - Finding the global (the lowest-energy) minimum; i.e., the most stable configuration of the nuclei

While the local molecular mechanics represents a standard procedure, the global one is still *in statu nascendi*.

- Any of the potential energy minima can be approximated by a paraboloid. Then, for N nuclei, we obtain $3N-6$ *normal modes* (i.e., harmonic and having the same phase) of the molecular vibrations. This represents important information about the molecule, because it is sufficient to calculate the IR and Raman spectra frequencies (cf. p. e17). Each of the normal modes makes all the atoms move, but some atoms may move more than others. It often happens that a certain mode is dominated by the vibration of a particular bond or functional group and therefore the corresponding frequency is *characteristic* for this bond or functional group, which may be very useful in chemical analysis.
- Molecular mechanics does not involve atomic kinetic energy, *molecular dynamics* (MD) does. MD represents a method of solving the Newton equations of motion[69] for all the atoms of the system. The forces acting on each atom at a given configuration of the nuclei are computed (from the potential energy V assumed to be known[70]) as $\boldsymbol{F}_j = -\nabla_j V$ for atoms $j = 1, 2, \ldots, N$. Now that the forces are known, we calculate the acceleration vector, and from that, the velocities and the new positions of the atoms. The system starts to evolve, as time goes on. Important ingredients of the MD procedure are:
  - Choice of starting conformation
  - Choice of starting velocities
  - Thermalization at a given temperature (with velocity adjustments to fulfill the appropriate Maxwell-Boltzmann distribution)
  - Harvesting the system trajectory
  - Conclusions derived from the trajectory.
- In MD (also in the other techniques listed below), there is the possibility of applying a sequence (protocol) of cooling and heating intervals in order to achieve a low-energy configuration of the nuclei (*simulated annealing*). The method is very useful and straightforward to apply.
- Besides MD, there are other useful techniques describing the motion of the system:
  - Langevin dynamics that allows taking into account the surrounding solvent ("at virtually no cost")
  - Monte Carlo dynamics is a powerful technique basing on drawing and then accepting/rejecting random configurations by using the *Metropolis criterion*. The criterion says that if the energy of the new configuration is lower, the configuration is accepted, and if it is higher, it is accepted with a certain probability.
  - Car-Parrinello dynamics allows for the electronic structure to be changed "*in flight*," when the nuclei move.
  - Cellular automata is a technique of general importance, which divides the total system in cells. Each cell is characterized by its state representing a vector with its components being Boolean variables. There are

[69] In some cases, we refer to this as *integration*.

[70] Usually, it is a force field.

propagation rules that change the state, as time goes on, and collision rules, which solve conflicts of the propagation rules. Both types of rules have a local character. Cellular automata evolution may have some features in common with thermodynamic equilibria.

Main Concepts, New Terms

angular momenta addition (p. 343)
autocorrelation (p. 366)
Boolean variables (p. 382)
Car-Parrinello algorithm (p. 377)
cellular automaton (p. 381)
characteristic frequency (p. 359)
cooling protocol (p. 370)
entropy (p. 353)
force field (p. 345)
free energy (p. 353)
global minimum (p. 352)
global optimization (p. 354)
Jacobi coordinate system (p. 341)
kinetic minimum (p. 353)
Langevin dynamics (p. 371)
Lennard-Jones potential (p. 347)
Metropolis algorithm (p. 374)
molecular dynamics (p. 364)
molecular mechanics (p. 349)
Monte Carlo dynamics (p. 371)
Monte Carlo with Replica Exchange (p. 375)
normal modes (p. 355)
rovibrational spectrum (p. 340)
simulated annealing (p. 370)
spatial correlation (p. 366)
thermalization (p. 367)
thermodynamic minimum (p. 353)
time correlation (p. 366)
torsional potential (p. 349)

From the Research Front

The number of atoms taken into account in MD nowadays may reach a million. The real problem is not the size of the system, but rather its complexity and the wealth of possible structures, with their too large number to be investigated. Some problems may be simplified by considering a quantum-mechanical part in the details and a classical part described by Newton equations. Another important problem is to predict the 3-D structure of proteins, starting from the available amino acid sequence. Every two years beginning in 1994, CASP (Critical Assessment of techniques for protein Structure Prediction) has been organized in California. CASP is a kind of scientific competition, in which theoretical laboratories (knowing only the amino acid sequence) make blind predictions about 3-D protein structures about to be determined in experimental laboratories. Most of the theoretical methods are based on the similarity of the sequence to a sequence from the Protein Data Bank of the 3-D structures, only some of the methods are related to chemical physics. Fig. 7.17 shows an example of the latter.

Ad Futurum

The maximum size of the systems investigated by the MM and MD methods will increase systematically to several million atoms in the near future. A critical problem will be the choice of the system to be studied, as well as the question to be asked. Very probably non-equilibrium systems will become more and more important in such areas as concerning impact physics, properties of materials subject to various stresses, explosions, self-organization (see Chapter 13), and, most important, chemical reactions. At the same time, the importance of MD simulations of micro-tools of dimensions of tens of thousands Å will very probably increase.

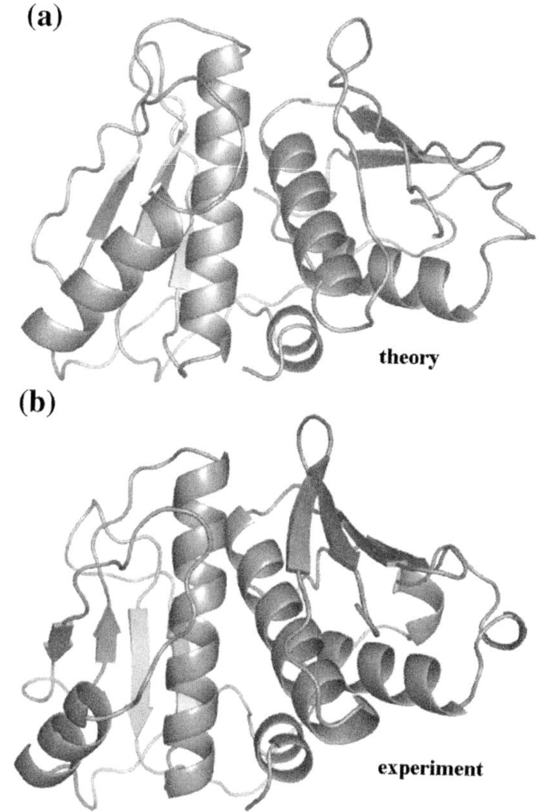

Fig. 7.17. One of the target proteins in the 2004 CASP6 competition. The 3-D structure (in ribbon representation) was obtained for the putative nitroreductase, one of the 1877 proteins of the bacterium *Thermotoga maritima*, which lives in geothermal marine sediments. The energy expression that was used in the theoretical calculations took into account the physical interactions (such as hydrogen bonds, hydrophobic interactions, etc.; see Chapter 13), as well as an empirical potential deduced from representative proteins' experimental structures deposited in the Brookhaven Protein Data Bank (no bias toward the target protein). The molecule represents a chain of 206 amino acids; i.e., about 3000 heavy atoms. Both theory (CASP6 blind prediction) and experiment (carried out within CASP6 as well) give the target molecule containing five α-helices and two β- pleated sheets (wide arrows). These secondary structure elements interact and form the unique (native) tertiary structure, which is able to perform its biological function. (a) predicted by A. Kolinski and K. Bujnicki by the Monte Carlo method, and (b) determined experimentally by X-ray diffraction. Both structures in atomic resolution differ (rms) by 2.9 Å. *Reproduced courtesy of Professor Andrzej Koliński.*

Additional Literature

A. R. Leach, *Molecular Modelling. Principles and Applications*, Longman, Prentice Hall, Upper Saddle River, New Jersey, (2001).

This book is considered the "bible" of theoretical simulations.

M. P. Allen and D. J. Tildesley, *Computer Simulations of Liquids*, Oxford Science Publications, Clarendon Press, Oxford (1987).

This book offers a more theoretical take on this subject.

Questions

1. The energy $V(\mathbf{R})$ as a potential energy for the motion of the nuclei in the adiabatic approximation:
   a. allows to identify the most stable configuration of the nuclei as such a point of the configurational space, for which $\nabla V(\mathbf{R}) = \mathbf{0}$
   b. differs for two isotopomers
   c. when in the Born-Oppenheimer approximation gives the same $V(\mathbf{R})$ for two isotopomers
   d. the gradient $\nabla V(\mathbf{R})$ calculated for a stable configuration $\mathbf{R}$ of the nuclei is equal $\mathbf{0}$

2. The symbols in the Hamiltonian $\hat{H} = -\frac{\hbar^2}{2\mu R^2}\frac{d}{dR}R^2\frac{d}{dR} + \frac{\hat{l}^2}{2\mu R^2} + \frac{\hat{j}^2}{2\mu_{AB} r_{eq}^2} + V$ for a triatomic system C...AB mean:
   a. V stands for the ground-state electronic energy as a function of the configuration of the nuclei
   b. $R =$ the CA distance
   c. $r_{eq} =$ the AB distance
   d. $\hat{j}^2$ means the operator of the square of the angular momentum of C with respect to the center of mass of AB

3. A force field:
   a. represents an approximate ground state electronic energy as a function of the configuration of the nuclei
   b. represents an approximation for the total energy of the molecular system
   c. allows to compute the approximate forces acting on each nucleus at a given nuclear configuration
   d. means an electric field created by the molecule.

4. The normal mode frequencies for a molecule:
   a. pertain to a particular minimum of the potential energy (they will be different for another minimum)
   b. are calculated assuming that the vibrational amplitudes are small
   c. take into account a small anharmonicity
   d. are different for different isotopomers

5. The equation for the normal modes $(\mathbf{A}-\omega_k^2\mathbf{1})\mathbf{L}_k = \mathbf{0}$ ($\mathbf{M}$ is a diagonal matrix of atom masses):
   a. if $\mathbf{A}$ is computed for a point, for which the gradient of the potential energy is zero, it may happen that ω_k is imaginary
   b. if $\mathbf{A} = \mathbf{M}^{-\frac{1}{2}}\mathbf{V}''\mathbf{M}^{-\frac{1}{2}}$, where $\mathbf{V}''$ stands for the matrix of second derivatives computed for a minimum of the potential energy, $\omega_k \geq 0$ for all k
   c. the vectors $\mathbf{L}_k$ represent the columns of $\mathbf{A}$
   d. $Tr(\mathbf{V}'') < 0$, computed at a point where $\nabla V = \mathbf{0}$, means that the point is a saddle point

6. One of the following sets of the wave numbers (cm^{-1}) in an IR spectrum may correspond to vibrations of the following bonds: C–H, C–C, C=C:
   a. 2900, 1650, 800
   b. 800, 2900, 1650
   c. 1650, 800, 2900
   d. 2900, 800, 1650

7. In the simulated annealing method in molecular mechanics:
   a. one adapts the velocity of particles to the Maxwell distribution for a given temperature (that varies during the simulation)
   b. the goal is to find the most stable structure
   c. one carries out computation with slowly and monotonically decreasing temperature
   d. one alternatively increases and decreases the temperature

8. In the Metropolis algorithm used in the Monte Carlo method (for temperature T), a new configuration of the nuclei is accepted
   a. only if its energy is lower than the energy of the current configuration
   b. always
   c. always, if its energy is lower than the energy of the current one
   d. sometimes, even if its energy is higher than the energy of the current configuration
9. In the Langevin dynamics, the atoms of the solvent molecules:
   a. represent a source of stochastic forces acting on the molecule under consideration
   b. are responsible for a friction felt by moving atoms of the molecule under consideration
   c. are treated differently than the atoms of the molecule under consideration
   d. are treated in the same way as the atoms of the molecule under consideration
10. In the Car-Parrinello dynamics:
   a. the nuclei move according to the Newton dynamics, while for the electrons, one solves the Schrödinger equation for each configuration of the nuclei
   b. the electronic charge distribution adjusts to the current positions of the nuclei, while the position of the nuclei changes due to the electronic charge distribution
   c. each electron is considered as having the mass of 1 a.u.
   d. when the nuclei move, the electronic charge distribution changes in such a way as to decrease the electronic energy at any position of the nuclei

Answers

1b,c,d, 2a,c, 3a,c, 4a,b,d, 5a,b, 6d, 7a,b,d, 8c,d, 9a,b,c, 10b,d

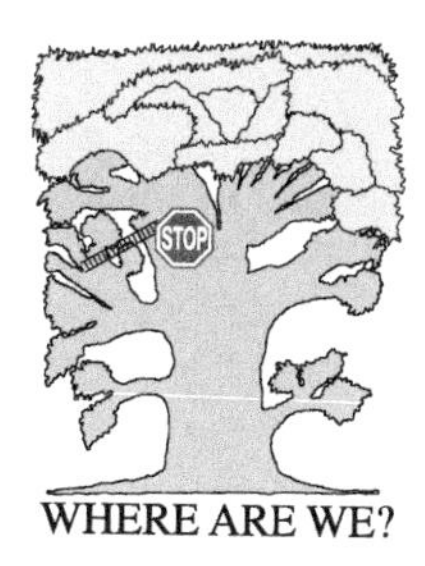

WHERE ARE WE?

Orbital Model of Electronic Motion in Atoms and Molecules

"Everything should be made as simple as possible, but not simpler."
Albert Einstein

Where Are We?

We are in the upper part of the main trunk of the TREE, most important for chemists.

An Example

What is the electronic structure of atoms? How do atoms interact in a molecule? Two *neutral* moieties (say, hydrogen atoms) attract each other with a large force *of a similar order of magnitude* to the Coulombic forces between two ions. This is quite surprising. What pulls these neutral objects toward one another? These questions are at the foundations of chemistry.

What Is It All About?

Ideas of Quantum Chemistry, Second Edition. http://dx.doi.org/10.1016/B978-0-444-59436-5.00008-8

The Born-Oppenheimer (or adiabatic) approximation is the central point of this book (note its position in the TREE). Thanks to the approximation, we can consider *separately* two *coupled* problems concerning molecules:

- The motion of the electrons at fixed positions of the nuclei (to obtain the electronic energy)
- The motion of nuclei in the potential representing the electronic energy of the molecule (see Chapter 7).

From now on, we will concentrate on the *motion of the electrons at fixed positions of the nuclei* (the Born-Oppenheimer approximation, p. 269).

To solve the corresponding Eq. (6.8), we have at our disposal the variational and the perturbation methods. The latter should have a reasonable starting point (i.e., an unperturbed system). This is not the case in the problem that we want to consider at the moment. Thus, only the variational method remains. If so, a class of the trial functions should be proposed. In this chapter, the trial wave function will have a very specific form, bearing significant importance for the theory. We mean here the so-called Slater determinant, which is composed of molecular orbitals. At a certain level of approximation, each molecular orbital is a "parking place" for two electrons. We will now learn on how to get the optimum molecular orbitals (using the Hartree-Fock method). Despite some quite complex formulas, which will appear below, the main idea behind them is extremely simple. It can be expressed in the following way.

Let us consider a traffic scenario where the cars (electrons) move past fixed positions of buildings (nuclei). The motion of the cars is very complex (as it is for the electrons) and therefore, the problem is extremely difficult. How may such a motion be described in an approximate way? To describe such a complex motion, one may use the so-called mean field approximation (paying the price of lower quality). In the mean field approximation method, we focus on the motion of *one car only*, considering its motion in such way that the car avoids *those streets that are usually most jammed.* In this chapter, we will treat the electrons in a similar manner (leaving the difficulties of considering the correlation of the motions of the electrons to Chapter 10). Now, the electrons will not feel the true electric field of the other electrons (as it should be in a precise approach), but rather their *mean* electric field (i.e., averaged over their motions).

Translating it into the quantum mechanical language, the underlying assumptions of the mean field method for the N identical particles (here: electrons) are as follows:

- There is a certain "*effective*" *one-particle* operator $\hat{F}(i)$ of an identical mathematical form for all particles $i = 1, 2, \ldots N$, which has the eigenfunctions ϕ_k, i.e., $\hat{F}\phi_k = \varepsilon_k \phi_k$
- $\left\langle \Psi | \hat{H} \Psi \right\rangle \approx \left\langle \tilde{\Psi} | \hat{H}^{ef} \tilde{\Psi} \right\rangle$, where $\tilde{\Psi}$ (normalized) is a wave function that approximates the exact wave function Ψ for the total system, $\hat{H}$ is the electronic Hamiltonian (in the clamped nuclei approximation, as discussed in Chapter 6), and $\hat{H}^{ef} = \sum_{i=1}^{N} \hat{F}(i)$. In such a case, the eigenvalue equation $\hat{H}^{ef} \Pi_{i=1}^{N} \phi_i(i) = E_0 \Pi_{i=1}^{N} \phi_i(i)$ holds, and the approximate total energy is equal to $E_0 = \sum_{i=1}^{N} \varepsilon_k$, as if the particles were independent.

Any mean field method needs to solve two problems:

- How should $\tilde{\Psi}$ be constructed using N eigenfunctions ϕ_k?
- What is the form of the one-particle effective operator $\hat{F}$?

These questions will be answered in this chapter.

Such effectively independent, yet interacting particles, are called quasiparticles, or as we sometimes used to say, bare particles dressed up by the interaction with others.

It is worth remembering that the mean field method is known by several different names in chemistry:

- One-determinant approximation
- One-electron approximation
- One-particle approximation
- Molecular orbital method
- Independent-particle approximation
- Mean field approximation
- Hartree-Fock method
- Self-consistent field method (as regards practical solutions).

It will be shown how the mean field method implies that milestone of chemistry: the periodic table of chemical elements.

Next, we will endeavor to understand *why* two atoms create a chemical bond, and also what affects the ionization energy and the electron affinity of a molecule.

Then, still within the molecular orbital scheme, we will show how we can reach a localized description of a molecule, with chemical bonds between *some* atoms, with the inner electronic shells, and the lone electronic pairs. The last terms are elements of a rich and very useful language commonly used by chemists.

Why Is This Important?

Contemporary quantum chemistry uses better methods than the mean field, as described in this chapter. We will get know them in Chapters 10 and 11. Yet all these methods *start from the mean field approximation*, and in most cases, they only perform cosmetic changes in energy and in electron distribution. For example, the methods described here yield about 99% of the total energy of a system.[1] There is one more reason why this chapter is important. Methods beyond the one-electron approximation are computationally very time-consuming (hence they may be applied only to small systems), while the molecular orbital approach is the "*daily bread*" of quantum chemistry. It is a sort of standard method, and the standards have to be learned.

What Is Needed?

- Postulates of quantum chemistry (Chapter 1)
- Operator algebra, Hermitian operators (see Appendix B available at booksite.elsevier.com/978-0-444-59436-5, p. e7)
- Complete set of functions (Chapter 1)
- Hilbert space (see Appendix B available at booksite.elsevier.com/978-0-444-59436-5, p. e7, recommended)
- Determinants (see Appendix A available at booksite.elsevier.com/978-0-444-59436-5, p. e1, necessary)
- Slater-Condon rules (see Appendix M available at booksite.elsevier.com/978-0-444-59436-5, p. e109, only the results are needed)
- Lagrange multipliers (see Appendix N available at booksite.elsevier.com/978-0-444-59436-5, p. e121)
- Mulliken population analysis (see Appendix S available at booksite.elsevier.com/978-0-444-59436-5, p. e143, occasionally used)

Classical Works

This chapter deals with the basic theory explaining the electronic structure of atoms and molecules. This is why we begin with Dmitri Ivanovich Mendeleev, who discovered in 1865, when writing his textbook *Osnovy Khimii* (*Principles of Chemistry*), St. Petersburg, Tovarishchestvo Obshchestvennaya Polza, 1869–71, his famous periodic table of elements, which is one of the greatest human achievements. ★ Gilbert Newton Lewis, in the paper "*The atom and the molecule,*" *J. Amer. Chem. Soc., 38,* 762 (1916) and Walter Kossel, in the article "*Über die Molekülbildung als Frage des Atombaus*" published in *Annalen der Physik, 49,* 229 (1916), introduced such important algorithmic theoretical tools as the octet rule and stressed the importance of the noble gas electronic configurations. ★ As soon as quantum mechanics was formulated in 1926, Douglas R. Hartree published several papers in the *Proceedings of the Cambridge Philosophical Society. 24,* 89 (1927); *24,* 111 (1927); *26,* 89 (1928); entitled "*The wave mechanics of an atom with a non-Coulomb central field,*" containing the computations for atoms such large as Rb and Cl. These were *self-consistent ab initio*[2] computations, and the wave function was assumed to be the *product* of spinorbitals. ★ The LCAO approximation (for the solid state) was introduced by Felix Bloch in his Ph.D. thesis "*Über die Quantenmechanik der Elektronen in Kristallgittern,*" University of Leipzig, 1928, and three years later, Erich Hückel used this method to describe the first molecule (benzene) in a publication "*Quantentheoretische Beitrage zum Benzolproblem. I. Die Elektronenkonfiguration des Benzols,*" which appeared in *Zeitschrift für Physik, 70,* 203 (1931). ★ Vladimir Fock introduced the antisymmetrization of the spinorbital product in his publication "*Näherungsmethode zur Lösung des quantenmechanischen Mehrkörperproblems*" in *Zeitschrift für Physik, 61,* 126 (1930), and *ibid. 62,* 795 (1930). ★ John Slater proposed the idea of the multi-configurational wave function ("*Cohesion in monovalent metals,*" *Phys. Rev., 35,* 509 (1930)). ★

---

[1] In physics and chemistry, we are seldom interested in the total energy. The energy differences of various states are important. Sometimes such precision is not enough, but the result speaks for itself.

[2] That is, they were derived from the first principles of (non-relativistic) quantum mechanics. Note that these scientists worked incredibly fast (with no help from e-mail or computers).

The Hartree-Fock method in the LCAO approximation was formulated by Clemens C.J. Roothaan in his work "*New developments in molecular orbital theory,*" published in the *Rev. Mod. Phys., 23,* 69 (1951), and, independently, by George G. Hall in a paper called "*The molecular orbital theory of chemical valency,*" in *Proc. R. Soc. (London), A205,* 541 (1951). ★ The physical interpretation of the orbital energies in the Hartree-Fock method was given by Tjalling C. Koopmans in his *only* quantum chemical paper "*On the assignment of wave functions and eigenvalues to the individual electron of an atom,*" published in *Physica, 1,* 104 (1934). ★ The first localized orbitals (for the methane molecule) were computed by Charles A. Coulson despite the difficulties of wartime (*Trans. Faraday Soc. 38,* 433 (1942)). ★ Hideo Fukutome, first in an article "*Spin Density Wave and Charge Transfer Wave in Long Conjugated Molecules*" in *Progress in Theoretical Physics*, 40 (1968) 998, and then, in several following papers, analyzed general solutions for the Hartree-Fock equations from the symmetry viewpoint, and showed exactly eight classes of such solutions.

In the previous chapter, the motion of the nuclei was considered. In the Born-Oppenheimer approximation (Chapter 6), the motion of the nuclei takes place in the potential, which is the electronic energy of a system (being a function of the nuclei position, $\boldsymbol{R}$, in the configurational space). The electronic energy $E_k^0(\boldsymbol{R})$ is an eigenvalue given in Eq. 6.8 (adapted to the polyatomic case, hence $R \rightarrow \boldsymbol{R}$): $\hat{H}_0\psi_k(\boldsymbol{r};\boldsymbol{R}) = E_k^0(\boldsymbol{R})\psi_k(\boldsymbol{r};\boldsymbol{R})$. We will now deal exclusively with this equation; i.e., we will consider the electronic motion at fixed positions of the nuclei (clamped nuclei). Thus, our goal is twofold: we are interested in what the electronic structure looks like and in how the electronic energy depends on the positions of the nuclei.[3]

Any theoretical method applicable to molecules may be also used for atoms (albeit very accurate wave functions, even for simple atoms, are not easy to calculate).[4] In fact, for atoms, we know the solutions quite well only in the mean field approximation (i.e., the atomic orbitals). Such orbitals play an important role as building blocks of many-electron wave functions.

8.1 Hartree-Fock Method–A Bird's-Eye View

Douglas R. Hartree (1897–1958) was born and died in Cambridge, U.K. He was a British mathematician and physicist, professor at Manchester University, and then professor of mathematical physics at Cambridge. Until 1921, his interest was in the development of numerical methods for anti-aircraft artillery (he had some experience from World War I), but a lecture by Niels Bohr has completely changed his career. Hartree immediately started investigating atoms. He used the atomic wave function in the form of the spinorbital *product.* Hartree learned to use machines to solve differential

[3] In the previous chapter, the ground-state electronic energy $E_0^0(\boldsymbol{R})$ was denoted as $V(\boldsymbol{R})$.

[4] If an atom is considered in the Born-Oppenheimer approximation, the problem is even simpler, and the electronic equation also holds; we can then take, e.g., $\boldsymbol{R} = \boldsymbol{0}$. People still try to compute correlated wave functions (i.e., beyond the mean field approximation; see Chapter 10) for heavier atoms. Besides, relativistic effects (see Chapter 3) play increasingly important roles for such atoms. Starting with magnesium, they are larger than the correlation corrections. Fortunately, the relativistic corrections for atoms are largest for the inner electronic shells, which are least important for chemists.

equations while in Boston, and then he built one for himself at Cambridge. The machine was invented by Lord Kelvin, and constructed by Vannevar Bush in the United States. The machine integrated equations using a circle which rolled on a rotating disk. Later, the first electronic computer, ENIAC was used, and Hartree was asked to come and help to compute missile trajectories.

An excerpt from *Solid State and Molecular Theory*, Wiley, London, 1975 by John C. Slater:

"Douglas Hartree was very distinctly of the matter-of-fact habit of thought that I found most congenial. The hand-waving magical type of scientist regarded him as a 'mere computer.' Yet he made a much greater contribution to our knowledge of the behavior of real atoms than most of them did. And while he limited himself to atoms, his demonstration of the power of the self-consistent field for atoms is what has led to the development of that method for molecules and solids as well."

Vladimir A. Fock (1898–1974), Russian physicist, professor at Leningrad University (Saint Petersburg), led investigations on quantum mechanics, gravity theory, general relativity theory, and in 1930, while explaining atomic spectra, invented the *antisymmetrization* of the spinorbitals product.

Before introducing the detailed formalism of the Hartree-Fock method, let us look at its principal features. It will help us to understand our mathematical goal.

First of all, the positions of the nuclei are frozen (Born-Oppenheimer approximation) and then we focus on the wave function of N *electrons*. Once we want to move nuclei, we need to repeat the procedure from the beginning (for the new position of the nuclei).

8.1.1 Spinorbitals as the One-Electron Building Blocks

Although this comparison is not precise, the electronic wave function for a molecule is built of segments, as a house is constructed from bricks.

The electronic wave function of a molecule containing N electrons depends on $3N$ Cartesian coordinates of the electrons and on their N spin coordinates (for each electron, its $\sigma = \frac{1}{2}$ or $-\frac{1}{2}$). Thus, it is a function of position in $4N$–dimensional space. This function will be created out of simple "*bricks*"; i.e., *molecular spinorbitals*. Each of those will be a function of the coordinates of *one* electron only: three Cartesian coordinates and one spin coordinate (cf., Chapter 1). A spinorbital is therefore a function of the coordinates in the 4-D space,[5] and in the most general case, a normalized spinorbital reads as (see Fig. 8.1):

$$\phi_i(\boldsymbol{r}, \sigma) = \varphi_{i1}(\boldsymbol{r})\alpha(\sigma) + \varphi_{i2}(\boldsymbol{r})\beta(\sigma), \tag{8.1}$$

[5] The analogy of house and bricks fails here because both the house and the bricks come from the same 3-D space.

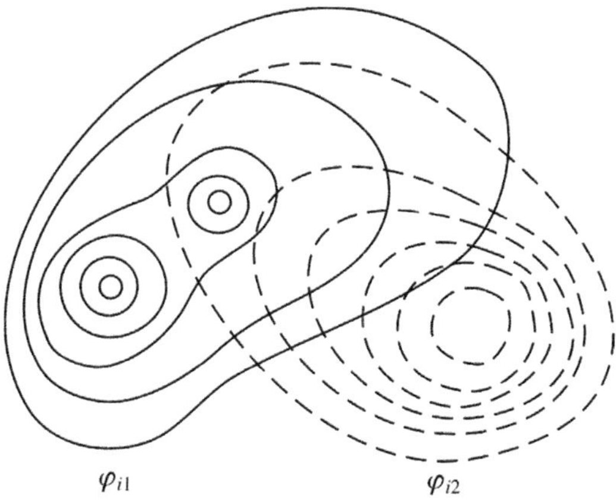

Fig. 8.1. According to Eq. (8.1), a spinorbital is a mixture of α and β orbital components: $\varphi_{i1}(\boldsymbol{r})$ and $\varphi_{i1}(\boldsymbol{r})$, respectively. This image shows two sections of such a spinorbital (z denotes the Cartesian axis perpendicular to the plane of the page): section $z = 0, \sigma = \frac{1}{2}$ (solid isolines) and section $z = 0, \sigma = -\frac{1}{2}$ (dashed isolines). In practical applications, most often a restricted form of spinorbitals is used: either $\varphi_{i1} = 0$ or $\varphi_{i2} = 0$; i.e., a spinorbital is taken as an orbital part times spin function α or β.

where the orbital components φ_{i1} and φ_{i2} (square-integrable functions) that depend on position $\boldsymbol{r}$ of the electron can adopt *complex* values, while the spin functions α and β, which depend on the spin coordinate σ, are defined in Chapter 1, p. 28.

In the vast majority of quantum mechanical calculations, the spinorbital ϕ_i is a *real* function, and φ_{i1} and φ_{i2} are such that either $\varphi_{i1} = 0$ or $\varphi_{i2} = 0$.

Yet for the time being we do not introduce any significant[6] *restrictions for the spinorbitals.* Spinorbital ϕ_i will adopt different complex values for various spatial coordinates, as well as for a given value[7] of the spin coordinate σ.

8.1.2 Variables

Thus, the variables, which the wave function depends on, are as follows:

x_1, y_1, z_1, σ_1 or briefly 1,

x_2, y_2, z_2, σ_2 or briefly 2,

....

x_N, y_N, z_N, σ_N or briefly N,

where x_i, y_i, z_i are the Cartesian coordinates and σ_i is the spin coordinate of electron i.

[6] The normalization condition does not reduce the generality of the approach.

[7] That is, we put $\sigma = \frac{1}{2}$ or $\sigma = -\frac{1}{2}$.

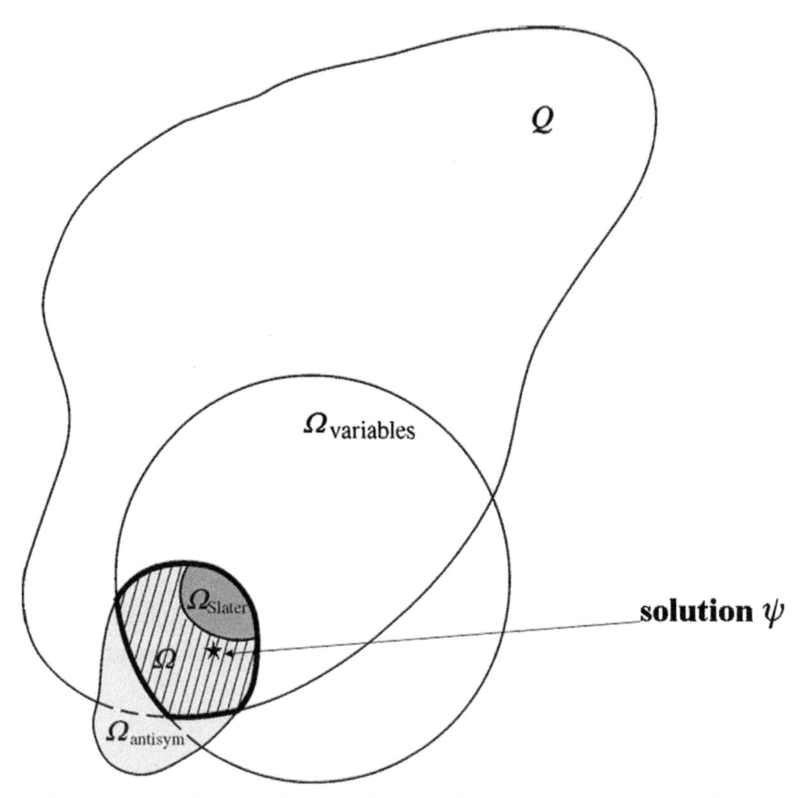

Fig. 8.2. Diagram of the sets among which the solution ψ of the Schrödinger equation is sought. The Q set is the one of all square-integrable functions, $\Omega_{variables}$ is the set of the functions with variables as those of the solution of the Schrödinger equation, ψ, and $\Omega_{antisym}$ is the set of the functions that are antisymmetric with respect to the exchange of coordinates of any two electrons. The solutions of the Schrödinger equation, ψ, will be sought in the common part of these three sets: $\psi \in \Omega = Q \cap \Omega_{variables} \cap \Omega_{antisym}$. The Ω_{Slater} represents the set of single Slater determinants built of normalizable spinorbitals. The exact wave function always belongs to $\psi \in \Omega - \Omega_{Slater}$.

An exact wave function $\psi(x_1, y_1, z_1, \sigma_1, x_2, y_2, z_2, \sigma_2, \ldots) = \psi(1, 2, \ldots, N)$ belongs (see Fig. 8.2) to the set Ω, which is the common part of the following sets:

- Set Q of all square-integrable functions
- Set $\Omega_{variables}$ of all the functions dependent on the abovementioned variables
- Set $\Omega_{antisym}$ of all the functions that are antisymmetric with respect to the mutual exchange of the coordinates of any two electrons (p. 34):

$$\psi \in \Omega = Q \cap \Omega_{variables} \cap \Omega_{antisym}.$$

8.1.3 Slater Determinant–An Antisymmetric Stamp

There should be something in the theory that assures us that if we renumber the electrons, no theoretical prediction will change. The postulate of the *antisymmetric* character of the wave function with respect to the *exchange* of the coordinates of *any* two electrons, certainly ensures this (Chapter 1, p. 33). The solution of the Schrödinger equation for a given stationary state of interest should be sought among *such* functions.

John C. Slater (1901–1976), American physicist, for 30 years a professor and dean at the Physics Department of the Massachusetts Institute of Technology, then at the University of Florida, Gainesville, where he participated in the Quantum Theory Project. His youth was in the stormy period of the intense development of quantum mechanics, and he participated vividly in it. For example, in 1926–1932, he published articles on the ground state of the helium atom, the screening constants (Slater orbitals), the antisymmetrization of the wave function (Slater determinant), and the algorithm for calculating the integrals (the Slater-Condon rules).

A Slater determinant is a function of the coordinates of N electrons, which *automatically* belongs to Ω:

$$\psi = \frac{1}{\sqrt{N!}} \begin{vmatrix} \phi_1(1) & \phi_1(2) & \dots & \phi_1(N) \\ \phi_2(1) & \phi_2(2) & \dots & \phi_2(N) \\ \dots & \dots & \dots & \dots \\ \phi_N(1) & \phi_N(2) & \dots & \phi_N(N) \end{vmatrix}, \tag{8.2}$$

where ϕ_i are the orthonormal[8] one-electron functions (i.e., molecular spinorbitals).[9] The Slater determinants form a subset $\Omega_{Slater} \subset \Omega$.

A Slater determinant carries two important attributes of the exact wave function:

- Suppose that we want to calculate the probability density that two electrons with the same spin coordinate σ are in the same place (i.e., that these two electrons have *all* their coordinates, spatial and spin ones, identical). If so, then the two columns of the abovementioned determinant are identical. And this means that the determinant becomes equal to zero.[10] From this, and from the continuity of the wave function, we may conclude that

> electrons of the same spin cannot approach each other.

- Let us now imagine two electrons with opposite values of their spin coordinate σ. If these two electrons take the same position in space, the Slater determinant will not vanish because in general, there is nothing that forces $\phi_i(1)$ be equal to $\phi_i(2)$, when $1 \equiv \left(\boldsymbol{r}_1, \sigma = \frac{1}{2}\right)$ and

[8] It is *most often* so, and then the factor standing in front of the determinant ensures the normalization. The spinorbitals could be *non-normalized* (but if they are to describe a stationary state, they should be square-integrable). They also do not need to be mutually orthogonal, but certainly they *need to be linearly independent*. Any attempt to insert linearly dependent functions in the determinant will have a "tragic outcome"–we will get 0. It comes from the properties of the determinant (if a row is a linear combination of the others, then the determinant is zero). It also follows that if we have a set of non-orthogonal spinorbitals in a Slater determinant, we could orthogonalize them by making the appropriate linear combinations. This would multiply the original Slater determinant by an irrelevant constant. This is why *it is no loss of generality to require the spinorbitals to be orthonormal.*

[9] In the theory of the atomic nucleus, the determinant wave function for the nucleons (fermions) is also used.

[10] Indeed, this is why we exist. Two objects built out of fermions (e.g., electrons) cannot occupy the same position in space. If it were not so, our bodies would sink into the ground.

$2 \equiv \left(\boldsymbol{r}_1, \sigma = -\frac{1}{2}\right)$ for $i = 1, 2, \ldots$ From this non-vanishing, and from the continuity of the wave function (which means a non-zero probability for having both electrons close in space), we conclude that

electrons of opposite spins can approach each other.

8.1.4 What Is the Hartree-Fock Method All About?

The Hartree-Fock method is a *variational* one (p. 232); i.e., it uses the variational wave function ψ in the form of a *single Slater determinant* and minimizes the mean value of the Hamiltonian $\varepsilon = \frac{\langle \psi | \hat{H} \psi \rangle}{\langle \psi | \psi \rangle}$ producing the Hartree-Fock energy $\varepsilon_{\min} = E_{HF} = \frac{\langle \psi_{HF} | \hat{H} \psi_{HF} \rangle}{\langle \psi_{HF} | \psi_{HF} \rangle}$.

The Slater determinant is an antisymmetric function, but an antisymmetric function does not necessarily need to take the shape of a Slater determinant.

Taking the variational wave function in the form of one determinant means an automatic limitation to the subset Ω_{Slater} for searching for the optimum wave function. In fact, we should search the optimum wave function in the set Ω. Thus, it is an *approximation* for the solution of the Schrödinger equation, with no chance of representing the exact result.

Why are Slater determinants used so willingly? There are two reasons for this:

- A determinant is a kind of "*stamp*." Whatever you put inside, the result (if not zero) is antisymmetric *by definition*; i.e., it automatically satisfies one of the postulates of quantum mechanics.
- It is constructed out of simple "*bricks*"–the one-electron functions (spinorbitals).

The Slater determinants built out of the complete set of spinorbitals do form the complete set.

Because of this, the true wave function can take the form of a linear combination of the determinants (we will discuss this later in this book, in Chapter 10).[11]

[11] But never represents a single determinant because the Hamiltonian of a molecule is not the sum of the effective Hamiltonians for the individual electrons.

AN ANALOGY–THE COUPLED HARMONIC OSCILLATORS (1 of 5): The derivation of the Fock equation given below looks more complex than it really is. We decided to go in parallel and illustrate all steps of the Fock equation derivation in a much simpler case of the two coupled harmonic (bosonic, 1-D) oscillators. These illustrations will be given as the inserts like the present one. Thus, reading only the inserts in the text, the reader is able to catch the very essence of what is being explained in the main text.
The Hamiltonian for the two oscillators will be given by $\hat{H} = \hat{T} + \hat{V}$, where $\hat{T} = -\frac{\hbar^2}{2m_1}\frac{\partial^2}{\partial x_1^2} - \frac{\hbar^2}{2m_2}\frac{\partial^2}{\partial x_2^2}$ and $V = \frac{1}{2}kx_1^2 + \frac{1}{2}kx_2^2 + \lambda x_1^4 x_2^4$, with $\lambda x_1^4 x_2^4$ as the coupling term. Considering the bosonic nature of the particles (the wave function has to be symmetric; see Chapter 1), we will use $\psi = \phi(1)\phi(2)$ as a variational function, where ϕ is a normalized spinorbital. This represents a restriction analogous to taking a single Slater determinant because an exact wave function should not be $\psi = \phi(1)\phi(2)$, but rather $\Sigma_{ij}\phi_i(1)\phi_j(2)$.

8.2 Toward the Optimal Spinorbitals and the Fock Equation

8.2.1 Dirac Notation for Integrals

The integrals over the spatial *and* spin coordinates (ϕ are the spinorbitals, φ - the orbitals) in the *Dirac notation* will be denoted with angle brackets $\langle\rangle$ ($\hat{h}$ denotes a one-electron operator and r_{12} – the distance between electrons 1 and 2), as follows:
for the one-electron integrals,

$$\left\langle i|\hat{h}|j\right\rangle \equiv \sum_{\sigma_1}\int dx_1dy_1dz_1\phi_i^*(1)\hat{h}\phi_j(1), \tag{8.3}$$

and for the two-electron integrals,

$$\langle ij|kl\rangle \equiv \sum_{\sigma_1}\sum_{\sigma_2}\int dx_1dy_1dz_1\int dx_2dy_2dz_2\phi_i^*(1)\phi_j^*(2)\frac{1}{r_{12}}\phi_k(1)\phi_l(2). \tag{8.4}$$

The integrals over the spatial coordinates (only) will be denoted by round brackets (), for the one-electron integrals:

$$\left(i|\hat{h}|j\right) \equiv \int dx_1dy_1dz_1\varphi_i^*(1)\hat{h}(1)\varphi_j(1), \tag{8.5}$$

and for the two-electron integrals:

$$(ij|kl) \equiv \int dx_1dy_1dz_1\int dx_2dy_2dz_2\varphi_i^*(1)\varphi_j^*(2)\frac{1}{r_{12}}\varphi_k(1)\varphi_l(2). \tag{8.6}$$

This is called Dirac notation (of the integrals).[12]

8.2.2 Energy Functional to Be Minimized

Applying the first Slater–Condon rule,[13] we get the following equation for *the mean value of the Hamiltonian* (without a constant nuclear repulsion) calculated using the normalized *Slater one-determinant function* ψ; i.e., the energy functional $E[\psi]$:

$$E[\psi] = \langle\psi|\hat{H}|\psi\rangle = \sum_{i=1}^{N}\langle i|\hat{h}|i\rangle + \frac{1}{2}\sum_{i,j=1}^{N}(\langle ij|ij\rangle - \langle ij|ji\rangle), \tag{8.7}$$

where the indices symbolize the spinorbitals, and the symbol $\hat{h}$

$$\hat{h}(1) = -\frac{1}{2}\Delta_1 - \sum_{a=1}^{M}\frac{Z_a}{r_{a1}} \tag{8.8}$$

is the one-electron operator (in atomic units) of the kinetic energy of the electron plus the operator of the nucleus-electron attraction (there are M nuclei).

AN ANALOGY–THE COUPLED HARMONIC OSCILLATORS (continued, 2 of 5): The expression for the mean value of the Hamiltonian takes the form $E[\phi] = \left\langle\psi|\hat{H}\psi\right\rangle = \left\langle\phi(1)\phi(2)|\left(\hat{h}(1)+\hat{h}(2)\right)\phi(1)\phi(2)\right\rangle + \lambda\left\langle\phi(1)\phi(2)|x_1^4x_2^4\,\phi(1)\phi(2)\right\rangle = \left\langle\phi(1)\phi(2)|\hat{h}(1)\,\phi(1)\phi(2)\right\rangle + \left\langle\phi(1)\phi(2)|\hat{h}(2)\phi(1)\phi(2)\right\rangle + \lambda\left\langle\phi(1)|x_1^4\,\phi(1)\right\rangle\left\langle\phi(2)|x_2^4\,\phi(2)\right\rangle = \left\langle\phi(1)|\hat{h}(1)\,\phi(1)\right\rangle + \left\langle\phi(2)|\hat{h}(2)\phi(2)\right\rangle + \lambda\left\langle\phi(1)|x_1^4\,\phi(1)\right\rangle\left\langle\phi(2)|x_2^4\,\phi(2)\right\rangle = 2\left\langle\phi|\hat{h}\phi\right\rangle + \lambda\left\langle\phi|x^4\phi\right\rangle^2$, where one-particle operator $\hat{h}(i) = -\frac{\hbar^2}{2m_i}\frac{\partial^2}{\partial x_i^2} + \frac{1}{2}kx_i^2$.

[12] Sometimes one uses *Coulomb notation* $[(ij|kl)_{Dirac} \equiv (ik|jl)_{Coulomb}$, also $\langle ij|kl\rangle_{Dirac} \equiv \langle ik|jl\rangle_{Coulomb}]$, which emphasizes the physical interpretation of the two electron integral as the energy of the Coulombic interaction of two charge distributions $\varphi_i^*(1)\varphi_k(1)$ for electron 1 and $\varphi_j^*(2)\varphi_l(2)$ for electron 2. Dirac notation for the two-electron integrals emphasizes the two-electron functions "*bra*" and "*ket*" from the general Dirac notation (p. 20). *In this book, we will consequently use Dirac notation* (both for integrals using spinorbitals, and for those using orbitals, the difference being emphasized by the type of bracket).

[13] See Appendix M available at booksite.elsevier.com/978-0-444-59436-5, p. e109; look at this rule on p. e119 (you may leave out its derivation).

8.2.3 Energy Minimization with Constraints

We would like to find such spinorbitals ("*the best ones*"), that *any* change in them leads to an *increase* in energy $E[\psi]$. But the changes of the spinorbitals need to be such that the above formula still holds, and it would hold only by assuming the *orthonormality* of the spinorbitals. This means that there are some constraints for the spinorbitals:

$$\langle i|j\rangle - \delta_{ij} = 0, \tag{8.9}$$

for $i, j = 1, 2, \ldots N$.

Thus we seek the *conditional minimum*. We will find it using the Lagrange multipliers method (see Appendix N available at booksite.elsevier.com/978-0-444-59436-5, p. e121). In this method, the equations of the constraints multiplied by the Lagrange multipliers are added (or subtracted, does not matter) to the original function that is to be minimized. Then we minimize the function as if the constraints did not exist.

We do the same for the functionals. The necessary condition for the minimum is that the variation[14] of $E - \sum_{ij} L_{ij}(\langle i|j\rangle - \delta_{ij})$ equals zero (the numbers L_{ij} denote the Lagrange multipliers to be found).

The variation of a functional is defined as the *linear* part of the functional change coming from a change in the function which is its argument.

Variation is an analog of the differential (the differential is just the linear part of the function's change). Thus, we calculate the linear part of a change (variation):

$$\delta(E - \sum_{ij} L_{ij}\langle i|j\rangle) = 0 \tag{8.10}$$

using the (as-yet) undetermined Lagrange multipliers L_{ij}, and we set the variation equal to zero.[15]

---

[14] However, this is not a sufficient condition because the vanishing of the differential for certain values of independent variables happens not only for minima, but also for maxima and saddle points (stationary points).

[15] Note, that $\delta(\delta_{ij}) = 0$.

AN ANALOGY–THE COUPLED HARMONIC OSCILLATORS (continued, 3 of 5): The change of E, because of the variation $\delta\phi^*$, is $E[\phi + \delta\phi] - E[\phi] = 2\left\langle\phi + \delta\phi|\hat{h}\phi\right\rangle + \lambda\left\langle\phi + \delta\phi|x^4\phi\right\rangle^2 - \left[2\left\langle\phi|\hat{h}\phi\right\rangle + \lambda\left\langle\phi|x^4\phi\right\rangle^2\right] = 2\left\langle\phi|\hat{h}\phi\right\rangle + 2\left\langle\delta\phi|\hat{h}\phi\right\rangle + \lambda\left\langle\phi|x^4\phi\right\rangle^2 + 2\lambda\left\langle\delta\phi|x^4\phi\right\rangle\left\langle\phi|x^4\phi\right\rangle + \lambda\left\langle\delta\phi|x^4\phi\right\rangle^2 - \left[2\left\langle\phi|\hat{h}\phi\right\rangle + \lambda\left\langle\phi|x^4\phi\right\rangle^2\right] = 2\left\langle\delta\phi|\hat{h}\phi\right\rangle + 2\lambda\left\langle\delta\phi|x^4\phi\right\rangle\left\langle\phi|x^4\phi\right\rangle + \lambda\left\langle\delta\phi|x^4\phi\right\rangle^2$. The linear part in $\delta\phi$ of the energy change (i.e., the variation) is, therefore, equal to $\delta E = 2\left\langle\delta\phi|\hat{h}\phi\right\rangle + 2\lambda\left\langle\delta\phi|x^4\phi\right\rangle\left\langle\phi|x^4\phi\right\rangle$. The variation $\delta\phi^*$, however, has to ensure the normalization of ϕ; i.e., $\langle\phi|\phi\rangle = 1$. After multiplying by the unknown Lagrange multiplier chosen as 2ε, we get the extremum condition $\delta(E - 2\varepsilon\langle\phi|\phi\rangle) = 0$; i.e., $2\left\langle\delta\phi|\hat{h}\phi\right\rangle + 2\lambda\left\langle\delta\phi|x^4\phi\right\rangle\left\langle\phi|x^4\phi\right\rangle - 2\varepsilon\langle\delta\phi|\phi\rangle = 0$.

The Stationarity Condition for the Energy Functional

It is sufficient to vary *only the function's complex conjugate to the spinorbitals* or only the spinorbitals (cf., p. 234), yet the result is always the same. We decide the first instance of this.

Substituting $\phi_i^* \to \phi_i^* + \delta\phi_i^*$ in Eq. (8.7), and retaining only linear terms in $\delta\phi_i^*$ to be inserted into Eq. (8.10), the variation takes the form (the symbols δi^* and δj^* mean $\delta\phi_i^*$ and $\delta\phi_j^*$)

$$\sum_{i=1}^{N}\left(\langle\delta i|\hat{h}|i\rangle + \frac{1}{2}\sum_j\left(\langle\delta i, j|ij\rangle + \langle i, \delta j|ij\rangle - \langle\delta i, j|ji\rangle - \langle i, \delta j|ji\rangle - 2L_{ij}\langle\delta i|j\rangle\right)\right) = 0. \tag{8.11}$$

Now we will express this in the form:

$$\sum_i\langle\delta i|\ldots\rangle = 0.$$

Since the δi^* may be arbitrary, the equation $|\ldots\rangle = 0$ (called the *Euler equation* in variational calculus) results. This will be our next goal.

Noticing that the sum indices and the numbering of electrons in the integrals are arbitrary, we have the following equalities:

$$\sum_{ij}\langle i, \delta j|ij\rangle = \sum_{ij}\langle j, \delta i|ji\rangle = \sum_{ij}\langle\delta i, j|ij\rangle,$$

$$\sum_{ij}\langle i, \delta j|ji\rangle = \sum_{ij}\langle j, \delta i|ij\rangle = \sum_{ij}\langle\delta i, j|ji\rangle,$$

and after substitution in the expression for the variation, we get

$$\sum_i\left(\langle\delta i|\hat{h}|i\rangle + \frac{1}{2}\sum_j\left(\langle\delta i, j|ij\rangle + \langle\delta i, j|ij\rangle - \langle\delta i, j|ji\rangle - \langle\delta i, j|ji\rangle - 2L_{ij}\langle\delta i|j\rangle\right)\right) = 0. \tag{8.12}$$

Let us rewrite this equation in the following manner ($\langle \delta i| \equiv \langle \delta\phi_i(1)|$):

$$\sum_i \Big\langle \delta\phi_i(1)\Big|\Big[\hat{h}\phi_i(1)+\sum_j \Big(\int d\tau_2 \frac{1}{r_{12}}\phi_j^*(2)\phi_j(2)\phi_i(1)$$

$$- \int d\tau_2 \frac{1}{r_{12}}\phi_j^*(2)\phi_i(2)\phi_j(1) - L_{ij}\phi_j(1)\Big)\Big]\Big\rangle_1 = 0, \tag{8.13}$$

where $\langle \delta\phi_i(1)|\ldots\rangle_1$ means integration over spatial coordinates of electron 1 and summation over its spin coordinate, $d\tau_2$ refers to the spatial coordinate integration and spin coordinate summing for electron 2. The above must be true for *any* $\langle \delta\phi_i(1)| \equiv \delta\phi_i^*$, which means that *each individual term* in the square parentheses needs to be equal to zero:

$$\hat{h}\phi_i(1) + \sum_j \Big(\int d\tau_2 \frac{1}{r_{12}}\phi_j^*(2)\phi_j(2)\cdot\phi_i(1) - \int d\tau_2 \frac{1}{r_{12}}\phi_j^*(2)\phi_i(2)\cdot\phi_j(1)\Big) = \sum_j L_{ij}\phi_j(1). \tag{8.14}$$

The Coulombic and Exchange Operators

Let us introduce the following linear operators:

a) *Two Coulombic operators: the total operator* $\hat{J}(1)$ and *the spinorbital operator* $\hat{J}_j(1)$, defined *via* their action on an arbitrary function $u(1)$ of the coordinates of electron 1, as follows:

$$\hat{J}(1)u(1) = \sum_j \hat{J}_j(1)u(1) \tag{8.15}$$

$$\hat{J}_j(1)u(1) = \int d\tau_2 \frac{1}{r_{12}}\phi_j^*(2)\phi_j(2)u(1) \tag{8.16}$$

b) Similarly, *two exchange operators: the total operator* $\hat{K}(1)$ and *the spinorbital operator* $\hat{K}_j(1)$

$$\hat{K}(1)u(1) = \sum_j \hat{K}_j(1)u(1) \tag{8.17}$$

$$\hat{K}_j(1)u(1) = \int d\tau_2 \frac{1}{r_{12}}\phi_j^*(2)u(2)\phi_j(1). \tag{8.18}$$

Then Eq. (8.14) takes the form

$$\left(\hat{h}(1) + \hat{J}(1) - \hat{K}(1)\right)\phi_i(1) = \sum_j L_{ij}\phi_j(1). \tag{8.19}$$

The equation is handy and concise except for one thing. It would be even better if the right side were proportional to $\phi_i(1)$ instead of being a linear combination of all the spinorbitals. In such a case, the equation would be similar to the eigenvalue problem and we would find it quite satisfactory. It would be similar but not identical, since the operators $\hat{J}$ and $\hat{K}$ include the *sought* spinorbitals ϕ_i. Because of this, the equation would be called the *pseudo-eigenvalue problem.*

8.2.4 Slater Determinant Subject to a Unitary Transformation

How can we help? Let us notice that we do not care too much about the spinorbitals themselves because these are *by-products* of the method that gives the optimum mean value of the Hamiltonian, and the corresponding N-electron wave function. We can choose some other spinorbitals, such that the mean value of the Hamiltonian, so long as the wave function do not change and the Lagrange multipliers L_{ij} are diagonal. Is this at all possible? Let us see.

Let us imagine the linear transformation of spinorbitals ϕ_i; i.e., in matrix notation:

$$\boldsymbol{\phi}' = \boldsymbol{A}\boldsymbol{\phi}, \tag{8.20}$$

where $\boldsymbol{\phi}$ and $\boldsymbol{\phi}'$ are vertical vectors containing components ϕ_i. A vertical vector is uncomfortable for typography, in contrast to its transposition (a horizontal vector), and it is easier to write the transposed vector: $\boldsymbol{\phi}'^T = \left[\phi_1', \ \phi_2', \ \ldots, \ \phi_N'\right]$ and $\boldsymbol{\phi}^T = [\phi_1, \ \phi_2, \ \ldots, \ \phi_N]$. If we construct the determinant built of spinorbitals $\boldsymbol{\phi}'$ and not of $\boldsymbol{\phi}$, an interesting chain of transformations will result:

$$\frac{1}{\sqrt{N!}}\begin{vmatrix} \phi_1'(1) & \phi_1'(2) & \ldots & \phi_1'(N) \\ \phi_2'(1) & \phi_2'(2) & \ldots & \phi_2'(N) \\ \ldots & \ldots & \ldots & \ldots \\ \phi_N'(1) & \phi_N'(2) & \ldots & \phi_N'(N) \end{vmatrix} = \frac{1}{\sqrt{N!}}\begin{vmatrix} \sum_i A_{1i}\phi_i(1) & \ldots & \sum_i A_{1i}\phi_i(N) \\ \sum_i A_{2i}\phi_i(1) & \ldots & \sum_i A_{2i}\phi_i(N) \\ \ldots & \ldots & \ldots \\ \sum_i A_{Ni}\phi_i(1) & \ldots & \sum_i A_{Ni}\phi_i(N) \end{vmatrix} \tag{8.21}$$

$$= \frac{1}{\sqrt{N!}}\det\left\{\boldsymbol{A}\begin{bmatrix} \phi_1(1) & \phi_1(2) & \ldots & \phi_1(N) \\ \phi_2(1) & \phi_2(2) & \ldots & \phi_2(N) \\ \ldots & \ldots & \ldots & \ldots \\ \phi_N(1) & \phi_N(2) & \ldots & \phi_N(N) \end{bmatrix}\right\}$$

$$= \det\boldsymbol{A} \cdot \frac{1}{\sqrt{N!}}\begin{vmatrix} \phi_1(1) & \phi_1(2) & \ldots & \phi_1(N) \\ \phi_2(1) & \phi_2(2) & \ldots & \phi_2(N) \\ \ldots & \ldots & \ldots & \ldots \\ \phi_N(1) & \phi_N(2) & \ldots & \phi_N(N) \end{vmatrix}. \tag{8.22}$$

Therefore, we have obtained our initial Slater determinant multiplied by a *number:* det $\boldsymbol{A}$. Thus, provided that det $\boldsymbol{A}$ is not zero,[16]

the new wave function would provide the same mean value of the Hamiltonian.

The only problem from such a transformation is the loss of the normalization of the wave function. Yet we may even preserve the normalization. Let us choose such a matrix $\boldsymbol{A}$, that $|\det \boldsymbol{A}| = 1$. This condition will hold if $\boldsymbol{A} = \boldsymbol{U}$, where $\boldsymbol{U}$ is a *unitary* matrix (when $\boldsymbol{U}$ is real, we call $\boldsymbol{U}$ an *orthogonal* transformation).[17] This means that

if a unitary transformation $\boldsymbol{U}$ is performed on the orthonormal spinorbitals, then the new spinorbitals ϕ' are also orthonormal.

This is why a unitary transformation is said to represent a rotation in the Hilbert space: the mutually orthogonal and perpendicular vectors do not lose these features upon rotation.[18] This can be verified by a direct calculation:

$$\begin{aligned}\left\langle \phi_i'(1)|\phi_j'(1)\right\rangle &= \left\langle \sum_r U_{ir}\phi_r(1)|\sum_s U_{js}\phi_s(1)\right\rangle \\ &= \sum_{rs} U_{ir}^* U_{js} \left\langle \phi_r(1)|\phi_s(1)\right\rangle = \sum_{rs} U_{ir}^* U_{js}\delta_{rs} \\ &= \sum_r U_{ir}^* U_{jr} = \delta_{ij}.\end{aligned}$$

Thus, in the case of a unitary transformation, even the normalization of the total one-determinant wave function is preserved; at worst, the phase χ of this function will change (while $\exp(i\chi) = \det \boldsymbol{U}$), and this factor does not change either $|\psi|^2$ or the mean value of the operators.

16 The $\boldsymbol{A}$ transformation thus cannot be singular (see the Appendix B available atbooksite.elsevier.com/978-0- 978-0-444-59436-5, p. e7).

17 For a unitary transformation, $\boldsymbol{U}\boldsymbol{U}^\dagger = \boldsymbol{U}^\dagger\boldsymbol{U} = \mathbf{1}$. The matrix $\boldsymbol{U}^\dagger$ arises from $\boldsymbol{U}$ *via* the exchange of rows and columns (this does not influence the value of the determinant), and *via* the complex conjugation of all elements (and this gives $\det \boldsymbol{U}^\dagger = (\det \boldsymbol{U})^*$). Finally, since $(\det \boldsymbol{U})(\det \boldsymbol{U}^\dagger) = 1$, we have $|\det \boldsymbol{U}| = 1$.

18 Just as three fingers held at right angles do not stop being the same length (normalization), and they continue to be orthogonal after rotation.

8.2.5 The $\hat{J}$ and $\hat{K}$ Operators Are Invariant

How does the Coulombic operator change upon a unitary transformation of the spinorbitals? Let us see the following:

$$\begin{aligned}
\hat{J}(1)'\chi(1) &= \int d\tau_2 \frac{1}{r_{12}} \sum_j \phi_j'^*(2)\phi_j'(2)\chi(1) \\
&= \int d\tau_2 \frac{1}{r_{12}} \sum_j \sum_r U_{jr}^* \phi_r^*(2) \sum_s U_{js}\phi_s(2)\chi(1) \\
&= \int d\tau_2 \frac{1}{r_{12}} \sum_{r,s} \left(\sum_j U_{js} U_{jr}^* \right) \phi_r^*(2)\phi_s(2)\chi(1) \\
&= \int d\tau_2 \frac{1}{r_{12}} \sum_{r,s} \left(\sum_j U_{rj}^\dagger U_{js} \right) \phi_r^*(2)\phi_s(2)\chi(1) \\
&= \int d\tau_2 \frac{1}{r_{12}} \sum_{r,s} \delta_{sr} \phi_r^*(2)\phi_s(2)\chi(1) \\
&= \int d\tau_2 \frac{1}{r_{12}} \sum_r \phi_r^*(2)\phi_r(2)\chi(1) = \hat{J}(1)\chi(1).
\end{aligned}$$

The operator $\hat{J}(1)'$ proves to be *identical* to the operator $\hat{J}(1)$. Similarly, we may prove *the invariance of the operator* K.

> The operators $\hat{J}$ and $\hat{K}$ are invariant with respect to any unitary transformation of the spinorbitals. In conclusion, while deriving the new spinorbitals from a unitary transformation of the old ones, we do not need to worry about $\hat{J}$ and $\hat{K}$ since they remain the same.

8.2.6 Diagonalization of the Lagrange Multipliers

Equation (8.19) may be written in matrix form as follows:

$$[\hat{h}(1) + \hat{J}(1) - \hat{K}(1)]\boldsymbol{\phi}(1) = \boldsymbol{L}\boldsymbol{\phi}(1), \tag{8.23}$$

where $\boldsymbol{\phi}$ is a column of spinorbitals. Transforming $\boldsymbol{\phi} = \boldsymbol{U}\boldsymbol{\phi}'$ and multiplying the last equation by $\boldsymbol{U}^\dagger$ (where $\boldsymbol{U}$ is a unitary matrix), we obtain

$$\boldsymbol{U}^\dagger[\hat{h}(1) + \hat{J}(1) - \hat{K}(1)]\boldsymbol{U}\boldsymbol{\phi}(1)' = \boldsymbol{U}^\dagger\boldsymbol{L}\boldsymbol{U}\boldsymbol{\phi}(1)', \tag{8.24}$$

because $\hat{J}$ and $\hat{K}$ did not change upon the transformation.

The $\boldsymbol{U}$ matrix *can* be chosen such that $\boldsymbol{U}^\dagger \boldsymbol{L}\boldsymbol{U}$ represents the *diagonal* matrix.

Its diagonal elements[19] will now be denoted as ε_i. Because $\hat{h}(1) + \hat{J}(1) - \hat{K}(1)$ is a linear operator, we get the following equation:

$$\boldsymbol{U}^\dagger \boldsymbol{U}\big(\hat{h}(1) + \hat{J}(1) - \hat{K}(1)\big)\boldsymbol{\phi}(1)' = \boldsymbol{U}^\dagger \boldsymbol{L}\boldsymbol{U}\boldsymbol{\phi}(1)' \tag{8.25}$$

or, alternatively,

$$\big(\hat{h}(1) + \hat{J}(1) - \hat{K}(1)\big)\boldsymbol{\phi}(1)' = \boldsymbol{\varepsilon}\boldsymbol{\phi}(1)', \tag{8.26}$$

where $\varepsilon_{ij} = \varepsilon_i \delta_{ij}$.

8.2.7 Optimal Spinorbitals Are Solutions of the Fock Equation (General Hartree-Fock Method)

We leave out the "*prime*" to simplify the notation[20] and write *the Fock equation* for a single spinorbital:

The Fock Equation in the General Hartree-Fock (GHF) Method:

$$\hat{F}(1)\phi_i(1) = \varepsilon_i \phi_i(1), \tag{8.27}$$

where the Fock operator $\hat{F}$ is

$$\hat{F}(1) = \hat{h}(1) + \hat{J}(1) - \hat{K}(1). \tag{8.28}$$

These ϕ_i are called *canonical spinorbitals* and are the solution of the Fock equation, and ε_i is the *orbital energy* corresponding to the spinorbital ϕ_i. It is indicated in brackets that both the *Fock operator* and the molecular spinorbital depend on the coordinates of one electron only (represented by electron 1).

[19] Such diagonalization is possible because $\boldsymbol{L}$ is a Hermitian matrix (i.e., $\boldsymbol{L}^\dagger = \boldsymbol{L}$), and each Hermitian matrix may be diagonalized *via* the transformation $\boldsymbol{U}^\dagger \boldsymbol{L}\boldsymbol{U}$ with the unitary matrix $\boldsymbol{U}$. Matrix $\boldsymbol{L}$ is indeed Hermitian. It is clear when we write the complex conjugate of the variation $\delta(E - \sum_{ij} L_{ij}\langle i|j\rangle) = 0$. This gives $\delta\left(E - \sum_{ij} L^*_{ij}\langle j|i\rangle\right) = 0$, because E is real, and after the change of the summation indices, $\delta\left(E - \sum_{ij} L^*_{ji}\langle i|j\rangle\right) = 0$. Thus, $L_{ij} = L^*_{ji}$, i.e., $\boldsymbol{L} = \boldsymbol{L}^\dagger$.

[20] This means that we finally forget about ϕ' (we pretend that they have never appeared), and we will deal only with such ϕ as correspond to the *diagonal* matrix of the Lagrange multipliers.

AN ANALOGY–THE COUPLED HARMONIC OSCILLATORS (continued, 4 of 5): The equation for the conditional extremum may be rewritten as $2\left\langle\delta\phi|\left[\hat{h}+\lambda\bar{x}^4x^4-\varepsilon\right]\phi\right\rangle=0$, where $\bar{x}^4=\langle\phi|x^4\phi\rangle$ is a *number*. This gives (remember that $\delta\phi^*$ is arbitrary) the Euler equation $\left[\hat{h}+\lambda\bar{x}^4x^4-\varepsilon\right]\phi=0$; i.e., the analog of the Fock [Eq. (8.27)]. $\hat{F}\phi=\varepsilon\phi$ with the operator $\hat{F}=\hat{h}+\lambda\bar{x}^4x^4$. Let us emphasize that the operator $\hat{F}$ is a one-particle operator, *via* the notation $\hat{F}(1)\,\phi(1)=\varepsilon\phi(1)$, while $\hat{F}(1)=\hat{h}(1)+\lambda\bar{x}^4x_1^4$. The term $\lambda x_1^4x_2^4$ is analogous to the term $\frac{1}{r_{12}}$, the term $\lambda\bar{x}^4x_1^4\phi(1)$ may be rewritten as $\int dx_2\lambda x_1^4x_2^4\phi^*(2)\phi(2)\phi(1)$, which for the coupled oscillators might be written as $\hat{J}_{osc}(1)\phi(1)$, where $\hat{J}_{osc}(1)=\int dx_2\lambda x_1^4x_2^4\phi^*(2)\phi(2)$ operator (with the interaction operator $\lambda x_1^4x_2^4$ instead of $\frac{1}{r_{12}}$ for electrons) is analogous to $\hat{J}(1)$. There will be no term corresponding to $\hat{K}(1)$ since the latter results from the antisymmetry of the variational wave function (the Slater determinant), while the bosonic function is a symmetric product $\phi(1)\phi(2)$ of two "*spinorbitals*" ϕ and we do not have such an effect. Also, there is no unitary transformation appearing because we have a single spinorbital here, not a set of N spinorbitals ϕ_i.

8.2.8 "Unrestricted" Hartree-Fock (UHF) Method

One may limit the GHF method by forcing some restrictions on the form of the GHF spinorbitals. A special case of the GHF method is known in textbooks as the *unrestricted Hartree-Fock method* (UHF). Despite its name, UHF is not a fully unrestricted method (as the GHF is). In the UHF, we assume that, [cf., Eq. (8.1)]:

- Orbital components φ_{i1} and φ_{i2} are *real.*
- *There is no mixing of the spin functions* α and β; i.e., either $\varphi_{i1}=0$ and $\varphi_{i2}\neq 0$ or $\varphi_{i1}\neq 0$ and $\varphi_{i2}=0$.

The UHF method is called sometimes the DODS method, which means Different Orbitals for Different Spins. In the UHF (or DODS) method, each spinorbital has its own orbital energy.

8.2.9 The Closed-Shell Systems and the Restricted Hartree-Fock (RHF) Method

Double Occupation of the Orbitals and the Pauli Exclusion Principle

When the number of electrons is even, the spinorbitals are usually formed out of orbitals in a very easy [and simplified with respect to Eq. (8.1)] manner,[21] by multiplication of each orbital by the spin functions α or β:

[21] It is not necessary, but it is quite comfortable.

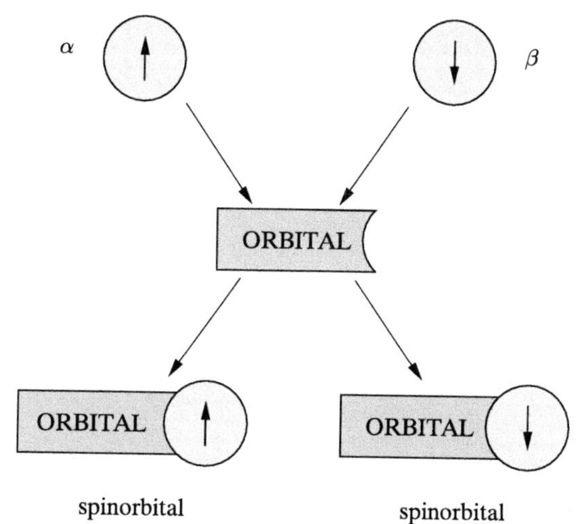

Fig. 8.3. Construction (scheme) of a spinorbital in the RHF method (i.e., a function x, y, z, σ) as a *product* of an orbital (a function of x, y, z) and one of the two spin functions $\alpha(\sigma)$ or $\beta(\sigma)$.

$$\phi_{2i-1}(\boldsymbol{r}, \sigma) = \varphi_i(\boldsymbol{r})\alpha(\sigma) \tag{8.29}$$

$$\phi_{2i}(\boldsymbol{r}, \sigma) = \varphi_i(\boldsymbol{r})\beta(\sigma), \tag{8.30}$$

$$i = 1, 2, \ldots \frac{N}{2}, \tag{8.31}$$

where, as it can be clearly seen, there are half as many occupied orbitals φ as occupied spinorbitals ϕ (*occupation means that a given spinorbital appears in the Slater determinant*[22]) (see Fig. 8.3). Thus, we introduce an artificial *restriction* for spinorbitals (some of the consequences will be described on p. 437). This is why the method is called the Restricted Hartree-Fock (RHF) method. Nothing forces us to do this; the criterion is simplicity. The problem of whether we lose the Hartree-Fock energy by doing this will be discussed on p. 441.

There are as many spinorbitals as electrons, and there can be a maximum of two electrons per orbital.

If we wished to occupy a given orbital with more than two electrons, we would need once again to use the spin function α or β when constructing the spinorbitals (i.e., repeating spinorbitals). This would imply two identical rows in the Slater determinant, and the wave function would equal zero. This cannot be accepted. The above rule of maximum double occupation is often called the *Pauli exclusion principle.*[23] Such a formulation of the Pauli exclusion principle

[22] When the Slater determinant is written, the electrons lose their identity–they are no longer distinguishable.

[23] From *Solid State and Molecular Theory*, Wiley, London, 1975 by John Slater: "*I had a seminar about the work which I was doing over there–the only lecture of mine which happened to be in German. It has appeared that not*

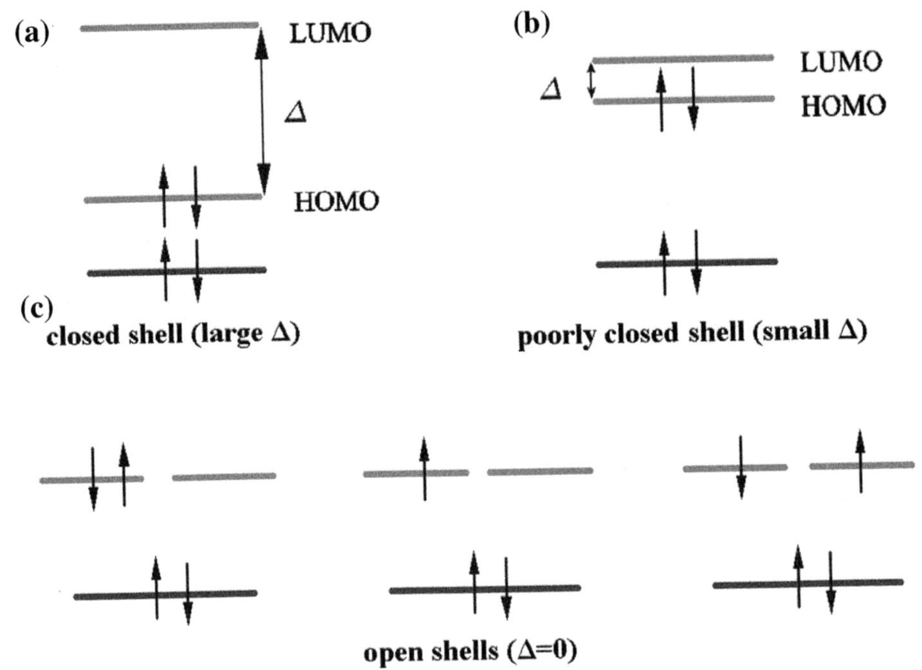

Fig. 8.4. Electron occupancy of orbitals and the concept of the HOMO and LUMO. The LUMO-HOMO energy difference is shown as Δ. (a) If Δ is large, we have the closed shell. (b) If $\Delta > 0$ is small, we have to do with a poorly closed shell. (c) $\Delta = 0$ means the open shell.

requires two concepts: the postulate of the antisymmetrization of the electronic wave function, p. 29, and double orbital occupancy. The first of these is of fundamental importance, while the second is of a technical nature.[24]

We often assume the double occupancy of orbitals within what is called the *closed shell.* The latter term has an approximate character (Figs. 8.4 and 8.5). It means that for the studied system, there is a large energy difference between HOMO and LUMO orbital energies.

> HOMO is the Highest Occupied Molecular Orbital, and LUMO is the Lowest Unoccupied Molecular Orbital. The unoccupied molecular orbitals are called *virtual orbitals.*

only Heisenberg, Hund, Debye, and young Hungarian Ph.D. student Edward Teller were present, but also Wigner, Pauli, Rudolf Peierls and Fritz London, all of them on their way to winter holidays. Pauli, of course, behaved in agreement with the common opinion about him, and disturbed my lecture saying that 'he had not understood a single word out of it,' but Heisenberg has helped me to explain the problem. (...) Pauli was extremely bound to his own way of thinking, similar to Bohr, who did not believe in the existence of photons. Pauli was a warriorlike man, a kind of dictator ...".

24 The concept of orbitals, occupied by electron pairs, exists only in the mean field method. We will abandon this idea in the future, and the Pauli exclusion principle will be understood in its generic form as a postulate (see Chapter 1) of the antisymmetry of the electronic wave function.

A Closed Shell

A closed shell means that the HOMO is doubly occupied, as are all the orbitals that have equal or lower energy. The occupancy is such that the mathematical form of the Slater determinant does not depend on the spatial orientation of the x-, y-, or z-axis. Using group theory nomenclature (see Appendix C available at booksite.elsevier.com/978-0-444-59436-5), this function transforms according to *fully symmetric irreducible representation* of the symmetry group of the electronic Hamiltonian.

If a shell is not closed, it is called "*open*."[25] We assume that there is a unique assignment for which molecular spinorbitals[26] within a closed shell are occupied in the ground state. The concept of the closed shell is approximate because it is not clear what it means when we say that the HOMO-LUMO energy distance[27] is large or small.[28]

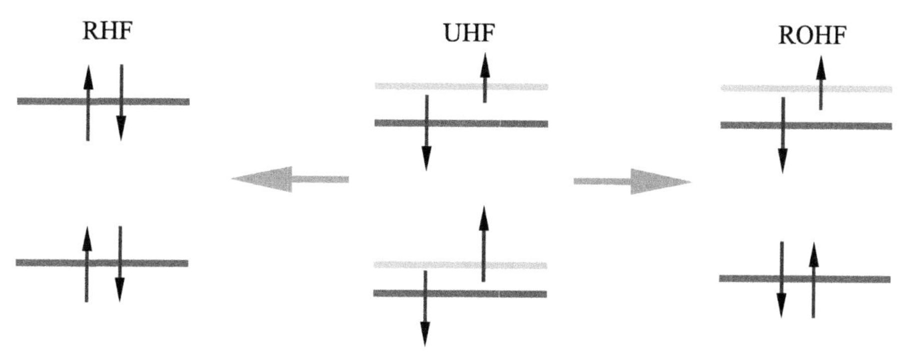

Fig. 8.5. Some restrictions imposed on the GHF method in computational practice: the RHF, the UHF, and the ROHF (Restricted Open Shell) methods. (a) RHF: we force the same (real) orbitals for the electron pair (opposite spins) producing a pair of spinorbitals by using the same orbital. (b) UHF: we relieve this restriction for orbitals (still being real). (c) ROHF: we keep the double occupancy for inner shells as for the RHF method, while for the valence shell, we use the UHF-type splitting of orbitals.

[25] Sometimes we use the term *semi-closed shell* if it is half-occupied by the electrons and we are interested in the state bearing maximum spin. In this case, the Slater determinant is a very good approximation. The reasons for this is, of course, the uniqueness of electron assignment to various spinorbitals. *If there is no uniqueness (as in the carbon atom), then the single-determinant approximation cannot be accepted.*

[26] The adjective *molecular* is suggested even for calculations for an atom. In a correct theory of electronic structure, the number of nuclei present in the system should not play any role. Thus, from the point of view of the computational machinery, an atom is just a molecule with one nucleus.

[27] The decision to occupy only the lowest-energy MOs (so-called *Aufbau Prinzip*; a name left over from the German origins of quantum mechanics) is accepted under the *assumption* that the total energy differences are sufficiently well approximated by the differences in the orbital energies.

[28] This is so unless the distance is zero. The helium atom, with the two electrons occupying the $1s$ orbital (HOMO), is a $1s^2$ shell of impressive "*closure*" because the LUMO-HOMO energy difference calculated in a high-quality basis set (6-31G**; see p. 431) of atomic orbitals is of the order of 62 eV. On the other hand, the HOMO-LUMO distance is zero for the carbon atom because in the ground state, 6 electrons occupy the $1s$, $2s$, $2p_x$, $2p_y$ and $2p_z$ orbitals. There is room for 10 electrons, and we only have 6. Hence, the occupation (configuration) in the ground

We need to notice that HOMO and LUMO have somewhat different meanings. As will be shown on p. 465, $-\varepsilon_{HOMO}$ represents an approximate ionization energy i.e., binding energy of an electron interacting with the $(N-1)$ - electron system, while $-\varepsilon_{LUMO}$ is an approximate electron affinity energy; i.e., energy of an electron interacting with the N - electron system.

The Fock Equation for Closed Shells

The Fock equations for a closed shell (using the RHF method) can be derived in a very similar way as with the GHF method. This means that we perform the following steps:

1. We write down the expression for the mean value of the Hamiltonian (as a functional of the *orbitals*, the summation extends over all the *occupied orbitals*[29] (there are $N/2$ of them, as will be recalled by the upper limit denoted by MO); see p. 419: $E = 2\sum_i^{\text{MO}}\left(i|\hat{h}|i\right) + \sum_{i,j}^{\text{MO}}[2(ij|ij) - (ij|ji)]$.
2. We seek the conditional minimum of this functional (Lagrange multipliers method) allowing for the variation of the *orbitals* that takes their orthonormality into account $\delta E = 2\sum_i^{\text{MO}}\left(\delta i|\hat{h}|i\right) + \sum_{i,j}^{\text{MO}}[2(\delta ij|ij) - (\delta ij|ji) + 2(i\delta j|ij) - (i\delta j|ji)] - \sum_{i,j}^{\text{MO}} L'_{ij}(\delta i|j) = 0$.
3. We derive the Euler equation for this problem from $(\delta i|\ldots) = 0$. In fact, it is the Fock equation expressed in *orbitals*.[30]

$$\hat{\mathcal{F}}(1)\varphi_i(1) = \varepsilon_i\varphi_i(1), \tag{8.32}$$

where φ are the *orbitals*. *The Fock operator* is defined for the closed shell as

$$\hat{\mathcal{F}}(1) = \hat{h}(1) + 2\hat{\mathcal{J}}(1) - \hat{\mathcal{K}}(1), \tag{8.33}$$

where the first term [$\hat{h}$, see Eq. (8.8)] is the sum of the kinetic energy operator of electron 1 and the operator of the interaction of this electron with the nuclei in the molecule, the next two terms (i.e., Coulombic $\hat{\mathcal{J}}$ and exchange $\hat{\mathcal{K}}$ operators) are connected with the potential energy of the

---

state is $1s^2 2s^2 2p^2$. Thus, both HOMO and LUMO are the $2p$ orbitals, with zero energy difference. If we asked for a single sentence describing why carbon compounds play a dominant role in nature, it should be emphasized that for carbon atoms, the HOMO-LUMO distance is equal to zero (and that the orbital levels ε_{2s} and ε_{2p} are close in energy).

On the other hand, the beryllium atom is an example of a closed shell, which is not very tightly closed. Four electrons are in the lowest-lying configuration $1s^2 2s^2$, but the orbital level $2p$ (LUMO) is relatively close to $2s$ (HOMO) (10 eV for the 6-31G** basis set is not a small gap, yet it amounts to much less than that of the helium atom).

[29] Not spinorbitals.

[30] After a suitable unitary transformation of orbitals, analogous to what we have done in the case of GHF.

interaction of electron 1 with all electrons in the system, and they are defined (slightly differently than before for $\hat{J}$ and $\hat{K}$ operators[31]) *via* the action on any function χ of the position of electron 1:

$$2\hat{\mathcal{J}}(1)\chi(1) = \sum_{i=1}^{\mathrm{MO}} 2\hat{\mathcal{J}}_i(1)\chi(1) = \sum_{i=1}^{\mathrm{MO}} 2\int \mathrm{d}V_2 \frac{1}{r_{12}}\varphi_i^*(2)\varphi_i(2)\chi(1)$$
$$\equiv 2\sum_i^{\mathrm{MO}} \int \mathrm{d}V_2 \frac{1}{r_{12}}\varphi_i^*(2)\varphi_i(2)\chi(1) \tag{8.34}$$

$$\hat{\mathcal{K}}(1)\chi(1) = \sum_{i=1}^{\mathrm{MO}} \hat{\mathcal{K}}_i(1)\chi(1) = \sum_{i=1}^{\mathrm{MO}} \int \mathrm{d}V_2 \frac{1}{r_{12}}\varphi_i^*(2)\chi(2)\varphi_i(1)$$
$$\equiv \sum_i^{\mathrm{MO}} \int \mathrm{d}V_2 \frac{1}{r_{12}}\varphi_i^*(2)\chi(2)\varphi_i(1), \tag{8.35}$$

where integration is now exclusively over the spatial coordinates[32] of electron 2. Factor 2 multiplying the Coulombic operator results (as the reader presumably guessed) from the *double* occupation of the orbitals.

Interpretation of the Coulombic Operator

The Coulombic operator is nothing else but calculation of the Coulombic potential (with the opposite sign as created by all the *electrons*; Fig. 8.6) at the position of electron 1. Indeed, such a potential coming from an electron occupying molecular orbital φ_i is equal to

$$\int \frac{\rho_i(2)}{r_{12}}\mathrm{d}V_2, \tag{8.36}$$

where $\rho_i(2) = \varphi_i(2)^*\varphi_i(2)$ is the probability density of finding electron 2 described by orbital φ_i. If we take into account that the orbital φ_i is occupied by *two* electrons, and that the number of the doubly occupied molecular orbitals is $N/2$, then the electrostatic potential calculated at the position of the electron 1 is

$$\int \frac{\sum_i^{\mathrm{MO}} 2\rho_i(2)}{r_{12}}\mathrm{d}V_2 = \sum_i^{MO} 2\hat{\mathcal{J}}_i = 2\hat{\mathcal{J}}(1).$$

The same expression also means the interaction energy (in a.u.) of two charges: 1 (pointlike elementary charge -1) and 2, represented by a diffused cloud of charge density distribution $\rho(2) = \sum_i^{\mathrm{MO}} 2\rho_i(2)$ carrying N elementary charges -1 because $-\int \rho(2)d\mathrm{V}_2 = -2\sum_i 1 = N(-1)$.

[31] This occurs because we have orbitals here, and not spinorbitals.

[32] Simply, the summation over the spin coordinates has already been done when deriving the equation for the mean value of the Hamiltonian.

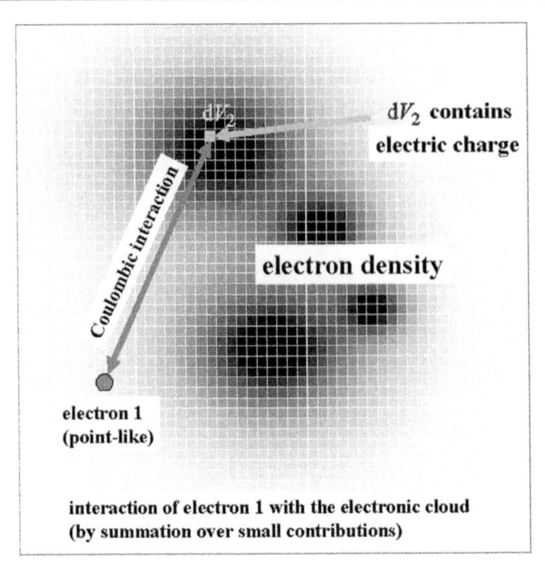

Fig. 8.6. Pointlike electron 1 interacts with the total electron density (shown as an electron cloud with density $\rho = \sum_i^{\mathrm{MO}} 2\rho_i(2)$). To compute the interaction energy, the total electron density is chopped into small cubes. The interaction energy of electron 1 with one of these cubes of volume dV_2 containing charge $-\sum_i^{\mathrm{MO}} 2\rho_i(2)dV_2$ is calculated according to the Coulomb law: charge × charge divided by their distance: $\frac{-1\times(-1)\sum_i^{\mathrm{MO}} 2\rho_i(2)dV_2}{r_{12}}$ or, alternatively, as charge -1 times electric potential produced by a single cube at electron 1. The summation over all cubes gives $\int \frac{\sum_i^{\mathrm{MO}} 2\rho_i(2)}{r_{12}} dV_2 = 2\hat{\mathcal{J}}(1)$.

Integration in the formula for the operator $\hat{\mathcal{J}}$ is a consequence of the approximation of independent particles.

This approximation means that, in the Hartree-Fock method, the electrons do not move in the electric field of the other pointlike electrons, but in the *mean static field of all the electrons* represented by electron cloud $\rho = \sum_i^{\mathrm{MO}} 2\rho_i(2)$. This is similar to a driver (one of the electrons) in Paris not using the position of other cars, but a map showing only the traffic intensity *via* the probability density cloud. The driver would then have a diffuse image of other vehicles and could not satisfactorily optimize his car's position with respect to other cars (which means higher energy for the molecule under study).

The Mean Field

This is typical for all the *mean field* methods. In these methods, *instead of watching the motion of other objects in detail, we average* these motions, and the problem simplifies (obviously, we pay the price of reduced quality).

This trick is general, ingenious, and worth remembering.[33]

Coulombic Self-Interaction

There is a problem with this, though. From what we have said, it follows that the electron 1 uses the "*maps*" of *total* electron density (i.e., including its own contribution to the density).[34] This looks strange, though. Let us take a closer look, maybe we have missed something in our reasoning. Note first of all that the repulsion of electron 1 (occupying, say, orbital φ_k) with the electrons, which is visible in the Fock operator, reads as $\left(\varphi_k | \left(2\hat{\mathcal{J}} - \hat{\mathcal{K}}\right) \varphi_k\right)$ and not as $\left(\varphi_k | \left(2\hat{\mathcal{J}}\right) \varphi_k\right)$. Let us write it down in more detail:

$$\begin{aligned}
\left(\varphi_k | \left(2\hat{\mathcal{J}} - \hat{\mathcal{K}}\right) \varphi_k\right) &= \int \mathrm{d}V_1 \, |\varphi_k(1)|^2 \sum_{i=1}^{\mathrm{MO}} 2 \int \mathrm{d}V_2 \frac{1}{r_{12}} \varphi_i^*(2)\varphi_i(2) \\
&\quad - \sum_{i=1}^{\mathrm{MO}} \int \mathrm{d}V_1 \varphi_k(1)^* \varphi_i(1) \int \mathrm{d}V_2 \frac{1}{r_{12}} \varphi_i^*(2)\varphi_k(2) \\
&= \int \mathrm{d}V_1 \, |\varphi_k(1)|^2 \sum_{i=1}^{\mathrm{MO}} 2 \int \mathrm{d}V_2 \frac{1}{r_{12}} \varphi_i^*(2)\varphi_i(2) \\
&\quad - \int \mathrm{d}V_1 \varphi_k(1)^* \varphi_k(1) \int \mathrm{d}V_2 \frac{1}{r_{12}} \varphi_k^*(2)\varphi_k(2) \\
&\quad - \sum_{i(\neq k)}^{\mathrm{MO}} \int \mathrm{d}V_1 \varphi_k(1)^* \varphi_i(1) \int \mathrm{d}V_2 \frac{1}{r_{12}} \varphi_i^*(2)\varphi_k(2) \\
&= \iint \mathrm{d}V_1 \mathrm{d}V_2 \frac{1}{r_{12}} \rho_k(1)[\rho(2) - \rho_k(2)] - \sum_{i(\neq k)}^{\mathrm{MO}} (ki|ik),
\end{aligned}$$

where $\rho_k = |\varphi_k(1)|^2$ (i.e., the distribution of electron 1) interacts electrostatically with all the *other* electrons[35] (i.e., with the distribution $[\rho(2) - \rho_k(2)]$, with ρ denoting the total electron density $\rho = \sum_{i=1}^{\mathrm{MO}} 2\,|\varphi_i|^2$ and $-\rho_k$ excluding from it the self-interaction energy of the electron in question). Thus, the Coulombic and exchange operators together ensure that an electron interacts electrostatically with other electrons, not with itself.

[33] We use it every day, although we do not call it a mean field approach. Indeed, if we say: *"I will visit my aunt at noon, because it is easier to travel out of rush hours,"* or *"I avoid driving through the center of town, because of the traffic jams,"* in practice we are using the mean field method. We average the motions of all citizens (including ourselves!) and we get a "*map*" (temporal or spatial), which allows us optimize *our own* motion. The motion of our fellow-citizens *disappears*, and we obtain a *one-body* problem.

[34] Exactly as happens with real city traffic maps.

[35] The fact that the integration variables pertain to electron 2 is meaningless, it is just a definite integration and the name of the variable does not count at all.

AN ANALOGY–THE COUPLED HARMONIC OSCILLATORS (continued, 5 of 5):
The Fock operator is: $\hat{F}(1) = \hat{h}(1) + \lambda \bar{x}^4 x_1^4$. It is now clear what the mean field approximation really is: the two-particle problem is reduced to a *single*-particle one (denoted as number 1), and the influence of the second particle, which appears as a result of the coupling term $\lambda x_1^4 x_2^4$, is *averaged over its positions* $\bar{x}^4 = \langle \phi | x^4 \phi \rangle = \langle \phi(2) | x_2^4 \phi(2) \rangle$. We see a similar effect in the Hartree-Fock problem for molecules: a single electron denoted by 1 interacts with all other electrons through the operators $2\hat{\mathcal{J}} - \hat{\mathcal{K}}$, in which we have an integration (averaging) over positions of all these electrons. So, we have essentially the same mean-field picture for the electronic system and for the two coupled oscillators.

Here, we end the illustration of the Hartree-Fock procedure by the two coupled harmonic oscillators.

Electrons with Parallel Spins Repel Less

There is also an exchange remainder $-\sum_{i(\neq k)}^{\text{MO}} (ki|ik)$, which is just a by-product of the antisymmetrization of the wave function (i.e., the Slater determinant), which tells us that in the Hartree-Fock picture, electrons of the same spin functions[36] repel ... less. Can this really be so? As shown at the beginning of this chapter, two electrons of the same spin cannot occupy the same point in space, and therefore (from the continuity of the wave function) they avoid each other. It is as if they repelled each other because of the Pauli exclusion principle, *in addition* to their Coulombic repulsion. Is there something wrong in our result, then? No, everything is all right. The necessary antisymmetric character of the wave function says simply that the same spins should keep electrons apart. However, when the electrons described by the same spin functions keep apart, this obviously means that *their Coulombic repulsion is weaker than that of electrons of opposite spins*. This is what the negative term $-\sum_{i(\neq k)}^{\text{MO}} (ki|ik)$ really means.

Hartree Method

The exchange operator represents a (non-intuitive) result of the antisymmetrization postulate for the total wave function (Chapter 1) and it has no classical interpretation. If the variational wave function were the *product* of the spinorbitals[37] (Douglas Hartree did this in the beginning of quantum chemistry),

$$\phi_1(1)\phi_2(2)\phi_3(3)\ldots\phi_N(N),$$

[36] When deriving the total energy expression (see Appendix M available at booksite.elsevier.com/978-0-444-59436-5), only those exchange terms survived, which correspond to the parallel spins of the interacting electrons. Note also that for real orbitals (as in the RHF method), every exchange contribution $(ki|ik) \equiv \int dV_1 \varphi_k(1)\varphi_i(1) \int dV_2 \frac{1}{r_{12}} \varphi_i(2)\varphi_k(2)$ means a repulsion because this is a self-interaction of the cloud $\varphi_k \varphi_i$.

[37] Such a function is not legal–it does not fulfill the antisymmetrization postulate. This illegal character (caused by a lack of the Pauli exclusion principle) would sometimes give unfortunate consequences: e.g., more than two electrons would occupy the 1s orbital, etc.

then we would get the corresponding Euler equation, which in this case is called the *Hartree equation*:

$$\hat{F}_{Hartree}(1)\phi_i(1) = \varepsilon_i\phi_i(1)$$

$$\hat{F}_{Hartree}(1) = \hat{h}(1) + \sum_{j(\neq i)}^{N} \hat{J}_j(1),$$

where $\hat{F}_{Hartree}$ corresponds to the Fock operator. Note that there is no self-interaction there.

8.2.10 Iterative Solution: The Self-Consistent Field Method

The following is a typical technique of solving the Fock equation.

First, we must handle the problem that in order to solve the Fock equation, we should know ... its solution. Indeed, the Fock equation is not an eigenvalue problem, but a pseudo-eigenvalue problem because the Fock operator depends on the *solutions* (which obviously are unknown). So, in the Fock equation, we do not know anything: all three quantities $\hat{\mathcal{F}}$, φ_i, ε_i that constitute the equation are unknown. Regardless of how strange all this might seem, we deal with this situation quite easily using an iterative approach because of the structure of $\hat{\mathcal{F}}$. This is called the *self-consistent field method (SCF)*. In this method (shown in Fig. 8.7), we do the following:

- Assume at the beginning (zeroth iteration) a certain shape of molecular orbitals.[38]

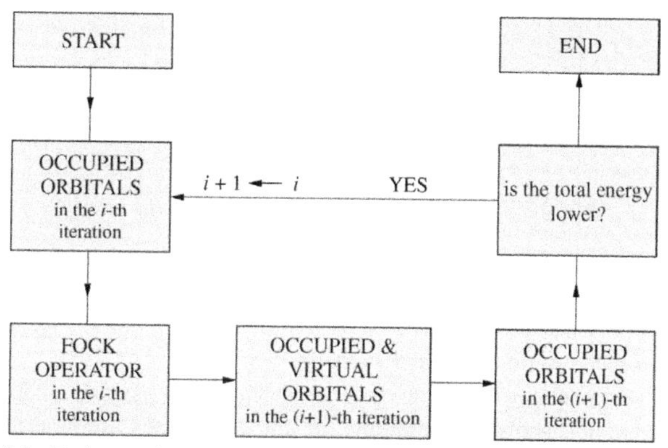

Fig. 8.7. Iterative solution of the Fock equation (the SCF). We do all of the following:

— Start from any set of occupied orbitals (zeroth iteration).
— Insert them into the Fock operator.
— Solve the Fock equation.
— Obtain the molecular orbitals of the first approximation.
— Choose those of the lowest energy as the occupied ones, and if your criterion of the total energy is not satisfied, repeat the procedure.

[38] These are usually the any-sort "*orbitals*," although recently, because of the *direct SCF* idea (we calculate the integrals whenever they are needed; i.e., at each iteration), an effort is made to save computational time per

- Introduce these orbitals to the Fock operator, thus obtaining a sort of "caricature" of it (the zero-order Fock operator).
- Solve the eigenvalue problem using the above "Fock operator" and get the molecular orbitals of the first iteration.
- Repeat the process until the shape of the orbitals does not change in the next iteration; i.e., until the Fock equations are solved.[39]

8.3 Total Energy in the Hartree-Fock Method

In Appendix M available at booksite.elsevier.com/978-0-444-59436-5, p. e109, we derived the following expressions for the mean value of the Hamiltonian using the normalized determinant (without a constant additive term for the nuclear repulsion energy V_{nn}, SMO means summation over the spinorbitals $i = 1, \ldots, N$; in the RHF method, and the MO summation limit means summation over the orbitals $i = 1, \ldots, N/2$):

$$E'_{HF} = \sum_{i}^{\mathrm{SMO}} \langle i|\hat{h}|i\rangle + \frac{1}{2}\sum_{i,j=1}^{\mathrm{SMO}} [\langle ij|ij\rangle - \langle ij|ji\rangle] \equiv \sum_{i}^{\mathrm{SMO}} h_{ii} + \frac{1}{2}\sum_{i,j=1}^{\mathrm{SMO}} [J_{ij} - K_{ij}]. \tag{8.37}$$

If double occupancy is assumed (i.e., the flexibility of the variational wave function is *restricted*), we may transform this expression in the following way:

$$E'_{RHF}(\text{double occupancy}) = \sum_{i}^{\mathrm{MO}} \left(\langle i\alpha|\hat{h}|i\alpha\rangle + \langle i\beta|\hat{h}|i\beta\rangle\right)$$
$$+\frac{1}{2}\sum_{i}^{\mathrm{MO}}\sum_{j}^{\mathrm{SMO}}[\langle i\alpha, j|i\alpha, j\rangle - \langle i\alpha, j|j, i\alpha\rangle + \langle i\beta, j|i\beta, j\rangle - \langle i\beta, j|ji\beta\rangle] = 2\sum_{i}^{\mathrm{MO}} \langle i|\hat{h}|i\rangle$$

---

iteration and therefore to provide as excellent a starting function as possible. We may obtain it *via* an initial calculation with some two-electron integrals neglected.

[39] Using our driver analogy, we may say that at the beginning, the driver has *false* maps of the probability density (thus, the system energy is high, and in our analogy, the car repair costs are large). The next iterations (repair costs effectively teach all the drivers) improve the map, the final energy decreases, and at the very end, we get the best map possible. The mean energy is the lowest possible (within the mean field method). A further energy reduction is only possible beyond the Hartree-Fock approximation; i.e., outside of the mean field method, which for the drivers means not believing maps, but their own eyes. A suspicious person (scientist) should be careful, because our solution *may depend on the starting point used*; i.e., from the initial, completely arbitrary orbitals. Besides, the iteration process does not necessarily need to be convergent. But it appears that the solutions in the Hartree-Fock method are usually independent of the zeroth-order MOs, and convergence problems are very rare. This is surprising. This situation is much worse for better-quality computations, where the AOs of small exponents are included (*diffuse orbitals*). Then we truly meet the problem already described (p. 352) of searching for the global energy minimum among a multitude of local ones.

$$+\frac{1}{2}\sum_i^{\mathrm{MO}}\sum_j^{\mathrm{MO}}[\langle i\alpha, j\alpha|i\alpha, j\alpha\rangle + \langle i\alpha, j\beta|i\alpha, j\beta\rangle - \langle i\alpha, j\alpha|j\alpha, i\alpha\rangle - \langle i\alpha, j\beta|j\beta, i\alpha\rangle$$

$$+\langle i\beta, j\alpha|i\beta, j\alpha\rangle + \langle i\beta, j\beta|i\beta, j\beta\rangle - \langle i\beta, j\alpha|j\alpha, i\beta\rangle - \langle i\beta, j\beta|j\beta, i\beta\rangle]$$

$$= 2\sum_i^{\mathrm{MO}}(i|\hat{h}|i) + \frac{1}{2}\sum_i^{\mathrm{MO}}\sum_j^{\mathrm{MO}}[4(ij|ij) - 2(ij|ji)]$$

$$= 2\sum_i^{\mathrm{MO}}(i|\hat{h}|i) + \sum_i^{\mathrm{MO}}\sum_j^{\mathrm{MO}}[2(ij|ij) - (ij|ji)].$$

This finally gives

$$E'_{RHF} = 2\sum_i^{\mathrm{MO}} h_{ii} + \sum_{i,j}^{\mathrm{MO}}[2\mathcal{J}_{ij} - \mathcal{K}_{ij}], \tag{8.38}$$

where $\mathcal{J}_{ij} \equiv \left(ij|ij\right)$, $\mathcal{K}_{ij} \equiv \left(ij|ji\right)$, see p. 413.

Both Eqs. (8.37) and (8.38) may in general give different results, because in the first, no double occupancy is assumed (we will discuss this further on p. 441).

Given the equality $\langle i|\hat{h}|i\rangle = (i|\hat{h}|i)$, these integrals are written here as h_{ii}. The *Coulombic and exchange* integrals defined in spinorbitals are denoted J_{ij} and K_{ij} and defined in orbitals as $\mathcal{J}_{ij}$ and $\mathcal{K}_{ij}$.

The additive constant corresponding to the internuclear repulsion (it is constant, since the nuclei positions are frozen) is

$$V_{nn} = \sum_{a<b}\frac{Z_a Z_b}{R_{ab}}, \tag{8.39}$$

which has not been introduced in the electronic Hamiltonian $\hat{H}_0$, and thus, the full Hartree-Fock energy is

$$E_{RHF} = E'_{RHF} + V_{nn}. \tag{8.40}$$

Note that the mean value of the electronic repulsion energy in our system is[40]

$$V_{ee} = \left\langle \psi_{HF}\left|\sum_{i<j}\frac{1}{r_{ij}}\right|\psi_{HF}\right\rangle = \sum_{i,j}^{\mathrm{MO}}[2(ij|ij) - (ij|ji)] = \sum_{i,j}^{\mathrm{MO}}[2\mathcal{J}_{ij} - \mathcal{K}_{ij}]. \tag{8.41}$$

It is desirable (interpretation purposes) to include the orbital energies in the formulas derived. Let us recall that the orbital energy ε_i is the mean value of the Fock operator for orbital i; i.e., the energy of an effective electron described by this orbital.[41] Based on Eqs. (8.33)–(8.35), this

[40] Recall that $\langle\psi_{RHF}|\hat{H}|\psi_{RHF}\rangle = E_{RHF}$ and V_{ee} is, therefore, the Coulombic interaction of electrons.

[41] $\left(\varphi_i|\hat{\mathcal{F}}\varphi_i\right) = \varepsilon_i$ because $\hat{\mathcal{F}}\varphi_i = \varepsilon_i\varphi_i$.

can be expressed as (i stands for the molecular orbital)

$$\varepsilon_i = h_{ii} + \sum_j^{MO} [2\mathcal{J}_{ij} - \mathcal{K}_{ij}], \tag{8.42}$$

and this in turn gives an elegant expression for the Hartree-Fock electronic energy:

$$E'_{RHF} = \sum_i^{MO} [h_{ii} + \varepsilon_i]. \tag{8.43}$$

From Eqs. (8.38), (8.41) and (8.42), the total electronic energy may be expressed as

$$E'_{RHF} = \sum_{i=1}^{MO} 2\varepsilon_i - V_{ee}. \tag{8.44}$$

It can be seen that the total electronic energy (i.e., E'_{RHF}) is not the sum of the orbital energies of electrons $\sum_i 2\varepsilon_i$.

And we would already expect full additivity, since the electrons in the Hartree-Fock method are treated as independent. Yet "*independent*" does not mean "*non-interacting.*" The reason for the non-additivity is that for *each* electron, we need to calculate its effective interaction with *all* the electrons, hence we would get too much repulsion.[42] Of course, the total energy, and not the sum of the orbital energies, is the most valuable. Yet in many quantum chemical problems, we interpret orbital energy lowering as energetically profitable. And it turns out that such an interpretation has an approximate justification. Works by Fraga, Politzer, and Ruedenberg[43] show that at the equilibrium geometry of a molecule, the formula

$$E_{RHF} = E'_{RHF} + V_{nn} \approx \frac{3}{2} \sum_{i=1}^{MO} 2\varepsilon_i, \tag{8.45}$$

works with 2%–4% precision, and even better results may be obtained by taking a factor of 1.55 instead of $\frac{3}{2}$.

8.4 Computational Technique: Atomic Orbitals as Building Blocks of the Molecular Wave Function

One of most powerful methods of computational analysis is to represent a function to be sought as a linear combination of some predefined set of functions (basis set). Fig. 8.8 shows the

[42] For example, the interaction of electron 5 and electron 7 is calculated *twice*: as the interaction $\frac{1}{r_{57}}$ and $\frac{1}{r_{75}}$.

[43] S. Fraga, *Theor. Chim. Acta, 2,* 406 (1964); P. Politzer, *J. Chem. Phys., 64,* 4239 (1976); K. Ruedenberg, *J. Chem. Phys., 66,* 375 (1977).

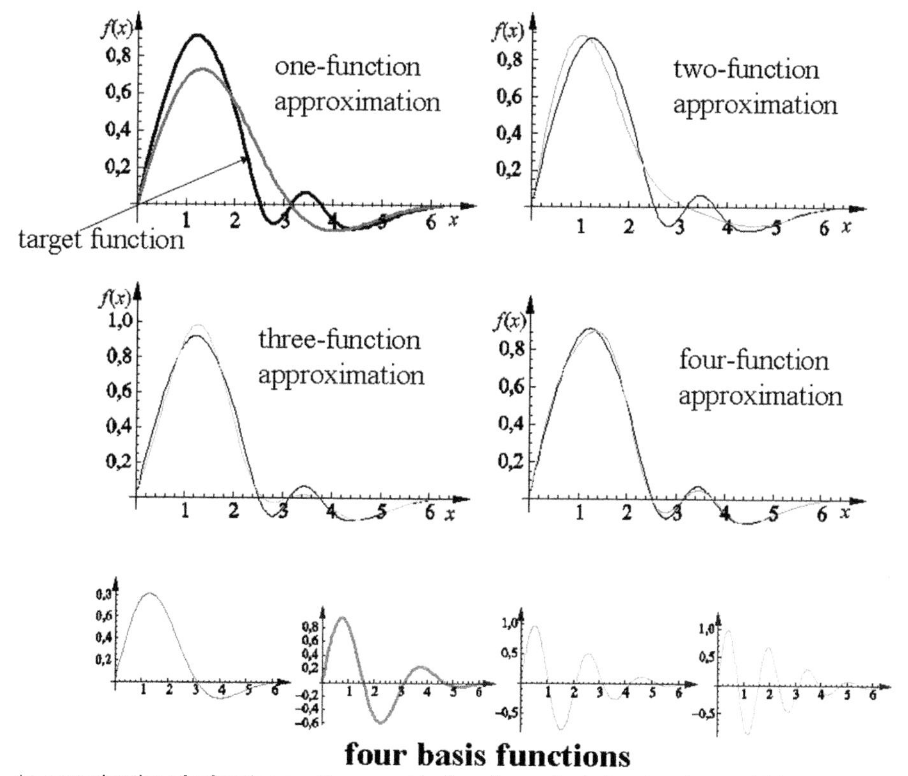

Fig. 8.8. An approximation of a function as a linear combination of some basis functions (one variable case). The first (larger) figures show how a function (black solid line) is approximated by, consecutively: one, two, three, and four basis functions (shown individually as small inserts at the bottom). As one can see, the approximation gets better and better with the increase of the number of the expansion functions. In an ideal situation (which is never reached in practice), this number would be infinite.

efficiency of such an idea in a step-by-step way (in the case of a smooth function of a single variable). In quantum chemistry we use atomic orbitals as the basis functions.

Atomic Orbital (AO)
means a function

$$g(\mathbf{r}) = f(x, y, z)\exp(-\zeta r^n),$$

where $f(x, y, z)$ is a polynomial and $n = 1, 2$. Such an atomic orbital is localized (centered) around $(0, 0, 0)$. The larger the exponent $\zeta > 0$, the more effective is this centering.

For $n = 1$, we have what is called *Slater Type Orbitals (STOs)*, and for $n = 2$, *Gaussian Type Orbitals (GTOs)*.

We have to be careful because the term *atomic orbital* is used in quantum chemistry with a double meaning. These are (1) the Hartree-Fock orbitals for a particular atom, or (2) functions localized in the space about a given center.[44] The role of the AOs is to provide a complete set of functions; i.e., a suitable linear combination of the AOs should in principle be able to approximate any continuous and square-integrable function.

8.4.1 Centering of the Atomic Orbital

If a complete set of the orbitals were at our disposal, then all the AOs might be centered around a *single* point.

> It is more economic, however, to allow using the incomplete set and the possibility of the AOs centered in various points of space.

Atomic orbital $g(\boldsymbol{r})$ may be shifted by a vector $\boldsymbol{A}$ in space [translation operation $\hat{\mathcal{U}}(\boldsymbol{A})$; see Chapter 2] to result in the new function $\hat{\mathcal{U}}(\boldsymbol{A})g(\boldsymbol{r}) = g(\hat{T}^{-1}(\boldsymbol{A})\,\boldsymbol{r}) = g(\hat{T}(-\boldsymbol{A})\,\boldsymbol{r}) = g(\boldsymbol{r}-\boldsymbol{A})$, because $\hat{T}^{-1}(\boldsymbol{A}) = \hat{T}(-\boldsymbol{A})$. Hence, the orbital centered at a given point (indicated by a vector $\boldsymbol{A}$) is (Fig. 8.9):

$$g(\boldsymbol{r}-\boldsymbol{A}) = f(x-A_x, y-A_y, z-A_z)\exp[-\zeta\,|\boldsymbol{r}-\boldsymbol{A}|^n]. \tag{8.46}$$

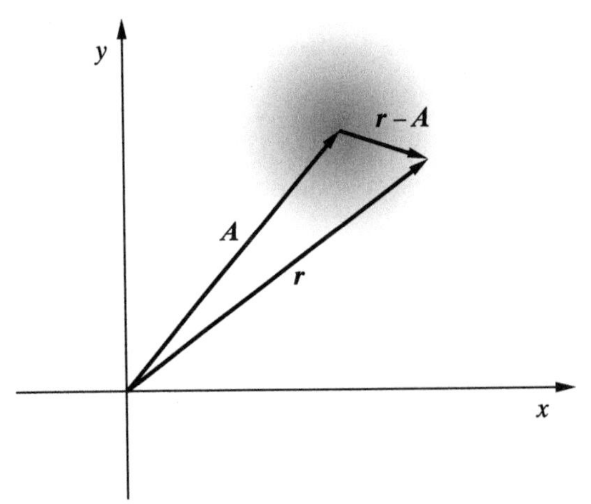

Fig. 8.9. The centering of $g(\boldsymbol{r})$ at the point shown by vector $\boldsymbol{A}$ means the creation of the orbital $g(\boldsymbol{r}-\boldsymbol{A})$. A linear combination of such orbitals can describe any smooth function of the position in space, of any degree of complexity.

---

[44] Atomic orbitals (the first meaning) may be thought of as expressed through linear combinations of atomic orbitals in the second meaning. The AOs may be centered on the nuclei (common practice), but they can also be centered off the nuclei.

8.4.2 Slater-Type Orbitals (STOs)

The Slater-type orbitals[45]:

$$\chi_{STO,nlm}\left(r,\theta,\phi\right)=Nr^{n-1}\exp\left(-\zeta r\right)Y_l^m\left(\theta,\phi\right), \tag{8.47}$$

(N stands for a suitable normalization constant) differ from the atomic orbitals of the hydrogen atom (see p. 202). The first difference is that the radial part is simplified in the STOs (the radial part of a STO has no nodes). The second difference is in the orbital exponent, which has no constraint except[46] $\zeta > 0$.

The STOs have a great advantage: they decay with distance from the center in a similar way to the "true" orbitals; let us recall the exponential vanishing of the hydrogen atom orbitals (see Chapter 4).[47] STOs would be fine, but finally we have to compute a large number of the integrals needed.[48] And here is a real problem: Since the Hamiltonian contains the electron-electron interactions, integrals appear with, in general, four atomic orbitals (of different centers). These integrals are difficult to calculate, and therefore consume an excessive amount of computer time.

8.4.3 Gaussian-Type Orbitals (GTOs)

If the exponent in Eq. (8.46) is equal to $n=2$, we are dealing with GTOs (N' is a normalization constant):

$$\chi_{GTO,nlm}\left(r,\theta,\phi\right)=N'r^{n-1}\exp\left(-\zeta r^2\right)Y_l^m\left(\theta,\phi\right). \tag{8.48}$$

The most important among them are $1s$-type orbitals, given for an arbitrary center $\boldsymbol{R}_p$:

$$\chi_p\equiv G_p(\boldsymbol{r};\alpha_p,\boldsymbol{R}_p)=\left(\frac{2\alpha_p}{\pi}\right)^{\frac{3}{4}}\exp\left(-\alpha_p|\boldsymbol{r}-\boldsymbol{R}_p|^2\right), \tag{8.49}$$

---

[45] We will distinguish two similar terms here: Slater-type orbitals and Slater orbitals. The latter term is reserved for special Slater-type orbitals, in which the exponent is easily computed by considering the effect of the screening of nucleus by the internal electronic shells. The screening coefficient is calculated according to the Slater rules; see p. 451.

[46] Otherwise, the orbital would not be square-integrable.

[47] It has been proved that *each* of the Hartree-Fock orbitals has *the same* asymptotic dependence on the distance from the molecule (N.C. Handy, M.T. Marron, and H.J. Silverstone, *Phys. Rev.*, *180,* 45 (1969)); i.e., $const\cdot\exp(-\sqrt{-2\varepsilon_{\max}}r)$, where $\varepsilon_{\max}$ is the orbital energy of HOMO. Earlier, people thought the orbitals decay as $\exp(-\sqrt{-2\varepsilon_i}r)$, where ε_i is the orbital energy expressed in atomic units. The last formula, as is easy to prove, holds for the atomic orbitals of hydrogen atom (see p. 201). R. Ahlrichs, M. Hoffmann-Ostenhoff, T. Hoffmann-Ostenhoff, and J.D. Morgan III, *Phys. Rev.*, *A23,* 2106 (1981), have shown that at a long distance r from an atom or a molecule, the square root of the *ideal* electron density satisfies the inequality: $\sqrt{\rho}\le C\left(1+r\right)^{\frac{(Z-N+1)}{\sqrt{2\varepsilon}}-1}\exp\left[-\left(2\varepsilon r\right)\right]$, where ε is the first *ionization potential*, Z is the sum of the nuclear charges, N is the number of electrons, and $C>0$ is a constant.

[48] The number of necessary integrals may reach billions.

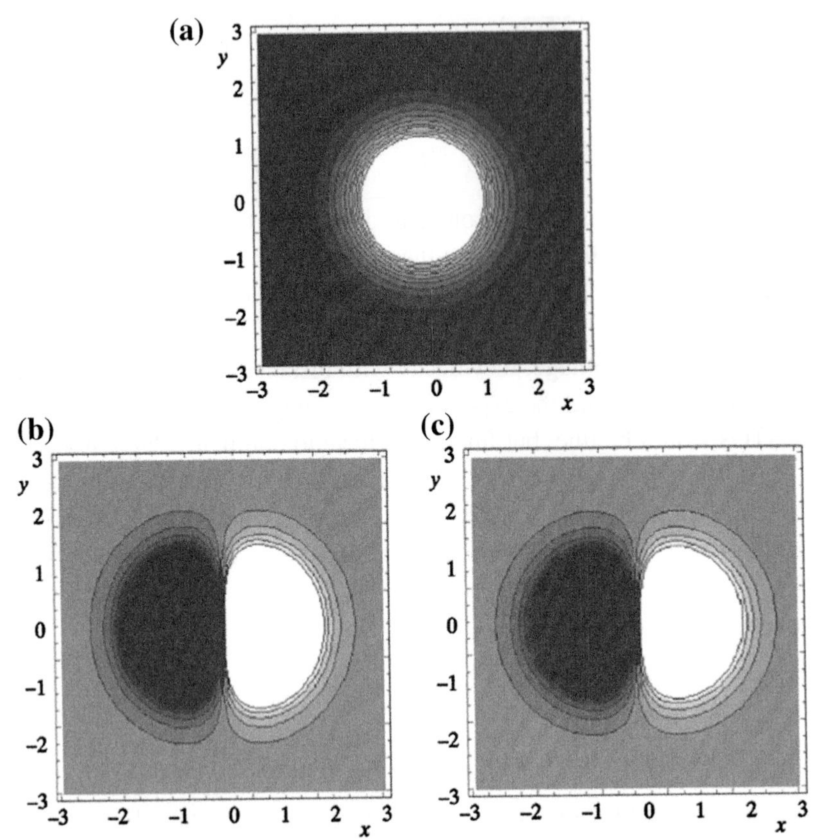

Fig. 8.10. Two spherically symmetric GTOs of the "1*s*-*type*" $G(\mathbf{r}; 1, \mathbf{0})$ (a) These are used to form the difference orbital (b): $G(\mathbf{r}; 1, -0.5\mathbf{i}) - G(\mathbf{r}; 1, +0.5\mathbf{i})$, where $\mathbf{i}$ is the unit vector along the x-axis. For comparison (c), the Gaussian-type p_x orbital is shown: $xG(\mathbf{r}; 1, \mathbf{0})$. It can be seen that the spherical orbitals may indeed simulate the $2p$ ones. Similarly, they can model the spatial functions of arbitrary complexity.

where α_p is the orbital exponent. Why are $1s$-type orbitals so important? Because we may construct "everything" (even s, p, d-like orbitals) out of them using proper linear combinations. For example, the difference of two $1s$ orbitals, centered at $(a, 0, 0)$ and $(-a, 0, 0)$, is similar to the $2p_x$ orbital (Fig. 8.10).

> The most important reason for the great progress of quantum chemistry in recent years is replacing the STOs, formerly used, by GTOs as the expansion functions.

Orbital Size

Each orbital extends to infinity, and it is impossible to measure its extension using a ruler. Still, the α_p coefficient may allow comparison of the sizes of various orbitals. And the quantity

$$\rho_p = (\alpha_p)^{-\frac{1}{2}} \tag{8.50}$$

may be called (which is certainly an exaggeration) the *orbital radius* of the orbital χ_p, because[49]

$$\int_0^{\rho_p}\int_0^{\pi}\int_0^{2\pi} \chi_p^2 dV = 4\pi \int_0^{\rho_p} \chi_p^2 r^2 dr = 0.74, \tag{8.51}$$

where the integration over r goes through the inside of a sphere of radius ρ_p. This gives us an idea about the part of space in which the orbital has an important amplitude. For example, the $1s$ hydrogen atom orbital can be approximated as a linear combination of three $1s$ GTOs (here centered on the origin of the coordinate system; this popular approximation is abbreviated to STO-3G)[50]:

$$1s \approx 0.64767\ G_1(\boldsymbol{r}; 0.151374, \boldsymbol{0}) + 0.40789\ G_2(\boldsymbol{r}; 0.681277, \boldsymbol{0}) + 0.07048\ G_3(\boldsymbol{r}; 4.50038, \boldsymbol{0}), \tag{8.52}$$

which corresponds to the following radii ρ of the three GTOs: 2.57, 1.21, and 0.47 a.u.

Product of GTOs

The GTOs have an outstanding feature (along with the square dependence in the exponent), which decides about their importance in quantum chemistry.

> The product of two Gaussian-type $1s$ orbitals (even if they have different centers) is a single Gaussian-type $1s$ orbital.[51]

The case of GTOs other than $1s$ does not give any trouble, but the result is slightly different. The product of the exponential factors is, of course, the $1s$-type GTO, shown above. The polynomials of x, y, z standing in both GTOs multiplied by each other [recall the dependence of

[49] See, e.g., I.S. Gradshteyn, and I.M. Ryzhik, *Table of Integrals, Series, and Products*, Academic Press, Orlando (1980), formula 3.381.

[50] S. Huzinaga, *J. Chem. Phys., 42*, 1293 (1965).

[51] Let us take two (not normalized) GTOs $1s$: $\exp(-a(\boldsymbol{r}-\boldsymbol{A})^2)$ and $\exp(-b(\boldsymbol{r}-\boldsymbol{B})^2)$, the first centered on the point shown by vector $\boldsymbol{A}$, the second centered on the point shown by vector $\boldsymbol{B}$. It will be shown that their product is the GTO:

$$\begin{aligned} \exp(-a(\boldsymbol{r}-\boldsymbol{A})^2)\exp(-b(\boldsymbol{r}-\boldsymbol{B})^2) &= N\exp(-c(\boldsymbol{r}-\boldsymbol{C})^2), \\ \text{with parameters } c &= a+b, \\ \boldsymbol{C} &= (a\boldsymbol{A}+b\boldsymbol{B})/(a+b), \\ N &= \exp\left[-\frac{ab}{a+b}(\boldsymbol{A}-\boldsymbol{B})^2\right]. \end{aligned}$$

Vector $\boldsymbol{C}$ shows the center of the new GTO. It is identical to the center of mass position, where the role of mass is played by the orbital exponents a and b.
Here is the proof:

the polynomial on the orbital centering; see Eq. (8.46)], can always be presented as a certain polynomial of x', y', z' taken versus the new center $\boldsymbol{C}$. Hence, in the general case,

> the product of any two GTOs is a linear combination of GTOs.

Integrals

If somebody wanted to perform[52] quantum chemical calculations by themselves, they would immediately face integrals to compute, the simplest among them being the $1s$-type. Expressions for these integrals are given in Appendix P available atbooksite.elsevier.com/978-0-444-59436-5 on p. e131.

Left side =

$$= \exp(-ar^2 + 2a\boldsymbol{rA} - aA^2 - br^2 + 2b\boldsymbol{rB} - bB^2)$$
$$= \exp(-(a+b)r^2 + 2\boldsymbol{r}(a\boldsymbol{A} + b\boldsymbol{B}))\exp[-(aA^2 + bB^2)]$$
$$= \exp(-cr^2 + 2c\boldsymbol{C}\boldsymbol{r})\exp[-(aA^2 + bB^2)]$$

Right side =

$$= N\exp(-c(\boldsymbol{r} - \boldsymbol{C})^2) = N\exp[-c(r^2 - 2\boldsymbol{Cr} + C^2)]$$
$$= \text{Left side},$$

if $N = \exp(cC^2 - aA^2 - bB^2)$.

It is instructive to transform the expression for N, which is a kind of amplitude of the GTO originating from the multiplication of two GTOs. So,

$$N = \exp[(a+b)C^2 - aA^2 - bB^2] = \exp\left(\frac{(a^2A^2 + b^2B^2 + 2ab\boldsymbol{AB})}{(a+b)} - aA^2 - bB^2\right)$$
$$= \exp\left(\frac{1}{a+b}(a^2A^2 + b^2B^2 + 2ab\boldsymbol{AB} - a^2A^2 - abA^2 - b^2B^2 - abB^2)\right)$$
$$= \exp\left(\frac{1}{a+b}(2ab\boldsymbol{A}B - abA^2 - abB^2)\right)$$
$$= \exp\left(\frac{ab}{a+b}(2\boldsymbol{AB} - A^2 - B^2)\right) = \exp\left(\frac{-ab}{a+b}(\boldsymbol{A} - \boldsymbol{B})^2\right).$$

This is what we wanted to show.
It is seen that if $\boldsymbol{A} = \boldsymbol{B}$, then amplitude N is equal to 1 and the GTO with the $a+b$ exponent results (as it should). The amplitude N strongly depends on the distance $|\boldsymbol{A} - \boldsymbol{B}|$ between two centers. If the distance is large, then N is very small, which gives the product of two distant GTOs as practically zero (in agreement with common sense). It is also clear that if we multiply two strongly contracted GTOs ($a, b \gg 1$) of different centers, the "*GTO–product*" is again small. Indeed, let us take $a = b$ as an example. We get $N = \exp\{[-a/2][\boldsymbol{A} - \boldsymbol{B}]^2\}$.

[52] That is, independent of existing commercial programs, which only require the knowledge of how to push a few buttons.

8.4.4 Linear Combination of the Atomic Orbitals (LCAO) Method

Algebraic Approximation

Usually we apply the SCF approach with the linear combination of the atomic orbitals (LCAO) method; this is then known as the LCAO MO and SCF LCAO MO.[53] In the SCF LCAO MO method, each molecular orbital is presented as a linear combination of atomic orbitals χ_s

$$\varphi_i(1) = \sum_s^M c_{si}\chi_s(1), \tag{8.53}$$

where the symbol (1) emphasizes that each of the atomic orbitals, and the resulting molecular orbital, depend on the spatial coordinates of *one* electron only (say, electron 1). The coefficients c_{si} are called the *LCAO coefficients*. The STOs and GTOs are important only in the context of the LCAO.

The approximation, in which the molecular orbitals are expressed as linear combinations of the atomic orbitals is also called the *algebraic approximation*.[54]

Erich Hückel (1896–1980), German physicist and professor at the universities in Stuttgart and Marburg, student of Bohr and Debye. Erich Hückel, presumably inspired by his brother Walter, an eminent organic chemist, created a simplified version of the Hartree-Fock method, which played a major role in linking the quantum theory with chemistry. Even today, although this tool is extremely simplistic and has been superceded by numerous and much better computational schemes, Hückel theory is valued as an initial insight into the electronic structure of some categories of molecules and solids.

A curious piece of trivia: these people liked to amuse themselves with little rhymes.

Felix Bloch has translated a poem by Walter Hückel from German to English. It does not look like a great poem, but it deals with the famous Erwin (Schrödinger) and his mysterious function ψ:

Erwin with his ψ can do
Calculations quite a few.
But one thing has not been seen,
Just what does ψ really mean

[53] This English abbreviation turned out to be helpful for Polish quantum chemists in communist times (as specialists in "MO methods", MO standing for the mighty "citizen police" which included the secret police). It was independently used by Professors Wiktor Kemula (University of Warsaw) and Kazimierz Gumiński (Jagiellonian University). A young coworker of Prof. Gumiński complained that despite much effort, he still could not get the official registered address in Kraków, required for employment at the university. The professor wrote a letter to the officials, and asked his coworker to deliver it in person. The reaction was immediate: *"Why didn't you show this to us earlier?!"*

[54] It was introduced in solid-state theory by Felix Bloch (his biography is on p. 512), and used in chemistry for the first time by Hückel.

Why is it so useful? Imagine we do not have such a tool at our disposal. Then we are confronted with defining a function that depends on the position in 3-D space and has a quite complex shape. If we want to do this accurately, we should provide the function values at many points in space, say for a large lattice with a huge number of nodes, and the memory of our PC will not be able to handle it. Also, in such an approach, one would not make use of the fact that the function is smooth. We find our way through by using atomic orbitals. For example, even if we wrote that a molecular orbital is in fact *a single* atomic orbital (we can determine the latter by giving *only four* numbers: three coordinates of the orbital center and the orbital exponent), although very primitive, this would carry a lot of physical intuition (truth...): (1) the spatial distribution of the probability of finding the electron is concentrated in some small region of space, (2) the function decays exponentially when we go away from this region, etc.

"Blocks" of molecular orbitals φ_i are constructed out of "primary building blocks" – the one-electron functions χ_r (in the jargon called *atomic orbitals*), which are required to fill two basic conditions:

- They need to be square-integrable.
- They need to form the complete set, i.e., "everything" can be constructed from this set (any smooth square-integrable function of x, y, z).

In addition, there are several *practical* conditions:

- They should be *effective*; i.e., each single function should include a part of the physics of the problem (position in space, decay rate while going to ∞, etc.).
- They should be "*flexible*"; i.e., their parameters should influence their shape to a large extent.
- The resulting *integrals should be easily computable* (numerically and/or analytically); see p. 426.

In *computational practice*, unfortunately, we fulfill the second set of conditions only to some extent: the set of orbitals taken into calculations (i.e., *the basis set*) is always limited because computing time requires money, etc. In some calculations for crystals, we also remove the first set of conditions (we often use *plane waves*: $\exp(i\boldsymbol{k}\cdot\boldsymbol{r})$, and these are not square-integrable).

We could construct a molecular orbital of any complexity *exclusively* using a linear combination of the orbitals $g(\boldsymbol{r}) = \exp(-\zeta\,|\boldsymbol{r}-\boldsymbol{A}|^n)$ with different ζ and $\boldsymbol{A}$; i.e., the $f(x, y, z) = const$, known as the $1s$ orbitals. We could do it also even for a Beethoven bust in a kind of "hole-repairing" (plastering-like) procedure by building Beethoven from a mist of very many $1s$ orbitals similarly as a sculptor would make it from clay.[55] But why we do not do it like this in practice? The reason is simple: the number of atomic orbitals that we would have to include in the calculations would be too large. Instead, chemists allow for higher-order polynomials $f(x, y, z)$. This makes for more efficient "plastering," because instead of spherically sym-

[55] Frost derived the method of FSGO in a paper "*Floating spherical Gaussian orbitals,*" A.A. Frost, *J. Chem. Phys.*, *47*, 3707 (1967); i.e., GTOs of variably chosen positions. Their number is truly minimal – equal to the number of occupied MOs.

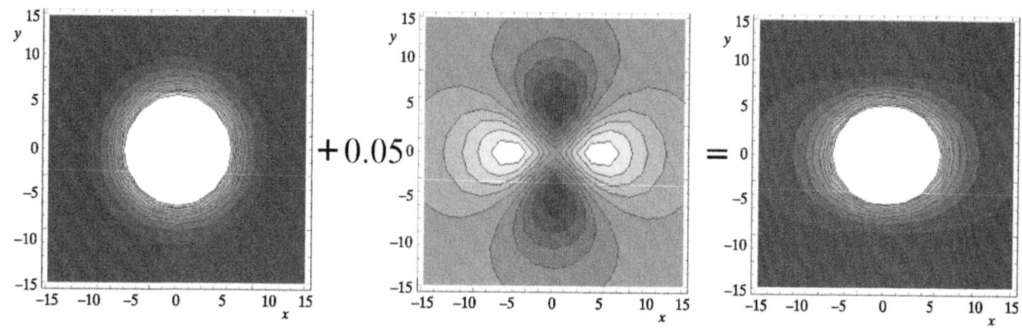

Fig. 8.11. An example of function modeling by a linear combination of AOs. If a tiny admixture of the $3d_{x^2-y^2}$ function is added to the spherically symmetric $1s$ orbital (a football ball, both with 0.5 orbital exponent). We will get shrinking in one direction, and elongation in the other (the dimension in the third direction is unchanged); i.e., a flattened rugby ball. In this case, the tiny admixture means 0.05. If the admixture were of the $2p$ type, the ball would look more like an egg. As we see, nearly everything can be simulated like this. This is the essence of the LCAO method.

metric objects ($1s$), we can use orbitals $g(\mathbf{r})$ of virtually any shape (*via* an admixture of the $p, d, f, \ldots$ functions that introduce more and more complicated patterns of wave function's increasing or decreasing). For example, the way that a rugby-ball shaped orbital can be achieved is shown in Fig. 8.11.

If in Fig. 8.12, we take five LCAOs and provide a reasonable choice of their centers, the exponents, and the weights of the functions, we will get quite a good approximation of the ideal orbital. We account for the advantages as follows: instead of providing a huge number of function values at the grid nodes, we master the function using only $5 \times 5 = 25$ numbers.[56]

Interpretation of LCAO

The idea of LCAO MO is motivated by the fact that the molecular orbital should consist of spatial sections (atomic orbitals), because in a molecule in the vicinity of a given atom, an electron should be described by an atomic orbital of *this* atom. The essence of the LCAO approach is just the connection (unification) of such sections. But only some AOs are important in practice. This means that the main effort of constructing MOs is connected to precise shaping and polishing, by inclusion of more and more of the necessary AOs.[57]

Effectiveness of AOs Mixing

When could we expect that two normalized AOs will have comparable LCAO coefficients in a low-energy MO? Two rules hold [both can be deduced immediately from Eq. (D.1)] for the *mixing effectiveness* of the AOs, obtained from numerical experience:

[56] Three coordinates of the center, the exponent, and the coefficient standing at AO altogether give five parameters per one AO.

[57] This plays the role of the filling mass because we aim for a beautiful (i.e., ideal from the point of view of the variational method) shape for the MOs.

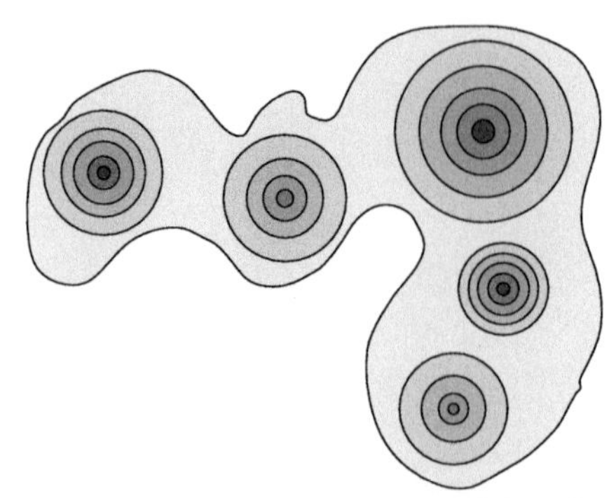

Fig. 8.12. The concept of a MO as a LCAO, a section view. From the point of view of mathematics, it is an expansion in a series of a complete set of functions. From the viewpoint of physics, it is just recognizing that when an electron is close to nucleus a, it should behave in a similar way as that required by the atomic orbital of atom a. From the point of view of a bricklayer, it represents the construction of a large building from soft and mutually interpenetrating bricks.

Effectiveness of AO Mixing

- AOs must correspond to comparable energies (in the meaning of the mean value of the Fock operator).
- AOs must have a large overlap integral.

Let us see what we obtain as the orbital energies[58] (in a.u.) for several important atoms:

	$1s$	$2s$	$2p$	$3s$	$3p$
H	**−0.5**	–	–	–	–
C	−11.34	**−0.71**	**−0.41**	–	–
N	−15.67	−0.96	**−0.51**	–	–
O	−20.68	−1.25	**−0.62**	–	–
F	−26.38	−1.57	**−0.73**	–	–
Cl	−104.88	−10.61	−8.07	−1.07	**−0.51**

Now, which orbitals will mix effectively when forming methane? The hydrogen atom offers the $1s$ orbital with energy −0.5. As we can see from the table, there is no possibility of effectively mixing the carbon $1s$ orbital, while $2s$ and $2p$ are very good candidates. Note that

[58] J.B. Mann, "*Atomic structure calculations. I. Hartree-Fock energy results for the elements H through Lr,*" Report LA-3690 (Los Alamos National Laboratory, 1967).

the orbital energies of all the outermost (the so-called valence) orbitals are similar for all the elements (highlighted in bold in the table), and therefore they are able to mix effectively (i.e., to lower energy by forming chemical bonds). This is why chemistry is mainly the science of outer shell orbitals.

The Mathematical Meaning of LCAO

From a mathematical point of view, Eq. (8.53) represents an expansion of an unknown function φ_i in a series of the known functions χ_r, which belong to a certain complete set, thus M should equal ∞. In real life (cf. Fig. 8.8), we need to truncate this series; i.e., use some limited M.

8.4.5 Basis Sets of Atomic Orbitals

Basis Set

The set of the AOs $\{\chi_r\}$ used in the LCAO expansion is called a *basis set.*

The choice of the basis set functions χ (the incomplete set) is one of the most important *practical* (numerical) problems of quantum chemistry. Yet, because it is of a technical character, we will just limit ourselves to a few remarks.

Although atomic functions do not need to be *atomic* orbitals (e.g., they may be placed between nuclei), in most cases, they are centered directly on the nuclei of the atoms belonging to the molecule under consideration. If M is small (in the less precise calculations), the Slater atomic orbitals discussed above are often used as the expansion functions χ_r; for larger M (in more accurate calculations), the relation between χ_r and the orbitals of the isolated atoms is lost, and χ_r are chosen based on the numerical experience gathered from the literature.[59]

8.4.6 The Hartree-Fock-Roothaan Method (SCF LCAO MO)

The Hartree-Fock equations are nonlinear differential-integral equations that can be solved by appropriate numerical methods. For example, in the case of atoms and diatomics,

[59] For those who are interested in such problems, we recommend the chapter by S. Wilson, "*Basis sets,*" in the book "*Ab initio Methods in Quantum Chemistry,*" ed. by K.P. Lawley, 1987, p. 439. In fact, this knowledge is a little magical. Certain notations describing the quality of basis sets are in common use. For example, the symbol 6-31G* means that the basis set uses GTOs (G), the hyphen divides two electronic shells (here K and L, see p. 448). The K shell is described by a single atomic orbital, which is a certain linear combination (a "*contracted orbital*") of six GTOs of the $1s$ type, and the two digits, 31, pertain to the L shell and denote two contracted orbitals for each valence orbital ($2s, 2p_x, 2p_y, 2p_z$), one of these contains three GTOs, the other one GTO (the latter is called "*contracted,*" with a bit of exaggeration). The starlet corresponds to d functions used additionally in the description of the L shell (called polarization functions).

Clemens C.J. Roothaan (b. 1918), American physicist and professor at the University of Chicago. He became interested in this topic after recognizing that in the literature, people write about the effective one-electron operator, but he could not find its mathematical expression. He derived the expression within the SCF LCAO MO procedure.

George G. Hall (b. 1925), Irish physicist and professor of mathematics at the University of Nottingham. His scientific achievements are connected to localized orbitals, ionization potentials, perturbation theory, solvation, and chemical reactions.

the orbitals may be obtained in a numerical form.[60] High accuracy at long distances from the nuclei is their great advantage. However, the method is prohibitively difficult to apply to larger systems.

A solution is the use of the LCAO MO method (algebraization of the Fock equations). It leads to simplification of the computational scheme of the Hartree-Fock method.[61] In the SCF LCAO MO method, the Fock equations (complicated differential-integral equations) are solved in a very simple way. From Eqs. (8.53) and (8.32), we have

$$\hat{\mathcal{F}} \sum_s c_{si} \chi_s = \varepsilon_i \sum_s c_{si} \chi_s. \tag{8.54}$$

Making the scalar product with χ_r for $r = 1, 2, \ldots, M$, we obtain (the symbols h_{rs} and $\mathcal{F}_{rs}$ are the matrix elements of the corresponding operators in the AO basis set)

$$\sum_s (\mathcal{F}_{rs} - \varepsilon_i S_{rs}) c_{si} = 0. \tag{8.55}$$

This is equivalent to the *Roothaan matrix equation*[62]:

$$\boldsymbol{\mathcal{F}} \boldsymbol{c} = \boldsymbol{S} \boldsymbol{c} \boldsymbol{\varepsilon}, \tag{8.56}$$

60 J. Kobus, *Adv. Quantum Chem, 28,* 1 (1997).

61 The LCAO approximation was introduced to the Hartree-Fock method, independently, by C.C.J. Roothaan and G.G. Hall.

62 On the left side: $L = \sum_s \mathcal{F}_{rs} c_{si}$; on the right side: $P = \sum_{s,l} S_{rs} c_{sl} \varepsilon_{li} = \sum_{s,l} S_{rs} c_{sl} \delta_{li} \varepsilon_i = \sum_s S_{rs} c_{si} \varepsilon_i$. Comparison of both sides of the equation gives the desired result.

where $\boldsymbol{S}$ is the matrix of the overlap integrals $\langle \chi_r | \chi_s \rangle$ involving the AOs, $\boldsymbol{\varepsilon}$ is the diagonal matrix of the orbital energies[63] ε_i, and $\boldsymbol{\mathcal{F}}$ is the Fock operator matrix. Each of these matrices is square (of the order M). $\boldsymbol{\mathcal{F}}$ depends on $\boldsymbol{c}$ (which is why it is a *pseudo*-eigenvalue equation).

If the LCAO expansion is introduced to the expression for the total energy, Eq. (8.43) gives ($\varepsilon_i = (i|\hat{\mathcal{F}}|i)$):

$$E'_{HF} = \sum_i [h_{ii} + (i|\hat{\mathcal{F}}|i)] = \sum_{i=1}^{MO} \sum_{rs} c^*_{ri} c_{si} [(r|\hat{h}|s) + (r|\hat{\mathcal{F}}|s)] \equiv \frac{1}{2} \sum_{rs} P_{sr} [h_{rs} + \mathcal{F}_{rs}], \tag{8.57}$$

where $\boldsymbol{P}$ in the RHF method is called the *bond-order matrix*,

$$P_{sr} = 2 \sum_j^{MO} c^*_{rj} c_{sj},$$

and the summation goes over all the occupied MOs. In consequence, a useful expression for the total energy in the HF method may be written as

$$E_{RHF} = \frac{1}{2} \sum_{rs}^{AO} P_{sr} (h_{rs} + \mathcal{F}_{rs}) + \sum_{a<b} \frac{Z_a Z_b}{R_{ab}}, \tag{8.58}$$

where the first summation goes over the atomic orbitals (AO). For completeness, we also give the expression for $\mathcal{F}_{rs}$:

$$\mathcal{F}_{rs} = (r|\hat{h} + 2\hat{\mathcal{J}} - \hat{\mathcal{K}}|s) \tag{8.59}$$

$$= h_{rs} + \sum_i^{MO} [2(ri|si) - (ri|is)] = h_{rs} + \sum_i^{MO} \sum_{pq}^{AO} c^*_{pi} c_{qi} [2(rp|sq) - (rp|qs)] \tag{8.60}$$

$$= h_{rs} + \sum_{pq}^{AO} P_{qp} \left[(rp|sq) - \frac{1}{2}(rp|qs) \right], \tag{8.61}$$

where i is the index of a MO, and r and s denote the AOs.

The SCF Solution

The Hartree-Fock-Roothaan matrix equation is solved iteratively as follows:

1. We assume an initial $\boldsymbol{c}$ matrix (i.e., also an initial $\boldsymbol{P}$ matrix; often in the zeroth iteration, we put $\boldsymbol{P} = \boldsymbol{0}$, as if there were no electron repulsion).
2. We find the $\boldsymbol{\mathcal{F}}$ matrix using matrix $\boldsymbol{P}$.

[63] In fact, they are some approximations to them. Their values approach the orbital energies, when the basis set of AOs gets closer to the complete basis set.

3. We solve the Hartree-Fock-Roothaan equation (see Appendix L available at booksite.elsevier.com/978-0-444-59436-5, p. e107) and obtain the M MOs, we choose the $N/2$ occupied orbitals (those of the lowest energy).
4. We obtain a new $\boldsymbol{c}$ matrix, and then a new $\boldsymbol{P}$, etc.
5. We go back to step 1.

The iterations are terminated when the total RHF energy (the more liberal approach) or the coefficients $\boldsymbol{c}$ (the less liberal one) change less than the assumed threshold values. Both these criteria (ideally fulfilled) may be considered as a sign that the output orbitals are already *self-consistent*. Practically, these are never the exact solutions of the Fock equations, because a *limited* number of AOs was used, while expansion to the complete set requires the use of an infinite number of AOs (the total energy in such a case would be called the *Hartree-Fock limit energy*).

After finding the MOs (hence, also the HF function) in the SCF LCAO MO approximation, we may calculate the total energy of the molecule as the mean value of its Hamiltonian. We need only the *occupied orbitals,* and not the *virtual* ones for this calculation.

> The Hartree-Fock method only takes care of the total energy *and completely ignores the virtual orbitals*, which may be considered as a kind of by-product.

8.4.7 Some Practical Problems

Size of the AO Basis Set

> **Number of MOs**
> The number of MOs obtained from the SCF procedure is always equal to the number of the AOs used. Each MO consists of various contributions of the same basis set of AOs. The apparent exception is when *due to symmetry*, the coefficients at some AOs are equal to zero.

For double occupancy, M needs to be larger or equal to $N/2$. Typically, we are forced to use large basis sets ($M \gg N/2$), and then along the occupied orbitals, we get $M - N/2$ unoccupied orbitals (*virtual orbitals*). Of course, we should aim at the highest-quality MOs (i.e., how close they are to the solutions of the Fock equations), and avoiding large M (computational effort is proportional to M^4), but in practice, a better basis set often means a larger M. The variational principle implies the ordering of the total energy values obtained in different approximations (Fig. 8.13).

It is required that the largest possible basis set be used (mathematics: we approach the complete set), but we may also ask if a basis set dimension may be *decreased* freely (economy!).

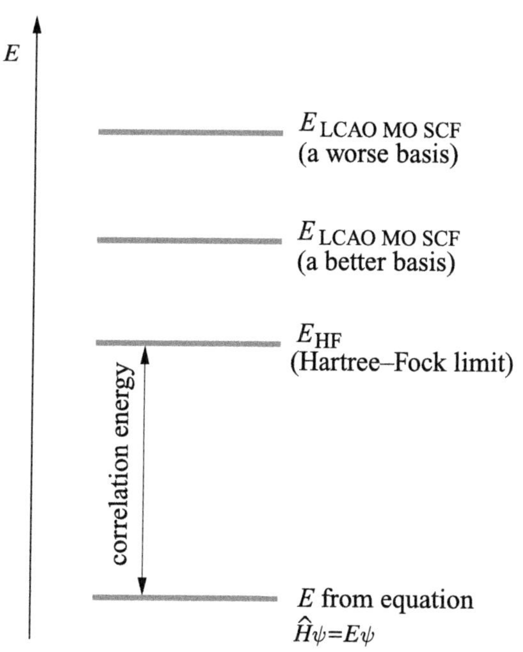

Fig. 8.13. The Hartree-Fock method is variational. The better the wave function, the lower the mean value of the Hamiltonian. An extension of the AO basis set (i.e., adding new AOs) *has to* lower the energy, and the ideal solution of the Fock equations gives the "*Hartree-Fock limit.*" The ground-state eigenvalue of the Hamiltonian is *always* lower than the HF limit because the Hartree-Fock method is able to produce only an approximation to the solution of the Schrödinger equation.

Of course, the answer is no. The absolute limit M is equal to half the number of the electrons, because only then can we create N spinorbitals and write the Slater determinant. However, in quantum chemistry (rather misleadingly), we call the minimum basis set the basis set resulting from an inner shell and valence orbitals in the corresponding atoms. For example, the minimum basis set for a water molecule is $1s$, $2s$ and three $2p$ orbitals of oxygen and two $1s$ orbitals of hydrogen atoms, which is seven AOs in total (while the truly minimal basis would contain only $10/2 = 5$ AOs).

Flip-Flop

The M MOs result from each iteration. We order them using the increasing orbital energy ε criterion, and then we use the $N/2$ orbitals of the lowest orbital energy in the Slater determinant; we call this the occupation of MOs by electrons. We might ask why we make the lowest-lying MOs occupied? *The variational principle does not hold for orbital energies.* And yet we do this (not trying all possible occupations), and only very rarely we get into trouble. The most frequent problem is that the criterion of orbital energy leads to the occupation of one set of MOs in odd iterations, and another set of MOs in even ones (typically both sets differ by including/excluding one of the two MOs that are neighbors on the energy scale) and the energy resulting from the

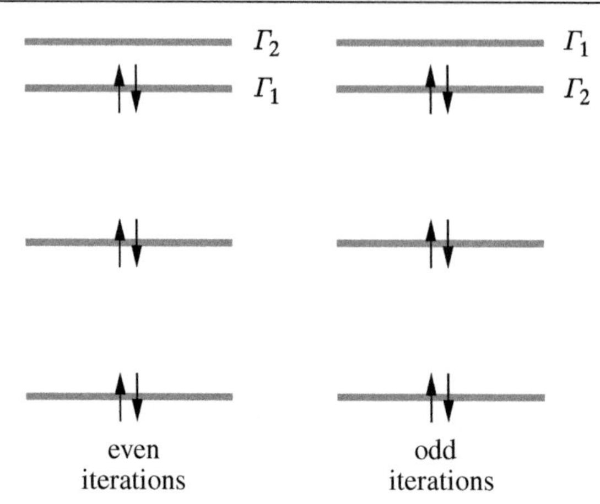

Fig. 8.14. A difficult case for the SCF method ("*flip-flop*"). We are sure that the orbitals exchange in subsequent iterations, because they differ in symmetry (Γ_1, Γ_2).

odd iterations is different from that of the even ones (a "*flip-flop*" behaviour). Such behavior of the Hartree–Fock method is indeed annoying[64] (Fig. 8.14).

Dilemmas of the AOs Centering

Returning to the total energy issue, we should recall that in order to decrease the total energy, we may move the nuclei (so far frozen during the HF procedure). This is called the *geometry optimization.* Practically all calculations should be repeated for each nuclear geometry during such optimization.[65] And there is one more subtlety. As was said before, the AOs are most often centered on the nuclei. When the nuclei are moved, the question arises whether a nucleus should pull its AOs to a new place, or not.[66] If *not*, then this "*slipping off*" the nuclei will significantly increase the energy (independent of, whether the geometry is improved or not). If *yes*, then in fact we use different basis sets for each geometry. Hence, in each case, we search for the solution in a slightly different space (because it is spanned by another basis set). People use the second approach. It is worth notifying that the problem would disappear if the basis set of AOs were complete.

The problem of AO centering is a bit shameful in quantum chemistry. Let us consider the LCAO approximation and a real molecule such as Na_2CO_3. As mentioned above, the LCAO

[64] There are methods for mastering this circus by using the matrix $\boldsymbol{P}$ in the kth iteration, not taken from the previous iteration (as usual), but as a certain linear combination of $\boldsymbol{P}$ from the $k-1$ and $k-2$ iterations. When the contribution of $\boldsymbol{P}$ from the $k-2$ iteration is large, in comparison with that from the $k-1$ iteration, it corresponds to a gentle attempt at quietening a nervous stallion.

[65] Let us take an example of CH_4. First, we set any starting geometry, say, a square-like planar. Now, we try to change the configuration to make it out-of-plane (the energy goes down). Taking the HCH angles as all equal (tetrahedral configuration) once more lowers the total energy computed. Putting all the CH bonds of equal length gives even lower energy. Finally, by trying different CH bond lengths, we arrive at the optimum geometry (for a given AO basis set). In practice, such geometry changes are made automatically by computing the gradient of total energy.

[66] Even if the AOs were off the nuclei, we would have the same dilemma.

functions have to form a complete set. But which functions are we talking about? Since they have to form a complete set, they may be chosen as the eigenfunctions of a certain Hermitian operator (e.g., the energy operator for the 3-D harmonic oscillator or the energy operator for the hydrogen atom or the Fock operator for the uranium atom). We decide, and we are free to choose. In addition to this freedom, we add another freedom, that of centering. Where should the eigenfunctions (of the oscillator or the hydrogen or uranium atom) of the complete set be centered (i.e., positioned in space)? Since it is the complete set, *each way of centering is good by definition*. It really looks like this if we hold to the principles.

But in practical calculations, we never have the complete set at our disposal. We always need to limit it to a certain finite number of functions, and it does not represent any complete set. We usually try to squeeze the best results from the available time and money. How do we do that? We apply our physical intuition to the problem, believing that it will pay off. First of all, intuition suggests the use of functions for some atom that is present in the molecule, and not those of the harmonic oscillator, or the hydrogen or uranium atom, which are absent from our molecule. And here we encounter another problem: *Which* atom is meant, because we have Na, C, and O in Na_2CO_3? It appears that

> the solution close to optimum is to take as a basis set *the beginnings* of several complete sets, each of them centered on one of the atoms.

So we could center the $1s$, $2s$, $2p$, and $3s$ orbitals on both Na atoms, and the $1s$, $2s$, and $2p$ set on the C and O atoms.[67]

8.5 Back to the Basics

8.5.1 When Does the RHF Method Fail?

The reason for any Hartree-Fock method failure can be only one thing: the wave function is approximated as a *single* determinant. All possible catastrophes come from this, and we might even deduce when the Hartree–Fock method is not appropriate for description of a particular real system. First, let us ask when a single determinant would be OK? Well, out of all Slater determinants that could be constructed from a certain spinorbital basis set, only its energy (i.e., the mean value of the Hamiltonian for this determinant) would be close to the true energy of

[67] This is nearly everything, except for a small paradox: if we are moderately poor (reasonable but not extensive basis sets), then our results will be good, but if we became rich (and performed high-quality computations using very large basis sets for each atom), then we would get into trouble. This would come from the fact that our basis set starts to look like six distinct, complete sets. Well, that looks too good, doesn't it? We have an *overcomplete* set, and trouble must result from this. The overcompleteness means that any orbital from one set is already representable as a linear combination of another complete set. You would see strange things when trying to diagonalize the Fock matrix, and that won't do. You can be sure that you would be begging to be less rich.

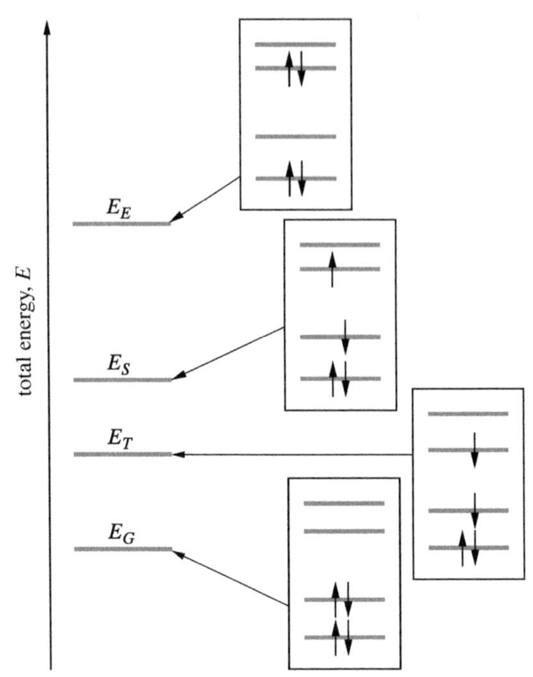

Fig. 8.15. In exact theory, there is no such a thing as molecular orbitals. In such a theory, we would only deal with the many-electron states and the corresponding energies of the molecule (left). If, nevertheless, we decided to *stick to the one-electron approximation*, we would have the MOs and the corresponding orbital energies (right). These one-electron energy levels can be occupied by electrons (0, 1, or 2) in various ways (the meaning of the occupation is given on p. 409), and a many-electron wave function (a Slater determinant) corresponds to each occupation. This function gives a certain mean value of the Hamiltonian (i.e., the total energy of the molecule.) *In this way (in many cases), one value of the total energy of the molecule corresponds to a diagram of orbital occupation*. The case of the S and T states is somewhat more complex than the one shown here, and we will come back to it on p. 460.

the molecule. In such a case, only this determinant would matter in the linear combination of determinants,[68] and the others would have negligible coefficients. This could be so[69] if the energies of the occupied orbitals were much lower than those of the virtual ones. Indeed, various electronic states of different total energies may be approximately formed while the orbitals scheme is occupied by electrons, and if the virtual levels are at high energies, the total energy calculated from the "*excited determinant*" (replacement: occupied spinorbital → virtual spinorbital) would also be high (Fig. 8.15).

In other words, the danger for the RHF method is when the energy difference between HOMO and LUMO is small. For example, RHF completely fails to describe metals properly, because $\Delta = 0$ there.[70] When the HOMO–LUMO gap is small, always expect bad results.

[68] The Slater determinants form the complete set, as discussed on p. 398. In the configuration interaction method (which will be described in Chapter 10), the electronic wave function is expanded using Slater determinants.

[69] We shift here from the total energy to the one-electron energy (i.e., to the orbital picture).

[70] It shows up as strange behavior of the total energy per metal atom, which exhibits poorly decaying oscillations with an increasing number of atoms. In addition, the exchange interactions notorious for fast (exponential) decay as calculated by the Hartree-Fock method are of a long-range character (see Chapter 9).

Incorrect Description of Dissociation by the RHF Method

Another example is provided by the H_2 molecule at long internuclear distances. In the simplest LCAO MO approach, two electrons are described by the *bonding orbital* (χ_a and χ_b are $1s$ orbitals centered on the H nuclei, a and b, respectively; both obtained by using the symmetry requirement):

$$\varphi_{bond} = \frac{1}{\sqrt{2(1+S)}}(\chi_a + \chi_b), \tag{8.62}$$

but there is another orbital, an *antibonding* one

$$\varphi_{antibond} = \frac{1}{\sqrt{2(1-S)}}(\chi_a - \chi_b). \tag{8.63}$$

These names stem from the respective energies, which are obtained if we accept that the molecular orbitals satisfy a sort of "Schrödinger equation" using an effective Hamiltonian (say, an analog of the Fock operator): $\hat{H}_{ef}\varphi = E\varphi$ and, after introducing notation, the overlap integral $S = (\chi_a|\chi_b)$, $H_{aa} = (\chi_a|\hat{H}_{ef}\chi_a)$, *the resonance integral*[71] $H_{ab} = H_{ba} = (\chi_a|\hat{H}_{ef}\chi_b) < 0$. For the bonding orbital, we have $E_{bond} = \left(\varphi_{bond}|\hat{H}_{ef}\varphi_{bond}\right) = \frac{1}{2(1+S)}\left[H_{aa} + H_{bb} + 2H_{ab}\right]$ which gives

$$E_{bond} = \frac{H_{aa} + H_{ab}}{1+S} < H_{aa},$$

and for the antibonding orbital (similar derivation),

$$E_{antibond} = \frac{H_{aa} - H_{ab}}{1-S} > H_{aa}.$$

The resonance integral H_{ab}, and the overlap integral S, decay exponentially when the internuclear distance R increases.

Incorrect Dissociation Limit of the Hydrogen Molecule

Thus, we have obtained the *quasi-degeneracy* (a near degeneracy of two orbitals) for long distances, while we need to occupy *only one* of these orbitals (bonding one) in the RHF method. The antibonding orbital is treated as virtual, and as such, is *completely ignored.* However, as a matter of fact, for long distances R, it corresponds to the same energy as the bonding energy.

We have to pay for such a huge drawback. And the RHF method pays, for its result significantly deviates (Fig. 8.16) from the exact energy for large R values (tending to the energy of the two isolated hydrogen atoms). This effect is known as an *incorrect dissociation* of a molecule in the RHF method (here exemplified by the hydrogen molecule). The failure may be explained in several ways, and we have presented one point of view here.

[71] This integral is negative. Its sign is what decides the energy effect of the chemical bond formation (because H_{aa} is nearly equal to the energy of an electron in the H atom; i.e., $-\frac{1}{2}$ a.u.).

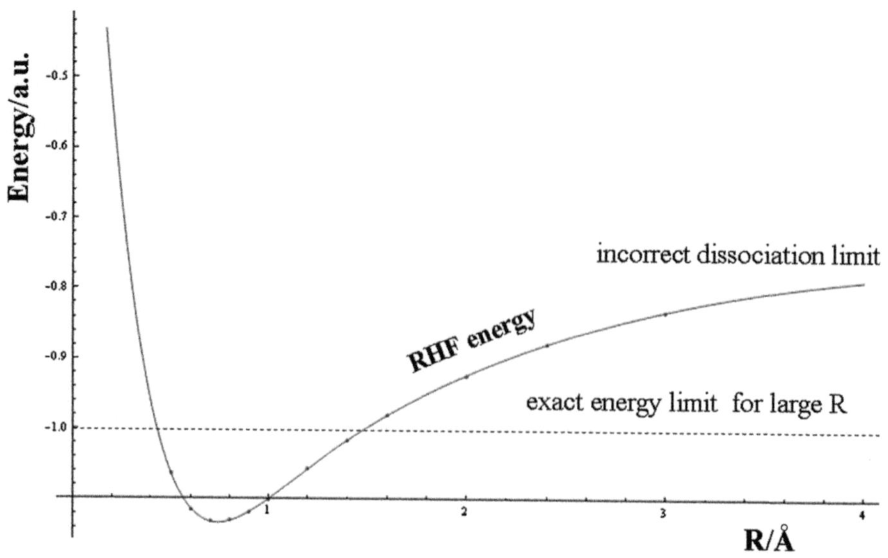

Fig. 8.16. Incorrect dissociation of H_2 in the MO (RHF) method. The wave function, in the form of one Slater determinant, leads to dissociation products, which are neither atoms nor ions (while they should be two ground-state hydrogen atoms with energy $2E_H = -1$ a.u.). The Hartree-Fock computations have been carried out (program GAUSSIAN) for the internuclear distances shown as points by using the atomic basis set known as $6-31G(d,p)$. The energy of an isolated hydrogen atom (E_H) has been calculated in the corresponding single-atom basis set. The curve displayed has been created as a multi-exponential approximation by using the least-mean-square method. Computed in such a way, the minimum of the curve corresponds to the internuclear distance $R = 1.385$ a.u. (the Hartree–Fock limit is 1.370 a.u., while the solution of the Schrödinger equation in the Born-Oppenheimer approximation gives 1.401 a.u.). The corresponding energy equals -1.131 a.u., while the Hartree–Fock limit is -1.134 a.u., and the solution of the Schrödinger equation within the Born-Oppenheimer approximation gives -1.174 a.u.

8.5.2 Fukutome Classes

Symmetry Dilemmas and the Fock Operator

We have derived the general Hartree-Fock method (GHF, p. 407) providing completely free variations for the spinorbitals taken from Eq. (8.1). As a result, the Fock equation [Eq. (8.27)] was derived.

We then decided to limit the spinorbital variations *via our own* condition of the double occupancy of the molecular orbitals as the real functions. This has led to the RHF method and to the Fock equation in the form of Eq. (8.32).

The Hartree-Fock method is a complex (nonlinear) machinery. Do the HF solutions have any symmetry features related to those of the Hamiltonian? This question may be asked both for the GHF method and for any *spinorbital constraints* (e.g., the RHF constraints). The following problems may be addressed:

- Do the output orbitals belong to the irreducible representations of the symmetry group (see Appendix C available at booksite.elsevier.com/978-0-444-59436-5 on p. e17) of the Hamiltonian? Or, if we set the nuclei in the configuration corresponding to symmetry group G, will the canonical orbitals transform according to some irreducible representations of the G group? Or, stated in yet another way, does the Fock operator exhibits the symmetry typical of the G group?[72]

- Does the same apply to electron density?
- Is the Hartree-Fock determinant an eigenfunction of the $\hat{S}^2$ and $\hat{S}_z$ operators?
- Is the probability density of finding a $\sigma = \frac{1}{2}$ electron equal to the probability density of finding a $\sigma = -\frac{1}{2}$ electron at any point of space?

Instabilities in the Hartree-Fock Method

The above mentioned questions are connected to the *stability of the solutions.* The HF solution is stable if any change of the spinorbitals leads to a *higher* energy. We may put certain conditions for spinorbital changes. *Relaxing the condition of double occupancy may take various forms*; e.g., the paired orbitals may be equal but *complex*, or all orbitals may be *different* real functions, or we may admit them as different complex functions, among other options. Could the energy increase along with this gradual orbital constraints removal? No, an energy increase is impossible, of course, because of the variational principle, the energy might, however, remain constant or decrease.

The general answer to this question (the character of the energy change) cannot be given since it depends on the molecule under study, interatomic distances, the AO basis set, etc. However, as shown by Fukutome[73] by a group theory analysis, there are exactly eight situations that *may* occur. Each of these leads to a characteristic shape of the set of occupied orbitals, and they are all given in Table 8.1. We may pass the borders between these eight classes of GHF method solutions while changing various parameters.

Example. RHF/UHF Triplet Instability

May the UHF method give lower energy for the hydrogen molecule than the RHF procedure? Let us take the RHF function:

$$\psi_{RHF} = \frac{1}{\sqrt{2}} \begin{vmatrix} \phi_1(1) & \phi_1(2) \\ \phi_2(1) & \phi_2(2) \end{vmatrix},$$

where both spinorbitals have a *common real orbital part* φ : $\phi_1 = \varphi\alpha, \phi_2 = \varphi\beta$.

Now we allow for a *diversification* of the orbital part (keeping the functions *real*: i.e., staying within the Fukutome class 4, usually called UHF in quantum chemistry) for both spinorbitals.

[72] It has been shown that the translational symmetry in the RHF solution for polymers is broken and that the symmetry of the electron density distribution in polymers exhibits a unit cell twice as long as that of the nuclear pattern [Bond-Order Alternating Solution (BOAS); J. Paldus and J. Čižek, *J. Polym. Sci., Part C, 29,* 199 (1970); also J.-M. André, J. Delhalle, J.G. Fripiat, G. Hennico, J.-L. Calais, and L. Piela, *J. Mol. Struct. (Theochem), 179,* 393 (1988)]. The BOAS represents a feature related to the Jahn-Teller effect in molecules and to the Peierls effect in the solid state (see Chapter 9).

[73] A series of papers by H. Fukutome starts with the article in *Prog. Theor. Phys., 40,* 998 (1968), and the review article *Int. J. Quantum Chem., 20,* 955 (1981). I recommend a beautiful paper by J.-L. Calais, *Adv. Quantum Chem., 17,* 225 (1985).

Table 8.1. Fukutome classes [for φ_{i1} and φ_{i2}; see Eq. (8.1)].

Class	Orbital Components $\begin{bmatrix} \varphi_{11} & \varphi_{21} & \cdots & \varphi_{N1} \\ \varphi_{12} & \varphi_{22} & \cdots & \varphi_{N2} \end{bmatrix}$	Remarks	Eigenfunction
1	$\begin{bmatrix} \varphi_1 & 0 & \varphi_2 & 0 & \cdots & \varphi_{N/2} & 0 \\ 0 & \varphi_1 & 0 & \varphi_2 & \cdots & 0 & \varphi_{N/2} \end{bmatrix}$	φ_i; real	RHF, $\hat{S}^2$, $\hat{S}_z$
2	$\begin{bmatrix} \varphi_1 & 0 & \varphi_2 & 0 & \cdots & \varphi_{N/2} & 0 \\ 0 & \varphi_1 & 0 & \varphi_2 & \cdots & 0 & \varphi_{N/2} \end{bmatrix}$	φ_i; complex	$\hat{S}^2$, $\hat{S}_z$
3	$\begin{bmatrix} \varphi_1 & 0 & \varphi_2 & 0 & \cdots & \varphi_{N/2} & 0 \\ 0 & \varphi_1^* & 0 & \varphi_2^* & \cdots & 0 & \varphi_{N/2}^* \end{bmatrix}$	φ_i; complex	$\hat{S}_z$
4	$\begin{bmatrix} \varphi_1 & 0 & \varphi_2 & 0 & \cdots & \varphi_{N/2} & 0 \\ 0 & \chi_1 & 0 & \chi_2 & \cdots & 0 & \chi_{N/2} \end{bmatrix}$	φ, χ; real	UHF, $\hat{S}_z$
5	$\begin{bmatrix} \varphi_1 & 0 & \varphi_2 & 0 & \cdots & \varphi_{N/2} & 0 \\ 0 & \chi_1 & 0 & \chi_2 & \cdots & 0 & \chi_{N/2} \end{bmatrix}$	φ, χ; complex	$\hat{S}_z$
6	$\begin{bmatrix} \varphi_1 & \chi_1 & \varphi_2 & \chi_2 & \cdots & \varphi_{N/2} & \chi_{N/2} \\ -\chi_1^* & \varphi_1^* & -\chi_2^* & \varphi_2^* & \cdots & -\chi_{N/2}^* & \varphi_{N/2}^* \end{bmatrix}$	φ, χ; complex	
7	$\begin{bmatrix} \varphi_1 & \chi_1 & \varphi_2 & \chi_2 & \cdots & \varphi_{N/2} & \chi_{N/2} \\ \tau_1 & \kappa_1 & \tau_2 & \kappa_2 & \cdots & \tau_{N/2} & \kappa_{N/2} \end{bmatrix}$	$\varphi, \chi, \tau, \kappa$; real	
8	$\begin{bmatrix} \varphi_1 & \chi_1 & \varphi_2 & \chi_2 & \cdots & \varphi_{N/2} & \chi_{N/2} \\ \tau_1 & \kappa_1 & \tau_2 & \kappa_2 & \cdots & \tau_{N/2} & \kappa_{N/2} \end{bmatrix}$	$\varphi, \chi, \tau, \kappa$; complex	

We proceed slowly from the closed-shell situation, using as the orthonormal spinorbitals:

$$\phi_1' = N_-(\varphi - \delta)\alpha,$$
$$\phi_2' = N_+(\varphi + \delta)\beta,$$

where δ is a small real correction to the φ function, and N_+ and N_- are the normalization factors.[74] The electrons hate each other (Coulomb law) and may thank us for giving them separate apartments[75]: $\varphi + \delta$ and $\varphi - \delta$. We will worry about the particular mathematical shape of δ in a minute. For the time being, though, let us see what happens to the UHF function:

$$\psi_{UHF} = \frac{1}{\sqrt{2}} \begin{vmatrix} \phi_1'(1) & \phi_1'(2) \\ \phi_2'(1) & \phi_2'(2) \end{vmatrix}$$
$$= \frac{1}{\sqrt{2}} N_- \begin{vmatrix} [\varphi(1) - \delta(1)]\alpha(1) & [\varphi(2) - \delta(2)]\alpha(2) \\ \phi_2'(1) & \phi_2'(2) \end{vmatrix}$$
$$= \frac{1}{\sqrt{2}} N_- \left\{ \begin{vmatrix} \varphi(1)\alpha(1) & \varphi(2)\alpha(2) \\ \phi_2'(1) & \phi_2'(2) \end{vmatrix} - \begin{vmatrix} \delta(1)\alpha(1) & \delta(2)\alpha(2) \\ \phi_2'(1) & \phi_2'(2) \end{vmatrix} \right\}$$

[74] Such a form is not fully equivalent to the UHF method, in which a general form of real orbitals is allowed, not just $\varphi \pm \delta$.

[75] Well, this is not quite the case because the apartments overlap.

$$= \frac{1}{\sqrt{2}} N_+ N_- \left\{ \begin{vmatrix} \varphi(1)\alpha(1) & \varphi(2)\alpha(2) \\ [\varphi(1)+\delta(1)]\beta(1) & [\varphi(2)+\delta(2)]\beta(2) \end{vmatrix} - \begin{vmatrix} \delta(1)\alpha(1) & \delta(2)\alpha(2) \\ [\varphi(1)+\delta(1)]\beta(1) & [\varphi(2)+\delta(2)]\beta(2) \end{vmatrix} \right\}$$

$$= \frac{1}{\sqrt{2}} N_+ N_- \left\{ \begin{vmatrix} \varphi(1)\alpha(1) & \varphi(2)\alpha(2) \\ \varphi(1)\beta(1) & \varphi(2)\beta(2) \end{vmatrix} + \begin{vmatrix} \varphi(1)\alpha(1) & \varphi(2)\alpha(2) \\ \delta(1)\beta(1) & \delta(2)\beta(2) \end{vmatrix} - \begin{vmatrix} \delta(1)\alpha(1) & \delta(2)\alpha(2) \\ \varphi(1)\beta(1) & \varphi(2)\beta(2) \end{vmatrix} - \begin{vmatrix} \delta(1)\alpha(1) & \delta(2)\alpha(2) \\ \delta(1)\beta(1) & \delta(2)\beta(2) \end{vmatrix} \right\}$$

$$= N_+ N_- \psi_{RHF} + \frac{1}{\sqrt{2}} N_+ N_- \left\{ \left[\varphi(1)\delta(2) - \varphi(2)\delta(1) \right] \left[\alpha(1)\beta(2) + \alpha(2)\beta(1) \right] \right\}$$

$$- \frac{1}{\sqrt{2}} N_+ N_- \begin{vmatrix} \delta(1)\alpha(1) & \delta(2)\alpha(2) \\ \delta(1)\beta(1) & \delta(2)\beta(2) \end{vmatrix}.$$

The first and last functions are singlets ($S_z = 0,\ S = 0$), while the second function represents a *triplet state* ($S_z = 0,\ S = 1$); see Appendix Q available at booksite.elsevier.com/978-0-444-59436-5 on p. e133. Thus, a small diversification of the orbital functions leads to some *triplet (second term) and singlet (third term) admixtures* to the original singlet function $N_+ N_- \psi_{RHF}$ (called *triplet contamination,* generally *spin contamination*). The former is proportional to δ and the latter to δ^2. Now the total wave function is no longer an eigenfunction of the $\hat{S}^2$ operator. How is this possible? If one electron has a spin coordinate of $\frac{1}{2}$ and the second one of $-\frac{1}{2}$, aren't they paired? Well, not necessarily, because one of the triplet functions (which describes the *parallel* configuration of both spins[76]) is $\left[\alpha(1)\beta(2) + \alpha(2)\beta(1)\right]$.

Is the resulting UHF energy (calculated for such a function) lower than the corresponding RHF energy (calculated for ψ_{RHF}); i.e., is the RHF solution unstable toward UHF-type spinorbitals changes (no. 4 in the table of Fukutome classes)?

It depends on a particular situation. Earlier, we have promised to consider what the δ function should look like for the hydrogen molecule. In the RHF method, both electrons occupy the same molecular orbital φ. If we ensured within the UHF method that whenever one electron is close to nucleus a, the second one prefers to be closer to b, this would happily be accepted by the electrons, since they repel each other (the mean value of the Hamiltonian would decrease, this is welcome). Taking $\delta = \varepsilon\tilde{\varphi}$ (where $\tilde{\varphi}$ is the antibonding orbital, and $\varepsilon > 0$ is a small coefficient) would have such consequences. Indeed, the sum $\varphi + \delta = \varphi + \varepsilon\tilde{\varphi}$ takes larger absolute value preferentially at one of the nuclei[77] (Fig. 8.17). Since both orbitals correspond to electrons with opposite spins, there will be some net spin on each of the nuclei.

[76] To call them parallel is an exaggeration, since they form an angle of 70.5^0 (see Chapter 1, p. 33), but this is customary in physics and chemistry.

[77] In our example, the approximate bonding orbital is $\varphi = \frac{1}{\sqrt{2}}(1s_a + 1s_b)$, and $\tilde{\varphi} = \frac{1}{\sqrt{2}}(1s_a - 1s_b)$, hence $\varphi + \varepsilon\tilde{\varphi} = \frac{1}{\sqrt{2}}[(1+\varepsilon)\,1s_a + (1-\varepsilon)\,1s_b]$, while $\varphi - \varepsilon\tilde{\varphi} = \frac{1}{\sqrt{2}}[(1-\varepsilon)\,1s_a + (1+\varepsilon)\,1s_b]$. Thus, one of the new orbitals has a larger amplitude at nucleus a, while the other one is at nucleus b (as we had initially planned).

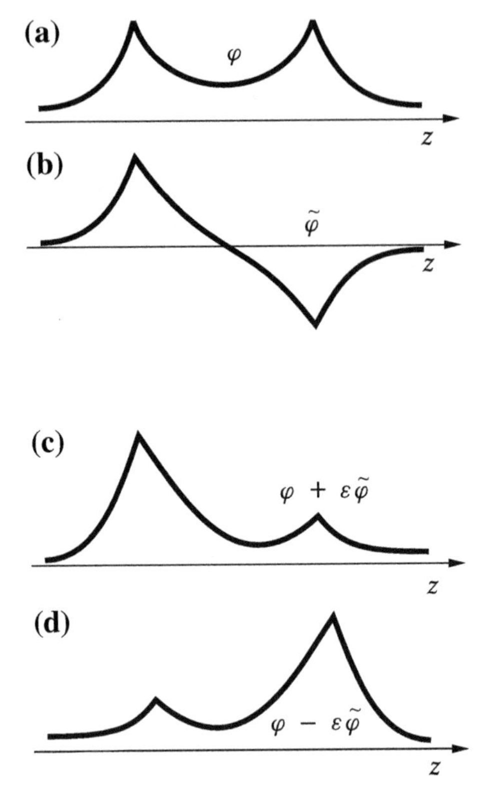

Fig. 8.17. The effect of mixing the bonding orbital φ (shown in panel a) with the antibonding orbital $\tilde{\varphi}$ (shown in panel b). A small admixture (c) of $\tilde{\varphi}$ to the orbital φ leads to an increase of the probability amplitude of the resulting orbital at the left nucleus, while a subtraction of $\tilde{\varphi}$ (d) leads to a larger probability amplitude of the resulting orbital at the right nucleus. Thus, it results in partial separation of the spins $\frac{1}{2}$ and $-\frac{1}{2}$.

A similar reasoning pertaining function $\varphi - \delta = \varphi - \varepsilon\tilde{\varphi}$ results in opposite preferences for the nuclei. Such a particular UHF method, which uses virtual orbitals $\tilde{\varphi}$ to change RHF orbitals, is called the *AMO approach.*[78]

Now,

$$\psi_{UHF} = N_+N_-\psi_{RHF} + \frac{1}{\sqrt{2}}N_+N_-\varepsilon\left\{\left[\varphi(1)\tilde{\varphi}(2) - \varphi(2)\tilde{\varphi}(1)\right]\left[\alpha(1)\beta(2) + \alpha(2)\beta(1)\right]\right\}$$
$$-\frac{1}{\sqrt{2}}N_+N_-\varepsilon^2\begin{vmatrix}\tilde{\varphi}(1)\alpha(1) & \tilde{\varphi}(2)\alpha(2)\\ \tilde{\varphi}(1)\beta(1) & \tilde{\varphi}(2)\beta(2)\end{vmatrix} = N_+N_-\left[\psi_{RHF} + \varepsilon\sqrt{2}\psi_{T'} - \varepsilon^2\psi_E\right],$$

where the following notation is used for normalized functions: ψ_{RHF} for the ground state of the energy E_{RHF}, $\psi_{T''}$ for the triplet state of the energy E_T, and ψ_E for the singlet state with a doubly occupied antibonding orbital that corresponds to the energy E_E ("*doubly excited*").

[78] *Alternant Molecular Orbitals*; P.-O. Löwdin, *Symp. Mol. Phys., Nikko (Tokyo Maruzen)*, (1954), p. 13; also R. Pauncz, *Alternant Molecular Orbitals*, Saunders, Philadelphia (1967).

Let us calculate the mean value of the Hamiltonian using the ψ_{UHF} function. Because of the orthogonality of the spin functions (remember that the Hamiltonian is independent of spin), we have $\left\langle \psi_{RHF} | \hat{H} \psi_{T''} \right\rangle = \langle \psi_{RHF} | \psi_{T''} \rangle = 0$, and we obtain (with accuracy up to ε^2 terms):

Per-Olov Löwdin (1916–2000), Swedish chemist and physicist and a professor at the University of Uppsala (Sweden), founder and professor of the Quantum Theory Project at Gainesville University (Florida). He was very active in organizing the scientific life of the international quantum chemistry community.

$$
\begin{aligned}
\bar{E}_{UHF} &\approx \frac{\left\langle \psi_{RHF} | \hat{H} \psi_{RHF} \right\rangle + 2\varepsilon^2 \left\langle \psi_{T''} | \hat{H} \psi_{T''} \right\rangle - 2\varepsilon^2 \left\langle \psi_{RHF} | \hat{H} \psi_E \right\rangle}{\langle \psi_{RHF} | \psi_{RHF} \rangle + 2\varepsilon^2 \langle \psi_{T''} | \psi_{T''} \rangle} \\
&= \frac{E_{RHF} + 2\varepsilon^2 E_T - 2\varepsilon^2 (\varphi\varphi|\tilde{\varphi}\tilde{\varphi})}{1 + 2\varepsilon^2} \approx E_{RHF} + 2\varepsilon^2 \left[(E_T - E_{RHF}) - (\varphi\varphi|\tilde{\varphi}\tilde{\varphi}) \right],
\end{aligned}
$$

where the Taylor expansion and the III Slater-Condon rule have been used (p. e107): $\left\langle \psi_{RHF} | \hat{H} \psi_E \right\rangle = (\varphi\varphi|\tilde{\varphi}\tilde{\varphi}) > 0$. The last integral is greater than zero because it corresponds to the Coulombic self-repulsion of a certain charge distribution.

It is now clear that if the singly-excited triplet state $\psi_{T''}$ is of high energy as compared to the ground state ψ_{RHF}, then the spatial diversification of the opposite spin electrons (connected with the stabilization of $-2\varepsilon^2(\varphi\varphi|\tilde{\varphi}\tilde{\varphi})$) will not pay. But if the E_T is close to the ground state energy, then the *total energy will decrease upon the addition of the triplet state*; i.e., the RHF solutions will be unstable toward the UHF (or AMO)–type change of the orbitals.

This is the picture we obtain in numerical calculations for the hydrogen molecule (Fig. 8.18). At short distances between the atoms (up to 2.30 a.u.), the interaction is strong and the triplet state is of high energy. Then the variational principle does not allow the triplet state to contribute to the ground state and the UHF and the RHF give the same result. But beyond the 2.30 a.u. internuclear distance, the triplet admixture results in a small stabilization of the ground state and the UHF energy is lower than the RHF. For very long distances (when the energy difference between the singlet and triplet states is very small), the energy gain associated with the triplet component is very large.

We can see from Fig. 8.18b the drama occurring at $R = 2.30$ a.u. for the mean value of the $\hat{S}^2$ operator. For $R < 2.30$ a.u. the wave function preserves the singlet character, for larger R the triplet addition increases fast, and at $R = \infty$ the mean value of the square of the total spin $\hat{S}^2$ is equal to 1, i.e., half-way between the $S(S+1) = 0$ result for the singlet ($S = 0$) and the $S(S+1) = 2$ result for the triplet ($S = 1$), since the UHF determinant is exactly 50%:50% singlet : triplet mixture. Thus, *one determinant (UHF) is able to describe properly the dissociation of the hydrogen molecule in its ground state (singlet), but at the expense of a large spin contamination (triplet admixture).*

RESULTS OF THE HARTREE-FOCK METHOD

8.6 Mendeleev Periodic Table

8.6.1 Similar to the Hydrogen Atom—The Orbital Model of an Atom

Dmitri Ivanovich Mendeleev (1834–1907), Russian chemist, professor at the University in Petersburg, and later controller of the Russian Standards Bureau of Weights and Measures (after he was expelled from the university by the tsarist powers for supporting a student protest). He was born in Tobolsk, as the youngest of 14 children of a headmaster. In 1859, young Mendeleev went to Paris and Heidelberg on a tsarist scholarship, where he worked with Robert Bunsen and Gustav Kirchhoff. After getting his Ph.D.in 1865, he became at 32 professor of chemistry at the University in St. Petersburg. Since he had no good textbook, he started to write his own (*Principles of Chemistry*). This is when he discovered one of the major human generalizations (1869): the Periodic Table of chemical elements. In 1905 he was nominated for the Nobel Prize, but lost by one vote to Henri Moissan, the discoverer of fluorine. The Swedish Royal Academy thus lost *its* chance, because in a year or so Mendeleev was dead. Many scientists have had similar intuition as had Mendeleev, but it was Mendeleev who completed the project, who organized the known elements in the table, and who predicted the existence of unknown elements. The following example shows how difficult it was for science to accept the periodic table. In 1864, John Newlands presented to The Royal Society in London his work showing similarities of the light elements, occurring for each eighth element with increasing atomic mass. The president of the meeting, quite amused by these considerations, suggested: "*Haven't you tried to organize them according to the alphabetic order of their names?*"

The Hartree-Fock method gives an approximate wave function for the atom of *any* chemical element from the Mendeleev periodic table *(orbital picture)*. The Hartree-Fock method stands behind the *orbital model of atoms*. The model says essentially that a single electron configuration can describe the atom to an accuracy that in most cases satisfies chemists. To tell the truth, the orbital model is in principle false,[79] but it is remarkable that nevertheless, the conclusions drawn from it agree with experimental results, at least qualitatively. It is exciting that

> the electronic structure of all elements can be generated to a reasonable accuracy using the *Aufbau Prinzip*; i.e., a certain scheme of filling the atomic orbitals of the hydrogen atom.

[79] Because the contributions of other Slater determinants (configurations) is not negligible (see Chapter 10).

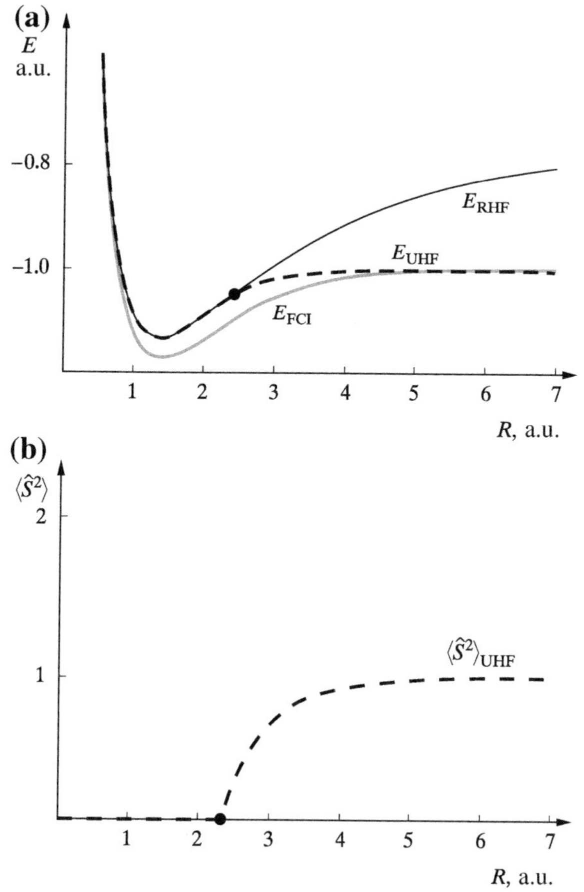

Fig. 8.18. (a) The mean value of the Hamiltonian (E) calculated by the RHF and UHF methods (by T. Helgaker, P. Jørgensen, and J. Olsen, "Molecular Electronic Structure Theory", Wiley, Chichester, 2000). The lowest curve (E_{FCI}) corresponds to the accurate result (called the full configuration interaction method; see Chapter 10). (b) The mean value of the $\hat{S}^2$ operator calculated by the RHF and UHF methods. The energies $E_{RHF}(R)$ and $E_{UHF}(R)$ are identical for internuclear distances $R < 2.30$ a.u. For larger R values, the two curves separate, and the RHF method gives an incorrect description of the dissociation limit, while the UHF method still gives a correct dissociation limit. For $R < 2.30$ a.u., the RHF and UHF wave functions are identical, and they correspond to a singlet, while for $R > 2.30$, the UHF wave function has a triplet contamination.

Thus, the simple and robust orbital model serves chemistry as a "work horse." Let us take some examples. All the atoms are build on a similar principle. A nodeless, spherically symmetric atomic orbital of the lowest orbital energy is called $1s$, the second lowest (and also the spherically symmetric, one-radial node) is called $2s$, etc. Therefore, when filling orbital energy states by electrons, some electronic shells are formed: K($1s^2$), L($2s^2 2p^6$), ..., where the maximum for shell orbital occupation by electrons is shown.

The very foundations of a richness around us (its basic building blocks being atoms in the Mendeleev periodic table) result from a very simple idea: that the proton and electron form a stable system called the hydrogen atom.

8.6.2 Shells and Subshells

The larger the atomic number, the more complex the electronic structure. For neutral atoms, the following occupation scheme applies (sign $<$ relates to the orbital energy):

Aufbau Prinzip

The *Aufbau Prinzip* relies on the following scheme of orbital energies (in ascending order).

	Orbital energy **Direction of occupying subshells** $\Longrightarrow\Downarrow$				**[Noble gas atoms]**
n					
1	$1s$				$[He(2)] = 1s^2$
2	$2s$			$2p$	$[Ne(10)] = [He(2)]2s^2 2p^6$
3	$3s$			$3p$	$[Ar(18)] = [Ne(10)]3s^2 3p^6$
4	$4s$		$3d$	$4p$	$[Kr(36)] = [Ar(18)]4s^2 3d^{10} 4p^6$
5	$5s$		$4d$	$5p$	$[Xe(54)] = [Kr(36)]5s^2 4d^{10} 5p^6$
6	$6s$	$4f$	$5d$	$6p$	$[Rn(86)] = [Xe(54)]6s^2 4f^{14} 5d^{10} 6p^6$
7	$7s$	$5f$	$6d$	$7p$	$[Uuo(118)] = [Rn(86)]7s^2 5f^{14} 6d^{10} 7p^6$

This sequence of the orbital energies may be viewed as a result of the Restricted Open-Shell Hartree-Fock (ROHF, p. 411) calculations for atoms. As one can see, the sequence differs from that for the hydrogen atom orbital energies. The main reason for this is that unlike for the hydrogen atom, for other atoms for a given principal quantum number n (counting atomic *shells*: K for $n = 1$, L for $n = 2$, M for $n = 3$, etc.), the orbital energies increase with the quantum number l, which define *subshells*: $s, p, d, \ldots$ for $l = 0, 1, 2, \ldots$ This in turn reflects the fact an electron in such an atom does not see a point nucleus. Instead, it sees the point nucleus surrounded by an electron cloud; i.e., with the nuclear charge screened by an electron cloud, something like a huge non-point-like nucleus. This leads to such an "anomaly" that the $3d$ orbital energy is higher than that of $4s$, etc.

However, we cannot expect that all nuances of atomic stabilities and of the ions corresponding to them might be deduced from a single simple rule like the *Aufbau Prinzip*, that would replace the hard work of solving the Schrödinger equation (plus also the relativistic effects; see Chapter 3) individually for each particular system.

> Increasing the nuclear charge of an atom (together with its number of electrons) leads to the consecutive occupation by electrons of the electronic shells and subshells of higher and higher energy. This produces a quasi-periodicity (sometimes called *periodicity* in chemistry) of the valence shells, and as a consequence, a quasi-periodicity of all chemical and physical properties of the elements (reflected in the Mendeleev periodic table).

Example 1. ***Noble gases.*** The atoms He, Ne, Ar, Kr, Xe, and Rn have a remarkable feature: they all exhibit the full occupancy of the electronic shells. According to the discussion on p. 430, what chemistry is all about is the outermost occupied orbitals (constituting the *valence shell*) that participate in forming chemical bonds. To have a chemical bond an atom has to offer or receive an electron. However,

> the noble gases have the highest ionization potential among all chemical elements and the zero electron affinity. This confirms the common chemical knowledge that the noble gases do not form chemical bonds.

One has to remember, however, that even the closed shells of the noble gases can be opened either in extreme physical conditions (like pressure) or by using aggressive compounds that are able to detach an electron from them.

Example 2. ***Alkali metals.*** The atoms Li, Na, K, Rb, Cs, Fr have the following dominant electronic configurations (the inner shells have been abbreviated by reporting the corresponding noble gas atom configuration):

	Inner shells	Valence configuration
Li	[He]	$2s^1$
Na	[Ne]	$3s^1$
K	[Ar]	$4s^1$
Rb	[Kr]	$5s^1$
Cs	[Xe]	$6s^1$
Fr	[Rn]	$7s^1$

Since the valence shell decides about chemistry, no wonder the elements Li, Na, K, Rb, Cs, and Fr exhibit similar chemical and physical properties. Let us take any property we want (e.g., what will we get if the element is thrown into water?). Lithium is a metal that reacts slowly with water, producing a colorless basic solution and hydrogen gas. Sodium is a metallic substance, and with water produces a very dangerous spectacle (wild dancing flames): it reacts rapidly with water to form a colorless basic solution and hydrogen gas. The other alkali metals are even more dangerous. Potassium is a metal as well, and it reacts very rapidly with water, giving a colorless basic solution and hydrogen gas. Rubidium is a metal that reacts very rapidly with water, producing a colorless basic solution and hydrogen gas. Cesium metal reacts rapidly with water. The result is a colorless solution and hydrogen gas. Francium is very scarce and expensive, and probably no one has tried its reaction with water. *However, we may expect, with very high probability, that if the reaction were made, it would be faster than the reaction with cesium, and a basic solution would be produced.*

However, maybe all elements react rapidly with water to form a colorless basic solution and hydrogen gas? Well, this is not true. The noble gases do not. They only dissolve in water, without any accompanying chemical reaction. They seem to be for the water structure just inert balls of increasing size. No wonder then that (at 293 K) one obtains the following monotonic sequence of their solubilities: 8.61 (He) < 10.5 (Ne) < 33.6 (Ar) < 59.4 (Kr) < 108.1 (Xe) < 230 (Rn) cm^3/kg.

Example 3. ***Halogens.*** Let us see whether there are other families. Let us concentrate on atoms that have p^5 as the outermost configuration. Using our scheme of orbital energies, we produce the following configurations with this property:

- $[He]2s^22p^5$ with 9 electrons (i.e., F),
- $[Ne]3s^23p^5$) with 17 electrons (i.e., Cl),
- $[Ar]4s^23d^{10}4p^5$) with 35, which corresponds to Br,
- $[Kr]4d^{10}, 5p^5$ with 53 electrons, which is iodine,
- $[Xe]6s^24f^{14}5d^{10}6p^5$ means 85 electrons (i.e., astatine, or At).

Are these elements similar? What happens to halogens when they make contact with water? Maybe they react very rapidly, with water producing a colorless basic solution and hydrogen gas, as with the alkali metals, or do they just dissolve in water like the noble gases? Let us see.

Fluorine reacts with water to produce oxygen, O_2, and ozone, O_3. This is strange in comparison with alkali metals. Next, chlorine reacts with water to produce hypochlorite, OCl^-. Bromine and iodine behave similarly, producing hypobromite OBr^- and hypoiodite OI^-. Nothing is known about the reaction of astatine with water. Apart from the exceptional behavior of fluorine,[80] there is no doubt that we have a family of elements. This family is different from the noble gases and from the alkali metals.

Thus, the families show evidence that elements differ widely among families, but much less within a family, with rather small (and often monotonic) changes within it. This is what (quasi) periodicity of the Mendeleev periodic table is all about. The families are called *groups* (usually columns) in the Mendeleev table.

The Mendeleev table represents more than just a grid of information–it is a kind of compass in chemistry. Instead of having a wilderness where all the elements exhibit their unique physical and chemical properties as *deus ex machina*, we obtain the *understanding* that the animals are in a zoo, and are not unrelated, that there are some families, which follow from similar structures and occupancies of the outer electronic shells. Moreover, it became clear for Mendeleev that there were cages in the zoo waiting for animals yet to be discovered. The animals could have been described in detail *before they were actually found by experimentation.* This periodicity pertains not only to the chemical and physical properties of elements, but also to all parameters that appear in theory and are related to atoms, molecules, and crystals.

[80] For light elements, the details of the electronic configuration play a more important role. For example, hydrogen may also be treated as an alkali metal, but its properties differ widely from the properties of the other members of this family.

8.6.3 Educated Guess of Atomic Orbitals–The Slater Rules

John Slater, by analyzing what his young coworkers were bringing from the computer room (as he wrote: *"when the boys were computing"*), noticed that he can quite easily *predict* the approximate shape of the orbitals without any calculation. According to his rules, it is enough to introduce a screening in the STO's exponent in order to get a rough idea of realistic AOs for a particular atom. Slater proposed the STOs with $\zeta = \frac{Z-\sigma}{n}$, where Z stands for the nuclear charge, σ tells us how other electrons screen (i.e., effectively diminish) the charge of the nucleus for an electron "sitting on" the analyzed STO, and n is the principal quantum number, the same as that in the *Aufbau Prinzip*.[81]

We focus on the electron occupying the orbital in question (that for which we are going to find ζ), and we try to estimate what the electron "sees." The electron sees that the nucleus charge is screened by its fellow electrons. The Slater rules are as follows:

- Write down the electronic configuration of the atom by grouping its orbitals in the following way: $[1s][2s2p][3s3p][3d]\ldots$
- Electrons from the groups to the right of this sequence give zero contribution.
- The electrons in the same group contribute 0.35 each, except the $[1s]$ fellow electron (if we consider $1s$ orbital), which contributes 0.30.
- For an electron in a $[nsnp]$ group, each electron in the $n-1$ group contributes 0.85, for lower groups (more on the left side) each contributes 1.0 and for the $[nd]$ or $[nf]$ groups, all electrons in the groups to the left contribute 1.0.

Example: Carbon atom

Configuration in groups $[1s^2][2s^2 2p^2]$. There will be two σ's: $\sigma_{1s} = 0.30$, $\sigma_{2s} = \sigma_{2p} = 3 \cdot 0.35 + 2 \cdot 0.85 = 2.75$. Hence, $\zeta_{1s} = \frac{6-0.30}{1} = 5.70$, $\zeta_{2s} = \zeta_{2p} = \frac{6-2.75}{2} = 1.625$, which means the following AOs $1s_C = N_{1s} \exp(-5.70r)$, $2s_C = N_{2s} r \exp(-1.625r)$, $2p_{x,C} = N_{2p} x \exp(-1.625r)$, $2p_{y,C} = N_{2p} y \exp(-1.625r)$, $2p_{z,C} = N_{2p} z \exp(-1.625r)$. The advantage of such estimations is that prior to any computation we have already an idea what kind of AOs we should expect from more accurate approaches.

8.7 The Nature of the Chemical Bond

As shown on p. 439, the MO method explains the nature of the chemical bond *via* the argument that the orbital energy in the molecule is lower than that in the isolated atom. But why is this so? Which interactions decide bond formation? Do they have their origin in quantum or simply in classical mechanics?

To answer these questions, we will analyze the simplest case: chemical bonding in molecular ion H_2^+. It seems that quantum mechanics is not required here: we deal with one repulsion and two attractions. No wonder there is bonding, since the net effect is one attraction. But the same

[81] For $n \geq 4$, some modifications have been designed.

applies, however, to the dissociated system (the hydrogen atom and the proton). Thus, the story is becoming more subtle.

8.7.1 The Simplest Chemical Bond: H_2^+ in the MO Picture

Let us analyze chemical bonding as viewed by the poor version of the MO method (only two 1*s* hydrogen atom orbitals are used in the LCAO expansion; see Appendix R available at booksite.elsevier.com/978-0-444-59436-5 on p. e137). Much can be seen because this is such a simple version. The mean kinetic energy of the (only) electron of H_2^+, residing on the bonding MO $\varphi = [2(1+S)]^{-1/2}(a+b)$, is given as ($a$ and b denote the atomic 1*s* orbitals centered, respectively, on the a and b nuclei)

$$\bar{T} \equiv (\varphi|\hat{T}\varphi) = \frac{T_{aa}+T_{ab}}{1+S}, \tag{8.64}$$

where S is the overlap integral $S = (a|b)$, and

$$T_{aa} = (a|-\frac{1}{2}\Delta|a) = T_{bb},$$
$$T_{ab} = (a|-\frac{1}{2}\Delta|b) = T_{ba}.$$

The non-interacting hydrogen atom and the proton have the mean kinetic energy of the electron equal to T_{aa}. The kinetic energy change is, thus,

$$\Delta T = \bar{T} - T_{aa} = \frac{T_{ab} - ST_{aa}}{1+S}. \tag{8.65}$$

Let us note (recall the a and b functions are the eigenfunctions of the hydrogen atom Hamiltonian), that $T_{ab} = E_H S - V_{ab,b}$ and $T_{aa} = E_H - V_{aa,a}$, where E_H is the ground state energy of the H atom,[82] and

$$V_{ab,b} = V_{ab,a} = -\left(a|\frac{1}{r_b}|b\right),$$
$$V_{aa,a} = -\left(a|\frac{1}{r_a}|a\right).$$

Now, ΔT can be presented as

$$\Delta T = -\frac{V_{ab,a} - SV_{aa,a}}{1+S}, \tag{8.66}$$

because the terms with E_H cancel each other out. In this way, the change in kinetic energy of the electron when a molecule is formed may be *formally* presented as the integrals describing

[82] For example, $T_{ab} = (a|-\frac{1}{2}\Delta|b) = (a|-\frac{1}{2}\Delta - \frac{1}{r_b} + \frac{1}{r_b}|b) = E_H S + (a|\frac{1}{r_b}|b) = E_H S - V_{ab,b}$.

the potential energy. Using the results of Appendix R available at booksite.elsevier.com/978-0-444-59436-5, p. e137, we have $-V_{ab,a} + SV_{aa,a} = (1 + R)\exp(-R) - S = S - \frac{R^2}{3}\exp(-R) - S = -\frac{R^2}{3}\exp(-R) \leq 0$. This means $\Delta T \leq 0$; i.e.,

the kinetic energy change stabilizes the molecule.

This agrees with our intuition that an electron in the molecule has more space ("larger box," see p. 163), and the energy levels in the box (potential energy is zero in the box, so we mean kinetic energy) decrease, when the box dimension increases. This example shows that some abstract problems that can be solved exactly (here the particle in the box), serve as a beacon for more complex problems.

Now let us calculate the change in the mean potential energy. The mean potential energy of the electron (the nucleus-nucleus interaction will be added later) equals

$$\bar{V} = (\varphi|V|\varphi) = (\varphi|-\frac{1}{r_a} - \frac{1}{r_b}|\varphi) = \frac{(V_{aa,a} + V_{aa,b} + 2V_{ab,a})}{1+S} \tag{8.67}$$

while in the hydrogen atom, it was equal to $V_{aa,a}$. The difference, ΔV, is

$$\Delta V = \frac{(-SV_{aa,a} + 2V_{ab,a} + V_{aa,b})}{1+S}. \tag{8.68}$$

We can see that when the change in *total* electronic energy ($\Delta E_{el} = \Delta T + \Delta V$) is calculated, some kinetic energy terms will cancel the corresponding potential energy terms, and *potential energy will dominate during bond formation*:

$$\Delta E_{el} = \frac{V_{ab,a} + V_{aa,b}}{1+S}. \tag{8.69}$$

To obtain the change, ΔE, in the total energy of the system during bond formation, we have to add the term $1/R$ describing the nuclear repulsion:

$$\Delta E = \frac{V_{ab,a}}{1+S} + \frac{V_{aa,b}}{1+S} + \frac{1}{R}. \tag{8.70}$$

This formula is identical (because $V_{ab,a} = V_{ab,b}$) to the difference in orbital energies in the molecule H_2^+ and in the hydrogen atom, as given in Appendix R available at booksite.elsevier.com/978-0-444-59436-5 on p. e137.

Eq. (8.70) can be easily interpreted. Let us first consider the electron density described by the φ orbital: $\varphi^2 = [2(1+S)]^{-1}(a^2 + b^2 + 2ab)$. Let us note that the density can be divided into the part ρ_a close to nucleus a, ρ_b close to nucleus b, and ρ_{ab} concentrated in the bonding region[83]

$$\varphi^2 = \rho_a + \rho_b + \rho_{ab}, \tag{8.71}$$

[83] We will see it later in detail.

where $\rho_a = [2(1+S)]^{-1}a^2$, $\rho_b = [2(1+S)]^{-1}b^2$, $\rho_{ab} = [(1+S)]^{-1}ab$. It can be seen[84] that the charge associated with ρ_a is $[-2(1+S)]^{-1}$, the charge connected with the nucleus b is the same, and the overlap charge ρ_{ab} is $-S/(1+S)$. Their sum gives $-2/[2(1+S)]-2S/[2(1+S)] = -1$ (the unit electronic charge). The formula for ΔE may also be written as (we use symmetry: the nuclei are identical, and the a and b orbitals differ only in their centers only):

$$\Delta E = \frac{V_{ab,a}}{[2(1+S)]} + \frac{V_{ab,b}}{[2(1+S)]} + \frac{V_{aa,b}}{[2(1+S)]} + \frac{V_{bb,a}}{[2(1+S)]} + \frac{1}{R}. \tag{8.72}$$

Now it is clear that this formula exactly describes the following Coulombic interactions (Fig. 8.19a,b):

- Of the electron cloud from the a atom (with density $\frac{1}{2}\rho_{ab}$) with the nuclei a and b (the first two terms of the expression)
- Of the electron cloud of density ρ_a with the b nucleus (third term)
- Of the electron cloud of density ρ_b with the a nucleus (fourth term)
- Of the a and b nuclei (fifth term).

If we consider *classically* a proton approaching a hydrogen atom, the only terms for the total interaction energy are (see Fig. 8.19c):

$$\Delta E_{class} = V_{aa,b} + \frac{1}{R}. \tag{8.73}$$

The difference between ΔE and ΔE_{class} only originates from the difference in electron density calculated quantum mechanically and classically; cf. Fig. 8.19. The ΔE_{class} is a *weak* interaction (especially for long distances), and tends to $+\infty$ for small R, because[85] of the $1/R$ term. This can be understood because ΔE_{class} is the difference between two Coulombic interactions: of a point charge with a spherical charge cloud, and of the respective two point charges (called *penetration energy*). ΔE contains two more terms in comparison with ΔE_{class}: $V_{ab,a}/[2(1+S)]$ and $V_{ab,b}/[2(1+S)]$, and both decrease exponentially to $V_{aa,a} = -1$ a.u., when R decreases to zero. Thus, these terms are not important for long distances, stabilize the molecule for intermediate distances (and provide the main contribution to the chemical bond energy), and are dominated by the $1/R$ repulsion for small distances.

In the quantum case, for the electron charge cloud connected with the a nucleus, a^2 is decreased by a charge of $S/(1+S)$, *which shifts to the halfway point toward nucleus b*. In the classical case, there is no charge shift–the whole charge is close to a. In both cases, there is the nucleus-nucleus and the nucleus-electron interaction. The first is identical, but the latter is completely different in both cases. Yet even in the latter interaction, there is something in common: the interaction of the nucleus with the major part of the electron cloud, with charge

[84] This happens after the integration of ρ_a.

[85] $V_{aa,a}$ is finite.

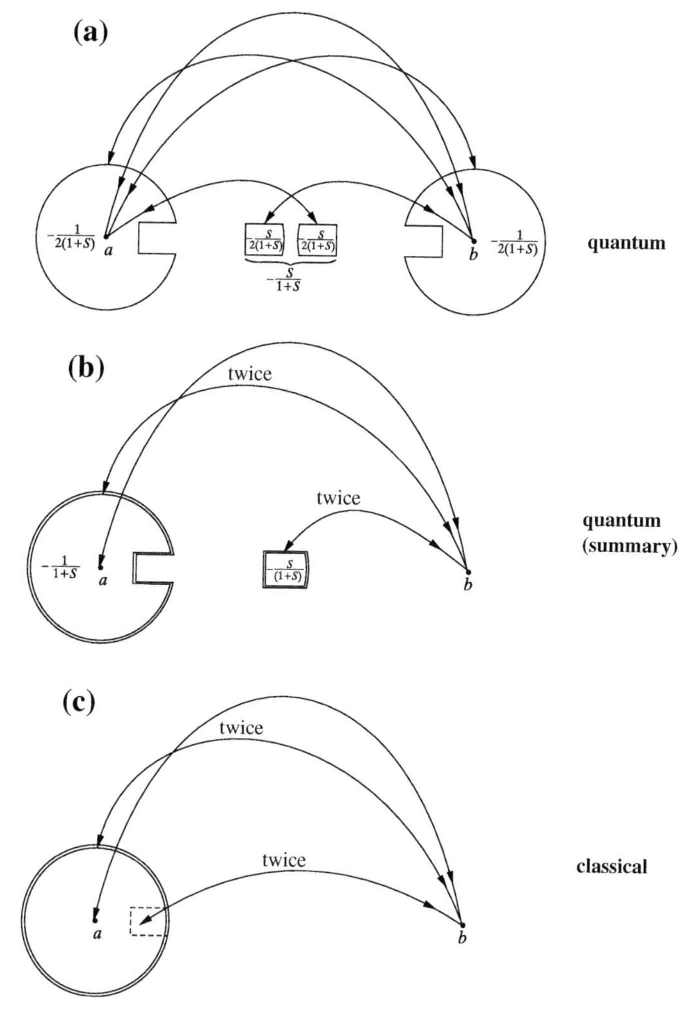

Fig. 8.19. The nature of the chemical bond in the H_2^+ molecule [schematic interpretation according to Eq. (8.72)]:

(a) *The quantum picture of the interaction.* The total electron density $\varphi^2 = \rho_a + \rho_b + \rho_{ab}$, consists of three electronic clouds $\rho_a = [2(1+S)]^{-1}a^2$ bearing the $-\frac{1}{2(1+S)}$ charge (in a.u.) concentrated close to the a nucleus, a similar cloud $\rho_b = [2(1+S)]^{-1}b^2$ concentrated close to the b nucleus, and the rest (the total charge is -1) $\rho_{ab} = [(1+S)]^{-1}ab$ bearing the charge of $-2\frac{S}{2(1+S)}$, concentrated in the bond. The losses of the charge on the a and b atoms have been shown schematically, since the charge in the middle of the bond originates from these losses. The interactions have been denoted by arrows: there are all kinds of interactions of the fragments of one atom with the fragments of the second one.

(b) *The quantum picture–summary* (we will need it in just a moment). This scheme is similar to (a), but it has been emphasized that the attraction of ρ_a by nucleus b is the same as the attraction of ρ_b by nucleus a; hence, they were both presented as one interaction of nucleus b with charge of $-2\rho_a$ at a (hence the double contour line). In this way, two of the interaction arrows have disappeared compared to (a).

(c) *The classical picture of the interaction between the hydrogen atom and a proton.* The proton (nucleus b) interacts with the electron of the a atom, bearing the charge of $-1 = -2\frac{1}{2(1+S)} - 2\frac{S}{2(1+S)}$ and with nucleus a. Such division of the electronic charge indicates that it consists of two fragments ρ_a [as in (b)] and of two fragments of the $-\frac{S}{2(1+S)}$ charge [i.e., similar to (b), but *centered in another way*]. A major difference compared to (b) is that in the classical picture nucleus b interacts with two *quite distant* electronic charges (put in the vicinity of nucleus a), while in the quantum picture [schemes (a) and (b)], the same charges attract at short distance.

$-[1-S/(1+S)] = -1/(1+S)$. The difference in the cases is the interaction with the remaining part of the electron cloud[86]: the charge $-S/(1+S)$.

In the classical view, this cloud is located close to distant nucleus a, while in the quantum view, it is distributed within the bond. The latter is much better for bonding. This interaction, of the (negative) electron cloud ρ_{ab} within the bond with the positive nuclei, stabilizes the chemical bond.

Fig. 8.19 shows an idea of the quantum mechanical nature of the chemical bond by using some particular *schematically drawn* electronic clouds, their interaction favoring the quantum mechanical picture over the classical one. However, there is nothing that prevents us *from showing the same clouds in a realistic way*, as Fig. 8.20 demonstrates.

Note that the critically important charge density distribution ρ_{ab} has a very unusual shape. It represents something similar to an electron density rod (the figure shows its section only) that connects the two nuclei, its density being the largest and *constant* within the section of the straight line connecting the two nuclei (an extremely striking shape[87]) and decaying very fast beyond.

8.7.2 Can We See a Chemical Bond?

If a substance forms crystals, it may be subjected to X-ray analysis. Such an analysis is quite exceptional, since it is one of very few techniques (which also include neutronography and nuclear magnetic resonance spectroscopy), which can show atomic positions in space. More precisely, the X-ray analysis shows electronic density maps because the radiation sees electrons, not nuclei. The inverse is true in neutronography. If we have the results of X-ray and neutron scattering, we can subtract the electron density of atoms (positions shown by neutron scattering) from the electron density of the molecular crystal (shown by X-ray scattering). This difference would be a result of the chemical bonding (and, to a smaller extent, of the intermolecular

---

[86] This simple interpretation gets more complex when further effects are considered, such as contributions to energy due to the polarization of the spherically symmetric atomic orbitals or the exponent dependence of the $1s$ orbitals (i.e., the dimensions of these orbitals) on the internuclear distance. When there are several factors at play (some positive, some negative) and when the final result is of the order of a single component, then *we* decide which component carries responsibility for the outcome. The situation is similar to that in Parliament, when two MPs from a small party are blamed for the result of a vote (the party may be called the "balancing party"), while perhaps 200 others who also voted in a similar manner are left alone.

[87] For points along the section ($0 \leq x \leq R$), one gets $\rho_{ab}(x,0,0) = \frac{1}{1+S}a(x,0,0)b(x,0,0) = \frac{1}{1+S}\frac{1}{\pi}\exp(-|x|)\exp(-|x-R|) = \frac{1}{1+S}\frac{1}{\pi}\exp(-x)\exp(x-R) = \frac{1}{1+S}\frac{1}{\pi}\exp(-x+x-R) = \frac{1}{1+S}\frac{1}{\pi}\exp(-R)$, which is a number that is independent of x. In Chapter 11, we will detect a chemical bond as a "*rope*" formed by the total electron density distribution (the Bader analysis), but the rod we find here as distinguishing the classical and the quantum situations represents only a part of this "*rope*" from Chapter 11.

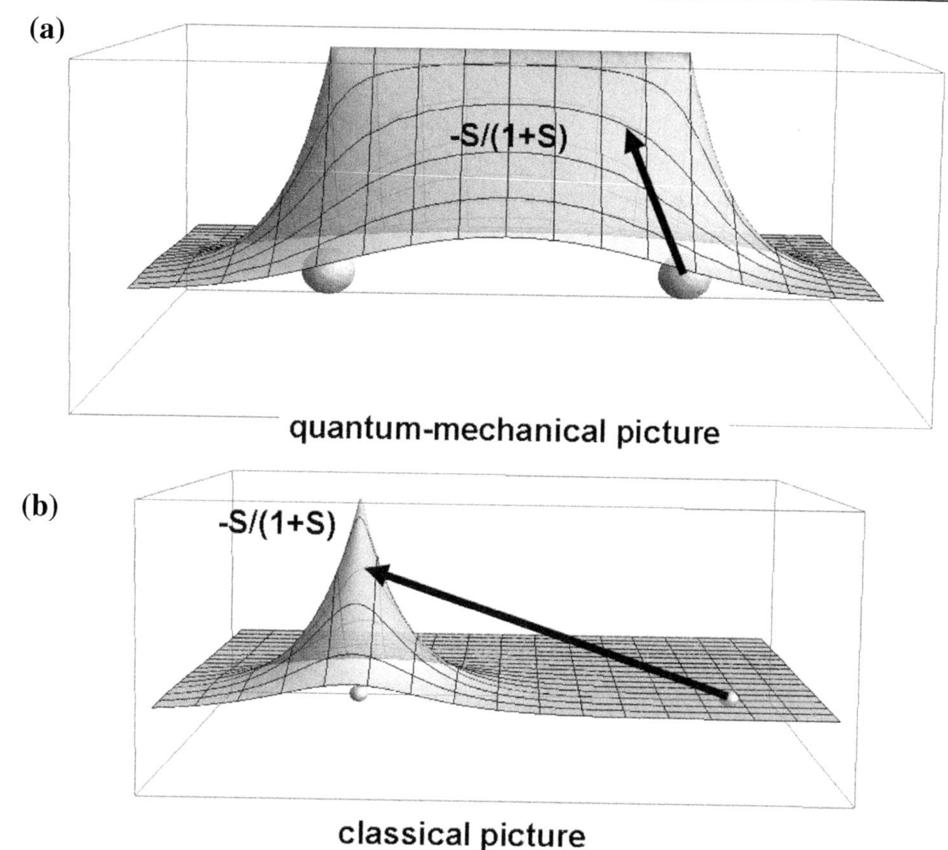

Fig. 8.20. The same reasoning is used as in Fig. 8.19 , but this time, the key charge distributions are drawn in a realistic way instead of a schematic diagram. The sections $z = 0$ of the two crucial electronic charge distributions are drawn. (a) In the quantum mechanical picture, what decides about chemical bonding is an electron cloud $-\frac{1}{1+S} 1s_a \cdot 1s_b \equiv -\frac{1}{1+S} ab$ that contains the charge $-\frac{S}{1+S}$ [S stands for the overlap integral $(a|b)$] and has a uniform density along the bond. This is why its attraction with the nucleus b (of charge $+1$), which is in the immediate neighborhood, is strong. (b) In the classical picture, the cloud $-\frac{S}{1+S}\left(1s_a\right)^2$, corresponding to the same total charge $-\frac{S}{1+S}$ as before (the scale used is changed), is close to the nucleus a, so its attraction with the nucleus b is far weaker. Thus, the atoms bind so strongly due to the quantum nature of the electron involved.

interactions). This method is called X-N or X-Ray minus Neutron Diffraction.[88] Hence, differential maps of the crystal are possible, where we can see the shape of the "additional" electron density at the chemical bond, or the shape of the electron deficit (negative density) in places where the interaction is antibonding.[89]

[88] There is also a pure X-ray version of this method. It uses the fact that the X-ray reflections obtained at large scattering angles see only the spherically symmetric part of the atomic electron density, similarly to that which we obtain from neutron scattering.

[89] R. Boese, *Chemie in unserer Zeit*, 23(1989)77, D. Cremer, E. Kraka, *Angew. Chem.*, 96(1984)612.

From the differential maps, we can estimate the following:

1) The strength of a chemical bond *via* the value of the positive electron density at the bond
2) The deviation of the bond electron density (perpendicular intersection) from the cylindrical symmetry, which gives information on the π character of the chemical bond
3) The shift of the maximum electron density toward one of the atoms, which indicates the *polarization* of the bond
4) The shift of the maximum electron density away from the straight line connecting the two nuclei, which indicates bent (banana-like) bonding.

This opens up new possibilities for comparing theoretical calculations with experimental data.

8.8 Excitation Energy, Ionization Potential, and Electron Affinity (RHF Approach)

8.8.1 Approximate Energies of Electronic States

Let us consider (within the RHF scheme) the simplest two-electron closed-shell system with both electrons occupying the same orbital φ_1. The Slater determinant, called ψ_G (G standing for the *ground state*) is built from two spinorbitals $\phi_1 = \varphi_1\alpha$ and $\phi_2 = \varphi_1\beta$. We also have the virtual orbital φ_2, corresponding to orbital energy ε_2, and we may form two other spinorbitals from it. We are now interested in the energies of *all the possible excited states* that can be formed from this pair of orbitals. These states will be represented as Slater determinants, built from φ_1 and φ_2 orbitals with the appropriate electron occupancy. We will also assume that excitations do not deform the φ orbitals (which is, of course, only partially true). Now all possible states may be listed by occupation of the ε_1 and ε_2 orbital levels, as shown in Table 8.2.

E is a doubly excited electronic state, and T and T' are two of three possible triplet states of the same energy. If we require that any state should be an eigenfunction of the $\hat{S}^2$ operator (it also needs to be an eigenfunction of $\hat{S}_z$, but this condition is fortunately fulfilled by all the functions listed above), it appears that only ψ_1 and ψ_2 are illegal. However, their combinations, shown here:

$$\psi_S = \frac{1}{\sqrt{2}}(\psi_1 - \psi_2) \tag{8.74}$$

$$\psi_{T''} = \frac{1}{\sqrt{2}}(\psi_1 + \psi_2). \tag{8.75}$$

Table 8.2. All possible occupations of levels ε_1 and ε_2.

Level| Function	ψ_G	ψ_T	$\psi_{T'}$	ψ_1	ψ_2	ψ_E
ε_2	–	α	β	β	α	$\alpha\beta$
ε_1	$\alpha\beta$	α	β	α	β	–

are legal. The first describes the singlet state, and the second is the triplet state (the third function, missing from the complete triplet set).[90] This may be easily checked by inserting the spinorbitals into the determinants, then expanding the determinants, and separating the spin part. For ψ_S, the spin part is typical for the singlet, $\frac{1}{\sqrt{2}}[\alpha(1)\beta(2) - \alpha(2)\beta(1)]$, and for T, T', and T'', the spin parts are, respectively, $\alpha(1)\alpha(2)$, $\beta(1)\beta(2)$, and $\frac{1}{\sqrt{2}}[\alpha(1)\beta(2) + \alpha(2)\beta(1)]$. This is expected for triplet functions with components of total spin equal to 1, -1, and 0, respectively (see Appendix Q available at booksite.elsevier.com/978-0-444-59436-5).

Now let us calculate the mean values of the Hamiltonian using the states mentioned above. Here, we will use the Slater–Condon rules (see the diagram on p. e119), which when expressed in orbitals[91] produce in the MO representation:

$$E_G = 2h_{11} + \mathcal{J}_{11}, \tag{8.76}$$

$$E_T = h_{11} + h_{22} + \mathcal{J}_{12} - \mathcal{K}_{12} \tag{8.77}$$

(for all three components of the triplet);

$$E_S = h_{11} + h_{22} + \mathcal{J}_{12} + \mathcal{K}_{12}, \tag{8.78}$$

$$E_E = 2h_{22} + \mathcal{J}_{22}, \tag{8.79}$$

where $h_{ii} = (\varphi_i|\hat{h}|\varphi_i)$, and $\hat{h}$ is a one-electron operator, the same as that appearing in the Slater-Condon rules, and *explicitly* shown on p. 400, $\mathcal{J}_{ij}$ and $\mathcal{K}_{ij}$ are the two-electron integrals (Coulombic and exchange): $\mathcal{J}_{ij} = (ij|ij)$ and $\mathcal{K}_{ij} = (ij|ji)$.

The orbital energies of a molecule (calculated for the state with the doubly occupied φ_1 orbital) are

$$\varepsilon_i = (\varphi_i|\hat{\mathcal{F}}|\varphi_i) = (\varphi_i|\hat{h} + 2\hat{\mathcal{J}} - \hat{\mathcal{K}}|\varphi_i). \tag{8.80}$$

Thus, we get

$$\varepsilon_1 = h_{11} + \mathcal{J}_{11}, \tag{8.81}$$

$$\varepsilon_2 = h_{22} + 2\mathcal{J}_{12} - \mathcal{K}_{12}. \tag{8.82}$$

Now, the energies of the electronic states can be expressed in terms of the orbital energies:

$$E_G = 2\varepsilon_1 - \mathcal{J}_{11}, \tag{8.83}$$

$$E_T = \varepsilon_1 + \varepsilon_2 - \mathcal{J}_{11} - \mathcal{J}_{12} \tag{8.84}$$

(for the ground singlet state and for the three triplet components of the common energy E_T). The distinguished role of φ_1 (in E_T) may be surprising (since the electrons reside on φ_1 and

[90] Let us make a convention, that in the Slater determinant $\frac{1}{\sqrt{2}}\det|\phi_1(1)\phi_2(2)|$, the spinorbitals are organized according to increasing orbital energy, because then the signs in Eqs. (8.74) and (8.75) are valid.

[91] For E_G, the derivation of the final formula is given on p. 419 (E'_{RHF}). The other derivations are simpler.

φ_2), but φ_1 is indeed distinguished, because the ε_i values are derived from the Hartree–Fock problem with the *only* occupied orbital φ_1. So we get

$$E_S = \varepsilon_1 + \varepsilon_2 - \mathcal{J}_{11} - \mathcal{J}_{12} + 2\mathcal{K}_{12}, \quad (8.85)$$

$$E_E = 2\varepsilon_2 + \mathcal{J}_{22} - 4\mathcal{J}_{12} + 2\mathcal{K}_{12}. \quad (8.86)$$

Now it is time to draw some conclusions.

8.8.2 Singlet or Triplet Excitation?

Aleksander Jabłoński (1898–1980), Polish theoretical physicist and professor at the John Casimirus University in Vilnius, then at the Nicolaus Copernicus University in Toruń. He studied photoluminescence problems. *Courtesy of Nicolaus Copernicus University, Poland*

The Jabłoński diagram plays an important role in molecular spectroscopy (Fig. 8.21). It shows three energy levels: the ground state (G), the first excited singlet state (S), and the metastable in-between state. Later, researchers identified this metastable state as the lowest triplet (T).

Let us compute the energy difference between the singlet and triplet states:

$$E_T - E_S = -2\mathcal{K}_{12} < 0. \quad (8.87)$$

This inequality says that

a molecule always has lower energy in the excited triplet state than in the excited singlet state (both states resulting from the use of the same orbitals),

because $\mathcal{K}_{12} = (\varphi_1(1)\varphi_2(2)|\frac{1}{r_{12}}|\varphi_2(1)\varphi_1(2))$ is always positive, being the interaction of two identical charge distributions (interpretation of an integral, real functions assumed). This rule holds firmly for the energy of the two lowest (singlet and triplet) excited states.

8.8.3 Hund's Rules

The difference between the energies of the ground and triplet states is

$$E_T - E_G = (\varepsilon_2 - \varepsilon_1) - \mathcal{J}_{12}. \quad (8.88)$$

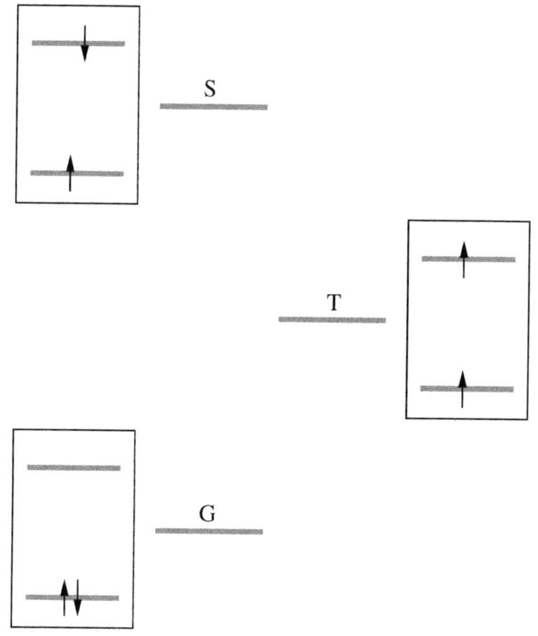

Fig. 8.21. The Jabłoński diagram. The ground state is labeled G. The energy of the singlet excited state (S) is *higher* than the energy of the corresponding triplet state (T; that results from use of the same orbitals).

This result has a simple interpretation. The excitation of a single electron (to the triplet state) costs some energy $(\varepsilon_2 - \varepsilon_1)$, but (since $\mathcal{J}_{12} = (12|12) = \int dV_1 dV_2 |\varphi_1(1)|^2 \frac{1}{r_{12}} |\varphi_2(2)|^2 > 0$) there is also an energy gain $(-\mathcal{J}_{12})$ connected with the removal of the (mutually repulsing) electrons from the "common apartment" (orbital φ_1) to the two separate "apartments" (φ_1 and φ_2). Apartment φ_2 is admittedly on a higher floor $(\varepsilon_2 > \varepsilon_1)$, but if $\varepsilon_2 - \varepsilon_1$ is small, then it may still pay to move.

Friedrich Hermann Hund (1896–1997), professor of theoretical physics at the universities in Jena, Frankfurt am Main, and finally Göttingen, where in his youth he had worked with Born and Franck. He applied quantum theory to atoms, ions, and molecules and discovered his famous empirical rule in 1925. Considered by many as one of the founders of quantum chemistry. (biography, written in German: *Intern. J. Quantum Chem., S11,* 6 (1977)).

In the limiting case, if $\varepsilon_2 - \varepsilon_1 = 0$, the system prefers to put electrons in separate orbitals and with the same spins (the empirical Hund rule; see Fig. 8.22).

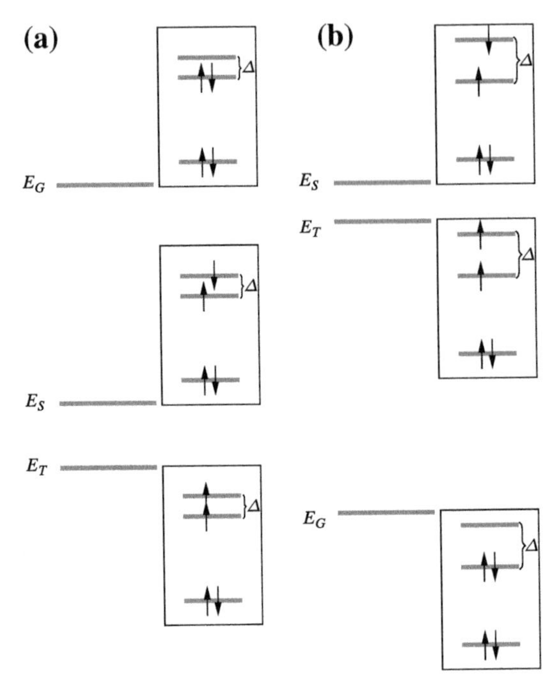

Fig. 8.22. Hund's rule. Energy of each configuration (left) corresponds to an electron occupation of the orbital energy levels (shown in boxes). Two electrons of the highest orbital energy face a dilemma (two upper diagrams): Is it better to occupy a common apartment on the lower floor (...but electrons do not like each other); or is it better for one of them (fortunately, they are not distinguishable...) to make a sacrifice and move to the upper-floor apartment (then they can avoid each other)? If the upper floor is not too high in the energy scale (small Δ, shown in a), then the electrons prefer the second case: each of them occupies a separate apartment and they feel best having their spins parallel (triplet state). But when the upper-floor energy is very high (large Δ, as shown in panel b), then both electrons are forced to live in the same apartment, and in that case, they are forced to have antiparallel spins. Hund's rule pertains to the first case in its extreme form ($\Delta = 0$). When there are several orbitals of the same energy and there are many possibilities of their occupation, then the state with the lowest energy is such that the electrons go each to a separate orbital, and the alignment of their spins is "*parallel*" (see p. 31).

8.8.4 Hund's Rules for the Atomic Terms

The question whether electron pairing is energetically favorable is most delicate for atoms. In the atomic case, one has to do with quite a lot of possible electronic configurations ("*occupancies*"), and the problem is which of them better describes the reality (i.e., which of them is of the lowest energy).

Hund discovered that this can be determined by using some empirical rules (now known as *Hund's rules*): Here they are

- Hund's first rule: the lowest energy corresponds to that configuration, which corresponds to the maximum of the spin angular momentum $|\mathbf{S}|$, where $|\mathbf{S}|^2 = S(S+1)\hbar^2$, with S either an integer or a half-integer. A large value of S requires a "same-spin situation," which is only possible for different orbitals for valence electrons; i.e., a single occupation of the orbitals (low electronic repulsion).

- Hund's second rule: if several such configurations come into play, the lowest energy corresponds to that one, which has the largest orbital angular momentum $|\mathbf{L}|$, with $|\mathbf{L}|^2 = L(L+1)\hbar^2$, where $L = 0, 1, 2, \ldots$
- Hund's third rule is a relativistic correction to the first two rules, introducing a splitting of the terms given by the previous rules. The energy operator (Hamiltonian) commutes with the square of the total angular momentum $\mathbf{J} = \mathbf{L} + \mathbf{S}$, and therefore, the energy levels depend rather on the *total* momentum $|\mathbf{J}|^2 = J(J+1)\hbar^2$. This means that they depend on the mutual orientation of **L** and **S** (this is a relativistic effect due to the spin-orbit coupling in the Hamiltonian). The vectors **L** and **S** add in quantum mechanics in a specific way (see Chapter 7, p. 343): one has $J = |L+S|, |L+S-1|, \ldots |L-S|$. *The III Hund's rule says that if the shell is less than half-filled, the lowest energy corresponds to* $J = |L-S|$, *while if more than half-filled to* $J = |L+S|$.

Hund's rules not only allow one to identify the lowest energy level [or term $^{2S+1}[L]_J$, where we use the symbol $[L] = S(\text{for } L=0),\ P(\text{for } L=1),\ D(\text{for } L=2), \ldots$] but they enable one to give their sequence on the energy scale.

Example. Closed Shells–A Neon Atom

We write first the neon electronic configuration: $1s^2 2s^2 2p^6$, with a convention that we choose as atomic orbitals the ones that represent the eigenfunctions of the $\hat{L}_z$ operator (in this case, $2p_0, 2p_1, 2p_{-1}$, not $2p_x, 2p_y, 2p_z$). Hund's rules are about the orbital angular momentum and spin angular momentum. Let us calculate the z component of the total *orbital* angular momentum as a sum of the corresponding components for individual electrons: $L_z = 2 \cdot 0\hbar + 2 \cdot 0\hbar + 2 \cdot (-1)\hbar + 2 \cdot 0\hbar + +2 \cdot (+1)\hbar = 0$. The only possibility for the total orbital angular momentum to have *only* this value of the z component is that $L = 0$. The term, therefore, has to have the symbol $[L] = S$ (this is a remnant of a misleading quantum tradition: do not confuse this symbol with the spin quantum number S). Now, about the z component of the *spin* angular momentum as a sum of contributions of the individual electrons, the following is true: $S_z = \left(\frac{1}{2} - \frac{1}{2}\right)\hbar + \left(\frac{1}{2} - \frac{1}{2}\right)\hbar + 3 \cdot \left(\frac{1}{2} - \frac{1}{2}\right)\hbar = 0$. This can happen only if $S = 0$ (this time S means spin). The multiplicity of the term is $2S+1 = 1$, and therefore, the term symbol $^{2S+1}[L]$ becomes ^1S (we say "*singlet S*"). Since $\mathbf{L} = \mathbf{0}$ and $\mathbf{S} = \mathbf{0}$, they may add up only to $\mathbf{J} = \mathbf{0}$, which means that $J = 0$. Therefore, the full term symbol should be $^1S_0$, but to avoid the banality, we write it always as ^1S. This term is the only one possible for the electronic configuration $1s^2 2s^2 2p^6$, as well as for any other closed electronic shell:

Any closed shell gives zero contribution to the orbital angular momentum **L** and to the spin angular momentum **S**. This makes possible, when determining the terms, just to ignore any closed-shell contribution.

Example. The Lowest-Energy Term for the Carbon Atom

The carbon atom electronic configuration reads as $1s^2 2s^2 2p^2$. Since the closed shells $1s^2 2s^2$ do not contribute anything, we just ignore them. What matters is the configuration $2p^2$. Three orbitals $2p$ allow for six spinorbitals, occupied by two electrons only. The number of possible Slater determinants is, therefore, $\binom{6}{2} = 15$. The lowest-energy Slater determinant will be determined by using Hund's rules to find the lowest-energy term. To this end, we draw a box diagram of orbital occupations, with the convention that an up or a down arrow stands for an electron with the α or β spin function, respectively. The boxes in the diagram (each one corresponding to an atomic orbital with quantum numbers n, l, m) begin by the one with maximum value of m and continue in descending order. Then we begin the occupation of the boxes by the electrons (maximum two electrons–of opposite spins–per box). Our aim is to find the lowest-energy term by using Hund's rules. We begin by the *I Hund's rule* and this means maximizing of S (same spins). To get this, we place electrons in separate boxes, but to satisfy also the *II Hund's rule*, we begin to do this from the left side of the diagram:

$$\begin{array}{ccc} 2p_1 & 2p_0 & 2p_{-1} \\ \uparrow & \uparrow & \end{array}.$$

We get $M_S = \frac{1}{2} + \frac{1}{2} = 1$ (such a projection implies $S = 1$) and $M_L = 1 + 0 = 1$ (this projection implies $L = 1$).

Hence, the term indicated by the first two Hund's rules that explains these projections is ^3P. This is what the non-relativistic approach gives as the ground state.

However, the relativistic effects will split this degenerate energy level into three separate levels according to three possible mutual orientations of the vectors **L** and **S**, leading to $J = |L+S|\ldots$, $J = |L-S|$ i.e., to $J = 2, 1, 0$. Therefore, we get three terms: $^3P_2, ^3P_1, ^3P_0$, and the *III Hund's rule* predicts the following energy sequence:

$$^3P_0 <^3 P_1 <^3 P_2.$$

This sequence is valid for all atoms having the $2p^2$ configuration: C, Si, Ge, Sn, Pb, with increasing splitting in this series (from about 43 cm^{-1} for C to about 10650 cm^{-1} for Pb[92]).

[92] This reminds us about a general rule: the heavier, the atom, the more important the relativistic effects are.

8.8.5 Ionization Potential and Electron Affinity (Koopmans's Theorem)

The ionization potential of the molecule M is defined as the minimum energy needed for an electron to detach from the molecule. The electron affinity energy of the molecule M is defined as the minimum energy for an electron detachment from M^-. Let us assume again, naively, that *during these operations, the molecular orbitals and the orbital energies do not undergo any changes*. In fact, of course, everything changes, and the computations should be repeated for each system separately (the same applies in the previous section for excitations).

In our two-electron system, which is a model of any closed-shell molecule, the electron removal leaves the molecule with one electron only, and its energy has to be

$$E_+ = h_{11}. \tag{8.89}$$

However,

$$h_{11} = \varepsilon_1 - \mathcal{J}_{11}. \tag{8.90}$$

This formula looks like trouble. After the ionization, there is only a single electron in the molecule, while here, some electron–electron repulsion (integral $\mathcal{J}$) appears! But everything is fine because we still use the two-electron problem as a reference, and ε_1 relates to the two-electron problem, in which $\varepsilon_1 = h_{11} + \mathcal{J}_{11}$.

Hence,

Ionization Energy

The ionization energy is equal to the negative of the orbital energy of an electron:

$$E_+ - E_G = -\varepsilon_1. \tag{8.91}$$

To calculate the electron affinity energy, we need to consider a determinant as large as 3×3, but this proves easy if the useful Slater-Condon rules (see Appendix M available at booksite.elsevier.com/978-0-444-59436-5) are applied. Rule I gives (we write everything using the ROHF spinorbitals, then note that the three spinorbitals are derived from two orbitals, and then sum over the spin variables):

$$E_- = 2h_{11} + h_{22} + \mathcal{J}_{11} + 2\mathcal{J}_{12} - \mathcal{K}_{12}, \tag{8.92}$$

and introducing the orbital energies, we get

$$E_- = 2\varepsilon_1 + \varepsilon_2 - \mathcal{J}_{11}, \tag{8.93}$$

which gives

$$E_- - E_G = \varepsilon_2. \tag{8.94}$$

Hence,

Electron Affinity
The electron affinity is the difference of the energies of the system without an electron and that of the anion, $E_G - E_- = -\varepsilon_2$. It is equal approximately to the negative energy of the virtual orbital on which the electron lands (if $\varepsilon_2 < 0$ attaching an electron means energy lowering).

A Comment on Koopmans's Theorem

Tjalling Charles Koopmans (1910–1985), American econometrist of Dutch origin and professor at Yale University, introduced mathematical procedures of linear programming to economics, and received the Nobel Prize in 1975 *"for work on the theory of optimum allocation of resources."*

The MO approximation represents a rough approximation to reality. So is Koopmans's theorem, which proves to be poorly satisfied for most molecules. But these approximations are often used for practical purposes. This is illustrated by a certain quantitative relationship, derived by Grochala et al.[93]

The authors noted that a very simple relationship holds surprisingly well for the equilibrium *bond lengths* R of four objects: the ground state M_0 of the closed shell molecule, its excited triplet state M_T, its radical–cation $M^{+\cdot}$, and radical–anion $M^{-\cdot}$:

$$R(M_T) = R(M^{-\cdot}) + R(M^{+\cdot}) - R(M_0). \tag{8.95}$$

The cyclobutadiene in the triplet state has the square symmetry (the two adjacent C–C bonds of equal length), while in the ground state the molecule is rectangular. Therefore, if one takes Eq. (8.95) for the first C–C bond and then for the second one, we should get the same length in the triplet state (all values calculated by the DFT method; see Chapter 11). Let us insert the bond lengths for the first C–C bond:

$$1.378\ \text{Å} + 1.363\ \text{Å} - 1.318\ \text{Å} = 1.423\ \text{Å},$$

while for the second C–C bond:

$$1.501\ \text{Å} + 1.489\ \text{Å} - 1.565\ \text{Å} = 1.425\ \text{Å},$$

both values are very close to what has been calculated independently for the triplet state: 1.426 Å.

The above relationship is similar to that pertaining to the corresponding energies:

$$E(M_T) = E(M^{-\cdot}) + E(M^{+\cdot}) - E(M_0),$$

[93] W. Grochala, A.C. Albrecht, and R. Hoffmann, *J. Phys. Chem. A, 104,* 2195 (2000).

which may be deduced, basing on certain approximations, from Koopmans's theorem[94] or from the Schrödinger equation, while neglecting the two-electron operators. The difference between these two expressions is substantial: the latter holds for the four species at *the same* nuclear geometry, while the former describes the geometry *changes* for the "relaxed" species.[95] The first equation proved to be satisfied for a variety of molecules: ethylene, cyclobutadiene, divinylbenzene, diphenylacetylene, *trans*–N_2H_2, CO, CN^-, N_2, and NO^+. It is not yet clear if it would hold beyond the one-electron approximation, or for experimental bond lengths (these are usually missing, especially for polyatomic molecules).

8.9 Toward Chemical Picture–Localization of MOs

The canonical MOs derived from the RHF method are usually delocalized over the whole molecule; i.e., their amplitudes are in general nonzero for all atoms in the molecule. This applies, however, mainly to high-energy MOs, which exhibit a similar AO amplitude for most atoms. Yet the canonical MOs of the inner shells are usually very well localized. The canonical MOs are occupied, as usual, by putting two electrons on each low-lying orbital (the Pauli exclusion principle).

> The picture obtained is in contrast to chemical intuition, which indicates that the electron pairs are localized within the chemical bonds, free electron pairs, and inner atomic shells. The picture, which agrees with intuition, may be obtained after the localization of the MOs.

The localization is based on making new orbitals to be linear combinations of the canonical MOs, a fully legal procedure. Then, the determinantal wave function, as shown on p. 404, expressed in the new spinorbitals, takes the form $\psi' = (det\boldsymbol{A})\psi$ and the total energy will remain unchanged. If linear transformation applied ($\boldsymbol{A}$) is an orthogonal transformation (i.e., $\boldsymbol{A}^T\boldsymbol{A} = \mathbf{1}$), or a unitary one ($\boldsymbol{A}^\dagger\boldsymbol{A} = \mathbf{1}$), the new MOs preserve orthonormality (like the canonical ones), as shown on p. 405. We emphasize that we can make *any non-singular*[96] *linear transformation* $\boldsymbol{A}$, not only orthogonal or unitary ones. This means something important, namely,

[94] Let us check it using the formulas derived by us:

$$\begin{aligned} E(M_T) &= \varepsilon_1 + \varepsilon_2 - \mathcal{J}_{11} - \mathcal{J}_{12}, \text{ and } E(M^{-\cdot}) + E(M^{+\cdot}) - E(M_0) \\ &= \left[2\varepsilon_1 + \varepsilon_2 - \mathcal{J}_{11}\right] + \left[\varepsilon_1 - \mathcal{J}_{11}\right] - \left[2\varepsilon_1 - \mathcal{J}_{11}\right] \\ &= \varepsilon_2 + \varepsilon_1 - \mathcal{J}_{11}. \end{aligned}$$

The equality is obtained after neglecting $\mathcal{J}_{12}$, as compared to $\mathcal{J}_{11}$.

[95] If we assume that a geometry change in these states induces an energy increase that is proportional to the square of the change, and that the curvature of all these parabolas is identical, then the above relationship would be easily proved. The problem is that these states have significantly different force constants, and the curvature of parabolas strongly varies among them.

[96] For any singular matrix, $\det \boldsymbol{A} = 0$, and this should not be allowed (as discussed on p. 404).

the solution in the Hartree-Fock method depends on the *space* spanned by the occupied orbitals (i.e., on the set of all linear combinations that can be formed from the occupied MOs), and not on some particular set of the molecular orbitals. The new orbitals do not satisfy the Fock equation (8.32); these are satisfied by canonical orbitals only.

The localized orbitals (being some other orthonormal basis set in the space spanned by the canonical orbitals) satisfy the Fock equation (8.19) with the off-diagonal Lagrange multipliers.

8.9.1 Can a Chemical Bond Be Defined in a Polyatomic Molecule?

Unfortunately, the view to which chemists get used (i.e., the chemical bonds between pairs of atoms, lone electron pairs, and inner shells) can be derived in an infinite number of ways (because of the arbitrariness of transformation $\boldsymbol{A}$), and in each case, the effects of localization vary. Hence,

we cannot uniquely define the chemical bond in a polyatomic molecule.

It is not a drama, however, because what really matters is the probability density; i.e., the square of the complex modulus of the *total* many-electron wave function. The concept of the (localized or delocalized) molecular orbitals represents simply an attempt to divide this total density into various spatially separated although overlapping parts, each belonging to a single MO. It is similar to dividing an apple into N equal parts. The freedom of such a division is unlimited. For example, we could envisage that each part would have the dimension of the apple ("*delocalized orbitals*"), or an apple would be simply cut axially, horizontally, concentrically, etc. into N equal parts, forming an analog of the localized orbitals. Yet each time, the full apple could be reconstructed from these parts.

As we will soon convince ourselves, the problem of defining a chemical bond in a polyatomic molecule is not so hopeless as it looks now, because various methods lead to essentially the same results.

Now let us consider some practical methods of localization. There are two categories: internal and external.[97] In the external localization methods we plan where the future MOs will be localized, and the localization procedure only slightly alters our plans. This is in contrast with the internal methods, where certain general conditions are imposed that induce automatically localization of the orbitals.

[97] Like medicines.

8.9.2 The External Localization Methods

Projection Method

This is an amazing method,[98] in which *we* first construct some *arbitrary*[99] (but linearly independent[100]) orbitals χ_i of the bonds, lone pairs, and the inner shells, the total number of these being equal to the number of the occupied MOs. Now let us project them on the space of the occupied RHF molecular orbitals $\{\varphi_j\}$ using the projection operator $\hat{P}$:

$$\hat{P}\chi_i \equiv \left(\sum_j^{MO} |\varphi_j\rangle\langle\varphi_j|\right)\chi_i. \tag{8.96}$$

The projection operator is used to create the new orbitals:

$$\varphi_i' = \sum_j^{MO} \langle\varphi_j|\chi_i\rangle\varphi_j. \tag{8.97}$$

The new orbitals φ_i', as linearly independent combinations of the occupied canonical orbitals φ_j, span the space of the canonical occupied HF orbitals $\{\varphi_j\}$. They are generally non-orthogonal, but we may orthogonalize them by applying the Löwdin orthogonalization procedure (symmetric orthogonalization; see Appendix I available at booksite.elsevier.com/978-0-444-59436-5).

Do the final localized orbitals depend on the starting χ_i in the projection method? The answer[101] is shown in Table 8.3. The influence is small.

Table 8.3. Influence of the initial approximation on the final localized MOs in the projection method of localization (the LCAO coefficients for the CH_3F molecule).

Function χ for the CF Bond				The localized orbital of the CF bond				
2s(C)	2p(C)	2s(F)	2p(F)	2s(C)	2p(C)	2s(F)	2p(F)	1s(H)
0.300	0.536	0.000	−0.615	0.410	0.496	−0.123	−0.654	−0.079
0.285	0.510	0.000	−0.643	0.410	0.496	−0.131	−0.655	−0.079
0.272	0.487	0.000	−0.669	0.410	0.496	−0.138	−0.656	−0.079
0.260	0.464	0.000	−0.692	0.410	0.496	−0.144	−0.656	−0.079
0.237	0.425	0.000	−0.730	0.410	0.496	−0.156	−0.658	−0.079

[98] A. Meunier, B. Levy, and G. Bertier, *Theoret. Chim. Acta, 29,* 49 (1973).
[99] This is the beauty of the projection method.
[100] A linear dependence cannot be allowed. If this happens, then we need to change the set of functions χ_i.
[101] B. Lévy, P. Millié, J. Ridard, and J. Vinh, *J. Electr. Spectr., 4,* 13 (1974).

8.9.3 The Internal Localization Methods

Ruedenberg Method: The Maximum Interaction Energy of the Electrons Occupying a MO

The basic concept of this method was given by Lennard-Jones and Pople[102] and applied by Edmiston and Ruedenberg.[103] It may be easily shown that for a given geometry of the molecule, the functional $\sum_{i,j=1}^{MO} \mathcal{J}_{ij}$ is invariant with respect to *any* unitary transformation of the orbitals:

$$\sum_{i,j=1}^{MO} \mathcal{J}_{ij} = const. \tag{8.98}$$

The proof is very simple and similar to the one on p. 406, where we derived the invariance of the Coulombic and exchange operators in the Hartree–Fock method.

This further implies that

> *maximization* of $\sum_{i=1}^{MO} \mathcal{J}_{ii}$ means making MOs small (the Ruedenberg localization criterion)

is at the same time equivalent to the *minimization* of the off-diagonal elements:

$$\sum_{i<j}^{MO} \mathcal{J}_{ij}. \tag{8.99}$$

This means that to localize the molecular orbitals, we try to put them as far apart as possible in space, because then their repulsion will be least.

Similarly, we can prove another invariance:

$$\sum_{i,j=1}^{MO} \mathcal{K}_{ij} = const'. \tag{8.100}$$

It may be also expressed in another way, given that $\sum_{i,j}^{MO} \mathcal{K}_{ij} = const' = \sum_{i}^{MO} \mathcal{K}_{ii} + 2\sum_{i<j}^{MO} \mathcal{K}_{ij} = \sum_{i}^{MO} \mathcal{J}_{ii} + 2\sum_{i<,j}^{MO} \mathcal{K}_{ij}$. Since we maximize the $\sum_{i}^{MO} \mathcal{J}_{ii}$, then simultaneously,

> *we minimize* the sum of the exchange contributions:
>
> $$\sum_{i<j}^{MO} \mathcal{K}_{ij}. \tag{8.101}$$

[102] J.E. Lennard-Jones and J.A. Pople, *Proc. Roy. Soc. (London), A202*, 166 (1950).

[103] C. Edmiston and K. Ruedenberg, *Rev. Modern Phys., 34*, 457 (1962).

Boys Method: The Minimum Distance Between Electrons Occupying an MO

In this method,[104] we search such a unitary transformation of the orbitals which minimizes the functional[105]

$$\sum_i^{MO} \left(ii\left|r_{12}^2\right|ii\right), \tag{8.102}$$

where the symbol $\left(ii\left|r_{12}^2\right|ii\right)$ denotes an integral similar to $\mathcal{J}_{ii} = (ii|ii)$, but instead of the $1/r_{12}$ operator, we have r_{12}^2. Functional Eq. (8.102) is invariant with respect to any unitary transformation of the molecular orbitals.[106] Since the integral $\left(ii\left|r_{12}^2\right|ii\right)$ represents the definition of the mean square of the distance between two electrons described by $\varphi_i(1)\varphi_i(2)$, the Boys criterion means that we try to obtain the localized orbitals as small as possible (small orbital *dimensions*); i.e., localized in some small volume in space.[107] The detailed technique of localization will be given in a moment.

8.9.4 Examples of Localization

Despite the freedom of the localization criterion choice, the results are usually similar. The orbitals of the CC and CH bonds in ethane, obtained by various approaches, are shown in Fig. 8.23 and Table 8.4.

Let us discuss the table some more. First, note the similarity of the results of various localization methods. The methods are different and the starting points are different, and yet we get almost the same result in the end. It is both striking and important that

> the results of various localizations are similar to one another, and in practical terms (not theoretically), we can speak of the unique definition of a chemical bond in a polyatomic molecule.

---

[104] S.F. Boys, in *Quantum Theory of Atoms, Molecules, and the Solid State* , P.O. Löwdin, ed., Acad. Press, New York (1966), p. 253.

[105] Minimization of the interelectronic distance (Boys method) is in fact similar in concept to the maximization of the Coulombic interaction of two electrons in the same orbital (Ruedenberg method).

[106] We need to represent the orbitals as components of a vector, the double sum as two scalar products of such vectors, then transform the orbitals, and show that the matrix transformation in the integrand results in a unit matrix.

[107] The integrals [Eq. (8.102)] are trivial. Indeed, using the Pythagorean theorem, we finally get simple one-electron integrals of the following type:

$$\begin{aligned}\left(i(1)i(2)|(x_2-x_1)^2|i(1)i(2)\right) &= \left(i(2)\left|x_2^2\right|i(2)\right) + \left(i(1)\left|x_1^2\right|i(1)\right)\\ &\quad -2\left(i(1)|x_1|i(1)\right)\left(i(2)|x_2|i(2)\right)\\ &= 2\left(i|x^2|i\right) - 2\left(i|x|i\right)^2.\end{aligned}$$

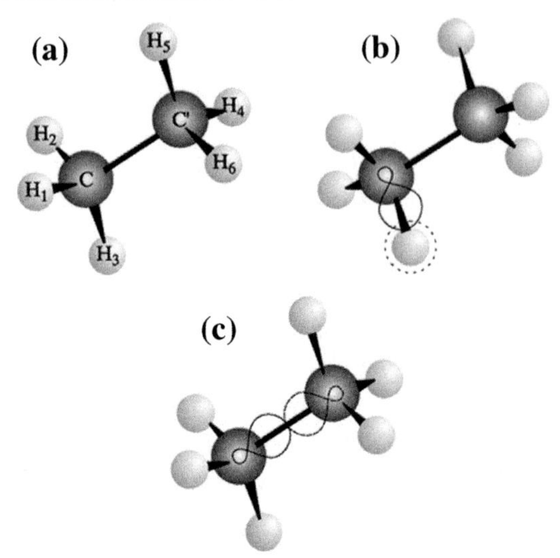

Fig. 8.23. The ethane molecule in the antiperiplanar configuration (a). The localized orbital of the CH bond (b) and the localized orbital of the CC bond (c). The carbon atom hybrid forming the CH bond is quite similar to the hybrid forming the CC bond.

Table 8.4. Localized orbitals for the ethane molecule[a] calculated by using three different localization methods.[b] The LCAO coefficients are shown for the non-equivalent atomic orbitals only. The z-axis is along the CC' bond.

	The Projection Method	Minimum Distance Method	Maximum Repulsion Method
		CC′ bond	
$1s$(C)	−0.0494	−0.1010	−0.0476
$2s$(C)	0.3446	0.3520	0.3505
$2p_z$(C)	0.4797	0.4752	0.4750
$1s$(H)	−0.0759	−0.0727	−0.0735
		CH bond	
$1s$(C)	−0.0513	−0.1024	−0.0485
$2s$(C)	0.3397	0.3373	0.3371
$2p_z$(C)	−0.1676	−0.1714	−0.1709
$2p_x$(C)	0.4715	0.4715	0.4715
$1s$(C′)	0.0073	0.0081	0.0044
$2s$(C′)	−0.0521	−0.0544	−0.054
$2p_z$(C′)	−0.0472	−0.0503	−0.0507
$2p_x$(C′)	−0.0082	−0.0082	−0.0082
$1s$(H1)	0.5383	0.5395	0.5387
$1s$(H2)	−0.0942	−0.0930	−0.0938
$1s$(H3)	−0.0942	−0.0930	−0.0938
$1s$(H4)	0.0580	0.0584	0.0586
$1s$(H5)	−0.0340	−0.0336	−0.0344
$1s$(H6)	−0.0340	−0.0336	−0.0344

[a] In its antiperiplanar conformation

[b] P. Millié, B. Lévy, and G. Berthier, in: *Localization and Delocalization in Quantum Chemistry*, ed. O. Chalvet, R. Daudel, S. Diner, and J.P. Malrieu, Reidel Publish. Co., Dordrecht (1975)

Nobody would reject the statement that a human body is composed of the head, the hands, the legs, etc. Yet a purist (i.e., a theoretician) might get into trouble defining a hand (where does it end up?). Therefore, purists would claim that it is impossible to define a hand, and as a consequence, there is no such a thing as a hand–it simply does not exist. This situation is quite similar to the definition of the chemical bonding between two atoms in a polyatomic molecule.

Note that the localized orbitals are concentrated mainly in *one* particular bond between two atoms. For example, in the CC bond orbital, the coefficients at the $1s$ orbitals of the hydrogen atom are small (-0.08); i.e., what really counts belongs to the carbon atoms. Similarly, the $2s$ and $2p$ orbitals of *one* carbon atom and the $1s$ orbital of *one* (the closest) of the hydrogen atoms, dominate the CH bond orbital. Of course, localization is never complete. The oscillating "tails" of the localized orbital may be found even at distant atoms. They assure the mutual orthogonality of the localized orbitals.

8.9.5 Localization in Practice–Computational Technique

Let us take as an example the maximization of the electron interaction within all the orbitals (Ruedenberg method):

$$I = \sum_{i}^{MO} \mathcal{J}_{ii} = \sum_{i}^{MO} (ii|ii). \tag{8.103}$$

Suppose that we want to make an orthogonal transformation (i.e., a rotation in the Hilbert space, see Appendix B available at booksite.elsevier.com/978-0-444-59436-5) of (so far only two) orbitals[108]: $|i\rangle$ and $|j\rangle$, in order to maximize I. The rotation (an orthogonal transformation, which preserves the orthonormality of the orbitals) can be written as

$$\left|i'(\vartheta)\right\rangle = |i\rangle \cos\vartheta + |j\rangle \sin\vartheta,$$
$$\left|j'(\vartheta)\right\rangle = -\,|i\rangle \sin\vartheta + |j\rangle \cos\vartheta,$$

where ϑ is an angle measuring the rotation (we are going to find the optimum angle ϑ). The contribution from the changed orbitals to I, is

$$I(\vartheta) = (i'i'|i'i') + (j'j'|j'j'). \tag{8.104}$$

Then[109]

$$I(\vartheta) = I(0)\Big(1-\frac{1}{2}\sin^2 2\vartheta\Big)+\big(2(ii|jj)+(ij|ij)\big)\sin^2 2\vartheta+\big((ii|ij)-(jj|ij)\big)\sin 4\vartheta, \tag{8.105}$$

where $I(0) = \big(ii|ii\big) + \big(jj|jj\big)$ is the contribution of the orbitals before their rotation.

[108] The procedure is an iterative one. First, we rotate one pair of orbitals, then we choose another pair and make another rotation, etc., until the next rotations do not introduce anything new.

[109] Derivation of this formula is simple, taking only one page.

Requesting that $\frac{dI(\vartheta)}{d\vartheta} = 0$, we easily get the condition for optimum $\vartheta = \vartheta_{opt}$:

$$-2I(0)\sin 2\vartheta_{opt}\cos 2\vartheta_{opt} + \big(2(ii|jj) + (ij|ij)\big)4\sin 2\vartheta_{opt}\cos 2\vartheta_{opt}$$
$$+\big((ii|ij) - (jj|ij)\big)4\cos 4\vartheta_{opt} = 0, \tag{8.106}$$

and hence

$$\mathrm{tg}(4\vartheta_{opt}) = 2\frac{(ij|jj) - (ii|ij)}{2(ii|jj) + (ij|ij) - \frac{1}{2}I(0)}. \tag{8.107}$$

The operation described here needs to be performed for all pairs of orbitals, and then repeated (iterations) until the numerator vanishes for each pair; i.e.,

$$(ij|jj) - (ii|ij) = 0. \tag{8.108}$$

Thus, the value of the numerator for each pair of orbitals is the criterion for whether a rotation is necessary for this pair or not. The matrix of the full orthogonal transformation represents the product of the matrices of these successive rotations.

The same technique of successive 2×2 rotations applies to other localization criteria.

8.9.6 The Chemical Bonds of σ, π, δ Symmetry

Localization of the MOs leads to the orbitals corresponding to chemical bonds (as well as lone pairs and inner shells). In the case of a bond orbital, a given localized MO is in practice dominated by the AOs of *two atoms only*–those that create the bond.[110] According to the discussion on p. 430, the larger the overlap integral of the AOs, the stronger the bonding. The energy of a molecule is most effectively decreased if the AOs are oriented in such a way as to maximize their overlap integral (see Fig. 8.24). We will now analyze the kind and the mutual orientation of these AOs.

As shown in Fig. 8.25, the orbitals σ, π, δ (either canonical or not) have the following features:

- The σ- type orbital *has no nodal plane containing the bond axis.*
- The π- type orbital has *one such nodal plane.*
- The δ- type orbital has *two such nodal planes.*

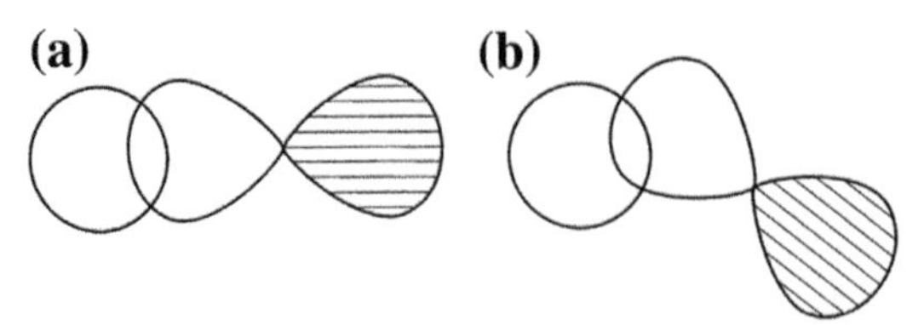

Fig. 8.24. Maximization of the AO overlap requests position (a), while position (b) is less preferred.

[110] That is, they have the largest absolute values of LCAO coefficients.

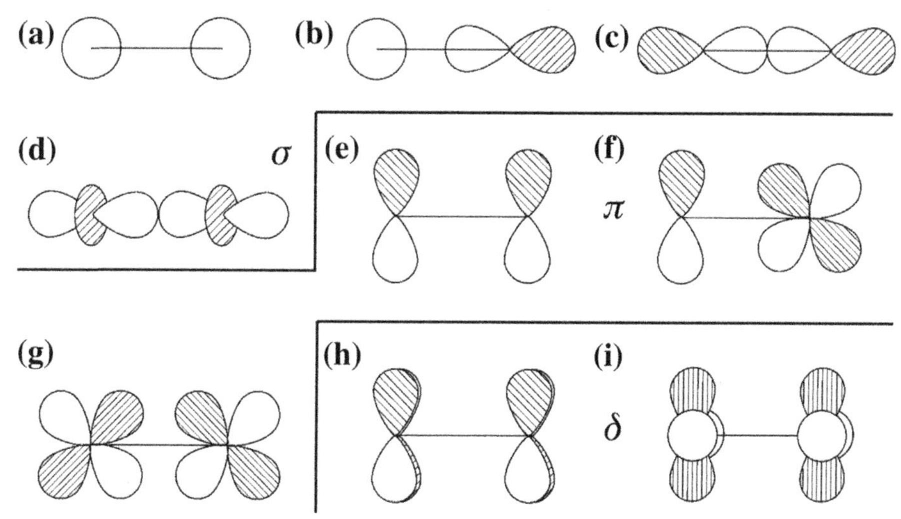

Fig. 8.25. Symmetry of the MOs results from the mutual arrangement of those AOs of both atoms, which have the largest LCAO coefficients. Panels (a) through (d) show the σ type bonds, panels (e) through (g) show the π type bonds, and panels (h) and (i) show the δ type bonds. The σ bond orbitals have no nodal plane (containing the nuclei), the π orbitals have one such plane, and the δ ones have two such planes.

If the z-axis is set as the bond axis, and the x-axis is set in the plane of the figure, then the cases (b-i) correspond (compare Chapter 4) to the overlap of the following AOs: (b) s with p_z, (c) p_z with p_z, (d) $3d_{3z^2-r^2}$ with $3d_{3z^2-r^2}$, (e) p_x with p_x, (f) p_x with $3d_{xz}$, (g) $3d_{xz}$ with $3d_{xz}$, (h) $3d_{xy}$ with $3d_{xy}$, and (i) $3d_{x^2-y^2}$ with $3d_{x^2-y^2}$. The images show the atomic orbitals that correspond to the *bonding* MOs. To get the corresponding *antibonding* MOs, we need to change the sign of *one* of the two AOs.

If a MO is antibonding, then a little star (tradition) is added to its symbol (e.g., σ^*, π^*, etc.). Usually, we also give the orbital quantum number (in order of increasing energy); e.g., 1σ, 2σ, ... etc. For homonuclear diatomics, additional notation is used (Fig. 8.26) showing the main atomic orbitals participating in the MO; e.g., $\sigma 1s = 1s_a + 1s_b$, $\sigma^* 1s = 1s_a - 1s_b$, $\sigma 2s = 2s_a + 2s_b$, $\sigma^* 2s = 2s_a - 2s_b$, etc.

The very fact that the π and δ molecular orbitals have a value of zero at the positions of the nuclei (the region most important for lowering the potential energy of electrons) suggests that they are bound to have a higher energy than the σ ones, and they do.

8.9.7 Electron Pair Dimensions and the Foundations of Chemistry

What are the dimensions of the electron pairs described by the localized MOs, and how do you define such dimensions? All orbitals extend to infinity, so you cannot measure them using a ruler, but some may be more diffuse than others. It also depends on the molecule itself, the role of a given MO in the molecular electronic structure (the bonding orbital, lone electron pair, or the inner shell), the influence of neighboring atoms, etc. These are fascinating problems, and the issue is at the heart of structural studies of chemistry.

Several concepts may be given to calculate the abovementioned dimensions of the molecular orbitals. For instance, we may take the integrals $\left(ii\left|r_{12}^2\right|ii\right) \equiv \left\langle r^2\right\rangle$ calculated within the Boys

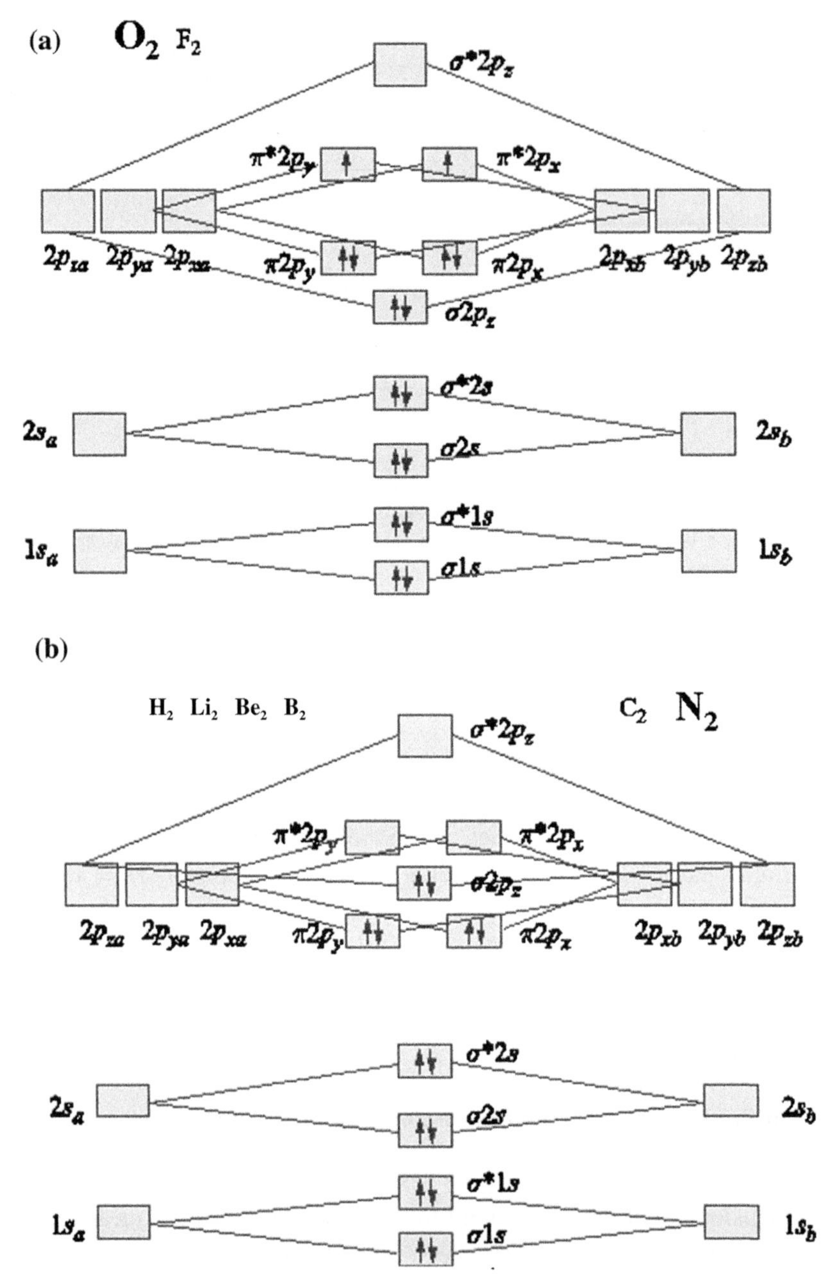

Fig. 8.26. Scheme of the bonding and antibonding MOs in homonuclear diatomics from H_2 through F_2. This scheme is better understood when you recall the rules of effective mixing of AOs discussed on p. 429. All the orbital energies become lower in this series (due to increasing of the nuclear charge), but lowering of the bonding π orbitals leads to changing the *order* of the orbital energies, when going from N_2 to F_2. This is why we get two sequences of orbital energies (schematically) for the molecules (b) from H_2 through N_2 (electron configuration is shown for N_2) and (a) for O_2 and F_2 (electron configuration is shown for O_2).

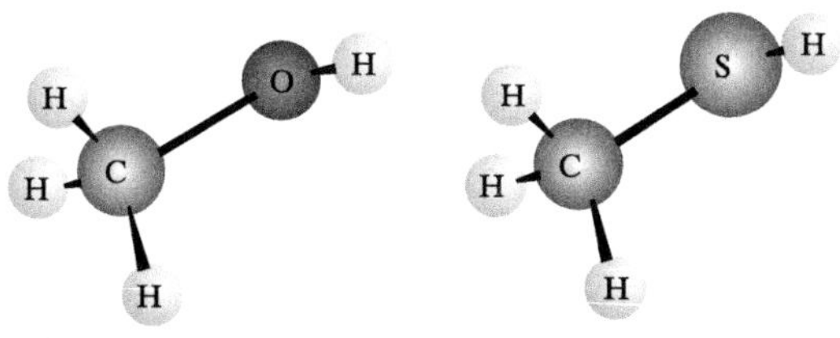

Fig. 8.27. Methanol (CH_3OH) and mercaptan (CH_3SH).

Table 8.5. The Dimensions of the electron pairs; i.e., $\sqrt{\langle r^2\rangle}$ (a.u.) for CH_3OH and CH_3SH according to Csizmadia.[a] "*Core*" means the 1*s* orbital of the atom indicated.

CH_3OH		CH_3SH	
core O	0.270	core S (1*s*)	0.148
core C	0.353	core C	0.357
		S (L shell)	0.489
			0.489
			0.483
			0.484
CO	1.499	CS	2.031
CH_1[b]	1.576	CH_1[b]	1.606
$CH_{2,3}$[b]	1.576	$CH_{2,3}$[b]	1.589
OH	1.353	SH	1.847
lone $pair_{1,2}$[c]	1.293	lone $pair_{1,2}$[c]	1.886

[a] I.G. Csizmadia, in *Localization and Delocalization in Quantum Chemistry*, ed. by O. Chalvet and R. Daudel, D. Reidel Publ. Co., Dordrecht (1975).
[b] Different electron pair dimensions originate from their different positions vs the OH or SH group.
[c] There are two lone pairs in the molecule.

localization procedure, and use them to estimate the square of the dimension of the (normalized) molecular orbital φ_i. Indeed, $\langle r^2\rangle$ is the mean value of the interelectronic distance for a two-electron state $\varphi_i(1)\varphi_i(2)$, and $\rho_i(\text{Boys}) = \sqrt{\langle r^2\rangle}$ may be viewed as a measure of the φ_i orbital dimension. Alternatively, we may do a similar thing using the Ruedenberg method, by noting that the Coulombic integral $\mathcal{J}_{ii}$, calculated in atomic units, is nothing more than the mean value of the *inverse of the distance* between two electrons described by the φ_i orbital. In this case, the dimension of the φ_i orbital may be proposed as $\rho_i(\text{Ruedenberg}) = \frac{1}{\mathcal{J}_{ii}}$. Below, the calculations are reported, in which the concept of $\rho_i(\text{Boys})$ is used. Let us compare the results for CH_3OH and CH_3SH (Fig. 8.27) in order to see what makes these two molecules so different.[111]

Interesting features of both molecules can be deduced from Table 8.5. The most fundamental aspect of the molecular structure is whether, speaking formally, the same chemical bonds (say, the CH ones) are indeed similar for both molecules. A purist approach says that each molecule

[111] Only those who have carried out experiments in person with mercaptan, or who have had neighbors (even distant ones) involved in such experiments, understand how important the difference between the OH and SH bonds really is. In view of the theoretical results reported, I am sure they also appreciate the blessing of theoretical work. According to the *Guinness Book of Records*, CH_3SH (mercaptan) is the smelliest substance in the Universe.

is a new world, and thus, these are two different bonds by definition. Yet chemical intuition says that some *local* interactions (in the vicinity of a given bond) should mainly influence the bonding. If such local interactions are similar, the bonds should turn out similar as well. Of course, the purist approach is formally right, but the local interactions turn out to be rather small. If chemists desperately clung to purist theory, they would know some 0.01% or so of what they now know about molecules. *It is of fundamental importance for chemistry that we do not study particular cases, case by case, but derive general rules.* Strictly speaking, these rules are false from the very beginning, for they are valid only to a certain extent, but they enable chemists to understand, to operate, and to be efficient. Otherwise, there would be no chemistry at all.

The periodicity of chemical elements discovered by Mendeleev is another great idea in chemistry. It has its source in the shell structure of atoms. This means, for example, that the compounds of sulfur with hydrogen should be similar to the compounds of oxygen with hydrogen, because sulfur and oxygen have analogous electronic configuration of the valence electrons (i.e., those of the highest energies), and they differ mainly in the inner shells (O: $[He]2s^2 2p^4$ as compared to S: $[Ne]3s^2 3p^4$).

Looking at Table 8.5, you can see the following:

1. The dimension of the electron lone pair localized on the $1s$ orbital of the sulfur atom is twice as small as the dimension of a similar pair of the $1s$ orbital of the oxygen atom. This is nothing special. The electrons occupying the $1s$ orbital of S experience a strong electric field of the nucleus charged +16, while the charge of the O nucleus is only +8. Let us note that the core of the carbon atom is even larger, because it is controlled by an even less charged nucleus[112] (+6).
2. The dimension of the electron pair of the $1s$ orbital of the carbon atom (core C) for CH_3OH is very similar to that of the corresponding orbital for CH_3SH (0.353 vs 0.357).[113] This means that the influence of the S atom (as compared to the oxygen atom) on the $1s$ orbital of the *neighboring atom* is small. The *local character of the interactions* is thus the most decisive.
3. The influence of the S and O atoms on the CH bonds of the methyl group is only slightly larger. For example, in CH_3OH, one of the CH bond localized orbitals has the dimension of 1.576 a.u., while in CH_3SH, the dimension is 1.606 a.u.
4. The three CH bonds in methanol are very similar to each other (the numbers in Table 8.5 are identical), yet only two of them are strictly equivalent due to symmetry. It is even more interesting that the CH bonds in CH_3SH are also similar to them, although the differences

[112] Let us check what is given for the Slater orbitals of the C,O,S atoms. The nuclear charge has to be diminished by 0.3, so these three cases should be $Z - \sigma = 5.70, 7.70, 15.70$, respectively. The mean value of the nucleus-electron distance can be easily computed as $\sqrt{\langle 1s|r^2|1s\rangle} = \sqrt{\frac{Z^3}{\pi}\int_0^\infty r^4 \exp(-2Zr)dr \int_0^\pi \sin\theta d\theta \int_0^{2\pi} d\phi} = \sqrt{\frac{Z^3}{\pi}\cdot 4!\,(2Z)^{-5}\cdot 2\cdot 2\pi} = \frac{\sqrt{3}}{Z}$. Our results are 0.30, 0.22, 0.11, whereas the reported, more accurate data are 0.35, 0.27, 0.15 a.u., respectively.

[113] Even these small changes may be detected experimentally by removal of electrons from the molecules by monochromatic X-ray radiation and subsequent measurement of the kinetic energy of the removed electrons. Those that were more strongly bound run slower.

between the various CH bonds of mercaptan, and between the corresponding CH bonds in methanol and mercaptan, are clearer. So, *even despite the different atomic environment, the chemical bond preserves its principal and individual features.*

5. This may sound like a banality, but it bears stating anyway: CH_3SH differs from CH_3OH in that the O atom is replaced by the S atom. No wonder then that large differences in the close vicinity of the O and S atoms are easily noticeable. The dimensions of the electron pairs at the S atom (lone pairs and the SH and CS bonds) are always larger than the corresponding pair at the O atom. The differences are at the 30% level. The sulfur atom is simply larger than the oxygen atom, indicating that the electrons are more loosely bound when we go down within a given group in the periodic table.

These conclusions are instructive and strongly encouraging, because *we see a locality in chemistry, and therefore chemistry is easier than it might be* (e.g., CH bonds have similar properties in two different molecules). On the other hand, we may play a subtle game with local differences on purpose by making suitable chemical substitutions. In this way, we have the possibility of *tuning the chemical and physical properties of materials, which is of prime importance in practical applications.*

8.9.8 Hybridization or Mixing One-Center AOs

The localized orbitals may serve to illustrate the idea of a *hybrid atomic orbital.* A given localized orbital φ of a bond represents a linear combination of the atomic orbitals of *mainly two atoms*–the partners that form the chemical bond (say a and b). If so, then (for each localized bond orbital), all the atomic orbitals of atom a may be added together with their specific LCAO coefficients,[114] and the same can be done for atom b. These two sums represent two *normalized hybrid atomic orbitals* χ_a and χ_b multiplied by the resulting coefficients c_a and c_b and together form the approximate[115] bond orbital:

$$\varphi \approx c_a h_a + c_b h_b,$$

with the corresponding LCAO expansions:

$$h_a = \sum_{j \in a} c_{ji} \chi_j,$$

$$h_b = \sum_{j \in b} c_{ji} \chi_j.$$

[114] That serve to express the localized orbital through the atomic basis set.

[115] The "*tails*" of the localized orbital (i.e., its amplitudes on other atoms) have been neglected.

Such a definition of the hybrid orbitals is not unique, since the localized orbitals used are also not unique. However, as shown above, this ambiguity is of secondary importance. The advantages of such an approach to hybridization are as follows:

- It can be determined for any configuration of the nuclei; e.g., for the tetrahedral as well as for any other configuration of CH_4, etc.
- The definition is applicable at any LCAO basis set used.
- It gives a clear message that *all* the atoms in a molecule are hybridized (why not?); e.g., the carbon atom in the methane molecule, as well as all the hydrogen atoms. The only difference between these two hybridizations is that the χ_a for the carbon atom does not resemble any of the χ_j in $\sum_{j\in a} c_{ji}\chi_j$ (because of comparable values[116] of $|c_{ji}|$ meaning an effective mixing of the atomic orbitals), while the χ_b for the hydrogen atom is *dominated by a single atomic orbital* $1s_b$, what should be understood as a lack of hybridization.[117]

How will the hybridization in the optimized geometry of methane look? Well, among five doubly occupied localized molecular orbitals, four[118] protrude from the carbon nucleus toward one of the hydrogens (four hydrogens form a regular tetrahedron) and will have only some marginal amplitudes on the three other hydrogens. If we neglect these "tails" on the other atoms and the contributions of the atomic orbitals other than $2s$ and $2p$ (i.e., their c_{ji}) of the carbon atom (also eliminating from the MO the $1s$ orbital of the partner hydrogen atom), we obtain the following normalized hybrid carbon orbitals:

$$h_i = \frac{1}{\sqrt{1+\lambda_i^2}}\left[(2s) + \lambda_i(2p_i)\right],$$

for $i = 1, 2, 3, 4$ denoting the four directions of p_i and therefore of the h_i. If we force the four hybrids to be equivalent, then this means $\lambda_i = \lambda$. Forcing the hybrids to be mutually orthogonal,[119]

$$\langle h_i|h_j\rangle = \frac{1}{1+\lambda^2}\left[1+\lambda^2\langle 2p_i|2p_j\rangle\right] = \frac{1}{1+\lambda^2}\left[1+\lambda^2\cos\theta_{ij}\right] = 0,$$

we obtain as the $2s$ and $2p$ mixing ratio,

$$\lambda = \sqrt{\frac{-1}{\cos\theta_{ij}}}. \tag{8.109}$$

---

116 These correspond mainly to $2s_a$ and $2p_a$, which have the highest values of the LCAO coefficients.

117 The reason why the carbon atom (and some other atoms such as N, O, etc.) is effectively hybridized, while the hydrogen atom not, is that the $2s$ and $2p$ orbital energy levels in those atoms are close in the energy scale, while the energy difference between the $1s$ hydrogen orbital energy and higher-energy hydrogen orbitals is larger.

118 The fifth will be composed mainly of the $1s$ carbon orbital.

119 "Orthogonal" also means "absolutely independent."

sp^3 Hybridization (Tetrahedral)

Since, for the tetrahedral configuration $\theta_{ij} = 109^0 28'$, from Eq. (8.109), $\cos 109^0 28' = -\frac{1}{3}$ and, therefore, $\lambda_i = \sqrt{3}$. Therefore, the orthogonal hybrids on the carbon atom (Figs. 8.28 and 8.29a) read as

$$h_i \left(sp^3\right) = \frac{1}{2}\left[(2s) + \sqrt{3}(2p_i)\right],$$

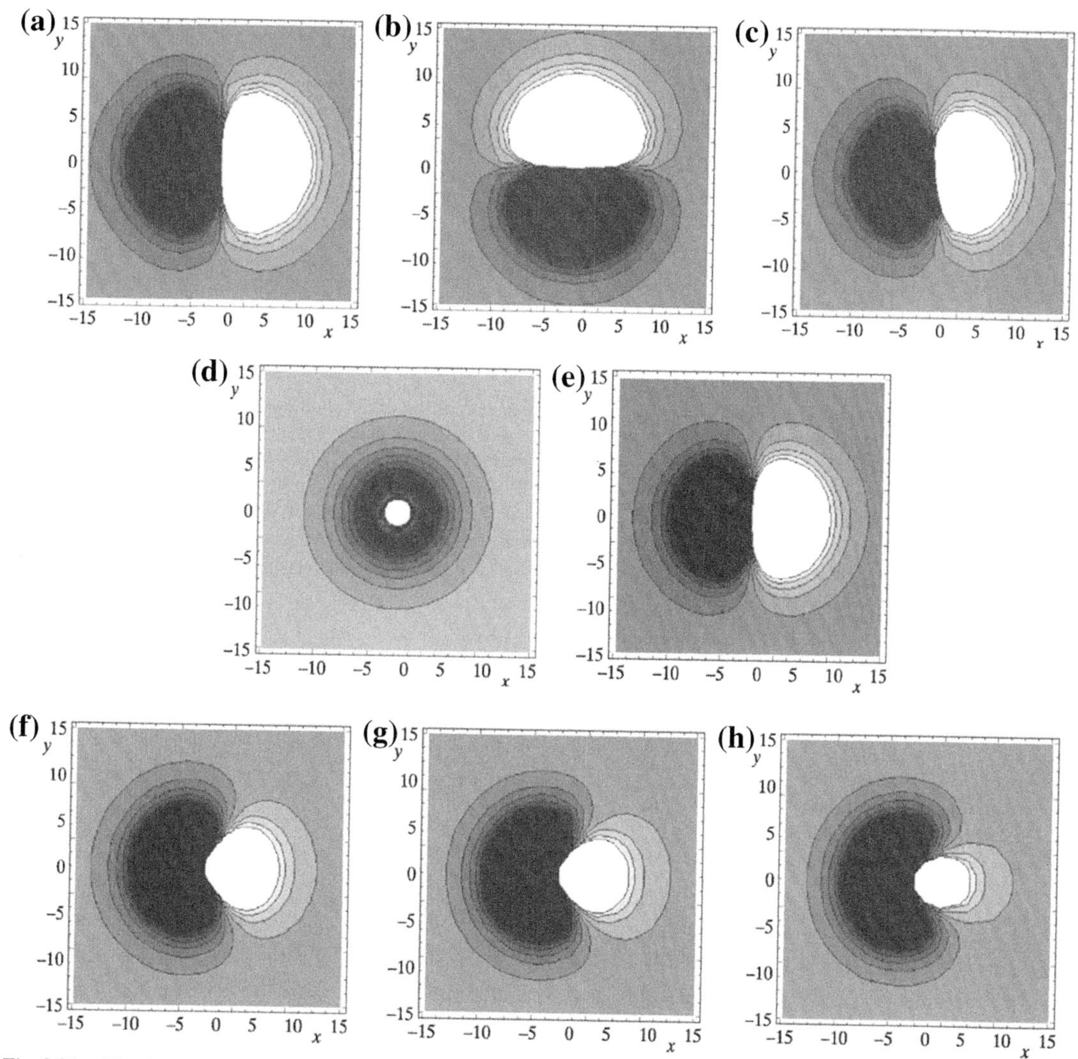

Fig. 8.28. The Slater-type orbitals shown as contours of the section at $z = 0$. The background corresponds to the zero value of the orbital, the darker regions to the negative, the brighter to the positive value of the orbital. (a) $2p_x$ and (b) $2p_y$, and their linear combination (c) equal to $\cos 5^0 2p_x + \sin 5^0 2p_y$, which is also a $2p$ orbital, but rotated by 5^0 with respect to the $2p_x$ orbital. In panels (d) and (e), we show the normalized $2s$ and $2p$ orbitals, which will now be mixed in various proportions: (f) the $1:1$ ratio, i.e., the sp hybridization, (g) the $1:\sqrt{2}$ ratio, i.e., the sp^2 hybridization, and (h) the $1:\sqrt{3}$ ratio, i.e., the sp^3 hybridization.

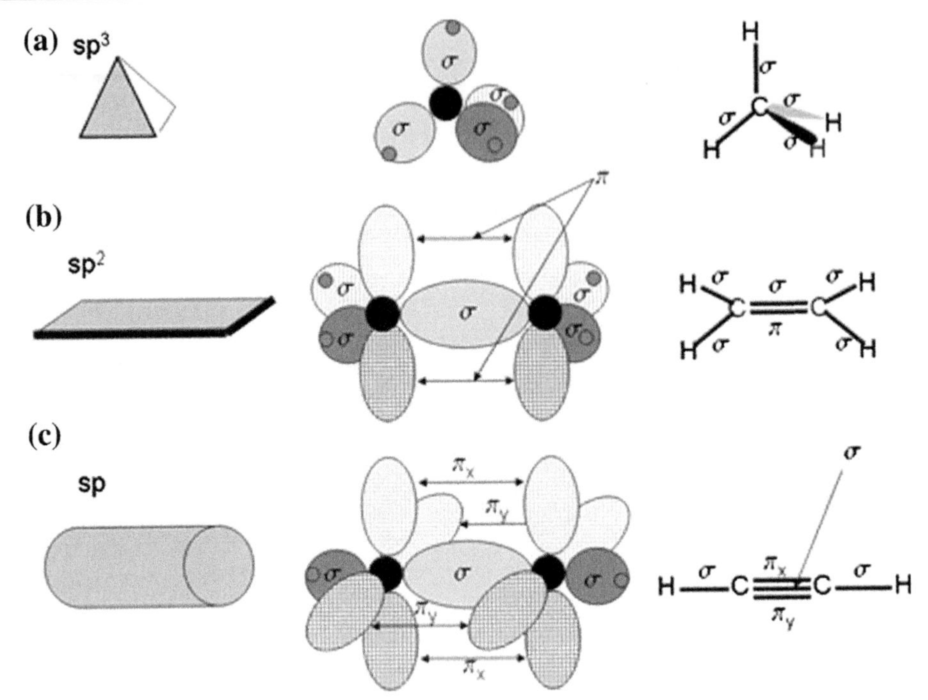

Fig. 8.29. (a) The sp^3 hybridization in the methane molecule in its tetrahedral equilibrium geometry [that corresponds to the minimum of the ground-state electronic energy $E_0^0(\boldsymbol{R})$; see p. 270]. There are four doubly occupied CH localized MOs and one that is essentially the doubly occupied $1s$ carbon atomic orbital. Each of the CH MOs (of the *nearly* cylindrical symmetry) is composed mainly of the carbon hybrid and the hydrogen $1s$ atomic orbital. The figure shows a scheme of the four carbon hybrids called the *sp^3 hybrids.* (b) An example of the nearly perfect sp^2 hybridization of the carbon atoms in the ethylene (C_2H_4), which is perfectly planar in its ground electronic state (D_{2h} symmetry). Such a hybridization is only approximate because the CCH angle has to differ from the HCH angle, both slightly deviate from 120^0. The localized MOs are the following (occupied by altogether 16 electrons):

— Two essentially $1s$ carbon orbitals
— Four CH orbitals and one CC orbital having the *nearly* cylindrical symmetry (i.e., σ type)
— One bond orbital being antisymmetric with respect to the reflection in the molecular plane (i.e., of the π symmetry)

(c) An example of the sp hybridization: the acetylene molecule. The Hartree-Fock geometry optimization gives the lowest-energy linear configuration: HCCH. The localization gives seven localized molecular orbitals:

— Two of them are essentially the $1s$ carbon orbitals.
— Two represent the cylindrical CH orbitals (σ).
— One cylindrical CC σ orbital.
— Two CC orbitals that are of π symmetry (perpendicular to each other).

where $2s$ and $2p_i$ are the normalized carbon atomic orbitals,[120] with i denoting the direction of the hybrid, one of the four directions from the carbon atom toward the tetrahedrally located hydrogen atoms, Fig. 8.29a.[121]

[120] Say, the Slater Type Orbitals (STOs), as discussed on p. 423.

[121] Such orientation of the (normalized) $2p_i$ may be achieved by the following choices (just look at the vortices of a cube with the carbon atom at its center and the four directions forming the tetrahedron):

sp^2 Hybridization (Trigonal)

If we tried to find the lowest-energy configuration of ethylene (C_2H_4), it would correspond to a planar structure (Fig. 8.29b) of D_{2h} symmetry. After analyzing the localized molecular orbitals, it would turn out that three hybrids protrude from each carbon nucleus, their directions lying in the molecular plane (say, xy). These hybrids form angles very close to 120^0.

For the trigonal hybridization (i.e., pure sp^2 hybridization, with the $\theta_{ij} = 120^0$ angles), we obtain from Eq. (8.109) $\lambda = \sqrt{2}$, and, therefore, the three orthogonal normalized sp^2 hybrids are

$$h_i\left(sp^2\right) = \frac{1}{\sqrt{3}}\left[(2s) + \sqrt{2}(2p_i)\right],$$

where the directions $i = 1, 2, 3$ form the mercedes logo on a plane.

sp Hybridization (Digonal)

The sp digonal hybridization is said to occur in acetylene: HCCH, which, after optimization of the Hartree-Fock energy, corresponds to the linear symmetric configuration. According to this explanation, each carbon atom exposes two hybrids (Fig. 8.29c): one toward its carbon and one toward its hydrogen partner. These hybrids use the two carbon $2s$ and the two carbon $2p_z$, and together with the two $1s$ orbitals of the hydrogens, form the two HC σ bonds and one CC σ bond. This means that each carbon atom has two electrons left, which occupy its $2p_x$ and $2p_y$ orbitals (perpendicular to the molecular axis). The $2p_x$ orbitals of the two carbon atoms form the doubly occupied π_x bonding localized molecular orbital and the same happens to the $2p_y$ orbitals. In this way, the carbon atoms form the C≡C triple bond composed of one σ and two π (i.e., π_x and π_y) bonds.

The angle between the two equivalent orthonormal hybrids should be $\theta_{ij} = 180^0$, then the mixing ratio will be determined by $\lambda = 1$. Two such hybrids are, therefore,[122] $h_i(sp) = \frac{1}{\sqrt{2}}[(2s) + (2p_i)]$, and making the two opposite directions explicit: $h_1(sp) = \frac{1}{\sqrt{2}}[(2s) + (2p_z)]$ and $h_2(sp) = \frac{1}{\sqrt{2}}[(2s) - (2p_z)]$.

$$2p_1 = \frac{1}{\sqrt{3}}\left(2p_x + 2p_y + 2p_z\right),$$
$$2p_2 = \frac{1}{\sqrt{3}}\left(2p_x - 2p_y - 2p_z\right),$$
$$2p_3 = \frac{1}{\sqrt{3}}\left(-2p_x + 2p_y - 2p_z\right),$$
$$2p_4 = \frac{1}{\sqrt{3}}\left(-2p_x - 2p_y + 2p_z\right).$$

The normalization of the above functions is obvious, since the $2p_x, 2p_y, 2p_z$ are orthogonal.

[122] This cannot be exact (cf., the ethylene case), because the two hybrids must not be equivalent. One corresponds to the CC, and the other to the CH bond.

Is Hybridization of Any Value?

The general chemistry textbook descriptions of hybridization for methane, ethylene, and acetylene usually start from the electronic configuration of the carbon atom: $1s^2 2s^2 2p^2$. Then it is said that, according to valence bond theory (VB; see Chapter 10), this configuration predicts CH_2 as the carbon hydride (bivalent carbon atom) with the CH bonds forming the right angle.[123] This differs very much from the way the methane molecule looks in reality (regular tetrahedron and tetravalent carbon). If the carbon atom were excited (this might happen at the expense of future energy gains, known as "*promotion*") then the configuration might look like $1s^2 2s^1 2p_x^1 2p_y^1 2p_z^1$. The textbooks usually go directly to the mixing of the valence atomic orbitals $2s, 2p_x 2p_y 2p_z$ to form four equivalent sp^3 hybrids, which lead directly to the tetrahedral hydride: the methane. Note, however, that we would draw the conclusion that the $1s^2 2s^1 2p_x^1 2p_y^1 2p_z^1$ configuration leads to four *non*-equivalent CH bonds in the CH_4 hydride.[124] Only equivalent mixing (hybridization) gives the correct picture. When aiming at ethylene or acetylene, the reasoning changes because some orbitals are left without mixing. We assume sp^2 (one orbital left) or sp (two orbitals left) hybridizations, respectively, which leads to the correct compounds and (almost) correct structures. *It looks as if when we know what we are aiming for, we decide what mixes and what does not.* This seems to be a bit unfair.

Example. Water Molecule

Let us carry out the Hartree-Fock calculations for the water molecule.[125] We focus on a subsequent calculation of the localized molecular orbitals and get five doubly occupied molecular orbitals, as shown in Fig. 8.30.

The lowest mean value of the orbital energy[126] corresponds to an orbital that is small and strongly localized on the oxygen nucleus, we identify it as practically $1s$ orbital of oxygen, Fig. 8.30a. Next, we have two bond orbitals, Figs. Fig. 8.30b,c, each of them looking at first sight like a hybrid of the oxygen atom oriented toward a given hydrogen, but it is a bit misleading because it absorbed also the $1s$ orbital of the corresponding hydrogen atom. Then we find something that seems embarrassing, Figs. 8.30(d,e), the last two localized molecular orbitals–the two lone pairs: one of them (d), with its axis within the plane of the molecule, looks as an sp^2 orbital of oxygen, while the second one (e) is orthogonal to the molecular plane and represents nothing but a pure $2p$ orbital of oxygen! Is it simply nonsense? Maybe the computer made a mistake? Well, it may be that the reader studied too literally the popular literature, where always the two lone pairs in the water molecule protruded right from the oxygen as in

[123] Because $2p^2$ means, say, $2p_x^1 2p_y^1$, and these singly occupied atomic orbitals form the two CH bonds with two $1s$ hydrogen orbitals.

[124] Three CH bonds would form right angles (because of $2p_x^1, 2p_y^1, 2p_z^1$), but one CH bond (formed by $2s^1$ together with the corresponding $1s$ hydrogen orbital) would have a quite different character. This contradicts what we get from experiments.

[125] For example, by using the web server Webmo.com (Gaussian program).

[126] The localized orbitals are not the eigenfunctions of the Fock operator.

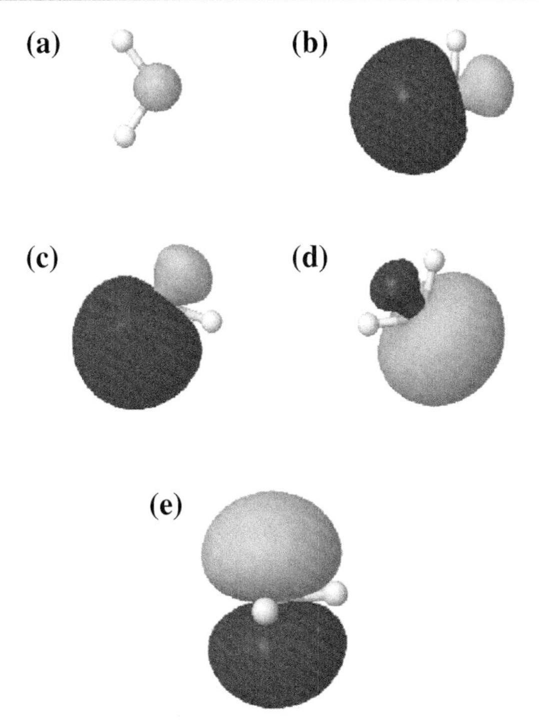

Fig. 8.30. A shock therapy. The localized molecular orbitals of water molecule. (a) An orbital, which is strongly localized on the oxygen nucleus and nearly spherical (practically $1s$ of the oxygen atom). (b) and (c) are the OH bond orbitals, one for each OH bond. (d) The first of the two lone pair orbitals is a hybrid with its axis within the molecular plane. (e) The second of the two lone pair orbitals (orthogonal to the plane of the molecule) is a pure $2p$ orbital of oxygen. It is clear that a sum of the last two orbitals leads to a hybrid, which has to go up off the plane. On the other hand, a subtraction of these orbitals brings a twin hybrid but down from the plane. In this way, we recover (in a fully correct way) the equivalent lone pairs of water molecules seen in textbooks.

a tetrahedral configuration. Everything is all right. If the reader added or subtracted the above two orbitals (which is a perfectly legal operation), he would get the nearly tetrahedrally oriented sp^3 lone pair orbitals known from textbooks. They are no better than the ones obtained in our calculations. Well, maybe they are a bit nicer because they will look similar. Both sets of the orbitals lead to the same total electronic density and the same total energy.

Example. Methane Molecule

Let us check how important the role of hybridization is in the formation of chemical bonds in methane. Let us imagine four scientists performing Hartree-Fock computations for methane in its tetrahedral configuration[127] of nuclei. They use four LCAO basis sets. Professor A believes that in this situation, it is important to remember sp^3 hybridization and uses the following basis set:

A:$1s_{H1}, 1s_{H2}, 1s_{H3}, 1s_{H4}, 1s_C, h_1\left(sp^3\right), h_2\left(sp^3\right), h_3\left(sp^3\right), h_4\left(sp^3\right)$.

Student B did not read anything about hybridization and just uses the common orbitals:

B:$1s_{H1}, 1s_{H2}, 1s_{H3}, 1s_{H4}, 1s_C, 2s_C, 2p_{x,C}, 2p_{y,C}, 2p_{z,C}$.

[127] Or any other one.

Students C and D are not the brightest–they have mixed up the hybridization for methane with that for ethylene and acetylene and used the following basis sets:

C: $1s_{H1}, 1s_{H2}, 1s_{H3}, 1s_{H4}, 1s_C, 2p_{x,C}, h_1\left(sp^2\right), h_2\left(sp^2\right), h_3\left(sp^2\right)$.

D: $1s_{H1}, 1s_{H2}, 1s_{H3}, 1s_{H4}, 1s_C, 2p_{x,C}, 2p_{y,C}, h_1\left(sp\right), h_2\left(sp\right)$.

Which of these scientists will obtain the lowest total energy (i.e., the best approximation to the wave function)?

Well, we could perform these calculations, but it is a waste of time. Indeed, each of the scientists used different basis sets, but they all used *the same space* spanned by the AOs.[128] This is because all these hybrids are linear combinations of the orbitals of student B. All the scientists are bound to obtain the same total energy, the same molecular orbitals,[129] and the same orbital energies.

Hybridization Is Useful Before the Calculations Are Performed

Is hybridization a useless concept, then? No, it is not. It serves as a first indicator (when calculations are not yet performed) of what happens to a local atomic electronic structure, if the atomic configuration is tetrahedral, trigonal, etc. For example, the trigonal hybrids describe the main features of the electronic configuration in the benzene molecule (see Fig. 8.31).

Let us take the slightly more complicated example of a molecule that is of great importance in biology (see Fig. 8.32).

It is important to remember that we always start from some chemical intuition[130] and use the structural formula given in Fig. 8.32a. Most often, we do not even consider other possibilities (isomers), like those shown in Fig. 8.32b. Now, we try to imagine what kind of local electronic structure we have around the particular atoms. Let us start from the methyl (i.e., $-CH_3$) functional groups. Of course, such a group resembles methane, except that one carbon hybrid extends to another atom (not hydrogen). Thus, we expect hybridization over there close to sp^3 one (with all consequences; i.e., angles, etc.). Next, we have the carbon atom that is believed[131] to make the double bond with the oxygen atom. The double bond means

[128] Any of the AOs (and, therefore, also of the MOs) of a person can be expressed as a linear combination of the AOs of another person.

[129] Although the LCAO coefficients will be different, of course, because the expansion functions are different. The orbital plots will be the same.

[130] This is based on the vast experience of chemists.

[131] Here, we rely on the concept of what is known as the *valency* of atoms; i.e., the number of bonds that a given atom is able to form with its neighbors. The valency is equal to the number of valence electrons or valence holes; e.g., the valency of the carbon atom is four (because its electron configuration is $K2s^22p^2$, four valence electrons), of the oxygen atom is two (because its electron configuration is $K2s^22p^4$, two valence holes). An element may have several valencies because of the possible opening of several electronic shells.

Note that we are making several assumptions based on chemical intuition or knowledge. The reason is that we want to go quickly without performing any computations. *This ambiguity disappears, if we make real computations (e.g., using the Hartree-Fock method). Then the chemical bonds, hybrids, and other items are obtained as a result of the computations.*

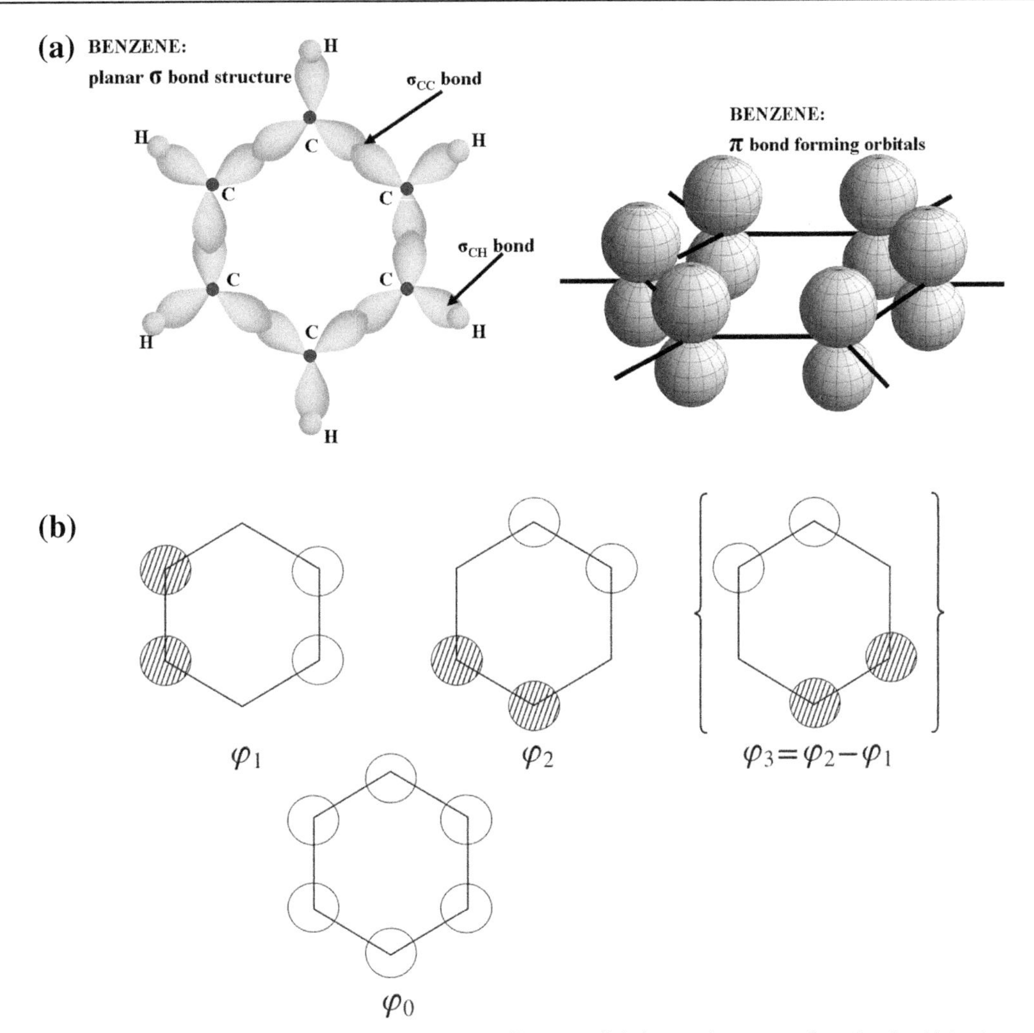

Fig. 8.31. The benzene molecule. The hybridization concept allows us to link the actual geometry of a molecule with its electronic structure (a). The sp^2 hybrids of the six carbon atoms form the six σ CC bonds, and the structure is planar. Each carbon atom thus uses two out of its three sp^2 hybrids; the third one lying in the same plane protrudes toward a hydrogen atom and forms the σ CH bond. In this way, each carbon atom uses its three valence electrons. The fourth one resides on the $2p$ orbital that is perpendicular to the molecular plane. The six $2p$ orbitals form six π molecular orbitals, out of which three are doubly occupied and three are empty (b). The doubly occupied ones are shown in panel (b). The φ_0 of the lowest energy is an all-in-phase linear combination of the $2p$ atomic orbitals (only their upper lobes are shown). The φ_1 and φ_2 correspond to higher energy and to the same energy, and have a single node (apart from the node plane of the AOs). The φ_3 orbital that apparently completes all combinations of single-node molecular orbitals is redundant (that is why it is in parentheses), because the orbital represents a linear combination of the φ_1 and φ_2.

an ethylene-like situation; i.e., both atoms should have hybridizations similar to sp^2. Let us begin from the oxygen atom (Fig. 8.32c). The sp^2 means three hybrids (planar configuration) protruding from the O atom. One of them will certainly bind to a similar one protruding from the carbon atom (OC σ bond); it therefore needs only a single electron from the oxygen. The

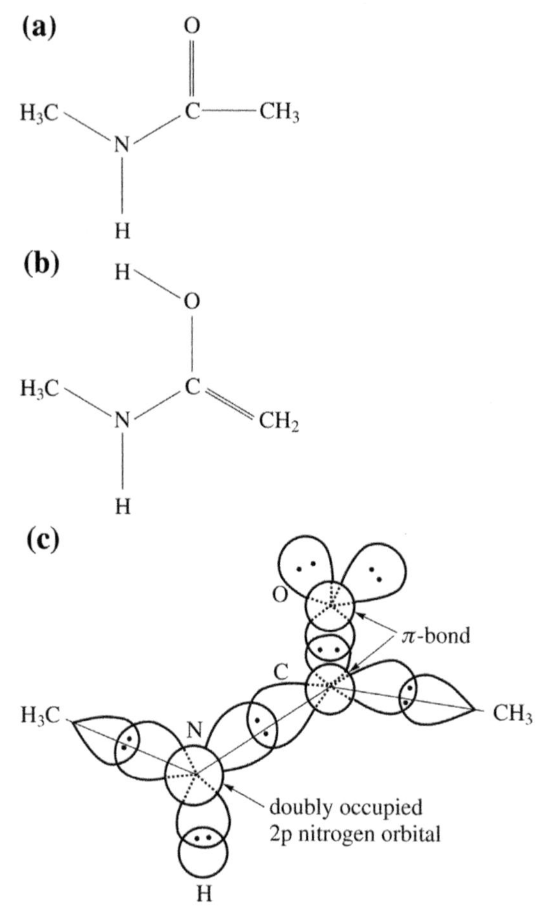

Fig. 8.32. How does the hybridization concept help? This shows the all-important (protein) example of the peptide bond. (a) We assume a certain pattern of the chemical bonds (this choice is knowledge based) ignoring other possibilities, such as the isomers shown in (b). Apart from the methyl groups (they have the familiar tetrahedral configuration), the molecule is planar. Usually in chemistry, knowing the geometry, we make a conjecture pertaining to the hybridization of particular atoms. This leads to the electron count for each atom: the electrons left are supposed to participate in bonds with other atoms. In the example shown, the sp^2 hybridization is assumed for the central carbon and for the nitrogen and oxygen atoms (c). A π bonding interaction of the nitrogen, carbon, and oxygen should therefore stabilize the planarity of the system, which is indeed an experimental fact.

oxygen atom has six valence electrons, so there remain five more to think of. Four of them will occupy the other two hybrids protruding into space (nothing to bind; they are lone pairs). Hence, there is one electron left. This is very good because it will participate in the OC π bond. Let us go to the partner carbon atom. It is supposed to make a double bond with the oxygen. Hence, it is reasonable to ascribe to it an ethylene-like hybridization as well. Out of four valence electrons for carbon, two are already used up by the σ and π CO bonds. Two other sp^2 hybrids remain, which of course accommodate the two electrons and therefore are able to make two σ bonds: one with $-CH_3$ and one with the nitrogen atom. Then we go to the nitrogen atom. It has three substituents in most cases in the (almost) planar configuration (we know this from experiment). To make the analysis simple, we assume an sp^2 ideal hybridization. The nitrogen atom has five valence electrons. Three of them will form the σ NC, NH, and N–CH_3 bonds. Note that although the configuration at N is assumed to be planar, this plane may not coincide with the analogous plane on the carbon atom. Finally, we predict the last two valence electrons of the nitrogen will occupy the $2p$ orbital perpendicular to the plane determined by the substituents of the nitrogen. Note that the $2p$ orbital could overlap (making a bonding effect) with the analogous $2p$ orbital of the carbon atom *provided that the two planes will coincide*. This is why we could expect the planarity of the C–N bond. This bond plays a prominent role in proteins because it is responsible for making the chain of amino acid residues (known as the *amide bond*). It is an experimental fact that *deviations of the amide bond from planarity are very small*.

The value of the analyses as that given above is limited to qualitative predictions. Of course, computations would give us a much more precise picture of the molecule. In such computations, the orbitals would be more precise, or would not be present at all, because, to tell the truth, there is no such thing as orbitals. We badly need to interpret the numbers, to communicate them to others in a understandable way, to say whether we understand these numbers or they are totally unexpected. Reasoning like this has a great value as part of our understanding of chemistry, of speaking about chemistry, and of predicting and discussing the structures. This is why we need hybridization. Moreover, if our calculations were performed within the VB method (in its simplest formulation; the details of the method will be explained in Chapter 10), then the lowest energy would be obtained by Professor A (who assumed the sp^3 hybridization), because the energy gain over there is very much connected to the overlap of the atomic orbitals forming the basis, and the overlap with the $1s$ hydrogen orbitals is the best for the basis set of Professor A. The other people would get high total energies because of poor overlap of their atomic orbitals with the $1s$ hydrogen orbitals.

8.10 A Minimal Model of a Molecule

It is easy to agree that our world is complex. It would be great, however, to understand how it operates. At least sometimes, answers look more and more complex as we go from crude to more and more accurate theories. Therefore, we would like to consider a simpler world (a model of our real world), that:

- Would work with good accuracy; i.e., would resemble the real world quite well.
- Would be based on such simple rules that we can understand it in detail.

We could explain these rules to any interested parties. Not only could we predict a lot for a molecular system, but we ourselves could be confident that we understand *most* of chemistry because it is based on *several simple rules*. Moreover, why worry about details? Most often, we want just to grasp the essence of the problem. On top of that, if this essence were free, only sometimes would we be interested in a more detailed (and expensive) picture.

Is this utopia, or can such a model of chemistry be built?

Well, it seems that theoretical chemistry nowadays offers such a model describing chemical structures.

The model is based on the following basic simplifications of the real world:

- The *non-relativistic approach*; i.e., the speed of light is assumed to be infinite, which leads to the Schrödinger equation (Chapter 2).
- The Born-Oppenheimer approximation (Chapter 6) that separates the motion of the nuclei from the motion of the electrons. This approximation allows us to introduce the concept of a *3-D structure* of the molecule: the heavy nuclear framework of the molecule kept together by "*electronic glue*" moves (translation), and at the same time rotates in space.
- The mean-field approximation of this chapter offers us the *orbital model* of the electronic structure of molecules within the RHF approach. In this picture, the electrons are described by the doubly occupied molecular orbitals. Localization of the orbitals gives the doubly occupied *inner shell, lone pair, and bond* MOs. The first and second are sitting on atoms, and the third on chemical bonds. Not all atoms are bound with all, but instead the *molecule has a pattern of chemical bonds.*
- These bonds are traditionally and formally represented by graphs suggesting a *single* (e.g., C–H); *double* (e.g., C=C), or *triple* (e.g., C ≡ C), although some intermediate situations usually take place. The total number of these formal bonds of a given atom is equal to its valency. The idea of valency helps a lot in selecting the chemical bond pattern, which afterwards may be checked against experiment (e.g., bond distances).[132] In most cases, a single bond is of the σ type, a double one is composed of one σ and one π, and a triple bond means one σ and two π bonds (cf., p. 474).
- The minimal model of a molecule may explain most of the chemical reactions, if besides the closed-shell configuration (double occupancy of the molecular orbitals, including HOMO), we consider excited configurations corresponding to electron transfer(s) from the HOMO to LUMO orbital (see Chapter 14).
- The bonds behave very much like *springs of a certain strength and length*,[133] and therefore, apart from the translational and rotational motion, the atoms vibrate about their equilibrium positions.[134] As to the structural problems (not chemical reactions), these vibrations may be treated as harmonic.

[132] For some molecules, this procedure is not unique; i.e., several chemical bond patterns may be conceived (called sometimes *resonance structures or mesomeric forms;* cf. the valence bond method in Chapter 10). In such cases, the real electronic structure corresponds to an averaging of all of them (in space or in time).

[133] Both depend first of all of the elements making the bond; also, a single bond is the weakest and longest, and the triple is the strongest and shortest.

[134] The model of molecule visualized in virtually all popular computer programs shows spherical atoms and chemical bonds as shining rods connecting them. First of all, atoms are not spherical, as is revealed by Bader analysis (p. 667) or atomic multipole representations (see Appendix S available at booksite.elsevier.com/978-0-444-59436-5). Second, a chemical bond resembles more a "*rope*" (higher values) of electronic density than a cylindrical rod. The rope is not quite straight and is slimmest at a critical point (see p. 671). Moreover, the rope, when cut

- The 3-D shape of simple chemical structures can be correctly predicted using the Hartree-Fock model. The main features of this 3-D structure can be also predicted (without any calculation) by using the concept of the *minimum repulsion energy of the electron pairs*. Within the molecular orbital model, such repulsion is given by Eq. (8.99).

8.11 Valence Shell Electron Pair Repulsion (VSEPR) Algorithm[135]

Is it possible to predict (in a minute, without any quantum chemical calculations) the shape of a molecule corresponding to the lowest energy, or more exactly the configuration of ligands around a central atom? It turns out that often[136] such a goal is feasible, and moreover, the rules (the VSEPR algorithm) behind such predictions are very simple:

- The VSEPR algorithm starts from choosing the central atom (C) in the molecule and defines the rest as a set of its ligands (L).[137]
- The key step of the VSEPR is to calculate an integer number (N) of electrons assigned to the valence shell of the central atom according to the formula: $N = n_C + n_L + n_{ion}$, where n_C stands for the number of the valence electrons of the central atom itself, n_L denotes the number of electrons offered by the ligands L, while n_{ion} is an obvious correction for the number of electrons if one considers an ion, not a neutral molecule.
  - Calculation of n_C. This number is calculated from the position of the central atom in the Mendeleev periodic table (n_C is the group number): for alkali metals, $n_C = 1$; for alkali earth metals, $n_C = 2$; for analogs of boron, $n_C = 3$; for analogs of carbon, $n_C = 4$; for analogs of nitrogen, $n_C = 5$; for analogs of oxygen, $n_C = 6$; for halogens, $n_C = 7$; for noble gases, $n_C = 8$ (except helium, for which $n_C = 2$).
  - Calculation of n_L. The integer n_L is a sum of integer contributions from all individual ligands. Each ligand bound to the central atom by a single bond contributes 1. The ligands O, S, CH_2, NH (i.e., those bound to the central atom by a double bond) contribute 0, and the ligands N, CH (i.e., those bound to the central atom by a triple bond) contribute -1.
  - Calculation of n_{ion}. If the total system is a cation of charge $+n|e|$, the number $n_{ion} = -n$; for an anion of charge $-n|e|$, the number $n_{ion} = +n$.
- One calculates *the number of the valence electronic pairs assigned to the central atom* as[138] $P = \frac{N}{2}$.

---

perpendicularly, has a circular cross section for pure σ bonds, and an oval cross section for the double bond σ and π (cf. Fig. 11.1).

[135] The VSEPR algorithm comes from Ronald Gillespie and Ronald Nyholm [R.J. Gillespie and R.S. Nyholm, Quart. Rev. Chem. Soc., 11, 339 (1957)].

[136] The VSEPR works correctly for small molecules almost without exceptions; for larger molecules, one meets difficulties of two types: there may be problems with heavy atoms and the VSEPR is unable to solve the conformational problems (i.e., to show the lowest-energy conformer).

[137] As the central atom can be chosen any atom of the molecule, one may also repeat the VSEPR procedure for the consecutive choices of the central atom and check whether the outputs are consistent.

[138] If N is an odd number (a rare case), one rounds off the number to $(N + 1)$.

- Distribution of the electron pairs found around the central atom: the pairs are distributed on a surface of a sphere in such a way as to have their repulsion be as small as possible (the largest distances between them).
- Distribution of the ligands around the central atom (i.e., the shape of the molecule): each ligand is assigned to one of the electron pairs. They are called the ligand pairs; the other ones will play the role of the lone electron pairs. If the number of ligands equals P, the structure is determined. If the number of ligands is larger than P, there is an error in our calculations! If the number of the ligands is smaller than P, we have a problem of several structures possible. *There comes the last VSEPR rule: two lone pairs repel stronger than a lone pair–ligand pair and even stronger than the ligand pair–ligand pair.*[139] This sequence leads to a unique VSEPR-predicted structure, all they are listed in Fig. 8.33 for various P and n_L.

Example. Water

Let us first try the VSEPR model. For the oxygen atom as the central one, we have $N = n_C + n_L + n_{ion} = 6+2+0 = 8$. Hence, the number of electronic pairs $P = N/2 = 4$. The largest distances between the pairs will be assured by the tetrahedral configuration (Fig. 8.33; the angles equal $109^0 28'$). Any assignment of the ligands (H) gives, of course, the same bent HOH structure. Recalling the strong repulsion between lone pairs, we can predict the HOH angle to be smaller than $109^0 28'$. This is a good guess because the experimental HOH angle is equal to 104.5^0.

Let us check now what the minimal model gives. The model (STO 6-31G** basis set, optimized geometry) predicts correctly that we are dealing with two *equivalent* OH bonds, because both bonds have the same length[140] $R_{OH} = 0.943$ Å, which is quite close to what can be deduced from the microwave spectroscopy[141]: $R_{OH} = 0.957$ Å. The minimal model used predicts also that the molecule is bent (C_{2v} symmetry), the HOH angle being 106.0^0 (the Hartree-Fock limit; i.e., calculated with the complete AO set is 105.3^0). *These results are quite typical for the minimal model: it is able to predict the bond lengths typically*[142] *to the accuracy of about* 0.01 Å, *and the bond-bond angles with the accuracy of about* 1^0. There is no H–H bond in H_2O at this geometry.[143]

---

[139] Note that at a given geometry, a minimization of the electron pair repulsion given by Eq. (8.99) means localization of the MOs. If, however, a change of geometry is considered (which is the crux of the VSEPR), a smaller electron repulsion [i.e., a smaller *const* in Eq. (8.98)] stabilizes the structure. For small variation of angles L-C-L, one may expect a marginal change of the self-repulsion; i.e., $\sum_i^{MO} \mathcal{J}_{ii} = \sum_i^{MO} (ii|ii)$, because each term depends uniquely on a localized orbital on C or on an individual C-L, not involving any two ligands. The repulsion is expected to vary for *different* localized orbitals because their distance will vary. The larger distance, the smaller the value of Eqs. (8.98) and (8.99). This may be seen as a theoretical hint for validity of the VSEPR.

[140] The VSEPR model does not provide information about interatomic distances.

[141] The microwave spectroscopy determines the moments of inertia.

[142] We mean here molecules with well-closed electronic shells.

[143] This is a common fact of chemistry, but if you are wondering how you know this, then it turns out you have to quote your teachers. As to a more serious argument, this conclusion can be drawn either from an electronic population analysis described in Appendix S available at booksite.elsevier.com/978-0-444-59436-5 or by performing the Bader analysis (as described in Chapter 11). In the first case, we would get the positive (and equivalent) populations between atoms O and H, which will result from a net bonding interaction, while the population between H and H

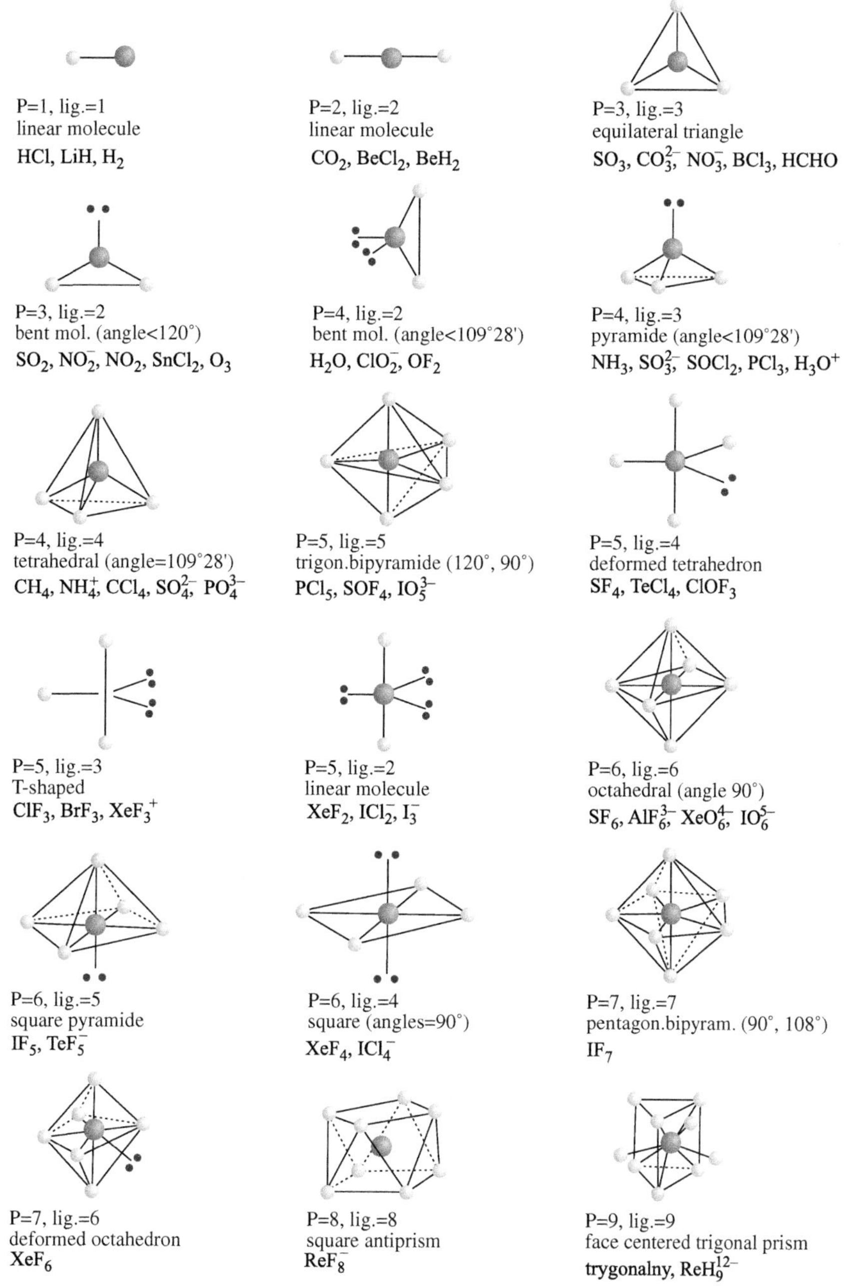

Fig. 8.33. The most stable structures according to VSEPR. The structures are assigned by assuming that the lone pair–lone pair repulsion is larger than the lone pair–bonding pair repulsion, the latter still larger than the bonding pair–bonding pair repulsion.

---

will be negative, which indicates an antibonding interaction. The Bader analysis would reveal a kind of electronic density "*rope*" between O and H and the absence of such a "*rope*" for H and H. No doubt, the H–H bond will appear, if for any reason the HOH angle is forced to be small.

The minimal model predicts three harmonic modes for the water molecule (Fig. 7.9 on p. 361): antisymmetric stretching vibration with the wave number 4264 cm^{-1}, symmetric stretching vibration 4147 cm^{-1}, and the bending vibration 1770 cm^{-1}. It is less easy to say how to compare these numbers to the experimental absorption lines observed in spectra. What an experimentalist measures is related to the vibrational excitations from the ground state $v = 0$ to the excited state with $v = 1$, but of an anharmonic oscillator. One may deduce what the corresponding frequencies would be if the potential energy well were parabolic, they will be equal: 3942, 3832, 1648 cm^{-1}. *This also is typical: the minimal model with its harmonic approximation gives the frequencies that are higher by 7–8% than the experimental values.*[144]

Example. Ozone

This simple molecule is composed of three oxygen atoms (O_3), which at first sight appear to play the same role. The concept of chemical valence (valence equals two) seems to confirm this idea, allowing a structure of an equilateral triangle of single bonds. What does a more serious approach offer? One has to solve the Schrödinger equation (even at the level of the Hartree-Fock method). Before we explore this, let us turn first to what the VSEPR algorithm predicts. We choose one of the oxygen atoms as a "central" one; the two other oxygen atoms will be the ligands. We count it as follows: $N = n_C + n_L + n_{ion} = 6 + 2 \cdot 0 + 0 = 6$. Hence, the number of the electronic pairs is $P = 3$. The repulsion principle (Fig. 8.33) gives the configuration in the form of the mercedes sign (at a 120^0 angle). We assign two pairs to the ligands, and one lone pair is left. Therefore, the VSEPR predicts the O-O-O angle to be a bit smaller than 120^0 (due to a stronger lone pair–ligand pair repulsion), which excludes the three oxygen atoms to be equivalent.

Let us see what the Hartree-Fock method has to say about this.[145] Let us start from an equilateral triangle configuration. We optimize the geometry and get an equilateral angle ($R_{OO} = 1.373$ Å) with the energy −224.245 a.u. Well, it may be that this is a saddle point. We calculate the normal modes, and all three frequencies turn out to be real, so it is a true minimum. But this can't be–the VSEPR must have made a mistake. The VSEPR algorithm is certainly primitive, so it is no wonder that it is sometimes misleading. Well, before we put the VSEPR away, let us perform a Hartree-Fock calculation (with geometry optimization), this time starting from the configuration predicted by the VSEPR. To our amazement, we find another stable configuration: an isosceles triangle (C_{2v} symmetry) with the energy −224.261 a.u.[146], which is *lower* than that computed before. The experiment confirms the VSEPR and our last result: symmetry C_{2v}. The side of the isosceles triangle is $R_{OO} = 1.204$ Å (the experimental

[144] This systematic error of theoretical description is quite often "*corrected*" by a proper scaling of the results. If anharmonicity becomes very large (like in case of vibronic coupling), the error may be much larger.

[145] There are 24 electrons in the system, which is quite a lot for quantum mechanical methods. Meanwhile, the VSEPR algorithm is *insensitive* to the number of electrons.

[146] The computed harmonic vibrational mode frequencies are 1537, 1453 and 849 cm^{-1}, which proves we have to do with a true minimum.

value is[147] 1.272 Å), whereas the largest OO distance is computed as $R_{OO} = 2.076$ Å. The computed angle is 119^0, and the experiment gives 116.78^0, while the VSEPR predicted "*less than* 120^0." A triumph of the simple VSEPR algorithm, even more amazing since obtained in a quite subtle situation of two non-equivalent and competing minima on the hypersurface of the ground-state electronic energy.

Example. I_3^- Anion

Our experience with H_2O and O_3 up to now suggests (hypothesis) that if one has three atoms, one gets the C_{2v} symmetry of the molecule. We may check our hypothesis in the case of I_3^-, the product of dissolving iodine in a solution of potassium iodide.

Well, what does the VSEPR say about the structure? We count (for an iodine atom taken as a central one): $N = n_C + n_L + n_{ion} = 7 + 2 \cdot 1 + 1 = 10$. We get $P = 5$, which means a configuration of a trigonal bipyramide (Fig. 8.33). In the center of the bipyramid, one has the central iodine atom, but where are the ligands? There are three possibilities: two iodine atoms occupy two axial positions (an I-I-I angle equal to 180^0); one of them occupies an axial position, while the other an equatorial position (an I-I-I angle equal to 90^0); or both are equatorial (an I-I-I angle equal to 120^0). Figure 8.33, which summarizes the result of the simple VSEPR assumption about the sequence of the strength of the electron pair repulsion, indicates unambiguously the first possibility. Thus, the VSEPR predicts a linear configuration–exactly what one finds in experiments.

What about this says the minimal quantum mechanical model? Let us optimize the geometry[148] starting from several distinct configurations of the nuclei.[149] We get the energy -20552.059 a.u., always corresponding to a symmetric linear configuration with the I-I distance equal to 2.94 Å. The experimental I-I distance is about 2.90 Å (and changes a bit depending on the accompanying cation in a salt). Therefore, the Hartree-Fock calculations (as well as the VSEPR) gave the structure confirmed by the experiment.

Example. The Infernal[150] ClF_3

The VSEPR gives (we choose chlorine as the central atom): $N = n_C + n_L + n_{ion} = 7 + 3 \cdot 1 + 0 = 10$. So, we have $P = 5$, and this means again a trigonal bipyramide (Fig. 8.33). The three chlorine atoms may be in the following configurations: (1) two axial and one equatorial; (2) one axial and two equatorial; (3) three equatorial. The VSEPR is able to distinguish these three situations (Fig. 8.33) and indicates case #1 has the lowest energy. This, however, means

---

[147] T. Tanaka, Y. Morino, *Spectroscopy, 33,* 538 (1970).

[148] 160 electrons is quite a lot (not for the VSEPR algorithm though). This is the reason why the calculations are carried out within the minimal basis set (STO-3G).

[149] This result is obtained from a starting point corresponding to a linear symmetric structure. A start from a bent structure (the angle about 120^0) gave the same linear configuration. A start from a strongly bent structure (the angle equal to 90^0) led to a dissociation into $I^- + I_2$ (of higher energy).

[150] When in contact with many organic substances, it catches fire spontaneously and has many other vicious surprises.

a very strange molecule: planar and . . . T-shaped. The roof of the T is formed by F–Cl–F lying on the straight line, while there is additional fluorine atom protruding from the chlorine. Does it have any sense at all? Let us optimize the geometry within the Hartree-Fock method.[151] We do get a T-shaped molecule; only the roof deviates from the straight line, but not too much: $\measuredangle F - Cl - F = 172.5^0$ (the experimental value is 175.0^0). The two fluorine atoms (of the roof of T) are equivalent (the net charge computed by the Mulliken population analysis is -0.515, the distance $r_{ClF} = 1.67$ Å, and its experimental value is 1.70 Å). The third fluorine atom differs: its net charge is -0.308, and $r_{ClF} = 1.58$ Å, the experimental value is 1.60 Å. When the optimization starts from configurations #2 or #3, it ends up again in the strange configuration #1, which suggests that these structures (2 and 3) are simply unstable. This underlines once more the success of the VSEPR, for a very unusual molecular shape.

8.12 The Isolobal Analogy

What is the electronic structure of the $Fe_2(CO)_8$ molecule? This is not easy to say with no background. Could there be some multiple bonds between iron atoms of the σ, π, or δ character? Well, the isolobal analogy will allow us to say the following: in fact, $Fe_2(CO)_8$ is an ethylene-like molecule, with all consequences concerning the iron-iron bond (double), the planarity of $(CO)_2Fe{=}Fe(CO)_2$ fragment, etc. Also, we will be able to predict that the $Fe(CO)_4$ moiety resembles ("*is isolobal with*") the CH_2, that it will bind to CH_2, forming the $H_2C{=}Fe(CO)_4$, and can just replace CH_2 in many other instances in organic chemistry.

Now comes a very short story about the isolobal analogy idea.[152]

The concept of localized molecular orbitals not only brought the quantum chemical approach much closer to everyday chemists' reasoning, but also enabled them to see chemistry in a kind of holistic perspective, beyond any useless details.[153] It became evident that to form a chemical structure (e.g., a single chemical bond) it is sufficient to fulfill necessary conditions concerning AOs, like for instance to have two particular atoms each offering a hybridized orbital protruding in space (of similar energy, "*a lobe*") and occupied by a single electron. The details of how the hybridized AO looks are not important. It may have either sp, sp^2, sp^3 orbital or e.g., dsp^3, d^2sp^3, etc.[154] Such outer-most crucial orbitals are known as *frontier orbitals* (they may be identified with the HOMO or LUMO molecular orbitals).

[151] We use the $6-31G(d)$ basis set.

[152] R. Hoffmann, *Nobel Lecture*, December 8, 1981.

[153] Stefan Banach: "*A good mathematician sees analogies among theorems, an outstanding one sees the analogies among theories, and a genius sees the analogies among analogies.*"

[154] To see what such a hybrid is, look at Fig. 4.24 on p. 210. If you add to what you see there (it is a linear combination of $3d_{z^2-x^2}$ and $3d_{z^2-y^2}$) $3s$ and $3p_z$ orbitals, you increase one of the lobes and diminish the other one along z. The result is a lobe protruding toward $+z$.

The isolobal analogy introduced by Roald Hoffmann says that

> the electronic structure (typically of inorganic d-electron complexes) can be rationalized by reducing it to the structures made by much simpler organic radicals like CH_3, CH_2, CH, with *similar* frontier orbitals, and therefore similar chemical properties. More formally, two fragments (or molecules) are said to be isolobal if they have the same number of the frontier orbitals of the same symmetry and electron occupation. So, the isolobal molecular fragments will bind similarly and for similar reasons, while the isolobal molecules will have similar electronic structures.

Let us consider some examples.

In Fig. 8.34, we see several isolobal chemical objects. Let us consider the methane molecule (upper-left corner). By detaching consecutive hydrogen atoms, one moves down the first column and the result are the radicals: CH_3, CH_2, CH *with one, two, and three mono-occupied lobes (all in the tetrahedral configuration)*, respectively. Now, consider the CrL_6 complex of chromium and six ligands L. Each ligand binds to the chromium atom because the chromium offers low-energy empty orbitals,[155] while ligands (like CO, NH_3, etc.) provide an electron pair each. The CrL_6 complex has, therefore, the electronic configuration $4s^2 3d^4 L^{12}$ (i.e., 18 valence electrons), which means the very stable Lewis-type closed shell of the krypton atom, $[Ar(18)]4s^2 3d^{10} 4p^6$, see p. 448. The next column begins by $Mn(CO)_3Cp$, where Cp means the cyclopentadienyl ligand offering 5 electrons of the π type. The electron count gives $4s^2 3d^5 (CO)^6\ Cp^5$, again the stable system of 18 valence electrons.

Now we will go down the columns (except the first column already discussed), beginning with CrL_6. Going toward the second row, we remove L with *two* electrons, but at the same time replace the central atom by the next one in the periodic table (i.e., by manganese). Therefore, the extra electron from the manganese atom (when compared to chromium) enters the lobe, which is thus singly occupied (see Fig. 8.34). The electron configuration is $4s^2 3d^4 (\text{lobe})^1 L^{10}$, which means 17 electrons or one electron hole in the 18-electron shell. We do the same in the third column (with the tetrahedral complex containing Cp). These compounds are analogous (isolobal) to the CH_3 radical. Going down once more (third row), we create two singly-occupied lobes by the same type of "*alchemical transmutation*": removing L with two electrons, but increasing the central atom atomic number by 1 (two holes in the 18-electron shell, isolobal with CH_2 group). Finally, we repeat this for the fourth row, receiving the three singly-occupied lobes (three electron holes, isolobal with CH group).

The philosophy behind the isolobal analogy is to see complex molecular structures as being interrelated and systemized through a simple idea. It allows to predict the essence of the

[155] In this case, six d^2sp^3-type hybrid lobes protrude toward the six corners of the tetragonal bipyramid.

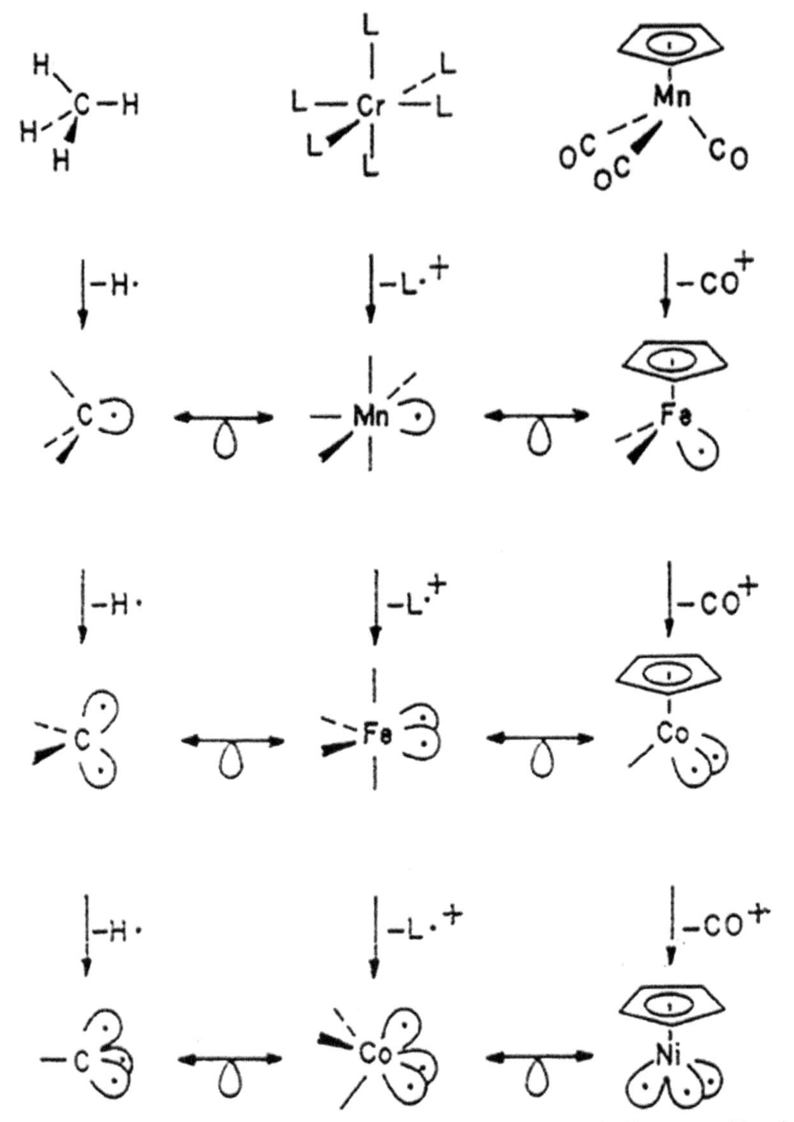

Fig. 8.34. The isolobal analogy between molecular fragments (symbolized by a double arrow with a loop). All species shown are electrically neutral and are derived conceptually from the first row that contains the "*generic*" compounds: CH_4, CrL_6, $Mn(CO)_3(C_5H_4)$, where L are ligands that offer an electron pair each. The second row is a result of an alchemical transmutation: we remove an electron pair with a ligand L, but compensate for it by increasing the atomic number of the central atom by 1 (its additional electron enters the empty lobe). In such a way, all species in the second row have an orbital lobe carrying a single electron and each pair of them is isolobal. Similarly, in the third row, all species share the same two-lobe structure (all are mutually isolobal), while in the fourth row, we have the same three-lobe isolobal structures.

electronic structure just from basic knowledge of much simpler molecules, and go across the often too-detailed quantum mechanical description, in other words, to understand, why the results *should be* like that. Understanding has direct consequences in chemical synthesis: the isolobal analogy helped a lot in producing new organometallic materials.

Summary

- The Hartree-Fock procedure represents a *variational* method. The variational function takes the form of *a single Slater determinant* ψ built of orthonormal molecular spinorbitals: $\psi = \frac{1}{\sqrt{N!}} \begin{vmatrix} \phi_1(1) & \phi_1(2) & \dots & \phi_1(N) \\ \phi_2(1) & \phi_2(2) & \dots & \phi_2(N) \\ \dots & \dots & \dots & \dots \\ \phi_N(1) & \phi_N(2) & \dots & \phi_N(N) \end{vmatrix}$.
- A molecular spinorbital $\phi_i(1)$ is a one-electron function of the coordinates x_1, y_1, z_1, σ_1. In the RHF method, it is the product $\varphi_i(x_1, y_1, z_1)\alpha(\sigma_1)$ or $\varphi_i(x_1, y_1, z_1)\beta(\sigma_1)$ of a real *molecular orbital* $\varphi_i(x_1, y_1, z_1)$ and of the *spin function* $\alpha(\sigma_1)$ or $\beta(\sigma_1)$, respectively. In the general HF method (GHF), a spinorbital is a complex function, which depends both on $\alpha(\sigma_1)$ and $\beta(\sigma_1)$. The UHF method uses, instead, real orbitals, which are all different and are multiplied either by α or β ("*different orbitals for different spins*").
- Minimization of the mean value of the Hamiltonian, $\varepsilon = \frac{\langle \psi | \hat{H} \psi \rangle}{\langle \psi | \psi \rangle}$, with respect to the orthonormal spinorbitals ϕ_i (GHF) leads to equations for *optimum spinorbitals* (Fock equations): $\hat{F}(1)\phi_i(1) = \varepsilon_i \phi_i(1)$, where the Fock operator $\hat{F}$ is $\hat{F}(1) = \hat{h}(1) + \hat{J}(1) - \hat{K}(1)$, the Coulombic operator is defined by $\hat{J}(1)u(1) = \sum_j \hat{J}_j(1)u(1)$ and $\hat{J}_j(1)u(1) = \int d\tau_2 \frac{1}{r_{12}} \phi_j^*(2)\phi_j(2)u(1)$, and the exchange operator by $\hat{K}(1)u(1) = \sum_j \hat{K}_j(1)u(1)$ and $\hat{K}_j(1)u(1) = \int d\tau_2 \frac{1}{r_{12}} \phi_j^*(2)u(2)\phi_j(1)$.
- In the Restricted Hartree-Fock method (RHF) for closed shell systems, we assume *double occupancy of orbitals*; i.e., we form two spinorbitals out of each MO (by multiplying either by α or β).
- The Fock equations are solved by an iterative approach (with *an arbitrary* starting point) and as a result, we obtain approximations to the following:
  - The total energy
  - The wave function (the optimum Slater determinant)
  - The canonical molecular orbitals (spinorbitals)
  - The orbital energies.
- Use of the LCAO expansion leads to the Hartree-Fock-Roothaan equations $\boldsymbol{Fc} = \boldsymbol{Sc\varepsilon}$. Our job, then, is to find the LCAO coefficients $\boldsymbol{c}$. This is achieved by transforming the matrix equation to the form of the eigenvalue problem, and to diagonalize the corresponding Hermitian matrix. The canonical MOs obtained are linear combinations of the atomic orbitals. The lowest-energy orbitals are occupied by electrons, while those of higher energy are called *virtual* and are left empty.
- Using the H_2^+ and H_2 examples, we found that a chemical bond results from a quantum effect of an electron density flow toward the bond region. This results from a superposition of atomic orbitals due to the variational principle.
- In the simplest MO picture:
  - The *excited triplet state has lower energy than the corresponding excited singlet state*
  - In the case of orbital degeneracy, the system prefers *parallel electron spins* (Hund's rule)
  - The ionization energy is equal to the negative of the *orbital energy of the removed* electron. The electron affinity is equal to the negative of the *orbital energy corresponding to the virtual orbital accommodating the added* electron (Koopmans's theorem).
- The canonical MOs for closed-shell systems (the RHF method) may be transformed to orbitals localized in the chemical bonds, lone pairs, and inner shells.
- There are many methods of localization. The most important ones are: the projection method, the method of minimum distance between two electrons from the same orbital (Boys approach), and the method of maximum interaction of electrons from the same orbital (Ruedenberg approach).
- Different localization methods lead to sets of localized MOs which are slightly different but their general shape is similar.

- The MOs (localized as well as canonical) can be classified as to the number of nodal surfaces going through the nuclei. A σ bond orbital has no nodal surface at all, a π bond orbital has a single nodal surface, and a δ bond orbital has two such surfaces.
- The localization allows comparison of the molecular fragments of different molecules. It appears that the features of the MO localized on the AB bond relatively weakly depend on the molecule in which this bond is found. This is a strong argument and a true source of experimental tactics in chemistry, which is to tune the properties of particular atoms by changing their neighborhood in a controlled way.
- Localization may serve to determine hybrids.
- In everyday practice, chemists often use a minimal model of molecules that enables them to compare the geometry and vibrational frequencies with experiment to the accuracy of about 0.01 Å for bond lengths and about 1^0 for bond angles. This model assumes that the speed of light is infinite (non-relativistic effects only), the Born-Oppenheimer approximation is valid (i.e., the molecule has a 3-D structure), the nuclei are bound by chemical bonds and vibrate in a harmonic way, the molecule moves (translation) and rotates as a whole in space.
- In many cases, we can successfully predict the 3-D structure of a molecule by using a very simple tool: the Valence Shell Electron Pair Repulsion (VSEPR) algorithm.

Main Concepts, New Terms

From the Research Front

John Pople (1925–2004), British mathematician and one of the founders of modern quantum chemistry. His childhood was during time war in England (every day 25 mile train journeys, sometimes under bombing). He came from a lower middle class family (drapers and farmers), but his parents were ambitious for the future of their children. At the age of 12, John developed an intense interest in mathematics. He entered Cambridge University after receiving a special scholarship. Pople made important contributions to theoretical chemistry. To cite a few: proposing semiempirical methods–the famous PPP method for π electron systems, the once very popular CNDO approach for all-valence calculations, and finally the monumental joint work on GAUSSIAN, a system of programs that constitutes one of most important computational tools for quantum chemists. Pople received the Nobel Prize in 1998 *"for his development of computational methods in quantum chemistry"* sharing it with Walter Kohn.

The Hartree-Fock method belongs to a narrow two- to three- member class of standard methods of quantum chemistry. It is the source of basic information about the electronic ground state of a molecule. It also allows its geometry optimization. At present, the available computational codes limit the calculations to the systems built of several hundreds of atoms. Moreover, the programs allow calculations to be made by clicking the mouse. The Hartree-Fock method is always at their core. The GAUSSIAN is one of the best known programs. It is the result of many years of coding by a team of quantum chemists working under John Pople. Pople was given the Nobel Prize in 1998, mainly for this achievement. To get a flavor of the kind of data needed, I provide below a typical data set necessary for GAUSSIAN to perform the Hartree-Fock computations for the water molecule:

```
#HF/STO-3G opt freq pop
water, the STO-3G basis set
0 1
O
H1 1 r12
H2 1 r12 2 a213
r12 = 0.96
a213 = 104.5
```

The explanatory comments for this program, line by line, follow:

- #HF/STO-3G opt freq pop is a command that informs GAUSSIAN that the computations are of the Hartree-Fock (HF) type, that the basis set used is of the STO-3G type (each STO is expanded into three GTOs), that we want to optimize geometry (opt), compute the harmonic vibrational frequencies (freq), and perform the charge population analysis for the atoms (known as Mulliken population analysis; see Appendix S available at booksite.elsevier.com/978-0-444-59436-5, p. e143).

- Just a comment line.
- 0 1 means that the total charge of the system is equal to 0, and the singlet state is to be computed (1).
- O means that the first atom in the list is oxygen.
- H1 1 r12 means that the second atom in the list is hydrogen (named H1), distant from the first atom by r12.
- H2 1 r12 2 a213 means that the third atom in the list is hydrogen (named H2), distant from atom number 1 by r12, and forming the 2-1-3 angle equal to a213.
- r12 = 0.96 is a starting OH bond length in Å.
- a213 = 104.5 is a starting angle in degrees.

Similar inputs are needed for other molecules. The initial geometry is to some extent arbitrary, and therefore in fact it cannot be considered as real input data. The only true information is the number and charge (kind) of the nuclei, the total molecular charge (i.e., we know how many electrons are in the system), and the multiplicity of the electronic state to be computed. The basis set issue (STO-3G) is purely technical and gives information about the quality of the results.

Ad Futurum

Along with the development of computational techniques, and with the progress in the domain of electronic correlation, the importance of the HF method as a source of information about total energy or total electron density will probably decrease. Simply, much larger molecules (beyond the HF level) will be within the reach of future computers. Yet HF calculations still will be carried out, and their results will be carefully analyzed. There are at least two reasons for this:

- HF calculations are most often the necessary step before more precise computations are performed
- HF computations result in the MO model: the MOs and the orbital energies scheme ("*minimal model*"), they provide the *conceptual framework for the molecule.* It is the sort of model that may be discussed, thought of, and used to search for explanation of physical and chemical phenomena. So far, such a possibility does not exist for advanced methods, where often we obtain very good results, but it is extremely difficult to get an idea *why* they agree so well with experiments.[156]

Additional Literature

A. Szabo and N.S. Ostlund, *Modern Quantum Chemistry,* McGraw-Hill, New York, 1989, pp. 108–231.
Excellent book.

T. Helgaker, P. Jørgensen, and J. Olsen, *Molecular Electronic Structure Theory*, Wiley, Chichester, 2000, pp. 433–513.
This book is a contemporary compendium of computational quantum chemistry.

O. Chalvet, R. Daudel, S. Diner, and J.P. Malrieu, Eds., *Localization and Delocalization in Quantum Chemistry*, D. Reidel Publish. Co., Dordrecht, 1975.
A set of the very interesting articles by the leading quantum chemists of that time.

[156] The fact of solving the Schrödinger equation, unfortunately does not instruct us on the nature of physical phenomena in most cases.

Questions

1. The Hartree-Fock method for a system with N electrons (closed-shell case) leads to a wave function
   a. that satisfies the Schrödinger equation
   b. in the form of a Slater determinant, which in the Hilbert space is the closest to the solution of the Schrödinger equation
   c. in the form of such a Slater determinant, which gives the lowest mean value of the electronic Hamiltonian
   d. in the form of an antisymmetrized product of the spinorbitals, each satisfying the Fock equation
2. The canonical orbitals:
   a. represent a minimal basis set of the atomic orbitals of the atoms present in the molecule
   b. give the lowest mean value of the Hamiltonian among all possible orbitals when inserted into the Slater determinant
   c. satisfy the Fock equation
   d. are either the core orbitals or bond orbitals or lone pair orbitals
3. The localized molecular orbitals
   a. are mutually orthogonal
   b. give the lowest mean value of the Hamiltonian of all the Slater determinants possible when inserted into the Slater determinant
   c. give the canonical orbitals when transformed by a particular unitary transformation
   d. are localized on individual atoms of the molecule
4. Localization of molecular orbitals
   a. is made for determining which molecular orbitals are bonding, antibonding and non-bonding
   b. may lead to different sets of localized orbitals
   c. lowers the mean value of the Hamiltonian
   d. enables one to obtain the orbitals of individual chemical bonds, of the lone pairs, and of the atomic core orbitals
5. The orbital energy
   a. for each MO, multiplied by its occupancy and summed up over all MOs, equals the total HF electronic energy
   b. is equal to the mean value of the Fock operator with the corresponding MO
   c. is a sum of the energies of the two electrons occupying the MO
   d. is equal to the mean value of the energy per one electron of the molecule
6. The Fock operator contains the following operators:
   a. of the kinetic energy of the nuclei
   b. of the electrostatic attraction of the electron pairs by the nuclei
   c. the Coulomb operator of a repulsion of a point-like electron with the charge density distribution of all electrons
   d. the electrostatic attraction electron-nucleus
7. In the LCAO MO method, each molecular orbital represents:
   a. a function of position of an electron in 3-D space
   b. a function that depends on the coordinates of the two electrons occupying this molecular orbital
   c. a function of class Q
   d. a linear combination of atomic orbitals (that belong to the atomic basis set chosen)
8. In the HF method (closed-shell case, U = a sum of the orbital energies of the doubly occupied molecular orbitals, V_{nn} stands for the repulsion of the nuclei) the total energy is
   a. lower than $2U + V_{nn}$

b. is equal $2U$
c. equal to $2U + V_{nn}$
d. lower than $2U - \frac{1}{2}V_{ee} + V_{nn}$

9. A comparison of the RHF and the UHF methods
   a. in the UHF method, one always gets a spin contamination of the wave function
   b. the Slater determinant in the RHF method represents an eigenfunction of the operator of the square of the spin angular momentum
   c. $E_{UHF} < E_{RHF}$
   d. $E_{UHF} \leq E_{RHF}$, and both functions are eigenfunctions of the z component of the total spin of the electrons
10. The RHF method for the hydrogen molecule (R is the internuclear distance)
   a. gives wrong results for large R, because a HOMO - LUMO quasi-degeneracy
   b. gives a correct description of the dissociation producing two ground-state hydrogen atoms
   c. gives wrong products of dissociation because of a too large difference between the bonding and antibonding energy levels
   d. does not take into account any correlation of motion of the two electrons

Answer

1c,d, 2b,c, 3a,b,c, 4b,d, 5b, 6c,d, 7a,c,d, 8a,d, 9b,d, 10a,d

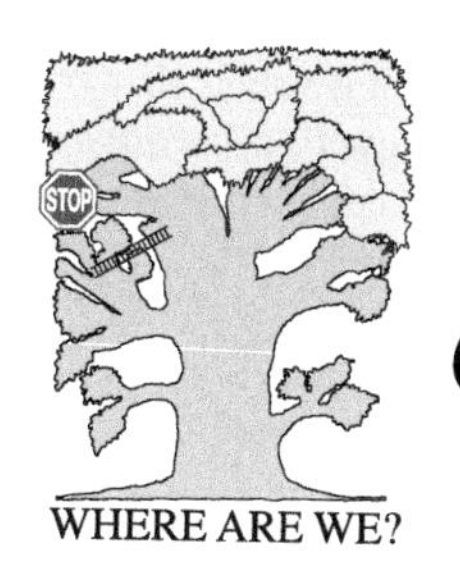

WHERE ARE WE?

Orbital Model of Electronic Motion in Periodic Systems

"Beauty of style and harmony and grace and good rhythm depend on simplicity."

Plato

Where Are We?

We are on the upper-left branch of the TREE.

An Example

Polyacetylene[1] represents a *practically* infinite polymeric chain[2]: ⋯–CH=CH–CH=CH–CH=CH–CH=CH–⋯. There is no such thing in nature as a truly infinite system. Yet, if we examine larger and larger portions of a homogeneous material, we discover the idea that such quantities as energy per stoichiometric unit, electron excitation energy, vibrational frequencies, etc. depend less and less on system size. This means that a boundary-region (polymer ends, crystal surface) contribution to

Herman Staudinger (1881–1965), German polymer chemist and professor at the University of Freiburg, received the Nobel Prize in 1953 *"for his discoveries in the field of macromolecular chemistry."* However strange it may sound now, the concept of polymers was unthinkable in chemistry as late as 1926.

It will be encouraging for Ph.D. students to read that a professor advised Staudinger in the late 1920s: *"Dear colleague, leave the concept of large molecules well alone: organic molecules with a molecular weight above 5000 do not exist. Purify your products, such as rubber, then they will crystallise and prove to be lower molecular substances."*

[1] The discovery of conducting polymers (like polyacetylene) was honored with the Nobel Prize in 2000 for Hideki Shirakawa (who synthesized a crystalline form of polyacetylene), as well as Allan G.MacDiarmid and Allan J. Heeger, who increased its electric conductivity by 18 orders of magnitude by doping the crystal with some electron acceptors and donors. This incredible increase is probably the largest known to humanity in any domain of experimental sciences [H.Shirakawa, E.J. Louis, A.G. MacDiarmid, C.K. Chiang, and A.J. Heeger, *Chem.Soc.Chem.Commun.*, 578 (1977)].

[2] That is, a macromolecule. The concept of polymer was introduced to chemistry by Herman Staudinger.

Ideas of Quantum Chemistry, Second Edition. http://dx.doi.org/10.1016/B978-0-444-59436-5.00009-X

these quantities becomes negligible. Therefore, these quantities (known as *intensive*) attain limit values that are identical to those for an infinite system. *It pays to investigate the infinite system because we can use its translational symmetry to simplify its description.* Well, this is what this chapter is about.

What Is It All About?

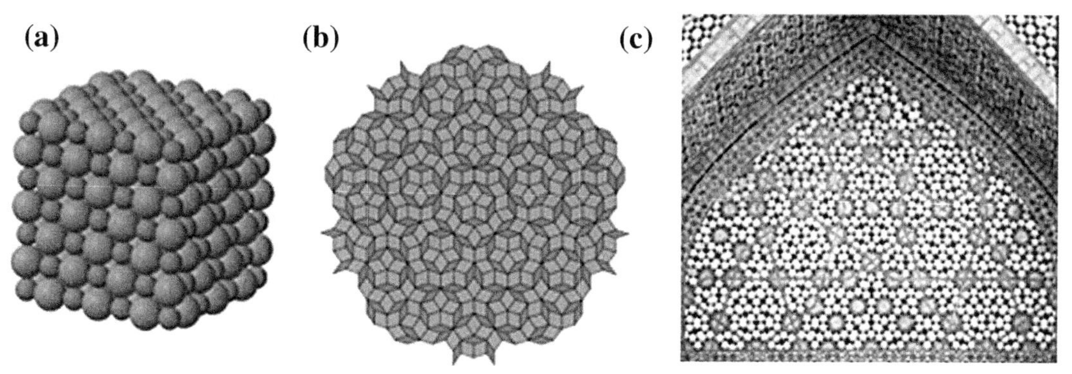

Fig. 9.1. Translational symmetry existing in crystals and its lack in quasi-crystals despite a perfect long-range order. (a) The translational symmetry in the NaCl crystal build of Na^+ and Cl^- ions; (b) the Penrose tiling as an example of a 2-D quasicrystal, without translational symmetry; (c) a medieval Arabian mosaic as an example of a long-range, non-translational order.

If a motif (e.g., a cluster of atoms) associated with a unit cell is *regularly translated along three different directions in space*, we obtain an infinite 3-D periodic structure (translational symmetry, Fig. 9.1a). In 2-D, this means that having a single (special) type of tiles (unit cells), we have been successful in tiling the *complete* 2-D space. One of the consequences of this is that the fivefold symmetry axes have to be absent in the atomic arrangements in such crystals. *It turned out that a vast majority of real crystals can be reliably modeled using this idea.* There were efforts in mathematics to design some non-translational complete tilings. In 1963, it was first shown that for number $N = \ldots 20000$ of square tiles, such a non-translational tiling is possible. This number has been gradually reduced, and in 1976, Roger Penrose proposed covering by $N = 2$ kinds of tiles; see Fig. 9.1b. Then, Daniel Shechtman discovered[3] that there are substances (known now as *quasicrystals*) that indeed show fivefold symmetry axes (also other translationally forbidden symmetry axes). There are also chaotic quasicrystals. The reason why the quasicrystals exist in nature is quite simple: some strong short-range interactions force unusual five-ligand complexes. It is remarkable that the discovery of quasicrystals has been preceded by ancient artists (Fig. 9.1c).

When applying the Hartree-Fock method to such periodic infinite objects, one usually exploits the translational symmetry of the system (e.g., in calculating integrals). It would be indeed prodigal to compute the integrals many times, the equality of which is *guaranteed* by translational symmetry. When translational symmetry is taken into account, the problem reduces to the calculation of the interaction of a single unit cell (with a reference labeled 0) with all other unit cells, the nearest neighbor cells being most important. The infinite size of the system is hidden in the plethora of points (to be taken into account) that is known as the First Brillouin Zone (FBZ). The FBZ represents a unit cell in what is called *inverse lattice* (associated with a given lattice reflecting the translational symmetry).

The electronic orbital energy becomes a function of the FBZ points and we obtain what is known as band structure of the energy levels. This band structure decides the electronic properties of the system (insulator, semiconductor, metal). We will also show how to carry out the mean field (Hartree-Fock) computations on infinite periodic systems. The calculations require infinite summations (interaction of the reference unit cell with the infinite crystal) to be made. This creates some mathematical problems, which will be also described in this chapter.

Why Is This Important?

This chapter is particularly important for those readers who are interested in solid-state physics and chemistry. Others may treat it as exotic, and if they decide they do not like exotic matter, they may skip this discussion and go to other chapters.

---

[3] He received the 2011 Nobel Prize in chemistry for "*the discovery of quasicrystals.*"

The properties of a polymer or a crystal sometimes differ widely from those of the atoms or molecules of which they are built. *The same* substance may form *different* periodic structures, which have *different* properties (e.g., graphite and diamond). The properties of periodic structures could be computed by extrapolation of the results obtained for larger and larger clusters of the atoms from which the substance is composed. This avenue, however, is non-economic. It is easier to carry out quantum mechanical calculations for an infinite system,[4] than for a large cluster.[5]

What Is Needed?

- Operator algebra (see Appendix B available at booksite.elsevier.com/978-0-444-59436-5, p. e7)
- Translation operator (see Appendix C available at booksite.elsevier.com/978-0-444-59436-5, p. e17)
- Hartree-Fock method (Chapter 8)
- Multipole expansion (see Appendix X available at booksite.elsevier.com/978-0-444-59436-5, p. e169, advised)
- Matrix diagonalization (see Appendix K available at booksite.elsevier.com/978-0-444-59436-5, p. e105, advised)

Classical Works

At the age of 23, Felix Bloch published an article called "*Über die Quantenmechanik der Elektronen in Kristallgittern*" in *Zeitschrift für Physik*, 52, 555 (1928) (only two years after Schrödinger's historic publication) on the translational symmetry of the wave function. This was also the first application of LCAO expansion. ★ In 1931, Leon Brillouin published a book entitled *Quantenstatistik* (Springer Verlag, Berlin), in which the author introduced some of the fundamental notions of band theory. ★ The first *ab initio* calculations for a polymer were made by Jean-Marie André in a paper "*Self-consistent field theory for the electronic structure of polymers*," published in the *Journal of Chemical Physics, 50*, 1536 (1969).

9.1 Primitive Lattice

Let us imagine an *infinite crystal*; e.g., a system that exhibits the *translational symmetry* of the charge distribution (nuclei and electrons). The translational symmetry will be fully determined by three (linearly independent) basis vectors[6]: $\boldsymbol{a}_1, \boldsymbol{a}_2$, and $\boldsymbol{a}_3$ having the property that $\boldsymbol{a}_i$ beginning at any atom extends to the corresponding identical atom located in the crystal. The lengths of the *basis* vectors $\boldsymbol{a}_1, \boldsymbol{a}_2$, and $\boldsymbol{a}_3$ are called the *lattice constants* along the three periodicity axes.[7]

There is a lot of such basis sets possible. Any choice of the basis vectors, is acceptable from the point of view of mathematics. For economic reasons, we choose one of the possible vector sets that give the *least volume parallelepiped*[8] with sides $\boldsymbol{a}_1$, $\boldsymbol{a}_2$, and $\boldsymbol{a}_3$. This parallelepiped

---

[4] The surface effects can be neglected, and the units the system is composed of turn out to be equivalent.

[5] Sometimes we may be interested in a particular cluster, not in an infinite system. Then it may turn out to be more economic to perform the calculations for the infinite system and use the results in computations for the clusters [e.g., R.A. Wheeler, L. Piela, and R. Hoffmann, *J. Am. Chem. Soc., 110*, 7302 (1988)].

[6] These are not necessarily perpendicular, though; they determine the periodicity axes.

[7] As shown on p. 440, a symmetry of the nuclear framework does not guarantee the same symmetry of the electronic charge distribution computed using a mean field method. We may have to cope with the period doubling as compared to the period of the nuclear framework (cf. BOAS, p. 441). If this happens, then we should choose lattice constants that ensure the periodicity of both nuclear and electron distributions.

[8] Yes, because the multiplicity of $\boldsymbol{a}_i$ would also lead to unit cells that, when repeated, would reproduce the whole crystal. We are, however, interested in the smallest unit cell.

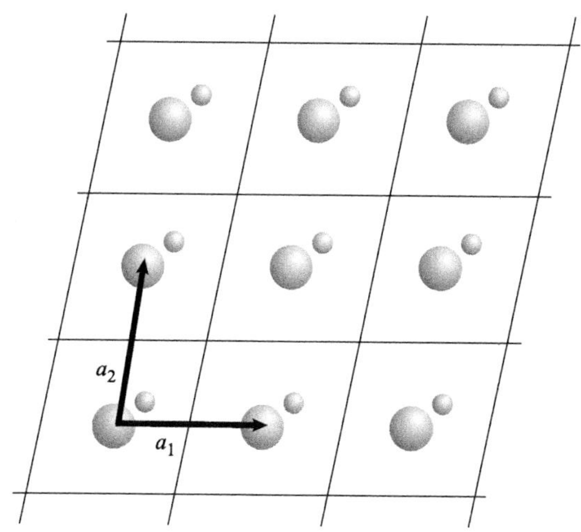

Fig. 9.2. Periodicity in 2-D. We choose the unit cell (the parallelogram with vectors $\boldsymbol{a}_1$ and $\boldsymbol{a}_2$) and its content (motif) in such a way as to reproduce the whole infinite crystal by repeating the unit cells through its translation vectors $\boldsymbol{R}_i = n_1\boldsymbol{a}_1 + n_2\boldsymbol{a}_2$ with integer n_1, n_2. In 3-D, instead of the parallelogram, we would have a parallelepiped, which would be repeated by translation vectors $\boldsymbol{R}_i = n_1\boldsymbol{a}_1 + n_2\boldsymbol{a}_2 + n_3\boldsymbol{a}_3$, with integer n_1, n_2, n_3.

(which is arbitrarily shifted in space;[9] see Fig. 9.2) represents *our* choice of the *unit cell*,[10] which together with its content (*motif*) is to be translationally repeated.[11]

Let us now introduce the *space of translation vectors* $\boldsymbol{R}_i = \sum_{j=1}^{3} n_{ij}\boldsymbol{a}_j$, where n_{ij} are *arbitrary integer numbers* (see Appendix B available at booksite.elsevier.com/978-0-444-59436-5, p. e7).

The points indicated by all the translation vectors ("*lattice vectors*") are called the *crystallographic lattice* or the *primitive lattice*, or simply the *lattice*.

Let us introduce the *translation operators* $\hat{T}(\boldsymbol{R}_i)$ defined as translations of a *function*, on which the operator acts, by vector $\boldsymbol{R}_i$ (cf. Chapter 2 and see Appendix C available at booksite.elsevier.com/978-0-444-59436-5 on p. e17):

$$\hat{T}(\boldsymbol{R}_i) f(\boldsymbol{r}) = f(\boldsymbol{r} - \boldsymbol{R}_i). \tag{9.1}$$

The function $f(\boldsymbol{r}) \equiv f(\boldsymbol{r} - \boldsymbol{0})$ is centered at the origin of the coordinate system, while the function $f(\boldsymbol{r} - \boldsymbol{R}_i)$ is centered on the point shown by vector $\boldsymbol{R}_i$.

9 The choice of the origin of the coordinate system is arbitrary, and the basis vectors are determined within the accuracy of an arbitrary translation.

10 An example of a jigsaw puzzle shows that other choices are possible as well. A particular choice may result from its convenience. This freedom will be discussed further on p. 516.

11 The motif can be ascribed to the unit cell (i.e., chosen) in many different ways, provided that after putting the cells together, we get the same original infinite crystal. Let me propose disregarding this problem for the time being (as well as the problem of the choice of the unit cell) and to think of the unit cell as a space-fixed *parallelepiped with the motif that has been enclosed in it*. We will come back to this complex problem at the end of this chapter.

The crystal periodicity is reflected by the following property of the potential energy V for an electron (where V depends on its position in the crystal):

$$V(\boldsymbol{r}) = V(\boldsymbol{r} - \boldsymbol{R}_i), \tag{9.2}$$

for any $\boldsymbol{R}_i$. The equation simply says that the infinite crystal looks exactly the same close to the origin O as to the point shown by any lattice vector $\boldsymbol{R}_i$.

It is easy to see that the operators $\hat{T}(\boldsymbol{R}_i)$ form a *group* (see Appendix C available at booksite.elsevier.com/978-0-444-59436-5, p. e17) with respect to their multiplication as the group operation.[12,13] In Chapter 2, it was shown that the Hamiltonian is invariant with respect to any translation of a molecule. For infinite systems, the proof looks the same for the kinetic energy operator, the invariance of V is guaranteed by Eq. (9.2). Therefore, the effective one-electron Hamiltonian commutes with any translation operator:

$$\hat{H}\hat{T}(\boldsymbol{R}_i) = \hat{T}(\boldsymbol{R}_i)\hat{H}.$$

9.2 Wave Vector

Since $\hat{T}(\boldsymbol{R}_i)$ commutes with the Hamiltonian, its eigenfunctions also represent the eigenfunctions of the translation operator[14] (cf. Chapter 2, p. 77, also see Appendix C available at booksite.elsevier.com/978-0-444-59436-5 on p. e17); i.e., in this case, $\hat{H}\psi = E\psi$ *and*

[12] Indeed, first a product of such operators represents a translational operator:

$$\hat{T}(\boldsymbol{R}_1)\hat{T}(\boldsymbol{R}_2)f(\boldsymbol{r}) = \hat{T}(\boldsymbol{R}_1)f(\boldsymbol{r} - \boldsymbol{R}_2) = f(\boldsymbol{r} - \boldsymbol{R}_1 - \boldsymbol{R}_2) = f(\boldsymbol{r} - (\boldsymbol{R}_1 + \boldsymbol{R}_2))$$
$$= \hat{T}(\boldsymbol{R}_1 + \boldsymbol{R}_2)f(\boldsymbol{r})$$

therefore:

$$\hat{T}(\boldsymbol{R}_1)\hat{T}(\boldsymbol{R}_2) = \hat{T}(\boldsymbol{R}_1 + \boldsymbol{R}_2). \tag{9.3}$$

The second requirement is to have a unity operator. This role is played by $\hat{T}(\boldsymbol{0})$, since

$$\hat{T}(\boldsymbol{0})f(\boldsymbol{r}) = f(\boldsymbol{r} + \boldsymbol{0}) = f(\boldsymbol{r}). \tag{9.4}$$

The third condition is the existence [for every $\hat{T}(\boldsymbol{R}_i)$] of the inverse operator, which in our case is $\hat{T}(-\boldsymbol{R}_i)$, because

$$\hat{T}(\boldsymbol{R}_i)\hat{T}(-\boldsymbol{R}_i) = \hat{T}(\boldsymbol{R}_i - \boldsymbol{R}_i) = \hat{T}(\boldsymbol{0}). \tag{9.5}$$

The group is Abelian (i.e., the operations commute), since

$$\hat{T}(\boldsymbol{R}_1)\hat{T}(\boldsymbol{R}_2) = \hat{T}(\boldsymbol{R}_1 + \boldsymbol{R}_2) = \hat{T}(\boldsymbol{R}_2 + \boldsymbol{R}_1) = \hat{T}(\boldsymbol{R}_2)\hat{T}(\boldsymbol{R}_1). \tag{9.6}$$

[13] Besides the translational group, the crystal may also exhibit what is called the *point group*, associated with rotations, reflections in planes, inversion, etc., and the *space group* that results from the translational group and the point group. In such cases, a smaller unit cell may be chosen because the whole crystal is reproduced not only by translations, but also by other symmetry operations. In this book, we will concentrate on the translational symmetry group only.

[14] The irreducible representations of an Abelian group are 1-D. In our case (translational group), this means that there is no degeneracy, and that an eigenfunction of the Hamiltonian is also an eigenfunction of all the translation operators.

$\hat{T}(\boldsymbol{R}_j)\psi(\boldsymbol{r}) = \psi(\boldsymbol{r} - \boldsymbol{R}_j) = \lambda_{\boldsymbol{R}_j}\psi(\boldsymbol{r})$. The symmetry of V requires the equality of the probability densities

$$|\psi(\boldsymbol{r} - \boldsymbol{R}_j)|^2 = |\psi(\boldsymbol{r})|^2 \tag{9.7}$$

for any lattice vector $\boldsymbol{R}_j$, which gives $|\lambda_{\boldsymbol{R}_j}|^2 = 1$. Therefore we may write

$$\lambda_{\boldsymbol{R}_j} = \exp(-i\theta_{\boldsymbol{R}_j}), \tag{9.8}$$

where $\theta_{\boldsymbol{R}_j}$ will be found in a moment.[15]

From the equation $\hat{T}(\boldsymbol{R}_j)\psi(\boldsymbol{r}) = \lambda_{\boldsymbol{R}_j}\psi(\boldsymbol{r})$, it follows that

$$\lambda_{\boldsymbol{R}_j}\lambda_{\boldsymbol{R}_l} = \lambda_{\boldsymbol{R}_j+\boldsymbol{R}_l}, \tag{9.9}$$

because

$$\hat{T}(\boldsymbol{R}_j + \boldsymbol{R}_l)\psi(\boldsymbol{r}) = \lambda_{\boldsymbol{R}_j+\boldsymbol{R}_l}\psi(\boldsymbol{r}). \tag{9.10}$$

On the other hand,

$$\begin{aligned}\hat{T}(\boldsymbol{R}_j + \boldsymbol{R}_l)\psi(\boldsymbol{r}) &= \hat{T}(\boldsymbol{R}_j)\hat{T}(\boldsymbol{R}_l)\psi(\boldsymbol{r}) = \lambda_{\boldsymbol{R}_l}\hat{T}(\boldsymbol{R}_j)\psi(\boldsymbol{r}) \\ &= \lambda_{\boldsymbol{R}_j}\lambda_{\boldsymbol{R}_l}\psi(\boldsymbol{r}).\end{aligned}$$

Since this relation has to be satisfied for any $\boldsymbol{R}_j$ and $\boldsymbol{R}_l$, it is therefore sufficient to have

$$\theta_{\boldsymbol{R}_j} = \boldsymbol{k} \cdot \boldsymbol{R}_j, \tag{9.11}$$

because a multiplication of λ by λ corresponds to adding the exponents, which results in adding vectors $\boldsymbol{R}$, which we need to have. The dot product $\boldsymbol{k} \cdot \boldsymbol{R}_j$ for simplicity will also be written as $\boldsymbol{k}\boldsymbol{R}_j$.

Conclusion:

The eigenfunctions of the one-electron Hamiltonian and the translation operators correspond to the following eigenvalues of the translation operator: $\lambda_{\boldsymbol{R}_j} = \exp(-i\boldsymbol{k}\boldsymbol{R}_j)$,

where the vector $\boldsymbol{k}$ characterizes the function, not the direction of $\boldsymbol{R}_j$. In other words, any one-electron wave function (crystal orbital), which is the eigenfunction of the one-electron Hamiltonian could be labelled by its corresponding vector $\boldsymbol{k}$; i.e., $\psi(\boldsymbol{r}) \rightarrow \psi_{\boldsymbol{k}}(\boldsymbol{r})$.

Bloch Theorem

The value of such a function in the point shifted by the vector $\boldsymbol{R}_j$ is equal to

$$\psi_{\boldsymbol{k}}(\boldsymbol{r} - \boldsymbol{R}_j) = \exp(-i\boldsymbol{k}\boldsymbol{R}_j)\psi_{\boldsymbol{k}}(\boldsymbol{r}) \tag{9.12}$$

[15] The exponent sign is arbitrary; we use "−" following a widely used convention.

Felix Bloch (1905–1983), American physicist of Swiss origin and professor at the Stanford University from 1936 to 1971. Bloch contributed to the electronic structure of metals, superconductivity, ferromagnetism, quantum electrodynamics, and the physics of neutrons. In 1946, independently from E.M. Purcell, he discovered the nuclear magnetic resonance effect.

Both scientists received the Nobel Prize in 1952 *"for the development of new methods for nuclear magnetic precision measurements and the discoveries in connection therewith."*

This relation represents a necessary condition to be fulfilled by the eigenfunctions for a perfect periodic structure (crystal, layer, and polymer). This equation differs widely from Eq. (9.2) for potential energy. Unlike potential energy, which does not change upon a lattice translation, the wave function undergoes a change of its phase acquiring the factor $\exp(-i\boldsymbol{k}\boldsymbol{R}_j)$.

Any linear combination of functions labeled by the same $\boldsymbol{k}$ represents an eigenfunction of any lattice translation operator and corresponds to the same $\boldsymbol{k}$. Indeed, from the linearity of the translation operator,

$$\begin{aligned}\hat{T}(\boldsymbol{R}_l)\big(c_1\phi_{\boldsymbol{k}}(\boldsymbol{r}) + c_2\psi_{\boldsymbol{k}}(\boldsymbol{r})\big) &= c_1\phi_{\boldsymbol{k}}(\boldsymbol{r}-\boldsymbol{R}_l) + c_2\psi_{\boldsymbol{k}}(\boldsymbol{r}-\boldsymbol{R}_l)\\ &= c_1\exp(-i\boldsymbol{k}\boldsymbol{R}_l)\phi_{\boldsymbol{k}}(\boldsymbol{r}) + c_2\exp(-i\boldsymbol{k}\boldsymbol{R}_l)\psi_{\boldsymbol{k}}(\boldsymbol{r})\\ &= \exp(-i\boldsymbol{k}\boldsymbol{R}_l)\big(c_1\phi_{\boldsymbol{k}}(\boldsymbol{r}) + c_2\psi_{\boldsymbol{k}}(\boldsymbol{r})\big).\end{aligned}$$

Let us construct the following function (called a *Bloch function*) from a function $\chi(\boldsymbol{r})$, that in future will play the role of an atomic orbital (in this case centered at the origin):

$$\phi(\boldsymbol{r}) = \sum_j \exp(i\boldsymbol{k}\boldsymbol{R}_j)\chi(\boldsymbol{r}-\boldsymbol{R}_j),$$

where the summation extends over all possible $\boldsymbol{R}_j$; i.e., over the whole crystal lattice. The function ϕ is *automatically* an eigenfunction of any translation operator and may be labeled by the index[16] $\boldsymbol{k}$.

Our function ϕ represents, therefore, an eigenfunction of the translation operator with the same eigenvalue as that corresponding to $\psi_{\boldsymbol{k}}$. In the following, very often $\psi_{\boldsymbol{k}}$ will be constructed as a linear combination of Bloch functions ϕ.

[16] Indeed, first

$$\begin{aligned}\hat{T}(\boldsymbol{R}_l)\phi(\boldsymbol{r}) &= \hat{T}(\boldsymbol{R}_l)\sum_j \exp(i\boldsymbol{k}\boldsymbol{R}_j)\chi(\boldsymbol{r}-\boldsymbol{R}_j) = \sum_j \exp(i\boldsymbol{k}\boldsymbol{R}_j)\hat{T}(\boldsymbol{R}_l)\chi(\boldsymbol{r}-\boldsymbol{R}_j)\\ &= \sum_j \exp(i\boldsymbol{k}\boldsymbol{R}_j)\chi(\boldsymbol{r}-\boldsymbol{R}_j-\boldsymbol{R}_l).\end{aligned}$$

Instead of the summation over $\boldsymbol{R}_j$, let us introduce a summation over $\boldsymbol{R}_{j'} = \boldsymbol{R}_j + \boldsymbol{R}_l$, which means an *identical* summation as before, but we begin to sum the term up from another point of the lattice. Then, we can write

A *Bloch function* is nothing but a symmetry orbital built from the functions $\chi(\boldsymbol{r} - \boldsymbol{R}_j)$.

A *symmetry orbital* is a linear combination of atomic orbitals that transforms according to an irreducible representation Γ of the symmetry group of the Hamiltonian (see Appendix C available at booksite.elsevier.com/978-0-444-59436-5). In order to obtain such a function, we may use the corresponding projection operator [see Eq. (C.13)].

There is also another way to construct a function $\phi_{\boldsymbol{k}}(\boldsymbol{r})$ of a given $\boldsymbol{k}$ from an auxiliary function $u(\boldsymbol{r})$ satisfying an equation similar to Eq. (9.2) for the potential V:

$$\widehat{T}(\boldsymbol{R}_i)u(\boldsymbol{r}) = u(\boldsymbol{r} - \boldsymbol{R}_i) = u(\boldsymbol{r}). \tag{9.13}$$

Then, $\phi_{\boldsymbol{k}}(\boldsymbol{r}) = \exp(i\boldsymbol{k}\boldsymbol{r})u(\boldsymbol{r})$. Indeed, let us check

$$\begin{aligned}\widehat{T}(\boldsymbol{R}_j)\phi_{\boldsymbol{k}}(\boldsymbol{r}) &= \widehat{T}(\boldsymbol{R}_j)\exp(i\boldsymbol{k}\boldsymbol{r})u(\boldsymbol{r}) = \exp(i\boldsymbol{k}(\boldsymbol{r} - \boldsymbol{R}_j))u(\boldsymbol{r} - \boldsymbol{R}_j) \\ &= \exp(-i\boldsymbol{k}\boldsymbol{R}_j)\phi_{\boldsymbol{k}}(\boldsymbol{r}),\end{aligned} \tag{9.14}$$

9.3 Inverse Lattice

Let us now construct the so-called *biorthogonal basis* $\boldsymbol{b}_1, \boldsymbol{b}_2, \boldsymbol{b}_3$ with respect to the basis vectors $\boldsymbol{a}_1, \boldsymbol{a}_2, \boldsymbol{a}_3$ of the primitive lattice, i.e., the vectors that satisfy the *biorthogonality relations*:

$$\boldsymbol{b}_i \boldsymbol{a}_j = 2\pi\delta_{ij}. \tag{9.15}$$

The vectors $\boldsymbol{b}_i$ can be expressed by the vectors $\boldsymbol{a}_i$ in the following way:

$$\boldsymbol{b}_i = 2\pi \sum_j \boldsymbol{a}_j \left(S^{-1}\right)_{ji}, \tag{9.16}$$

$$S_{ij} = \boldsymbol{a}_i \cdot \boldsymbol{a}_j. \tag{9.17}$$

The vectors $\boldsymbol{b}_1, \boldsymbol{b}_2$, and $\boldsymbol{b}_3$ form the basis of a lattice in a 3-D space. This lattice will be called the *inverse lattice*. The inverse lattice vectors are, therefore,

$$\boldsymbol{K}_j = \sum_{i=1}^{i=3} g_{ji}\boldsymbol{b}_i, \tag{9.18}$$

where g_{ij} represent arbitrary *integers*. We have $\boldsymbol{K}_j\boldsymbol{R}_i = 2\pi M_{ij}$, where M_{ij} are integers.

$$\begin{aligned}\sum_{j'}\exp(i\boldsymbol{k}(\boldsymbol{R}_{j'} - \boldsymbol{R}_l))\chi(\boldsymbol{r} - \boldsymbol{R}_{j'}) &= \exp(-i\boldsymbol{k}\boldsymbol{R}_l)\sum_{j'}\exp(i\boldsymbol{k}\boldsymbol{R}_{j'})\chi(\boldsymbol{r} - \boldsymbol{R}_{j'}) \\ &= \exp(-i\boldsymbol{k}\boldsymbol{R}_l)\phi(\boldsymbol{r}),\end{aligned}$$

which had to be proved.

Indeed,

$$K_j \cdot R_i = \sum_{l=1}^{l=3} g_{jl} b_l \cdot \sum_{k=1}^{k=3} n_{ik} a_k = \sum_{l=1}^{l=3} \sum_{k=1}^{k=3} n_{ik} g_{jl} b_l \cdot a_k \tag{9.19}$$

$$= \sum_{l=1}^{l=3} \sum_{k=1}^{k=3} n_{ik} g_{jl} (2\pi) \delta_{lk} = (2\pi) \sum_{l=1}^{l=3} n_{il} g_{jl} = 2\pi M_{ij}, \tag{9.20}$$

with n_{ik}, g_{jl}, and, therefore, M_{ij} being *integers*.

The inverse lattice is composed, therefore, from the isolated points indicated from the origin by the vectors $\boldsymbol{K}_j$. All the vectors that begin at the origin form the *inverse space*.

Examples

Let us see how we obtain the inverse lattice (1-D, 2-D, 3-D) in practice.

1-D

We have only a single biorthogonality relation: $\boldsymbol{b}_1 \boldsymbol{a}_1 = 2\pi$; i.e., after skipping the index $ba = 2\pi$. Because of the single dimension, we have to have $\boldsymbol{b} = \frac{2\pi}{a} \left(\frac{\boldsymbol{a}}{a}\right)$, where $|\boldsymbol{a}| \equiv a$. Therefore,

the vector $\boldsymbol{b}$ has length $\frac{2\pi}{a}$ and the same direction as $\boldsymbol{a}$.

2-D

With 2-D, we have to satisfy $\boldsymbol{b}_1 \boldsymbol{a}_1 = 2\pi, \boldsymbol{b}_2 \boldsymbol{a}_2 = 2\pi, \boldsymbol{b}_1 \boldsymbol{a}_2 = 0$, and $\boldsymbol{b}_2 \boldsymbol{a}_1 = 0$. This means that the game takes place within the plane determined by the lattice vectors $\boldsymbol{a}_1$ and $\boldsymbol{a}_2$. The vector $\boldsymbol{b}_1$ has to be perpendicular to $\boldsymbol{a}_2$, while $\boldsymbol{b}_2$ has to be perpendicular to $\boldsymbol{a}_1$, their directions as shown in Fig. 9.3 [each of the $\boldsymbol{b}$ vectors is a linear combination of $\boldsymbol{a}_1$ and $\boldsymbol{a}_2$, according to Eq. (9.16)].

3-D

In the 3-D case, the biorthogonality relations are equivalent to setting

$$\boldsymbol{b}_1 = \boldsymbol{a}_2 \times \boldsymbol{a}_3 \frac{2\pi}{V}, \tag{9.21}$$

$$\boldsymbol{b}_2 = \boldsymbol{a}_3 \times \boldsymbol{a}_1 \frac{2\pi}{V}, \tag{9.22}$$

and

$$\boldsymbol{b}_3 = \boldsymbol{a}_1 \times \boldsymbol{a}_2 \frac{2\pi}{V}, \tag{9.23}$$

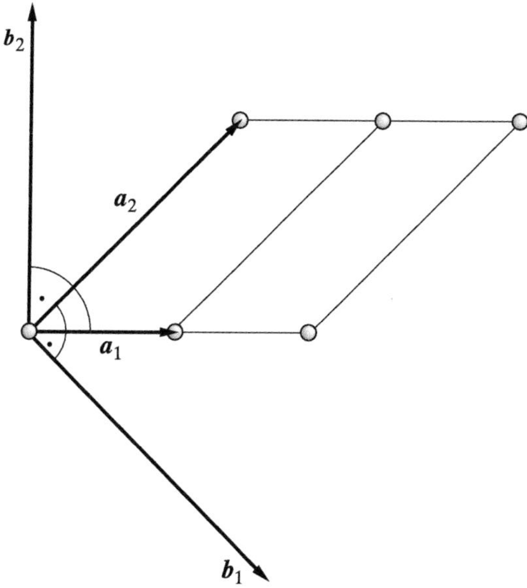

Fig. 9.3. Construction of the inverse lattice in 2-D. In order to satisfy the biorthogonality relations Eq. (9.15), the vector $\boldsymbol{b}_1$ has to be orthogonal to $\boldsymbol{a}_2$, while $\boldsymbol{b}_2$ must be perpendicular to $\boldsymbol{a}_1$. The lengths of the vectors $\boldsymbol{b}_1$ and $\boldsymbol{b}_2$ also follow from the biorthogonality relations: $\boldsymbol{b}_1 \cdot \boldsymbol{a}_1 = \boldsymbol{b}_2 \cdot \boldsymbol{a}_2 = 2\pi$.

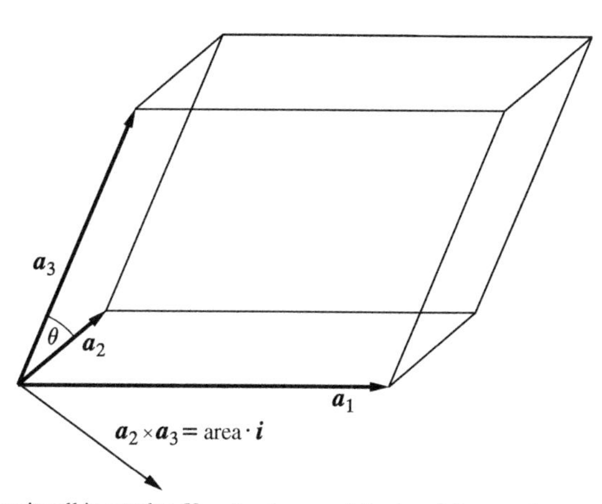

Fig. 9.4. The volume V of the unit cell is equal to $V = \boldsymbol{a}_1 \cdot (area\ of\ the\ base)\, \boldsymbol{i} = \boldsymbol{a}_1 \cdot (\boldsymbol{a}_2 \times \boldsymbol{a}_3)$; $\boldsymbol{i}$ is the unit vector orthogonal to the base.

where

$$V = \boldsymbol{a}_1 \cdot (\boldsymbol{a}_2 \times \boldsymbol{a}_3) \tag{9.24}$$

is the volume of the unit cell of the crystal (see Fig. 9.4).

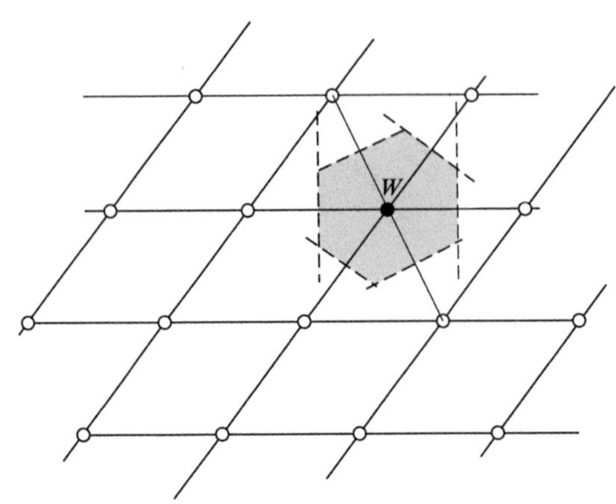

Fig. 9.5. Construction of the FBZ as a Wigner-Seitz unit cell of the inverse lattice in 2-D. The circles represent the nodes of the inverse lattice. We cut the lattice in the middle between the origin node W and all the other nodes (here, it turns out to be sufficient to take only the nearest and the next nearest neighbors) and remove all the sawed-off parts that do not contain W. Finally we obtain the FBZ in the form of a hexagon. The Wigner-Seitz unit cells (after performing all allowed translations in the inverse lattice) reproduce the complete inverse space.

9.4 First Brillouin Zone (FBZ)

Léon Nicolas Brillouin (1889–1969), French physicist, and professor at the Sorbonnne and College de France in Paris, and starting in 1941, at the University of Madison, Columbia University, Harvard University. His contributions included quantum mechanics and solid-state theory (he is one of the founders of electronic band theory). He introduced the notion of the FBZ in 1930.

As was remarked at the beginning of this chapter, the example of a jigsaw puzzle shows us that a parallelepiped unit cell is not the only choice. Now, we will profit from this extra freedom and will define the so-called *Wigner-Seitz unit cell.* Here is the prescription for how to construct it (Fig. 9.5): We focus on a node W, saw the crystal along the plane that dissects (symmetrically) the distance to a nearest-neighbor node, throw the part that *does not* contain W into the fireplace, and then repeat the procedure until we are left with a solid containing W. This solid represents the First Brillouin Zone (FBZ).

9.5 Properties of the FBZ

The vectors $\boldsymbol{k}$, which begin at the origin and end within the FBZ, label *all different* irreducible representations of the translational symmetry group.

Let us imagine two inverse space vectors $\boldsymbol{k}'$ and $\boldsymbol{k}''$ related by the equality $\boldsymbol{k}'' = \boldsymbol{k}' + \boldsymbol{K}_s$, where $\boldsymbol{K}_s$ stands for an inverse lattice vector. Taking into account the way the FBZ has been constructed, if one of them, (say $\boldsymbol{k}'$) indicates a point in the interior of the FBZ, then the second, $\boldsymbol{k}''$, "*protrudes*" outside the FBZ. Let us try to construct a Bloch function that corresponds to $\boldsymbol{k}''$:

$$\phi_{\boldsymbol{k}''} = \sum_j \exp(i\boldsymbol{k}''\boldsymbol{R}_j)\chi(\boldsymbol{r}-\boldsymbol{R}_j) = \sum_j \exp\left(i(\boldsymbol{k}'+\boldsymbol{K}_s)\boldsymbol{R}_j\right)\chi(\boldsymbol{r}-\boldsymbol{R}_j) \tag{9.25}$$

$$= \exp(i\boldsymbol{K}_s\boldsymbol{R}_j)\sum_j \exp(i\boldsymbol{k}'\boldsymbol{R}_j)\chi(\boldsymbol{r}-\boldsymbol{R}_j) \tag{9.26}$$

$$= \exp(i2\pi M_{sj})\sum_j \exp(i\boldsymbol{k}'\boldsymbol{R}_j)\chi(\boldsymbol{r}-\boldsymbol{R}_j) \tag{9.27}$$

$$= \sum_j \exp(i\boldsymbol{k}'\boldsymbol{R}_j)\chi(\boldsymbol{r}-\boldsymbol{R}_j) = \phi_{\boldsymbol{k}'}. \tag{9.28}$$

It turns out that our function ϕ does behave like corresponding to $\boldsymbol{k}'$. We say that the two vectors are *equivalent.*

Vector $\boldsymbol{k}$ outside the FBZ is always equivalent to a vector from inside the FBZ, while two vectors from inside of the FBZ are never equivalent. Therefore, if we are interested in electronic states (the irreducible representation of the translational group are labeled by $\boldsymbol{k}$ vectors), it is sufficient to limit ourselves to those $\boldsymbol{k}$ vectors that are enclosed in the FBZ.

9.6 A Few Words on Bloch Functions

9.6.1 Waves in 1-D

Let us take a closer look of a Bloch function corresponding to the vector $\boldsymbol{k}$:

$$\phi_{\boldsymbol{k}}(\boldsymbol{r}) = \sum_j \exp(i\boldsymbol{k}\boldsymbol{R}_j)\chi(\boldsymbol{r}-\boldsymbol{R}_j) \tag{9.29}$$

and limit ourselves to 1-D periodicity. In such a case, the wave vector $\boldsymbol{k}$ reduces to a *wave number* k, and the vectors $\boldsymbol{R}_j$ can all be written as $\boldsymbol{R}_j = aj\boldsymbol{z}$, where $\boldsymbol{z}$ stands for the unit vector along the periodicity axis, a means the lattice constant (i.e., the nearest-neighbor distance), while $j = 0, \pm 1, \pm 2, \ldots$ Let us assume that in the lattice nodes, we have hydrogen atoms with orbitals $\chi = 1s$. Therefore, in 1-D, we have

$$\phi_k(\boldsymbol{r}) = \sum_j \exp(ikja)\chi(\boldsymbol{r}-aj\boldsymbol{z}). \tag{9.30}$$

Let me stress that ϕ_k represents a function of position $\boldsymbol{r}$ in the 3-D space, only the periodicity has a 1-D character. The function is a linear combination of the hydrogen atom $1s$ orbitals. The coefficients of the linear combination depend exclusively on the value of k. Equation (9.28) tells us that the allowed $k \in \left(0, \frac{2\pi}{a}\right)$, or alternatively, $k \in \left(-\frac{\pi}{a}, \frac{\pi}{a}\right)$. If we exceed the FBZ *length* $\frac{2\pi}{a}$, then we would simply repeat the Bloch functions. For $k = 0$, we get

$$\phi_0 = \sum_j \exp(0)\chi(\mathbf{r} - ajz) = \sum_j \chi(\mathbf{r} - ajz); \tag{9.31}$$

i.e., simply a sum of the $1s$ orbitals. Such a sum has a large value on the nuclei, and close to a nucleus the function ϕ_0 will be delusively similar to its $1s$ orbital (Fig. 9.6a).

The function looks like a chain of buoys floating on a perfect water surface. If we ask whether ϕ_0 represents a wave, the answer could be, that if any, then its wavelength is ∞. What about $k = \frac{\pi}{a}$? In such a case:

$$\phi_{\frac{\pi}{a}}(\mathbf{r}) = \sum_j \exp(ij\pi)\chi(\mathbf{r} - ajz) = \sum_j (\cos\pi j + i sin\pi j)\chi(\mathbf{r} - ajz)$$
$$= \sum_j (-1)^j \chi(\mathbf{r} - ajz).$$

If we decide to draw the function in space, we would obtain Fig. 9.6b. When asked this time, we would answer that the wavelength is equal to $\lambda = 2a$, which, by the way, is equal to[17] $\frac{2\pi}{|k|}$. There is a problem. Does the wave correspond to $k = \frac{\pi}{a}$ or $k = -\frac{\pi}{a}$? It corresponds to *both* of them. Well, does it contradict the theorem that the FBZ contains all *different* states? No, it does not. Both functions are from the border of the FBZ; their k values differ by $\frac{\pi}{2a}$ (one of the inverse lattice vectors), and therefore both functions represent *the same state*.

Now, let us take $k = \frac{\pi}{2a}$. We obtain

$$\phi_k(\mathbf{r}) = \sum_j \exp\left(\frac{i\pi j}{2}\right)\chi(\mathbf{r} - ajz) = \sum_j \left(\cos\left(\frac{\pi j}{2}\right) + i\sin\left(\frac{\pi j}{2}\right)\right)\chi(\mathbf{r} - ajz), \tag{9.32}$$

with some coefficients being complex numbers. For $j = 0$, the coefficient is equal to 1; for $j = 1$, it equals i; for $j = 2$, it takes the value -1; for $j = 3$, it attains $-i$; for $j = 4$, it is again 1, and the values repeat periodically. This is depicted in Fig. 9.6c. If this time we ask whether we see any wave there, we have to answer that we do, because after the length $4a$, everything begins to repeat. Therefore, $\lambda = 4a$, and again $\frac{2\pi}{k} = \frac{2\pi}{\frac{\pi}{2a}}$. Everything is OK except that humans like pictures more than schemes. Can we help it somehow? Let us take a look of $\phi_k(\mathbf{r})$, which corresponds to $k = -\frac{\pi}{2a}$. We may easily convince ourselves that this situation corresponds to what we have in Fig. 9.6d.

Let us stress that $\phi_{-k} = \phi_k^*$ represents *another* complex wave. By adding and subtracting $\phi_k(\mathbf{r})$ and $\phi_{-k}(\mathbf{r})$, we receive the real functions, which can be plotted and that is all we need. By adding $\frac{1}{2}(\phi_k + \phi_{-k})$, we obtain

$$\frac{1}{2}(\phi_k + \phi_{-k}) = \sum_j \cos\left(\frac{\pi j}{2}\right)\chi(\mathbf{r} - ajz), \tag{9.33}$$

[17] In the preceding case, the formula $\lambda = \frac{2\pi}{k}$ also worked because it gave $\lambda = \infty$.

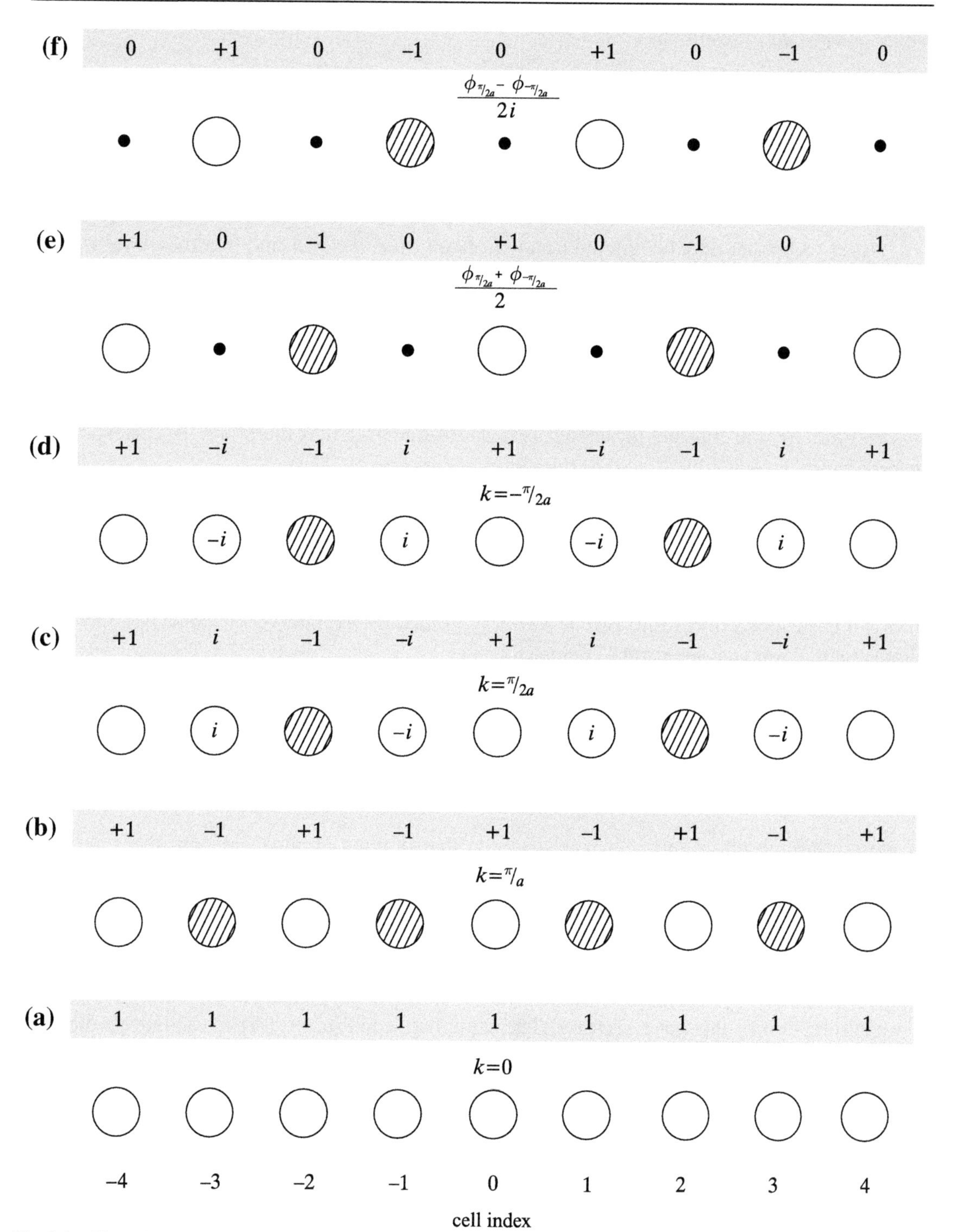

Fig. 9.6. Waves in 1-D. Shadowed (white) circles mean negative (positive) values of the function. Despite the fact that the waves are complex, in each of the cases (a)–(f), we are able to determine their wavelength.

while $\frac{1}{2i}(\phi_k - \phi_{-k})$ results in

$$\frac{1}{2i}(\phi_k - \phi_{-k}) = \sum_j \sin\left(\frac{\pi j}{2}\right) \chi(\boldsymbol{r} - ajz). \qquad (9.34)$$

Now, there is no problem with plotting the new functions (Fig. 9.6e,f).[18]

A similar technique may be applied to any k. Each time, we will find that the wave we see exhibits the wavelength $\lambda = \frac{2\pi}{k}$.

9.6.2 Waves in 2-D

Readers confident in their understanding of the wave vector concept may skip this subsection.

This time, we will consider the crystal as 2-D rectangular lattice; therefore, the corresponding inverse lattice is also 2-D, as well as the wave vectors $\boldsymbol{k} = (k_x, k_y)$.

Let us take first $\boldsymbol{k} = (0, 0)$. We immediately obtain $\phi_{\boldsymbol{k}}$ (shown in Fig. 9.7a), which corresponds to an infinite wavelength (again $\lambda = \frac{2\pi}{k}$), which looks like no wave at all.

Let us try $\boldsymbol{k} = \left(\frac{\pi}{a}, 0\right)$. The summation over j may be replaced by a double summation (indices m and n along the x- and y-axes, respectively); therefore, $\boldsymbol{R}_j = ma\boldsymbol{x} + nb\boldsymbol{y}$, where m and n correspond to the unit cell j, and a and b denote the lattice constants along the axes shown by the unit vectors $\boldsymbol{x}$ and $\boldsymbol{y}$. We have

$$\begin{aligned}\phi_{\boldsymbol{k}} &= \sum_{mn} \exp\left(i\left(k_x ma + k_y nb\right)\right) \chi(\boldsymbol{r} - ma\boldsymbol{x} - nb\boldsymbol{y}) \\ &= \sum_{mn} \exp(i\pi m) \chi(\boldsymbol{r} - ma\boldsymbol{x} - nb\boldsymbol{y}) = \sum_{mn} (-1)^m \chi(\boldsymbol{r} - ma\boldsymbol{x} - nb\boldsymbol{y}).\end{aligned}$$

If we go through all m and n, it is easily seen that moving along x, we will meet the signs $+1, -1, +1, -1, \ldots$, while moving along y, we have the same sign all the time. This will correspond to Fig. 9.7b.

This is a wave.

The wave fronts are oriented along y; i.e., the wave runs along the x-axis. Therefore, it runs in the direction of the wave vector $\boldsymbol{k}$. The same happened in the 1-D cases, but we did not express that explicitly: the wave moved along the (1-D) vector $\boldsymbol{k}$.

Exactly as before, the wavelength is equal to 2π divided by the length of $\boldsymbol{k}$. Since we are at the FBZ border, a wave with $-\boldsymbol{k}$ simply means the same wave as for $\boldsymbol{k}$.

[18] And what would happen if we took $k = \frac{\pi}{a}\frac{m}{n}$, with the integer $m < n$? We would again obtain a wave with the wavelength $\lambda = \frac{2\pi}{k}$; i.e., in this case, $\lambda = \frac{n}{m}2a$. It would be quite difficult to recognize such a wave computed at the lattice nodes because the closest wave maxima would be separated by $n2a$ and this length would have been covered by m wavelengths.

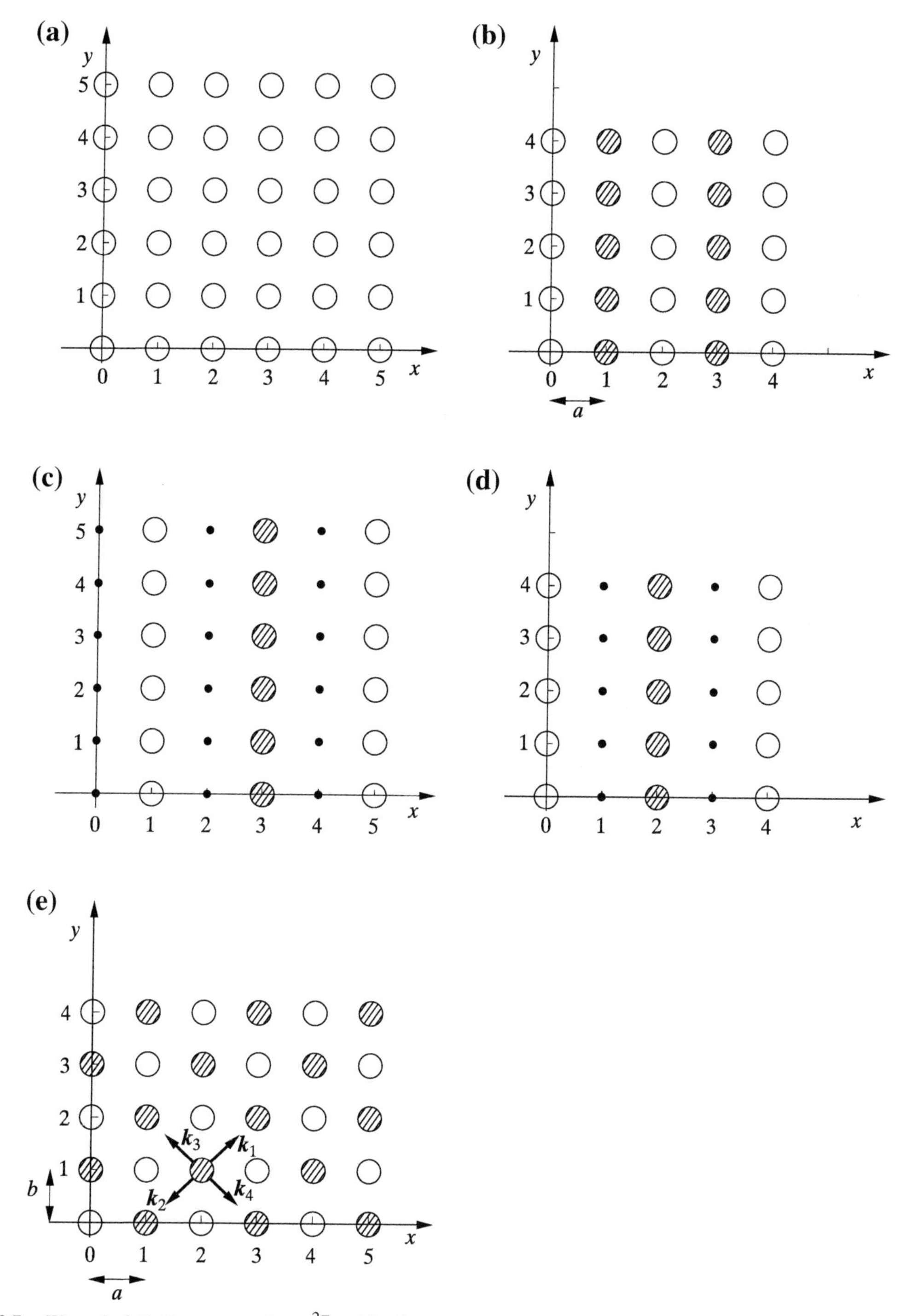

Fig. 9.7. Waves in 2-D. In any case, $\lambda = \frac{2\pi}{k}$, while the wave vector $\boldsymbol{k}$ points to the direction of the wave propagation. (a) $\boldsymbol{k} = (0, 0)$; (b) $\boldsymbol{k} = \left(\frac{\pi}{a}, 0\right)$; (c) $\boldsymbol{k} = \left(\frac{\pi}{2a}, 0\right)$, $\frac{1}{2i}\left(\phi_{\boldsymbol{k}} - \phi_{-\boldsymbol{k}}\right)$; (d) $\boldsymbol{k} = \left(\frac{\pi}{2a}, 0\right)$, $\frac{1}{2}\left(\phi_{\boldsymbol{k}} + \phi_{-\boldsymbol{k}}\right)$; (e) $\boldsymbol{k} = \left(\frac{\pi}{a}, \frac{\pi}{b}\right)$.

If we take $\boldsymbol{k} = \left[\frac{\pi}{2a}, 0\right]$, then

$$\phi_{\boldsymbol{k}} = \sum_{mn} \exp\left(i(\boldsymbol{k}_x ma + \boldsymbol{k}_y nb)\right)\chi(\boldsymbol{r} - ma\boldsymbol{x} - nb\boldsymbol{y})$$
$$= \sum_{mn} \exp\left(\frac{i\pi m}{2}\right)\chi(\boldsymbol{r} - ma\boldsymbol{x} - nb\boldsymbol{y}).$$

This case is very similar to that in 1-D for $k = \frac{\pi}{2a}$, when we look at the index m and $k = 0$, and when we take into account the index n. We may carry out the same trick with addition and subtraction, and immediately get Figs. 9.7c and d.

Is there any wave over there? Yes, there is. The wavelength equals $4a$ (i.e., $\lambda = \frac{2\pi}{k}$), and the wave is directed along vector $\boldsymbol{k}$. When making the figure, we also used the wave corresponding to $-\boldsymbol{k}$; therefore, neither the sum nor the difference correspond to $\boldsymbol{k}$ or $-\boldsymbol{k}$, but rather to both of them (we have two standing waves). The reader may guess the wavelength and direction of propagation for $\phi_{\boldsymbol{k}}$ corresponding to $\boldsymbol{k} = [0, \frac{\pi}{2b}]$.

Let us see what happens for $\boldsymbol{k} = [\frac{\pi}{a}, \frac{\pi}{b}]$. We obtain

$$\phi_{\boldsymbol{k}} = \sum_{mn} \exp\left(i(k_x ma + k_y nb\right)\chi(\boldsymbol{r} - ma\boldsymbol{x} - nb\boldsymbol{y})$$
$$= \sum_{mn} \exp\left(i(m\pi + n\pi)\right)\chi(\boldsymbol{r} - ma\boldsymbol{x} - nb\boldsymbol{y})$$
$$= \sum_{mn} (-1)^{m+n}\chi(\boldsymbol{r} - ma\boldsymbol{x} - nb\boldsymbol{y}),$$

which produces waves propagating along $\boldsymbol{k}$. And what about the wavelength? We obtain[19]

$$\lambda = \frac{2\pi}{\sqrt{(\frac{\pi}{a})^2 + (\frac{\pi}{b})^2}} = \frac{2ab}{\sqrt{a^2 + b^2}}. \tag{9.35}$$

In the last example, there is something that may worry us. As we can see, Fig. 9.7 corresponds not only to $\boldsymbol{k}_1 = (\frac{\pi}{a}, \frac{\pi}{b})$ and $\boldsymbol{k}_2 = (-\frac{\pi}{a}, -\frac{\pi}{b})$, which is understandable (as discussed above), but also to the wave with $\boldsymbol{k}_3 = (-\frac{\pi}{a}, \frac{\pi}{b})$ and to the wave evidently coupled to it–namely, with $\boldsymbol{k}_4 = (\frac{\pi}{a}, -\frac{\pi}{b})$. What is going on? Again, let us recall that we are on the FBZ border and this identity is natural because the vectors $\boldsymbol{k}_2$ and $\boldsymbol{k}_3$ as well as $\boldsymbol{k}_1$ and $\boldsymbol{k}_4$ differ by the inverse lattice vector $(0, \frac{2\pi}{b})$, which makes the two vectors equivalent.

[19] The formula can be easily verified in two limiting cases. The first corresponds to $a = b$. Then, $\lambda = a\sqrt{2}$, and this agrees with Fig. 9.7e. The second case is when $b = \infty$, which gives $\lambda = 2a$, exactly as in the 1-D case with $k = \frac{\pi}{a}$. This is what we expected.

9.7 Infinite Crystal as a Limit of a Cyclic System

9.7.1 Origin of the Band Structure

Let us consider the hydrogen atom in its ground state (cf. p. 201). The atom is described by the atomic orbital $1s$ and corresponds to energy -0.5 a.u. Let us now take two such atoms. We have two molecular orbitals: bonding and antibonding (cf. p. 439), which correspond, respectively, to energies a bit lower than -0.5 and a bit higher than -0.5 (this splitting is larger if the overlap of the atomic orbitals gets larger). We therefore have two energy levels, which stem directly from the $1s$ levels of the two hydrogen atoms. For three atoms, we would have three levels; for 10^{23} atoms, we would get 10^{23} energy levels that would be densely distributed along the energy scale but would not cover the whole scale. There will be a bunch of energy levels stemming from $1s$, i.e., an *energy band* of allowed electronic states. If we had an infinite chain of hydrogen atoms, there would be a band resulting from 1s levels, a band stemming from 2s, 2p, etc., the bands might be separated by *energy gaps*.

How dense would the distribution of the electronic levels be? Will the distribution be uniform? Answers to such questions are of prime importance for the electronic theory of crystals. It is always advisable to come to a conclusion by steps, starting from something as simple as possible, which we understand very well.

Fig. 9.8 shows how the energy level distribution looks for longer and larger rings (regular polygon) of hydrogen atoms. One of the important features of the distribution is that

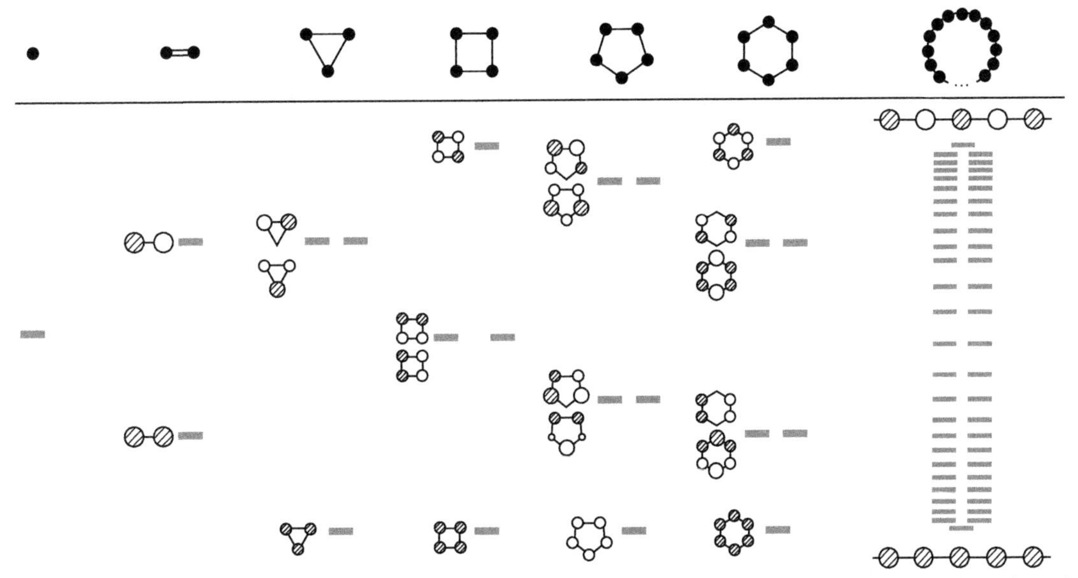

Fig. 9.8. Energy level distribution for a regular polygon built from hydrogen atoms. It is seen that the energy levels are located within an energy band and are closer to one another at the band edges. The center of the band is close to energy 0, taken as the binding energy in the isolated hydrogen atom (equal to -0.5 a.u.). Next to energy levels, the molecular orbitals are shown schematically (the shadowed circles mean negative values).

the levels extend over an energy interval and are more numerous for energy extremes.

How do the wave functions that correspond to higher and higher energy levels in a band look? Let us see the situation in the ring H_n molecules. Fig. 9.8 indicates that the rule is very simple. The number of nodes of the wave function increases by 1 when we go to the next level (higher in the energy scale).[20]

9.7.2 Born–von Kármán Condition in 1-D

How is it in the case of a crystal? Here we are confronted with the first difficulty. Which crystal, are we dealing with, and what shape is it? Should it be an ideal crystal–i.e., with perfectly ordered atoms? There is nothing like the perfect crystal in nature. For the sake of simplicity (as well as generality), let us assume, however, that our crystal *is* perfect indeed. Well, and now what about its surface (shape)? Even if we aimed at studying the surface of a crystal, the first step would be the infinite crystal (i.e., with no surface). This is the way that theoreticians always operate.[21]

One of the ingenious ideas in this direction is known as the *Born–von Kármán boundary conditions*. The idea is that instead of considering a crystal treated as a stick (let us consider the 1-D case) we treat it as a circle; i.e., the value of the wavefunction at one end of the stick has to be equal to the wavefunction value at the other end. In this way, we remove the problem of the crystal ends, and on top of that, all the unit cells become equivalent.

Theodore von Kármán (1881–1963), American physicist of Hungarian origin and director of the Guggenheim Aeronautical Laboratory at the California Institute of Technology in Pasadena. Professor von Kármán was also a founder of the NASA Jet Propulsion Laboratory and father of the concept of the first supersonic aeroplane. On the Hungarian stamp, one can see the famous "*Kármán vortex street*" behind an aeroplane. He was asked by the father of the young mathematical genius John

von Neumann to persuade him that the job of a mathematician is far less exciting than that of a banker. Theodore von Kármán did not accomplish this mission well (to the benefit of science).

The same may be done in 2-D and 3-D cases. We introduce usually the Born–von Kármán boundary conditions for a finite N and then go with N to ∞. After such a procedure is carried out, we are pretty sure that the solution we are going to obtain will not only be true for an infinite cycle, but also for the mass (bulk) of the infinite crystal. This stands to reason, provided

[20] They are bound to differ by the number of nodes because this ensures their mutual orthogonality (required for the eigenfunctions of a Hermitian operator).

[21] People say that when theoreticians attack the problem of stability of a table as a function of the number n of its legs, they do it in the following way. First, they start with $n = 0$, then they proceed with $n = 1$, then they go to $n = \infty$, and after that, they have no time to consider other values of n.

that the crystal surface does not influence the (deep) bulk properties at all.[22] In the ideal periodic case, we have to do with the cyclic translational symmetry group (see Appendix C available at booksite.elsevier.com/978-0-444-59436-5 on p. e17). The group is Abelian and, therefore, all the irreducible representations have dimension 1.

Let us assume that we have to do with N equidistant atoms located on a circle, the nearest-neighbor distance being a. From the Bloch theorem, Eq. (9.12), for the wave function ψ, we have

$$\psi(N) = \exp(-ikaN)\psi(0), \tag{9.36}$$

where we have assumed that the wave function ψ corresponds to the wave vector $\boldsymbol{k}$ (here, in 1D, to the wave number k), the translation has been carried out by Na, and as the argument of the function ψ we have (symbolically) used the number $(0, 1, 2, \ldots N-1)$ of the atom on which the function is computed.

The Born–von Kármán condition means

$$\psi(N) = \psi(0), \tag{9.37}$$

or

$$\exp(-ikaN) = 1 \tag{9.38}$$

From this, it follows that:

$$kaN = 2\pi J, \tag{9.39}$$

where $J = 0, \pm 1, \pm 2, \ldots$. This means that only *some* k are allowed; namely, $k = \frac{2\pi}{a}\frac{J}{N}$.

The Bloch functions take the form [cf. Eq. (9.29)]

$$\sum_j \exp(ikja)\chi_j, \tag{9.40}$$

where χ_j denotes a given atomic orbital (e.g., $1s$) centered on atom j. The summation over j in our case is finite, because we only have N atoms, $j = 0, 1, 2, \ldots, N-1$. Let us consider $J = 0, 1, 2, \ldots, N-1$ and the corresponding values of $k = \frac{2\pi}{a}\frac{J}{N}$. For each k, we have a Bloch function; altogether we have, therefore, N Bloch functions. Now, we may try to increase J and take $J = N$. The corresponding Bloch function may be written as

$$\sum_j \exp(i2\pi j)\chi_j = \sum_j \chi_j, \tag{9.41}$$

which turns out to be identical to the Bloch function with $k = 0$; i.e., with $J = 0$. We are reproducing what we already have. It is clear, therefore, that we have a set of those k, that form a *complete set of non-equivalent states*, they correspond to $J = 0, 1, 2, \ldots N-1$. It is also seen that if the limits of this set are shifted by the same integer (like e.g., $J = -3, -2, -1, 0, 1, 2, \ldots, N-4$), then we still have the same complete set of non-equivalent states. Staying for the time being with our primary choice of the set, we will get N values of $k \in$

[22] We circumvent the difficult problem of the crystal surface. The boundary (surface) problem is extremely important for obvious reasons: we usually *have to do with this, not with the bulk*. The existence of the surface leads to some specific, surface-related electronic states.

$\left[0, \frac{2\pi}{a}\frac{N-1}{N}\right]$; i.e., $k \in \left\{0, \frac{2\pi}{a}\frac{1}{N}, \frac{2\pi}{a}\frac{2}{N}, \ldots \frac{2\pi}{a}\frac{N-1}{N}\right\}$. *Those k values are equidistant.* When $N \to \infty$, then the section to be divided attains the length $\frac{2\pi}{a}$. Hence

the non-equivalent states (going with N to infinity) correspond to those k that are from section $\left[0, \frac{2\pi}{a}\right]$ or shifted section $\left[-\frac{\pi}{a}, +\frac{\pi}{a}\right]$, called the FBZ. From now on, we will adopt this last choice; i.e., $\left[-\frac{\pi}{a}, +\frac{\pi}{a}\right]$. We are allowed to make any shift, because, as we have shown, we keep the same non-equivalent values of k. The allowed k values are distributed *uniformly* within the FBZ. The number of the allowed k is equal to ∞ because $N = \infty$ (and the number of the allowed k is always equal to N).

9.7.3 k-Dependence of Orbital Energy

Note that the higher the energy of a molecular orbital (in our case, they are identical to the Bloch functions), the more nodes molecular orbitals have. Let us take the example of benzene ($N = 6$, cf. Fig. 9.8, this time for carbon atoms) and consider only those molecular orbitals that can be written as linear combinations of the carbon $2p_z$, where z is the axis orthogonal to the plane of the molecule. The wave vectors[23] $\left(k = \frac{2\pi}{a}\frac{J}{N}\right)$ may be chosen as corresponding to $J = 0, 1, 2, \ldots, 5$, or equivalently to $J = -3, -2, -1, 0, +1, +2$. It is seen that $J = 0$ gives a nodeless function,[24] $J = \pm 1$ lead to a pair of the Bloch functions with a single node, $J = \pm 2$ give a pair of the two-node functions, and finally $J = -3$ corresponds to a three-node function.

It has occasionally been remarked in this book (cf. e.g., Chapter 4), that increasing the number of nodes[25] results in higher energy. This rule becomes most transparent in the present case, see Fig. 9.8. A nodeless Bloch function means that all the contacts between the $2p$ orbitals being *π–bonding*, which results in *low energy*. A single node means introducing two nearest-neighbor *π–antibonding* interactions, and this causes an energy increase. Two nodes result in four antibonding interactions, and the energy goes up even more. Three nodes already give all the nearest-neighbor contacts of antibonding character, and the energy is the highest possible.

9.8 A Triple Role of the Wave Vector

As has already been said, the wave vector (in 1-D, 2-D and 3-D) plays several roles. Here they are:

1. The wave vector $\boldsymbol{k}$ tells us which type of plane wave arranged from certain objects (like atomic orbitals) we are concerned with. The direction of $\boldsymbol{k}$ is the propagation direction, and the wavelength is $\lambda = \frac{2\pi}{|\boldsymbol{k}|}$.

[23] In this case, this is a wave number.

[24] We disregard here the node that follows from the reflection in the molecular plane as being shared by all the molecular orbitals considered.

[25] That is, considering another wave function that has a larger number of nodes.

2. The wave vector may also be treated as a label for the irreducible representation of the translational group.

In other words, $\boldsymbol{k}$ determines which irreducible representation we are dealing with (see Appendix C available at booksite.elsevier.com/978-0-444-59436-5 on p. e17). This means that $\boldsymbol{k}$ tells us *which permitted rhythm* is exhibited by the coefficients at atomic orbitals in a particular Bloch function (i.e., ensuring that the square has the symmetry of the crystal). There are a lot of such rhythms; e.g., all the coefficients equal each other ($k = 0$), or one node introduced, two nodes, etc. The FBZ represents a set of such $\boldsymbol{k}$, which corresponds to *all possible rhythms*; i.e., non-equivalent Bloch functions.[26] In other words, the FBZ gives us all the possible symmetry orbitals that can be formed from an atomic orbital.

3. The longer the $\boldsymbol{k}$ is, the more nodes the Bloch function $\phi_{\boldsymbol{k}}$ has: $|\boldsymbol{k}| = 0$ means no nodes, while at the boundary of the FBZ, there is the maximum number of nodes.

9.9 Band Structure

9.9.1 Born–von Kármán Boundary Condition in 3-D

The Hamiltonian $\hat{H}$ that we have been talking about represents an effective, one-electron Hamiltonian, its form not yet given. From Chapter 8, we know that it may be taken as the Fock operator. A crystal represents nothing but a huge (quasi-infinite) molecule, and assuming the Born–von Kármán condition, a huge cyclic molecule.

This is how we will get the Hartree-Fock solution for the crystal–by preparing the Hartree-Fock solution for a cyclic molecule and then letting the number of unit cells N go to infinity.

Hence, let us take a large piece of crystal–a parallelepiped with the number of unit cells *in each of the periodicity directions* (i.e., along the three basis vectors) equals $2N+1$ (the reference cell 0, N cells on the right, N cells on the left). The particular number, $2N + 1$, is not very important, we have only to be sure that such a number is large. We assume that the Born–von Kármán condition is fulfilled. This means that we treat the crystal like a snake eating its tail, and this will happen on each of the three periodicity axes. This enables us to treat the translational group as a cyclic group, which gives an enormous simplification to our task (no end effects, all cells equivalent). The cyclic group of the lattice constants a, b, c implies that [cf. Eq. (9.38)]

$$\exp\left(ik_x a(2N+1)\right) = 1, \tag{9.42}$$

$$\exp\left(ik_y b(2N+1)\right) = 1, \tag{9.43}$$

[26] That is, linearly independent.

$$\exp\left(ik_z c(2N+1)\right) = 1, \tag{9.44}$$

which can be satisfied only for some special vectors $\boldsymbol{k} = (k_x, k_y, k_z)$:

$$k_x = \frac{2\pi}{a}\frac{J_x}{2N+1}, \tag{9.45}$$

$$k_y = \frac{2\pi}{b}\frac{J_y}{2N+1}, \tag{9.46}$$

$$k_z = \frac{2\pi}{c}\frac{J_z}{2N+1}, \tag{9.47}$$

with any of J_x, J_y, J_z taking $2N+1$ consecutive integer numbers. We may, for example, assume that $J_x, J_y, J_z \in \{-N, -N+1, \ldots, 0, 1, 2, \ldots, N\}$. Whatever N is, $\boldsymbol{k}$ will always satisfy

$$-\frac{\pi}{a} < k_x < \frac{\pi}{a}, \tag{9.48}$$

$$-\frac{\pi}{b} < k_y < \frac{\pi}{b}, \tag{9.49}$$

$$-\frac{\pi}{c} < k_z < \frac{\pi}{c}, \tag{9.50}$$

which is what we call the FBZ. Therefore, we may say that before letting $N \to \infty$,

the FBZ is filled with the allowed vectors $\boldsymbol{k}$ in a *grainlike way*; the number being equal to the number of unit cells; i.e., $(2N+1)^3$. Note that the distribution of the vectors allowed in the FBZ is *uniform*. This is ensured by the numbers J, which divide the axes $k_x, k_y,$ and k_z in the FBZ into equal pieces.

9.9.2 Crystal Orbitals from Bloch Functions (LCAO CO Method)

What we expect to obtain finally in the Hartree-Fock method for an infinite crystal are the molecular orbitals, which in this context will be called the *crystal orbitals (COs)*. As usual, we will plan to expand the CO as linear combinations of atomic orbitals (cf. 427). Which atomic orbitals? Well, those that we consider appropriate for a satisfactory description of the crystal[27]; e.g., the atomic orbitals of all the atoms of the crystal. We feel, however, that we are going to have a big problem trying to perform this task.

There will be a lot of atomic orbitals, and therefore also an astronomic number of integrals to compute (infinite for the infinite crystal), and there is nothing that can be done about that. On the other hand, if we begin such a hopeless task, the value of any integral would repeat an infinite number of times. This indicates a chance to simplify the problem. Indeed, we have not yet used the translational symmetry of the system.

[27] As for molecules.

If we are going to use the symmetry, then *we may create the Bloch functions representing the building blocks that guarantee the proper symmetry in advance*. Each Bloch function is built from an atomic orbital χ:

$$\phi_k = (2N+1)^{-\frac{3}{2}} \sum_j \exp(i\boldsymbol{k}\boldsymbol{R}_j)\chi(\boldsymbol{r}-\boldsymbol{R}_j). \tag{9.51}$$

The function is identical to that of Eq. (9.29), except it has a factor $(2N+1)^{-\frac{3}{2}}$, which makes the function approximately normalized.[28]

Any CO will be a linear combination of such Bloch functions, each corresponding to a given χ. This is equivalent to the LCAO expansion for molecular orbitals, the only difference is that we have cleverly preorganized the atomic orbitals (of one type) into symmetry orbitals (Bloch functions). Hence, it is indeed appropriate to call this approach as the LCAO CO method (*Linear Combination of Atomic Orbitals — Crystal Orbitals*), analogous to the LCAO MO (cf. p. 429). There is, however, a problem. Each CO should be a linear combination of the ϕ_k for various types of χ and for *various* $\boldsymbol{k}$. Only then would we have the full analogy: a molecular orbital is a linear combination of all the atomic orbitals belonging to the atomic basis set.[29]

It will be shown below that the situation is far better:

each CO corresponds to a single vector $\boldsymbol{k}$ from the FBZ and is a linear combination of the Bloch functions, each characterized by the same $\boldsymbol{k}$.

---

[28] The function without this factor is of class Q; i.e., normalizable for any finite N, but non-normalizable for $N=\infty$. The approximate normalization makes the function square integrable, even for $N=\infty$. Look at the following:

$$\begin{aligned}\langle\phi_k \mid \phi_k\rangle &= (2N+1)^{-3}\sum_j\sum_{j'}\exp\left(i\boldsymbol{k}(\boldsymbol{R}_j-\boldsymbol{R}_{j'})\right)\int \chi(\boldsymbol{r}-\boldsymbol{R}_j)\chi(\boldsymbol{r}-\boldsymbol{R}_{j'})d\tau\\ &= (2N+1)^{-3}\sum_j\sum_{j'}\exp\left(i\boldsymbol{k}(\boldsymbol{R}_j-\boldsymbol{R}_{j'})\right)\int \chi(\boldsymbol{r})\chi(\boldsymbol{r}-(\boldsymbol{R}_j-\boldsymbol{R}_{j'}))d\tau,\end{aligned}$$

because the integral does depend on a *relative* separation in space of the atomic orbitals. Further,

$$\langle\phi_k \mid \phi_k\rangle = \sum_j \exp(i\boldsymbol{k}\boldsymbol{R}_j)\int \chi(\boldsymbol{r})\chi(\boldsymbol{r}-\boldsymbol{R}_j)d\tau, \tag{9.52}$$

because we can replace a double summation over j and j' by a double summation over j and $j''=j-j'$ (both double summations exhaust all the lattice nodes), and the latter summation always gives the same independent of j; the number of such terms is equal to $(2N+1)^3$. Finally, we may write $\langle\phi_k \mid \phi_k\rangle = 1+$ various integrals. The largest of these integrals is the nearest-neighbor overlap integral of the functions χ. For normalized χ, each of these integrals represents a fraction of 1; additionally, the contributions for further neighbors decay exponentially (cf. p. e137). As a result, $\langle\phi_k \mid \phi_k\rangle$ is a number of the order of 1 or 2. This is what we have referred to as an *approximate normalization*.

[29] Indeed, for any $\boldsymbol{k}$, the number of distinct Bloch functions is equal to the number of atomic orbitals per unit cell. The number of allowed vectors, $\boldsymbol{k}$, is equal to the number of unit cells in the crystal. Hence, using the Bloch functions for all allowed $\boldsymbol{k}$ would be justified, and any CO would represent a linear combination of all the atomic orbitals of the crystal.

There are, however, only a few Bloch functions — their number is equal to the number of the atomic orbitals per unit cell (denoted by ω).[30]

It is easy to show that, indeed, we can limit ourselves to a single vector $\boldsymbol{k}$. Imagine that this is false, and our CO is a linear combination of all the Bloch functions with all possible $\boldsymbol{k}$. When, in the next step, we solve the orbital equation with the effective (i.e., Fock) Hamiltonian using the Ritz method, then we will end up computing the integrals $\langle \phi_{\boldsymbol{k}} \mid \hat{F}\phi_{\boldsymbol{k}'} \rangle$ and $\langle \phi_{\boldsymbol{k}} \mid \phi_{\boldsymbol{k}'} \rangle$. For $\boldsymbol{k} \neq \boldsymbol{k}'$, such integrals *equal zero* according to group theory (see Appendix C available at booksite.elsevier.com/978-0-444-59436-5 on p. e17), because $\hat{F}$ transforms according to the fully symmetric irreducible representation of the translational group,[31] while $\phi_{\boldsymbol{k}}$ and $\phi_{\boldsymbol{k}'}$ transform according to *different* irreducible representations.[32] Therefore, the secular determinant in the Ritz method will have a *block form* (see Appendix C available at booksite.elsevier.com/978-0-444-59436-5). The first block will correspond to the first $\boldsymbol{k}$, the second to the next $\boldsymbol{k}$, etc., where every block[33] would look as if in the Ritz method, we used the Bloch functions corresponding uniquely to that particular $\boldsymbol{k}$. Conclusion: since a CO has to be a wave with a *given* $\boldsymbol{k}$, let us construct it with Bloch functions, which already have just this type of behavior with respect to translation operators (i.e., have just this $\boldsymbol{k}$). This is fully analogous with the situation in molecules, if we used atomic symmetry orbitals.[34]

Thus, each vector $\boldsymbol{k}$ from the FBZ is associated with a crystal orbital, and therefore with a set of LCAO CO coefficients.

The number of such CO sets (each $\boldsymbol{k}$ being one set) *in principle* has to be equal to the number of unit cells (i.e., infinite).[35] The only profit we may expect could be associated with the hope that the computed quantities do not depend on $\boldsymbol{k}$ too much but will rather change smoothly when $\boldsymbol{k}$ changes. This is indeed what will happen, and *then a small number of vectors* $\boldsymbol{k}$ *will be used, and the quantities requiring other* $\boldsymbol{k}$ *will be computed by interpolation.*

Only a part of the computed COs will be occupied, and this depends on the orbital energy of a given CO, the number of electrons, and the corresponding $\boldsymbol{k}$, similar to what we saw for molecules.

---

[30] Our optimism pertains, of course, to taking a modest atomic basis set (small ω).

[31] Unit cells (by definition) are identical.

[32] Recall that $\boldsymbol{k}$ also has the meaning of the irreducible representation index (of the translational group).

[33] The whole problem can be split into independent problems for individual blocks.

[34] A symmetry atomic orbital (SAO) represents such linear combination of equivalent-by-symmetry AOs that transforms according to one of the irreducible representations of the symmetry group of the Hamiltonian. Then, when molecular orbitals (MOs) are formed in the LCAO MO procedure, any given MO is a linear combination of the SAOs belonging to a particular irreducible representation. For example, the water molecule exhibits a symmetry plane (σ) that is perpendicular to the plane of the molecule. A MO, which is symmetric with respect to σ does contain the SAO $1s_a + 1s_b$, but does not contain the SAO $1s_a - 1s_b$.

[35] Well, we cannot fool Mother Nature. Was there an infinite molecule (crystal) to be computed or not? Then the number of such sets of computations has to be infinite.

The set of SCF LCAO CO equations will be very similar to the set for the molecular orbital method (SCF LCAO MO). In principle, the only difference will be that, in the crystal case, we will consequently use symmetry orbitals (Bloch functions) instead of atomic orbitals.

That's it. The rest of this section is associated with several technical details accompanying the operation $N \to \infty$.

9.9.3 SCF LCAO CO Equations

Let us write down the SCF LCAO CO equations as if they corresponded to a large molecule (Bloch functions will be used instead of atomic orbitals). Then the nth CO may be written as

$$\psi_n(\boldsymbol{r}, \boldsymbol{k}) = \sum_q c_{qn}(\boldsymbol{k})\phi_q(\boldsymbol{r}, \boldsymbol{k}), \tag{9.53}$$

where ϕ_q is the Bloch function corresponding to the atomic orbital χ_q:

$$\phi_q(\boldsymbol{r}, \boldsymbol{k}) = (2N+1)^{-\frac{3}{2}} \sum_j \exp(i\boldsymbol{k}\boldsymbol{R}_j)\chi_q^j, \tag{9.54}$$

with $\chi_q^j \equiv \chi_q(\boldsymbol{r} - \boldsymbol{R}_j)$ (for $q = 1, 2, \ldots, \omega$).

The symbol χ_q^j means the qth atomic orbital (from the set that we prepared for the unit cell motif) located in the cell indicated by vector $\boldsymbol{R}_j$ (jth cell).

In the expression for ψ_n, we have taken into account that there is no reason whatsoever that the coefficients c were $\boldsymbol{k}$–independent, since the expansion functions ϕ depend on $\boldsymbol{k}$. This situation does not differ from what we encountered in the Hartree-Fock-Roothaan method (cf. p. 431), with one technical exception: instead of the atomic orbitals, we have symmetry orbitals (in our case Bloch functions).

The secular equations for the Fock operator will have, of course, the form of the Hartree and Fock-Roothaan equations (cf. Chapter 8, p. 431):

$$\sum_{q=1}^{\omega} c_{qn}[F_{pq} - \varepsilon_n S_{pq}] = 0$$

for $p = 1, 2, \ldots, \omega$,

where the usual notation has been applied. For the sake of simplicity, we have not highlighted the $\boldsymbol{k}$–dependence of c, F, and S. Whenever we decide to do this in the future, we will put it in the form $F_{pq}(\boldsymbol{k})$, $S_{pq}(\boldsymbol{k})$, etc. Of course, ε_n will become a function of $\boldsymbol{k}$, as will be stressed by the symbol $\varepsilon_n(\boldsymbol{k})$. Theoretically, the secular equation has to be solved for every $\boldsymbol{k}$ of the FBZ.

Therefore, despite the fact that the secular determinant is of rather low rank (ω), the infinity of the crystal forces us to solve this equation an infinite number (the number of vectors $\boldsymbol{k}$) of times. For the time being, though, do not worry too much.

9.9.4 Bandwidth

The number of secular equation solutions is equal to ω. Let us label them using index n. If we focus on one such solution and check whether $\varepsilon_n(\boldsymbol{k})$ and $\psi_n(\boldsymbol{r}, \boldsymbol{k})$ are sensitive to a tiny change of $\boldsymbol{k}$ within the FBZ, it turns out that $\varepsilon_n(\boldsymbol{k})$ and $\psi_n(\boldsymbol{r}, \boldsymbol{k})$ change smoothly. This may not be true when $\boldsymbol{k}$ passes through the border of the FBZ, however.

The function $\varepsilon_n(\boldsymbol{k})$ is called the nth electronic band.

If we traveled in the FBZ, starting from the origin and continuing along a straight line, then $\varepsilon_1, \varepsilon_2$, ...etc. would change as functions of $\boldsymbol{k}$ and create several energy bands. If $\varepsilon_n(\boldsymbol{k})$ changes very much during our travel over the FBZ, we would say that the nth band has large width or *dispersion*.

As it was shown on p. 523 for the hydrogen atoms, an energy band forms due to the bonding and antibonding effects, the energy splitting being of the order of the overlap integral between the nearest-neighbor $1s$ AOs. If instead of hydrogen atoms, we put a unit cell with a few atoms inside (motif), then the story is similar: the motif has some one-electron energy levels (orbital energies), putting together the unit cells makes changing these energy levels into energy bands, the number of levels in any band is equal to the number of unit cells, or the number of allowed $\boldsymbol{k}$ vectors in the FBZ.

The bandwidth is related to interactions among the unit cell contents, and is roughly proportional to the overlap integral between the orbitals of the interacting unit cells.

How do we plot the band structure? For the 1-D crystal, such as a periodic polymer, there is no problem: the number k (the wave vector $\boldsymbol{k}$) changes from $-\frac{\pi}{a}$ to $\frac{\pi}{a}$, we plot the function $\varepsilon_n(k)$. For each n, we have a single plot; e.g., for the hydrogen atom, the band ε_1 collects energies resulting from the $1s$ atomic orbital interacting with other atoms, similarly the band ε_2, which resulted from $2s$, etc. In the 3-D case, we usually choose a path in FBZ. We start from the point Γ, defined as $\boldsymbol{k} = \boldsymbol{0}$. Then, we continue to some points located on the faces and edges of the FBZ surface. It is impossible to go through the whole FBZ. The band structure in the 3-D case is usually shown by putting the described itinerary through the FBZ on the abscissa (Fig. 9.9), and $\varepsilon_n(\boldsymbol{k})$ on the ordinate. Fig. 9.9 shows an example of what we might obtain from such calculations.

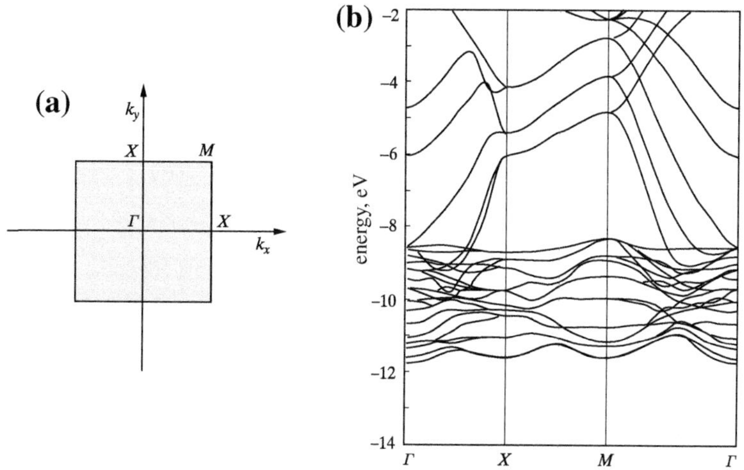

Fig. 9.9. (a) FBZ for four regular layers of nickel atoms (a crystal surface model); (b) the band structure for this system. We see that we cannot understand much: it is just a horribly irregular mess of lines. All the band structures look equally clumsy. Despite of this, from such a plot, we may determine the electrical and optical properties of the nickel slab. We will see later in this chapter why the bands have such a mysterious form.

9.9.5 Fermi Level and Energy Gap: Insulators, Metals, and Semiconductors

Insulators

How many electrons are in a crystal? The answer is simple: the infinite crystal contains an infinite number of electrons. But infinities are often different. The decider is the number of electrons per unit cell. Let us denote this number by n_0.

If this means a double occupation of the molecular orbitals of the unit cell, then the corresponding band in the crystal will also be fully occupied, because the number of energy levels in a band is equal to the number of unit cells, and each unit cell contributes two electrons from the above mentioned molecular orbital. The bands that come from the valence orbitals of the motif are called *valence bands*. Therefore,

> doubly occupied orbitals of the motif, related usually to the inner electronic shells, lead to fully occupied bands. Accordingly, singly occupied orbitals lead to bands that are *half–occupied*, while empty (virtual) orbitals lead to empty bands (unoccupied bands, also called conduction bands).

The highest-occupied crystal orbital is known as the Fermi level; it is equivalent to the HOMO of the crystal.[36] The two levels HOMO and LUMO are fundamental as always–they

[36] We sometimes find a thermodynamic definition of the Fermi level, but in this book, it will always be the highest-occupied crystal orbital.

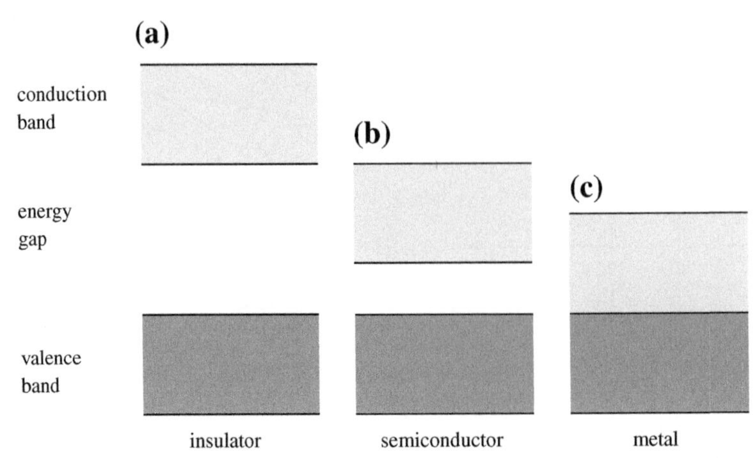

Fig. 9.10. Valence bands (highest occupied by electrons, dark gray) and conduction bands (empty, light gray). The electric properties of a crystal depend on the energy gap between them (i.e., HOMO-LUMO separation). A large gap (a) is typical for an insulator, a medium gap (b) means a semiconductor, and a zero gap (c) is typical of metals.

decide about the chemistry of the system (in our case, the chemical and physical properties of the crystal).

> The gap between the HOMO and LUMO of the crystal is the gap between the top of the occupied valence band and the bottom of the conduction band; see Fig. 9.10. When the gap is large we have to do with insulators.

Metals, 1-D Metals, and Peierls Distortion

A partially filled band may lead to the situation where the band gap equals zero.

> A metal is characterized by having empty levels (conduction band) immediately above doubly occupied valence ones.

Metals, because of the zero gap, are conductors of electric current.[37]

The conductivity of the metallic systems is typically orientation independent. In the last several decades, 2-D and 1-D metals (with anisotropy of conductivity) have been discovered. The latter are called *molecular wires* and may have unusual properties, but are difficult to prepare because they often undergo spontaneous dimerization of the lattice (known as the *Peierls transition*).

[37] When an electric field is applied to a crystal, its energy levels change. If the field is weak, then the changes may be computed by perturbation theory (treating the zero–field situation as the unperturbed one). This means that the perturbed states acquire some admixtures of the excited states (cf. Chapter 5). The lower the energy gap is, the more mixing takes place. For metallic systems (with a gap of zero), such perturbation theory certainly would not be applicable, but real excitation to the conduction band may take place.

As Fig. 9.11a shows, dimerization makes the bonding and antibonding effects stronger a little below and above the middle of the band ($k = \frac{\pi}{2a}$), whereas at the band edges ($k = 0$), the effect is almost zero (since dimerization makes the bonding or antibonding effects cancel within a pair of consecutive bonds). As a result, the degeneracy is removed in the middle of the band (Fig. 9.11b); i.e., a band gap appears and the system undergoes metal-insulator transition (Fig. 9.11c). This is why polyacetylene, instead of having all the CC bonds equivalent (Fig. 9.11d), which would make it a metal, exhibits alternation of bond lengths (Fig. 9.11e), and it becomes an insulator or semiconductor.

Rudolph Peierls (1907–1995), British physicist and professor at the universities of Birmingham and Oxford. Peierls participated in the Manhattan Project (which resulted in the atomic bomb) as the leader of the British group.

Peierls proposed the mechanism of the metal-insulator phase transition.

To a chemist, the Peierls transition is natural. The hydrogen atoms will not stay equidistant in a chain but will simply react and form hydrogen molecules; i.e., it will dimerize like lightning. Also, the polyacetylene will try to form π bonds by binding the carbon atoms in *pairs*. There is simply a shortage of electrons to keep *all* the CC bonds strong; in fact, *there are only enough for every second bond*, which means simply dimerization through creating π bonds. On the other hand, the Peierls transition may be described as similar to the Jahn-Teller effect: there is a degeneracy of the occupied and empty levels at the Fermi level, and it is therefore possible to lower the energy by removing the degeneracy through a distortion of geometry (i.e., dimerization).

Polyacetylene, becomes ionized after doping if the dopants are electron acceptors, or it receives extra electrons if the dopant represents an electron donor (symbolized by D^+ in Fig. 9.12). The perfect polyacetylene exhibits the bond alternation discussed above, but it may be that we have a defect that is associated with a region of "*changing rhythm*" (or "*phase*"): from[38] $(= - = - =)$ to $(- = - = -)$. Such a kink is sometimes described as a *soliton* wave (Fig. 9.12a,b); i.e., a "*solitary*" wave first observed in the 19th century in Scotland on a water channel, where it preserved its shape while moving over a distance of several kilometers. The soliton defects cause some new energy levels ("*solitonic levels*") to appear within the gap. These levels too form their own solitonic band.

Charged solitons may travel when subject to an electric field, and therefore the doped polyacetylene turns out to be a good conductor (organic metal).

In polyparaphenylene, soliton waves are impossible because the two phases (aromatic and quinoid, as shown in Fig. 9.12c) differ in energy (low-energy aromatic phase and high-energy quinoid phase). However, when the polymer is doped, a charged double defect (*bipolaron*, Fig. 9.12c) may form, and the defect may travel when an electric field is applied. Hence, the doped polyparaphenylene, similarly to the doped polyacetylene, is an organic metal.

[38] This possibility was first recognized by J.A. Pople and S.H. Walmsley, *Mol.Phys.*, *5*, 15 (1962), which was published 15 years before the experimental discovery of this effect.

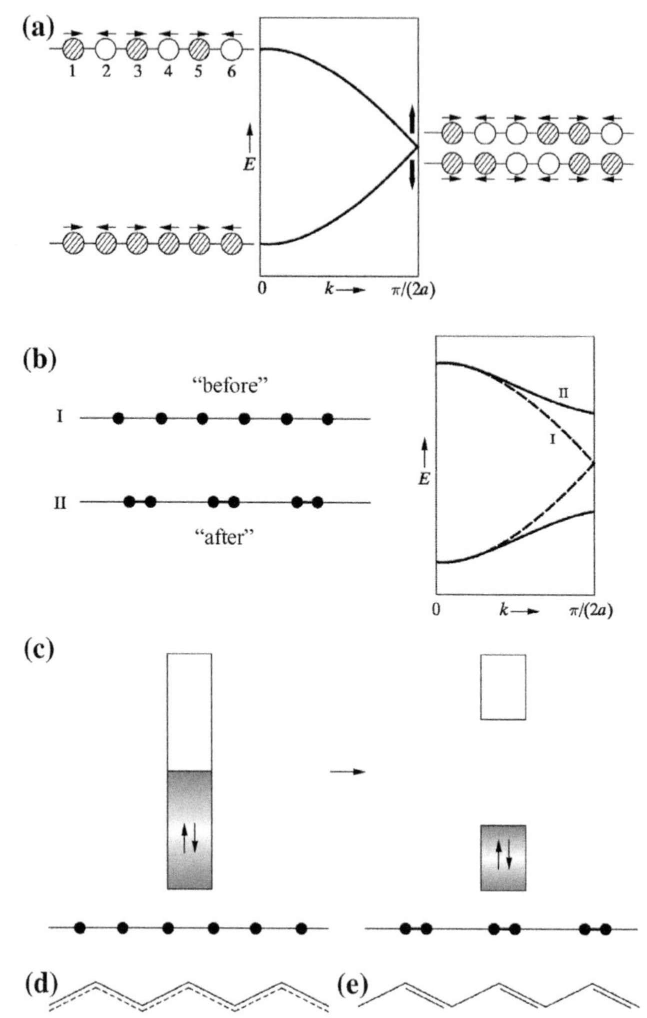

Fig. 9.11. The Peierls effect has the same origin as the Jahn-Teller effect in removing the electronic level degeneracy by distorting the system [H.A.Jahn and E.Teller, *Proc.Roy.Soc.A 161*, 220 (1937)]. The electrons occupy half the FBZ; i.e., $-\frac{\pi}{2a} \leq k \leq \frac{\pi}{2a}$, a standing for the nearest-neighbor distance. The band has been plotted on the assumption that the period is equal to $2a$; hence a characteristic back folding of the band (similar to the way that we would fold a sheet of paper with the band structure drawn, the period equaling a). (a) A lattice dimerization, shown by little arrows, amplifies the bonding and antibonding effects close to the middle of the FBZ; i.e., in the neighborhood of $k = \pm\frac{\pi}{2a}$. Close to $k = 0$, there is a cancellation of the opposite effects: bonding (bottom) and antibonding (top). (b) As a result, the degeneracy at $k = \frac{\pi}{2a}$ is removed and the band gap appears, which corresponds to lattice dimerization. (c) The system lowers its energy when undergoing metal-insulator or metal-semiconductor transition. (d) The polyacetylene chain, if forcing equivalence of all CC bonds, represents a metal. However, due to the Peierls effect, the system undergoes dimerization (e) and becomes an insulator.

Controlling the Metal Fermi Level–An Electrode

The Fermi level (i.e., HOMO level) is especially interesting in metal because there are ways to change its position on the energy scale. We may treat the metal as a container for electrons: we may pump the electrons into it or create the electron deficiency in it by using it as a cathode or anode, respectively. Having a tunable HOMO level, we decide if and when our reactant (i.e., the

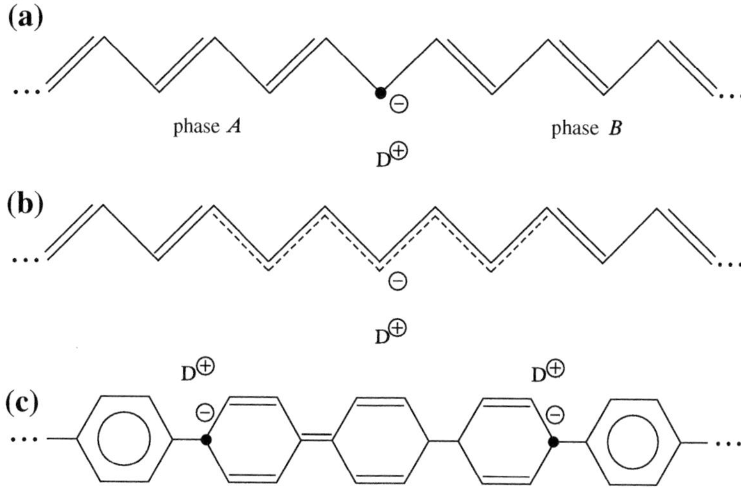

Fig. 9.12. Solitons and bipolarons as a model of electric conductivity in polymers. (a) Two phases of polyacetylene separated by a defect. Originally, the defect was associated to an unpaired electron, but when a donor, D, gave its electron to the chain, the defect became negatively charged. (b) The energy of such a defect is independent of its position in the chain (important for charge transportation), but in reality, the change of phase takes place in sections of about 15 CC bonds, not two bonds as (a) suggests. Such a situation is sometimes modeled by a nonlinear differential equation, which describes a soliton motion ("*solitary wave*") that represents the traveling phase boundary. (c) In the polyparaphenylene chain, two phases (low-energy aromatic and high-energy quinoid) are possible as well, but in this case, they are of different energies. Therefore, the energy of a single defect (aromatic structures-kink-quinoid structures) depends on its position in the chain (therefore, no charge transportation). However, a *double* defect with a (higher-energy) section of a quinoid structure has a position-independent energy, and when charged by dopants (*bipolaron*) can conduct electricity. The abovementioned polymers can be doped either by electron donors (e.g., arsenium, potassium) or electron acceptors (iodine), which results in a spectacular increase in their electric conductivity.

electrode) acts as an electron donor or electron acceptor. This opens new avenues such as polarography, when scanning the electrode potential results in consecutive electrode reactions occurring whenever the electrode Fermi level matches the LUMO of a particular substance present in the solution. Since the matching potentials are characteristic for the substances, this is a way of performing chemical identification with quantitative analysis.

Semiconductors

An intrinsic semiconductor exhibits a conduction band separated by a small energy gap (band gap) from the valence band (see Fig. 9.13a).

If the empty energy levels of the dopant are located just over the occupied band of an intrinsic semiconductor, the dopant may serve as an electron acceptor for the electrons from the occupied band (thus introducing its own conduction band), we have a *p-type semiconductor*, (Fig. 9.13b). If the dopant energy levels are occupied and located just under the conduction band, the dopant may serve as a *n-type semiconductor* (Fig. 9.13c).

Among these three fundamental classes of materials: insulators, metals, and semiconductors, the semiconductors are most versatile as to their properties and practical applications. The metals just conduct electricity, and the carriers of the electric current are electrons. The metals'

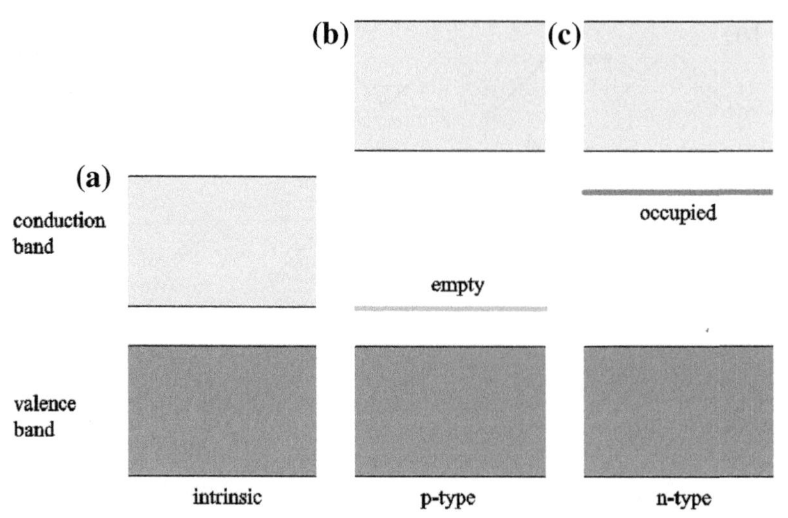

Fig. 9.13. Energy bands for semiconductors. (a) Intrinsic semiconductor (small gap), (b) p-type semiconductor (electron acceptor levels close to the occupied band), (c) n-type semiconductor (electron donor levels close to the conduction band).

conductivity spans only one order of magnitude. The insulators are useful only because they do not conduct electric current. In contrast to the metals and the insulators the conductivity of semiconductors can be controlled within many orders of magnitude (mainly by doping; i.e., admixture of other materials). The second extraordinary feature is that only in semiconductors, the conductivity can be tuned by using two types of the charge carriers: (negative) electrons and (positive) electron holes. The results of such tuning depends on temperature, light, and electric and magnetic fields. In contrast to metals and insulators, the semiconductors are able to emit visible light. All these features make it possible to tailor functional semiconductor devices to versatile electric and photonic properties. This is why in practically any electric or photonic equipment, there is a semiconductor device.

Additionally, the reasons why the organic metals and semiconductors are of practical interest are the versatility and tunability (precision) offered by organic chemistry, the easy processing typical of the plastic industry, the ability to literally bent the device without losing its properties, and last but not least, a low weight.

What kind of substances are semiconductors? Well, the most important class of them can be derived directly from a section of the Mendeleev periodic table (the first row shows the group number) Table 9.1. According to the Table we have: *IV-IV semiconductors*: the elemental

Table 9.1. A "*semiconductor section*" of the Mendeleev periodic table.

II	III	IV	V	VI
	B	C	N	O
	Al	Si	P	S
Zn	Ga	Ge	As	Se
Cd	In	Sn	Sb	Te

semiconductors C, Si, Ge as well as the compounds SiGe, SiC, *III-V semiconductors*: GaN, GaP, InP, InSb, etc., *II-IV semiconductors*: CdSe, CdS, CdTe, ZnO, ZnS, etc.

9.10 Solid-State Quantum Chemistry

A calculated band structure, with information about the position of the Fermi level, tell us a lot about the electric properties of the material being looked at here (insulator, semiconductor, and metal). They tell us also about basic optical properties; e.g., the band gap indicates what kind of absorption spectrum we may expect. We can calculate any measurable quantity because we have at our disposal the computed (approximate though) wave function.

However, despite this very precious information, which is present in the band structure, there is a little worry. When we look at any band structure, such as that shown in Fig. 9.9, the overwhelming feeling is a kind of despair. All band structures look similar–as just a tangle of plots. Some go up, some down, some stay unchanged, and some change their direction (seemingly for no reason). Can we understand this? What is the theory behind this band behavior?

9.10.1 Why Do Some Bands Go Up?

Let us take our beloved chain of hydrogen atoms in the 1*s* state, to which we already owe so much (Fig. 9.14).

When will the state of the chain have the lowest possible energy? Of course, when all the atoms interact in a bonding way and not in an antibonding way. This corresponds to Fig. 9.14a (no nodes of the wave function). When, in this situation, we introduce a single nearest-neighbor antibonding interaction, the energy will increase a bit (Fig. 9.14b). When two such interactions are introduced (Fig. 9.14c), the energy goes up even more, and the plot corresponds to two nodes. Finally, the highest-energy situation: all nearest-neighbor interactions are antibonding (maximum number of nodes), as shown in Fig. 9.14d. Let us recall that the wave vector was associated with the number of nodes. Hence, if k increases from zero to $\frac{\pi}{a}$, the energy increases from the energy corresponding to the nodeless wave function to the energy characteristic for

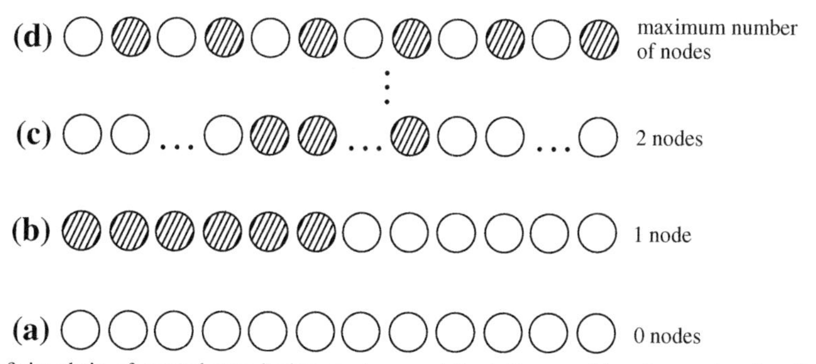

Fig. 9.14. The infinite chain of ground-state hydrogen atoms and the influence of bonding and antibonding effects. (a) All interactions are bonding; (b) introduction of a single node results in an energy increase; (c) two nodes increase the energy even more; (d) with the maximum number of nodes, the energy is at its maximum.

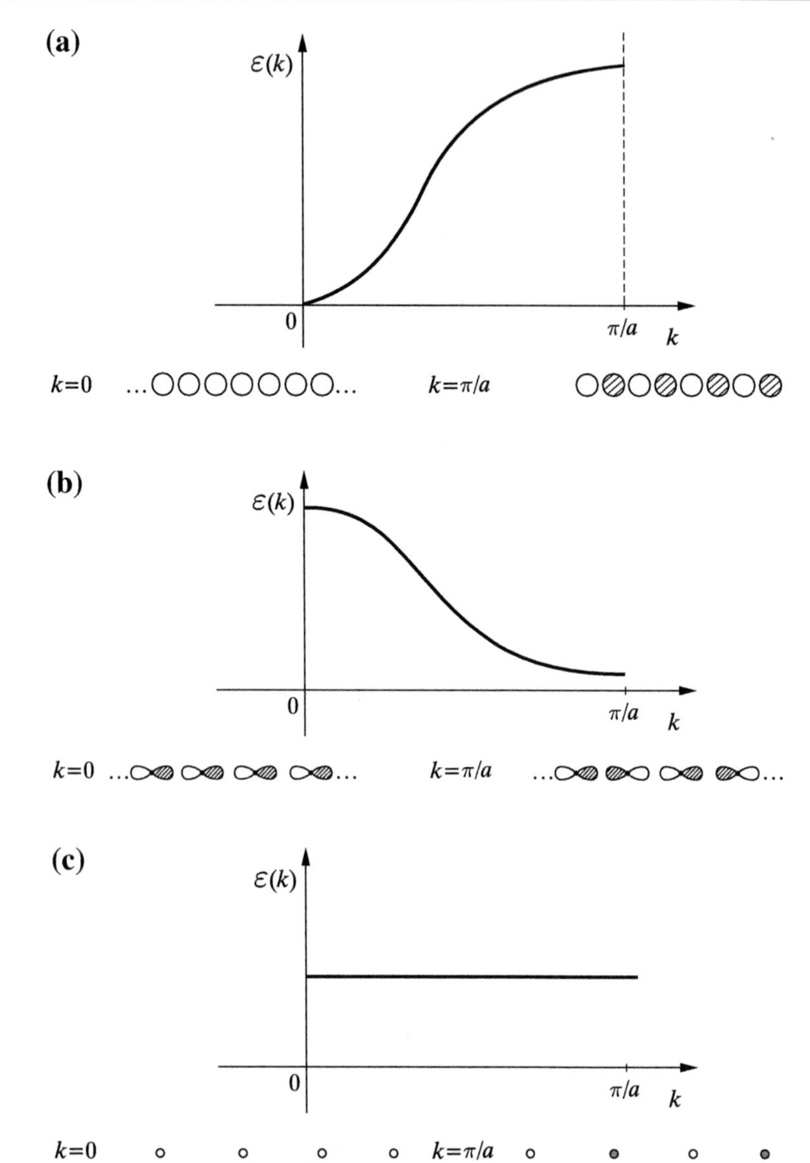

Fig. 9.15. Three typical band plots in the FBZ. (a) $1s$ orbitals. Increasing k is accompanied by an *increase* of the antibonding interactions, which is why the energy goes up. (b) $2p_z$ orbitals (z denotes the periodicity axis). Increasing k results in *decreasing* the number of antibonding interactions and the energy goes down. (c) Inner-shell orbitals. The overlap is small as it is; therefore, the band width is practically zero.

the maximum-node wave function. We understand, therefore, that some band plots are as they appear in Fig. 9.15a.

9.10.2 Why Do Some Bands Go Down?

Sometimes the bands go in the opposite direction: the lowest energy corresponds to $k = \frac{\pi}{a}$, the highest energy to $k = 0$. What happens over there? Let us once more take the hydrogen atom chain, but this time in the $2p_z$ state (z is the periodicity axis). Now, the Bloch function corresponding to $k = 0$ (i.e., a function that follows just from locating the orbitals $2p_z$ side by side), describes the highest-energy interaction — the *nearest-neighbor interactions are all antibonding*. Introduction of a node (increasing k) means a relief for the system–instead of one painful antibonding interaction, we get a soothing bonding one. The energy goes down. No wonder, therefore, some bands look like those shown in Fig. 9.15b.

9.10.3 Why Do Some Bands Stay Constant?

According to numerical rules, inner-shell atomic orbitals do not form effective linear combinations (crystal orbitals). Such orbitals have very large exponential coefficients and the resulting overlap integral, and therefore the band width (bonding vs antibonding effect), is negligible. This is why the nickel $1s$ orbitals (deep-energy level) result in a low-energy band of almost zero width (Fig. 9.15c); i.e., staying flat as a pancake all the time. Since they are always of very low energy, they are doubly occupied, and their plot is so boring that they are not even displayed (they are absent in Fig. 9.9).

9.10.4 More Complex Behavior Explainable–Examples

We understand, therefore, at least why some bands are monotonically going down, some going up, and others staying constant. In explaining these cases, we have assumed that a given CO is dominated by a single Bloch function. Other behaviors can be explained as well by detecting what kind of Bloch function *combination* we have in a given crystal orbital.

2-D Regular Lattice of the Hydrogen Atoms

Let us take a planar regular lattice of hydrogen atoms in their ground state.[39] Fig. 9.9 shows the FBZ of a similar lattice. We (arbitrarily) choose as the itinerary through the FBZ: $\Gamma - X - M - \Gamma$. From Fig. 9.7a, we easily deduce that the band energy for the point Γ has to be the lowest because it corresponds to all the interaction bonding. What will happen at the point X (Fig. 9.9a), which corresponds to $\boldsymbol{k} = (\pm\frac{\pi}{a}, 0)$ or $\boldsymbol{k} = (0, \pm\frac{\pi}{a})$? This situation is related to Fig. 9.7b. If we focus on any of the hydrogen atoms, it has four nearest-neighbor interactions: two bonding and two antibonding. This, to a good approximation, corresponds to the nonbonding situation (hydrogen atom ground-state energy), because the two effects nearly cancel out. Halfway between Γ and X, we go through the point that corresponds to Fig. 9.7c,d. For such a point, any hydrogen atom has two bonding and two nonbonding interactions; i.e., the energy is the average of the Γ and X energies. The point M is located in the corner of the FBZ, and corresponds to Fig. 9.7e. All the nearest-neighbor interactions are antibonding there, and the energy will be very high. We may, therefore, anticipate a band structure of the kind sketched in Fig. 9.16a, which was drawn

[39] A chemist's first thought would be that this could never stay this way, when the system is isolated. We are bound to observe the formation of hydrogen molecules.

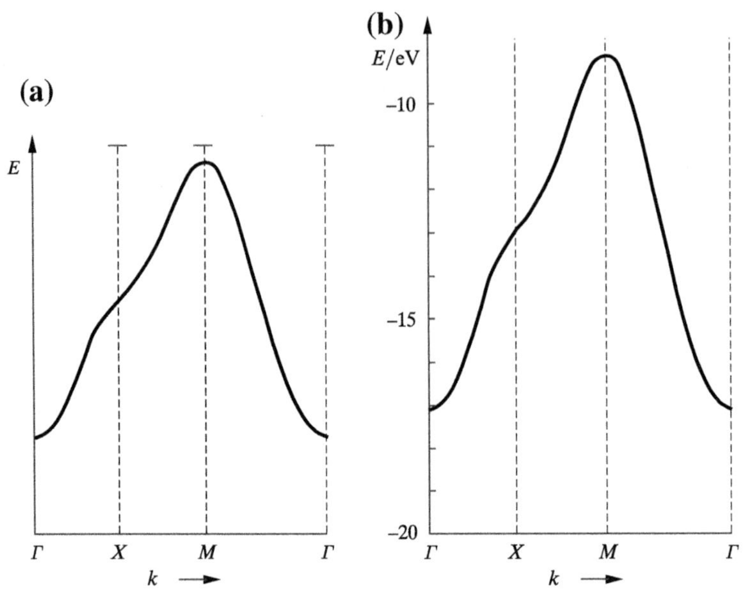

Fig. 9.16. (a) A Roald Hoffmann's sketch of the valence band for a regular planar lattice of ground-state hydrogen atoms and (b) the valence band as computed in the laboratory of Roald Hoffmann for a nearest-neighbor distance equal to 2 Å. The similarity of the two plots confirms that we are able, at least in some cases, to predict band structure.

to reflect the fact that the density of states for the band edges is the largest, and therefore the slope of the curves has to reflect this. Fig. 9.16b shows the results of the computations.[40] It is seen that, even very simple reasoning may rationalize the main features of band structure plots.

Trans-polyacetylene (Regular 1-D Polymer)

Polyacetylene already has quite a complex band structure, but as usual, the bands close to the Fermi level (valence bands) are the most important in chemistry and physics. All these bands are of the π type; i.e., their COs are antisymmetric with respect to the plane of the polymer. Fig. 9.17 shows how the valence bands are formed.[41] We can see that the principle is identical to that for the chain of the hydrogen atoms: the more nodes there are, the higher the energy is. The highest energy corresponds to the band edge.

The resulting band is only *half-filled* (metallic regime) because each of the carbon atoms offers one electron, and the number of COs is equal to the number of carbon atoms (each CO can accommodate two electrons). Therefore, the Peierls mechanism (Fig. 9.11) is bound to enter into play, and in the middle of the band, a gap will open. The system is, therefore, predicted

[40] R. Hoffmann, *Solids and Surfaces. A Chemist's View of Bonding in Extended Structures*, VCH Publishers, New York (1988).

[41] J.-M. André, J. Delhalle, and J.-L. Brédas, *Quantum Chemistry Aided Design of Organic Polymers*, World Scientific, Singapore (1991).

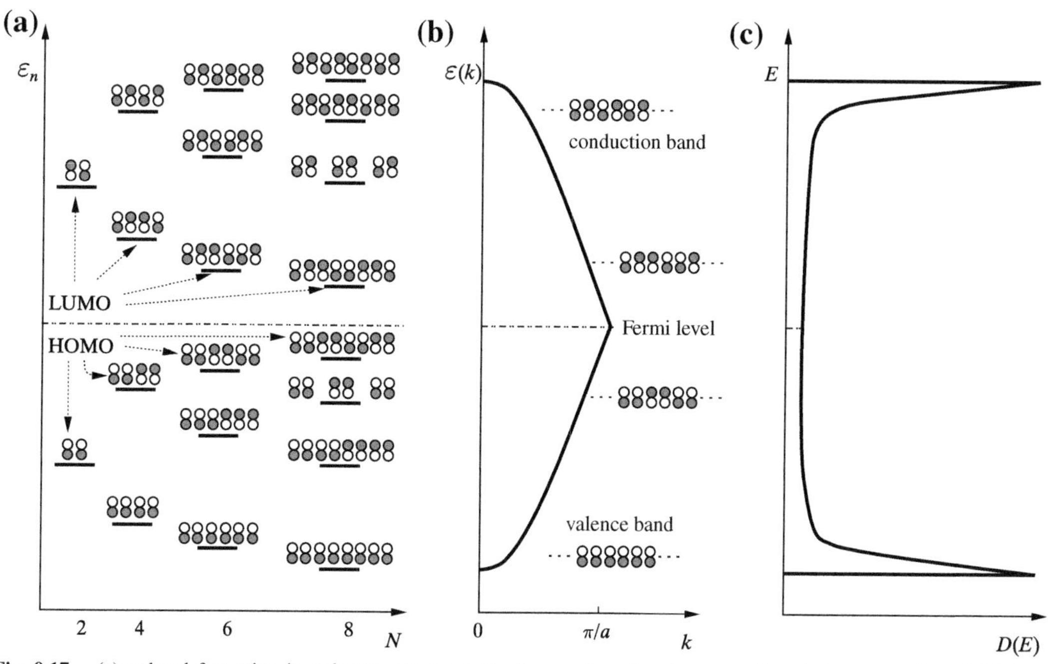

Fig. 9.17. (a) π-band formation in polyenes (N stands for the number of carbon atoms) with the assumption of CC bond equivalence (each has length $a/2$). For $N = \infty$, this gives the metallic solution (no Peierls effect). As we can see, the band formation principle is identical to that, which we have seen for hydrogen atoms. (b) Band structure. (c) Density of states $D(E)$; i.e., the number of states per energy unit at a given energy E. The density has maxima at the extremal points of the band. If we allowed the Peierls transtition, at $k = \pm\pi/a$ we would have a gap.

to be an insulator (or semiconductor) and indeed it is. It may change to a metal when doped. Fig. 9.17 shows a situation analogous to the case of a chain of the ground-state hydrogen atoms.

Polyparaphenylene

The extent to which the COs conform to the rule of increasing number of nodes with energy (or k) will be seen in the example of a planar conformation of polyparaphenylene[41]. On the left side of Fig. 9.18, we have the valence π-orbitals of benzene as follows:

- The lowest-energy orbital has a nodeless,[42] doubly occupied molecular orbital φ_1.
- Then, we have a doubly degenerate and fully occupied level with the corresponding orbitals, φ_2 and φ_3, each having a single node.
- Next, a similar double degenerate empty level with orbitals φ_4 and φ_5 (each with two nodes).
- Finally, we have the highest-energy empty three-node orbital φ_6.

Thus, even in the single monomer we have the rule fulfilled.

Binding phenyl rings by using CC σ bonds results in polyparaphenylene. Let us see what happens when the wave number k increases (the middle and the right side of Fig. 9.18). What counts now is how two *complete monomer orbitals* combine: in phase or out of phase. The

[42] Not counting the nodal plane of the nuclear framework.

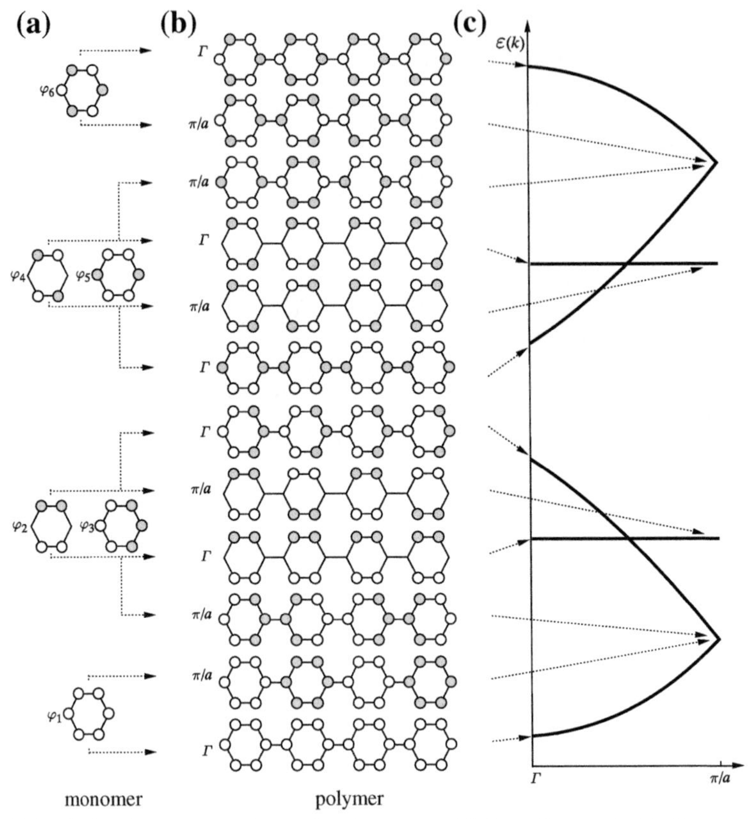

Fig. 9.18. Rationalizing the band structure of polyparaphenylene (π-bands). The COs (in center) built as in-phase or out-of-phase combinations of the benzene π molecular orbitals (left side). It is seen that energy of the COs for $k = 0$ and $k = \frac{\pi}{a}$ agree with the rule of an increasing number of nodes. A small bandwidth corresponds to small overlap integrals of the monomer orbitals.

lowest-energy π-orbitals of benzene (φ_1), arranged in phase ($k = 0$), give point Γ – the lowest energy in the polymer, while out-of-phase, point $k = \frac{\pi}{a}$ has the highest energy. At $k = \frac{\pi}{a}$, there is a degeneracy of this orbital and of φ_3 arranged out of phase. The degeneracy is quite interesting because, despite a superposition of the orbitals with the different number of nodes, the result corresponds to the same number of nodes. Note the extremely small dispersion of the band which results from the arrangement of φ_2. The figure shows that it is bound to be small because it is caused by the arrangement of two molecular orbitals that are farther away in space than those so far considered (the overlap results from the atomic orbitals separated by three bonds, and not by a single bond as it has been). We see a similar regularity in the conduction bands that correspond to the molecular orbitals φ_4, φ_5, and φ_6. The rule works here without any exception and results from the simple statement that a bonding superposition has a lower energy than the corresponding antibonding one.

Thus, when looking at the band structure for polyparaphenylene, we understand every detail of this tangle of bands.

A Stack of Pt(II) Square Planar Complexes

Let us try to predict[43] qualitatively (without making calculations) the band structure of a stack of platinum square planar complexes – typically $[Pt(CN^-)_4^{2-}]_\infty$. Consider the eclipsed configuration of all the monomeric units. Let us first simplify our task. Who likes cyanides? Let us throw them away and take something theoreticians really love: H^-. This isn't just laziness. If needed, we are able to make calculations for cyanides too, but to demonstrate that we really understand the machinery, we are always recommended to make the system as simple as possible (but not simpler). We suspect that the main role of CN^- is just to interact electrostatically, and H^- does this too (being much smaller). In reality, it turns out that what decides is the Pauli exclusion principle, rather than the ligand charge.[44]

The electronic dominant configuration of the platinum atom in its ground state is[45] $[Xe](4f^{14})5d^9\,6s^1$. As we can see, we have the xenon-like closed shell and also the full closed subshell $4f$. The orbital energies corresponding to these closed shells are much lower than the orbital energy of the hydrogen anion (with which they are to be combined). This is why they will not participate in the Pt-H bonds. Of course, they will contribute to the band structure, but this contribution will be trivial: flat bands (because of small overlap integrals) with energies very close to the energies characterizing the corresponding atomic orbitals. The Pt valence shell is, therefore, $5d^9\,6s^1\,6p^0$ for Pt^0, and $5d^8\,6s^0\,6p^0$ for Pt^{2+}, which we have in our stack. The corresponding orbital energies are shown on the left side of Fig. 9.19a.

Let us choose a Cartesian coordinate system with the origin on the platinum atom and the four ligands at equal distances on the x- and y-axes. In the Koopmans approximation (cf. Chapter 8, p. 465), an orbital energy represents the electron energy on a given orbital. We see that because the ligands are pushing (and the Pauli exclusion principle is operating), all the platinum atom orbital energies will go up (destabilization; in Fig. 9.19a, this shift is not shown; only a relative shift is given). The largest shift up will be undergone by the $5d_{x^2-y^2}$ orbital energy because the orbital lobes protrude right across to the ligands. Eight electrons of Pt^{2+} will therefore occupy four other d orbitals[46] ($5d_{xy}$, $5d_{xz}$, $5d_{yz}$, $5d_{3z^2-r^2}$), while $5d_{x^2-y^2}$

[43] We follow R.Hoffmann, *Solids and Surfaces. A Chemist's View of Bonding in Extended Structures*, VCH publishers, New York (1988).

[44] When studying complexes of Fe^{2+} and Co^{2+} (of planar and tetrahedral symmetry), I have got that splitting the d energy levels by negative point charges (simulating ligands) has been very ineffective even when making the negative charges excessively large and pushing them closer to the ion. In contrast to that, replacing the point charges by some closed-shell entities resulted in strong splitting.

[45] Xe denotes the xenon-like configuration of electrons.

[46] Of these four, the lowest-energy ones will correspond to the orbitals $5d_{xz}$, $5d_{yz}$, because their lobes just avoid the ligands. The last two orbitals $5d_{xy}$ and $5d_{3z^2-r^2} = 5d_{z^2-x^2} + 5d_{z^2-y^2}$ will go up somewhat in the energy scale (each to different extents), because they aim in part at the ligands. However, these splits will be smaller when compared to the fate of the orbital $5d_{x^2-y^2}$ and therefore, these levels have been shown in Fig. 9.19 as a single degenerate level.

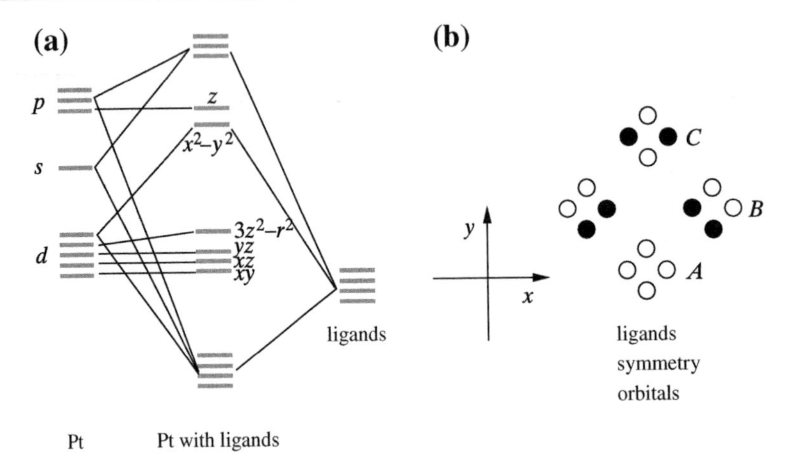

Fig. 9.19. Predicting the band structure of $\left(PtH_4^{2-}\right)_\infty$. (a) Monomer ($PtH_4^{2-}$) molecular orbitals built of the atomic orbitals of Pt^{2+} (the three p and five d Pt atomic orbitals correspond to two degenerate energy p and d levels) and four ligand ($L = H^-$) orbitals. One of the platinum orbitals ($5d_{x^2-y^2}$) corresponds to high energy because it protrudes right across to the ligands. The four ligand AOs, due to the long distance, practically do not overlap and are shown as a quadruply degenerate level. (b) The ligand symmetry orbitals form linear combinations with those of the metal.

will become LUMO. The four ligand atomic orbitals practically do not overlap (long distance), and this is why in Fig. 9.19a, they are depicted as a quadruply quasi-degenerate level. We organize them as the ligand symmetry orbitals, as shown in Fig. 9.19b: the nodeless orbital (A) and two single-node orbitals (B) corresponding to the same energy, and the two-node orbital (C). The effective linear combinations (cf. p. 429, what counts most is symmetry) are formed by the following pairs of orbitals: $6s$ with A, $6p_x$ and $6p_y$ with B, and the orbital $5d_{x^2-y^2}$ with C (in each case, we obtain the bonding and the antibonding orbital); the other platinum orbitals, $5d$ and $6p_z$, do not have partners of the appropriate symmetry (and therefore, their energy does not change). Thus, we obtain the energy-level diagram of the monomer in Fig. 9.19a.

Now, we form a stack of PtH_4^{2-} along the periodicity axis z. Let us form the Bloch functions (Fig. 9.20a) for each of the valence orbitals at two points of the FBZ: $k = 0$ and $k = \frac{\pi}{a}$. The results are given in Fig. 9.20b. Because of large overlap of the $6p_z$ orbitals with themselves, and $3d_{3z^2-r^2}$ also with themselves, these σ bands will have very large dispersions. The smallest dispersion will correspond to the $5d_{xy}$ band (as well as to the empty band $5d_{x^2-y^2}$), because the orbital lobes of $5d_{xy}$ (also of $5d_{x^2-y^2}$) are oriented perpendicularly to the periodicity axis. Two bands $5d_{xz}$ and $5d_{yz}$ have a common fate (i.e., the same plot) due to the symmetry, and a medium bandwidth (Fig. 9.20b). We predict, therefore, the band structure shown in Fig. 9.21a. It is to be compared with the calculated band structure for $\left(PtH_4^{2-}\right)_\infty$ (Fig. 9.21b). As we can see, the prediction turns out to be correct.

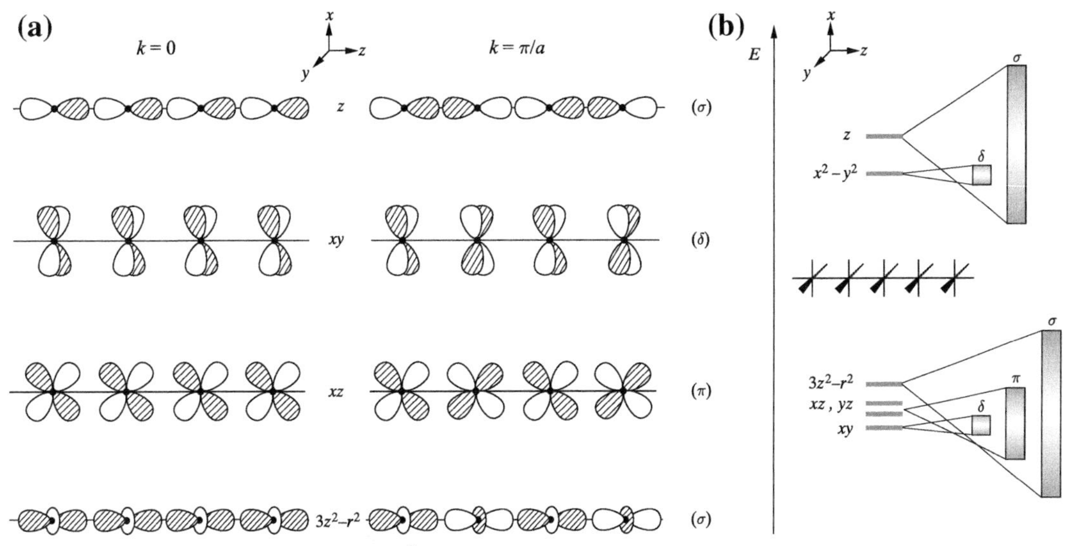

Fig. 9.20. Predicting the band structure of $(PtH_4^{2-})_\infty$. (a) The Bloch functions for $k = 0$ and $k = \frac{\pi}{a}$ corresponding to the atomic orbitals $6p_z$(σ type orbitals), $5d_{xy}$ (δ type orbitals), $5d_{xz}$ (π type orbitals, similarly for $5d_{yz}$), $5d_{3z^2-r^2}$ (σ type orbitals); (b) The bandwidth is very sensitive to the overlap of the atomic orbitals. The bandwidths in $(PtH_4^{2-})_\infty$ result from the overlap of the (PtH_4^{2-}) orbitals.

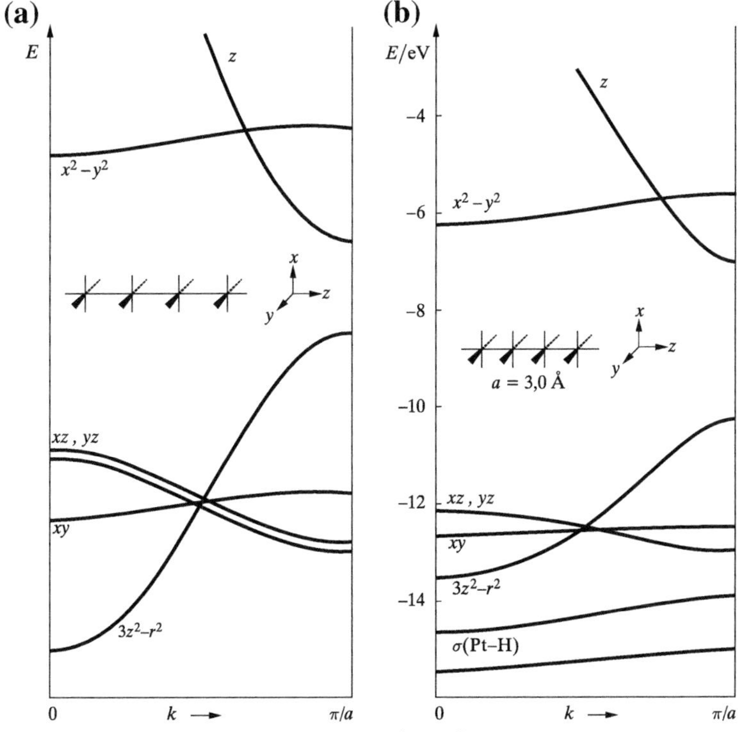

Fig. 9.21. Predicting (after Roald Hoffmann) the band structure of $\left(PtH_4^{2-}\right)_\infty$. (a) The predicted band structure; (b) the computed band structure (by Roald Hoffmann) for $a = 3$ Å.

9.11 The Hartree-Fock Method for Crystals

9.11.1 Secular Equation

What has been said previously about the Hartree-Fock method is only a sort of general theory. The time has now arrived to show how the method works in practice. We have to solve the Hartree-Fock-Roothaan equation (cf. Chapter 8, pp. 431 and 531).

The Fock matrix element is equal to (noting that $(\chi_p^j|\hat{F}\chi_q^{j'}) \equiv F_{pq}^{jj'}$ depends on the *difference*[47] between the vectors $\boldsymbol{R}_j$ and $\boldsymbol{R}_{j'}$)

$$F_{pq} = (2N+1)^{-3} \sum_{jj'} \exp\left(i\boldsymbol{k}(\boldsymbol{R}_j - \boldsymbol{R}'_j)\right)(\chi_p^{j'}|\hat{F}\chi_q^j) \tag{9.55}$$

$$= \sum_j \exp(i\boldsymbol{k}\boldsymbol{R}_j) F_{pq}^{0j}. \tag{9.56}$$

The same can be done with S_{pq}, and therefore, the Hartree-Fock-Roothaan secular equation (see p. 531) has the form:

$$\sum_{p=1}^{\omega} c_{pn}(\boldsymbol{k}) \Big(\sum_j \exp(i\boldsymbol{k}\boldsymbol{R}_j)\big(F_{pq}^{0j}(\boldsymbol{k}) - \varepsilon_n(\boldsymbol{k}) S_{pq}^{0j}(\boldsymbol{k})\big)\Big) = 0, \tag{9.57}$$

for $q = 1, 2, \ldots \omega$. The integral S_{pq} equals

$$S_{pq} = \sum_j \exp(i\boldsymbol{k}\boldsymbol{R}_j) S_{pq}^{0j}. \tag{9.58}$$

The summation goes over the lattice nodes and $S_{pq}^{0j} \equiv \left(\chi_p^0|\chi_q^j\right)$. In order to be explicit, let us see what is inside the Fock matrix elements $F_{pq}^{0j}(\boldsymbol{k})$. *We have* to find a dependence there on the Hartree-Fock-Roothaan solutions (determined by the coefficients c_{pn}), and more precisely on the bond order matrix.[48] Any CO, according to Eq. (9.53), has the form

$$\psi_n(\boldsymbol{r}, \boldsymbol{k}) = (2N+1)^{-\frac{3}{2}} \sum_q \sum_j c_{qn}(\boldsymbol{k}) \exp(i\boldsymbol{k}\boldsymbol{R}_j) \chi_q^j(\boldsymbol{r}), \tag{9.59}$$

where we promise to use such c_{qn} that ψ_n are normalized. For molecules, the bond order matrix element (for the atomic orbitals χ_p and χ_q) has been defined as $2\sum c_{pi}c_{qi}^*$ (the summation is over the doubly occupied orbitals), where the factor 2 results from the double occupation of the

[47] As a matter of fact, all depends on how distant and how oriented the unit cells j and j' are. We have used the fact that $\hat{F}$ exhibits the crystal symmetry.

[48] We have seen the same in the Hartree-Fock method for molecules, where the Coulomb and exchange operators depended on the solutions to the Fock equation, (cf. p. 412).

closed shell. We have exactly the same for the crystal, where we define the bond order matrix element corresponding to atomic orbitals χ_q^j and χ_p^l as

$$P_{pq}^{lj} = 2(2N+1)^{-3} \sum_{\text{occupied}} c_{pn}(\boldsymbol{k}) \exp(i\boldsymbol{k}\boldsymbol{R}_l) c_{qn}(\boldsymbol{k})^* \exp(-i\boldsymbol{k}\boldsymbol{R}_j) \tag{9.60}$$

where the summation goes over all the occupied COs (we assume double occupation, hence factor 2). This means that in the summation, we have to go over all the occupied bands (index n), and in each band over all allowed COs; i.e., all the allowed $\boldsymbol{k}$ vectors in the FBZ. Thus,

$$P_{pq}^{lj} = 2(2N+1)^{-3} \sum_n \sum_{\boldsymbol{k}}^{FBZ} c_{pn}(\boldsymbol{k}) c_{qn}(\boldsymbol{k})^* \exp\left(i\boldsymbol{k}(\boldsymbol{R}_l - \boldsymbol{R}_j)\right). \tag{9.61}$$

This definition of the $\boldsymbol{P}$ matrix is exactly what we should have for a large closed-shell molecule. The matrix element has to have four indices (instead of the two indices in the molecular case), because we have to describe the atomic orbitals indicating that atomic orbital p is from unit cell l, and atomic orbital q from unit cell j. It is easily seen that P_{pq}^{lj} depends on the *difference* $\boldsymbol{R}_l - \boldsymbol{R}_j$, not on the $\boldsymbol{R}_l, \boldsymbol{R}_j$ themselves. The reason for this is that in a crystal, everything is repeated, and the important thing to consider is the *relative* distances. Thus, the $\boldsymbol{P}$ matrix is determined by all the elements P_{pq}^{0j}.

9.11.2 Integration in the FBZ

There is a problem with $\boldsymbol{P}$: namely, it requires a summation over $\boldsymbol{k}$. We do not like this because the number of the permitted vectors $\boldsymbol{k}$ is huge for large N (and N has to be large because we are dealing with a crystal). We have to do something with it.

Let us try a small exercise. Imagine that we have to perform a summation $\sum_{\boldsymbol{k}} f(\boldsymbol{k})$, where f represents a smooth function in the FBZ. Let us denote the sum to be found by X. Let us multiply X by a small number $\Delta = \frac{V_{FBZ}}{(2N+1)^3}$, where V_{FBZ} stands for the FBZ volume:

$$X\Delta = \sum_{\boldsymbol{k}}^{FBZ} f(\boldsymbol{k})\Delta \tag{9.62}$$

In other words, we just cut the FBZ into tiny segments of volume Δ, their number equal to the number of the permitted $\boldsymbol{k}$. It is clear that if N is large (as it is in our case), then a very good approximation of $X\Delta$ would be

$$X\Delta = \int_{FBZ} f(\boldsymbol{k}) d\boldsymbol{k}. \tag{9.63}$$

Hence,

$$X = \frac{(2N+1)^3}{V_{FBZ}} \int_{FBZ} f(\boldsymbol{k}) d\boldsymbol{k}. \tag{9.64}$$

After applying this result to the bond order matrix, we obtain

$$P_{pq}^{lj} = \frac{2}{V_{FBZ}} \int^{FBZ} \sum_n c_{pn}(\boldsymbol{k}) c_{qn}(\boldsymbol{k})^* \exp\left(i\boldsymbol{k}(\boldsymbol{R}_l - \boldsymbol{R}_j)\right) d\boldsymbol{k}. \tag{9.65}$$

For a periodic *polymer* (in 1-D: $V_{FBZ} = \frac{2\pi}{a}$, $\Delta = \frac{V}{2N+1}$), we would have

$$P_{pq}^{lj} = \frac{a}{\pi} \int \sum_n c_{pn}(k) c_{qn}(k)^* \exp\left(ika(l-j)\right) dk. \tag{9.66}$$

9.11.3 Fock Matrix Elements

In full analogy with the formula on p. 427, we can express the Fock matrix elements by using the bond order matrix $\boldsymbol{P}$ for the crystal:

$$F_{pq}^{0j} = T_{pq}^{0j} - \sum_h \sum_u Z_u V_{pq}^{0j}(\boldsymbol{A}_u^h) + \sum_{hl} \sum_{rs} P_{sr}^{lh} \left(\begin{pmatrix} 0h & jl \\ pr & qs \end{pmatrix} - \frac{1}{2} \begin{pmatrix} 0h & lj \\ pr & sq \end{pmatrix} \right), \tag{9.67}$$

where $\boldsymbol{P}$ satisfies the normalization condition[49]

[49] The $\boldsymbol{P}$ matrix satisfies the normalization condition, which we obtain in the following way: As in the molecular case, the normalization of CO means

$$\begin{aligned} 1 &= \langle \psi_n(\boldsymbol{r},\boldsymbol{k}) \mid \psi_n(\boldsymbol{r},\boldsymbol{k}) \rangle \\ &= (2N+1)^{-3} \sum_{pq} \sum_{jl} c_{pn}(\boldsymbol{k})^* c_{qn}(\boldsymbol{k}) \exp[i\boldsymbol{k}(\boldsymbol{R}_j - \boldsymbol{R}_l)] S_{pq}^{lj} \\ &= (2N+1)^{-3} \sum_{pq} \sum_{jl} c_{pn}(\boldsymbol{k})^* c_{qn}(\boldsymbol{k}) \exp[i\boldsymbol{k}(\boldsymbol{R}_j - \boldsymbol{R}_l))] S_{pq}^{o(j-l)} \\ &= \sum_{pq} \sum_{j} c_{pn}(\boldsymbol{k})^* c_{qn}(\boldsymbol{k}) \exp(i\boldsymbol{k}\boldsymbol{R}_j) S_{pq}^{0j}. \end{aligned}$$

Now let us do the same for all the occupied COs and sum the results. On the left side, we sum just 1, so we obtain the number of doubly occupied COs; i.e., $n_0(2N+1)^3$, because n_0 denotes the number of doubly occupied bands. Further, in each band, we have in 3-D $(2N+1)^3$ allowed vectors $\boldsymbol{k}$. Therefore, we have

$$\begin{aligned} n_0(2N+1)^3 &= \sum_{pq} \sum_{j} \Big(\sum_n \sum_{\boldsymbol{k}}^{FBZ} c_{pn}(\boldsymbol{k})^* c_{qn}(\boldsymbol{k}) \exp(i\boldsymbol{k}\boldsymbol{R}_j) \Big) S_{pq}^{0j} \\ &= \sum_{pq} \sum_{j} \frac{1}{2} (2N+1)^3 P_{qp}^{j0} S_{pq}^{0j}, \end{aligned}$$

where from Eq. (9.61), after exchanging $p \leftrightarrow q$, $j \leftrightarrow l$, we get

$$P_{qp}^{jl} = 2(2N+1)^{-3} \sum_n \sum_{\boldsymbol{k}}^{FBZ} c_{qn}(\boldsymbol{k}) c_{pn}(\boldsymbol{k})^* \exp\left(i\boldsymbol{k}(\boldsymbol{R}_j - \boldsymbol{R}_l)\right).$$

$$\sum_j \sum_{pq} P_{qp}^{j0} S_{pq}^{0j} = 2n_0, \tag{9.68}$$

where $2n_0$ means the number of electrons in the unit cell.

The first term on the right side of Eq. (9.67) represents the kinetic energy matrix element

$$T_{pq}^{0j} = \left(\chi_p^0 | -\frac{1}{2}\Delta | \chi_q^j \right), \tag{9.69}$$

where the second term is a sum of matrix elements, each corresponding to the nuclear attraction of an electron and the nucleus of index u and charge Z_u in the unit cell h:

$$V_{pq}^{0j}(\boldsymbol{A}_u^h) = \left(\chi_p^0 | \frac{1}{|\boldsymbol{r} - \boldsymbol{A}_u^h|} | \chi_q^j \right), \tag{9.70}$$

where the upper index of χ denotes the cell number, the lower index is the number of the atomic orbital in a cell, the vector $\boldsymbol{A}_u^h$ indicates nucleus u (numbering within the unit cell) in unit cell h (from the coordinate system origin), and the third term is connected to the Coulombic operator (the first of two terms) and the exchange operator (the second of two terms). The summations over h and l go over the unit cells of the whole crystal, and therefore are very difficult and time consuming.

The definition of the two-electron integral,

$$\left(\begin{array}{c|c} 0h & jl \\ pr & qs \end{array} \right) = \int \mathrm{d}\boldsymbol{r}_1\, \mathrm{d}\boldsymbol{r}_2 \chi_p^0(\boldsymbol{r}_1)^* \chi_r^h(\boldsymbol{r}_2)^* \frac{1}{r_{12}} \chi_q^j(\boldsymbol{r}_1) \chi_s^l(\boldsymbol{r}_2), \tag{9.71}$$

is in full analogy to the notation of Chapter 8 and Appendix M available at booksite.elsevier.com/978-0-444-59436-5, p. e109.

9.11.4 Iterative Procedure (SCF LCAO CO)

To solve Eq. (9.57), one uses the SCF LCAO MO technique as applied for molecules (Chapter 8) and now adapted for crystals. This particular method will be called SCF LCAO CO, because the linear combinations (LC) of the symmetry AOs are used as the expansion functions for the

---

and then

$$P_{qp}^{j0} = 2(2N+1)^{-3} \sum_n \sum_{\boldsymbol{k}}^{FBZ} c_{qn}(\boldsymbol{k}) c_{pn}(\boldsymbol{k})^* \exp\left(i\boldsymbol{k}\boldsymbol{R}_j\right).$$

Hence,

$$\sum_{pq} \sum_j P_{qp}^{j0} S_{pq}^{0j} = 2n_0.$$

crystal orbitals (COs) in a self-consistent procedure (SCF - Self-Consistent Field) described below. How does the SCF LCAO CO method work?

1. First (zeroth iteration), we start from a guess[50] for $\boldsymbol{P}$.
2. Then, we calculate the elements F_{pq}^{0j} for all atomic orbitals p, q for unit cells $j = 0, 1, 2, \ldots$ $j_{\max}$. What is $j_{\max}$? The answer is certainly non-satisfactory: $j_{\max} = \infty$. In practice, however, we often take $j_{\max}$ as being of the order of a few cells; most often, we take[51] $j_{\max} = 1$.
3. For each $\boldsymbol{k}$ from the FBZ, we calculate the elements F_{pq} and S_{pq} of Eqs. (9.56) and (9.58), and then solve the secular equations within the Hartree-Fock-Roothaan procedure. This step requires diagonalization[52] (see Appendix K available at booksite.elsevier.com/978-0-444-59436-5, p. e105). As a result, for each $\boldsymbol{k}$ we obtain a set of coefficients c for the crystal orbitals and the energy eigenvalue $\varepsilon_n(\boldsymbol{k})$.
4. We repeat all this for the values of $\boldsymbol{k}$ covering in some optimal way (some recipes exist) the FBZ. We are then all set to carry out the numerical integration in the FBZ and we calculate an approximate matrix $\boldsymbol{P}$.
5. This enables us to calculate a new approximation to the matrix $\boldsymbol{F}$ and so on, until the procedure converges in a self-consistent way; i.e., produces $\boldsymbol{P}$ very close to that matrix $\boldsymbol{P}$, which has been inserted into the Fock matrix $\boldsymbol{F}$. In this way, we obtain the band structure $\varepsilon_n(\boldsymbol{k})$ and all the corresponding COs.

9.11.5 Total Energy

How do we calculate the total energy for an infinite crystal? We know the answer without calculating: $-\infty$. Indeed, since the energy represents an extensive quantity, for an infinite number of the unit cells, we get $-\infty$ because a single cell usually represents a bound state (negative energy). Therefore, the question has to be posed in another way.

How do we calculate the total energy *per unit cell*? This is a different question. Let us denote this quantity by E_T. Since a crystal only represents a *very* large molecule, we may use the expression for the total energy of a molecule. In the 3-D case, we get

$$(2N+1)^3 E_T = \frac{1}{2}\sum_{pq}\sum_{lj} P_{qp}^{jl}\left(h_{pq}^{lj} + F_{pq}^{lj}\right) + \frac{1}{2}\sum_{lj}\sum_{uv}{}' \frac{Z_u Z_v}{R_{uv}^{lj}}, \tag{9.72}$$

where the summation over p and q extends over the ω atomic orbitals that any unit cell offers, l and j tell us, in which cells these orbitals are located. The last term on the right side refers to the nuclear repulsion of all the nuclei in the crystal; u, v number the nuclei in a unit cell; while

[50] The result is presumed to be independent of this choice.

[51] This is the nearest-neighbor approximation. We encounter a similar problem *inside* the F_{pq}^{0j} because we have somehow truncated the summations over h and l. These problems will be discussed later in this chapter.

[52] Unlike the molecular case, this time the matrix to diagonalize is Hermitian, and is not necessarily symmetric. Methods of diagonalization exist for such matrices, and there is a guarantee that their eigenvalues are real.

l, j indicate the cells (a prime means that there is no contribution from the charge interaction with itself). Since the summations over l and j extend over the whole crystal,

$$(2N+1)^3 E_T = \frac{1}{2}(2N+1)^3 \sum_{pq}\sum_{j} P_{qp}^{j0}[h_{pq}^{0j} + F_{pq}^{0j}] + (2N+1)^3 \frac{1}{2} \sum_{j}\sum_{uv}{}' \frac{Z_u Z_v}{R_{uv}^{0j}}, \quad (9.73)$$

because each term has an equal contribution, and the number of such terms is equal to $(2N+1)^3$.

Therefore, the total energy per unit cell amounts to

$$E_T = \frac{1}{2}\sum_{j}\sum_{pq} P_{qp}^{j0}\left(h_{pq}^{0j} + F_{pq}^{0j}\right) + \frac{1}{2}\sum_{j}\sum_{u}\sum_{v}{}' \frac{Z_u Z_v}{R_{uv}^{0j}}. \quad (9.74)$$

The formula is correct, but we can easily see that we are to be confronted with some serious problems. E.g., the summation over nuclei represents a divergent series and we will get $+\infty$. *This problem appears only because we are dealing with an infinite system.* We have to manage the problem somehow.

9.12 Long-Range Interaction Problem

What is left to be clarified is the question of how to go from N to infinity.[53] It will be soon shown how dangerous this problem is.

We see from Eqs. (9.67) and (9.74) that thanks to the translational symmetry, we may treat each $\boldsymbol{k}$ separately, infinity continues to make us a little nervous. In the expression for F_{pq}^{0j}, we

[53] Let me tell you about my adventure with this problem, because I remember how as a student, I wanted to hear about struggles with understanding matter and ideas instead of some dry summaries.
The story began quite accidentally. In 1977, at the University of Namur (Belgium), Professor Joseph Delhalle asked the Ph.D. student Christian Demanet to perform a numerical test. The test consisted of taking a simple infinite polymer (the infinite chain ... LiH LiH LiH ... had been chosen), to use the simplest atomic basis set possible and to see what we should take as N, to obtain the Fock matrix with sufficient accuracy. Demanet first took $N = 1$, then $N = 2$, $N = 3$ — the Fock matrix changed all the time. He got impatient and took $N = 10$, $N = 15$ — but the matrix continued to change. Only when he used $N = 200$ did the Fock matrix elements stabilize within the accuracy of six significant figures. We could take $N = 200$ for an extremely poor basis set and for a few such tests, but never in good quality calculations because their cost would become astronomic. Even for the case in question, the computations had to be done overnight. In a casual discussion at the beginning of my six-week term at the University of Namur, Joseph Delhalle told me about the problem. He said also that in a recent paper, the Austrian scientists Alfred Karpfen and Peter Schuster also noted that the results depend strongly on the chosen value of N. They made a correction *after* the calculations with a small N had been performed. They added the dipole-dipole electrostatic interaction of the cell 0 with a few hundred neighboring cells, and as the dipole moment of a cell, they took the dipole moment of the isolated LiH molecule. As a result, the Fock matrix elements changed much less with N. This information made me think about implementing the multipole expansion right from the beginning in the self-consistent Hartree-Fock-Roothaan procedure for a polymer. Below you will see what has been done. The presented theory pertains to a regular polymer (a generalization to 2-D and 3-D is possible).

have a summation (over the whole infinite crystal) of the interactions of an electron with all the nuclei, and in the next term, a summation over the whole crystal of the electron-electron interactions. This is of course natural, because our system is infinite. The problem is, however, that both summations diverge: the first tends to $-\infty$, the second to $+\infty$. On top of this, to compute the bond order matrix $\boldsymbol{P}$, we have to perform another summation in Eq. (9.65) over the FBZ of the crystal. We have a similar, very unpleasant, situation in the total energy expression, where the first term tends to $-\infty$, while the nuclear repulsion term goes to $+\infty$.

The routine approach in the literature was to replace infinity by taking the first-neighbor interactions. This approach is quite understandable because any attempt to take further neighbors ends up with an exorbitant bill to pay.[54]

9.12.1 Fock Matrix Corrections

A first idea that we may think of is to separate carefully the long-range part of the Fock matrix elements and of the total energy from these quantities as calculated in a traditional way (i.e., by limiting the infinite-range interactions to those for the N neighbors on the left from cell 0 and N neighbors on the right of it). For the Fock matrix element, we would have

$$F_{pq}^{0j} = F_{pq}^{0j}(N) + C_{pq}^{0j}(N), \tag{9.75}$$

where $C_{pq}^{0j}(N)$ stands for the long-range correction, while $F_{pq}^{0j}(N)$ is calculated assuming interactions with the N right and N left neighbors of cell 0:

$$F_{pq}^{0j}(N) = T_{pq}^{0j} + \sum_{h=-N}^{h=+N} \left(-\sum_{u} Z_u V_{pq}^{0j}\left(\boldsymbol{A}_u^h\right) + \sum_{l=h-N}^{l=h+N} \sum_{rs} P_{sr}^{lh} \left(\left(\begin{matrix} 0h \\ pr \end{matrix} \middle| \begin{matrix} jl \\ qs \end{matrix} \right) - \frac{1}{2} \left(\begin{matrix} 0h \\ pr \end{matrix} \middle| \begin{matrix} lj \\ sq \end{matrix} \right) \right) \right) \tag{9.76}$$

$$C_{pq}^{0j}(N) = \sum_{h}^{\#} \left(-\sum_{u} Z_u V_{pq}^{0j}\left(\boldsymbol{A}_u^h\right) + \sum_{l=h-N}^{l=h+N} \sum_{rs} P_{sr}^{lh} \left(\begin{matrix} 0h \\ pr \end{matrix} \middle| \begin{matrix} jl \\ qs \end{matrix} \right) \right), \tag{9.77}$$

where the symbol $\sum_h^{\#}$ will mean a summation over all the unit cells *except* the section of unit cells with numbers $-N, -N+1, \ldots, 0, 1, \ldots N$; i.e., the neighborhood of cell 0 ("*short-range*"). The nuclear attraction integral[55]

$$V_{pq}^{0j}(\boldsymbol{A}_u^h) = (\chi_p^0 | \frac{1}{|\boldsymbol{r} - (\boldsymbol{A}_u + ha\boldsymbol{z})|} | \chi_q^j), \tag{9.78}$$

where the vector $\boldsymbol{A}_u$ shows the position of the nucleus u in cell 0, while $\boldsymbol{A}_u^h \equiv \boldsymbol{A}_u + ha\boldsymbol{z}$ points to the position of the equivalent nucleus in cell h ($\boldsymbol{z}$ denotes the unit vector along the periodicity axis).

[54] The number of two-electron integrals, which quantum chemistry positively dislikes, increases with the number of neighbors to take (N) and the atomic basis set size per unit cell (ω) as $N^3\omega^4$. Besides, the nearest-neighbors *are* indeed the most important.

[55] Without the minus sign in the definition, the name is not quite adequate.

The expression for $C_{pq}^{0j}(N)$ has a clear physical interpretation. The first term represents the interaction of the charge distribution $-\chi_p^0(1)^*\chi_q^j(1)$ (of electron 1, hence the sign $-$) with *all the nuclei*,[56] except those enclosed in the short-range region (i.e., extending from $-N$ to $+N$). The second term describes the interaction of the same electronic charge distribution with the *total electronic distribution* outside the short-range region. How do we see this? The integral $\left(\begin{smallmatrix}0h \\ pr\end{smallmatrix}\middle|\begin{smallmatrix}jl \\ qs\end{smallmatrix}\right)$ means the Coulombic interaction of the distribution under consideration $-\chi_p^0(1)^*\chi_q^j(1)$ with its partner-distribution $-\chi_r^h(2)^*\chi_s^l(2)$, doesn't it? This distribution is multiplied by P_{sr}^{lh} and then summed over all possible atomic orbitals r and s in cell h and its neighborhood (the sum over cells l from the neighborhood of cell h), which gives the total partner electronic distribution $-\sum_{l=h-N}^{l=h+N}\sum_{rs} P_{sr}^{lh}\chi_r^h(2)^*\chi_s^l(2)$. This, however, simply represents the electronic charge distribution of cell h. Indeed, the distribution, when integrated, gives [(just look at Eq. (9.68)] $-\sum_{l=h-N}^{l=h+N}\sum_{rs} P_{sr}^{lh}S_{rs}^{hl} = 2n_0$. Therefore, our electron distribution $-\chi_p^0(1)^*\chi_q^j(1)$ interacts electrostatically with the charge distribution of all cells except those enclosed in the short-range region, because Eq. (9.77) contains the summation over all cells h except the short-range region. Finally,

> the long-range correction to the Fock matrix elements $C_{pq}^{0j}(N)$ represents the Coulombic interaction of the charge distribution $-\chi_p^0(1)^*\chi_q^j(1)$ with all the unit cells (nuclei and electrons) from outside the short-range region.

In the $C_{pq}^{0j}(N)$ correction, in the summation over l, *we have neglected the exchange term* $-\frac{1}{2}\sum_h^{\#}\sum_{l=h-N}^{l=h+N}\sum_{rs} P_{sr}^{lh}\left(\begin{smallmatrix}0h \\ pr\end{smallmatrix}\middle|\begin{smallmatrix}lj \\ sq\end{smallmatrix}\right)$. The reason for this was that we have been *convinced* that P_{sr}^{0h} vanishes fast with h. Indeed, the largest integral in the summation over l is $-\frac{1}{2}\sum_h^{\#}\sum_{rs} P_{sr}^{0h}\left(\begin{smallmatrix}0h \\ pr\end{smallmatrix}\middle|\begin{smallmatrix}0j \\ sq\end{smallmatrix}\right)$. This term is supposed to be small not because of the integral $\left(\begin{smallmatrix}0h \\ pr\end{smallmatrix}\middle|\begin{smallmatrix}0j \\ sq\end{smallmatrix}\right)$, which can be quite important [e.g., $\left(\begin{smallmatrix}0h \\ pr\end{smallmatrix}\middle|\begin{smallmatrix}0,h-1 \\ pq\end{smallmatrix}\right)$], but because of P_{sr}^{0h}. We will come back to this problem later in this chapter.[57]

9.12.2 Total Energy Corrections

The total energy per unit cell could similarly be written as

$$E_T = E_T(N) + C_T(N), \tag{9.79}$$

where $E_T(N)$ means the total energy per unit cell as calculated by the traditional approach (i.e., with truncation of the infinite series on the N left and N right neighbors of the cell 0).

---

[56] Refer to the interpretation of the integral $-V_{pq}^{0j}(\boldsymbol{A}_u^h) = -\left(\chi_p^0(\boldsymbol{r})\middle|\frac{1}{|\boldsymbol{r}-\boldsymbol{A}_u^h|}\middle|\chi_q^j(\boldsymbol{r})\right)$.

[57] Matrix element P_{sr}^{0h} [i.e., the bond order contribution from the AO product: $\chi_s^0(1)\chi_r^h(1)^*$ pertaining to distant cells 0 and h] seems to be a small number. This will turn out to be delusive. We have to stress, however, that trouble will come only in some "*pathological*" situations.

The quantity $C_T(N)$ therefore represents the error (i.e., the long-range correction). The detailed formulas for $E_T(N)$ and $C_T(N)$ are the following:

$$E_T(N) = \frac{1}{2}\sum_{j=-N}^{j=+N}\sum_{pq} P_{qp}^{j0}\left(h_{pq}^{0j} + F_{pq}^{0j}(N)\right) + \frac{1}{2}\sum_{j=-N}^{j=+N}\sum_{u}\sum_{v}{}' \frac{Z_u Z_v}{R_{uv}^{0j}}, \tag{9.80}$$

$$C_T(N) = \frac{1}{2}\sum_{j}\sum_{pq} P_{qp}^{j0} C_{pq}^{0j}(N)$$
$$+\frac{1}{2}\sum_{h}^{\#}\left(\sum_{j}\sum_{pq} P_{qp}^{j0}\sum_{u}\left[-Z_u V_{pq}^{0j}(\mathbf{A}_u^h)\right] + \sum_{u}\sum_{v}{}' \frac{Z_u Z_v}{R_{uv}^{0h}}\right), \tag{9.81}$$

where from F_{pq}^{0j}, we have already separated its long-range contribution $C_{pq}^{0j}(N)$, so that $C_T(N)$ contains *all* the long-range corrections.

Equation (9.81) for $C_T(N)$ may be obtained *just by looking* at Eq. (9.80). The first term with $C_{pq}^{0j}(N)$ is evident,[58] it represents the Coulombic interaction of the *electronic distribution* [let us recall Eq. (9.68)] associated with cell 0 with *the whole polymer chain* except the short-range region. What, therefore, is yet to be added to $E_T(N)$? What it lacks is the Coulombic interaction of the *nuclei* of cell 0 with *the whole polymer chain*, except for the short-range region. Let us see whether we have it in Eq. (9.81). The last term means the Coulombic interaction of the nuclei of cell 0 with *all the nuclei of the polymer* except for the short-range region (and again we know why we have the factor $\frac{1}{2}$). What, therefore, is represented by the middle term[59]? It is clear, that it has to be (with the factor $\frac{1}{2}$) the Coulombic interaction of the *nuclei of cell* 0 with the total

[58] The factor $\frac{1}{2}$ may worry us a little. Why just $\frac{1}{2}$? Let us see. Imagine N *identical objects* $i = 0, 1, 2, \ldots N-1$ *playing identical roles in a system* (like our unit cells). We will be interested in the energy per object, E_T. The total energy may be written as (let us assume here pairwise interactions only)

$$NE_T = \sum_j E_j + \sum_{i<j} E_{ij},$$

where E_j and E_{ij} are, respectively, the isolated object energy and the pairwise interaction energy. Since the objects are identical, then

$$NE_T = NE_0 + \frac{1}{2}\sum_{i,j}{}' E_{ij} = NE_0 + \frac{1}{2}\sum_i\left(\sum_j{}' E_{ij}\right) = NE_0 + \frac{1}{2}N\left(\sum_j{}' E_{0j}\right),$$

where the prime means excluding self-interaction and the term in parentheses means the interaction of object 0 with all others. Finally,

$$E_T = E_0 + \frac{1}{2}\left(\sum_j{}' E_{0j}\right),$$

where we have the factor $\frac{1}{2}$ before the interaction of one of the objects with the rest of the system.

[59] As we can see, we have to sum (over j) to infinity the expressions h_{pq}^{0j}, which contain T_{pq}^{0j} [but these terms decay very fast with j and can all be taken into account in $E_T(N)$] and the long-range terms, the Coulombic interaction

electronic distribution outside the short-range region. We look at the middle term now. We have the minus sign, which is very good indeed, because we have to have an attraction. Further, we have the factor $\frac{1}{2}$, which is also OK; and then we have $\sum_h^{\#}$, which is perfect, because we expect a summation over only the long range, and finally, we have $\sum_j \sum_{pq} P_{qp}^{j0} \sum_u \left[-Z_u V_{pq}^{0j}(\mathbf{A}_u^h) \right]$ and we do not like this. This is the Coulombic interaction of the total *electronic distribution of cell* 0 with the *nuclei of the long-range region,* while we expected the interaction of the *nuclei* of cell 0 with the *electronic charge distribution of the long-range region*. What is going on? Everything is OK. Just count the interactions pairwise, and at each of them, reverse the locations of the interacting objects – the two interactions mean the same. Therefore,

the long-range correction to the total energy per cell $C_T(N)$ represents the Coulombic interaction of cell 0 with all the cells from outside the short-range region.

We are now all set to calculate the long-range corrections $C_{pq}^{0j}(N)$ and $C_T(N)$. It is important to realize that all the interactions to calculate pertain to objects that are *far away in space.*[60] This is what we have carefully prepared. This is the condition that enables us to apply the multipole expansion to each of the interactions (see Appendix X available at booksite.elsevier.com/978-0-444-59436-5).

9.12.3 Multipole Expansion Applied to the Fock Matrix

Let us first concentrate on $C_{pq}^{0j}(N)$. As seen from Eq. (9.77), there are two types of interactions to calculate: the nuclear attraction integrals $V_{pq}^{0j}(\mathbf{A}_u^h)$ and the electron repulsion integrals $\left(\begin{smallmatrix}0h \\ pr\end{smallmatrix}\middle|\begin{smallmatrix}jl \\ qs\end{smallmatrix}\right)$. In the second term, we may use the multipole expansion of $\frac{1}{r_{12}}$ given in Appendix X available

of the electronic charge distribution of cell 0 with the nuclei beyond the short-range region [the middle term in $C_T(N)$]. The argument about fast decay with j of the kinetic energy matrix elements mentioned before follows from the double differentiation with respect to the coordinates of the electron. Indeed, this results in another atomic orbital, but with the same center. This leads to the overlap integral of the atomic orbitals centered like those in $\chi_p^0\chi_q^j$. Such an integral decays exponentially with j.

[60] Let us check this. What objects we are talking about? Let us begin from $C_{pq}^{0j}(N)$. As it is seen from the formula, one of the interacting objects is the charge distribution of the first electron $\chi_p^0(1)^*\chi_q^j(1)$. The second object is the whole polymer except for the nuclei and electrons of the neighborhood of the cell 0. The charge distributions $\chi_p^0(1)^*\chi_q^j(1)$ with various j are always close to cell 0, because the orbital $\chi_p^0(1)$ is anchored at cell 0, and such a distribution decays exponentially when cell j goes away from cell 0. The fact that the nuclei with which the distribution $\chi_p^0(1)^*\chi_q^j(1)$ interacts are far apart is evident, but less evident is that the electrons with which the distribution interacts are also far away from cell 0. Let us have a closer look at the electron-electron interaction. The charge distribution of electron 2 is $\chi_r^h(2)^*\chi_s^l(2)$, and the summation over cells h excludes the neighborhood of cell 0. Hence, because of the exponential decay, there is a guarantee that the distribution $\chi_r^h(2)^*\chi_s^l(2)$ is bound to be close to cell h, if this distribution is to be of any significance. Therefore, the charge distribution $\chi_r^h(2)^*\chi_s^l(2)$ is certainly far away from cell 0.

Similar reasoning may be used for $C_T(N)$. The interacting objects are of the type $\chi_p^0(1)^*\chi_q^j(1)$ – i.e., always close to cell 0 – with the nuclei of cell h, and there is a guarantee that h is far from cell 0. The long distance of the interacting nuclei (second term) is evident.

at booksite.elsevier.com/978-0-444-59436-5 on p. e169. In the first term, we will do the same, but this time, one of the interacting particles will be the nucleus indicated by vector $\boldsymbol{A}_u^h$. The corresponding multipole expansion reads as (in a.u.; the nucleus u of the charge Z_u interacts with the electron of charge -1, $n_k = n_l = \infty$, $S = min(k, l)$):

$$-\frac{Z_u}{r_{u1}} = \sum_{k=0}^{n_k}\sum_{l=0}^{n_l}\sum_{m=-S}^{m=+S} A_{kl|m|} R^{-(k+l+1)} M_a^{(k,m)}(1)^* M_b^{(l,m)}(u), \tag{9.82}$$

where R stands for the distance between the origins of the coordinate system centered in cell 0 and the coordinate system in cell h, which, of course, is equal to $R = ha$. The multipole moment of electron 1, $M_a^{(k,m)}(1)$, reads as

$$M_a^{(k,m)}(1) = -r_a^k P_k^{|m|}(\cos\theta_{a1}) \exp(im\phi_{a1}), \tag{9.83}$$

while

$$M_b^{(l,m)}(u) = Z_u r_u^l P_l^{|m|}(\cos\theta_u) \exp(im\phi_u) \tag{9.84}$$

denotes the multipole moment of nucleus u computed in the coordinate system of the cell h. When this expansion, as well as the expansion for $\frac{1}{r_{12}}$, are inserted into Eq. (9.77) for $C_{pq}^{0j}(N)$, we obtain

$$C_{pq}^{0j}(N) = \sum_h^{\#}\sum_{k=0}^{n_k}\sum_{l=0}^{n_l}\sum_{m=-S}^{m=+S} A_{kl|m|} R^{-(k+l+1)} \Big((\chi_p^0|\hat{M}_a^{(k,m)}(1)^*|\chi_q^j)$$
$$\cdot\Big[\sum_u M_b^{(l,m)}(\boldsymbol{A}_u^h)\Big] + (\chi_p^0|\hat{M}_a^{(k,m)}(1)^*|\chi_q^j)$$
$$\cdot \sum_{l'=h-N}^{l'=h+N}\sum_{rs} P_{sr}^{l'h}(\chi_r^h|\hat{M}_b^{(l,m)}(2)|\chi_s^{l'}) \Big)$$
$$= \sum_h^{\#}\sum_{k=0}^{n_k}\sum_{l=0}^{n_l}\sum_{m=-S}^{m=+S} A_{kl|m|} R^{-(k+l+1)} (\chi_p^0|\hat{M}_a^{(k,m)}(1)^*|\chi_q^j)$$
$$\cdot\Big[\sum_u M_b^{(l,m)}(\boldsymbol{A}_u^h) + \sum_{l'=h-N}^{l'=h+N}\sum_{rs} P_{sr}^{l'h}(\chi_r^h|\hat{M}_b^{(l,m)}(2)|\chi_s^{l'})\Big].$$

Let us note that in the brackets, we have nothing but a multipole moment of *unit cell* h. Indeed, the first term represents the multipole moment of all the nuclei of cell h, while the second term is the multipole moment of electrons of unit cell h. The latter can be best seen if we recall the normalization condition [Eq. (9.68)]: $\sum_{l'=h-N}^{l'=h+N}\sum_{rs} P_{sr}^{l'h} S_{rs}^{hl'} = \sum_{l'=-N}^{l'=+N}\sum_{rs} P_{sr}^{l'0} S_{rs}^{0l'} = 2n_0$, with $2n_0$ denoting the number of electrons per cell. Hence, we can write

$$C_{pq}^{0j}(N) = \sum_h^{\#}\sum_{k=0}\sum_{l=0}\sum_{m=-S}^{m=+S} A_{kl|m|} R^{-(k+l+1)} (\chi_p^0|\hat{M}_a^{(k,m)}(1)^*|\chi_q^j) M^{(l,m)}(h), \tag{9.85}$$

where the multipole moment of cell h is given by

$$M^{(l,m)}(h) = \Big[\sum_u M_b^{(l,m)}(A_u^h) + \sum_{l'=h-N}^{l'=h+N} \sum_{rs} P_{sr}^{l'h}(\chi_r^h|\hat{M}_b^{(l,m)}(2)|\chi_s^{l'})\Big], \tag{9.86}$$

because the summation over u goes over the nuclei belonging to cell h, and the coordinate system b is anchored in cell h. Now is the time to say something most important.

Despite the fact that $M^{(l,m)}(h)$ depends formally on h, in reality, it is h–independent, because all the unit cells are identical.

Therefore, we may safely write that $M^{(l,m)}(h) = M^{(l,m)}$.

Now we will try to avoid a well-hidden trap, and then we will be all set to prepare ourselves to pick the fruit from our orchard. The trap is that $A_{kl|m|}$ depends on h. Why is this? Well, in the $A_{kl|m|}$, there is $(-1)^l$, while the corresponding $(-1)^k$ is absent; i.e., there is a thing that is associated with the 2^l–pole in the coordinate system b, and there is no an analogous expression for its partner, the 2^k-pole of coordinate system a. Remember, however (as discussed in Appendix X available at booksite.elsevier.com/978-0-444-59436-5), that the z-axis of each coordinate system has been chosen in such a way that a "shoots" toward b, and b *does not* shoot toward a. Therefore, the two coordinate systems are not equivalent, and hence one may have $(-1)^l$, and not $(-1)^k$. The coordinate system a is associated with cell 0, and the coordinate system b is connected to cell h. If $h > 0$, then it is true that a shoots to b, but if $h < 0$, their roles are exchanged. In such a case, in $A_{kl|m|}$, we should not put $(-1)^l$, but $(-1)^k$. If we do this, then in the summation over h in Eq. (9.85), the only dependence on h appears in a simple term $(ha)^{-(k+l+1)}$!

It appears, therefore, to be a possibility of exactly summing the electrostatic interaction along an infinite polymer chain.

Indeed, the sum

$$\sum_{h=1}^{\infty} h^{-(k+l+1)} = \zeta(k+l+1), \tag{9.87}$$

where $\zeta(n)$ stands for the *Riemann dzeta function*, which is known to a high accuracy and be available in mathematical tables.[61]

[61] For example, M. Abramovitz and I. Stegun, Eds, *Handbook of Mathematical Functions*, Dover, New York (1968), p. 811.

Georg Friedrich Bernhard Riemann (1826–1866), German mathematician and physicist and professor at the University of Göttingen. Nearly all his papers gave rise to a new mathematical theory. His life was full of personal tragedies; he lived only 40 years, but despite this, he made a giant contribution to mathematics, mainly in non-Euclidean geometries (his geometry plays an important role in the general theory of relativity), in the theory of integrals (Riemann integral), and in the theory of trigonometric series.

In his only paper on the number theory he considered the Riemann dzeta function and has shown its importance for understanding the distribution of prime numbers. He was the first to consider that physical reality may involve more than three or four dimensions.

The interactions of cell 0 with all the other cells are enclosed in this number. When this is inserted into $C_{pq}^{0j}(N)$, then we obtain

$$C_{pq}^{0j}(N) = \sum_{k=0}^{\infty}\sum_{l=0}^{\infty} U_{pq}^{0j(k,l)} \frac{\Delta_N^{(k+l+1)}}{a^{(k+l+1)}}, \tag{9.88}$$

where

$$U_{pq}^{0j(k,l)} = \sum_{m=-S}^{m=+S} (-1)^m[(-1)^k + (-1)^l] \frac{(k+l)!}{(k+|m|)!(l+|m|)!} M_{pq}^{0j(k,m)*} M^{(l,m)} \tag{9.89}$$

$$\Delta_N^{(n)} = \zeta(n) - \sum_{h=1}^{N} h^{-n}. \tag{9.90}$$

Note that the formula for $C_{pq}^{0j}(N)$ represents a sum of the multipole-multipole interactions. The formula also shows that

electrostatic interactions in a regular polymer come from a multipole-multipole interaction with different parities of the multipoles,

which can be seen from the term[62] $[(-1)^k + (-1)^l]$.

According to the discussion in Appendix X available at booksite.elsevier.com/978-0-444-59436-5, to preserve the invariance of the energy with respect to translation of the coordinate

[62] The term appears due to the previously discussed problem of "who shoots to whom" in the multipole expansion. What happens is that the interaction of an even (odd) multipole of cell 0 with an odd (even) multipole on the right side of the polymer cancels with a similar interaction with the left side. This is easy to understand. Let us imagine the multipoles as non-pointlike objects built of the appropriate point charges. We look along the periodicity axis. An even multipole has the same signs at both ends, while an odd one has opposite signs. Thus, when the even multipole is located in cell 0, and the odd one on its right side, this interaction will cancel exactly, with the interaction of the odd one located on the left side (at the same distance).

system, when computing $C_{pq}^{0j}(N)$, we have to add all the terms with $k+l+1 = const$, i.e.,

$$C_{pq}^{0j}(N) = \sum_{n=3,5,..}^{\infty} \frac{\Delta_N^{(n)}}{a^n} \sum_{l=1}^{n-1} U_{pq}^{0j(n-l-1,l)}. \tag{9.91}$$

The above expression is equivalent to Eq. (9.88), but it ensures automatically the translational invariance by taking into account all the necessary multipole-multipole interactions.[63]

What should we know, therefore, to compute the long-range correction $C_{pq}^{0j}(N)$ to the Fock matrix[64]? From Eq. (9.91), it is seen that one has to know how to calculate three numbers: $U_{pq}^{0j(k,l)}$, a^{-n} and $\Delta_N^{(n)}$. The equation for the first one is given in Table 9.2; the other two are trivial. Δ is easy to calculate knowing the Riemann ζ function (from tables): in fact, we have to calculate the multipole moments, and these are *one*-electron integrals (easy to calculate). Originally, before the multipole expansion method was designed, we also had a large number of *two*-electron integrals (expensive to calculate). Instead of overnight calculations, the computer time was reduced to about 1 *s* and the results were more accurate.

9.12.4 Multipole Expansion Applied to the Total Energy

As shown above, the long-range correction to the total energy means that the interaction of cell 0 with all the cells from the long-range region multiplied by $\frac{1}{2}$. The reasoning pertaining to its computation may be repeated exactly in the way we have shown in the previous subsection. However, we must remember a few differences:

- What interacts is not the charge distribution $\chi_p^{0*}\chi_q^j$, but the complete cell 0.
- The result has to be multiplied by $\frac{1}{2}$, for reasons discussed earlier.

Finally, we obtain

$$C_T(N) = \frac{1}{2} \sum_{k=0}^{\infty} \sum_{l=0}^{\infty} U_T^{(k,l)} \frac{\Delta_N^{(k+l+1)}}{a^{k+l+1}}, \tag{9.92}$$

where

$$U_T^{(k,l)} = \sum_{m=-S}^{m=+S} \left((-1)^k + (-1)^l\right) \frac{(k+l)!(-1)^m}{(k+|m|)!(l+|m|)!} M^{(k,m)*} M^{(l,m)}. \tag{9.93}$$

[63] Indeed, $\sum_{l=1}^{n-1} U_{pq}^{0j(n-l-1,l)} = U_{pq}^{0j(n-2,1)} + U_{pq}^{0j(n-3,2)} + \cdots + U_{pq}^{0j(0,n-1)}$; i.e., a review of all terms with $k+l+1 = n$ except $U_{pq}^{0j(n-1,0)}$. This term is absent because it requires calculation of $M^{(0,0)}$; i.e., of the *charge of the elementary cell*, which has to stay electrically neutral (otherwise the polymer falls apart). Therefore $M^{(0,0)} = 0$. Why, however, does the summation over n not simply represent $n = 1, 2, \ldots \infty$, but contains only odd values of n except $n = 1$? What would happen if we took $n = 1$? Look at Eq. (9.88). The value $n = 1$ requires $k = l = 0$. This leads to the "monopole–monopole" interaction, but this is 0, since the whole unit cell (and one of the multipoles is that of the unit cell) carries no charge. The summation in Eq. (9.91) does not contain any even values of n, because they would correspond to k and l of different parities, and such interactions (as we have shown before) are equal to 0. Therefore, indeed, Eq. (9.91) contains all the terms that are necessary.

[64] L. Piela and J. Delhalle, *Intern. J. Quantum Chem. 13*, 605 (1978).

Table 9.2. The quantities $U^{(k,l)}$ for $k+l<7$ necessary for calculating the most important long-range corrections to the Fock matrix elements $U_{pq}^{0j(k,l)}$ and to the total energy per unit cell $U_T^{(k,l)}$. The parentheses [] mean the corresponding Cartesian multipole moment. When computing the Fock matrix correction, the first multipole moment [] stands for the multipole moment of the charge distribution $\chi_p^0\chi_q^j$, the second, of the unit cell. For example, $U^{(0,2)}$ for the correction $C_{pq}^{0j}(N)$ is equal to $\left(\begin{smallmatrix}0 & j\\ p & q\end{smallmatrix}\right)\left(\sum_u Z_u\left(3z_u^2-r_u^2\right)-\sum_{l'=-N}^{l'=+N}\sum_{rs}P_{sr}^{l'0}\left(\chi_r^0|3z^2-r^2|\chi_s^{l'}\right)\right)$, while $U^{(0,2)}$ for $C_T(N)$ equals 0 because [1] means the *charge* of the unit cell, which is equal to zero. In the table, only values of U for $k\le l$ are given. If $l<k$, then the formula is the same, but the order of the moments is reversed.

n	$U^{(k,l)}, k+l+1=n$
3	$U^{(0,2)}=[1][3z^2-r^2]$
	$U^{(1,1)}=2[x][x]+2[y][y]-4[z][z]$
5	$U^{(0,4)}=\frac{1}{4}[1][35z^4-30z^2r^2+3r^4]$
	$U^{(1,3)}=4[z][3r^2z-5z^3]+3[x][5xz^2-r^2x]+3[y][5yz^2-r^2y]$
	$U^{(2,2)}=3[3z^2-r^2][3z^2-r^2]-24[xz][xz]-24[yz][yz]$
	$+\frac{3}{2}[x^2-y^2][x^2-y^2]+6[xy][xy]$
7	$U^{(0,6)}=\frac{1}{8}[1][231z^6-315z^4r^2+105z^2r^4-5r^6]$
	$U^{(1,5)}=-\frac{3}{2}[z][63z^5-70z^3r^2+15zr^4]+\frac{15}{4}[x][21z^4x-14z^2xr^2+xr^4]$
	$+\frac{15}{4}[y][21z^4y-14z^2yr^2+yr^4]$
	$U^{(2,4)}=\frac{15}{8}[3z^2-r^2][35z^4-30z^2r^2+3r^4]-30[xz][7z^3x-3xzr^2]$
	$-30[yz][7z^3y-3yzr^2]+\frac{15}{4}[x^2-y^2][7z^2(x^2-y^2)-r^2(x^2-y^2)]$
	$+15[xy][7z^2xy-xyr^2]$
	$U^{(3,3)}=-10[5z^3-3zr^2][5z^3-3zr^2]+\frac{45}{4}[5z^2x-xr^2][5z^2x-xr^2]$
	$+\frac{45}{4}[5z^2y-yr^2][5z^2y-yr^2]-45[zx^2-zy^2][zx^2-zy^2]$
	$-180[xyz][xyz]+\frac{5}{4}[x^3-3xy^2][x^3-3xy^2]+\frac{5}{4}[y^3-3x^2y][y^3-3x^2y]$

Let us note that (for the same reasons as before)

> the interaction of multipoles of different parities equals zero.

and this time we have to do with the interaction of the multipoles of complete cells. The quantities $U_T^{(k,l)}$ are given in Table 9.2.

Do the Fock Matrix Elements and the Total Energy per Cell Represent Finite Values?

If the Fock matrix elements were infinite, then we could not manage to carry out the Hartree-Fock-Roothaan self-consistent procedure. If E_T were infinite, the periodic system could not exist at all. It is, therefore, important to know when we can safely *model* an infinite system.

For any finite system, there is no problem: the results are always finite. The only danger, therefore, is summation to infinity ("*lattice sums*"), which always ends with the interaction of a part or whole unit cell with an infinite number of distant cells. Let us take such an example in the simplest case of a single atom per cell. Let us assume that the atoms interact by the Lennard-Jones pairwise potential (p. 347): $E = \varepsilon\left[\left(\frac{r_0}{r}\right)^{12} - 2\left(\frac{r_0}{r}\right)^{6}\right]$, where r means the interatomic distance, r_0 means the equilibrium distance, and ε is the depth of the potential well. Let us try to compute the lattice sum $\sum_j E_{0j}$, where E_{0j} means the interaction energy of the cells 0 and j. We see that, due to the form of the potential over long distances, what counts is uniquely the attractive term $-2\varepsilon\left(\frac{r_0}{r}\right)^6$. When we take such interactions that pertain to a sphere of the radius R (with the origin located on atom 0), each individual term (i.e., its absolute value) decreases with increasing R. This is very important, because when we have a 3-D lattice, the number of such interactions within the sphere *increases* as R^3. We see that the decay rate of the interactions will finally win and the lattice sum will converge. However, we can easily see that if the decay of the pairwise interaction energy were slower, then we might have had trouble calculating the lattice sum. For example, if, instead of the neutral Lennard-Jones atoms, we took ions of the same charge, the interaction energy would explode to ∞. It is evident, therefore, that for each periodic system, there are some conditions to be fulfilled if we want to have finite lattice sums.

These conditions are more severe for the Fock matrix elements because each of the terms represent the interaction of a *charge* with complete distant unit cells. The convergence depends on the asymptotic interaction energy of the potential. In the case of the multipole-multipole interaction, we know what the asymptotic behavior looks like: it is $R^{-(0+l+1)} = R^{-(l+1)}$, where R stands for the intercell distance. The lattice summation in a nD lattice ($n = 1, 2, 3$) gives the partial sum dependence on R as $\frac{R^n}{R^{l+1}} = R^{n-l-1}$. This means that[65]

in 1-D, the unit cell cannot have any nonzero net charge ($l = 0$); in 2-D, it cannot have a nonzero charge and dipole moment ($l = 1$); and in 3-D, it cannot have a nonzero charge, dipole moment, and quadrupole moment ($l = 2$).

9.13 Back to the Exchange Term

The long-range effects discussed so far result from the Coulomb interaction in the Fock equation for a regular polymer. There is, however, also an exchange contribution, which has been postponed in the long-range region. It is time now to reconsider this contribution. The exchange term in the Fock matrix element F_{pq}^{0j} had the following form [see Eq. (9.67)]:

$$-\frac{1}{2}\sum_{h,l}\sum_{rs} P_{sr}^{lh}\left({}_{pr}^{0h}\middle|{}_{sq}^{lj}\right) \tag{9.94}$$

and gave the following contribution to the total energy per unit cell:

$$E_{exch} = \sum_j E_{exch}(j), \tag{9.95}$$

[65] L.Z. Stolarczyk and L. Piela, *Intern.J.Quantum Chem.* 22, 911 (1982).

where the cell 0–cell j interaction has the form [see Eq. (9.81)]:

$$E_{exch}(j) = -\frac{1}{4}\sum_{h,l}\sum_{pqrs} P_{qp}^{j0} P_{sr}^{lh} \left({}^{0h}_{pr} \middle| {}^{lj}_{sq} \right). \tag{9.96}$$

It would be very nice to have the exchange contribution $E_{exch}(j)$ decaying fast, when j increases, because it could be enclosed in the short-range contribution. Do we have good prospects for this? The above formula shows (the integral) that the summation over l is *safe*: the contribution of those cells l that are far from cell 0 is negligible due to differential overlaps of type $\chi_p^0(1)^* \chi_s^l(1)$. The summation over cells h is safe as well (for the same reasons), because it is bound to be limited to the neighborhood of cell j (see the integral).

In contrast, the only guarantee of a satisfactory convergence of the sum over j is that *we hope* the matrix element P_{qp}^{j0} decays fast if j increases.

So far, exchange contributions have been neglected, and there has been an indication that suggested this was the right procedure. This was the magic word "*exchange*". All the experience of myself and my colleagues in intermolecular interactions whispers, "*This is surely a short-range type.*" In a manuscript by Sandor Suhai, I read that the exchange contribution is of a long-range type. To our astonishment, this turned out to be right (just a few numerical experiments). We have a long-range exchange. After analysis was performed, it turned out that

the long-range exchange interaction appears if and only if the system is metallic.

A metallic system is notorious for its HOMO–LUMO quasidegeneracy; therefore, we began to suspect that when the HOMO–LUMO gap decreases, the P_{qp}^{j0} coefficients do not decay with j.

Such things are most clearly seen when the simplest example is taken, and the hydrogen molecule at long internuclear distance is the simplest prototype of a metal. Indeed, this is a system with half-filled orbital energy levels when the LCAO MO method is applied (in the simplest case: two atomic orbitals). Note that, after subsequently adding two extra electrons, the resulting system (let us not worry too much that such a molecule could not exist!) would model an insulator; i.e., all the levels are doubly occupied.[66]

Analysis of these two cases convinces us that indeed our suspicions were justified. Here are the bond order matrices that we obtain in both cases (see Appendix S available at booksite.elsevier.com/978-0-444-59436-5, p. e143), S denotes the overlap integral of the $1s$ atomic orbitals of

[66] Of course, we could take two helium atoms. This also would be good. However, the first principle in research is "*In a single step, only change a single parameter, analyze the result, draw the conclusions, and take the second step.*"
Just in passing, a second principle also applies here. If we do not understand an effect, what should we do? Just divide the system in two parts and look where the effect persists. Keep dividing until the effect disappears. Take the simplest system in which the effect still exists, analyze the problem, understand it, and go back slowly to the original system (this is why we have H_2 and H_2^{2-} here).

atoms a and b):

$$\boldsymbol{P} = (1+S)^{-1}\begin{pmatrix} 1 & 1 \\ 1 & 1 \end{pmatrix} \quad \text{for } H_2 \tag{9.97}$$

$$\boldsymbol{P} = (1-S^2)^{-1}\begin{pmatrix} 1 & -S \\ -S & 1 \end{pmatrix} \quad \text{for } H_2^-. \tag{9.98}$$

We see[67] how profoundly these two cases differ in the off-diagonal elements (they are analogues of P_{qp}^{j0} for $j \neq 0$).

In the second case, the proportionality of P_{qp}^{j0} and S ensures an *exponential, therefore very fast, decay* if j tends to ∞. In the first case, there is no decay of P_{qp}^{j0} at all.

A detailed analysis for an infinite chain of hydrogen atoms ($\omega = 1$) leads to the following formula[68] for P_{qp}^{j0}:

$$P_{11}^{j0} = \frac{2}{\pi j}\sin\left(\frac{\pi j}{2}\right). \tag{9.99}$$

This means an extraordinarily slow decay of these elements (and therefore of the exchange contribution) with j. When the metallic regime is even slightly removed, the decay gets much, much faster.

This result shows that the long-range character of the exchange interactions does not exist in reality. It seems to represent an indication that the Hartree-Fock method fails in such a case.

9.14 Choice of Unit Cell

The concept of the unit cell has been important throughout this chapter. The unit cell represents an object that, when repeated by translations, gives an infinite crystal. In this simple definition, almost every word can be a trap.

Is this feasible? Is the choice unique? If not, then what are the differences among them? How is the motif connected to the unit cell choice? Is the motif unique? Which motifs may we consider?

As we have already noted, the choice of unit cell as well as of motif is not unique. This is easy to see. Indeed Fig. 9.22 shows that the unit cell and the motif can be chosen in many different and equivalent ways.

Moreover, there is no chance of telling in a responsible way which of the choices are reasonable and which are not. And it happens that in this particular case, we really have a plethora of choices. Putting no limits to our fantastic scenario, we may choose a unit cell in a particularly capricious way (see Figs. 9.22b and 9.23).

---

[67] L. Piela, J.-M. André, J.G. Fripiat, and J. Delhalle, *Chem. Phys. Lett. 77*, 143 (1981).

[68] I.I. Ukrainski, *Theor. Chim. Acta 38*, 139 (1975); $q = p = 1$ means that we have a single 1s hydrogen orbital per unit cell.

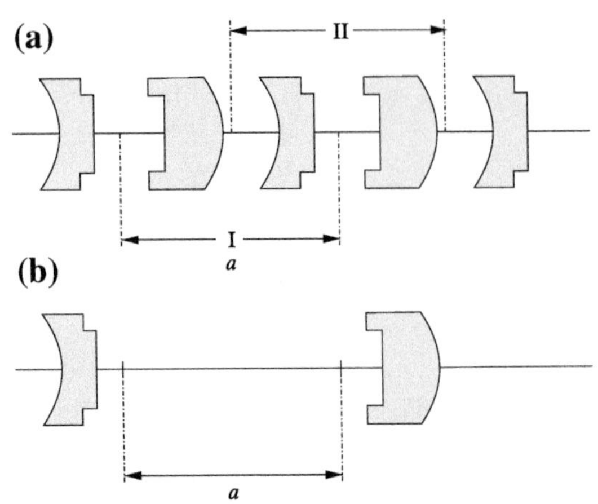

Fig. 9.22. Three of many possible choices of the unit cell motif. (a) choices I and II differ, but both look "reasonable"; (b) choice III might be called strange. Despite of this strangeness, choice III is as legal (from the point of view of mathematics) as I or II.

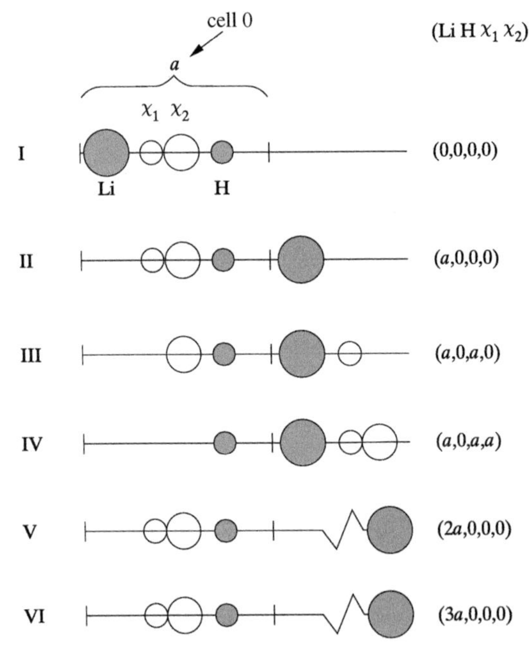

Fig. 9.23. Six different choices (I-VI) of unit cell content (motifs) for a linear chain $(\text{LiH})_\infty$. Each cell has the same length $a = 6.3676$ a.u. There are two nuclei: Li^{3+} and H^+ and two Gaussian doubly occupied $1s$ atomic orbitals (denoted by χ_1 and χ_2, with exponents 1.9815 and 0.1677, respectively) *per cell*. Motif I corresponds to a common sense situation: both nuclei and electron distribution determined by χ_1 and χ_2 are within the section $(0, a)$. The other motifs (II-VI), all corresponding to the same unit cell $(0, a)$ of length a are very strange. Each motif is characterized by the symbol (k, l, m, n), which means that the Li nucleus, H nucleus, χ_1 and χ_2 are shifted to the right by ka, la, ma, and na, respectively. *All the unit cells with their contents (motifs) are fully justified, equivalent from the mathematical point of view, and, therefore, "legal" from the point of view of physics.* Note that the nuclear framework and the electronic density corresponding to a cell are very different for all the choices.

Fig. 9.23 shows six different, fully legitimate, choices of motif associated with a unit cell in a 1-D "*polymer*" $(LiH)_\infty$. Each motif consists of the lithium nucleus, a proton, and an electronic charge distribution in the form of two Gaussian 1s orbitals that accommodate four electrons altogether. By repeating any of these motifs, we reconstitute the same original chain.

We may say there may be many legal choices of motif, but this is without any theoretical meaning because all the choices lead to the same infinite system. Well, this is true with respect to theory, but in practical applications, the choice of motif may be of prime importance. We can see this from Table 9.3, which corresponds to Fig. 9.23.

The results, without taking into account the long-range interactions, depend very strongly on the choice of unit cell motif.

Use of the multipole expansion greatly improves the results and, to very good accuracy, makes them *independent of the choice of unit cell motif* as it should be.

Table 9.3. Total energy per unit cell E_T in the "polymer" LiH as a function of unit cell definition (Fig. 9.23). For each choice of unit cells, this energy is computed in four ways: (1) without long-range forces (long range = 0); i.e., unit cell 0 interacts with $N = 6$ unit cells on its right-hand side and N unit cells on its left-hand-side (2), (3), (4) with the long range computed with multipole interactions up to the a^{-3}, a^{-5} and a^{-7} terms, respectively. The bold figures are exact. The corresponding dipole moment μ of the unit cell (in Debyes) is also given.

Unit Cell	Long Range	μ	$-E_T$
I	0	6.6432	**6.61**0869
	a^{-3}	6.6432	**6.6127946**92
	a^{-5}	6.6432	**6.6127946**87
	a^{-7}	6.6432	**6.612794674**
II	0	−41.878	**6.**524802885
	a^{-3}	−41.878	**6.61**2519674
	a^{-5}	−41.878	**6.61279**0564
	a^{-7}	−41.878	**6.6127946**04
III	0	−9.5305	**6.60**7730984
	a^{-3}	−9.5305	**6.61278**8446
	a^{-5}	−9.5305	**6.6127946**33
	a^{-7}	−9.5305	**6.61279467**3
IV	0	22.82	**6.5**7395630
	a^{-3}	22.82	**6.612**726254
	a^{-5}	22.82	**6.612793**807
	a^{-7}	22.82	**6.6127946**62
V	0	−90.399	6.148843431
	a^{-3}	−90.399	**6.60**7530384
	a^{-5}	−90.399	**6.61**2487745
	a^{-7}	−90.399	**6.6127**74317

Note that the larger the dipole moment of the unit cell, the worse the results with the short-range forces only. This is understandable because the first non-vanishing contribution in the multipole expansion is the dipole-dipole term (see Appendix X available at booksite.elsevier.com/978-0-444-59436-5). Note how considerably the unit cell dependence drops after this term is switched on (a^{-3}).

The conclusion is that in the standard (i.e., short-range) calculations, we should always choose the unit cell motif that corresponds to the smallest dipole moment. It seems, however, that such a motif is what everybody would choose using common sense.

9.14.1 Field Compensation Method

In a moment, we will unexpectedly find a very different conclusion. The logical chain of steps that led to it has, in my opinion, a didactic value, and contains a considerable amount of optimism. When this result was obtained by Leszek Stolarczyk and myself, we were stunned by its simplicity.

Is it possible to design a unit cell motif with a dipole moment of zero? This would be a great unit cell because its interaction with other cells would be weak and it would decay fast with intercellular distance. We could therefore compute the interaction of a few cells like this and the job would be done: we would have an accurate result at very low cost.

There is such a unit cell motif.

Imagine that we start from the concept of the unit cell with its motif (with lattice constant a). This motif is, of course, electrically neutral (otherwise, the total energy would be $+\infty$), and its dipole moment component along the periodicity axis is equal to μ. Let us put its symbol in the unit cell (see Fig. 9.24a).

Now let us add to the motif two extra (i.e., fictitious) pointlike opposite charges ($+q$ and $-q$), located on the periodicity axis and separated by a. The charges are chosen in such a way ($q = \frac{\mu}{a}$) that they alone give the dipole moment component along the periodicity axis equal to $-\mu$; Fig. 9.24b.

In this way, the new unit cell dipole moment (with the additional fictitious charges included) is equal to zero. Is this an acceptable choice of motif? Well, what does *acceptable* mean in this context? The only requirement is that by repeating the new motif with period a, we have to reconstruct the whole crystal. What will we get when repeating the new motif? Let us see (by taking a look at Fig. 9.24c).

We get the original periodic structure because the charges all along the polymer, except the boundaries, have cancelled each other out. Simply, the pair of charges $+q$ and $-q$, when located at a point, result in nothing.

In practice, we would like to repeat just a few neighboring unit cell motifs (a cluster) and then compute their interaction. In such case, we will observe the charge cancellation inside the cluster, but no cancellation on its boundaries ("*surface*").

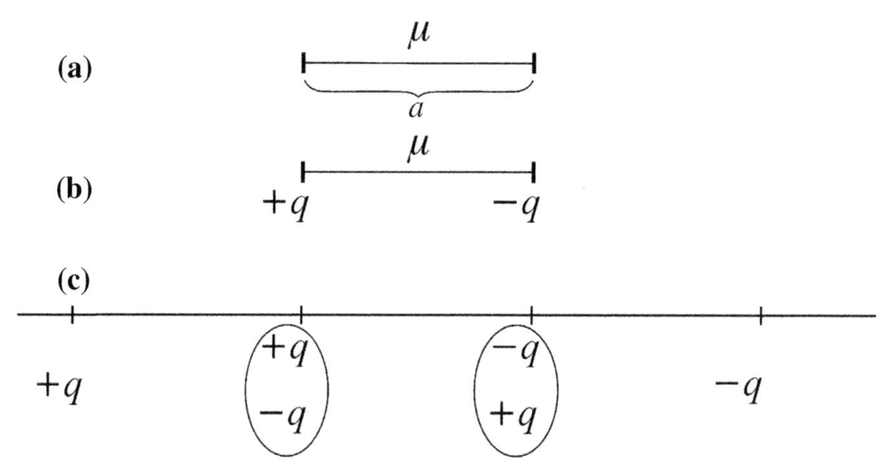

Fig. 9.24. Field compensation method. (a) The unit cell with length a and dipole moment $\mu > 0$. (b) The modified unit cell with additional fictitious charges ($|q| = \frac{\mu}{a}$) that cancel the dipole moment. (c) The modified unit cells (with $\mu' = 0$) give the original polymer when added together.

Therefore, we get a sort of point charge distribution at the boundaries.

If the boundary charges did not exist, it would correspond to the traditional calculations of the original unit cells without taking any long-range forces into account. The boundary charges, therefore, play the important role of replacing the electrostatic interaction with the rest of the *infinite* crystal, by the boundary charge interactions with the cluster ("*field compensation method*").

This is all. The consequences are simple.

Let us not only kill the dipole moment, but also other multipole moments of the unit cell content (up to a maximum moment). The resulting cell will be unable to interact electrostatically with anything. Therefore, interaction within a small cluster of such cells will give us an accurate energy per cell result.

This multipole killing (field compensation) may be carried out in several ways.[69]

Application of the method is extremely simple. Imagine unit cell 0 and its neighboring unit cells (a cluster). Such a cluster is sometimes treated as a molecule and its role is to represent a bulk crystal. This is a very expensive way to describe the bulk crystal properties, for the cluster surface atom ratio to the bulk atom is much higher than we would wish (the surface still playing an important role). What is lacking is the crystal field that will change the cluster properties. In the field compensation method, we do the same, but there are some fictitious charges at the cluster boundaries that take care of the crystal field. This enables us to use a smaller cluster than before (low cost) and still get the influence of the infinite crystal. The fictitious charges are treated the same way in computations as the nuclei (even if some of them are

[69] L. Piela and L.Z. Stolarczyk, *Chem.Phys.Letters 86*, 195 (1982).

negatively charged). However artificial it may seem, the results are far better when using the field compensation method than without it.[70]

9.14.2 The Symmetry of Subsystem Choice

The example described above raises an intriguing question, pertaining to our understanding of the relation between a part and the whole.

There are an infinite number of ways to reconstruct the same system from parts. *These ways are not equivalent in practical calculations* if, for any reason, we are unable to compute all the interactions in the system. However, if we have a theory (in our case the multipole method) that is able to compute the interactions,[71] including the long-range forces, then it turns out the final result is virtually independent of the choice of unit cell motif. This arbitrariness of choice of subsystem looks analogous to the arbitrariness of the choice of coordinate system. The final results do not depend on the coordinate system used, but still the numerical results (as well as the effort to get the solution) do.

The separation of the whole system into subsystems is of key importance to many physical approaches, but we rarely think of the freedom associated with the choice. For example, an atomic nucleus does not in general represent an elementary particle, and yet in quantum mechanical calculations, we treat it as a point particle, without an internal structure, and we can do this successfully.[72] Further, in the *Bogolyubov*[73]*transformation*, the Hamiltonian is represented by creation and annihilation operators, each being a linear combination of the creation and annihilation operators for electrons (described in Appendix U available at booksite.elsevier.com/978-0-444-59436-5, p. e153). The new operators also fulfill the anticommutation rules, only the Hamiltonian contains more additional terms than before (see Appendix U available at booksite.elsevier.com/978-0-444-59436-5). A particular Bogolyubov transformation may describe the creation and annihilation of quasi-particles, such as the electron hole (and others). We are dealing with the same physical system as before, but we look at it from a completely different point of view, by considering it as being composed of something else. Is there any theoretical (i.e., *serious*) reason for preferring one division into subsystems over another? Such a reason may be only of practical importance.[74] Any correct theory should give the same description of the total system *independently of subsystems that we decide to choose.*

[70] Using "negative nuclei" looked so strange that some colleagues doubted receiving anything reasonable from such a procedure.

[71] This is done with controlled accuracy; i.e., we still neglect the interactions of higher multipoles.

[72] This represents only a fragment of the storylike structure of science (cf. p. 67), one of its most intriguing features. It makes science work; otherwise, when considering the genetics of peas in biology, we would have to struggle with the quark theory of matter.

[73] Nicolai Nicolaevitch Bogolyubov (1909–1992) was a Russian physicist, director of the Dubna Nuclear Institute, and an outstanding theoretician.

[74] For example, at temperature $t < 0°$C, we may solve the equations of motion for N frozen water drops, and we may obtain reasonable dynamics of the system. At $t > 0°$C, obtaining such dynamics will be virtually impossible.

Symmetry with Respect to Division into Subsystems

The symmetry of *objects* is important for the description of *them*, and therefore may be viewed as being of limited interest. The symmetry of the laws of nature [i.e., of the theory that describes all objects (whether symmetric or not)] is much more important. This has been discussed in detail in Chapter 2 (cf. p. 68), but it seems that we did not list there a fundamental symmetry of any correct theory: the *symmetry with respect to the choice of subsystems. A correct theory has to describe the total system independently of what we decide to treat as subsystems.*

We will meet this problem once more in intermolecular interactions (in Chapter 13). However, in the periodic system, it has been possible to use, in computational practice, the symmetry described above.

Our problem resembles an excerpt from "*Dreams of a Final Theory*" by Steven Weinberg[75] pertaining to *gauge symmetry*: "*The symmetry underlying it has to do with changes in our point of view about the identity of the different types of elementary particle. Thus it is possible to have a particle wave function that is neither definitely an electron nor definitely a neutrino, until we look at it.*" Here we have freedom in the choice of subsystems as well, and a correct theory has to reconstitute the description of the whole system.

This is an intriguing problem.

Summary

- A crystal is often approximated by an infinite crystal *(primitive) lattice,* which leads to the concept of the *unit cell.* By translational repeating of a chosen atomic *motif* associated with a unit cell, we reconstruct the whole infinite crystal.
- The one-electron Hamiltonian is invariant with respect to translations by any lattice vector. Therefore, its eigenfunctions (crystal orbitals) are simultaneously eigenfunctions of the translation operators (Bloch theorem): $\hat{T}(\boldsymbol{R}_j)\phi_{\boldsymbol{k}}(\boldsymbol{r}) = \phi_{\boldsymbol{k}}(\boldsymbol{r}-\boldsymbol{R}_j) = \exp\left(-i\boldsymbol{k}\boldsymbol{R}_j\right)\phi_{\boldsymbol{k}}(\boldsymbol{r})$ and transform according to the irreducible representation of the translational group labeled by the *wave vector* $\boldsymbol{k}$.
- *Bloch functions* may be treated as atomic symmetry orbitals $\phi = \sum_j \exp\left(i\boldsymbol{k}\boldsymbol{R}_j\right)\chi\left(\boldsymbol{r}-\boldsymbol{R}_j\right)$ formed from the atomic orbital $\chi\left(\boldsymbol{r}\right)$. Their symmetry is determined by $\boldsymbol{k}$.
- The crystal lattice basis vectors allow the formation of the basis vectors of the *inverse lattice.* Linear combinations of them (with integer coefficients) determine the *inverse lattice* subject to translational symmetry.
- A special *(Wigner-Seitz)* unit cell of the inverse lattice is called the *First Brillouin Zone (FBZ).*
- The vectors $\boldsymbol{k}$ inside the FBZ label possible non-equivalent irreducible representations of the translational group.
- The wave vector plays a triple role:
  - It indicates the *direction of the wave*, which is an eigenfunction of $\hat{T}(\boldsymbol{R}_j)$ with eigenvalue $\exp(-i\boldsymbol{k}\boldsymbol{R}_j)$.
  - It labels the *irreducible representations* of the translational group.
  - The longer the wave vector $\boldsymbol{k}$, the more nodes the wave has.
- In order to neglect the crystal surface, we apply the *Born–von Kármán boundary condition*: "*Instead of a stick-like system, we take a circle.*"

[75] Pantheon Books, New York (1992), Chapter 6.

- In full analogy with molecules, we can formulate the SCF LCAO CO Hartree-Fock-Roothaan method (a CO instead of an MO). Each CO is characterized by a vector $\boldsymbol{k} \in$ FBZ and is a linear combination of the Bloch functions with the same $\boldsymbol{k}$.
- The orbital energy dependence on $\boldsymbol{k} \in$ FBZ is called the *energy band*. The stronger the intercell interaction, the wider the bandwidth (dispersion).
- Electrons occupy the *valence bands*, and the *conduction bands* are empty. The *Fermi level* is the HOMO energy of the crystal. If the HOMO-LUMO energy difference (*energy gap* between the valence and conduction bands) is zero, we have a *metal*; if it is large, we have an *insulator*; if it is medium, we have a *semiconductor*.
- Semiconductors may be intrinsic, or n–type (if the donor dopant levels are slightly below the conduction band), or p–type (if the acceptor dopant levels are slightly above the occupied band)
- Metals when cooled may undergo what is known as the *Peierls transition*, which denotes lattice dimerization and band gap formation. In this way, the system changes from a metal to a semiconductor or insulator. This transition corresponds to the Jahn-Teller effect in molecules.
- Polyacetylene is an example of a *Peierls transition* (dimerization), which results in shorter bonds (a little less multiple than double ones) and longer bonds (a little more multiple than single ones). Such a dimerization introduces the possibility of a defect separating two rhythms ("*phases*") of the bonds: from double-single to single-double. This defect can move within the chain, which may be described as a solitonic wave. The soliton may become charged, and in this case, participates in electric conduction (increasing it by many orders of magnitude).
- In polyparaphenylene, a soliton wave is not possible because the two phases, quinoid and aromatic, are not of the same energy, which excludes free motion. A double defect is possible though, a bipolaron. Such a defect represents a section of the quinoid structure (in the aromatic-like chain), at the end of which we have two unpaired electrons. The electrons, when paired with extra electrons from donor dopants or when removed by acceptor dopants, form a double ion (bipolaron), which may contribute to electric conductance.
- The band structure may be foreseen in simple cases and logically connected to the subsystem orbitals.
- To compute the Fock matrix elements or the total energy per cell, we have to calculate the interaction of cell 0 with all other cells.
- The interaction with neighboring cells is calculated without approximations, while that with distant cells uses multipole expansion. Multipole expansion applied to the electrostatic interaction gives accurate results, while the numerical effort is dramatically reduced.
- In some cases (metals), we meet long-range exchange interaction, which disappears as soon as the energy gap emerges. This indicates that the Hartree-Fock method is not applicable in this case.
- The choice of unit cell motif is irrelevant from the theoretical point of view, but leads to different numerical results when the long-range interactions are omitted. When taking into account the long-range interactions, the theory becomes independent of the division of the whole system into arbitrary motifs.

Main Concepts, New Terms

From the Research Front

The Hartree-Fock method for periodic systems nowadays represents a routine approach coded in several *ab initio* computer packages. We may analyze the total energy, its dependence on molecular conformation, the density of states, the atomic charges, etc. Also, calculations of first-order responses to the electric field (polymers are of interest for optoelectronics) have been successful in the past. However, nonlinear problems (like the second harmonic generation; see Chapter 12) still represent a challenge.

Ad Futurum

Probably there will soon be no problem in carrying out the Hartree-Fock or DFT (see Chapter 11) calculations, even for complex polymers and crystals. What will remain for a few decades is the very important problem of lowest-energy crystal packing and of solid-state reactions and phase transitions. Post-Hartree-Fock calculations will be more and more important taking into account electronic correlation effects. The real challenge will start in designing non-periodic materials, where the polymer backbone will serve as a molecular rack for installing some functions (transport, binding, releasing, signal transmitting). The functions will be expected to cooperate (*"intelligent materials,"* cf. Chapter 15).

Additional Literature

R. Hoffmann, *Solids and Surfaces. A Chemist's View of Bonding in Extended Structures*, VCH Publishers, New York (1988).

A masterpiece written by a Nobel Prize winner, one of the founders of solid-state quantum chemistry. Solid-state theory was traditionally the domain of physicists. Some concepts typical of chemistry, such as atomic orbitals, bonding and antibonding effects, chemical bonds, and localization of orbitals, were usually absent in such descriptions. They are highlighted in this book.

J.-M. André, J. Delhalle, and J.-L. Brédas, *Quantum Chemistry Aided Design of Organic Polymers*, World Scientific, Singapore, 1991.

A well-written book oriented mainly toward the response of polymers to the electric field.

Questions

1. Bloch theorem [$\phi_{\mathbf{k}}(\mathbf{r})$ stands for a crystal orbital (CO), $\mathbf{R}_j$ is a lattice vector, $\mathbf{k}$ is a wave vector]:
   a. pertains to the eigenvalue (corresponding to $\phi_{\mathbf{k}}$) of the translation operator by a lattice vector

b. $|\phi_{\mathbf{k}}(\mathbf{r})| = |\phi_{\mathbf{k}}(\mathbf{r} - \mathbf{R}_j)|$
c. $|\phi_{\mathbf{k}}(\mathbf{r})|^2$ exhibits the same symmetry as the potential energy $V(\mathbf{r})$
d. if $\phi_{\mathbf{k}}(\mathbf{r})$ corresponds to the wave vector $\mathbf{k}$, $\phi_{\mathbf{k}}(\mathbf{r} - \mathbf{R}_j) = \exp\left(-i\mathbf{k} \cdot \mathbf{R}_j\right) \phi_{\mathbf{k}}(\mathbf{r})$

2. The First Brillouin Zone (FBZ)
   a. means the smallest unit cell of a primitive lattice
   b. means a smallest motif to be repeated in a crystal
   c. does not contain in its inner part any pair of equivalent wave vectors
   d. the wave vectors that correspond to the surface of the FBZ may differ by an inverse lattice vector
3. A function $\phi_{\mathbf{k}}$ corresponding to the wave vector $\mathbf{k}$
   a. for $\mathbf{k} = \mathbf{0}$ the function $\phi_{\mathbf{k}}$, build of $1s$ atomic orbitals, does not have any nodal planes
   b. represents a wave with the front perpendicular to $\mathbf{k}$
   c. the larger $|\mathbf{k}|$ the greater the number of nodes
   d. has to be a crystal orbital (CO)
4. A crystal orbital (CO)
   a. represents any linear combination of the atomic orbitals of the atoms the crystal is composed of
   b. is characterized by its wave vector $\mathbf{k}$
   c. with $\mathbf{k} = \mathbf{0}$ corresponds to the lowest energy for a given electronic band
   d. two COs with $\mathbf{k}$ that differ by any inverse lattice vector are identical
5. An infinite polyacetylene chain
   a. represents a conductor
   b. exhibits an alternation of the CC bond lengths
   c. exhibits a nonzero energetic gap between the valence band and the conduction band
   d. conducts the electric current thanks to the solitonic defects that result from donor or acceptor doping
6. The band width (CO means a crystal orbital) increases if
   a. one goes from the COs that correspond to the inner electronic shells to the COs corresponding to valence electrons
   b. one increases a pressure
   c. the distance between atoms gets larger
   d. the atomic orbitals overlap more
7. A semiconductor
   a. exhibits a small energy gap
   b. has about a half of the conductivity of copper
   c. has the energy gap equal to zero
   d. conducts electricity, but only in one direction
8. The Fermi level
   a. represents an electronic energy level from which removing an electron needs the least energy
   b. has the energy corresponding to the HOMO orbital of the crystal
   c. means the mean energy of the occupied electronic states
   d. is the lowest energy of the conducting band
9. The dipole-quadrupole interaction per unit cell in a regular polymer is
   a. 0
   b. equal to the difference between the dipole-dipole and quadrupole-quadrupole interactions
   c. equal to the mean value o the dipole-dipole and quadrupole-quadrupole interactions
   d. a sum of the interaction of the dipole of the unit cell 0 with the quadrupoles from beyond the section of cells $-N, \dots N$

10. The unit cell dipole moment in a regular polymer
    a. is uniquely defined for an electrically neutral polymer
    b. does not depend on the position of the cell with respect to cell 0
    c. must be equal 0, otherwise the total dipole-dipole interaction energy would be equal ∞
    d. does depend on the choice of the unit cell

Answers

1a,b,c,d, 2c,d, 3a,b,c, 4b,d, 5b,c,d, 6a,b,d, 7a, 8a,b, 9a,d, 10b,d

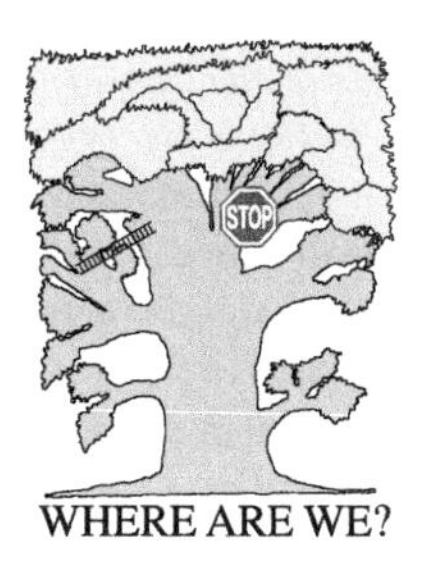

Correlation of the Electronic Motions

"God does not care about our mathematical difficulties, He integrates empirically."
Albert Einstein

Where Are We?

The main road on the trunk leads us to the right part of the crown of the tree.

An Example

As usual, let us consider the simplest example: the hydrogen molecule. The normalized Hartree-Fock determinant,

$$\psi_{RHF}(1,2) = \frac{1}{\sqrt{2}} \begin{vmatrix} \phi_1(1) & \phi_1(2) \\ \phi_2(1) & \phi_2(2) \end{vmatrix},$$

with double occupancy of the normalized molecular orbital $\varphi(\phi_1 = \varphi\alpha, \phi_2 = \varphi\beta)$, after expansion, immediately gives

$$\psi_{RHF}(1,2) = \varphi(1)\varphi(2)\frac{1}{\sqrt{2}}\{\alpha(1)\beta(2) - \beta(1)\alpha(2)\}.$$

The key quantity here is $|\psi_{RHF}(1,2)|^2$, since it tells us about the probability density of the occurrence of certain coordinates of the electrons. We will study the fundamental problem for the motion of electrons: whether the electrons react to their presence.

Let us ask a few very important questions. What is the probability density of occurrence of the situation when electron 1 occupies *different positions* in space *on the contour line* $\varphi = const$ and has spin coordinate $\sigma_1 = 1/2$ while electron 2 has spin coordinate $\sigma_2 = -1/2$, and its space coordinates are x_2, y_2, z_2 (*conditional probability*)?

We calculate

$$\begin{aligned} |\psi_{RHF}(1,2)|^2 &= \left[\varphi(1)\varphi(2)\frac{1}{\sqrt{2}}\{\alpha(\sigma_1)\beta(\sigma_2) - \beta(\sigma_1)\alpha(\sigma_2)\}\right]^2 \\ &= \left[const \times \varphi(x_2,y_2,z_2)\frac{1}{\sqrt{2}}\{\alpha(1/2)\beta(-1/2) - \beta(1/2)\alpha(-1/2)\}\right]^2 \\ &= \left[const \times \varphi(x_2,y_2,z_2)\frac{1}{\sqrt{2}}\{1 \times 1 - 0 \times 0\}\right]^2 \\ &= \frac{1}{2}(const)^2 \times \varphi^2(x_2,y_2,z_2) \end{aligned}$$

Ideas of Quantum Chemistry, Second Edition. http://dx.doi.org/10.1016/B978-0-444-59436-5.00010-6

Electron 1 changes its position on the contour line, but the distribution of the probability density of electron 2 (of the opposite spin) does not change a bit, although electron 2 should move away from its partner, since the electrons repel each other. Electron 2 is not afraid to approach electron 1. The latter can even touch electron 2, and it does not react at all. For such a deficiency, we have to pay through the high average value of the Hamiltonian (since there is a high average energy of the electron repulsion). The Hartree-Fock method, therefore, has an obvious shortcoming.

We now ask about the probability density of finding a situation in which we leave everything the same as before, but now electron 2 has spin coordinate $\sigma_2 = 1/2$ (so this is the situation where both electrons have identical projections of spin angular momentum[1]). What will the response to this change be of $\left|\psi_{RHF}(1,2)\right|^2$ as a function of the position of electron 2?

Again, we calculate

$$\left|\psi_{RHF}(1,2)\right|^2 = \left[const\ \varphi(x_2, y_2, z_2)\frac{1}{\sqrt{2}}\left\{\alpha\left(\frac{1}{2}\right)\beta\left(\frac{1}{2}\right) - \beta\left(\frac{1}{2}\right)\alpha\left(\frac{1}{2}\right)\right\}\right]^2$$
$$= \left[const\ \varphi(x_2, y_2, z_2)\frac{1}{\sqrt{2}}\{1 \times 0 - 0 \times 1\}\right]^2 = 0.$$

We ask about the distribution of the electron of the same spin. The answer is that this distribution is *everywhere equal to zero*; i.e., we do not find electron 2 with spin coordinate $\frac{1}{2}$ independent of the position of the electron 1 with spin coordinate $\frac{1}{2}$ (in whatever point on the contour line or beyond it).

The second conclusion can be accepted, since it follows from the pairing of the spins,[2] but the first conclusion is just absurd. Such nonsense is admitted by the Hartree-Fock method. In this chapter, we will ponder how can we introduce a correlation of electronic motions.

We define the electronic *correlation energy* as

$$E_{corel} = E - E_{RHF},$$

where E is the energy entering the Schrödinger equation,[3] and E_{RHF} is the Restricted Hartree-Fock energy.[4] One has to note that the Hartree-Fock procedure takes into account the Pauli exclusion principle, so it also considers the correlation of electrons of the same spin coordinate. Hence, the correlation energy E_{corel} is defined here with respect to the Hartree-Fock level of electron correlation.

What Is It All About?

The outline of the chapter is as follows:

- First, we will discuss the methods that explicitly (*via* the form of the suggested wave function) allow the electrons to control their mutual distance ("*a correlation of motions*").

---

1 We may ask: "*Why is this*?" After all, we consider a singlet state, hence the spin projections are opposite. We will not find the situation with parallel spin projections. But this is nothing to worry about. If, in fact, we are right, then we will get 0 as the density of the respective conditional probability. Let us see whether it will really be so.

2 This is ensured by the singlet form of the spin part of the function.

3 This is the rigorous nonrelativistic energy of the system in its ground state. This quantity is not available experimentally; we can *evaluate* it by subtraction of the calculated relativistic corrections from the energy of the total ionization of the system.

4 Usually, we define the correlation energy for the case of double occupancy of the molecular orbitals (the RHF method; see p. 394). In the case of open shells, especially when the multideterminantal description is required, the notion of correlation energy still remains to be defined. These problems will not be discussed in this book.

- In the second part of the chapter, the correlation will be less visible, since it will be accounted for by application of linear combinations of the Slater determinants. We will discuss the variational methods (VB, CI, MC SCF), and then the non-variational ones (CC, EOM-CC, MBPT).

In the previous chapter, we dealt with the description of electronic motion in the mean field approximation. Now *we use this approximation as a starting point toward methods that account for electron correlation.* Each of the methods considered in this chapter, when rigorously applied, should give an exact solution of the Schrödinger equation. Thus, this chapter will give us access to methods providing accurate solutions of the Schrödinger equation.

Why Is This Important?

Perhaps, in our theories, the electrons do not need to correlate their motion and the results will be still acceptable? Unfortunately, this is not so. The mean field method provides ca. 99% of the total energy of the system. This is certainly a lot, and in many cases, the mean field method gives very satisfactory results, but still falls short of treating several crucial problems correctly. For example,

- *Only through electron correlation* do the noble gas atoms attract each other in accordance with experiment (liquefaction of gases).
- According to the Hartree-Fock method, the F_2 molecule *does not exist* at all, whereas the fact is that it exists, and is doing quite well (bonding energy equal to 38 kcal/mol).[5]
- About *half* the interaction energy of large molecules (often of biological importance) calculated at the equilibrium distance originates purely from the correlation effect.
- The Restricted Hartree-Fock (RHF) method used to describe the dissociation of the chemical bond gives simply tragic results (cf. Chapter 8, p. 437), *qualitatively wrong*; on the other hand, the Unrestricted Hartree-Fock (UHF) method gives a qualitatively correct description.

We see that in many cases, electronic correlation must be taken into account.

What Is Needed?

- Operator algebra (see Appendix B available at booksite.elsevier.com/978-0-444-59436-5)
- Hartree-Fock method (Chapter 8)

[5] Yet this is not a strong bond. For example, the bonding energy of the H_2 molecule equals 104 kcal/mol, of the HF - 135 kcal/mol.

- Eigenvalue problem (see Appendix L available at booksite.elsevier.com/978-0-444-59436-5, p. e107)
- Variational method (Chapter 5)
- Perturbation theory (Chapter 5, recommended)
- Matrix diagonalization (see Appendix K available at booksite.elsevier.com/978-0-444-59436-5, p. e105, recommended)
- Second quantization (see Appendix U available at booksite.elsevier.com/978-0-444-59436-5, p. e153)

Classic Papers

The first calculations with electron correlation for molecules were performed by Walter Heitler and Fritz Wolfgang London in a paper called "*Wechselwirkung neutraler Atome und homöopolare Bindung nach der Quantenmechanik,*" published in *Zeitschrift für Physik, 44,* 455 (1927). The covalent bond (in the hydrogen molecule) could be correctly described only after the electron correlation was included. June 30, 1927, when Heitler and London submitted this paper, is the birth date of quantum chemistry. ★ The first calculations incorporating electron correlation in an atom (helium) were published by Egil Andersen Hylleraas in an article called "*Neue Berechnung der Energie des Heliums im Grundzustande, sowie des tiefsten Terms von Ortho-Helium,*" published in *Zeitschrift für Physik, 54,* 347 (1929). ★ Later, significantly more accurate results were obtained for the hydrogen molecule by Hubert M. James and Albert S. Coolidge in an article called "*The ground state of the hydrogen molecule,*" published in the *Journal of the Chemical Physics, 1,* 825 (1933), and a contemporary reference point for that molecule are several papers by Włodzimierz Kołos and Lutosław Wolniewicz, among which was an article entitled "*Potential energy curves for the $X^1\Sigma_g^+$, $B^3\Sigma_u^+$, $C^1\Pi_u$ states of the hydrogen molecule*" published in the *Journal of Chemical Physics, 43,* 2429 (1965). ★ Christian Møller and Milton S. Plesset in *Physical Review, 46,* 618 (1934), published a paper called "*Note on an approximation treatment for many-electron systems,*" where they presented a perturbational approach to electron correlation. ★ The first calculations with the Multi-configurational self-consistent field (MC SCF) method for atoms was published by Douglas R. Hartree, his father, William Hartree, and Bertha Swirles in a paper called "*Self-consistent field, including exchange and superposition of configurations, with some results for oxygen,*" *Philosophical Transactions of the Royal Society (London), A238,* 229 (1939), and the general MC SCF theory was presented by Roy McWeeny in a work called "*On the basis of orbital theories,*" *Proceedings of the Royal Society (London), A232,* 114 (1955). ★ As a classic paper in electronic correlation, we also recommend an article by Per-Olov Löwdin, "*Correlation problem in many-electron quantum mechanics,*" published in *Advances in Chemical Physics, 2,* 207 (1959). ★ The idea of the coupled cluster (CC) method was introduced by Fritz Coester in a paper in *Nuclear Physics, 7,* 421 (1958), entitled "*Bound states of a many-particle system.*" ★ Jiří Čížek introduced the (diagrammatic) CC method into electron correlation theory in a paper "*On the correlation problem in atomic and molecular systems. Calculation of wavefunction components in Ursell-type expansion using quantum-field theoretical methods,*" published in the *Journal of Chemical Physics, 45,* 4256 (1966). ★ The book "*Three Approaches to Electron Correlation in Atoms*" (Yale University Press, New Haven, CT, and London; 1970), edited by Oktay Sinanoğlu and Keith A. Brueckner, contains several reprints of the papers that cleared the path toward the CC method. ★ A derivation of the CC equations for interacting nucleons was presented by Herman Kümmel and Karl-Heinz Lührmann, *Nuclear Physics,* A191, 525 (1972), in a paper entitled "*Equations for linked clusters and the energy variational principle.*"

Size Consistency Requirement

The methods presented in this chapter will take into account the electronic correlation. A particular method may be a better or worse way to deal with this difficult problem. The better the solution, the more convincing its results are.

There is, however, one requirement that we believe to be a natural one for any method. Namely,

> any reliable method when applied to a system composed of very distant (i.e., non-interacting) subsystems should give the energy, which is a sum of the energies for the individual subsystems. A method having this feature is known as *size consistent.*[a]
>
> [a] The size consistency has some theoretical issues to be solved. One may define the subsystems and their distances in many different ways, some of them quite weird. For instance, one may consider all possible dissociation channels (with different products) with unclear electronic states to assume. Here, we consider the simplest cases: the closed-shell character of the total system and of the subsystems. Even this is not unique, however.

Before we consider other methods, let us check whether our fundamental method (i.e., the Hartree-Fock method) is size consistent or not.

Hartree-Fock Method

As shown on p. 417, the Hartree-Fock electronic energy reads as $E'_{HF} = \sum_i^{\text{SMO}} \langle i|\hat{h}|i\rangle + \frac{1}{2}\sum_{i,j=1}^{\text{SMO}}[\langle ij|ij\rangle - \langle ij|ji\rangle]$, while the total energy is $E_{HF} = E'_{HF} + V_{nn}$, where the last term represents a constant repulsion of the nuclei. When the intersubsystem distances are infinite (they are then non-interacting), one can divide the spinorbitals $|i\rangle$, $i = 1, 2, \ldots N$ into non-overlapping sets $i \in A, i \in B, i \in C, \ldots$, where $i \in A$ means the molecular spinorbital $|i\rangle$ is localized on the subsystem A and represents a Hartree-Fock spinorbital of molecule A, etc. Then, in the limit of large distances (symbolized by lim, V_B stands for the operator of the interaction of the nuclei of molecule B with an electron, while $\lim V_{nn} = \sum_A V_{nn,A}$, with $V_{nn,A}$ representing the nuclear repulsion within molecule A, and $E_{HF}(A)$ denotes the Hartree-Fock energy of molecule A):

$$\lim E_{HF} = \sum_i^{\text{SMO}} \lim\langle i|\hat{h}|i\rangle + \frac{1}{2}\lim\sum_{i,j=1}^{\text{SMO}}[\langle ij|ij\rangle - \langle ij|ji\rangle] + \lim V_{nn}$$

$$= \sum_A \left[\sum_{i\in A}^{\text{SMO}} \langle i|\hat{h}^A|i\rangle + \lim\sum_{i\in A}^{\text{SMO}}\langle i|\sum_{B\neq A} V_B|i\rangle\right] + \frac{1}{2}\sum_{i,j\in A}^{\text{SMO}}[\langle ij|ij\rangle - \langle ij|ji\rangle]$$

$$+\frac{1}{2}\lim \sum_{i\in A, j\in B}^{\text{SMO}} [\langle ij|ij\rangle - \langle ij|ji\rangle] + \sum_A V_{nn,A}$$

$$= \sum_A \left[\sum_{i\in A}^{\text{SMO}} \langle i|\hat{h}^A|i\rangle + 0 + \frac{1}{2} \sum_{i,j\in A}^{\text{SMO}} [\langle ij|ij\rangle - \langle ij|ji\rangle] + 0 + V_{nn,A} \right]$$

$$= \sum_A E_{HF}(A).$$

The zeros in the above formula appeared instead of the terms that vanish because of the Coulombic interaction of the objects that are farther and farther from one another. For example, in the mixed terms $\frac{1}{2}\sum_{i\in A, j\in B}^{\text{SMO}}[\langle ij|ij\rangle - \langle ij|ji\rangle]$, the spinorbitals $|i\rangle$ and $|j\rangle$ belong to different molecules, all integrals of the type $\langle ij|ij\rangle$ vanish because they correspond to the Coulomb interaction of electron 1, with the probability density distribution $\phi_i^*(1)\phi_i(1)$ in molecule A, and electron 2, with the distribution $\phi_j^*(2)\phi_j(2)$ centered on molecule B. Such an interaction vanishes as the inverse of the AB distance; i.e., it goes to zero in the limit under consideration. The integrals $\langle ij|ji\rangle$ vanish even faster because they correspond to the Coulombic interaction of $\phi_i^*(1)\phi_j(1)$ with $\phi_j^*(2)\phi_i(2)$ and each of these distributions itself vanishes exponentially if the distance AB goes to infinity. Hence, all the mixed terms tend to zero.

Thus,

> The Hartree-Fock method is size consistent.

* * *

We have learned, from the example given at the beginning of this chapter, that the "genetic defect" of the mean field methods is that they describe electrons that ignore the fact that they are close to or far from each other. For instance, in the two-electron case previously considered, where we established the coordinates of electron 1, electron 2 has a certain distribution of the probability density. *This distribution does not change when electron* 1 *moves to a different position.* This means that the electrons are not "*afraid*" to get close to each other, although they should, since when electrons are close, the energy increases (Fig. 10.1a,b).

*The explicitly correlated wave function (*which we will explain in a moment*) has the inter-electronic distance built in its mathematical form.* We may compare this to making the electrons wear spectacles.[6] Now they avoid each other. One of my students said that it would be the best if the electrons moved apart to infinity. Well, they cannot. They are attracted by the nucleus

[6] Of course, the methods described further also provide their own "*spectacles*" (otherwise, they would not give the solution of the Schrödinger equation), but the spectacles in the explicitly correlated functions are easier to construct with a small number of parameters.

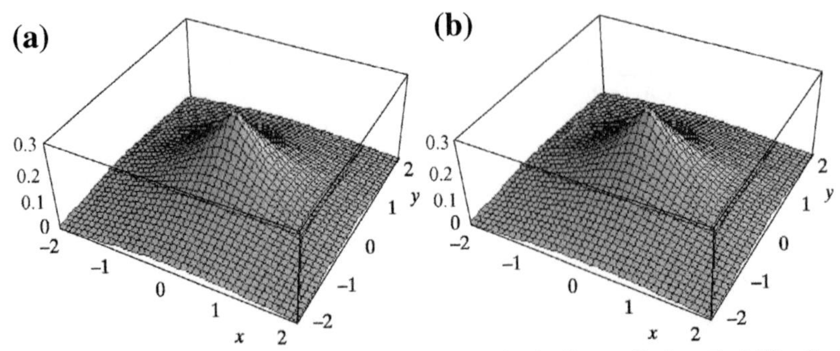

Fig. 10.1. Absence of electronic correlation in the helium atom as seen by the Hartree-Fock method. Visualization of the cross-section of the square of the wave function (probability density distribution) describing electron 2 within the plane xy, provided that electron 1 is located in a certain point in space: (a) at $(-1, 0, 0)$; b) at $(1, 0, 0)$. *Note that in both cases, the conditional probability density distributions of electron 2 are identical.* This means electron 2 does not react to the motion of electron 1; i.e., there is no correlation whatsoever of the electronic motions (when the total wave function is the Hartree-Fock one).

(energy gain), and, being close to it, must be close to each other too (energy loss). There is a compromise to achieve.

VARIATIONAL METHODS USING EXPLICITLY CORRELATED WAVE FUNCTION

10.1 Correlation Cusp Condition

Short distances are certainly the most important for the Coulombic interaction of two charges, although obviously the regions of configurational space connected with the long interelectronic distances are much larger. Thus, the region is not large, but important, within it the "collisions" take place. It turns out that the wave function calculated in the region of a collision must satisfy some very simple mathematical condition (called the *correlation cusp* condition). This is what we want to demonstrate. The derived formulas[7] are universal, and they apply to any pair of charged particles.

Let us consider *two* particles with charges q_i and q_j and masses m_i and m_j *separated from other particles.* This makes sense since simultaneous collisions of three or more particles occur very rarely compared to two-particle collisions. Let us introduce a Cartesian system of coordinates (say, in the middle of the beautiful market square in Brussels), so that the system

[7] T. Kato, *Commun. Pure Appl. Math. 10*, 151 (1957).

of two particles is described with six coordinates. Then (atomic units are used) the sum of the kinetic energy operators of the particles is

$$\hat{T} = -\frac{1}{2m_i}\Delta_i - \frac{1}{2m_j}\Delta_j. \tag{10.1}$$

Now we separate the motion of the center of mass of the two particles with position vectors $\boldsymbol{r}_i$ and $\boldsymbol{r}_j$. The center of mass in our coordinate system is indicated by the vector $\boldsymbol{R}_{CM} = (X_{CM}, Y_{CM}, Z_{CM})$:

$$\boldsymbol{R}_{CM} = \frac{m_i \boldsymbol{r}_i + m_j \boldsymbol{r}_j}{m_i + m_j} \tag{10.2}$$

Tosio Kato (1917–1999) was an outstanding Japanese physicist and mathematician. His studies at the University of Tokyo were interrupted by World War II. After the war, he got his Ph.D. at this university (his thesis was about convergence of the perturbational series), and obtained the title of professor in 1958.

In 1962, Kato became professor at the University of Berkeley, California. He admired the botanic garden there, knew a lot of Latin names of plants, and appreciated very much the Charles Linnaeus classification of plants.

Let us also introduce the total mass of the system $M = m_i + m_j$, the reduced mass of the two particles $\mu = \frac{m_i m_j}{m_i + m_j}$ and the vector of their relative positions $\boldsymbol{r} = \boldsymbol{r}_i - \boldsymbol{r}_j$. Introducing the three coordinates of the center of mass measured with respect to the market square in Brussels and the three coordinates x, y, and z, which are components of the vector $\boldsymbol{r}$, we get (see Appendix I available at booksite.elsevier.com/978-0-444-59436-5 on p. e93, example 1)

$$\hat{T} = -\frac{1}{2M}\Delta_{CM} - \frac{1}{2\mu}\Delta, \tag{10.3}$$

$$\Delta_{CM} = \frac{\partial^2}{\partial X_{CM}^2} + \frac{\partial^2}{\partial Y_{CM}^2} + \frac{\partial^2}{\partial Z_{CM}^2}, \tag{10.4}$$

$$\Delta = \frac{\partial^2}{\partial x^2} + \frac{\partial^2}{\partial y^2} + \frac{\partial^2}{\partial z^2}. \tag{10.5}$$

After this operation, the Schrödinger equation for the system is separated (as always in the case of two particles; see Appendix I available at booksite.elsevier.com/978-0-444-59436-5) into two equations: the first describing the motion of the center of mass (seen from Brussels) and the second describing the *relative* motion of the two particles (with Laplacian of x, y, z, and reduced mass μ). We are not interested in the first equation; the second one (Brussels-independent) is what we are after. Let us write down the Hamiltonian corresponding to the second equation:

$$\hat{H} = -\frac{1}{2\mu}\Delta + \frac{q_i q_j}{r}. \tag{10.6}$$

We are interested in how the wave function looks when the distance of the two particles r is getting very small. If r is small, it makes sense to expand the wave function in a power series[8] of $r: \psi = C_0 + C_1 r + C_2 r^2 + \cdots$ Let us calculate $\hat{H}\psi$ close to $r = 0$. The Laplacian expressed in the spherical coordinates represents the sum of three terms (see Appendix H available at booksite.elsevier.com/978-0-444-59436-5, p. e91, Eq. H.1): the first, which contains the differentiation with respect to r, and the remaining two, which contain the differentiation with respect to the angles θ and $\phi: \Delta = \frac{1}{r^2}\frac{\partial}{\partial r} r^2 \frac{\partial}{\partial r} +$ terms depending on θ and ϕ. Since we have assumed the function to be dependent on r only, upon the action of the Laplacian, only the first term gives a nonzero contribution.

We obtain

$$\hat{H}\psi = \left(-\frac{1}{2\mu}\Delta + \frac{q_i q_j}{r}\right)\psi \tag{10.7}$$

$$= -\frac{1}{2\mu}\left(\frac{1}{r^2}\frac{\partial}{\partial r} r^2 \frac{\partial}{\partial r} + \cdots\right)(C_0 + C_1 r + C_2 r^2 + \cdots) \tag{10.8}$$

$$+ \frac{q_i q_j}{r}(C_0 + C_1 r + C_2 r^2 + \cdots) \tag{10.9}$$

$$= 0 - \frac{1}{2\mu}\left(\frac{2C_1}{r} + 6C_2 + 12C_3 r + \cdots\right) \tag{10.10}$$

$$+ C_0\frac{q_i q_j}{r} + C_1 q_i q_j + C_2 q_i q_j r + \cdots \tag{10.11}$$

The wave function cannot go to infinity when r goes to zero, while in the above expression, we have two terms ($-\frac{1}{2\mu}\frac{2C_1}{r}$ and $C_0\frac{q_i q_j}{r}$), which would then "*explode*" to infinity.

These terms must cancel each other out.

Hence, we obtain

$$C_0 q_i q_j = \frac{C_1}{\mu}. \tag{10.12}$$

This condition is usually expressed in another way. We use the fact that $\psi(r = 0) = C_0$ and $\left(\frac{\partial \psi}{\partial r}\right)_{r=0} = C_1$ and obtain the cusp condition as follows:

[8] Assuming such a form, we exclude the possibility that the wave function goes to $\pm\infty$ for $r \to 0$. This must be so, since otherwise, either the respective probability would go to infinity or the operators would become non-Hermitian (cf. p. 80). Both possibilities are unacceptable. We covertly assumed also (to simplify our considerations) that the wave function does not depend on the angles θ and ϕ. This dependence can be accounted for by making the constants C_0, C_1, C_2 the functions of θ and ϕ. Then the final result still holds, but for the coefficients C_0 and C_1 averaged over θ and ϕ.

$$\left(\frac{\partial\psi}{\partial r}\right)_{r=0} = \mu q_i q_j \psi(r=0).$$

- The case of two electrons:
  Then $m_i = m_j = 1$; hence, $\mu = \frac{1}{2}$ and $q_i = q_j = -1$. We get the cusp condition for the collision of two electrons as
  $$\left(\frac{\partial\psi}{\partial r}\right)_{r=0} = \frac{1}{2}\psi(r=0)$$
  or (introducing variable $r = r_{12}$ together with particles' position vectors $\boldsymbol{r}_1$ and $\boldsymbol{r}_2$)

  the wave function should be of the form
  $\psi = \phi(\boldsymbol{r}_1, \boldsymbol{r}_2)\left[1 + \frac{1}{2}r_{12} + \cdots\right]$,

  where $+\cdots$ means higher powers of r_{12}.
- The nucleus-electron case:
  When one of the particles is a nucleus of charge Z, then $\mu \simeq 1$, and we get
  $$\left(\frac{\partial\psi}{\partial r}\right)_{r=0} = -Z\psi(r=0).$$

Thus

the correct wave function for the electron in the vicinity of a nucleus should have an expansion $\psi = const(1 - Zr_{a1} + \cdots)$, where r_{a1} replacing r is the distance from the nucleus.

Let us see how it is with the $1s$ function for the hydrogen-like atom (the nucleus has charge Z) expanded in a Taylor series in the neighborhood of $r = 0$. We have $1s = N\exp(-Zr) = N(1 - Zr + \cdots)$, which works.

The correlation cusp makes the wave function not differentiable at $r = 0$.

10.2 The Hylleraas CI Method

In 1929, two years after the birth of quantum chemistry, a paper by Egil Hylleraas[9] appeared, where, for the ground state of the helium atom, a trial variational function, containing the inter-

[9] E.A. Hylleraas, *Zeit. Phys., 54,* 347 (1929). Egil Andersen Hylleraas arrived in 1926 in Göttingen, Germany, to collaborate with Max Born. His professional experience was related to crystallography and to the optical properties of quartz. When one of the employees fell ill, Born told Hylleraas to continue his work on the helium atom in the context of the newly developed quantum mechanics. The helium atom problem had already been attacked by Albrecht Unsöld in 1927 using first-order perturbation theory, but Unsöld obtained the ionization potential equal

electronic distance explicitly, was applied. This was a brilliant idea, since it showed that already a small number of terms provide very good results. Even though no fundamental difficulties were encountered for larger atoms, the enormous numerical problems were prohibitive for atoms with larger numbers of electrons. In this case, the century-long progress means going from 2- to 10-electron systems.

In the Hylleraas-CI method,[10] the Hylleraas idea has been exploited when designing a method for larger systems. The electronic wave function is proposed as a linear combination of Slater determinants, and in front of each determinant $\Phi_i(1, 2, 3, \ldots, N)$, we insert, next to the variational coefficient c_i, correlational factors with some powers $(v, u, \ldots)$ of the interelectronic distances (r_{mn} between electron m and electron n, etc.):

$$\psi = \sum_i c_i \hat{A}\left[r_{mn}^{v_i} r_{kl}^{u_i} \cdots \Phi_i(1, 2, 3, \ldots, N)\right], \tag{10.13}$$

where $\hat{A}$ denotes an antisymmetrization operator (see Appendix M available at booksite.elsevier.com/978-0-444-59436-5, p. e109). If $v_i = u_i = 0$, we have the CI expansion: $\psi = \sum_i c_i \Phi_i$ (which we will discuss on p. 615). If $v_i \neq 0$ or $u_i \neq 0$, we include a variationally proper treatment of the appropriate distances r_{mn} or r_{kl}; i.e., correlation of the motions of the electrons m and n, or k and l, etc. The antisymmetrization operator ensures the requirement for symmetry of the wave function with respect to the exchange of the arbitrary two electrons. The method described was independently proposed in 1971 by Wiesław Woźnicki[11] and by Sims and Hagstrom.[12] The method of correlational factors has a nice feature, in that even a short expansion should give a very good total energy for the system, since we combine the power of the CI method with the great success of the explicitly correlated approaches. Unfortunately, the method has also a serious drawback. To make practical calculations, it is necessary to evaluate the integrals occurring in the variational method, and they are very difficult to calculate.[13]

---

to 20.41 eV, while the experimental value was equal to 24.59 eV. In the reported calculations (done on a recently installed calculator), Hylleraas obtained a value of 24.47 eV (cf. contemporary accuracy, p. 148).

[10] Here, CI stands for "*Configuration Interaction.*"

[11] W. Woźnicki, in *Theory of Electronic Shells in Atoms and Molecules* (A. Yutsis, ed.), Mintis, Vilnius (1971), p. 103.

[12] J.S. Sims and S.A. Hagstrom, *Phys. Rev. A4,* 908 (1971).

[13] It is enough to realize that, in the matrix element of the Hamiltonian containing two terms of the above expansion, we may find, e.g., a term $1/r_{12}$ (from the Hamiltonian) and r_{13} (from the factor in front of the determinant), as well as the product of six spinorbitals describing the electrons 1, 2, 3. Such integrals have to be computed, and the existing algorithms are inefficient.

10.3 Two-Electron Systems

10.3.1 The Harmonic Helium Atom

An unpleasant feature of the electron correlation is that we deal either with intuitive concepts or, if our colleagues want to help us, they bring wave functions with formulas so long as the distance from Cracow to Warsaw (or longer[14]) and say: look, this is what *really* happens. It would be good to analyze such formulas term by term, but this approach does not make sense because there are too many terms. Even the helium atom, when we write down the formula for its ground-state wave function, becomes a mysterious object. Correlation of motion of any element seems to be so difficult to grasp mathematically that we easily give up. A group of scientists published a paper in 1993 that has generated interest on this point. They obtained a rigorous solution of the Schrödinger equation (described in Chapter 4, p. 212), the only exact solution which has been obtained so far for correlational problems.[15]

Note that the exact wave function (its spatial part[16]) is a *geminal* (i.e., two-electron function).

$$\psi\left(\boldsymbol{r}_1,\boldsymbol{r}_2\right)=N\left(1+\frac{1}{2}r_{12}\right)e^{-\frac{1}{4}\left(r_1^2+r_2^2\right)}. \tag{10.14}$$

Let me be naive. Do we have two harmonic springs here? Yes, we do (see Fig. 4.26, p. 212). Then, let us treat them first as independent oscillators and take the *product* of the ground-state functions of both oscillators: $\exp\left[-\frac{1}{4}\left(r_1^2+r_2^2\right)\right]$. Well, it would be good to account for the cusp condition $\psi=\phi(\boldsymbol{r}_1,\boldsymbol{r}_2)\left[1+\frac{1}{2}r_{12}+\cdots\right]$ and take care of it, even in a naive way. Let us just implement the crucial correlation factor $\left(1+\frac{1}{2}r_{12}\right)$, *the simplest* that satisfies the cusp condition (see p. 587). It turns out that such a recipe leads to a *rigorous* wave function![17]

From Eq. (10.14), we see that for $r_1=r_2=const$ (in such a case, both electrons move on the surface of the sphere), the larger value of the function (and *eo ipso* of the probability) is obtained for *larger* r_{12}. This means that, it is most probable that the electrons prefer to occupy opposite sides of a nucleus. This is a practical manifestation of the existence of the Coulomb hole around electrons (i.e., the region of the reduced probability of finding a second electron):

[14] This is a very conservative estimate. Let us calculate–half jokingly. Writing down a single Slater determinant would easily take up 10 cm of space. The current world record amounts to several billion such determinants in the CI expansion (say, 3 billion). Now let us calculate: $10\,\text{cm}\times 3\times 10^9=3\times 10^{10}\,\text{cm}=3\times 10^8\,\text{m}=3\times 10^5\,\text{km}=300000\,\text{km}$. So, this not Warsaw to Cracow, but Earth to the Moon.

[15] S. Kais, D.R. Herschbach, N.C. Handy, C.W. Murray, and G.J. Laming, *J. Chem. Phys. 99,* 417 (1993).

[16] For one- and two-electron systems, the wave function is *a product* of the spatial and spin factors. A normalized spin factor for two-electron systems $\frac{1}{\sqrt{2}}\{\alpha(1)\beta(2)-\beta(1)\alpha(2)\}$ guarantees that the state in question is a singlet (see Appendix Q available at booksite.elsevier.com/978-0-444-59436-5 p. e133). Since we will only manipulate the spatial part of the wave function, the spin is the default. Since the total wave function has to be antisymmetric, and the spin function is antisymmetric, the spatial function should be symmetric–and it is.

[17] As a matter of fact, that is true only for a single force constant. Nevertheless, the unusual simplicity of that analytic formula is most surprising.

the electrons simply repel each other. They cannot move apart to infinity since both are held by the nucleus. The only thing they can do is to be close to the nucleus and to avoid each other–and this is what we observe in Eq. (10.14).

10.3.2 The James-Coolidge and Kołos-Wolniewicz Functions

One-electron problems are the simplest. For systems with *two* electrons,[18] we can apply certain mathematical tricks that allow very accurate results. We are going to talk about such calculations in a moment.

Kołos and Wolniewicz applied the Ritz variational method (see Chapter 5) to the hydrogen molecule with the following trial function:

$$\Psi = \frac{1}{\sqrt{2}}\left[\alpha(1)\beta(2) - \alpha(2)\beta(1)\right]\sum_{i}^{M} c_i\left(\Phi_i(1,2) + \Phi_i(2,1)\right),$$

$$\Phi_i(1,2) = \exp\left(-A\xi_1 - \bar{A}\xi_2\right)\xi_1^{n_i}\eta_1^{k_i}\xi_2^{m_i}\eta_2^{l_i}\left(\frac{2r_{12}}{R}\right)^{\mu_i}$$

$$\cdot\left(\exp\left(B\eta_1 + \bar{B}\eta_2\right) + (-1)^{k_i+l_i}\exp\left(-B\eta_1 - \bar{B}\eta_2\right)\right), \tag{10.15}$$

where the elliptic coordinates of the electrons with index $j = 1, 2$ are given by

$$\xi_j = \frac{r_{aj} + r_{bj}}{R}, \tag{10.16}$$

$$\eta_j = \frac{r_{aj} - r_{bj}}{R}, \tag{10.17}$$

where R denotes the internuclear distance, r_{aj} and r_{bj} are nucleus-electron distances (the nuclei are labeled by a, b), r_{12} is the (crucial to the method) interelectronic distance, $c_i, A, \bar{A}, B, \bar{B}$ are variational parameters, and n, k, l, m, μ are integers (smaller than selected limiting values).

The simplified form of this function with $A = \bar{A}$ and $B = \bar{B} = 0$ is the James-Coolidge[19] function, thanks to which these authors enjoyed the most accurate result for the hydrogen molecule for 27 years.

---

[18] For a larger number of electrons, it is much more difficult.

[19] H.M. James and A.S. Coolidge, *J. Chem. Phys., 1*, 825 (1933). Hubert M. James in the 1960s was professor at Purdue University.

Kołos and Roothaan,[20] and later on, Kołos and Wolniewicz,[21] Kołos and Rychlewski, and others,[22] applied longer and longer expansions (helped by the fact that computer technology was improving fast) up to M of the order of thousands; see Table 10.1.

Włodzimierz Kołos (1928–1996), Polish chemist and professor at the Warsaw University. His calculations on small molecules (with Roothaan, Wolniewicz, and Rychlewski) had an unprecedented accuracy in quantum chemistry. The Department of Chemistry of Warsaw University and the Polish Chemical Society established the Włodzimierz Kołos Medal accompanying a lecture (the first lecturers were Roald Hoffmann, Richard Bader, and Paul von Ragué Schleyer). In the Ochota quarter in Warsaw, there is a Włodzimierz Kołos Street. Lutosław Wolniewicz (born 1930), Polish physicist and professor at the Nicolaus Copernicus University in Toruń.

As can be seen from Table 10.1, there was a competition between theoreticians and the experimental laboratory of Herzberg. When, in 1964 Kołos and Wolniewicz obtained 36117.3 cm^{-1} (see Table 10.1) for the dissociation energy of the hydrogen molecule, quantum chemists held their breath. The experimental result of Herzberg and Monfils, obtained four years earlier (see Table 10.1), was *smaller*, and this seemed to contradict the variational principle (Chapter 5; i.e., as if the theoretical result were below the ground-state energy), the foundation of quantum mechanics. There were only three possibilities: either the theoretical or experimental results are wrong or quantum mechanics has internal inconsistency. Kołos and Wolniewicz increased the accuracy of their calculations in 1968 and excluded the first possibility. It soon turned out that the problem lay in the accuracy of the experiment.[23] When Herzberg increased the accuracy,

Gerhard Herzberg (1904–1999), Canadian chemist of German origin and professor at the National Research Council and at the University of Saskatchewan in Saskatoon and the University of Ottawa. The greatest spectroscopist of the 20th century. Herzberg laid the foundations of molecular spectroscopy, is author of the fundamental monograph on this subject, and received a Nobel prize in 1971 "*for his contribution to knowledge of the electronic structure and geometry of molecules, particularly free radicals.*"

[20] W. Kołos and C.C.J. Roothaan, *Rev. Modern Phys., 32,* 205 (1960).

[21] For the first time in quantum chemical calculations, relativistic corrections and corrections resulting from quantum electrodynamics were included. This accuracy was equivalent to hitting, from Earth, an object on the Moon the size of a car. These results are cited in nearly all textbooks on quantum chemistry to demonstrate that the theoretical calculations have a solid background.

[22] The description of these calculations is given in the review article by Piszczatowski et al. cited in Table 10.1

[23] At that time, Herzberg was hosting them in Canada and treated them to a homemade fruit liquor, which was considered by his coworkers to be absolutely exceptional. This is probably the best time to give the recipe for the exquisite drink, which is known in the circles of quantum chemists as "*kolosovka*":

Table 10.1. Dissociation energy of H_2 in the ground state (in cm^{-1}). Comparison of the results of theoretical calculations and experimental measurements. The figures in parentheses mean the error in units of the last digit reported. Bold numbers are used to indicate the values connected with the Herzberg-Kołos-Wolniewicz controversy.

Year	Author	Experiment	Theory
1926	Witmer	35000	
1927	Heitler-London		23100[a]
1933	James-Coolidge		36104[a]
1935	Beutler	36116(6)	
1960	Kołos-Roothaan		36113.5[a]
1960	Herzberg-Monfils	**36113.6(3)**	
1964	Kołos-Wolniewicz		**36117.3**[a]
1968	Kołos-Wolniewicz		**36117.4**[a]
1970	Herzberg	**36118.3**[c]	
1970	Stwalley	36118.6(5)	
1975	Kołos-Wolniewicz		36118.0
1978	Kołos-Rychlewski		36118.12[b]
1978	Bishop-Cheung		36117.92
1983	Wolniewicz		36118.01
1986	Kołos-Szalewicz-Monkhorst		36118.088
1991	McCormack-Eyler	36118.26(20)	
1992	Balakrishnan-Smith-Stoicheff	36118.11(8)	
1992	Kołos-Rychlewski		36118.049
1995	Wolniewicz		36118.069
2009	Piszczatowski et al.		36118.0695(10)[d]
2009	Liu et al.	36118.0696(4)	

[a]Obtained from calculated binding energy by subtracting the energy of zero vibrations.
[b]Obtained by treating the improvement of the binding energy as an additive correction to the dissociation energy.
[c]Upper bound.
[d]The references to the cited works can be found in the paper by K. Piszczatowski, G. Łach, M. Przybytek, J. Komasa, K. Pachucki, and B. Jeziorski, J.Chem.Theory and Comput., 5, 3039 (2009).

he obtained 36118.3 cm^{-1} as the dissociation energy (Table 10.1), which was then consistent with the variational principle.

The theoretical result of 2009 given in the table includes non-adiabatic, relativistic and quantum electrodynamic (QED) corrections. The relativistic and QED corrections have been calculated assuming the adiabatic approximation and, by taking into account all the terms up to $\left(\frac{1}{c}\right)^3$ and the leading term in the QED $\left(\frac{1}{c}\right)^4$ contribution, some effects never taken into account before for any molecule. To get an idea about the importance of the particular levels of theory, let me report their contributions to the H_2 dissociation energy (the number in parentheses means the error in the units of the last digit given). The $\left(\frac{1}{c}\right)^0$ contribution (i.e., the solution of the Schrödinger equation) gives 36118.7978(2) cm^{-1}, the $\left(\frac{1}{c}\right)^1$ is equal to zero, $\left(\frac{1}{c}\right)^2$ is the Breit

Pour a pint of pure spirit into a beaker. Hang an orange on a piece of gauze directly over the meniscus. Cover tightly and wait for two weeks. Then throw the orange away–there is nothing of value left in it–and turn your attention to the spirit. It should contain now all the flavors from the orange. Next, slowly pour some spring water into the beaker until the liquid becomes cloudy, and then some more spirit to make it clear again. Propose a toast to the future of quantum chemistry!

correction (see p. 145) and turned out to be $-0.5319(3)$ cm^{-1}, the QED (see p. 148) $\left(\frac{1}{c}\right)^3$ correction is $-0.1948(2)$ cm^{-1}, while the $\left(\frac{1}{c}\right)^4$ contribution is $-0.0016(8)$ cm^{-1}. We see that to obtain such agreement with the experimental value as shown in Table 10.1, one needs to include all the abovementioned corrections.

10.3.3 Neutrino Mass

Calculations like those above required unique software, especially in the context of the nonadiabatic effects included. Additional gains appeared unexpectedly when Kołos and others[24] initiated work aiming at explaining whether the electronic neutrino has a nonzero mass.[25] In order to interpret the expensive experiments, precise calculations were required for the β-decay of the tritium molecule as a function of the neutrino mass. The emission of the antineutrino (ν) in the process of β-decay:

$$T_2 \rightarrow HeT^+ + e + \nu$$

Alexandr Alexandrovitch Friedmann (1888–1925), Russian mathematician and physicist, in his article in *Zeit. Phys., 10,* 377 (1922), proved on the basis of Einstein's general theory of relativity that the curvature of the Universe must change, which became the basis of cosmological models of the expanding Universe. During World War I, Friedman was a pilot in the Russian army and made bombing raids over my beloved Przemyśl.

In one of his letters, he asked his friend, the eminent Russian mathematician Steklov, for advice about the integration of equations he derived to describe the trajectories of his bombs. Later, in a letter to Steklov dated February 28, 1915, he wrote: "*Recently I had an opportunity to verify my theory during a flight over Przemyśl, the bombs fell exactly in the places predicted by the theory. To get the final proof of my theory I intend to test it in flights during next few days.*"

More information can be found at http://www-groups.dcs.st-and.ac.uk/~history/Mathematicians/Friedmann.html.

should have consequences for the final quantum states of the HeT^+ molecule. To enable evaluation of the neutrino mass by the experimentalists Kołos et al. performed precise calculations of all possible final states of HeT^+ and presented them as a function of the hypothetical mass of the neutrino. There is such a large number of neutrinos in the Universe that if its mass exceeded a certain, even very small threshold value of the order of[26] 1 eV, the mass of the Universe would exceed the critical value predicted by Alexander Friedmann in his cosmological theory (based on

[24] W. Kołos, B. Jeziorski, H.J. Monkhorst, and K. Szalewicz, *Int. J. Quantum Chem.*, S19, 421 (1986).

[25] Neutrinos are stable fermions of spin $\frac{1}{2}$. Three types of neutrinos exist (each has its own antiparticle): electronic, muonic, and taonic. The neutrinos are created in the weak interactions (e.g., in β-decay) and do not participate either in the strong interactions, or in electromagnetic interactions. The latter feature expresses itself in an incredible ability to penetrate matter (e.g., crossing the Earth as though through a vacuum). The existence of the electronic neutrino was postulated in 1930 by Wolfgang Pauli and discovered in 1956 by F. Reines and C.L. Cowan; the muonic neutrino was discovered in 1962 by L. Lederman, M. Schwartz, and J. Steinberger.

[26] The mass of the elementary particle is given in the form of its energetic equivalent mc^2.

Einstein's general theory of relativity). This would mean that the currently occurring expansion of the Universe (discovered by Hubble) would finally stop, and its collapse would follow. If the neutrino mass turned out to be too small, then the Universe would continue its expansion. Thus, quantum chemical calculations for the HeT^+ molecule may turn out to be helpful in predicting our fate (unfortunately, being crushed or frozen). So far, the estimate of neutrino mass gives a value smaller than 1 eV, which indicates the expansion of the universe.[27]

Edwin Powell Hubble (1889–1953), American astronomer and explorer of galaxies, found in 1929 that the distance between galaxies is proportional to the infrared shift in their spectrum caused by the Doppler effect, which is consequently interpreted as expansion of the Universe. A surprise from recent astronomical studies is that the expansion is faster and faster (for unknown reasons).

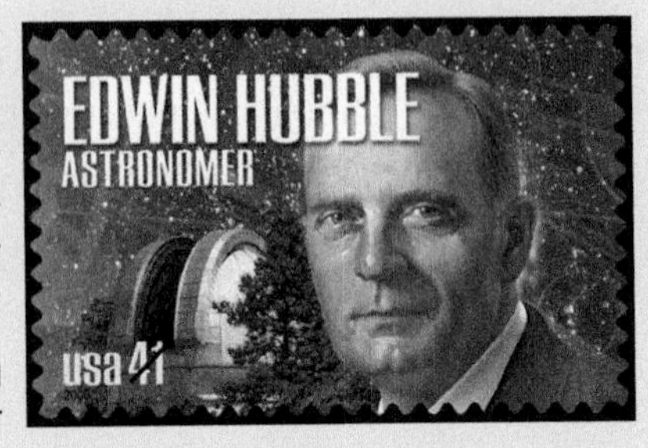

10.4 Exponentially Correlated Gaussian Functions

In 1960, Boys[28] and Singer[29] noticed that the functions that are products of Gaussian orbitals and correlational factors of Gaussian type, $\exp\left(-br_{ij}^2\right)$, where r_{ij} is the distance between electron i and electron j, generate relatively simple integrals in the quantum chemical calculations. A product of two Gaussian orbitals, with positions shown by the vectors $\boldsymbol{A}$, $\boldsymbol{B}$, and of an exponential correlation factor is called *an exponentially correlated Gaussian geminal*[30]:

$$g\left(\boldsymbol{r}_i, \boldsymbol{r}_j; \boldsymbol{A}, \boldsymbol{B}, a_1, a_2, b\right) = N e^{-a_1\left(\boldsymbol{r}_i - \boldsymbol{A}\right)^2} e^{-a_2\left(\boldsymbol{r}_j - \boldsymbol{B}\right)^2} e^{-br_{ij}^2}.$$

A geminal is an analog of an orbital–there is a one-electron function, and here is a two-electron one. A single geminal is very rarely used in computations,[31] we apply hundreds or even thousands of Gaussian geminals. When we want to find out the optimal positions $\boldsymbol{A}$, $\boldsymbol{B}$ and

---

27 At this moment, there are other candidates for contributing significantly to the mass of the Universe, mainly the mysterious "*dark matter.*" This constitutes the major part of the mass of the Universe. We know very little about it.
Recently, it turned out that neutrinos undergo what are called *oscillations*; e.g., an electronic neutrino travels from the Sun and on its way spontaneously changes to a muonic neutrino. The oscillations indicate that the mass of the neutrino is nonzero. According to current estimations, however, it is much smaller than the accuracy of the tritium experiments.

28 S.F. Boys, *Proc. Royal Soc. A258,* 402 (1960).

29 K. Singer, *Proc. Royal Soc. A258,* 412 (1960).

30 This is an attempt to go beyond the two-electron systems with the characteristic for the systems approach of James, Coolidge, Hylleraas, Kołos, Wolniewicz, and others.

31 Ludwik Adamowicz introduced an idea of the minimal basis of the Gaussian geminals [equal to the number of the electron pairs) and applied to the LiH and HF molecules, L. Adamowicz and A.J. Sadlej, *J. Chem. Phys., 69,* 3992 (1978).

the optimal exponents a and b in these thousands of geminals, it turns out that nothing certain is known about them, the $\boldsymbol{A}, \boldsymbol{B}$ positions are scattered chaotically[32]; in the $a > 0$ and $b > 0$ exponents, there is no regularity either. Nevertheless, the above formula for a single Gaussian geminal looks as if it suggested $b > 0$.

10.5 Electron Holes

10.5.1 Coulomb Hole (Correlation Hole)

It is always good to count "*on fingers*" to make sure that everything is all right. Let us see how a *single* Gaussian geminal describes the correlation of the electronic motion. Let us begin with the helium atom with the nucleus in the position $\boldsymbol{A} = \boldsymbol{B} = \boldsymbol{0}$. The geminal takes the form

$$g_{He} = Ne^{-a_1 r_1^2} e^{-a_1 r_2^2} e^{-b r_{12}^2}, \tag{10.18}$$

where $N > 0$ is a normalization factor. Let us assume[33] that electron 1 is at $(x_1, y_1, z_1) = (1, 0, 0)$. In such a situation, where does electron 2 prefer to be? We will discover this (Fig. 10.2) from the position of electron 2 for which g_{He} assumes the largest value.

Just to get an idea, let us try to restrict the motion of electron 2. For instance, let us demand that it moves only on the sphere of radius equal to 1 centered at the nucleus. So we insert $r_1 = r_2 = 1$. Then, $g_{He} = const \exp\left[-b r_{12}^2\right]$ and we will find out easily what electron 2 likes most. With $b > 0$, the latter factor tells us that what electron 2 likes best is just to sit on electron 1. Is it what the correlation is supposed to mean that one electron sits on the other? Here, we have rather an anticorrelation. Something is going wrong. According to this analysis, we should rather take the geminal of the form, e.g.:

$$g_{He} = Ne^{-a_1 r_1^2} e^{-a_1 r_2^2} \left[1 - e^{-b r_{12}^2}\right].$$

Now everything is qualitatively in order. When the interelectronic distance increases, the value of the g_{He} function also increases, which means that such a situation is *more* probable than that corresponding to a short distance. If the electrons become too agitated and begin to think that it would be better when their distance gets very large, they would be called to order by the factors $\exp\left[-a_1 r_1^2\right] \exp\left[-a_1 r_2^2\right]$. Indeed, in such a case, the distance between the nucleus and at least one of the electrons is long and the probability of such a situation is quenched by one or both exponential factors. For large r_{12} distances, the factor $\left[1 - \exp\left[-b r_{12}^2\right]\right]$ is practically equal to 1. This means that for large interelectronic distances, g_{He} is practically equal to $N \exp\left[-a_1 r_1^2\right] \exp\left[-a_1 r_2^2\right]$; i.e., to the product of the orbitals (no correlation of motions at long interelectronic distances and rightly so).

---

[32] The methods in which those positions are selected at random achieved a great success.

[33] We use atomic units.

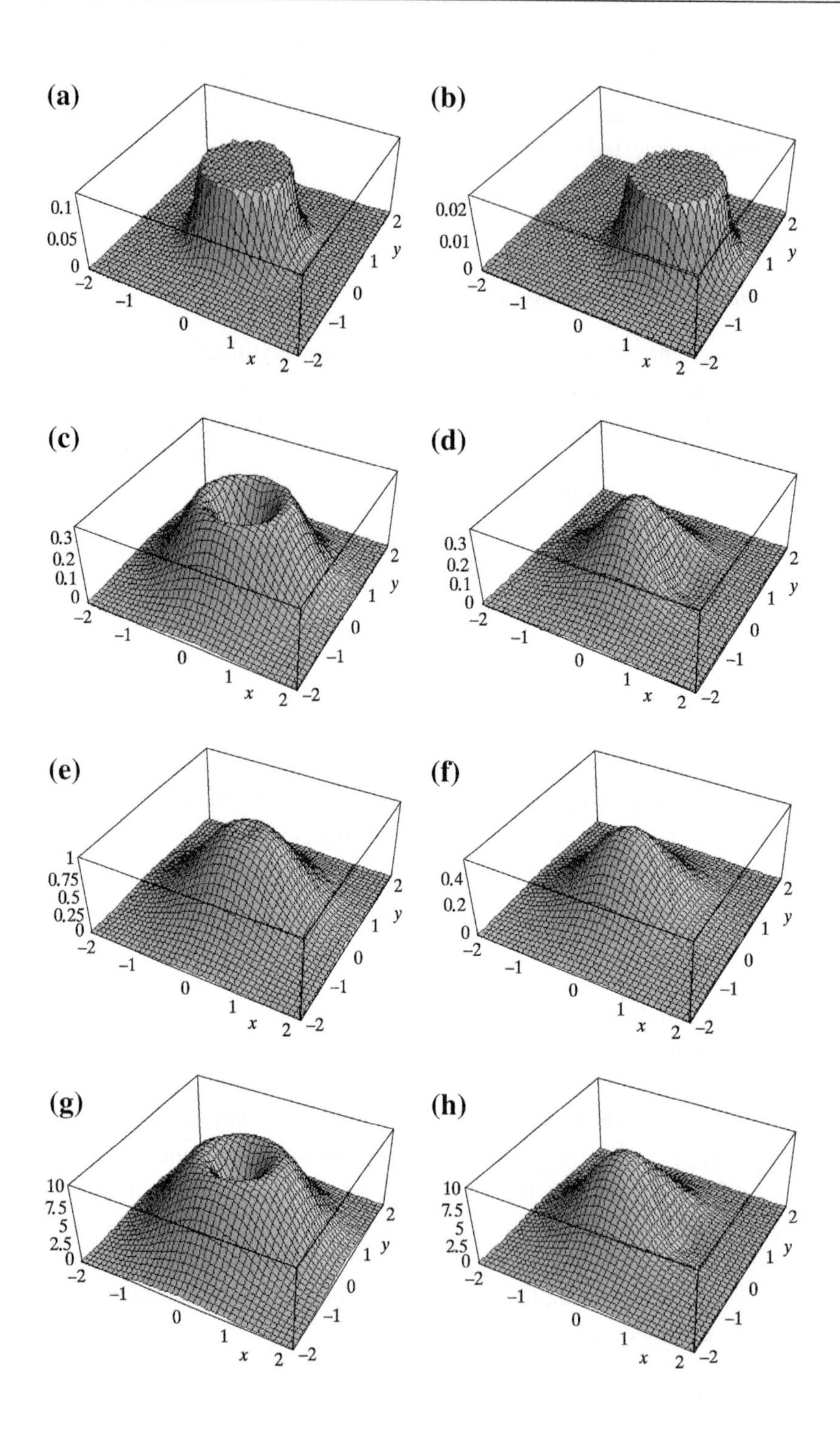
(a)
0.1
0.05
0
−2
−1
0
1
x
2
−2
−1
0
1
2
y
(b)
0.02
0.01
0
(c)
0.3
0.2
0.1
(d)
(e)
1
0.75
0.5
0.25
(f)
0.4
0.2
(g)
10
7.5
5
2.5
(h)

Around electron 1, there is a region of low probability of finding electron 2. This region is called the *Coulomb hole.*

The Gaussian geminals do not satisfy the correlation cusp condition (p. 587), because of factor $\exp(-br_{ij}^2)$. It is required (for simplicity, we write $r_{ij} = r$) that $\left(\frac{\partial g}{\partial r}\right)_{r=0} = \frac{1}{2}g(r=0)$, whereas the left side is equal to 0, while the right side $\frac{1}{2}N\exp[-a_1(\boldsymbol{r}_i - \boldsymbol{A})^2]\exp[-a_2(\boldsymbol{r}_j - \boldsymbol{B})^2]$ is not equal to zero. This is not a disqualifying feature, since the region of space in which this condition should be fulfilled is very small.

The area of application of this method is–for practical (computational) reasons–relatively small. The method of Gaussian geminals has been applied in unusually accurate calculations for three- and four-electron systems.[34]

10.5.2 Exchange Hole (Fermi Hole)

The mutual avoidance of electrons in the helium atom or in the hydrogen molecule is caused by Coulombic repulsion of electrons (described in the previous subsection). As we have shown in this chapter, in the Hartree-Fock method the Coulomb hole is absent, whereas methods that account for electron correlation generate such a hole. However, electrons avoid each other also for reasons other than their charge. The Pauli principle is another reason this occurs. One of the consequences is the fact that electrons with the same spin coordinate cannot reside in the same place; see p. 34. The continuity of the wave function implies that the probability density of them staying *in the vicinity* of each other is small; i.e.,

←

Fig. 10.2. Illustration of the correlation and anticorrelation of the electrons in the helium atom. Panels (a) and (b) present the machinery of the "*anticorrelation*" connected with the geminal $g_{He} = N\exp\left[-r_1^2\right]\exp\left[-r_2^2\right]\exp\left[-2r_{12}^2\right]$. In (a), electron 1 has a position (0, 0, 0), while (b) corresponds to electron 1 being at point (1, 0, 0) (cutting off the top parts of the plots is caused by graphical limitations, not by the physics of the problem). It can be seen that electron 2 *holds to electron 1*; i.e., it behaves in a completely unphysical manner (since the electrons repel each other). Panels (c) and (d) show how electron 2 will respond to such two positions of electron 1, if the wave function is described by the geminal $g_{He} = N\exp\left[-r_1^2\right]\exp\left[-r_2^2\right]\left[1-\exp\left[-2r_{12}^2\right]\right]$. In (c), we see that electron 2 runs away "with all its strength" (the hollow in the middle) from electron 1 placed at (0, 0, 0), we have correlation. Similarly, in (d), if electron 1 is in point (1, 0, 0), then it causes a slight depression for electron 2 in this position, we do have correlation. However, the graph is different than in case (c). This is understandable since the nucleus is all the time in the point (0, 0, 0). Panels (e) and (f) correspond to the same displacements of electron 1, but this time, the correlation function is equal to $\psi(\boldsymbol{r}_1, \boldsymbol{r}_2) = \left(1 + \frac{1}{2}r_{12}\right)\exp\left[-\left(r_1^2 + r_2^2\right)\right]$; i.e., it is similar to the wave function of the harmonic helium atom. It can be seen (particularly in panel e) that there *is* a correlation, although much less visible than in the previous examples. To amplify (artificially) the correlation effect, panels (g) and (h) show the same as (e) and (f), but for the function $\psi(\boldsymbol{r}_1, \boldsymbol{r}_2) = (1+25r_{12})\exp\left[-\left(r_1^2 + r_2^2\right)\right]$, which [unlike in (e) and (f)] does not satisfy the correlation cusp condition.

[34] W. Cencek, Ph.D. thesis, Adam Mickiewicz University, Poznań, 1993; also J. Rychlewski, W. Cencek, and J. Komasa, *Chem. Phys. Letters, 229,* 657 (1994); W. Cencek, and J. Rychlewski, *Chem. Phys. Letters, 320,* 549 (2000).

around the electron, there is a no-parking area for other electrons with the same spin coordinate (*known as the exchange, or Fermi hole*).

Let us see how such exchange holes arise. We will try to make the calculations as simple as possible.

We have shown above that the Hartree-Fock function does not include any electron correlation. We must admit, however, that we have come to this conclusion on the basis of the two-electron, closed-shell case. This is a special situation, since both electrons have *different* spin coordinates $\left(\sigma = \frac{1}{2} \text{ and } \sigma = -\frac{1}{2}\right)$. Is it really true that the Hartree-Fock function does not include any correlation of electronic motion?

We take the H_2^- molecule in the simplest formulation of the LCAO MO method.[35] We have three electrons. As a wave function, we will take the single (normalized) Hartree-Fock determinant (of the UHF type) with the following orthonormal spinorbitals occupied: $\phi_1 = \varphi_1\alpha, \phi_2 = \varphi_1\beta, \phi_3 = \varphi_2\alpha$:

$$\psi_{UHF}(1,2,3) = \frac{1}{\sqrt{3!}} \begin{vmatrix} \phi_1(1) & \phi_1(2) & \phi_1(3) \\ \phi_2(1) & \phi_2(2) & \phi_2(3) \\ \phi_3(1) & \phi_3(2) & \phi_3(3) \end{vmatrix}.$$

Example 1: The Great Escape

We are interested in electron 3 with electron 1 residing at nucleus a with space coordinates $(0,0,0)$ and with spin coordinate $\sigma_1 = \frac{1}{2}$ and with electron 2 located at nucleus b with coordinates $(R,0,0)$ and $\sigma_2 = -\frac{1}{2}$, whereas the electron 3 itself has spin coordinate $\sigma_3 = \frac{1}{2}$. The square of the absolute value of the function ψ_{UHF} calculated for these values depends on x_3, y_3, z_3 and represents the *conditional probability* density distribution for finding electron 3 (provided electrons 1 and 2 have the fixed coordinates given above and denoted by $1_0, 2_0$). So, let us calculate individual elements of the determinant $\psi_{UHF}(1_0, 2_0, 3)$, taking into account the properties of spin functions α and β (cf. p. 27):

$$\psi_{UHF}(1_0, 2_0, 3) = \frac{1}{\sqrt{3!}} \begin{vmatrix} \varphi_1(0,0,0) & 0 & \varphi_1(x_3,y_3,z_3) \\ 0 & \varphi_1(R,0,0) & 0 \\ \varphi_2(0,0,0) & 0 & \varphi_2(x_3,y_3,z_3) \end{vmatrix}.$$

Using the Laplace expansion (see Appendix A available at http://booksite.elsevier.com/978-0- 444-59436-5 on p. e1), we get

$$\psi_{UHF}(1_0, 2_0, 3) = \frac{1}{\sqrt{3!}}[\varphi_1(0,0,0)\varphi_1(R,0,0)\varphi_2(x_3,y_3,z_3) \\ - \varphi_1(x_3,y_3,z_3)\varphi_1(R,0,0)\varphi_2(0,0,0)].$$

[35] This involves two atomic orbitals only: $1s_a = \chi_a$ and $1s_b = \chi_b$, two molecular orbitals: [bonding $\varphi_1 = \frac{1}{\sqrt{2(1+S)}}(\chi_a + \chi_b)$], and antibonding [$\varphi_2 = \frac{1}{\sqrt{2(1-S)}}(\chi_a - \chi_b)$, cf. p. 437] and the overlap integral $S \equiv (\chi_a|\chi_b)$.

The plot of this function (the overlap integral S is included in normalization factors of the molecular orbitals) is given in Fig. 10.3.

Qualitatively, however, everything is clear even without the calculations. Due to the forms of the molecular orbitals (S is small) $\varphi_1(0,0,0) = \varphi_1(R,0,0) \approx \varphi_2(0,0,0) = \text{const}$, we get

$$\psi_{UHF}(1_0, 2_0, 3) \approx -\text{const}^2 \frac{1}{\sqrt{3}} \chi_b(3),$$

so the conditional probability of finding electron 3 is

$$\rho(3) \approx \frac{1}{3}\text{const}^4 \, [\chi_b(3)]^2. \tag{10.19}$$

We can see that *for some reason, electron 3 has chosen to be in the vicinity of nucleus b.* What scared it so much when we placed one electron on *each* nucleus? Electron 3 ran to be as far as possible from electron 1 residing on a. It hates electron 1 so much that it has just ignored the Coulomb repulsion, of electron 2 sitting on b, and jumped on it![36] What has happened?

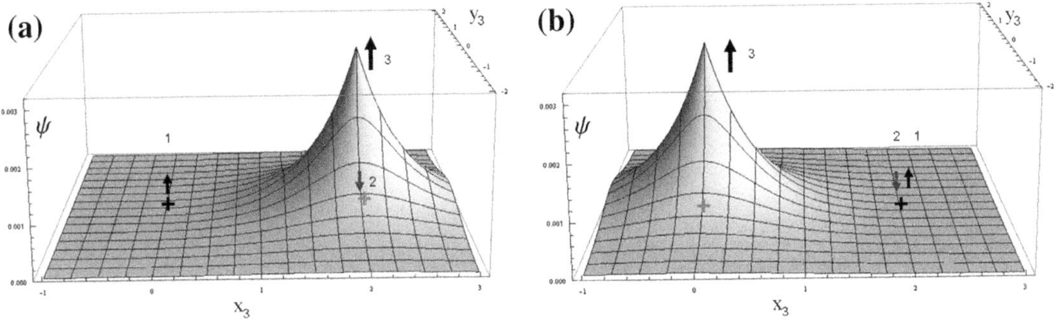

Fig. 10.3. Demonstration of the power of the Pauli exclusion principle, or the Fermi hole formation for the H_2^- molecule in the UHF model (p. 448, a wave function in the form of a single Slater determinant). The two protons (a and b), indicated by "+," occupy positions $(0, 0, 0)$ and $(2, 0, 0)$ in a.u., respectively. The space and spin coordinates (the latter shown as arrows) of electrons 1 and 2 [$\left(x_1, y_1, z_1, \sigma_1 = \frac{1}{2}\right)$ and $\left(x_2, y_2, z_2, \sigma_2 = -\frac{1}{2}\right)$, so they have opposite spins] as well as the spin coordinate of electron 3 ($\sigma_3 = \frac{1}{2}$, the same as the spin coordinate of electron 1) will be fixed at certain values: electron 2 will always sit on nucleus b, electron 1 will occupy some chosen positions on the x-axis (i.e., we keep $y_1 = 0, z_1 = 0$). In this way, we will have to work with a section $\psi(x_3, y_3, z_3)$ of the wave function, visualized in the figure by setting $z_3 = 0$. The square of the resulting function represents a conditional probability density of finding electron 3, if electrons 1 and 2 have the assigned coordinates. (a) Corresponds to example 1: electron 1 sits on nucleus a. Electron 3 runs away to the nucleus b, despite the fact that there is already electron 2! (b) Corresponds to example 2: electron 1 sits on nucleus b together with electron 2. Electron 3 runs away to the nucleus a. (c) Corresponds to example 3–a dilemma for electron 3: electron 1 sits in the middle between the nuclei. Electron 3 chooses the antibonding molecular orbital (c1), because it offers a node exactly at the position of electron 1 (with same spin), when squared it creates a Fermi hole (c2)! (d1) is an even tougher case: electron 1 sits at $\frac{1}{3}$ of the internuclear distance; so, what is electron 3 going to do? Electron 3 chooses such a combination of the bonding and of the antibonding molecular orbitals that creates a node (and a Fermi hole, d2) precisely at the position of electron 1 with the same spin. Clearly, with a single Slater determinant as the wave function, electrons with the same spin hate one another (Fermi hole), while electrons with the opposite spin just ignore each other (no Coulomb hole).

[36] In fact it does not see electron 2 (because of the one-determinantal wave function).

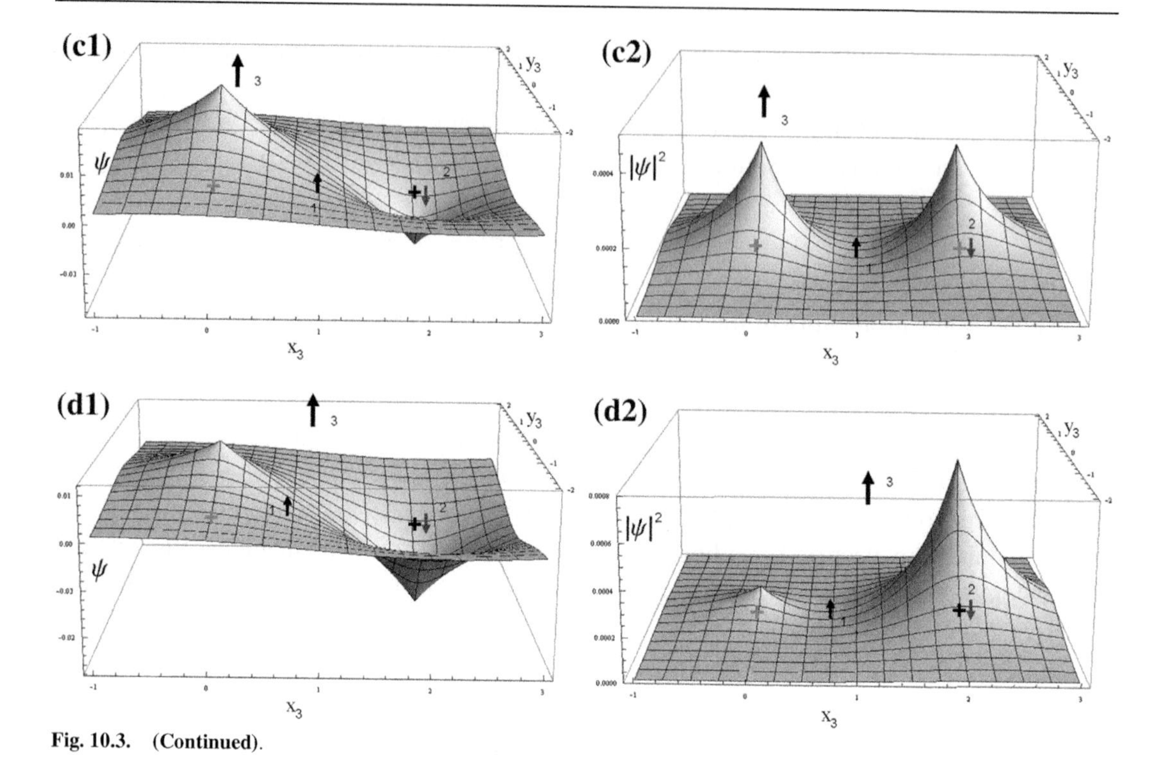

Fig. 10.3. (Continued).

Well, we have some suspicions. Electron 3 could have been scared only by the spin coordinate of electron 1, *the same as its own.*

This is just an indication of the exchange hole around each electron.

Example 2: Another Great Escape

Maybe electron 3 does not run away from anything, but simply always resides at nucleus b. Let us make sure of that by moving electron 1 to nucleus b (electron 2 is already sitting over there, but that does not matter). What, then, will electron 3 do? Let us see. We have electrons 1 and 2 at nucleus b with space coordinates $(R, 0, 0)$ and spin coordinates $\sigma_1 = \frac{1}{2}, \sigma_2 = -\frac{1}{2}$, whereas electron 3 has spin coordinate $\sigma_3 = \frac{1}{2}$. To calculate the conditional probability, we have to calculate the value of the wave function.

This time,

$$\psi_{UHF}(1_0, 2_0, 3) = \frac{1}{\sqrt{3!}} \begin{vmatrix} \varphi_1(R,0,0) & 0 & \varphi_1(x_3, y_3, z_3) \\ 0 & \varphi_1(R,0,0) & 0 \\ \varphi_2(R,0,0) & 0 & \varphi_2(x_3, y_3, z_3) \end{vmatrix} \approx \text{const}^2 \frac{1}{\sqrt{3}} \chi_a(3)$$

or

$$\rho(3) \approx \frac{1}{3} \text{const}^4 [\chi_a(3)]^2. \tag{10.20}$$

We see that electron 3 with spin coordinate $\sigma_3 = \frac{1}{2}$ runs in panic to nucleus a, because it is as scared of electron 1 with spin $\sigma_1 = \frac{1}{2}$ as the devil is of holy water.

Example 3: A Dilemma

And what would happen if we made the decision for electron 3 more difficult? Let us put electron 1 $\left(\sigma_1 = \frac{1}{2}\right)$ *in the center* of the molecule and electron 2 $\left(\sigma_2 = -\frac{1}{2}\right)$ as before, at nucleus b. According to what we think about the whole machinery, electron 3 (with $\sigma_3 = \frac{1}{2}$) should run away from electron 1 because both electrons have the same spin coordinates, and this is what they hate most. But *where* should it run? Will electron 3 select nucleus a or nucleus b? The nuclei do not look equivalent. There is an electron sitting at b, while the a center is empty. Maybe electron 3 will jump to a then? Well, the function analyzed is the Hartree-Fock type, electron 3 ignores the Coulomb hole (it does not see electron 2 sitting on b) and therefore, it will not prefer the empty nucleus a to sit at. It looks like electron 3 will treat both nuclei on the same basis. In the case of two atomic orbitals, electron 3 has a choice: either bonding orbital φ_1 or antibonding orbital φ_2 (either of these situations corresponds to equal electron densities on a and on b). Out of the two molecular orbitals, φ_2 looks much more attractive to electron 3, because it has a node[37] exactly, where electron 1 with its nasty spin is. This means that there is a chance for electron 3 to take care of the Fermi hole of electron 1: we predict that electron 3 will "choose" only φ_2. Let us check this step by step:

$$\begin{aligned}
\psi_{UHF}(1_0, 2_0, 3) &= \frac{1}{\sqrt{3!}}\begin{vmatrix} \varphi_1\left(\frac{R}{2},0,0\right) & 0 & \varphi_1(x_3,y_3,z_3) \\ 0 & \varphi_1(R,0,0) & 0 \\ \varphi_2\left(\frac{R}{2},0,0\right) & 0 & \varphi_2(x_3,y_3,z_3) \end{vmatrix} \\
&= \frac{1}{\sqrt{3!}}\begin{vmatrix} \varphi_1\left(\frac{R}{2},0,0\right) & 0 & \varphi_1(x_3,y_3,z_3) \\ 0 & \varphi_1(R,0,0) & 0 \\ 0 & 0 & \varphi_2(x_3,y_3,z_3) \end{vmatrix} \\
&= \frac{1}{\sqrt{3!}}\varphi_1\left(\frac{R}{2},0,0\right)\varphi_1(R,0,0)\varphi_2(x_3,y_3,z_3) = \text{const}_1\varphi_2(x_3,y_3,z_3).
\end{aligned}$$

And it does exactly that.

In Fig. 10.3, in panel (d1), we give an example with electron 1 at $\frac{1}{3}R$. The result is similar: a Fermi hole is precisely at the position of electron 1.

Which hole is more important: Coulomb or exchange? This question will be answered in Chapter 11.

[37] That is, low probability of finding electron 3 over there.

VARIATIONAL METHODS WITH SLATER DETERMINANTS

In all of these methods, the variational wave function will be sought in the form of a linear combination of Slater determinants. As we have seen a while ago, even a single Slater determinant assures a very serious avoiding of electrons with the same spin coordinate. Using a linear combination of Slater determinants means an automatic (based on variational principle) optimization of the exchange hole (Fermi hole).

What about the Coulomb hole? If this hole were also optimized, a way to the solution of the Schrödinger equation would open up. However, as we have carefully checked before, a single Slater determinant does not know anything about the Coulomb hole. If it does not know, then perhaps a linear combination of guys, each of them not knowing anything, will not do any better... Wrong! A linear combination of Slater determinants *is* able to describe the Coulomb hole.[38]

10.6 Static Electron Correlation

Some of these Slater determinants are necessary for fundamental reasons. For example, consider the carbon atom ground state, its (triplet) ground state corresponding to the $1s^2 2s^2 2p^2$ configuration. The configuration does not define *which* of the triply *degenerate* $2p$ orbitals have to be included in the Slater determinant. Any choice of the $2p$ orbitals will be therefore non-satisfactory: one is forced to go beyond a single Slater determinant. A similar situation occurs if an obvious *quasi-degeneracy* occurs, like for the hydrogen molecule at large distances (see Chapter 8). In such a case, we are also forced to include in calculations another Slater determinant. One may say that

> what is known as a *static correlation* represents an energy gain coming from considering in the wave function (in the form of a linear combination of Slater determinants) low-energy Slater determinants, which follow from occupying a set of degenerate or quasi-degenerate orbitals.

10.7 Dynamic Electron Correlation

The dynamic electron correlation means the rest of the correlation effect, beyond the static one. It corresponds also to occupying orbital energies, but not those related to the degeneracy or quasi-degeneracy of the ground state. As we see, the distinction between the static and the dynamic correlation is a bit arbitrary.

[38] Not all linear combinations of Slater determinants describe the Coulomb hole. Indeed, for example, a Hartree-Fock function in the LCAO MO approximation may be expanded in a series of Slater determinants (see Appendix A available at booksite.elsevier.com/978-0-444-59436-5) with the atomic orbitals, but no Coulomb hole is described by this function.

Example of Beryllium

Let us take a beryllium atom as an illustration. The beryllium atom has four electrons ($1s^2 2s^2$ configuration). Beryllium represents a tough case in quantum chemistry because the formally occupied $2s$ orbital energy is quite close to the formally unoccupied orbital energy of $2p$. In the present example, we will claim this as a dynamic correlation, but to tell the truth, it is just between the static and dynamic correlation. One may, therefore, suspect that the excited configurations $2s^1 2p^1$ and $2p^2$ will be close in energy scale to the ground-state configuration $2s^2$. There is, therefore, no legitimate arguments for neglecting these excited configurations in the wave function (what the Hartree-Fock method does). Since the Hartree-Fock method is poor in this case, this means the electronic correlation energy must be large for the beryllium atom.[39]

Why to worry then about the closed-shell electrons $1s^2$? Two of the electrons are bound very strongly ($1s^2$)–so strongly that we may treat them as passive observers that do not react to anything that may happen. Let us just ignore the inner shell[40] in such a way that we imagine an "effective nucleus of the pseudoatom" of beryllium as a genuine beryllium nucleus surrounded by the electronic cloud $1s^2$. The charge of this "nucleus" is $4 - 2 = 2$. Then, the ground-state Slater determinant for such a pseudoatom reads as

$$\psi_0 = \frac{1}{\sqrt{2!}} \begin{vmatrix} 2s(1)\alpha(1) & 2s(2)\alpha(2) \\ 2s(1)\beta(1) & 2s(2)\beta(2) \end{vmatrix}, \tag{10.21}$$

where we decide to approximate the function $2s$ as a normalized Slater orbital[41] ($\zeta > 0$):

$$2s = \sqrt{\frac{\zeta^5}{3\pi}}\, r \exp(-\zeta r).$$

Since the Hartree-Fock method looks to be a poor tool for beryllium, we propose a more reasonable wave function in the form of a linear combination of the ground-state configuration [Eq. (10.21)] and the configuration given by the following Slater determinant:

$$\psi_1 = \frac{1}{\sqrt{2!}} \begin{vmatrix} 2p_x(1)\alpha(1) & 2p_x(2)\alpha(2) \\ 2p_x(1)\beta(1) & 2p_x(2)\beta(2) \end{vmatrix}, \tag{10.22}$$

where just to keep things as simple as possible, we use the $2p_x$ orbital.

---

[39] This is why we took the beryllium atom and not just the helium atom, in which the energy difference between the orbital levels $1s$ and $2s$ is much larger (i.e., the correlation energy much smaller).

[40] The reasoning below may be repeated with the $1s^2$ shell included; the calculations will be a bit more complicated, but the final result very similar.

[41] Let us check whether the normalization coefficient is correct: $\int (2s)^2 dV = \frac{\zeta^5}{3\pi} \int r^2 \exp(-2\zeta r) dV = \frac{\zeta^5}{3\pi} \int_0^\infty r^4 \exp(-2\zeta r) dr \int_0^\pi \sin\theta d\theta \int_0^{2\pi} d\phi = \frac{4\pi\zeta^5}{3\pi} \int_0^\infty r^4 \exp(-2\zeta r) dr = \frac{4\pi\zeta^5}{3\pi} 4!(2\zeta)^{-5} = 1$, as it should be.

Such a function, being a linear combination of antisymmetric functions, is itself antisymmetric with respect to the electron exchange (as it should be–see Chapter 1). Just to grasp the essence of the problem, we omit all other excitations, including $2s^2 \rightarrow 2p^2$ with the orbitals $2p_y, 2p_z$ as well as the excitations of the type $2s^2 \rightarrow 2s^1 2p^1$. The latter excitation seems to require low energy, so it is potentially important. However, it will be shown later in this chapter that there are arguments for neglecting it (because of a weak coupling with the ground-state configuration). The x-axis has been highlighted by us (through taking $2p_x$ orbitals only) for purely didactic reasons, because soon we are going to frighten electron 2 by using electron 1 in certain points on the x-axis (therefore, this axis is expected to be the main direction of escaping for electron 2):

$$2p_x = \zeta \sqrt{\frac{\zeta^3}{\pi}} x \exp(-\zeta r) = \zeta x\,(2s).$$

The drastically simplified wave function reads, therefore, as

$$\psi = \psi_0 + \kappa \psi_1, \tag{10.23}$$

where κ stands for a coefficient to be determined, which measures how much of the $2p^2$ configuration has to be added to the $2s^2$ configuration in order to describe correctly the physical behavior of the electrons[42] (for example, this is forced by the variational method or by a perturbational approach, see Chapter 5). Let us use a perturbational approach, in which we assume ψ_0 as a unperturbed wave function. Eq. (5.24), p. 245 says, that with our current notation, the coefficient κ may be estimated as

$$\kappa = \frac{\left\langle \psi_1 | \hat{H}^{(1)} \psi_0 \right\rangle}{E_0 - E_1}, \tag{10.24}$$

where the energies E_0 and E_1 correspond to the ground-state configuration (ψ_0) and the excited-state configuration (ψ_1), while $\hat{H}^{(1)}$ stands for the perturbation. Right now, we have no idea what this perturbation is, but it is not necessary to know this since (see Chapter 5) $\left\langle \psi_1 | \hat{H}^{(1)} \psi_0 \right\rangle = \left\langle \psi_1 | (\hat{H} - \hat{H}^{(0)}) \psi_0 \right\rangle = \left\langle \psi_1 | \hat{H} \psi_0 \right\rangle - E_0 \langle \psi_1 | \psi_0 \rangle = \left\langle \psi_1 | \hat{H} \psi_0 \right\rangle - 0 = \left\langle \psi_1 | \hat{H} \psi_0 \right\rangle$, where $\hat{H}^{(0)} \psi_0 = E_0 \psi_0$ and $\langle \psi_1 | \psi_0 \rangle = 0$ (the latter because of the orthogonality of $2s$ and $2p_x$).

It is seen, therefore, that we have to do with a matrix element of the Hamiltonian calculated with two Slater determinants containing orthonormal spinorbitals: $2s\alpha, 2s\beta, 2p_x\alpha, 2p_x\beta$, the first two composing ψ_0, the last ones present in ψ_1. Hence, all necessary conditions are satisfied

[42] We are not intending to get a perfect description of the system because with such a trial function, there is no chance to solve the Schrödinger equation anyway. Rather, we are here to grasp a qualitative picture: will it be a Coulomb hole or not?

for operating the third Slater-Condon rule (see appendix M available at booksite.elsevier.com/978-0-444-59436-5, p. e109). We get

$$\left\langle\psi_1|\hat{H}\psi_0\right\rangle = \langle 2s\alpha 2s\beta|2p_x\alpha 2p_x\beta\rangle - \langle 2s\alpha 2s\beta|2p_x\beta 2p_x\alpha\rangle$$
$$= \langle 2s\alpha 2s\beta|2p_x\alpha 2p_x\beta\rangle - 0 = (2s2s|2p_x2p_x)$$
$$\equiv \int \left[2s(1)2p_x(1)\right]\frac{1}{r_{12}}\left[2s(2)2p_x(2)\right]\mathrm{d}V_1\mathrm{d}V_2 > 0.$$

We have got a key inequality,[43] because from Eq. (10.24) and $E_0 < E_1$, it follows that

$$\kappa < 0. \tag{10.25}$$

Our qualitative conclusions will depend only on the sign of κ, not on its particular value. Let us make a set of exercises listed below (all distances in a.u.), first with ψ_0, then with ψ_1, and finally with $\psi = \psi_0 + \kappa\psi_1$. In all of them, the following is true:

- The nucleus is immobilized at (0, 0, 0).
- Let us put electron 1, having the spin coordinate $\sigma_1 = \frac{1}{2}$, at $(-1, 0, 0)$.
- We will search the probability distribution of finding electron 2 with the spin coordinate $\sigma_2 = -\frac{1}{2}$.
- We will repeat the two last points with electron 1 at $(+1, 0, 0)$; i.e., on the opposite side of the nucleus and at the same electron-nucleus distance.
- We will compare the two probability distributions. If they were identical, there would be no correlation whatsoever; otherwise, there would be a correlation.

To this end, we will need three numbers to be calculated (the three numbers in parentheses represent x, y, and z):

$$2s(-1,0,0) = 2s(1,0,0) = \sqrt{\frac{\zeta^5}{3\pi}}\exp(-\zeta) \equiv A > 0,$$

$$2p_x(1,0,0) = \zeta\sqrt{\frac{\zeta^3}{\pi}}\exp(-\zeta) = B > 0,$$

$$2p_x(-1,0,0) \equiv \zeta\sqrt{\frac{\zeta^3}{\pi}}(-1)\exp(-\zeta) \equiv -B.$$

Function ψ_0.

We expand the determinant [Eq. (10.21)] for electron 1 being at position $(-1, 0, 0)$ and obtain a function of position of electron 2 in the form[44] $\frac{1}{\sqrt{2}}A \cdot 2s(2)$. Therefore, the (conditional)

[43] The inequality follows from evident repulsion of two identical electron clouds (of electron 1 and of electron 2), because they sit on top of each other.

[44] Only the diagonal elements of the Slater determinant are nonzero (the rest of elements vanish because of the spin functions), so we get the result right away.

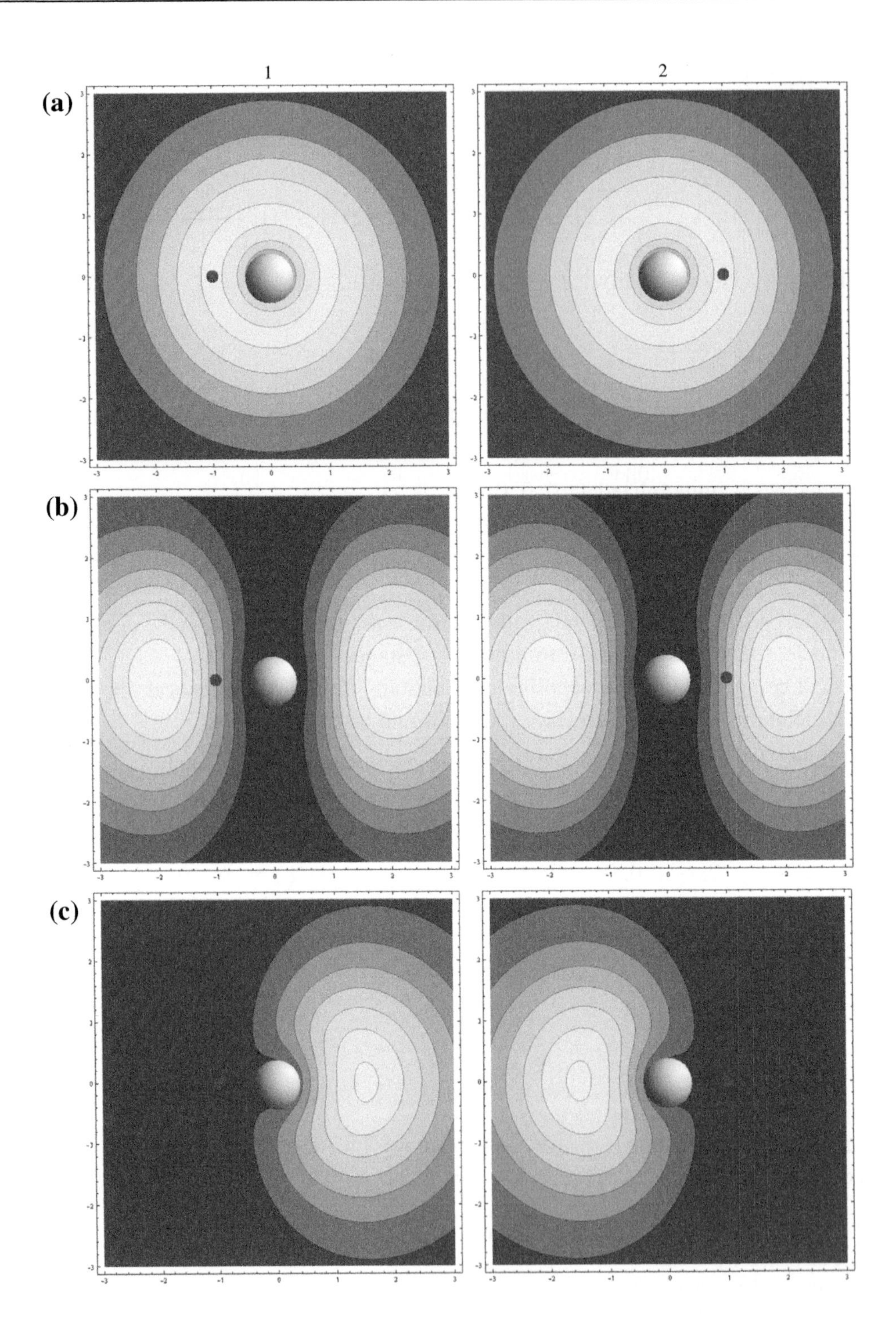
1
2
(a)
(b)
(c)

probability density distribution of electron 2 is $\frac{1}{2}A^2[2s(2)]^2$; (see Fig. 10.4a1). We repeat the same for position $(1, 0, 0)$ of electron 1 and get the identical result (see Fig. 10.4a2). Conclusion: there is no Coulomb hole for the ground-state Slater determinant. Well, this is what we should expect. However, it may be that this result depends on a type of Slater determinant. Let us take the Slater determinant ψ_1.

Function ψ_1.

Expanding [Eq. (10.22)] for a fixed position $(-1, 0, 0)$ of electron 1, one gets a function depending on the position of electron 2 in the form $\frac{1}{\sqrt{2}}(-B) \cdot 2p_x(2)$, and therefore the conditional probability of finding electron 2 is $\frac{1}{2}B^2[2p_x(2)]^2$ (Fig. 10.5b1). Repeating the same for position $(1, 0, 0)$ of electron 1, we obtain a function: $\frac{1}{\sqrt{2}}B \cdot 2p_x(2)$, but still we get the same probability distribution: $\frac{1}{2}B^2[2p_x(2)]^2$ (see Fig. 10.5b2). Once again, we obtain no Coulomb hole.

Function $\psi = \psi_0 + \kappa\psi_1$.

We calculate $\psi = \psi_0 + \kappa\psi_1$ for position $(-1, 0, 0)$ of electron 1 and we obtain a function of position of electron 2 in the form $\frac{1}{\sqrt{2}}A \cdot 2s(2) + \kappa\left[\frac{1}{\sqrt{2}}(-B) \cdot 2p_x(2)\right]$ with the corresponding conditional probability distribution of electron 2 as $\rho_-(2) = \frac{1}{2}A^2[2s(2)]^2 + \frac{1}{2}\kappa^2 B^2[2p_x(2)]^2 - \kappa AB \cdot 2s(2) \cdot 2p_x(2)$ (Fig. 10.4c1). When repeating the same for position $(1, 0, 0)$ of electron 1, we obtain a *different* result: $\frac{1}{\sqrt{2}}A \cdot 2s(2) + \kappa\left[\frac{1}{\sqrt{2}}B \cdot 2p_x(2)\right]$ and therefore a *different* probability distribution: $\rho_+(2) = \frac{1}{2}A^2[2s(2)]^2 + \frac{1}{2}\kappa^2 B^2[2p_x(2)]^2 + \kappa AB \cdot 2s(2) \cdot 2p_x(2)$ (see Fig. 10.4c2). So, there *is* a correlation of the electronic motion. It would be even better to have this correlation reasonable.[45] Panels (c1) and (c2) of Fig. 10.4 show that indeed, the correlation stands to reason: *the two electrons avoid one another, if electron 1 is on the left side, electron 2 is on the right side and vice versa.*

If we did not have the inequality [Eq. (10.25)], this conclusion could not be derived. For $\kappa > 0$, electron 2 would accompany electron 1 ("*anticorrelation*"), which means "a completely non-physical" behavior. For $\kappa = 0$ or $\kappa = \pm\infty$, there would be no correlation.[46] All, therefore,

←

Fig. 10.4. A single Slater determinant cannot describe any Coulomb correlation, but a linear combination of the Slater determinants can. The image shows the beryllium atom, with a pseudonucleus (of charge +2) shown as a large sphere in the center. All the images show the sections ($z = 0$) of the (conditional) probability density distribution of finding electron 2 (a, upper row–for the single Slater determinant ψ_0; b, second row–for the single Slater determinant ψ_1; c, bottom row–for a two-determinantal wave function $\psi = \psi_0 + \kappa\psi_1$), when electron 1, symbolized by a small sphere, resides at $(-1, 0, 0)$ (the left side has the symbol 1) or at $(1, 0, 0)$ (the right side has the symbol 2). Only in the case of the two-determinantal wave function $\psi = \psi_0 + \kappa\psi_1$, one obtains any difference between the probability distributions, when electron 1 occupies two positions: $(-1, 0, 0)$ and $(1, 0, 0)$. The values $\kappa < 0$ correspond to mutual avoiding of the two electrons (in such a case, the wave function takes into account the Coulomb hole), $\kappa = 0$ means mutual ignoring of the two electrons, $\kappa > 0$ would correspond to a very bad wave function, that describes the two electrons sticking one to the other. In order to highlight the correlation effect (purely didactic reasons), we took quite arbitrarily $\kappa = -0.7$ and $\zeta = 1$.

[45] A unreasonable *correlation* would be, for example, when the two electrons were sticking to each other.

[46] All these cases correspond to a single determinant ψ_0 (for $\kappa = 0$) and ψ_1 or $-\psi_1$ (for $\kappa = \pm\infty$).

depends on the coefficients of the linear combination of Slater determinants. This is the variational principle or the perturbational theory that takes care that the wave function was close to the solution of the Schrödinger equation for the ground state. This forces a physics-based description of the electronic correlation–in our case, $\kappa < 0$.

> A two-determinantal function $\psi = \psi_0 + \kappa\psi_1$ with $\kappa < 0$ can (in contrast with the single determinantal functions ψ_0 and ψ_1) approximate the effect of the dynamic correlation (Coulomb hole). Of course, a combination of many Slater determinants with appropriate coefficients can do it better.

10.8 Anticorrelation, or Do Electrons Stick Together in Some States?

What about electronic correlation in excited electronic states? Not much is known for excited states in general. In our case of Eq. (10.23), the Ritz variational method would give two solutions. One would be of lower energy corresponding to $\kappa < 0$ (this solution has been approximated by us using the perturbational approach). The second solution (the excited electronic state) will be of the form $\psi_{exc} = \psi_0 + \kappa'\psi_1$. In such a simple two-state model, the coefficient κ' can be found just from the (necessary) orthogonality of the two solutions: $\langle\psi_{exc}|\psi\rangle = \langle\psi_0 + \kappa'\psi_1|\psi_0 + \kappa\psi_1\rangle = 1 + \kappa\kappa'^* + \kappa'^*\langle\psi_1|\psi_0\rangle + \kappa\langle\psi_0|\psi_1\rangle = 1 + \kappa\kappa'^* = 0$.

Hence, $\kappa'^* = -\frac{1}{\kappa} > 0$. We have, therefore, $\kappa' > 0$ and it is quite intriguing that our excited state corresponds now to what we call here an "*anticorrelation.*" In the excited state, we got the two electrons sticking to each other. This result certainly cannot be thought as of general value for excited states. It is probable that in excited electronic states, the electronic correlation gets weaker, but according to what we have found in our two-state model, *some excited states might exhibit the electronic anticorrelation*! This indication may be less surprising than it sounds. For example, the hydrogen molecule has not only the covalent states, but also the excited states of ionic character (as we will discuss next). In the ionic states, the two electrons prefer to occupy the same space (still repelling each other), as if there were a kind of "attraction" between them.

Electrons Attract Themselves!

Do the electrons repel each other? Of course. Does this mean that the electrons try to be as far from themselves as possible? Yes, but the words "*as possible*" are important. What does that mean? Usually, this means a game between the electrons strongly attracted by a nucleus and their important repulsion through the Pauli exclusion principle (Fermi hole), together with much less important Coulomb repulsion (Coulomb hole).

Let us try to simplify the situation. First, let us remove the presence of the nuclei and see what electrons like without them. Then, while all the time keeping the Coulomb repulsion, we will either switch on the Fermi hole by considering the triplet states with the two electrons having opposite spins or switch off the Fermi hole by taking the singlet states of these two electrons.

Let us take a toy (a model): a circle of radius R (the potential energy within the circle set to zero, the infinity outside) and two electrons moving along the circle. Thus, there is no nuclei, only a circle with two electrons living in it. Independent of the singlet or triplet states considered, our common sense says that these two electrons will avoid each other; i.e., they *will prefer to be on the opposite sites of the circle*. Let us check whether this idea is true.

When you write down the corresponding Hamiltonian, it will depend on ϕ_1 and ϕ_2 (two position angles) and contain the kinetic energy operator of the two electrons plus the Coulombic repulsion of the electrons $\frac{e^2}{r_{12}}$. Now, we can introduce the center of mass angular coordinate (proportional to $\phi_1 + \phi_2$) and the relative coordinate $\phi = \phi_1 - \phi_2$. After exact separation of the center-of-mass motion, we get the Schrödinger equation for ϕ (μ is the reduced mass of the two electrons):

$$\left(-\frac{\hbar^2}{2\mu R^2}\frac{\partial^2}{\partial\phi^2} + \frac{e^2}{r_{12}}\right)\psi(\phi) = E\psi(\phi).$$

If the Coulombic repulsion were absent, the solutions would be $const, \exp(im\phi)$ and $\exp(-im\phi), m = 1, 2, \ldots$ which means the non-degenerate nodeless ground state and all other states doubly degenerate. In the future, we will use their combinations ($\sin m\phi$ and $\cos m\phi$) as the expansion functions for the wave function.

Now we reconsider the Coulombic repulsion. In fact, after separation is done, we may treat electron 1 as sitting all the time at $\phi = 0$ and electron 2 (with the coordinate ϕ) moving. The eigenfunctions for this problem are:

- The nodeless ground state ψ_0, which because of the Coulombic term, will not be a constant, but have a maximum at $\phi = 180^\circ$ (i.e., the farthest distance from electron 1). The spatial function is a symmetric function of ϕ, so this describes the singlet ground-state.
- The first excited state ψ_1 has one node, and this nodal line should be along a straight line: electron 1 and position $\phi = 180^\circ$, This function is antisymmetric with respect to exchange of the electrons ($\phi \to -\phi$), so this is the (lowest) triplet state. This state will be of low-energy, because it takes care of the Fermi hole, the wave function equal zero for electron 2 at the position of electron 1.
- The second excited state (ψ_2) will also have one node (recall the benzene π orbitals, or think about m and $-m$), but the nodal plane has to be orthogonal to that of ψ_1 (symmetric function; i.e., the first excited singlet). The function ψ_2 has to be orthogonal to ψ_0 and ψ_1. The orthogonality to ψ_0 means it has to have larger absolute amplitude at the position of electron 1 than on the opposite site ($\phi = 180^\circ$). So we see that already such a low-energy state as ψ_2 is of the kind that electron 2 prefers to be closer to electron 1.
- Similar phenomenon will appear for higher states.

The above description has a resemblance to the rigid dipolar rotator rotating in plane with a uniform electric field within this plane (with orientation $\phi = 0$); see p. 736. There is only one difference with respect to the problem of two electrons: the reason why the negative pole of the dipole hates to get the orientation $\phi = 0$ (which corresponds to electron 2 avoiding electron 1 at $\phi = 0$) is the uniform electric field and not the non-uniform electric field created by electron 1. This difference is of secondary importance. So, the very fact that there are experimental observations of what is known as low-field seeker dipole molecules (with the dipole moment *against* the electric field, see p.736) represents a strong indication that the same should happen here. So

> there will be excited states that describe electrons close to each other, as if they attracted themselves!

10.9 Valence Bond (VB) Method

10.9.1 Resonance Theory–Hydrogen Molecule

Slater determinants are usually constructed from *molecular* spinorbitals. If, instead, we use *atomic* spinorbitals and the Ritz variational method (Slater determinants as the expansion functions), we would get the most general formulation of the valence bond (VB) method. The beginning of VB theory goes back to papers by Heisenberg, the first application was made by Heitler and London, and later theory was generalized by Hurley, Lennard-Jones, and Pople.[47]

The essence of the VB method can be explained by an example. Let us take the hydrogen molecule with atomic spinorbitals of type $1s_a\alpha$ and $1s_b\beta$ (abbreviated as $a\alpha$ and $b\beta$) centered at two nuclei. Let us construct from them several (non-normalized) Slater determinants, for instance:

$$\psi_1 = \frac{1}{\sqrt{2}}\begin{vmatrix} a(1)\alpha(1) & a(2)\alpha(2) \\ b(1)\beta(1) & b(2)\beta(2) \end{vmatrix} = \frac{1}{\sqrt{2}}\left[a(1)\alpha(1)b(2)\beta(2) - a(2)\alpha(2)b(1)\beta(1)\right],$$

$$\psi_2 = \frac{1}{\sqrt{2}}\begin{vmatrix} a(1)\beta(1) & a(2)\beta(2) \\ a(1)\alpha(1) & a(2)\alpha(2) \end{vmatrix} = \frac{1}{\sqrt{2}}\left[a(1)\beta(1)b(2)\beta(2) - a(2)\beta(2)b(1)\alpha(1)\right],$$

$$\psi_3 = \frac{1}{\sqrt{2}}\begin{vmatrix} a(1)\alpha(1) & a(2)\alpha(2) \\ a(1)\beta(1) & a(2)\beta(2) \end{vmatrix} = \frac{1}{\sqrt{2}}\left[a(1)\alpha(1)a(2)\beta(2) - a(2)\alpha(2)a(1)\beta(1)\right]$$

$$= a(1)a(2) \cdot \frac{1}{\sqrt{2}}[\alpha(1)\beta(2) - \alpha(2)\beta(1) \equiv \psi_{H^-H^+}$$

$$\psi_4 = \frac{1}{\sqrt{2}}\begin{vmatrix} b(1)\alpha(1) & b(2)\alpha(2) \\ b(1)\beta(1) & b(2)\beta(2) \end{vmatrix} = b(1)b(2) \cdot \frac{1}{\sqrt{2}}[\alpha(1)\beta(2) - \alpha(2)\beta(1) \equiv \psi_{H^+H^-}.$$

The functions ψ_3, ψ_4 and the normalized difference $N_{HL}(\psi_1 - \psi_2) \equiv \psi_{HL}$ (N_{HL} is a normalization factor)

---

[47] W. Heisenberg, *Zeit. Phys., 38,* 411 (1926); *ibid., 39,* 499 (1926); *ibid. 41,* 239 (1927); W. Heitler and F. London, *Zeit. Phys., 44,* 455 (1927); A.C. Hurley, J.E. Lennard-Jones, and J.A. Pople, *Proc. Roy. Soc. London, A220,* 446 (1953).

HEITLER – LONDON FUNCTION (10.26)

$$\psi_{HL} = N_{HL}\left[a(1)b(2) + a(2)b(1)\right] \cdot \frac{1}{\sqrt{2}}\left[\alpha(1)\beta(2) - \alpha(2)\beta(1)\right] \quad (10.27)$$

are eigenfunctions of the operators $\hat{S}^2$ and $\hat{S}_z$ (cf. Appendix Q available at booksite.elsevier.com/978-0-444-59436-5, p. e133) corresponding to the singlet state. The functions ψ_3, ψ_4 for obvious reasons are called *ionic structures* (H^-H^+ and H^+H^-),[48] whereas the function ψ_{HL} is called a *Heitler-London function* or *a covalent structure*.[49]

The VB method relies on optimization of the expansion coefficients c in front of these structures in the Ritz procedure (p. 238):

$$\psi = c_{cov}\psi_{HL} + c_{ion1}\psi_{H^-H^+} + c_{ion2}\psi_{H^+H^-}. \quad (10.28)$$

Fritz Wolfgang London (1900–1954) was born in Breslau (now Wrocław) and studied in Bonn, Frankfurt, Göttingen, Munich (getting his Ph.D. at 21), and Paris. Later, he worked in Zurich, Rome, and Berlin. He escaped from Nazism to the United Kingdom, where he worked at Oxford University (1933–1936). In 1939, London emigrated to the United States, where he became professor of theoretical chemistry at Duke University in Durham, North Carolina.

Fritz London rendered great services to quantum chemistry. He laid the foundations of the theory of the *chemical* (covalent) bond and also introduced dispersion interactions, one of the most important *intermolecular* interactions. This is nearly all of what chemistry is about. He also worked in the field of superconductivity.

The covalent structure itself, ψ_{HL}, was one great success of Walter Heitler[50] and Fritz London. For the first time, the qualitatively correct description of the chemical bond was obtained. The crucial point turned out to be an inclusion, in addition to the product function $a(1)b(2)$, its counterpart *with exchanged electron numbers* $a(2)b(1)$, since the electrons *are*

---

[48] This is because both electrons reside at the same nucleus.

[49] This is because both electrons belong to the same extent to each of the nuclei.

[50] Walter Heitler (1904–1981) was a German chemist and professor at the University in Göttingen, and later he worked in Bristol and Zürich.

indistinguishable. If we expand the Hartree-Fock determinant with doubly occupied bonding orbital $a + b$, we would also obtain a certain linear combination of the three structures mentioned,[51] but with *the constant coefficients independent of the interatomic distance*:

$$\psi_{RHF} = N\left(\frac{1}{N_{HL}}\psi_{HL} + \psi_{H^-H^+} + \psi_{H^+H^-}\right). \quad (10.29)$$

This leads to a very bad description of the H_2 molecule at long internuclear distances with the Hartree-Fock method. Indeed, for long internuclear distances, the Heitler-London function should dominate, because it corresponds to the (correct) dissociation limit (two ground-state hydrogen atoms). The trouble is that with fixed coefficients, *the Hartree-Fock function overestimates the role of the ionic structure* for long interatomic distances. Fig. 10.5 shows that the Heitler-London function describes the electron correlation (Coulomb hole), whereas the Hartree-Fock function does not.

10.9.2 Resonance Theory–Polyatomic Case

The VB method was developed by Linus Pauling with the name of *theory of resonance*.

Linus Carl Pauling (1901–1994), American physicist and chemist; in the years 1931–1964, he was a professor at the California Institute of Technology in Pasadena; in 1967–1969, he was a professor at the University of California, San Diego; and from 1969–1974 at the Stanford University. He received the 1954 Nobel prize "*for his research into the nature of the chemical bond and its application to the elucidation of the structure of complex substances.*" In 1962, he received the Nobel peace prize. His major achievements are the development of the theory of chemical bond–the VB method (also called *resonance theory*), and determining the structure of one of the fundamental structural elements of proteins, the α–helix.

[51] Indeed, the normalized Hartree-Fock determinant [double occupation of the molecular orbital $\varphi_1 = \frac{1}{\sqrt{2(1+S)}}(a+b)$, where the overlap integral between the atomic orbitals $S = (a|b)$] can be rewritten as

$$\begin{aligned}\psi_{HF} &= \frac{1}{\sqrt{2!}}\begin{vmatrix}\varphi_1(1)\alpha(1) & \varphi_1(2)\alpha(2)\\ \varphi_1(1)\beta(1) & \varphi_1(2)\beta(2)\end{vmatrix}\\ &= \frac{1}{2(1+S)}\left[a(1)a(2)+b(1)b(2)+a(1)b(2)+a(2)b(1)\right]\frac{1}{\sqrt{2}}[\alpha(1)\beta(2)-\alpha(2)\beta(1)]\\ &= \frac{1}{2(1+S)}\left[\psi_{H^-H^+}+\psi_{H^+H^-}+\frac{1}{N_{HL}}\psi_{HL}\right].\end{aligned}$$

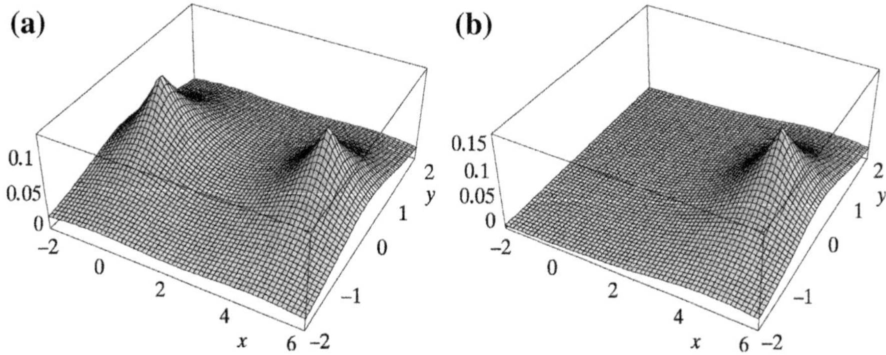

Fig. 10.5. Illustration of electron correlation in the hydrogen molecule. The nuclear positions are (0, 0, 0) and (4, 0, 0) in a.u. Slater orbitals of $1s$ type have an orbital exponent equal to 1. (a) Visualization of the xy cross-section of the wave function of electron 2, assuming that electron 1 resides on the nucleus (either the first or the second one), has spin coordinate $\sigma_1 = \frac{1}{2}$, whereas electron 2 has spin coordinate $\sigma_2 = -\frac{1}{2}$ and the total wave function is equal $\psi = N\{ab + ba + aa + bb\}\{\alpha\beta - \beta\alpha\}$; i.e., it is a Hartree-Fock function. The plot is the same independent of which nucleus electron 1 resides; i.e., we observe the *lack of any correlation* of the motions of electrons 1 and 2. If we assume the spins to be parallel $\left(\sigma_2 = \frac{1}{2}\right)$, the wave function vanishes. (b) A similar plot, but for the Heitler-London function $\psi_{HL} = N_{HL}[a(1)b(2) + a(2)b(1)]\frac{1}{\sqrt{2}}[\alpha(1)\beta(2) - \alpha(2)\beta(1)]$ and with electron 1 residing at nucleus (0, 0, 0). Electron 2 runs to the nucleus in position (4, 0, 0). We have the correlation of the electronic motion. If we assume parallel spins $\left(\sigma_2 = \frac{1}{2}\right)$, the wave function vanishes.

The method can be applied to all molecules, although a particularly useful field of applications of resonance theory can be found in the organic chemistry of aromatic systems. For example, the total electronic wave function of the benzene molecule is presented as a linear combination of resonance structures[52]:

$$\psi = \sum_I c_I \Phi_I, \tag{10.30}$$

and to each one (in addition to the mathematical form), a graph is assigned. For example, six π electrons can participate in the following "adventures" (forming covalent and ionic bonds).

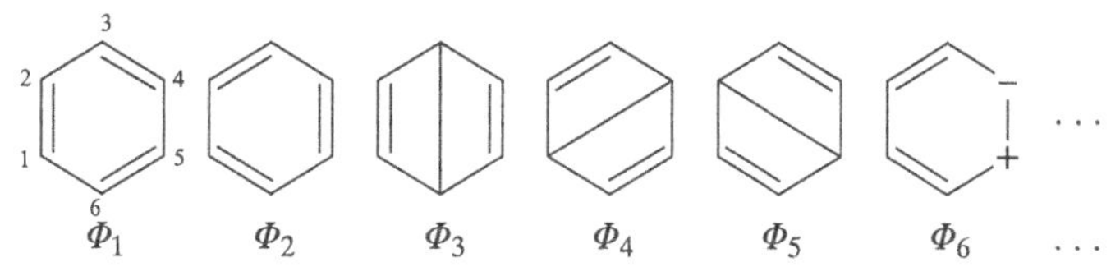

The first two structures are famous Kekulé structures, the next three are Dewar structures, and the sixth is an example of the possible mixed covalent-ionic structures. From these graphs, we may deduce which atomic orbitals (out of the $2p_z$ orbital of carbon atoms, z is perpendicular to the plane of the benzene ring) take part in the covalent bond (of the π type). As far as the mathematical form of the Φ_1 structure is concerned, we can write it as the antisymmetrized

[52] Similar to the original applications, we restrict ourselves to the π electrons and the σ electrons are treated as inactive in each structure, forming, among other things, the six C–C bonds.

(cf. antisymmetrization operator; p. e107) product of three Heitler-London functions (involving the proper pairs of $2p_z$ carbon atomic orbitals), the first for electrons 1, 2, the second for electrons 3, 4, and the third for 5, 6. Within the functions Φ_I, the ionic structures can also occur. The rules for writing the structures were not quite clear, and the electrons were located to some extent in an arbitrary manner, making the impression that it is up to theoretical chemists to use their imaginations and draw imaginary pictures, and next to translate them into mathematical form to obtain–after applying the variational method–an approximation to the wave function (and to the energy).

In fact, the problem is connected to the Ritz method and to expansion into the complete set of functions[53] (i.e., with a purely mathematical problem). Although it may seem very strange to students (fortunately), many people were threatened for supporting the theory of resonance. Scientists serving the totalitarian regime decided to attack Eq. (10.30). Why was this[54]? The Stalinists did not like the idea that "the sum of fictitious structures can describe reality." But wait! If some artificial functions could interfere with reality, then socialist realism loses to abstraction, a kolkhoz (collective farm) member to an intellectual, Lysenkoism to Mendelism,[55] gulags to the idea of freedom, and you are on the brink of disaster (if you are a Stalinist, that is).

---

[53] In principle, *they should* form the complete set, but even so, in practical calculations, we never deal with true complete sets.

[54] Of course, *the true* reason was not a convergence of a series in the Hilbert space, but their personal careers *at any price*. Totalitarian systems never have problems finding such "scientists." In chemistry, there was the danger of losing a job–and in biology, lives were actually at risk.

It is rather difficult to think about Joseph Stalin as a quantum chemist. He was, however, kept informed about the current situation in the group of people involved in carrying out summation in Eq. (10.30); i.e., working in the resonance theory. To encourage young people to value and protect the freedom they have, and to reflect on human nature, some exempts from the resolution adopted by the All Soviet Congress of Chemists of the Soviet Union are reported. The resolution pertains to the theory of resonance (after the disturbing and reflective book by S.E. Shnoll, *Gheroy i zladieyi rossiyskoy nauki* Kron-Press, Moscow, 1997, p. 297):

"*Dear Joseph Vissarionovich (Stalin),*

the participants of the All Soviet Congress send to you, the Great Leader and Teacher of all progressive mankind, our warm and cordial greetings. We Soviet chemists gathered together to decide, by means of broad and free discussion, the fundamental problems of the contemporary theory of the structure of molecules, want to express our deepest gratitude to you for the everyday attention you pay to Soviet science, particularly to chemistry. Our Soviet chemistry is developing in the Stalin era, which offers unlimited possibilities for the progress of science and industry. Your brilliant work in the field of linguistics put the tasks for still swifter progress in front of all scientists of our fatherland (...). Motivated by the resolutions of the Central Committee of the Bolshevik Communist Party concerning ideological matters and by your instructions, Comrade Stalin, the Soviet chemists wage war against the ideological concepts of bourgeois science. The lie of the so called "resonance theory" has been disclosed, and the remains of this idea will be thrown away from the Soviet chemistry. We wish you, our dear Leader and Teacher, good health and many, many years of famous life to the joy and happiness of the whole of progressive mankind (...)."

The events connected with the theory of resonance started in the autumn of 1950 at Moscow University. Quantum chemistry lecturers, Yakov Kivovitch Syrkin and Mirra Yefimovna Diatkina, were attacked. The accusation was about diffusion of the theory of resonance and was launched by former assistants of Syrkin. Since everything was in the hands of the professionals, Syrkin and Diatkina confessed guilty to each of the charges.

[55] Trofim Lysenko (1898–1976), Soviet scientist of enormous political influence, rejected the genetic laws of Mendel. In my seventh-grade biology textbook, virtually only his "theory" was mentioned. As a pupil, I recall wanting to learn this theory. It was impossible to find any information. With difficulty, I finally found something: acorns

Gregor Johann Mendel (1822–1884), modest Moravian monk, from 1843 a member of the Augustinian order at Brno from 1843 on (abbot beginning in 1868). His unusually precise and patient experiments with sweet peas of two colors and seeds of two degrees of smoothness, allowed him to formulate the principal laws of genetics. Only in 1900 were his fundamental results remembered, and since then, the rapid progress of contemporary genetics began.

10.10 Configuration Interaction (CI) Method

In the *configuration interaction* method,[56]

> the variational wave function is a linear combination of Slater determinants constructed from *molecular* spinorbitals, Eq. (10.30): $\psi = \sum_{I=0}^{M} c_I \Phi_I$.

In most cases, we are interested in the function ψ for *the electronic ground state of the system* (in addition, when solving the CI equations we also get approximations to the excited states with different values of the c_I coefficients).

Generally, we construct the Slater determinants Φ_I by placing electrons on the molecular spinorbitals obtained with the Hartree-Fock method,[57] in most cases, the set of determinants is also limited by imposing an upper bound for the orbital energy. In that case, the expansion in Eq. (10.30) is finite. The Slater determinants Φ_I are obtained by the replacement of occupied spinorbitals with virtual ones in the single Slater determinant, which is the Hartree-Fock function

should be placed in a hole in the ground in large numbers to permit something like the class struggle. The winner will be the strongest oak-tree, which is what we all want.

[56] This is also called the method of superposition of configurations or configuration mixing.

[57] In this method, we obtain M molecular orbitals; i.e., $2M$ molecular spinorbitals, where M is the number of atomic orbitals employed. The Hartree-Fock determinant Φ_0 is the best form of wave function so long as the electronic correlation is not important. The criterion of this "*goodness*" is the mean value of the Hamiltonian. If we want to include the electron correlation, we may think of another form of the 1-D function more suitable as the starting point. We do not change our definition of correlation energy; i.e., we consider the RHF energy as that which does not contain any correlation effects. For instance, we may ask which of the normalized single-determinant functions Φ is closest to the normalized exact function ψ. As a measure of this, we might use:

$$|\langle\psi|\Phi\rangle| = \text{maximum}. \tag{10.31}$$

The single determinantal function $\Phi = \Phi_B$, which fulfills the above condition, is called a Bruckner function (O. Sinanoğlu and K.A. Brueckner *Three Approaches to Electron Correlation in Atoms* Yale Univ. Press, New Haven and London, 1970).

(Φ_0; i.e., ψ_{RHF}) in most cases. When one spinorbital is replaced, the resulting determinant is called singly excited, when two it is doubly excited, etc.[58,59]

The virtual spinorbitals form an orthonormal basis in *the virtual space*. If we carry out any non-singular linear transformation (cf. p. 467) of virtual spinorbitals, each "new" n-tuply excited Slater determinant becomes a linear combination of all "old" n-tuply excited determinants and only n-tuply excited ones.[60] In particular, the unitary transformation would preserve the mutual orthogonality of the n-tuply excited determinantal functions.

Thus, the total wave function [Eq. (10.30)] is a linear combination of the *known* Slater determinants (we assume that the spinorbitals are always known) with *unknown c* coefficients.

The name of the CI methods refers to the linear combination of the configurations rather than to the Slater determinants.

A configuration (i.e., a *configuration state function*, or *CSF*) is a linear combination of determinants that is an eigenfunction of the operators: $\hat{S}^2$ and $\hat{S}_z$, and belongs to the proper irreducible representation of the symmetry group of the Hamiltonian. We say that this is a linear combination of the (spatial and spin) symmetry adapted determinants. Sometimes we refer to the spin-adapted configurations, which are eigenfunctions only of the $\hat{S}^2$ and $\hat{S}_z$ operators.

The particular terms in the CI expansion may refer to the respective CSFs or to the Slater determinants. Both versions lead to the same results, but using CSFs may be more efficient

---

[58] In the language of the second quantization (see Appendix U available at booksite.elsevier.com/978-0-444-59436-5, p. e153), the wave function in the CI method has the form (the Φ_0 function is a Slater determinant which does not necessarily need to be a Hartree-Fock determinant):

$$\psi = c_0\Phi_0 + \sum_{a,p} c_p^a \hat{p}^\dagger \hat{a}\Phi_0 + \sum_{a<b,p<q} c_{pq}^{ab} \hat{q}^\dagger \hat{p}^\dagger \hat{a}\hat{b}\Phi_0 + \text{higher excitations}, \tag{10.32}$$

where c are the expansion coefficients, the creation operators $\hat{q}^\dagger, \hat{p}^\dagger, \ldots$ refer to the virtual spinorbitals $\phi_p, \phi_q, \ldots$ and the annihilation operators $\hat{a}, \hat{b}, \ldots$ refer to occupied spinorbitals $\phi_a, \phi_b, \ldots$ (the operators are denoted with the same indices as spinorbitals but the former are equipped with hat symbols), and the inequalities satisfied by the summation indices ensure that the given Slater determinant occurs only once in the expansion.

[59] The Hilbert space corresponding to N electrons is the sum of the orthogonal subspaces $\Omega_n, n = 0, 1, 2, \ldots N$, which are spanned by the n-tuply excited (orthonormal) Slater determinants. Elements of the space Ω_n are all linear combinations of n-tuply excited Slater determinants. It does not mean, of course, that each element of this space is an n-tuply excited Slater determinant. For example, the sum of two doubly excited Slater determinants is a doubly excited Slater determinant only when one of the excitations is common to both determinants.

[60] Indeed, the Laplace expansion (see Appendix A available at booksite.elsevier.com/978-0-444-59436-5) along the row corresponding to the first new virtual spinorbital leads to the linear combination of the determinants containing new (*virtual, which means that the rank of excitation is not changed by this*) orbitals in this row. Continuing this procedure with the Slater determinants obtained, we finally get a linear combination of n-tuply excited Slater determinants expressed in old spinorbitals.

if we are looking for a wave function that transforms itself according to a single irreducible representation.

Next, this problem is reduced to the Ritz method (see Appendices L, p. e107 and K, p. e105), and subsequently to the secular equations $(\boldsymbol{H} - \epsilon \boldsymbol{S})\,\boldsymbol{c} = \boldsymbol{0}$. It is worth noting here that, e.g., the CI wave function for the ground state of the helium atom would be linear combinations of the determinants where the largest c coefficient occurs in front of the Φ_0 determinant constructed (say from the spinorbitals $1s\alpha$ and $1s\beta$), but the nonzero contribution would also come from the other determinants constructed from the $2s\alpha$ and $2s\beta$ spinorbitals (one of the doubly excited determinants). The CI wave functions for all states (ground and excited) are linear combinations of *the same Slater determinants*; they differ only in the c coefficients.

The state energies obtained from the solution of the secular equations always approach the exact values from above.

10.10.1 Brillouin Theorem

In the CI method, we have to calculate matrix elements H_{IJ} of the Hamiltonian.

The Brillouin theorem says that

$$\langle \Phi_0 | \hat{H} \Phi_1 \rangle = 0 \tag{10.33}$$

if Φ_0 is a solution of the Hartree-Fock problem ($\Phi_0 \equiv \psi_{RHF}$), and Φ_1 is a singly excited Slater determinant in which the spinorbital $\phi_{i'}$ is orthogonal to all spinorbitals used in Φ_0.

Proof:
From the second Slater-Condon rule (see Appendix M available at booksite.elsevier.com/ 978-0-444-59436-5 p. e109), we have

$$\langle \Phi_0 | \hat{H} \Phi_1 \rangle = \langle i | \hat{h} i' \rangle + \sum_j \left[\langle ij | i'j \rangle - \langle ij | ji' \rangle \right]. \tag{10.34}$$

On the other hand, considering the integral $\langle i | \hat{F} i' \rangle$, where $\hat{F}$ is a Fock operator, we obtain from 8.28 (using the definition of the Coulomb and exchange operators from p. 403):

$$\langle i | \hat{F} i' \rangle = \langle i | \hat{h} i' \rangle + \sum_j [\langle i | \hat{J}_j i' \rangle - \langle i | \hat{K}_j i' \rangle] = \langle i | \hat{h} i' \rangle + \sum_j [\langle ij | i'j \rangle - \langle ij | ji' \rangle] = \langle \Phi_0 | \hat{H} \Phi_1 \rangle.$$

From the Hermitian character of $\hat{F}$, it follows that

$$\langle i | \hat{F} i' \rangle = \langle \hat{F} i | i' \rangle = \varepsilon_i \delta_{ii'} = 0. \tag{10.35}$$

We have proved the theorem.

The Brillouin theorem is sometimes useful in discussion of the importance of particular terms in the CI expansion for the ground state.

10.10.2 Convergence of the CI Expansion

Increasing the number of expansion functions by adding a new function lowers the energy (due to the variational principle). It often happens that the inclusion of only two determinants gives qualitative improvement with respect to the Hartree-Fock method; however, when going further, the situation becomes more difficult. The convergence of the CI expansion is slow (i.e., to achieve a good approximation to the wave function), the number of determinants in the expansion must usually be large. Theoretically, the shape of the wave function ensures solution of the Schrödinger equation $H\psi = E\psi$, but in practice, we are *always* limited by the basis of the atomic orbitals employed.

> To obtain satisfactory results, we need to increase the number M of atomic orbitals in the basis. The number of molecular orbitals produced by the Hartree-Fock method is also equal to M, hence the number of spinorbitals is equal to $2M$. In this case, the number of all determinants is equal to $\binom{2M}{N}$, where N refers to the number of electrons.

10.10.3 Example of H_2O

We are interested in the ground state of the water molecule, which is a singlet state ($S = 0$, $M_S = 0$).

The minimal basis set, composed of seven atomic orbitals (two $1s$ orbitals of the hydrogen atoms, $1s$, $2s$, and three $2p$ orbitals of the oxygen atom), is considered too poor; therefore, we prefer what is called the *double dzeta basis*, which provides two functions with different exponents for each orbital of the minimal basis. This creates a basis of $M = 14$ atomic orbitals. There are 10 electrons, so $\binom{28}{10}$ gives 13 million Slater determinants. For a matrix of that size to be diagonalized is certainly impressive. Even more impressive is that we achieve only *an approximation* to the correlation energy which amounts to about 50% of the exact correlation energy,[61] since M is only equal to 14, but in principle, it should be equal to ∞. Nevertheless, for *comparative* purposes, we assume that the correlation energy obtained is 100%.

The simplest remedy is to get rid of some determinants in such a way that the correlation energy is not damaged. Which ones? Well, many of them correspond to the incorrect projection S_z of the total spin or the incorrect total spin S. For instance, we are interested in the singlet state (i.e., $S = 0$ and $S_z = 0$), but some determinants are built of spinorbitals containing exclusively α spin functions. This is a pure waste of resources, since the non-singlet functions

[61] We see here how vicious the dragon of electron correlation is.

do not make any contributions to the singlet state. When we remove these and other *incorrect* determinants, we obtain a smaller matrix to be diagonalized. The number of Slater determinants with $S_z = 0$ equals $\binom{M}{N/2}^2$. In our case, this makes slightly over 4 million determinants (instead of 13 million). What would happen if we diagonalized the huge original matrix anyway? Well, nothing would happen. There would be more work, but the computer would create *the block form*[62] from our enormous matrix, and each would correspond to the particular S^2 and S_z, *while the whole contribution to the correlation energy of the ground state comes from the block corresponding to* $S = 0$ and $S_z = 0$.

Let us continue throwing away determinants. This time, however, we have to make a compromise; i.e., some of the Slater determinants are arbitrarily considered not to be important (which will worsen the results, if they are rejected). Which of the determinants should be considered as not important? The general opinion in quantum chemistry is that the multiple excitations are less and less important (when the multiplicity increases). If we take only the singly, doubly, triply, and quadruply excited determinants, the number of determinants will reduce to 25000 and we will obtain 99% of the approximate correlation energy defined above. If we take the singly and doubly excited determinants only, there are only 360 of them, and 94% of the correlation effect is obtained. This is why this *CI Singles and Doubles (CISD)* method is used so often.

For larger molecules, this selection of determinants becomes too demanding, therefore we have to decide individually for each configuration: to include or reject it? The decision is made either on the basis of the perturbational estimate of the importance of the determinant[63] or by a test calculation with inclusion of the determinant in question (see Fig. 10.6).

To obtain good results, we need to include a large number of determinants (e.g., of the order of thousands, millions, or even billions). This means that contemporary quantum chemistry has made enormous technical progress.[64] This, however, is a sign, not of the strength of quantum chemistry, but of its weakness. What are we going to do with such a function? We may load it back into the computer and calculate all the properties of the system with high accuracy (although this cannot be guaranteed). To answer the question about why we obtained some particular numbers, we have to answer that we do not know–it is the computer that knows. This is a trap. It would be better to get, say, two Slater determinants, which describe the system to a reasonable approximation, and we can understand what is going on in the molecule.

---

[62] These square blocks would be easily noticed after proper ordering of the expansion functions.

[63] The perturbational estimate mentioned relies on the calculation of the weight of the determinant based on the first-order correction to the wave function in perturbation theory (p. 245). In such an estimate, the denominator contains the excitation energy evaluated as the difference in orbital energies between the Hartree-Fock determinant and the one in question. In the numerator, there is a respective matrix element of the Hamiltonian calculated with the help of the known Slater-Condon rules (see Appendix M available at booksite.elsevier.com/978-0-444-59436-5, p. e109).

[64] To meet such needs, quantum chemists have had to develop entirely new techniques of applied mathematics.

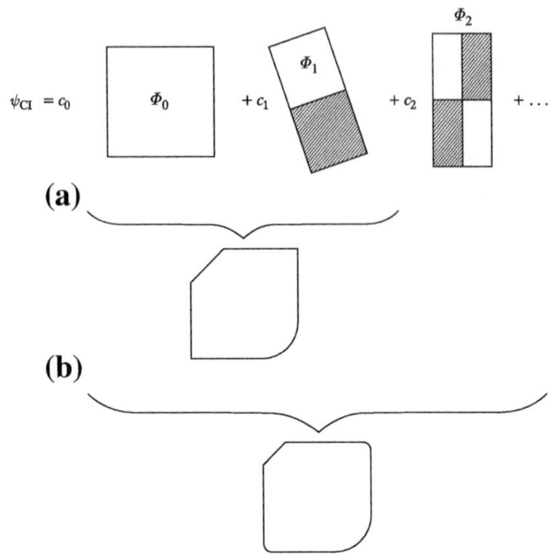

Fig. 10.6. Symbolic illustration of the principle of the CI method with one Slater determinant Φ_0 dominant in the ground state (this is a problem of the many electron wave functions so the picture cannot be understood literally). The purpose of this diagram is to emphasize a relatively small role of electronic correlation (more exactly, of what is known as the *dynamical correlation*; i.e., correlation of electronic motion). The function ψ_{CI} is a linear combination (the c coefficients) of the determinantal functions of different shapes in the many-electron Hilbert space. The shaded regions correspond to the negative sign of the function; the nodal surfaces of the added functions allow for the effective deformation of ψ_0 to have lower and lower average energy. (a) Since c_1 is small in comparison to c_0, the result of the addition of the first two terms is a slightly deformed ψ_0. (b) Similarly, the additional excitations just make cosmetic changes in the function (although they may substantially affect the quantities calculated with it).

10.10.4 Which Excitations Are Most Important?

The convergence can be particularly bad if we use the virtual spinorbitals obtained by the Hartree-Fock method. Not all excitations are equally important. It turns out that usually, although this is not a rule, low excitations dominate the ground-state wave function.[65] The single excitations *themselves* do not contribute anything to the ground-state *energy* (if the spinorbitals are generated with the Hartree-Fock method, then the Brillouin theorem mentioned above applies). *They are crucial, however, for excited states or in dipole moment calculations.* For the ground state, only when coupled to other types of excitation do they assume nonzero (although small) contribution. Indeed, if in the CI expansion we only use the Hartree-Fock determinant and the determinants corresponding to single excitations, then, due to the Brillouin theorem, the secular determinant would be factorized.[66] This factorization (Fig. 10.7) pertains to the single determinant corresponding to the Hartree-Fock function and to the determinants corresponding exclusively to single excitations. Since we are interested in the ground state, only the first determinant

[65] That is, those that require the lowest excitation energies. Later, a psychological mechanism began to work supported by economics: the *high-energy* excitations are numerous and, because of that, very expensive and they correspond to a high number of electrons excited. Due to this, a reasonable restriction for the number of configurations in the CI expansion is excitation rank. We will come back to this problem later.

[66] That is, it could be written in block form, which would separate the problem into several smaller subproblems.

$H =$

	HF	S	D	T	Q	...
HF	E_{HF}	0^a	III	0^b	0^b	0^b
S	0^a	block S	II	III	0^b	0^b
D	III	II	block D	II	III	0^b
T	0^b	III	II	block T	II	III
Q	0^b	0^b	III	II	block Q	II
⋮	0^b	0^b	0^b	III	II	block ...

Fig. 10.7. The block structure of the Hamiltonian matrix (***H***) is the result of the Slater-Condon rules (see Appendix M available at booksite.elsevier.com/978-0-444-59436-5, p. e119). S indicates single excitations, D indicates double excitations, T indicates triple excitations, and Q indicates quadruple excitations. (a) A block of zero values due to the Brillouin theorem. (b) A block of zero values due to the fourth Slater-Condon rule, (II) the nonzero block obtained according to the second and third Slater-Condon rules, (III) the nonzero block obtained according to the third Slater-Condon rule. All the nonzero blocks are sparse matrices dominated by zero values, which is important in the diagonalization process.

is of importance to us, and the result does not change whether we include or not a contribution coming from single excitations into the wave function.

Usually, performing CI calculations with the inclusion of all excitations (for the assumed value of M; i.e., the *full CI*), is not possible in practical calculations due to the extremely long expansion. We are forced to truncate the CI basis somewhere. It would be good to terminate it in such a way that all *essential* terms are retained. The problem with this, however, is determining what we mean by *essential. The most significant terms for the correlation energy come from the double excitations since these are the first excitations coupled to the Hartree-Fock function.* Smaller, although important, contributions come from other excitations (usually of low excitation rank). We certainly wish that it would be like this for large molecules. Nobody knows what the truth is.

10.10.5 Natural Orbitals (NOs)

The fastest convergence is achieved in the basis set of *natural* orbitals (NOs); i.e., when we construct spinorbitals with *these* orbitals and from them the Slater determinants. The NO is defined *a posteriori* in the following way. After carrying out the CI calculations, we construct the density ρ (see Appendix S available at booksite.elsevier.com/978-0-444-59436-5, p. e143) as follows:

$$\rho(1) = N\int \psi^*(1,2,3,\ldots N)\psi(1,2,3,\ldots,N)\mathrm{d}\tau_2\mathrm{d}\tau_3\cdots\mathrm{d}\tau_N$$
$$= \sum_{ij} D_{ji}\phi_i^*(1)\phi_j(1); \quad D_{ij} = D_{ji}^*, \tag{10.36}$$

where the summation runs over all the spinorbitals. By diagonalization of matrix $\boldsymbol{D}$ (a rotation in the Hilbert space spanned by the spinorbitals), we obtain the density expressed in the natural spinorbitals (NOs) transformed by the unitary transformation

$$\rho(1) = \sum_i (D_{diag})_{ii} \phi_i'^*(1)\phi_i'(1). \tag{10.37}$$

The most important ϕ'_i from the viewpoint of the correlation are the NOs with large *occupancies*; i.e., $(D_{diag})_{ii}$ values. Inclusion of only the most important ϕ'_i in the CI expansion creates a short and quite satisfactory wave function.[67]

10.10.6 Size Inconsistency of the CI Expansion

A truncated CI expansion has one unpleasant feature that affects the applicability of the method.

Let us imagine that we want to calculate the interaction energy of two beryllium atoms, and that we decide that to describe the beryllium atom, we have to include not only the $1s^2 2s^2$ configuration, but also the doubly excited one, $1s^2 2p^2$. In the case of beryllium, this is a very reasonable step, since both configurations have very close energies. Let us assume now that we calculate the wave function for *two* beryllium atoms. If we want this function to describe the system correctly, also at large interatomic distances, we have to make sure that the departing atoms have appropriate excitations at their disposition (i.e., in our case $1s^2 2p^2$ for each). To achieve this, we *must incorporate quadruple excitations into the method.*[68]

If we include quadruples, we have a chance to achieve (an approximate) size consistency; i.e., the energy will be proportional to good accuracy to the number of atoms, or else our results will not be size consistent.

Let us imagine 10 beryllium atoms. In order to have size consistency we need to include 20-fold excitations. This would be very expensive. We clearly see that, for many systems, the size consistency requires inclusion of multiple excitations. If we carried out CI calculations for all possible (for a given number of spinorbitals) excitations, such a CI method (i.e., *FCI*) would be size consistent.

10.11 Direct CI Method

We have already mentioned that the CI method converges slowly. Due to this, the Hamiltonian matrices and overlap integral matrices are sometimes so large that they cannot fit into the

[67] Approximate natural orbitals can also be obtained directly without performing the CI calculations.

[68] See J.A. Pople, R. Seeger, and R. Krishnan, *Intern. J. Quantum Chem. S11,* 149 (1977), as well as p. 47 of the book by P. Jørgensen and J. Simons *Second Quantization–Based Methods in Quantum Chemistry* Academic Press (1981).

computer memory. In practice, such a situation occurs in all high-quality calculations for small systems and in all calculations for medium and large systems. Even for quite large atomic orbital basis, the number of integrals is much smaller than the number of Slater determinants in the CI expansion.

Björn Roos[69] first noticed that to find the lowest eigenvalues and their eigenvectors, we do not need to store a huge $\boldsymbol{H}$ matrix in computer memory. Instead, we need to calculate the *residual vector* $\boldsymbol{\sigma} = (\boldsymbol{H} - E\boldsymbol{1})\boldsymbol{c}$, where $\boldsymbol{c}$ is a trial vector (defining the trial function in the variational method, p. 232). If $\boldsymbol{\sigma} = \boldsymbol{0}$, it means that the solution is found. Knowing $\boldsymbol{\sigma}$, we may find (on the basis of first-order perturbation theory) a slightly improved $\boldsymbol{c}$, etc. The product $\boldsymbol{Hc}$ can be obtained by going through the set of integrals and assigning to each a coefficient resulting from $\boldsymbol{H}$ and $\boldsymbol{c}$, and next adding the results to the new $\boldsymbol{c}$ vector. Then the procedure is repeated. Until 1971, CI calculations with 5000 configurations were considered a significant achievement. After Roos's paper, there was a leap of several orders of magnitude, bringing the number of configurations to the range of billions. For the computational method, this was a revolution.

10.12 Multireference CI Method

Usually in the CI expansion, the dominant determinant is Hartree-Fock. We construct the CI expansion, replacing the spinorbitals in this determinant (*single reference method*). We can easily imagine a situation in which taking one determinant is not justified, since the shell is not well closed (e.g., four hydrogen atoms). We already know that certain determinants (or, in other words: configurations) absolutely need be present ("*static correlation*") in the correct wave function. To be sure, we are the judges, deciding which is good or bad. This set of determinants is a basis in the *model space*.

In the single reference CI method, the model space (Fig. 10.8) is formed by a single Slater determinant. In the multireference CI method, the set of determinants constitute the model space. This time, the CI expansion is obtained by replacement of the spinorbitals participating in the model space by other virtual orbitals. We proceed further as in CI.

There is no end to the problems yet, since again, we have billions of possible excitations.[70] We do other tricks to survive in this situation. We may, for instance, get the idea not to excite

[69] B.O. Roos, *Chem. Phys. Letters, 15,* 153 (1972).

[70] There is another trouble too called *intruder states*; i.e., states that are of unexpectedly low energy. How could these states appear? First, the CI states known as "*front-door intruders*" appear if some important (low-energy) configurations were for some reason not included into the model space. Second, we may have the "*back-door intruder*" states, when the energy gap between the model space and the other configurations was too small (quasi-degeneracy), and some CI states became of low energy (enter the model space energy zone) even if they are not composed of the model space configurations.

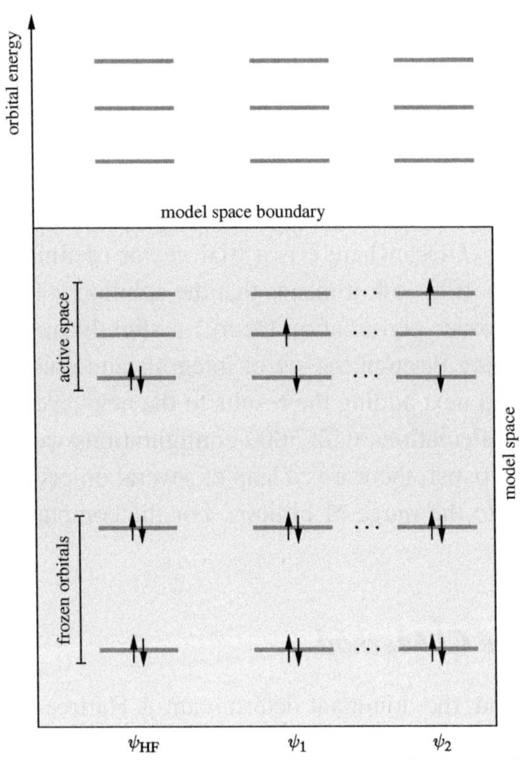

Fig. 10.8. Illustration of the model space in the multireference CI method used mainly in the situation when no single Slater determinant dominates the CI expansion. The orbital levels of the system are presented here. Part of them are occupied in all Slater determinants considered ("*frozen spinorbitals*"). Above them is a region of closely spaced orbital levels called *active space*. In the optimal case, a significantly large energy gap occurs between the latter and unoccupied levels lying higher. The model space is spanned by all or some of the Slater determinants obtained by various occupancies of the active space levels.

the inner-shell orbitals, since the numerical effort is serious, the lowering of the total energy can also be large, but the effect on the energy *differences* (this is what chemists are usually interested in) is negligible. We say that such orbitals are *frozen*. Some of the orbitals are kept doubly occupied in all Slater determinants but we optimize their shape. Such orbitals are called *inactive*. Finally, the orbitals of varied occupancy in different Slater determinants are called *active*. The frozen orbitals are, in our method, important spectators of the drama, the inactive orbitals contribute a little toward lowering the energy, but the most efficient work is done by the active orbitals.

10.13 Multiconfigurational Self-Consistent Field Method (MC SCF)

In the configuration interaction method, it is sometimes obvious that certain determinants of the CI expansion *must* contribute to the wave function if the latter is to correctly describe the system. For example, if we want to describe the system in which a bond is being broken (or is being formed), for its description, we need several determinants for sure (cf. the description of the dissociation of the hydrogen molecule on p. 437).

Why is this? In the case of the dissociation with which we are dealing here, there is a quasidegeneracy of the bonding and antibonding orbital of the bond in question; i.e., the approximate equality of their energies (the bond energy is of the order of the overlap integral and the latter goes to zero when the bond is being broken). The determinants, which can be constructed by various occupancies of these orbitals, have very close energies and, consequently, their contributions to the total wave function are of similar magnitude and *should be included* in the wave function.

In the multiconfigurational self-consistent field (MC SCF) method, as in CI, it is *up to us* to decide which set of determinants we consider sufficient for the description of the system.

> Each of the determinants is constructed from molecular spinorbitals that are not fixed (as in the CI method) but are modified in such a way as to have the total energy as low as possible.

The MC SCF method is the most general scheme of the methods that use a linear combination of Slater determinants as an approximation to the wave function. In the limiting case of the MC SCF, when the number of determinants is equal to 1, we have, of course, the Hartree-Fock method.

10.13.1 Classical MC SCF Approach

We will describe first the classical MC SCF approach, which is a variational method. As was mentioned, the wave function in this method has the form of a finite linear combination of Slater determinants Φ_I:

$$\psi = \sum_I d_I \Phi_I, \tag{10.38}$$

where d are variational coefficients.

> In the classical MC SCF method, we do the following:
>
> 1. Take a finite CI expansion (the Slater determinants and the orbitals for their construction are fixed)
> 2. Calculate the coefficients for the determinants by the Ritz method (the orbitals do not change)
> 3. Vary the LCAO coefficients in the orbitals at the fixed CI coefficients to obtain the best MOs
> 4. Return to point 1 until self-consistency is achieved

10.13.2 Unitary MC SCF Method

Another version of the MC SCF problem, *a unitary method* suggested by Lévy and Berthier[71] and later developed by Daalgaard and Jørgensen,[72] is gaining increasing importance. The eigenvalue problem does not appear in this method.

We need two mathematical facts to present the unitary MC SCF method. The first is a theorem: If $\hat{A}$ is a Hermitian operator (i.e., $\hat{A}^\dagger = \hat{A}$), then $\hat{U} = \exp(i\hat{A})$ is a unitary operator satisfying $\hat{U}^\dagger\hat{U} = 1$.

Let us see how $\hat{U}^\dagger$ looks:

$$\begin{aligned}\hat{U}^\dagger = \left(\exp(i\hat{A})\right)^\dagger &= \left(1 + i\hat{A} + \frac{1}{2!}(i\hat{A})^2 + \frac{1}{3!}(i\hat{A})^3 + \cdots\right)^\dagger \\ &= \left(1 + (-i)\hat{A}^\dagger + \frac{1}{2!}(-i\hat{A}^\dagger)^2 + \frac{1}{3!}(-i\hat{A}^\dagger)^3 + \cdots\right) \\ &= \left(1 + (-i)\hat{A} + \frac{1}{2!}(-i\hat{A})^2 + \frac{1}{3!}(-i\hat{A})^3 + \cdots\right) = \exp(-i\hat{A})\end{aligned}$$

Hence, $\hat{U}\hat{U}^\dagger = 1$; i.e., $\hat{U}$ is a unitary operator.[73]

Now we will look at the second mathematical fact, which is a commutator expansion:

$$e^{-\hat{A}}\hat{H}e^{\hat{A}} = \hat{H} + [\hat{H}, \hat{A}] + \frac{1}{2!}[[\hat{H}, \hat{A}], \hat{A}] + \frac{1}{3!}[[[\hat{H}, \hat{A}], \hat{A}], \hat{A}] + \cdots \tag{10.39}$$

This theorem can be proved by induction, expanding the exponential functions.

Now we are all set to describe the unitary method. We introduce two new operators:

$$\hat{\lambda} = \sum_{ij} \lambda_{ij}\hat{i}^\dagger\hat{j}, \tag{10.40}$$

where $\hat{i}^\dagger$ and $\hat{j}$ are the creation and annihilation operators, respectively, associated to spinorbitals i, j (see Appendix U available at booksite.elsevier.com/978-0-444-59436-5). Further,

$$\hat{S} = \sum_{IJ} S_{IJ}|\Phi_I\rangle\langle\Phi_J|. \tag{10.41}$$

---

[71] B. Lévy and G. Berthier, *Intern. J. Quantum Chem.*, *2*, 397 (1968).

[72] E. Dalgaard and P. Jørgensen, *J. Chem. Phys.*, *69*, 3833 (1978).

[73] Is an operator ($\hat{C}$) of multiplication by a constant Hermitian? Let us see: $\langle\varphi|\hat{C}\psi\rangle \stackrel{?}{=} \langle\hat{C}\varphi|\psi\rangle$; l.h.s = $\langle\varphi|c\psi\rangle = c\langle\varphi|\psi\rangle$; r.h.s. = $\langle c\varphi|\psi\rangle = c^*\langle\varphi|\psi\rangle$. l.h.s. = r.h.s., if $c = c^*$. An operator conjugate to c is, therefore, c^*. Further, if $\hat{B} = i\hat{A}$, what is a form of $\hat{B}^\dagger$? We have $\langle\hat{B}^\dagger\varphi|\psi\rangle = \langle\varphi\hat{B}|\psi\rangle$, then $\langle\varphi|i\hat{A}|\psi\rangle = \langle -i\hat{A}^\dagger\varphi|\psi\rangle$, and finally $\hat{B}^\dagger = -i\hat{A}^\dagger$.

We assume that λ_{ij} and S_{IJ} are elements of the Hermitian matrices[74] (their determination is the goal of the whole method), and Φ_I are determinants from the MC SCF expansion [Eq. (10.38)].

It can be seen that the $\hat{\lambda}$ operator replaces a single spinorbital in a Slater determinant and forms a linear combination of such modified determinantal functions; the $\hat{S}$ operator replaces such a combination with another. The "*knobs*" that control these changes are coefficients λ_{ij} and S_{IJ}.

We will need the unitary transformations $\exp(i\hat{\lambda})$ and $\exp(i\hat{S})$. They are very convenient, since when starting from some set of the orthonormal functions (spinorbitals or Slater determinants) and applying this transformation, we always retain the orthonormality of new spinorbitals (due to $\hat{\lambda}$) and of the linear combination of determinants (due to $\hat{S}$). This is an analogy to the rotation of the Cartesian coordinate system. It follows from the above equations that $\exp(i\hat{\lambda})$ modifies spinorbitals (i.e., operates in the one-electron space), and $\exp(i\hat{S})$ rotates the determinants in the space of many-electron functions.

Now we suggest the form of our variational function for the ground state:

$$|\tilde{0}\rangle = \exp(i\hat{\lambda})\exp(i\hat{S})|0\rangle, \tag{10.42}$$

where $|0\rangle$ denotes a starting combination of determinants with specific spinorbitals and the matrices $\boldsymbol{\lambda}$ and $\boldsymbol{S}$ contain the variational parameters. So, we modify the spinorbitals and change the coefficients in front of the determinants to obtain a new combination of the modified determinants, $|\tilde{0}\rangle$. The mean energy value for that function is[75]

$$E = \langle\tilde{0}|\hat{H}|\tilde{0}\rangle = \langle 0|\exp(-i\hat{S})\exp(-i\hat{\lambda})\hat{H}\exp(i\hat{\lambda})\exp(i\hat{S})|0\rangle, \tag{10.43}$$

Taking advantage of the commutator expansion [Eq. (10.39)], we have

$$E = \langle 0|\hat{H}|0\rangle - i\langle 0|[\hat{S}+\hat{\lambda}, \hat{H}]|0\rangle + \frac{1}{2}\langle 0|[\hat{S}, [\hat{H}, \hat{S}]]|0\rangle + \frac{1}{2}\langle 0|[\hat{\lambda}, [\hat{H}, \hat{\lambda}]]|0\rangle$$
$$+\langle 0|[\hat{S}, [\hat{H}, \hat{\lambda}]]|0\rangle + \cdots$$

It follows from the last equation that in order to calculate E, we have to know the result of the operation of $\hat{\lambda}$ on $|0\rangle$ (i.e., on the linear combination of determinants), which comes down to the operation of the creation and annihilation operators on the determinants, which is simple. It can also be seen that we need to apply the operator $\hat{S}$ to $|0\rangle$, but its definition shows that this

[74] Considering the matrix elements of the operators $\hat{\lambda}$ and $\hat{S}$, we would easily be convinced that both operators are also Hermitian.

[75] Here, we use the equality $[\exp(i\hat{A})]^{\dagger} = \exp(-i\hat{A})$.

is trivial. This expression[76] can now be optimized; i.e., the best Hermitian matrices $\boldsymbol{\lambda}$ and $\boldsymbol{S}$ can be selected. It is done in the same step (this distinguishes the current method from the classical one). Usually the calculations are carried out in the matrix form, neglecting the higher terms and retaining only the quadratic ones in $\hat{S}$ and $\hat{\lambda}$. Neglecting the higher terms is equivalent to allowing for very small rotations in Eq. (10.42), but instead we have a large number of rotations (iterative solution).[77]

The success of the method depends on the starting point. The latter strongly affects the energy and its hypersurface (in the space of the parameters of the matrices $\boldsymbol{\lambda}$ and $\boldsymbol{S}$) is very complicated, it has many local minima. This problem is not yet solved, but various procedures accelerating the convergence are applied; e.g., the new starting point is obtained by averaging the starting points of previous iterations. The method also has other problems, since the orbital rotations partially replace the rotation in the space of the Slater determinants (the rotations do not commute and are not independent). In consequence, linear dependencies may appear.

10.14 Complete Active Space SCF (CAS SCF) Method

An important special case of the MC SCF method is the complete active space SCF (CAS SCF) method of Roos, Taylor, and Siegbahn (see Fig. 10.9).[78] Let us assume that we are dealing with a closed-shell molecule. The RHF method (p. 394) provides the molecular orbitals and the orbital energies. From them, we select the low-energy orbitals. Part of them are *inactive*; i.e., they are doubly occupied in all determinants, but they are varied, which results in lowering the mean value of the Hamiltonian (some of the orbitals may be frozen–i.e., kept unchanged). These are the spinorbitals corresponding to the inner shells. The remaining spinorbitals belong to the *active space*. Now we consider all possible occupancies and excitations of the active spinorbitals (this is where the adjective *complete* comes from) to obtain the set of determinants in the expansion of the MC SCF. By taking all possible excitations within the active space, we achieve *a size consistency*; i.e., when dividing the system into subsystems and separating them (infinite distances) we obtain the sum of the energies calculated for each subsystem separately. By taking the complete set of excitations, we also determine that the results do not depend on any (non-singular) linear transformation of the molecular spinorbitals within the given subgroup

[76] The term with i gives a real number

$$i \cdot \langle 0|[\hat{S}+\hat{\lambda}, \hat{H}]|0\rangle = i \cdot \left(\langle(\hat{S}+\hat{\lambda})0|\hat{H}0\rangle - \langle\hat{H}0|(\hat{S}+\hat{\lambda})0\rangle\right) \rightarrow i \cdot (z - z^*) = i(2i\mathrm{Im}z) \in R,$$

where R is a set of real numbers.

[77] In the classical MC SCF method, when minimizing the energy with respect to the parameters, we use only linear terms in the expansion of the energy with respect to these parameters. In the unitary formulation, on the other hand, we use both linear and quadratic terms. This implies much better convergence of the unitary method.

[78] B.O. Roos and P.E.M. Siegbahn, in *Modern Theoretical Chemistry* vol. III, ed. H.F. Schaefer, Plenum Press, New York (1977); P.E.M. Siegbahn, *J. Chem. Phys., 70,* 5391 (1979); B.O. Roos, P.R. Taylor and P.E.M. Siegbahn, *Chem. Phys., 48,* 157 (1980).

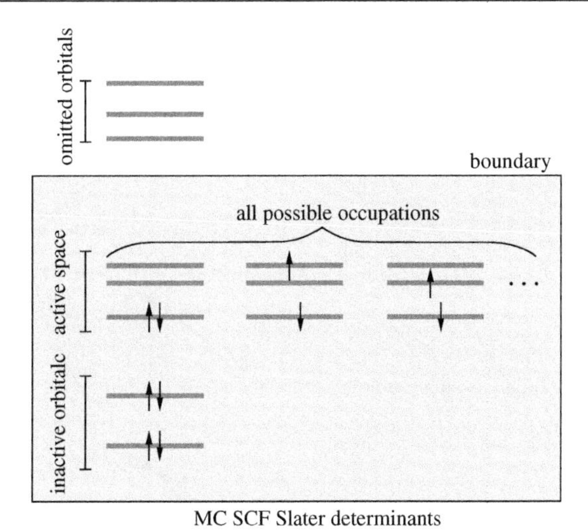

Fig. 10.9. CAS SCF, a method of construction of the Slater determinants in the MC SCF expansion. The inner-shell orbitals are usually inactive. From the active space + inactive spinorbitals, we create the complete set of possible Slater determinants to be used in the MC SCF calculations. The spinorbitals of the energy higher than a certain selected threshold are entirely ignored in the calculations.

of orbitals (i.e., within the inactive or active spinorbitals). This makes the result invariant with respect to the localization of the molecular orbitals.

NON-VARIATIONAL METHOD WITH SLATER DETERMINANTS

10.15 Coupled Cluster (CC) Method

The CC method is the most reliable one among quantum mechanical methods applied to chemistry today.

The problem of many-body correlation of motion of anything is extremely difficult and so far unresolved (e.g., weather forecasting). The problem of electron correlation also seemed to be hopelessly difficult. It still remains that way; however, it turns out that we can exploit a certain observation made by Sinanoğlu.[79] This author noticed that the major portion of the correlation is included through the introduction of correlation within electron pairs, next through pair-pair interactions, then pair-pair-pair interactions, etc. The canonical molecular spinorbitals, which we can use, are in principle delocalized over the whole molecule, but practically the delocalization is not so large. Even in the case of canonical spinorbitals, and certainly when using localized molecular spinorbitals, we can think about an electron excitation as a transfer of an electron

[79] O. Sinanoğlu and K.A. Brueckner, *Three Approaches to Electron Correlation in Atoms* Yale University Press, New Haven and London (1970).

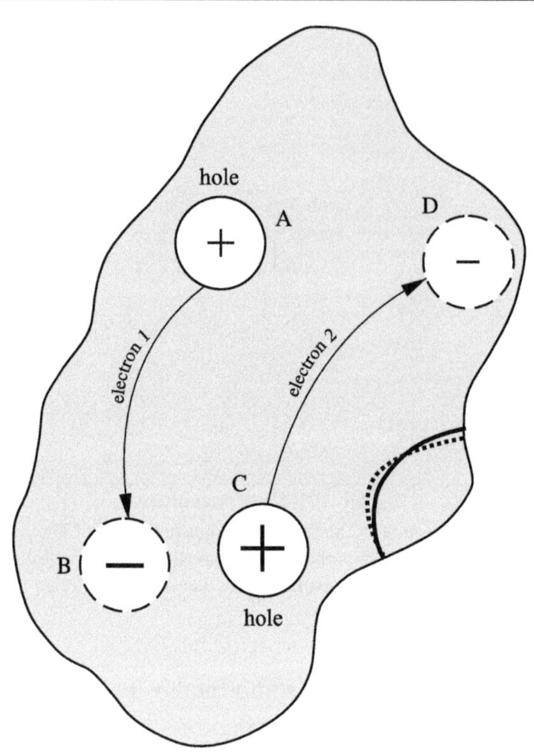

Fig. 10.10. In order to include the electron correlation, the wave function should somehow reflect the fact that electrons avoid each other. Electron 1 jumping from A (an orbital) to B (another orbital) should make electron 2 escape from C (close to B) to D (close to A). This is the very essence of electron correlation. The other orbitals play a role of spectators. However, the spectators change upon the excitations described above. These changes are performed by allowing their own excitations (symbolized by changing from the solid line to the dashed line on the right side). This is how triple, quadruple, and higher excitations emerge and contribute to electronic correlation.

from one place in the molecule to another. Inclusion of the correlation of electronic motion represents, in the language of electron excitations, the following philosophy: when electron 1 jumps from an orbital localized in place A to an orbital localized in place B, it would be good from the point of view of the variational principle if electron 2 jumped from the orbital localized at C to the orbital localized at D (see Fig. 10.10).

The importance of a given double excitation depends on the energy connected with the electron relocation and the arrangement of points A,B,C,D. Yet this simplistic reasoning suggests single excitations do not carry any correlation (this is confirmed by the Brillouin theorem) and this is why their role is very small in the ground state. Moreover, it also suggests that double excitations should be very important.

10.15.1 Wave and Cluster Operators

We start by introducing a special Slater determinant, the *reference determinant* (called the *vacuum state*, which can be the Hartree-Fock determinant) Φ_0, and we write that the exact wave function for the ground state is

$$\psi = \exp(\hat{T})\Phi_0 \tag{10.44}$$

where $\exp(\hat{T})$ is *a wave operator*, and $\hat{T}$ itself is *a cluster operator*. In the CC method, *an intermediate normalization*[80] of the function ψ is assumed; i.e.,

$$\langle \psi | \Phi_0 \rangle = 1.$$

Equation (10.44) represents a very ambitious task. It assumes that we will find an operator $\hat{T}$ such that the wave operator ($e^{\hat{T}}$), as with the touch of a wizard's wand, will make an ideal solution of the Schrödinger equation from the Hartree-Fock function. The formula with $\exp(\hat{T})$ is *an Ansatz*. The charming sounding word *Ansatz*[81] can be translated as an arrangement or order, but in mathematics, the term refers to the construction assumed.

In the research literature, we use the argument that the wave operator ensures the size consistency of the CC. According to this reasoning, for an infinite distance between molecules A and B, both ψ and Φ_0 functions can be expressed in the form of the product of the wave functions for A and B. When the cluster operator is assumed to be of the form (obvious for infinitely separated systems) $\hat{T} = \hat{T}_A + \hat{T}_B$, then the exponential form of the wave operator $\exp(\hat{T}_A + \hat{T}_B)$ ensures a desired form of the product of the wave function $[\exp(\hat{T}_A + \hat{T}_B)]\Phi_0 = \exp \hat{T}_A \exp \hat{T}_B \Phi_0$. *If we took a finite CI expansion:* $(\hat{T}_A + \hat{T}_B)\Phi_0$, *then we would not get the product but the sum which is incorrect.* In this reasoning, there is an error, since due to the Pauli principle (antisymmetry of the wave function with respect to the electron exchange), over long distances, neither the function ψ nor the function Φ_0 is the product of the functions for the subsystems.[82] Although the reasoning is not quite correct, the conclusion is correct, as will be shown at the end of the description of the CC method shortly.

The CC method is automatically size consistent.

As a cluster operator $\hat{T}$, we assume a sum of the excitation operators (see Appendix U available at booksite.elsevier.com/978-0-444-59436-5):

$$\hat{T} = \hat{T}_1 + \hat{T}_2 + \hat{T}_3 + \cdots + \hat{T}_{l_{\max}}, \tag{10.45}$$

where

$$\hat{T}_1 = \sum_{a,r} t_a^r \hat{r}^\dagger \hat{a} \tag{10.46}$$

[80] It contributes significantly to the numerical efficiency of the method.

[81] This word has survived in the literature in its original German form.

[82] For instance, the RHF function for the hydrogen molecule is not a product function for long distances; see p. 610.

is an operator for single excitations,

$$\hat{T}_2 = \frac{1}{4} \sum_{\substack{ab \\ rs}} t_{ab}^{rs} \hat{s}^\dagger \hat{r}^\dagger \hat{a} \hat{b}, \tag{10.47}$$

is an operator for double excitations, etc. The subscript $l = 1, 2, \ldots, l_{\max}$ in $\hat{T}_l$ indicates the rank of the excitations involved (with respect to the vacuum state). The symbols $a, b, \ldots$ refer to the spinorbitals occupied in Φ_0, and $p, q, r, s, \ldots$ refer to the virtual ones, and

> t represents *amplitudes* (i.e., the numbers whose determination is the goal of the CC method). The rest of this chapter will be devoted to the problem of how to obtain these miraculous amplitudes.

In the CC method, we want to obtain correct results with the assumption that $l_{\max}$ of Eq. (10.45) *is relatively small* (usually $2 \div 5$). If $l_{\max}$ were equal to N (i.e., to the number of electrons), then the CC method would be identical to the full (usually unfeasible) CI method.

10.15.2 Relationship Between CI and CC Methods

Obviously, there is a relation between the CI and CC methods. For instance, if we write $\exp(\hat{T})\Phi_0$ in such a way as to resemble the CI expansion

$$\begin{aligned} \exp(\hat{T})\Phi_0 &= \left[1 + (\hat{T}_1 + \hat{T}_2 + \hat{T}_3 + \cdots) + \frac{1}{2}(\hat{T}_1 + \hat{T}_2 + \hat{T}_3 + \cdots)^2 + \cdots \right] \Phi_0 \\ &= (1 + \hat{C}_1 + \hat{C}_2 + \hat{C}_3 + \cdots)\Phi_0, \end{aligned} \tag{10.48}$$

the operators $\hat{C}_i$ (index i denoting the excitation rank: $i = 1$ for singles, $i = 2$ for double, etc.), pertaining to the CI method, have the following structure:

$$\begin{aligned} \hat{C}_1 &= \hat{T}_1 \\ \hat{C}_2 &= \hat{T}_2 + \frac{1}{2!}\hat{T}_1^2 \\ \hat{C}_3 &= \hat{T}_3 + \frac{1}{3!}\hat{T}_1^3 + \hat{T}_1\hat{T}_2 \\ \hat{C}_4 &= \hat{T}_4 + \frac{1}{4!}\hat{T}_1^4 + \frac{1}{2!}\hat{T}_2^2 + \hat{T}_3\hat{T}_1 + \frac{1}{2!}\hat{T}_1^2\hat{T}_2 \end{aligned} \tag{10.49}$$

$$\cdots \tag{10.50}$$

We see that the multiple excitations $\hat{C}_l$ result from mathematically distinct terms; e.g., $\hat{C}_3$ is composed of triple excitations $\hat{T}_3$, $\hat{T}_1^3$, and $\hat{T}_1\hat{T}_2$.

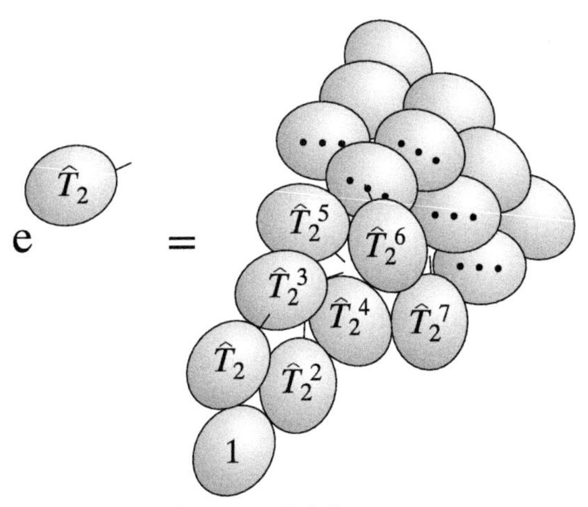

Fig. 10.11. Why such a name? An artistic impression on *coupled clusters*.

On the basis of current numerical experience,[83] we believe that, within the excitation of a given rank, the contributions coming from the correlational interactions of the electron pairs are the most important; e.g., within C_4, the $\frac{1}{2!}\hat{T}_2^2$ excitations containing the product of amplitudes for two electron pairs are the most important, $\hat{T}_4$ (which contains the amplitudes of quadruple excitations) is of little importance, since they correspond to the coupling of the motions of four electrons, and the terms $\hat{T}_1^4$, $\hat{T}_3\hat{T}_1$ and $\hat{T}_1^2\hat{T}_2$ can be made small by using the MC SCF orbitals. Contemporary quantum chemists use diagrammatic language following Richard Feynman. The point is that the mathematical terms (the energy contributions) appearing in CC theory can be translated one by one into the figures according to certain rules. It turns out that it is much easier to think in terms of diagrams than to speak about the mathematical formulae or to write them out. The coupled cluster method, terminated at $\hat{T}_2$ in the cluster operator automatically includes $\hat{T}_2^2$, etc. We may see in it some resemblance to a group of something (excitations), or in other words to a cluster (see Fig. 10.11).

10.15.3 Solution of the CC Equations

The strategy of the CC method is the following: first, we make a decision with respect to l_{max} in the cluster expansion 10.45 ($l_{\max}$ should be small[84]).

The exact wave function $\exp(\hat{T})\Phi_0$ satisfies the Schrödinger equation; i.e.,

$$\hat{H}\exp(\hat{T})\Phi_0 = E\exp(\hat{T})\Phi_0, \tag{10.51}$$

which, after operating from the left with $\exp(-\hat{T})$ gives

$$\exp(-\hat{T})\hat{H}\exp(\hat{T})\Phi_0 = E\Phi_0 \tag{10.52}$$

[83] This is a contribution by Oktay Sinanoğlu; O. Sinanoğlu, and K.A. Brueckner (ed.), *Three Approaches to Electron Correlation in Atoms* Yale University Press, New Haven and London (1970).

[84] Only then is the method cost-effective.

The $\exp(-\hat{T})\hat{H}\exp(\hat{T})$ operator can be expressed in terms of the commutators [see Eq. (10.39)][85]:

$$e^{-\hat{T}}\hat{H}e^{\hat{T}} = \hat{H} + [\hat{H},\hat{T}] + \frac{1}{2!}[[\hat{H},\hat{T}],\hat{T}] + \frac{1}{3!}[[[\hat{H},\hat{T}],\hat{T}],\hat{T}] + \frac{1}{4!}[[[[\hat{H},\hat{T}],\hat{T}],\hat{T}],\hat{T}]. \tag{10.53}$$

The expansion of Eq. (10.53) is *finite* (justification can be only diagrammatic) since in the Hamiltonian $\hat{H}$, we have only two-particle interactions.

Multiplying Eq. (10.52) from the left by the function $\langle^{mn\ldots}_{ab\ldots}|$ representing the determinant obtained from the vacuum state by the action of the excitation operator with the annihilators $\hat{a}, \hat{b}, \ldots$ and creators $\hat{n}^{\dagger}, \hat{m}^{\dagger}, \ldots$ and integrating, we obtain one equation for each function used[86]:

$$\langle^{mn\ldots}_{ab\ldots}|\exp(-\hat{T})\hat{H}\exp(\hat{T})|\Phi_0\rangle = 0, \tag{10.54}$$

where we have zero on the right side due to the orthogonality. The Slater determinants $|^{mn}_{ab}\rangle$ represent all excitations from Φ_0 resulting from the given cluster expansion $\hat{T} = \hat{T}_1 + \hat{T}_2 \cdots + \hat{T}_{l_{\max}}$. This is the fundamental equation of the CC method. For such a set of excited configurations the number of CC equations is equal to the number of the amplitudes sought.

$t^{mn\ldots}_{ab\ldots}$ are unknown quantities; i.e., amplitudes determining the $\hat{T}_l$, and, consequently, the wave operator [Eq. (10.44)] and wave function for the ground state $\psi = \psi_0$. The equations that we get in the CC method are *nonlinear*

since the ts occur at higher powers than the first [which can be seen from Eq. (10.54) that the highest power of t is 4], which, on one hand, requires much more demanding and capricious (than linear ones) numerical procedures, and, on the other, contributes to the greater efficiency of the method. The number of such equations often exceeds 100000 or a million.[87] These equations are solved iteratively assuming certain starting amplitudes t and iterating the equations until self-consistency.

We hope that in such a procedure, an approximation to the ground-state wave function is obtained, although sometimes an unfortunate starting point may lead to some excited state.[88]

[85] It is straightforward to demonstrate the correctness of the first few terms by expanding the wave operator in the Taylor series.

[86] Therefore, the number of equations is equal to the number of the amplitudes t to be determined.

[87] This refers to calculations with $\hat{T} = \hat{T}_2$ for ca. 10 occupied orbitals (for instance, 2 water molecules) and 150 virtual orbitals. These are not calculations for large systems.

[88] The first complete analysis of all CC solutions was performed by K. Jankowski and K. Kowalski, *Phys. Rev. Letters, 81,* 1195 (1998); *J. Chem. Phys., 110,* 37, 93 (1999); *ibid.* 111, 2940, 2952 (1999). Recapitulation can

We usually use as a starting point that which is obtained from the linear version (reduced to obtain a linearity) of the CC method. We will write down these equations as $t_{ab}^{mn} = \ldots$ various powers of all t amplitudes. First, we neglect the nonlinear terms, which represents the initial approximation. The amplitudes are substituted into the right side and we iterate until self-consistency. When all the amplitudes are found, then we obtain the energy E by projecting Eq. (10.54) against Φ_0 function instead of $|_{ab}^{mn}\rangle$:

$$E = \langle \Phi_0 | e^{-\hat{T}} \hat{H} e^{\hat{T}} \Phi_0 \rangle. \tag{10.55}$$

The Non-variational Character of the Method

The operator $(e^{-T})^\dagger$, conjugate to e^{-T}, is $e^{-T^\dagger}$; i.e., the energy

$$E = \langle e^{-\hat{T}^\dagger} \Phi_0 | \hat{H} e^{\hat{T}} \Phi_0 \rangle \tag{10.56}$$

does not represent the mean value of the Hamiltonian. Hence, the CC method is not variational. If we multiplied Eq. (10.51) from the left by $e^{\hat{T}^\dagger}$, we would obtain the variational character of E:

$$E = \frac{\langle \Phi_0 | e^{\hat{T}^\dagger} \hat{H} e^{\hat{T}} \Phi_0 \rangle}{\langle \Phi_0 | e^{\hat{T}^\dagger} e^{\hat{T}} \Phi_0 \rangle} = \frac{\langle e^{\hat{T}} \Phi_0 | \hat{H} | e^{\hat{T}} \Phi_0 \rangle}{\langle e^{\hat{T}} \Phi_0 | e^{\hat{T}} \Phi_0 \rangle}. \tag{10.57}$$

However, it would not be possible to apply the commutator expansion and instead of the four terms in Eq. (10.53) we would have an infinite number. Thus, the non-variational CC method benefits from the very economical condition of the intermediate normalization. For this reason, we prefer the non-variational approach.

10.15.4 Example: CC with Double Excitations

How does the CC machinery work? Let us show it for a relatively simple case, $\hat{T} = \hat{T}_2$. Equation (10.54), written without the commutator expansion, takes the form

$$\langle_{ab}^{mn} | e^{-\hat{T}_2} \hat{H} e^{\hat{T}_2} \Phi_0 \rangle = 0. \tag{10.58}$$

Taking advantage of the commutator expansion, we have

$$\begin{aligned}
\langle_{ab}^{mn} | e^{-\hat{T}_2} \hat{H} e^{\hat{T}_2} \Phi_0 \rangle &= \langle_{ab}^{mn} | \left(1 - \hat{T}_2 + \frac{1}{2}\hat{T}_2^2 + \ldots\right) \hat{H} \left(1 + \hat{T}_2 + \frac{1}{2}\hat{T}_2^2 + \ldots\right) \Phi_0 \rangle \\
&= \langle_{ab}^{mn} | \hat{H} \Phi_0 \rangle + \langle_{ab}^{mn} | \hat{H} \hat{T}_2 \Phi_0 \rangle + \frac{1}{2} \langle_{ab}^{mn} | \hat{H} \hat{T}_2^2 \Phi_0 \rangle \\
&\quad - \langle_{ab}^{mn} | \hat{T}_2 \hat{H} \Phi_0 \rangle - \langle_{ab}^{mn} | \hat{T}_2 \hat{H} \hat{T}_2 \Phi_0 \rangle + A = 0.
\end{aligned}$$

be found in K. Jankowski, K. Kowalski, I. Grabowski, and H.J. Monkhorst, *Intern. J. Quantum Chem.*, *95*, 483 (1999).

However,

$$A = -\frac{1}{2}\langle{}^{mn}_{ab}|\hat{T}_2\hat{H}\hat{T}_2^2\Phi_0\rangle + \frac{1}{2}\langle{}^{mn}_{ab}|\hat{T}_2^2\hat{H}\Phi_0\rangle + \frac{1}{2}\langle{}^{mn}_{ab}|\hat{T}_2^2\hat{H}\hat{T}_2\Phi_0\rangle + \frac{1}{4}\langle{}^{mn}_{ab}|\hat{T}_2^2\hat{H}\hat{T}_2^2\Phi_0\rangle = 0.$$

The last equality follows from the fact that each term is equal to zero. The first vanishes since both determinants differ by four *excitations*. Indeed, $\langle\left(\hat{T}_2^\dagger\right)^{mn}_{ab}|$ denotes a double *deexcitation*[89] of the doubly excited function (i.e., something proportional to $\langle\Phi_0|$). For similar reasons (too strong deexcitations give zero), the remaining terms in A also vanish. As a result, we need to solve the equation

$$\langle{}^{mn}_{ab}|\hat{H}\Phi_0\rangle + \langle{}^{mn}_{ab}|\hat{H}\hat{T}_2\Phi_0\rangle + \frac{1}{2}\langle{}^{mn}_{ab}|\hat{H}\hat{T}_2^2\Phi_0\rangle - \langle{}^{mn}_{ab}|\hat{T}_2\hat{H}\Phi_0\rangle - \langle{}^{mn}_{ab}|\hat{T}_2\hat{H}\hat{T}_2\Phi_0\rangle = 0.$$

After *several days*[90] of algebraic manipulations, we get the equations for the t amplitudes (for each t_{ab}^{mn} amplitude, there is one equation):

$$\begin{aligned}\left(\varepsilon_m + \varepsilon_n - \varepsilon_a - \varepsilon_b\right) t_{ab}^{mn} = \langle mn|ab\rangle &- \sum_{p>q}\langle mn|pq\rangle t_{ab}^{pq} - \sum_{c>d}\langle cd|ab\rangle t_{cd}^{mn} \\ &+ \sum_{c,p}\left[\langle cn|bp\rangle t_{ac}^{mp} - \langle cm|bp\rangle t_{ac}^{np} - \langle cn|ap\rangle t_{bc}^{mp} + \langle cm|ap\rangle t_{bc}^{np}\right]\end{aligned} \quad (10.59)$$

$$\begin{aligned}&+ \sum_{c>d,p>q}\langle cd|pq\rangle\left[t_{ab}^{pq}t_{cd}^{mn} - 2\left(t_{ab}^{mp}t_{cd}^{nq} + t_{ab}^{nq}t_{cd}^{mp}\right)\right. \\ &\left. - 2\left(t_{ac}^{mn}t_{bd}^{pq} + t_{ac}^{pq}t_{bd}^{mn}\right) + 4\left(t_{ac}^{mp}t_{bd}^{nq} + t_{ac}^{nq}t_{bd}^{mp}\right)\right].\end{aligned} \quad (10.60)$$

It can be seen that the last expression includes the term independent of t, the linear terms, and the quadratic terms.

How can we find the ts that satisfy Eq. (10.60)? We do it with the help of the iterative method. First, we substitute zeros for all ts on the right side of the equation. Thus, from the left side, the first approximation to t_{ab}^{mn} is[91] $t_{ab}^{mn} \cong \frac{\langle mn|ab\rangle}{(\varepsilon_m+\varepsilon_n-\varepsilon_a-\varepsilon_b)}$. We have now an estimate of each amplitude, so we are making progress. The approximation to t obtained in this way is substituted into the right side to evaluate the left side, and so forth. Finally, we achieve a self-consistency of the iterative process and obtain the CC wave function for the ground state of our system. With the amplitudes, we calculate the energy of the system with Eq. (10.55).

This is how the CCD (the CC with double excitations in the cluster operator) works from the practical viewpoint. It is more efficient when the initial amplitudes are taken from a short CI

[89] This is the opposite of excitation.

[90] Students – more courage!

[91] As we see, we would have trouble if $(\varepsilon_m + \varepsilon_n - \varepsilon_a - \varepsilon_b)$ is close to 0 (quasidegeneracy of the vacuum state with some other state), because then $t_{ab}^{mn} \to \infty$.

expansion,[92] with subsequent linearization (as above) of terms containing the initial (known) amplitudes.

The computational cost of the CCD and CCSD (singles and doubles) methods scales as N^6, where N is a number of molecular orbitals (occupied and virtual[93]), whereas the analogous cost of the CCSDT (singles, doubles, triples) method requires N^8 scaling. This means that, if we increase the orbital basis twice, the increase in the computational cost of the CCSDT method will be four times larger than that of the CCSD scheme. This is a lot, and because of this widespread popularity, it has been gained for the CCSD(T) method, which only partly uses the triple excitations.

10.15.5 Size Consistency of the CC Method

The size consistency of the CC method can be proved on the basis of Eqs. (10.52) and (10.54). Let us assume that the system dissociates into two[94] non-interacting subsystems A and B (i.e., at infinite distance). Then the orbitals can be also divided into two separable (mutually orthogonal) subsets. We will show[95] that the cluster amplitudes, having *mixed* indices (from the first and second groups of orbitals), are equal to 0.

Let us note first that, for infinite distance, the Hamiltonian $\hat{H} = \hat{H}_A + \hat{H}_B$. In such a situation, the wave operator can be expressed as

$$\hat{T} = \hat{T}_A + \hat{T}_B + \hat{T}_{AB}, \tag{10.61}$$

where $\hat{T}_A, \hat{T}_B, \hat{T}_{AB}$ include the operators corresponding to spinorbitals from the subsystems A, B and from the system AB, respectively. Of course, in this situation, we have the following commutation condition:

$$[\hat{H}_A, \hat{T}_B] = [\hat{H}_B, \hat{T}_A] = 0. \tag{10.62}$$

Then, owing to the commutator expansion in Eq. (10.53), we obtain:

$$e^{-\hat{T}}(\hat{H}_A + \hat{H}_B)e^{\hat{T}} = e^{-\hat{T}_A}\hat{H}_A e^{\hat{T}_A} + e^{-\hat{T}_B}\hat{H}_B e^{\hat{T}_B} + O(\hat{T}_{AB}), \tag{10.63}$$

where $O(\hat{T}_{AB})$ denotes the linear and higher terms in $\hat{T}_{AB}$. Substituting this into Eq. (10.54) with *bra* $\langle$mixed$|$ vector representing mixed excitation, we observe that the first two terms on

[92] The configuration interaction method with inclusion of single and double excitations only:
CCD: J.A. Pople, R. Krishnan, H.B. Schlegel, and J.S. Binkley, *Intern. J. Quantum Chem., S14,* 545 (1978); R.J. Bartlett and G.D. Purvis III, *Intern. J. Quantum Chem. S14,* 561 (1978).
CCSD: G.D. Purvis III, *J. Chem. Phys., 76,* 1910 (1982).

[93] These estimations are valid for the same relative increase of the number of occupied and virtual orbitals, as it is, e.g., for going from a molecule to its dimer. In the case of calculations for the same molecule, but two atomic basis sets (that differ in size) the cost increases only as N^4.

[94] This can be generalized to many non-interacting subsystems.

[95] B. Jeziorski, J. Paldus, and P. Jankowski, *Intern. J. Quantum Chem., 56,* 129 (1995).

the right side of the last equation give zero. It means that we get the equation

$$\left\langle \text{mixed}|O(\hat{T}_{AB})\Phi_0 \right\rangle = 0, \tag{10.64}$$

which, due to the linear term in $O(\hat{T}_{AB})$, is fulfilled by $\hat{T}_{AB} = 0$. Conclusion: for the infinite distance between the subsystems, we do not have mixed amplitudes and the energy of the AB system is bound to be the sum of the energies of subsystem A and subsystem B (size consistency).

10.16 Equation-of-Motion Coupled Cluster (EOM-CC) Method

The CC method is used to calculate the ground-state energy and wave function. What about the excited states? This is a task for the equation-of-motion coupled cluster (EOM-CC) method, the primary goal being not the excited states themselves, but the excitation energies with respect to the ground state.

10.16.1 Similarity Transformation

Let us note that for the Schrödinger equation $\hat{H}\psi = E\psi$, we can perform an interesting sequence of transformations based on the wave operator $e^{\hat{T}}$:

$$e^{-\hat{T}}\hat{H}\psi = Ee^{-\hat{T}}\psi$$
$$e^{-\hat{T}}\hat{H}e^{\hat{T}}e^{-\hat{T}}\psi = Ee^{-\hat{T}}\psi.$$

We obtain the eigenvalue equation again, but for the *similarity transformed Hamiltonian*[96]

$$\hat{\mathcal{H}}\bar{\psi} = E\bar{\psi},$$

where $\hat{\mathcal{H}} = e^{-\hat{T}}\hat{H}e^{\hat{T}}$, $\bar{\psi} = e^{-\hat{T}}\psi$, and the energy E does not change at all after this transformation. This result will be very useful in a moment.

10.16.2 Derivation of the EOM-CC Equations

As the reference function in the EOM-CC method, we take the CC wave function for the ground state:

$$\psi_0 = \exp{(\hat{T})}\Phi_0, \tag{10.65}$$

where Φ_0 is usually a Hartree-Fock determinant. Now, we define the operator $\hat{U}_k$, which ("*EOM-CC Ansatz*") performs a miracle: from the wave function of the ground state ψ_0, it creates the

[96] In contrast to the Hamiltonian $\hat{H}$, the similarly transformed Hamiltonian does not represent a Hermitian operator. Moreover, it contains not only the one- and two-electron terms, as it does in $\hat{H}$, but also all other many-electron operators up to the total number of electrons in the system.

wave function ψ_k for the kth excited state of the system:

$$\psi_k = \hat{U}_k \psi_0.$$

The operators $\hat{U}_k$ change the coefficients in front of the configurations (see p. 616). The operators $\hat{U}_k$ are [unlike the wave operator $\exp(\hat{T})$] linear with respect to the excitations; i.e., the excitation amplitudes occur there in first powers. For the case of the single and double excitations (EOM-CCSD), we have $\hat{T}$ in the form of the sum of single and double excitations:

$$\hat{T} = \hat{T}_1 + \hat{T}_2$$

and

$$\hat{U}_k = \hat{U}_{k,0} + \hat{U}_{k,1} + \hat{U}_{k,2},$$

where the task for the $\hat{U}_{k,0}$ operator is to change the coefficient in front of the function Φ_0 to that appropriate to the $|k\rangle$ function, the role of the operators $\hat{U}_{k,1}, \hat{U}_{k,2}$ is an appropriate modification of the coefficients in front of the singly and doubly excited configurations. These tasks are done by the excitation operators with τ amplitudes (they have to be distinguished from the amplitudes of the CC method):

$$\hat{U}_{k,0} = \tau_0(k)$$
$$\hat{U}_{k,1} = \sum_{a,p} \tau_a^p(k)\hat{p}^\dagger \hat{a}$$
$$\hat{U}_{k,2} = \sum_{a,b,p,q} \tau_{ab}^{pq}(k)\hat{q}^\dagger \hat{p}^\dagger \hat{a}\hat{b},$$

where the amplitudes $\tau(k)$ are numbers that are the targets of the EOM-CC method. The amplitudes give the wave function ψ_k and the energy E_k.

We write down the Schrödinger equation for the excited state:

$$\hat{H}\psi_k = E_k \psi_k.$$

Now we substitute the EOM-CC *Ansatz*:

$$\hat{H}\hat{U}_k \psi_0 = E_k \hat{U}_k \psi_0,$$

and from the definition of the CC wave operator, we get[97]

$$\hat{H}\hat{U}_k \exp(\hat{T})\Phi_0 = E_k \hat{U}_k \exp(\hat{T})\Phi_0.$$

[97] By neglecting higher than single and double excitations, the equation represents an approximation.

Due to the missing *deexcitation* part (i.e., that which lowers the excitation rank, such as from doubles to singles) the operators $\hat{U}_k$ and $\hat{T}$ commute[98]; hence, the operators $\hat{U}_k$ and $\exp(\hat{T})$ also commute:

$$\hat{U}_k \exp(\hat{T}) = \exp(\hat{T})\hat{U}_k.$$

Substituting this, we have:

$$\hat{H} \exp(\hat{T})\hat{U}_k\Phi_0 = E_k \exp(\hat{T})\hat{U}_k\Phi_0$$

and multiplying from the left with $\exp(-\hat{T})$, we get:

$$[\exp(-\hat{T})\hat{H} \exp(\hat{T})]\hat{U}_k\Phi_0 = E_k\hat{U}_k\Phi_0$$

or, introducing the similarity transformed Hamiltonian,

$$\hat{\mathcal{H}} = e^{-\hat{T}}\hat{H}e^{\hat{T}},$$

we obtain

$$\hat{\mathcal{H}}\hat{U}_k\Phi_0 = E_k\hat{U}_k\Phi_0.$$

From the last equation, we will subtract the CC equation for the ground state:

$$[\exp(-\hat{T})\hat{H} \exp(\hat{T})]\Phi_0 = E_0\Phi_0.$$

Multiplying it from the left with $\hat{U}_k$ (i.e., $\hat{U}_k\hat{\mathcal{H}}\Phi_0 = E_0\hat{U}_k\Phi_0$), we get

$$\hat{\mathcal{H}}\hat{U}_k\Phi_0 - \hat{U}_k\hat{\mathcal{H}}\Phi_0 = E_k\hat{U}_k\Phi_0 - E_0\hat{U}_k\Phi_0.$$

Finally, we obtain an important result:

$$[\hat{\mathcal{H}}, \hat{U}_k]\Phi_0 = \left(E_k - E_0\right)\hat{U}_k\Phi_0.$$

The operator $\hat{U}_k$ contains the sought amplitudes $\tau(k)$.

We find them in a similar manner as in the CC method. For that purpose, we make a scalar product of the left and right sides of that equation with each excitation $|^{mn...}_{ab...}\rangle$ used in $\hat{U}_k$. We get the set of the EOM-CC equations whose number is equal to the number of sought amplitudes plus one more equation due to the normalization condition of ψ_k. The unknown parameters are amplitudes and the excitation energies $E_k - E_0$:

$$\left\langle \begin{matrix} mn\ldots \\ ab\ldots \end{matrix} \middle| [\hat{\mathcal{H}}, \hat{U}_k] \middle| \Phi_0 \right\rangle = \left(E_k - E_0\right) \left\langle \begin{matrix} mn\ldots \\ ab\ldots \end{matrix} \middle| \hat{U}_k \middle| \Phi_0 \right\rangle,$$

[98] If $\hat{U}_k$ contains true excitations, then it does not matter whether excitations are performed by $\hat{U}_k\hat{T}$ or $\hat{T}\hat{U}_k$ (commutation), because both $\hat{U}_k$ and $\hat{T}$ mean going up in the energy scale. If, however, $\hat{U}_k$ contains deexcitations, then it may happen that there is an attempt in $\hat{T}\hat{U}_k$ to deexcite the ground-state wave function–that makes 0, whereas $\hat{U}_k\hat{T}$ may be still OK because the excitations in $\hat{T}$ may be more important than the deexcitations in $\hat{U}_k$.

Once we solve these equations, the problem is over.

It is important that the excitations $|^{mn\ldots}_{ab\ldots}\rangle$ used in $\hat{U}_k$ include not only the regular singles and doubles, and the function with no excitation[99] (i.e., the function Φ_0), but also the states with different numbers of electrons (i.e., with the ionized states or the states with extra electrons). It turned out that the last possibility offers an intriguing way of determining a particular electronic state starting from several distinct points of view. Indeed, one may carry out the EOMCC computations for a given state (with N electrons) starting first from function $\Phi_0(1, 2, \ldots N)$, then repeating the calculations with different functions $\Phi_0(1, 2, \ldots N - M)$, where $M = \pm 1, \pm 2, \ldots$ and compare the results. As shown by Kucharski and Musiał[100] such a possibility is especially fruitful if $\Phi_0(1, 2, \ldots N)$ were a very bad approximation to the ground-state wave function e.g., in case of dissociation of a chemical bond. This approach may offer an elegant avenue to circumvent the serious problem of bond dissociation.

10.17 Many-body Perturbation Theory (MBPT)

The majority of routine calculations in quantum chemistry are done with variational methods (mainly the Hartree-Fock scheme). If we consider post-Hartree-Fock calculations, then non-variational [CCSD, CCSD(T)] and perturbational approaches (including MBPT) take the lead. The perturbational methods are based on the simple idea that the system in slightly modified conditions is similar to that before the perturbation is applied (cf. p. 240).

In the formalism of perturbation theory, knowing the unperturbed system and the perturbation allows us to provide successive corrections to obtain the solution of the perturbed system. Thus, for instance, the energy of the perturbed system is the energy of the unperturbed system plus the first-order correction, plus the second-order correction, plus..., etc. If the perturbation is small, then we *hope*[101] that the series is convergent; even then, however, there is no guarantee that the series converges fast.

10.17.1 Unperturbed Hamiltonian

In the perturbational approach (cf. 232) to the electron correlation, the Hartree-Fock function, Φ_0, is treated as the zero-order approximation to the true ground-state wave function; i.e., $\Phi_0 = \psi_0^{(0)}$. Thus, the Hartree-Fock wave function stands at the starting point, while the goal is the exact ground-state electronic wave function.

---

[99] More precisely, to get only the excitation energy we do not need the coefficient next to Φ_0.

[100] S. Kucharski and M. Musiał, *Proc. Conference HITY*, Krakow, Poland, 2011.

[101] Not much is known concerning the convergence of series occurring in quantum chemistry. Commonly, only a few perturbational corrections are computed.

In the majority of cases, this is a reasonable approximation, since the Hartree-Fock method usually provides as much as 98 to 99% of the total energy.[102] A Slater determinant Φ_I is constructed from the spinorbitals obeying the Fock equation. How do we construct the operator for which the Slater determinant is an eigenfunction? We will find out in a moment that this operator is the sum of the Fock operators (cf. Appendix U available at booksite.elsevier.com/978-0-444-59436-5):

$$\hat{H}^{(0)} = \sum_i \hat{F}(i) = \sum_i \epsilon_i \hat{\imath}^\dagger \hat{\imath}. \tag{10.66}$$

Indeed,

$$\hat{H}^{(0)} \Phi_I = \sum_i \epsilon_i \hat{\imath}^\dagger \hat{\imath} \cdot \Phi_I = \left(\sum_i \epsilon_i \right) \cdot \Phi_I, \tag{10.67}$$

since the annihilation of one spinorbital in the determinant and the creation of the same spinorbital leaves the determinant unchanged. This is so on the condition that the spinorbital ϕ_i *is present* in $\psi_0^{(0)}$.

> The eigenvalue of $\hat{H}_0 = \sum_i \epsilon_i \hat{\imath}^\dagger \hat{\imath}$ is always the sum of the orbital energies corresponding to all spinorbitals in the Slater determinant Φ_I.

This means that the sum of several determinants, each built from a different (in the sense of the orbital energies) set of spinorbitals, is not an eigenfunction of $\hat{H}^{(0)}$.

10.17.2 Perturbation Theory–Slightly Different Presentation

We have to solve the Schrödinger equation for the ground state[103] $\hat{H}\psi_0 = E\psi_0$, with $\hat{H} = \hat{H}^{(0)} + \hat{H}^{(1)}$, where $\hat{H}^{(0)}$ denotes the unperturbed Hamiltonian given by Eq. (10.66), and $\hat{H}^{(1)}$ is a perturbation operator. The eigenfunctions and the eigenvalues of $\hat{H}^{(0)}$ are given by Eq. (10.67), but remembering the perturbation theory formulas, we will denote the Slater determinants as $\Phi_I \equiv \psi_I^{(0)}$.

For the ground state, we expand the energy E_0 and the wave function ψ_0 in a power series[104]: we put $\lambda\hat{H}^{(1)}$ instead of $\hat{H}^{(1)}$ in the Hamiltonian and expand the energy and the wave function in a power series[105] with respect to λ:

---

[102] Sometimes, as we know, the method fails; and then the perturbation theory based on the Hartree-Fock starting point is a risky business, since the perturbation is very large.

[103] We use the notation from Chapter 5.

[104] This is an old trick of perturbation theory equivalent to saying that the shape of a bridge loaded with a car is the shape of the bridge without the car, plus the deformation proportional to the mass of the car, plus the deformation proportional to the square of the mass of the car, etc. This works if the bridge is solid and the car is light (the perturbation is small).

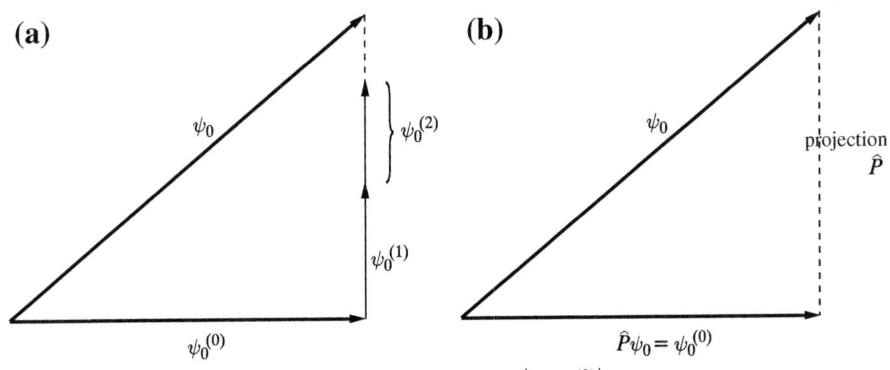

Fig. 10.12. Pictorial presentation of the intermediate normalization (a) $\left\langle \psi_0 | \psi_0^{(0)} \right\rangle = 1$ and (b) the projection onto the axis $\psi_0^{(0)}$ in the Hilbert space using the operator $\hat{P} = |\psi_0^{(0)}\rangle\langle\psi_0^{(0)}|$. Here, $\psi_0^{(n)}$, $n = 1, 2$ represents a correction of the nth order to the ground-state wave function. The picture can only be simplistic and schematic: the orthogonality of $\psi_0^{(n)}$ to ψ_0 is shown correctly, but the apparent parallelism of $\psi_0^{(1)}$ and $\psi_0^{(2)}$ is false.

$$E_0 = E_0^{(0)} + \lambda E_0^{(1)} + \lambda^2 E_0^{(2)} + \cdots, \tag{10.68}$$

$$\psi_0 = \psi_0^{(0)} + \lambda \psi_0^{(1)} + \lambda^2 \psi_0^{(2)} + \cdots \tag{10.69}$$

The Schrödinger equation does not force the normalization of the function. It is convenient to use the *intermediate normalization* (Fig. 10.12a); i.e., to require that $\langle \psi_0 | \psi_0^{(0)} \rangle = 1$.

This means that the (non-normalized) ψ_0 must include the normalized function of zeroth order $\psi_0^{(0)}$ *and, possibly, something orthogonal to it.*

10.17.3 MBPT Machinery–Part 1: Energy Equation

Let us write $\hat{H}\psi_0$ as $\hat{H}\psi_0 = (\hat{H}^{(0)} + \hat{H}^{(1)})\psi_0$, or, in another way, as $\hat{H}^{(1)}\psi_0 = (\hat{H} - \hat{H}^{(0)})\psi_0$. Multiplying this equation by $\psi_0^{(0)}$ and integrating, we get (taking advantage of the intermediate normalization)

$$\langle \psi_0^{(0)} | \hat{H}^{(1)} \psi_0 \rangle = \langle \psi_0^{(0)} | (\hat{H} - \hat{H}^{(0)}) \psi_0 \rangle = E_0 \langle \psi_0^{(0)} | \psi_0 \rangle - \langle \psi_0^{(0)} | \hat{H}^{(0)} \psi_0 \rangle = E_0 - E_0^{(0)} = \Delta E_0. \tag{10.70}$$

Thus,

$$\Delta E_0 = \langle \psi_0^{(0)} | \hat{H}^{(1)} \psi_0 \rangle. \tag{10.71}$$

[105] So we assume that the respective functions are analytic in the vicinity of $\lambda = 0$.

Reduced Resolvent or the "Almost" Inverse of $(E_0^{(0)} - \hat{H}^{(0)})$

Let us define several useful quantities–we need to get familiar with them now–which will introduce a certain elegance into our final equations.

Let the first be *a projection operator* on the ground-state zeroth order function:

$$\hat{P} = |\psi_0^{(0)}\rangle\langle\psi_0^{(0)}|. \tag{10.72}$$

This means that $\hat{P}\chi$ is, within accuracy to a constant, equal to either $\psi_0^{(0)}$ or zero for an arbitrary function χ. Indeed, if χ is expressed as a linear combination of the eigenfunctions $\psi_n^{(0)}$ (these functions form an orthonormal complete set as eigenfunctions of the Hermitian operator)

$$\chi = \sum_n c_n \psi_n^{(0)}, \tag{10.73}$$

then (Fig. 10.12b)

$$\hat{P}\chi = \sum_n c_n \hat{P}\psi_n^{(0)} = \sum_n c_n |\psi_0^{(0)}\rangle\langle\psi_0^{(0)}|\psi_n^{(0)}\rangle = \sum_n c_n \delta_{0n} \psi_0^{(0)} = c_0 \psi_0^{(0)}. \tag{10.74}$$

Let us now introduce another projection operator:

$$\hat{Q} = 1 - \hat{P} = \sum_{n=1}^{\infty} |\psi_n^{(0)}\rangle\langle\psi_n^{(0)}| \tag{10.75}$$

on the space orthogonal to $\psi_0^{(0)}$. Obviously, $\hat{P}^2 = \hat{P}$ and $\hat{Q}^2 = \hat{Q}$. The latter holds since $\hat{Q}^2 = (1 - \hat{P})^2 = 1 - 2\hat{P} + \hat{P}^2 = 1 - \hat{P} = \hat{Q}$.

Now we define *a reduced resolvent*

$$\hat{R}_0 = \sum_{n=1}^{\infty} \frac{|\psi_n^{(0)}\rangle\langle\psi_n^{(0)}|}{E_0^{(0)} - E_n^{(0)}}. \tag{10.76}$$

The definition says that the reduced resolvent represents an operator that from an arbitrary vector ϕ of the Hilbert space, takes the following actions:

- Cuts out its components along all the unit (i.e., normalized) basis vectors $\psi_n^{(0)}$ *except* $\psi_0^{(0)}$
- Weighs the projections by the factor $\frac{1}{E_0^{(0)} - E_n^{(0)}}$, so they become less and less important for higher and higher energy states
- Adds all the weighed vectors together.

We easily obtain[106]

$$\hat{R}_0\left(E_0^{(0)}-\hat{H}^{(0)}\right)=\left(E_0^{(0)}-\hat{H}^{(0)}\right)\hat{R}_0=\hat{Q}. \tag{10.77}$$

For functions ϕ orthogonal to $\psi_0^{(0)}$ (i.e., satisfying $\hat{Q}\phi=\phi$), the action of the operator $\hat{R}_0$ is identical to that of the operator $(E_0^{(0)}-\hat{H}^{(0)})^{-1}$. $\hat{R}_0$ does not represent the inverse of $(E_0^{(0)}-\hat{H}^{(0)})$, however, because for $\phi=\psi_0^{(0)}$, we get $\hat{R}_0(E_0^{(0)}-\hat{H}^{(0)})\phi=0$, and not the unchanged ϕ.

10.17.4 MBPT Machinery–Part 2: Wave Function Equation

Our goal now will be to present the Schrödinger equation in a different form. Let us first write it down as follows:

$$(E_0-\hat{H}^{(0)})\psi_0=\hat{H}^{(1)}\psi_0. \tag{10.78}$$

We aim at having $(E_0^{(0)}-\hat{H}^{(0)})\psi_0$ on the left side. Let us add $(E_0^{(0)}-E_0)\psi_0$ to both sides of that equation to obtain

$$\left(E_0^{(0)}-\hat{H}^{(0)}\right)\psi_0=\left(E_0^{(0)}-E_0+\hat{H}^{(1)}\right)\psi_0. \tag{10.79}$$

Let us now operate on both sides of this equation with the reduced resolvent $\hat{R}_0$:

$$\hat{R}_0\left(E_0^{(0)}-\hat{H}^{(0)}\right)\psi_0=\hat{R}_0\left(E_0^{(0)}-E_0+\hat{H}^{(1)}\right)\psi_0. \tag{10.80}$$

---

[106] Let us make sure of this:

$$\begin{aligned}\hat{R}_0\left(E_0^{(0)}-\hat{H}^{(0)}\right)\phi&=\sum_{n=1}^{\infty}\left(E_0^{(0)}-E_n^{(0)}\right)^{-1}|\psi_n^{(0)}\rangle\langle\psi_n^{(0)}|\left(E_0^{(0)}-\hat{H}^{(0)}\right)|\phi\rangle\\&=\sum_{n=1}^{\infty}\left(E_0^{(0)}-E_n^{(0)}\right)^{-1}\left(E_0^{(0)}-E_n^{(0)}\right)|\psi_n^{(0)}\rangle\langle\psi_n^{(0)}|\phi\rangle\\&=\sum_{n=1}^{\infty}|\psi_n^{(0)}\rangle\langle\psi_n^{(0)}|\phi\rangle=\hat{Q}\phi.\end{aligned}$$

Let us now operate on the same function with the operator $(E_0^{(0)}-\hat{H}^{(0)})\hat{R}_0$ (i.e., the operators are in reverse order):

$$\begin{aligned}\left(E_0^{(0)}-\hat{H}^{(0)}\right)\hat{R}_0\phi&=\left(E_0^{(0)}-\hat{H}^{(0)}\right)\sum_{n=1}^{\infty}\left(E_0^{(0)}-E_n^{(0)}\right)^{-1}|\psi_n^{(0)}\rangle\langle\psi_n^{(0)}|\phi\rangle\\&=\sum_{n=1}^{\infty}\left(E_0^{(0)}-E_n^{(0)}\right)^{-1}\left(E_0^{(0)}-\hat{H}^{(0)}\right)|\psi_n^{(0)}\rangle\langle\psi_n^{(0)}|\phi\rangle\\&=\sum_{n=1}^{\infty}|\psi_n^{(0)}\rangle\langle\psi_n^{(0)}|\phi\rangle=\hat{Q}\phi.\end{aligned}$$

On the left side, we have $\hat{Q}\psi_0$ [as follows from Eq. (10.77)], but $\hat{Q}\psi_0 = (1 - \hat{P})\psi_0 = \psi_0 - |\psi_0^{(0)}\rangle\langle\psi_0^{(0)}|\psi_0\rangle = \psi_0 - \psi_0^{(0)}$, due to the intermediate normalization. As a result, the equation takes the form

$$\psi_0 - \psi_0^{(0)} = \hat{R}_0 \left(E_0^{(0)} - E_0 + \hat{H}^{(1)}\right) \psi_0. \tag{10.81}$$

Thus, we obtain

$$\psi_0 = \psi_0^{(0)} + \hat{R}_0 \left(E_0^{(0)} - E_0 + \hat{H}^{(1)}\right) \psi_0. \tag{10.82}$$

At the same time, based on the expression for ΔE in perturbation theory (Eq. (10.71)), we have

$$E_0 = E_0^{(0)} + \left\langle \psi_0^{(0)} | \hat{H}^{(1)} \psi_0 \right\rangle. \tag{10.83}$$

These are the equations of the many body perturbation theory, in which the exact wave function and energy are expressed in terms of the unperturbed functions and energies plus certain corrections. The problem is that, as can be seen, these corrections involve the unknown function and unknown energy.

Let us not despair in this situation, but try to apply an iterative technique. First, substitute for ψ_0 on the right side of Eq. (10.82) that which most resembles ψ_0; i.e., $\psi_0^{(0)}$. We obtain

$$\psi_0 \cong \psi_0^{(0)} + \hat{R}_0 \left(E_0^{(0)} - E_0 + \hat{H}^{(1)}\right) \psi_0^{(0)}, \tag{10.84}$$

and then the new approximation to ψ_0 should again be plugged into the right side and this procedure is continued until convergence. It can be seen that the successive terms form a series (let us hope that it is convergent).

$$\psi_0 = \sum_{n=0}^{\infty} \left[\hat{R}_0 \left(E_0^{(0)} - E_0 + \hat{H}^{(1)}\right)\right]^n \psi_0^{(0)}. \tag{10.85}$$

Now only known quantities occur on the right side except for E_0, the exact energy. Let us pretend that its value is known and insert into the energy expression [Eq. (10.83)] the function ψ_0:

$$E_0 = E_0^{(0)} + \left\langle \psi_0^{(0)} | \hat{H}^{(1)} \psi_0 \right\rangle = E_0^{(0)} + \left\langle \psi_0^{(0)} | \hat{H}^{(1)} \sum_{n=0}^{M} \left[\hat{R}_0 \left(E_0^{(0)} - E_0 + \hat{H^{(1)}} \right) \right]^n | \psi_0^{(0)} \right\rangle . \tag{10.86}$$

Let us go back to our problem: we want to have E_0 on the left side of the last equation, while - for the time being - E_0 occurs on the right sides of both equations. To exit the situation, we will treat E_0 occurring on the right side as a parameter manipulated in such a way as to obtain equality in both of these equations. We may do it in two ways. One leads to the Brillouin-Wigner perturbation theory, the other to the Rayleigh-Schrödinger perturbation theory.

10.17.5 Brillouin-Wigner Perturbation Theory

Let us decide first at what $n = M$ we terminate the series; i.e., to what order of perturbation theory the calculations will be carried out. Say that $M = 3$. Let us now take any reasonable value[107] as a parameter of E_0. We insert this value into the right side of Eq. (10.86) for E_0 and calculate the left side (i.e., E_0). Then let us again insert the new E_0 into the right side and continue in this way until self-consistency [i.e., until Eq. (10.86) is satisfied]. After E_0 is known, we go to Eq. (10.85) and compute ψ_0 (through a certain order–e.g., M).

Brillouin-Wigner perturbation theory has, as seen, the somewhat unpleasant feature that successive corrections to the wave function depend on the M assumed at the beginning.

We may suspect[108] – and this is true – that *the Brillouin-Wigner perturbation theory is not size consistent.*

10.17.6 Rayleigh-Schrödinger Perturbation Theory

As an alternative to Brillouin-Wigner perturbation theory, we may consider Rayleigh-Schrödinger perturbation theory, which *is size consistent*. In this method, the total energy is computed in a stepwise manner:

$$E_0 = \sum_{k=0}^{\infty} E_0^{(k)} \tag{10.87}$$

in such a way that first we calculate the first-order correction $E_0^{(1)}$ [i.e., of the order of $\hat{H}^{(1)}$], then the second-order correction, $E_0^{(2)}$ [i.e., of the order of $(\hat{H}^{(1)})^2$], etc. If we insert into the

[107] A "unreasonable" value will lead to numerical instabilities. *Then* we will learn that it was unreasonable to take it.

[108] This is due to the iterative procedure.

right side of Eqs. (10.85) and (10.86) the expansion $E_0 = \sum_{k=0}^{\infty} E_0^{(k)}$ and then, by applying the usual perturbation theory argument, we equalize the terms of the same order and get

for $n = 0$:

$$E_0^{(1)} = \left\langle \psi_0^{(0)} | \hat{H}^{(1)} \psi_0^{(0)} \right\rangle, \tag{10.88}$$

for $n = 1$:

$$E^{(2)} = \left\langle \psi_0^{(0)} | \hat{H}^{(1)} \hat{R}_0 \left(E_0^{(0)} - E_0 + \hat{H}^{(1)} \right) \psi_0^{(0)} \right\rangle = \left\langle \psi_0^{(0)} | \hat{H}^{(1)} \hat{R}_0 \hat{H}^{(1)} \psi_0^{(0)} \right\rangle, \tag{10.89}$$

since $\hat{R}_0 \psi_0^{(0)} = 0$;
for $n = 2$:
$E^{(3)} =$ the third-order terms from the expression:

$$\left\langle \psi_0^{(0)} | \hat{H}^{(1)} \left[\hat{R}_0 \left(E_0^{(0)} - E_0^{(0)} - E_0^{(1)} - E_0^{(2)} - \cdots + \hat{H}^{(1)} \right) \right]^2 \psi_0^{(0)} \right\rangle$$
$$= \left\langle \psi_0^{(0)} | \hat{H}^{(1)} \hat{R}_0 \left(-E_0^{(1)} - E_0^{(2)} - \cdots + \hat{H}^{(1)} \right) \hat{R}_0 \left(-E_0^{(1)} - E_0^{(2)} - \cdots + \hat{H}^{(1)} \right) \psi_0^{(0)} \right\rangle$$

and the only terms of the third order are:

$$E^{(3)} = \left\langle \psi_0^{(0)} | \hat{H}^{(1)} \hat{R}_0 \hat{H}^{(1)} \hat{R}_0 \hat{H}^{(1)} \psi_0^{(0)} \right\rangle - E_0^{(1)} \left\langle \psi_0^{(0)} | \hat{H}^{(1)} R_0^2 \hat{H}^{(1)} \psi_0^{(0)} \right\rangle, \tag{10.90}$$

etc.

Unfortunately, we cannot give a general expression for the kth correction to the energy although we can give an algorithm for the construction of such an expression.[109] Rayleigh-Schrödinger perturbation theory (unlike the Brillouin-Wigner approach) has the nice feature that the corrections of the particular orders are independent of the maximum order chosen.

10.18 Møller-Plesset Version of Rayleigh-Schrödinger Perturbation Theory

Let us consider the case of a closed shell.[110] In the Møller-Plesset perturbation theory, we assume as $\hat{H}^{(0)}$ the sum of the Hartree-Fock operators [from the RHF method; see Eq. (10.66)], and $\psi_0^{(0)} = \psi_{RHF}$, i.e.:

$$\hat{H}^{(0)} = \sum_i^N \hat{F}(i) = \sum_i^{\infty} \epsilon_i i^\dagger i,$$
$$\hat{H}^{(0)} \psi_{RHF} = E_0^{(0)} \psi_{RHF}, \tag{10.91}$$

[109] J. Paldus and J. Čížek, *Adv. Quantum Chem.*, 9, 105 (1975).

[110] Møller–Plesset perturbation theory also has its multireference formulation when the function Φ_0 is a linear combination of determinants [K. Woliński, P. Pulay, *J. Chem. Phys.*, 90, 3647 (1989)].

$$E_0^{(0)} = \sum_i \epsilon_i \tag{10.92}$$

(the last summation is over spinorbitals occupied in the RHF function); hence, the perturbation, known in the literature as *a fluctuation potential*, is equal to

$$\hat{H}^{(1)} = \hat{H} - \hat{H}^{(0)}. \tag{10.93}$$

For such a perturbation, we may carry out calculations through a given order n: we have a sequence of approximations MPn. A very popular method relies on the inclusion of the perturbational corrections to the energy through the second order (known as the *MP2 method*) and through the fourth order (MP4).

10.18.1 Expression for MP2 Energy

What is the expression for the total energy in the MP2 method?

Let us note first that, when calculating the mean value of the Hamiltonian in the standard Hartree-Fock method, we automatically obtain the sum of the zeroth-order energies $\sum_i \epsilon_i$ and the first-order correction to the energy $\left\langle \psi_{RHF} | \hat{H}^{(1)} \psi_{RHF} \right\rangle$. Indeed, $E_{RHF} = \left\langle \psi_{RHF} | \hat{H} \psi_{RHF} \right\rangle$ $= \left\langle \psi_{RHF} | (\hat{H}^{(0)} + \hat{H}^{(1)}) \psi_{RHF} \right\rangle = (\sum_i \epsilon_i) + \left\langle \psi_{RHF} | \hat{H}^{(1)} \psi_{RHF} \right\rangle$. So what is left to be done (in the MP2 approach) is the addition of the second-order correction to the energy (p. 245, the prime in the summation symbol indicates that the term making the denominator equal to zero is omitted), where, as the complete set of functions, we assume the Slater determinants $\psi_k^{(0)}$ corresponding to the energy $E_k^{(0)}$ (they are generated by various spinorbital occupancies):

$$E_{MP2} = E_{RHF} + \sum_k{}' \frac{\left| \left\langle \psi_k^{(0)} | \hat{H}^{(1)} \psi_{RHF} \right\rangle \right|^2}{E_0^{(0)} - E_k^{(0)}} = E_{RHF} + \sum_k{}' \frac{\left| \left\langle \psi_k^{(0)} | \hat{H} \psi_{RHF} \right\rangle \right|^2}{E_0^{(0)} - E_k^{(0)}}, \tag{10.94}$$

the last equality holds because ψ_{RHF} is an eigenfunction of $\hat{H}^{(0)}$, and $\psi_k^{(0)}$ and ψ_{RHF} are orthogonal. It can be seen that among possible functions $\psi_k^{(0)}$, we may ignore all but doubly excited ones. Why? This is for two reasons:

- The single excitations give $\left\langle \psi_k^{(0)} | \hat{H} \psi_{RHF} \right\rangle = 0$ due to the Brillouin theorem.
- The triple and higher excitations differ by more than two excitations from the functions ψ_{RHF} and, due to the fourth Slater-Condon rule (see Appendix M available at booksite.elsevier.com/978-0-444-59436-5 p. e119), give a contribution equal to 0.

In such a case, we take as the functions $\psi_k^{(0)}$ only doubly excited Slater determinants ψ_{ab}^{pq}, which means that we replace the occupied spinorbitals: $a \to p, b \to q$, and, to avoid repetitions, $a < b$, $p < q$. These functions are eigenfunctions of $\hat{H}^{(0)}$ with the eigenvalues being the sum of the respective orbital energies [see Eq. (10.67)]. Thus, using the third Slater-Condon rule, we obtain the energy correct through the second order:

$$E_{MP2} = E_{RHF} + \sum_{a<b,p<q} \frac{|\langle ab|pq\rangle - \langle ab|qp\rangle|^2}{\varepsilon_a + \varepsilon_b - \varepsilon_p - \varepsilon_q}; \tag{10.95}$$

hence, the MP2 scheme viewed as an approximation to the correlation energy gives[111]

$$E_{corel} \approx E_{MP2} - E_{RHF} = \sum_{a<b,p<q} \frac{|\langle ab|pq\rangle - \langle ab|qp\rangle|^2}{\varepsilon_a + \varepsilon_b - \varepsilon_p - \varepsilon_q}. \tag{10.96}$$

Well, how effective is the MP method in computing the electron correlation? Fig. 10.13 shows a comparison of the RHF, MP2, MP3, and CISD (in this case, equivalent to CI) methods applied to the hydrogen molecule for several values of the internuclear distance R. The results

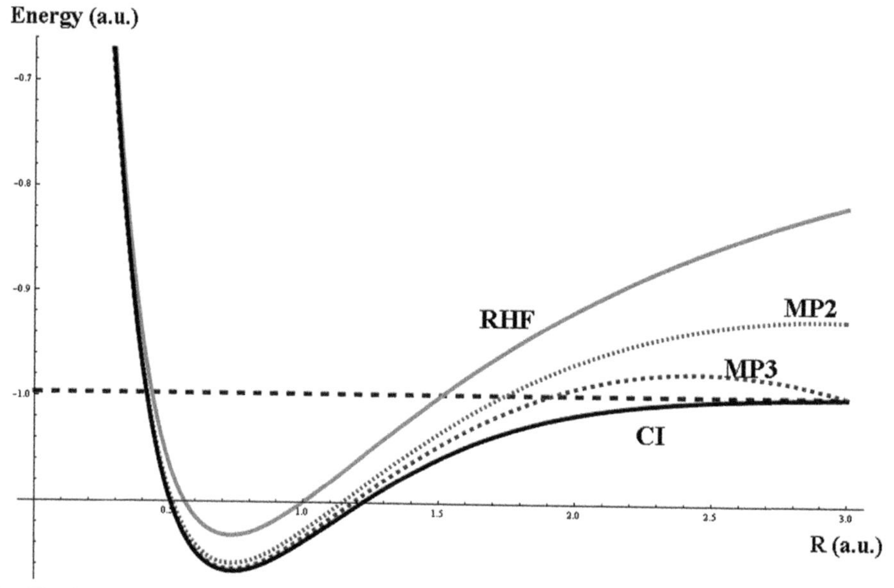

Fig. 10.13. The electronic energy of the hydrogen molecule as a function of the internuclear distance R. The energy is computed by using the RHF (gray solid line), MP2 (lighter dotted line), MP3 (darker dotted line) and CI (black solid line). The energy of the two isolated hydrogen atoms is shown as a horizontal dashed line. The computations have been carried out by using the Gaussian program with a standard basis of atomic orbitals 6-311G(d,p). Energies and distances are given in a.u.

[111] The MP2 method usually gives satisfactory results (e.g., the frequencies of the normal modes). There are indications, however, that the deformations of the molecule connected with some vibrations strongly affecting the electron correlation (vibronic coupling) create too severe a test for the method–the error may amount to 30 to 40% for frequencies of the order of hundreds of cm^{-1}, as has been shown by D. Michalska, W. Zierkiewicz, D.C. Bieńko, W. Wojciechowski, and T. Zeegers-Huyskens, *J. Phys. Chem., A105,* 8734 (2001).

of CI are better than those of the Restricted Hartree-Fock method (RHF)–a feature guaranteed by the variational principle. As one can see, the RHF method indicates quite accurately the position of the minimum, although it makes there a clearly visible error in energy. In contrast to this, for large R, the method creates a kind of disaster. The duty of the perturbational MP2 and MP3 methods is to improve the RHF mess by adding some corrections. This difficult job is done very well for distances R close to the minimum. The duty is, however, too demanding for large internuclear distances, although even there the improvement is important, especially for the MP3 method.

10.18.2 Is the MP2 Method Size Consistent?

Let us see. From Eq. (10.96), we have $E_{MP2} = E_{HF} + \sum'_{a<b,p<q} \frac{|\langle ab|pq\rangle - \langle ab|qp\rangle|^2}{\varepsilon_a+\varepsilon_b-\varepsilon_p-\varepsilon_q}$. On the right side, the E_{HF} energy *is* size consistent, as it was shown at the beginning of this chapter. It is therefore sufficient to prove that the second term is also size consistent. For separated subsystems, the excitations $a \to p$ and $b \to q$ must correspond to the spinorbitals a and p belonging to the same molecule (and represent the Hartree-Fock orbitals for the subsystems). The same can be said for the spinorbitals b and q. We have, therefore (lim denotes the limit corresponding to all distances among the subsystems equal to infinity, and $E_{RHF}(A)$ stands for the Hartree-Fock energy of molecule A),

$$\begin{aligned}
\lim E_{MP2} &= \sum_A E_{RHF}(A) + \lim \sum'_{a<b,p<q} \frac{|\langle ab|pq\rangle - \langle ab|qp\rangle|^2}{\varepsilon_a+\varepsilon_b-\varepsilon_p-\varepsilon_q} \\
&= \sum_A E_{RHF}(A) + \sum_A \sum'_{a,b,p,q\in A} \frac{|\langle ab|pq\rangle - \langle ab|qp\rangle|^2}{\varepsilon_a+\varepsilon_b-\varepsilon_p-\varepsilon_q} \\
&\quad + \sum_{A<B} \sum'_{a,p\in A,b,q\in B} \lim \frac{|\langle ab|pq\rangle - \langle ab|qp\rangle|^2}{\varepsilon_a+\varepsilon_b-\varepsilon_p-\varepsilon_q} \\
&= \sum_A E_{MP2}(A) + 0,
\end{aligned}$$

because in the last term, the integral $\langle ab|pq\rangle$ vanishes as $\frac{1}{R_{AB}}$, while the integral $\langle ab|qp\rangle$ vanishes even faster (exponentially, because of the overlap of spinorbitals belonging to different molecules).

The result obtained means that

> the MP2 method is size consistent.

Example

The proofs of the size consistency should be reflected by numerical results in practical applications. Let us perform some routine calculations for two helium atoms[112] by using the HF, MP2, CISD, and CCSD methods. We perform the calculations for a single helium atom, and then for two separated helium atoms, but with the internuclear distance so large that there are serious grounds for rejecting any suspicion about their significant mutual interaction. Then, we will see whether the energy for the two atoms is twice the energy of a single atom (as it should be for size consistency). Well, how to decide about such a safe distance? A helium atom is an object of the diameter of about 2 Å (in a simple and naive view). The distance of about 30 Å should be sufficiently large to have the interaction energy negligible. The numerical results are collected in the following table:

	2 He	He_2 ($R = 30$ Å)
HF	**−5.7103209**	**−5.7103209**
MP2	**−5.7327211**	**−5.7327211**
CISD	**−5.7403243**	**−5.740**1954
CCSD	**−5.7403243**	**−5.7403243**

The numbers given confirm the theoretical considerations. The numbers in the second column (twice the energy of the isolated helium atom) and the third column (the energy of the two distant atoms) are identical to eight significant figures (shown in bold) for the HF, MP2, and CCSD methods. In contrast to that, according to what we know about the CI method, the CISD method is size inconsistent (the difference is on the fifth significant figure).

10.18.3 Convergence of the Møller-Plesset Perturbation Series

Does the Møller-Plesset perturbational series converge? Very often this question can be considered surrealist, since *most frequently* we carry out calculations through the second, third, and–at most–fourth order of perturbation theory. Such calculations usually give a satisfactory description of the physical quantities considered and we do not think about going to high orders requiring major computational effort. There were, however, scientists interested to see how fast the convergence is if very high orders are included (MPn) for $n < 45$. And there was a surprise (see Fig. 10.14).

[112] One may use, for example, the public domain www.webmo offering several quantum chemistry programs; we use here the Gaussian program with the atomic orbital basis set $6-31G(d)$.

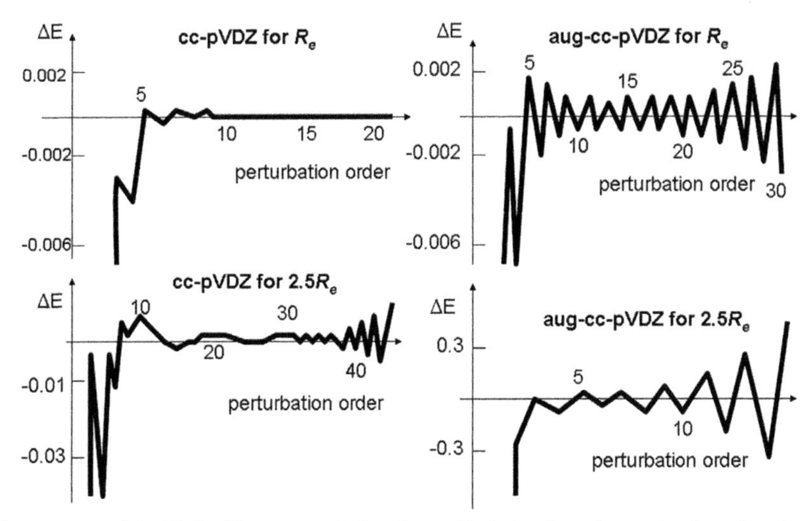

Fig. 10.14. Convergence of the Møller-Plesset perturbation theory (deviation from the exact value, given in a.u.) for the HF molecule as a function of the basis set used (cc-pVDZ and augmented cc-pVDZ) and assumed bond length, R_e denotes the HF equilibrium distance (following T. Helgaker, P. Jørgensen, and J. Olsen, *Molecular Electronic-Structure Theory* Wiley, Chichester, 2000, p. 780, Fig. 14.6). *Courtesy of the authors.*

It is true that the first few orders of the MP perturbation theory give reasonably good results, but later, the accuracy of the MP calculations gets worse. A lot depends on the atomic orbital basis set adopted and wealthy people (using the augmented basis sets, which is much more rare) encounter some difficulties, whereas poor ones (modest basis sets) do not. Moreover, for long bond lengths (2.5 of the equilibrium distance R_e), the MPn performance is worse. For high orders, the procedure is heading for a catastrophe[113] of the kind already described on p. 249. The reason for this is the highly excited and diffuse states used as the expansion functions.[114]

10.18.4 Special Status of Double Excitations

In Møller-Plesset perturbation theory, $\Delta E = E_0 - E_0^{(0)} = E_0 - E_{RHF} - E_0^{(0)} + E_{RHF} = E_{corel} + \left(E_{RHF} - E_0^{(0)}\right)$. On the other hand,[115] $\Delta E = E_0 - E_0^{(0)} = \left\langle \psi_0^{(0)} | \hat{H} \psi_0 \right\rangle - E_0^{(0)}$.

The function ψ_0 can be expanded in Slater determinants of various excitation rank (we use intermediate normalization): $\psi_0 = \psi_0^{(0)} + excitations$. Then, by equalizing the two expressions for ΔE obtained above, we have

[113] This is so except for the smaller basis set and the equilibrium bond length, but the problem has been studied up to $n = 21$.

[114] An analysis of this problem is given in T. Helgaker, P. Jørgensen, and J. Olsen, Molecular Electronic-Structure Theory, Wiley, Chichester (2000), p. 769.

[115] In this instance, we take advantage of the intermediate normalization $\left\langle \psi_0^{(0)} | \psi_0 \right\rangle = 1$ and $\left\langle \psi_0^{(0)} | \psi_0^{(0)} \right\rangle = 1$ and the fact that ψ_0 is an eigenfunction of $\hat{H}$.

$E_{corel} + E_{RHF} = \left\langle \psi_0^{(0)} | \hat{H} \psi_0 \right\rangle = \left\langle \psi_0^{(0)} | \hat{H} (\psi_0^{(0)} + excitations) \right\rangle = E_{RHF} + \left\langle \psi_0^{(0)} | \hat{H} \right.$ $\left. (excitations) \right\rangle$; hence

$$E_{corel} = \left\langle \psi_0^{(0)} | \hat{H} (excitations) \right\rangle . \tag{10.97}$$

The Slater-Condon rules (see Appendix M available at booksite.elsevier.com/978-0-444-59436-5, p. e109) show immediately that the only excitations that give nonzero contributions are the single and double excitations. Moreover, taking advantage of the Brillouin theorem, we obtain single excitation contributions exactly equal zero. So we get the result that

> the exact correlation energy can be obtained using only that part of a formula for the configuration interaction wave function ψ_{CI} that contains exclusively double excitations: $E_{corel} = \left\langle \psi_0^{(0)} | \hat{H} (double\ excitations\ only) \right\rangle$.

The problem, however, lies in the fact that these doubly excited determinants are equipped with coefficients obtained in the *full* CI method (i.e., with all possible excitations). How is this? We should draw attention to the fact that, in deriving the formula for ΔE, intermediate normalization is used. If someone gave us the normalized FCI wave functions as a Christmas gift,[116] then the coefficients occurring in the formula for ΔE would not be the double excitation coefficients in the FCI function. We would have to denormalize this function to have the coefficient for the Hartree-Fock determinant equal to 1. We cannot do this without knowledge of the coefficients for higher excitations.

It is as if somebody said: the treasure is hidden in our room, but to find it, you have to solve a very difficult problem in the kingdom of Far Far Away. Imagine a compass that leads you unerringly to that place in our room where the treasure is hidden. Perhaps a functional exists whose minimization would provide us directly with the solution, but we do not know it yet.[117]

Summary

- In the Hartree-Fock method, *electrons of opposite spins do not correlate their motion*[118] which is an absurd situation (in contrast to when electrons of the same spins avoid each other, which is reasonable). In many cases (like the F_2 molecule, description of dissociation of chemical bonds, or interaction of atoms and non-polar molecules), this leads to wrong results. In this chapter, we have learned about the methods that do take into account a correlation of electronic motions.

---

[116] Dreams...

[117] It looks like the work by H. Nakatsuji, *Phys. Rev.* A, *14*, 41 (1976), and M. Nooijen, *Phys. Rev. Letters, 84,* 2108 (2000) go in this direction.

[118] Note, however, that they repel each other (mean field) as if they were electron clouds.

Variational Methods Using Explicitly Correlated Wave Function

- Rely on using in the variational method a trial function that contains the explicit distance between the electrons. This improves the results significantly, but requires evaluation of *very* complex integrals.
- The correlation cusp condition, $\left(\frac{\partial\psi}{\partial r}\right)_{r=0} = \mu q_i q_j \psi(r=0)$ can be derived, where r is the distance of two particles with charges q_i and q_j, and μ is the reduced mass of the particles. This condition helps to determine the correct form of the wave function ψ. For instance, for the two electrons, the correct wave function has to satisfy (in a.u.): $\left(\frac{\partial\psi}{\partial r}\right)_{r=0} = \frac{1}{2}\psi(r=0)$.
- The family of variational methods with explicitly correlated functions includes the Hylleraas method, the Hylleraas CI method, the James-Coolidge and the Kołos-Wolniewicz approaches, as well as a method with exponentially correlated Gaussians. The method of explicitly correlated functions is very successful for two-, three-, and four-electron systems. For larger systems, due to the excessive number of complicated integrals, variational calculations are not yet feasible.

Variational Methods with Slater Determinants

- The configuration interaction (CI) approach is *a Ritz method* (see Chapter 5), which uses the expansion in terms of *known* Slater determinants. These determinants are constructed from the molecular spinorbitals (usually occupied and virtual ones) produced by the Hartree-Fock method.
- Full CI expansion usually contains an enormous number of terms and is not feasible. Therefore, the CI expansion must be truncated somewhere. Usually, we truncate it at a certain maximum rank of excitations with respect to the Hartree-Fock determinant (i.e., the Slater determinants corresponding to single, double, or up to some maximal excitations are included).
- Truncated (limited) CI expansion *is not size consistent*; i.e., the energy of the system of non-interacting objects is not equal to the sum of the energies of the individual objects (calculated separately with the same truncation pattern).
- The multiconfiguration self-consistent field (MC SCF) method is similar to the CI scheme, but we vary not only *the coefficients in front of the Slater determinants, but also the Slater determinants themselves* (changing the analytical form of the orbitals in them). We have learned about two versions: the classic one (where we optimize alternatively coefficients of Slater determinants and the orbitals) and a unitary one (where we optimize *simultaneously* the determinantal coefficients and orbitals).
- The complete active space self-consistent field (CAS SCF) method is a special case of the MC SCF approach and relies on the *selection* of a set of spinorbitals (usually separated energetically from others) and on construction from them of all possible Slater determinants within the MC SCF scheme. Usually, low-energy spinorbitals are *inactive* during this procedure; i.e., they all occur in *each* Slater determinant (and are either frozen or allowed to vary).

Non-variational Method Based on Slater Determinants

- The coupled-cluster (CC) method is an attempt to find such an expansion of the wave function in terms of the Slater determinants, which would preserve size consistency. In this method, the wave function for the electronic ground state is obtained as a result of the operation of the wave operator $\exp(\hat{T})$ on the Hartree-Fock function (this *ensures* size consistency). The wave operator $\exp(\hat{T})$ contains the cluster operator $\hat{T}$, which is defined as the sum of the operators for the l-tuple excitations, $\hat{T}_l$ up to a certain maximum $l = l_{\max}$. Each $\hat{T}_l$ operator is the sum of the operators each responsible for *a particular l-tuple excitation* multiplied by its *amplitude t*. The aim of the CC method is to find the t values since they determine the wave function and energy. The method generates nonlinear equations with respect to unknown t amplitudes. The CC method usually provides very good results.

- The equation-of-motion coupled-cluster (EOM-CC) method is based on the CC wave function obtained for the ground state and is designed to provide the electronic excitation energies and the corresponding excited-state wave functions.
- The many-body perturbation theory (MBPT) method is a perturbation theory in which the unperturbed system is usually described by a single Slater determinant. We obtain two basic equations of the MBPT approach for the ground-state wave function: $\psi_0 = \psi_0^{(0)} + \hat{R}_0 \left(E_0^{(0)} - E_0 + \hat{H}^{(1)} \right) \psi_0$ and $E_0 = E_0^{(0)} + \left\langle \psi_0^{(0)} | \hat{H}^{(1)} \psi_0 \right\rangle$, where $\psi_0^{(0)}$ is usually the Hartree-Fock function, $E_0^{(0)}$ the sum of the orbital energies, $\hat{H}^{(1)} = \hat{H} - \hat{H}^{(0)}$ is the fluctuation potential, and $\hat{R}_0$ the reduced resolvent (i.e., the "*almost*" inverse of the operator $E_0^{(0)} - \hat{H}^{(0)}$). These equations are solved in an iterative manner. Depending on the iterative procedure chosen, we obtain either the Brillouin-Wigner or the Rayleigh-Schrödinger perturbation theory. The latter is applied in the Møller-Plesset method.
- *One of the basic computational methods for the correlation energy is the MP2 method*, which gives the result correct through the second order of the Rayleigh-Schrödinger perturbation theory (with respect to energy).

Main Concepts, New Terms

From the Research Front

The computational cost in the Hartree-Fock method scales with the size N of the atomic orbital basis set as N^4 and, while using devices similar to *direct* CI, even[119] as N^3. However, after making the Hartree-Fock computations, we perform more and more frequently calculations of the electronic correlation. The main approaches used to this end are the MP2 method, the CC method with single and double excitations in $\hat{T}$ and partial inclusion of triple ones [the so-called CCSD(T) approach]. The CC method has been generalized for important cases involving chemical bond breaking.[120] The state of the art in CC theory currently includes the full CCSDTQP model, which incorporates into the cluster expansion all the operators through pentuple excitations.[121] The formulas in these formalisms become monstrous to such an extent, that scientists desperately invented an "anti-weapon": first, automatic (computer-based) derivation of the formulas is used, followed by automatic coding of the derived formulas into executable programs (usually using the Fortran). In such an approach we do not need to see our formulas...

The computational cost of the CCSD scheme scales as N^6. The computational strategy often adopted relies on obtaining the optimum geometry of the system with a less sophisticated method (e.g., Hartree-Fock) and, subsequently, calculating the wave function for that geometry with a more sophisticated approach (e.g., the MP2 that scales as N^5, MP4 or CCSD(T) scaling as N^7). In the next chapter, we will learn about the density functional theory (DFT), which represents an alternative to the above-mentioned methods.

Recoupling Quantum Chemistry with Nuclear Forces

The CC method has been designed first in the field of nuclear physics. This fact, however, had no consequences until recent years, since the numerical procedure has been judged by the community as untractable. Only because the quantum chemist Jiří Čížek accidentally looked up a nuclear physics journal, the idea diffused to quantum chemistry community and after some spectacular developments turned out to become the most successful in studying atoms and molecules. It turned out, however, that the idea went back to nuclear physics from quantum chemistry. The quantum chemistry CC technique has been applied to compute the energy levels for nucleons in several nuclei with much higher precision, than it was possible before.[122]

Nakatsuji Strategy

Hiroshi Nakatsuji looked at the Schrödinger equation from an unexpected side.[123] He wrote two equations:

$$\left\langle \delta\psi | (\hat{H} - E)\psi \right\rangle = 0, \tag{10.98}$$

$$\left\langle \psi | (\hat{H} - E)^2 \psi \right\rangle = 0 \tag{10.99}$$

and asked: what is their relation to the Schrödinger equation $(\hat{H} - E)\psi = 0$.

[119] This reduction is caused mainly by a preselection of the two-electron integrals. The preselection allows us to estimate the value of the integral without its computation and to reject the large number of integrals of values close to zero.

[120] P. Piecuch, M. Włoch, 123, 224105 (2005).

[121] M. Musiał S.A. Kucharski, and R.J. Bartlett, *J. Chem. Phys., 116,* 4382 (2002).

[122] M. Włoch, D.J. Dean, J.R. Gour, M. Hjorth-Jensen, K. Kowalski, T. Papenbrock, and P. Piecuch, *Phys. Rev. Letters, 94,* 212501 (2005).

[123] H. Nakatsuji, *J. Chem. Phys., 113,* 2949 (2000).

Hiroshi Nakatsuji, professor at Kyoto University, Japan, then professor at Quantum Chemistry Research Institute, Kyoto. When visiting Warsaw, he presented me his ingenious way of solving the Schrödinger equation. I was deeply impressed and said: "*You are a mathematician I presume?*". Professor Nakatsuji: "*No, I am just an organic chemist!*"

Note, that Eq. (10.98) follows from minimizing the functional $\left\langle \psi | \hat{H} \psi \right\rangle$ under normalization constraint[124] ($\langle \psi | \psi \rangle = 1$) of the trial function ψ. This is the essence of the variational method described in Chapter 5. Satisfaction of Eq. (10.98) may happen *either* because ψ fulfills the Schrödinger equation, *or*, at ψ not satisfying the Schrödinger equation, but optimal within the variational method restricted to a class of variations[125] $\delta\psi$. Anyway, if ψ satisfies Eq. (10.98), it does not necessarily represent a solution to the Schrödinger equation, it does with no restrictions imposed on $\delta\psi$.

Eq. (10.99) has a different status: it is satisfied only for the solution ψ of the Schrödinger equation.[126] Unfortunately, it contains the square of the Hamiltonian. This seems to hint that difficult integrals will be calculated in the future, but for the time being, we are going forward courageously.

Imagine, that the variation of ψ in Eq. (10.98) was chosen to have a very special form:

$$\delta\psi = (\hat{H} - E)\psi \cdot \delta C, \tag{10.100}$$

where C is a variational parameter in ψ. Then, from Eq. (10.98), we have the precious Eq. (10.99):

$$\left\langle (\hat{H} - E)\psi | (\hat{H} - E)\psi \right\rangle \cdot \delta C^* = 0,$$

and in such a case,[127] $(\hat{H} - E)\psi = 0$ (solution of the Schrödinger equation). It is seen, therefore, that the right side of Eq. (10.100) in a sense "forces" correct structure of the wave function, and hopefully this also takes place when we take an approximation instead of the exact (and unknown) energy E. Having this in mind, let us construct a variational function satisfying Eq. (10.100). But how do we get this? Well, let us begin an iterational game with functions ($n = 0, 1, 2, \ldots$ numbers the iterations, $\delta\psi$ represents an analog of $\psi_{n+1} - \psi_n$, we define $\bar{E}_n \equiv \left\langle \psi_n | \hat{H} \psi_n \right\rangle$) as

$$\psi_{n+1} = \left[1 + C_n(\hat{H} - \bar{E}_n)\right] \psi_n. \tag{10.101}$$

We start from an arbitrary function ψ_0, and in each iteration, we determine variationally the value of the coefficient C_n. We hope the procedure converges; i.e., what we get as the left side is the function inserted into the right side. If this happens, we achieve the satisfaction of

$$\psi = \left[1 + C(\hat{H} - E)\right] \psi, \tag{10.102}$$

where we have removed the lower indices because they do not matter at convergence. For $C \neq 0$, this means the achievement of our aim; i.e., $(\hat{H} - E)\psi = 0$.

As it turned out, this recipe needs some corrections when applied in practical calculations. In order to be able to calculate the integrals $\bar{E}_n = \left\langle \psi_n | \hat{H} \psi_n \right\rangle$ safely,[128] Nakatsuji considered what is known as the scaled Schrödinger

[124] A conditional minimum can be found by using the Lagrange multipliers method, as described in Appendix N available at booksite.elsevier.com/978-0-444-59436-5.

[125] If no restriction is imposed, the function found satisfies the Schrödinger equation.

[126] Indeed, $\left\langle \psi | (\hat{H} - E)^2 \psi \right\rangle = \left\langle \left(\hat{H} - E\right) \psi | (\hat{H} - E)\psi \right\rangle = ||(\hat{H} - E)\psi||^2 = 0$, where $||(\hat{H} - E)\psi||$ is the vector length. The latter equals 0 only if all the components of the vector equal 0. This means that in any point of space, we have $(\hat{H} - E)\psi = 0$.

[127] This happens because of the arbitrariness of δC.

[128] We have to calculate the mean values of higher and higher powers of the Hamiltonian. These integrals are notorious for diverging.

equation[129]:

$$g(\hat{H} - E)\psi = 0, \tag{10.103}$$

instead of the original one (satisfied by the same ψ), where the arbitrary function (of the electronic coordinates) g does not commute with the Hamiltonian, must be positive everywhere, except points of singularity, but even approaching a singularity, it has to be $\lim gV \neq 0$. Thus, the philosophy behind function g is to destroy the "singularity character in singularities" and, at the same time not to destroy the precious information about these singular points, present in the potential energy V. Several possibilities have been tested (e.g., $g = \frac{1}{-V_{ne}+V_{ee}}$ or $g = -\frac{1}{V_{ne}V_{ee}}$, etc.), where V_{ne} and V_{ee} are the Coulomb potential energy of the electron-nucleus and electron-electron interactions.[130]

The results witness about great effectiveness of this iterative method. For example,[131] in a little more than 20 iterations, the Schrödinger equation was practically solved (with nearly 100% of the correlation energy within finite basis sets) for molecules HCHO, CH_3F, HCN, CO_2, and C_2H_4. Analytical calculations[132] for H_2 within four to six iterations gave the electronic energy (at the equilibrium distance) with 15 significant figures (independently of several tested starting functions ψ_0). Similar calculations for the helium atom gave an accuracy of over 40 digits.[133]

No doubt, Nakatsuji's idea does represent not only a fresh look at the quantum theory, but it also has a significant practical power. It remains to learn what the complicated final form of the wave function is telling us. This, however, pertains also to wave functions produced by many other methods.

Ad Futurum

Experimental chemistry is focused, in most cases, on molecules of a *larger* size than those for which fair calculations with correlation are possible. However, after thorough analysis of the situation, it turns out that the cost of the calculations does not necessarily increase very fast with the size of a molecule.[134] Employing localized molecular orbitals and using the multipole expansion (see Appendix X available at booksite.elsevier.com/978-0-444-59436-5) of the integrals involving the orbitals separated in space causes, for elongated molecules, the cost of the post-Hartree-Fock calculations to scale linearly with the size of a molecule.[135] It can be expected that if the methods described in this chapter are to survive in practical applications, such a step has to be made.

There is one more problem, which will probably be faced by quantum chemistry when moving to larger molecules containing heteroatoms. Nearly all the methods, including electron correlation, described so far (with the exception of the explicitly correlated functions) are based on the silent and pretty "obvious" assumption, that the higher the excitation we consider, the higher the configuration energy we get. This assumption seems to be satisfied so far, but the molecules considered were always small, and the method has usually been limited to a small number of excited electrons. This assumption can be challenged in certain cases.[136] The multiple excitations in large molecules containing easily polarizable fragments can result in electron transfers that cause energetically favorable strong electrostatic interactions ("*mnemonic effect*"[137]) that lower the energy of the configuration. The reduction can be large enough to make the energy of the *formally* multiply excited determinant close to that of the Hartree-Fock

[129] H. Nakatsuji, *Phys. Rev. Lett.*, 93, 30403 (2004). In this reference Nakatsuji's standard method is described.

[130] The integration difficulty can be circumvented also by considering satisfaction (in points of space) of the Schrödinger equation in the form $\frac{\hat{H}\psi}{\psi}$ = const as described in H. Nakatsuji, H. Nakashima, Y. Kurokawa, and A. Ishikawa, *Phys. Rev. Lett.*, *99*, 240402 (2007).

[131] H. Nakatsuji, *Bull. Chem. Soc. Japan, 78*, 1705 (2005).

[132] Iterations result in a (nested) analytical form of the wave function.

[133] H. Nakashima, and H. Nakatsuji, *J. Chem. Phys.*, *127*, 224104 (2007).

[134] H.-J. Werner, *J. Chem. Phys.*, *104*, 6286 (1996).

[135] See e.g., W. Li, P. Piecuch, J.R. Gour, S. Li, *J. Chem. Phys.*, *131*, 114109 (2009).

[136] There are exceptions though; see A. Jagielska, and L. Piela, *J. Chem. Phys.*, 112, 2579 (2000).

[137] L.Z. Stolarczyk and L. Piela, *Chem. Phys. Letters*, *85*, 451 (1984).

determinant. Therefore, it should be taken into account on the same footing as Hartree-Fock. This is rather unfeasible for the methods discussed above.

The explicitly correlated functions have a built-in adjustable and efficient basic mechanism accounting for the correlation within the interacting electronic pair. The mechanism is based on the obvious thing: the electrons should avoid each other.[138]

Let us imagine the CH_4 molecule and look at it from the viewpoint of localized orbitals. With the method of explicitly correlated geminal functions for bonds, we would succeed in making the electrons avoid each other within the same bond. And what should happen if the center of gravity of the electron pair of one of the bonds shifts toward the carbon atom? The centers of gravity of the electron pairs of the remaining three bonds should move away along the CH bonds. The wave function must be designed in such a way that it accounts for this. In current theories, this effect is either deeply hidden or entirely neglected. A similar effect may happen in a polymer chain. One of the natural correlations of electronic motions should be a shift of electron pairs of all bonds in the same phase. As a highly many-electron effect the latter is neglected in current theories. However, the purely correlational Axilrod-Teller effect in the case of linear configuration, discussed in Chapter 13 (three-body dispersion interaction in the third order of perturbation theory), suggests clearly that the correlated motion of many electrons should occur.

Additional Literature

A. Szabo and N. S. Ostlund, *Modern Quantum Chemistry*, McGraw-Hill, New York (1989).

This classical book gives a detailed and crystal clear description of most important methods used in quantum chemistry.

T. Helgaker, P. Jørgensen, and J. Olsen, *Molecular Electronic Structure Theory*, Wiley, Chichester (2000).

Practical information on the various methods accounting for electron correlation presented in a clear and competent manner.

Questions

1. Hartree-Fock method
   a. describes the electrons with their positions being completely independent
   b. introduces the correlation of motion of electrons with the same spin coordinate
   c. does not introduce any correlation of motion of electrons with the opposite spin coordinates
   d. ignores the Coulomb hole, but takes care of the Fermi hole
2. The ground state of helium atom in the Hartree-Fock method:
   a. if one electron is on the nucleus, the probability of finding the second one in a small volume dV is also the largest on the nucleus
   b. if electron 1 is on one side of the nucleus, electron 2 is easiest to find on the nucleus
   c. if both electrons are at the same distance from the nucleus, it is equally easy to find them in the same point as in two points opposite to each other with respect to the nucleus
   d. if both electrons are at the same distance from the nucleus, they will tend to be on the opposite sides of the nucleus
3. The CI method truncated at double excitations gives energy E_{BeBe} for two beryllium atoms at large distance R. In calculations by using this method:
   a. if $R \longrightarrow \infty$, there will be $E_{BeBe} = 2E_{Be}$

[138] In special conditions, one electron can follow the other, forming a Cooper pair. The Cooper pairs are responsible for the mechanism of superconductivity. This will be a fascinating field of research for chemist-engineered materials in the future.

b. if $R \longrightarrow \infty$, one will obtain $E_{BeBe} - 2E_{Be} = const \neq 0$
c. if $R \longrightarrow \infty$, one will get $E_{BeBe} = 2E_{Be}$, but under condition that the CI calculation for the individual beryllium atom was limited to double excitations
d. the result obtained contains an error coming from the size inconsistency

4. The CC method (with the cluster operator truncated at double excitations) gives energy E_{BeBe} for two beryllium atoms at very large internuclear distance R, and the energy E_{Be} for a single beryllium atom. In the calculations using this method:

a. if $R \longrightarrow \infty$, there will be $E_{BeBe} = 2E_{Be}$
b. if $R \longrightarrow \infty$, one will obtain $E_{BeBe} - 2E_{Be} = const \neq 0$
c. if $R \longrightarrow \infty$, one will get $E_{BeBe} = 2E_{Be}$, but under condition that the CC calculation for the individual beryllium atom was limited to single excitations
d. the result obtained contains an error coming from the size inconsistency

5. The cusp condition for collision of two charged particles (μ means the reduced mass, all quantities in a.u.):

a. follows from the requirement that a wave function cannot acquire infinite values
b. for an electron and an atomic nucleus of charge Z reads as $\left(\frac{\partial \psi}{\partial r}\right)_{r=0} = Z\psi(r=0)$
c. for two electrons: $\left(\frac{\partial \psi}{\partial r}\right)_{r=0} = \frac{1}{2}\psi(r=0)$
d. for any two particles with charges q_1 and q_2 : $\left(\frac{\partial \psi}{\partial r}\right)_{r=0} = \mu q_1 q_2 \psi(r=0)$

6. The wave function $\psi(\mathbf{r}_1, \mathbf{r}_2) = N\left(1 + \frac{1}{2}r_{12}\right)\exp\left[-\frac{1}{4}\left(r_1^2 + r_2^2\right)\right]$ (N stands for the normalization constant, $\mathbf{r}_1$ and $\mathbf{r}_2$ denote the radius vectors for two electrons, respectively, r_{12} means their distance) represents:

a. an exact wave function for harmonium ("*harmonic helium atom*") with the force constant equal to $\frac{1}{4}$
b. an orbital occupied by electrons 1 and 2
c. a product of two orbitals
d. a geminal that takes into account the Coulomb hole

7. A helium atom with an approximate wave function (see question 6): $\psi(\mathbf{r}_1, \mathbf{r}_2) = N\left(1 + \frac{1}{2}r_{12}\right)\exp\left[-\frac{1}{4}\left(r_1^2 + r_2^2\right)\right]$. From this function it follows

a. if the nucleus-electron distance is the same for the two electrons, the electrons will have a tendency to be more often on the opposite sides of the nucleus
b. that finding the electrons at the same point in space is more probable for smaller nucleus-electron distances
c. it takes into account the Fermi hole
d. that the electrons are always on the opposite sides of the nucleus

8. An intermediate normalization of the wave function ψ_0 and the normalized function $\psi_0^{(0)}$ means that:

a. $\left\langle \psi_0 | \psi_0 - \psi_0^{(0)} \right\rangle = 1$
b. the Hilbert space vector ψ_0 is composed of the unit vector $\psi_0^{(0)}$ plus some vectors that are orthogonal to $\psi_0^{(0)}$
c. $\langle \psi_0 | \psi_0 \rangle \neq 1$
d. $\left\langle \psi_0 | \psi_0^{(0)} \right\rangle = 1$

9. The Møller-Plesset method, known as MP2:

a. is equivalent to the Ritz variational method (CI procedure) with the double excitations only
b. is based on the perturbational approach with the Hartree-Fock wave function as the unperturbed
c. represents a perturbational approach with calculation of the electronic energy up to the second order; the zeroth order plus the first order energies gives the Hartree-Fock energy

d. in this method the zeroth-order electronic energy represents a sum of the orbital energies of all spinorbitals present in the Hartree-Fock Slater determinant.

10. To calculate the electronic correlation energy
    a. it is sufficient to carry out calculations within the Hartree-Fock method, and then to perform the full CI computation
    b. it is sufficient to know the Hartree-Fock energy and all ionization potentials for the system
    c. one has to use an explicitly correlated variational wave function
    d. it is sufficient to know a wave function expansion containing only the double excitations, but with their CI coefficients obtained in presence of all excitations

Answers

1b,c,d, 2a,b,c, 3b,d, 4a, 5a,c,d, 6a,d, 7a,b, 8b,c,d, 9b,c,d, 10a,d

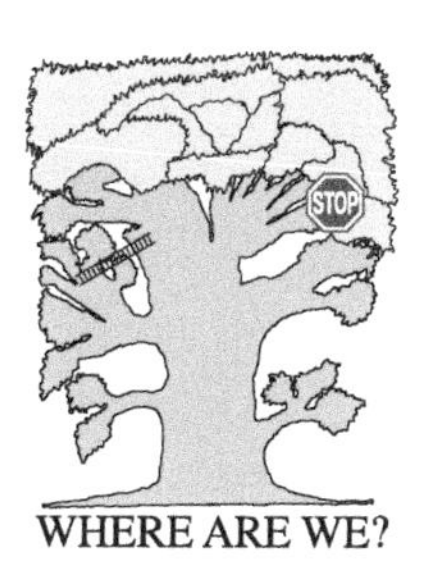

CHAPTER 11

Chasing Correlation Dragon: Density Functional Theory (DFT)

"As I observe, meditate and pray
Are we like clouds on a summer's day?"
Ruth Oliver, "Clouds"

Where Are We?

We are on an upper-right-side branch of the TREE.

An Example

A metal represents a system that is very difficult to describe using the quantum chemistry methods given so far. The Restricted Hartree-Fock (RHF) method here offers a very bad, if not pathological, approximation (cf. Chapter 9, p. 555), because the HOMO-LUMO gap equals zero in metal. The methods based on the Slater determinants (CI, MC SCF, CC, etc., as discussed in Chapter 10) are ruled out as involving a giant number of excited configurations to be taken into account because of the continuum of the occupied and virtual energy levels (see Chapter 9). Meanwhile, in the past, some properties of metals could be obtained, from simple theories that assumed that the electrons in a metal behave similarly to a homogeneous *electron gas* (also known as *jellium*), and the nuclear charge (to make the whole system neutral) has been treated as smeared out uniformly in the metal volume. Something physically important has to be captured in such theories.

What Is It All About?

Ideas of Quantum Chemistry, Second Edition. http://dx.doi.org/10.1016/B978-0-444-59436-5.00011-8

The preceding chapter showed how difficult it is to calculate correlation energy. Basically, there are two approaches: either to follow configuration interaction type methods (CI, MC SCF, CC, etc.), or to go in the direction of explicitly correlated functions. The first means a barrier of more and more numerous excited configurations to be taken into account, while the second involves very tedious and time-consuming integrals. In both cases, we know the Hamiltonian and fight for a satisfactory wave function (often using the variational principle, as discussed in Chapter 5). It turns out that there is also a third direction (presented in this chapter) that does not regard configurations (except a single special one) and does not have the bottleneck of difficult integrals. Instead, we have the kind of wave function in the form of a single Slater determinant, but we have a serious problem of defining the proper Hamiltonian.

The ultimate goal of the density functional theory (DFT) method is the calculation of the total energy of the system and the ground-state electron density distribution without using the wave function of the system.

Why Is This Important?

The DFT calculations (despite taking electronic correlation into account) are not expensive; their cost is comparable to that of the Hartree-Fock method. Therefore, the same computer power allows us to explore much larger molecules than with other post-Hartree-Fock (correlation) methods.

What Is Needed?

- The Hartree-Fock method (Chapter 8)
- The perturbational method (Chapter 5, advised)
- Lagrange multipliers (see Appendix N available at booksite.elsevier.com/978-0-444-59436-5, p. e121, advised)

Classic Works

The idea of treating electrons in metal as an electron gas was conceived in 1900, independently, by Lord Kelvin[1] and by Paul Drude.[2] ★ The concept explained the electrical conductivity of metals, and was then used by Llewellyn Hilleth Thomas in "*The calculation of atomic fields*," published in *Proceedings of the Cambridge Philosophical Society*, *23*, 542 (1926), as well as by Enrico Fermi in "*A statistical method for the determination of some atomic properties and the application of this method to the theory of the periodic system of elements*," in *Zeitschrift für Physik*, *48*, 73 (1928). They (independently) calculated the electronic kinetic energy per unit volume of the electron gas (this is known as the *kinetic energy density*) as a function of the local electron density ρ. ★ In 1930, Paul Adrien Maurice Dirac presented a similar result in "*Note on the exchange phenomena in the Thomas atom*," in *Proceedings of the Cambridge Philosophical Society*, *26*, 376 (1930), for the *exchange energy* as a function of ρ. ★ In a classic paper "*A simplification of the Hartree-Fock method*," published in *Physical Review*, *81*, 385 (1951), John Slater showed that the Hartree-Fock method applied to metals gives the exchange energy density proportional to $\rho^{\frac{1}{3}}$. ★ For classical positions, specialists often use a book by Pál Gombas *Die statistische Theorie des Atoms und ihre Anwendungen*, Springer Verlag, Wien (1948). ★ The contemporary theory was born in 1964–1965, when two fundamental works appeared: Pierre Hohenberg and Walter Kohn published "*Inhomogeneous electron gas*," in *Physical Review*, *136*, B864 (1964); and Walter Kohn and Lu J.Sham published "*Self-consistent equations including exchange and correlation effects*" in *Physical Review*, *A140*, 1133 (1965). ★ Mel Levy, in the article "*Electron densities in search of Hamiltonians*," published in *Physical Review*, *A26*, 1200 (1982), proved that the variational principle in quantum chemistry can be equivalently presented as a minimization of the Hohenberg-Kohn functional that depends on the electron density ρ. ★ Richard F.W. Bader wrote a book on mathematical analysis of the electronic density called *Atoms in Molecules. A Quantum Theory*, Clarendon Press, Oxford (1994), that enabled chemists to look at molecules in a synthetic way, independent of the level of theory that has been used to describe it. ★ Erich Runge and Eberhard K.U. Gross in "*Density-functional theory for time-dependent systems*," published in *Phys. Rev. Lett.*, *52*, 997 (1984), have extended the Hohenberg-Kohn-Sham formalism to time domain.

11.1 Electronic Density–The Superstar

In the DFT method, we begin with the Born–Oppenheimer approximation, which allows us to obtain the electronic wave function corresponding to fixed positions of the nuclei. We will be interested in the ground-state of the system.

Let us introduce a notion of the *first-order density matrix*:[3]

$$\rho(\mathbf{r};\mathbf{r}') = \rho_\alpha(\mathbf{r};\mathbf{r}') + \rho_\beta(\mathbf{r};\mathbf{r}'), \tag{11.1}$$

which we define as follows (α stands for the spin coordinate $\sigma = \frac{1}{2}$, and β means $\sigma = -\frac{1}{2}$):

$$\rho_\sigma(\mathbf{r};\mathbf{r}') = N \int \mathrm{d}\tau_2 \mathrm{d}\tau_3 \ldots \mathrm{d}\tau_N \Psi^*(\mathbf{r}', \sigma, \mathbf{r}_2, \sigma_2, \ldots, \mathbf{r}_N, \sigma_N)\Psi(\mathbf{r}, \sigma, \mathbf{r}_2, \sigma_2, \ldots, \mathbf{r}_N, \sigma_N). \tag{11.2}$$

[1] Lord Kelvin was born William Thomson (1824–1907), British physicist and mathematician, professor at the University of Glasgow. His main contributions are in thermodynamics (the second law, internal energy), theory of electric oscillations, theory of potentials, elasticity, hydrodynamics, etc. His great achievements were honored by the title of Lord Kelvin in 1892.

[2] Paul Drude (1863–1906) was a German physicist and professor at the universities in Leipzig, Giessen, and Berlin.

[3] The "indices" of this "matrix element" are $\mathbf{r}$ and $\mathbf{r}'$.

Thus, we integrate $N\Psi^*\Psi$ over all electron coordinates except electron number 1 (and, just to preserve an additional mathematical freedom, we assign *two distinct positions*: $\mathbf{r}$ and $\mathbf{r}'$ for electron 1).

The key quantity in this chapter will be the diagonal element of $\rho(\mathbf{r};\mathbf{r}')$ and $\rho_\sigma(\mathbf{r};\mathbf{r}')$ i.e., $\rho(\mathbf{r};\mathbf{r}) \equiv \rho(\mathbf{r})$ and $\rho_\sigma(\mathbf{r};\mathbf{r}) \equiv \rho_\sigma(\mathbf{r})$, respectively, where obviously

$$\rho(\mathbf{r}) = \rho_\alpha(\mathbf{r}) + \rho_\beta(\mathbf{r}). \tag{11.3}$$

which is an observable physical quantity.

The wave function Ψ is antisymmetric with respect to the exchange of the coordinates of any two electrons, and, therefore $|\Psi|^2$ is symmetric with respect to such an exchange. Hence, the definition of ρ is independent of the label of the electron we do not integrate over. According to this definition,

ρ represents nothing else but the density of the electron cloud carrying N electrons, because (integration over the whole 3D space)

$$\int \rho(\boldsymbol{r})\mathrm{d}\boldsymbol{r} = N. \tag{11.4}$$

Therefore, the electron density distribution $\rho(\boldsymbol{r})$ is given for a point $\boldsymbol{r}$ in the units: the number of electrons per volume unit. Since $\rho(\boldsymbol{r})$ represents an integral of a non-negative integrand, $\rho(\boldsymbol{r})$ is always non-negative. Let us check that ρ may be also defined as the mean value of the *electron density operator* $\hat{\rho}(\boldsymbol{r}) = \sum_{i=1}^N \delta(\boldsymbol{r}_i - \boldsymbol{r})$, a sum of the Dirac delta operators (cf. Appendix E available at booksite.elsevier.com/978-0-444-59436-5 on p. e69) for individual electrons at position $\boldsymbol{r}$:

$$\langle \Psi \mid \hat{\rho}\Psi\rangle = \langle \Psi \mid \left(\sum_{i=1}^N \delta(\boldsymbol{r}_i - \boldsymbol{r})\right)\Psi\rangle = \sum_{i=1}^N \langle \Psi \mid \left(\delta(\boldsymbol{r}_i - \boldsymbol{r})\right)\Psi\rangle = \rho(\boldsymbol{r}). \tag{11.5}$$

Indeed, each of the integrals in the summation is equal to[4] $\rho(\boldsymbol{r})/N$, the summation over i gives N; therefore, we obtain $\rho(\boldsymbol{r})$.

If the function Ψ is taken as a normalized Slater determinant built of N spinorbitals ϕ_i, from the I rule of Slater-Condon (see Appendix M available at http://booksite.elsevier.com/978-0-444-59436-5, we replace there $\hat{h}$ with δ) for $\langle \Psi \mid (\sum_{i=1}^N \delta(\boldsymbol{r}_i - \boldsymbol{r}))\Psi\rangle$,

[4] Please remember $\langle | \rangle$ means integration over space coordinates and summation over spin coordinates.

we obtain[5]

$$\rho(\boldsymbol{r}) = \langle\phi_1(1)\,|\delta(\boldsymbol{r}_1-\boldsymbol{r})\phi_1(1)\rangle_1 + \langle\phi_2(1)|\,\delta(\boldsymbol{r}_1-\boldsymbol{r})\phi_2(1)\rangle_1$$
$$+\ldots\langle\phi_N(1)\,|\delta(\boldsymbol{r}_1-\boldsymbol{r})\phi_N(1)\rangle_1$$
$$=\sum_{i=1}^{N}\sum_{\sigma=-\frac{1}{2},+\frac{1}{2}}|\phi_i(\boldsymbol{r},\sigma)|^2=\sum_{i=1}^{N}\sum_{\sigma}\left|\phi_i\left(\boldsymbol{r},\frac{1}{2}\right)\right|^2+\sum_{i=1}^{N}\sum_{\sigma}\left|\phi_i\left(\boldsymbol{r},-\frac{1}{2}\right)\right|^2$$
$$\equiv\rho_\alpha(\boldsymbol{r})+\rho_\beta(\boldsymbol{r}). \tag{11.6}$$

In $\rho_\alpha(\boldsymbol{r})$, we have only those spinorbitals ϕ_i, which have the spin function α, similarly in $\rho_\beta(\boldsymbol{r})$ we have those with the spin function β.

If additionally we assume the double occupancy of the molecular orbitals, we have

$$\rho(\boldsymbol{r})=\sum_{i=1}^{N}\sum_{\sigma}|\phi_i(\boldsymbol{r},\sigma)|^2=\sum_{i=1}^{N/2}\sum_{\sigma}|\varphi_i(\boldsymbol{r})\,\alpha(\sigma)|^2+\sum_{i=1}^{N/2}\sum_{\sigma}|\varphi_i(\boldsymbol{r})\,\beta(\sigma)|^2$$
$$=\sum_{i=1}^{N/2}2\,|\varphi_i(\boldsymbol{r})|^2,$$

where φ_i stand for the molecular orbitals. We see that when we admit the open shells, we have

in the one-determinantal approximation,

$$\rho(\boldsymbol{r})=\sum_i n_i\,|\varphi_i(\boldsymbol{r})|^2, \tag{11.7}$$

with $n_i = 0, 1, 2$ denoting orbital occupancy in the Slater determinant.

11.2 Electron Density Distributions- Bader Analysis

11.2.1 Overall Shape of ρ

Imagine an electron cloud with a charge distribution[6] that carries the charge of N electrons. Unlike a storm cloud, the electron cloud does not change in time (stationary state), but it has density $\rho(\boldsymbol{r})$ that changes in space (similar to the storm cloud). Inside the cloud, the nuclei are located. The function $\rho(\boldsymbol{r})$ exhibits non-analytical behavior (discontinuity of its gradient) at the positions of the nuclei, which results from the poles ($-\infty$) of the potential energy at these

[5] After renaming the electron coordinates in the integrals, the integration is over the spatial and spin coordinates of electron 1. This expression is invariant with respect to any unitary transformation of the molecular orbitals; cf. Chapter 8.

[6] This is similar to a storm cloud in the sky.

positions. Recall the shape of the 1s wave function for the hydrogen-like atom (see Fig. 4.20), it has a spike at $r = 0$. In Chapter 10, it was shown that the correct electronic wave function has to satisfy the cusp condition in the neighborhood of each of the nuclei, where ρ changes as $\exp(-2Zr)$ (p. 205). This condition results in spikes of $\rho(\boldsymbol{r})$ exactly at the positions of the nuclei (see Fig. 11.1a). How sharp the spike is depends on the charge of the nucleus Z: an

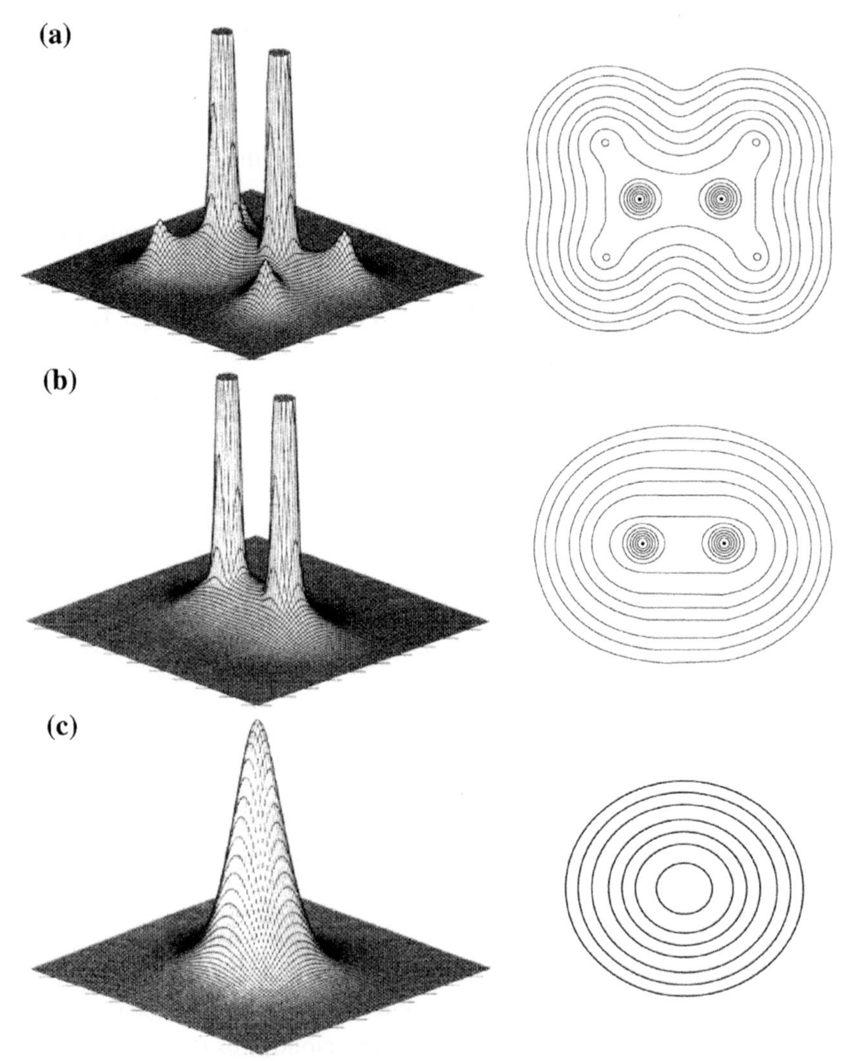

Fig. 11.1. Electron density ρ for the planar ethylene molecule shown in three cross sections. $\int \rho(\boldsymbol{r})d\boldsymbol{r} = 16$, the number of electrons in the molecule. Panel (a) shows the cross section *within the molecular plane*. The positions of the nuclei can be easily recognized by the "*spikes*" of ρ (obviously much more pronounced for the carbon atoms than for the hydrogens atoms), their charges can be computed from the slope of ρ. Panel (b) shows the cross section *along the CC bond perpendicular to the molecular plane*; therefore, only the maxima at the positions of the carbon nuclei are visible. Panel (c) is the cross section *perpendicular to the molecular plane and intersecting the CC bond* (through its center). It is seen that ρ decays monotonically with the distance from the bond center. Most interesting, however, is that the cross section resembles an ellipse rather than a circle. Note that we do not see any separate σ or π densities. This is what the concept of π bond is all about, just to reflect the bond cross section ellipticity. R.F.W. Bader, T.T. Nguyen-Dang, and Y.Tal, *Rep. Progr. Phys.*, *44*, 893 (1981); *courtesy of Richard Bader.*

infinitesimal deviation from the position of the nucleus (p. 586)[7] has to be accompanied by such a decreasing of the density[8] that $\frac{\partial \rho}{\partial r}/\rho = -2Z$.

Thus, because of the Coulombic interactions, the electrons will concentrate close to the nuclei, and therefore, we will have maxima of ρ right on them. At great distances from the nuclei, the density ρ will decay to practically zero with the asymptotics $\exp[-2\sqrt{2I}r]$, where I being the first ionization potential. Further details will be of great interest–for example, are there any concentrations of ρ besides the nuclei, such as in the regions *between* nuclei? If yes, will it happen for every pair of nuclei, or for some pairs only? This is of obvious importance for chemistry, which deals with the idea of chemical bonds between atoms and a model of the molecule as the nuclei kept together by a chemical bond pattern.

11.2.2 Critical Points

For analysis of any smooth function, including the electronic density as a function of the position in space, the *critical* (or *stationary*) *points* are defined as those for which we see the vanishing of the gradient

$$\nabla \rho = \mathbf{0}.$$

These are maxima, minima, and saddle points. If we start from an arbitrary point and follow the direction of $\nabla \rho$, we end up at a maximum of ρ. Its position may correspond to any of the nuclei or to a non-nuclear concentration distribution (Fig. 11.2). Formally, positions of the nuclei are not the stationary points because $\nabla \rho$ has a discontinuity here connected to the cusp condition (see Chapter 10, p. 585), but the largest maxima correspond to the positions of the nuclei. Maxima may appear not only at the positions of the nuclei, but also elsewhere[9] (*non-nuclear attractors*, (Fig. 11.2a). The compact set of starting points which converge in this way

[7] If nonzero size nuclei were considered, the cusps would be rounded (within the size of the nuclei), the discontinuity of the gradient would be removed, and regular maxima would be observed.

[8] It has been shown [P.D. Walker and P.G. Mezey, *J.Am. Chem. Soc.*, *116*, 12022 (1994)] that despite the non-analytical character of ρ (because of the spikes), the function ρ has the following remarkable property: *if we know* ρ even in the smallest volume, this determines ρ in the whole space. A by-product of this theorem is of interest to chemists. Namely, this means that a functional group in two different molecules or in two conformations of the same molecule cannot have an identical ρ characteristic for it. If it had, from ρ in its neighborhood we would be able to reproduce the whole density distribution $\rho(\boldsymbol{r})$, but for which of the molecules or conformers? Therefore, by *reductio ad absurdum*, we have the result: it is impossible to define (with all details) the notion of a functional group in chemistry. This is analogous to the conclusion drawn in Chapter 8 about the impossibility of a rigorous definition of a chemical bond (p. 468). This also shows that chemistry and physics (relying on mathematical approaches) profit very much, and further, are heavily based on some ideas that mathematics destroys in a second. Nevertheless, without these ideas, natural sciences would lose their generality, efficiency, and beauty.

[9] For example, imagine a few dipoles with their positive poles oriented toward a point in space. If the dipole moments exceed some value, it may turn out that around this point, there will be a concentration of electron density with a maximum there. This is what happens in certain dipoles, in which an electron is far away from the nuclear framework (sometimes as far as 50 Å) and keeps following the positive pole of the dipole ("*a dipole-bound electron*") when the dipole rotates in space; see, e.g., J. Smets, D.M.A. Smith, Y. Elkadi, and L. Adamowicz, *Pol. J. Chem.*, *72*, 1615 (1998).

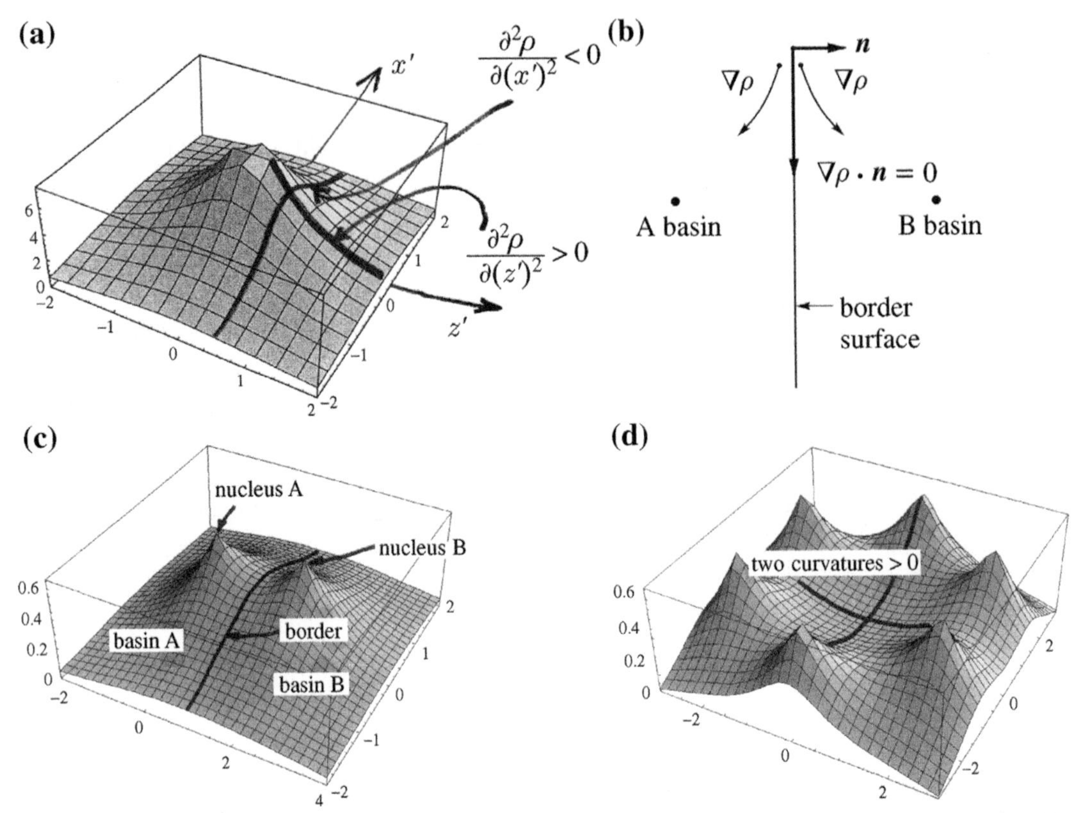

Fig. 11.2. How does the electronic density change in space? Panel (a) illustrates the non-nuclear attractor (maximum of ρ). Note that we can tell the signs of some second derivatives (curvatures) computed at the intersection of black lines (slope), the radial curvature $\frac{\partial^2\rho}{\partial(z')^2}$ is positive, while the two lateral ones (only one of them: $\frac{\partial^2\rho}{\partial(x')^2}$ is shown) are negative. If for the function shown, the curvatures were computed at the maximum, all three curvatures would be negative. Panel (b) shows the idea of the border surface separating two basins of ρ corresponding to two nuclei: A and B. Right at the border between the two basins, the force lines of $\nabla\rho$ diverge: if you take a step left from the border, you end up in nucleus A, and if you take a step right, you get into the basin of B. Just at the border, you have to have $\nabla\rho \cdot \boldsymbol{n} = 0$ because the two vectors: $\nabla\rho$ and $\boldsymbol{n}$ are perpendicular. (c) The same as panel (b) showing additionally the density function for chemical bond AB. The border is shown as a black line. Two of three curvatures are negative (one of them shown), the third one is positive. Panel (d) illustrates the electronic density distribution in benzene. In the middle of the ring, two curvatures are positive (shown), and the third curvature is negative (not shown). If the curvatures were computed in the center of the fullerene (not shown), all three curvatures would be positive (because the electron density increases when going out of the center).

(i.e., following $\nabla\rho$) to the same maximum is called the *basin of attraction of this maximum*, and the position of the maximum is known as an *attractor*. We have therefore the nuclear and non-nuclear attractors and basins. A basin has its neighbor-basins, and the border between the basins represents a surface satisfying $\nabla\rho \cdot \boldsymbol{n} = 0$, where $\boldsymbol{n}$ is a unit vector perpendicular to the surface (Fig. 11.2b).

In order to tell whether a particular critical point represents a maximum (non-nuclear attractor), a minimum or a saddle point we have to calculate at this point the Hessian; i.e., the matrix of the second derivatives: $\left\{\frac{\partial^2\rho}{\partial\xi_i\partial\xi_j}\right\}$, where $\xi_1 = x$, $\xi_2 = y$, $\xi_3 = z$. Now, the stationary point is used as the origin of a local Cartesian coordinate system, which will be rotated in such a way as to obtain the Hessian matrix (computed in the rotated coordinate system) diagonal. This means that the rotation has been performed in such a way that the axes of the new local coordinate system are collinear with the principal axes of a quadratic function that approximates ρ in the neighborhood of the stationary point (this rotation is achieved simply by diagonalization of the Hessian $\left\{\frac{\partial^2 \rho}{\partial \xi_i \partial \xi_j}\right\}$; see Appendix K available at booksite.elsevier.com/978-0-444-59436-5). The diagonalization gives three eigenvalues. We have the following possibilities (the case when the Hessian matrix is singular will be considered later on):

- *All three eigenvalues are negative–we have a maximum of* ρ (non-nuclear attractor; Fig. 11.2a).
- *All three eigenvalues are positive–we have a minimum of* ρ. The minimum appears when we have a cavity; e.g., in the center of fullerene. When we leave this point, independent of the direction of this motion, the electron density increases.
- *Two eigenvalues are positive, one is negative–we have a first-order saddle point of* ρ. The center of the benzene ring may serve as an example (Fig. 11.2d). If we leave this point in the molecular plane in any of the two independent directions, ρ increases (thus, a minimum of ρ within the plane, the two eigenvalues positive), but when leaving perpendicularly to the plane, the electronic density decreases (thus a maximum of ρ along the axis, the negative eigenvalue).
- *One eigenvalue is positive, while two are negative–we have a second-order saddle point of* ρ. It is a very important case, because this is what happens at any covalent chemical bond (Figs. 11.1 and 11.2c). In the region between *some*[10] nuclei of a polyatomic molecule, we may have such a critical point. When we go perpendicularly to the bond in any of the two possible directions, ρ decreases (a maximum within the plane, two eigenvalues negative), while going toward any of the two nuclei, ρ increases (to achieve maxima at the nuclei; a minimum along one direction; i.e., one eigenvalue positive). The critical point needs not be located along the straight line going through the nuclei ("*banana*" bonds are possible), and its location may be closer to one of the nuclei (polarization). Thus, the nuclei are connected by a kind of electronic density "rope" (most dense at its core and decaying outside) extending from one nucleus to the other along (in general) a curved

René Thom (1923–2002), French mathematician, professor at the Université de Strasbourg, and founder of catastrophe theory (1966). The theory analyzes abrupt changes of functions (change of the number and character of stationary points) upon changing some parameters. In 1958, René Thom received the Fields Medal, the highest distinction for a mathematician.

[10] Only *some* pairs of atoms correspond to chemical bonds.

Richard Bader (1931–2012), Canadian chemist and professor at McMaster University in Canada. After earning his Ph.D. at the Massachussets Institute of Technology, he won an international fellowship to study at Cambridge University in England under Christopher Longuet-Higgins. At their first meeting, Bader was given the titles of two books with the instruction, "*When you have read these books, maybe we can talk again*". From these books, Bader learned about theories of electron density.

From that time on, he became convinced that electron density was the quantity of prime importance for the theory. *Photo reproduced courtesy of Richard Bader.*

line, having a single critical point on it. Its cross section for some bonds is circular, while for others it is elliptic-like.[11]

- Some of the eigenvalues may equal zero. The set of parameters (like the internuclear distance) at which $\det\left\{\frac{\partial^2\rho}{\partial\xi_i\partial\xi_j}\right\}=0$ (corresponding to an eigenvalue equal to 0) is called the *catastrophe set*. Calculations have shown that when the two nuclei separate, the rope elongates and *suddenly*, at a certain internuclear distance, it breaks down (this corresponds to zeroing out one of the eigenvalues). Thus, the catastrophe theory of René Thom turns out to be instrumental in chemistry.

11.2.3 Laplacian of the Electronic Density as a "Magnifying Glass"

Fig. 11.3 shows the functions $f(x)$, $f' = \frac{df}{dx}$ and $f'' = \frac{d^2f}{dx^2}$, where $f(x)$ is a function with a visible maximum at $x = 0$ and a hump close to $x = 0.9$. The hump is hardly visible–it is so small that there is no local maximum of $f(x)$ over there. Such a function resembles somewhat the electron density decay for an atom, when we go off the nucleus (position of the maximum[12]).

We may say that $-\frac{d^2f}{dx^2}$ can detect some subtle features of the $f(x)$ plot and gives maxima where the original function $f(x)$ has only almost invisible humps.

There is a similar situation with the function $-\Delta\rho(x, y, z) = -\left(\frac{\partial^2\rho}{\partial x^2} + \frac{\partial^2\rho}{\partial y^2} + \frac{\partial^2\rho}{\partial z^2}\right)$, except that here, we have three Cartesian coordinates. The way that we choose the directions of the Cartesian axes is irrelevant because at any point in space, $-\Delta\rho(x, y, z)$ does not depend on such a choice. Indeed, the coordinate systems, which we may choose, differ by an orthogonal transformation, which is peculiar because it does leave the *trace* of the Hessian invariant.

Imagine now ρ of an atom decaying with the distance to the nucleus as $f(x)$, similar to the decay of a smoke cloud (Fig. 11.4a), dense in the center and vanishing outward. Let us

[11] All the details may be computed nowadays by using quantum mechanical methods, often the most demanding ones (with the electronic correlation included). Contemporary crystallography is able to measure the same quantities in some excellent X-ray experiments. Therefore, the physicochemical methods are able to indicate precisely which atoms are involved in a chemical bond, is it strong or not, is it straight or curved ("rope-like"), what is the thickness of the "rope", has it a cylindrical or oval cross section (connected to its σ or π character), etc. A good review is available in T.S. Koritsanszky, P. Coppens, *Chem. Rev.*, *101*, 1583 (2001).

[12] Well, there is no cusp, so we have a nonzero size of the nucleus and/or Gaussian type orbitals used.

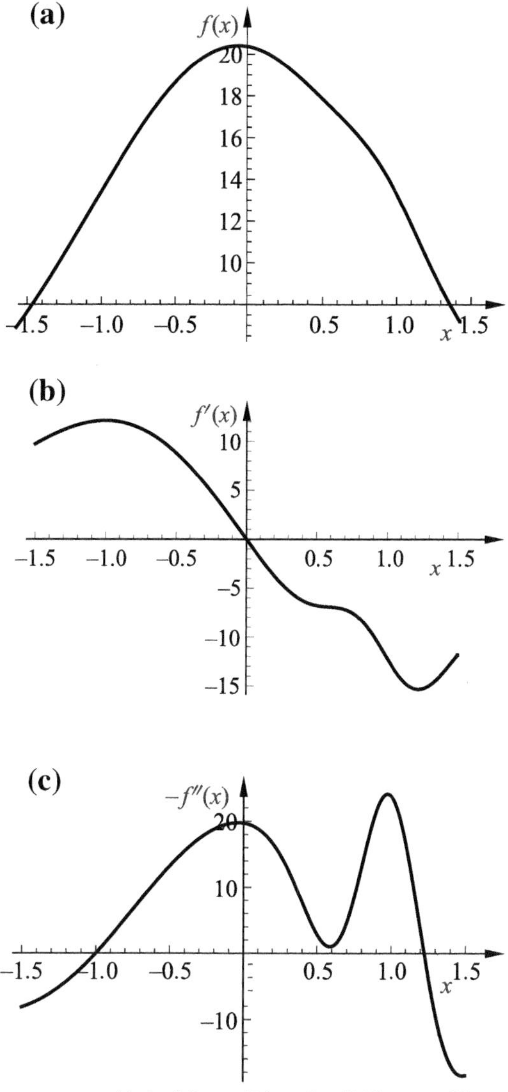

Fig. 11.3. The Laplacian $-\Delta\rho$ represents a kind of "*magnifying glass*." Here, we illustrate this in a 1-D case: instead of $-\Delta\rho(x, y, z)$, we have $-f''(x) \equiv -\frac{d^2 f}{dx^2}$. (a) A function $f(x)$ with a single maximum. One can see a small asymmetry of the function resulting from a hardly visible hump on the right side. (b) The first derivative $f'(x)$. (c) the plot of $-f''(x)$ shows two maxima. One of them (at $x = 0$) indicates the maximum of f, and the second one (close to $x = 1$) makes the small hump of $f(x)$ clearly visible.

calculate the Hessian at every point along the radius. It is easy to calculate $\Delta\rho(x, y, z)$ simply by summing up the diagonal terms of the Hessian. If we diagonalized the Hessian (i.e., rotated the axes in a particular way), its eigenvalues would correspond to the curvatures of the sections of ρ along the new coordinate axes (x', y', z'):

- *The section along the radius* (say z'). This curvature (see also Fig. 11.2a) is expected to be large and positive since this is the direction ρ exponentially decays.

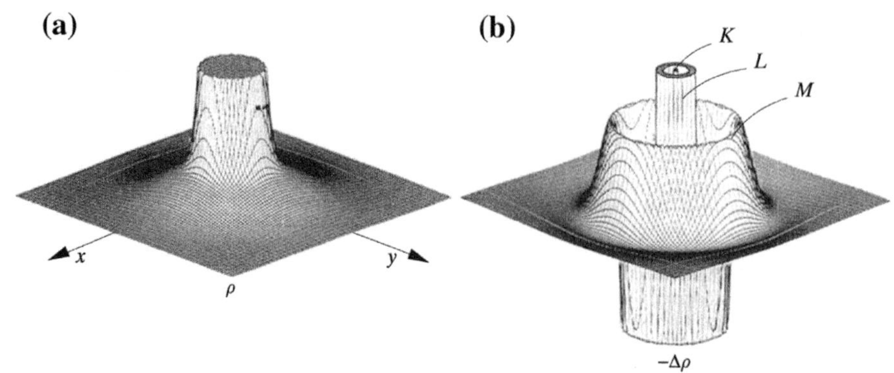

Fig. 11.4. A cross section of ρ (a) as well as a cross section of $-\Delta\rho$ (b) for the argon atom. The three humps (b) correspond to the K,L and M electron shells (p. 447). R.F.W. Bader, *Atoms in Molecules. A Quantum Theory*, Clarendon Press, Oxford (1994), *courtesy of Richard Bader.*

- *Two other sections: along x' and along y'* (only the first of them is shown in Fig. 11.2a). These sections at a given radius mean cutting perpendicularly to the radius, and whether looking along x' or along y', we see the same: a larger value at the radius and a decay outside; i.e., both eigenvalues are negative.

Fig. 11.4 displays ρ and $-\Delta\rho$ for the argon atom. Despite an apparent lack of any internal structure of the function ρ (left), the function $-\Delta\rho$ detected three concentrations of charge similar to the hump of the function $f(x)$. We may say that $-\Delta\rho(x, y, z)$ plays the role of a "magnifying glass": these are the K,L,M shells of the argon atom, seen very clearly.

Fig. 11.5 shows $-\Delta\rho$ for the systems N_2, Ar_2, and F_2. The figure highlights the shell character of the electronic structure of each of the atoms.[13] Fig. 11.2c shows that the electronic density is the greatest along the bond and drops outside in *each of the two orthogonal* directions. If, however, we went along the bond approaching any of the nuclei, the *density would increase.* This means that there is a saddle point of the second order because one eigenvalue of the Hessian is positive and two negative.

If there were no covalent bond at all (non-bonded atoms or ionic bond: no electron density "rope" connecting the nuclei), the last two values would be zero, and this means that $-\Delta\rho < 0$. Thus, if it happens that for a bond $-\Delta\rho > 0$, this means a large perpendicular contribution; i.e., a strong, "rope-like" covalent bond.

For the N_2 molecule, we have a large value of $-\Delta\rho > 0$ between the nuclei, which means an electronic charge concentrated in a strong bond. Therefore, the nuclei have a dilemma: whether

[13] Note that the nitrogen and the fluorine have two shells (K and L), while the argon atom has three shells (K,L,M); cf. Chapter 8.

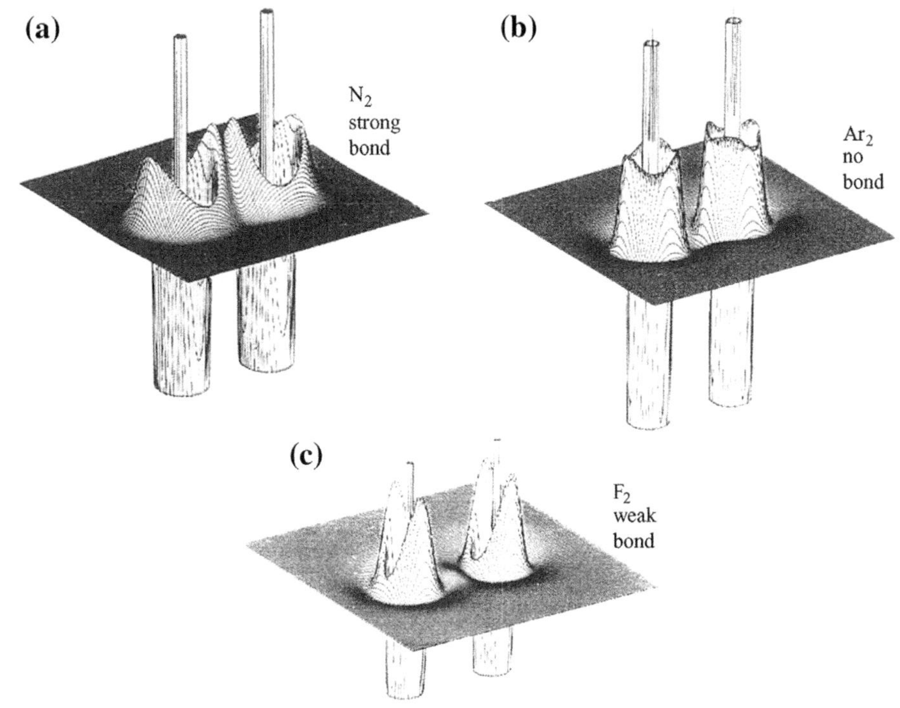

Fig. 11.5. A cross section of the quantity $-\Delta\rho$ for N_2, Ar_2 and F_2. We will focus now on the $-\Delta\rho$ value, computed in the middle of the internuclear distance. (a) We can see that for N_2, the value of $-\Delta\rho > 0$ (chemical bond), (b) for Ar-Ar, $-\Delta\rho < 0$ (no chemical bond), and (c) a very small positive $-\Delta\rho$ for F_2 (weak chemical bond). R.F.W. Bader, *Atoms in Molecules. A Quantum Theory*, Clarendon Press, Oxford (1994), *courtesy of Richard Bader.*

to run off, because they repel each other, or to run only a little, because there is such a beautiful negative charge in the middle of the bond (here, the nuclei choose the second possibility). This dilemma is absent in the Ar_2 system (Fig. 11.5b): the electronic charge runs off the middle of the bond, and the nuclei get uncovered and run off. The molecule F_2 sticks together but not very strongly–just look at the internuclear region: $-\Delta\rho$ is quite low there.[14]

11.3 Two important Hohenberg-Kohn theorems

11.3.1 Correlation Dragon Resides in Electron Density: Equivalence of Ψ_0 and ρ_0

Hohenberg and Kohn proved in 1964 an interesting theorem.[15]

> The ground-state electronic density $\rho_0(\boldsymbol{r})$ and the ground-state wave function Ψ_0 can be used alternatively as full descriptions of the ground state of the system.

[14] We see now why the F_2 molecule does not represent an easy task for the Hartree-Fock method (see Chapter 8; the method indicated that the molecule does not exist).

[15] P. Hohenberg and W. Kohn, *Phys. Rev.*, *136*, B864 (1964).

Walter Kohn (b. 1923), American physicist of Austrian origin and professor at the University of California – Santa Barbara. His conviction about the primary role the electronic density plays, led him to fundamental theoretical discoveries. Kohn shared the Nobel Prize with John A. Pople in 1998, receiving it *"for his development of the density-functional theory."*

This theorem is sometimes proved in a special way. Imagine that somebody gave us $\rho_0(\boldsymbol{r})$ without a single word of explanation. We have no idea which system it corresponds to. First, we calculate $\int \rho_0(\boldsymbol{r})d\boldsymbol{r}$, where the integration goes over the whole space. This gives a natural number N, which is the number of electrons in the system. We did not know it, but now we do. Next, we investigate the function $\rho_0(\boldsymbol{r})$, looking point by point at its values. We are searching for the "spikes" (cusps), because every cusp tells us where a nucleus is.[16] After this is done, we know all the positions of the nuclei. Now, we concentrate on each of the nuclei and look how fast the density drops when leaving the nucleus. The calculated slope has to be equal to a negative even number: $-2Z$ (see p. 447), and Z gives us the charge of the nucleus. Thus, we have deduced the composition of our system. Now we are in a position to write down the Hamiltonian for the system and *solve the Schrödinger equation*. After that, we know the ground-state wave function.

We started, therefore, from $\rho_0(\boldsymbol{r})$, and we got the ground-state wave function Ψ_0. According to Eqs. (11.1) and (11.2), from the wave function by integration, we obtain the density distribution $\rho_0(\boldsymbol{r})$. Hence, $\rho_0(\boldsymbol{r})$ contains the same precise information about the system as Ψ_0.

Thus, if we know ρ_0, we also know[17] Ψ_0, and, if we know Ψ_0, we also know ρ_0.[18]

The proof we carried out pertains only to the case when the *external potential* (everything except the interelectronic interaction) acting on the electrons stems from the nuclei.

[16] $\rho(\boldsymbol{r})$ represents a cloud similar to those that float in the sky. This "spike", therefore, means simply a high density in the cloud.

[17] And all the excited states wave functions as well! This is an intriguing conclusion, supported by experts; see W. Koch and M.C. Holthausen, *A Chemist's Guide to Density Functional Theory*, 2d ed., Wiley, Weinheim, 2001. On p. 59, it says "*the DFT is usually termed a ground state theory. The reason for this is not that the ground state density does not contain the information on the excited states–it actually does!–but because no practical way to extract this information is known so far.*"

[18] The theorem just proved shines in its simplicity. People thought that the wave function, usually a very complicated mathematical object (that depends on $3N$ space and N spin coordinates) is indispensable for computing the properties of the system. Moreover, the larger the system, the worse the difficulties in calculating it (recall Chapter 10 with billions of excitations, nonlinear parameters, etc.). Besides, how can we interpret such a complex object? This is a horribly complex problem. And it turns out that everything about the system just sits in $\rho(\boldsymbol{r})$, a function of position in our well-known 3-D space. It turns out that information about nuclei is hidden in such a simple object. This seems trivial (cusps), but it also includes much more subtle information about how electrons avoid each other due to Coulombic repulsion and the Pauli exclusion principle.

The Hohenberg-Kohn theorem can be proved for an arbitrary external potential–this property of the density is called the *v-representability*. The arbitrariness mentioned above is necessary in order to define in future the functionals for more general densities (than for isolated molecules). We will need that generality when introducing the functional derivatives (p. 584) in which $\rho(\boldsymbol{r})$ has to result from any external potential (or to be a *v-representable density*). Also, we will be interested in a *non-Coulombic* potential corresponding to the harmonic helium atom (cf. harmonium, p. 589) to see how exact the DFT method is. We may imagine ρ, which is *not* v-representable; e.g., discontinuous (in one, two, or even in every point like the Dirichlet function). The density distributions that are not v-representable are out of our field of interest.

11.3.2 A Secret of the Correlation Dragon: The Existence of Energy Functional Minimized by ρ_0

Hohenberg and Kohn also proved an analog of the variational principle (p. 232):

Hohenberg-Kohn Theorem:
For a given number of electrons (the integral over ρ equals N) and external potential v, there *exists* a functional of ρ, denoted by $E_v^{\text{HK}}[\rho]$, for which the following variational principle is satisfied:

$$E_v^{\text{HK}}[\rho] \geq E_v^{\text{HK}}[\rho_0] = E_0,$$

where ρ_0 stands for the (ideal) ground-state electronic density distribution corresponding to the ground state energy E_0.

We will prove this theorem using the variational principle in a way shown first by Levy.[19] The variational principle states that

$$E_0 = \min\langle\Psi \mid \hat{H} \mid \Psi\rangle,$$

where we search among the wave functions Ψ normalized to 1 and describing N electrons.

This minimization may be carried out in two steps, Fig. 11.6:

$$E_0 = \min_{\rho,\int \rho dV=N} \min_{\Psi\to\rho}\langle\Psi \mid \hat{T} + U + V \mid \Psi\rangle, \tag{11.8}$$

where $\hat{T}, U, V$ represent the kinetic energy, the electron repulsion, and the electron-nuclei attraction operators, respectively, for all the N electrons of our system (the hat in operators will be omitted if the operator has a multiplicative character).

The two minimization steps have the following meanings:

- The internal minimization is performed at the condition labeled as "$\Psi \to \rho$", which means minimization of the integral among the N-electron functions that are normalized to 1, and

[19] M. Levy, *Phys. Rev. A*, *26*, 1200 (1982).

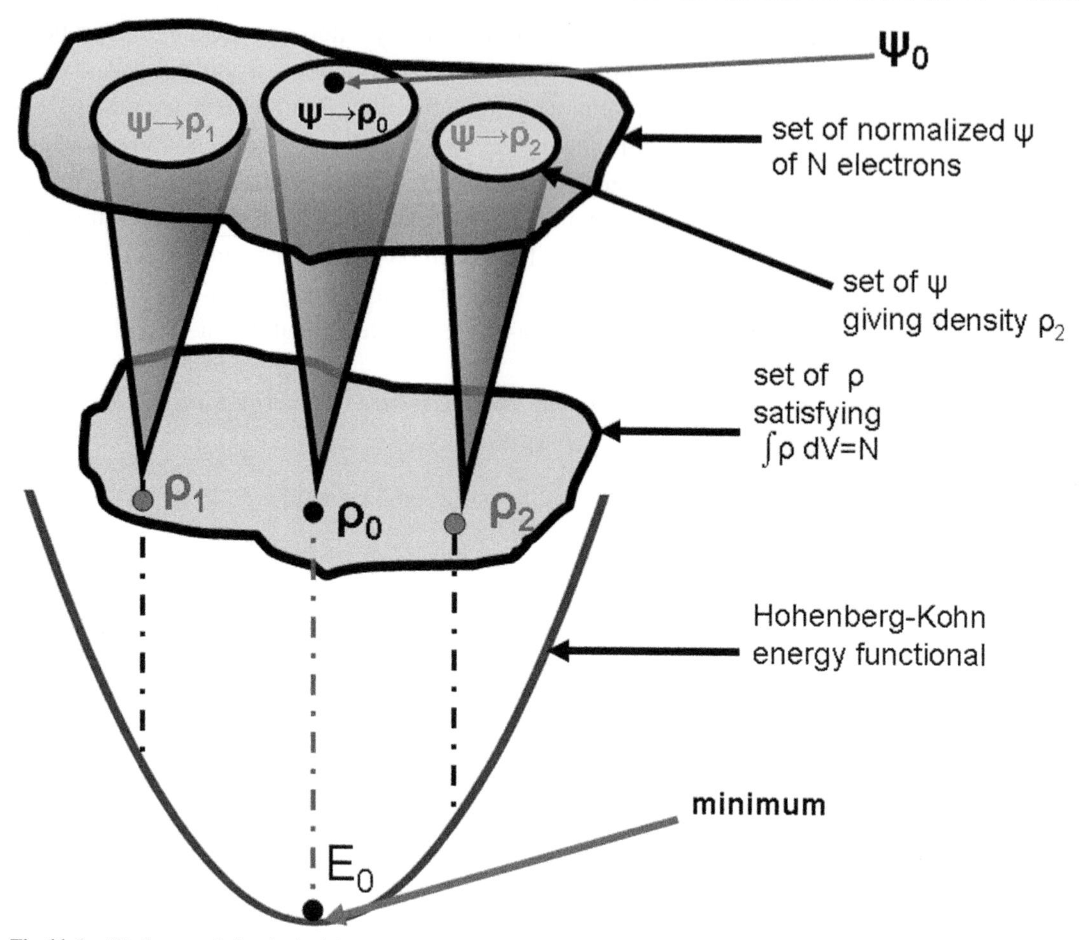

Fig. 11.6. The Levy variational principle (scheme). The task of the internal minimization is that at a *given fixed* density distribution ρ carrying N electrons, you must choose among those normalized functions Ψ, *that all produce* ρ (we will denote this by the symbol "$\Psi \rightarrow \rho$"); *such a function that minimizes* $\langle \Psi \mid \hat{T} + U \mid \Psi \rangle$ of Eq. (11.12). In the upper part of the figure, three sets of such functions Ψ are shown: one set gives ρ_1, the second ρ_0, and the third ρ_2. The external minimization symbolized by "$\rho, \int \rho dV = N$" chooses among all possible electron distributions ρ (that correspond to N electrons, shown in the center part of the figure) such a distribution $\rho = \rho_0$, that gives the lowest value (the ground state energy E_0, see the bottom part of the figure) of the Hohenberg-Kohn functional E_v^{HK}; i.e., $E_0 = \min_{\rho, \int \rho dV = N} E_v^{\mathrm{HK}}$. Note that among the functions Ψ that give ρ_0, there is the exact ground-state wave function Ψ_0.

any of them giving a *fixed* density distribution ρ "carrying" N electrons (the minimum attained at $\Psi = \Psi_{\min}$). As a result of this minimization, we obtain a functional of ρ given as $\min_{\Psi \rightarrow \rho} \langle \Psi \mid \hat{T} + U + V \mid \Psi \rangle = \langle \Psi_{\min}(\rho) \mid \hat{T} + U + V \mid \Psi_{\min}(\rho) \rangle$, because $\langle \Psi_{\min} \mid \hat{T} + U + V \mid \Psi_{\min} \rangle$ depends what we have taken as ρ.

- In the external minimization symbolized by $\rho, \int \rho dV = N$, we go over all the density distributions ρ that integrate to N (i.e., describe N electrons), and we choose that $\rho = \rho_0$ which minimizes the functional $\langle \Psi_{\min}(\rho) \mid \hat{T} + U + V \mid \Psi_{\min}(\rho) \rangle$. According to the variational principle (p. 233), this minimum is bound to be the exact ground-state energy E_0, while ρ_0 is the exact ground-state density distribution.

Therefore, both minimizations do the same as the variational principle.

The External Potential

It is easy to show that $\langle \Psi \mid V\Psi \rangle$ may be expressed as an integral involving the density distribution ρ instead of Ψ. Indeed, since

$$V = \sum_{i=1}^{N} v(\boldsymbol{r}_i), \quad \text{where} \quad v(\boldsymbol{r}_i) = \sum_A -\frac{Z_A}{|\boldsymbol{r}_i - \boldsymbol{r}_A|}, \tag{11.9}$$

then in each of the resulting integrals $\langle \Psi \mid v(\boldsymbol{r}_i)\Psi \rangle$, we may carry out the integration over all the electrons except the i-th one, and for this single one, we sum over its spin coordinate. It is easy to see that every such term (their number is N) gives *the same result* $\langle \Psi \mid v(\boldsymbol{r}_i)\Psi \rangle = \frac{1}{N}\int v(\boldsymbol{r})\rho(\boldsymbol{r})d\boldsymbol{r}$, because the electrons are indistinguishable (this is why we omit the index i). Because of this, we will get

$$\langle \Psi \mid V\Psi \rangle = \int v(\boldsymbol{r})\rho(\boldsymbol{r})d\boldsymbol{r}. \tag{11.10}$$

Therefore, the Levy minimization may be written as

$$E_0 = \min_{\rho, \int \rho dV = N} \left\{ \int v(\boldsymbol{r})\rho(\boldsymbol{r})d\boldsymbol{r} + \min_{\Psi\to\rho} \langle \Psi \mid (\hat{T} + U)\Psi \rangle \right\}. \tag{11.11}$$

The Universal Potential

At this point, we define the auxiliary functional[20] F^{HK}:

$$F^{HK}[\rho] = \min_{\Psi\to\rho} \langle \Psi \mid (\hat{T} + U)\Psi \rangle \equiv \langle \Psi_{min}(\rho) \mid (\hat{T} + U)\Psi_{min}(\rho) \rangle, \tag{11.12}$$

where Ψ_{min} stands for a normalized function which has been chosen among those that produce a given ρ, and makes the smallest value of $\langle \Psi \mid \hat{T} + U \mid \Psi \rangle$. This functional is often called *universal*, because it does not depend on any external potential–rather, it pertains solely to interacting electrons only.

[20] A functional is always defined in a domain (in this case a domain of the allowed ρs). How do allowed ρs look? Here are the conditions to fulfill: (a) $\rho \geq 0$ (b) $\int \rho dV = N$ (c) $\nabla \rho^{1/2}$ square-integrable. Among these conditions, we do not find any that would require the existence of such an antisymmetric Ψ of N electrons that would correspond [in the sense of Eq. (11.2)] to the density ρ under consideration (this is known as N-representability). It turns out that such a requirement is not needed, since it was proved by Thomas Gilbert (the proof may be found in the book by R.G. Parr and W. Yang *Density Functional Theory of Atoms and Molecules*, Oxford University Press, New York (1989), that every ρ, that satisfies the above conditions is N-representable because it corresponds to at least one antisymmetric N-electron Ψ.

The Hohenberg-Kohn Potential

In the DFT, we define the crucial

Hohenberg-Kohn functional $E_v^{\mathrm{HK}}[\rho]$ as

$$E_v^{\mathrm{HK}}[\rho] = \int v(\boldsymbol{r})\rho(\boldsymbol{r})\mathrm{d}\boldsymbol{r} + F^{\mathrm{HK}}[\rho], \tag{11.13}$$

and the minimum of this functional is the ground-state energy

$$E_0 = \min_{\rho, \int \rho dV = N} E_v^{\mathrm{HK}}[\rho], \tag{11.14}$$

while ρ that minimizes $E_v^{\mathrm{HK}}[\rho]$ represents the exact ground-state density distribution ρ_0 (see Fig. 11.6). Each ρ corresponds to at least one antisymmetric electronic wave function (the $N-$representability mentioned above), and there is no better wave function than the ground state, which, of course, corresponds to the density distribution ρ_0. This is why we have:

Hohenberg-Kohn Functional:
The Hohenberg-Kohn functional $E_v^{\mathrm{HK}}[\rho]$ attains minimum $E_0 = E_v^{\mathrm{HK}}[\rho_0]$ for the ideal density distribution. Now our job will be to find out what mathematical form the functional could have. And here we meet the basic problem of the DFT method: nobody has so far been able to give such a formula. The best that has been achieved to date are some approximations. These approximations, however, are so good that they begin to supply results that satisfy chemists.

Therefore, when the question is posed: "*Is it possible to construct a quantum theory, in which the basic concept would be electronic density?*", we have to answer: "*Yes, it is.*" This answer, however, has only an existential value ("*Yes, there exists*"). We have no information about how such a theory could be constructed.

An indication may come from the concept of the wave function. In order to proceed toward the abovementioned unknown functional, we will focus on the ingenious idea of a *fictitious Kohn-Sham system of non-interacting electrons.*

11.4 The Kohn-Sham Equations

11.4.1 A Kohn-Sham System of Non-interacting Electrons

Let us consider an electron subject to some "*external*" potential $v(\boldsymbol{r})$; *for example* coming from the Coulombic interaction with the nuclei (with charges Z_A in a.u. and positions $\boldsymbol{r}_A$):

$$v(\mathbf{r}) = \sum_A -\frac{Z_A}{|\mathbf{r} - \mathbf{r}_A|}. \tag{11.15}$$

In this system, we have N electrons, which also interact by Coulombic forces among themselves. All these interactions produce the ground-state electronic density distribution ρ_0 (ideal; i.e., that we obtain from the exact, 100% correlated wave function). Now let us consider

Fictitious Kohn-Sham System:
the fictitious Kohn-Sham system of N model electrons (fermions), that *do not interact* at all (as if their charge equaled zero), but instead of the interaction with the nuclei, they are subject to an external potential $v_0(\mathbf{r})$ so ingeniously tailored that ρ does not change; i.e., we still have the ideal ground-state electronic density $\rho = \rho_0$.

Let us assume for a while that we have found such a wonder potential $v_0(\mathbf{r})$. We will worry later about how to find it in reality. Now we assume that the problem has been solved. Can we find ρ_0? Of course, we can. Since the Kohn-Sham electrons do not interact between themselves, we have only to solve the one-electron equation (with the wonder v_0)

$$\left(-\frac{1}{2}\Delta + v_0\right)\phi_i = \varepsilon_i \phi_i, \tag{11.16}$$

where ϕ_i are the solutions - some spinorbitals, of course, called the *Kohn-Sham spinorbitals*.[21]

The total wave function is a Slater determinant, which in our situation should be called the *Kohn-Sham determinant* instead. The electronic density distribution of such a system is given by Eq. (11.7) and the density distribution ρ_0 means *exact*; i.e., correlated 100% (thanks to the "*wonder*" and unknown operator v_0).

11.4.2 Chasing the Correlation Dragon into an Unknown Part of the Total Energy

Let us try to write down a general expression for the electronic ground-state energy of the system under consideration. Obviously, we have to have in it the kinetic energy of the electrons, their interaction with the nuclei, and their repulsion among themselves. However, in the DFT

[21] If the electrons do not interact, the corresponding wave function can be taken as a *product* of the spinorbitals for individual electrons. Such a function for electrons is not antisymmetric, and, therefore, is "illegal". Taking the *Kohn-Sham determinant* (instead of the product) helps because it is antisymmetric and represents an eigenfunction of the total Hamiltonian of the fictitious system [i.e., the sum of the one-electron operators given in Eq. (11.16)]. This is easy to show because a determinant represents a sum of products of the spinorbitals, the products differing only by permutation of electrons. If the total Hamiltonian of the fictitious system acts on such a sum, each term (product) is its eigenfunction, and each eigenvalue amounts to $\sum_{i=1}^{N} \varepsilon_i$; i.e., it is the same for each product. Hence, the Kohn-Sham determinant represents an eigenfunction of the fictitious system. Scientists compared the Kohn-Sham orbitals with the canonical Hartree-Fock orbitals with great interest. It turns out that the differences are small.

approach, we write the following:

$$E = T_0 + \int v(\boldsymbol{r})\rho(\boldsymbol{r})\mathrm{d}\boldsymbol{r} + J[\rho] + E_{\mathrm{xc}}[\rho], \tag{11.17}$$

where:

- Instead of the electronic kinetic energy of the system, we write down the electronic kinetic energy *of the fictitious Kohn-Sham system of (non-interacting) electrons* T_0 (recall the Slater-Condon rules, discussed on p. e119):

$$T_0 = -\frac{1}{2}\sum_{i=1}^{N}\langle\phi_i \mid \Delta\phi_i\rangle. \tag{11.18}$$

- Next, there is the correct electron-nuclei interaction (or other external potential) term: $\int v(\boldsymbol{r})\rho(\boldsymbol{r})\mathrm{d}\boldsymbol{r}$.
- Then, there is an interaction of the electron cloud with itself[22]:

$$J[\rho] = \frac{1}{2}\iint \frac{\rho(\boldsymbol{r}_1)\rho(\boldsymbol{r}_2)}{|\boldsymbol{r}_1 - \boldsymbol{r}_2|}\mathrm{d}\boldsymbol{r}_1\mathrm{d}\boldsymbol{r}_2. \tag{11.19}$$

No doubt such an idea looks reasonable, for the energy expression should contain an interaction of the electron cloud with itself (because the electrons repel each other). However, there is a trap in this concept–a malady hidden in $J[\rho]$. The illness is seen best if one considers the simplest system: the hydrogen atom ground state. In Eq. (11.19), we have an interelectronic self-repulsion, which actually does not exist because we have only one electron. So, whatever reasonable remedy is to be designed in the future, it should reduce this unwanted *self-interaction* in the hydrogen atom to zero. The problem is not limited, of course, to the hydrogen atom. When taking $J[\rho]$, *an electron is interacting with itself, and this self-interaction has to be somehow excluded from* $J[\rho]$ *by introducing a correction.* Two electrons repel each other electrostatically, and therefore, around each of them there has to exist a kind of no-parking zone for the other one (a "*Coulomb hole*"; cf. p. 595). Also, a no-parking zone results because electrons of the

[22] How can we compute the Coulombic interaction within a storm cloud exhibiting certain charge distribution ρ? At first sight, it looks like a difficult problem, but remember that we know how to calculate the Coulombic interaction of two *point* charges. Let us divide the whole cloud into tiny cubes, each with volume $\mathrm{d}V$. The cube that is pointed by the vector $\boldsymbol{r}_1$ contains a tiny charge $\rho(\boldsymbol{r}_1)\mathrm{d}V \equiv \rho(\boldsymbol{r}_1)\mathrm{d}\boldsymbol{r}_1$. We know that when calculating the Coulombic interaction of two such cubes, we have to write $\frac{\rho(\boldsymbol{r}_1)\rho(\boldsymbol{r}_2)}{|\boldsymbol{r}_1-\boldsymbol{r}_2|}\mathrm{d}\boldsymbol{r}_1\mathrm{d}\boldsymbol{r}_2$. This has to be summed over all possible positions of the first and the second cube: $\iint \frac{\rho(\boldsymbol{r}_1)\rho(\boldsymbol{r}_2)}{|\boldsymbol{r}_1-\boldsymbol{r}_2|}\mathrm{d}\boldsymbol{r}_1\mathrm{d}\boldsymbol{r}_2$, but in this way each interaction is computed twice, whereas they represent parts of the same cloud. Hence, the final self-interaction of the storm cloud is $\frac{1}{2}\iint \frac{\rho(\boldsymbol{r}_1)\rho(\boldsymbol{r}_2)}{|\boldsymbol{r}_1-\boldsymbol{r}_2|}\mathrm{d}\boldsymbol{r}_1\mathrm{d}\boldsymbol{r}_2$. The expression for the self-interaction of the electron cloud is the same.

same spin coordinate hate one another[23] ("*exchange*", or "*Fermi hole*"; cf. p. 597). The integral J does not take such a correlation of motions into account.

Thus, we have written a few terms and we do not know what to write next. Well,

in the DFT, in the expression for E, we write in Eq.(11.17) the lacking remainder as E_{xc}, and we call it *exchange-correlation energy* (label x stands for "*exchange*", c is for "*correlation*") and declare, courageously, that we will manage somehow to get it.

The above formula represents a definition of the *exchange-correlation energy*, although it is rather a strange definition–it requires us to know E. We should not forget that in E_{xc}, a correction to the kinetic energy also must be included (besides the exchange and correlation effects) that takes into account that kinetic energy has to be calculated for the true (i.e., interacting) electrons, not for the non-interacting Kohn-Sham ones. The next question is connected to what kind of mathematical form E_{xc} might have. Let us assume for the time being that we have no problem with this mathematical form. For now, we will establish a relation between our wonder external potential v_0 and our mysterious E_{xc}, both quantities performing miracles, but not known.

11.4.3 Derivation of the Kohn-Sham Equations

Now we will make a variation of E; i.e., we will find the linear effect of changing E due to a variation of the spinorbitals (and therefore also of the density). We make a spinorbital variation denoted by $\delta\phi_i$ (as discussed in p. 402, it is justified to vary either ϕ_i or ϕ_i^*, the result is the same: we choose, therefore, $\delta\phi_i^*$) and see what effect it will have on E, keeping only the linear term. We have [see Eq. (11.6)]

$$\phi_i^* \rightarrow \phi_i^* + \delta\phi_i^* \tag{11.20}$$

$$\rho \rightarrow \rho + \delta\rho \tag{11.21}$$

$$\delta\rho\left(\boldsymbol{r}\right) = \sum_{\sigma}\sum_{i=1}^{N} \delta\phi_i^*\left(\boldsymbol{r},\sigma\right)\phi_i\left(\boldsymbol{r},\sigma\right). \tag{11.22}$$

We insert the right sides of the above expressions into E, and identify the variation; i.e., the linear part of the change of E. The variations of the individual terms of E look like (note that the symbol $\langle | \rangle$ stands for an integral over space coordinates and a summation over the spin coordinates, as discussed on p. 399):

$$\delta T_0 = -\frac{1}{2}\sum_{i=1}^{N}\langle\delta\phi_i \mid \Delta\phi_i\rangle \tag{11.23}$$

[23] A correlated density and a non-correlated density differ in that in the correlated one, we have smaller values in the high-density regions, because the holes make the overcrowding of space by electrons less probable.

$$\delta \int v\rho \mathrm{d}\boldsymbol{r} = \int v\delta\rho \mathrm{d}\boldsymbol{r} = \sum_{i=1}^{N} \langle \delta\phi_i \mid v\phi_i \rangle \tag{11.24}$$

$$\begin{aligned}
\delta J &= \frac{1}{2}\left[\int \frac{\rho(\boldsymbol{r}_1)\delta\rho(\boldsymbol{r}_2)}{|\boldsymbol{r}_1-\boldsymbol{r}_2|}\mathrm{d}\boldsymbol{r}_1\mathrm{d}\boldsymbol{r}_2 + \int \frac{\delta\rho(\boldsymbol{r}_1)\rho(\boldsymbol{r}_2)}{|\boldsymbol{r}_1-\boldsymbol{r}_2|}\mathrm{d}\boldsymbol{r}_1\mathrm{d}\boldsymbol{r}_2\right] \\
&= \int \frac{\rho(\boldsymbol{r}_2)\delta\rho(\boldsymbol{r}_1)}{|\boldsymbol{r}_1-\boldsymbol{r}_2|}\mathrm{d}\boldsymbol{r}_1\mathrm{d}\boldsymbol{r}_2 \\
&= \sum_{i=1}^{N}\int \sum_{\sigma_1} \delta\phi_i^*(\boldsymbol{r}_1,\sigma_1)\,\phi_i(\boldsymbol{r}_1,\sigma_1)\,\frac{\rho(\boldsymbol{r}_2)}{|\boldsymbol{r}_1-\boldsymbol{r}_2|}\mathrm{d}\boldsymbol{r}_1\mathrm{d}\boldsymbol{r}_2 \\
&= \sum_{i=1}^{N}\sum_{\sigma_1}\int_1 \mathrm{d}\boldsymbol{r}_1\delta\phi_i^*(\boldsymbol{r}_1,\sigma_1)\,\phi_i(\boldsymbol{r}_1,\sigma_1)\int_2 \frac{\rho(\boldsymbol{r}_2)}{|\boldsymbol{r}_1-\boldsymbol{r}_2|}\mathrm{d}\boldsymbol{r}_2 \\
&= \sum_{i=1}^{N}\sum_{\sigma_1}\int_1 \mathrm{d}\boldsymbol{r}_1\delta\phi_i^*(\boldsymbol{r}_1,\sigma_1)\,\phi_i(\boldsymbol{r}_1,\sigma_1)\int_2 \frac{\sum_j\sum_{\sigma_2}\phi_j^*(\boldsymbol{r}_2,\sigma_2)\,\phi_j(\boldsymbol{r}_2,\sigma_2)}{|\boldsymbol{r}_1-\boldsymbol{r}_2|}\mathrm{d}\boldsymbol{r}_2 \\
&= \sum_{i,j=1}^{N} \langle \delta\phi_i(\boldsymbol{r}_1,\sigma_1)|\hat{J}_j(\boldsymbol{r}_1)\phi_i(\boldsymbol{r}_1,\sigma_1)\rangle_1,
\end{aligned} \tag{11.25}$$

where $\langle\ldots\mid\ldots\rangle_1$ means integration over spatial coordinates and the summation over the spin coordinate of electron 1 ($\int_1$ means the integration only), with the Coulomb operator $\hat{J}_j$ associated with the spinorbital ϕ_j

$$\hat{J}_j(\boldsymbol{r}_1) = \sum_{\sigma_2}\int \frac{\phi_j(\boldsymbol{r}_2,\sigma_2)^*\phi_j(\boldsymbol{r}_2,\sigma_2)}{|\boldsymbol{r}_1-\boldsymbol{r}_2|}\mathrm{d}\boldsymbol{r}_2. \tag{11.26}$$

Finally, we come to the variation of E_{xc}; i.e., δE_{xc}. We are in a quite difficult situation because we do not know the mathematical dependence of the functional E_{xc} on ρ, and therefore also on $\delta\phi_i^*$. Nevertheless, we somehow have to get the linear part of E_{xc} (i.e., the variation).

A change of functional F (due to $f \to f+\delta f$) contains a part linear in δf denoted by δF, plus some higher powers[24] of δf denoted by $O((\delta f)^2)$:

$$F[f+\delta f] - F[f] = \delta F + O((\delta f)^2). \tag{11.27}$$

The δF is defined through the *functional derivative*[25] (Fig. 11.7) of F with respect to the function f (denoted by $\frac{\delta F[f]}{\delta f(x)}$), for a single variable x:

$$\delta F = \int_a^b \mathrm{d}x \frac{\delta F[f]}{\delta f(x)}\delta f(x). \tag{11.28}$$

[24] If δf is very small, the higher terms are negligible.

[25] The functional derivative itself is a functional of f and a function of x (just for the sake of simplicity). An example of a functional derivative may be found in Eq. (11.25), when looking at $\delta J = \int \frac{\rho(\boldsymbol{r}_2)\delta\rho(\boldsymbol{r}_1)}{|\boldsymbol{r}_1-\boldsymbol{r}_2|}\mathrm{d}\boldsymbol{r}_1\mathrm{d}\boldsymbol{r}_2 = \int \mathrm{d}\boldsymbol{r}_1\{\int \mathrm{d}\boldsymbol{r}_2\frac{\rho(\boldsymbol{r}_2)}{|\boldsymbol{r}_1-\boldsymbol{r}_2|}\}\delta\rho(\boldsymbol{r}_1)$. Indeed, as we can see from Eq. (11.28) $\int \mathrm{d}\boldsymbol{r}_2\frac{\rho(\boldsymbol{r}_2)}{|\boldsymbol{r}_1-\boldsymbol{r}_2|} \equiv \frac{\delta J[\rho]}{\delta\rho(\boldsymbol{r}_1)}$, which is a 3-D equivalent of $\frac{\delta F[f]}{\delta f(x)}$. Note that $\int \mathrm{d}\boldsymbol{r}_2\frac{\rho(\boldsymbol{r}_2)}{|\boldsymbol{r}_1-\boldsymbol{r}_2|}$ is a functional of ρ and a function of $\boldsymbol{r}_1$.

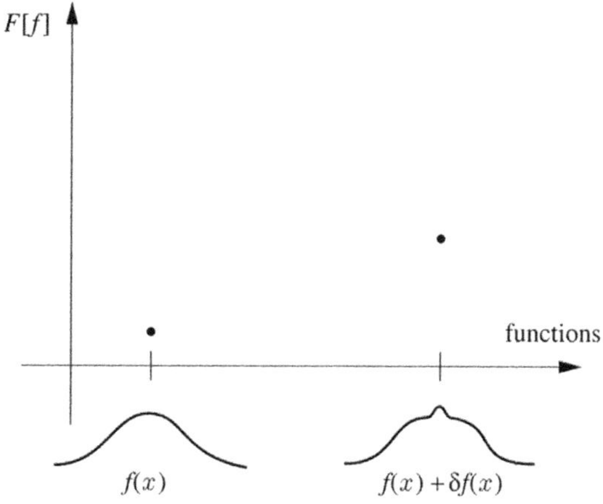

Fig. 11.7. A scheme showing what a functional derivative is about. The ordinate represents the values of a functional $F[f]$, while each point of the horizontal axis represents a function $f(x)$. The functional $F[f]$ depends, of course, on details of the function $f(x)$. If we consider a *small local change* of $f(x)$, this change may result in a large change of F, and then the derivative $\frac{\delta F}{\delta f}$ is large, or in a small change of F, and then the derivative $\frac{\delta F}{\delta f}$ is small (this depends on the particular functional).

Indeed, in this case, we obtain as δE_{xc}:

$$\delta E_{xc} = \int d\boldsymbol{r} \frac{\delta E_{xc}}{\delta \rho(\boldsymbol{r})} \delta \rho(\boldsymbol{r}) = \sum_{i=1}^{N} \langle \delta \phi_i | \frac{\delta E_{xc}}{\delta \rho} \phi_i \rangle. \tag{11.29}$$

Therefore, a unknown quantity E_{xc} is replaced by another unknown quantity $\frac{\delta E_{xc}}{\delta \rho}$, but there is profit from this: the functional derivative enables us to write an equation for spinorbitals. The variations of the spinorbitals are not arbitrary in this formula – they have to satisfy the orthonormality conditions [because our formulas such as Eq. (11.6), are valid only for such spinorbitals] for $i, j = 1, \ldots N$, which gives

$$\langle \delta \phi_i \mid \phi_j \rangle = 0 \quad \text{for} \quad i, j = 1, 2, \ldots N. \tag{11.30}$$

Let us multiply each of the results of Eq. (11.30) by a Lagrange multiplier ε_{ij}, add them together, then subtract from the variation δE and write the result as equal to zero[26] (in the minimum, we have $\delta E = 0$). We obtain

$$\delta E - \sum_{i,j}^{N} \varepsilon_{ij} \langle \delta \phi_i \mid \phi_j \rangle = 0 \tag{11.31}$$

or (note that $\langle \delta \phi_i \mid \phi_j \rangle_1$; i.e., integration over electron 1 is equal to $\langle \delta \phi_i \mid \phi_j \rangle$)

$$\sum_{i=1}^{N} \left\langle \delta \phi_i \mid \left\{ \left[-\frac{1}{2}\Delta + v + \sum_{j=1}^{N} \hat{J}_j + \frac{\delta E_{xc}}{\delta \rho} \right] \phi_i - \sum_{j=1}^{N} \varepsilon_{ij} \phi_j \right\} \right\rangle_1 = 0. \tag{11.32}$$

[26] See Appendix N available at booksite.elsevier.com/978-0-444-59436-5, p. e121 explains why such a procedure corresponds to minimization with constraints.

After inserting the Lagrange multipliers, the variations of ϕ_i^* are already *independent*, and the only possibility to have zero on the right side is that every individual ket $|\rangle$ is zero (Euler equation; cf. p. e122):

$$\left\{-\frac{1}{2}\Delta + v + v_{\text{coul}} + v_{\text{xc}}\right\}\phi_i = \sum_{j=1}^{N} \varepsilon_{ij}\phi_j, \tag{11.33}$$

$$v_{\text{coul}}(\boldsymbol{r}) \equiv \sum_{j=1}^{N} \hat{J}_j(\boldsymbol{r}), \tag{11.34}$$

$$v_{\text{xc}}(\boldsymbol{r}) \equiv \frac{\delta E_{\text{xc}}}{\delta\rho(\boldsymbol{r})}. \tag{11.35}$$

It would be good now to get rid of the non-diagonal Lagrange multipliers in order to obtain a beautiful one-electron equation analogous to the Fock equation. To this end, we need the operator in the curly brackets in Eq. (11.33) to be invariant with respect to an arbitrary unitary transformation of the spinorbitals. The sum of the Coulomb operators (v_{coul}) is invariant, as has been demonstrated on p. 406. As to the unknown functional derivative $\delta E_{xc}/\delta\rho$ (i.e., potential v_{xc}), its invariance follows from the fact that it is a functional of ρ [and ρ of Eq. (11.6) is invariant]. Finally, after applying such a unitary transformation that diagonalizes the matrix of ε_{ij}, we obtain the Kohn-Sham equation ($\varepsilon_{ii} \equiv \varepsilon_i$):

Kohn-Sham Equation

$$\left\{-\frac{1}{2}\Delta + v + v_{\text{coul}} + v_{\text{xc}}\right\}\phi_i = \varepsilon_i\phi_i. \tag{11.36}$$

The equation is analogous to the Fock equation (p. 407)[27]. We solve the Kohn-Sham equation by an *iterative method.* We start from any zero-iteration orbitals. This enables us to calculate a zero approximation to ρ, and then the zero approximations to the operators v_{coul} and v_{xc} [in a moment, we will see how to compute E_{xc}, and then, using Eq. (11.35), we obtain v_{xc}]. The solution to the Kohn-Sham equation gives new orbitals and new ρ. The procedure is then repeated until consistency is achieved.

Hence, finally, we "*know*" what the wonder operator v_0 looks like:

$$v_0 = v + v_{\text{coul}} + v_{\text{xc}}. \tag{11.37}$$

[27] There is a difference in notation: the one-electron operator $\hat{h}$ and the Coulomb operator $\hat{J}$ from the Fock equation are now replaced by $-\frac{1}{2}\Delta + v \equiv \hat{h}$ and $\hat{J} \equiv v_{coul}$. There is, however, a serious difference: instead of the exchange operator $-\hat{K}$ in the Fock equation, we have here the exchange-correlation potential v_{xc}.

As in the Hartree-Fock method, there is no problem with v_{coul}, but a serious difficulty arises with the exchange-correlation operator v_{xc}, or (equivalent) with the energy E_{xc}. The second Hohenberg–Kohn theorem says that the functional $E_v^{HK}[\rho]$ exists, but it does not guarantee that it is simple. For now, we will worry about this potential, but we will go ahead anyway.

Kohn-Sham Equations with Spin Polarization

Before searching for v_{xc}, let us generalize the Kohn-Sham formalism and use Eq. (11.3) for splitting ρ into the α and β spin functions. If these contributions are not equal (even for some $\boldsymbol{r}$), we will have a *spin polarization*. In order to reformulate the equations, we consider two non-interacting fictitious electron systems: one described by the spin functions α, and the other by functions β, with the corresponding density distributions $\rho_\alpha(\boldsymbol{r})$ and $\rho_\beta(\boldsymbol{r})$ exactly equal to ρ_α and ρ_β, respectively, in the (original) interacting system. Then, we obtain two coupled[28] Kohn-Sham equations, for $\sigma = \alpha$ and $\sigma = \beta$, with potential v_0 that depends on the spin coordinate σ:

$$v_0^\sigma = v + v_{\text{coul}} + v_{\text{xc}}^\sigma. \tag{11.38}$$

The situation is analogous to the unrestricted Hartree-Fock (UHF) method, cf. p. 408.

This extension of the DFT is known as *spin density functional theory (SDFT)*.

11.5 Trying to Guess the Appearance of the Correlation Dragon

We now approach the point where we promised to write down the mysterious exchange-correlation energy. Well, truthfully and straightforwardly: we do not know the analytical form of this quantity. Nobody knows what the exchange-correlation is–there are only guesses. The number of formulas will be almost unlimited, as is usual with guesses.[29] Let us take the simplest ones to show the essence of the procedure.

11.5.1 Local Density Approximation (LDA)

The electrons in a molecule are in a very complex situation because they not only interact among themselves, but also with the nuclei. However, a simpler system has been elaborated theoretically for years: a homogeneous gas model in a box[30], or an electrically neutral system (the nuclear charge is smeared out uniformly). It does not represent the simplest system to study, but it turns out that theory is able to determine (exactly) some of its properties. For example, it has been deduced how E_{xc} depends on ρ, and even how it depends on ρ_α and ρ_β. Since the gas

---

[28] Through the common operator v_{coul}, a functional of $\rho_\alpha + \rho_\beta$, and through v_{xc} because the latter is in general a functional of both, ρ_α and ρ_β.

[29] Some detailed formulas are reported in the book by J.B. Foresman and A. Frisch, *Exploring Chemistry with Electronic Structure Methods*, Gaussian, Pittsburgh, 1996, str.272.

[30] This gas model has periodic boundary conditions. This is a common trick to avoid the surface problem. We consider a box having a property such that if something goes out through one wall, it enters through the opposite wall (cf. p. 524).

is homogeneous and the volume of the box is known, then we could easily work out how the E_{xc} per unit volume depends on these quantities.

Then, the reasoning described next.[31]

The electronic density distribution in a molecule is certainly inhomogeneous, but locally (within a small volume) we may assume its homogeneity. Then, if someone asks about the exchange-correlation energy contribution from this small volume, we would say that in principle, we do not know, but to a good approximation the contribution could be calculated as a product of the small volume and the exchange-correlation energy density from the homogeneous gas theory (with the electronic gas density as calculated inside the small volume).

Thus, everything is decided locally: we have a sum of contributions from each infinitesimally small element of the electron cloud with the corresponding density. This is why it is called the *local density approximation* (*LDA*, when the ρ dependence is used) or the *local spin density approximation* (*LSDA*, when the ρ_α and ρ_β dependencies are exploited).

11.5.2 Non-local Approximation (NLDA)

Gradient Expansion Approximation (GEA)

There are approximations that go beyond the LDA. They consider that the dependence $E_{xc}[\rho]$ may be *non-local*; i.e., E_{xc} may depend on ρ at a given point (locality), but also on ρ nearby (non-locality). When we are at a point, what happens further off depends not only on ρ at that point, but also the gradient of ρ at the point, etc.[32] This is how the idea of the gradient expansion approximation (GEA) appeared

$$E_{xc}^{GEA} = E_{xc}^{LSDA} + \int B_{xc}(\rho_\alpha, \rho_\beta, \nabla\rho_\alpha, \nabla\rho_\beta)d\mathbf{r}, \tag{11.39}$$

where the exchange-correlation function B_{xc} is carefully selected as a function of ρ_α, ρ_β, and their gradients, in order to maximize the successes of the theory/experiment comparison. However, this recipe was not so simple, and some strange unexplained discrepancies were still taking place.

Perdew-Wang Functional (PW91)

A breakthrough in the quality of results is represented by the following proposition of Perdew and Wang:

$$E_{xc}^{PW91} = \int f(\rho_\alpha, \rho_\beta, \nabla\rho_\alpha, \nabla\rho_\beta)d\mathbf{r}, \tag{11.40}$$

[31] W. Kohn and L.J. Sham, *Phys. Rev.*, *140*, A1133 (1965).

[32] As in a Taylor series, then we may need not only the gradient, but also the Laplacian, etc.

where the function f of ρ_α, ρ_β and their gradients has been tailored in an ingenious way. It sounds unclear, but it will be shown below that their approximation used some fundamental properties and this enabled them *without introducing any parameters* to achieve a much better agreement between the theory and experiment.

The Famous B3LYP Hybrid Functional

The B3LYP approach belongs to the *hybrid (i.e., mixed) approximations* for the exchange-correlation functional. The approximation is famous because it gives very good results and, therefore, is extremely popular. So far so good, but there is a danger of Babylon-type science.[33] It seems like a witch's brew for the B3LYP exchange-correlation potential E_{xc}: *take the exchange-correlation energy from the LSDA method (a unit), add a pinch* (0.20 *unit) of the difference between the Hartree-Fock exchange energy*[34] E_x^{KS} *and the LSDA* E_x^{LSDA}. *Then, mix well* 0.72 *unit of Becke exchange potential* E_x^{B88} *which includes the 1988 correction, then add* 0.81 *unit of the Lee-Young-Parr correlation potential* E_c^{LYP}. You will like this homeopathic magic potion most (a "*hybrid*") if you conclude by putting in 0.19 unit of the Vosko-Wilk-Nusair potential[35] E_c^{VWN}:

$$E_{xc} = E_{xc}^{LSDA} + 0.20\left(E_x^{HF} - E_x^{LSDA}\right) + 0.72 E_x^{B88} + 0.81 E_c^{LYP} + 0.19 E_c^{VWN}. \quad (11.41)$$

If you do it this way, satisfaction is (almost) guaranteed, and your results will agree very well with the experiment.

11.5.3 The Approximate Character of the DFT vs. the Apparent Rigor of Ab Initio Computations

There are lots of exchange-correlation potentials in the literature. There is an impression that their authors worried most about theory/experiment agreement. We can hardly admire this kind of science, but the alternative (i.e., the practice of *ab initio* methods with the intact and "*holy*" Hamiltonian operator) has its own disadvantages. This is because finally we have to choose a given atomic basis set, and this influences the results. It is true that we have the variational principle at our disposal, and it is possible to tell which result is more accurate. But more and more often in quantum chemistry, we use some non-variational methods (cf. Chapter 10). Besides, the Hamiltonian holiness disappears when the theory becomes relativistic (cf. Chapter 3).

Everybody would like to have agreement with experiments, and it is no wonder people tinker with the exchange-correlation enigma. This tinkering, however, is by no means arbitrary. There are some serious physical restraints with it, which will be shown shortly.

---

[33] The Chaldean priests working on "Babylonian science" paid attention to making their small formulas efficient. The ancient Greeks (to whom contemporary science owes so much) favored crystal clear reasoning.

[34] In fact, this is Kohn-Sham exchange energy [see Eq. (11.72)], because the Slater determinant wave function used to calculate it is the Kohn-Sham determinant, not the Hartree-Fock one.

[35] S.H. Vosko, L. Wilk, and M. Nusair, *Can. J. Phys.*, *58*,1200 (1980).

11.6 On the Physical Justification for the Exchange-Correlation Energy

Now we are going to introduce several useful concepts, such as the electron pair distribution function and the electron hole (in a more formal way than we did in Chapter 10, p. 597), etc.

11.6.1 The Electron Pair Distribution Function

From the N-electron wave function, we may compute what is called the *electron pair correlation function* $\Pi(\boldsymbol{r}_1, \boldsymbol{r}_2)$–in short, a pair function defined as[36]

$$\Pi(\boldsymbol{r}_1, \boldsymbol{r}_2) = N(N-1) \sum_{\sigma_1, \sigma_2} \int |\Psi|^2 d\tau_3 d\tau_4 \ldots d\tau_N \tag{11.42}$$

where the summation over spin coordinates pertains to all electrons (for the electrons 3, 4, ... N, the summation is hidden in the integrals over $d\tau$), while the integration is over the space coordinates of the electrons 3, 4, ... N.

The function $\Pi(\boldsymbol{r}_1, \boldsymbol{r}_2)$ measures the probability density of finding one electron at the point indicated by $\boldsymbol{r}_1$ and another at $\boldsymbol{r}_2$, and tells us how the motions of two electrons are correlated. If Π were a *product* of two functions $\rho_1(\boldsymbol{r}_1) > 0$ and $\rho_2(\boldsymbol{r}_2) > 0$, then this motion is not correlated (because the probability of two events represents a product of the probabilities for each of the events only for *independent; i.e., uncorrelated events*).

Note that [see Eqs. (11.1) and (11.2) on p. 665]

$$\int \Pi(\boldsymbol{r}_1, \boldsymbol{r}_2) dV_2 = N(N-1) \sum_{\sigma_1} \int d\tau_2 \int |\Psi|^2 d\tau_3 d\tau_4 \ldots d\tau_N = (N-1)\rho(\boldsymbol{r}_1) \tag{11.43}$$

and

$$\iint \Pi(\boldsymbol{r}_1, \boldsymbol{r}_2) dV_1 dV_2 = (N-1) \int \rho(\boldsymbol{r}_1) dV_1 = N(N-1). \tag{11.44}$$

Function Π appears in a natural way, when we compute the mean value of the total electronic repulsion $\langle \Psi \mid U \mid \Psi \rangle$ with the Coulomb operator $U = \sum_{i<j}^{N} \frac{1}{r_{ij}}$ and a normalized N-electron wave function Ψ. Indeed, we have ("*prime*" in the summation corresponds to omitting the diagonal term)

$$\langle \Psi \mid U\Psi \rangle = \frac{1}{2} \sum_{i,j=1}^{N}{}' \langle \Psi \mid \frac{1}{r_{ij}} \Psi \rangle$$

[36] The function represents the diagonal element of the *two-particle electron density matrix*: $\Gamma(\boldsymbol{r}_1, \boldsymbol{r}_2; \boldsymbol{r}_1', \boldsymbol{r}_2') = N(N-1) \sum_{\text{all } \sigma} \int \Psi^* (\boldsymbol{r}_1'\sigma_1, \boldsymbol{r}_2', \sigma_2, \boldsymbol{r}_3, \sigma_3, \ldots, \boldsymbol{r}_N, \sigma_N) \Psi (\boldsymbol{r}_1, \sigma_1, \boldsymbol{r}_2, \sigma_2, \boldsymbol{r}_3, \sigma_3, \ldots, \boldsymbol{r}_N, \sigma_N) d\boldsymbol{r}_3 d\boldsymbol{r}_4 \cdots d\boldsymbol{r}_N$, $\Pi(\boldsymbol{r}_1, \boldsymbol{r}_2) \equiv \Gamma(\boldsymbol{r}_1, \boldsymbol{r}_2; \boldsymbol{r}_1, \boldsymbol{r}_2)$.

$$
\begin{aligned}
&= \frac{1}{2}\sum_{i,j=1}^{N}{}' \left\{ \sum_{\sigma_i,\sigma_j} \int \mathrm{d}\boldsymbol{r}_i \mathrm{d}\boldsymbol{r}_j \frac{1}{r_{ij}} \int |\Psi|^2 \frac{\mathrm{d}\tau_1 \mathrm{d}\tau_2 \ldots \mathrm{d}\tau_N}{\mathrm{d}\tau_i \mathrm{d}\tau_j} \right\} \\
&= \frac{1}{2}\sum_{i,j=1}^{N}{}' \int \mathrm{d}\boldsymbol{r}_i \mathrm{d}\boldsymbol{r}_j \frac{1}{r_{ij}} \frac{1}{N(N-1)} \Pi(\boldsymbol{r}_i, \boldsymbol{r}_j) \\
&= \frac{1}{2}\frac{1}{N(N-1)} \sum_{i,j=1}^{N}{}' \int \mathrm{d}\boldsymbol{r}_1 \mathrm{d}\boldsymbol{r}_2 \frac{1}{r_{12}} \Pi(\boldsymbol{r}_1, \boldsymbol{r}_2) \\
&= \frac{1}{2}\frac{1}{N(N-1)} \int \mathrm{d}\boldsymbol{r}_1 \mathrm{d}\boldsymbol{r}_2 \frac{\Pi(\boldsymbol{r}_1, \boldsymbol{r}_2)}{r_{12}} \sum_{i,j=1}^{N}{}' 1 \\
&= \frac{1}{2} \int \mathrm{d}\boldsymbol{r}_1 \mathrm{d}\boldsymbol{r}_2 \frac{\Pi(\boldsymbol{r}_1, \boldsymbol{r}_2)}{r_{12}}.
\end{aligned} \tag{11.45}
$$

We will need this result in a moment. We see that to determine the contribution of the electron repulsions to the total energy, we need the two-electron function Π. The first Hohenberg-Kohn theorem tells us that it is sufficient to know something simpler (namely, the electronic density ρ). How can we reconcile these two demands?

> The further DFT story will pertain to the question: how can we change the potential in order to replace Π by ρ?

11.6.2 *Adiabatic Connection: From What Is Known Towards the Target*

To begin, let us write two Hamiltonians that are certainly very important for our goal: the first is the total Hamiltonian of our system (of course, with the Coulombic electron-electron interactions U). Let us denote the operator as $H(\lambda = 1)$, we use the abbreviation $v(\boldsymbol{r}_i) \equiv v(i)$:

$$
\hat{H}(\lambda = 1) = \sum_{i=1}^{N} \left[-\frac{1}{2}\Delta_i + v(i) \right] + U. \tag{11.46}
$$

The second Hamiltonian $H(\lambda = 0)$ pertains to the Kohn-Sham fictitious system of the *non-interacting* electrons (it contains our wonder v_0, which we solemnly promise to search for, and the kinetic energy operator and nothing else):

$$
\hat{H}(\lambda = 0) = \sum_{i=1}^{N} \left[-\frac{1}{2}\Delta_i + v_0(i) \right]. \tag{11.47}
$$

We will try to connect these two important systems by generating some intermediate Hamiltonians $\hat{H}(\lambda)$ for λ *intermediate* between 0 and 1:

$$\hat{H}(\lambda) = \sum_{i=1}^{N} \left[-\frac{1}{2}\Delta_i + v_\lambda(i) \right] + U(\lambda), \tag{11.48}$$

where

$$U(\lambda) = \lambda \sum_{i<j}^{N} \frac{1}{r_{ij}}.$$

Our electrons are not real electrons for intermediate values of λ; rather, each electron carries the electric charge $\sqrt{\lambda}$.

The intermediate Hamiltonian $\hat{H}(\lambda)$ contains a mysterious v_λ, which generates the exact density distribution ρ that corresponds to the Hamiltonian $\hat{H}(\lambda = 1)$ i.e., with all interactions in place. The same exact ρ corresponds to $\hat{H}(\lambda = 0)$.

We have, therefore, the ambition to go from the $\lambda = 0$ situation to the $\lambda = 1$ situation, all the time guaranteeing that the antisymmetric ground-state eigenfunction of $\hat{H}(\lambda)$ for any λ gives *the same electron density distribution* ρ*, the ideal (exact)*. The way chosen represents a kind of "path of life" for us, because by sticking to it, we do not lose the most precious of our treasures: the ideal density distribution ρ. We will call this path the *adiabatic connection* because all the time, we will adjust the correction computed to our actual position on the path.

Our goal will be the total energy $E(\lambda = 1)$. The adiabatic transition will be carried out in tiny steps. We will start with $E(\lambda = 0)$, and end up with $E(\lambda = 1)$:

$$E(\lambda = 1) = E(\lambda = 0) + \int_0^1 E'(\lambda)\mathrm{d}\lambda, \tag{11.49}$$

where the increments $\mathrm{d}E(\lambda) = E'(\lambda)\mathrm{d}\lambda$ will be calculated as the first-order perturbation energy correction, Eq. (5.20). The first-order correction is sufficient, because we are going to apply only infinitesimally small λ increments.[37] Each time, when λ changes from λ to $\lambda + \mathrm{d}\lambda$, the situation at λ [i.e., the Hamiltonian $\hat{H}(\lambda)$ and the wave function $\Psi(\lambda)$] will be treated as unperturbed. What, therefore, does the perturbation operator look like? Well, when we go from λ to $\lambda+\mathrm{d}\lambda$, the Hamiltonian changes by $\hat{H}^{(1)}(\lambda) = \mathrm{d}\hat{H}(\lambda)$. Then, the first-order perturbation correction to the energy given by (5.20), represents the mean value of $\mathrm{d}\hat{H}(\lambda)$ with the unperturbed function $\Psi(\lambda)$:

$$\mathrm{d}E(\lambda) = \langle \Psi(\lambda) | \mathrm{d}\hat{H}(\lambda)\Psi(\lambda) \rangle, \tag{11.50}$$

[37] λ plays a different role here than the perturbational parameter λ on p. 241.

where in $\mathrm{d}\hat{H}$ we only have a change of v_λ and of $U(\lambda)$ due to the change of λ:

$$\mathrm{d}\hat{H}(\lambda) = \sum_{i=1}^{N} \mathrm{d}v_\lambda(i) + \mathrm{d}\lambda \sum_{i<j}^{N} \frac{1}{r_{ij}}. \tag{11.51}$$

Note that we have succeeded in writing such a simple formula, *because the kinetic energy operator stays unchanged all the time (it does not depend on* λ*)*. Let us insert this into the first-order correction to the energy in order to get $\mathrm{d}E(\lambda)$ and use Eqs. (11.10) and (11.45):

$$\mathrm{d}E(\lambda) = \langle \Psi(\lambda)|\mathrm{d}\hat{H}(\lambda)\Psi(\lambda)\rangle = \int \rho(\boldsymbol{r})\mathrm{d}v_\lambda(\boldsymbol{r})\mathrm{d}\boldsymbol{r} + \frac{1}{2}\mathrm{d}\lambda \iint \mathrm{d}\boldsymbol{r}_1\mathrm{d}\boldsymbol{r}_2 \frac{\Pi_\lambda(\boldsymbol{r}_1, \boldsymbol{r}_2)}{r_{12}}. \tag{11.52}$$

In the last formula, we introduced a function Π_λ that is an analog of the pair function Π but pertains to the electrons carrying the charge $-\sqrt{\lambda}$ [we have used Eq. (11.45), noting that we have a λ-dependent wave function $\Psi(\lambda)$].

In order to go from $E(\lambda = 0)$ to $E(\lambda = 1)$, it is sufficient just to integrate this expression from 0 to 1 over λ (this corresponds to the infinitesimally small increments of λ as mentioned before). Note that (by definition) ρ *does not depend on* λ, which is of fundamental importance in the success of the integration $\int \rho(\boldsymbol{r})\mathrm{d}v_\lambda(\boldsymbol{r})\mathrm{d}\boldsymbol{r}$ and gives the result

$$E(\lambda = 1) - E(\lambda = 0) = \int \rho(\boldsymbol{r})\{v - v_0\}(\boldsymbol{r})\mathrm{d}\boldsymbol{r} + \frac{1}{2}\int_0^1 \mathrm{d}\lambda \iint \mathrm{d}\boldsymbol{r}_1\mathrm{d}\boldsymbol{r}_2 \frac{\Pi_\lambda(\boldsymbol{r}_1, \boldsymbol{r}_2)}{r_{12}}. \tag{11.53}$$

The energy for $\lambda = 0$; i.e., for the non-interacting electrons in an unknown external potential v_0 will be written as [cf. Eqs. (11.16) and (11.18)]:

$$E(\lambda = 0) = \sum_i \varepsilon_i = T_0 + \int \rho(\boldsymbol{r})v_0(\boldsymbol{r})\mathrm{d}\boldsymbol{r}. \tag{11.54}$$

Inserting this into Eq. (11.53) we obtain $E(\lambda = 1)$; i.e., the energy of our original system:

$$E(\lambda = 1) = T_0 + \int \rho(\boldsymbol{r})v(\boldsymbol{r})\mathrm{d}\boldsymbol{r} + \frac{1}{2}\int_0^1 \mathrm{d}\lambda \iint \mathrm{d}\boldsymbol{r}_1\mathrm{d}\boldsymbol{r}_2 \frac{\Pi_\lambda(\boldsymbol{r}_1, \boldsymbol{r}_2)}{r_{12}}. \tag{11.55}$$

Note, that according to Eq. (11.43), we get $\int \Pi_\lambda(\boldsymbol{r}_1, \boldsymbol{r}_2)\mathrm{d}\boldsymbol{r}_2 = (N-1)\rho(\boldsymbol{r}_1)$, because $\rho(\boldsymbol{r}_1)$ does not depend on λ due to the nature of our adiabatic transformation.

The expression for $E(\lambda = 1)$ may be simplified by introducing the pair distribution function Π_{aver} which is the $\Pi_\lambda(\boldsymbol{r}_1, \boldsymbol{r}_2)$ averaged over $\lambda = [0, 1]$:

$$\Pi_{aver}(\boldsymbol{r}_1, \boldsymbol{r}_2) \equiv \int_0^1 \Pi_\lambda(\boldsymbol{r}_1, \boldsymbol{r}_2)\mathrm{d}\lambda. \tag{11.56}$$

Here also (we will use this result in a moment),

$$\int \Pi_{aver}(\boldsymbol{r}_1, \boldsymbol{r}_2) d\boldsymbol{r}_2 = \int_0^1 \int \Pi_\lambda(\boldsymbol{r}_1, \boldsymbol{r}_2) d\lambda d\boldsymbol{r}_2 = (N-1)\rho(\boldsymbol{r}_1) \int_0^1 d\lambda = (N-1)\rho(\boldsymbol{r}_1). \tag{11.57}$$

Finally, we obtain the following expression for

the total energy E:

$$E(\lambda = 1) = T_0 + \int \rho(\boldsymbol{r}) v(\boldsymbol{r}) d\boldsymbol{r} + \frac{1}{2} \iint d\boldsymbol{r}_1 d\boldsymbol{r}_2 \frac{\Pi_{aver}(\boldsymbol{r}_1, \boldsymbol{r}_2)}{r_{12}}. \tag{11.58}$$

Note that this equation is similar to the total energy expression appearing in traditional quantum chemistry[a] (without repulsion of the nuclei),

$$E = T + \int \rho(\boldsymbol{r}) v(\boldsymbol{r}) d\boldsymbol{r} + \frac{1}{2} \iint d\boldsymbol{r}_1 d\boldsymbol{r}_2 \frac{\Pi(\boldsymbol{r}_1, \boldsymbol{r}_2)}{r_{12}}. \tag{11.59}$$

[a] It is evident from the mean value of the total Hamiltonian [taking into account the mean value of the electron-electron repulsion, Eqs. (11.10) and (11.45)].

As we can see, the DFT total energy expression, instead of the mean kinetic energy of the fully interacting electrons T, contains T_0; i.e., the mean kinetic energy of the non-interacting (Kohn-Sham) electrons.[38] We pay a price, however, which is that we need to compute the function Π_{aver} somehow. But note that the correlation energy dragon has been driven into the problem of finding a two-electron function Π_{aver}.

11.6.3 Exchange-Correlation Energy vs. Π_{aver}

What is the relation between Π_{aver} and the exchange-correlation energy E_{xc} introduced earlier? We find that immediately, comparing the total energy given in Eqs. (11.17) and (11.19), and now in Eq. (11.58). It is seen that the exchange-correlation energy is as follows:

$$E_{xc} = \frac{1}{2} \iint d\boldsymbol{r}_1 d\boldsymbol{r}_2 \frac{1}{r_{12}} \{\Pi_{aver}(\boldsymbol{r}_1, \boldsymbol{r}_2) - \rho(\boldsymbol{r}_1)\rho(\boldsymbol{r}_2)\}. \tag{11.60}$$

[38] As a matter of fact, the whole Kohn-Sham formalism with the fictitious system of non-interacting electrons has been designed precisely for this reason.

The energy looks as if it were a potential energy, but it implicitly incorporates (in Π_{aver}) the kinetic energy correction for changing the electron non-interacting system to the electron–interacting system.

Now let us try to get some information about the integrand (i.e., Π_{aver}), by introducing the notion of the electron hole.

11.6.4 The Correlation Dragon Hides in the Exchange-Correlation Hole

Electrons do not like each other, which manifests itself in Coulombic repulsion. On top of that, two electrons having the same spin coordinates hate each other (Pauli exclusion principle) and also try to get out of each other's way. This has been analyzed in Chapter 10, p. 597. We should highlight these features because both concepts are basic and simple.

Let us introduce the definition of the exchange-correlation hole h_{xc} as satisfying the equation

$$\Pi_{\text{aver}}(\boldsymbol{r}_1, \boldsymbol{r}_2) = \rho(\boldsymbol{r}_1)\rho(\boldsymbol{r}_2) + \rho(\boldsymbol{r}_1)h_{\text{xc}}(\boldsymbol{r}_1; \boldsymbol{r}_2). \tag{11.61}$$

Thus, in view of Eqs. (11.59) and (11.45), we have the electron repulsion energy

$$\frac{1}{2}\iint d\boldsymbol{r}_1 d\boldsymbol{r}_2 \frac{\Pi_{\text{aver}}(\boldsymbol{r}_1, \boldsymbol{r}_2)}{r_{12}} = \frac{1}{2}\iint d\boldsymbol{r}_1 d\boldsymbol{r}_2 \frac{\rho(\boldsymbol{r}_1)\rho(\boldsymbol{r}_2)}{r_{12}} + \frac{1}{2}\iint d\boldsymbol{r}_1 d\boldsymbol{r}_2 \frac{\rho(\boldsymbol{r}_1)h_{\text{xc}}(\boldsymbol{r}_1; \boldsymbol{r}_2)}{r_{12}}. \tag{11.62}$$

as the self-interaction of the electron cloud of the density distribution $\rho(\boldsymbol{r})$ Eq. (11.19), plus a correction $\frac{1}{2}\iint d\boldsymbol{r}_1 d\boldsymbol{r}_2 \frac{\rho(\boldsymbol{r}_1)h_{\text{xc}}(\boldsymbol{r}_1;\boldsymbol{r}_2)}{r_{12}}$, which takes into account all necessary interactions; i.e., our complete correlation dragon is certainly hidden in the unknown hole function $h_{\text{xc}}(\boldsymbol{r}_1; \boldsymbol{r}_2)$. Note that the hole charge distribution integrates over $\boldsymbol{r}_2$ to the charge -1 irrespectively of the position $\boldsymbol{r}_1$ of the electron 1. Indeed, integrating Eq. (11.61) over $\boldsymbol{r}_2$ and using Eq. (11.43), we get[39]

$$\int d\boldsymbol{r}_2 h_{\text{xc}}(\boldsymbol{r}_1; \boldsymbol{r}_2) = -1. \tag{11.63}$$

11.6.5 Electron Holes in Spin Resolution

First, we will decompose the function Π_{aver} into the components related to the spin functions[40] of electrons 1 and 2; $\alpha\alpha, \alpha\beta, \beta\alpha, and \beta\beta$:

$$\Pi_{\text{aver}} = \Pi^{\alpha\alpha}_{\text{aver}} + \Pi^{\alpha\beta}_{\text{aver}} + \Pi^{\beta\alpha}_{\text{aver}} + \Pi^{\beta\beta}_{\text{aver}}, \tag{11.64}$$

---

[39] $\int d\boldsymbol{r}_2 h_{\text{xc}}(\boldsymbol{r}_1; \boldsymbol{r}_2) = \int d\boldsymbol{r}_2 h_{\text{xc}}(\boldsymbol{r}_1; \boldsymbol{r}_2) = \int d\boldsymbol{r}_2 \frac{\Pi_{\text{aver}}(\boldsymbol{r}_1,\boldsymbol{r}_2) - \rho(\boldsymbol{r}_1)\rho(\boldsymbol{r}_2)}{\rho(\boldsymbol{r}_1)} = \frac{1}{\rho(\boldsymbol{r}_1)}\int d\boldsymbol{r}_2 \left(\Pi_{\text{aver}}(\boldsymbol{r}_1, \boldsymbol{r}_2) - \rho(\boldsymbol{r}_1)\rho(\boldsymbol{r}_2)\right)$
$= \frac{1}{\rho(\boldsymbol{r}_1)}[(N-1)\,\rho(\boldsymbol{r}_1) - N\rho(\boldsymbol{r}_1)] = -1.$

[40] Such a decomposition follows from Eq. (11.42). We average all the contributions $\Pi^{\sigma\sigma'}$ separately and obtain the formula.

where $\Pi_{\text{aver}}^{\alpha\beta} dV_1 dV_2$ represents a measure of the probability density[41] that two electrons are in their small boxes, indicated by the vectors $\boldsymbol{r}_1$ and $\boldsymbol{r}_2$; the boxes have the volumes dV_1 and dV_2; and the electrons are described by the spin functions α and β (the other components of Π_{aver} are defined in a similar way). Since $\rho = \rho_\alpha + \rho_\beta$, the exchange-correlation energy can be written as[42]

$$E_{\text{xc}} = \frac{1}{2} \sum_{\sigma\sigma'} \iint d\boldsymbol{r}_1 d\boldsymbol{r}_2 \frac{\Pi_{\text{aver}}^{\sigma\sigma'}(\boldsymbol{r}_1, \boldsymbol{r}_2) - \rho_\sigma(\boldsymbol{r}_1)\rho_{\sigma'}(\boldsymbol{r}_2)}{r_{12}}, \tag{11.65}$$

where the summation goes over the spin coordinates. It is seen that

a nonzero value of E_{xc} tells us whether the behavior of electrons deviates from their *independence* (the latter is described by the product of the probability densities; i.e., the second term in the numerator). This means that E_{xc} has to contain the electron-electron correlation resulting from Coulombic interaction and their avoidance from the Pauli exclusion principle.

By using the abbreviation for the exchange-correlation hole

$$h_{\text{xc}}^{\sigma\sigma'}(\boldsymbol{r}_1, \boldsymbol{r}_2) \equiv \frac{\Pi_{\text{aver}}^{\sigma\sigma'}(\boldsymbol{r}_1, \boldsymbol{r}_2) - \rho_\sigma(\boldsymbol{r}_1)\rho_{\sigma'}(\boldsymbol{r}_2)}{\rho_\sigma(\boldsymbol{r}_1)},$$

we obtain

$$E_{\text{xc}} = \frac{1}{2} \sum_{\sigma\sigma'} \int d\boldsymbol{r}_1 \int d\boldsymbol{r}_2 \frac{\rho_\sigma(\boldsymbol{r}_1)}{r_{12}} h_{\text{xc}}^{\sigma\sigma'}(\boldsymbol{r}_1, \boldsymbol{r}_2). \tag{11.66}$$

The final expression for the exchange-correlation hole is

Exchange-Correlation Hole

$$h_{\text{xc}}^{\sigma\sigma'}(\boldsymbol{r}_1, \boldsymbol{r}_2) = \frac{\Pi_{\text{aver}}^{\sigma\sigma'}(\boldsymbol{r}_1, \boldsymbol{r}_2)}{\rho_\sigma(\boldsymbol{r}_1)} - \rho_{\sigma'}(\boldsymbol{r}_2). \tag{11.67}$$

The hole pertains to that part of the pair distribution function that is inexplicable by a product-like dependence. Since a product function describes independent electrons, the hole function grasps the "*intentional*" avoidance of the two electrons.

We have, therefore, four exchange-correlation holes: $h_{\text{xc}}^{\alpha\alpha}, h_{\text{xc}}^{\alpha\beta}, h_{\text{xc}}^{\beta\alpha}, h_{\text{xc}}^{\beta\beta}$.

---

[41] Here, the probability density is λ-averaged.

[42] Indeed, $E_{\text{xc}} = \frac{1}{2} \iint d\boldsymbol{r}_1 d\boldsymbol{r}_2 \frac{\Pi_{\text{aver}}(\boldsymbol{r}_1,\boldsymbol{r}_2) - \rho(\boldsymbol{r}_1)\rho(\boldsymbol{r}_2)}{r_{12}} = \frac{1}{2} \iint d\boldsymbol{r}_1 d\boldsymbol{r}_2 \frac{\sum_{\sigma\sigma'} \Pi_{\text{aver}}^{\sigma\sigma'}(\boldsymbol{r}_1,\boldsymbol{r}_2) - (\sum_\sigma \rho_\sigma(\boldsymbol{r}_1))(\sum_{\sigma'} \rho_{\sigma'}(\boldsymbol{r}_2))}{r_{12}} = \frac{1}{2} \sum_{\sigma\sigma'} \iint d\boldsymbol{r}_1 d\boldsymbol{r}_2 \frac{\Pi_{\text{aver}}^{\sigma\sigma'}(\boldsymbol{r}_1,\boldsymbol{r}_2) - \rho_\sigma(\boldsymbol{r}_1)\rho_{\sigma'}(\boldsymbol{r}_2)}{r_{12}}$.

11.6.6 The Dragon's Ultimate Hideout: The Correlation Hole

Dividing the Exchange-Correlation Hole into the Exchange Hole and the Correlation Hole

The restrictions introduced come from the Pauli exclusion principle (cf. Slater determinant), and hence have been related to the exchange energy. So far, no restriction has appeared that would stem from the Coulombic interactions of electrons.[43] This made people think of differentiating the holes into two contributions: exchange hole h_{x} and correlation hole h_{c} (called the *Coulombic hole*). Let us begin with a formal division of the exchange-correlation energy into the exchange and the correlation parts:

Exchange-Correlation Energy

$$E_{\text{xc}} = E_{\text{x}} + E_{\text{c}}, \tag{11.68}$$

and *we will say that we know, what the exchange part is.*

The DFT exchange energy (E_x) is calculated in the same way as in the Hartree-Fock method, but with the Kohn-Sham determinant. The correlation energy E_{c} represents just a rest.

This is the same strategy of chasing the electronic correlation dragon into a hole–this time into the correlation hole. When we do not know a quantity, we write down what we know plus a remainder. And the dragon with 100 heads sits in it. Because of this division, the Kohn–Sham equation will contain the sum of the exchange and correlation potentials instead of v_{xc}:

$$v_{\text{xc}} = v_{\text{x}} + v_{\text{c}}, \tag{11.69}$$

with

$$v_{\text{x}} \equiv \frac{\delta E_{\text{x}}}{\delta \rho}, \tag{11.70}$$

$$v_{\text{c}} \equiv \frac{\delta E_{\text{c}}}{\delta \rho}. \tag{11.71}$$

Let us recall what the Hartree-Fock exchange energy[44] looks like [(Chapter 8, Eq. (8.38)]. The Kohn-Sham exchange energy looks the same, of course, except that the spinorbitals are now Kohn-Sham, not Hartree-Fock. Therefore, we have the exchange energy E_{x} as (the sum is

[43] This is the role of the Hamiltonian.

[44] This is the one that appeared from the exchange operator (i.e., containing the exchange integrals).

over the molecular spinorbitals[45])

$$\begin{aligned} E_{\mathrm{X}} &= -\frac{1}{2}\sum_{i,j=1}^{\mathrm{SMO}} K_{ij} = -\frac{1}{2}\sum_{i,j=1}^{\mathrm{SMO}} \langle ij \mid ji \rangle \\ &= -\frac{1}{2}\sum_{\sigma}\int \frac{\left\{\sum_{i=1}^{N}\phi_i^*(1)\phi_i(2)\right\}\left\{\sum_{j=1}^{N}\phi_j^*(2)\phi_j(1)\right\}}{r_{12}}\mathrm{d}\boldsymbol{r}_1\mathrm{d}\boldsymbol{r}_2 \\ &= -\frac{1}{2}\sum_{\sigma}\int \frac{|\rho_\sigma(\boldsymbol{r}_1;\boldsymbol{r}_2)|^2}{r_{12}}\mathrm{d}\boldsymbol{r}_1\mathrm{d}\boldsymbol{r}_2, \end{aligned} \tag{11.72}$$

where

$$\rho_\sigma(\boldsymbol{r}_1;\boldsymbol{r}_2) \equiv \sum_{i=1}^{N}\phi_i(\boldsymbol{r}_1,\sigma)\phi_i^*(\boldsymbol{r}_2,\sigma) \tag{11.73}$$

represents the *one-particle* density matrix for the σ subsystem [Eq. (11.2)], and ρ_σ is obtained from the Kohn-Sham determinant. Note that density $\rho_\sigma(\boldsymbol{r})$ is its diagonal; i.e., $\rho_\sigma(\boldsymbol{r}) \equiv \rho_\sigma(\boldsymbol{r};\boldsymbol{r})$.

The above may be incorporated into the exchange energy E_{X}, equal to

$$E_{\mathrm{X}} = \frac{1}{2}\sum_{\sigma\sigma'}\iint \mathrm{d}\boldsymbol{r}_1\mathrm{d}\boldsymbol{r}_2\frac{\rho_\sigma(\boldsymbol{r}_1)}{r_{12}}h_{\mathrm{X}}^{\sigma\sigma'}(\boldsymbol{r}_1,\boldsymbol{r}_2), \tag{11.74}$$

if the exchange hole (also known as the *Fermi hole*) h is proposed as

$$h_{\mathrm{X}}^{\sigma\sigma'}(\boldsymbol{r}_1,\boldsymbol{r}_2) = \delta_{\sigma\sigma'}\left\{-\frac{|\rho_\sigma(\boldsymbol{r}_1;\boldsymbol{r}_2)|^2}{\rho_\sigma(\boldsymbol{r}_1)}\right\}. \tag{11.75}$$

It is seen that the exchange hole is negative everywhere[46] and diagonal in the spin index. Let us integrate the exchange hole over $\boldsymbol{r}_2$ for an arbitrary position of electron 1. First, we have

$$\begin{aligned} |\rho_\sigma(\boldsymbol{r}_1;\boldsymbol{r}_2)|^2 &= \sum_i \phi_i^*(\boldsymbol{r}_2,\sigma)\phi_i(\boldsymbol{r}_1,\sigma)\sum_j \phi_j^*(\boldsymbol{r}_1,\sigma)\phi_j(\boldsymbol{r}_2,\sigma) \\ &= \sum_{ij}\phi_i^*(\boldsymbol{r}_2,\sigma)\phi_i(\boldsymbol{r}_1,\sigma)\phi_j^*(\boldsymbol{r}_1,\sigma)\phi_j(\boldsymbol{r}_2,\sigma). \end{aligned}$$

[45] Note that spinorbital i has to have the same spin function as spinorbital j (otherwise, $K_{ij} = 0$).
[46] This has its origin in the minus sign before the exchange integrals in the total energy expression.

Then, the integration gives[47]

$$\int h_{\mathrm{x}}^{\sigma\sigma'}(\boldsymbol{r}_1, \boldsymbol{r}_2)\mathrm{d}\boldsymbol{r}_2 = -\delta_{\sigma\sigma'}. \tag{11.76}$$

Therefore,

the exchange hole $h_{\mathrm{x}}^{\sigma\sigma}(\boldsymbol{r}_1, \boldsymbol{r}_2)$ is negative everywhere and when integrated over $\boldsymbol{r}_2$ at any position $\boldsymbol{r}_1$ of electron 1 gives -1; i.e., exactly the charge of one electron is expelled from the space around electron 1.

What, therefore, the correlation hole look like? According to the philosophy of dragon chasing it is the rest

$$h_{\mathrm{xc}}^{\sigma\sigma'} = h_{\mathrm{x}}^{\sigma\sigma'} + h_{\mathrm{c}}^{\sigma\sigma'}. \tag{11.77}$$

The correlation energy from Eq. (11.68) therefore has the form:

$$E_{\mathrm{c}} = \frac{1}{2}\sum_{\sigma\sigma'}\iint \mathrm{d}\boldsymbol{r}_1\mathrm{d}\boldsymbol{r}_2\frac{\rho_\sigma(\boldsymbol{r}_1)}{r_{12}}h_{\mathrm{c}}^{\sigma\sigma'}(\boldsymbol{r}_1, \boldsymbol{r}_2). \tag{11.78}$$

Since the exchange hole has already fulfilled the boundary conditions of Eqs. (11.63) through (11.76) forced by the Pauli exclusion principle, the correlation hole satisfies a simple boundary condition

$$\int h_{\mathrm{c}}^{\sigma\sigma'}(\boldsymbol{r}_1, \boldsymbol{r}_2)\mathrm{d}\boldsymbol{r}_2 = 0. \tag{11.79}$$

Thus, the correlation hole means that electron 1 is pushing electron 2 (i.e., other electrons) off, but this means only that the pushed electrons are moved further out.

The dragon of electronic correlation has been chased into the correlation hole. Numerical experience turns out to conclude (below an example will be given) that

the exchange energy E_{x} is much more important than the correlation energy E_{c} and, therefore, scientists managed to replace the terrible exchange correlation dragon to a tiny beast hiding in the correlation hole (to be found).

[47] Indeed, $\int h_{\mathrm{x}}^{\sigma\sigma'}(\boldsymbol{r}_1, \boldsymbol{r}_2)\mathrm{d}\boldsymbol{r}_2 = -\delta_{\sigma\sigma'}\frac{1}{\rho_\sigma(\boldsymbol{r}_1)}\int |\rho_\sigma(\boldsymbol{r}_1; \boldsymbol{r}_2)|^2\,\mathrm{d}\boldsymbol{r}_2 = -\delta_{\sigma\sigma'}\frac{1}{\rho_\sigma(\boldsymbol{r}_1)}\sum_{ij}\phi_i(\boldsymbol{r}_1,\sigma)\phi_j^*(\boldsymbol{r}_1,\sigma)$ $\int \phi_i^*(\boldsymbol{r}_2,\sigma)\phi_j(\boldsymbol{r}_2,\sigma)\mathrm{d}\boldsymbol{r}_2 = -\delta_{\sigma\sigma'}\frac{1}{\rho_\sigma(\boldsymbol{r}_1)}\sum_{ij}\phi_i(\boldsymbol{r}_1,\sigma)\phi_j^*(\boldsymbol{r}_1,\sigma)\delta_{ij} = -\delta_{\sigma\sigma'}\frac{1}{\rho_\sigma(\boldsymbol{r}_1)}\sum_i \phi_i(\boldsymbol{r}_1,\sigma)\phi_i^*(\boldsymbol{r}_1,\sigma) =$ $-\delta_{\sigma\sigma'}\frac{1}{\rho_\sigma(\boldsymbol{r}_1)}\rho_\sigma(\boldsymbol{r}_1) = -\delta_{\sigma\sigma'}.$

11.6.7 Physical Grounds for the DFT Functionals

LDA

The LDA is not as primitive as it looks. The electron density distribution for the homogeneous gas model satisfies the Pauli exclusion principle and, therefore, this approximation gives the Fermi holes that fulfill the boundary conditions with Eqs. (11.63), (11.76) and (11.79). The LDA is often used because it is rather inexpensive, while still giving a reasonable geometry of molecules and vibrational frequencies.[48] The quantities that the LDA fails to reproduce are the binding energies[49], ionization potentials, and the intermolecular dispersion interaction.

The Perdew-Wang Functional (PW91)

Perdew noted a really dangerous feature in an innocent and reasonable-looking GEA potential. It turned out that in contrast to the LDA, the boundary conditions for the electron holes were not satisfied. For example, the exchange hole was not negative everywhere, as Eq. (11.75) requires. Perdew and Wang corrected this deficiency in a way similar to that of Alexander the Great, when he cut the Gordian knot. They tailored the formula for E_{xc} in such a way as to change the positive values of the function to zero, while the peripheral parts of the exchange holes were cut to force the boundary conditions to be satisfied anyway. The authors noted an important improvement in the results.

The Functional B3LYP

It was noted that the LDA and even GEA models systematically give too large chemical bond energies. On the other hand, it was known that the Hartree-Fock method is notorious for making the bonds too weak. What are we to do? Well, just mix the two types of potential and hope to have an improvement with respect to any of the models. Recall Eq. (11.56) for Π_{aver}, where the averaging extended from $\lambda = 0$ to $\lambda = 1$. The contribution to the integral for λ close to 0 comes from the situations similar to the fictitious model of non-interacting particles, where the wave function has the form of the Kohn-Sham determinant. Therefore, those contributions contain the exchange energy E_{x} corresponding to such a determinant. We may conclude that a contribution from the Kohn-Sham exchange energy $E_{\mathrm{x}}^{\mathrm{HF}}$ might look quite natural.[50] This is what the B3LYP method does, Eq. (11.41). Of course, it is not possible to justify the particular proportions of the B3LYP ingredients. Such things are justified only by their success.[51]

---

[48] Some colleagues of mine sometimes add a malicious remark that the frequencies are so good that they even take into account the anharmonicity of the potential.

[49] The average error in a series of molecules may even be of the order of 40 kcal/mol; this is a lot, since the chemical bond energy is of the order of about 100 kcal/mol.

[50] The symbol HF pertains to Kohn-Sham rather than to Hartree-Fock.

[51] The same sentiment applies in herbal therapy.

11.7 Visualization of Electron Pairs: Electron Localization Function (ELF)

One of the central ideas of general chemistry is the notion of an electron pair (i.e., two electrons of opposite spins that occupy a certain region of space). Understanding chemistry means knowing the role of these electron pairs in individual molecules (which is directly related to their structure) and what may happen to them when two molecules are in contact (chemical reactions). Where in a molecule do electron pairs prefer to be? This is the role of the ELF, which may be seen as an idea of visualization that helps chemists to elaborate what is known as *chemical intuition* ("*understanding*"), an important qualitative generalization that supports any practical chemist's action such as planning chemical synthesis.

In Chapter 10 (p. 597) and in this chapter, we were dealing with the Fermi hole that characterized quantitatively the strength of the Pauli exclusion principle: two electrons with the same spin coordinate avoid each other. We have given several examples showing that a (probe) electron with the same spin as a reference electron tries to be as far as possible from the latter one–a very strong effect. And what about electrons of opposite spins? Well, they are not subject to this restriction and can approach each other, but not too close, because of the Coulomb interaction. As a result, in molecules (and an atom as well) *we have to do with a shell-like electronic structure*: an electron pair (while keeping a reasonable electron-electron distance in it) may profit from occupying a domain very close to the nuclei. There is no future there for any other electron or electron pair, because of the Pauli exclusion principle. Other electron pairs have to occupy separately other domains in space.

The strength of the Pauli exclusion principle will certainly depend on the position in space with respect to the nuclear framework. Testing this strength is our goal now.

Let us take a reference electron at position $\mathbf{r}$ in a global coordinate system (Fig. 11.8a) and try to approach it with a "*probe electron*" of the same spin coordinate, shown by the radius vector $\mathbf{r}+\mathbf{r}_p$ (i.e., the probe electron would have the radius vector $\mathbf{r}_p$, when seen from the reference electron shown by $\mathbf{r}$). We will consider only such $\mathbf{r}_p$ that ensure that the probe electron be enclosed around the reference one in a sphere of radius R_p, i.e., $r_p \leq R_p$. The key function is the Fermi hole function, $h_x^{\sigma\sigma}(\mathbf{r},\mathbf{r}+\mathbf{r}_p)$ of Eq. (11.75) on p. 698. We are interested in what fraction of the probe electron is outside the abovementioned sphere. For small r_p, one can certainly write the following Taylor expansion about point $\mathbf{r}$:

$$h_x^{\sigma\sigma}(\mathbf{r},\mathbf{r}+\mathbf{r}_p) = -\rho_\sigma(\mathbf{r}) + (\nabla h_x^{\sigma\sigma})_{\mathbf{r}_p=\mathbf{0}} \cdot \mathbf{r}_p + C(\mathbf{r})\ \ r_p^2 + \ldots \tag{11.80}$$

Since function $h_x^{\sigma\sigma}(\mathbf{r},\mathbf{r}+\mathbf{r}_p)$ has a minimum for $\mathbf{r}_p=\mathbf{0}$ (the most improbable scenario[52]: two electrons of the same spin coordinate would sit one on top of the other), we see a vanishing of the gradient: $(\nabla h_x^{\sigma\sigma})_{\mathbf{r}_p=\mathbf{0}} = \mathbf{0}$. As to the last term shown, instead of the usual second derivatives calculated at the minimum, we simplified things by putting a rotation-averaged

[52] This is true, but only for those positions $\mathbf{r}$ of the reference electron for which the density $\rho(\mathbf{r})$ is not too small. For $\mathbf{r}$ belonging to peripheries of the molecule, a substantial exchange hole cannot be dug out at $\mathbf{r}$ because the ground there is shallow. In such a case, the exchange hole "*stays behind*" $\mathbf{r}$, in the region of appreciable values of ρ.

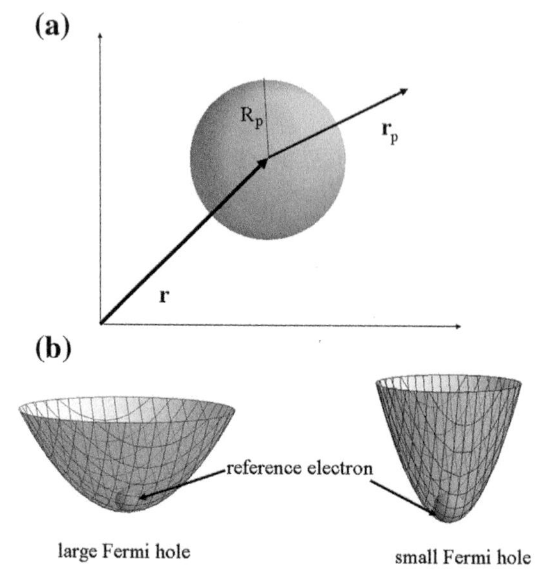

Fig. 11.8. A gear to test the power of the Pauli exclusion principle (Fermi hole). (a) The reference electron with the radius vector $\mathbf{r}$ and probe electron with the radius vector $\mathbf{r}+\mathbf{r}_p$ (both with the same spin coordinate). From the sphere shown, an electron of the same spin as that of the reference one is expelled. Therefore, the same sphere is a residence for two electrons with opposite spins (electron pair); (b) Two parabolic Fermi holes–a result of the Pauli exclusion principle. In each hole, the reference electron is shown (represented by a small ball). A narrow well means a large value of $C(\mathbf{r})$ and therefore a small value of ELF, which means a small propensity to host an electron pair. In contrast to that, a wide well corresponds to a large propensity to home there an electron pair.

constant $C(\mathbf{r}) > 0$. Truncating in Eq. (11.80) all terms beyond quadratic ones, we get

$$h_x^{\sigma\sigma}(\mathbf{r}, \mathbf{r}+\mathbf{r}_p) = -\rho_\sigma(\mathbf{r}) + C(\mathbf{r}) \ \ r_p^2. \tag{11.81}$$

Thus, $h_x^{\sigma\sigma}(\mathbf{r}, \mathbf{r}+\mathbf{r}_p)$ is nothing but a paraboloidal well with the minimum equal to $-\rho_\sigma(\mathbf{r})$ at position $\mathbf{r}$ of the reference electron. The well is controlled by the value of $C(\mathbf{r})$: for small $C(\mathbf{r})$ the well is wide, for large $C(\mathbf{r})$ the well is narrow (Fig.11.8b).

What fraction of the probe electron charge is expelled from the sphere of radius R_p? This can be calculated from the hole function by integrating it over all possible θ_p, ϕ_p and $r_p < R_p$):

$$\int d\mathbf{r}_p h_x^{\sigma\sigma}(\mathbf{r}, \mathbf{r}+\mathbf{r}_p) = -\rho_\sigma(\mathbf{r}) \int d\mathbf{r}_p + C(\mathbf{r}) \int r_p^2 d\mathbf{r}_p =$$

$$-\rho_\sigma(\mathbf{r})\frac{4}{3}\pi R_p^3 + C(\mathbf{r}) \int dr_p d\theta_p d\phi_p r_p^2 r_p^2 \sin\theta_p = -\rho_\sigma(\mathbf{r})\frac{4}{3}\pi R_p^3 + C(\mathbf{r})\frac{4\pi R_p^5}{5}.$$

Now we have to decide what to choose as R_p, if the reference electron has position $\mathbf{r}$. It is reasonable to make R_p dependent on position in space because the Fermi hole should be created easier (i.e., it would be larger) for small values of the electron density $\rho_\sigma(\mathbf{r})$, and harder for larger values. Quite arbitrarily, we choose such a function $R_p(\mathbf{r})$ that the first term on the right side satisfies:

$$-\rho_\sigma(\mathbf{r})\frac{4}{3}\pi R_p^3 = -1, \tag{11.82}$$

which means that close to the point shown by vector $\mathbf{r}$, the volume of the sphere of radius R_p is equal to the mean volume per single electron of the spin coordinate σ in the uniform electron gas of density $\rho_\sigma(\mathbf{r})$. If we have the closed-shell case ($\rho_\alpha(\mathbf{r}) = \rho_\beta(\mathbf{r})$), the same volume also contains an electron of the opposite spin. This means that such a volume contains on average a complete *electron pair*. As a consequence, using Eq. (11.82), one may write

$$C(\mathbf{r})\frac{4\pi R_p^5}{5} \sim C(\mathbf{r})\left(\rho_\sigma(\mathbf{r})\right)^{-\frac{5}{3}}.$$

Becke and Edgecombe used this expression to construct a function ELF($\mathbf{r}$) that reflects a tendency for an electron pair[53] to reside at point $\mathbf{r}$ (a large value of the ELF for a strong tendency, a small one for a weak tendency):

$$\text{ELF}(\mathbf{r}) = \frac{1}{1 + \kappa C(\mathbf{r})\left(\rho_\sigma(\mathbf{r})\right)^{-\frac{5}{3}}},$$

where $\kappa > 0$ represents an arbitrary scaling constant. Since $\infty > C(\mathbf{r}) > 0$, at any κ, we have $0 \leq \text{ELF}(\mathbf{r}) \leq 1$.

A large ELF($\mathbf{r}$) value) at position $\mathbf{r}$ means that a large Fermi hole is there, or a lot of space for an electron pair.

ELF($\mathbf{r}$) represents a function in a 3-D space. How can we visualize such a function inside a molecule? Well, one way is to look at an "*iso-ELF*" surface. But which one–because we have to decide among the ELF values ranging from 0 to 1? There is a general problem with isosurfaces because one has to choose an ELF value that returns an interesting information. A unfortunate value may give a useless result.[54]

Let us take a series of diatomic molecules: F_2, Cl_2, Br_2, whose electronic structure is believed to be well known in chemistry. Saying "well known" in reality means that whoever we have met in the past, when asked about it, said that there is a single covalent bond in all of these molecules. This book has already discussed about the VSEPR algorithm (see Chapter 8), which also predicts on each atom a tetrahedral configuration of the three lone pairs and a single chemical bond with the partner. If one were interested in the electron density coming from these three lone pairs, one would discover that instead of a "tripod-like" density, we would get an object with *cylindrical* symmetry. This would reflect the fact that the tripod's legs can be positioned anywhere on the ring with the center on the atom-atom axis, and perpendicular to the axis.

Let us begin with Br_2. The above conviction seems to be confirmed by the obtained ELF($\mathbf{r}$) function (see Fig. 11.9a). Indeed, as one can see, the ELF isosurface shows two tori, each behind

[53] A.D. Becke and K.E. Edgecombe, *J. Chem. Phys.*, *92*, 5397 (1990).

[54] For example, a section of Himalaya mountains at 10 000 m altitude brings a function that is zero everywhere.

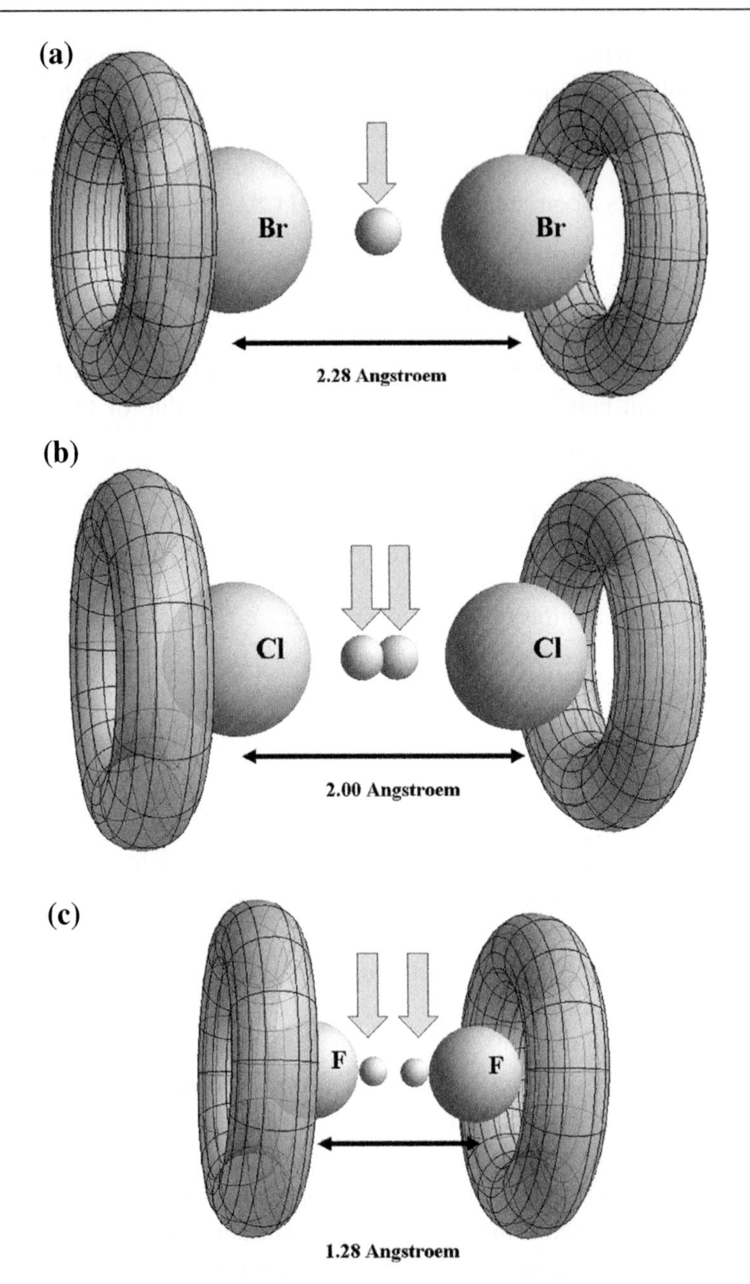

Fig. 11.9. A scheme composed of some selected isosurfaces of ELF (r) for (a) Br_2; (b) Cl_2; (c) F_2. In all cases, we see the peripheral tori, conserving the cylindrical symmetry of the system and corresponding to the electronic lone pairs (three for each atom, together with the bond they form a nearly tetrahedral configuration). The iso-ELF islands shown by the arrows correspond to the regions with higher probability density for finding that electron pair which is responsible for the chemical bond. The cases of F_2 and Cl_2 surprisingly show two such islands, while in case of Br_2, we have a single island of the largest tendency to find an electron pair.

the corresponding bromine atom, as could be expected for the electron density coming from the three corresponding atomic lone pairs, and each conserving the cylindrical symmetry. What about the bond electron pair? Well we see (Fig. 11.9a) that right in the middle of the $Br - Br$ distance, there is a preferred place for an electron pair, also conserving the cylindrical symmetry of the total system.

This gives the impression that ELF(**r**) tells us[55] what every freshman knows either from teachers, or from Professor Gilbert Lewis, or from the VSEPR algorithm. What could this student and ourselves expect from Cl_2 and F_2? Obviously, the same! And yet the ELF(**r**) has a surprise for us! It turns out (Fig. 11.9b, c) that the ELF procedure shows *two* regions for the bond electron pair. Why? Well, there is an indication. Let us recall (say, from the valence bond method, discussed in Chapter 10), that among important VB structures is the covalent one and two ionic structures. In the case of Cl_2, they would be the Heitler-London function describing the covalent bond Cl-Cl and the two ionic structures corresponding to Cl^+Cl^- and Cl^-Cl^+, respectively. Such ionic structures are necessary for a reliable description of the molecule at finite internuclear distances.[56] In a particular ionic structure, one electron is shifted toward one of the atoms; i.e., such a structure breaks the symmetry. However, the presence of the two such structures (of equal weight, for a homopolar molecule) restores this symmetry. One may say that for Cl_2 and F_2, there is a large fluctuation of the bond electron-pair position that strengthens the bond–a *charge-shift bonding*, CS). Therefore,

> besides the covalent bonds (like in Br-Br), the ionic bonds (as in NaCl; see Chapter 6), the polar bonds (like in C-H), there are the CS bonds, the bonds with fluctuating position of the bonding electron pair.

The concept of the CS bonds as some distinct kind of chemical bonds comes from independent theoretical considerations[57]*, but also seems to find its confirmation in a specialized visualization tool, which is in fact what the ELF idea really provides.*

11.8 The DFT Excited States

Ground States for a Given Symmetry

The DFT is usually considered as a ground-state theory. One should, however, remember that the exact ground-state electron density ρ_0 contains information about all the excited states (remember the discussion on p. 235). Well, the problem is that we do not know yet how to extract this information from ρ_0. Some of the excited states are the lowest-energy states belonging to a

[55] Using strange shapes, colors, shading, and even reflexes of light on them, which shamelessly play the role of making us naively believe that all these things are real.

[56] For infinite distance, they do not count.

[57] S. Shaik, D. Danovich, B. Silvi, D. L. Lauvergnat, and P.C. Hiberty, *Chem. Eur. J.*, *11*, 6358 (2005).

given irreducible representation of the symmetry group of the Hamiltonian. In such cases, forcing the proper symmetry of the Kohn-Sham orbitals leads to the solution for the corresponding excited state. Thus, these states are excited ones, but formally they can be treated as the ground states (in the corresponding irreducible representation).

Time-Dependent DFT

Is it possible to detect excited states by exciting the ground state? Well, there is a promising path showing how to do it.[58] From Chapter 2, we know that this requires the time-dependent periodic perturbation $-\hat{\mu} \cdot \mathcal{E} \exp(\pm i\omega t)$ of frequency ω, where $\hat{\mu}$ denotes the dipole moment operator of the system, and $\mathcal{E}$ is the electric field amplitude. Such a theory is valid under the assumption that the perturbation is relatively small and the electronic states of the isolated molecule are still relevant. In view of that, we consider only a linear response of the system to the perturbation. Let us focus on the dipole moment of the system as a function of ω. It turns out that at certain values, $\omega = \omega_{0k} = \frac{E_k^{(0)} - E_0^{(0)}}{\hbar}$; for $E_k^{(0)}$ denoting the energy of the k–th excited state ($E_0^{(0)}$ is the ground state energy), one has an abrupt change of the mean value of the dipole moment. In fact, this means that for $\omega = \omega_{0k}$, the ω-dependent polarizability (the dipole moment change is proportional to the polarizability) goes to infinity (i.e., has a pole). By detecting these poles[59] we are able to calculate the excited states in the DFT within the accuracy of a few tenths of eV.

11.9 The Hunted Correlation Dragon Before Our Eyes

The DFT method has a long history behind it, which began with Thomas, Dirac, Fermi, etc. At the beginning, the successes were quite modest (the electron gas theory, known as the $X\alpha$ *method*). Real success came after a publication by Jan Andzelm and Erich Wimmer.[60] The DFT method, offering results at a correlated level for a wide spectrum of physical quantities, turned out to be roughly as inexpensive as the Hartree-Fock procedure–this is the most sensational feature of the method.

---

[58] E. Runge and E.K.U. Gross, *Phys. Rev. Lett.*, *52*, 997 (1984); M.E. Casida, "*Time-dependent density functional response theory for molecules*," in *Recent Advances in Density Functional Methods, Part 1*, D.P. Chong, ed., World Scientific, Singapore, 1995.

[59] As a first guess, it may serve the orbital energy differences from the ground-state theory.

[60] J. Andzelm and E. Wimmer, *J. Chem. Phys.*, *96*, 1280 (1992). (Jan was my Ph.D. student.) In the paper by A. Scheiner, J. Baker, J. Andzelm, *J. Comp. Chem.*, *18*, 775 (1997) the reader will find an interesting comparison of the methods used. One of the advantages (or deficiencies) of the DFT methods is that they offer a wide variety of basis functions (in contrast to the *ab initio* methods, where Gaussian basis sets rule), recommended for some particular problems to be solved. For example, in electronics (Si,Ge) the plane wave $\exp(ikr)$ expansion is a preferred choice. On the other hand, these functions are not advised for catalysis phenomena with rare earth atoms. The Gaussian basis sets in the DFT had a temporary advantage (in the 1990s) over others, because the standard Gaussian programs offered analytically computed gradients (for optimization of the geometry). Now this is also offered by many DFT methodologies.

We Have a Beacon–Exact Electron Density Distribution of Harmonium

Hohenberg and Kohn proved their famous theorem on the existence of the energy functional, but nobody was able to give the functional for any system. All the DFT efforts are directed toward elaborating such a potential, and the only criterion of whether a model is any good is comparison with experimental results. However, it turned out that there is a system for which every detail of the DFT can be verified. Uniquely, the dragon may be driven out of the hole, and we may fearlessly and with impunity analyze all the details of its anatomy. The system is a bit artificial–it is the harmonic helium atom (harmonium) discussed on p. 212, in which the two electrons attract the nucleus by a harmonic force, while repelling each other by Coulombic interaction. For some selected force constants k (e.g., for $k = \frac{1}{4}$), the Schrödinger equation *can be solved analytically*. The wave function is extremely simple; see p. 589. The electron density (normalized to 2) is computed as

$$\rho_0(\boldsymbol{r}) = 2N_0^2 \mathrm{e}^{-\frac{1}{2}r^2} \left\{ \left(\frac{\pi}{2}\right)^{\frac{1}{2}} \left[\frac{7}{4} + \frac{1}{4}r^2 + \left(r + \frac{1}{r}\right) \mathrm{erf}\left(\frac{r}{\sqrt{2}}\right) \right] + \mathrm{e}^{-\frac{1}{2}r^2} \right\}, \tag{11.83}$$

where erf is the error function, $\mathrm{erf}\left(z\right) = \frac{2}{\sqrt{\pi}} \int_0^z \exp\left(-u^2\right) \mathrm{d}u$, and

$$N_0^2 = \frac{\pi^{\frac{3}{2}}}{(8 + 5\sqrt{\pi})}. \tag{11.84}$$

We should look at the $\rho_0(\boldsymbol{r})$ with great interest – it is a unique occasion, it is probable you will never see again an *exact* result. The formula is not only exact, but on top of that, it is simple. Kais et al. compare the exact results with two DFT methods: the BLYP (a version of B3LYP) and the Becke-Perdew (BP) method.[61]

Because of the factor $\exp\left(-0.5r^2\right)$, the density distribution ρ is concentrated on the nucleus.[62] The authors compare this density distribution with the corresponding Hartree-Fock density (appropriate for the potential used), and even with the density distribution related to the hydrogen-like atom (after neglecting $1/r_{12}$ in the Hamiltonian, the wave function becomes an antisymmetrized product of the two hydrogen-like orbitals). In the latter case, the electrons do not see each other[63], and the corresponding density distribution is too concentrated on the nucleus. As soon as the term $1/r_{12}$ is restored, the electrons immediately move apart, and ρ on the nucleus decreases by about 30%. The second result is also interesting: the Hartree-Fock density is very close to ideal–it is almost the same curve.[64]

---

[61] The detailed references to these methods are given in S. Kais, D.R. Herschbach, N.C. Handy, C.W. Murray, and G.J. Laming, *J. Chem. Phys.*, *99*, 417 (1993).

[62] This is as it should be.

[63] This is so even in the sense of the mean field (as it is in the Hartree-Fock method).

[64] This is why the HF method is able to give 99.6% of the total energy. Nevertheless, in some cases, this may not be a sufficient accuracy.

Table 11.1. Harmonium (harmonic helium atom). Comparison of the components (a.u.) of the total energy $E[\rho_0]$ calculated by the HF, BLYP, and BP methods with the exact values (row KS; exact Kohn-Sham solution).[a]

	$E[\rho_0]$	$T_0[\rho_0]$	$\int v\rho_0 d\mathbf{r}$	$J[\rho_0]$	$E_x[\rho_0]$	$E_c[\rho_0]$
KS	**2.0000**	**0.6352**	**0.8881**	**1.032**	**−0.5160**	**−0.0393**
HF	2.0392	0.6318	0.8925	1.030	−0.5150	0
BLYP	2.0172	0.6313	0.8933	1.029	−0.5016	−0.0351
BP	1.9985	0.6327	0.8926	1.028	−0.5012	−0.0538

[a] The row KS with the bold digits corresponds to the exact result.

Total Energy Components

It turns out that in the case analyzed (and so far only in this case), we can calculate the *exact* total energy E [Eq. (11.17)], "wonder" potential v_0 that in the Kohn-Sham model gives the exact density distribution ρ [Eq. (11.83)], exchange potential v_x and correlation potential v_c [Eqs. (11.70) and (11.71)].[65] Let us begin from the total energy.

In the second row of Table 11.1 labeled KS (for Kohn-Sham), the exact total energy is reported ($E[\rho_0] = 2.0000$ a.u.) and its components (bold figures) calculated according to Eqs. (11.10), (11.17)–(11.19), (11.68), and (11.72). The exact correlation energy E_c is calculated as the difference between the exact total energy and the listed components. Thus, $T_0[\rho_0]$ stands for the kinetic energy of the non-interacting electrons, $\int v\rho_0 d\mathbf{r}$ means the electron-nucleus attraction (which is positive because the harmonic potential is positive), and $J[\rho_0]$ represents the self-interaction energy of ρ_0. According to Eq. (11.19), and taking into account ρ_0 (i.e., twice a square of the orbital), we obtain $J[\rho_0] = 2\mathcal{J}_{11}$ with the Coulombic integral $\mathcal{J}_{11}$. On the other hand, the exchange energy is given by Eq. (11.72): $E_x = -\frac{1}{2}\sum_{i,j=1}^{\text{SMO}} K_{ij}$, and after summing over the spin coordinates, we obtain the exchange energy $E_x = -\mathcal{K}_{11} = -\mathcal{J}_{11}$. We see such a relation between J and E_x in the second row (KS[66]). The other rows report already various approximations computed by HF, BLYP, and BP, each of which gives its own Kohn-Sham spinorbitals and its own approximation of the density distribution ρ_0. This density distribution was used for the calculation of the components of the total energy within each approximate method. Of course, the Hartree-Fock method (third row) gave 0 for the correlation energy, because there is no correlation in it except that which follows from the Pauli exclusion principle taken into account in the exchange energy (cf. Chapter 10).

It is remarkable that all the methods are doing quite well. The BLYP gives the total energy with an error of 0.87%–twice as small as the Hartree-Fock method, while the BP functional missed by as little as 0.08%. The total energy components are a bit worse, which proves that a

[65] These potentials as functions of ρ or r.

[66] Only for spin-compensated two-electron systems, we have $E_x[\rho] = -\frac{1}{2}J[\rho_0]$ and therefore, $v_x = \frac{\delta E_x}{\delta\rho}$ can be calculated analytically. In all other cases, although E_x can be easily evaluated (knowing orbitals), the calculation of v_x is very difficult and costly (it can only be done numerically). In the present two-electron case, v_x^{HF} is a multiplicative operator rather than an integral operator.

certain cancellation of errors occurs among the energy components, which improves the value of the total energy. The KS kinetic energy T_0 amounts to 0.6352, while that calculated as the mean value of the kinetic energy operator (of two electrons) is a bit larger, 0.6644–the rest is in the exchange-correlation energy.[67]

Exact "Wonder" v_0 Potential–The Correlation Dragon Is Finally Caught

Fig. 11.10 shows a unique thing, our long-awaited "wonder," as well as exact potential v_0 as a function of r, and alternatively as a function of $\rho^{\frac{1}{3}}$. We look at it with great curiosity. The exact $v_0(r)$ represents a monotonic function increasing with r and represents a modification (influence of the second electron) of the external potential v , we see that v_0 is shifted *upward* with respect to v, because the electron repulsion is effectively included. As we can see, the best approximate potential is the Hartree-Fock one.

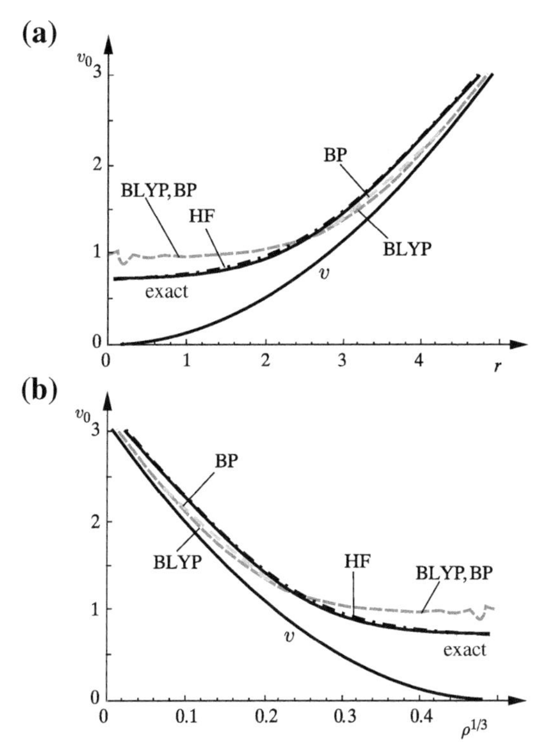

Fig. 11.10. Efficiency analysis of various DFT methods and comparison with the exact theory for the harmonium (with force constant $k = \frac{1}{4}$) according to Kais et al. Panel (a) shows one-electron effective potential $v_0 = v + v_{coul} + v_{xc}$, with external potential $v = \frac{1}{2}kr^2$. Panel (b) presents the same quantities as functions of $\rho^{\frac{1}{3}}$. The solid line corresponds to the exact results. The symbol HF pertains to the Fock potential (for the harmonic helium atom, · — ·—), and the symbols BLYP (— — —) and BP (===) stand for two popular DFT methods.

[67] We have described this before.

Exchange Potential

As to exchange potential v_X (Fig. 11.11), it has to be negative–and indeed it is. How are the BLYP and BP exchange potentials doing? Their plots are very close to each other and go almost parallel to the exact exchange potential for most values of r; i.e., they are very good (any additive constant in any potential energy does not count). For small r, both DFT potentials undergo some strange vibration. This region (high density) is surely the most difficult to describe, and no wonder that simple formulas cannot accurately describe the exact electronic density distribution (Fig.11.11).

Measuring the Correlation Dragon: It Is a Small Beast

The correlation potential v_C turns out (Fig. 11.12) to correspond to forces $10 - 20$ *times smaller* than those typical for exchange potential v_X (just look at the corresponding slopes). This is an important message because, as the reader may remember, at the very end, we tried to push the dragon into the correlation hole v_C and, as we see now, we have succeeded. The dragon of the correlation energy turned out to be a small beast.

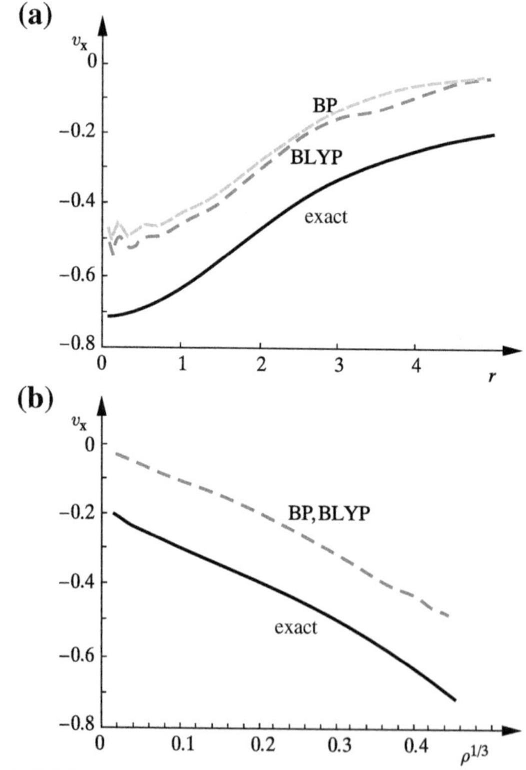

Fig. 11.11. Exchange potential. Efficiency analysis of various DFT methods and comparison with the exact theory for the harmonium (with the force constant $k = \frac{1}{4}$) according to Kais et al. Panel (a) shows exchange potential v_X as a function of the radius r, and Panel (b) uses a function of the density distribution ρ. The notation of Fig. 11.10 is used. It is seen that both DFT potentials produce plots that differ by nearly a constant from the exact potential (it is, therefore, an almost exact potential). The two DFT methods exhibit some non-physical oscillations for small r.

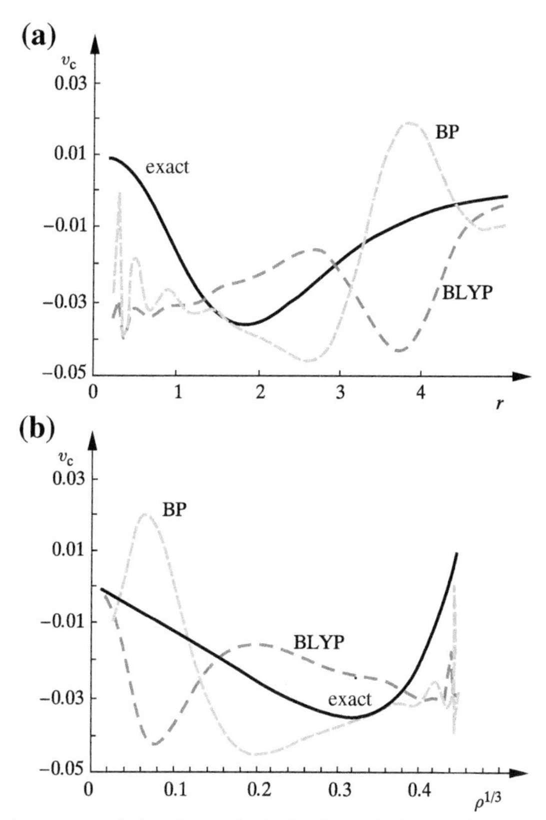

Fig. 11.12. The long-chased electron correlation dragon is finally found in its correlation hole, and we have an exceptional opportunity to see what it looks like. Correlation potential–efficiency analysis of various DFT methods and comparison with the exact theory for the harmonic helium atom (with the force constant $k = \frac{1}{4}$) according to Kais et al. Panel (a) shows correlation potential v_C (which is less important than the exchange potential) as a function of the radius r (a) and of density ρ (b). The same notation is used as in Fig. 11.10. The DFT potentials produce plots that differ widely from the exact correlation potential.

The exact potential represents a smooth hooklike curve. The BLYP and BP correlation plots twine loosely like eels round about the exact curve[68], and, for small r, exhibit some vibration similar to that for v_X. It is most impressive that the BLYP and BP curves twine as if they were in counterphase, which suggests that, if added, they might produce good[69] results.

Conclusion

The harmonic helium atom represents an instructive example that pertains to medium electronic densities. It seems that the dragon of the correlation energy does not have hundreds of heads and is of quite good character, although it remains a bit unpredictable.

The results of various DFT versions are generally quite effective, although this comes from a cancellation of errors. Nevertheless, great progress has been made. At present, many chemists

[68] The deviations are very large.

[69] Such temptations give birth to Babylon-type science.

prefer the DFT method (economy and accuracy) than to getting stuck at the barrier of the configuration interaction excitations. And yet the method can hardly be called *ab initio*, since the exchange-correlation potential is tailored in a somewhat too practical manner.

Summary

- The main theoretical concept of the DFT method is the electronic density distribution $\rho(\boldsymbol{r}) = N\sum_{\sigma_1=\frac{1}{2}}^{-\frac{1}{2}} \int d\tau_2 d\tau_3 \ldots d\tau_N \mid \Psi(\boldsymbol{r},\sigma_1,\boldsymbol{r}_2,\sigma_2,\ldots,\boldsymbol{r}_n,\sigma_N)|^2$, where $\boldsymbol{r}$ indicates a point in 3-D space, and the sum is over all the spin coordinates of N electrons, while the integration is over the space coordinates of $N-1$ electrons. For example, within the molecular orbital (RHF) approximation $\rho = \sum_i 2\left|\varphi_i\left(\boldsymbol{r}\right)\right|^2$ is the sum of the squares of all the molecular orbitals multiplied by their occupation number. The electronic density distribution ρ is a function of the position in 3-D space.
- ρ carries a lot of information. The density ρ exhibits maxima at nuclei (with a discontinuity of the gradient because of the cusp condition, as discussed on p. 585). The Bader analysis is based on identification of the critical (stationary) points of ρ (i.e., those for which $\nabla\rho = \mathbf{0}$); for each of them, the Hessian is computed (the second derivatives matrix). Diagonalization of the Hessian tells us whether the critical point corresponds to a maximum of ρ (non-nuclear attractor[70]), a minimum (e.g., cavities), a first-order saddle point (e.g., a ring center), or a second-order saddle point (chemical bond).
- The DFT relies on the two Hohenberg-Kohn theorems:
  - The *ground-state electronic density distribution* (ρ_0) *contains the same information as the ground-state wave function*(Ψ_0). Therefore, instead of a complex mathematical object (the ground-state wave function Ψ_0) depending on $4N$-variables, we have a much simpler object (ρ_0) that depends on three variables (Cartesian coordinates) only.
  - A total energy functional of ρ *exists* that attains its minimum at $\rho = \rho_0$. This mysterious functional is not yet known.
- Kohn and Sham presented the concept of a system with *non-interacting electrons,* subject to some "*wonder*" external field $v_0(\boldsymbol{r})$ (instead of that of the nuclei), such that the resulting *density ρ remains identical to the exact ground-state density distribution* ρ_0. This *fictitious* system of electrons plays a very important role in the DFT.
- Since the Kohn-Sham electrons do not interact, their wave function represents a single Slater determinant (called the *Kohn-Sham determinant*).
- We write the total energy expression $E = T_0 + \int v(\boldsymbol{r})\rho(\boldsymbol{r})d\boldsymbol{r} + J[\rho] + E_{xc}[\rho]$, which contains:
  - The kinetic energy of the *non-interacting* electrons (T_0)
  - The potential energy of the electron-nuclei interaction ($\int v(\boldsymbol{r})\rho(\boldsymbol{r})d\boldsymbol{r}$)
  - The Coulombic electron-electron self-interaction energy ($J[\rho]$)
  - The remainder E_{xc} i.e., the unknown exchange-correlation energy
- Using the single-determinant Kohn-Sham wave function (which gives the exact ρ_0), we vary the Kohn-Sham spinorbitals in order to find the minimum of the energy E.
- We are immediately confronted with the problem of how to find the unknown exchange-correlation energy E_{xc}, which is replaced by an unknown exchange-correlation potential in the form of a functional derivative $v_{xc} \equiv \frac{\delta E_{xc}}{\delta\rho}$. We obtain the Kohn-Sham equation (resembling the Fock equation) $\{-\frac{1}{2}\Delta + v_0\}\phi_i = \varepsilon_i\phi_i$, where "wonder-potential" $v_0 = v + v_{coul} + v_{xc}$, v_{coul} stands for the sum of the usual Coulombic operators (as in the Hartree-Fock method[71], but built from the Kohn-Sham spinorbitals), and v_{xc} is an exchange-correlation potential to be found.

[70] The maxima on the nuclei are excluded from the analysis because of the discontinuity of $\nabla\rho$ mentioned above.

[71] It is, in fact, $\frac{\delta J[\rho]}{\delta\rho}$.

- The main problem now resides in the nature of $E_{\rm xc}$ (and $v_{\rm xc}$). A variety of practical guesses that we are forced to make begin here.
- The simplest guess is the local density approximation (LDA). We assume that $E_{\rm xc}$ can be summed up from the contributions of all the points in space, and that the individual contribution depends only on ρ computed at this point. Now, the key question is: *What does this dependence $E_{\rm xc}[\rho]$ look like?* The LDA answers this question by using the following approximation: each point $\boldsymbol{r}$ in the 3D space contributes to $E_{\rm xc}$ depending on the computed value of $\rho(\boldsymbol{r})$ *as if it were a homogeneous gas of uniform density* ρ, where the dependence $E_{\rm xc}[\rho]$ is *exactly* known.
- There are also more complex $E_{\rm xc}[\rho]$ functionals that go beyond the local approximation. They not only use the local value of ρ, but sometimes also $\nabla\rho$ (gradient approximation).
- In each of these choices, there is a lot of ambiguity. This, however, is restricted by some physical requirements.
- The requirements are related to the *electron pair distribution function* $\Pi(\boldsymbol{r}_1,\boldsymbol{r}_2) = N(N-1)\sum_{\text{all }\sigma_i}\int|\Psi|^2 d\boldsymbol{r}_3 d\boldsymbol{r}_4\ldots d\boldsymbol{r}_N$, which takes into account that the two electrons, shown by $\boldsymbol{r}_1$ and $\boldsymbol{r}_2$, avoid each other.
- First-order perturbation theory leads to the exact expression for the total energy E as $E = T_0 + \int\rho(\boldsymbol{r})v(\boldsymbol{r})d\boldsymbol{r} + \frac{1}{2}\iint d\boldsymbol{r}_1 d\boldsymbol{r}_2 \frac{\Pi_{\rm aver}(\boldsymbol{r}_1,\boldsymbol{r}_2)}{r_{12}}$, where $\Pi_{\rm aver}(\boldsymbol{r}_1,\boldsymbol{r}_2) = \int_0^1 \Pi_\lambda(\boldsymbol{r}_1,\boldsymbol{r}_2)d\lambda$, with the parameter $0 \le \lambda \le 1$ being instrumental when transforming the system of *non-interacting* electrons ($\lambda = 0$, Kohn-Sham model) into the system of *fully interacting* ones ($\lambda = 1$), and *all the while preserving the exact density distribution* ρ. Unfortunately, the function $\Pi_\lambda(\boldsymbol{r}_1,\boldsymbol{r}_2)$ remains unknown.
- The function $\Pi_\lambda(\boldsymbol{r}_1,\boldsymbol{r}_2)$ serves to define the electron hole functions, which will tell us where electron 2 prefers to be, if electron 1 occupies the position $\boldsymbol{r}_1$. The exchange-correlation energy is related to the $\Pi_{\rm aver}^{\sigma\sigma'}$ function by $E_{\rm xc} = \frac{1}{2}\sum_{\sigma\sigma'}\iint d\boldsymbol{r}_1 d\boldsymbol{r}_2 \frac{\Pi_{\rm aver}^{\sigma\sigma'}(\boldsymbol{r}_1,\boldsymbol{r}_2) - \rho_\sigma(\boldsymbol{r}_1)\rho_{\sigma'}(\boldsymbol{r}_2)}{r_{12}}$, where the sum is over the spin coordinate σ of electron 1 and spin coordinate σ' of electron 2, with the decomposition $\Pi_{\rm aver} = \Pi_{\rm aver}^{\alpha\alpha} + \Pi_{\rm aver}^{\alpha\beta} + \Pi_{\rm aver}^{\beta\alpha} + \Pi_{\rm aver}^{\beta\beta}$. For example, the number $\Pi_{\rm aver}^{\alpha\beta} dV_1 dV_2$ is proportional to the probability of finding simultaneously an electron with the spin function α in the volume dV_1 located at $\boldsymbol{r}_1$, another electron with the spin function β in the volume dV_2 located at $\boldsymbol{r}_2$, etc.
- The definition of the exchange-correlation hole function $h_{\rm xc}^{\sigma\sigma'}(\boldsymbol{r}_1,\boldsymbol{r}_2)$ is as follows: $E_{\rm xc} = \frac{1}{2}\sum_{\sigma\sigma'}\int d\boldsymbol{r}_1\int d\boldsymbol{r}_2 \frac{\rho_\sigma(\boldsymbol{r}_1)}{r_{12}} h_{\rm xc}^{\sigma\sigma'}(\boldsymbol{r}_1,\boldsymbol{r}_2)$, which is equivalent to setting $h_{\rm xc}^{\sigma\sigma'}(\boldsymbol{r}_1,\boldsymbol{r}_2) = \frac{\Pi_{\rm aver}^{\sigma\sigma'}(\boldsymbol{r}_1,\boldsymbol{r}_2)}{\rho_\sigma(\boldsymbol{r}_1)} - \rho_{\sigma'}(\boldsymbol{r}_2)$. This means that the hole function is related to that part of the pair distribution function that indicates the *avoidance of the two electrons* [i.e., beyond their independent motion described by the *product* of the densities $\rho_\sigma(\boldsymbol{r}_1)\rho_{\sigma'}(\boldsymbol{r}_2)$].
- Due to the antisymmetry requirement for the wave function (see Chapter 1), the holes have to satisfy some general (integral) conditions. The electrons with parallel spins have to avoid each other: $\int h_{\rm xc}^{\alpha\alpha}(\boldsymbol{r}_1,\boldsymbol{r}_2)d\boldsymbol{r}_2 = \int h_{\rm xc}^{\beta\beta}(\boldsymbol{r}_1,\boldsymbol{r}_2)d\boldsymbol{r}_2 = -1$ (one electron disappears from the neighborhood of the other), while the electrons with opposite spins are not influenced by the Pauli exclusion principle: $\int h_{\rm xc}^{\alpha\beta}(\boldsymbol{r}_1,\boldsymbol{r}_2)d\boldsymbol{r}_2 = \int h_{\rm xc}^{\beta\alpha}(\boldsymbol{r}_1,\boldsymbol{r}_2)d\boldsymbol{r}_2 = 0$.
- The exchange correlation hole is a sum of the exchange hole and the correlation hole: $h_{\rm xc}^{\sigma\sigma'} = h_{\rm x}^{\sigma\sigma'} + h_{\rm c}^{\sigma\sigma'}$, where the exchange hole follows in a simple way from the Kohn-Sham determinant (and is therefore supposed to be known). Then, we have to guess the correlation holes. *All the correlation* holes have to satisfy the condition $\int h_{\rm c}^{\sigma\sigma'}(\boldsymbol{r}_1,\boldsymbol{r}_2)d\boldsymbol{r}_2 = 0$, which means only that the average has to be zero, but that says nothing about the particular form of $h_{\rm c}^{\sigma\sigma'}(\boldsymbol{r}_1,\boldsymbol{r}_2)$. The only thing sure is that close to the origin, the function $h_{\rm c}^{\sigma\sigma'}$ has to be negative, and, therefore, for longer distances, it has to be positive.
- The popular approximations (e.g., LDA, PW91) in general satisfy the integral conditions for the holes.
- The hybrid approximations (e.g., B3LYP)–i.e., such a linear combination of the potentials that will assure good agreement with experimental results–become more and more popular.
- The DFT models can be tested when applied to exactly solvable problems with electronic correlation (like the harmonium, as discussed in Chapter 4). It turns out that despite the exchange and correlation DFT potentials deviating from the exact ones, the total energy is quite accurate.

- There is a possibility in DFT to calculate the excitation energies. This is possible within the time-dependent DFT. In this formulation, one is looking at the frequency-dependent polarizabilities in the system subject to the electric field perturbation of frequency ω. The polarizabilities have poles at $\hbar\omega$ equal to an excitation energy.

Main Concepts, New Terms

adiabatic connection (p. 691)
attractor (p. 570)
Bader analysis (p. 712)
basin (p. 670)
B3LYP functional (p. 700)
catastrophe set (p. 672)
correlation hole (p. 697)
critical (stationary) points (p. 669)
CS, charge shift bond (p. 705)
density matrix (p. 665)
electron gas (p. 706)
electronic density (p. 665)
electron pair distribution (p. 690)
ELF, Electron Localization Function (p. 701)
exchange-correlation energy (p. 683)
exchange-correlation hole (p. 696)
exchange hole (p. 697)
exchange-correlation potential (p. 689)
Fermi hole (p. 698)
gradient approximation, NLDA (GEA) (p. 688)
Hohenberg-Kohn functional (p. 680)
Hohenberg-Kohn theorem (p. 675)
holes (p. 695)
hybrid approximations, NLDA (p. 689)
Kohn-Sham equation (p. 680)
Kohn-Sham system (p. 680)
Levy minimization (p. 679)
local density approximation, LDA (p. 687)
non-nuclear attractor (p. 670)
one-particle density matrix (p. 698)
Perdew-Wang functional (p. 688)
self-interaction energy (p. 708)
spin polarization (p. 687)
Time-Dependent DFT (p. 706)
v-representability (p. 677)

From the Research Front

Computer technology has been revolutionary–and not only because computers are fast. Much more important is that each programmer uses the full experience of his predecessors and easily "*stands on the shoulders of giants.*" The computer era has made an unprecedented transfer of the most advanced theoretical tools from the finest scientists to practically everybody. Experimentalists often investigate large molecules. If there is a method like DFT, which gives answers to their vital questions in a shorter time than the *ab initio* methods, they will not hesitate and choose the DFT, even if the method is of semi-empirical type. Something like this happens now. Nowadays, the DFT procedure is applicable to systems with hundreds of atoms.

The DFT method also is developing fast in the conceptual sense[72]; e.g., the theory of reactivity ("*charge sensitivity analysis*"[73]) has been derived, which established a link between the intermolecular electron transfer and the charge density changes in atomic resolution. For systems in magnetic fields, current DFT was developed.[74] Relativistic effects[75] and time-dependent phenomena[76] are included in some versions of the theory.

[72] See, e.g., P. Geerlings, F. De Proft, and W. Langenaeker, *Chem. Rev.*, *103*, 1793 (2003).

[73] R.F. Nalewajski and J. Korchowiec, *Charge Sensitivity Approach to Electronic Structure and Chemical Reactivity* World Scientific, Singapore, 1997; R.F. Nalewajski, J. Korchowiec, and A. Michalak, "*Reactivity criteria in charge sensitivity analysis*," *Topics in Current Chemistry*, *183*, 25 (1996); R.F. Nalewajski, "*Charge sensitivities of molecules and their fragments*," *Rev. Mod. Quant. Chem.*, K. D. Sen, ed., World Scientific, Singapore (2002)1071; R.F. Nalewajski and R.G. Parr, *Proc. Natl. Acad. Sci. USA*, *97*, 8879 (2000).

[74] G. Vignale and M. Rasolt, *Phys. Rev. Letters*, *59*, 2360 (1987); *Phys.Rev.B*, *37*, 10685 (1988).

[75] A.K. Rajagopal and J. Callaway, *Phys. Rev. B*, *7*, 1912 (1973); A.H. MacDonald, S.H. Vosko, *J. Phys. C*, *12*, 2977 (1979).

[76] E. Runge and E.K.U. Gross, *Phys. Rev. Lett. 52*, 997 (1984); R. van Leeuwen, *Phys. Rev. Lett. 82*, 3863 (1999).

Ad Futurum

The DFT will be, of course, further elaborated in the future. There are already investigations under way that will allow us to calculate the dispersion energy.[77] The impetus will probably be directed toward such methods as the density matrix functional theory (DMFT) proposed by Levy[78], and currently being developed by Jerzy Ciosłowski[79] The idea is to abandon $\rho(\boldsymbol{r})$ as the central quantity, and instead use the one-particle density matrix$\rho(\boldsymbol{r}';\boldsymbol{r})$ of Eqs. (11.1) and (11.2).

The method has the advantage that we are not forced to introduce the non-interacting Kohn-Sham electrons, because the mean value of the electron kinetic energy may be expressed directly by the new quantity (this follows from the definition):

$$T = -\frac{1}{2}\int \mathrm{d}\boldsymbol{r} \quad [\Delta_{\boldsymbol{r}}\rho(\boldsymbol{r};\boldsymbol{r}')]|_{\boldsymbol{r}'=\boldsymbol{r}},$$

where the symbol $|_{\boldsymbol{r}'=\boldsymbol{r}}$ means replacing $\boldsymbol{r}'$ by $\boldsymbol{r}$ after the result $\Delta_{\boldsymbol{r}}\rho(\boldsymbol{r};\boldsymbol{r}')$ is ready. Thus, in the DMFT exchange-correlation, we have no kinetic energy left.

The success of the DFT approach will probably make the traditional *ab initio* procedures faster, up to the development of methods with linear scaling (with the number of electrons for long molecules). The massively parallel "*computer farms*," with 2000 processors currently to millions expected to come soon, will saturate most demands of experimental chemistry. The results will be calculated fast; it will be much more difficult to define an interesting target to compute.

We will have an efficient hybrid potential, say, of the B3LYP5PW2013/2014-type. There remains, however, a problem that already appears in laboratories. A colleague delivers a lecture and proposes a hybrid B3LYP6PW2013update[80], which is more effective for aromatic molecules. What will these two scientists talk about? It is very good that the computer understands all this, but what about the scientists? In my opinion, science will move into such areas as planning new materials and new molecular phenomena (cf. Chapter 15) with the programs mentioned above as tools.

Additional Literature

W. Koch and M. C. Holthausen, *A Chemist's Guide to Density Functional Theory*, New York, Wiley-VCH (2000).

A very good and clear book. It contains the theory and, in the second half, a description of the DFT reliability when calculating various physical and chemical properties.

R.H. Dreizler and E.K.U. Gross, *Density Functional Theory*, Springer, Berlin (1990).

A rigorous book on DFT for specialists in the field.

R.G. Parr and W. Yang, *Density Functional Theory of Atoms and Molecules*, Oxford University Press, Oxford (1989).

The classic textbook on DFT for chemists.

[77] W. Kohn, Y. Meir, and D. Makarov, *Phys. Rev. Lett.*, *80*, 4153 (1998); E. Hult, H. Rydberg, B.I. Lundqvist, and D.C. Langreth, *Phys. Rev. B*, *59*, 4708 (1999); J. Ciosłowski and K. Pernal, *J. Chem. Phys.*, *116*, 4802 (2002).

[78] M. Levy, *Proc. Nat. Acad. Sci.(USA)*, *76*, 6062 (1979).

[79] J. Ciosłowski and K. Pernal, *J. Chem. Phys.*, *111*, 3396 (1999); J. Ciosłowski and K. Pernal, *Phys. Rev. A*, *61*, 34503 (2000); J. Ciosłowski, P. Ziesche, and K. Pernal, *Phys. Rev. B*, *63*, 205105 (2001); J. Ciosłowski and K. Pernal, *J. Chem. Phys.*, *115*, 5784 (2001); J. Ciosłowski, P. Ziesche, and K. Pernal, *J. Chem. Phys.*, *115*, 8725 (2001).

[80] The same pertains to the traditional methods. Somebody operating billions of the expansion functions meets a colleague using even more functions. It would be a real pity if we changed into experts ("*This is what we are paid for...*") knowing which particular BLYP is good for calculating interatomic distances, which for charge distribution, etc.

A.D. Becke, in *Modern Electronic Structure Theory. Part II*, D.R. Yarkony, ed., World Scientific, p. 1022.
An excellent and comprehensible introduction into DFT written by a renowned expert in the field.

J. Andzelm and E. Wimmer, *J. Chem. Phys.*, *96*, 1280 (1992).
A competent presentation of DFT technique introduced by the authors.

R.F.W. Bader, *Atoms in Molecules. A Quantum Theory*, Clarendon Press, Oxford (1994).
An excellent and comprehensive book written by the founder of the atoms-in-molecule approach.

Questions

1. In Bader analysis, in the critical point of the charge density for a covalent bond,
   a. the value of the density is positive
   b. the electronic density Hessian has precisely two negative eigenvalues
   c. all three components of the electron density gradient are equal zero
   d. we are always in the middle of the distance between the two nuclei
2. In Bader analysis, the electronic density Hessian calculated at the center of the benzene ring (of D_{6h} symmetry):
   a. has exactly one positive eigenvalue
   b. has exactly two positive eigenvalues
   c. all three components of the electron density gradient are equal zero
   d. the trace of the Hessian depends on the Cartesian coordinate system chosen.
3. Hohenberg and Kohn (ρ stands for the electron density, $\rho = \rho_0$ corresponds to the ideal ground-state electronic density, E_0 is the ground state energy)
   a. gave the functional $E^{HK}[\rho]$ exhibiting a minimum that corresponds to the density ρ_0
   b. have proved that having ρ_0 one can obtain the ground state wave function
   c. have proved that from E_0 one can obtain ρ_0
   d. have proved that there exists an energy functional $E^{HK}[\rho] \geq E^{HK}[\rho_0] = E_0$
4. The Kohn-Sham system of electrons (ρ stands for the electron density, $\rho = \rho_0$ corresponds to the ideal ground-state electronic density, E_0 is the ground state energy)
   a. represents N electrons leading to the Hartree-Fock electronic density
   b. represents a system of N non-interacting electrons that give the same electronic density $\rho = \rho_0$ as the electronic density of the fully interacting system
   c. is described by N spinorbitals, each being a solution of a one-electron equation
   d. leads to the Slater determinant corresponding to the electronic density ρ_0
5. In the LDA approximation (E_{xc} stands for the exchange–correlation energy),
   a. the uniform electron gas represents a system of N electrons in a box of volume V with the periodic boundary conditions
   b. the uniform electron gas represents a system of N electrons in a box of volume V with the boundary condition of the wave function vanishing at the border of the box
   c. the uniform electron gas represents a system of N electrons in a box of volume V with the periodic boundary conditions and the uniform distribution of the nuclear matter (to get the electrically neutral system)
   d. $E_{xc}[\rho]$ for molecules is such a functional of ρ that for a fixed ρ the value of E_{xc} is equal to the known value E_{xc}^{gas} for the uniform electron gas corresponding to the same density
6. The DFT hybrid approximation (E_{xc} stands for the exchange–correlation energy):
   a. means using a linear combination of atomic hybrid orbitals in the expansion of the Kohn-Sham molecular orbitals
   b. the B3LYP method belongs to the hybrid approximations

c. the hybrid approximations represent in fact some semi-empirical methods
d. one uses as E_{xc} a linear combination of E_{xc} expressions stemming from several DFT functionals and from the Hartree-Fock method

7. The exchange-correlation energy E_{xc} in the Kohn-Sham method
   a. contains a part of the electron kinetic energy
   b. effectively takes into account the Coulomb hole and the Fermi hole
   c. depends on a particular DFT functional
   d. is equal zero.
8. The exchange-correlation hole functions satisfy
   a. $\int h_{\text{xc}}^{\alpha\beta}(\mathbf{r}_1, \mathbf{r}_2)d\mathbf{r}_2 = 0$ and $\int h_{\text{xc}}^{\beta\beta}(\mathbf{r}_1, \mathbf{r}_2)d\mathbf{r}_2 = -1$
   b. $\int h_{\text{xc}}^{\alpha\beta}(\mathbf{r}_1, \mathbf{r}_2)d\mathbf{r}_2 = -1$ and $\int h_{\text{xc}}^{\beta\beta}(\mathbf{r}_1, \mathbf{r}_2)d\mathbf{r}_2 = -1$
   c. $\int h_{\text{xc}}^{\alpha\beta}(\mathbf{r}_1, \mathbf{r}_2)d\mathbf{r}_2 = 0$ and $h_{\text{xc}}^{\beta\beta}(\mathbf{r}, \mathbf{r}) = -\rho_{\beta}(\mathbf{r})$
   d $h_{\text{xc}}^{\alpha\beta}(\mathbf{r}_1, \mathbf{r}_2) = 0$ and $h_{\text{xc}}^{\alpha\alpha}(\mathbf{r}_1, \mathbf{r}_2) = -1$
9. The DFT exchange energy E_x
   a. is more important than the correlation energy
   b. $E_x < 0$
   c. is calculated using the exchange Hartree-Fock expression, but with the Kohn-Sham orbitals
   d. must be a repulsion for He...He and attraction for H...H
10. The Kohn-Sham DFT method
    a. is able to describe the interaction of the two argon atoms
    b. is as time-consuming as the Hartree-Fock method
    c. does not take into account the electron correlation, because it uses a one-determinantal wave function
    d. which would contain a correct exchange-correlation potential, would describe the dispersion interaction

Answers

1a,b,c, 2b,c, 3b,d, 4b,c,d, 5c,d, 6b,c,d, 7a,b,c, 8a,c, 9a,b,c, 10b,d

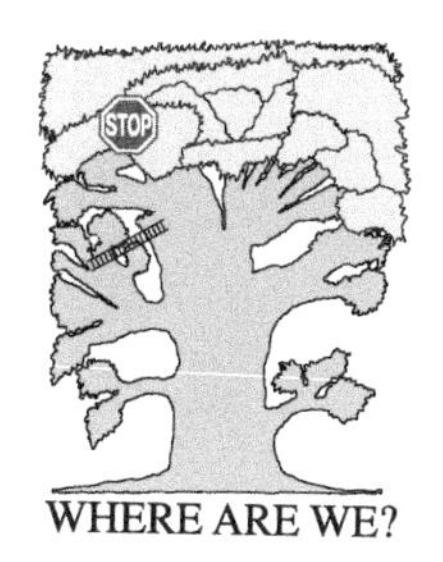

The Molecule Subject to the Electric or Magnetic Field

"For the time being I was not aware of, but soon I have experienced by myself, how dangerous for our ship approaching the Magnetic Mountain was. (...) Not only the anchor, the iron trunks, the knives, spoons and other objects, but also the nails, that have been used to join the boards of the ship, jumped off just by themselves..."

Bolesław Leśmian "Adventures of the Sailor Sindbad"

Where Are We?

We are in the crown of the TREE (left side)

An Example

How does a molecule react to an applied electric field? How do we calculate the changes that undergoes? In some materials, there is a strange phenomenon: a monochromatic *red* laser light beam enters a transparent substance and leaves the specimen as a *blue* beam. Why?

Let's look at another example, this time involving a magnetic field. We apply some long wavelength electromagnetic field to a specimen. We do not see any absorption whatsoever. However, if, in addition, we apply a static magnetic field that gradually increases in intensity, we observe absorption at some intensities. If we analyze the magnetic field values corresponding to the absorption, then they cluster into some mysterious groups that depend on the chemical composition of the specimen. Why is this?

What Is It All About?

The properties of a substance with and without an external electric field *differ*. The problem is how to compute the molecular properties in the electric field from the properties of the isolated molecule and the characteristics of the applied field. Molecules also react upon application of a magnetic field, which changes the internal electric currents and modifies the local magnetic field. A nucleus may be treated as a small magnet that reacts to the local magnetic field that it encounters. This local field depends not only on the external magnetic field, but also on those from other nuclei and on the electronic structure in the vicinity. This produces some energy levels in the spin system, with transitions leading to the nuclear magnetic resonance (NMR) phenomenon, which has wide applications in chemistry, physics, and medicine.

Ideas of Quantum Chemistry, Second Edition. http://dx.doi.org/10.1016/B978-0-444-59436-5.00012-X

The following topics will be described in this chapter.

Why Is This Important?

There is no such thing as an isolated molecule, since any molecule interacts with its neighborhood. In most cases, this is the electric field of another molecule, which represents the only information about the external world for the molecule. The source of the electric field (another molecule or a technical equipment) is of no importance. *Any molecule will respond to the electric field, but some will respond dramatically, while others may respond quite weakly.* The way they respond is of prime importance in technical and scientific applications.

The molecular electronic structure does not respond to a change in orientation of the nuclear magnetic moments because the corresponding perturbation is too small. On the other hand, the molecular electronic structure influences the subtle energetics of interaction of the nuclear spin magnetic moments, and these effects may be recorded in the NMR spectrum. This is of great practical importance because it means that *we have in the molecule under study a system of sounds (nuclear spins) that characterizes the electronic structure almost without perturbing it.*

What Is Needed?

- Perturbation theory (Chapter 5)
- Variational method (Chapter 5, advised)
- Harmonic oscillator and rigid rotator (Chapter 4, advised)
- Breit Hamiltonian (Chapter 3, advised)
- Appendix S available at booksite.elsevier.com/978-0-444-59436-5, p. e143 (advised)
- Appendix G available at booksite.elsevier.com/978-0-444-59436-5, p. e81 (necessary for magnetic properties)
- Appendix M available at booksite.elsevier.com/978-0-444-59436-5, p. e109 (advised)
- Appendix W available at booksite.elsevier.com/978-0-444-59436-5, p. e163 (advised)

Classical Works

Peter Debye, as early as 1921, predicted in "*Molekularkräfte und ihre Elektrische Deutung*," *Physikalische Zeitschrift*, *22*, 302 (1921), that a non-polar gas or liquid of molecules with a nonzero quadrupole moment, when subject to a non-homogeneous electric field, will exhibit the birefringence phenomenon due to the orientation of the quadrupoles in the electric field gradient. ★ A book by John Hasbrouck Van Vleck called *Electric and Magnetic Susceptibilities* Oxford University Press (1932) represented enormous progress. ★ The theorem that forces acting on nuclei result from classical interactions with electron density (computed by a quantum mechanical method) was first proved by Hans Gustav Adolf Hellmann in the world's first textbook of quantum chemistry, *Einführung in die Quantenchemie* Deuticke, Leipzig und Wien[1] (1937), p. 285; and then, independently, by Richard Feynman in "*Forces in molecules*," published in *Physical Review*, *56*, 340 (1939). ★ The first idea of nuclear magnetic resonance (NMR) came from a Dutch scholar named Cornelis Jacobus Gorter in "*Negative result in an attempt to detect nuclear spins*," in *Physica*, *3*, 995 (1936). ★ The first *electron* paramagnetic resonance (EPR) measurement was carried out by Evgeniy Zavoiski from Kazan University (USSR) , and he reported his results in "*Spin-magnetic resonance in paramagnetics*," published in *Journal of Physics (USSR)*, *9*, 245, 447 (1945). ★ The first NMR absorption experiment was performed by Edward M. Purcell, Henry C. Torrey, and Robert V. Pound and published in "*Resonance absorption by nuclear magnetic moments in a solid*," which appeared in *Physical Review*, *69*, 37 (1946), while the first correct explanation of nuclear spin-spin coupling (through the chemical bond) was given by Norman F. Ramsey and Edward M. Purcell in "*Interactions between nuclear spins*

John Hasbrouck Van Vleck (1899–1980), American physicist and professor at the University of Minnesota, received the Nobel Prize in 1977 for "*fundamental theoretical investigations of the electronic structure of magnetic and disordered systems.*"

[1] A Russian edition of Hellmann's book had appeared a few months earlier, but that version does not contain the theorem.

in molecules," published in *Physical Review*, *85*, 143 (1952). ★ The first successful experiment in nonlinear optics with frequency doubling was reported by Peter A. Franken, Alan E. Hill, Wilbur C. Peters, and Gabriel Weinreich in "*Generation of optical harmonics*," published in *Physical Review Letters*, *7*, 118 (1961). ★ Hendrik F. Hameka's book *Advanced Quantum Chemistry. Theory of Interactions between Molecules and Electromagnetic Fields*" (1965), Reading, MA is also considered a classic work.

12.1 Hellmann-Feynman Theorem

Let us assume that a system with Hamiltonian $\hat{H}$ is in a *stationary state* described by the (normalized) function ψ. Now let us begin to do a little tinkering with the Hamiltonian by introducing a parameter P. So we have $\hat{H}(P)$, and assume that we may change the parameter smoothly. For example, as the parameter P, we may take the electric field intensity, or, if we assume the Born-Oppenheimer approximation, then as P, we may take a nuclear coordinate.[2] If we change P in the Hamiltonian $\hat{H}(P)$, then its eigenfunctions and eigenvalues become functions of P.

The Hellmann-Feynman theorem pertains to the rate of the change[3] of $E(P)$:

Hans Gustav Adolf Hellmann (1903–1938), German physicist and one of the pioneers of quantum chemistry. He contributed to the theory of dielectric susceptibility, theory of spin, chemical bond theory (semiempirical calculations, also the virial theorem and the role of kinetic energy), intermolecular interactions theory, electronic affinity. Hellmann wrote the world's first textbook of quantum chemistry *Vviedieniye v kvantovuyu khimiyu*, a few months later published in Leipzig as *Einführung in die Quantenchemie*. In 1933, Hellmann presented his habilitation thesis at the Veterinary College of Hannover. As part of the paperwork he filled out a form in which, according to the recent Nazi requirement, he wrote that his wife was of Jewish origin. The German ministry rejected the habilitation. The situation grew more and more dangerous (many students of the school were active Nazis), and the Hellmanns decided to emigrate. Since his wife was born in the Ukraine, they chose the Eastern route. Hellmann obtained a position at the Karpov Institute of Physical Chemistry in Moscow as a theoretical group leader. A leader of another group, the Communist Party First Secretary of the Institute (Hellmann's colleague and a co-author of one of his papers) A.A. Zukhovitskiy as well as the former First Secretary, leader of the Heterogenic Catalysis Group Mikhail Tiomkin, denounced Hellmann to the institution, later called the KGB, which soon arrested him. Years later, an investigation protocol was found in the KGB archives, with material about Hellmann's spying written by somebody else, but with Hellmann's signature. This was a common result of such "*investigations.*" On May 16, 1938, Albert Einstein, and on May 18, three other Nobel Prize recipients, Irene Joliot-Curie, Frederick Joliot-Curie, and Jean-Baptiste Perrin, asked Stalin for mercy for Hellmann. Stalin ignored the eminent scholars' supplication, and on May 29, 1938 Hans Hellmann was executed by firing squad. After W.H.E. Schwarz et al., *Bunsen-Magazin*, (1999) 10, 60. Portrait reproduced from a painting by Tatiana Livschitz, courtesy of Professor Eugen Schwarz.

---

[2] Recall that in the adiabatic approximation, the electronic Hamiltonian depends parametrically on the nuclear coordinates (discussed in Chapter 6). Then $E(P)$ corresponds to $E_k^0(R)$ from Eq. (6.8).

[3] We may define $\left(\frac{\partial \hat{H}}{\partial P}\right)_{P=P_0}$ as an operator, being a limit when $P \to P_0$ of the operator sequence $\frac{\hat{H}(P)-\hat{H}(P_0)}{P-P_0}$.

Richard Philips Feynman (1919–1988), American physicist and for many years professor at the California Institute of Technology. His father was his first informal teacher of physics, who taught him the extremely important skill of independent thinking. Feynman studied at the Massachusetts Institute of Technology, then at Princeton University, where he earned his Ph.D. under the supervision of John Archibald Wheeler.

In 1945–1950, Feynman served as a professor at Cornell University. A paper plate thrown in the air by a student in the Cornell cafe was the first impulse for Feynman to think about creating a new version of quantum electrodynamics. For this achievement, Feynman received the Nobel Prize in 1965; cf. p. 14.

Feynman was a genius who contributed to several branches of physics (superfluidity, weak interactions, quantum computers, and nanotechnology). His textbook *The Feynman Lectures on Physics* is considered an unchallenged achievement in academic literature. Several of his books became best-sellers. Feynman was famous for his unconventional, straightforward, and crystal-clear thinking, and for his courage and humor. Curiosity and courage made possible his investigations of the ancient Maya calendar, ant habits, and his activity in painting and music.

From John Slater's autobiography *"Solid State and Molecular Theory"*, London, Wiley, (1975):

"The theorem known as the Hellmann-Feynman theorem, stating that the force on a nucleus can be rigorously calculated by electrostatics (...), remained, as far as I was concerned, only a surmise for several years. Somehow, I missed the fact that Hellmann, in Germany, proved it rigorously in 1936, and when a very bright undergraduate turned up in 1938–1939 wanting a topic for a bachelor's thesis, I suggested to him that he see if it could be proved. He come back very promptly with a proof. Since he was Richard Feynman (...), it is not surprising that he produced his proof without trouble."

Hellmann-Feynman Theorem:

$$\frac{\partial E}{\partial P} = \langle \psi | \frac{\partial \hat{H}}{\partial P} | \psi \rangle. \tag{12.1}$$

The proof is simple. The differentiation with respect to P of the integrand in $E = \langle \psi | H | \psi \rangle$ gives

$$\begin{aligned} \frac{\partial E}{\partial P} &= \langle \frac{\partial \psi}{\partial P} | \hat{H} \psi \rangle + \langle \psi | \frac{\partial \hat{H}}{\partial P} \psi \rangle + \langle \psi | \hat{H} \frac{\partial \psi}{\partial P} \rangle \\ &= E \left(\langle \frac{\partial \psi}{\partial P} | \psi \rangle + \langle \psi | \frac{\partial \psi}{\partial P} \rangle \right) + \langle \psi | \frac{\partial \hat{H}}{\partial P} \psi \rangle = \langle \psi | \frac{\partial \hat{H}}{\partial P} \psi \rangle, \end{aligned} \tag{12.2}$$

because the expression in parentheses is equal to zero (we have profited from the fact that $\hat{H}$ is a Hermitian and that ψ represents its eigenfunction[4]). Indeed, differentiating $\langle \psi | \psi \rangle = 1$, we have

$$0 = \langle \frac{\partial \psi}{\partial P} | \psi \rangle + \langle \psi | \frac{\partial \psi}{\partial P} \rangle, \tag{12.3}$$

which completes the proof.

[4] If, instead of the exact eigenfunction, we use an approximate function ψ, then the theorem would have to be modified. In such a case, we have to take into account the terms $\langle \frac{\partial \psi}{\partial P} | \hat{H} | \psi \rangle + \langle \psi | \hat{H} | \frac{\partial \psi}{\partial P} \rangle$.

Soon we will use the Hellmann-Feynman theorem to compute the molecular response to an electric field[5].

ELECTRIC PHENOMENA

12.2 The Molecule Immobilized in an Electric Field

The Homogeneous Electric Field

The electric field intensity $\mathcal{E}$ at a point represents the force acting on a unit positive point charge (probe charge): $\mathcal{E} = -\nabla V$, where V stands for the electric field potential energy at this point.[6] When the potential decreases linearly in space (Fig. 12.1a), the electric field intensity is constant (Fig. 12.1b,c). If at such a potential, we shift the probe charge from a to $a + x$ $(x > 0)$, then the potential energy will *decrease* by $V(a + x) - V(a) = -\mathcal{E}x < 0$. Similarly, the potential energy of a stone will decrease after sliding the stone downhill.

If, instead of a unit charge, we shift the charge Q, then the energy will be lower by $-\mathcal{E}Qx$. It is seen that if we change the *direction* of the shift or the *sign* of the probe charge, then the

5 In case P is a nuclear coordinate (say, x coordinate of the nucleus C, denoted by X_C), and E stands for the potential energy for the motion of the nuclei [cf. Chapter 6, the quantity corresponds to E_0^0 of Eq. (6.8)], the quantity $-\frac{\partial E}{\partial P} = F_{X_C}$ represents the x component of the force acting on the nucleus. The Helmann-Feynman theorem says that this component can be computed as the mean value of the derivative of the Hamiltonian with respect to the parameter P. Since the electronic Hamiltonian reads

$$\hat{H}_0 = -\frac{1}{2}\sum_i \Delta_i + V$$

$$V = -\sum_A \sum_i \frac{Z_A}{r_{Ai}} + \sum_{i<j} \frac{1}{r_{ij}} + \sum_{A<B} \frac{Z_A Z_B}{R_{AB}},$$

then, after differentiating, we have

$$\frac{\partial \hat{H}_0}{\partial X_C} = \frac{\partial V}{\partial X_C} = \sum_i \frac{Z_C}{(r_{Ci})^3}(X_C - x_i) - \sum_{B(\neq C)} \frac{Z_C Z_B}{(R_{BC})^3}(X_C - X_B).$$

Therefore,

$$F_{X_C} = -\langle\psi|\frac{\partial \hat{H}}{\partial P}|\psi\rangle = Z_C\left[\int d\boldsymbol{r}_1 \rho(1)\frac{x_1 - X_C}{(r_{C1})^3} - \sum_{B(\neq C)} \frac{Z_B}{(R_{BC})^3}(X_B - X_C)\right],$$

where $\rho(1)$ stands for the electronic density defined in Chapter 11, Eqs. (11.1) and (11.2).
The last term can be easily calculated from the positions of the nuclei, the first term requires calculation of the one-electron integrals. Note that the resulting formula says that the forces acting on the nuclei follow from the *classical* Coulomb interaction involving the electronic density ρ, even if the electronic density has been (and must be) computed from quantum mechanics.

6 We see that two potential functions that differ by a constant will give the same forces; i.e., will describe identical physical phenomena (this is why this constant is arbitrary).

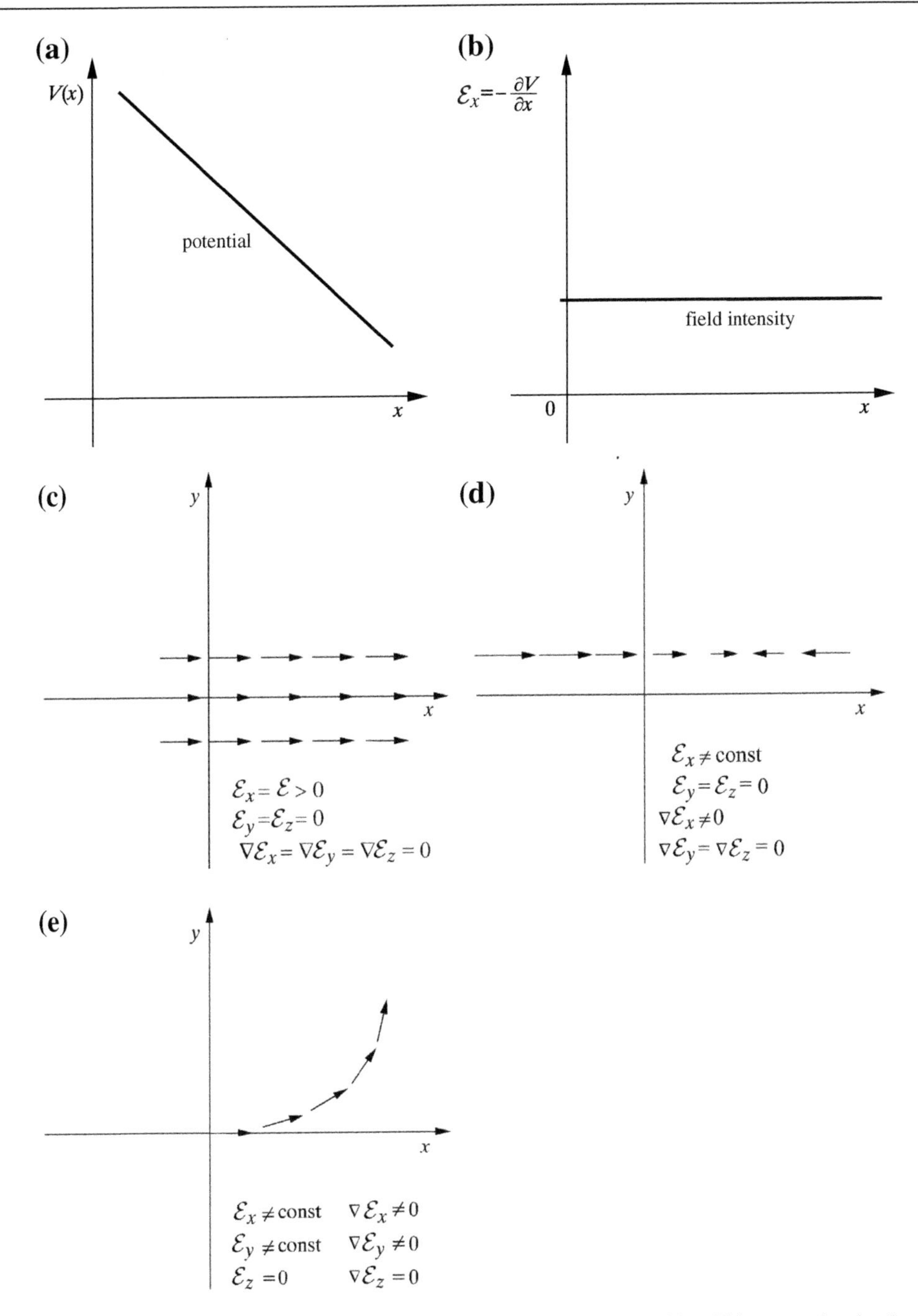

Fig. 12.1. Recalling the electric field properties. (a) 1-D: the potential V decreases with x. This means that the electric field intensity $\boldsymbol{\mathcal{E}}$ is constant; i.e., the field is uniform (b). 3-D: (c) Uniform electric field $\boldsymbol{\mathcal{E}} = (\mathcal{E}, 0, 0)$. (d) Inhomogeneous electric field $\boldsymbol{\mathcal{E}} = (\mathcal{E}(x), 0, 0)$. (e) Inhomogeneous electric field $\boldsymbol{\mathcal{E}} = (\mathcal{E}_x(x, y), \mathcal{E}_y(x, y), 0)$.

energy will go *up* (in case of the stone, we may change only the direction). Therefore,

> the change of the potential energy in a homogeneous electric field $\boldsymbol{\mathcal{E}}$ when shifting charge Q by vector $\boldsymbol{r}$ is equal to $\Delta E = -Q\boldsymbol{\mathcal{E}} \cdot \boldsymbol{r}$.

12.2.1 The Electric Field as a Perturbation

The Non-Homogeneous Field at a Slightly Shifted Point

Imagine a Cartesian coordinate system in 3-D space and a non-homogeneous electric field (Fig. 12.1d,e) in it $\boldsymbol{\mathcal{E}} = [\mathcal{E}_x(x, y, z), \mathcal{E}_y(x, y, z), \mathcal{E}_z(x, y, z)]$.

Assume that the electric field vector $\boldsymbol{\mathcal{E}}(\boldsymbol{r}_0)$ is measured at a point indicated by the vector $\boldsymbol{r}_0$. What will we measure at a point shifted by a small vector $\boldsymbol{r} = (x, y, z)$ with respect to $\boldsymbol{r}_0$? The components of the electric field intensity represent smooth functions in space and this is why we may compute the electric field from the Taylor expansion; for each of the components $\mathcal{E}_x, \mathcal{E}_y, \mathcal{E}_z$ separately, all the derivatives are computed at point $\boldsymbol{r}_0$ (see Fig. 12.2), indices $q, q', q'' = x, y, z$:

$$
\begin{aligned}
\mathcal{E}_x &= +\left(\frac{\partial \mathcal{E}_x}{\partial x}\right)_0 x + \left(\frac{\partial \mathcal{E}_x}{\partial y}\right)_0 y + \left(\frac{\partial \mathcal{E}_x}{\partial z}\right)_0 z \\
&\quad + \frac{1}{2}\left(\frac{\partial^2 \mathcal{E}_x}{\partial x^2}\right)_0 x^2 + \frac{1}{2}\left(\frac{\partial^2 \mathcal{E}_x}{\partial x \partial y}\right)_0 xy + \frac{1}{2}\left(\frac{\partial^2 \mathcal{E}_x}{\partial x \partial z}\right)_0 xz \\
&\quad + \frac{1}{2}\left(\frac{\partial^2 \mathcal{E}_x}{\partial y \partial x}\right)_0 yx + \frac{1}{2}\left(\frac{\partial^2 \mathcal{E}_x}{\partial y^2}\right)_0 y^2 + \frac{1}{2}\left(\frac{\partial^2 \mathcal{E}_x}{\partial y \partial z}\right)_0 yz \\
&\quad + \frac{1}{2}\left(\frac{\partial^2 \mathcal{E}_x}{\partial z \partial x}\right)_0 zx + \frac{1}{2}\left(\frac{\partial^2 \mathcal{E}_x}{\partial z \partial y}\right)_0 zy + \frac{1}{2}\left(\frac{\partial^2 \mathcal{E}_x}{\partial z^2}\right)_0 z^2 + \cdots \\
&= \mathcal{E}_{x,0} + \sum_q \left(\frac{\partial \mathcal{E}_x}{\partial q}\right)_0 q + \frac{1}{2}\sum_{q,q'} \left(\frac{\partial^2 \mathcal{E}_x}{\partial q \partial q'}\right)_0 qq' + \cdots
\end{aligned}
$$

and similarly:

$$
\mathcal{E}_y = \mathcal{E}_{y,0} + \sum_q \left(\frac{\partial \mathcal{E}_y}{\partial q}\right)_0 q + \frac{1}{2}\sum_{q,q'} \left(\frac{\partial^2 \mathcal{E}_y}{\partial q \partial q'}\right)_0 qq' + \cdots
$$

$$
\mathcal{E}_z = \mathcal{E}_{z,0} + \sum_q \left(\frac{\partial \mathcal{E}_z}{\partial q}\right)_0 q + \frac{1}{2}\sum_{q,q'} \left(\frac{\partial^2 \mathcal{E}_z}{\partial q \partial q'}\right)_0 qq' + \cdots
$$

Energy Gain due to a Shift of the Electric Charge Q

These two electric field intensities (at points $\boldsymbol{r}_0$ and $\boldsymbol{r}_0 + \boldsymbol{r}$) have been calculated in order to consider the energy gain associated with the shift $\boldsymbol{r}$ of the electric point charge Q. Similar to the 1-D case just considered, we have the energy gain $\Delta E = -Q\boldsymbol{\mathcal{E}} \cdot \boldsymbol{r}$. There is only one problem:

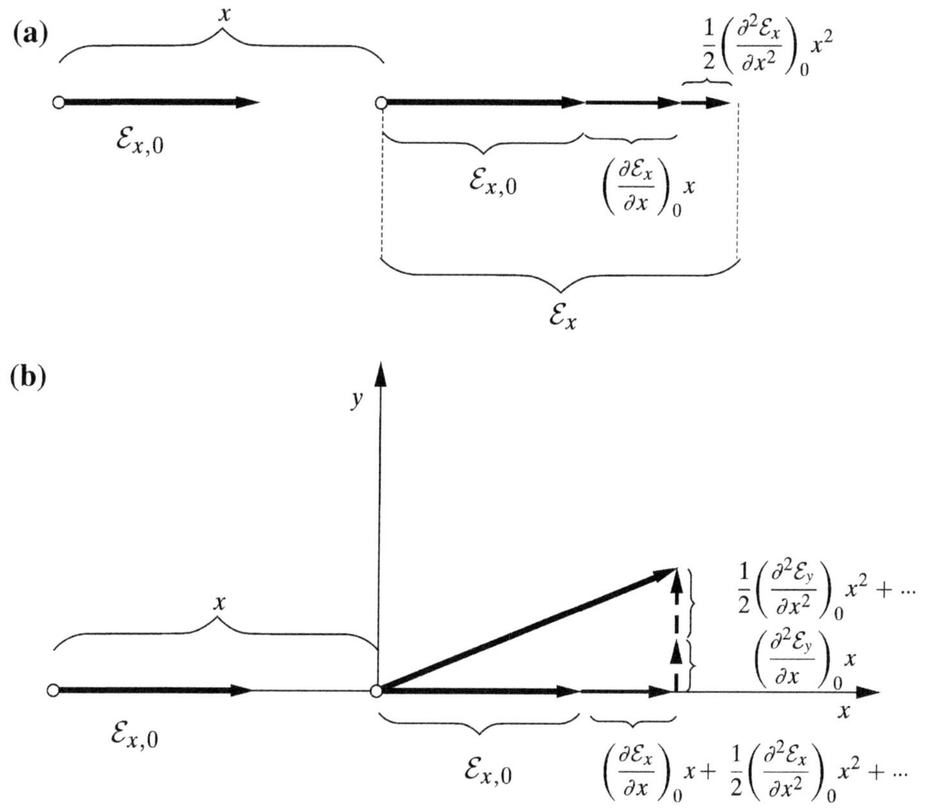

Fig. 12.2. The electric field computed at point $x \ll 1$ from its value (and the values of its derivatives) at point 0. (a) 1-D case; (b) 2-D case.

which of the two electric field intensities is to be inserted into the formula? Since the vector $\boldsymbol{r} = \boldsymbol{i}x + \boldsymbol{j}y + \boldsymbol{k}z$ is small ($\boldsymbol{i}, \boldsymbol{j}, \boldsymbol{k}$ stand for unit vectors corresponding to axes x, y, z, respectively), we may insert, e.g., the mean value of $\boldsymbol{\mathcal{E}}\left(\boldsymbol{r}_0\right)$ and $\boldsymbol{\mathcal{E}}\left(\boldsymbol{r}_0 + \boldsymbol{r}\right)$. We quickly get the following:

$$
\begin{aligned}
\Delta E &= -Q\boldsymbol{\mathcal{E}} \cdot \boldsymbol{r} = -Q\frac{1}{2}\left[\boldsymbol{\mathcal{E}}\left(\boldsymbol{r}_0\right) + \boldsymbol{\mathcal{E}}\left(\boldsymbol{r}_0 + \boldsymbol{r}\right)\right]\boldsymbol{r} \\
&= -\frac{1}{2}Q\left[\boldsymbol{i}\left(\mathcal{E}_{x,0} + \mathcal{E}_x\right) + \boldsymbol{j}\left(\mathcal{E}_{y,0} + \mathcal{E}_y\right) + \boldsymbol{k}\left(\mathcal{E}_{z,0} + \mathcal{E}_z\right)\right]\left(\boldsymbol{i}x + \boldsymbol{j}y + \boldsymbol{k}z\right) \\
&= -\mathcal{E}_{x,0}Qx - \mathcal{E}_{y,0}Qy - \mathcal{E}_{z,0}Qz \\
&\quad - Q\frac{1}{2}\sum_q \left(\frac{\partial\mathcal{E}_x}{\partial q}\right)_0 qx - Q\frac{1}{4}\sum_{q,q'}\left(\frac{\partial^2\mathcal{E}_x}{\partial q\partial q'}\right)_0 qq'x \\
&\quad - Q\frac{1}{2}\sum_q \left(\frac{\partial\mathcal{E}_y}{\partial q}\right)_0 qy - Q\frac{1}{4}\sum_{q,q'}\left(\frac{\partial^2\mathcal{E}_y}{\partial q\partial q'}\right)_0 qq'y
\end{aligned}
$$

$$-Q\frac{1}{2}\sum_{q}\left(\frac{\partial\mathcal{E}_z}{\partial q}\right)_0 qz - Q\frac{1}{4}\sum_{q,q'}\left(\frac{\partial^2\mathcal{E}_z}{\partial q\partial q'}\right)_0 qq'z + \cdots,$$

$$= -\sum_{q}\mathcal{E}_{q,0}\tilde{\mu}_q - \frac{1}{2}\sum_{q,q'}\left(\frac{\partial\mathcal{E}_q}{\partial q'}\right)_0 \tilde{\Theta}_{qq'} - \frac{1}{4}\sum_{q,q',q''}\left(\frac{\partial^2\mathcal{E}_q}{\partial q'\partial q''}\right)_0 \tilde{\Omega}_{qq'q''} + \cdots, \quad (12.4)$$

where "$+\cdots$" denotes higher-order terms, while $\tilde{\mu}_q = Qq$, $\tilde{\Theta}_{qq'} = Qqq'$, $\tilde{\Omega}_{qq'q''} = Qqq'q''$, ... represent the components of the successive *electric moments* of a particle with electric charge Q pointed by the vector $\boldsymbol{r}_0 + \boldsymbol{r}$ and calculated within the coordinate system located at $\boldsymbol{r}_0$. For example, $\tilde{\mu}_x = Qx$, $\tilde{\Theta}_{xy} = Qxy$, $\tilde{\Omega}_{xzz} = Qxz^2$, etc.

Traceless Multipole Moments

The components of such moments in general are not independent. The three components of the dipole moment are indeed independent, but among the quadrupole components, we have the obvious relations $\tilde{\Theta}_{qq'} = \tilde{\Theta}_{q'q}$ from their definition, which reduces the number of independent components from 9 to 6. This, however, is not all. From the Maxwell equations (see Appendix G available at booksite.elsevier.com/978-0-444-59436-5, p. e81), we obtain the *Laplace equation* $\Delta V = 0$ (Δ means the Laplacian), which is valid for points without electric charges. Since $\boldsymbol{\mathcal{E}} = -\nabla V$, and therefore $-\nabla\boldsymbol{\mathcal{E}} = \Delta V$, we obtain

$$\nabla\boldsymbol{\mathcal{E}} = \sum_{q}\frac{\partial\mathcal{E}_q}{\partial q} = 0. \quad (12.5)$$

Thus, in the energy expression $-\frac{1}{2}\sum_{q,q'}\left(\frac{\partial\mathcal{E}_q}{\partial q'}\right)_0\tilde{\Theta}_{qq'}$ of Eq. (12.4), the quantities $\tilde{\Theta}_{qq'}$ are not independent, since we have to satisfy the condition in Eq. (12.5).

We have, therefore, only 5 independent moments that are quadratic in coordinates. For the same reasons, we have only 7 (among 27) independent moments with the third power of coordinates. Indeed, 10 original components $\Omega_{q,q',q''}$, with $(q, q', q'') = xxx, yxx, yyx, yyy, zxx, zxy, zzx, zyy, zzy, zzz$, correspond to all permutational non-equivalent moments. We have, however, three relations that these components have to satisfy. They correspond to the three equations, each obtained from the differentiation of Eq. (12.5) over x, y, z, respectively. This results in only seven *independent* components[7] $\Omega_{q,q',q''}$.

These relations between moments can be taken into account (adding to the energy expression the zeros resulting from the Laplace equation (12.5)) and we may introduce what are known

[7] In Appendix X available at booksite.elsevier.com/978-0-444-59436-5 on p. e169, the definition of the polar coordinate-based multipole moments is reported. The number of independent components of such moments is equal to the number of independent Cartesian components and equals $(2l + 1)$ for $l = 0, 1, 2, \ldots$ with the consecutive l pertaining, respectively, to the monopole (or charge) ($2l + 1 = 1$), dipole (3), quadrupole (5), octupole (7), etc. (in agreement with what we have found a while before for the particular moments).

as the *traceless Cartesian multipole moments*[8] (the symbol without tilde), which may be chosen in the following way:

$$\mu_q \equiv \tilde{\mu}_q, \tag{12.6}$$

$$\Theta_{qq'} \equiv \frac{1}{2}\left[3\tilde{\Theta}_{qq'} - \delta_{qq'}\sum_q \tilde{\Theta}_{qq}\right], \tag{12.7}$$

The adjective *traceless* results from relations of the type $\mathrm{Tr}\Theta = \sum_q \Theta_{qq} = 0$, etc.

Then the expression for the energy contribution changes to (check that both expressions are identical after using the Laplace formula)

$$\Delta E = -\sum_q \mathcal{E}_{q,0}\mu_q - \frac{1}{3}\sum_{q,q'}\left(\frac{\partial \mathcal{E}_q}{\partial q'}\right)_0 \Theta_{qq'} - \cdots \tag{12.8}$$

Most often, we first compute the moments and then use them to calculate the traceless multipole moments (cf. Table on p. 562).

System of Charges in a Non-Homogeneous Electric Field

Since we are interested in constructing the *perturbation operator* that is to be added to the Hamiltonian, from now on, according to the postulates of quantum mechanics (Chapter 1), we will treat the coordinates x, y, z in Eq. (12.8) as *operators* of multiplication by just x, y, z. In addition, we would like to treat many charged particles, not just one, because we want to consider molecules. To this end, we will sum up all the above expressions, computed for each charged particle, separately. As a result, the Hamiltonian for the total system (nuclei and electrons) in the electric field $\mathcal{E}$ represents the Hamiltonian of the system without field ($\hat{H}^{(0)}$) and the perturbation ($\hat{H}^{(1)}$):

$$\hat{H} = \hat{H}^{(0)} + \hat{H}^{(1)}, \tag{12.9}$$

where

$$\hat{H}^{(1)} = -\sum_q \hat{\mu}_q \mathcal{E}_q - \frac{1}{3}\sum_{qq'} \hat{\Theta}_{qq'}\mathcal{E}_{qq'} \cdots \tag{12.10}$$

with the convention

$$\mathcal{E}_{qq'} \equiv \frac{\partial \mathcal{E}_q}{\partial q'},$$

[8] The reader will find the corresponding formulas in the article by A.D. Buckingham, *Advan. Chem. Phys.*, *12*, 107 (1967); or by A.J. Sadlej, "*Introduction to the theory of intermolecular interactions*," *Lund's Theoretical Chemistry Lecture Notes*, Lund (1990).

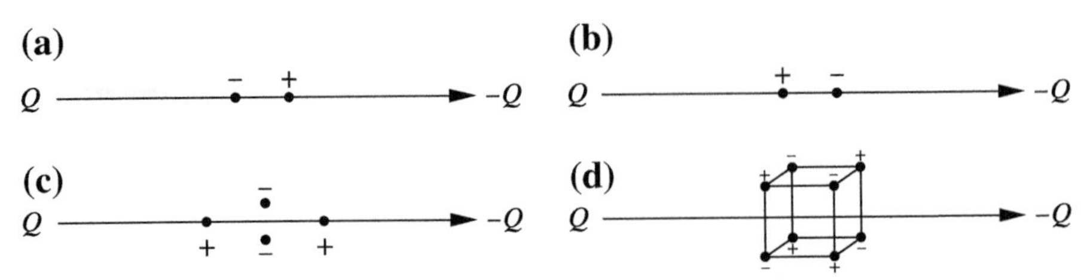

Fig. 12.3. Explanation of why a dipole moment interacts with the electric field intensity, a quadrupole moment with its gradient, while the octupole moment does not interact either with the first or with the second. The external electric field is produced by two *distant* electric charges $Q > 0$ and $-Q$ (for long distances between them, the field in the central region between the charges resembles a homogeneous field) and interacts with an object (a dipole, a quadrupole, etc.) located in the central region. A favorable orientation of the object corresponds to the lowest interaction energy with Q and $-Q$. Panel (a) Shows such a low-energy situation for a dipole: the charge "+" protrudes toward $-Q$, while the charge "−" protrudes toward Q. Panel (b) Corresponds to the opposite situation, energetically unfavourable. As we can see, the interaction energy of the dipole with the electric field *differentiates* these two situations. Now, let us locate a quadrupole in the middle (c). Let us imagine that a neutral point object has just split into four point charges (of the same absolute value). The system lowers its energy by the "−" charges going off the axis because they have increased their distance from the charge $-Q$, but at the same time, the system energy has increased by the same amount since the charges went off the symmetrically located charge $+Q$. What about the "+" charges? The splitting of the "++" charges leads to an energy gain for the right-side "+" charge, because it approached $-Q$, and went off the charge $+Q$, but the left-side "+" charge gives the opposite energy effect. All together, the net result is zero. Conclusion: *the quadrupole does not interact with the homogeneous electric field.* Now, let us imagine an inhomogeneous field having a nonzero gradient along the axis (e.g., both Q charges differ by their absolute values). There will be no energy difference for the "−" charges, but one of the "+" charges will be attracted more strongly than the other. Therefore, the *quadrupole interacts with the field gradient.* We may foresee that the quadrupole will align with its longer axis along the field. Panel (d) Shows an octupole (all charges have the same absolute value). Indeed, the total charge, all the components of the dipole as well as of quadrupole moment, are equal to zero, but the octupole (eight charges in the vertices of a cube) is nonzero. Such an octupole does not interact with a homogeneous electric field (because the right and left sides of the cube do not gain anything when interacting), it also does not interact with the field gradient (because each of the abovementioned sides of the cube is composed of two plus and two minus charges—what the first ones gain the second ones lose).

where the field component and its derivatives are computed at a given point ($\boldsymbol{r}_0$) (e.g., in the center of mass of the molecule), while $\hat{\mu}_q$, $\hat{\Theta}_{qq'}$, ... denote the operators of the components of the traceless Cartesian multipole moments of the total system; i.e., of the molecule.[9] How can we imagine multipole moments? We may associate a given multipole moment with a simple object that exhibits a nonzero value for this particular moment, but all lower multipole moments equal zero.[10] Some of these objects are shown in Fig. 12.3, located between two charges Q and $-Q$ producing an external field. Note that the multipole moment names (*di*pole, *quadru*pole, *octu*pole) indicate the *number* of the point charges from which the objects are built.

Equation (12.10) means that if the system exhibits nonzero multipole moments (before any interaction or due to the interaction), they will interact with the external electric field: the dipole with the electric field intensity, the quadrupole with its gradient, etc. Fig. 12.3 shows why this happens.

[9] This is also calculated with respect to this point. This means that if the molecule is large, then $\boldsymbol{r}$ may become dangerously large. In such a case, as a consequence, the series [Eq. (12.8)] may converge slowly.

[10] *Higher* moments in general will be nonzero.

12.2.2 The Homogeneous Electric Field

In case of a *homogeneous* external electric field, the contribution to $\hat{H}^{(1)}$ comes from the first term in Eq. (12.10):

$$\hat{H} = \hat{H}^{(0)} + \hat{H}^{(1)} = \hat{H}^{(0)} - \hat{\mu}_x \mathcal{E}_x - \hat{\mu}_y \mathcal{E}_y - \hat{\mu}_z \mathcal{E}_z = \hat{H}^{(0)} - \hat{\boldsymbol{\mu}} \cdot \boldsymbol{\mathcal{E}}, \tag{12.11}$$

where the dipole moment operator $\hat{\boldsymbol{\mu}}$ has the form:

$$\hat{\boldsymbol{\mu}} = \sum_i \boldsymbol{r}_i Q_i, \tag{12.12}$$

with the vector $\boldsymbol{r}_i$ indicating the particle i of charge Q_i.

Hence,

$$\frac{\partial \hat{H}}{\partial \mathcal{E}_q} = -\hat{\mu}_q. \tag{12.13}$$

From this, it follows that

$$\langle \psi | \frac{\partial \hat{H}}{\partial \mathcal{E}_q} \psi \rangle = -\langle \psi | \hat{\mu}_q \psi \rangle = -\mu_q, \tag{12.14}$$

where μ_q is the expectation value of the qth component of the dipole moment.

From the Hellmann-Feynman theorem, we have

$$\langle \psi | \frac{\partial \hat{H}}{\partial \mathcal{E}_q} \psi \rangle = \frac{\partial E}{\partial \mathcal{E}_q}; \tag{12.15}$$

therefore

$$\frac{\partial E}{\partial \mathcal{E}_q} = -\mu_q. \tag{12.16}$$

On the other hand, in the case of a *weak electric field* $\boldsymbol{\mathcal{E}}$, we certainly may write the Taylor expansion as

$$E(\boldsymbol{\mathcal{E}}) = E^{(0)} + \sum_q \left(\frac{\partial E}{\partial \mathcal{E}_q} \right)_{\boldsymbol{\mathcal{E}}=\mathbf{0}} \mathcal{E}_q + \frac{1}{2!} \sum_{q,q'} \left(\frac{\partial^2 E}{\partial \mathcal{E}_q \partial \mathcal{E}_{q'}} \right)_{\boldsymbol{\mathcal{E}}=\mathbf{0}} \mathcal{E}_q \mathcal{E}_{q'}$$
$$+ \frac{1}{3!} \sum_{q,q',q''} \left(\frac{\partial^3 E}{\partial \mathcal{E}_q \partial \mathcal{E}_{q'} \partial \mathcal{E}_{q''}} \right)_{\boldsymbol{\mathcal{E}}=\mathbf{0}} \mathcal{E}_q \mathcal{E}_{q'} \mathcal{E}_{q''} + \cdots, \tag{12.17}$$

where $E^{(0)}$ stands for the energy of the unperturbed molecule.

Linear and Nonlinear Responses to a Homogeneous Electric Field

Comparing Eqs. (12.16) and (12.17) we get

$$\frac{\partial E}{\partial \mathcal{E}_q} = -\mu_q = \left(\frac{\partial E}{\partial \mathcal{E}_q}\right)_{\boldsymbol{\mathcal{E}}=\mathbf{0}} + \sum_{q'} \left(\frac{\partial^2 E}{\partial \mathcal{E}_q \partial \mathcal{E}_{q'}}\right)_{\boldsymbol{\mathcal{E}}=\mathbf{0}} \mathcal{E}_{q'} + \frac{1}{2}\sum_{q'} \left(\frac{\partial^3 E}{\partial \mathcal{E}_q \partial \mathcal{E}_{q'} \partial \mathcal{E}_{q''}}\right)_{\boldsymbol{\mathcal{E}}=\mathbf{0}} \mathcal{E}_{q'}\mathcal{E}_{q''} \ldots \tag{12.18}$$

or replacing the derivatives by their equivalents (permanent dipole moment, molecular polarizability, and hyperpolarizabilities)

$$\mu_q = \mu_{0q} + \sum_{q'} \alpha_{qq'}\mathcal{E}_{q'} + \frac{1}{2}\sum_{q'q''} \beta_{qq'q''}\mathcal{E}_{q'}\mathcal{E}_{q''} + \cdots \tag{12.19}$$

The meaning of the formula for μ_q is clear: in addition to the permanent dipole moment $\boldsymbol{\mu}_0$ of the isolated molecule, we have its modification [i.e., an induced dipole moment, which consists of the *linear* part in the field $(\sum_{q'} \alpha_{qq'}\mathcal{E}_{q'})$ and of the *nonlinear* part $(\frac{1}{2}\sum_{q'q''} \beta_{qq'q''}\mathcal{E}_{q'}\mathcal{E}_{q''} + \cdots)$]. The quantities that characterize the molecule: vector $\boldsymbol{\mu}_0$ and tensors $\boldsymbol{\alpha}$, $\boldsymbol{\beta}$, ... are of key importance. By comparing Eq. (12.18) with Eq. (12.19), we have the following relations:

the permanent (field-independent) dipole moment of the molecule (component q):

$$\mu_{0q} = -\left(\frac{\partial E}{\partial \mathcal{E}_q}\right)_{\boldsymbol{\mathcal{E}}=\mathbf{0}}, \tag{12.20}$$

the total dipole moment (field-dependent):

$$\mu_q = -\left(\frac{\partial E}{\partial \mathcal{E}_q}\right), \tag{12.21}$$

the component qq' of the *dipole polarizability* tensor:

$$\alpha_{qq'} = -\left(\frac{\partial^2 E}{\partial \mathcal{E}_q \partial \mathcal{E}_{q'}}\right)_{\boldsymbol{\mathcal{E}}=\mathbf{0}} = \left(\frac{\partial \mu_q}{\partial \mathcal{E}_{q'}}\right)_{\boldsymbol{\mathcal{E}}=\mathbf{0}}, \tag{12.22}$$

the component $qq'q''$ of the *dipole hyperpolarizability* tensor:

$$\beta_{qq'q''} = -\left(\frac{\partial^3 E}{\partial \mathcal{E}_q \partial \mathcal{E}_{q'} \partial \mathcal{E}_{q''}}\right)_{\boldsymbol{\mathcal{E}}=\mathbf{0}}. \tag{12.23}$$

Next, we would obtain higher-order dipole hyperpolarizabilities ($\gamma, \ldots$), which will contribute to the characteristics of the way the molecule is polarized when subject to a weak electric field.

The Homogeneous Field: Dipole Polarizability and Dipole Hyperpolarizabilities

When using the definition of $\boldsymbol{\mu}, \boldsymbol{\alpha}, \boldsymbol{\beta}, \boldsymbol{\gamma}$ from Eq. (12.17), we have the following expression for the energy of the molecule in the electric field:

$$E(\boldsymbol{\mathcal{E}}) = E^{(0)} - \sum_{q} \mu_{0q}\mathcal{E}_q - \frac{1}{2}\sum_{qq'} \alpha_{qq'}\mathcal{E}_q\mathcal{E}_{q'} - \frac{1}{3!}\sum_{qq'q''} \beta_{qq'q''}\mathcal{E}_q\mathcal{E}_{q'}\mathcal{E}_{q''} - \frac{1}{4!}\sum_{qq'q''q'''} \gamma_{qq'q''q'''}\mathcal{E}_q\mathcal{E}_{q'}\mathcal{E}_{q''}\mathcal{E}_{q'''} \ldots \quad (12.24)$$

Due to the homogeneous character of the electric field, this formula pertains exclusively to the interaction of the molecular *dipole* (the permanent dipole plus the induced linear and nonlinear response) with the electric field.

As seen from Eq. (12.19), the induced dipole moment with the components $\mu_q - \mu_{0q}$ may have a different direction from the applied electric field (due to the tensor character of the polarizability and hyperpolarizabilities). This is quite understandable because the electrons will move in a direction that will represent a compromise between the direction the electric field forces them to move in and the direction where the polarization of the molecule is easiest (Fig. 12.4).

It is seen from Eqs. (12.19) and (12.22) that

- As a second derivative of a continuous function E, the polarizability represents a symmetric tensor ($\alpha_{qq'} = \alpha_{q'q}$).
- The polarizability characterizes this part of the *induced dipole moment, which is proportional to the field.*
- If non-diagonal components of the polarizability tensor are nonzero, then the charge flow direction within the molecule will differ from the direction of the field. This would happen when the electric field forced the electrons to flow into empty space, while they had a "*highway*" to travel along some chemical bonds (cf. Fig. 12.4).
- If a molecule is symmetric with respect to the plane $q = 0$ (say, $z = 0$), then all the (hyper)polarizabilities with odd numbers of the indices q, are equal to zero (cf. Fig. 12.4). It has to be like this because otherwise, a change of the electric field component from $\mathcal{E}_z$ to $-\mathcal{E}_z$ would cause a change in energy [see Eq. (12.24)], which is impossible because the molecule is symmetric with respect to the plane $z = 0$.

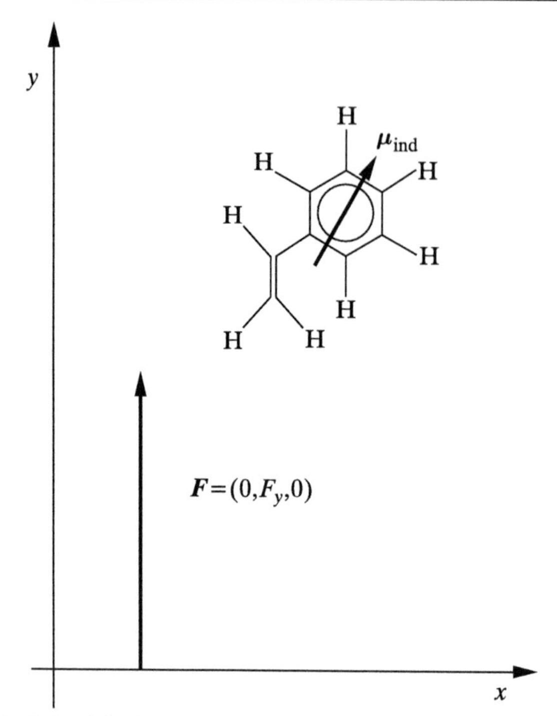

Fig. 12.4. The direction of the induced dipole moment may differ from the direction of the electric field applied (due to the tensor character of the polarizability and hyperpolarizabilities). Example: the vinyl molecule in a planar conformation. Assume the following Cartesian coordinate system: x (horizontal in the Figure plane), y (vertical in the Figure plane) and z (perpendicular to the Figure plane), and the external electric field: $\boldsymbol{\mathcal{E}} = \left(0, \mathcal{E}_y, 0\right)$. The component x of the induced dipole moment is equal to [within the accuracy of linear terms, Eq. (12.19)] $\mu_{ind,x} = \mu_x - \mu_{0x} \approx \alpha_{xy}\mathcal{E}_y, \mu_{ind,y} \approx \alpha_{yy}\mathcal{E}_y, \mu_{ind,z} \approx \alpha_{zy}\mathcal{E}_y$. Due to the symmetry plane $z = 0$ of the molecule (cf. p. 704) $\alpha_{zy} = \alpha_{zx} = 0$, and similarly for the hyperpolarizabilities, we have $\mu_{ind,z} = 0$. As we can see, despite the field having its x component equal to zero, the induced dipole moment x component does not equal to zero ($\mu_{ind,x} \neq 0$).

- The dipole *hyper*polarizabilities (β and higher-order) are very important because if we limited ourselves to the first two terms of Eq. (12.19) containing only μ_{0q} and $\alpha_{qq'}$ (i.e., neglecting β and higher hyperpolarizabilities), the molecule would be equally easy to polarize in two opposite directions.[11] This is why, for a molecule with a center of inversion, all *odd* dipole hyperpolarizabilities (i.e., with an odd number of indices q) have to equal zero because the invariance of the energy with respect to the inversion will be preserved that way. If the molecule does not exhibit an inversion center, the nonzero odd dipole hyperpolarizabilities ensure that polarization of the molecule depends, in general, on whether we change the direction of the electric field vector to the opposite. This is how it should be. Why do the electrons move to the same extent toward an electron donor (on one end of the molecule) and to an electron acceptor (on the other end)?

[11] According to Eq. (12.19), the absolute value of the q component of the *induced dipole moment* $\boldsymbol{\mu}_{ind} = \boldsymbol{\mu} - \boldsymbol{\mu}_0$ would be identical for $\mathcal{E}_q$, as well as for $-\mathcal{E}_q$.

Does the Dipole Moment Really Exist?

Now, let us complicate things. What is $\boldsymbol{\mu}_0$? We used to say that it is the dipole moment of the molecule: $\boldsymbol{\mu}_0 = \langle\psi|\hat{\boldsymbol{\mu}}_0\psi\rangle$. Unfortunately, no molecule has a nonzero dipole moment in any of its stationary states. This follows from the invariance of the Hamiltonian with respect to the inversion operation and was described on p. 72. The mean value of the dipole moment operator is bound to be zero since $|\psi|^2$ is symmetric, while the dipole moment operator itself is antisymmetric with respect to the inversion. Thus, for *any* molecule[12], $\mu_{0q} = 0$ for $q = x, y, z$.

Is this strange? No, not at all. The reason is the rotational part of the wave function (cf. p. 272). This is quite natural. Have you ever tried to figure out why the hydrogen atom does not exhibit a dipole moment despite having two opposite poles: that of the proton and of the electron? The reason is the same. The electron in its ground state is described by the $1s$ orbital, which does not prefer any direction and the dipole moment integral for the hydrogen atom gives zero.[13] Evidently, we have got into trouble.

But this trouble disappears after the Born-Oppenheimer approximation (the clamped nuclei approximation, cf. p. 268) is used; i.e., if we hold the *molecule fixed in space*. In such a case, the molecule *has* the dipole moment, and this dipole moment is to be inserted into formulas as $\boldsymbol{\mu}_0$, and then we may calculate the polarizability, hyperpolarizabilities, etc. (see p. 72). But what do we do when we do not apply the Born-Oppenheimer approximation? Yet, in experiments, we do not use the Born-Oppenheimer approximation (or any other one). We have to allow the molecule to rotate and then the dipole moment $\boldsymbol{\mu}_0$ disappears. Well, that is not quite true since the space is no longer isotropic. There is a chance to measure a dipole moment. What do we measure, then?

It is always good to see things working in a simple model, and simple models resulting in exact solutions of the Schrödinger equation were described in Chapter 4. A good model for our rotating molecule may be the rigid rotator with a dipole moment (a charge Q on one mass and $-Q$ on the other).[14] The Hamiltonian remains, in principle, the same as for the rigid rotator because we have to add a *constant* $-\frac{Q^2}{R}$ to the potential energy, which does not change anything. Thus, the ground state wave function is Y_0^0 = const as before, which tells us that *every* orientation of the rigid dipolar rotor in space is *equally probable*.

In a homogeneous electric field, such a wave function will not be a constant but will have a single maximum for the electric field direction (and the minimum for the opposite direction). For any field intensity, this will correspond to the state of the lowest energy. This is natural because a dipole should have a tendency to align along the orientation of the electric field.

What will happen if the rotator were in one of its excited states? Well, we can guess. A strong electric field will stop the rotation in order to align the rotator along the field and make

[12] It is common sense that the HF molecule has a nonzero dipole moment. Common knowledge says that when an electric field is applied, the HF dipole gets aligned along the electric field vector. Does this happen at any field, no matter how small? This would be an incredible scenario. No, the picture has to be more complex.

[13] The same is in any excited stationary state because $\int d\boldsymbol{r}\, x\, |\psi_{nlm}|^2 = 0$.

[14] This moment, therefore, has a constant length.

from the rotator a kind of oscillator that allows only its vibrations about the field direction. In the first approximation, this can be viewed as a harmonic oscillator, and hence the equidistant distribution of the energy levels.[15]

The above reasoning suggests that the larger the field intensity, the lower the energy of dipolar molecules. If this happened, the molecules would seek regions with the highest electric field. We will see in a while that this is the case, but not always.[16]

A Surprise at Excited States

It has been shown in experiments that dipole molecules (*even the same* as high field-seeking ones) may sometimes seek low-field ones; i.e., they may be expelled from the electric field.[17] How is it possible?

Let us study these things in more detail. Imagine a free rigid rotator–a model of a dipolar molecule, where for the sake of simplicity, we assume now it is rotating within a plane. After separation of the center of mass (cf. Appendix I available at booksite.elsevier.com/978-0-444-59436-5, p. e93), one has to solve a problem for a single particle (with the reduced mass μ). It is like if one of the particles, having the negative charge $-Q$, resided in the origin, while the second particle, with the charge Q, moved around in a circle of radius R , where R stands for the length of the rigid rotator. The only variable is angle ϕ. This problem has been solved[18] in Chapter 4, p. 167. The expression for the energy (after inserting $L = 2\pi R$), reads as[19]

$$E_J = -\frac{Q^2}{R} + J^2 \frac{\hbar^2}{2\mu R^2} = \text{const} + J^2 \frac{\hbar^2}{2\mu R^2}, \tag{12.25}$$

[15] The perturbation for small angle θ can be written as

$$\hat{H}^{(1)} = -\hat{\mu} \cdot \mathcal{E} = -RQ\mathcal{E}\cos\phi \simeq -\mathcal{E}RQ\left(1 - \frac{1}{2}\phi^2\right) = \text{const} + \frac{1}{2}\mathcal{E}RQ\phi^2 = \text{const} + \frac{1}{2}k\phi^2,$$

which corresponds to a harmonic dependence on ϕ with the force constant $k = \mathcal{E}RQ$.

[16] Such molecules are known as the "*high-field seekers.*" Basing on molecular dynamics, it is possible to predict [H.J. Loesch and B. Scheel, *Phys. Rev. Letters*, *85*, 2709 (2000)] what happens in the following situation. Suppose that we have a steel cylinder with a metal wire along its axis. There is a voltage difference applied to the cylinder and the wire resulting in an inhomogeneous electric field, the highest field being at the wire. A molecular beam of polar molecules (like NaCl, NaBr, NaI) when injected on one side of the cylinder begins to orbit in a helix-like motion about the wire. It is also possible to join the ends of the cylinder (making torus) and forming a closed trajectory of the beam. Such devices might serve in the future as reservoirs of molecules in a given quantum state.

[17] These are known as "*low-field seekers.*"

[18] The moving particle was an electron, but it does not matter. For the dipole, there will be the electrostatic interaction of the two charges, but this interaction is constant (since R is a constant) and therefore irrelevant.

[19] In a more formal derivation, we write down first the Hamiltonian for a dipole rotator (two point masses with charges Q and $-Q$ and distance R): $H^{(0)} = -\frac{\hbar^2}{2m_1}\Delta_1 - \frac{\hbar^2}{2m_2}\Delta_2 + \text{const}$, $\text{const} = -\frac{Q^2}{R}$. Next, we separate the center of mass motion (see Appendix I available at booksite.elsevier.com/978-0-444-59436-5, p. e93) and get the Schrödinger equation with the Hamiltonian $\hat{H}^{(0)} = -\frac{\hbar^2}{2\mu}\Delta - \frac{Q^2}{R}$, that describes the relative motion of the two particles. After expressing Δ in spherical coordinates (see Appendix R available at booksite.elsevier.com/978-0-444-59436-5, p. e137) and putting $R = \text{const}$, and $\theta = \pi/2$ (rotations about the z-axis only), we get $\hat{H}^{(0)} = -\frac{\hbar^2}{2\mu R^2}\frac{\partial^2}{\partial\phi^2} + \text{const}$. The corresponding eigenfunctions are $\Phi_J(\phi) = \frac{1}{\sqrt{2\pi}}\exp(iJ\phi)$, $J = 0, \pm 1, \pm 2, \ldots$, with the eigenvalues given by Eq. (12.25).

for $J = 0, \pm 1, \pm 2, \ldots$ We have, therefore, the ground state corresponding to $J = 0$ and the doubly degenerate excited states with $J = \pm 1, \pm 2, \ldots$, corresponding to the wave functions $\Phi_J(\phi) = \frac{1}{\sqrt{2\pi}} \exp(iJ\phi)$. One of the excited state wave functions, $\Phi_J(\phi)$, $J > 0$, corresponds to the rotation that increases ϕ, the second one, $\Phi_{-J}(\phi)$, describes the rotation in the opposite direction. Note that in each of these states, the probability density of finding the rotating particle is uniform $|\Phi_J(\phi)|^2 = |\Phi_{-J}(\phi)|^2 = \frac{1}{2\pi} \neq f(\phi)$; i.e., the rotational motion is uniform (the wave functions are perfectly delocalized).

In the external electric field $\mathcal{E}$ the Hamiltonian reads as [see Eq. (12.11]

$$\hat{H} = \hat{H}^{(0)} - \hat{\boldsymbol{\mu}} \cdot \boldsymbol{\mathcal{E}}. \tag{12.26}$$

Note that a uniform electric field along the $\phi = 0$ direction has to cause the degeneracy to be lifted. Indeed, for a very strong field, the wave functions and the energies have to be similar to those of the harmonic oscillator (the dipole will oscillate about the $\phi = 0$ direction), which means non-degeneracy. How does such a transition from the rotator (the degenerate levels) to the oscillator (the non-degenerate levels) may look like?

There are two opposite effects manifesting themselves in every excited state: a free rotation of the particle ($E_J = J^2 \frac{\hbar^2}{2\mu R^2}$) is modified by the interaction of the dipole with the field that tends to stop the rotation and to orient the dipole along the field with the maximum energy gain (for a perfect alignment) $-\mu\mathcal{E} = -\mathcal{E}QR$. The ratio of these two tendencies can be characterized by parameter $\gamma = \frac{\mathcal{E}}{J^2}$.

Let us increase γ to some medium values (a stronger field, in our calculations $\frac{\hbar^2}{2\mu R^2} = 0.00005$ a.u. and $\mathcal{E}QR = 0.001$ a.u.) and consider the low energy levels. We obtain a symmetric (with respect to $\phi \to -\phi$) nodeless ground state wave function ψ_0, but unlike that for the free rotator, *showing a maximum amplitude for the orientation of the field* ($\phi = 0$). Thus, in this state the dipole has a propensity for orientation along the electric field. Next energy levels correspond to two wave functions (stemming from the non-perturbed functions with $J = \pm 1$), with a single nodal plane each. The lower level corresponds to the antisymmetric wave function ψ_1, with the nodal plane going through points $\phi = 0$ and $\phi = 180°$ and resembling very much the first excited state of the harmonic oscillator. Thus, in this state the dipolar rotator is oscillating about the direction of the field. The second one-node state, ψ_2, is symmetric and has maximum $|\psi_2|$ on the side *opposite* to the field! The reason is that all these functions have to be orthogonal.[20] The nodal plane of ψ_2 when orthogonal to that of ψ_1 keeps $\langle\psi_1|\psi_2\rangle = 0$. However, to make

The same result can be obtained equivalently by postulating that an integer number ($|J|$) of the de Broglie wavelengths (λ) should match the $2\pi R$ distance: $2\pi R = |J|\lambda$, where the quantum number $J = 0, \pm 1, \pm 2, \ldots$ From the de Broglie relation $\lambda = \frac{h}{p}$ (p stands for the momentum of the moving particle of mass μ), one gets $p = |J|\frac{\hbar}{R}$. Hence, the total energy (being the kinetic energy only) is $E = \frac{\mu v^2}{2} = \frac{p^2}{2\mu} = J^2 \frac{\hbar^2}{2\mu R^2}$, as in Eq. (12.25).

[20] The wave functions represent the eigenfunctions of a Hermitian operator and correspond to different energies in a non-zero field. Such functions must be orthogonal (Appendix B available at booksite.elsevier.com/ 978-0-444-59436-5, p.e7).

$\langle\psi_0|\psi_2\rangle = 0$, one has to have $|\psi_2|$ small on the side of the field and *large on the side opposite to the field*!

We will come to similar conclusions for excited states.

For very large γ (very strong electric fields), the electric field will overcome the rotational kinetic energy for the states up to some large J: the ground state, as well as the excited states of both kinds described above will lower their energies due to the overwhelming influence of the electric field. These states will be localized close to the direction ($\phi = 0$), all resembling the harmonic oscillator wave functions.

> When in a non-homogeneous field, the molecules in such states will seek the stronger field domains to lower their energies. The higher excited states still will be localized about the direction opposite to the field. The molecules in such states will be the low-field seekers; they will be expelled outside the field.

The orientation of an electric dipole opposite to the electric field seems counterintuitive, but it is not. We will use an analogy to explain this. Children like to use a swing, which is nothing but a rotator in the gravitational field. If there were no such field (say, on a solid spaceship), the swing would move around at a certain speed in one of two possible directions. The gravitational field (similarly as the electric field acting on the moving positive charge) forces the swing position to prefer the down direction. This means children like the high-field-seeking states of the swing. However, besides the children, there are acrobats who manage not only to make almost free rotation (a delocalized state), but also to get a state of much higher energy, as shown in Fig. 12.5.

The acrobat, after exceeding some kinetic energy is able to *spend more time being oriented opposite to the field*, than along the field! In theoretician's reasoning there is a possibility to lower the energy of such a state by going off the gravitational field. This is therefore an analog of the low-field seeker dipole molecule.

12.2.3 The Non-Homogeneous Electric Field: Multipole Polarizabilities and Hyperpolarizabilities

Let us come back to the non-rotating (immobilized) molecules.

The formula $\mu_q = \mu_{0q} + \sum_{q'} \alpha_{qq'}\mathcal{E}_{q'} + \frac{1}{2}\sum_{q'q''} \beta_{qq'q''}\mathcal{E}_{q'}\mathcal{E}_{q''} + \cdots$ pertains to the polarizabilities and hyperpolarizabilities in a *homogeneous* electric field. The polarizability $\alpha_{qq'}$ characterizes a *linear* response of the molecular *dipole* moment to the electric field, the hyperpolarizability $\beta_{qq'q''}$ and the higher ones characterize the corresponding *nonlinear* response of the molecular *dipole* moment. However, a change of the molecular charge distribution contains more information than just that offered by the induced dipole moment. For a *non-homogeneous* electric field, the energy expression changes because besides the dipole moment, higher multipole moments (permanent as well as induced) come into play (see Fig. 12.3). Using the Hamiltonian equation (12.9) with the perturbation equation (12.10), which corresponds to a molecule

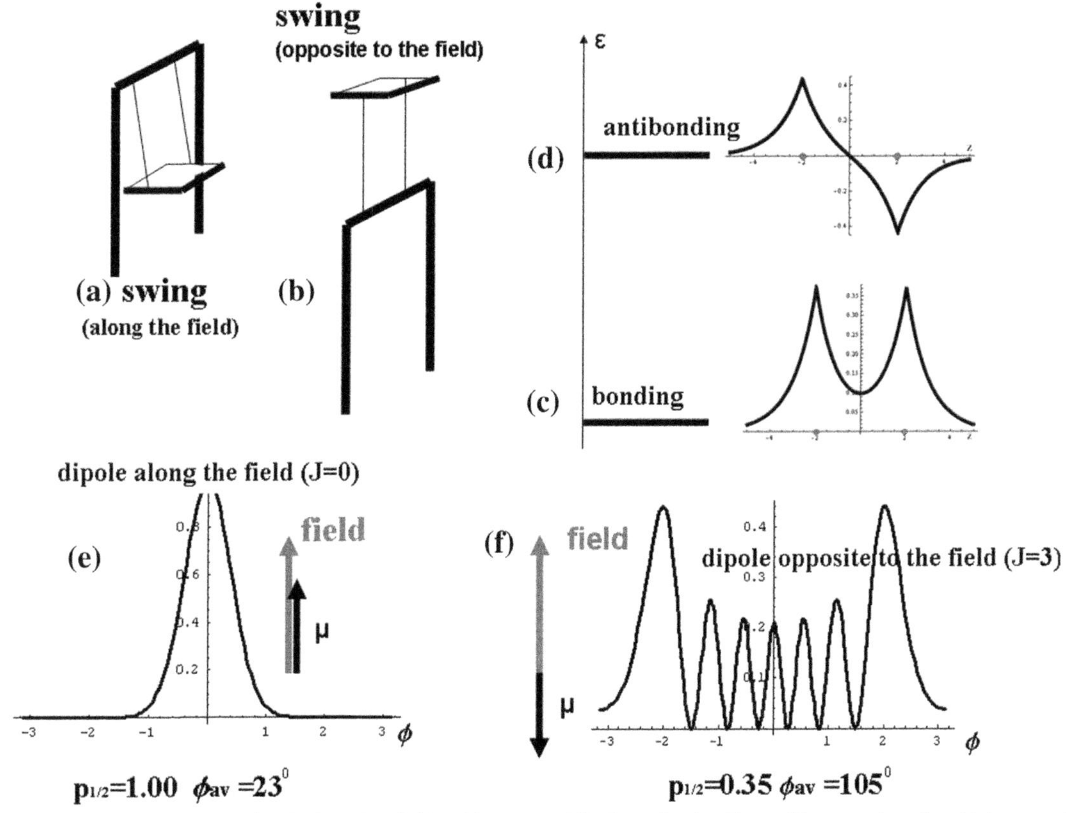

Fig. 12.5. Analogies of the high-field and low-field seeking states of dipolar molecules. Two stable states of a swing: (a) downward (a large stability); (b) upward (marginal stability); (c,d) two electronic states of the molecular ion H_2^+; (c) the ground state corresponding to the stable molecule (the bonding orbital); (d) an excited state that corresponds to a dissociating molecule (the antibonding orbital); (e,f) the probability density of a given orientation ϕ of a dipole in the electric field (corresponding to $\phi = 0$) in two particular states. The quantity $p_{\frac{1}{2}}$, stands for the calculated probability that the dipole-field angle $\phi \in (-90°, 90°)$, ϕ_{av} means the mean value of $|\phi|$; (e) in the ground rotational state ($J = 0$) $\phi_{\mathrm{av}} = 23°$ i.e., the dipole prefers to be oriented along the field, in the excited rotational state ($J = 3$), the expected orientation is $\phi_{\mathrm{av}} = 105°$; i.e., mostly opposite to the field. These data correspond to $\frac{\hbar^2}{2\mu R^2} = 0.00005$ a.u. and $\mathcal{E}QR = 0.001$ a.u.

immersed in a non-homogeneous electric field, we obtain the following energy expression from the Hellmann-Feynman theorem [Eqs. (12.15) and (12.17)]:

$$E\left(\mathcal{E}\right) = E^{(0)} + E_{\mu} + E_{\Theta} + E_{\mu-\Theta} + \cdots , \tag{12.27}$$

where besides the unperturbed energy $E^{(0)}$ of the molecule, we have the following:

- The dipole-field interaction energy E_{μ} (including the permanent and induced dipole, these terms appeared earlier for the homogeneous field):

$$E_{\mu} = -\left[\sum_{q} \mu_{0q}\mathcal{E}_q + \frac{1}{2}\sum_{qq'} \alpha_{qq'}\mathcal{E}_q\mathcal{E}_{q'} + \frac{1}{6}\sum_{q,q',q''} \beta_{q,q',q''}\mathcal{E}_q\mathcal{E}_{q'}\mathcal{E}_{q''} \ldots\right]. \tag{12.28}$$

- Next, the terms that pertain to the *non-homogeneity* of the electric field: the energy E_Θ of the interaction of the field gradient with the quadrupole moment (the permanent one Θ, the first term, and of the induced one; C stands for the *quadrupole polarizability*, and then, in the terms denoted by "$+\cdots$," there are the nonlinear responses with *quadrupole hyperpolarizabilities*):

$$E_\Theta = -\left[\frac{1}{3}\sum_{qq'}\Theta_{qq'}\mathcal{E}_{qq'} + \frac{1}{6}\sum_{qq'q''q'''} C_{qq'q''q'''}\mathcal{E}_{qq'}\mathcal{E}_{q''q'''} + \cdots\right]. \tag{12.29}$$

- The dipole-quadrupole cross term $E_{\mu-\Theta}$:

$$E_{\mu-\Theta} = -\left[\frac{1}{3}\sum_{q,q',q''} A_{q,q'q''}\mathcal{E}_q\mathcal{E}_{q'q''} + \frac{1}{6}\sum_{q,q',q'',q'''} B_{qq',q''q'''}\mathcal{E}_q\mathcal{E}_{q'}\mathcal{E}_{q''q'''}\right], \tag{12.30}$$

  and

- The interaction of higher multipoles (permanent as well as induced: first, the octupole Ω with the corresponding *octupole polarizabilities and hyperpolarizabilities*, etc.) with the higher derivatives of electric field together with the corresponding cross terms denoted as: $+\cdots$

12.3 How to Calculate the Dipole Moment

The dipole moment in normalized state $|n\rangle$ is to be calculated (according to the postulates of quantum mechanics, as discussed in Chapter 1; the Born-Oppenheimer approximation is assumed) as the mean value $\boldsymbol{\mu} = \langle n|\hat{\boldsymbol{\mu}}|n\rangle$ of the dipole moment operator[21]

$$\hat{\boldsymbol{\mu}} = -\sum_i \boldsymbol{r}_i + \sum_A Z_A \boldsymbol{R}_A, \tag{12.31}$$

where $\boldsymbol{r}_i$ are the vectors indicating the electrons and $\boldsymbol{R}_A$ shows nucleus A with the charge Z_A (in a.u.).

12.3.1 Coordinate System Dependence

The dipole moment operator and the dipole moment itself do not depend on the choice of the origin of the coordinate system *only for a neutral molecule*. When two coordinate systems differ by translation $\boldsymbol{R}$, then, in general, we may obtain two different results:

$$\hat{\boldsymbol{\mu}} = \sum_i Q_i \boldsymbol{r}_i$$

$$\hat{\boldsymbol{\mu}}' = \sum_i Q_i \boldsymbol{r}_i' = \sum_i Q_i(\boldsymbol{r}_i + \boldsymbol{R}) = \hat{\boldsymbol{\mu}} + \sum_i Q_i \boldsymbol{R} = \hat{\boldsymbol{\mu}} + \boldsymbol{R}\sum_i Q_i. \tag{12.32}$$

[21] As is seen, this is an operator having x, y, and z components in a chosen coordinate system, and each of its components means a *multiplication by the corresponding coordinates and electric charges.*

It is seen that $\hat{\boldsymbol{\mu}}' = \hat{\boldsymbol{\mu}}$, only if $\sum_i Q_i = 0$ (i.e., for a neutral system).[22]

This represents a special case of the theorem, saying that the lowest non-vanishing multipole moment does not depend on the choice of coordinate system; all others may depend on that choice.

12.3.2 Hartree-Fock Approximation

In order to show how we calculate the dipole moment in practice, let us use the Hartree-Fock approximation. Using the normalized Slater determinant $|\Phi_0\rangle$, we have as the Hartree-Fock approximation to the dipole moment:

$$\boldsymbol{\mu} = \langle\Phi_0| - \sum_i \boldsymbol{r}_i + \sum_A Z_A \boldsymbol{R}_A |\Phi_0\rangle = \langle\Phi_0| - \sum_i \boldsymbol{r}_i |\Phi_0\rangle + \langle\Phi_0| \sum_A Z_A \boldsymbol{R}_A |\Phi_0\rangle = \boldsymbol{\mu}_{el} + \boldsymbol{\mu}_{nucl}, \tag{12.33}$$

where the integration goes over the electronic coordinates. The dipole moment of the nuclei $\boldsymbol{\mu}_{nucl} = \sum_A Z_A \boldsymbol{R}_A$ is very easy to compute because, in the Born-Oppenheimer approximation, the nuclei occupy some fixed positions in space. The electronic component of the dipole moment $\boldsymbol{\mu}_{el} = \langle\Phi_0| - \sum_i \boldsymbol{r}_i |\Phi_0\rangle$, according to the Slater-Condon rules (rule I, see Appendix M available at booksite.elsevier.com/978-0-444-59436-5 on p. e109), amounts to $\boldsymbol{\mu}_{el} = -\sum_i n_i (\varphi_i | \boldsymbol{r}_i \varphi_i)$, where n_i stands for the occupation number of the orbital φ_i (let us assume double occupation; i.e., $n_i = 2$). After the LCAO expansion is applied ($\varphi_i = \sum_j c_{ji} \chi_j$) and combining the coefficients c_{ji} into the bond order matrix (see p. 433) $\boldsymbol{P}$, we have

$$\boldsymbol{\mu}_{el} = -\sum_{kl} P_{lk} (\chi_k | \boldsymbol{r} | \chi_l). \tag{12.34}$$

This is all we can say in principle about calculation of the dipole moment in the Hartree-Fock approximation. The rest belongs to the technical side. We choose a coordinate system and calculate all the integrals of type $(\chi_k | \boldsymbol{r} \chi_l)$; i.e., $(\chi_k | x \chi_l)$, $(\chi_k | y \chi_l)$, $(\chi_k | z \chi_l)$. The bond order matrix $\boldsymbol{P}$ is just a by-product of the Hartree-Fock procedure.

12.3.3 Atomic and Bond Dipoles

It is interesting that within the Hartree-Fock model, the total dipole moment can be decomposed into atomic and pairwise contributions:

$$\boldsymbol{\mu}_{el} = -\sum_A \sum_{k \in A} \sum_{l \in A} P_{lk} (\chi_k | \boldsymbol{r} | \chi_l) - \sum_A \sum_{k \in A} \sum_{B \neq A} \sum_{l \in B} P_{lk} (\chi_k | \boldsymbol{r} | \chi_l), \tag{12.35}$$

[22] If you ever have to debug a computer program that calculates the dipole moment, then remember that there is a simple and elegant test at your disposal that is based on the above theorem. You just make two runs of the program for a neutral system each time using a different coordinate system (the two systems differing by a translation). The two results have to be identical.

where we assume that the atomic orbital centers (A, B) correspond to the nuclei.[23] To this end, we construct the vectors, which indicate from the origin the nuclei and the centers of any pair of them. If the two atomic orbitals k and l belong to *the same* atom, then we insert $\boldsymbol{r} = \boldsymbol{R}_A + \boldsymbol{r}_A$, where $\boldsymbol{R}_A$ shows the atom (nucleus) A from the origin, and $\boldsymbol{r}_A$ indicates the electron from the local origin centered on A. If k and l belong to *different* atoms, then $\boldsymbol{r} = \boldsymbol{R}_{AB} + \boldsymbol{r}_{AB}$, where $\boldsymbol{R}_{AB}$ indicates the center of the AB section and $\boldsymbol{r}_{AB}$ represents the position of the electron with respect to this center. Then,

$$\begin{aligned}\boldsymbol{\mu}_{el} = &-\sum_A \boldsymbol{R}_A \sum_{k\in A}\sum_{l\in A} S_{kl}P_{lk} - \sum_A\sum_{k\in A}\sum_{l\in A} P_{lk}(\chi_k|\boldsymbol{r}_A|\chi_l) \\ &-\sum_A\sum_{B\neq A} \boldsymbol{R}_{AB} \sum_{k\in A}\sum_{l\in B} S_{kl}P_{lk} - \sum_A\sum_{k\in A}\sum_{B\neq A}\sum_{l\in B} P_{lk}(\chi_k|\boldsymbol{r}_{AB}|\chi_l).\end{aligned} \tag{12.36}$$

After adding the dipole moment of the nuclei, we obtain

$$\boldsymbol{\mu} = \sum_A \boldsymbol{\mu}_A + \sum_A\sum_{B\neq A} \boldsymbol{\mu}_{AB}, \tag{12.37}$$

where

$$\begin{aligned}\boldsymbol{\mu}_A &= \boldsymbol{R}_A\Big(Z_A - \sum_{k\in A}\sum_{l\in A} S_{kl}P_{lk}\Big) - \sum_{k\in A}\sum_{l\in A} P_{lk}(\chi_k|\boldsymbol{r}_A|\chi_l) \\ \boldsymbol{\mu}_{AB} &= -\boldsymbol{R}_{AB}\sum_{k\in A}\sum_{l\in B} S_{kl}P_{lk} - \sum_{k\in A}\sum_{l\in B} P_{lk}(\chi_k|\boldsymbol{r}_{AB}|\chi_l).\end{aligned}$$

We therefore have a quite interesting result[24]:

> The molecular dipole moment can be represented as the sum of the individual atomic dipole moments and the pairwise atomic dipole contributions.

The P_{lk} is large, when k and l belong to the atoms forming a *chemical bond* (if compared to two non-bonded atoms; see Appendix S available at booksite.elsevier.com/978-0- 444-59436-5, p. e143); therefore, the dipole moments related to *pairs* of atoms come practically uniquely from *chemical bonds*. The contribution of the lone pairs of the atom A is hidden in the second term of $\boldsymbol{\mu}_A$ and may be quite large (cf. Appendix T available at booksite.elsevier.com/978-0-444-59436-5 on p. e149).

---

[23] We use the LCAO notation in the form: $\varphi_i = \sum_A \sum_{k\in A} c_{ki}\chi_k$.

[24] This does not represent a unique partitioning, only the total dipole moment should remain the same. For example, the individual atomic contributions include the lone pairs, which otherwise could be counted as a separate lone pair contribution.

12.3.4 Within the ZDO Approximation

In several semi-empirical methods of quantum chemistry (e.g., in the Hückel method), we assume the Zero Differential Overlap (ZDO) approximation; i.e., that $\chi_k \chi_l \approx (\chi_k)^2 \delta_{kl}$ and hence the second terms in $\boldsymbol{\mu}_A$, as well as in $\boldsymbol{\mu}_{AB}$, are equal to zero[25], and therefore

$$\boldsymbol{\mu} = \sum_A \boldsymbol{R}_A \left(Z_A - \sum_{k \in A} P_{kk} \right) = \sum_A \boldsymbol{R}_A Q_A, \tag{12.38}$$

where $Q_A = (Z_A - \sum_{k \in A} P_{kk})$ represents the net electric charge of the atom[26] A. This result is extremely simple: the dipole moment comes only from the atomic charges.

12.4 How to Calculate the Dipole Polarizability

We have a formal expression [Eq. (12.24)] involving the dipole polarizability, but we need to calculate this expansion to be able to write the formula for $\alpha_{qq'}$.

12.4.1 Sum Over States Method (SOS)

Perturbation theory gives the energy of the ground state $|0\rangle$ in a weak electric field as (the sum of the zeroth, first and second-order energies[27]; see Chapter 5):

$$E\left(\boldsymbol{\mathcal{E}}\right) = E^{(0)} + \langle 0|\hat{H}^{(1)}|0\rangle + \sum_n{}' \frac{|\langle 0|\hat{H}^{(1)}|n\rangle|^2}{E_0^{(0)} - E_n^{(0)}} + \cdots \tag{12.39}$$

If we assume a homogeneous electric field [see Eq. (12.11)], the perturbation is equal to $\hat{H}^{(1)} = -\hat{\boldsymbol{\mu}} \cdot \boldsymbol{\mathcal{E}}$, and we obtain

$$E = E^{(0)} - \langle 0|\hat{\boldsymbol{\mu}}|0\rangle \cdot \boldsymbol{\mathcal{E}} + \sum_n{}' \frac{[\langle 0|\hat{\boldsymbol{\mu}}|n\rangle \cdot \boldsymbol{\mathcal{E}}][\langle n|\hat{\boldsymbol{\mu}}|0\rangle \cdot \boldsymbol{\mathcal{E}}]}{E^{(0)} - E_n^{(0)}} + \cdots \tag{12.40}$$

The first term represents the energy of the unperturbed molecule, and the second term is a correction for the interaction of the permanent dipole moment with the field. The next term already takes into account that not only the permanent dipole moment but also an *induced moment* interact with the electric field; Eq. (12.19):

$$\sum_n{}' \frac{[\langle 0|\hat{\boldsymbol{\mu}}|n\rangle \cdot \boldsymbol{\mathcal{E}}][\langle n|\hat{\boldsymbol{\mu}}|0\rangle \cdot \boldsymbol{\mathcal{E}}]}{E^{(0)} - E_n^{(0)}} = -\frac{1}{2} \sum_{qq'} \alpha_{qq'} \mathcal{E}_q \mathcal{E}_{q'}, \tag{12.41}$$

---

[25] The second term in $\boldsymbol{\mu}_A$ equals zero because the integrands $\chi_k^2 x, \chi_k^2 y, \chi_k^2 z$ are all antisymmetric with respect to transformation of the coordinate system $x \to -x, y \to -y, z \to -z$.

[26] The molecule stays neutral. Indeed, $\sum_A \sum_{k \in A} P_{kk} = \sum_A \sum_{k \in A} \sum_i n_i c_{ki}^* c_{ki} = \sum_A \sum_{k \in A} \sum_i n_i |c_{ki}|^2 \approx \sum_i n_i = N$, where we consequently used the ZDO approximation.

[27] Prime in the summation means that the 0th state is excluded.

where the component qq' of the polarizability is equal to

$$\alpha_{qq'} = 2\sum_n{}' \frac{\langle 0|\hat{\mu}_q|n\rangle\langle n|\hat{\mu}_{q'}|0\rangle}{\Delta_n}, \tag{12.42}$$

where $\Delta_n = E_n^{(0)} - E^{(0)}$. The polarizability has the dimension of a volume.[28]

Similarly, we may obtain the perturbational expressions for the dipole, quadrupole, octupole hyperpolarizabilities, etc. For example, the ground-state dipole hyperpolarizability β_0 has the form (the $qq'q''$ component, with the prime meaning that the ground state is omitted, and we skip the derivation):

$$\beta_{qq'q''} = \sum_{n,m}{}' \frac{\langle 0|\hat{\mu}_q|n\rangle\langle n|\hat{\mu}_{q'}|m\rangle\langle m|\mu_{q''}|0\rangle}{\Delta_n\Delta_m} - \langle 0|\mu_q|0\rangle \sum_n{}' \frac{\langle 0|\hat{\mu}_{q'}|n\rangle\langle n|\hat{\mu}_{q''}|0\rangle}{(\Delta_n)^2}. \tag{12.43}$$

A problem with the SOS method is its slow convergence and the fact that whenever the expansion functions do not cover the energy continuum, the result is incomplete.

Example 1. ***The Hydrogen Atom in an Electric Field–Perturbational Approach***

An atom or molecule, when located in electric field, undergoes a *deformation*. We will show this in detail, taking the example of the hydrogen atom.

First, let us introduce a Cartesian coordinate system, within which the whole event will be described. Let the electric field be directed toward your right; i.e., it has the form $\boldsymbol{\mathcal{E}} = (\mathcal{E}, 0, 0)$, with a constant $\mathcal{E} > 0$. The positive value of $\mathcal{E}$ means, according to the definition of electric field intensity, that a positive unit charge would move along $\boldsymbol{\mathcal{E}}$ (i.e., from left to right). Thus, the anode is on your left and the cathode on your right.

We will consider a weak electric field; therefore, perturbation theory is applicable, which means just small corrections to the unperturbed situation. In our case, the first-order correction to the wave function [Eq. (5.24)], will be expanded in the series of hydrogen atomic orbitals (they form the complete set[29], cf. Chapter 5):

$$\psi_0^{(1)} = \sum_{k(\neq 0)} \frac{\left\langle \psi_k^{(0)}|\hat{H}^{(1)}|1s\right\rangle}{E_0^{(0)} - E_k^{(0)}} \psi_k^{(0)}, \tag{12.44}$$

where $\psi_k^{(0)} \equiv |nlm\rangle$ and $E_k^{(0)} = -\frac{1}{2n^2}$ denote the orbitals and energies (in a.u.) of the isolated hydrogen atom, respectively (the unperturbed state $|nlm\rangle = |100\rangle = 1s$); and $\hat{H}^{(1)}$ is the perturbation, which for a homogeneous electric field has the form $\hat{H}^{(1)} = -\hat{\boldsymbol{\mu}} \cdot \boldsymbol{\mathcal{E}} = -\hat{\mu}_x\mathcal{E}$,

[28] Because μ^2 has the dimension of charge^2 × length^2, and Δ_n has the dimension of energy; for example, in Coulombic energy: charge^2/length.

[29] Still, they do not span the continuum.

with $\hat{\mu}_x$ standing for the dipole moment operator (its x-component). The operator, according to Eq. (12.31), represents the sum of products: charge (in our case of the electron or proton) times the x-coordinate of the corresponding particle (let us denote them x and X, respectively): $\hat{\mu}_x = -x + X$, where the atomic units have been assumed. To keep the expression as simple as possible, let us locate the proton at the origin of the coordinate system[30]; i.e., $X = 0$. Finally, $\hat{H}^{(1)} = x\mathcal{E}$. Thus the perturbation $\hat{H}^{(1)}$ is simply proportional to the x-coordinate of the electron.

In order not to work in vain, let us first check which unperturbed states k will contribute to the summation on the right side of Eq. (12.44). The ground state ($k = 0$; i.e., the $1s$ orbital) is excluded by the perturbation theory. Next, $k = 1, 2, 3, 4$ denote the orbitals $2s, 2p_x, 2p_y, 2p_z$. The contribution of the $2s$ is equal to zero because $\langle 2s|\hat{H}^{(1)}|1s\rangle = 0$, due to the antisymmetry of the integrand with respect to reflection $x \to -x$ ($\hat{H}^{(1)}$ changes its sign, while the orbitals $1s$ and $2s$ do not). A similar argument excludes the $2p_y$ and $2p_z$ orbitals. Hence, for the time being, we have only a single candidate[31] $2p_x$. This time, the integral is not zero, and we will calculate it shortly. If the candidates from the next shell ($n = 3$) are considered, similarly, the only nonzero contribution comes from $3p_x$. We will, however, stop the calculation at $n = 2$ because our goal is only to show how the whole machinery works. Thus, we need to calculate $\frac{\langle 2p_x|\hat{H}^{(1)}|1s\rangle}{E_0^{(0)} - E_1^{(0)}} = \frac{\langle 2p_x|x|1s\rangle}{E_0^{(0)} - E_1^{(0)}}\mathcal{E}$. The denominator is equal to $-1/2 + 1/8 = -3/8$ a.u. Calculation of the integral (a fast exercise for students[32]) gives 0.7449 a.u. At $\mathcal{E} = 0.001$ a.u., we obtain the coefficient -0.001986 at the normalized orbital $2p_x$ in the first-order correction to the wave function. The negative value of the coefficient means that the orbital $-0.001986(2p_x)$ has its positive lobe oriented leftward.[33] The small absolute value of the coefficient results in such a tiny modification of the $1s$ orbital after the electric field is applied, that it will be practically invisible. In order to make the deformation visible, let us use $\mathcal{E} = 0.1$ a.u. Then, the admixture of $2p_x$ is equal to $-0.1986(2p_x)$; i.e., an approximate wave function of the hydrogen atom has the form $1s - 0.19862p_x$. Fig. 12.6 shows the unperturbed and perturbed $1s$ orbital. As seen, the deformation makes an egg shape of the wave function (from a spherical one), and the

[30] The proton might be located anywhere. The result does not depend on this choice because the perturbation operators will differ by a constant. This, however, means that the nominator $\langle \psi_k^{(0)}|\hat{H}^{(1)}|1s\rangle$ in the formula will remain unchanged because $\langle \psi_k^{(0)}|1s\rangle = 0$.

[31] Note how fast our computation of the integrals proceeds. The main job (zero or not zero—that is the question) is done by the group theory.

[32] From p. 206, we have $\langle 2p_x|x|1s\rangle = \frac{1}{4\pi\sqrt{2}}\int_0^\infty dr r^4 \exp\left(-\frac{3}{2}r\right)\int_0^\pi d\theta \sin^3\theta \int_0^{2\pi} d\phi \cos^2\phi = \frac{1}{4\pi\sqrt{2}}4!\left(\frac{3}{2}\right)^{-5}\frac{4}{3}\pi = 0.7449$, where we have used the formula $\int_0^\infty x^n \exp(-\alpha x)\,dx = n!\alpha^{-(n+1)}$ to calculate the integral over r.

[33] $2p_x \equiv x \times$ the positive spherically symmetric factor means that the positive lobe of the $2p_x$ orbital is on your right (i.e., on the positive part of the x-axis).

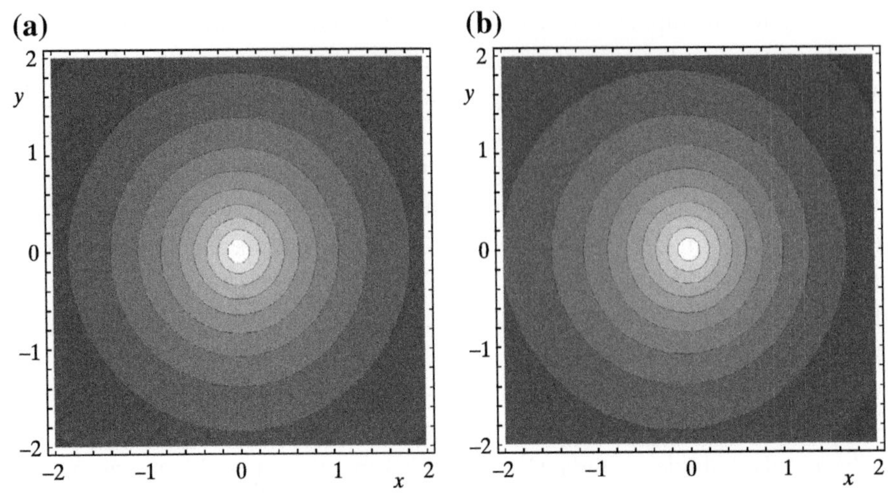

Fig. 12.6. Polarization of the hydrogen atom in an electric field. The wave functions for (a) the unperturbed atom (b) the atom in the electric field (a.u.) $\mathcal{E} = (0.1, 0, 0)$ are shown. As we can see, there are differences in the corresponding electronic density distributions: in the second case, the wave function is deformed toward the anode (i.e., leftward). Note that the wave function is less deformed in the region close to the nucleus than in its left or right neighborhood. This is a consequence of the fact that the deformation is made by the $-0.1986(2p_x)$ function. Its main role is to subtract on the right and add on the left, and the smallest changes are at the nucleus because $2p_x$ has its node there.

electron is pulled toward the anode.[34] This is what we expected. Higher expansion functions $(3p_x, 4p_x, \ldots)$ would change the shape of the wave function only a little.

Just in passing, we may calculate a crude approximation to the dipole polarizability α_{xx}. From Eq. (12.42), we have

$$\alpha_{xx} \cong \frac{16}{3} \langle 2p_x | x(1s) \rangle^2 = \frac{16}{3}(0.7449)^2 = 2.96 \text{ a.u.}$$

The exact (non-relativistic) result is $\alpha_{xx} = 4.5$ a.u. This shows that the number that we have received is somewhat off, but after recalling that only a single expansion function has been used (instead of infinity of them), we should be quite happy with our result.[35]

12.4.2 Finite Field Method

One may solve the Schrödinger equation, including the term $-\hat{\boldsymbol{\mu}} \cdot \boldsymbol{\mathcal{E}}$, in the Hamiltonian. The solution is valid, then, for this particular $\boldsymbol{\mathcal{E}}$. This procedure is known as the *finite field method.*

[34] This "*pulling*" results from adding together $1s$ and (with a negative coefficient) $2p_x$; i.e., we decrease the probability amplitude on the right side of the nucleus, and increase it on the left side.

[35] Such a situation is quite typical in the practice of quantum chemistry: the first terms of expansions give a lot, while the next ones give less and less, the total result approaching its limit with more and more pain. Note that in the present case, all terms are of the same sign, and we obtain better and better approximations when the expansion becomes longer and longer.

Example 2. ***Hydrogen Atom in Electric Field – The Variational Approach***

The polarizability of the hydrogen atom also may be computed by using the variational method (Chapter 5), in which the variational wave function $\psi = \chi_1 + c\chi_2$ is the $\chi_1 \equiv 1s$ plus an admixture (this is controlled by a variational parameter) of the p type orbital χ_2 with a certain exponential coefficient ζ (Ritz method of Chapter 5), see Appendix V available at booksite.elsevier.com/978-0-444-59436-5, Eq. (V.1). As it is seen from Eq. (V.4), if χ_2 is taken as the $2p_x$ orbital (i.e., $\zeta = \frac{1}{2}$), we obtain $\alpha_{xx} = 2.96$ a.u., the same number that we have already obtained by the perturbational method. However, if we take $\zeta = 1$ (i.e., the same as in hydrogenic orbital $1s$), we will obtain $\alpha_{xx} = 4$ a.u. This is a substantial improvement.

Is it possible to obtain an even better result with the variational function ψ? Yes, it is. If we use the finite field method (with the electric field equalling $\mathcal{E} = 0.01$ a.u.), we will obtain[36] the minimum of E of Eq. (V.3) as corresponding to $\zeta_{opt} = 0.797224$. If we insert $\zeta = \zeta_{opt}$ into Eq. (V.4), we will obtain 4.475 a.u., which is only 0.5% off the exact result. This nearly perfect result is computed with a single correction function[37].

Sadlej Relation–Electric Field Variant Orbitals

In order to compute accurate values of $E(\mathcal{E})$ extended LCAO expansions have to be used. Andrzej Sadlej[38] noticed that this huge numerical task in fact only takes into account a very simple effect: just a kind of *shift*[39] of the electronic charge distribution toward the anode. Since the atomic orbitals are usually centered on the nuclei and the electronic charge distribution shifts, to compensate for this using the on-nuclei atomic orbitals requires monstrous and expensive LCAO expansions.

In LCAO calculations nowadays, we most often use Gaussian-type orbitals (GTOs; see Chapter 8). They are rarely thought of as representing wave functions of the harmonic oscillator (cf. Chapter 4), which they really do.[40] Sadlej became interested in what would happen if an electron described by a GTO were subject to the electric field $\mathcal{E}$.

Sadlej noticed that the GTO will change in a similar way as the wave functions of a charged harmonic oscillator in an electric field do, the later however simply shift, Fig. 12.7a.

[36] You may use *Mathematica* and the command FindMinimum[E, {ζ, 1}] to do this.

[37] This success means that sometimes long expansions in the Ritz method may result from unfortunate choice of the expansion functions.

[38] A.J. Sadlej, *Chem.Phys.Letters*, *47*, 50 (1977); A.J. Sadlej, *Acta Phys. Polon. A*, *53*, 297 (1978).

[39] This shift occurs with a deformation.

[40] At least, they do if they represent the $1s$ GTOs.

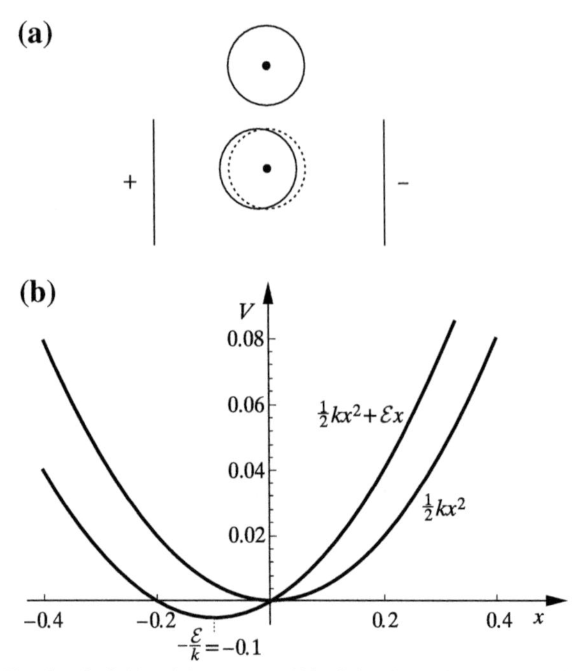

Fig. 12.7. Sadlej relation. The electric field mainly causes a shift of the electronic charge distribution toward the anode (a). A GTO represents the eigenfunction of a harmonic oscillator. Suppose that an electron oscillates in a parabolic potential energy well (with the force constant k). In this situation, a homogeneous electric field $\mathcal{E}$ corresponds to the perturbation $\mathcal{E}x$, that *conserves the harmonicity with unchanged force constant k* (b).

Indeed, this can be shown as follows. The Schrödinger equation for the harmonic oscillator (here, for an electron with $m = 1$ in a.u., its position is x) without any electric field is given on p. 186. According to the example of the hydrogen atom in an electric field, the Schrödinger equation for an electron oscillating in homogeneous electric field $\mathcal{E} > 0$ takes the form:

$$\left(-\frac{1}{2}\frac{d^2}{dx^2} + \frac{1}{2}kx^2 + \mathcal{E}x\right)\psi(x, \mathcal{E}) = E(\mathcal{E})\psi(x, \mathcal{E}). \tag{12.45}$$

Now, let us find constants a and b, such that

$$\frac{1}{2}kx^2 + \mathcal{E}x = \frac{1}{2}k(x-a)^2 + b. \tag{12.46}$$

We immediately get $a = -\mathcal{E}/k$, $b = -\frac{1}{2}ka^2$. The constant b is completely irrelevant, since it only shifts the zero on the energy scale. Thus,

> the solution to a charged harmonic oscillator (oscillating electron) in a homogeneous electric field represents the same function as without the field, but shifted by $-\frac{\mathcal{E}}{k}$.

Andrzej Jerzy Sadlej (1941–2010), Polish quantum chemist, and the only person I know who wrote a scientific book while still a undergraduate (*Elementary Methods of Quantum Chemistry*, PWN, Warsaw, 1966), the first book on quantum chemistry written in the Polish language. After half a century, the book still holds up due to its competence. The book that you are reading now owes very much to friendship of the author with Andrzej. He did not expect any acknowledgment–his only concern was science. Andrzej Sadlej made several important contributions to quantum chemistry. Among others he proposed the first rigorous two-component Dirac theory of the relativistic electron.

Indeed, inserting $x' = x + \frac{\mathcal{E}}{k}$ leads to $\mathrm{d}/\mathrm{d}x = \mathrm{d}/\mathrm{d}x'$ and $\mathrm{d}^2/\mathrm{d}x^2 = \mathrm{d}^2/\mathrm{d}x'^2$ which gives a similar Schrödinger equation except that the harmonic potential is shifted. Therefore, the solution to the equation can be written as simply a zero-field solution $\psi(x') = \psi(x + \frac{\mathcal{E}}{k})$ shifted by $-\frac{\mathcal{E}}{k}$. This is quite understandable because the operation only means adding to the parabolic potential energy $kx^2/2$ a term proportional to x; i.e., a parabola potential again (though it is a displaced one; see Fig. 12.7b).

To see how this displacement depends on the GTO *exponent*, let us recall its relation to the harmonic oscillator force constant k (cf. p. 186). The harmonic oscillator eigenfunction corresponds to a GTO with an exponent equal to $\alpha/2$, where $\alpha^2 = k$ (in a.u.). Therefore, if we have a GTO with exponent equal to A, this means the corresponding harmonic oscillator has the force constant $k = 4A^2$. Now, if the homogeneous electric field $\mathcal{E}$ is switched on, the center of this atomic orbital has to move by $\Delta(A) = -\mathcal{E}/k = -\frac{1}{4}\mathcal{E}/A^2$. This means that all the atomic orbitals have to move opposite to the applied electric field (as expected), and the displacement of the orbital is small, if its exponent is large, and *vice versa*. Also, if the atomic electron charge distribution results from several GTOs (as in the LCAO expansion), it deforms in the electric field in such a way that the diffuse orbitals shift more, while the compact ones (with large exponents) shift only a little. All together, this does not mean just a simple shift of the electronic charge density, but instead its shift accompanied by a deformation. On the other hand, we may simply optimize the GTO positions within the finite field Hartree-Fock method and check whether the corresponding shifts $\Delta_{opt}(A)$ indeed follow the Sadlej relation.[41]

[41] We have tacitly assumed that in the unperturbed molecule, the atomic orbitals occupy optimal positions. This assumption may sometimes cause trouble. If the centers of the atomic orbitals in an isolated molecule are non-optimized, we may end up with a kind of antipolarizability: we apply the electric field and, when the atomic orbital centers are optimized, the electron cloud moves opposite to the way we expect. This is possible only because in such a case, the orbital centers mainly follow the strong intramolecular electric field, rather than the much weaker external field $\mathcal{E}$ (J.M. André, J. Delhalle, J.G. Fripiat, G. Hennico, and L. Piela, *Intern. J. Quantum Chem.*, *22S*, 665 (1988)).

It turns out that the relation $\Delta_{opt}(A) \sim -E/A^2$ is satisfied to a good level of accuracy[42], *despite the fact that the potential energy in an atom does not represent that of a harmonic oscillator.*

The Electrostatic Catastrophe of the Theory

There is a serious problem in finite field theory. If even the *weakest* homogeneous electric field is applied and a very good basis set is used, we are *bound* to have some kind of catastrophe. It's a nasty word, but unfortunately it accurately reflects a mathematical horror that we are going to be exposed to after adding to the Hamiltonian operator $\hat{H}^{(1)} = x\mathcal{E}$ with an electric field (here, x symbolizes the component of the dipole moment).[43] The problem is that this operator is *unbound*; i.e., for a normalized trial function ϕ, the integral $\langle\phi|\hat{H}^{(1)}\phi\rangle$ may attain ∞ or $-\infty$. Indeed, by gradually shifting the function toward the negative values of the x-axis, we obtain more and more negative values of the integral, and for $x = -\infty$, we get $\langle\phi|\hat{H}^{(1)}\phi\rangle = -\infty$. In other words,

> when using atomic orbitals centered far from the nuclei in the region of the negative x (or allowing optimization of the orbital centers with the field switched on), we will lower the energy to $-\infty$ (i.e., catastrophe). This is quite understandable because such a system (electrons separated from the nuclei and shifted far away along the x-axis) has a huge dipole moment and therefore *very low energy*.

Suppose that calculations for a molecule in an electric field $\mathcal{E}$ are carried out. According to the Sadlej relation, we shift the corresponding atomic orbitals proportionally to $\eta\mathcal{E}/A^2$, with $\eta < 0$, and the energy goes down. Around $\eta = -\frac{1}{4}$, which according to Sadlej corresponds to optimal shifts[44], we may expect the lowest energy, then, for larger $|\eta|$, the energy has to go up. What if we continue to increase (Fig. 12.8) the shift parameter $|\eta|$?

The energy increase will continue only up to some critical value of η. Then, according to the discussion above, the energy will fall to $-\infty$ (i.e., to a catastrophe). Thus, the energy curve exhibits a barrier (Fig. 12.8), that is related to the basis set quality (its "*saturation*"): a poor basis means a high barrier, while the ideal basis (i.e., the complete basis set) gives no barrier at all. It just falls into the abyss with the polarizability going to infinity, etc. Therefore, rather paradoxically, reliable values of polarizability are to be obtained using a medium-quality basis set. An improvement of the basis will lead to worse results.[45]

This pertains to variational calculations. What about the perturbational method? In the first- and second-order corrections to the energy, the formulas contain the zero-order approximation

[42] This is how the *electric-field–variant orbitals* (EFVOs) were born.

[43] The most dramatic form of the problem would appear if the finite field method were combined with the numerical solution of the Schrödinger or Fock equation.

[44] They are optimal for a parabolic potential.

[45] Once more, we make this point; wealth does not necessarily improve life.

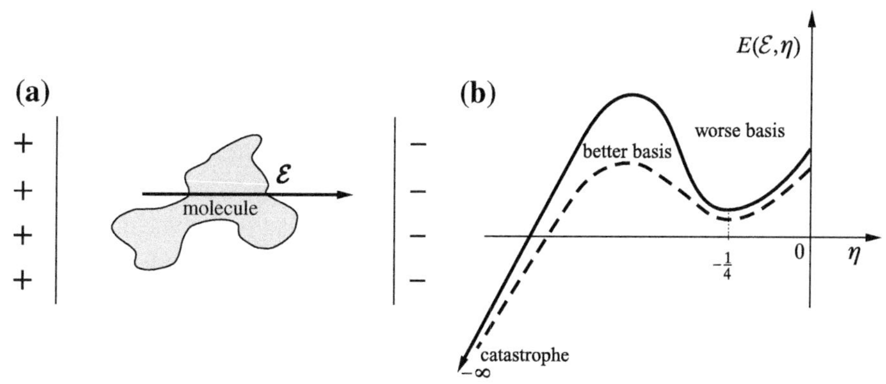

Fig. 12.8. A molecule in a homogeneous electric field (a). In (b), η is a parameter describing the shift of the Gaussian atomic orbitals along the electric field, with $\eta = 0$ showing the centering on the nuclei. The total energy $E(\mathcal{E}, x)$ is a function of the electric field intensity $\mathcal{E}$ and the basis set shift parameter η. Optimization of η gives a result close to the Sadlej value $\eta = \frac{1}{4}$. Larger absolute η values first lead to an increase of E, but then end up in a decrease toward a catastrophe: $lim_{x\to-\infty} E(\mathcal{E}, x) = -\infty$.

to the wave function $\psi_0^{(0)}$; e.g., $E^{(2)} = \langle\psi_0^{(0)}|\hat{H}^{(1)}\psi_0^{(1)}\rangle$. If the origin of the coordinate system is located on the molecule, then the exponential decay of $\psi_0^{(0)}$ forces the first-order correction to the wave function $\psi_0^{(1)}$ to be localized close to the origin; otherwise, it would tend to zero through the shifting toward the negative values of x (this prevents the integral diverging to $-\infty$). However, the third-order correction to the energy contains the term $\langle\psi_0^{(1)}|\hat{H}^{(1)}\psi_0^{(1)}\rangle$, which may already go to $-\infty$. Hence, the perturbation theory also carries the seed of future electrostatic catastrophe.

12.4.3 What Is Going on at Higher Electric Fields?

Polarization

The theory described so far is applicable only when the electric field intensity is small. Such a field can mainly polarize (a small deformation) the electronic charge distribution. More fascinating phenomena begin when the electric field gets stronger.

Deformation

Of course, the equilibrium configurations of the molecule with and without an electric field differ. In a simple case, say the HCl molecule, the HCl distance increases. It has to increase since the cathode pulls the hydrogen atom and repels the chlorine atom, while the anode does the opposite. In more complex cases, like a flexible molecule, the field may change its conformation. This means that the polarizability results both from the electron cloud deformation and the displacement of the nuclei. It turns out that the latter effect (called *vibrational polarization*) is of great importance.[46]

[46] J.-M. André and B. Champagne, in "*Conjugated Oligomers, Polymers, and Dendrimers: From Polyacetylene to DNA*", J.L. Brédas (Ed.), Bibliothéque Scientifique Francqui, De Boeck Université, 1999, p. 349.

Dissociation

When the electric field gets stronger, the molecule may dissociate into ions. To this end, the external electric field intensity has to become comparable to the electric field produced by the molecule itself in its neighborhood. The intramolecular electric fields are huge, and the intermolecular ones are weaker but also very large–of the order of 10^8 V/m, much larger than those offered by current technical installations. No wonder, then, that the molecules may interact to such an extent that they may even undergo chemical reactions. When the interaction is weaker, the electric fields produced by molecules may lead to intermolecular complexes. Many beautiful examples of this may be found in biochemistry (see Chapters 13 and 15). A strong external electric field applied to a crystal may cause a cascade of processes; e.g., the so-called *displacive phase transitions*, when sudden displacements of atoms occur, and a new crystal structure appears.

Destruction

A sufficiently strong electric field will destroy the molecules through their ionization. The resulting ions accelerate in the field, collide with the molecules, and ionize them even more (these phenomena are accompanied by light emission, as in vacuum tubes). Such processes may lead to the final decomposition of the system (plasma) with the electrons and the nuclei finally reaching the anode and cathode. We will have a vacuum.

Creation

Let us keep increasing the electric field applied to the vacuum. Will anything interesting happen? From Chapter 3, we know that when huge electric field intensities are applied (of the order of the electric field intensity in the vicinity of a proton, which is infeasible for the time being), then *the particles and antiparticles will leap off the vacuum*. The vacuum is not just nothing.

12.5 A Molecule in an Oscillating Electric Field

Constant and Oscillating Components

A nonzero hyperpolarizability indicates a *nonlinear* response (the dipole moment proportional to the second and higher powers of the field intensity). This may mean an "inflated" reaction to the applied field, a highly desired feature for contemporary optoelectronic materials. One such reaction is the second- and third-harmonic generation (SHG and THG, respectively), where the light of frequency ω generates in a material light with frequencies 2ω and 3ω, respectively. A simple statement about why this may happen is shown below.[47]

[47] The problem of how the polarizability changes as a function of inducing wave frequency is described in detail in J. Olsen and P. Jørgensen, *J. Chem. Phys.*, *82*, 3235 (1985).

Let us imagine a molecule immobilized in a laboratory coordinate system (as in an oriented crystal). Let us switch on a homogeneous electric field $\boldsymbol{\mathcal{E}}$, which has two components, a static component $\boldsymbol{\mathcal{E}}^0$ and an oscillating one $\boldsymbol{\mathcal{E}}^\omega$ with frequency ω:

$$\boldsymbol{\mathcal{E}} = \boldsymbol{\mathcal{E}}^0 + \boldsymbol{\mathcal{E}}^\omega \cos\left(\omega t\right). \tag{12.47}$$

We may imagine various experiments here: the steady field along x, y, or z and a light beam polarized along x, y, or z, we may also vary ω for each beam, etc. Such choices lead to a rich set of nonlinear optical phenomena.[48] What will the reaction of the molecule be in such an experiment? Let us see.[49]

Induced Dipole Moment

The total dipole moment of the molecule (i.e., the permanent moment $\boldsymbol{\mu}_0$ plus the induced moment $\boldsymbol{\mu}_{ind}$) will depend on time because $\boldsymbol{\mu}_{ind}$ does:

$$\mu_q(t) = \mu_{0,q} + \mu_{ind,q}, \tag{12.48}$$

$$\mu_{ind,q}(t) = \sum_{q'} \alpha_{qq'}\mathcal{E}_{q'} + \frac{1}{2}\sum_{q'q''} \beta_{qq'q''}\mathcal{E}_{q'}\mathcal{E}_{q''} + \frac{1}{6}\sum_{q',q'',q'''} \gamma_{qq'q''q'''}\mathcal{E}_{q'}\mathcal{E}_{q''}\mathcal{E}_{q'''} + \cdots \tag{12.49}$$

Therefore, if we insert $\mathcal{E}_q = \mathcal{E}_q^0 + \mathcal{E}_q^\omega \cos\left(\omega t\right)$ as the electric field component for $q = x, y, z$, we obtain

$$\begin{aligned}\mu_q\left(t\right) = {} & \mu_{0,q} + \sum_{q'} \alpha_{qq'}\left[\mathcal{E}_{q'}^0 + \mathcal{E}_{q'}^\omega \cos\left(\omega t\right)\right] \\ & + \frac{1}{2}\sum_{q'q''} \beta_{qq'q''}\left[\mathcal{E}_{q'}^0 + \mathcal{E}_{q'}^\omega \cos\left(\omega t\right)\right] \times \left[\mathcal{E}_{q''}^0 + \mathcal{E}_{q''}^\omega \cos\left(\omega t\right)\right] \\ & + \frac{1}{6}\sum_{q',q'',q'''} \gamma_{qq'q''q'''}\left[\mathcal{E}_{q'}^0 + \mathcal{E}_{q'}^\omega \cos\left(\omega t\right)\right]\left[\mathcal{E}_{q''}^0 + \mathcal{E}_{q''}^\omega \cos\left(\omega t\right)\right] \\ & \times \left[\mathcal{E}_{q'''}^0 + \mathcal{E}_{q'''}^\omega \cos\left(\omega t\right)\right] + \cdots\end{aligned} \tag{12.50}$$

[48] S. Kielich, *Molecular nonlinear optics* Warszawa-Poznań, PWN (1977).

[49] For the sake of simplicity, we have used the same frequency and the same phases for the light polarized along x, y, and z.

SHG and THG Harmonic Generation

After multiplication and simple trigonometry, we have

$$\mu_q(t) = \mu_{\omega=0,q} + \mu_{\omega,q} \cos \omega t + \mu_{2\omega,q} \cos\left(2\omega t\right) + \mu_{3\omega,q} \cos\left(3\omega t\right), \qquad (12.51)$$

where the amplitudes μ corresponding to the coordinate $q \in x, y, z$ and to the particular resulting frequencies $0, \omega, 2\omega, 3\omega$ have the form given below. The polarizabilities and hyperpolarizabilities depend on the frequency ω and the direction of the incident light waves. According to the convention, a given (hyper) polarizability, such as $\gamma_{qq'q''q'''}(-3\omega; \omega, \omega, \omega)$, is characterized by the frequencies ω corresponding to the three directions x, y, and z of the incident light polarization (preceded by the negative Fourier frequency of the term, -3ω, which symbolizes the photon energy conservation law). Some of the symbols [e.g., $\gamma_{qq'q''q'''}(-\omega; \omega, -\omega, \omega)$] after a semicolon have negative values, which means a partial (as in $\gamma_{qq'q''q'''}(-\omega; \omega, -\omega, \omega)$) or complete (as in $\beta_{q,q',q''}(0; -\omega, \omega)$) cancellation of the intensity of the oscillating electric field. The formulas for the amplitudes are:

$$\begin{aligned}
\mu_{\omega=0,q} &= \mu_{0,q} + \sum_{q'} \alpha_{qq'}\left(0; 0\right) \mathcal{E}^0_{q'} + \frac{1}{2} \sum_{q',q''} \beta_{qq'q''}(0; 0, 0)\mathcal{E}^0_{q'}\mathcal{E}^0_{q''} \\
&+ \frac{1}{6} \sum_{q',q'',q'''} \gamma_{qq'q''q'''}(0; 0, 0, 0)\mathcal{E}^0_{q'}\mathcal{E}^0_{q''}\mathcal{E}^0_{q'''} \\
&+ \frac{1}{4} \sum_{q'q''} \beta_{q,q',q''}(0; -\omega, \omega)\mathcal{E}^\omega_{q'}\mathcal{E}^\omega_{q''} + \frac{1}{4} \sum_{q,q',q'',q'''} \gamma_{qq'q''q'''}(0; 0, -\omega, \omega)\mathcal{E}^0_{q'}\mathcal{E}^\omega_{q''}\mathcal{E}^\omega_{q'''},
\end{aligned}$$

$$\begin{aligned}
\mu_{\omega,q} &= \sum_{q'} \alpha_{qq'}\left(-\omega; \omega\right) \mathcal{E}^\omega_{q'} + \sum_{q',q''} \beta_{qq'q''}(-\omega; \omega, 0)\mathcal{E}^\omega_{q'}\mathcal{E}^0_{q''} \\
&+ \frac{1}{2} \sum_{q,q',q'',q'''} \gamma_{qq'q''q'''}(-\omega; \omega, 0, 0)\mathcal{E}^\omega_{q'}\mathcal{E}^0_{q''}\mathcal{E}^0_{q'''} \\
&+ \frac{1}{8} \sum \sum_{q,q',q'',q'''} \gamma_{qq'q''q'''}(-\omega; \omega, -\omega, \omega)\mathcal{E}^\omega_{q'}\mathcal{E}^\omega_{q''}\mathcal{E}^\omega_{q'''},
\end{aligned}$$

$$\begin{aligned}
\mu_{2\omega,q} &= \frac{1}{4} \sum_{q'q''} \beta_{q,q',q''}(-2\omega; \omega, \omega)\mathcal{E}^\omega_{q'}\mathcal{E}^\omega_{q''} \\
&+ \frac{1}{4} \sum \sum_{q,q',q'',q'''} \gamma_{qq'q''q'''}(-2\omega; \omega, \omega, 0)\mathcal{E}^\omega_{q'}\mathcal{E}^\omega_{q''}\mathcal{E}^0_{q'''},
\end{aligned} \qquad (12.52)$$

$$\mu_{3\omega,q} = \frac{1}{24} \sum_{q,q',q'',q'''} \gamma_{qq'q''q'''}(-3\omega; \omega, \omega, \omega)\mathcal{E}^\omega_{q'}\mathcal{E}^\omega_{q''}\mathcal{E}^\omega_{q'''}. \qquad (12.53)$$

We see the following:

- An oscillating electric field may result in a non-oscillating dipole moment related to the hyperpolarizabilities $\beta_{q,q',q''}(0; -\omega, \omega)$ and $\gamma_{qq'q''q'''}(0; 0, -\omega, \omega)$, which manifests as an electric potential difference on two opposite crystal faces.
- The dipole moment oscillates with the basic frequency ω of the incident light and in addition, with two other frequencies: the second (2ω) and third (3ω) harmonics (SHG and THG, respectively). This is supported by experiments (mentioned in the example at the beginning of the chapter); applying incident light of frequency ω, we obtain emitted light with frequencies[50] 2ω and 3ω.

Note that to generate a large SHG, the material has to have large values of the hyperpolarizabilities β and γ. The THG needs a large γ. In both cases, a strong laser electric field is a must. The SHG and THG therefore require support from the theoretical side: we are looking for high hyperpolarizability materials and quantum mechanical calculations *before* an expensive organic synthesis is done.[51]

MAGNETIC PHENOMENA

The electric and magnetic fields (both of them related by the Maxwell equations, see Appendix G available at booksite.elsevier.com/978-0-444-59436-5) interact differently with matter, which is highlighted in Fig. 12.9, where the electron trajectories in both fields are shown. They are totally different, the trajectory in the magnetic field has a circle-like character, while in the electric field, it is a parabola. This is why the description of magnetic properties differs so much from that of electric properties.

12.6 Magnetic Dipole Moments of Elementary Particles

12.6.1 Electron

An elementary particle, besides its orbital angular momentum, may also have internal angular momentum, or spin; cf. p. 29. In Chapter 3, the Dirac theory led to a relation between the *spin* angular momentum $\boldsymbol{s}$ of the electron and its *dipole magnetic moment* $\boldsymbol{M}_{spin,el}$ [Eq. (3.63), p. 134]:

$$\boldsymbol{M}_{spin,el} = \gamma_{el}\boldsymbol{s},$$

[50] This experiment was first carried out by P.A. Franken, A.E. Hill, C.W. Peters, and G. Weinreich, *Phys. Rev. Letters*, 7, 118 (1961).

[51] In molecular crystals, it is not sufficient that particular molecules have high values of hyperpolarizability. What counts is the hyperpolarizability of the crystal unit cell.

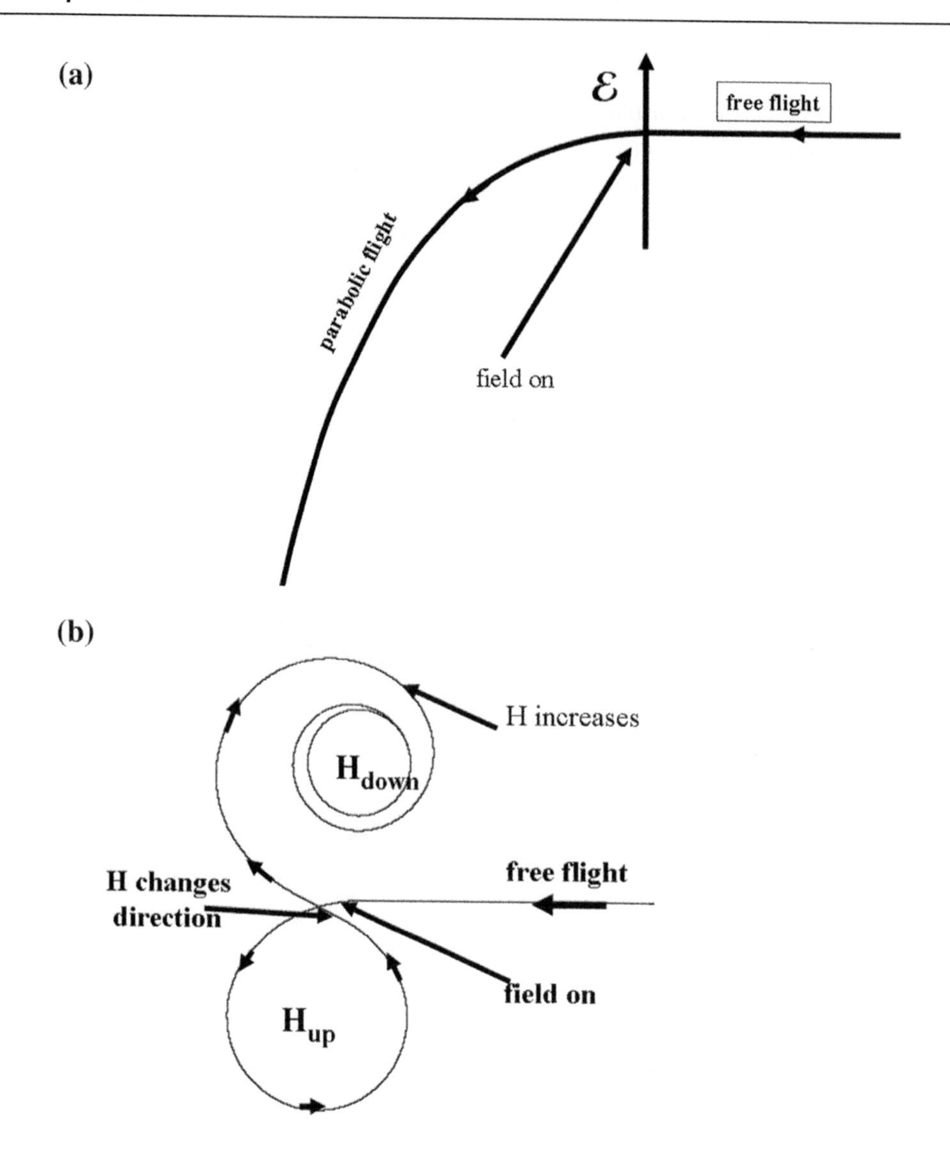

Fig. 12.9. Dramatic differences between electron trajectories in the electric and magnetic fields. (a) The uniform electric field $\mathcal{E}$ and the parabolic electron trajectory; (b) first, the magnetic field is switched off, then it is set on (H_{up}), being perpendicular to the picture and oriented toward the reader. After the electron made a single loop, the orientation of the magnetic field has been inverted (H_{down}) and its intensity kept increasing. The latter caused the electron to make loops with smaller and smaller radius.

with the *gyromagnetic coefficient*[52]

$$\gamma_{el} = -2\frac{\mu_B}{\hbar},$$

[52] The word *gyromagnetic* is derived from Greek word *gyros,* or circle; it is believed that a circular motion of a charged particle is related to the resulting magnetic moment.

where the Bohr magneton (m_0 is the electronic rest mass)

$$\mu_B = \frac{e\hbar}{2m_0c}.$$

The *gyromagnetic factor* is twice as large as that appearing in the relation between the electron *orbital* angular momentum $\boldsymbol{L}$ and the associated magnetic dipole moment:

$$\boldsymbol{M}_{orb,el} = -\frac{\mu_B}{\hbar}\boldsymbol{L}. \tag{12.54}$$

Quantum electrodynamics explains this effect much more precisely, predicting the factor very close to the experimental value[53] 2.0023193043737, known with the breathtaking accuracy level of ± 0.0000000000082.

12.6.2 Nucleus

Let us stay within the Dirac theory, as pertaining to a single elementary particle. If, instead of an electron, we take a *nucleus* of charge $+Ze$ and atomic mass[54] M, then we *would presume* (after insertion into the above formulas) the gyromagnetic factor should be $\gamma = -2\frac{\mu_{\text{nucl}}}{\hbar} = -2\frac{\frac{(-Ze)}{2Mm_Hc}\hbar}{\hbar} = 2\frac{Z}{M}\frac{e\hbar}{2m_Hc} = 2\frac{Z}{M}\frac{\mu_N}{\hbar}$, where $\mu_N = \frac{e\hbar}{2m_Hc}$ (m_H denoting the proton mass) is known as the *nuclear magneton*.[55] For a proton ($Z = 1, M = 1$), we would have $\gamma_p = 2\mu_N/\hbar$, whereas the experimental value[56] is $\gamma_p = 5.59\mu_N/\hbar$. What is going on? In both cases, we have a single elementary particle (electron or proton), both with a spin quantum number equal to $\frac{1}{2}$. We might expect that nothing special will happen for the proton, and only the mass ratio and charge will make a difference. Instead, we see that Dirac theory does pertain to the electron, but not to the nuclei. Visibly, the proton is more complex than the electron. We see that even the simplest nucleus has internal machinery, which results in the observed strange deviation. There are lots of quarks in the proton (three valence quarks and a sea of virtual quarks together with the gluons, etc.). The proton and electron polarize the vacuum differently and this results in different gyromagnetic factors. Other nuclei exhibit even stranger properties. Sometimes we even have negative gyromagnetic coefficients. In such a case, their magnetic moment is the opposite of the spin angular momentum. The complex nature of the internal machinery of the nuclei and vacuum polarization lead to the observed gyromagnetic coefficients.[57] Science has

[53] R.S. Van Dyck Jr., P.B. Schwinberg, and H.G. Dehmelt, *Phys. Rev. Letters*, *59*, 26 (1990).

[54] This is a unitless quantity.

[55] This is ca. 1840 times smaller than the Bohr magneton (for the electron).

[56] Also, the gyromagnetic factor for an electron is expected to be ca. 1840 times larger than that for a proton. This means that a proton is expected to create a magnetic field that is ca. 1840 times weaker than the field created by an electron.

[57] The relation between spin and magnetic moment is as mysterious as that between the magnetic moment and charge of a particle (the spin is associated with a rotation, while the magnetic moment is associated with a rotation of a

had some success here (e.g., for leptons[58]), but for nuclei, the situation is worse. This is why we are simply forced to take this into account in this book[59] and treat the *spin magnetic moments* of the nuclei as the experimental data:

$$\boldsymbol{M}_A = \gamma_A \boldsymbol{I}_A, \tag{12.55}$$

where $\boldsymbol{I}_A$ represents the spin angular momentum of the nucleus A.

12.6.3 Dipole Moment in the Field

12.6.3.1 Electric Field

The problem of an electric dipole μ rotating in an electric field was described on p. 735. When the field is switched off (cf. p. 198), the ground state is non-degenerate ($J = M = 0$) and represents a constant, while the excited states are all degenerate. After an electric field ($\mathcal{E}$) is switched on, the ground-state wave function deforms in such a way as to prefer the alignment of the rotating dipole moment along the field, and, for the excited states, the degeneracy is lifted. Since we may always use a complete set of rigid rotator wave functions (at zero field), this means the deformed wave functions *have to be linear combinations of the wave functions corresponding to different* J.

12.6.3.2 Magnetic Field

Imagine a spinning top like the ones children like to play with. If you make it spin (with angular momentum $\boldsymbol{I}$) and leave it in space without any interaction, then due to the fact that space is isotropic, its angular momentum will stay constant (i.e., the top will rotate about its axis with a constant speed and the axis will not move with respect to distant stars, as shown in Fig. 12.10a).

The situation changes if a homogeneous vector field (e.g., a magnetic field) is switched on. Now, the space is no longer isotropic and the vector of the angular momentum is no longer conserved. However, the conservation law *for the projection of the angular momentum on the direction of the field* is still valid. This means that the top makes a precession about the field axis because this is what keeps the projection constant (see Fig. 12.10b). The magnetic dipole moment $\boldsymbol{M} = \gamma \boldsymbol{I}$ in the magnetic field $\boldsymbol{H} = (0, 0, H)$, $H > 0$ has as many stationary states as is the number of possible projections of the spin angular momentum on the field direction. From Chapter 1, we know that this number is $2I + 1$, where I is the spin quantum number of

charged object) or its mass. A neutron has spin equal to $\frac{1}{2}$ and magnetic moment similar to that of a proton despite there being zero electric charge. The neutrino has no charge, nearly zero mass and magnetic moment, and still spin equal to $\frac{1}{2}$.

[58] And what about the "*heavier brothers*" of the electron, the muon and taon (cf. p. 327)? For the muon, the coefficient in the gyromagnetic factor (2.0023318920) is similar to that of the electron (2.0023193043737), just a bit larger and agrees equally well with experimental results. For the taon, we have only a theoretical result, a little larger than the two other "*brothers*." Thus, the whole lepton family hopefully behaves in a similar way.

[59] This is done with a suitable respect for nature's complexity.

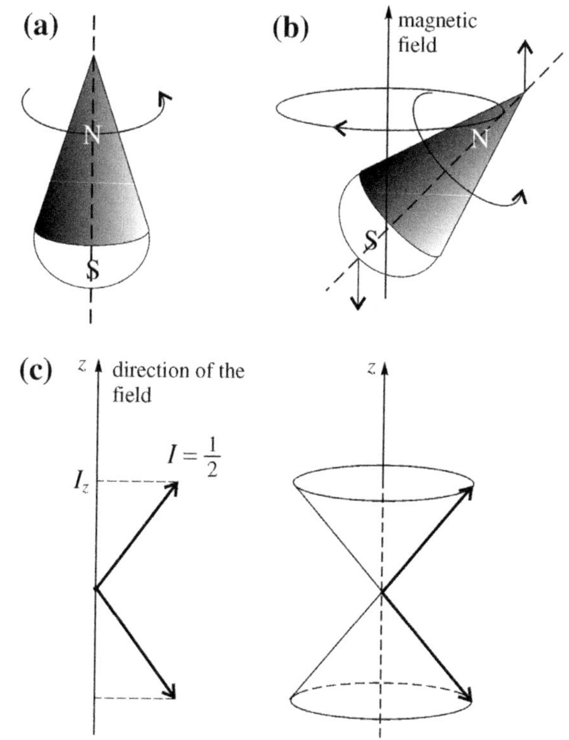

Fig. 12.10. Classical and quantum tops in space. (a) The space is isotropic and therefore the classical top preserves its angular momentum; i.e., its axis does not move with respect to distant stars and the top rotates about its axis with a constant speed. This behavior is used in the gyroscopes that help to orient a spaceship with respect to distant stars. (b) The same top in a homogeneous vector field. The space is no longer isotropic, and therefore the total angular momentum is no longer preserved. The projection of the total momentum on the field direction is still preserved. This is achieved by the precession of the top axis about the direction of the field. (c) A quantum top; i.e., an elementary particle with spin quantum number $I = \frac{1}{2}$ in the magnetic field. The projection I_z of its spin $\boldsymbol{I}$ is quantized: $I_z = m_I \hbar$ with $m_I = -\frac{1}{2}, +\frac{1}{2}$ and, therefore, we have two energy eigenstates that correspond to two precession cones, directed up and down.

the particle (e.g., for a proton: $I = \frac{1}{2}$). The projections (Fig. 12.10c) are equal to $m_I \hbar$ with $m_I = -I, -I+1, \ldots 0, \ldots + I$. Therefore,

the energy levels in the magnetic field are equal to

$$E_{m_I} = -\gamma m_I \hbar H. \tag{12.56}$$

Note that the energy level splitting is proportional to the magnetic field intensity; see Fig. 12.11.

If a nucleus has $I = \frac{1}{2}$, then the energy difference ΔE between the two states in a magnetic field H (one with $m_I = -\frac{1}{2}$ and the other one with $m_I = \frac{1}{2}$), equals $\Delta E = 2 \times \frac{1}{2}\gamma \hbar H =$

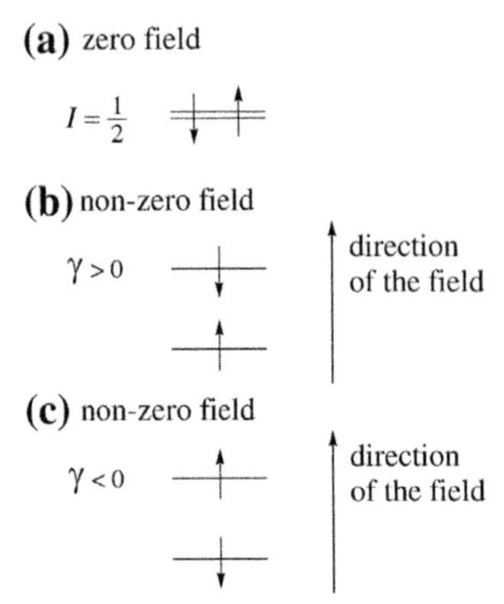

Fig. 12.11. Energy levels in magnetic field $\boldsymbol{H} = (0, 0, H)$ for a nucleus with spin angular momentum $\boldsymbol{I}$ corresponding to spin quantum number $I = \frac{1}{2}$. The magnetic dipole moment equals $\boldsymbol{M} = \gamma \boldsymbol{I}$. (a) At the zero field, the level is doubly degenerate. (b) For $\gamma > 0$ (e.g., a proton), $\boldsymbol{I}$ and $\boldsymbol{M}$ have the same direction. In a nonzero magnetic field, the energy equals $E = -\boldsymbol{M} \cdot \boldsymbol{H} = -M_Z H = -\gamma m_I \hbar H$, where $m_I = \pm\frac{1}{2}$. Thus, the degeneracy is lifted: the state with $m_I = \frac{1}{2}$; i.e., with the positive projection of $\boldsymbol{I}$ on direction of the magnetic field has lower energy. (c) For $\gamma < 0$, $\boldsymbol{I}$ and $\boldsymbol{M}$ have the opposite direction. The state with $m_I = \frac{1}{2}$; i.e., has higher energy.

$\gamma \hbar H$, and

$$\Delta E = h\nu_L, \tag{12.57}$$

where the Larmor[60] frequency is defined as

$$\nu_L = \frac{\gamma H}{2\pi}. \tag{12.58}$$

We see (Fig. 12.11) that for nuclei with $\gamma > 0$, lower energy corresponds to $m_I = \frac{1}{2}$; i.e., to the spin moment along the field (forming an angle $\theta = 54°44'$ with the magnetic field vector; see p. 28).

Note that

> there is a difference between the energy levels of the electric dipole moment in an electric field and the levels of the magnetic dipole in a magnetic field. The difference is that for the magnetic dipole of an elementary particle, the states do not have admixtures from the other I values (which is given by nature), while for the electric dipole, there are admixtures from states with other values of J.

This suggests that we may also expect such admixtures in the magnetic field. Anyway, it is at least true if the particle is complex. For example, the singlet state ($S = 0$) of the hydrogen

[60] Named after Joseph Larmor (1857–1942), Irish physicist and professor at Cambridge University who highlighted the concept of precession in atomic physics.

molecule gets an admixture of the triplet state ($S = 1$) in the magnetic field, because the spin magnetic moments of both electrons tend to align parallel to the field.

12.7 NMR Spectra–Transitions Between the Nuclear Quantum States

Is there any possibility of making the nuclear spin flip from one quantum state to another? Yes. Evidently, we have to create distinct energy levels corresponding to different spin projections; i.e., to switch the magnetic field on (Figs. 12.11 and 12.12a). After the electromagnetic field is applied and its frequency matches the energy level difference, *the system* absorbs the energy. It looks as if *a nucleus* absorbs the energy and changes its quantum state. In a genuine NMR experiment, the electromagnetic frequency is fixed (radio wavelengths) and the specimen is scanned

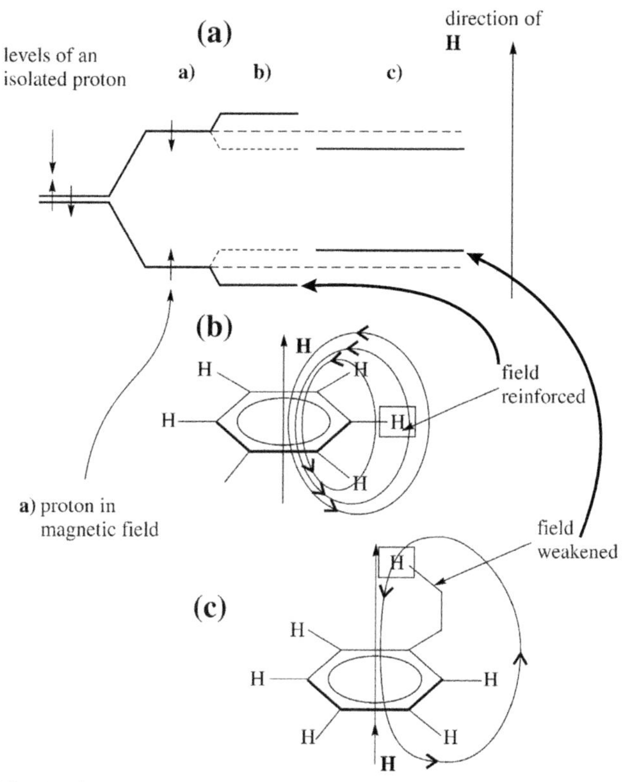

Fig. 12.12. Proton's shielding by the electronic structure. (a) The energy levels of an isolated proton in magnetic field; (b) the energy levels of the proton of the benzene ring (no nuclear spin interaction is assumed). The most mobile π electrons of benzene (which may be treated as a conducting circular wire) move around the benzene ring in response to the external magnetic field (perpendicular to the plane), thus producing an induced magnetic field. The latter one (when considered along the ring's six-fold axis) opposes to the external magnetic field, but *at the position of the proton actually leads to an additional increase in the magnetic field felt by the proton*. This is why the figure (b, upper part) shows the increasing of the energy level difference due to the *electron shielding effect*. (c) The energy levels of another proton (located along the ring's axis) in a similar molecule. This proton feels a local magnetic field that is decreased with respect to the external one (due to the induction effect).

by a variable magnetic field. At some particular field values, the energy difference matches the electromagnetic frequency and the transition (nuclear magnetic resonance) is observed.

The magnetic field a particular nucleus feels differs from the external magnetic field applied because the electronic structure in which the nucleus is immersed in makes its own contribution (see Fig. 12.12b and c). Also, the nuclear spins interact by creating their own magnetic fields.

We have not yet considered these effects in the non-relativistic Hamiltonian (2.1) on p. 67 (e.g., no spin-spin or spin-field interactions). The effects that we are now dealing with are so small–on the order of 10^{-11} kcal/mole–that they are of no importance for most applications, including UV-VIS, IR, Raman spectra, electronic structure, chemical reactions, intermolecular interactions, etc. This time, however, the situation is different: we are going to study very subtle interactions using the NMR technique, which aims precisely at the energy levels that result from spin-spin and spin- magnetic field interactions. Even if these effects are very small, they can be observed. Therefore,

we have to consider more exact Hamiltonians. First, we have to introduce the following:

- The interaction of our system with the electromagnetic field.
- Next, we will consider the influence of the electronic structure on the magnetic field acting on the nuclei.
- Finally, the nuclear magnetic moment interaction ("*coupling*") will be considered.

12.8 Hamiltonian of the System in the Electromagnetic Field

The non-relativistic Hamiltonian[61] $\hat{H}$ of the system of N particles (the jth particle having mass m_j and charge q_j) moving in an external electromagnetic field with vector potential $\boldsymbol{A}$ and scalar potential ϕ may be written as[62]

$$\hat{H} = \sum_{j=1} \left[\frac{1}{2m_j} \left(\hat{\boldsymbol{p}}_j - \frac{q_j}{c} \boldsymbol{A}_j \right)^2 + q_j \phi_j \right] + \hat{V}, \tag{12.59}$$

---

[61] To describe the interactions of the spin magnetic moments, this Hamiltonian will soon be supplemented by the relativistic terms from the Breit Hamiltonian (p. 147).

[62] To obtain this equation, we may use Eq. (3.34) as the starting point, which together with $E = mc^2$, gives with the accuracy of the first two terms the expression $E = m_0 c^2 + \frac{p^2}{2m_0}$. In the electromagnetic field, after introducing the vector and scalar potentials for particle of charge q, we have to replace E by $E - q\phi$, and $\boldsymbol{p}$ by $\left(\boldsymbol{p} - \frac{q}{c}\boldsymbol{A}\right)$. Then, after shifting the zero of the energy by $m_0 c^2$, the energy operator for a single particle reads as $\frac{1}{2m}\left(\hat{\boldsymbol{p}} - \frac{q}{c}\boldsymbol{A}\right)^2 + q\phi$, where $\boldsymbol{A}$ and ϕ are the values of the corresponding potentials at the position of the particle. For many particles, we sum these contributions up and add the interparticle interaction potential (V). This is what we wanted to obtain [H. Hameka, *Advanced Quantum Chemistry* Addison-Wesley, Reading, MA (1965), p. 40].

where $\hat{V}$ stands for the "*internal*" potential coming from the mutual interactions of the particles, and A_j and ϕ_j denote the external vector[63] and scalar potentials A and ϕ, respectively, calculated at the position of particle j.

12.8.1 Choice of the Vector and Scalar Potentials

In Appendix G available at booksite.elsevier.com/978-0-444-59436-5 on p. e81, it was shown that there is a certain arbitrariness in the choice of both potentials, which leaves the physics of the system unchanged. If, for a homogeneous magnetic field $\boldsymbol{H}$, we choose the vector potential at the point indicated by $\boldsymbol{r} = \left(x, y, z\right)$ as (Eq. (G.14)) $\boldsymbol{A}\left(\boldsymbol{r}\right) = \frac{1}{2}\left[\boldsymbol{H} \times \boldsymbol{r}\right]$, then, as shown in Appendix G available at booksite.elsevier.com/978-0-444-59436-5, we will satisfy the Maxwell equations, and in addition obtain the commonly used relation (Eq. (G.13)) $\mathrm{div}\boldsymbol{A} \equiv \nabla\boldsymbol{A} = 0$, known as the *Coulombic gauge*. In this way, the origin of the coordinate system ($\boldsymbol{r} = \boldsymbol{0}$) was chosen as the origin of the vector potential (which need not be a rule).

Because $\boldsymbol{\mathcal{E}} = \boldsymbol{0}$ and $\boldsymbol{A}$ is time-independent $\phi = const$ (p. e81), which of course means also that $\phi_j = const$, as an additive constant, may simply be eliminated from the Hamiltonian equation (12.59).

12.8.2 Refinement of the Hamiltonian

Let us assume the Born-Oppenheimer approximation (p. 269). Thus, the nuclei occupy some fixed positions in space, and in the electronic Hamiltonian equation (12.59), we have the electronic charges $q_j = -e$ and masses $m_j = m_0 = m$ (we skip the subscript 0 for the rest mass of the electron). Now, let us refine the Hamiltonian by adding the interaction of the particle magnetic moments (of the electrons and nuclei; the moments result from the orbital motion of the electrons, as well as from the spin of each particle) with themselves and with the external magnetic field. We have, therefore, a refined Hamiltonian of the system, the particular terms of the Hamiltonian correspond[64] to the relevant terms of the Breit Hamiltonian[65] (p. 147)

$$\hat{\mathcal{H}} = \hat{\mathcal{H}}_1 + \hat{\mathcal{H}}_2 + \hat{\mathcal{H}}_3 + \hat{\mathcal{H}}_4, \tag{12.60}$$

[63] Note that the presence of the magnetic field (and therefore of $\boldsymbol{A}$) makes it as if the charged particle moves faster on one side of the vector potential origin and slower on the opposite side.

[64] All the terms used in the theory of magnetic susceptibilities and the Fermi contact term can be derived from classical electrodynamics.

[65] However, it does not correspond to all of them. As we will see later, the NMR experimental spectra are described by using for each nucleus what is known as the *shielding constant* (related to the shielding of the nucleus by the electron cloud) and the internuclear coupling constants. The shielding and coupling constants enter in a specific way into the energy expression. Only those terms are included in the Hamiltonian that give nonzero contributions to these quantities.

where (δ stands for the Dirac delta function, see Appendix E available at booksite.elsevier.com/978-0-444-59436-5, and the spins have been replaced by the corresponding operators)

$$\hat{\mathcal{H}}_1 = \sum_{j=1}^{N} \frac{1}{2m}\left(\hat{\boldsymbol{p}}_j + \frac{e}{c}\boldsymbol{A}_j\right)^2 + \boldsymbol{V} + \hat{H}_{SH} + \hat{H}_{IH} + \hat{H}_{LS} + \hat{H}_{SS} + \hat{H}_{LL}, \tag{12.61}$$

$$\hat{\mathcal{H}}_2 = \gamma_{el} \sum_{j=1}^{N} \sum_{A} \gamma_A \left[\frac{\hat{\boldsymbol{s}}_j \cdot \hat{\boldsymbol{I}}_A}{r_{Aj}^3} - 3\frac{\left(\hat{\boldsymbol{s}}_j \cdot \boldsymbol{r}_{Aj}\right)\left(\hat{\boldsymbol{I}}_A \cdot \boldsymbol{r}_{Aj}\right)}{r_{Aj}^5} \right], \tag{12.62}$$

$$\hat{\mathcal{H}}_3 = -\gamma_{el}\frac{8\pi}{3} \sum_{j=1}^{N} \sum_{A} \gamma_A \delta\left(\boldsymbol{r}_{Aj}\right) \hat{\boldsymbol{s}}_j \cdot \hat{\boldsymbol{I}}_A, \tag{12.63}$$

$$\hat{\mathcal{H}}_4 = \sum_{A<B} \gamma_A \gamma_B \left[\frac{\hat{\boldsymbol{I}}_A \cdot \hat{\boldsymbol{I}}_B}{R_{AB}^3} - 3\frac{\left(\hat{\boldsymbol{I}}_A \cdot \boldsymbol{R}_{AB}\right)\left(\hat{\boldsymbol{I}}_B \cdot \boldsymbol{R}_{AB}\right)}{R_{AB}^5} \right], \tag{12.64}$$

where, in the global coordinate system, the internuclear distance means the length of the vector $\boldsymbol{R}_{AB} = \boldsymbol{R}_B - \boldsymbol{R}_A$, while the electron-nucleus distance (of the electron j with nucleus A) will be the length of $\boldsymbol{r}_{Aj} = \boldsymbol{r}_j - \boldsymbol{R}_A$. These terms have the following meaning:

- In the term $\hat{\mathcal{H}}_1$, besides the kinetic energy operator in the external magnetic field [with vector potential $\boldsymbol{A}$, with the convention $\boldsymbol{A}_j \equiv \boldsymbol{A}\left(\boldsymbol{r}_j\right)$] given by $\sum_{j=1}^{N} \frac{1}{2m}\left(\hat{\boldsymbol{p}}_j + \frac{e}{c}\boldsymbol{A}_j\right)^2$, we have the Coulomb potential V of the interaction of all the charged particles. Next, we have the following:

Pieter Zeeman (1865–1943), Dutch physicist and professor at the University of Amsterdam. He became interested in the influence of a magnetic field on molecular spectra and discovered a field-induced splitting of the absorption lines in 1896. He shared the Nobel Prize with Hendrik Lorentz *"for their researches into the influence of magnetism upon radiation phenomena"* in 1902. The Zeeman splitting of star spectra allows us to determine the value of the magnetic field of the star at the moment the light was emitted.

- The interaction of the spin magnetic moments of the electrons ($\hat{H}_{SH}$) and of the nuclei ($\hat{H}_{IH}$) with the field $\boldsymbol{H}$. These terms come from the first part of the term $\hat{H}_6$ of the Breit Hamiltonian, and represent the simple Zeeman terms:

$-\hat{\boldsymbol{\mu}} \cdot \boldsymbol{H}$, where $\hat{\boldsymbol{\mu}}$ is the magnetic moment operator of the corresponding particle. Why, together with $\hat{H}_{SH} + \hat{H}_{IH}$, do we not have in $\hat{\mathcal{H}}_1$ the term $\hat{H}_{LH}$; i.e., the interaction of the electron orbital magnetic moment with the field? It would be so nice to have the full set of terms: the spin and the orbital magnetic moments interacting with the field. Everything is fine though, such a term is hidden in the mixed term resulting from $\frac{1}{2m}\left(\hat{\boldsymbol{p}}_j + \frac{e}{c}\boldsymbol{A}_j\right)^2$. Indeed, we get the corresponding *Zeeman term* from the transformation $\frac{e}{mc}\hat{\boldsymbol{p}}_j \cdot \boldsymbol{A}_j = \frac{e}{mc}\boldsymbol{A}_j \cdot \hat{\boldsymbol{p}}_j = \frac{e}{2mc}(\boldsymbol{H} \times \boldsymbol{r}_j) \cdot$

$\hat{\boldsymbol{p}}_j = \frac{e}{2mc}\boldsymbol{H}\cdot(\boldsymbol{r}_j\times\hat{\boldsymbol{p}}_j) = \frac{e}{2mc}\boldsymbol{H}\cdot\hat{\boldsymbol{L}}_j = -\boldsymbol{H}\cdot(-\frac{e}{2mc}\hat{\boldsymbol{L}}_j) = -\boldsymbol{H}\cdot\boldsymbol{M}_{orb,el}(j)$, where $\boldsymbol{M}_{orb,el}(j)$ is, according to the definition of Eq. (12.54), the orbital magnetic moment of the electron j.

- The electronic spin-orbit terms ($\hat{H}_{LS}$); i.e., the corresponding magnetic dipole moment interactions; related to the term $\hat{H}_3$ in the Breit Hamiltonian
- The electronic spin-spin terms ($\hat{H}_{SS}$); i.e., the corresponding spin magnetic moment interactions, related to the term $\hat{H}_5$ in the Breit Hamiltonian
- The electronic orbit-orbit terms ($\hat{H}_{LL}$); i.e., the electronic orbital magnetic dipole interactions (corresponding to the term $\hat{H}_2$ in the Breit Hamiltonian)

- The terms $\hat{\mathcal{H}}_2, \hat{\mathcal{H}}_3, \hat{\mathcal{H}}_4$ (crucial for the NMR experiment) correspond to the magnetic "*dipole-dipole*" interaction involving *nuclear* spins (the term $\hat{H}_5$ of the Breit Hamiltonian): the classical electronic spin – nuclear spin interaction ($\hat{\mathcal{H}}_2$) plus the corresponding Fermi contact term[66] ($\hat{\mathcal{H}}_3$) and the classical interaction of the nuclear spin magnetic dipoles ($\hat{\mathcal{H}}_4$), this time without the contact term, because the nuclei are kept at long distances by the chemical bond framework.[67]

The magnetic dipole moment (of a nucleus or electron) "feels" the magnetic field acting on it through the vector potential $\boldsymbol{A}_j$ at the particle's position $\boldsymbol{r}_j$. This $\boldsymbol{A}_j$ is composed of the external field vector potential $\frac{1}{2}\left[\boldsymbol{H}\times\left(\boldsymbol{r}_j-\boldsymbol{R}\right)\right]$(i.e., associated with the external magnetic field[68] $\boldsymbol{H}$), the individual vector potentials coming from the magnetic dipoles of the nuclei[69] $\sum_A \gamma_A \frac{\boldsymbol{I}_A\times\boldsymbol{r}_{Aj}}{r_{Aj}^3}$ (and having their origins on the individual nuclei) and the vector potential $\boldsymbol{A}_{el}(\boldsymbol{r}_j)$ coming from the orbital and spin magnetic moments of all the electrons:

$$\boldsymbol{A}_j \equiv \boldsymbol{A}\left(\boldsymbol{r}_j\right) = \frac{1}{2}\left[\boldsymbol{H}\times\boldsymbol{r}_{0j}\right] + \sum_A \gamma_A \frac{\boldsymbol{I}_A\times\boldsymbol{r}_{Aj}}{r_{Aj}^3} + \boldsymbol{A}_{el}\left(\boldsymbol{r}_j\right), \tag{12.65}$$

where

$$\boldsymbol{r}_{0j} = \boldsymbol{r}_j - \boldsymbol{R}. \tag{12.66}$$

For closed-shell systems (the majority of molecules) the vector potential $\boldsymbol{A}_{el}$ may be neglected [i.e., $\boldsymbol{A}_{el}\left(\boldsymbol{r}_j\right)\cong\boldsymbol{0}$], because the magnetic fields of the electrons cancel out for a closed-shell molecule (singlet state).

[66] Let us take the example of the hydrogen atom in its ground state. Just note that the highest probability of finding the electron described by the orbital 1*s* is on the proton. The electron and the proton have spin magnetic moments that necessarily interact after they coincide. This effect is certainly something other than just the dipole-dipole interaction, which as usual describes the magnetic interaction for long distances. We have to have a correction for very short distances–this is the Fermi contact term.

[67] And atomic electronic shell structure.

[68] The vector $\boldsymbol{R}$ indicates the origin of the external magnetic field $\boldsymbol{H}$ vector potential from the global coordinate system (cf. Appendix G available at booksite.elsevier.com/978-0-444-59436-5 and the commentary there related to the choice of origin).

[69] Recalling the force lines of a magnet, we see that the magnetic field vector $\boldsymbol{H}$ produced by the nuclear magnetic moment $\gamma A\boldsymbol{I}_A$ should reside within the plane of $\boldsymbol{r}_{Aj}$ and $\gamma_A\boldsymbol{I}_A$. This means that $\boldsymbol{A}$ has to be orthogonal to the plane. This is ensured by $\boldsymbol{A}_j$ being proportional to $\gamma_A\boldsymbol{I}_A\times\boldsymbol{r}_{Aj}$.

Rearranging Terms

When such a vector potential $\boldsymbol{A}$ is inserted into $\hat{\mathcal{H}}_1$ (just patiently make the square of the content of the parentheses) we immediately get

$$\hat{\mathcal{H}} = \hat{\mathcal{H}}_0 + \hat{\mathcal{H}}^{(1)}, \tag{12.67}$$

where $\hat{\mathcal{H}}_0$ is the usual non-relativistic Hamiltonian for the isolated system:

$$\hat{\mathcal{H}}_0 = -\sum_j \frac{\hbar^2}{2m}\Delta_j + \hat{V}, \tag{12.68}$$

$$\hat{\mathcal{H}}^{(1)} = \sum_k^{11} \hat{B}_k, \tag{12.69}$$

while *a few* minutes of a careful calligraphy leads to the result[70]

$$\hat{B}_1 = \frac{e^2}{2mc^2}\sum_{A,B}\sum_j \gamma_A\gamma_B \frac{\hat{\boldsymbol{I}}_A \times \boldsymbol{r}_{Aj}}{r_{Aj}^3}\frac{\hat{\boldsymbol{I}}_B \times \boldsymbol{r}_{Bj}}{r_{Bj}^3}, \tag{12.70}$$

$$\hat{B}_2 = \frac{e^2}{8mc^2}\sum_j \left(\boldsymbol{H} \times \boldsymbol{r}_{0j}\right)\cdot\left(\boldsymbol{H} \times \boldsymbol{r}_{0j}\right), \tag{12.71}$$

$$\hat{B}_3 = -\frac{i\hbar e}{mc}\sum_A\sum_j \gamma_A \nabla_j \cdot \frac{\hat{\boldsymbol{I}}_A \times \boldsymbol{r}_{Aj}}{r_{Aj}^3}, \tag{12.72}$$

$$\hat{B}_4 = -\frac{i\hbar e}{2mc}\sum_j \nabla_j \cdot \left(\boldsymbol{H} \times \boldsymbol{r}_{0j}\right), \tag{12.73}$$

$$\hat{B}_5 = \frac{e^2}{2mc^2}\sum_A\sum_j \gamma_A \left(\boldsymbol{H} \times \boldsymbol{r}_{0j}\right)\cdot \frac{\hat{\boldsymbol{I}}_A \times \boldsymbol{r}_{Aj}}{r_{Aj}^3}, \tag{12.74}$$

$$\hat{B}_6 = \hat{\mathcal{H}}_2 = \gamma_{el}\sum_{j=1}^N\sum_A \gamma_A \left[\frac{\hat{\boldsymbol{s}}_j \cdot \hat{\boldsymbol{I}}_A}{r_{Aj}^3} - 3\frac{\left(\hat{\boldsymbol{s}}_j \cdot \boldsymbol{r}_{Aj}\right)\left(\hat{\boldsymbol{I}}_A \cdot \boldsymbol{r}_{Aj}\right)}{r_{Aj}^5}\right], \tag{12.75}$$

$$\hat{B}_7 = \hat{\mathcal{H}}_3 = -\gamma_{el}\frac{8\pi}{3}\sum_{j=1}\sum_A \gamma_A \delta\left(\boldsymbol{r}_{Aj}\right)\hat{\boldsymbol{s}}_j \cdot \hat{\boldsymbol{I}}_A, \tag{12.76}$$

$$\hat{B}_8 = \hat{H}_{SH} = -\gamma_{el}\sum_j \hat{\boldsymbol{s}}_j \cdot \boldsymbol{H}, \tag{12.77}$$

[70] The operators $\hat{B}_3$ and $\hat{B}_4$ contain the nabla (differentiation) operators. It is worth noting that this differentiation pertains to *everything on the right side of the nabla, including any function on which* $\hat{B}_3$ and $\hat{B}_4$ *operators will act.*

$$\hat{B}_9 = \hat{\mathcal{H}}_4 = \sum_{A<B} \gamma_A \gamma_B \left[\frac{\hat{\boldsymbol{I}}_A \cdot \hat{\boldsymbol{I}}_B}{R_{AB}^3} - 3 \frac{\left(\hat{\boldsymbol{I}}_A \cdot \boldsymbol{R}_{AB}\right)\left(\hat{\boldsymbol{I}}_B \cdot \boldsymbol{R}_{AB}\right)}{R_{AB}^5} \right], \tag{12.78}$$

$$\hat{B}_{10} = \hat{H}_{IH} = -\sum_A \gamma_A \hat{\boldsymbol{I}}_A \cdot \boldsymbol{H}, \tag{12.79}$$

$$\hat{B}_{11} = \hat{H}_{LS} + \hat{H}_{SS} + \hat{H}_{LL}. \tag{12.80}$$

We are just approaching the coupling of our theory with the NMR experiment. To this end, let us first define an empirical Hamiltonian, which serves in the NMR experiment to find what are known as the *nuclear shielding constants* and the *spin-spin coupling constants*. Then we will come back to the perturbation $\hat{\mathcal{H}}^{(1)}$.

12.9 Effective NMR Hamiltonian

NMR spectroscopy[71] means recording the electromagnetic wave absorption by a system of interacting nuclear magnetic dipole moments.[72] It is important to note that the energy differences detectable by contemporary NMR equipment are of the order of 10^{-13} a.u., while the breaking of a chemical bond corresponds to about 10^{-1} a.u. This is why

> all possible changes of the spin state of a system of nuclei do not change the chemical properties of the molecule. This is really what we could only dream of: we have something like observatory stations (the nuclear spins) that are able to detect tiny chemical bond details.

As will be seen in a moment, to reproduce NMR spectra, we need an effective and rotation-averaged Hamiltonian that describes the interaction of the nuclear magnetic moments with the magnetic field and with themselves.

12.9.1 Signal Averaging

NMR experiments usually pertain to the recording of the radio-wave radiation coming from a liquid specimen (which can take many hours). Therefore, we obtain a static (time-averaged) record, which involves various kinds of averaging:

- Over the rotations of any single molecule that contributes to the signal (we assume that each dipole keeps the same orientation in space when the molecule is rotating). These rotations can be free or restrained.
- Over all the molecules present in the specimen.
- Over the vibrations of the molecule (including internal rotations).

[71] The first successful experiment of this kind was described by E.M. Purcell, H.C. Torrey, and R.V. Pound, *Phys. Rev.*, *69*, 37 (1946).

[72] The wavelengths used in the NMR technique are of the order of meters (radio frequencies).

12.9.2 Empirical Hamiltonian

The empirical NMR Hamiltonian contains some parameters that take into account the electronic cloud structure in which the nuclei are immersed. *These NMR parameters will represent our target.*

Now, let us proceed in this direction.

To interpret the NMR data, it is sufficient to consider an *effective* Hamiltonian (containing explicitly only the nuclear magnetic moments, the electron coordinates are absent and the electronic structure enters only implicitly through some interaction parameters). In the matrix notation, we have

$$\hat{\mathcal{H}} = -\sum_{A} \gamma_A \boldsymbol{H}^T \left(\mathbf{1} - \boldsymbol{\sigma}_A\right) \boldsymbol{I}_A + \sum_{A<B} \gamma_A \gamma_B \{\boldsymbol{I}_A^T \left(\boldsymbol{D}_{AB} + \boldsymbol{K}_{AB}\right) \boldsymbol{I}_B\}, \tag{12.81}$$

where $\boldsymbol{I}_C \equiv \left(I_{C,x}, I_{C,y}, I_{C,z}\right)^T$ stands for the spin angular momentum of the nucleus C, while $\boldsymbol{\sigma}_A, \boldsymbol{D}_{AB}, \boldsymbol{K}_{AB}$ denote the symmetric square matrices *(three-dimensional tensors):*

- $\boldsymbol{\sigma}_A$ is a *shielding constant* tensor of the nucleus A. Due to this shielding, nucleus A feels a *local field* $\boldsymbol{H}_{loc} = \left(\mathbf{1} - \boldsymbol{\sigma}_A\right)\boldsymbol{H} = \boldsymbol{H} - \boldsymbol{\sigma}_A \boldsymbol{H}$ instead of the external field $\boldsymbol{H}$ applied (due to the tensor character of $\boldsymbol{\sigma}_A$, the vectors $\boldsymbol{H}_{loc}$ and $\boldsymbol{H}$ may differ by their length and direction). The formula assumes that the shielding is proportional to the external magnetic field intensity that causes the shielding. Thus, the first term in the Hamiltonian $\hat{\mathcal{H}}$ may also be written as $-\sum_A \gamma_A \boldsymbol{H}_{loc}^T \boldsymbol{I}_A$.
- $\boldsymbol{D}_{AB}$ is the 3×3 matrix describing the (direct) *dipole-dipole interaction through space* defined above.
- $\boldsymbol{K}_{AB}$ is also a 3×3 matrix that takes into account that two magnetic dipoles also interact through the framework of the chemical bonds or hydrogen bonds that separate them. This is known as the *reduced spin-spin intermediate coupling tensor.*

Without Electrons...

Let us imagine, just for fun, removing all the electrons from the molecule (and keep them safely in a drawer), while the nuclei still reside in their fixed positions in space. The Hamiltonian would consist of two types of term:

- *The Zeeman term*: interaction of the nuclear magnetic moments with the external magnetic field (the nuclear analog of the first term in $\hat{H}_6$ of the Breit Hamiltonian; p. 147) $-\sum_A \boldsymbol{H} \cdot \hat{\boldsymbol{M}}_A = -\sum_A \gamma_A \boldsymbol{H} \cdot \hat{\boldsymbol{I}}_A$
- The "*through space*" dipole-dipole nuclear magnetic moment interaction (the nuclear analog of the $\hat{H}_5$ term in the Breit Hamiltonian) $\sum_{A<B} \gamma_A \gamma_B [\hat{\boldsymbol{I}}_A \cdot (\boldsymbol{D}_{AB} \hat{\boldsymbol{I}}_B)]$:

$$\boldsymbol{D}_{AB} = \frac{\boldsymbol{i} \cdot \boldsymbol{j}}{R_{AB}^3} - 3\frac{(\boldsymbol{i} \cdot \boldsymbol{R}_{AB})(\boldsymbol{j} \cdot \boldsymbol{R}_{AB})}{R_{AB}^5},$$

where $\boldsymbol{i}, \boldsymbol{j}$ denote the unit vectors along the x-, y-, *and* z-axes; e.g., $(\boldsymbol{D}_{AB})_{xx} = \frac{1}{R_{AB}^3} - 3\frac{(R_{AB,x})^2}{R_{AB}^5}$, $(\boldsymbol{D}_{AB})_{xy} = -3\frac{R_{AB,x}R_{AB,y}}{R_{AB}^5}$, etc. with $\boldsymbol{R}_{AB}$ denoting the vector pointing nucleus B from nucleus A (of length R_{AB}).

Rotations Average out the Dipole-Dipole Interaction

What would happen if we rotated the molecule? In the theory of NMR, there are a lot of notions stemming from classical electrodynamics. In the isolated molecule, the total angular momentum has to be conserved (this follows from the isotropic properties of space). The total angular momentum comes, not only from the particles' orbital motion, but also from their spin contributions. The empirical (non-fundamental) conservation law pertains to the total spin angular momentum alone (cf. p. 76), as well as the individual spins separately. The spin magnetic moments are oriented in space, and this orientation results from the history of the molecule and may be different in each molecule of the substance. These spin states are non-stationary. The *stationary states correspond to some definite values of the square of the total spin of the nuclei and of the spin projection on a chosen axis*. According to quantum mechanics (Chapter 1), only these values are to be measured. For example, in the hydrogen molecule, there are two stationary nuclear spin states: one with parallel spins (ortho-hydrogen) and the other with antiparallel (para-hydrogen). Then we may assume that the hydrogen molecule has two "*nuclear gyroscopes*" that keep pointing them in the same direction in space when the molecule rotates (Fig. 12.13).

Let us see what will happen if we average the interaction of two magnetic dipole moments (the formula for the interaction of two dipoles will be derived in Chapter 13, p. 815): $E_{dip-dip} = \frac{\boldsymbol{M}_A \cdot \boldsymbol{M}_B}{R_{AB}^3} - 3\frac{(\boldsymbol{M}_A \cdot \boldsymbol{R}_{AB})(\boldsymbol{M}_B \cdot \boldsymbol{R}_{AB})}{R_{AB}^5}$. Assume (without losing the generality of the problem) that $\boldsymbol{M}_A$ resides at the origin of a polar coordinate system and has a constant direction along the z-axis, while the dipole $\boldsymbol{M}_B$ just moves on the sphere of the radius R_{AB} around $\boldsymbol{M}_A$ (all orientations are equally probable), the $\boldsymbol{M}_B$ vector preserving the same direction in space $(\theta, \phi) = (u, 0)$ all the time. Now, let us calculate the average value of $E_{dip-dip}$ with respect to all possible positions of $\boldsymbol{M}_B$ on the sphere:

$$
\begin{aligned}
\bar{E}_{dip-dip} &= \frac{1}{4\pi}\int_0^{\pi} d\theta \sin\theta \int_0^{2\pi} d\phi E_{dip-dip} \\
&= \frac{1}{4\pi}\int_0^{\pi} d\theta \sin\theta \int_0^{2\pi} d\phi \left[\frac{1}{R_{AB}^3}\boldsymbol{M}_A \cdot \boldsymbol{M}_B - \frac{3}{R_{AB}^5}(\boldsymbol{M}_A \cdot \boldsymbol{R}_{AB})(\boldsymbol{M}_B \cdot \boldsymbol{R}_{AB})\right] \\
&= \frac{M_A M_B}{4\pi R_{AB}^3}\int_0^{\pi} d\theta \sin\theta \int_0^{2\pi} d\phi \left[\cos u - 3\cos\theta \cos(\theta - u)\right] \\
&= \frac{M_A M_B}{2R_{AB}^3}\int_0^{\pi} d\theta \sin\theta \left[\cos u - 3\cos\theta \cos(\theta - u)\right] \\
&= \frac{M_A M_B}{R_{AB}^3}\left\{\cos u - \frac{3}{2}\int_0^{\pi} d\theta \sin\theta \cos\theta \left[\cos\theta \cos u + \sin\theta \sin u\right]\right\}
\end{aligned}
$$

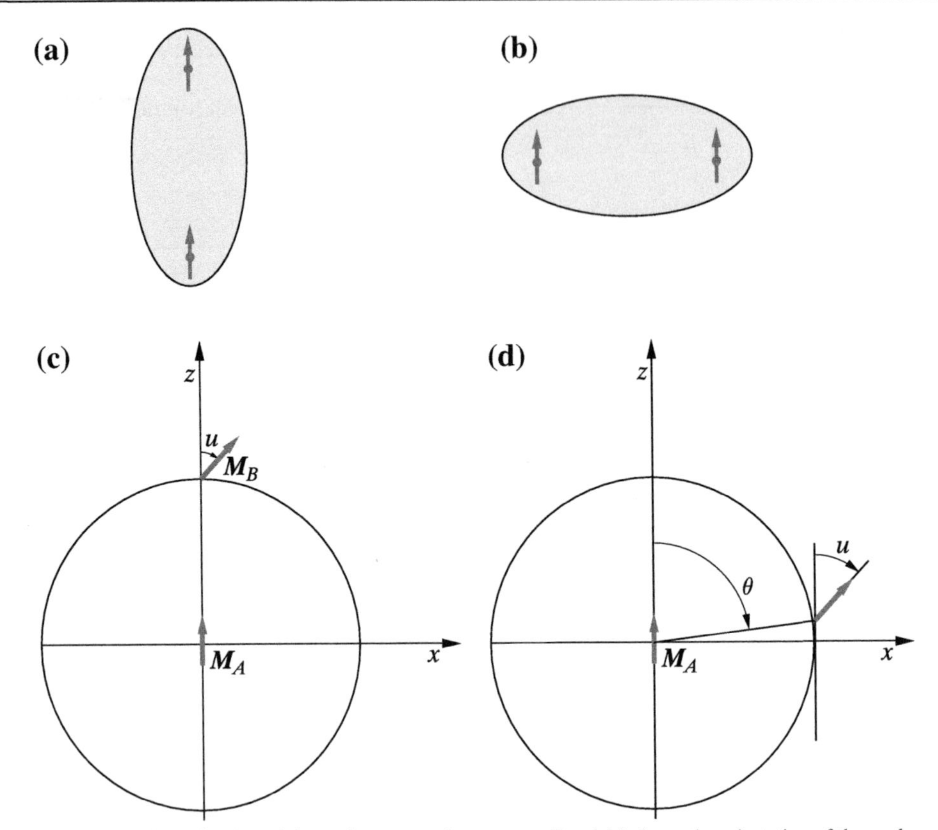

Fig. 12.13. Rotation of a molecule and the nuclear magnetic moments. Panel (a) shows the orientation of the nuclear magnetic moments in the orthohydrogen at the perpendicular configuration of the nuclei. Panel (b) shows the same, but the molecule is oriented horizontally. In the theory of NMR, we assume (in a classical way), that the motion of the molecule does not influence the orientation of both nuclear magnetic moments (c) averaging the dipole-dipole interaction over all possible orientations. Let us immobilize the magnetic moment $\boldsymbol{M}_A$ along the z-axis, the magnetic moment $\boldsymbol{M}_B$ will move on the sphere of radius 1 both moments still keeping the same direction in space $(\theta, \phi) = (u, 0)$. Panel (d) shows one of these configurations. Averaging over all possible orientations gives zero.

$$= \frac{M_A M_B}{R_{AB}^3} \left\{ \cos u - \frac{3}{2} \left[\cos u \cdot \frac{2}{3} + \sin u \cdot 0 \right] \right\} = 0. \tag{12.82}$$

Thus, the averaging gave 0 *regardless of the radius* R_{AB} *and of the angle u between the two dipoles. This result was obtained when assuming the orientations of both dipoles do not change (the abovementioned "gyroscopes"), and that all angles* θ and ϕ are equally probable.

Averaging over Molecular Rotations

An NMR experiment requires long recording times. This means that each molecule, when rotating freely (gas or liquid[73]) with respect to the NMR apparatus, acquires all possible orientations with equal probability. The equipment will detect an average signal. This is why the proposed effective Hamiltonian has to be averaged over the rotations. As we have shown, such an

[73] This is not the case in the solid state.

averaging causes the mean dipole-dipole interaction (containing $\boldsymbol{D}_{AB}$) to be equal to zero. If we assume that the external magnetic field is along the z-axis, then the averaged Hamiltonian reads as

$$\hat{\mathcal{H}}_{av} = -\sum_{A} \gamma_A \left(1 - \sigma_A\right) H_z \hat{I}_{A,z} + \sum_{A<B} \gamma_A \gamma_B K_{AB} (\hat{\boldsymbol{I}}_A \cdot \hat{\boldsymbol{I}}_B), \tag{12.83}$$

where $\sigma_A = \frac{1}{3}\left(\sigma_{A,xx} + \sigma_{A,yy} + \sigma_{A,zz}\right) = \frac{1}{3} Tr \boldsymbol{\sigma}_A,\ K_{AB} = \frac{1}{3} Tr \boldsymbol{K}_{AB}$.

This Hamiltonian is at the basis of NMR spectra interpretation. An experimentalist adjusts σ_A for all the magnetic nuclei and K_{AB} for all their interactions, in order to reproduce the observed spectrum. Any theory of NMR spectra should explain the values of these parameters.

Adding the Electrons–Why the Nuclear Spin Interaction Does not Average Out to Zero

We know already why $\boldsymbol{D}_{AB}$ averages out to zero, but why isn't this true for $\boldsymbol{K}_{AB}$?

Ramsey and Purcell[74] explained this by what is known as the *spin induction mechanism*, described in Fig. 12.14. Spin induction results that, in the averaging of $\boldsymbol{K}_{AB}$, the spin-spin configurations have different weights than in the averaging of $\boldsymbol{D}_{AB}$. This effect is due to the chemical bonds because it makes a difference if the correlating electrons have their spins oriented parallel or perpendicular to the bond line.

Norman F. Ramsey (b. 1915), American physicist and professor at the University of Illinois and Columbia University, and then from 1947 on, at Harvard University. He is, first of all, an outstanding experimentalist in the domain of NMR measurements in molecular jets, but his "*hobby*" is theoretical physics. Ramsey carried out the first accurate measurement of the neutron magnetic moment and gave a lower-bound theoretical estimation to its dipole moment. In 1989 he received the Nobel Prize *"for the invention of the separated oscillatory fields method and its use in the hydrogen maser and other atomic clocks."*

Where does such an effect appear in quantum chemistry? One of the main candidates may be the term $\hat{H}_3$ (the Fermi contact term in the Breit Hamiltonian; p. 147) which couples the orbital motion of the electrons with their spin magnetic moments. This is a relativistic effect (hence it is very small) and therefore, the rotational averaging leaves only a small value of K_{AB}.

Edwards Mills Purcell (1912–1997), American physicist and professor at the Massachusetts Institute of Technology and Harvard University. His main domains were relaxation phenomena and magnetic properties in low temperatures. He received in 1952 the Nobel Prize, together with Felix Bloch, *"for their development of new methods for nuclear magnetic precision measurements and discoveries in connection therewith."*

[74] N.F. Ramsey and E.M. Purcell, *Phys. Rev.*, *85*, 143 (1952).

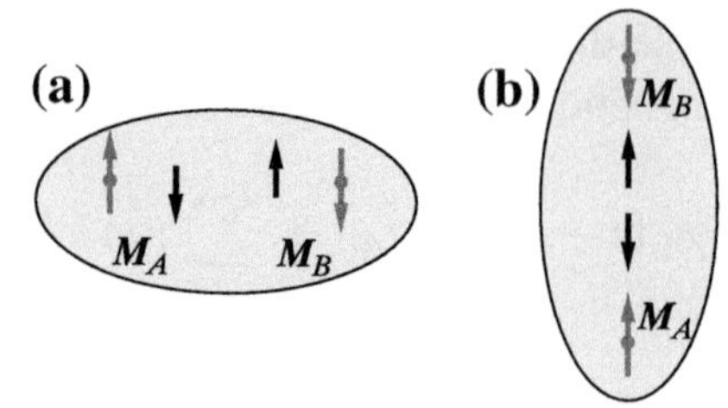

Fig. 12.14. The nuclear spin-spin coupling (Fermi contact) mechanism through chemical bond AB. The electrons repel each other and therefore correlate their motion (cf. p. 589). This is why, when one of them is close to nucleus A, the second prefers to run off to nucleus B. An electron close to A, will exhibit a tendency (i.e., the corresponding energy will be lower than in the opposite case) to have a spin antiparallel to the spin of A. The second electron, close to B, must have opposite spin to its partner, and therefore will exhibit a tendency to have its spin *the same as that of nucleus* A. We may say that the second electron exposes the spin of nucleus A right at the position of the nucleus B. Such a mechanism gives a much stronger magnetic dipole interaction than that through empty space. Panel (a) shows a favorable configuration of nuclear and electron spins, all perpendicular to the bond. Panel (b) shows the same situation after the molecule is rotated by 90°. The electronic correlation energy will obviously differ in these two orientations of the molecule, and this results in different averaging than in the case of the interaction through space.

12.9.3 Nuclear Spin Energy Levels

Calculating the mean value of the Hamiltonian from equation (12.83), we obtain the energy of the nuclear spins in the magnetic field:

$$E = -\sum_A (1-\sigma_A)\,\gamma_A H m_{I,A}\hbar + \sum_{A<B} \gamma_A\gamma_B K_{AB} \left\langle \hat{\boldsymbol{I}}_A \cdot \hat{\boldsymbol{I}}_B \right\rangle,$$

where $\left\langle \hat{\boldsymbol{I}}_A \cdot \hat{\boldsymbol{I}}_B \right\rangle$ is the mean value of the scalar product of the two spins calculated by using their spin functions. This expression can be simplified by the following transformation:

$$\begin{aligned} E &= -\sum_A (1-\sigma_A)\,\gamma_A H m_{I,A}\hbar + \sum_{A<B} \gamma_A\gamma_B K_{AB} \left\langle \hat{I}_{A,x}\hat{I}_{B,x} + \hat{I}_{A,y}\hat{I}_{B,y} + \hat{I}_{A,z}\hat{I}_{B,z} \right\rangle \\ &= -\sum_A (1-\sigma_A)\,\gamma_A H m_{I,A}\hbar + \sum_{A<B} \gamma_A\gamma_B K_{AB} \left(0\cdot 0 + 0\cdot 0 + \hbar^2 m_{I,A} m_{I,B}\right), \end{aligned}$$

because the mean values of $\hat{I}_{C,x}$ and $\hat{I}_{C,y}$ calculated for the spin functions of nucleus C both equal 0 (for the α or β functions describing a nucleus with $I_C = \frac{1}{2}$; see Chapter 1). Therefore,

the energy becomes a function of the magnetic spin quantum numbers $m_{I,C}$ for all the nuclei with a nonzero spin $\boldsymbol{I}_C$:

$$E\left(m_{I,A}, m_{I,B}, \ldots\right) = -\hbar H \sum_A (1-\sigma_A)\,\gamma_A m_{I,A} + \sum_{A<B} h J_{AB} m_{I,A} m_{I,B}, \qquad (12.84)$$

where the commonly used nuclear spin-spin *coupling constant* is defined as

$$J_{AB} \equiv \frac{\hbar}{2\pi} \gamma_A \gamma_B K_{AB}. \qquad (12.85)$$

Note that since hJ_{AB} has the dimension of the energy, then J_{AB} itself is a frequency and may be expressed in hertz.

Due to the presence of the rest of the molecule (electron shielding), the Larmor frequency $\nu_A = \frac{H\gamma_A}{2\pi}(1-\sigma_A)$ is changed by $-\sigma_A \frac{H\gamma_A}{2\pi}$ with respect to the Larmor frequency $\frac{H\gamma_A}{2\pi}$ for the isolated proton. Such changes are usually expressed (as parts per million (ppm)[75]) by the *chemical shift* δ_A:

$$\delta_A = \frac{\nu_A - \nu_{ref}}{\nu_{ref}} \cdot 10^6 = \frac{\sigma_{ref} - \sigma_A}{\sigma_{ref}} \cdot 10^6, \tag{12.86}$$

where ν_{ref} is the Larmor frequency for a reference nucleus [for protons, this means by convention the proton Larmor frequency in tetramethylsilane, $Si(CH_3)_4$]. The chemical shifts (unlike the Larmor frequencies) are independent of the magnetic field applied.

Example. ***The carbon nucleus in an external magnetic field.*** We consider a single carbon ^{13}C nucleus (spin quantum number $I_C = \frac{1}{2}$) in a molecule with non-magnetic other nuclei.

As seen from Eq. (12.84), such a nucleus in magnetic field H has two energy levels (for $m_{I,C} = \pm\frac{1}{2}$; see Fig. 12.15a):

$$E(m_{I,C}) = -\hbar H (1-\sigma_C)\gamma_C m_{I,C},$$

where the shielding constant σ_C characterizes the vicinity of the nucleus. For the isolated nucleus, $\sigma_C = 0$.

Example. ***The methane molecule*** $^{13}CH_4$ ***in magnetic field H.*** This time, there is an additional magnetic field coming from four equivalent protons, each having $I_H = \frac{1}{2}$. The energy levels of the carbon magnetic spin result from the magnetic field and from the $m_{I,H}$'s of the protons according to Eq. (12.84); see Fig. 12.15. The resonance of the ^{13}C nucleus means transition between energy levels that correspond to $m_{I,C} = \pm\frac{1}{2}$ and all the m_{I,H_i} being constant.[76] Thus, the lower level corresponds to

$$E_+(m_{I,H_1}, m_{I,H_2}, m_{I,H_3}, m_{I,H_4}) = -\frac{\hbar}{2} H (1-\sigma_C)\gamma_C + \frac{h}{2}\frac{1}{} J_{CH}(m_{I,H_1} + m_{I,H_2} + m_{I,H_3} + m_{I,H_4}).$$

At the higher level, we have the energy

$$E_-(m_{I,H_1}, m_{I,H_2}, m_{I,H_3}, m_{I,H_4}) = \frac{\hbar}{2} H (1-\sigma_C)\gamma_C - \frac{h}{2}\frac{1}{} J_{CH}(m_{I,H_1} + m_{I,H_2} + m_{I,H_3} + m_{I,H_4}).$$

Since $m_{I,H_i} = \pm\frac{1}{2}$, then each of the levels $E_\pm$ will be split into five levels (see Fig. 12.15):

[75] This means that the chemical shift (unitless quantity) has to be multiplied by 10^{-6} to obtain $\frac{\nu_A - \nu_{ref}}{\nu_{ref}}$.

[76] The NMR selection rule for a given nucleus says that the single nucleus undergoes a flip.

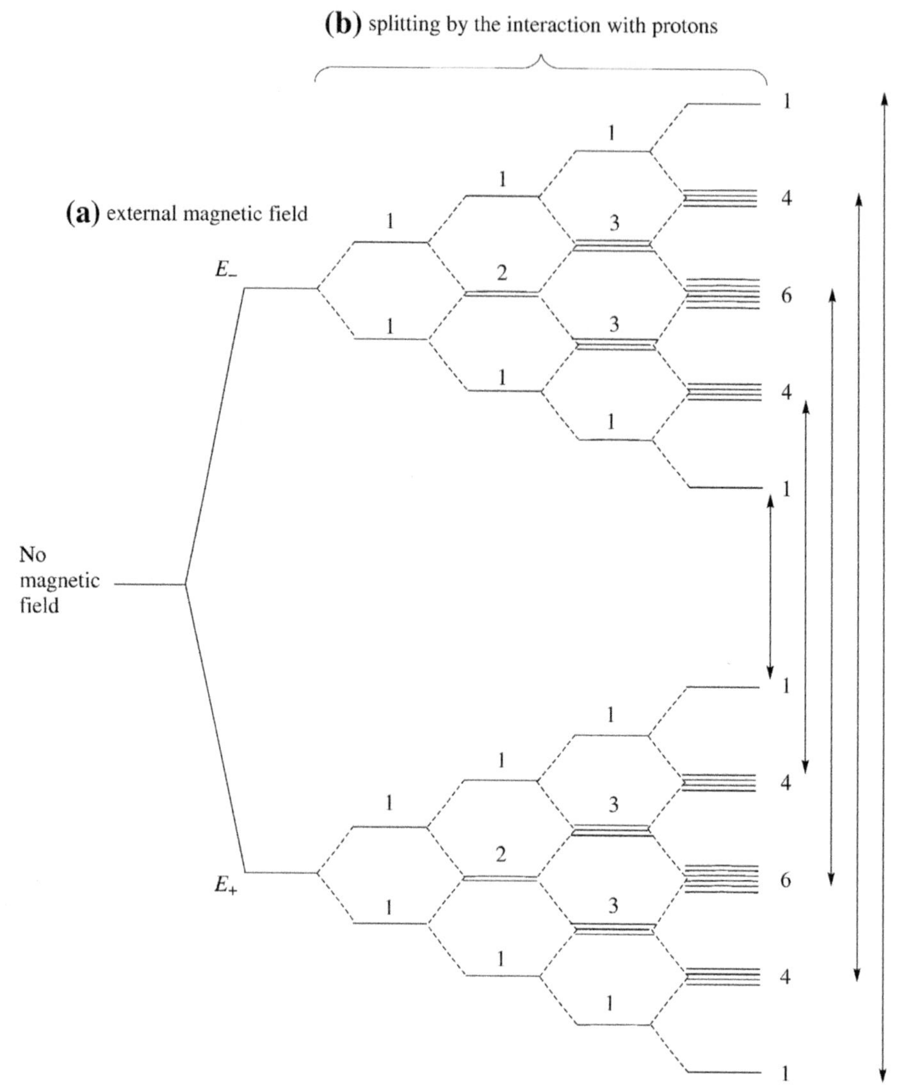

Fig. 12.15. The energy levels of the ^{13}C magnetic moment in an external magnetic field and in the methane molecule. (a) The spin energy levels of the ^{13}C atom in an external magnetic field; (b) additional interaction of the ^{13}C spin with the four equivalent proton magnetic moments switched on. As we can see, the energy levels in each branch follow the Pascal triangle rule. The splits within the branch come from the coupling of the nuclei and are field-independent. The E_+ and E_- energies are field-dependent: increasing field means a tuning of the separation between the energy levels. The resonance takes place when the field-dependent energy difference matches the energy of the electromagnetic field quanta. The NMR selection rule means that only the indicated transitions take place. Since the energy split due to the coupling of the nuclei is very small, the levels E_+ are equally occupied, and therefore, the NMR intensities satisfy the ratio: 1 : 4 : 6 : 4 : 1.

- A non-degenerate level arising from all $m_{I,H_i} = \frac{1}{2}$
- A quadruply degenerate level that comes from all $m_{I,H_i} = \frac{1}{2}$, except one equal to $-\frac{1}{2}$ (there are four positions of this one)
- A sextuply degenerate level that results from two $m_{I,H_i} = \frac{1}{2}$ and two $m_{I,H_i} = -\frac{1}{2}$ (six ways of achieving this)

- A quadruply degenerate level that comes from all $m_{I,H_i} = -\frac{1}{2}$, except one that equals $\frac{1}{2}$ (there are four positions of this one)
- A non-degenerate level arising from all $m_{I,H_i} = -\frac{1}{2}$

Example. ***Nuclear resonances in C_2H_5OH***

Our result may be generalized for n equivalent protons, which interact with a nuclear spin in an external magnetic field. The proton magnetic moment may be aligned either along or opposite to the external magnetic field. The number of ways for k moments aligned along and $n-k$ moments aligned opposite is $\binom{n}{k}$. For the previous case (methane) with $n = 4$ and for $k = 0, 1, 2, 3, 4$, the numbers of equivalent positions are 1,4,6,4,1, which leads to the degeneracy of the nuclear energy levels of the carbon nuclear momentum, as shown in Fig. 12.15.

What would happen if in a single molecule, one had several groups of equivalent nuclei? By the way, what does *equivalent* mean in this context? For example, how many equivalent protons do we have in a molecule like C_2H_5OH? The chemists' way of writing this formula suggests that we know something special about one of these protons. It turns out that this peculiarity comes from binding to the oxygen atom, while other hydrogen atoms are bound to carbon atoms. These other protons form two groups, which is reflected in a more detailed formula: CH_3–CH_2–OH. Up to now, we have discussed the non-equivalence, but what about equivalence? Are the three protons in the methyl group CH_3- equivalent for a chemist? If we take into account the conformational states, we see several conformers possible, but in none of these conformations are the three protons equivalent, although *the roles played by the three protons in the whole set of the conformations are identical*! The situation gets better if one recalls that NMR experiments take a long time and pertain to many molecules in solution. Therefore, every molecule is able to visit all the conformations (for sufficiently high temperatures), including those resulting from rotations of the CH_3- group about the C-C bond. There is no good reason to think that the three protons of the methyl group are non-equivalent in the NMR experiment, and the same applies to the two protons of the $-CH_2-$ group.[77]

The group of n_1-equivalent protons modifies the magnetic field felt by a nucleus (not belonging to the group), which will undergo a resonance transition. Due to the same coupling constants of this particular nucleus with all the nuclei of the group, a splitting of the NMR signal to $n_1 + 1$ signals occurs. For instance, in the case of the ^{13}C resonance in methane, we got $n_1 + 1 = 4 + 1 = 5$ signals.

[77] In the case of rigid molecules, such an averaging does not take place. In such a case, there appears to be a subtle difference between the concepts of the chemical and magnetic equivalence of two nuclei: a and a'. One has the chemical equivalence of a and a', if the two nuclei have the same neighborhood. If a and a' are chemically non-equivalent, they are also non-equivalent magnetically. However, *two chemically equivalent nuclei may turn out to be magnetically non-equivalent.* It will happen, when one finds at least one nucleus (excluding a and a'), that its nuclear coupling constant with a differs from that for a'. As an example may serve two protons in $H_2C = CF_2$. Indeed, let us take one of the fluorine nuclei. One of the protons corresponds to the *cis*, and the other to the *trans* coupling with this particular nucleus.

Possibly, another group of n_2-equivalent protons interacts with the nucleus differently (another value of the coupling constant) and causes its own splitting of each of the previously described signals into $n_2 + 1$ signals.

There is one more problem. What will be if the nucleus, the resonance of which we are considering, is equivalent with some other nuclei? This may happen in two cases. In the first case, the equivalent protons belong to the same molecule, and in the second one, they belong to different molecules. This second case happens always in macroscopic sample used in the NMR and the result is that the NMR signal is stronger if the concentration is larger.[78] The NMR installation does not know anything about our concept of a molecule. Therefore,

> in the case of the equivalent nuclei from the same molecule, one may treat all these nuclei as a "collective" single nucleus undergoing the resonance with the intensity multiplied by the number of the equivalent nuclei.

Such an effect looks natural if one recalls that the NMR electromagnetic waves correspond to radio frequencies. This means a long wave, which is unable to distinguish the equivalent nuclei distributed in a tiny section of space.

The NMR spectra may be quite complicated, especially when the chemical shifts turn out to be less important than the spin-spin coupling. In this example, the opposite is true (such spectra are known as the *first-order NMR spectra*), and the situation is easier. What, therefore, one should expect as the NMR spectrum of the CH_3–CH_2–OH molecule? Well, roughly speaking, we expect three signals shifted with respect to what one gets for the commonly used tetramethylsilane reference (internal standard) signal: one for the protons of the CH_3 group, one for those of the CH_2 group, and one for the proton of the OH group. Also, we expect that the shift of the methyl group signal should be the smallest one (because of the similarity to the reference), the shift for the methylene group should be larger (less similar to the reference), and finally, the OH signal should differ very much from the reference (resulting in the largest shift). The experiment confirms this rough estimation (see Fig. 12.16a). In addition, it is reasonable to expect the corresponding intensities to satisfy the proportion $3:2:1$ (proportionality to the number of the protons). Again, the experiment seems to show something like that (when considering the area under the peaks), besides the fact that we see some very complex splitting of each of the three signals (see Fig. 12.16a).

We are not satisfied by the above rule of thumb, and we want to discover why each signal has such a complex structure. We suspect that this has to do with the interaction with the neighbors of the resonating nucleus. Let us start from the collective nucleus H_3 from the methyl group. It interacts with a group of two equivalent protons (methylene group) separated from it by three bonds and with a single proton of the OH group separated by the four bonds. Therefore, first of all, one may expect the resonance of the collective nucleus H_3 from the CH_3 group split into

[78] It is also broader, because every molecule, even if chemically identical, has a somewhat different geometry.

(a)

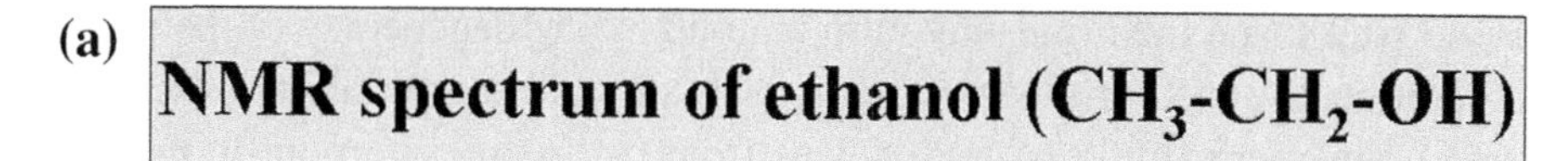

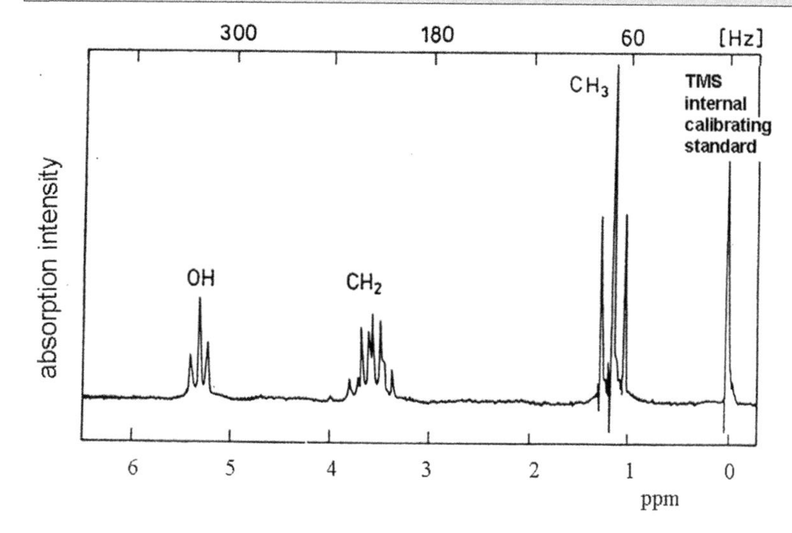

(b)

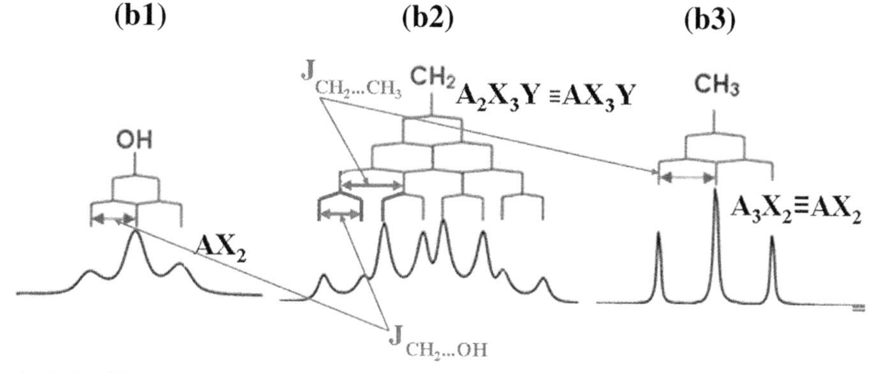

Fig. 12.16. Analysis of the proton magnetic resonance spectrum for the liquid ethanol. (a) The NMR experimental spectrum. From the right side: the tetramethylsilane (TMS) signal–the reference chemical shift taken as $\delta = 0$, at $\delta \approx 1$ ppm the triplet signal from the protons of the CH_3 group, at about 3.5 ppm a multiplet from the CH_2 group, at $\delta \approx 5.3$ ppm there is a triplet coming from the resonance of the proton from the hydroxyl group. (b) Rationalization of the experimental spectrum. (b1) The hydroxyl group proton interacts magnetically with the protons of the CH_2 group. This splits the hydroxyl proton signal into three signals with the intensity ratio 1 : 2 : 1 (there is no splitting visible from the CH_3 group because these protons are too far away). (b2) The explanation of the CH_2 multiplet signal goes in two steps. First, the splitting from the interaction with the CH_3 group is taken into account (the coupling constant $J_{CH_2-CH_3}$), resulting in a quartet with the intensities 1 : 3 : 3 : 1. Next, each of the resulting signals is split into two lines due to the interaction with the hydroxyl proton (with a smaller spin-spin coupling constant J_{CH_2-OH} than before, the intensity ratio is 1 : 1). (b3) The signal from the collective nucleus of the CH_3 group is split by the interaction with the methylene group protons into three signals with the spin-spin coupling constant $J_{CH_2-CH_3}$ (the interaction with the hydroxyl group proton is too weak to be visible). The nomenclature used: $A_nX_mY_p$ denotes the signal of the collective nucleus A (n equivalent nuclei) split by m equivalent protons X and p equivalent protons Y (non-equivalent to X).

$n_{CH_2} + 1 = 3$ peaks, and their intensity ratio (caused by the degeneracy of the energy levels) $1:2:1$. Additionally, each of the resulting signals should be split into two lines due to two possible spin orientations of the proton nucleus from the hydroxyl group. In the experiment, we see the first splitting (Fig. 12.16b3), but not the second one. This is because of too-weak sensitivity of the NMR equipment used, simply the hydroxyl proton is too far away and the splitting is small.

Now, what about $-CH_2-$? The signal from this "collective proton" is split by the methyl group protons into $n_{CH_3} + 1 = 4$ lines with the intensity ratio $1:3:3:1$ (the separation between them should be identical to that found for CH_3, because the two kinds of protons interact with the same coupling constant $J_1 = 7.2\ Hz$). In addition, each of these signals will be split into two lines due to the interaction with the hydroxyl proton (the corresponding coupling constant is $J_2 = 5.1\ Hz$). As a result, one gets a quite complex multiplet of signals, its structure fully rationalized by the above coupling constants (see Fig. 12.16b2).

Finally, we will consider the proton from the hydroxyl group. Its signal will be split into $n_{CH_2} + 1 = 3$ lines with the intensities $1:2:1$ (Fig. 12.16b1) because of two protons from the methylene group. The influence of the distant methyl group is not visible in the spectrum of this accuracy. The chemical shift dominates and is the largest for the protons of this molecule. It turns out that the chemical shift of this proton is sensitive to the details of the intermolecular interactions for this particular proton participates in the hydrogen bonds and other interactions, including possible chemical reactions.[79]

> We have, therefore, some important information: the NMR spectrum may serve for identification of chemical interactions within the molecules, as well as of intermolecular interactions.

12.10 The Ramsey Theory of the NMR Chemical Shift

An external magnetic field $\boldsymbol{H}$ or/and the magnetic field produced by the nuclear magnetic dipole moments $\boldsymbol{M}_1, \boldsymbol{M}_2, \boldsymbol{M}_3 \ldots$ certainly represent an extremely weak perturbation to the molecule, and therefore, the perturbational methods described in Chapter 5 seem to be a perfect choice.

[79] My NMR colleagues told me that the chemical shift of this proton may change very much, even changing the order of the multiplets observed. The chemical shift depends on such things as whether the ethanol is freshly made or not, how long it has been kept in the bottle, whether there are some traces of water in it or not, etc. This proton looks so crazy, that one may think it is responsible for the widely known extraordinary properties of the ethanol molecule. Is the rest of the molecule innocent? Certainly not. We have protons in organic and inorganic acids and nothing comes out of it. Maybe we have to have a OH functional group? No, because water itself would act much stronger than the ethanol. Maybe the proton is innocent, and the C_2H_5- group should be blamed? No, because the C_2H_5-H does not act like that. It seems one should have a hydroxyl group very close to a bulky hydrophobic group. Indeed, the CH_3- group makes an effect similar to that of C_2H_5-, except that in addition, it hurts the eyes and even may kill the drinker.

We decide to apply the theory through the second order. Such an effect is composed of two parts:

- The first-order correction (*the diamagnetic contribution*)
- The second-order correction (*the paramagnetic contribution*)

The corresponding energy change due to the perturbation $\hat{\mathcal{H}}^{(1)}$ from Eq. (12.69) (prime means that $k = 0$; i.e., the ground state is excluded from the summation):

$$\Delta E = E_0^{(1)} + E_0^{(2)} = \left\langle \psi_0^{(0)} | \hat{\mathcal{H}}^{(1)} \psi_0^{(0)} \right\rangle + \sum_k{}' \frac{\left\langle \psi_0^{(0)} | \hat{\mathcal{H}}^{(1)} \psi_k^{(0)} \right\rangle \left\langle \psi_k^{(0)} | \hat{\mathcal{H}}^{(1)} \psi_0^{(0)} \right\rangle}{E_0^{(0)} - E_k^{(0)}}. \tag{12.87}$$

12.10.1 Shielding Constants

In the equation (12.83) for Hamiltonian, the shielding constants occur in the term $\boldsymbol{I}_A \cdot \boldsymbol{H}$. The perturbation operator $\hat{H}^{(1)}$ contains a lot of terms, but most of them, when inserted into the above formula, are unable to produce terms that behave like $\boldsymbol{I}_A \cdot \boldsymbol{H}$. Only some very particular terms could produce such a dot product dependence. A minute of reflection leads directly to $\hat{B}_3$, $\hat{B}_4$, $\hat{B}_5$ and $\hat{B}_{10}$ as the only terms of the Hamiltonian that have any chance of producing the dot product form.[80] Therefore, using the definition of the reduced resolvent $\hat{R}_0$ of Eq. (10.76), we have[81]

$$\begin{aligned} \Delta E = E_0^{(1)} + E_0^{(2)} &= \left\langle \psi_0^{(0)} | (\hat{B}_{10} + \hat{B}_5) \psi_0^{(0)} \right\rangle \\ &+ \left\langle \psi_0^{(0)} | \left(\hat{B}_3 \hat{R}_0 \hat{B}_4 + \hat{B}_4 \hat{R}_0 \hat{B}_3 \right) \psi_0^{(0)} \right\rangle. \end{aligned} \tag{12.88}$$

After averaging the formula over rotations and extracting the proper term proportional to $\boldsymbol{I}_A$ and $\boldsymbol{H}$ (details given in Appendix W available at booksite.elsevier.com/978-0-444-59436-5,

[80] There is an elegant way to single out the only necessary B_i that give a contribution to the energy proportional to the product $x_i x_j$ (no higher terms included), where x_i and x_j stand for some components of the magnetic field intensity $\boldsymbol{H}$ and/or of the nuclear spin $\boldsymbol{I}_A$ (that cause perturbation of the molecule). As to the first-order correction ("*diamagnetic*"), we calculate the second derivative $\left(\frac{\partial^2 \hat{\mathcal{H}}^{(1)}}{\partial x_i \partial x_i} \right)_{\boldsymbol{H}=\boldsymbol{0}, \boldsymbol{I}_i=\boldsymbol{0}}$ of the Hamiltonian $\hat{\mathcal{H}}^{(1)}$ with respect to the components of $\boldsymbol{H}$ or/and $\boldsymbol{I}_A$, afterward inserting $\boldsymbol{H} = \boldsymbol{0}$ and $\boldsymbol{I}_A = \boldsymbol{0}$ (i.e., calculating the derivative at zero perturbation). Then the diamagnetic correction to the energy is $\left\langle \psi_0^{(0)} | \left(\frac{\partial^2 \hat{\mathcal{H}}^{(1)}}{\partial x_i \partial x_i} \right)_{\boldsymbol{H}=\boldsymbol{0}, \boldsymbol{I}_i=\boldsymbol{0}} \psi_0^{(0)} \right\rangle$. As to the second-order correction ("*paramagnetic*"), we calculate the first derivatives: $\left(\frac{\partial \hat{\mathcal{H}}^{(1)}}{\partial x_i} \right)_{\boldsymbol{H}=\boldsymbol{0}, \boldsymbol{I}_i=\boldsymbol{0}}$ and $\left(\frac{\partial \hat{\mathcal{H}}^{(1)}}{\partial x_j} \right)_{\boldsymbol{H}=\boldsymbol{0}, \boldsymbol{I}_i=\boldsymbol{0}}$ and, therefore, the contribution to the energy is $\sum_k' \frac{\left\langle \psi_0^{(0)} | \left(\frac{\partial \hat{\mathcal{H}}^{(1)}}{\partial x_i} \right)_{\boldsymbol{H}=\boldsymbol{0}, \boldsymbol{I}_i=\boldsymbol{0}} \psi_k^{(0)} \right\rangle \left\langle \psi_k^{(0)} | \left(\frac{\partial \hat{\mathcal{H}}^{(1)}}{\partial x_j} \right)_{\boldsymbol{H}=\boldsymbol{0}, \boldsymbol{I}_i=\boldsymbol{0}} \psi_0^{(0)} \right\rangle}{E_0^{(0)} - E_k^{(0)}}$.

[81] Note that whenever the reduced resolvent appears in a formula, infinite summation over unperturbed states is involved.

p. e163) we obtain the following as the shielding constant of the nucleus A:

$$\sigma_A = \frac{e^2}{3mc^2}\left\langle \psi_0^{(0)} \middle| \sum_j (\boldsymbol{r}_{0j} \cdot \boldsymbol{r}_{Aj}) \frac{1}{r_{Aj}^3} \psi_0^{(0)} \right\rangle$$

$$- \frac{e^2}{6m^2c^2}\left\langle \psi_0^{(0)} \middle| \left[\left(\sum_j \frac{\hat{\boldsymbol{L}}_{Aj}}{r_{Aj}^3} \right) \hat{R}_0 \left(\sum_j \hat{\boldsymbol{L}}_{0j} \right) \right.\right.$$

$$\left.\left. + \left(\sum_j \hat{\boldsymbol{L}}_{0j} \right) \hat{R}_0 \left(\sum_j \frac{\hat{\boldsymbol{L}}_{Aj}}{r_{Aj}^3} \right) \right] \psi_0^{(0)} \right\rangle, \tag{12.89}$$

where

$$\hat{\boldsymbol{L}}_{Aj} = -i\hbar \left(\boldsymbol{r}_{Aj} \times \nabla_j \right) \tag{12.90}$$

and

$$\hat{\boldsymbol{L}}_{0j} = -i\hbar \left(\boldsymbol{r}_{0j} \times \nabla_j \right) \tag{12.91}$$

stand for the angular momenta operators for the electron j calculated with respect to the position of nucleus A and with respect to the origin of vector potential $\boldsymbol{A}$, respectively.

12.10.2 Diamagnetic and Paramagnetic Contributions

The result [Eq. (12.89)] has been obtained in two parts:

$$\sigma_A = \sigma_A^{\text{dia}} + \sigma_A^{\text{para}}, \tag{12.92}$$

called the *diamagnetic contribution*,

$$\sigma_A^{\text{dia}} = \frac{e^2}{3mc^2}\left\langle \psi_0^{(0)} \middle| \sum_j (\boldsymbol{r}_{0j} \cdot \boldsymbol{r}_{Aj}) \frac{1}{r_{Aj}^3} \psi_0^{(0)} \right\rangle$$

and the *paramagnetic contribution*,

$$\sigma_A^{\text{para}} = -\frac{e^2}{6m^2c^2}\left\langle \psi_0^{(0)} \middle| \left[\left(\sum_j \frac{\hat{\boldsymbol{L}}_{Aj}}{r_{Aj}^3} \right) \hat{R}_0 \left(\sum_j \hat{\boldsymbol{L}}_{0j} \right) + \left(\sum_j \hat{\boldsymbol{L}}_{0j} \right) \hat{R}_0 \left(\sum_j \frac{\hat{\boldsymbol{L}}_{Aj}}{r_{Aj}^3} \right) \right] \psi_0^{(0)} \right\rangle.$$

Each of these contributions looks suspicious. Indeed, the diamagnetic contribution explicitly depends on the choice of origin $\boldsymbol{R}$ of vector potential $\boldsymbol{A}$ through $\boldsymbol{r}_{0j} = \boldsymbol{r}_j - \boldsymbol{R}$; see Eq. (12.66). Similarly, the paramagnetic contribution also depends on this choice through $\hat{\boldsymbol{L}}_{0j}$ and Eq. (12.66). We have already stressed the *practical* importance of the choice of $\boldsymbol{R}$ in Appendix G available at booksite.elsevier.com/978-0-444-59436-5. Since both contributions depend on the choice, they cannot have any physical significance separately.

Is it possible that the *sum* of the two contributions is invariant with respect to choice of $\boldsymbol{R}$? Yes, it is! The invariance has, fortunately, been proved.[82] This is good because any measurable quantity cannot depend on arbitrary choice of the origin of the coordinate system.

12.11 The Ramsey Theory of the NMR Spin-Spin Coupling Constants

We will apply the same philosophy to calculate the nuclear coupling constant. Taking into account the Hamiltonian $\hat{\mathcal{H}}^{(1)}$ from Eq. (12.69), p. 766, we note that the only terms in $\hat{\mathcal{H}}^{(1)}$ that have the chance to contribute to the NMR coupling constants [see Eq. (12.84)] are

$$\begin{aligned}\Delta E = E_0^{(1)} + E_0^{(2)} &= \left\langle \psi_0^{(0)} | (\hat{B}_1 + \hat{B}_9)\psi_0^{(0)} \right\rangle \\ &+ \left\langle \psi_0^{(0)} | \left(\hat{B}_3 + \hat{B}_6 + \hat{B}_7\right) \hat{R}_0 | \left(\hat{B}_3 + \hat{B}_6 + \hat{B}_7\right) \psi_0^{(0)} \right\rangle \\ &= E_{\text{dia}} + E_{\text{para}},\end{aligned} \tag{12.93}$$

because we are looking for terms that could result in the scalar product of the nuclear magnetic moments. The first term is the diamagnetic contribution (E_{dia}), and the second one is the paramagnetic contribution (E_{para}).

12.11.1 Diamagnetic Contributions

There are two diamagnetic contributions in the total diamagnetic effect $\left\langle \psi_0^{(0)} | (\hat{B}_1 + \hat{B}_9)\psi_0^{(0)} \right\rangle$:

- The $\left\langle \psi_0^{(0)} | \hat{B}_9\, \psi_0^{(0)} \right\rangle$ term simply represents the $\sum_{A<B} \gamma_A \gamma_B \boldsymbol{I}_A^T \boldsymbol{D}_{AB} \boldsymbol{I}_B$ contribution of Eq. (12.81); i.e., the *direct ("through space") nuclear spin-spin interaction.* This calculation does not require anything except summation over spin-spin terms. However, *as has been shown, averaging over free rotations of the molecule in the specimen renders this term equal to zero.*
- The $\left\langle \psi_0^{(0)} | \hat{B}_1 \psi_0^{(0)} \right\rangle$ term can be transformed in the following way:

$$\begin{aligned}\left\langle \psi_0^{(0)} | \hat{B}_1 \psi_0^{(0)} \right\rangle &= \left\langle \psi_0^{(0)} | \left(\frac{e^2}{2mc^2} \sum_{A,B} \sum_j \gamma_A \gamma_B \frac{\boldsymbol{I}_A \times \boldsymbol{r}_{Aj}}{r_{Aj}^3} \frac{\boldsymbol{I}_B \times \boldsymbol{r}_{Bj}}{r_{Bj}^3} \right) \psi_0^{(0)} \right\rangle \\ &= \frac{e^2}{2mc^2} \sum_{A,B} \sum_j \gamma_A \gamma_B \left\langle \psi_0^{(0)} | \frac{(\boldsymbol{I}_A \times \boldsymbol{r}_{Aj}) \cdot (\boldsymbol{I}_B \times \boldsymbol{r}_{Bj})}{r_{Aj}^3 r_{Bj}^3} \psi_0^{(0)} \right\rangle.\end{aligned}$$

[82] A.Abragam, *The Principles of Nuclear Magnetism* Clarendon Press, Oxford (1961).

Now, note that $(\boldsymbol{A} \times \boldsymbol{B}) \cdot \boldsymbol{C} = \boldsymbol{A} \cdot (\boldsymbol{B} \times \boldsymbol{C})$. Taking $\boldsymbol{A} = \boldsymbol{I}_A, \boldsymbol{B} = \boldsymbol{r}_{Aj}, \boldsymbol{C} = \boldsymbol{I}_B \times \boldsymbol{r}_{Bj}$ we first have the following:

$$\left\langle \psi_0^{(0)} | \hat{B}_1 \psi_0^{(0)} \right\rangle = \frac{e^2}{2mc^2} \sum_{A,B} \sum_j \gamma_A \gamma_B \boldsymbol{I}_A \cdot \left\langle \psi_0^{(0)} | \frac{\boldsymbol{r}_{Aj} \times (\boldsymbol{I}_B \times \boldsymbol{r}_{Bj})}{r_{Aj}^3 r_{Bj}^3} \psi_0^{(0)} \right\rangle.$$

Recalling that $\boldsymbol{A} \times (\boldsymbol{B} \times \boldsymbol{C}) = \boldsymbol{B}(\boldsymbol{A} \cdot \boldsymbol{C}) - \boldsymbol{C}(\boldsymbol{A} \cdot \boldsymbol{B})$ this term (called the *diamagnetic spin-orbit contribution, DSO*[83]) reads as

$$E_{\mathrm{DSO}} = \frac{e^2}{2mc^2} \sum_{A,B} \sum_j \gamma_A \gamma_B \boldsymbol{I}_A \cdot \left[\boldsymbol{I}_B \left\langle \psi_0^{(0)} | \frac{\boldsymbol{r}_{Aj} \cdot \boldsymbol{r}_{Bj}}{r_{Aj}^3 r_{Bj}^3} \psi_0^{(0)} \right\rangle - \left\langle \psi_0^{(0)} | \boldsymbol{r}_{Bj} \frac{\boldsymbol{r}_{Aj} \cdot \boldsymbol{I}_B}{r_{Aj}^3 r_{Bj}^3} \psi_0^{(0)} \right\rangle \right].$$

We see that we need to calculate some integrals with mono-electronic operators, which is an easy task.

12.11.2 Paramagnetic Contributions

The paramagnetic contribution E_{para} to the energy can be split into several terms:

$$\begin{aligned} E_{\mathrm{para}} &= \left\langle \psi_0^{(0)} | \left(\hat{B}_3 + \hat{B}_6 + \hat{B}_7 \right) \hat{R}_0 \left(\hat{B}_3 + \hat{B}_6 + \hat{B}_7 \right) \psi_0^{(0)} \right\rangle \\ &= E_{\mathrm{PSO}} + E_{\mathrm{SD}} + E_{\mathrm{FC}} + \text{mixed terms}, \end{aligned}$$

where we have

- The *paramagnetic spin-orbit contribution*:

$$E_{\mathrm{PSO}} = \left\langle \psi_0^{(0)} | \hat{B}_3 \hat{R}_0 \hat{B}_3 \psi_0^{(0)} \right\rangle,$$

with $\hat{B}_3$ meaning the interaction between the *nuclear spin* magnetic moment and the magnetic moment resulting from the *electronic angular momenta* of the individual electrons in an atom.

- The *spin-dipole contribution*

$$E_{\mathrm{SD}} = \left\langle \psi_0^{(0)} | \hat{B}_6 \hat{R}_0 \hat{B}_6 \psi_0^{(0)} \right\rangle,$$

which describes the interaction energy of the magnetic spin dipoles: the nuclear with the electronic dipole,

- The *Fermi contact interaction*

$$E_{\mathrm{FC}} = \left\langle \psi_0^{(0)} | \hat{B}_7 \hat{R}_0 \hat{B}_7 \psi_0^{(0)} \right\rangle,$$

which is related to the electronic spin–nuclear spin interaction with zero distance between them.

[83] The name comes, of course, from the nuclear spin-electronic orbit interaction.

- The mixed terms contain $\left\langle\psi_0^{(0)}|\hat{B}_i\hat{R}_0\hat{B}_j\psi_0^{(0)}\right\rangle$ for $i, j = 3, 6, 7$ and $i \neq j$. These terms are either exactly zero or (in most cases, not always) small.[84]

12.11.3 Coupling Constants

The energy contributions have to be averaged over rotations of the molecule and the coupling constants are to be extracted from the resulting formulas. How this is performed is shown in Appendix W available at booksite.elsevier.com/978-0-444-59436-5 on p. e163.

Using this result, the nuclear spin-spin coupling constant is calculated as the sum of the diamagnetic and paramagnetic contributions:

$$J_{AB} = J_{AB}^{\text{dia}} + J_{AB}^{\text{para}}, \tag{12.94}$$

$$J_{AB}^{\text{dia}} \equiv J_{AB}^{\text{DSO}}, \tag{12.95}$$

$$J_{AB}^{\text{para}} = J_{AB}^{\text{PSO}} + J_{AB}^{\text{SD}} + J_{AB}^{\text{FC}} + J_{AB}^{\text{mixed}}, \tag{12.96}$$

where the particular contributions to the coupling constant are[85]

$$J_{AB}^{\text{DSO}} = \frac{e^2\hbar}{3\pi mc^2}\gamma_A\gamma_B\sum_j\left\langle\psi_0^{(0)}|\frac{\boldsymbol{r}_{Aj}\cdot\boldsymbol{r}_{Bj}}{r_{Aj}^3 r_{Bj}^3}\psi_0^{(0)}\right\rangle,$$

$$J_{AB}^{\text{PSO}} = \frac{1}{3\pi}\hbar\left(\frac{e}{mc}\right)^2\gamma_A\gamma_B\sum_{j,l}\left\langle\psi_0^{(0)}|\hat{\boldsymbol{L}}_{Aj}\hat{R}_0\hat{\boldsymbol{L}}_{Bl}\psi_0^{(0)}\right\rangle,$$

---

[84] Let us consider all cross terms. First, let us check that $\left\langle\psi_0^{(0)}|\hat{B}_3\hat{R}_0\hat{B}_6\psi_0^{(0)}\right\rangle = \left\langle\psi_0^{(0)}|\hat{B}_6\hat{R}_0\hat{B}_3\psi_0^{(0)}\right\rangle = \left\langle\psi_0^{(0)}|\hat{B}_3\hat{R}_0\hat{B}_7\psi_0^{(0)}\right\rangle = \left\langle\psi_0^{(0)}|\hat{B}_7\hat{R}_0\hat{B}_3\psi_0^{(0)}\right\rangle = 0$. Note that $\hat{B}_6$ and $\hat{B}_7$ both contain electron spin operators, while $\hat{B}_3$ does not. Let us assume that, as it is usually the case, $\psi_0^{(0)}$ is a singlet function. Recalling Eq. (10.76) this implies that, because of $\left\langle\psi_0^{(0)}|\hat{B}_3\psi_k^{(0)}\right\rangle$ to survive, the function $\psi_k^{(0)}$ has to have the same multiplicity as $\psi_0^{(0)}$. This, however, kills the other factors: $\left\langle\psi_0^{(0)}|\hat{B}_6\psi_k^{(0)}\right\rangle$ and $\left\langle\psi_0^{(0)}|\hat{B}_7\psi_k^{(0)}\right\rangle$ terms describing the magnetic interaction of nuclei with exactly the same role played by electrons with α and β spins. Thus, the products $\left\langle\psi_0^{(0)}|\hat{B}_3\psi_k^{(0)}\right\rangle\left\langle\psi_k^{(0)}|\hat{B}_6\psi_0^{(0)}\right\rangle$ and $\left\langle\psi_0^{(0)}|\hat{B}_3\psi_k^{(0)}\right\rangle\left\langle\psi_k^{(0)}|\hat{B}_7\psi_0^{(0)}\right\rangle$ are zero.

The mixed term $\left\langle\psi_0^{(0)}|\hat{B}_6\hat{R}_0\hat{B}_7\psi_0^{(0)}\right\rangle$ vanishes for the isotropic electron cloud around the nucleus, because in the product $\left\langle\psi_0^{(0)}|\hat{B}_6\psi_k^{(0)}\right\rangle\left\langle\psi_k^{(0)}|\hat{B}_7\psi_0^{(0)}\right\rangle$, the Fermi term $\left\langle\psi_k^{(0)}|\hat{B}_7\psi_0^{(0)}\right\rangle$ survives, if $\psi_0^{(0)}\psi_k^{(0)}$ calculated at the nucleus is nonzero. However, this kills, $\left\langle\psi_0^{(0)}|\hat{B}_6\psi_k^{(0)}\right\rangle$, because for $\psi_0^{(0)}\psi_k^{(0)} \neq 0$ (which as a rule comes from a $1s$ orbital, this is isotropic situation), the electron-nucleus dipole-dipole magnetic interaction averages to zero when different positions of the electron are considered. For non-isotropic cases, this mixed contribution can be of importance.

[85] The empirical Hamiltonian equation (12.84) contains only the $A > B$ contributions, so the factor 2 appears in J.

$$J_{AB}^{SD} = \frac{1}{3\pi}\hbar\gamma_{el}^2\gamma_A\gamma_B \sum_{j,l=1}^{N}$$
$$\left\langle \psi_0^{(0)} | \left[\frac{\hat{\boldsymbol{s}}_j}{r_{Aj}^3} - 3\frac{\left(\hat{\boldsymbol{s}}_j \cdot \boldsymbol{r}_{Aj}\right)\boldsymbol{r}_{Aj}}{r_{Aj}^5} \right] \hat{R}_0 \left[\frac{\hat{\boldsymbol{s}}_l}{r_{Bl}^3} - 3\frac{\left(\hat{\boldsymbol{s}}_l \cdot \boldsymbol{r}_{Bl}\right)\left(\boldsymbol{r}_{Bl}\right)}{r_{Bl}^5} \right] \psi_0^{(0)} \right\rangle,$$

$$J_{AB}^{FC} = \frac{1}{3\pi}\hbar\left(\frac{8\pi}{3}\right)^2 \gamma_{el}^2\gamma_A\gamma_B \sum_{j,l=1} \left\langle \psi_0^{(0)} | \delta\left(\boldsymbol{r}_{Aj}\right)\hat{\boldsymbol{s}}_j \hat{R}_0 \delta\left(\boldsymbol{r}_{Bl}\right)\hat{\boldsymbol{s}}_l \psi_0^{(0)} \right\rangle.$$

Thus,

the nuclear spin magnetic moments are coupled *via* their magnetic interaction with the *electronic* magnetic moments as follows:

- $J_{AB}^{DSO} + J_{AB}^{PSO}$ results from the electronic *orbital* magnetic dipole moments.
- $J_{AB}^{SD} + J_{AB}^{FC}$ corresponds to such interactions with the electronic *spin* magnetic dipole moments.

As to the integrals involved, the Fermi contact contribution J_{AB}^{FC} (just the value of the wave function at the nucleus position) is the easiest to compute. Assuming that $\psi_k^{(0)}$ states are Slater determinants, the diamagnetic spin-orbit contribution J_{AB}^{DSO} requires some (easy) one-electron integrals of the type $\left\langle \psi_1 | \frac{x_{Aj}}{r_{Aj}^3} \psi_2 \right\rangle$; the paramagnetic spin-orbit contribution J_{AB}^{PSO} needs some one-electron integrals involving $\hat{\boldsymbol{L}}_{Aj}$ operators, which require differentiation of the orbitals; the spin-dipole contribution J_{AB}^{SD} leads to some simple one-electron integrals, but handling the spin operators is needed (see p. 28), as is the case for J_{AB}^{FC}. All the formulas require an infinite summation over states (due to the presence of $\hat{R}_0$), which is very tedious. This is why, in contemporary computational technique, some other approaches, mainly what is called *propagator theory*, are used.[86]

12.11.4 The Fermi Contact Coupling Mechanism

There are no simple rules, but usually the most important contribution to J_{AB} comes from the Fermi contact term (J_{AB}^{FC}), the next most important is paramagnetic spin-orbit term J_{AB}^{PSO}, other terms, including the mixed contributions J_{AB}^{mixed}, are of little importance. Let us consider the Fermi contact coupling mechanism between two protons through a single bond (the coupling constant J_{AB} denoted as ${}^1J_{HH}$). The proton and the electron close to it prefer to have opposite spins, Fig. 12.17. Then the other electron of the bond (being closer to the other nucleus) shows the other nucleus the spin of first nucleus, so the second nucleus prefers to have the opposite

[86] J. Linderberg and Y. Öhrn, *Propagators in Quantum Chemistry* 2d ed., John Wiley & Sons, Ltd. (2004).

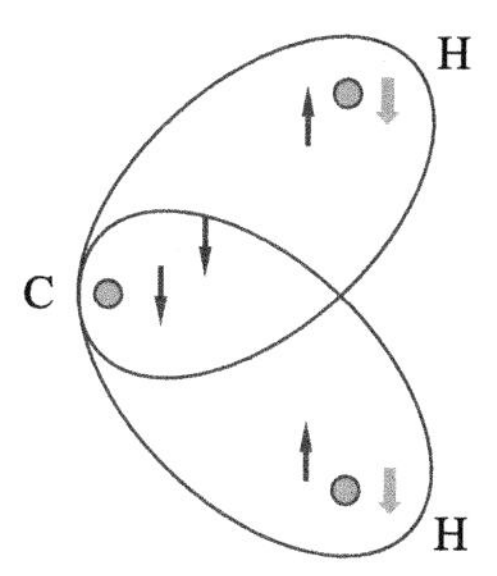

Fig. 12.17. Is the proton-proton coupling constant through two bonds (H–C–H), i.e., ${}^2J_{HH}$, positive or negative? Recall that ${}^1J_{HH} > 0$ (shown in Fig. 12.14a), where the induction mechanism is described. The interaction through two bonds depends on what happens at the central carbon atom: are the spins of the two electrons there (one from each bond C–H) parallel or antiparallel? Hund's rule suggests they prefer to be parallel. This means that the situation with the two proton spins parallel is more energetically favorable, and this means ${}^2J_{HH} < 0$. This rule of thumb may fail when the carbon atom participates in multiple bonds, as in ethylene. For more information, see the section "*From the Research Front,*" later in this chapter.

spin with respect to the first nucleus. According to Eq. (12.84), since $m_{I,A}m_{I,B} < 0$, this means $J_{AB} \equiv {}^1J_{HH} > 0$. What about ${}^2J_{HH}$? This time, to have a through-bond interaction we have to have a central atom, like carbon ${}^{12}C$ (i.e., with zero magnetic moment), Fig. 12.17. The key point now is what happens at the central atom: whether it is preferable to have on it two parallel or two antiparallel electron spins? We do not know, but we have a suggestion. Hund's rule says that, in case of orbital degeneracy (in our case, this corresponds to two *equivalent* C-H bonds), the electrons prefer to have parallel spins. This suggests that the two distant proton spins have a negative coupling constant (i.e., ${}^2J_{HH} < 0$), which is indeed the case. The same argument suggests that ${}^3J_{HH} > 0$, etc.[87]

12.12 Gauge-Invariant Atomic Orbitals (GIAOs)

The coupling constants in practical applications may depend on the choice of vector and scalar potentials. The arbitrariness in the choice of the potentials A and ϕ ("*gauge choice*") does not represent any problem for an *atom* because it is natural to locate the origin (related to the formula G.12 on p. e83) on the nucleus. The same reasoning, however, determines that there is a serious problem for a molecule, because even though any choice is equally justified, this justification is only theoretical, not practical. Should the origin be chosen at the center of mass, at the center of the electron cloud, halfway between them, or at another point? *An unfortunate (although mathematically fully justified) choice of the vector potential origin would lead to correct results,*

[87] Thus, although calculation of the coupling constants is certainly complex, we have in mind a simple model of the nuclear spin-spin interaction that seems to work. We love such models because they enable us to predict numbers knowing other numbers, or to predict new phenomena. This gives the impression that we understand what happens. This is by no means true. What the electrons are doing and how the spin magnetic moments interact is too complicated, but nevertheless, we may suspect the main principles of the game. Such models help us to discuss things with others, to communicate some conjectures, to verify them, and to get more and more confidence in ourselves. This continues until one day, something goes wrong. Then we try to understand why it happened. This *may* require a revision of our model (i.e., a new model, etc).

but only after calculating and summing up all the contributions to infinity, including application of the complete set of atomic orbitals. These requirements are too demanding.

12.12.1 London Orbitals

Atomic orbitals are used in quantum chemistry as the building blocks of many-electron functions (cf. p. 423). Where to center the orbitals sometimes presents a serious problem. On top of this, in the case of a magnetic field, there is additionally the abovementioned arbitrariness of choice of the vector potential origin. A remedy to the second problem was found by Fritz London[88] in the form of atomic orbitals that depend explicitly on the vector potential applied. Each atomic orbital $\chi\left(\boldsymbol{r}-\boldsymbol{R}_C\right)$ centered on nucleus C (with position shown by vector $\boldsymbol{R}_C$) and describing an electron pointed by vector $\boldsymbol{r}$, is replaced by the *London orbital* in the following way:

London Atomic Orbitals

$$\chi_L\left(\boldsymbol{r}-\boldsymbol{R}_C;\boldsymbol{A}_C\right)=\exp\left(-i\boldsymbol{A}_C\cdot\boldsymbol{r}\right)\chi\left(\boldsymbol{r}-\boldsymbol{R}_C\right), \tag{12.97}$$

where $\boldsymbol{A}_C$ stands for the value of vector field $\boldsymbol{A}$ at nucleus C, and $\boldsymbol{A}$ corresponds to the origin O according to formula G. 12 on p. e83, where $\boldsymbol{H}$ denotes the intensity of a homogeneous magnetic field (no contribution from the magnetic field created by the nuclei, etc.).

As seen, the London orbitals are not invariant with respect to the choice of vector potential origin;

e.g., with respect to shifting the origin of the coordinate system in Eq. (G.14) by vector $\boldsymbol{R}$:

$$\boldsymbol{A}'\left(\boldsymbol{r}\right)=\frac{1}{2}\left[\boldsymbol{H}\times(\boldsymbol{r}-\boldsymbol{R})\right]=\boldsymbol{A}(\boldsymbol{r})-\frac{1}{2}\left[\boldsymbol{H}\times\boldsymbol{R}\right]. \tag{12.98}$$

Indeed,

$$\begin{aligned}\chi_L\left(\boldsymbol{r}-\boldsymbol{R}_C;\boldsymbol{A}'_C\right)&=\exp\left(-i\boldsymbol{A}'_C\cdot\boldsymbol{r}\right)\chi\left(\boldsymbol{r}-\boldsymbol{R}_C\right)\\&=\exp\left(-i\boldsymbol{A}_C\cdot\boldsymbol{r}\right)\exp\left(i\frac{1}{2}\left[\boldsymbol{H}\times\boldsymbol{R}\right]\cdot\boldsymbol{r}\right)\chi\left(\boldsymbol{r}-\boldsymbol{R}_C\right)\\&=\exp\left(i\frac{1}{2}\left[\boldsymbol{H}\times\boldsymbol{R}\right]\cdot\boldsymbol{r}\right)\chi_L\left(\boldsymbol{r}-\boldsymbol{R}_C;\boldsymbol{A}_C\right)\neq\chi_L\left(\boldsymbol{r}-\boldsymbol{R}_C;\boldsymbol{A}_C\right).\end{aligned}$$

Despite this property, the London orbitals are also known as gauge-invariant atomic orbitals (GIAOs).

[88] F. London, *J. Phys. Radium*, 8, 397 (1937).

12.12.2 Integrals Are Invariant

Let us calculate the overlap integral S between two London orbitals centered at points C and D. After shifting the origin of the coordinate system in Eq. (G.14) by vector $\boldsymbol{R}$, we get

$$S = \left\langle \chi_{L,1}\left(\boldsymbol{r} - \boldsymbol{R}_C; \boldsymbol{A}'_C\right) | \chi_{L,2}\left(\boldsymbol{r} - \boldsymbol{R}_D; \boldsymbol{A}'_D\right)\right\rangle = \left\langle \exp\left(-i\boldsymbol{A}'_C \cdot \boldsymbol{r}\right)\chi_1 | \exp\left(-i\boldsymbol{A}'_D \cdot \boldsymbol{r}\right)\chi_2\right\rangle$$
$$= \left\langle \chi_1 | \exp\left(-i(\boldsymbol{A}'_D - \boldsymbol{A}'_C)\cdot \boldsymbol{r}\right)\chi_2\right\rangle;$$

i.e., *the result is independent of* $\boldsymbol{R}$. It turns out[89] that all the integrals needed: those of kinetic energy, nuclear attraction, and electron repulsion (cf. Appendix P available at booksite.elsevier.com/978-0-444-59436-5 on p. e131) are invariant with respect to an arbitrary shift of the origin of vector potential $\boldsymbol{A}$.

This means that when we use the London orbitals, the results do not depend on the choice of vector potential origin.

Summary

- The Hellmann-Feynman theorem tells us about the rate at which the energy changes when we change parameter P in the Hamiltonian (e.g., the electric field). This rate is $\frac{\partial E}{\partial P} = \langle \psi(P) | \frac{\partial \hat{H}}{\partial P} | \psi(P)\rangle$, where $\psi(P)$ means the *exact* normalized solution to the Schrödinger equation [with energy $E(P)$] at value P of the parameter.

Electric Phenomena

- When a molecule is located in a non-homogeneous electric field, the perturbation operator has the form $\hat{H}^{(1)} = -\sum_q \hat{\mu}_q \mathcal{E}_q - \frac{1}{3}\sum_{qq'} \hat{\Theta}_{qq'}\mathcal{E}_{qq'} \ldots$, where $\mathcal{E}_q$ for $q = x, y, z$ denote the electric field components along the corresponding axes of a Cartesian coordinate system, $\mathcal{E}_{qq'}$ stands for the q' component of the gradient of $\mathcal{E}_q$, while $\hat{\mu}_q, \hat{\Theta}_{qq'}$ stand for the operators of the corresponding components of the dipole and quadrupole moments. In a homogeneous electric field ($\mathcal{E}_{qq'} = 0$), this reduces to $\hat{H}^{(1)} = -\sum_q \hat{\mu}_q \mathcal{E}_q$.
- After using the last expression in the Hellmann-Feynman theorem, we obtain the dependence of the dipole moment components on the (weak) field intensity: $\mu_q = \mu_{0q} + \sum_{q'} \alpha_{qq'}\mathcal{E}_{q'} + \frac{1}{2}\sum_{q'q''} \beta_{qq'q''}\mathcal{E}_{q'}\mathcal{E}_{q''} + \cdots$, where μ_{0q} stands for the component corresponding to the isolated molecule, $\alpha_{qq'}$ denotes the q, q' component of the (dipole) polarizability tensor, $\beta_{qq'q''}$ is the corresponding component of the (dipole) first hyperpolarizability tensor, etc. The quantities $\mu_{0q}, \alpha_{qq'}, \beta_{qq'q''}$ in a given Cartesian coordinate system characterize the *isolated* molecule (no electric field) and represent the target of the calculation methods.
- Reversing the electric field direction may in general give different absolute values of the induced dipole moment, only because of nonzero hyperpolarizability $\beta_{qq'q''}$ and higher-order hyperpolarizabilities.
- In a non-homogeneous field, we have the following interactions:
  - Of the *permanent dipole moment* of the molecule with the electric field $-\boldsymbol{\mu}_0\boldsymbol{\mathcal{E}}$
  - Of the *induced dipole moment* proportional to the field ($\sum_{q'} \alpha_{qq'}\mathcal{E}_{q'}$), with the field plus higher-order terms proportional to higher powers of the field intensity involving *dipole hyperpolarizabilities*

[89] T. Helgaker and P. Jørgensen, *J. Chem. Phys.*, 95, 2595 (1991).

  - Of the *permanent quadrupole moment* $\Theta_{qq'}$ of the molecule with the field gradient: $-\frac{1}{3}\sum_{qq'}\Theta_{qq'}\mathcal{E}_{qq'}$
  - Of the *induced quadrupole moment* proportional to the field gradient with the field gradient ($-\frac{1}{6}\sum_{qq'q''q'''} C_{qq'q''q'''}\mathcal{E}_{qq'}\mathcal{E}_{q''q'''}$, the quantity C is called the *quadrupole polarizability*) + higher-order terms containing *quadrupole hyperpolarizabilities*.
  - Higher multipole interactions
- In the LCAO MO approximation, the dipole moment of the molecule can be decomposed into the sum of the atomic dipole moments and the dipole moments of the atomic pairs.
- The dipole polarizability may be computed by the following:
  - The *Sum over States method* (SOS), which is based on second-order correction to the energy in the perturbational approach
  - The *finite field method*; e.g., a variational approach in which the interaction with a weak homogeneous electric field is included in the Hamiltonian. The components of the polarizability are computed as the second derivatives of the energy with respect to the corresponding field components (the derivatives are calculated at the zero field). In practical calculations within the LCAO MO approximation, we often use the Sadlej relation that connects the shift of a Gaussian atomic orbital with its exponent and the electric field intensity.
- In laser fields, we may obtain a series of nonlinear effects (proportional to higher powers of field intensity), including the doubling and tripling of the incident light frequency.

Magnetic Phenomena

- An elementary particle has a magnetic dipole moment $\boldsymbol{M}$ proportional to its spin angular momentum $\boldsymbol{I}$; i.e., $\boldsymbol{M} = \gamma \boldsymbol{I}$, where γ stands for what is called the *gyromagnetic factor* (which is characteristic of the kind of particle)
- The magnetic dipole of a particle with spin $\boldsymbol{I}$ (corresponding to spin quantum number I) in homogeneous magnetic field $\boldsymbol{H}$ has $2I + 1$ energy states $E_{m_I} = -\gamma m_I \hbar H$, where $m_I = -I, -I+1, \ldots, +I$. Thus, the energy is proportional to H.
- The Hamiltonian of a system in an electromagnetic field has the form $\hat{H} = \sum_{j=1}[\frac{1}{2m_j}\left(\hat{\boldsymbol{p}}_j - \frac{q_i}{c}\boldsymbol{A}_j\right)^2 + q_j\phi_j] + \hat{V}$, where $\boldsymbol{A}$ and ϕ denote the vector and scalar fields (both are functions of position in the 3-D space, and here, they are calculated at particle j) that characterize the external electromagnetic field.
- $\boldsymbol{A}$ and ϕ potentials contain, in principle (see Appendix G available at booksite.elsevier.com/978-0-444-59436-5), the same information as the magnetic and electric field $\boldsymbol{H}$ and $\boldsymbol{\mathcal{E}}$. There is an arbitrariness in the choice of $\boldsymbol{A}$ and ϕ.
- In order to calculate the energy states of a system of nuclei (detectable in NMR spectroscopy), we have to use the Hamiltonian $\hat{H}$ given above supplemented by the interaction of all magnetic moments, related to the orbital and spin of the electrons and the nuclei.
- The refinement is based on classical electrodynamics and the usual quantum mechanical rules for forming operators (Chapter 1) or, alternatively, on the relativistic Breit Hamiltonian (p. 156). This is how we get the Hamiltonian equation (12.67), which contains the usual non-relativistic Hamiltonian plus the perturbation equation [Eq. (12.69)] with a number of terms (p. 766).
- NMR experimentalists use an empirical Hamiltonian [Eq. (12.83)], in which they have the interaction of the nuclear spin magnetic moments with the magnetic field (the Zeeman effect), the latter weakened by the shielding of the nuclei by the electrons plus the dot products of the nuclear magnetic moments weighted by the coupling constants. The experiment gives both the shielding (σ_A) and the coupling (J_{AB}) constants.
- Nuclear spin coupling takes place through the induction mechanism in the chemical bond (cf. Fig. 12.14). Of key importance for this induction is a high electron density at the position of the nuclei (the so-called Fermi contact term; see Fig. 12.14).
- The perturbational theory of shielding and coupling constants was given by Ramsey. According to the theory, each quantity consists of diamagnetic and paramagnetic contributions. The diamagnetic term is easy to calculate, while the paramagnetic one is more demanding.

- Each of the contributions individually depends on the choice of the origin of the vector potential $\boldsymbol{A}$, while their sum (i.e., the shielding constant), is invariant with respect to this choice.
- The London atomic orbitals $\chi_L = \exp\left(-i\boldsymbol{A}_C \cdot \boldsymbol{r}\right) \chi(\boldsymbol{r} - \boldsymbol{R}_C)$ used in calculations for a molecule in a magnetic field depend explicitly on that field, through the value $\boldsymbol{A}_C$ of the vector potential $\boldsymbol{A}$ calculated at the center $\boldsymbol{R}_C$ of the usual atomic orbital $\chi(\boldsymbol{r} - \boldsymbol{R}_C)$.
- The most important feature of London orbitals is that all the integrals appearing in calculations are invariant with respect to the origin of the vector potential. This is why results obtained using London orbitals are also independent of that choice.

Main Concepts, New Terms

atomic dipoles (p. 741)
Bohr magneton (p. 757)
bond dipoles (p. 741)
Cartesian multipole moments (p. 729)
chemical shift (p. 773)
coupling constant (pp. 768, 781)
coupling mechanism (p. 784)
diamagnetic effect (p. 780)
diamagnetic spin-orbit contribution (p. 782)
dipole hyperpolarizability (p. 732)
dipole, quadrupole, octupole moments (p. 728)
dipole polarizability (p. 732)
direct spin-spin interaction (p. 781)
empirical NMR Hamiltonian (p. 768)
Fermi contact contribution (p. 782)
finite field method (p. 746)
GIAO (p. 786)
gyromagnetic factor (p. 757)
Hellmann-Feynman theorem (p. 722)
homogeneous electric field (p. 731)
intermediate spin-spin coupling (p. 768)
linear response (p. 732)
local field (p. 719)
London orbitals (p. 786)
magnetic dipole (p. 755)
Maxwell equations (p. e81)
multipole hyperpolarizability (p. 738)
multipole moments (p. 729)
multipole polarizability (p. 738)
NMR (p. 768)
NMR Hamiltonian (p. 768)
non-homogeneous electric field (p. 729)
nonlinear response (p. 733)
nuclear magneton (p. 757)
oscillating electric field (p. 752)
paramagnetic effect (p. 780)
paramagnetic spin-orbit (p. 782)
Ramsey theory (p. 778)
Sadlej relation (p. 747)
second/third harmonic generation (SHG/THG) (p. 754)
shielding constants (p. 768)
spin-dipole contribution (p. 782)
spin magnetic moment (p. 758)
Sum over States (SOS) method (p. 743)
traceless multipole moments (p. 728)
ZDO (p. 743)

From the Research Front

The electric dipole (hyper)polarizabilities are not easy to calculate, for the following reasons:

- The Sum over States method (SOS) converges slowly; i.e., a huge number of states have to be taken into account, including those belonging to a continuum.
- The finite field method requires a large quantity of atomic orbitals with small exponents (they describe the lion's share of the electron cloud deformation), although, being diffuse, they do not contribute much to the minimized energy (and lowering the energy is the only indicator that tells us whether a particular function is important or not).

More and more often in their experiments, chemists investigate *large* molecules. Such large objects cannot be described by "*global*" polarizabilities and hyperpolarizabilities (except perhaps optical properties, where the wave length is often much larger than size of molecule). How such large molecules function (interacting with other molecules) depends first of all on their local properties. We have to replace such characteristics by new ones offering

Table 12.1. Comparison of theoretical and experimental shielding constants. The shielding constant σ_A (unitless quantity) is (as usual) expressed in ppm; i.e., the number given has to be multiplied by 10^{-6} to obtain σ_A of Eq. (12.84).

	CH_4		NH_3		H_2O		HF	
Method[a]	σ_C	σ_H	σ_N	σ_H	σ_O	σ_H	σ_F	σ_H
Hartree-Fock	194.8	31.7	262.3	31.7	328.1	30.7	413.6	28.4
MP2	201.0	31.4	276.5	31.4	346.1	30.7	424.2	28.9
MP4	198.6	31.5	269.9	31.6	337.5	30.9	418.7	29.1
CCSD(T)	198.9	31.6	270.7	31.6	337.9	30.9	418.6	29.2
CASSCF	200.4	31.19	269.6	31.02	335.3	30.21	419.6	28.49
Experiment[b]	**198.7**	**30.61**	**264.54**	**31.2**	**344.0**	**30.052**	**410**	**28.5**

[a]The Hartree-Fock, MP2, MP4 results are calculated in J. Gauss, Chem. Phys. Letters, 229, 198 (1994); the CCSD(T) in J. Gauss, J.F. Stanton, J. Chem. Phys., 104, 2574 (1996), and the CASSCF in K. Ruud, T. Helgaker, R. Kobayashi, P. Jørgensen, K.L. Bak, and H.J. Jensen, J. Chem. Phys., 100, 8178 (1994). For a discussion of the Hartree-Fock method, see Chapter 8; for the other methods mentioned here, see Chapter 10.
[b]The references to the corresponding experimental papers are given in T. Helgaker, M. Jaszuński, and K. Ruud, Chem. Rev., 99, 293 (1999). The experimental error is estimated for σ_H in ammonia as ± 1.0, for σ_O as ± 17.2, for σ_H in water as ± 0.015, for σ_F as ± 6, and for σ_H in hydrogen fluoride as ± 0.2.

Table 12.2. Comparison of theoretical and experimental spin-spin coupling constants $^nJ_{AB}$ for ethylene (n denotes the number of separating bonds), in Hz. For a discussion of the methods used see Chapter 10.

	Spin-spin coupling constants J_{AB} for ethylene, in Hertz					
Method[a]	$^1J_{CC}$	$^1J_{CH}$	$^2J_{CH}$	$^2J_{HH}$	$^3J_{HH\text{-cis}}$	$^3J_{HH\text{-trans}}$
MC SCF	71.9	146.6	−3.0	−2.7	10.9	18.1
EOM-CCSD	70.1	153.23	−2.95	0.44	11.57	17.80
Experiment	**67.457**	**156.302**	**−2.403**	**2.394**	**11.657**	**19.015**

[a]All references to experimental and theoretical results are in T. Helgaker, M. Jaszuński, and K. Ruud, Chem. Rev., 99, 293 (1999).

atomic resolution, similar to those proposed in the techniques of Stone or Sokalski (p. e143), where individual atoms are characterized by their multipole moments, polarizabilities, etc.

Not long ago, the shielding and especially spin-spin coupling constants were very hard to calculate with reasonable accuracy. Nowadays, these quantities are computed routinely using commercial software with atomic London orbitals (or other than GIAO basis sets).

The current possibilities of the theory in predicting the nuclear shielding constants and the nuclear spin-spin coupling constants are shown in the Tables 12.1 and 12.2.

Note that the accuracy of the theoretical results for shielding constants is nearly the same as that of experimental results. As to the spin-spin coupling constants, the theoretical results are only slightly away from experimental values.

Ad Futurum

It seems that the SOS method will gradually fall out of favor. The finite field method (in the electric field responses) will become more and more important, due to its simplicity. It remains, however, to solve the problem of how to process the information that we get from such computations and translate it into the abovementioned local characteristics of the molecule.

Contemporary numerical methods allow routine calculation of polarizability. It is difficult with the hyperpolarizabilities that are much more sensitive to the quality of the atomic basis set used. The hyperpolarizabilities relate to nonlinear properties, which are in high demand in new materials for technological applications.

Such problems as the dependence of the molecular spectra and of the molecular conformations and structure on the external electric field (created by our equipment or by a neighboring molecule) will probably become more and more important. This may pertain especially the femto-second spectroscopy, where the laser electric fields are very strong.

The theory of the molecular response to an electric field and the theory of the molecular response to a magnetic field look, despite some similarities, as if they were "from another story". One of the reasons is that the electric field response can be described by solving the Schrödinger equation, while that corresponding to the magnetic field is based inherently on relativistic effects, much less investigated beside some quite simple examples. Another reason may be the scale difference: the electric effects are much larger than the magnetic ones.

However, the theory for the interaction of matter with the electromagnetic field has to be coherent. The finite field method, so gloriously successful in electric field effects, is in the "stone age" stage for magnetic field effects. The propagator methods[90] look the most promising, these allow for easier calculation of NMR parameters than the sum-over-states methods.

Additional Literature

A.D. Buckingham, *Advan. Chem. Phys.*, *12*, 107 (1967).

A classical paper on molecules in a static or periodic electric field.

H.F. Hameka, *Advanced Quantum Chemistry. Theory of Interactions Between Molecules and Electromagnetic Field*, Addison-Wesley, Reading, MA (1965).

This is a first-class book, although it presents the state of the art before the *ab initio* methods for calculating the magnetic properties of molecules.

T. Helgaker, M. Jaszuński, and K. Ruud, *Chem. Rev.*, *99*, 293 (1999).

A review article on the magnetic properties of molecules (NMR) with presentation of suitable contemporary theoretical methods.

Questions

1. The Hellmann-Feynman theorem says that $\frac{\partial E}{\partial P} = \langle\psi|\frac{\partial \hat{H}}{\partial P}|\psi\rangle$ (where $\hat{H}$ is the Hamiltonian that depends on parameter P). This is true, when
   a. ψ stands for the ground-state wave function
   b. ψ is of class Q
   c. ψ is the exact Hartree-Fock function
   d. ψ is an eigenfunction of $\hat{H}$
2. In the expression for the energy of a molecule in electric field,
   a. a component of the quadrupole moment is multiplied by the gradient of the field
   b. a component of the dipole moment is multiplied by the intensity of the field
   c. the dipole hyperpolarizability represents a coefficient at the third power of the electric field intensity
   d. the dipole polarizability represents a coefficient at the square of the electric field intensity
3. A non-polar molecule (but having a nonzero quadrupole moment) in the field with a nonzero gradient
   a. will interact with the field
   b. will orient in such a way as to have the dipole moment along the electric field
   c. will orient in such a way as to have the longer axis of the quadrupole along the field
   d. will orient in such a way as to have the longer axis of the quadrupole along the gradient of the field

[90] J. Linderberg and Y. Öhrn, *Propagators in Quantum Chemistry* 2nd ed., John Wiley & Sons, Ltd. (2004).

4. The second harmonic generation in a uniform electric field depends on the following molecular property
   a. dipole hyperpolarizability β
   b. quadrupole and octupole polarizability
   c. octupole hyperpolarizability
   d. dipole hyperpolarizability γ
5. In some variational calculations for a molecule without electric field the positions of the atomic orbitals have been optimized. In the finite field variational method a small shift of a certain Gaussian atomic orbital off a nucleus
   a. will always increase the energy
   b. will always decrease energy, while at a larger shift the energy will increase
   c. may decrease the energy, if the shift is opposite to the electric field
   d. will decrease the energy, if the sufficiently small shift pertains to all atomic orbitals and is opposite to the electric field
6. The magnetic moment **M** (of an elementary particle)
   a. is the same as its spin angular momentum
   b. has the same length for the electron and for the proton
   c. interacts with uniform magnetic field **H**, and the interaction energy is equal $\frac{1}{2}\mathbf{H}\cdot\mathbf{M}$
   d. interacts with uniform magnetic field **H**, and the interaction energy is equal $-\mathbf{H}\cdot\mathbf{M}$
7. The hydrogen molecule in its electronic singlet state
   a. when in magnetic field acquires an admixture of the triplet electronic state
   b. may be of two kinds depending on the singlet or triplet state of its nuclei
   c. has the electron spins forming the angle 180°
   d. must have the opposite nuclear spins
8. The vector potential **A** of the electromagnetic field corresponds to the uniform magnetic field **H**. Then
   a. $\mathbf{A} = \mathrm{rot}\mathbf{H}$
   b. **A** may be chosen in such a way as to satisfy the Maxwell equation $\mathbf{H} = \mathrm{rot}\mathbf{A}$
   c. **A** also represents a uniform vector field
   d. **A** and $\mathbf{A} - \nabla \exp(-x^2)$ give the same magnetic field **H**
9. The shielding constant for a nucleus consists of the diamagnetic and paramagnetic parts.
   a. the diamagnetic part depends on the choice of the origin of the vector potential **A**
   b. the paramagnetic part depends on the choice of the origin of the vector potential **A**
   c. as physical effects no of them can depend on the choice of the origin of the vector potential **A**
   d. the sum of these parts does not depend on the choice of the origin of the vector potential **A**
10. The value of the London orbital $\chi_L(\mathbf{r}-\mathbf{R};\mathbf{A})$ calculated at the point indicated by vector **r** depends on
    a. the magnetic field at point **R**
    b. the vector potential at point **r**
    c. the vector potential at point $\mathbf{r}-\mathbf{R}$
    d. the vector potential **A** at the point indicated by **R**

Answers

1a,d, 2a,b,c,d, 3a,d, 4a,d, 5c,d, 6d, 7a,b,c, 8b,d, 9a,b,d, 10d

CHAPTER 13

Intermolecular Interactions

"Remember when discoursing about water, to adduce first experience, then reason."
from notes by Leonardo da Vinci (1452–1519)

Where Are We?

We are in the crown of the TREE.

An Example

Look at a snowflake on your hand. Why does such a fascinating, regular structure exist? Why does it change to a few drops of water once it sits on your skin for a second? Why does do water molecules stick together? Visibly, they attract each other *for some reason*. The interaction must not be very strong, however, since the snowflake transforms so easily and then the water drops evaporate (without destroying the water molecules, though).

What Is It All About?

Ideas of Quantum Chemistry, Second Edition. http://dx.doi.org/10.1016/B978-0-444-59436-5.00013-1

Chapter 8 dealt with the question of why atoms form molecules. Electrons and nuclei attract each other, and this results in almost exact electrical neutralization of matter. Despite this, the molecules formed interact for the following reasons:

- Two atoms or molecules cannot occupy the same space.
- Electrons and nuclei in an atom or molecule may still interact with their counterparts in other atoms or molecules. This chapter will tell us about the reason for and details of these interactions.

Why Is This Important?

What is the most important fact that humanity ever learned about matter? According to Richard Feynman, it is "*The world is built of atoms, which repel each other at short distances and attract at longer ones.*" If the intermolecular interactions were suddenly switched off, the world would disintegrate in about a femtosecond–i.e., that is in a single period of atomic vibration (the atoms simply would not come back when shifted from their equilibrium positions). Soon after, everything would evaporate and a sphere of gas, the remainder of the Earth would be held by gravitational forces. Isn't that enough?

What Is Needed?

- Perturbation theory (Chapter 5)
- Variational method (Chapter 5, recommended)
- see Appendix X available at booksite.elsevier.com/978-0-444-59436-5, p. e169
- Many-Body Perturbation Theory (MBPT) (Chapter 10, p. 644)
- Reduced resolvent (Chapter 10, p. 644)
- see Appendix Y available at booksite.elsevier.com/978-0-444-59436-5, p. e183 (recommended)
- see Appendix T available at booksite.elsevier.com/978-0-444-59436-5 (mentioned)

Classical Works

The important subject of intermolecular interactions was recognized and studied very early. The idea that the cohesion of matter stems from the interaction of small indivisible particles ("*atoms*") comes from Democritus. ★ An idea similar to that cited by Feynman was first stated clearly by the Croat scientist Rudjer Bošković in "*Theoria Philosophiae naturalis*," Venice, 1763. ★ Padé approximants were first proposed in the Ph.D. thesis of Henri Padé entitled "*Sur la représentation approchée d'une fonction pour des fractions rationnelles*," which was published in *Annales des Sciences d'Ecole Normale Superieure, Suppl.[3], 9*, 1 (1892). ★ The role of intermolecular interactions was highlighted in

Democritus of Abdera (ca.460 B.C.–ca.370 B.C.), Greek philosopher who formulated the first atomic theory. Its traces go further back in time, but this is Democritus who produced a much more elaborated picture. According to him, nature represents a constant motion of indivisible and permanent particles (atoms), whose interactions result in various materials. *It turned out after almost 25 centuries that this hypothesis was basically correct!* All the written works of Democritus have been lost, but his ideas continued to have an important impact on science for centuries.

Rudjer Josip Bosković (1711–1787), Croat physicist, mathematician, and astronomer and philosopher from Dubrovnik. Because of his expertise in statics and mechanics he was chosen to repair such masterpieces as the dome of St Peter's Basilica.

Johannes Diderik van der Waals (1837–1923), Dutch physicist and professor at the University of Amsterdam. His research topic was the influence of intermolecular forces on the properties of gases (equation of state of the real gas, 1873) and liquids. In 1910, van der Waals received a Nobel Prize "*for his work on the equation of state for gases and liquids.*" He is known also for introducing what is now called the van der Waals forces, which stem from the ubiquitous intermolecular interactions.

the work of Johannes Diderik van der Waals, especially in "*Die Kontinuität des gasformigen und flüssigen Zustandes,*" Barth, Leipzig (1899, 1900). From that time on, intermolecular interactions are often called *van der Waals interactions.* ★ Determination of parenthood of the hydrogen bond idea is not an easy task. By all reasonable indications, the idea comes from the work "*Über Haupt- und Nebenvalenzen und die Constitution der Ammoniumverbindungen,*" in *Liebig's Annalen der Chemie*, *322*, 261 (1902), which was written by the Swiss organic chemist Alfred Werner, the father of coordination chemistry and 1913 Nobel Prize winner. ★ The concept of ionic radii was first proposed by Linus Pauling in "*The sizes of ions and the structure of ionic crystals,*" *J. Amer. Chem. Soc.*, *49*, 765 (1927). ★ The quantum mechanical explanation of intermolecular forces, including the ubiquitous dispersion interactions, was given by Fritz London in "*Zur Theorie und Systematik der Molekularkräfte,*" *Zeitschrift für Physik*, *63*, 245 (1930); and in "*Über einige Eigenschaften und Anwendungen der Molekularkräfte*" from *Zeitschrift für Physikalische Chemie (B)*, *11*, 222 (1930). ★ The hydrophobic effect was first highlighted by Walter Kauzmann in a paper called "*Some factors in the interpretation of protein denaturation,*" in *Adv. Protein Chem.*, *14*, 1 (1959), The effect was further elaborated by George Nemethy, Harold Scheraga, Frank Stillinger, and David Chandler, among others. ★ Resonance interactions were first described by Robert S. Mulliken in an article called "*The interaction of differently excited like atoms at large distances,*" in *Phys. Rev.*, *120*, 1674 (1960). ★ Bogumił Jeziorski and Włodzimierz Kołos extended the existing theory of intermolecular forces to intermediate distances ["*On the symmetry forcing in the perturbation theory of weak intermolecular interactions,*" *Intl. J. Quantum Chem.*, 12 Suppl.*1*, 91 (1977)].

INTERMOLECULAR INTERACTIONS (THEORY)

There are two principal methods of calculating the intermolecular interactions: the supermolecular method and the perturbational method. Both assume the Born-Oppenheimer approximation.

We may all agree about what the total system under consideration should be.[1] Any idea of interaction poses one fundamental question: what kind of objects interact? The answer represents

[1] Even this is a matter of compromise. We (quite arbitrarily) cut the system out of the Universe and say that it does not interact with the rest of the Universe.

our arbitrary decision and has profound consequences. Even if we may be able to describe perfectly the total system for a large spectrum of our choices, the work needed may depend critically on the choice of the interacting objects.

13.1 Idea of the Rigid Interaction Energy

The configuration of the nuclei of the total system can be defined by a set of coordinates given by vector $\mathbf{R}$. We divide the whole system into the interacting subsystems ("molecules"): $A, B, C, \ldots$ with their *internal* geometries (configurations of the nuclei) defined by $\mathbf{R}_A, \mathbf{R}_B, \mathbf{R}_C, \ldots$ and the fixed numbers of electrons $N_A, N_B, , N_C, \ldots$, respectively. The rest of the coordinates ("external") that determine the intermolecular distances and the orientations of the molecules in a global coordinate system will be denoted as $\mathbf{R}_{ex}$:

$$\text{all coordinates} = \text{external} + \text{internal}.$$

Let us define the *rigid interaction energy* at the configuration $\mathbf{R}$ of the nuclei as

$$E_{int}\left(\mathbf{R}_{ex}; \mathbf{R}_A, \mathbf{R}_B, \mathbf{R}_C, \ldots\right) = E_{ABC\ldots}(\mathbf{R}) - [E_A\left(\mathbf{R}_A\right) + E_B\left(\mathbf{R}_B\right) + E_C\left(\mathbf{R}_C\right) + \cdots], \quad (13.1)$$

where the notation $E_{int}\left(\mathbf{R}_{ex}; \mathbf{R}_A, \mathbf{R}_B, \mathbf{R}_C, \ldots\right)$ means that E_{int} is a function of $\mathbf{R}_{ex}$ and depends parametrically on $\mathbf{R}_A, \mathbf{R}_B, \mathbf{R}_C, \ldots$.; i.e., $\mathbf{R}_A, \mathbf{R}_B, \mathbf{R}_C, \ldots$. *are fixed.* $E_{ABC\ldots}$ is the ground-state electronic energy [corresponding to $E_0^{(0)}$ from Eq. (6.21)] of the total system, and $E_A\left(\mathbf{R}_A\right), E_B\left(\mathbf{R}_B\right), E_C\left(\mathbf{R}_C\right), \ldots$ are the electronic energies of the n subsystems (molecules), calculated *at the same positions of the nuclei* as those in the total system.

This definition implies that the interaction energy represents just a theoretical concept, not a measurable quantity. Indeed, we calculate $E_{int}(\mathbf{R})$ for any geometry $\mathbf{R}$ that we wish to consider, which may have nothing to do with the optimized geometry of the total system or the geometry of the isolated molecules. We will see in a while, that there are also other reasons, why the E_{int} cannot be measured.

13.2 Idea of the Internal Relaxation

One may modify the concept of the rigid interaction energy [see Eq. (13.1)] by allowing relaxation of the individual molecules at a fixed $\mathbf{R}_{ex}$. This seems to be physically appealing since chemists often think of molecules as approaching each other, and during such an approach, the molecules *reorient and change* due to the interaction. This is not shown explicitly when calculating the rigid interaction energy, because we see the relaxation of the *electronic* structures of the interacting molecules, but we keep the nuclear frameworks of the individual molecules unrelaxed. The geometry of the individual molecules should change when they approach.

The internally relaxed interaction energy $E_{int}^{relax}\left(\mathbf{R}_{ex}\right)$ may be defined by

$$E_{int}^{relax}\left(\mathbf{R}_{ex}\right) = \min_{\mathbf{R}_A, \mathbf{R}_B, \mathbf{R}_C \ldots} E_{ABC\ldots}(\mathbf{R}) - [E_A\left(\mathbf{R}_A^0\right) + E_B\left(\mathbf{R}_B^0\right) + E_C\left(\mathbf{R}_C^0\right) + \cdots], \quad (13.2)$$

where $\min_{\mathbf{R}_A,\mathbf{R}_B,\mathbf{R}_C...} E_{ABC...}(\mathbf{R})$ means that the minimum of $E_{ABC...}$ is achieved by simultaneous optimization of the internal coordinates (i.e., internal geometries) $\mathbf{R}_A, \mathbf{R}_B, \mathbf{R}_C \ldots$ at all the other coordinates ($\mathbf{R}_{ex}$) fixed. In the second term are the electronic energies of the isolated molecules with the optimized geometries $\mathbf{R}_A^0, \mathbf{R}_B^0, \mathbf{R}_C^0, \ldots$ Thus, in general, both terms in Eq. (13.2) differ from those of Eq. (13.1).

After minimization $[\min_{\mathbf{R}_A,\mathbf{R}_B,\mathbf{R}_C...} E_{ABC...}(\mathbf{R})]$, we get the total electronic energy as a function of $\mathbf{R}_{ex}$, with the molecules $A, B, C, \ldots$ distorted by the interaction (their geometry differing from $\mathbf{R}_A^0, \mathbf{R}_B^0, \mathbf{R}_C^0, \ldots$).[2]

The relaxed interaction energy idea may be extended still further by also allowing energy optimization within a subset of the $\mathbf{R}_{ex}$ coordinates. For example, one of the choices may allow optimization of the rotational degrees of freedom of each molecule (while still keeping other degrees of freedom in $\mathbf{R}_{ex}$ fixed).

13.3 Interacting Subsystems

13.3.1 Natural Division

Although the notion of interaction energy is of great practical value, its theoretical meaning is a little unclear. Right at the beginning, we have a question: interaction *of what*? We view the system *as being composed of particular subsystems* that once isolated, have to be put together.

For instance, the supersystem

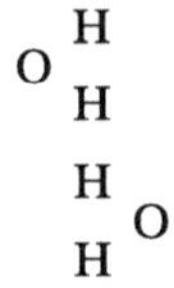

may be considered as two interacting water molecules, but even then, we still are not certain whether the two molecules correspond to (I) or to (II):

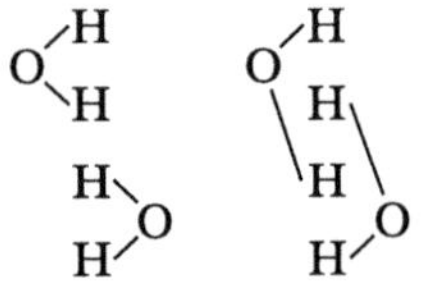

I II

In addition, the system might be considered to be composed of a hydrogen molecule interacting with two OH radicals:

[2] The minimization performed may be the subject of the multiple minima problem (see Chapter 7). In such a case, one may get a non-unique, though still physically meaningful, $E_{int}^{relax}(\mathbf{R}_{ex})$ function, depending on which of multiple minima has been achieved. This physical meaning is not complete since one has to take into account the zero vibration energy (which we will do shortly).

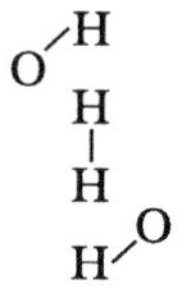

O–H
H
|
H
H–O

and the process repeats again.

The choice of subsystems is of no importance from the point of view of mathematics, but it is of crucial importance from the point of view of calculations in theoretical chemistry.

The particular choice of subsystem should depend on the kind of experiment with which we wish to compare our calculations. For example:

- We are interested in the interaction of water molecules when studying water evaporation or freezing.
- We are interested in the interaction of atoms and ions that exist in the system when heating water to 1000 °C.

Let us stress that in any case of choosing subsystems, we are forced to single out *particular atoms* belonging to subsystems[3] A and B. It is not sufficient to define the kind of molecules participating in the interaction (see our two first examples).

If we are dividing a system into n subsystems in two ways (I and II), and we obtain $|E_{int}|_I < |E_{int}|_{II}$, division I will be more *natural* than division II.

13.3.2 What Is Most Natural?

Which division is *most* natural? We do not have any experience in answering this question. But why should we have any difficulties? Isn't it sufficient to consider all possible divisions and to choose the one which requires the lowest energy? Unfortunately, this is not so obvious. Let us consider two widely separated water molecules (shown in Fig. 13.1a).

Right between the molecules (e.g., in the middle of the OO separation), we place two point charges $q > 0$ and $-q$; i.e., we place nothing, since the charges cancel each other out (please compare a similar trick on p. 570). So we just have two water molecules. Now we start our game. We say that the charges are *real*: one belongs to one of the molecules and the other to the second (see Fig.13.1b). The charge q could be anything, but we want to use it for a very special purpose: to construct the two subsystems in a *more natural* way than just two water molecules. It is interesting that after the choice is made, any of the subsystems has a *lower* energy than that of isolated water, since the molecules are oriented in such a way as to attract each other. This means that the value of q can be chosen from an interval that makes the choice of subsystems more natural. For a certain $q = q_{opt}$, we would obtain as the interaction energy of the new

[3] This means that the interaction energy idea is among the classical concepts. In a quantum system, particles of the same kind are indistinguishable. A quantum system does not allow us to separate a part from the system. Despite this, the interaction energy idea is important and useful.

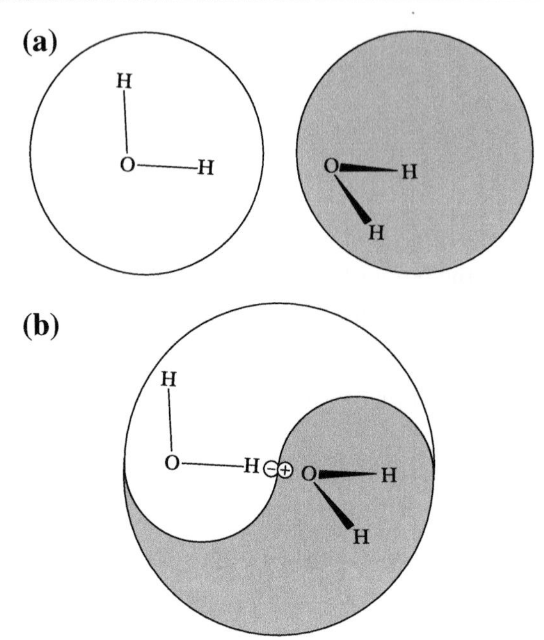

Fig. 13.1. The part-entity relationship, showing two distinct ways of dividing the $(H_2O)_2$ system into subsystems. Division (a) is traditional. The interacting objects are two isolated water molecules and the interaction energy is equal to about -5 kcal/mol (attraction). Division (b) is more subtle. For a certain point in space of charge zero, we wish to treat it as being composed of two fictitious charges $q > 0$ and $-q$, and one of the charges is ascribed to the first molecule and the other to the second molecule. In this way, two *new subsystems* are defined, each of them composed of the water molecule *and* the corresponding point charge. The value of q may be chosen in such a way as to produce the interaction energy of the new subsystems close to 0. Therefore, this is a more natural choice of subsystems than the traditional one. The total interaction energy of the two water molecules is now absorbed within the interactions of the fictitious point charges with "their" water molecules. Each of the point charges takes over the interaction of "its" water molecule with the rest of the Universe. Hence, I have permitted myself (with the necessary *poetic license*) to use the *yin* and *yang* symbols–the two basic elements of ancient Chinese philosophy.

subsystems: $E'_{int} = 0$. This certainly would be the most natural choice,[4] with the "*dressed*" water molecules not seeing each other.[5]

Later in this chapter, we will not use any fictitious charges.

13.4 Binding Energy

Interaction energy can be calculated at any configuration $\boldsymbol{R}$ of the nuclei. We may ask whether any "privileged" configuration exists with respect to the interaction energy. This was the subject of Chapter 7, and there it turned out that the electronic energy may have many minima (equilibria) as a function of $\boldsymbol{R}$. For each of these configurations, we may define the *binding energy with respect*

---

4 Although this choice is not unique, since the charges could be chosen at different points in space, and we could also use point multipoles, etc.

5 This alludes to the elementary particles being "*dressed*" by interactions (as discussed in the section "*What Is It All About?*" in Chapter 8). It is worth noting that we have to superpose the subsystems *first* (then the fictitious charges disappear), and *then* calculate the interaction energy of the water molecules that are deformed by the charges.

to a particular dissociation channel as the difference of the corresponding interaction energies (all subsystems at the optimal positions $\boldsymbol{R}_{opt(j)}$ of the nuclei with respect to the electronic energy E_j):

$$E_{bind} = E_{ABC...}\left(\boldsymbol{R}_{opt(tot)}\right) - \sum_{j=A,B,C,...} E_j\left(\boldsymbol{R}_{opt(j)}\right) = \min_{\boldsymbol{R}_{ex}} E_{int}^{relax}\left(\mathbf{R}_{ex}\right) \tag{13.3}$$

At a given configuration $\boldsymbol{R}_{opt(tot)}$, we usually have many dissociation channels differing by the possible products and their quantum-mechanical states.

13.5 Dissociation Energy

The calculated interaction energy of Eq. (13.1), as well as the binding energies are only theoretical quantities and cannot be measured. The measurable quantity is the *dissociation energy*

$$E_{diss} = E_{bind} + \left[\Delta E_{0,tot} - \sum_{j=A,B,C,...} \Delta E_{0,j}\right], \tag{13.4}$$

where $\Delta E_{0,tot}$ stands for what is known as the *zero vibration energy* of the total system (cf. p. 364) at the *equilibrium geometry* $\boldsymbol{R}_{opt(tot)}$, and $\Delta E_{0,j}$ for $j = A, B, C, \ldots$ representing the zero vibration energies for the subsystems. In the harmonic approximation, $\Delta E_{0,tot} = \frac{1}{2}\sum_i h\nu_{i,tot}$ and $\Delta E_{0,j} = \frac{1}{2}\sum_i h\nu_{i,j}, \ldots$ at their equilibrium geometries $\boldsymbol{R}_{opt(j)}$.

Let us stress that the formula $\frac{1}{2}h\nu$ for the zero vibration energy is related to the harmonic approximation.[6] Generally,

> the zero vibration energy has to be determined as the difference between the lowest vibrational energy level and the energy of the bottom of the potential well.

13.6 Dissociation Barrier

If a molecule receives dissociation energy, it is most often a sufficient condition for its dissociation (see Fig.13.2a). Sometimes, however, the energy is too low, and the reason is that there is an energy barrier to be overcome (see Fig. 13.2b). This dissociation barrier may be high and the system stable even if the dissociation products have (much) lower energy. The catenans and rotaxans shown in Fig. 13.2c may serve as examples.

The energy necessary to overcome the barrier from the trap side is equal to

$$E_{bar1} = E_{\#} - \left(E_{\min} + \frac{1}{2}\sum_j h\nu_j\right),$$

[6] Note that ν is well defined in this formula.

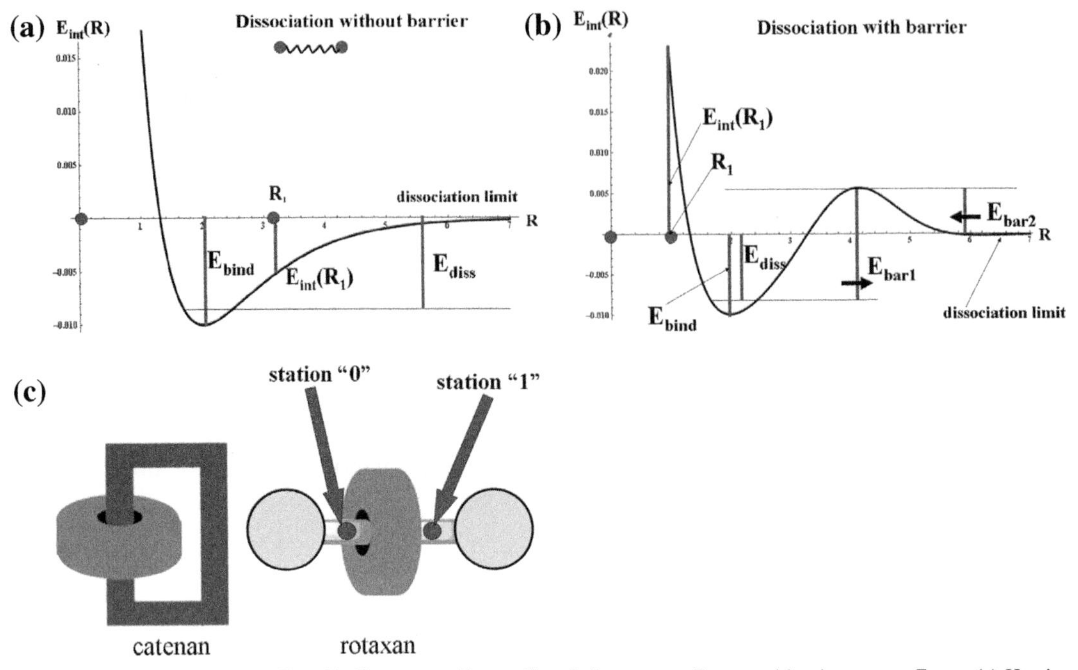

Fig. 13.2. Interaction energy E_{int}, binding energy E_{bind}, dissociation energy E_{diss}, and barrier energy E_{bar1}. (a) Here's a common situation: the interaction energy of two atoms or molecules (circles) represents a simple function of their distance R. (b) Here is a more complex situation: there is a barrier for dissociation of the complex (E_{bar1}) and a barrier (of height E_{bar2}) for approaching the atoms. (c) Two parts of the molecule are interlocked. In such a situation, like a catenan (two interlocked rings) and a rotaxan (a ring moving along a wire with two stable positions), the E_{bar1} energy is very large.

where E_{min} is the energy of the bottom of the well, $E_{\#}$ represents the barrier top energy, and $\frac{1}{2}\sum_j h\nu_j$ is the zero vibration energy of the well.

13.7 Supermolecular Approach

In the supermolecular method, the interaction energy is calculated from its definition [Eq. (13.1)] using any reliable method of the electronic energy calculation. For the sake of brevity, we will consider the interaction of only two subsystems: A and B.

13.7.1 Accuracy Should Be the Same

There is a problem, though. The trouble is that we are unable to solve the Schrödinger equation exactly, either for AB, for A, or for B. We have to use approximations. If so, we have to worry about *the same accuracy* of calculation for AB, A, and B. From this fact, we may expect that

> in determining E_{AB}, as well as E_A and E_B, the same theoretical method is preferred because any method introduces its own systematic error, and we may hope these errors will cancel at least partially in the above formula.

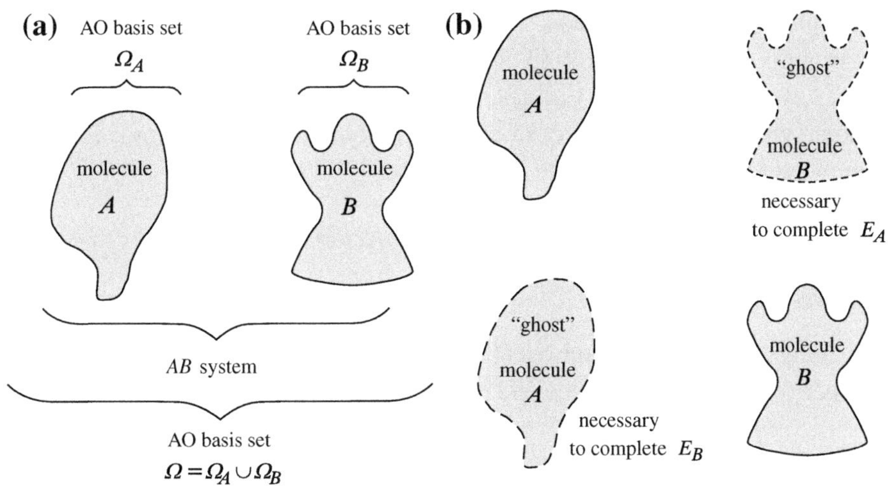

Fig. 13.3. (a) The BSSE problem. Each of the molecules offers its own atomic orbitals to the total basis set $\Omega = \Omega_A \cup \Omega_B$. Panel(b) illustrates the counter-poise method, in which the calculations for a single subsystem are performed within the full atomic basis set Ω: the orbitals centered on it and what are called *ghost orbitals* centered on the partner.

This problem is already encountered at the stage of basis set choice. For example, suppose that we have decided to carry out the calculations within the Hartree-Fock method in the LCAO MO approximation. The same method has to be used for AB, A, and B. However, what does this really mean? Suppose we use the following protocol:

1. Consider the atomic basis set Ω, which consists of the atomic orbitals centered on the nuclei of A (set Ω_A) and on the nuclei of B (set Ω_B); i.e., $\Omega = \Omega_A \cup \Omega_B$.
2. Calculate E_{AB} using Ω, E_A using Ω_A, and E_B using Ω_B (see Fig. 13.3a).

Apparently everything looks logical, but *we did not use the same method* when calculating the energies of AB, A, and B. The basis set used was different depending what we wanted to calculate. Thus, it seems more appropriate to calculate all three quantities using the same basis set Ω.

13.7.2 Basis Set Superposition Error (BSSE) . . . and the Remedy

Such an approach is supported by the following reasoning. When the calculations for E_{AB} are performed within the basis set Ω, we not only calculate implicitly the interaction energy, but also we allow the individual subsystems to lower their energy–for no physical reason whatsoever–and this reduction corresponds to the $\Omega - \Omega_A$ for A or $\Omega - \Omega_B$ for B, Fig. 13.3a. Conclusion: by subtracting from E_{AB} the energies (E_A calculated with Ω_A and E_B with Ω_B), we are left not only with the interaction energy (as should be), but also with an unwanted and non-physical extra term (an error) connected with the artificial lowering of the subsystems' energies, when calculating E_{AB}. This error is called the *Basis Set Superposition Error (BSSE)*.

To remove the BSSE, we may consider the use of the basis set Ω not only for E_{AB}, but also for E_A and E_B. This procedure, called the *counter-poise method*, was first introduced by Boys and

Bernardi.[7] Application of the full basis set Ω when calculating E_A results in the wave function of A containing not only its own atomic orbitals, but also the atomic orbitals of the ("*absent*") partner B, the "*ghost orbitals*" (see Fig. 13.3b). As a by-product, the charge density of A would exhibit broken symmetry with respect to the symmetry of A itself (if any); for example, the helium atom would have a small dipole moment, etc.

13.7.3 Good and Bad News About the Supermolecular Method

Two Deficiencies

When performing the subtraction in Eqs. (13.1) or (13.2), we obtain a number representing the interaction energy at a certain distance and orientation of the two subsystems.

> The resulting E_{int} has two disadvantages: it is less precise than E_A and E_B, and it does not tell us anything about why the particular value is obtained.

The first disadvantage could be compared (following Coulson[8]) to weighing a captain's hat by first weighing the ship with the captain wearing his hat and the ship with the captain without his hat (Fig. 13.4).

Formally, everything is perfect, but there is a cancellation of significative digits in E_{AB} and $(E_A + E_B)$, which may lead to a very poor interaction energy.

The second deficiency deals with the fact that the interaction energy obtained is just a number, and we have no idea why the number is of such magnitude.[9]

Both deficiencies will be removed in the perturbational approach to intermolecular interaction. Then, the interaction energy will be calculated directly and we will be able to tell which physical contributions it consists of.

Fig. 13.4. In the supermolecular method, we subtract two large numbers that differ only slightly and lose accuracy in this way. In order not to obtain a result like 240 kg or so, we have to have at our disposal a very accurate method of weighing things.

7 S.F. Boys and F. Bernardi, *Mol.Phys.*, *19*, 553 (1970).

8 C.A. Coulson, *Valence* Oxford University Press (1952).

9 The severity of this problem can be diminished by analyzing the supramolecular interaction energy expression (using molecular orbitals of A and B) and identifying the physically distinguishable terms by the kind of molecular integrals of which they are composed [K. Kitaura and K. Morokuma, *Intern.J.Quantum Chem.*, *10*, 325 (1976)].

A big advantage of the supermolecular method is its applicability at any intermolecular distance (i.e., regardless of how strong the interaction is).

13.8 Perturbational Approach

13.8.1 Intermolecular Distance–What Does It Mean?

What is the distance (in kilometers) between the Polish and German populations, or what does the distance between two buses mean? Because of the nonzero dimensions of both objects, it is difficult to tell what a reasonable distance could be, and any measure of it will be arbitrary. It is the same story with molecules. Until now, we did not need a notion for the intermolecular distance–the positions of the nuclei were sufficient. At the beginning, we need only an infinite distance, and therefore, in principle, any definition will be acceptable. Later, however, we will be forced to specify the intermolecular distance (cf., p. 812 and Appendix X available at booksite.elsevier.com/978-0-444-59436-5 on p. e169). The final numerical values should not depend on this choice, but intermediary results could depend on it. It will turn out that despite the existing arbitrariness, we will prefer those definitions that are based upon the center of charge distance or similar.

13.8.2 Polarization Approximation (Two Molecules)

According to the Rayleigh-Schrödinger perturbation theory (discussed in Chapter 5), the unperturbed Hamiltonian $\hat{H}^{(0)}$ is a sum of the isolated molecules' Hamiltonians:

$$\hat{H}^{(0)} = \hat{H}_A + \hat{H}_B.$$

Following quantum theory tradition, in this chapter the symbol for the perturbation operator will be changed (when compared to Chapter 5) as follows: $\hat{H}^{(1)} \equiv V$.

Despite the fact that we may also formulate the perturbation theory for excited states, we will assume that we are dealing with the ground state (and denote it by subscript "0"). In what is called the *polarization approximation*, the zeroth-order wave function will be taken as a product:

$$\psi_0^{(0)} = \psi_{A,0}\psi_{B,0}, \tag{13.5}$$

where $\psi_{A,0}$ and $\psi_{B,0}$ are the exact ground state wave functions for the isolated molecules A and B, respectively; i.e.,

$$\hat{H}_A\psi_{A,0} = E_{A,0}\psi_{A,0},$$
$$\hat{H}_B\psi_{B,0} = E_{B,0}\psi_{B,0}.$$

First, for the large separation of the two molecules, the electrons of molecule A are *distinguishable* from the electrons of molecule B. We have to stress the classical flavor of this approximation. Second, we assume that the exact wave functions of both isolated molecules[10] $\psi_{A,0}$ and $\psi_{B,0}$ are at our disposal.

Of course, function $\psi_0^{(0)}$ is only an approximation of the exact wave function of the total system. Intuition tells us that this approximation is probably very good because we presume that the perturbation is small and the product function $\psi_0^{(0)} = \psi_{A,0}\,\psi_{B,0}$ is an exact wave function for the *non-interacting* system.

The chosen $\psi_0^{(0)}$ has a wonderful feature–namely, it represents an eigenfunction of the $\hat{H}^{(0)}$ operator, as is required by the Rayleigh–Schrödinger perturbation theory (see Chapter 5).

The function has also an unpleasant feature: it differs from the exact wave function by symmetry. For example, it is easy to see that

the function $\psi_0^{(0)}$ *is not* antisymmetric with respect to the electron exchanges between molecules, while the exact function has to be antisymmetric with respect to any exchange of electron labels.

This deficiency exists for any intermolecular distance.[11] We will soon pay a high price for this.

---

[10] We will eliminate an additional complication that sometimes may occur. The nth state of the two non-interacting molecules comes, of course, from *some* states of the isolated molecules A and B. It may happen (most often when the two molecules are identical), that two different sets of the states give the same energy $E_n^{(0)}$. Typically, this may happen upon the exchange of excitations of both molecules. Then, $\psi_n^{(0)}$ has to be taken as a linear combination of these two possibilities, which leads to profound changes of the formulas with respect to the usual cases. Such an effect is called the *resonance interaction* [R.S. Mulliken, *Phys.Rev.*, *120*, 1674 (1960)]. The resulting interaction decays with the distance as R^{-3} (i.e., quite slowly), making possible an excitation energy transfer over long distances between the interacting molecules. The resonance interaction turns out to be very important (e.g., in biology). An interested reader may find more in the review article by J.O. Hirschfelder and W.J. Meath, *Advan.Chem.Phys.*, *12*, 3 (1967).

[11] We may say that the range of the Pauli principle is infinity. If somebody paints some electrons green and others red (we do this in the perturbational method), they are in "no man's land," between the classical and quantum worlds. Since the wave function $\psi_0^{(0)}$ does not have the proper symmetry, the corresponding operator $\hat{H}^{(0)} = \hat{H}_A + \hat{H}_B$ is just a mathematical object not quite appropriate to the total system under study.

First-order Effect: Electrostatic Energy

The first-order correction (see Chapter 5, p. 244),

$$E_0^{(1)} \equiv E_{\text{elst}} \equiv E_{pol}^{(1)} = \left\langle \psi_0^{(0)} | V \psi_0^{(0)} \right\rangle, \tag{13.6}$$

represents what iscalled the *electrostatic interaction energy* (E_{elst}). To stress that E_{elst} is the first-order correction to the energy in the polarization approximation, the quantity will be alternatively denoted by $E_{pol}^{(1)}$. The electrostatic energy represents the Coulombic interaction of two "frozen" charge distributions corresponding to the isolated molecules A and B, because it is the mean value of the Coulombic interaction energy operator V calculated with the wave function $\psi_0^{(0)}$ being the product of the wave functions of the isolated molecules $\psi_0^{(0)} = \psi_{A,0}\psi_{B,0}$.

Second-order Energy: Induction and Dispersion Energies

The second-order energy in the polarization approximation approach can be expressed in a slightly different way.

The nth state of the total system at long intermolecular distances corresponds to some states n_A and n_B of the individual molecules; i.e.,

$$\psi_n^{(0)} = \psi_{A,n_A}\psi_{B,n_B} \tag{13.7}$$

and[a]

$$E_n^{(0)} = E_{A,n_A} + E_{B,n_B}. \tag{13.8}$$

[a] Also, we exclude the resonance interaction in this case.

Using this assumption, the second-order correction to the ground-state energy (we assume that $n = 0$ and $\psi_0^{(0)} = \psi_{A,0}\psi_{B,0}$) can be expressed as (see Chapter 5, p. 244)

$$E_0^{(2)} = \sum_{n_A}\sum_{n_B}{}' \frac{|\langle \psi_{A,n_A}\psi_{B,n_B} | V \psi_{A,0}\psi_{B,0}\rangle|^2}{(E_{A,0} - E_{A,n_A}) + (E_{B,0} - E_{B,n_B})}, \tag{13.9}$$

where "*prime*" in the summation means excluding $(n_A, n_B) = (0, 0)$. The quantity $E_0^{(2)}$ can be divided in the following way:

$$E_0^{(2)} = \sum_{n_A}\sum_{n_B}{}' \cdots = \sum_{(n_A=0, n_B\neq 0)} \cdots + \sum_{(n_A\neq 0, n_B=0)} \cdots + \sum_{(n_A\neq 0, n_B\neq 0)} \cdots \tag{13.10}$$

Let us construct a matrix $\mathbf{A}$ (of infinite dimension) composed of the element $A_{00} = 0$ and the other elements calculated from the formula

$$A_{n_A,n_B} = \frac{|\langle \psi_{A,n_A}\psi_{B,n_B}|V\psi_{A,0}\psi_{B,0}\rangle|^2}{(E_{A,0} - E_{A,n_A}) + (E_{B,0} - E_{B,n_B})} \tag{13.11}$$

and divide it into the following parts:

n_B ↓ \ n_A →	0	1 2 3 4 5 ...
0	0	II
1 2 3 4 5 ⋮	I	III

The quantity $E_0^{(2)}$ is a sum of all the elements of $\mathbf{A}$. This summation will be carried out in three steps. First, the sum of all the elements of column 0 (part I, $n_A = 0$) represents the *induction energy* associated with forcing a change in the charge distribution of the molecule B by the charge distribution of the isolated ("frozen") molecule A. Second, the sum of all the elements of row 0 (part II, $n_B = 0$) has a similar meaning, but the roles of the molecules are interchanged. Finally, the sum of all the elements of the "interior" of the matrix (part III, n_A and n_B not equal to zero) represents the *dispersion energy*. Therefore,

$$E_0^{(2)} = \underset{\text{I}}{E_{ind}(A \to B)} + \underset{\text{II}}{E_{ind}(B \to A)} + \underset{\text{III}}{E_{disp}}, \tag{13.12}$$

where

$$E_{ind}(A \to B) = \sum_{n_B}{}' \frac{|\langle \psi_{A,0}\psi_{B,n_B}|V\psi_{A,0}\psi_{B,0}\rangle|^2}{(E_{B,0} - E_{B,n_B})},$$

$$E_{ind}(B \to A) = \sum_{n_A}{}' \frac{|\langle \psi_{A,n_A}\psi_{B,0}|V\psi_{A,0}\psi_{B,0}\rangle|^2}{(E_{A,0} - E_{A,n_A})},$$

$$E_{ind} = E_{ind}(A \to B) + E_{ind}(B \to A),$$

$$E_{disp} = \sum_{n_A}{}' \sum_{n_B}{}' \frac{|\langle \psi_{A,n_A}\psi_{B,n_B}|V\psi_{A,0}\psi_{B,0}\rangle|^2}{(E_{A,0} - E_{A,n_A}) + (E_{B,0} - E_{B,n_B})}. \tag{13.13}$$

The electrostatic and the induction interactions are visualized in Fig. 13.5.

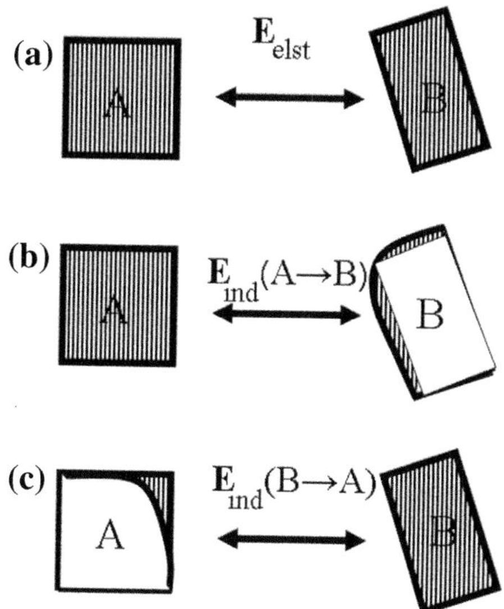

Fig. 13.5. The essence of the electrostatic and of the induction interactions (a schematic visualization). (a) the electrostatic energy ($E_{\text{elst}} \equiv E_0^{(1)} \equiv E_{\text{pol}}^{(1)}$) represents the classical Coulombic interaction of the "frozen" charge distributions of molecule A and of molecule B, the same as those of the isolated molecules. (b) The induction energy consists of two contributions. The first one, $E_{\text{ind}}(A \to B)$, means a modification of the electrostatic energy by allowing a polarization of the molecule B by the frozen (i.e., unperturbed) molecule A. (c) The second contribution to the induction energy, $E_{\text{ind}}(B \to A)$, corresponds to the exchange of the roles of the molecules.

What Do These Formulas Tell Us?

One thing has to be made clear. In Eq. (13.13), we sometimes see arguments for the interacting molecules undergoing excitations. We have to recall, however, that we are interested in the ground state of the total system the whole time, and we calculate its energy and wave function. The excited state wave functions appearing in the formulas are only a consequence of the fact that the first-order correction to the wave function is expanded in a complete basis set chosen deliberately as $\{\psi_n^{(0)}\}$. If we took another basis set (e.g., the wave functions of another isoelectronic molecule), we would obtain the same numerical results [although Eq. (13.13) will not hold], but the argument would be removed. From the mathematical point of view, the very essence of perturbation theory is a small deformation of the starting $\psi_0^{(0)}$ function. This tiny deformation is the target of the expansion in the basis set $\{\psi_n^{(0)}\}$. In other words, the perturbation theory is just a cosmetic effect with respect to $\psi_0^{(0)}$: add a small hump here (Fig. 13.6), subtract a small function there, etc. Therefore, the presence of the excited wave functions in the formulae is not an argument for observing some physical excitations. We may say that the mathematical procedure took what we have prepared for it, and we have prepared excited states.

This does not mean that the energy eigenvalues of the molecule have no influence on its induction or dispersion interactions with other molecules.[12] However, this is a different story.

[12] The smaller the gap between the ground and excited states of the molecule, the larger the polarizability; see Chapter 12.

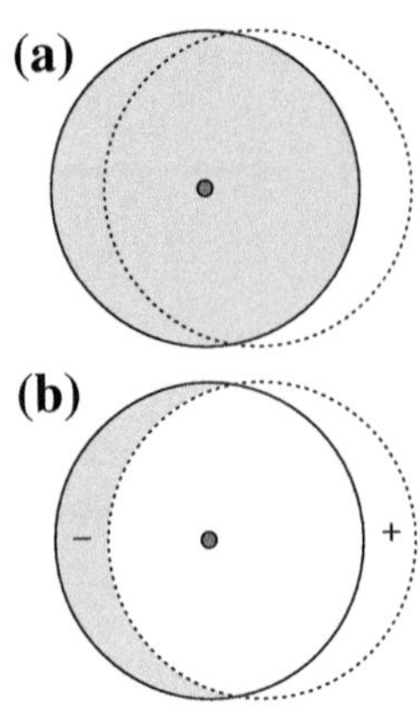

Fig. 13.6. A perturbation of the wave function is a *small* correction. Panel (a) shows in a schematic way how a wave function (solid line) can be transformed into a function that is shifted off the nucleus (dotted line). The function representing the correction added is shown schematically in Panel (b). Please note the function has symmetry of a p orbital.

It has to do with whether the small deformation that we have been talking about depends on the energy eigenvalue spectrum of the individual molecules.

> The denominators in the expressions for the induction and dispersion energies suggest that the lower excitation energies of the molecules, the larger their deformation, induction, and dispersion energy.

13.8.3 Intermolecular Interaction: Physical Interpretation

Now we would like to recommend the reader to study the multipole expansion concept (see Appendix X available at booksite.elsevier.com/978-0-444-59436-5 on p. e169, also cf. Chapter 12, p. 728).

> The essence of the multipole expansion is a replacement of the trouble-making Coulombic interaction of two pointlike particles (one from molecule A, and the other from molecule B) by an infinite sum of easily calculable interactions of what are called *multipoles*. Each interaction term has in the denominator (instead of r_{12}) an integer power of the intermolecular distance (R) between the origins of the two coordinate systems localized in the individual molecules.

In other words, multipole expansion describes the intermolecular interaction of two non-spherically symmetric, distant objects by the "interaction" of deviations (multipoles) from spherical symmetry.

To prepare ourselves for the application of the multipole expansion, let us introduce two Cartesian coordinate systems with x- and y-axes in one system parallel to the corresponding axes in the other system, and with the z-axes collinear (see Fig.X.1 on p. e170). One of the systems is connected to molecule A, the other one to molecule B, and the distance between the origins is R.[13]

The operator V of the interaction energy of two molecules may be written as

$$V = -\sum_j \sum_a \frac{Z_a}{r_{aj}} - \sum_i \sum_b \frac{Z_b}{r_{bi}} + \sum_{ij} \frac{1}{r_{ij}} + \sum_a \sum_b \frac{Z_a Z_b}{R_{ab}}, \tag{13.14}$$

where we have used the convention that the summations over i and a correspond to all electrons and nuclei of molecule A, respectively, and over j and b to molecule B. Since the molecules are assumed to be distant, we have a *practical* guarantee that the interacting particles are distant too. In V, many terms with inverse interparticle distance are present. For any such term, we may write the corresponding multipole expansion (see Appendix X available at booksite.elsevier.com/978-0-444-59436-5, p. e169, $s = \min(k, l)$):

$$-\frac{Z_a}{r_{aj}} = \sum_{k=0} \sum_{l=0} \sum_{m=-s}^{m=s} A_{kl|m|} R^{-(k+l+1)} \hat{M}_A^{(k,m)}(a)^* \hat{M}_B^{(l,m)}(j),$$

$$-\frac{Z_b}{r_{bi}} = \sum_{k=0} \sum_{l=0} \sum_{m=-s}^{m=s} A_{kl|m|} R^{-(k+l+1)} \hat{M}_A^{(k,m)}(i)^* \hat{M}_B^{(l,m)}(b),$$

$$\frac{1}{r_{ij}} = \sum_{k=0} \sum_{l=0} \sum_{m=-s}^{m=s} A_{kl|m|} R^{-(k+l+1)} \hat{M}_A^{(k,m)}(i)^* \hat{M}_B^{(l,m)}(j),$$

$$\frac{Z_a Z_b}{R_{ab}} = \sum_{k=0} \sum_{l=0} \sum_{m=-s}^{m=s} A_{kl|m|} R^{-(k+l+1)} \hat{M}_A^{(k,m)}(a)^* \hat{M}_B^{(l,m)}(b),$$

where

$$A_{kl|m|} = (-1)^{l+m} \frac{(k+l)!}{(k+|m|)!(l+|m|)!}, \tag{13.15}$$

[13] A sufficient condition for the multipole expansion convergence is a separation of the charge distributions of both molecules such that they could be enclosed in two non-penetrating spheres located at the origins of the two coordinate systems. This condition cannot be fulfilled with molecules because their electronic charge density distribution extends to infinity. The consequences of this are described in Appendix X available at booksite.elsevier.com/978-0-444-59436-5. However, the better the sphere condition is fulfilled (by a proper choice of the origins), the more effective are the first terms of the multipole expansion in describing the interaction energy.

The fact that we use closed sets (like the spheres) in the theory witnesses that in the polarization approximation, we are in a "no man's land" between the quantum and classical worlds.

and the multipole moment $M_C^{(k,m)}(n)$ pertains to particle n and is calculated in "its" coordinate system $C \in \{A, B\}$. For example,

$$\hat{M}_A^{(k,m)}(a) = Z_a R_a^k P_k^{|m|}(\cos\theta_a)\exp(im\phi_a), \tag{13.16}$$

where R_a, θ_a, ϕ_a are the polar coordinates of nucleus a (with charge Z_a) of molecule A taken in the coordinate system of molecule A. When all such expansions are inserted into the formula for V, we may perform the following chain of transformations:

$$\begin{aligned}
V &= -\sum_j\sum_a \frac{Z_a}{r_{aj}} - \sum_i\sum_b \frac{Z_b}{r_{bi}} + \sum_{ij}\frac{1}{r_{ij}} + \sum_a\sum_b \frac{Z_a Z_b}{R_{ab}} \\
&\cong \sum_j\sum_a\sum_{k=0}\sum_{l=0}\sum_{m=-s}^{m=s} A_{kl|m|} R^{-(k+l+1)} \hat{M}_A^{(k,m)}(a)^* \hat{M}_B^{(l,m)}(j) \\
&+ \sum_i\sum_b\sum_{k=0}\sum_{l=0}\sum_{m=-s}^{m=s} A_{kl|m|} R^{-(k+l+1)} \hat{M}_A^{(k,m)}(i)^* \hat{M}_B^{(l,m)}(b) \\
&+ \sum_{ij}\sum_{k=0}\sum_{l=0}\sum_{m=-s}^{m=s} A_{kl|m|} R^{-(k+l+1)} \hat{M}_A^{(k,m)}(i)^* \hat{M}_B^{(l,m)}(j) \\
&+ \sum_a\sum_b\sum_{k=0}\sum_{l=0}\sum_{m=-s}^{m=s} A_{kl|m|} R^{-(k+l+1)} \hat{M}_A^{(k,m)}(a)^* \hat{M}_B^{(l,m)}(b) \\
&= \sum_{k=0}\sum_{l=0}\sum_{m=-s}^{m=s} A_{kl|m|} R^{-(k+l+1)} \left\{ \left[\sum_a \hat{M}_A^{(k,m)}(a)\right]^* \left[\sum_j \hat{M}_B^{(l,m)}(j)\right] \right. \\
&+ \left[\sum_i \hat{M}_A^{(k,m)}(i)\right]^* \left[\sum_b \hat{M}_B^{(l,m)}(b)\right] + \left[\sum_i \hat{M}_A^{(k,m)}(i)\right]^* \left[\sum_j \hat{M}_B^{(l,m)}(j)\right] \\
&\left. + \left[\sum_a \hat{M}_A^{(k,m)}(a)\right]^* \left[\sum_b \hat{M}_B^{(l,m)}(b)\right] \right\} \\
&= \sum_{k=0}\sum_{l=0}\sum_{m=-s}^{m=s} A_{kl|m|} R^{-(k+l+1)} \left[\sum_a \hat{M}_A^{(k,m)}(a) \right. \\
&\left. + \sum_i \hat{M}_A^{(k,m)}(i)\right]^* \left[\sum_b \hat{M}_B^{(l,m)}(b) + \sum_j \hat{M}_B^{(l,m)}(j)\right] \\
&= \sum_{k=0}\sum_{l=0}\sum_{m=-s}^{m=s} A_{kl|m|} R^{-(k+l+1)} \hat{M}_A^{(k,m)*} \hat{M}_B^{(l,m)}.
\end{aligned} \tag{13.17}$$

In the final square brackets of Eq. (13.17), we can recognize the multipole moment operators for the molecules calculated in their coordinate systems:

$$\hat{M}_A^{(k,m)} = \sum_a \hat{M}_A^{(k,m)}(a) + \sum_i \hat{M}_A^{(k,m)}(i),$$

$$\hat{M}_B^{(l,m)} = \sum_b \hat{M}_B^{(l,m)}(b) + \sum_j \hat{M}_B^{(l,m)}(j).$$

This takes the form of a single multipole expansion, but this time the multipole moment operators correspond to *entire molecules.*

Using the table of multipoles (p. e174), we may easily write down the multipole operators for the individual molecules. The lowest moment is the net charge (monopole) of the molecules:

$$\hat{M}_A^{(0,0)} = q_A = Z_A - N_A,$$
$$\hat{M}_B^{(0,0)} = q_B = Z_B - N_B,$$

where Z_A is the sum of all the nuclear charges of molecule A, and N_A is its number of electrons (similarly for B). The next moment is $\hat{M}_A^{(1,0)}$, which is a component of the dipole operator equal to

$$\hat{M}_A^{(1,0)} = -\sum_i z_i + \sum_a Z_a z_a, \tag{13.18}$$

where z denotes the z-coordinates of the corresponding particles measured in the coordinate system A (Z denotes the nuclear charge). Similarly, we could very easily write other multipole moments, and the operator V takes the form (see Appendix X available at booksite.elsevier.com/978-0-444-59436-5)

$$V = \frac{q_A q_B}{R} - R^{-2}(q_A \hat{\mu}_{Bz} - q_B \hat{\mu}_{Az}) + R^{-3}(\hat{\mu}_{Ax}\hat{\mu}_{Bx} + \hat{\mu}_{Ay}\hat{\mu}_{By} - 2\hat{\mu}_{Az}\hat{\mu}_{Bz})$$
$$+ R^{-3}(q_A \hat{Q}_{B,z^2} + q_B \hat{Q}_{A,z^2}) + \cdots,$$

where the monopole q_A is the net charge of molecule A,

$$\hat{\mu}_{Ax} = -\sum_i x_i + \sum_a Z_a x_a,$$

$$\hat{Q}_{A,z^2} = -\sum_i \frac{1}{2}\left(3z_i^2 - r_i^2\right) + \sum_a Z_a \frac{1}{2}\left(3z_a^2 - R_a^2\right),$$

and A means that all these moments are measured in coordinate system A. The other quantities have similar definitions and are easy to derive.[14]

13.8.4 Electrostatic Energy in the Multipole Representation Plus the Penetration Energy

Electrostatic energy (p. 807) represents the first-order correction in polarization perturbational theory and is the mean value of V with the product wave function $\psi_0^{(0)} = \psi_{A,0}\psi_{B,0}$. Since now we have the multipole representation of V, we may insert it into Eq. (13.6).

Let us stress for the sake of clarity that V is an *operator* that contains the *operators* of the molecular multipole moments, and that the integration is, as usual, carried out over the x, y, z, σ coordinates of all electrons (the nuclei have positions fixed in space according to the Born–Oppenheimer approximation); i.e., over the coordinates of electrons 1, 2, 3, etc. Since in the polarization approximation, we know perfectly well which electrons belong to molecule A (*"they are painted green"*), and which belong to B (*"they are painted red"*), we perform the integration separately over the electrons of molecule A and those of molecule B. We have a comfortable situation now because every term in V represents a *product* of an operator depending on the coordinates of the electrons belonging to A and of an operator depending on the coordinates of the electrons of molecule B. This (together with the fact that in the integral, we have a *product* of $|\psi_{A,0}|^2$ and $|\psi_{B,0}|^2$) results in a product of two integrals: one over the electronic coordinates of A and the other one over the electronic coordinates of B. This is the reason why we like multipoles so much.

Therefore,

> the expression for $E_0^{(1)} = E_{elst}$ *formally* has to be of exactly the same form as the multipole representation of V, the only difference being that in V, we have the molecular multipole *operators*, whereas in E_{elst}, we have the molecular multipoles themselves as the *mean values* of the corresponding molecular multipole operators in the ground state (the index "0" has been omitted on the right side).

However, the operator V from Eq. (13.14) and the operator in the multipole form [Eq. (13.17)] are equivalent only when the multipole form converges. It does when the interacting objects are non-overlapping, which is not the case. The electronic charge distributions penetrate and this causes a small difference (*penetration energy* E_{penetr}) between the E_{elst} values calculated with and without the multipole expansion. The penetration energy vanishes very fast with

[14] There is one thing that may bother us here–namely, that $\hat{\mu}_{Bz}$ and $\hat{\mu}_{Az}$ appear in the charge-dipole interaction terms with opposite signs, so they are not on equal footing. The reason is that the two coordinate systems are also not on equal footing because the z-coordinate of the coordinate system A points to B, whereas the opposite is not true (see Appendix X available at booksite.elsevier.com/978-0-444-59436-5).

intermolecular distance R (cf. Appendix R available at booksite.elsevier.com/978-0-444-59436-5, p. e137):

$$E_{\text{elst}} = E_{multipol} + E_{penetr}, \tag{13.19}$$

where $E_{multipol}$ contains all the terms of the multipole expansion

$$E_{multipol} = \frac{q_A q_B}{R} - R^{-2}(q_A \mu_{Bz} - q_B \mu_{Az}) + R^{-3}(\mu_{Ax}\mu_{Bx} + \mu_{Ay}\mu_{By} - 2\mu_{Az}\mu_{Bz}) + R^{-3}(q_A Q_{B,z^2} + q_B Q_{A,z^2}) + \cdots$$

The molecular multipoles are

$$q_A = \langle \psi_{A,0} | -\sum_i 1 + \sum_a Z_a | \psi_{A,0} \rangle = \left(-\sum_i 1 + \sum_a Z_a \right) \langle \psi_{A,0} | \psi_{A,0} \rangle = \sum_a Z_a - N_A$$

$= \text{the same as operator } q_A,$

$$\mu_{Ax} = \langle \psi_{A,0} | \hat{\mu}_{Ax} \psi_{A,0} \rangle = \langle \psi_{A,0} | -\sum_i x_i + \sum_a Z_a x_a | \psi_{A,0} \rangle \tag{13.20}$$

and a similar thing happens with the other multipoles.

Since the multipoles in the formula for $E_{multipol}$ pertain to the isolated molecules, we may say that the electrostatic interaction represents the interaction of the permanent multipoles of both molecules.

Dipole-Dipole

The above multipole expansion also represents a useful source for the expressions for particular multipole-multipole interactions.

Let us take as an example the important case of the dipole-dipole interaction.

From the above formulas, the dipole-dipole interaction $E_{dip-dip} = \frac{1}{R^3}(\mu_{Ax}\mu_{Bx} + \mu_{Ay}\mu_{By} - 2\mu_{Az}\mu_{Bz})$ reads also as

$$E_{dip-dip} = \frac{1}{R^3}\left(\boldsymbol{\mu}_{A\perp} \cdot \boldsymbol{\mu}_{B\perp} - 2\boldsymbol{\mu}_{A\parallel} \cdot \boldsymbol{\mu}_{B\parallel}\right), \tag{13.21}$$

where $\perp$ and $\parallel$ mean that we are dealing with the vector components perpendicular to the axis connecting the two point dipoles and parallel to this axis, respectively.

This is a short, easy-to-memorize formula, and we might be completely satisfied in using it, *provided that we always remember the particular coordinate system used for its derivation.*

Taking into account our coordinate system, the vector pointing the coordinate system origin a from b is $\boldsymbol{R} = (0, 0, R)$. Then we can express $E_{dip-dip}$ in a very useful form *independent of any choice of coordinate system* (cf., e.g., pp. 147, 764):

Dipole-dipole Interaction:

$$E_{dip-dip} = \frac{\boldsymbol{\mu}_A \cdot \boldsymbol{\mu}_B}{R^3} - 3\frac{(\boldsymbol{\mu}_A \cdot \boldsymbol{R})(\boldsymbol{\mu}_B \cdot \boldsymbol{R})}{R^5}. \tag{13.22}$$

This form of the dipole-dipole interaction was used in Chapters 3 and 12.

It is important to understand the energies of the ubiquitous dipole-dipole interaction. It is sufficient to consider the two dipoles having orientation within a plane formed by $\boldsymbol{\mu}_A$ and the vector $\boldsymbol{R}$, connecting both point dipoles. If the dipole moments and the vector $\boldsymbol{R}$ do not form the common plane, we represent one of the dipole moments (say, $\boldsymbol{\mu}_B$) as a sum of two components: one forming the common plane with $\boldsymbol{\mu}_A$ and $\boldsymbol{R}$, and the second one orthogonal to this plane: $\boldsymbol{\mu}_B = \boldsymbol{\mu}_{B\parallel} + \boldsymbol{\mu}_{B\perp}$. The dipole-dipole interaction is a sum of the dipole-dipole interactions with these two components, but the interaction of $\boldsymbol{\mu}_A$ with $\boldsymbol{\mu}_{B\perp}$ is zero. Indeed, we have $E_{dip-dip} = \frac{\boldsymbol{\mu}_A \cdot (\boldsymbol{\mu}_{B\parallel}+\boldsymbol{\mu}_{B\perp})}{R^3} - 3\frac{(\boldsymbol{\mu}_A \cdot \boldsymbol{R})(\boldsymbol{\mu}_{B\parallel}+\boldsymbol{\mu}_{B\perp})\cdot \boldsymbol{R}}{R^5} = \frac{\boldsymbol{\mu}_A \cdot \boldsymbol{\mu}_{B\perp}}{R^3} + \frac{\boldsymbol{\mu}_A \cdot \boldsymbol{\mu}_{B\parallel}}{R^3} - 3\frac{(\boldsymbol{\mu}_A \cdot \boldsymbol{R})(\boldsymbol{\mu}_{B\perp} \cdot \boldsymbol{R})}{R^5} - 3\frac{(\boldsymbol{\mu}_A \cdot \boldsymbol{R})(\boldsymbol{\mu}_{B\parallel} \cdot \boldsymbol{R})}{R^5} = 0 + \frac{\boldsymbol{\mu}_A \cdot \boldsymbol{\mu}_{B\parallel}}{R^3} + 0 - 3\frac{(\boldsymbol{\mu}_A \cdot \boldsymbol{R})(\boldsymbol{\mu}_{B\parallel} \cdot \boldsymbol{R})}{R^5} = \frac{\boldsymbol{\mu}_A \cdot \boldsymbol{\mu}_{B\parallel}}{R^3} - 3\frac{(\boldsymbol{\mu}_A \cdot \boldsymbol{R})(\boldsymbol{\mu}_{B\parallel} \cdot \boldsymbol{R})}{R^5}$.

Therefore,

as to the dipole-dipole interaction, whatever is nonzero pertains to the situation where the two dipole moments and $\boldsymbol{R}$ form the common plane.

Hence, the orientation of these dipole moments with respect to an axis (chosen as $\boldsymbol{R}$) can be characterized by the angles θ_A and θ_B, which the vectors form with the axis. The dipole-dipole formula [Eq. (13.21)] gives the following in this case:

$$E_{dip-dip} = \frac{\mu_A \mu_B}{R^3}\left(\sin\theta_A \sin\theta_B - 2\cos\theta_A \cos\theta_B\right). \tag{13.23}$$

Fig. 13.7 shows function $E_{dip-dip}\left(\theta_A, \theta_B\right)$. As one can see, the strongest dipole-dipole attraction corresponds to the two dipoles aligned like this $\rightarrow\rightarrow$ ($\theta_A = \theta_B = 0$) or $\leftarrow\leftarrow$($\theta_A = \theta_B = 180^0$). The worst energetically are the orientations $\rightarrow\leftarrow$($\theta_A = 0, \theta_B = 180^0$) and $\leftarrow\rightarrow$ $\theta_A = 180^0, \theta_B = 0$). The lateral interaction corresponding to the antiparallel configurations $\uparrow\downarrow$($\theta_A = -\theta_B = 90^0$) or $\downarrow\uparrow$($\theta_A = -\theta_B = -90^0$) also corresponds to attraction, but it does

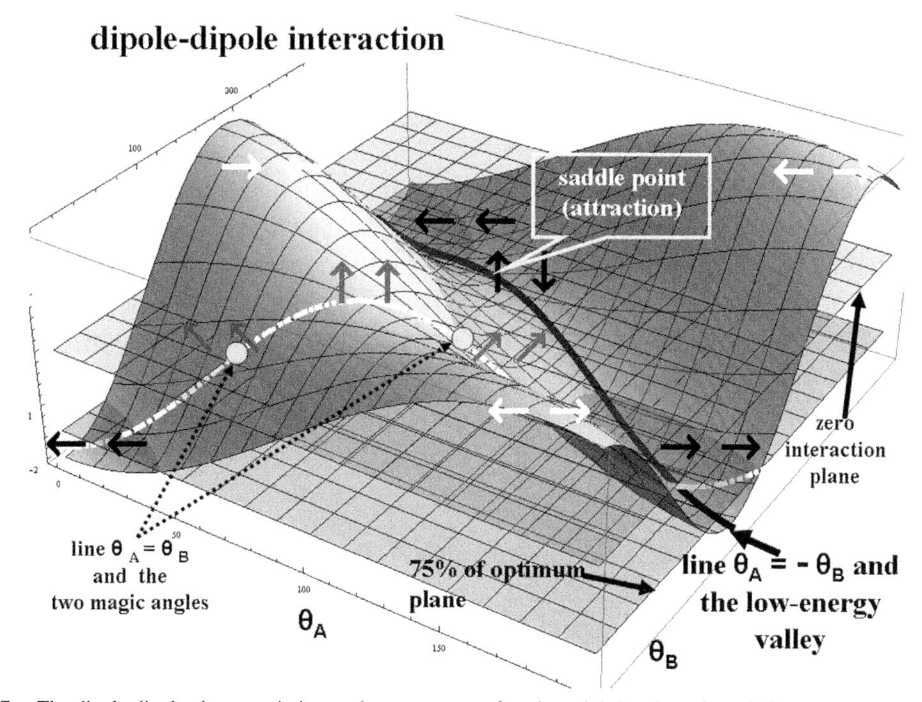

Fig. 13.7. The dipole-dipole electrostatic interaction energy as a function of their orientation within a plane (for convenience in $\frac{\mu_A\mu_B}{R^3}$ energy units, R stands for the distance between these two pointlike dipoles). The angles θ_A and θ_B measure the angular deviation of the dipoles from the connecting axis. The upper plane shown corresponds to the interaction energy equal to 0. The lowest energy is -2 and corresponds to two collinear parallel orientations (shown by the arrows $\longrightarrow \ \longrightarrow$ and $\longleftarrow \ \longleftarrow$). When moving along the valley (the central part of the figure, solid black line), one can pass from one to the other of these configurations (via the configuration $\uparrow\downarrow$ of energy -1), all the time having the dipole-dipole attraction. Therefore, if some other interactions (such as a steric hindrance) tried to destabilize the optimum dipole-dipole collinear configuration, an interesting low-energy compromise would be possible (see the lower plane)–just rotate the two dipoles starting from the configuration $\longrightarrow \ \longrightarrow$ toward the antiparallel configuration $\uparrow\downarrow$ (i.e., putting $\theta_A = -\theta_B \neq 0$ instead of $\theta_A = \theta_B = 0$ and arriving at configuration $\nearrow \ \searrow$). In the two points labeled by the white circle (the white line shown corresponds to $\theta_A = \theta_B$), both dipoles form with the axis what is known as the magic angle θ_{mag} (the corresponding configurations are $\nwarrow\nwarrow$ and $\nearrow\nearrow$), and their interaction vanishes. The energy maxima (equal to $+2$) correspond to the configurations labeled $\longrightarrow \ \longleftarrow$ and $\longleftarrow \ \longrightarrow$.

not represent any minimum; these are two saddle points. The lateral parallel configurations $\uparrow\uparrow$ or $\downarrow\downarrow$ mean repulsion of the absolute value opposite to the $\uparrow\downarrow$ attraction.

There is a lot of configurations giving zero dipole-dipole interaction [shown in Fig.13.7 as intersection of the $E_{dip-dip}(\theta_A, \theta_B)$ function with the upper plane]. However, if one forces the dipoles to be parallel ($\theta_A = \theta_B = \theta$), the zero energy will correspond only to the particular θ angle known as magic angle[15] θ_{mag} corresponding to the $\nearrow\nearrow$ (or $\nwarrow\nwarrow$) configuration.

[15] This angle plays an important role in the solid-state nuclear magnetic resonance measurements. We have $E_{dip-dip} = \frac{\mu_A\mu_B}{R^3}\left(\sin^2\theta_{\text{mag}} - 2\cos^2\theta_{\text{mag}}\right) = \frac{\mu_A\mu_B}{R^3}\left(1 - 3\cos^2\theta_{\text{mag}}\right) = 0$; from this, we get $\theta_{\text{mag}} = \arccos\frac{1}{\sqrt{3}} = 54.74^0$.

In the central part of Fig.13.7 is a deep valley connecting the most important configurations of two dipoles - two global minima with energy equal to -2 (for the configurations $\rightarrow\rightarrow$ and $\leftarrow\leftarrow$) separated by the saddle point corresponding to $\uparrow\downarrow$ or $\downarrow\uparrow$.

> If interactions other than dipole-dipole tried to distort the parallel collinear configuration of the interacting dipoles, it could be done without increasing the dipole-dipole electrostatic energy too much, provided that the distortion would correspond to the opposite angular deviations of both dipoles with respect to the **R** axis (i.e., the dipoles would be tilted in opposite directions, with the dipole orientations moved toward the saddle point configuration $\uparrow\downarrow$; see Fig. 13.7). In such a way, one gets an easy-to-achieve compromise, which is common in experimental dipole-dipole configurations: $\nearrow\ \searrow$.

Is the Electrostatic Interaction Important?

Electrostatic interaction can be attractive or repulsive. For example, in the electrostatic interaction Na^+ i Cl^-, the main role will be played by the charge-charge interaction, which is negative and therefore represents attraction, while for $Na^+\ldots Na^+$, the electrostatic energy will be positive (repulsion). For neutral molecules, the electrostatic interaction may depend on their *orientation* to such an extent that the sign may change. This is an exceptional feature that is peculiar only to electrostatic interaction.

When the distance R is small when compared to the size of the interacting subsystems, multipole expansion gives bad results. To overcome this, the total charge distribution may be divided into *atomic* segments (see Appendix S available at booksite.elsevier.com/978-0-444-59436-5). Each atom would carry its charge and other multipoles, and the electrostatic energy would be the sum of the atom-atom contributions, any of which would represent a series similar[16] to $E_0^{(1)}$.

Example:

Let us calculate (by using the Hartree-Fock method) the rigid interaction energy of two hydrogen fluoride molecules at the fixed F...F distance equal to 5 Å (each molecule has the optimum length equal to 0.911 Å). We obtain the following results[17] [in the last column, the dipole-dipole interaction energy computed from Eq. (13.23) is displayed, all energies in a.u.]:

Configuration	E_{int}	$E_{dip-dip}$
$\rightarrow\rightarrow$ collinear parallel	−0,00143	−0,00143
$\rightarrow\leftarrow$ collinear antiparallel	+0,00320	+0,00143
$\uparrow\uparrow$ parallel dipoles	+0,00071	+0,00071
$\uparrow\downarrow$ antiparallel dipoles	−0,00071	−0,00071

[16] A.J. Stone, *Chem.Phys.Lett.*, *83*, 233 (1981); A.J. Stone and M. Alderton, *Mol.Phys.*, *56*, 1047 (1985); W.A. Sokalski and R. Poirier, *Chem.Phys.Lett. 98*, 86 (1983); W.A. Sokalski and A. Sawaryn, *J.Chem.Phys.*, *87*, 526 (1987).

[17] We take the standard basis set 6-31G(d), for which the $H-F$ bond length of the individual molecule is optimal. The results are not corrected for the BSSE.

It is seen that, except for the collinear antiparallel case, the electrostatic dipole-dipole interaction dominates the other interactions contributing to the Hartree-Fock result. A lot of molecular integrals, so many terms to calculate, the SCF iterative method included, but the situation is correctly described by a primitive dipole-dipole term. The collinear antiparallel configuration turned out to be too difficult a case here, the discrepancy is large. This is because, even if the intermolecular distance $F \ldots F$ is kept constant, in reality the molecules are too close in this particular case; e.g., the $H \ldots H$ distance is about 3 Å only. In such a situation, the game involves not only the interaction of the point dipoles, but also other interactions like quadrupole-dipole, quadrupole-quadrupole and other electrostatic contributions, and non-electrostatic interactions (like e.g., the valence repulsion, induction, etc.). Therefore, one may expect this trouble.

Example: Dipole-Internal Field Interactions in Proteins

In the above example, all of the dipoles corresponded to the entire interacting molecule. One may consider, however, the molecular dipole moment as a sum of the dipole moments distributed on all atoms and bonds of the molecule [see Eq. (12.37) on p. 742], and then to calculate the *intermolecular* interaction as a sum of the interactions of such distributed dipoles. This may suggest that the same concept may be applied to the *intramolecular* electrostatic dipole-dipole interactions.

Here, we give an example of this idea. The native conformation is important, because many proteins assure their biological activity only in such a conformation. It is currently assumed that the native conformation corresponds to the lowest-energy conformational state [i.e., to the global minimum of energy (cf., p. 353)]. A hypothesis has been presented[18] that at the energy minimum, each peptide bond (-HN-CO-) of the molecule is oriented quite well along the local electric field created by the rest of the molecule.[19] The hypothesis has been tested considering a polyalanine oligomer in the α-helical conformation. It turned out that indeed, except of the very ends of the α-helix, the deviation from the alignment was of the order of a few degrees only. Later on extensive investigations of the intramolecular electric field in the native conformations of proteins have been performed.[20] The conclusion was similar: the peptide unit dipoles tend to align along the local electric field.

Reality or Fantasy?

In principle, this part of the discussion (about electrostatic interactions) may be considered completed. I am tempted, however, to enter some "obvious" subjects, which will turn out to lead us far from the usual track of intermolecular interactions.

[18] L. Piela and H.A. Scheraga, *Biopolymers, 26, S*33 (1987).

[19] Each atom of the molecule had an assigned electric charge according to a force field used.

[20] D. Ripoll, J.A. Vila, and H.A. Scheraga, *Proc.Natl Acad.Sci, USA, 102*, 7559 (2005).

Let us consider the Coulomb interaction of two point charges q_1 on molecule A and q_2 on molecule B, both charges separated by distance r:

$$E_{elst} = \frac{q_1 q_2}{r}. \tag{13.24}$$

This is an outstanding formula for the following reasons:

- First, we have the amazing power of r with the *exact* value -1.
- Second, change of a charge *sign* does not make any profound changes in the formula, except the change of *sign* of the interaction energy.
- Third, the formula is bound to be false (it has to be only an approximation), since instantaneous interaction is assumed, whereas the interaction has to have time to travel between the interacting objects and during that time, the objects change their distance (see Chapter 3, p. 146).

From these remarks follow some apparently obvious observations, that E_{elst} is invariant with respect to the following operations:

- $q_1' = -q_1 \quad q_2' = -q_2$ (charge conjugation, Chapter 2, 2.1.8)
  - $q_1' = q_2 \quad q_2' = q_1$ (*exchange* of charge positions)
  - $q_1' = -q_2 \quad q_2' = -q_1$ (charge conjugation *and* exchange of charge positions)

These invariance relations, when treated literally and rigorously, are not of particular usefulness in theoretical chemistry. They may, however, open new possibilities when considered as some limiting cases. Chemical reaction mechanisms very often involve the interaction of molecular ions. Suppose that we have a particular reaction mechanism. Now, let us make the charge conjugation of all the objects involved in the reaction (this would require the change of matter to antimatter). This will preserve the reaction mechanism. We cannot do such changes in chemistry. However, we may think of some *other molecular systems*, which have similar geometry but opposite overall charge pattern (a "*counter pattern*"). The new reaction has a chance to run in a similar direction as before. This concept is parallel to the idea of *Umpolung* (considering inversing of polarity of the species involved in reaction) functioning in organic chemistry. It seems that nobody has looked from that point of view at all known reaction mechanisms.[21]

13.8.5 Induction Energy in the Multipole Representation

The induction energy contribution consists of two parts: $E_{ind}(A \rightarrow B)$ and $E_{ind}(B \rightarrow A)$ or, respectively, the polarization energy of molecule B in the electric field of the unperturbed molecule A and *vice versa*.

[21] The author is aware of only a single example of such a pair of counter patterns: the Friedl-Crafts reaction and what is called the *vicarious nucleophilic substitution*, discovered by Mieczysław Mąkosza [M. Mąkosza and A. Kwast, *J.Phys.Org.Chem.*, *11*, 341(1998)].

The goal of the present section is to take apart the induction mechanism by showing its multipole components. If we insert the multipole representation of V into the induction energy $E_{\text{ind}}(A \to B)$, then

$$E_{\text{ind}}(A \to B) = \sum_{n_B}{}' \frac{|\langle \psi_{A,0}\psi_{B,n_B}|V\psi_{A,0}\psi_{B,0}\rangle|^2}{E_{B,0}-E_{B,n_B}} = \sum_{n_B}{}' \frac{1}{E_{B,0}-E_{B,n_B}}$$

$$\{|R^{-1}q_A \cdot 0 - R^{-2}q_A\langle\psi_{B,n_B}|\hat{\mu}_{Bz}\psi_{B,0}\rangle + R^{-2}\cdot 0$$

$$+R^{-3}[\mu_{Ax}\langle\psi_{B,n_B}|\hat{\mu}_{Bx}\psi_{B,0}\rangle + \mu_{Ay}\langle\psi_{B,n_B}|\hat{\mu}_{By}\psi_{B,0}\rangle$$

$$-2\mu_{Az}\langle\psi_{B,n_B}|\hat{\mu}_{Bz}\psi_{B,0}\rangle] + \cdots|\}^2$$

$$= \sum_{n_B}{}' \frac{1}{E_{B,0}-E_{B,n_B}} \{| - R^{-2}q_A\langle\psi_{B,n_B}|\hat{\mu}_{Bz}\psi_{B,0}\rangle +$$

$$R^{-3}[\mu_{Ax}\langle\psi_{B,n_B}|\hat{\mu}_{Bx}\psi_{B,0}\rangle + \mu_{Ay}\langle\psi_{B,n_B}|\hat{\mu}_{By}\psi_{B,0}\rangle$$

$$-2\mu_{Az}\langle\psi_{B,n_B}|\hat{\mu}_{Bz}\psi_{B,0}\rangle] + \cdots|\}^2$$

$$= -\frac{1}{2}\frac{1}{R^4}q_A^2\alpha_{B,zz} + \cdots,$$

where the following is true:

- The zeros appearing in the first part of the derivation come from the orthogonality of the eigenstates of the isolated molecule B.
- The symbol "+..." stands for higher powers of R^{-1}.
- $\alpha_{B,zz}$ represents the zz component of the dipole polarizability tensor of the molecule B, which absorbed the summation over the excited states of B according to definition 12.42.

A Molecule in the Electric Field of Another Molecule

Note that $\frac{1}{R^4}q_A^2$ represents the square of the electric field intensity $\mathcal{E}_z(A \to B) = \frac{q_A}{R^2}$ measured on molecule B and created by the net charge of molecule A. Therefore, we have $E_{\text{ind}}(A \to B) = -\frac{1}{2}\alpha_{A,zz}\, \mathcal{E}_z^2(A \to B) + \cdots$ according to formula 12.24 describing the molecule in an electric field. For molecule B, its partner, molecule A (and *vice versa*...), represents an external world creating the electric field, and molecule B has to behave as described in Chapter 12. The net charge of A created the electric field $\mathcal{E}_z(A \to B)$ on molecule B, which, as a consequence, induced on B a dipole moment $\mu_{B,\text{ind}} = \alpha_{B,zz}\, \mathcal{E}_z(A \to B)$ according to formula 12.19. This is associated with the interaction energy term $\frac{1}{2}\alpha_{B,zz}\, \mathcal{E}_z^2(A \to B)$, see Eq. (12.24).

There is, however, a small problem. Why is the induced moment proportional only to the net charge of molecule A? This would be absurd. Molecule B does not know anything about multipoles of molecule A, it only knows about the *local* electric field that acts on it and has to react to that field by a suitable polarization. Everything is all right, though. The rest of the problem is in the formula for $E_{\text{ind}}(A \to B)$. So far, we have analyzed the electric field on B

coming from the net charge of A, but the other terms of the formula will give contributions to the electric field coming from *all other* multipole moments of A. Then, the response of B will pertain to the total electric field created by "frozen" A on B, as it should be. A similar story applies to $E_{int}(B \to A)$.

13.8.6 Dispersion Energy in the Multipole Representation

After inserting V in the multipole representation into the expression for the dispersion energy, we obtain

$$E_{\text{disp}} = \sum_{n_A}{}' \sum_{n_B}{}' \frac{1}{(E_{A,0} - E_{A,n_A}) + (E_{B,0} - E_{B,n_B})} |R^{-1} q_A q_B \cdot 0 \cdot 0 - R^{-2} q_A \cdot 0 \cdot (\mu_{Bz})_{n_B,0}$$
$$- R^{-2} q_B \cdot 0 \cdot (\mu_{Az})_{n_A,0} + R^{-3} [(\mu_{Ax})_{n_A,0} (\mu_{Bx})_{n_B,0} + (\mu_{Ay})_{n_A,0} (\mu_{By})_{n_B,0}$$
$$- 2 (\mu_{Az})_{n_A,0} (\mu_{Bz})_{n_B,0}] + \cdots|^2$$
$$= \sum_{n_A}{}' \sum_{n_B}{}' \frac{|R^{-3} [(\mu_{Ax})_{n_A,0} (\mu_{Bx})_{n_B,0} + (\mu_{Ay})_{n_A,0} (\mu_{By})_{n_B,0} - 2 (\mu_{Az})_{n_A,0} (\mu_{Bz})_{n_B,0}] + \cdots|^2}{(E_{A,0} - E_{A,n_A}) + (E_{B,0} - E_{B,n_B})},$$

where $(\mu_{Ax})_{n_A,0} = \langle \psi_{A,n_A} | \hat{\mu}_{Ax} \psi_{A,0} \rangle$, $(\mu_{Bx})_{n_B,0} = \langle \psi_{B,n_B} | \hat{\mu}_{Bx} \psi_{B,0} \rangle$ and similarly for the y and z dipole moment components. The zeros in the first part of the equality chain come from the orthogonality of the eigenstates of each of the molecules.

The square in the formula pertains to all terms. The other terms, not shown in the formula, have the powers of R^{-1} be higher than R^{-3}.

Hence, if we squared the total expression in the numerator, the most important term would be the dipole-dipole contribution with the asymptotic R^{-6} distance dependence.

As we can see, Eq. (13.13), its calculation requires *double* electronic excitations (one on the first, the other one on the second interacting molecules), and these already belong to the correlation effect (cf. p. 649).

The dispersion interaction is a pure correlation effect, and therefore, the methods used in a supermolecular approach, which do not take into account the electronic correlation (such as the Hartree-Fock method), are unable to produce any nonzero dispersion contributions.

13.8.7 Dispersion Energy Model–Calculation on Fingers

Where does this physical effect come from?

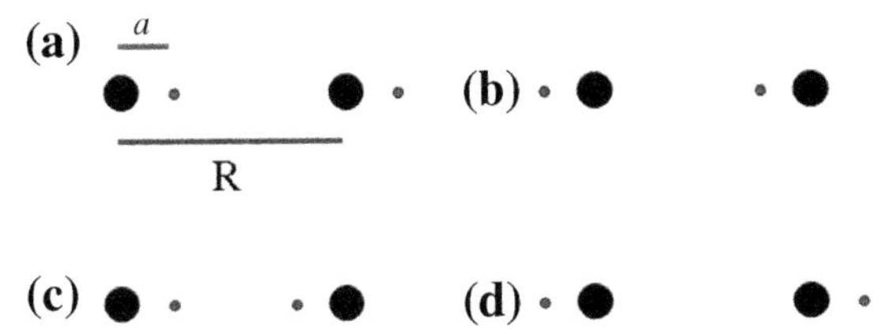

Fig. 13.8. Dispersion energy origin shown using a primitive model of two interacting hydrogen atoms. The nuclei (black circles) occupy some fixed positions in space (at distance R), while each of the electrons (gray) can occupy only two positions: on the left and on the right of the corresponding nucleus (at distance a from it). A popular explanation for the dispersion interaction is that, due to electron repulsion: situations (a) and (b) occur more often than situation (c), and this is why the dispersion interaction represents a net attraction of dipoles. The positions of the electrons that correspond to (a) and (b) represent two favorable instantaneous dipole-instantaneous dipole interaction, while (c) corresponds to a non-favorable instantaneous dipole-instantaneous dipole interaction. A trouble with this explanation is that there is also the possibility of having electrons far apart, as in (d). This most favorable situation (the longest distance between the electrons) means, however, repulsion of the resulting dipoles. It may be shown, though, that the net result (dispersion interaction) is still an attraction, as it should be.

We will try to catch the very essence of the dispersion interaction. As always, we will try to be as simple as possible in order to construct the model, which still contains the very reason that dispersion interactions happen.

- Well, why to complicate things and consider molecules? Do atoms interact *via* dispersion interaction? Yes. Let us take atoms, then.
- Is it essential to have more than two electrons? No.
- Let us take, therefore, two hydrogen atoms, each in its ground state (i.e., $1s$ state), and at a long internuclear distance R.
- Is it of importance for seeing the phenomenon to have the "full size," 3-D hydrogen atoms? No, because the effect comes from the electron correlation and such correlation may happen even for the toy 1-D hydrogen atoms, with the electrons moving along the internuclear axis only.
- Is it essential indeed to have such motion? No, we may still simplify things and give only the possibility of correlating *two positions* for each of the two electrons (along the internuclear axis, so our toy will be not only 1-D, but also "granular"): on the left side and on the right side (see Fig. 13.8), the fixed electron-proton distance being $a \ll R$.

Now, let us calculate the interaction energy of the two "toy hydrogen atoms" at large distances R by using the dipole-dipole interactions for all four possible situations $i = 1, 2, 3, 4$ from Eq. (13.22), assuming the local coordinate systems on the protons. In the total potential energy, there is a common contribution, identical in all the four situations: the interaction within the individual atoms: $-2/a$ [the remainder is the interaction energy $E_{int}(i)$]. The potential energy resulting from the Coulomb interactions of the particles, therefore, is

$$V(i) = -2/a + E_{int}(i),$$

where the first term on the right side is twice (two "hydrogen atoms") the electron-nucleus interaction, and the second one is given as follows:

Situation, i	Fig.13.8	Interaction Energy $E_{int}(i)$
1	a	$-2\frac{\mu^2}{R^3}$
2	b	$-2\frac{\mu^2}{R^3}$
3	c	$+2\frac{\mu^2}{R^3}$
4	d	$+2\frac{\mu^2}{R^3}$

with $\boldsymbol{\mu} = (0, 0, \pm a)$ for electrons 1 and 2 according to Eq. (13.18), and $\mu \equiv a$ in a.u. Note, that if we assume the same probability for each situation, the interatomic interaction energy per situation would be zero; i.e., $\frac{1}{4}\sum_i E_{int}(i) = 0$. These situations, however, have different probabilities (p_i), because the electrons repel each other, and the total energy depends where they actually are. *Note that the probabilities should be different only because of the electron correlation.* If we could guess somehow these probabilities $p_i, i = 1, 2, 3, 4$, then we could calculate the mean interaction energy of our model 1-D atoms as $\bar{E}_{int} = \sum_i p_i E_{int}(i)$. In this way, we could see whether it corresponds to net attraction ($\bar{E}_{int} < 0$) or repulsion ($\bar{E}_{int} > 0$), which is most interesting for us. Well, but how do we calculate them[22]?

We may suspect that for the ground state (we are interested in the ground state of our system), the lower the potential energy $V(i)$, the higher the probability density p_i. This is what happens for the harmonic oscillator, for the Morse oscillator, for the hydrogen-like atom, etc. It looks as a general rule. Is there any tip that could help us work out what such a dependence might be? If you do not know where to begin, then think of the harmonic oscillator model as a starting point! This is what people usually do as a first guess. As seen from Eq. (4.22), the ground-state wave function for the harmonic oscillator may be written as $\psi_0 = A\exp[-BV(x)]$, where $B > 0$ and $V(x)$ stands for the potential energy for the harmonic oscillator. Therefore, the probability density changes as $\rho = A^2\exp[-2BV(x)]$. Interesting...Now let us assume that a similar thing happens[23] for the probabilities p_i of finding the electrons 1 and 2 and they may be reasonably estimated as $p_i = N'A^2\exp[-2BV(i)]$, where $V(i) = -2/a + E_{int}(i)$ plays a role of potential energy. Finally, $p_i = N\exp[-2BE_{int}(i)]$, and $N = 1/\sum_i \exp[-2BE_{int}(i)]$ is the normalization constant assuring that in our model, $\sum_i p_i = 1$. For long distances R [small

[22] In principle, we could look at what people have calculated in the most sophisticated calculations for the hydrogen molecule at a large R, and assign the p_i as the squares of the wave function value for the corresponding four positions of both electrons. Since these wave functions are awfully complex and do not contribute anything qualitatively different, we disregard this path with no regrets.

[23] This is like having the electron attached to the nucleus by a harmonic spring (instead of having Coulombic attraction).

$E_{int}(i)]$, we may expand this expression in a Taylor series and obtain $p_i = \frac{[1-2BE_{int}(i)]}{\sum_j \exp[-2BE_{int}(j)]} \approx \frac{1-2BE_{int}(i)}{\sum_j (1-2BE_{int}(j)+...)} = \frac{1-2BE_{int}(i)}{4-2B\cdot 0+\sum_j \frac{1}{2}[2BE_{int}(j)]^2+...} \approx \frac{1}{4} - \frac{B}{2}E_{int}(i)$, where the Taylor series has been truncated to the accuracy of the linear terms in the interaction. Then, the mean interaction energy is

$$\bar{E}_{int} = \sum_i p_i E_{int}(i) \approx \sum_i \left[\frac{1}{4} - \frac{B}{2}E_{int}(i)\right] E_{int}(i)$$

$$= \frac{1}{4}\sum_i E_{int}(i) - \frac{B}{2}\sum_i [E_{int}(i)]^2 = 0 - \frac{B}{2}\frac{16\mu^4}{R^6} = -8B\frac{\mu^4}{R^6} < 0.$$

The approximations that we have made were extremely crude, but despite this, we were able to grasp three important features of the correct dispersion energy: that it corresponds to attractive interaction, that it vanishes with distance as R^{-6}, and that it is proportional to the fourth power of the instantaneous dipole moment. What really contributed to the success of the above reasoning is the dipole-dipole character of the interaction,[a] not the use of a harmonic model.

[a] In addition to the rule: the lower the potential energy $V(x)$, the higher the value of the ground-state wave function $\psi(x)$ and the higher the corresponding probability density.

Examples:

The electrostatic interaction energy of two molecules can be calculated from Eq. (13.6). However, it is very important for a chemist to be able to predict the main features of the electrostatic interaction *without any calculation* at all, based on some general rules. This will create chemical intuition or chemical common sense so important in planning, performing, and understanding experiments.

How do we recognize that a particular multipole-multipole interaction represents attraction or repulsion? First, we replace the molecules by their lowest nonzero multipoles represented by point charges; e.g., ions by + or −, dipolar molecules by +−, quadrupoles by $+\overset{-}{-}+$, etc. In order to do this, we have to know which atoms are electronegative and which electropositive.[24] After doing this, we replace the two molecules by the multipoles. If the nearest-neighbor charges in the two multipoles have opposite signs, the multipoles attract each other; otherwise, they repel. (Fig. 13.9).

[24] This is common knowledge in chemistry and is derived from experiments as well as from quantum mechanical calculations. The later provide the partial atomic charges from what is called population analysis (see Appendix S available at booksite.elsevier.com/978-0-444-59436-5). Despite its non-uniqueness, it would satisfy our needs. A unique and elegant method of calculation of atomic partial charges is related to the Bader analysis described on p. 669.

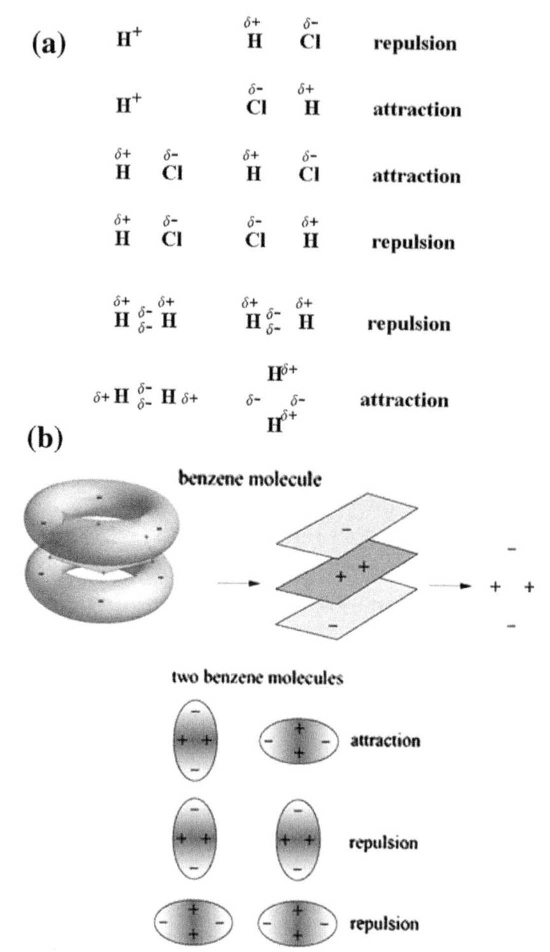

Fig. 13.9. For sufficiently large intermolecular separations, the interaction of the lowest non-vanishing multipoles dominates. Whether this is an attraction or repulsion can be recognized by representing the molecular charge distributions by non-pointlike multipoles (clusters of point charges). If such multipoles point to each other by point charges of the opposite (same) sign, then the electrostatic interaction of the molecules is attraction (repulsion). (a) A few examples of simple molecules and the atomic partial charges; (b) even the interaction of the two benzene molecules obeys this rule: in the face-to-face configuration they repel, while they attract each other in the perpendicular configuration.

The data of Table 13.1 were obtained assuming a long intermolecular distance and molecular orientations as shown in the table.

In composing Table 13.1, some helpful information has been used:

- *Induction and dispersion energies always represent attraction, except in some special cases when they are zero.* These special cases are obvious; e.g., it is impossible to induce some changes on molecule B if molecule A does not have any nonzero permanent multipoles. Also, the dispersion energy is zero if an interacting subsystem has no electrons.

Table 13.1. The asymptotic interaction energy (proportional to R^{-n}, the table gives the exponent n) of two molecules in their electronic ground states. For each pair of molecules or atoms, a short characteristic of their electrostatic, induction, and dispersion interactions is given. It consists of n and the sign (in parentheses) of the corresponding interaction type: the " $-$ " sign means attraction, the " $+$ " sign means repulsion, and 0 corresponds to the absence of such an interaction.

System	Electrostatic	Induction	Dispersion
He ... He	0	0	6(−)
He ··· H^+	0	4(−)	0
He ··· HCl	0	6(−)	6(−)
H^+ ··· HCl	2(+)	4(−)	0
HCl ··· ClH	3(+)	6(−)	6(−)
HCl ··· HCl	3(−)	6(−)	6(−)
H–H ··· He	0	8(−)	6(−)
H–H ··· H–H	5(+)	8(−)	6(−)
$^{H}_{H}$ ··· H-H	5(−)	8(−)	6(−)
$^{H}_{H}$O ··· H O_H	3(−)	6(−)	6(−)
$^{H}_{H}$O ··· H O$^{H}_{H}$	3(+)	6(−)	6(−)

- *Electrostatic energy is nonzero, if both interacting molecules have some nonzero permanent multipoles.*
- *Electrostatic energy is negative (positive), if the lowest non-vanishing multipoles of the interacting partners attract (repel) themselves.*[25]
- The dispersion energy always decays as R^{-6}
- The electrostatic energy vanishes as $R^{-(k+l+1)}$, where the 2^k–pole and 2^l–pole represent the lowest non-vanishing multipoles of the interacting subsystems
- The induction energy vanishes as $R^{-2(k+2)}$, where the 2^k–pole is the lower of the two lowest nonzero permanent multipoles of the molecules A and B. The formula is easy to understand if we take into account that the lowest *induced* multipole is always a dipole ($l = 1$), and that the induction effect is of the second order (hence 2 in the exponent).

13.9 Symmetry Adapted Perturbation Theories (SAPT)

The SAPT approach applies to intermediate intermolecular separations, when the electron clouds of both molecules overlap to such an extent, that the following occurs:

- The polarization approximation [i.e., ignoring the Pauli principle (p. 805)], becomes a very poor approximation.
- The multipole expansion becomes invalid.

[25] This statement is true for sufficiently long distances.

13.9.1 Polarization Approximation Is Illegal

The polarization approximation zero-order wave function $\psi_{A,0}\psi_{B,0}$ will be deprived of the privilege of being the unperturbed function $\psi_0^{(0)}$ in a perturbation theory. Since it will still play an important role in the theory, let us denote it by $\varphi^{(0)} = \psi_{A,0}\psi_{B,0}$.

The polarization approximation seems to have (at first glimpse) a very strong foundation because at long intermolecular distances R, the zero-order *energy* is close to the exact one. The trouble is, however, that a similar statement is not true for the zero-order *wave function* $\varphi^{(0)}$ and for the exact wave function at any intermolecular distance (even at infinity).

Let us take an example of two ground-state hydrogen atoms. The polarization approximation zero-order wave function

$$\varphi^{(0)}(1,2) = 1s_a(1)\alpha(1)\, 1s_b(2)\beta(2), \tag{13.25}$$

where the spin functions have been introduced (the Pauli principle is ignored.[26])

This function is neither symmetric (since $\varphi^{(0)}(1,2) \neq \varphi^{(0)}(2,1)$) nor antisymmetric (since $\varphi^{(0)}(1,2) \neq -\varphi^{(0)}(2,1)$), and therefore is "*illegal*" and in principle not acceptable.

13.9.2 Constructing a Symmetry Adapted Function

In the Born-Oppenheimer approximation, the electronic ground-state wave function of H_2 has to be the eigenfunction of the nuclear inversion symmetry operator $\hat{I}$ interchanging nuclei a and b (see Appendix C available at booksite.elsevier.com/978-0-444-59436-5). Since $\hat{I}^2 = 1$, the eigenvalues can be either -1 (called u symmetry) or $+1$ (g symmetry).[27] The ground state is of g symmetry, so the projection operator $\frac{1}{2}(1+\hat{I})$ will take care of that (it says: make a 50-50 combination of a function and its counterpart coming from the exchange of nuclei a and b).[28] On top of this, the wave function has to fulfill the Pauli exclusion principle, which we will ensure by the antisymmetrizer $\hat{A}$ (cf., p. e109), which, when acting, gives either an antisymmetric function or zero. Altogether, the proper symmetry will be ensured by projecting $\varphi^{(0)}$ using the idempotent projection operator:

$$\hat{\mathcal{A}} = \frac{1}{2}(1+\hat{I})\hat{A}. \tag{13.26}$$

[26] This is the essence of the polarization approximation.

[27] The symbols come from German: g stands for *gerade* (even) and u stands for *ungerade* (odd).

[28] We ignore the proton spins here.

We obtain as a projection $\hat{\mathcal{A}}\varphi^{(0)}$ of $\varphi^{(0)}$ (in case of no g or u symmetry we will use $\hat{\mathcal{A}} \equiv \hat{A}$):

$$\begin{aligned}
\frac{1}{2}(1+\hat{I})\hat{A}\varphi^{(0)} &= \frac{1}{2!}\frac{1}{2}(1+\hat{I})\sum_P (-1)^p \hat{P}[1s_a(1)\alpha(1)1s_b(2)\beta(2)] \\
&= \frac{1}{4}(1+\hat{I})[1s_a(1)\alpha(1)1s_b(2)\beta(2) - 1s_a(2)\alpha(2)1s_b(1)\beta(1)] \\
&= \frac{1}{4}[1s_a(1)\alpha(1)1s_b(2)\beta(2) - 1s_a(2)\alpha(2)1s_b(1)\beta(1) + 1s_b(1)\alpha(1)1s_a(2)\beta(2) \\
&\quad - 1s_b(2)\alpha(2)1s_a(1)\beta(1)] \\
&= \frac{1}{2\sqrt{2}}[1s_a(1)1s_b(2) + 1s_a(2)1s_b(1)]\left\{\frac{1}{\sqrt{2}}[\alpha(1)\beta(2) - \alpha(2)\beta(1)]\right\}.
\end{aligned}$$

This result is proportional to the Heitler-London wave function from p. 611, where its important role in chemistry was highlighted:

$$\psi_{HL} \equiv \psi_0^{(0)} = N[1s_a(1)1s_b(2) + 1s_a(2)1s_b(1)]\left\{\frac{1}{\sqrt{2}}[\alpha(1)\beta(2) - \alpha(2)\beta(1)]\right\}, \tag{13.27}$$

where N is the normalization constant.

The function has the same symmetry as the exact solution to the Schrödinger equation (antisymmetric with respect to the exchange of electrons and symmetric with respect to the exchange of protons). It is easy to calculate[29] that normalization of $\psi^{(0)}$ means $N = 2[(1+S^2)]^{-1/2}$, where S stands for the overlap integral of the atomic orbitals $1s_a$ and $1s_b$.

13.9.3 The Perturbation Is Always Large in Polarization Approximation

Let us check (see Appendix B available at booksite.elsevier.com/978-0-444-59436-5) how distant functions $\varphi^{(0)}$ and $\psi^{(0)}$ are in the Hilbert space (they are both normalized; i.e., they are unit vectors in the Hilbert space). We will calculate the norm of difference $\varphi^{(0)} - \psi_0^{(0)}$. If the

[29] $\int |\psi_{HL}|^2 d\tau_1 d\tau_2 = |N|^2(2+2S^2)\left\{\sum_{\sigma_1}\sum_{\sigma_2}\frac{1}{2}\left[\alpha(1)\beta(2) - \alpha(2)\beta(1)\right]^2\right\} = |N|^2 2(1+S^2)\frac{1}{2}(1+1-2\cdot 0) = 1$

Hence, $N = \frac{1}{\sqrt{2(1+S^2)}}$. Shortly, we will need function $\psi_0^{(0)}$ with the intermediate normalization with respect to $\varphi^{(0)}$; i.e., satisfying $\langle\psi_0^{(0)}|\varphi^{(0)}\rangle = 1$ instead of $\langle\psi_0^{(0)}|\psi_0^{(0)}\rangle = 1$. Then N will be different, and equal to $\langle\varphi^{(0)}|\hat{\mathcal{A}}\varphi^{(0)}\rangle^{-1}$.

norm were small, then the two functions would be close in the Hilbert space. Let us see:

$$||\varphi^{(0)} - \psi_0^{(0)}|| \equiv \left[\int \left(\varphi^{(0)} - \psi_0^{(0)}\right)^* \left(\varphi^{(0)} - \psi_0^{(0)}\right) \mathrm{d}\tau\right]^{\frac{1}{2}}$$

$$= \left[1 + 1 - 2\int \varphi^{(0)}\psi_0^{(0)} \mathrm{d}\tau\right]^{\frac{1}{2}} = \left\{2 - 2\int \left[1s_a(1)\alpha(1)1s_b(2)\beta(2)\right]\right.$$

$$\left. N\left[1s_a(1)1s_b(2) + 1s_a(2)1s_b(1)\right]\left\{\frac{1}{\sqrt{2}}\left[\alpha(1)\beta(2) - \alpha(2)\beta(1)\right]\right\} \mathrm{d}\tau\right\}^{\frac{1}{2}}$$

$$= \left\{2 - N\sqrt{2}\int \left[1s_a(1)1s_b(2)\right]\left[1s_a(1)1s_b(2) + 1s_a(2)1s_b(1)\right] \mathrm{d}\mathbf{r}_1 \mathrm{d}\mathbf{r}_2\right\}^{\frac{1}{2}}$$

$$= \left\{2 - \frac{1}{\sqrt{1+S^2}}(1+S^2)\right\}^{\frac{1}{2}} = \{2 - \sqrt{1+S^2}\}^{1/2},$$

where we have assumed that the functions are real. When $R \to \infty$, then $S \to 0$ and

$$\lim_{R\to\infty} ||\varphi^{(0)} - \psi_0^{(0)}|| = 1 \neq 0. \tag{13.28}$$

It is, therefore, clear that the Heitler-London wave function differs considerably and that this huge difference *does not vanish* when $R \to \infty$.

The two normalized functions $\varphi^{(0)}$ and $\psi_0^{(0)}$ represent two unit vectors in the Hilbert space; see Fig. 13.10. The scalar product of the two unit vectors $\langle\varphi^{(0)}|\psi_0^{(0)}\rangle$ is equal to $\cos\theta$. Let us calculate this angle $\theta_{\lim}$, which corresponds to R tending to ∞. The quantity $\lim_{R\to\infty} ||\varphi^{(0)} - \psi_0^{(0)}||^2 = \lim_{R\to\infty}\int(\varphi^{(0)} - \psi_0^{(0)})^*(\varphi^{(0)} - \psi_0^{(0)})\mathrm{d}\tau = \lim_{R\to\infty}\left[2 - 2\cos\theta\right] = 1$. Hence, $\cos\theta_{\lim} = \frac{1}{2}$, and therefore $\theta_{\lim} = 60^0$. This means that the three unit vectors, $\varphi^{(0)}$, $\psi_0^{(0)}$ and $\varphi^{(0)} - \psi_0^{(0)}$ for $R \to \infty$, form an equilateral triangle, and therefore $\varphi^{(0)}$ represents a highly "handicapped" function, which lacks about a half with respect to a function of the proper symmetry.[30] This is certainly bad news.

Therefore, the perturbation V has to be treated as *always large*, because it is responsible for a huge wave function change: from the unperturbed one of bad symmetry to the exact one of the correct symmetry.

[30] In Appendix Y available at booksite.elsevier.com/978-0-444-59436-5, p. e183, we show how the charge distribution changes when the Pauli exclusion principle is forced by a proper projection of the $\varphi^{(0)}$ wave function.

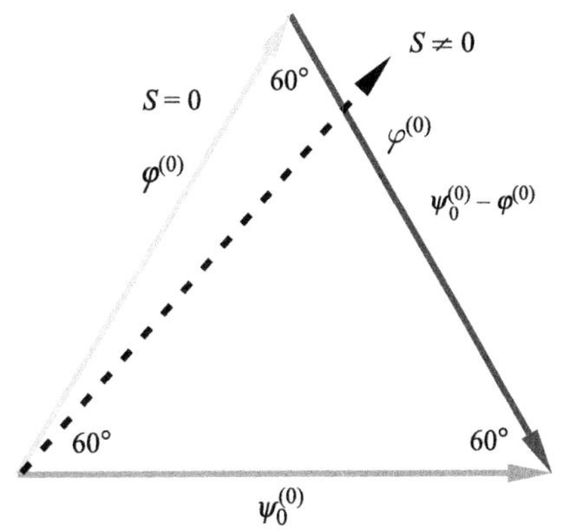

Fig. 13.10. A view from the Hilbert space showing that the polarization approximation is very bad, even for the infinite intersystem distance. The normalized functions $\varphi^{(0)}$ and $\psi_0^{(0)}$ for the hydrogen molecule are unit vectors belonging to the Hilbert space. The functions differ widely at any intermolecular distance R. For $S = 0$ (i.e., for long internuclear distances), the difference $\psi_0^{(0)} - \varphi^{(0)}$ represents a vector of the Hilbert space having the length 1. Therefore, for $R = \infty$, the three vectors $\varphi^{(0)}$, $\psi_0^{(0)}$, and $\psi_0^{(0)} - \varphi^{(0)}$ form an equilateral angle. For shorter distances, the angle between $\varphi^{(0)}$ and $\psi_0^{(0)}$ becomes smaller than 60^0.

In contrast to this, there would be no problem at all with the vanishing of the $||\psi_0^{(0)} - \psi_0||$ as $R \to \infty$, *where* ψ_0 *represents an exact ground-state solution of the Schrödinger equation. Indeed,* $\psi_0^{(0)}$ *correctly describes the dissociation of the molecule into two hydrogen atoms (both in the 1s state), as well as both functions have the same symmetry for all interatomic distances.* Therefore,

> the Heitler-London wave function represents a good approximation to the exact function for long (and we hope medium) intermolecular distances. Unfortunately, it is not the eigenfunction of the $\hat{H}^{(0)}$, and therefore, we cannot construct the usual Rayleigh-Schrödinger perturbation theory.

And this is the second piece of bad news here.

13.9.4 Iterative Scheme of SAPT

We now have two issues: either to construct another zero-order Hamiltonian, for which the $\psi_0^{(0)}$ function would be an eigenfunction (then the perturbation would be small and the Rayleigh-Schrödinger perturbation theory might be applied), or to abandon any Rayleigh-Schrödinger perturbation scheme and replace it by something else. The first of these possibilities was developed intensively in many laboratories. The approach had the deficiency that the operators appearing in the theories depended explicitly on the basis set used, and therefore, there was no guarantee that a basis independent theory exists.

The second possibility relies on an iterative solution of the Schrödinger equation, forcing the proper symmetry of the intermediate functions. The method was proposed by Bogumił Jeziorski and Włodzimierz Kołos.

Claude Bloch was probably the first to write the Schrödinger equation in the form shown in formulas[31] 10.82 and 10.71. Let us recall them in a notation adapted to the present situation:

$$\psi_0 = \varphi^{(0)} + \hat{R}_0(E_0^{(0)} - E_0 + V)\psi_0$$
$$E_0 = E_0^{(0)} + \langle\varphi^{(0)}|V\psi_0\rangle,$$

where we assume that $\varphi^{(0)}$ satisfies

$$\hat{H}^{(0)}\varphi^{(0)} = E_0^{(0)}\varphi^{(0)},$$

with the eigenvalues of the unperturbed Hamiltonian $\hat{H}^{(0)} = \hat{H}_A + \hat{H}_B$ given as the sum of the energies of the isolated molecules A and B:

$$E_0^{(0)} = E_{A,0} + E_{B,0},$$

and ψ_0 is the exact ground-state solution to the Schrödinger equation with the total non-relativistic Hamiltonian $\hat{H}$ of the system:

$$\hat{H}\psi_0 = E_0\psi_0.$$

We focus our attention on the difference $\mathcal{E}_0$ between E_0, which is our target, and $E_0^{(0)}$, which is at our disposal as the unperturbed energy. We may write the Bloch equations in a form exposing the interaction energy $\mathcal{E}_0 = E_0 - E_0^{(0)}$:

$$\psi_0 = \varphi^{(0)} + \hat{R}_0(-\mathcal{E}_0 + V)\psi_0$$
$$\mathcal{E}_0 = \langle\varphi^{(0)}|V\psi_0\rangle,$$

the equations are valid for intermediate normalization $\langle\varphi^{(0)}|\psi_0\rangle = 1$. This system of equations for $\mathcal{E}_0$ and ψ_0 might be solved by an iterative method:

Iterative Scheme:

$$\psi_0(n) = \varphi^{(0)} + \hat{R}_0[-\mathcal{E}(n) + V]\psi_0(n-1) \tag{13.29}$$

$$\mathcal{E}_0(n) = \langle\varphi^{(0)}|V\psi_0(n-1)\rangle, n = 1, 2, 3, \ldots, \tag{13.30}$$

where n is the *iteration* number.

In practice, everything depends on the starting point chosen, how many iterations have to be done and whether the convergence will be achieved. One of the most beautiful features of iterative schemes is that despite the fact that the freedom is usually very large, we are able to reach the limit; i.e., to get $||\psi_0(n+1) - \psi_0(n)|| < \varepsilon$ and $|\mathcal{E}_0(n+1) - \mathcal{E}_0(n)| < \varepsilon$ for $n > n_0$ and for an arbitrarily small $\varepsilon > 0$.

[31] C. Bloch, *Nucl.Phys.*, 6, 329 (1958).

Polarization Scheme Reproduced

We may start in the zeroth iteration with $\psi_0(0) = \varphi^{(0)}$. From the second of the iterative equations, we get right away $\mathcal{E}_0(1) = \langle\varphi^{(0)}|V\varphi^{(0)}\rangle \equiv E^{(1)}_{\text{pol}}$, which is the interaction energy accurate up to the first order of the polarization perturbational scheme. After inserting this result and $\psi_0(0) = \varphi^{(0)}$ into the first equation, one obtains (note that $\hat{R}_0\varphi^{(0)} = 0$) $\psi_0(1) = \varphi^{(0)} + \hat{R}_0(V - \langle\varphi^{(0)}|V\varphi^{(0)}\rangle)\varphi^{(0)} = \varphi^{(0)} + \hat{R}_0 V\varphi^{(0)}$. This is a sum of the unperturbed wave function and of the first-order correction, Eq. (5.24) on p. 245, exactly what we have obtained in the polarization approximation. The second iteration gives

$$\begin{aligned}\mathcal{E}_0(2) &= \left\langle\varphi^{(0)}|V\psi_0(1)\right\rangle = \left\langle\varphi^{(0)}|V\left(\varphi^{(0)} + \hat{R}_0 V\varphi^{(0)}\right)\right\rangle = E^{(1)}_{\text{pol}} + \left\langle\varphi^{(0)}|V\hat{R}_0 V\varphi^{(0)}\right\rangle \\ &= E^{(1)}_{\text{pol}} + E^{(2)}_{\text{pol}}\end{aligned}$$

because the second term represents nothing but the second-order energy correction in the polarization approximation.

When repeating the above iterative scheme and grouping the individual terms according to the powers of V, at each turn, we obtain the exact expression appearing in the Rayleigh-Schrödinger polarization approximation (discussed in Chapter 5), plus some higher-order terms.

It is worth noting that $\mathcal{E}_0(n)$ is the *sum* of all corrections of the Rayleigh-Schrödinger *up to the* nth order with respect to V, plus some higher-order terms. For large R, the quantity $\mathcal{E}_0(n)$ is an arbitrarily good approximation of the exact interaction energy.

Of course, the rate at which the iterative procedure converges depends very much on the starting point chosen. From this point of view, the start from $\psi_0(0) = \varphi^{(0)}$ is particularly unfortunate because the remaining (roughly) 50% of the *wave function* has to be restored by the hard work of the perturbational series (high-order corrections are needed). This will be especially pronounced for long intermolecular distances, where the exchange interaction energy will not be reproduced in any finite order.

Murrell-Shaw and Musher-Amos (MS-MA) Perturbation Theory Reproduced

A much more promising starting point in Eq. (13.29) seems to be $\psi_0(0) = \psi_0^{(0)}$, because the symmetry of the wave function is already correct and the function itself represents an exact solution at $R = \infty$. For convenience, the intermediate normalization is used (see p. 241) $\left\langle\varphi^{(0)}|\psi_0^{(0)}\right\rangle = 1$; i.e., $\psi_0^{(0)} = N\hat{A}\varphi^{(0)}$ with $N = \langle\varphi^{(0)}|\hat{A}\varphi^{(0)}\rangle^{-1}$. The first iteration of Eqs. (13.29) and (13.30) gives the first-order correction to the energy (which we split into the polarization part and the rest):

$$\mathcal{E}_0(1) = N\left\langle \varphi^{(0)} | V\hat{\mathcal{A}}\varphi^{(0)} \right\rangle = E^{(1)}_{\text{pol}} + E^{(1)}_{\text{exch}}, \tag{13.31}$$
$$E^{(1)}_{\text{pol}} \equiv E_{\text{elst}} = \left\langle \varphi^{(0)} | V\varphi^{(0)} \right\rangle.$$

We have obtained the electrostatic energy that was already known, plus a correction $E^{(1)}_{\text{exch}}$ which we will discuss shortly.

The first-iteration wave function will be obtained in the following way. First, we will use the commutation relation $\hat{\mathcal{A}}\hat{H} = \hat{H}\hat{\mathcal{A}}$ or

$$\hat{\mathcal{A}}(\hat{H}^{(0)} + V) = (\hat{H}_0 + V)\hat{\mathcal{A}}. \tag{13.32}$$

Of course,

$$\hat{\mathcal{A}}\left(\hat{H}^{(0)} - E^{(0)}_0 + V\right) = \left(\hat{H}^{(0)} - E^{(0)}_0 + V\right)\hat{\mathcal{A}}, \tag{13.33}$$

which gives[32] $V\hat{\mathcal{A}} - \hat{\mathcal{A}}V = [\hat{\mathcal{A}}, \hat{H}^{(0)} - E^{(0)}_0]$, as well as $(V - \mathcal{E}_1)\hat{\mathcal{A}} = \hat{\mathcal{A}}(V - \mathcal{E}_1) + [\hat{\mathcal{A}}, \hat{H}^0 - E^{(0)}_0]$. Now we are ready to use Eq. (13.30) with $n = 1$:

$$\begin{aligned}
\psi_0(1) &= \varphi^{(0)} + \hat{R}_0(V - \mathcal{E}_0(1))\psi^{(0)}_0 = \varphi^{(0)} + N\hat{R}_0(V - \mathcal{E}_0(1))\hat{\mathcal{A}}\varphi^{(0)} \\
&= \varphi^{(0)} + N\hat{R}_0\left\{\hat{\mathcal{A}}(V - \mathcal{E}_0(1)) + \hat{\mathcal{A}}\left(\hat{H}^{(0)} - E^{(0)}_0\right) - \left(\hat{H}^{(0)} - E^{(0)}_0\right)\hat{\mathcal{A}}\right\}\varphi^{(0)} \\
&= \varphi^{(0)} + N\hat{R}_0\hat{\mathcal{A}}\left(V - \mathcal{E}_0(1)\right)\varphi^{(0)} + N\hat{R}_0\hat{\mathcal{A}}\left(\hat{H}^{(0)} - E^{(0)}_0\right)\varphi^{(0)} \\
&\quad - N\hat{R}_0\left(\hat{H}^{(0)} - E^{(0)}_0\right)\hat{\mathcal{A}}\varphi^{(0)}.
\end{aligned}$$

The third term is equal to 0 because $\varphi^{(0)}$ is an eigenfunction of $\hat{H}^{(0)}$ with an eigenvalue $E^{(0)}_0$. The fourth term may be transformed by decomposing $\hat{\mathcal{A}}\varphi^{(0)}$ into the vector (in the Hilbert space) parallel to $\varphi^{(0)}$ or $\langle \hat{\mathcal{A}}\varphi^{(0)} | \varphi^{(0)} \rangle \varphi^{(0)}$ and the vector orthogonal to $\varphi^{(0)}$, or $(1 - |\varphi^{(0)}\rangle\langle\varphi^{(0)}|)\mathcal{A}\varphi^{(0)}$. The result of $\hat{R}_0(\hat{H}^{(0)} - E^{(0)}_0)$ acting on the first vector is zero (p. 644), while the second vector gives $(1 - |\varphi^{(0)}\rangle\langle\varphi^{(0)}|)\mathcal{A}\varphi^{(0)}$. This gives as the first iteration ground-state wave function $\psi_0(1)$:

$$\begin{aligned}
\psi_0(1) &= \varphi^{(0)} + N\hat{R}_0\hat{\mathcal{A}}(V - \mathcal{E}_0(1))\varphi^{(0)} + N\hat{\mathcal{A}}\varphi^{(0)} - N\langle\varphi^{(0)}|\hat{\mathcal{A}}\varphi^{(0)}\rangle\varphi^{(0)} \\
&= \frac{\hat{\mathcal{A}}\varphi^{(0)}}{\langle\varphi^{(0)}|\hat{\mathcal{A}}\varphi^{(0)}\rangle} + N\hat{R}_0\hat{\mathcal{A}}(V - \mathcal{E}_0(1))\varphi^{(0)} = \hat{\mathcal{B}}\varphi^{(0)} - N\hat{R}_0\hat{\mathcal{A}}(\mathcal{E}_0(1) - V)\varphi^{(0)},
\end{aligned}$$

where

$$\hat{\mathcal{B}}\varphi^{(0)} = \frac{\hat{\mathcal{A}}\varphi^{(0)}}{\langle\varphi^{(0)}|\hat{\mathcal{A}}\varphi^{(0)}\rangle}. \tag{13.34}$$

[32] Let us stress *in passing* that the left side is of the first order in V, while the right side is of the zeroth order. Therefore, in SAPT, the order is not a well-defined quantity, its role is taken over by the iteration number.

After inserting $\psi_0(1)$ into the iterative scheme [Eq. (13.29)] with $n = 2$, we obtain the second-iteration energy:

$$\mathcal{E}_0(2) = \langle \varphi^{(0)} | V \psi_0(1) \rangle = \frac{\langle \varphi^{(0)} | V \hat{\mathcal{A}} \varphi^{(0)} \rangle}{\langle \varphi^{(0)} | \hat{\mathcal{A}} \varphi^{(0)} \rangle} - N \langle \varphi^{(0)} | V \hat{R}_0 \hat{\mathcal{A}} [\mathcal{E}_0(1) - V] \varphi^{(0)} \rangle. \qquad (13.35)$$

These equations are identical to the corresponding corrections in perturbation theory derived by Murrell and Shaw[33] and by Musher and Amos[34] (MS–MA).

13.9.5 Symmetry Forcing

Finally, there is good news. It turns out that we may formulate a general iterative scheme that can produce various procedures, only some of them were known in the literature. In addition, the scheme has been designed by my nearest-neighbor colleagues (Jeziorski and Kołos). This scheme reads as

$$\psi_0(n) = \varphi^{(0)} + \hat{R}_0[-\mathcal{E}_0(n) + V]\hat{\mathcal{F}}\psi_0(n-1)$$
$$\mathcal{E}_0(n) = \langle \varphi^{(0)} | V \hat{\mathcal{G}} \psi_0(n-1) \rangle$$

where in Eqs. (13.29) and (13.30), we have inserted operators $\hat{\mathcal{F}}$ and $\hat{\mathcal{G}}$, which have to fulfill the obvious condition

$$\hat{\mathcal{F}}\psi_0 = \hat{\mathcal{G}}\psi_0 = \psi_0, \qquad (13.36)$$

where ψ_0 is the solution to the Schrödinger equation.

Why Force the Symmetry?
At the end of the iterative scheme (convergence), the insertion of the operators $\hat{\mathcal{F}}$ and $\hat{\mathcal{G}}$ has no effect at all, but *before that*, their presence may be crucial for the numerical convergence. This is the goal of symmetry forcing.

This method of generating perturbation theories has been called by the authors the *symmetry forcing* method in SAPT, Table 13.2.

[33] J.N. Murrell and G. Shaw, *J.Chem.Phys.*, *46*, 1768 (1867).
[34] J.I. Musher and A.T. Amos, *Phys.Rev.*, *164*, 31 (1967).

Table 13.2. Symmetry forcing in various perturbation schemes of the SAPT compared to the polarization approximation. The operator $\hat{\mathcal{B}}$ is defined by: $\hat{\mathcal{B}}\chi = \hat{\mathcal{A}}\chi/\langle\varphi^{(0)}|\hat{\mathcal{A}}\chi\rangle$.

Perturbation Scheme	$\psi_0(0)$	$\hat{\mathcal{F}}$	$\hat{\mathcal{G}}$
Polarization	$\varphi^{(0)}$	1	1
Symmetrized polarization	$\varphi^{(0)}$	1	$\hat{\mathcal{B}}$
MS–MA	$\hat{\mathcal{B}}\varphi^{(0)}$	1	1
Jeziorski-Kołos scheme[a]	$\hat{\mathcal{B}}\varphi^{(0)}$	$\hat{\mathcal{A}}$	1
EL–HAV[b]	$\hat{\mathcal{B}}\varphi^{(0)}$	$\hat{\mathcal{A}}$	$\hat{\mathcal{B}}$

[a] B. Jeziorski and W. Kołos, *Int.J.Quantum Chem.12*, 91 (1977).

[b] Eisenschitz–London and Hirschfelder–van der Avoird perturbation theory: R. Eisenschitz and F. London, *Zeit.Phys. 60*, 491 (1930); J.O. Hirschfelder, *Chem.Phys.Letters, 1*, 363 (1967); A. van der Avoird, *J.Chem.Phys. 47*, 3649 (1967).

Polarization Collapse Removed

The corrections obtained in SAPT differ from those of the polarization perturbational method.

To show the relation between the results of the two approaches, let us first introduce some new quantities. The first is an idempotent antisymmetrizer:

$$\hat{\mathcal{A}} = C\hat{\mathcal{A}}^A\hat{\mathcal{A}}^B(1+\hat{P}),$$

with $C = \frac{N_A!N_B!}{(N_A+N_B)!}$, where $\hat{\mathcal{A}}^A$, $\hat{\mathcal{A}}^B$ are idempotent antisymmetrizers for molecules A and B, each molecule contributing N_A and N_B electrons. Permutation operator $\hat{P}$ contains all the electron exchanges between molecules A and B:

$$\hat{P} = \hat{P}_{AB} + \hat{P}'$$
$$\hat{P}_{AB} = -\sum_{i\in A}\sum_{j\in B}\hat{P}_{ij},$$

with $\hat{P}_{AB}$ denoting the single exchanges only, and $\hat{P}'$ the rest of the permutations (i.e., the double, triple, etc. exchanges). Let us stress that $\varphi^{(0)} = \psi_{A,0}\psi_{B,0}$ represents a product of two antisymmetric functions,[35] and therefore, $\hat{\mathcal{A}}\varphi^{(0)} = C(1+\hat{P}_{AB}+\hat{P}')\psi_{A,0}\psi_{B,0}$. Taking into account the operator $\hat{P}$ in $\langle\varphi^{(0)}|V\hat{\mathcal{A}}\varphi^{(0)}\rangle$ and $\langle\varphi^{(0)}|\hat{\mathcal{A}}\varphi^{(0)}\rangle$ produces

$$E^{(1)} = \frac{\langle\psi_{A,0}\psi_{B,0}|V\psi_{A,0}\psi_{B,0}\rangle + \langle\psi_{A,0}\psi_{B,0}|V\hat{P}_{AB}\psi_{A,0}\psi_{B,0}\rangle + O(S^4)}{1+\langle\psi_{A,0}\psi_{B,0}|\hat{P}_{AB}\psi_{A,0}\psi_{B,0}\rangle + O(S^4)}, \quad (13.37)$$

[35] The product itself does not have this symmetry.

and therefore,

in the polarization approximation,

$$E_{\text{pol}}^{(1)} \equiv E_{\text{elst}} = \langle \varphi^{(0)} | V \varphi^{(0)} \rangle, \tag{13.38}$$

while in SAPT,

$$E^{(1)} = \frac{\langle \varphi^{(0)} | V \hat{\mathcal{A}} \varphi^{(0)} \rangle}{\langle \varphi^{(0)} | \hat{\mathcal{A}} \varphi^{(0)} \rangle}, \tag{13.39}$$

$$E^{(1)} = E_{\text{pol}}^{(1)} + E_{\text{exch}}^{(1)}, \tag{13.40}$$

where *the exchange interaction* in the first-order perturbation theory is

$$E_{\text{exch}}^{(1)} = \langle \psi_{A,0}\psi_{B,0} | V \hat{P}_{AB} \psi_{A,0}\psi_{B,0} \rangle - \langle \psi_{A,0}\psi_{B,0} | V \psi_{A,0}\psi_{B,0} \rangle \langle \psi_{A,0}\psi_{B,0} | \hat{P}_{AB} \psi_{A,0}\psi_{B,0} \rangle + O(S^4). \tag{13.41}$$

In the most commonly encountered interaction of closed shell molecules, the $E_{\text{exch}}^{(1)}$ term represents the *valence repulsion*.

The symbol $O(S^4)$ stands for all the terms that vanish with the fourth power of the overlap integrals or faster.[36] The valence repulsion already appears in the first order of the perturbation theory (besides the electrostatic energy $E_{\text{pol}}^{(1)}$) as a result of the Pauli exclusion principle.[37]

We have gained a remarkable thing, which may be seen by taking the example of two interacting subsystems: Na^+ and Cl^-. In the polarization approximation, the electrostatic, induction, and dispersion contributions to the interaction energy are negative, the total energy will go down, and we will soon have a catastrophe: both subsystems would occupy the same place in space and according to the energy calculated (that could attain even $-\infty$, as shown in Fig. 13.11), the system would be extremely happy. This is absurd.

If this were true, we could not exist. Indeed, sitting safely in a chair, we have an equilibrium of the gravitational force and..., well, and what? First of all, the force coming from the valence repulsion that we are talking about. It is claimed sometimes that quantum effects are peculiar to small objects (electrons, nuclei, atoms, molecules) and are visible when dealing with such

[36] This means that we also take into account such a decay in situations other than in overlap integrals; e.g., $\langle 1s_a 1s_b | 1s_b 1s_a \rangle$ is of the order S^2, where $S = \langle 1s_a | 1s_b \rangle$. Thus, the criterion is the differential overlap rather than the overlap integral.

[37] Here is an intriguing idea: the polarization approximation should be an extremely good approximation for the interaction of a molecule with an antimolecule (built from antimatter). Indeed, in the molecule we have electrons in the antimolecule positrons, and no antisymmetrization (between the systems) is needed. Therefore, a product wave function should be a very good starting point. No valence repulsion will appear, and the two molecules will penetrate like ghosts. "*Soon after the annihilation takes place and the system disappears*".

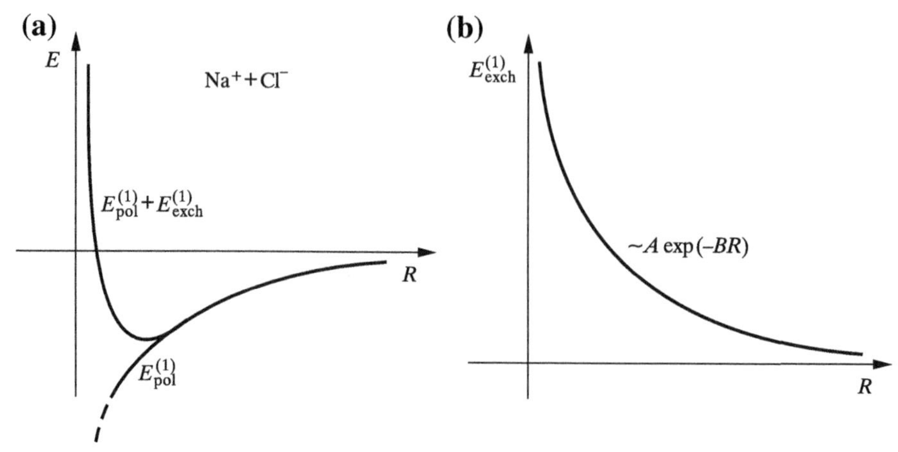

Fig. 13.11. Interaction energy of Na^+ and Cl^- (scheme). The polarization approximation gives an absurdity for small separations: the subsystems attract very strongly (mainly because of the electrostatic interaction), while they have had to repel very strongly. (a) The absurdity is removed when the valence repulsion is taken into account. Panel(b) shows the valence repulsion alone modeled by the term $A \exp(-BR)$, where A and B are positive constants.

particles. We see, however, that we owe even the ability to sit in a chair to the Pauli exclusion principle (a quantum effect).

> Valence repulsion removes the absurdity of the polarization approximation, which made the collapse of the two subsystems possible.

13.9.6 A Link to the Variational Method–The Heitler-London Interaction Energy

Since the $\hat{A}\varphi^{(0)}$ wave function is a good approximation of the exact ground state wave function at high values of R, we may calculate what is called the *Heitler-London interaction energy* (E_{int}^{HL}) as the mean value of the total (electronic) Hamiltonian minus the energies of the isolated subsystems:

$$E_{int}^{HL} = \frac{\langle \hat{A}\varphi^{(0)} | \hat{H} \hat{A}\varphi^{(0)} \rangle}{\langle \hat{A}\varphi^{(0)} | \hat{A}\varphi^{(0)} \rangle} - (E_{A,0} + E_{B,0}).$$

This expression may be transformed in the following way:

$$\begin{aligned} E_{int}^{HL} &= \frac{\langle \varphi^{(0)} | \hat{H} \hat{A}\varphi^{(0)} \rangle}{\langle \varphi^{(0)} | \hat{A}\varphi^{(0)} \rangle} - (E_{A,0} + E_{B,0}) \\ &= \frac{\langle \varphi^{(0)} | \hat{H}^{(0)} \hat{A}\varphi^{(0)} \rangle + \langle \varphi^{(0)} | V \hat{A}\varphi^{(0)} \rangle}{\langle \varphi^{(0)} | \hat{A}\varphi^{(0)} \rangle} - (E_{A,0} + E_{B,0}) \end{aligned}$$

$$= \frac{(E_{A,0} + E_{B,0})\langle \varphi^{(0)} | \hat{\mathcal{A}} \varphi^{(0)} \rangle + \langle \varphi^{(0)} | V \hat{\mathcal{A}} \varphi^{(0)} \rangle}{\langle \varphi^{(0)} | \hat{\mathcal{A}} \varphi^{(0)} \rangle} - (E_{A,0} + E_{B,0})$$

$$= \frac{\langle \varphi^{(0)} | V \hat{\mathcal{A}} \varphi^{(0)} \rangle}{\langle \varphi^{(0)} | \hat{\mathcal{A}} \varphi^{(0)} \rangle}.$$

Therefore, the Heitler-London interaction energy is equal to the first-order SAPT energy

$$E_{int}^{HL} = E^{(1)}.$$

13.9.7 Summary: The Main Contributions to the Interaction Energy

From the first two iterations ($n = 2$) of the SAPT scheme [Eqs. (13.29) and (13.30), p. 832] we got the energy $\mathcal{E}_0(2)$, which now will be written in a somewhat different form:

$$\mathcal{E}_0(2) = E_{\text{elst}} + E_{\text{exch}}^{(1)} - \frac{\langle \varphi^{(0)} | V \hat{R}_0 \hat{\mathcal{A}} [\mathcal{E}_0(1) - V] \varphi^{(0)} \rangle}{\langle \varphi^{(0)} | \hat{\mathcal{A}} \varphi^{(0)} \rangle}. \tag{13.42}$$

We recognize on the right side (the first two terms) the complete first-order contribution; i.e., the electrostatic energy (E_{elst}) and the valence repulsion energy ($E_{\text{exch}}^{(1)}$). From the definition of the induction and dispersion energies [Eqs. (13.13) on p. 808], as well as from the reduced resolvent $\hat{R}_0$ of Eq. (10.76) on p. 644, (applied here to the individual molecules), one may write

$$\hat{R}_0 = \hat{R}_{0,\text{ind}} + \hat{R}_{0,\text{disp}}, \tag{13.43}$$

where the induction part of the resolvent

$$\hat{R}_{0,\text{ind}} = \hat{R}_{0,\text{ind}(A \to B)} + \hat{R}_{0,\text{ind}(B \to A)}, \tag{13.44}$$

corresponds to deforming B by A and *vice versa*

$$\hat{R}_{0,\text{ind}(A \to B)} = \sum_{n_B (\neq 0)} \frac{|\psi_{A,0} \psi_{B,n_B}\rangle \langle \psi_{A,0} \psi_{B,n_B}|}{E_{B,0} - E_{B,n_B}}, \tag{13.45}$$

$$\hat{R}_{0,\text{ind}(B \to A)} = \sum_{n_A (\neq 0)} \frac{|\psi_{A,n_A} \psi_{B,0}\rangle \langle \psi_{A,n_A} \psi_{B,0}|}{E_{A,0} - E_{A,n_A}}, \tag{13.46}$$

while the dispersion part of the resolvent reads as

$$\hat{R}_{0,\text{disp}} = \sum_{(n_A, n_B) \neq (0,0)} \frac{|\psi_{A,n_A} \psi_{B,n_B}\rangle \langle \psi_{A,n_A} \psi_{B,n_B}|}{(E_{A,0} - E_{A,n_A}) + (E_{B,0} - E_{B,n_B})}. \tag{13.47}$$

Let us insert these formulas into Eq. (13.42). We get (because $\hat{R}_{0,\text{ind}} \varphi^{(0)} = \hat{R}_{0,\text{ind}} \psi_{A,0} \psi_{B,0} = 0$, and, for similar reasons, also $\hat{R}_{0,\text{disp}} \varphi^{(0)} = 0$)

$$\mathcal{E}_0(2) - \left(E_{\text{elst}} + E_{\text{exch}}^{(1)} \right) = \tag{13.48}$$

$$\frac{\langle \varphi^{(0)} | V \hat{R}_{0,\text{ind}} \hat{\mathcal{A}} V \varphi^{(0)} \rangle}{\langle \varphi^{(0)} | \hat{\mathcal{A}} \varphi^{(0)} \rangle} + \frac{\langle \varphi^{(0)} | V \hat{R}_{0,\text{disp}} \hat{\mathcal{A}} V \varphi^{(0)} \rangle}{\langle \varphi^{(0)} | \hat{\mathcal{A}} \varphi^{(0)} \rangle} = \tag{13.49}$$

$$\left(E_{\text{ind}} + E^{(2)}_{\text{ind-exch}} \right) + \left(E_{\text{disp}} + E^{(2)}_{\text{disp-exch}} \right). \tag{13.50}$$

In this way, we have introduced

- *The induction-exchange energy* $E^{(2)}_{\text{ind-exch}}$ representing a modification of the induction energy, known from the polarization approximation (E_{ind})

$$E^{(2)}_{\text{ind-exch}} = \frac{\langle \psi_{A,0}\psi_{B,0} | V \hat{R}_{0,\text{ind}} \hat{\mathcal{A}} V \psi_{A,0}\psi_{B,0} \rangle}{\langle \psi_{A,0}\psi_{B,0} | \hat{\mathcal{A}} \left(\psi_{A,0}\psi_{B,0} \right) \rangle} - E_{\text{ind}}.$$

Note that after using the definition of the antisymmetrization operator $\hat{\mathcal{A}}$ [Eq. (M.1) on p. e109] as well as after applying the Taylor expansion, one gets as the first term: $\langle \psi_{A,0}\psi_{B,0} | V \hat{R}_{0,\text{ind}} V \psi_{A,0}\psi_{B,0} \rangle \equiv E_{\text{ind}}$, which cancels the second term of the last equation. The other terms are

$$E^{(2)}_{\text{ind-exch}} = \langle \psi_{A,0}\psi_{B,0} | V \hat{R}_{0,\text{ind}} \hat{P}_{AB} V \psi_{A,0}\psi_{B,0} \rangle \tag{13.51}$$

$$- \langle \psi_{A,0}\psi_{B,0} | V \hat{R}_{0,\text{ind}} V \psi_{A,0}\psi_{B,0} \rangle \langle \psi_{A,0}\psi_{B,0} | \hat{P}_{AB} \left(\psi_{A,0}\psi_{B,0} \right) \rangle \tag{13.52}$$

$$+ O(S^4). \tag{13.53}$$

Therefore, they represent the corrections for the induction energy due to the forcing of the Pauli exclusion principle (for they appear as a result of the antisymmetrization) and are said to take care of the electron exchanges between the two molecules (due to the single electron exchanges from the permutation operator $\hat{P}_{AB}$, the double exchanges, etc.; see p. 836).[38]

- *The dispersion-exchange energy* $E^{(2)}_{\text{disp-exch}}$, which is a modification of the dispersion energy defined in the polarization perturbation theory (E_{disp})

$$E^{(2)}_{\text{disp-exch}} = \frac{\langle \psi_{A,0}\psi_{B,0} | V \hat{R}_{0,\text{disp}} \hat{\mathcal{A}} V \psi_{A,0}\psi_{B,0} \rangle}{\langle \psi_{A,0}\psi_{B,0} | \hat{\mathcal{A}} \left(\psi_{A,0}\psi_{B,0} \right) \rangle} - E_{\text{disp}}. \tag{13.54}$$

After representing $\hat{\mathcal{A}}$ by permutation operators and application of the Taylor expansion to the inverse of the denominator, the first term $\langle \psi_{A,0}\psi_{B,0} | V \hat{R}_{0,\text{disp}} V \psi_{A,0}\psi_{B,0} \rangle \equiv E_{\text{disp}}$ cancels out the second term on the right side. The other terms, as in the case of $E^{(2)}_{\text{ind-exch}}$, result from

[38] One more thing should be mentioned at this point. Because $\hat{R}_{0,\text{ind}} = \hat{R}_{0,\text{ind}(A\to B)} + \hat{R}_{0,\text{ind}(B\to A)}$ one may split $E^{(2)}_{\text{ind-exch}}$ into a part associated with a modification of polarization (due to the valence repulsion) of B by A and a similar part for polarization modification of A by B.

The electron permutations lead to the integrals that decay with R similarly as powers of the overlap integrals; i.e., the above modifications decay very fast with increasing R.

the Pauli exclusion principle:

$$E^{(2)}_{\text{disp–exch}} = \langle \psi_{A,0}\psi_{B,0}|V\hat{R}_{0,\text{disp}}\hat{P}_{AB}V\psi_{A,0}\psi_{B,0}\rangle \tag{13.55}$$

$$-\langle \psi_{A,0}\psi_{B,0}|V\hat{R}_{0,\text{disp}}V\psi_{A,0}\psi_{B,0}\rangle\langle \psi_{A,0}\psi_{B,0}|\hat{P}_{AB}\left(\psi_{A,0}\psi_{B,0}\right)\rangle \tag{13.56}$$

$$+O(S^4) \tag{13.57}$$

In the SAPT scheme, which takes into account the overlapping of the electron clouds of both interacting molecules, one obtains the terms known from the polarization approximation (no antisymmetrization) plus some important modifications.

In the first order, due to the Pauli exclusion principle (i.e., to the antisymmetrization) the electrostatic energy E_{elst} that is known from the polarization approximation is supplemented by the valence repulsion energy $E^{(1)}_{\text{exch}}$.

In the second order, the Pauli deformation (cf. Appendix Y available at booksite.elsevier.com/978-0-444-59436-5 on p. e183) of the electronic density in the AB complex results in exchange-based modifications ($E^{(2)}_{\text{ind–exch}}$ and $E^{(2)}_{\text{disp–exch}}$) of the induction and dispersion interactions (E_{ind} and E_{disp}) that are known from the polarization perturbation theory.

One may, therefore, suspect that a restriction of polarization and of the electron correlation (the essence of the dispersion energy) due to the Pauli exclusion principle should result in a smaller energy gain, and therefore, $E^{(2)}_{\text{ind–exch}}$ and $E^{(2)}_{\text{disp–exch}}$ should represent positive (i.e., destabilizing) energy effects. There is no theoretical evidence of this, but the numerical results seem to confirm the above conjecture; see Table 13.3.

Table 13.3. The SAPT contributions to the interaction energy calculated by using the DFT electronic density for Ne_2, $(H_2O)_2$ and $(C_6H_6)_2$. Numerical experience: the values for $E^{(2)}_{\text{ind–exch}}$ and $E^{(2)}_{\text{disp–exch}}$ are always positive!

Contribution	$Ne_2[\text{cm}^{-1}]$[a]	$(H_2O)_2\left[\frac{\text{kcal}}{\text{mol}}\right]$[a]	$(C_6H_6)_2\left[\frac{\text{kcal}}{\text{mol}}\right]$[b]
E_{elst}	−6.24	−7.17	−0.01
$E^{(1)}_{exch}$	26.40	5.12	3.82
E_{ind}	−4.76	−2.22	−1.30
$E^{(2)}_{ind-exch}$	4.88	1.13	1.07
E_{disp}	−46.23	−2.15	−5.77
$E^{(2)}_{disp-exch}$	1.85	0.36	0.51
E_{int}	−22.65	−4.50	−1.67

[a] A.J. Misquitta and K. Szalewicz, *J.Chem.Phys.*, *122*, 214109 (2005).

[b] S. Tsuzuki and H.P.Lüthi, *J.Chem.Phys.*, *114*, 3949 (2001).

13.10 Convergence Problems and Padé Approximants

A very slow convergence (of "pathological" character) is bound to appear. To see why, let us recall the Heitler-London function (p. 521), which for very large R approximates the exact wave function very well. Since the function is so accurate, and therefore indispensable for a good description, an effective iterative process is expected to reproduce it in the first few iterations. Of course, the Heitler-London wave function can be represented as a linear combination of the functions belonging to a complete basis set, because any complete basis set can do it. In particular, a complete basis set can be formed from the products of two functions belonging to complete one-electron basis sets[39]: the first one, describing electron 1, represents an orbital centered on the nucleus a, while the second function (of electron 2) is an orbital centered on b. There is, however, a difficulty of a purely numerical character: the second half of the Heitler-London wave function corresponds to electron 1 at b, and electron 2 at a, *just opposite to what describe the functions belonging to the complete set.* This would result in prohibitively long expansions (related to a large number of iterations and a high order of the perturbation theory) necessary for the theory reproduced the second half of the Heitler-London function. This effort is similar to expanding the function $1s_b$ (i.e., localized far away off the center a) as a linear combination of the functions $1s_a, 2s_a, 2p_{xa}, 2p_{xa}, 2p_{ya}, 2p_{za}, \ldots$ It can be done in principle, but in practice, it requires prohibitively long expansions, which should ensure a perfect destructive interference (of the wave functions) at a and, at the same time, a perfect constructive interference at b.

Another kind of convergence problems is related to a too late operation of the Pauli exclusion principle. This pertains also to SAPT. Why? Look at Table 13.2. One of the perturbational schemes given there (namely, the symmetrized polarization approximation) is based on calculation of the wave function, exactly as in the polarization approximation scheme, but just before calculation of the corrections to the energy, the polarization wave function is projected on the antisymmetrized space. This procedure is bound to have trouble. The system excessively changes its charge distribution without paying any attention to the Pauli exclusion principle (thus allowing it to polarize itself in a non-physical way–this may be described as *overpolarization*), and then the result has to be modified in order to fulfill a principle[40] (the Pauli principle).

This became evident after a study called the *Pauli blockade.*[41] It was shown that if the Pauli exclusion principle is not obeyed, the electrons of the subsystem A can flow, without any penalty and totally non-physically, to the low-energy orbitals of B. This may lead to occupation of that orbital by four electrons (!), whereas the Pauli principle admits only a maximum of a double occupation.

[39] The hydrogen orbitals form a complete set, because they are the eigenfunctions of a Hermitian operator (Hamiltonian).

[40] This is similar to allowing all plants grow as they want and, just after harvesting everything, select the wheat alone. We cannot expect much from such an agriculture.

[41] M. Gutowski and L. Piela, *Mol.Phys.*, *64*, 337 (1988).

Thus, any realistic deformation of the electron clouds has to take into account simultaneously the exchange interaction (valence repulsion, or the Pauli principle). Because of this, we have introduced what is called the *deformation-exchange interaction energy*:

$$E_{\text{def-exch}} = \mathcal{E}_0(2) - (E_{\text{elst}} + E_{\text{disp}}). \tag{13.58}$$

Now about making convergence faster. Any perturbational correction carries information. Summing up (this is the way we calculate the total effect), these corrections means a certain processing of the information. We may ask an amazing question: *is there any possibility of taking the same corrections and squeezing out more information*[42] *than just making the sum*?

In 1892, Henri Padé[43] wrote his doctoral dissertation in mathematics and presented some fascinating results.

For a power series,

$$f(x) = \sum_{j=0}^{\infty} a_j x^j, \tag{13.59}$$

we may define a *Padé approximant* $[L/M]$ as the ratio of two polynomials:

$$[L/M] = \frac{P_L(x)}{Q_M(x)}, \tag{13.60}$$

where $P_L(x)$ is a polynomial of at most the Lth order, while $Q_M(x)$ is a polynomial of the Mth order. The coefficients of the polynomials P_L and Q_M will be determined by the following condition:

$$f(x) - [L/M] = \text{terms of higher order than } x^{L+M}. \tag{13.61}$$

In other words,

the first $L + M$ terms of the Taylor expansion for a function $f(x)$ and, for its Padé approximant, are identical.

[42] That is, a more accurate result.

[43] H. Padé, *Ann.Sci.Ecole Norm.Sup.Suppl.[3]*, *9*, 1 (1892).

Since the numerator and denominator of the approximant can be harmlessly multiplied by any nonzero number, we may set, without losing anything, the following normalization condition:

$$Q_M(0) = 1. \tag{13.62}$$

Let us assume also that $P_L(x)$ and $Q_M(x)$ do not have any common factor.

If we now write the polynomials as

$$P_L(x) = p_0 + p_1 x + p_2 x^2 + \cdots p_L x^L$$
$$Q_M(x) = 1 + q_1 x + q_2 x^2 + \cdots q_M x^M,$$

then multiplying Eq. (13.60) by Q and forcing the coefficients at the same powers of x to be equal, we obtain the following system of equations for the unknowns p_i and q_i (there are $L + M + 1$ of them; the number of equations is the same):

$$\begin{aligned}
&a_0 &&= p_0\\
&a_1 + a_0 q_1 &&= p_1\\
&a_2 + a_1 q_1 + a_0 q_2 &&= p_2\\
&a_L + a_{L-1} q_1 + \cdots a_0 q_L &&= p_L\\
&a_{L+1} + a_L q_1 + \cdots a_{L-M+1} q_M &&= 0\\
&\vdots\\
&a_{L+M} + a_{L+M-1} q_1 + \cdots a_L q_M = 0.
\end{aligned} \tag{13.63}$$

Note that the sum of the subscripts in either term is a constant integer within the range $[0, L + M]$, which is connected to the abovementioned equal powers of x.

Example 1. (***Mathematical***)

The method is best illustrated in action. Let us take a function

$$f(x) = \frac{1}{\sqrt{1 - x}} \tag{13.64}$$

and then expand f in a Taylor series:

$$f(x) = 1 + \frac{1}{2}x + \frac{3}{8}x^2 + \frac{5}{16}x^3 + \frac{35}{128}x^4 + \cdots \tag{13.65}$$

Therefore, $a_0 = 1$; $a_1 = \frac{1}{2}$; $a_2 = \frac{3}{8}$; $a_3 = \frac{5}{16}$; $a_4 = \frac{35}{128}$. Now forget for the moment that these coefficients came from the Taylor expansion of $f(x)$. Many other functions may have the same *beginning* of the Taylor series. Let us calculate some partial sums of the right side of Eq. (13.65):

Approx. $f(\frac{1}{2})$	Sum up to the nth term
$n=1$	1.00000
$n=2$	1.25000
$n=3$	1.34375
$n=4$	1.38281
$n=5$	1.39990

We see that the Taylor series "works very hard"; and approaching $f(\frac{1}{2}) = \sqrt{2} = 1.414213562$, it succeeds, but not without a painful effort.

Now let us check how one of the simplest Padé approximants (namely, [1/1]), performs the same job. By definition,

$$\frac{(p_0 + p_1 x)}{(1 + q_1 x)}. \tag{13.66}$$

Solving Eq. (13.63) gives as the approximant[44]:

$$\frac{\left(1 - \frac{1}{4}x\right)}{\left(1 - \frac{3}{4}x\right)}. \tag{13.67}$$

Let us stress that information contained in the power series [Eq. (13.59)] has been limited to a_0, a_1, and a_2 (all other coefficients *have not been used*). For $x = \frac{1}{2}$, the Padé approximant [1/1] has the value

$$\frac{\left(1 - \frac{1}{4}\frac{1}{2}\right)}{\left(1 - \frac{3}{4}\frac{1}{2}\right)} = \frac{7}{5} = 1.4, \tag{13.68}$$

which is *more effective* than the painful efforts of the Taylor series, which used a coefficients up to a_4 (this gave 1.39990). To be fair, we have to compare the Taylor series result that used only a_0, a_1, a_2 and this gives only 1.34375. Therefore, the approximant failed by 0.01, while the Taylor series failed by 0.07. The Padé approximant [2/2] has the form

$$[2, 2] = \frac{\left(1 - \frac{3}{4}x + \frac{1}{16}x^2\right)}{\left(1 - \frac{5}{4}x + \frac{5}{16}x^2\right)}. \tag{13.69}$$

For $x = \frac{1}{2}$, its value is equal to $\frac{41}{29} = 1.414$, which means an accuracy of 10^{-4}, while *without Padé approximants, but using the same information contained in the coefficients, we get an accuracy that is two orders of magnitude worse.*

Our procedure did not have information that the function expanded is $(1-x)^{-\frac{1}{2}}$ for we gave the first five terms of the Taylor expansion only. Despite this, the procedure determined with high accuracy what will give higher terms of the expansion.

---

[44] Indeed, $L = M = 1$, and therefore the equations for the coefficients p and q are the following: $p_0 = 1$, $\frac{1}{2} + q_1 = p_1$, $\frac{3}{8} + \frac{1}{2}q_1 = 0$. This gives the solution: $p_0 = 1, q_1 = -\frac{3}{4}, p_1 = -\frac{1}{4}$.

Example 2. (*Quantum Mechanical*)

This is not the end of the story yet. The reader will see shortly some things which will be even stranger. Perturbation theory also represents a power series (with respect to λ) with coefficients that are energy corrections. If perturbation is small, the corrections are small as well; in principle, the higher the perturbation order the smaller perturbational corrections are. As a result, a partial sum of a few low-order corrections usually gives sufficient accuracy. However, the higher the order, the more difficult are the corrections to calculate. Therefore, we may ask if there is any possibility of obtaining good results and at a low price by using the Padé approximants. In Table 13.4, the results of a perturbational study of H_2^+ by Jeziorski et al. are collected.[45]

For $R = 12.5$ a.u., we see that the approximants had a very difficult task to do. First of all they "recognized" the series limit, already about $2L + 1 = 17$. Before that, they have been less effective than the original series. It has to be stressed, however, that they "recognized" it *extremely* well (see $2L + 1 = 21$). In contrast to this, the (traditional) partial sums ceased to improve when L increased. This means that either the partial sum series converges[46] to a false limit or it converges to the correct limit, but does it extremely slowly. We see from the variational result (the error is calculated with respect to this) that the convergence *is* false.[47] If the variational result had not been known, we would say that the series has already converged. However, the Padé approximants said: "*no, this is a false convergence*," and they were right.

For $R = 3.0$ a.u. (see Table 13.4), the original series represents a real tragedy. For this distance, the perturbation is too large and the perturbational series just evidently *diverges*.

Table 13.4. Convergence of the MS–MA perturbational series for the hydrogen atom in the field of a proton ($2p\sigma_u$ state) for internuclear distance R (a.u.). The Table reports the error (in %) for the sum of the original perturbational series and for the Padé $[L+1, L]$ approximant. The error is calculated with respect to the variational method (i.e., the best for the basis set used).

$2L+1$	$R = 12.5$		$R = 3.0$	
	pert. series	$[L+1, L]$	**pert. series**	$[L+1, L]$
3	0.287968	0.321460	0.265189	0.265736
5	0.080973	−0.303293	0.552202	−1.768582
7	0.012785	−0.003388	0.948070	0.184829
9	−0.000596	−0.004147	1.597343	0.003259
11	−0.003351	−0.004090	2.686945	0.002699
13	−0.003932	−0.004088	4.520280	0.000464
15	−0.004056	−0.004210	7.606607	0.000009
17	−0.004084	−0.001779	12.803908	0.000007
19	−0.004090	0.000337	21.558604	−0.000002
21	−0.004092	−0.000003	36.309897	0.000001

[45] B. Jeziorski, K. Szalewicz and M. Jaszuński, *Chem.Phys.Letters*, *61*, 391(1979).

[46] We have only numerics as an argument though. The described case represents an example of vicious behavior when summing up a series: starting from a certain order of the perturbational series, the improvement becomes extremely slow (we see the "*false*" convergence), although we are far from the true limit.

[47] At least, this is so as we see it numerically.

The greater our effort, the greater the error of our result. The error is equal to 13% for $2L+1 = 17$, then it goes to 22% for $2L + 1 = 19$, and finally, it attains 36% for $2L + 1 = 21$. Despite of these hopeless results, it turns out that the problem is easy for the Padé approximants.[48] They were already much better for $L = 3$.

Why Are the Padé Approximants so Effective?

The apparent garbage produced by the perturbational series for $R = 3.0$ a.u. represented for the Padé approximants precise information that the absurd perturbational corrections pertain the energy of the $\ldots 2p\sigma_u$ state of the hydrogen atom in the electric field of the proton. Why does this happen? Visibly low-order perturbational corrections, even if absolutely crazy, somehow carry information about the physics of the problem. The convergence properties of the Rayleigh-Schrödinger perturbation theory depend critically on the poles of the function approximated (see the discussion on p. 250). A pole cannot be described by any power series (as happens in perturbation theories), whereas the Padé approximants have poles built in the very essence of their construction (the denominator as a polynomial). This is why they may fit so well to the nature of the problems under study.[49]

13.11 Non-additivity of Intermolecular Interactions

13.11.1 Interaction Energy Represents the Non-additivity of the Total Energy

The *total* energy of interacting molecules is not an additive quantity; i.e., it does not represent the sum of the energies of the isolated molecules. The reason for this non-additivity is the interaction energy.

First of all, the interaction energy requires the declaration of which fragments of the total system we are treating as (interacting) molecules (see the beginning of this chapter). The only real system is the total system, not these subsystems. Therefore, the subsystems can be chosen in many different ways.

If the theory were exact, the total system could be described at any such choice (cf. p. 570). Only the supermolecular theory that is invariant with respect to such choices.[50] The perturbation theory so far has no such powerful feature (this problem is not even raised in the literature), because it requires the intramolecular and intermolecular interactions to be treated on the same footing.

---

[48] Similar findings are reported in T.M. Perrine, R.K. Chaudhuri, and K.F. Freed, *Intern.J.Quantum Chem.*, *105*, 18 (2005).

[49] There are cases, however, where Padé approximants may fail in a spectacular way.

[50] However, this occurs for rather trivial reasons; i.e., interaction energy represents a by-product of the method. The main goal is the total energy, which by definition is independent of the choice of subsystems.

13.11.2 Many-body Expansion of the Rigid Interaction Energy

A next question could be: *is the interaction energy pair-wise additive*? That is,

is the interaction energy a sum of *pairwise* interactions?

If this were true, it would be sufficient to calculate all possible interactions of pairs of molecules in the configuration identical to that they have in the total system,[51] and our problem would be solved.

For the time being, let us take the example of a stone, a shoe, and a foot. The owner of the foot will certainly remember what the three-body interaction is, while nothing special happens when you put a stone into the shoe, or your foot into the shoe, or a small stone on your foot (two-body interactions). The molecules behave similarly:

their molecular interactions are not pairwise additive.

In the case of three interacting molecules, there is an effect of a strictly three-body character, *which cannot be reduced to the two-body interactions*. Similarly, for larger numbers of molecules, there is a nonzero four-body effect, because all cannot be calculated as two- and three-body interactions, etc.

In what is called the *many-body expansion* for N molecules $A_1, A_2, \ldots, A_N$ the (rigid) interaction energy $E_{int}(A_1 A_2 \ldots A_N)$, i.e., the difference between the total energy $E_{A_1 A_2 \ldots A_N}$ and the sum of the energies of the isolated molecules[52] E_{A_i}, can be represented as a series of m-body interaction terms $\Delta E\left(m, N\right), m = 2, 3, \ldots, N$:

$$E_{int} = E_{A_1 A_2 \ldots A_N} - \sum_{i=1}^{N} E_{A_i} = \sum_{i>j}^{N} \Delta E_{A_i A_j}\left(2, N\right) + \sum_{i>j>k}^{N} \Delta E_{A_i A_j A_k}\left(3, N\right) + \cdots \Delta E_{A_1 A_2 \ldots A_N}\left(N, N\right). \tag{13.70}$$

The $\Delta E\left(m, N\right)$ contribution to the interaction energy of N molecules ($m \leq N$) represents the sum of the interactions of m molecules (all possible combinations of m molecules among N molecules keeping their configurations fixed as in the total system) inexplicable by the interactions of $m' < m$ molecules.

[51] This would be much less expensive than the calculation for the total system.

[52] With the same internal configuration as that in the total system and with the fixed number of electrons.

Now let's consider one more question. Should we really stay with the idea of the rigid interaction energy? For instance, we may be interested in how the conformation of the AB complex changes in the presence of molecule C. This is also a three-body interaction. Such dilemmas have not yet been solved in the literature.

Example. Three Helium Atoms

Let us take the helium atoms in the equilateral triangle configuration with the side length $R = 2.52$ Å. This value corresponds to 90% of the double helium atom van der Waals radius[53] (1.40 Å). The MP2 calculation, gives the energy for the three atoms equal to $E_{He_3} = -8.5986551$ a.u., for a pair of the atoms: $E_{He_2} = -5.7325785$ a.u., and for a single atom: $E_{He} = -2.8663605$ a.u. Therefore, the interaction energy of a single atomic pair is equal to[54] $\Delta E_{\mathrm{He\,He}}(2.3) = 0.0001425$ a.u., three such interactions give what is known as the two-body contribution (or the two-body effect) $\Delta E_{\text{two-body}} = \sum_{i<j}^{3} \Delta E_{A_i A_j}(2, 3) = 0.0001425 \cdot 3 = 0.0004275$ a.u. The three-body effect represents the rest; i.e., it is calculated as the difference between E_{He_3} and the sum of the one-body and two-body effects $\Delta E_{\mathrm{He\,He\,He}}(3.3) \equiv \Delta E_{\text{three-body}} = E_{He_3} - E_{He} \cdot 3 - \Delta E_{\text{two-body}} = -0.000001$ a.u. Therefore, we get the ratio $\left|\frac{\Delta E_{\text{three-body}}}{\Delta E_{\text{two-body}}}\right| = 0.26\%$. This means the three-body effect is quite small.

Example. Locality or Communication

Do two helium atoms sense each other when separated by a bridge molecule? In the other words, do the intermolecular interactions have local character or not? By *local character*, we mean that their influence is limited to, say, the nearest-neighbor atoms.[55] Let us check this by considering a bridge molecule (two cases: butane and butadiene) between two helium atoms. The first helium atom pushes the terminal carbon atom of the butane (the distance $R_{C\,He} = 1.5$ Å, the attack is perpendicular to the CC bond). The question is whether the second helium atom in the same position but at the other end of the molecule will feel that pushing or not (*the positions of all the nuclei do not change,* since we are within the Born-Oppenheimer approximation)? This means that we are interested in a three-body effect.

Let us calculate the three-body interaction for He...butane...He and for He...butadiene ...He. A large three-body effect would mean a communication between the two helium atoms through the electronic structure of the bridge. For the butane bridge, we get $\left|\frac{\Delta E_{\text{three-body}}}{\Delta E_{\text{two-body}}}\right| = 0.02\%$, and for the butadiene bridge, we obtain $\left|\frac{\Delta E_{\text{three-body}}}{\Delta E_{\text{two-body}}}\right| = 0.53\%$. Thus, we see the locality, but it is quite remarkable that there is an order of magnitude larger helium-helium communication in the case of butadiene (the distance between the terminal carbon atoms is similar in both cases:

[53] The notion of the van der Waals radius of an atom (which will be introduced a bit later), even if very useful, is arbitrary. The abovementioned 90% is set for didactic purposes in order to make the effect clearly visible.

[54] The description of the MP2 method is on p. 649. In the calculations reported here, the 6-31G(d) basis set is used, and there is no BSSE correction.

[55] The intermolecular interactions in general are certainly non-local in the abovementioned sense. As an example, look at the allosteric effect: when one ligand binds to a protein, it may help (or inhibit) the binding of another ligand that is far away. Crucially, there are conformational transformations, which are absent in our example.

3.7 Å for butadiene and 3.9 Å for butane), which is known for having the conjugated double and single CC bonds: = − =, while butane has only the single CC bonds: − − −. It seems that here we are touching one of the aspects of the all-important difference in chemistry between the labile π electron structure (butadiene) and the stiff σ electron structure (butane).

13.11.3 What Is Additive, and What Is Not?

Already vast experience has been accumulated in this area, and some generalizations are possible.[56,57] For three argon atoms in an equilibrium configuration, the three-body term is of the order of 1%. It should be noted, however, that in the argon *crystal*, there are a lot of three-body interactions and the three-body effect increases to about 7% . On the other hand, for liquid water, the three-body effect is of the order of 20% and the higher contributions are about 5%. Three-body effects are sometimes able to determine the crystal structure and have significant influence on the physical properties of the system close to a phase transition ("*critical region*").[58] In the case of the interaction of metal atoms, the non-additivity is much larger than that for the noble gases, and the three-body effects may attain a few tens of percentage points.

This is important information since the force fields widely used in molecular mechanics (see p. 345) are based almost exclusively on effective pairwise interactions (neglecting the three- and more-body contributions).[59]

Although the intermolecular interactions are non-additive, we may ask whether individual contributions to the interaction energy (electrostatic, induction, dispersion, or valence repulsion) are additive?

Let us begin from the electrostatic interaction.

13.11.4 Additivity of the Electrostatic Interaction

Suppose that we have three molecules A, B, and C, and intermolecular distances are long. Therefore, it is possible to use the polarization perturbation theory in a very similar way to that presented in the case of two molecules (p. 805). In this approach, the unperturbed Hamiltonian $\hat{H}^{(0)}$ represents the sum of the Hamiltonians for the isolated molecules A, B, and C. Let us change the abbreviations a little bit to be more concise for the case of three molecules. A *product* function $\psi_{A,n_A}\psi_{B,n_B}\psi_{C,n_C}$ will be denoted by $|n_A n_B n_C\rangle = |n_A\rangle\,|n_B\rangle\,|n_C\rangle$, where n_A, n_B, n_C $(= 0, 1, 2, \ldots)$ stand for the quantum numbers corresponding to the orthonormal wave functions

[56] V. Lotrich and K. Szalewicz, *Phys.Rev.Letters*, *79*, 1301(1997).

[57] In quantum chemistry, this almost always means a numerical convergence; i.e., a fast decay of individual contributions.

[58] R. Bukowski and K. Szalewicz, *J.Chem.Phys.*, *114*, 9518 (2001).

[59] That is, the effectivity of a force field relies on choice of interaction parameters such that the experimental data are reproduced (in such a way that the parameters implicitly contain part of the higher-order terms).

for the molecules A,B,C, respectively. The functions $|n_A n_B n_C\rangle = |n_A\rangle |n_B\rangle |n_C\rangle$ are the eigenfunctions of $\hat{H}^{(0)}$:

$$\hat{H}^{(0)} |n_A n_B n_C\rangle = [E_A(n_A) + E_B(n_B) + E_C(n_C)] |n_A n_B n_C\rangle,$$

which is analogous to Eqs. (13.7) and (13.8).

The perturbation is equal to $\hat{H} - \hat{H}^{(0)} = V = V_{AB} + V_{BC} + V_{AC}$, where the operators V_{XY} contain all the Coulomb interaction operators involving the nuclei and electrons of molecule X and those of molecule Y.

Let us recall that the electrostatic interaction energy $E_{\text{elst}}(ABC)$ of the ground state ($n_A = 0, n_B = 0, n_C = 0$) molecules is defined as the first-order correction to the wave function in the polarization approximation perturbation theory[60]:

$$E_{\text{pol}}^{(1)} \equiv E_{\text{elst}}(ABC) = \langle 0_A 0_B 0_C | V | 0_A 0_B 0_C \rangle = \langle 0_A 0_B 0_C | V_{AB} + V_{BC} + V_{AC} | 0_A 0_B 0_C \rangle,$$

where the quantum numbers 000 have been supplied (maybe because of my excessive caution) by the redundant and self-explanatory indices $(0_A, 0_B, 0_C)$.

The integration in the last formula goes over the coordinates of all electrons. In the polarization approximation, the electrons can be unambiguously divided into three groups: those belonging to A, to B, and to C. Because the zero-order wave function $|0_A 0_B 0_C\rangle$ represents a product $|0_A\rangle |0_B\rangle |0_C\rangle$, the integration over the electron coordinates of one molecule can be easily performed and yields

$$E_{\text{elst}}(A, B, C) = \langle 0_A 0_B | \hat{V}_{AB} | 0_A 0_B \rangle + \langle 0_B 0_C | \hat{V}_{BC} | 0_B 0_C \rangle + \langle 0_A 0_C | \hat{V}_{AC} | 0_A 0_C \rangle,$$

where, in the first term, the integration was performed over the electrons of C, in the second over the electrons of A, and in the third over those of B.

Now, let us look at the last formula. We easily see that the individual terms simply represent the electrostatic interaction energies of pairs of molecules: AB, BC, and AC, that we would obtain in the perturbational theory (within the polarization approximation) for the interaction of AB, BC, and AC, respectively. Conclusion:

> the electrostatic interaction is pairwise additive.

13.11.5 Exchange Non-additivity

What about the exchange contribution? This contribution *does not exist in the polarization approximation*. It appears only in SAPT in pure form in the first-order energy correction and

[60] The $E_{\text{elst}}(ABC)$ term in SAPT represents only part of the first-order correction to the energy (the rest being the valence repulsion).

coupled to other effects in higher-order energy corrections.[61] The exchange interaction is difficult to interpret because it appears as a result of the antisymmetry of the wave function (Pauli exclusion principle). The antisymmetry is forced by the postulates of quantum mechanics (p. 34), and its immediate consequence is that the probability density of finding two electrons with the same spin and space coordinates is equal to zero.

Pauli Deformation

The Pauli exclusion principle leads to a *deformation* of the wave functions describing the two molecules (by projecting the productlike wave function by the antisymmetrizer $\hat{A}$) with respect to the productlike wave function. The Pauli deformation (cf. Appendix Y available at booksite.elsevier.com/978-0-444-59436-5) appears already in the zeroth order of perturbation theory, whereas in the polarization approximation, the deformation of the wave function appears in the first order and is not related to the Pauli exclusion principle.

The antisymmetrizer pertains to the permutation symmetry of the wave function with respect to the coordinates of all electrons and therefore is different for a pair of molecules and for a system of three molecules.

The expression for the three-body non-additivity of the valence repulsion[62] is given by Eq. (13.42), based on Eq. (13.39) of the first-order correction in SAPT[63] and Eq. (13.70) of the three-body contribution:

$$E^{(1)}_{\text{exch},ABC} = N_{ABC}\langle 0_A 0_B 0_C | \hat{V}_{AB} + \hat{V}_{BC} + \hat{V}_{AC} | \hat{A}_{ABC}(0_A 0_B 0_C)\rangle - \sum_{(XY)=(AB),(AC),(BC)} N_{XY}\langle 0_X 0_Y | \hat{V}_{XY} | \hat{A}_{XY}(0_X 0_Y)\rangle, \tag{13.71}$$

where $N_{ABC}\hat{A}_{ABC}\,|0_A 0_B 0_C\rangle$ and $N_{AB}\hat{A}_{AB}\,|0_A 0_B\rangle$, and so forth represent the normalized (N_{ABC}, etc. are the normalization coefficients) antisymmetrized productlike wave function of the systems ABC, AB, etc. The antisymmetrizer $\hat{A}_{ABC}$ pertains to subsystems A, B, and C; similarly, $\hat{A}_{AB}$ pertains to A and B, etc., all antisymmetrizers containing only the intersystem electron exchanges and the summation goes over all pairs of molecules.

There is no chance of proving that the exchange interaction is additive; i.e., that Eq. (13.71) gives 0. Let us consider the simplest possible example: each molecule has only a single electron: $|0_A(1)0_B(2)0_C(3)\rangle$. The operator $\hat{A}_{ABC}$ makes (besides other terms) the following permutation: $\hat{A}_{ABC}\,|0_A(1)0_B(2)0_C(3)\rangle = \cdots - \frac{1}{(N_A+N_B+N_C)!}\,|0_A(3)0_B(2)0_C(1)\rangle + \cdots$, which according to

[61] Such terms are bound to appear. For instance, the induction effect is connected to deformation of the electron density distribution. The interaction (electrostatic, exchange, dispersive, etc.) of such a deformed object will change with respect to that of the isolated object. The coupling terms take care of this change.

[62] B. Jeziorski, M. Bulski, and L. Piela, *Intern.J.Quantum Chem.*, *10*, 281 (1976).

[63] Because, as we have already proved, the rest (i.e., the electrostatic energy) is an additive quantity.

Eq. (13.71) leads to the integral $-\frac{1}{(N_A+N_B+N_C)!} N_{ABC} \langle 0_A(1)0_B(2)0_C(3)| \frac{1}{r_{12}} |0_A(3)0_B(2)0_C(1)\rangle$ $= -\frac{1}{(N_A+N_B+N_C)!} N_{ABC} \langle 0_A(1)0_B(2)| \frac{1}{r_{12}} |0_B(2)0_C(1)\rangle$, involving the wave functions centered on A, B and C. This means that the term belongs to the *three-body* effect.

The permutation operators of which the $\hat{A}_{ABC}$ operator is composed correspond to the identity permutation[64] as well as to the exchange of one or more electrons between the interacting subsystems: $\hat{A}_{ABC} = 1+$ single exchanges $+$ double exchanges$+\cdots$

It is easy to demonstrate[65] that

> *the larger the number of electrons exchanged, the less important such exchanges are* because the resulting contributions would be proportional to higher and higher powers of the overlap integrals (S), which are small as a rule.

SE Mechanism[66]

The smallest nonzero number of electron exchanges in $\hat{A}_{ABC}$ equals 1. Such an exchange may take place only between two molecules–eg., AB.[67] This results in terms proportional to S^2 in the three-body expression, where S means the overlap integral between the molecular orbitals of A and B. The third molecule *does not participate in the electron exchanges, but is not just a spectator in the interaction.* If it were, the interaction would not be three-body (see Fig. 13.12).

> **SE Mechanism**
> Molecule C interacts electrostatically with the *Pauli deformation of molecules A and B (i.e., with the multipoles that represent the deformation).* Such a mixed interaction is called the *SE mechanism.*

When the double electron exchanges are switched on, we would obtain part of the three-body effect of the order of S^4. Since S is usually of the order of 10^{-2}, this contribution is expected to be small, although caution is advised, because the *number* of such terms is much more important.

---

[64] The operator reproduces the polarization approximation expressions in SAPT.

[65] The procedure is as follows:

1. First,we write down the exact expression for the first-order exchange non-additivity.
2. Then, we expand the expression in the Taylor series with respect to those terms that arise from all electron exchanges except the identity permutation.
3. Next, we see that the exchange non-additivity expression contains terms of the order of S^2 and higher, where S stands for the overlap integrals between the orbitals of the interacting molecules.

[66] In this expression, SE stands for single exchange.

[67] After that, we have to consider AC and BC.

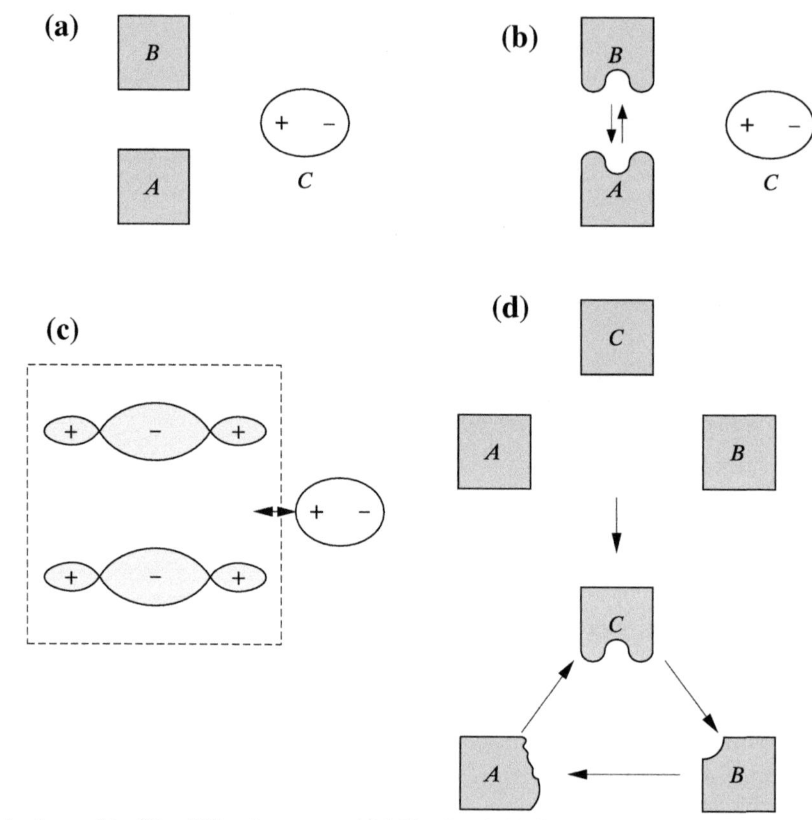

Fig. 13.12. A scheme of the SE and TE exchange non-additivities. Panels (a), (b), and (c) show the SE mechanism. (a) Three non-interacting molecules (schematic representation of electron densities). (b) Pauli deformation of molecules A and B. (c) Electrostatic interaction of the Pauli deformation resulting from single electron exchanges between A and B with the dipole moment of C. (d) The TE mechanism: molecules A and B exchange an electron with the mediation of molecule C. All molecular electron density distributions undergo Pauli deformation.

TE Mechanism[68]

Is there any contribution of the order of S^3? Yes. The antisymmetrizer $\hat{A}_{ABC}$ is able to make the single electron exchange between, say, A and B, but by mediation of C. The situation is schematically depicted in Fig. 13.12.

> **TE Mechanism**
> A molecule is involved in a single exchange with another molecule by mediation of a third one.

[68] In this expression, TE stands for triple exchange.

Exchange interaction is of short-range character

The Pauli deformation has a local (i.e., short-range) character; see Appendix Y available at booksite.elsevier.com/978-0-444-59436-5. Let us imagine that molecule B is very long and the configuration corresponds to A B C. When C is far from A, the three-body effect is extremely small because almost everything in the interaction is of the two-body character. If molecule C approaches A and has some nonzero, low-order multipoles (e.g., a charge), then it may interact by the SE mechanism even from far away. Both mechanisms (SE and TE) operate only at short BA distances.

The exchange interaction is non-additive, but the effects pertain to the contact region of both molecules. The Pauli exclusion principle does not have any finite range in space; i.e., after being introduced, it has serious implications for the wave function even at infinite intermolecular distances (cf. p. 829). Despite this, it always leads to the differential overlap of atomic orbitals (as in overlap or exchange integrals), which decays exponentially with increasing intermolecular distance. This may not be true for the SE mechanism, which has a partly long-range character and some caution is needed, when considering its spatial range.

13.11.6 Induction Non-additivity

The non-additivity of the intermolecular interaction results mainly from the non-additivity of the induction contribution.

How do we convince ourselves about the non-additivity? This is very easy. It will be sufficient to write the expression for the induction energy for the case of three molecules and to see whether it gives the sum of the pairwise induction interactions. Before we do this, let us write the formula for the total second-order energy correction (similar to the case of two molecules on p. 807):

$$E^{(2)}(ABC) = \sum_{n_A,n_B,n_C}{}' \frac{|\langle n_A n_B n_C|V|0_A 0_B 0_C\rangle|^2}{[E_A(0_A) - E_A(n_A)] + [E_B(0_B) - E_B(n_B)] + [E_C(0_C) - E_C(n_C)]}. \tag{13.72}$$

According to perturbation theory, the term with all the indices equal to zero has to be omitted in the above expression (symbolized by $\sum'$). It is much better like this because otherwise, the denominator would "cause explosion". Further, the terms with all nonzero indices are equal to zero. Indeed, let us recall that V is the sum of the Coulomb potentials corresponding to all three *pairs* of the three molecules. This is the reason why it is easy to perform the integration over the electron coordinates of the third molecule (not involved in the pair). A similar operation has been performed for the electrostatic interaction. This time, however, the integration makes the term equal to zero because of the orthogonality of the ground and excited states of the third molecule. All this leads to the conclusion that to have a nonzero term in the summation, there has *to be one index or two indices of zero value* among the three indices. Let us perform the summation in two stages: all the terms with only-two-zeros (or a single nonzero) indices will make a contribution to $E_{ind}(ABC)$, while all the terms with only-one-zero (or two nonzero) indices will sum to $E_{disp}(ABC)$:

$$E^{(2)}(ABC) = E_{ind}(ABC) + E_{disp}(ABC), \tag{13.73}$$

where the first term represents the *induction energy*:

$$E_{ind}(ABC) = E_{ind}\left(AB \to C\right) + E_{ind}\left(AC \to B\right) + E_{ind}\left(BC \to A\right),$$

where
$E_{ind}\left(BC \to A\right) \equiv \sum_{n_A \neq 0} \frac{|\langle n_A 0_B 0_C|V|0_A 0_B 0_C\rangle|^2}{E_A(0_A)-E_A(n_A)}$ means that the "frozen" molecules B and C *acting together* polarize molecule A, etc. The second term in Eq. (13.73) represents the *dispersion energy* (this will be considered later in this chapter).

For the time being, let us consider the induction energy $E_{ind}(ABC)$. Writing V as the sum of the Coulomb interactions of the pairs of molecules, we have

$$E_{ind}\left(BC \to A\right) = \sum_{n_A \neq 0} \frac{|\langle n_A 0_B 0_C|V_{AB} + V_{BC} + V_{AC}|0_A 0_B 0_C\rangle|^2}{E_A(0_A) - E_A(n_A)}$$
$$= \sum_{n_A \neq 0} \frac{|\langle n_A 0_B|V_{AB}|0_A 0_B\rangle + \langle n_A 0_C|V_{AC}|0_A 0_C\rangle|^2}{E_A(0_A) - E_A(n_A)}.$$

Look at the square in the numerator. The induction non-additivity arises because of this. If the square were equal to the sum of squares of the two components, the total expression shown explicitly would be equal to the induction energy corresponding to the polarization of A by the frozen charge distribution of B plus a similar term corresponding to the polarization of A by C (i.e., the polarization occurring *separately*). Together with the other such terms in $E_{ind}\left(AB \to C\right) + E_{ind}\left(AC \to B\right)$, *we would obtain the additivity* of the induction energy $E_{ind}\left(ABC\right)$. However, besides the sum of squares, we also have the mixed terms. They will produce the non-additivity of the induction energy:

$$E_{ind}(ABC) = E_{ind}(AB) + E_{ind}(BC) + E_{ind}(AC) + \Delta_{ind}(ABC). \tag{13.74}$$

Thus, we obtain the following expression for the *induction non-additivity* $\Delta_{ind}(ABC)$:

$$\Delta_{ind}(ABC) = 2\mathrm{Re} \sum_{n_A \neq 0} \frac{\langle n_A 0_B|V_{AB}|0_A 0_B\rangle \langle n_A 0_C|V_{AC}|0_A 0_C\rangle}{[E_A(0_A) - E_A(n_A)]} + \cdots, \tag{13.75}$$

where the term displayed explicitly is the non-additivity of $E_{\text{ind}}\left(BC \rightarrow A\right)$ and "$+\cdots$" stands for the non-additivities of $E_{\text{ind}}\left(AB \rightarrow C\right) + E_{\text{ind}}\left(AC \rightarrow B\right)$.

Example. Induction Non-additivity

To show that the induction interaction of two molecules depends on the presence of the third molecule, let us consider the system shown in Fig. 13.13.

Let molecule B be placed halfway between A^+ and C^+; thus, the configuration of the system is: A^+.......B.........C^+, with long distances between the subsystems. In such a situation, the total interaction energy is practically represented by the induction contribution plus the constant electrostatic repulsion of A^+ and C^+. Is the three-body term (induction non-additivity) large? We will show soon that this term is large and positive (destabilizing). Since the electric field intensities nearly cancel within molecule B, then despite the high polarizability of the latter, the induction energy will be small. On the contrary, the opposite is true when considering two-body interaction energies. Indeed, A^+ polarizes B very strongly, C^+ does the same, resulting in high stabilization due to high two-body induction energy. Since the total effect is nearly zero, the induction non-additivity is bound to be a large positive number.[69]

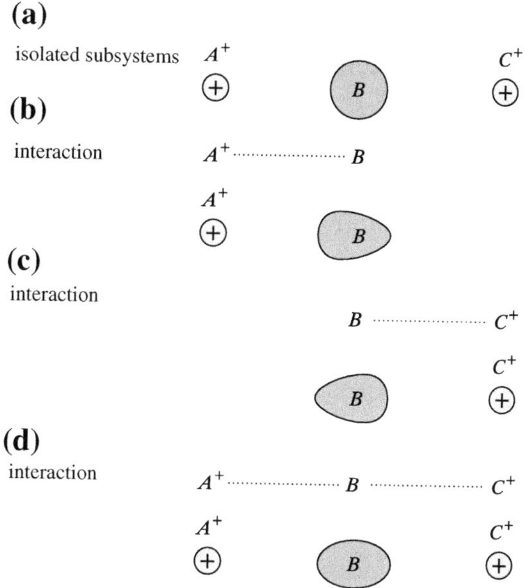

Fig. 13.13. The induction interaction may produce a large non-additivity. (a) Two distant non-polarizable cations: A^+, C^+, and a small, polarizable neutral molecule B placed exactly in the middle between AC. (b) The two-body induction interaction A^+B, a strong polarization. (c) The two-body induction interaction BC^+, a strong polarization. (d) The two cations polarize molecule B. Their electric field vectors cancel each other out in the middle of B and give a small electric field intensity within B (a weak polarization).

[69] If the intermolecular distances were small, B were not in the middle of AC or molecule B were of large spatial dimension, the strength of our conclusion would diminish.

Self-consistency and Polarization Catastrophe

The second-order induction effects pertain to polarization by the charge distributions corresponding to the isolated molecules. However, the induced multipoles introduce a change in the electric field and in this way contribute to further changes in charge distribution. These effects already belong to the third[70] and higher orders of perturbation theory.

It is therefore evident that a two-body interaction model cannot manage the induction interaction energy. This is because we have to ensure that any subsystem (e.g., A), has to experience polarization in an electric field which is the *vector sum* of the electric fields from all its partner subsystems (B,C,...) and calculated at the position of A. The calculated induced dipole moment of A (we focus on the lowest multipole) creates the electric field that produces some changes in the dipole moments of B,C,..., which in turn change the electric field acting on all the molecules, including A. The circle closes and the polarization procedure has to be performed until self-consistency is reached. This can often be done, although such a simplified interaction model does not allow for geometry optimization, which may lead to a *polarization catastrophe* ending up with induction energy equal to $-\infty$ (due to excessive approach and lack of the Pauli blockade described on p. 840).

13.11.7 Additivity of the Second-order Dispersion Energy

The dispersion energy is a second-order correction, [see Eq. (13.13)] that gives the formula for interaction of two molecules. For *three* molecules, we obtain the following formula for the dispersion part of the second-order effect (see the discussion on p. 855):

$$E_{\text{disp}}(ABC) = \sum_{(n_A,n_B)\neq(0_A,0_B)} \frac{|\langle n_A n_B 0_C | V_{AB} + V_{BC} + V_{AC} | 0_A 0_B 0_C\rangle|^2}{[E_A(0_A) - E_A(n_A)] + [E_B(0_B) - E_B(n_B)]} + \cdots,$$

where $+\cdots$ denotes analogous terms with summations over n_A, n_C as well as n_B, n_C. Of the three integrals in the numerator, only the first one will survive, since the other vanish due to the integration over the coordinates of the electrons of molecule Z not involved in the interaction V_{XY}:

$$E_{\text{disp}}(ABC) = \sum_{(n_A,n_B)\neq(0_A,0_B)} \frac{|\langle n_A n_B 0_C | V_{AB} | 0_A 0_B 0_C\rangle + 0 + 0|^2}{\left[E_A(0_A) - E_A(n_A)\right] + \left[E_B(0_B) - E_B(n_B)\right]} + \sum_{(n_A,n_C)\neq(0_A,0_C)} \cdots + \sum_{(n_B,n_C)\neq(0_B,0_C)} \cdots$$

[70] Each of the induced multipoles is proportional to V, their interaction introduces another V; all together, this gives a term proportional to VVV; i.e., indeed of the third order.

In the first term, we can integrate over the coordinates of C. Then the first term displayed in the above formula turns out to be the dispersion interaction of A and B:

$$E_{\text{disp}}(ABC) = \sum_{(n_A,n_B)\neq(0,0)} \frac{|\langle n_A n_B|V_{AB}|0_A 0_B\rangle|^2}{[E_A(0_A)-E_A(n_A)]+[E_B(0_B)-E_B(n_B)]} + \sum_{(n_A,n_C)\neq(0_A,0_C)} \cdots + \sum_{(n_B,n_C)\neq(0_B,0_C)} \cdots = E_{\text{disp}}(AB)+E_{\text{disp}}(AC)+E_{\text{disp}}(BC).$$

Thus, we have proved that

the dispersion interaction (second-order of the perturbation theory) *is additive*:

$$E_{\text{disp}}(ABC) = E_{\text{disp}}(AB)+E_{\text{disp}}(AC)+E_{\text{disp}}(BC).$$

13.11.8 Non-additivity of the Third-order Dispersion Interaction

One of the third-order energy terms represents a correction to the dispersion energy. The correction, as shown by Axilrod and Teller,[71] has a three-body character. The part connected to the interaction of three distant instantaneous dipoles on A, B and C reads as

$$E^{(3)}_{\text{disp}} = 3C^{(3)}_{ddd}\frac{1+3\cos\theta_A\cos\theta_B\cos\theta_C}{R^3_{AB}R^3_{AC}R^3_{BC}}, \tag{13.76}$$

where R_{XY} and θ_X denote the sides and the angles of the ABC triangle, and $C^{(3)}_{ddd} > 0$ represents a constant. The formula shows that

when the ABC system is in a linear configuration, the dispersion contribution is negative (i.e., *stabilizing*), while the equilateral triangle configuration corresponds to a *destabilization*.

[71] B.M. Axilrod and E. Teller, *J.Chem.Phys.*, 11299 (1943). The general anisotropic many-body intermolecular potentials for a system of N molecules have been derived by P. Piecuch, *Mol. Phys.*, 59 (1986)10671085.

ENGINEERING OF INTERMOLECULAR INTERACTIONS

13.12 Idea of Molecular Surface

13.12.1 Van der Waals Atomic Radii

It would be of practical importance to know how close two molecules can be to each other. We will not entertain this question too seriously, though, because this problem cannot have an elegant solution: it depends on the direction of the approach and the atoms involved, as well as how strongly the two molecules collide. Searching for the effective radii of atoms would be nonsense, if the valence repulsion were not a sort of "soft wall" or if the atom sizes were very sensitive to molecular details. Fortunately, it turns out that an atom, despite different roles played in molecules, can be characterized by its approximate radius, called the *van der Waals radius*. The radius may be determined in a naive but quite effective way. For example, we may approach two HF molecules like that : H-F...F-H, axially with the fluorine atoms heading on, then find the distance[72] R_{FF} at which the interaction energy is equal to, say, 5 kcals/mol. The proposed fluorine atom radius would be $r_F = \frac{R_{FF}}{2}$. A similar procedure may be repeated with two HCl molecules with the resulting r_{Cl}. Now, let us consider an axial complex H-F....Cl-H with the intermolecular distance corresponding to 5 kcals/mol. What F...Cl distance are we expecting? We expect something close to $r_F + r_{Cl}$. It turns out that we are about right. This is why the atomic van der Waals radius concept is so attractive from the practical point of view.

13.12.2 Definition of Molecular Surface

We may construct a superposition of atomic van der Waals spheres. This defines what is called the *van der Waals surface* of the molecule,[73] or a molecular shape–which is a concept of great importance and of the same arbitrariness as the radii themselves.

In a similar way, we may define *ionic radii*[74] to reproduce the ion packing in ionic crystals, as well as *covalent radii* to foresee chemical bond lengths.

13.12.3 Confining Molecular Space–The Nanovessels

Molecules at long distances interact through the mediation of the electric fields created by them. The valence repulsion is of a different character, since it results from the Pauli exclusion

---

[72] This may be done using a reliable quantum mechanical method.

[73] The van der Waals surface of a molecule may sometimes be very complex; e.g., a molecule may have two or more surfaces (like fullerenes).

[74] This concept was introduced by Pauling based on crystallographic data [L. Pauling, *J.Amer.Chem.Soc.*, *49*, 765 (1927)].

principle, and may be interpreted as an energy penalty for an attempt by electrons of the same spin coordinate to occupy the same space (cf. Chapter 1 and p. 597).

Luty and Eckhardt[75] have highlighted the role of pushing one molecule by another. Let us imagine an atomic probe, such as a helium atom. The pushing by the probe deforms the molecular electronic wave function (Pauli deformation), but the motion of the electrons is accompanied by the motion of the nuclei. Both motions may lead to dramatic events. For example, we may wonder how an explosive reaction takes place. The spire hitting the material in its metastable chemical state is similar to the helium atom probe pushing a molecule. Due to the pushing, the molecule distorts to such an extent that the HOMO-LUMO separation vanishes and the system rolls down (see Chapter 14) to a deep potential energy minimum on the corresponding potential energy hypersurface. The HOMO-LUMO gap closing takes place within the reaction barrier. Since the total energy is conserved, the large reaction net energy gain goes to highly excited vibrational states (in the classical approximation corresponding to large-amplitude vibrations). The amplitude may be sufficiently large to ensure the pushing of the next molecules in the neighborhood and a chain reaction starts with exponential growth.

Now imagine a lot of atomic probes confining the space (like a cage or template) available to a molecule under study. In such a case, the molecule will behave differently from a free one. For example, consider the following:

- A protein molecule, when confined, *may fold to another conformation.*
- Some photochemical reactions that require a space for the rearrangement of molecular fragments *will not occur*, if the space is not accessible.
- In a restricted space, some other chemical reactions will take place (new chemistry - *chemistry in "nanovessels"*).
- Some unstable molecules may *become stable* when enclosed in a nanovessel.

These are fascinating and little-explored topics.

13.12.4 Molecular Surface Under High Pressure

Chemical reactions in nanovessels, which may run in another way than without them, suggest that the atomic structure of materials may change under increasing pressure. Suppose that we are dealing with a molecular crystal and we are studying what happens after applying isotropic pressure.

The transformations after increasing pressure are characterized by the following rules of thumb:

- The weaker the interatomic bond (whether chemical or of the van der Waals character) is, the more its length changes. As a manifestation of this, the pressure after passing a threshold

[75] T. Luty and C.J. Eckhardt, in *Reactivity of Molecular Solids*, eds.E. Boldyreva and V. Boldyrev, Wiley, New Jersey, 1999, p.51.

value may change the packing of the molecules, leaving the molecules themselves virtually unchanged (*displacive phase transitions*).

- Then, at higher pressure, the chemical bonds, often of mixed character (i.e., ionic and covalent polarized bonds) become more and more covalent. This is because the overlap integrals increase very much due to shorter interatomic distances (exponential increase) and as a consequence, the covalent bond character becomes stronger (cf. p. 437).
- For even larger pressures, the atomic coordination numbers begin to increase. This corresponds to formation of new chemical bonds, which make the atoms to be closer in space. In order to make the new bonds, the atoms, when squeezed, open their closed core electronic shells (e.g., under sufficiently high pressures, even the noble gas atoms open their electronic shells; cf. Chapter 14). Some new phase transitions are accompanying these changes, known as *reconstructive phase transitions*.
- At even higher pressures, the crystal structure becomes the close-packed type, which means a further minimization of the crystal's empty spaces.
- Empirical observation: the atoms when under high pressure behave similarly as the atoms belonging to the same group and the next period of the Mendeleev table do at lower pressures. For example, silicon under high pressure behaves as germanium under lower pressure, etc.

13.13 Decisive Forces

13.13.1 Distinguished Role of the Valence Repulsion and Electrostatic Interaction

The valence repulsion plays the role of a hard wall (covered by a "soft blanket") that forbids the closed-shell molecules to approach too closely (this represents a very important factor, since those molecules that do not fit together receive an energy penalty).

The electrostatic contribution plays a prominent role in the intermolecular interaction, since the electrostatic forces already operate effectively at long intermolecular distances (their range may be reduced in polar solvents).

The induction and dispersion contributions, even if they are sometimes larger than the electrostatic interaction, usually play a less important role. This is because only the electrostatics may change the sign of the energy contribution when the molecules reorient, thus playing the pivotal role in creating structure.

The induction and dispersion contributions are negative (almost independent of the mutual orientation of the molecules), and we may say, as a rule of thumb, that their role is to stabilize the structure already determined by the valence repulsion and the electrostatics.

13.13.2 Hydrogen Bond

Among the electrostatic interactions, the most important are those having a strong dependence on orientation. The most representative of them are the *hydrogen bonds (also known as hydrogen bridges) X-H ...Y, where an electronegative atom X plays the role of a proton donor, while an electronegative atom Y – plays the role of a proton acceptor (see Fig. 13.14). Most often, the hydrogen bond X-H ...Y deviates only a little from linearity.* Additionally, the XY separation usually falls into a narrow range: 1.7–2.0 Å; i.e., it is to a large extent independent of X and Y. The hydrogen bond features are unique because of the extraordinary properties of the hydrogen atom in the hydrogen bridge. This is the only atom that occasionally may attain the partial charge equal to 0.45 a.u., which means that it represents a nucleus devoid to a large extent of the electron density. This is one of the reasons why the hydrogen bond is strong when compared to other types of intermolecular interactions.

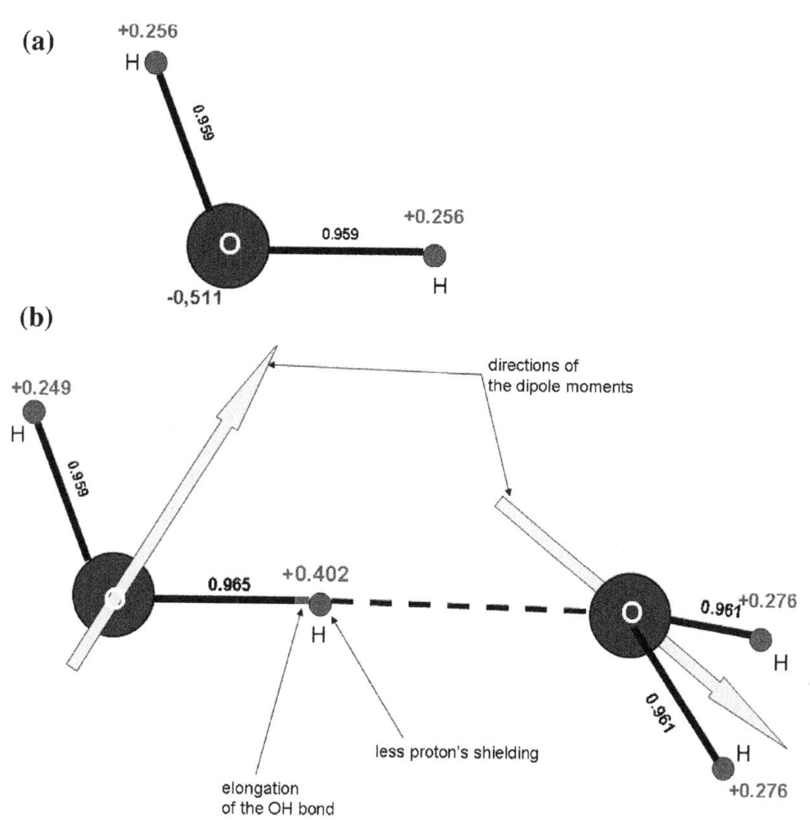

Fig. 13.14. The hydrogen bond in the water dimer described by the MP2 method. The optimal configuration (a) of an isolated water molecule and (b) of the complex of two water molecules. The numbers displayed at the nuclei represent their Mulliken net charges (in a.u.), at the bonds–their lengths in Å. Conclusion: the hydrogen bridge $O - H \ldots O$ is almost linear and the bridge hydrogen atom is devoid of some 40% of its electronic cloud (exceptional case in chemistry).

Example. Hydrogen Bond in the Water Dimer

This is the most important hydrogen bond on Earth, because liquid water means a prerequisite of life. Moreover, since it may be thought as being representative of other hydrogen bonds, we treat the present example as highlighting the main consequences of formation of a generic hydrogen bond. We will perform the quantum mechanical calculations (of medium quality) within the Born-Oppenheimer approximation, including the electron correlation and optimizing the system geometry.[76]

In Fig. 13.14a, the optimized geometry of the isolated water molecule is shown ($r_{OH} = 0.959$ Å, $\measuredangle$ HOH $= 103.4^0$), as well as the net Mulliken charges on the individual atoms[77] ($Q_O = -0.511$, $Q_H = 0.256$). All these numbers change, when two water molecules form the hydrogen bond (Fig. 13.14b). We see the following changes:

- The complex has C_{2v} symmetry with the proton donor molecule and the oxygen atom of the proton acceptor molecule within the symmetry plane (the book's page).
- The optimized oxygen-oxygen distance turns out to be $R_{OO} = 2.91$ Å, the hydrogen bond is *almost linear* ($\measuredangle$OH...O $= 176.7^0$), and the optimized O-H bond length (of the hydrogen bridge) is $r_{OH,donor} = 0.965$ with a tiny upward deviation of the proton from the linearity of the OH...O bridge (the effect of repulsion with the protons of the acceptor). *This very common quasi-linearity suggests the electrostatics is important in forming hydrogen bonds.* If it were true, we should see this also from the relative position of the molecular dipoles. At first sight, we might be disappointed though, because the dipole moments are almost perpendicular. However, the two dipole moments deviate in the opposite directions from the collinear configuration with the deviation angle about $\pm 45^0$ (this means that they are nearly orthogonal). As seen in Fig. 13.7, despite such a large deviation, the electrostatics is still favorable.[78]
- We do not see any significant charge transfer between the molecules. Indeed, the Mulliken population analysis shows only a tiny $0.02e$ electron transfer from the proton donor to the proton acceptor.
- The donor molecule is perturbed more than the acceptor one. This is seen first of all from the Mulliken net charges on hydrogen atoms and from the deformation of geometry (see Fig. 13.14). One sees *a remarkable positive charge on the bridge hydrogen atom*

[76] The Møller-Plesset method, which is accurate up to the second order, is used (the MP2 method is described in Chapter 10), and the basis set is 6–311G(d,p). This ensures that the results will be of a decent quality.

[77] Cf. Appendix S available at booksite.elsevier.com/978-0-444-59436-5 on p. e143. The charges are calculated using the Hartree-Fock method that precedes the MP2 perturbational calculation.

[78] Let us calculate from Eq. (13.23): $E_{dip-dip} = \frac{\mu_A^2}{R^3}\left(\sin\theta_A \sin\theta_B - 2\cos\theta_A \cos\theta_B\right) = \frac{\mu_A^2}{R^3}\left(\frac{1}{\sqrt{2}}\left(-\frac{1}{\sqrt{2}}\right) - 2\frac{1}{\sqrt{2}}\frac{1}{\sqrt{2}}\right) = \frac{\mu_A^2}{R^3}\left(-\frac{3}{2}\right)$, as compared with the optimal value $-2\frac{\mu_A^2}{R^3}$. The loss of the dipole-dipole electrostatics is of the order of 25%. The energy loss must be more than compensated by other attractive interactions.

Table 13.5. Energy contributions to the interaction energy $E_{\rm int}$ in the system HO-H...OH_2 (hydrogen bond) calculated[a] within the SAPT method: electrostatic energy $E_{\rm elst}$, valence repulsion energy $E^{(1)}_{\rm exch}$, induction energy $E_{\rm ind}$ and dispersion energy $E_{\rm disp}$ for three O...O distances: equilibrium distance $R_{eq} = 3.00$ and two distances a little larger: medium 3.70 Å and large 4.76 Å.

Contributions to E_{int} (in kcal/mol)				
$R_{\rm OO}$(Å)	$E_{\rm elst}$	$E^{(1)}_{\rm exch}$	$E_{\rm ind}$	$E_{\rm disp}$
3.00	−7.12	4.90	−1.63	−1.54
3.70	−2.79	0.30	−0.18	−0.31
4.76	−1.12	0.00	−0.02	−0.05

[a] B. Jeziorski and M. van Hemert, *Mol.Phys.*, *31*, 713 (1976).

$Q_{{\rm H},donor} = 0.402$; no other nucleus in the whole chemistry is as poor in electrons. This hydrogen bond effect is confirmed in the NMR recordings (low shielding constant).

- The calculated binding energy[79] is $E_{bind} = E^{MP2} - 2E^{MP2}_{\rm H_2O} = -6.0\frac{kcal}{mol}$. *This order of magnitude of the binding energy, about* 20 *times smaller than the energy of chemical bonds, is typical for hydrogen bridges.*

Example. Water-Water Dimer–Perturbational Approach

Let us take once more the example of two water molecules to show the dominant role of electrostatics in the hydrogen bond.

As you can see from Table13.5, while at the equilibrium distance $R_{\rm OO} = 3.00$ Å, all the contributions are of importance (although the electrostatics dominates). All the contributions except electrostatics diminish considerably after increasing separation by only about 0.70 Å. For the largest separation ($R_{\rm OO} = 4.76$), the electrostatics dominates by far. This is why the hydrogen bond is said to have a mainly electrostatic character.[80]

13.13.3 Coordination Interaction

Coordination interaction appears if an electronic pair of one subsystem (electron donor) lowers its energy by interacting[81] with an electron acceptor offering an empty orbital; e.g., a cation (acceptor) interacts with an atom or atoms (donor) offering lone electronic pairs. This may be also seen as a special kind of electrostatic interaction.[82] Fig. 13.15a shows a derivative of porphyrin, and Fig. 13.15b a *cryptand* (the name comes from the ritual of eternal depositing in a crypt), both compounds offering lone pairs for the interaction with a cation.

[79] No BSSE correction is included.

[80] It has been proved that covalent structures (cf. p. 610) also contribute to the properties of the hydrogen bond, but their role decreases dramatically when the molecules move apart.

[81] In this interaction, a molecular orbital is formed.

[82] A lone pair has a large dipole moment (see Appendix T available at booksite.elsevier.com/978-0-444-59436-5), which interacts with the positive charge of the acceptor.

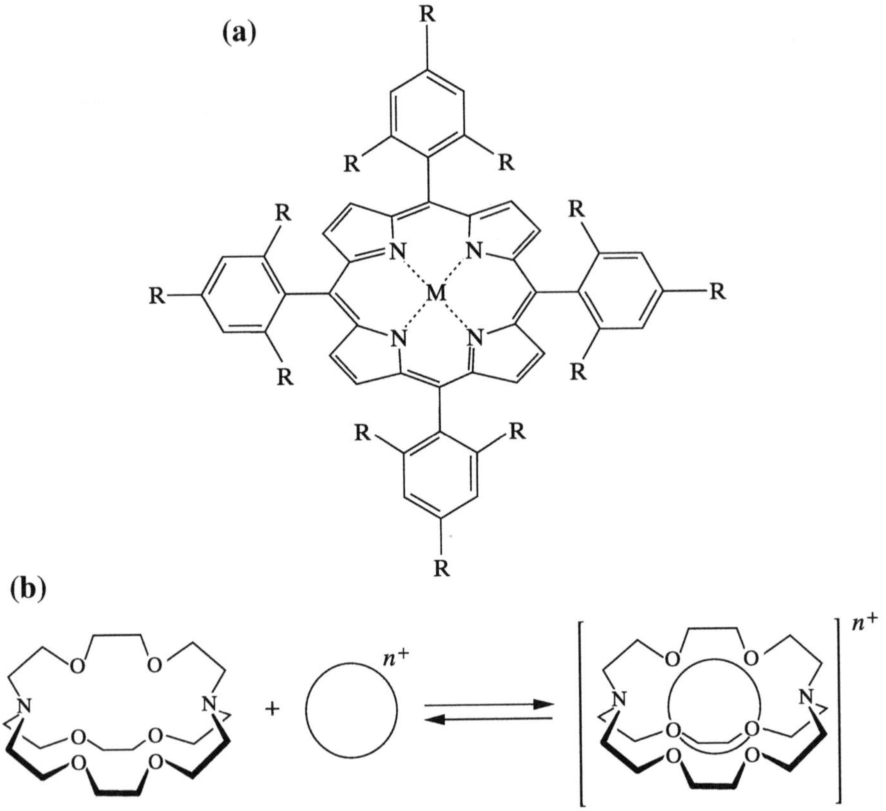

(a)

(b)

Fig. 13.15. A cation fits (a) the porphyrin ring or (b) the cryptand.

When concentrating on the ligands, we can see that in principle, they represent a negatively charged cavity (lone pairs) waiting for a monoatomic cation with *dimensions of a certain range only*. The interaction of such a cation with the ligand would be exceptionally large (selectivity of the interaction).

Let us consider a water solution containing ions: Li^+, Na^+, K^+, Rb^+, and Cs^+. After adding the abovementioned cryptand and after the equilibrium state is attained (ions/cryptand, ions/water and cryptand/water solvation), only for K^+ will the equilibrium be shifted toward the K^+/cryptand complex. For the other ions, the equilibrium will be shifted toward their association with water molecules, not the cryptand.[83] This is remarkable information.

[83] J.-M. Lehn, *Supramolecular Chemistry*, Institute of Physical Chemistry Publications (1993), p. 88: the equilibrium constants of the ion/cryptand association reactions are: for Li^+, Na^+, K^+, Rb^+, Cs^+ (only the order of magnitude is given): 10^2, 10^7, 10^{10}, 10^8, 10^4, respectively. As seen, the cryptand's cavity only fits well to the potassium cation.

We are able to selectively extract objects of some particular shape and dimensions.

13.13.4 Hydrophobic Effect

This is quite a peculiar type of interaction, which appears mainly (but not only) in water solutions.[84] The hydrophobic interaction does not represent any particular new interaction (beyond those we have already considered), because at least potentially, they could be explained by the electrostatic, induction, dispersion, valence repulsion, and other interactions already discussed.

The problem, however, may be seen from a different point of view. The basic interactions have been derived as if operating in a vacuum. However, in a medium, the molecules interact with one another through the mediation of other molecules, including those of the solvent. In particular, a water medium creates an elastic network of the hydrogen bonds[85] that surround the hydrophobic moieties, like hydrocarbon molecules, trying to expel them from the solvent[86] and, therefore, pushing together (i.e., minimizing the hole in the hydrogen bond network). This imitates their mutual attraction, resulting in the formation of a sort of "*oil drop.*"

We may say in a rather simplistic way that hydrophobic molecules aggregate not because they attract particularly strongly, but because water strongly prefers them to be out of its hydrogen bond net structure.

Hydrophobic interactions have a complex character and are not yet fully understood. The interaction depends strongly on the size of the hydrophobic entities. For small sizes, such as two methane molecules in water, the hydrophobic interaction is small, increasing considerably for larger synthons. The hydrophobic effects become especially important for what is called the *amphiphilic macromolecules* with their van der Waals surfaces differing in character (hydrophobic/hydrophilic). The amphiphilic molecules, an example shown in Fig. 13.16, are able to *self-organize*, forming structures up to the nanometer scale (*nanostructures*).

Fig. 13.17 shows an example of the hierarchic (multilevel) character of a molecular architecture:

[84] W. Kauzmann, *Advan.Protein Chem.*, *14*, 1(1959). A contemporary theory is given in K. Lum, D. Chandler, and J.D. Weeks, *J.Phys.Chem.*, *103*, 4570 (1999).

[85] M.N. Rodnikova, *J.Phys.Chem.(Russ.)*, *67*, 275 (1993).

[86] Hydrophobic interactions involve not only the molecules on which we focus our attention, but also, to an important extent, the water molecules of the solvent.

The idea of solvent-dependent interactions is a general and fascinating topic of research. Imagine the interaction of solutes in mercury, in liquid gallium, in liquid sodium, in a highly polarizable organic solvent, etc. Due to the peculiarities of these solvents, we will have different chemistry going on in them.

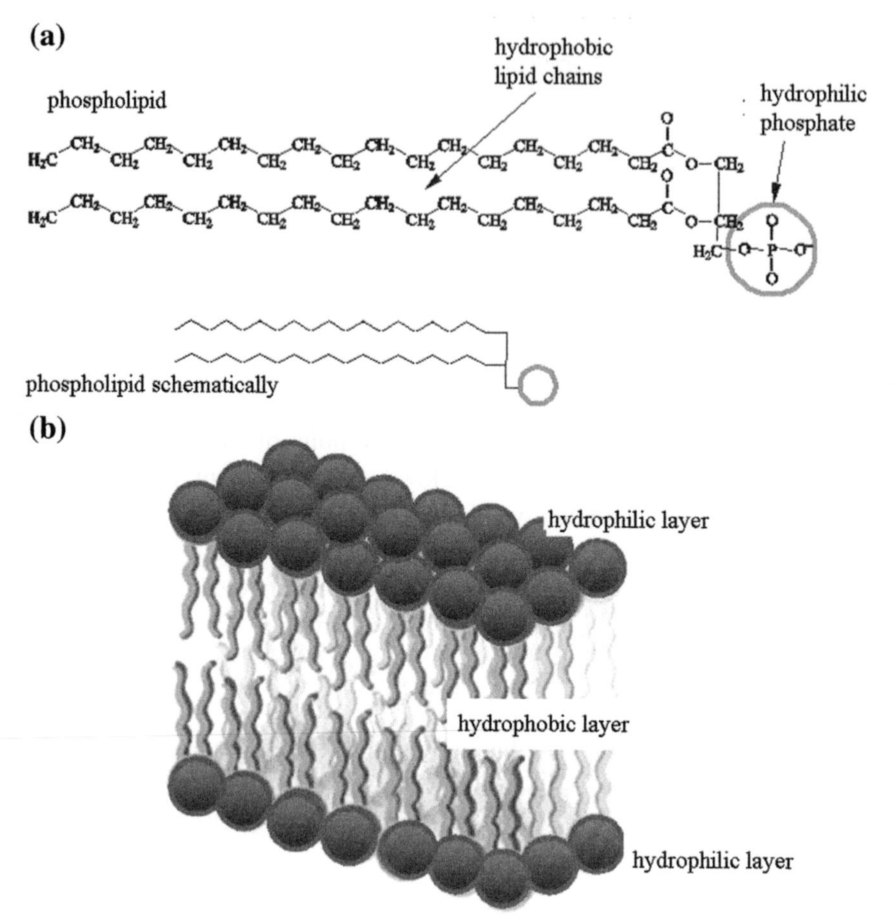

Fig. 13.16. An amphiphilic molecule [i.e., two contradictions (hydrophobicity and hydrophilicity)] side by side and their consequences in water. (a) A phospholipid with two hydrophobic aliphatic "*tails*" and a hydrophilic phosphate "*head*" (its chemical structure given), also shown schematically. (b) The hydrophobic effect in water leads to formation of a lipid bilayer structure, while the hydrophilic heads are exposed to the bulk water (due to a strong hydration effect). The lipid bilayer plays in biology the role of the cell wall.

- The chemical binding of the amino acids into the oligopeptides is the first ("stiff") level (known as *primary structure*).
- The second ("*soft*") level corresponds to a network of hydrogen bonds responsible for forming the α-helical conformation of each of the two oligopeptides (*secondary structure*).
- The third level (*tertiary structure*) corresponds to packing the two α-helices through an extremely effective hydrophobic interaction, the *leucine zipper*. Two α-helices form a very stable structure[87] winding up around each other and thus forming a kind of a superhelix, the so-called coiled-coil, due to the hydrophobic leucine zipper.[88]

[87] B. Tripet, L. Yu, D.L. Bautista, W.Y. Wong, T.R. Irvin, and R.S. Hodges, *Prot.Engin.*, *9*, 1029 (1996).

[88] Leucine may be called the "*flagship*" of the hydrophobic amino acids, although this is not the best compliment for a hydrophobe.

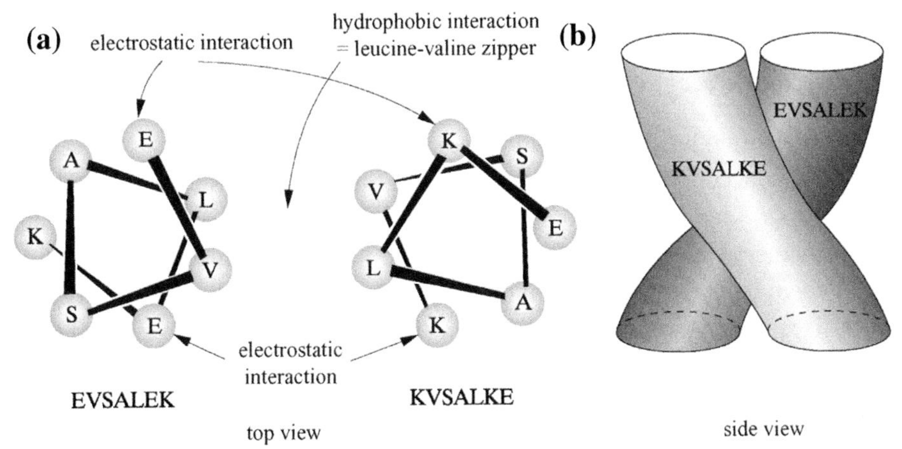

Fig. 13.17. An example of superhelix formation in the case of two oligopeptide chains(a): $(EVSALEK)_n$ with $(KVSALKE)_n$, with E standing for the glutamic acid, V for valine, S for serine, A for alanine, L for leucine, and K for lysine. This is an example of a multilevel molecular architecture. First, each of the two oligopeptide chains form α-helices, which then form a strong hydrophobic complex due to a perfect matching of the hydrophobic residues (leucine and valine of one of the α-helices with valine and leucine of the second one, called the leucine zipper, b). The complex is made stronger additionally by two salt bridges (COO^- and NH_3^+ electrostatic interaction) involving pairs of glutamic acid (E) and lysine (K). The resulting complex (b) is so strong that it serves in analytical chemistry for the separation of some proteins.

The molecular architecture described above was first planned by a chemist. The system fulfilled all the points of the plan and self-organized in a spontaneous process.[89]

When the hydrophobic moieties in water are free to move, they usually form a separate phase like the oil phase above the water phase or an oil film on the surface of water. If this freedom is limited like in the amphiphilic molecules, a compromise is achieved in which the hydrophobic moieties keep together, while the hydrophilic ones are exposed to water (e.g., lipid bilayer or micelle). Finally, when the positions of the hydrophobic amino acids are constrained by the polypeptide backbone, they have the tendency to be buried as close to the center of the globular protein as possible, while the hydrophilic amino acid residues are exposed to the surrounding water.

[89] One day, I said to my friend Leszek Stolarczyk: "*If those organic chemists wanted, they could synthesize anything you might dream of. They are even able to cook up in their flasks a molecule composed of the carbon atoms that would form the shape of a cavalry man on his horse.*" Leszek answered: "*Of course! And the cavalry man would have a little sabre, made of iron atoms.*"

13.14 Construction Principles

13.14.1 Molecular Recognition–Synthons

Organic molecules often have quite a few donor and acceptor substituents. These names may pertain to donating/accepting functional group or groups. Sometimes a particular side of a molecule displays a system of donors and acceptors. Such a system "awaiting" interaction with a complementary object is called a *synthon* (the notion introduced by the Indian scholar Desiraju[90]). Crown ethers, therefore, contain the synthons that can recognize a narrow class of cations (with sizes within a certain range). In Fig. 13.18, we show another example of synthons based on hydrogen bonds X–H...Y. Due to the particular geometry of the molecules, as well as to the abovementioned weak dependence of the XY distance on X and Y, both synthons are complementary.

The example just reported is of immense importance because it pertains to guanine (G), cytosine (C), adenine (A), and thymine (T). Thanks to these two pairs of synthons (GC and AT), we exist, because the G, C, A, and T represent the four letters that are sufficient to write the book of life, word by word, in a single molecule of DNA. The words, the sentences, and the chapters of this book decide the majority of the very essence of your (and my) personality. The whole DNA strand may be considered as a large single synthon. The synthon has its important counterpart which fits the DNA perfectly because of the complementarity. The molecular machine that synthesizes this counterpart molecule (a "negative") is the polymerase, a wonderful molecule (that you will learn more about in Chapter 15). Any error in this complementarity results in a mutation.[91]

Fig. 13.18. The hydrogen bond and Nature's all-important hydrogen bond-based synthons. (a) A hydrogen bond is almost linear; (b) the synthon of adenine (A) fits the synthon of thymine (T) forming a complementary AT pair (two hydrogen bonds involved) of these nucleobases, and the guanine synthon (G) fits the cytosine synthon (C) forming the complementary GC pair (three hydrogen bonds involved); (c) a section of the DNA double-helix shown in the form of a synthon-synthon interaction scheme. A single DNA thread represents a polymer; the monomer units composed of the sugar (five-carbon) rings (deoxyribose) and of the phosphate groups PO_4^{3-}, thus being a polyanion (the negative charge is compensated in the solution by the corresponding number of cations). Each deoxyribose offers an important substituent from the set of the nucleobases A,T,G,C. In order to form the double-helix, the second DNA thread must have the complementary bases to those of the first one. In this way, the two threads represent two complementary polysynthons bound by the hydrogen bonds. (d) A protein represents a chain-like polymer, the monomers being the peptide bonds – NH-CO –, one of them highlighted in the figure. The peptide bonds are bound together through the bridging carbon atoms (known as the C_α carbons): $-(\text{NH-CO-C}_\alpha \text{ HR}_i)_N-$, where N is usually of the order of hundreds, and the functional groups $R_i, i = 1, 2, \ldots, 20$ stand for the aminoacid side chains (Nature has 20 kinds of these side chains). Two structural motifs dominate in a protein 3-D structure, both created by the hydrogen bond-based self-recognition of the synthons within the protein molecule (one peptide bond is highlighted). (e) The α-helical motif, in which a system of the intramolecular hydrogen bonds NH ...OC stabilizes a helical structure. (f) The β-sheet motif, based as well on the intramolecular hydrogen bonds NH ...OC, but formed by a lateral approaching of the peptide β-strands.

[90] G.R. Desiraju, *Crystal Engineering, The Design of Organic Solids* Elsevier, Amsterdam, (1989).

[91] Any mutation represents a potential or real danger, as well as a chance for evolution.

13.14.2 "Key-and-Lock," Template-like, and "Hand-Glove" Synthon Interactions

The spectrum of the energy levels of a molecule represents something like its fingerprint. An energy level corresponds to certain electronic, vibrational, and rotational states (Chapter 6). Different *electronic* states[92] may be viewed as representing different chemical bond patterns.

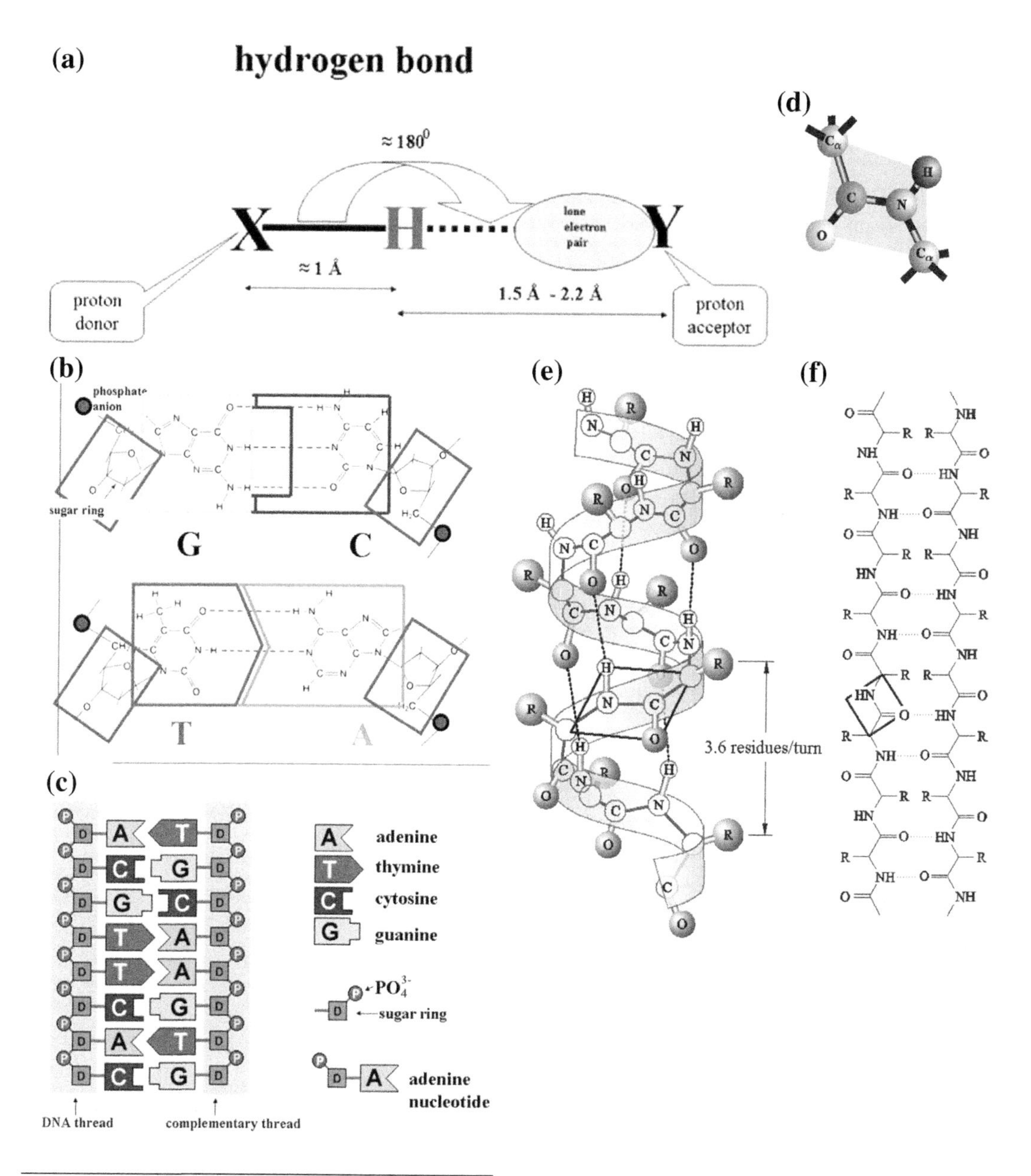

92 In the Born-Oppenheimer approximation, each corresponds to a potential energy hypersurface, PES.

Different *vibrational* states[93] form series, each series for an energy well on the PES (Potential Energy Surface). The energy level pattern is completed by the rotational states of the molecule as a whole. Since the electronic excitations are of the highest energy, the PES of the ground electronic state is most important. For larger molecules, such a PES is characterized by a lot of potential energy wells corresponding to the conformational states. If the bottoms of the excited conformational wells are of high energy (with respect to the lowest-energy well, as shown in Fig. 13.19a), then the molecule in its ground state may be called rigid because high energy is needed to change the molecular conformation.

If such rigid molecules A and B match perfectly each other, this corresponds to the *key-lock* type of molecular recognition. To match, the interacting molecules sometimes only need to orient properly in space when approaching one another and then dock (the AT or GC pairs may serve as an example). This key-lock concept of Fischer from 100 years ago (concerning enzyme-substrate interaction) is considered as the foundation of supramolecular chemistry – the chemistry that deals with the complementarity and matching of molecules.

One of the molecules, if rigid enough, may serve as a *template* for another molecule, which is flexible (see Fig. 13.19b). Finally two flexible molecules (Fig. 13.19c) may pay an energy penalty for acquiring higher-energy conformations, but ones that lead to a very strong interaction of the molecules in the *hand-glove* type of molecular recognition.

Fig. 13.20 shows the far-reaching consequences of a apparently minor change in the molecular structure (replacement of 16 hydrogen atoms by fluorines). Before the replacement, the molecule has a conelike shape, while after it the cone has to open. This means that the molecule has a concave part and resembles a plate. What happens next represents a direct consequence of this:

- In the first case, the conelike molecules associate laterally, which has to end up in a spherical structure similar to a micelle. The spheres pack into a cubic structure (as in NaCl).
- In the second case, the "plates" match with each other and one gets stacks of them. The most effective packing is of the stack-to-stack type, leading to the hexagonal liquid crystal packing.

Still another variation of the hand-glove interaction comes into play, when during the approach, a new type of synthon appears, and the synthons match afterward. For example, in the Hodges superhelical structure (Fig. 13.17), only after formation of the α-helices does it turn out that the leucine and valine side chains of one helix match perfectly similar synthons of the second helix (the leucine-zipper).

[93] The states include internal rotations, such as those of the methyl group.

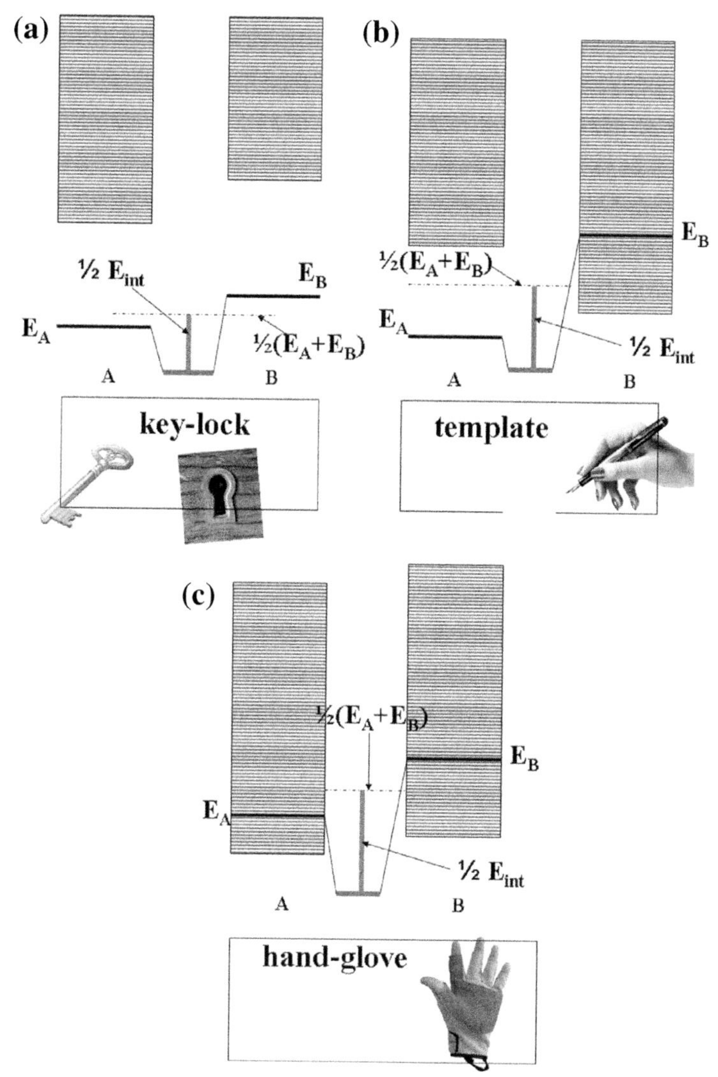

Fig. 13.19. The key-lock, template, and hand-glove interactions. Any molecule may be characterized by a spectrum of its energy levels. (a) In the key-lock type interaction of two rigid molecules A and B, their low-energy conformational states are separated from the quasi-continuum high-energy conformational states (including possibly those of some excited electronic states) by an energy gap, which is generally different for A and B. Due to the effective synthon interactions, the energy per molecule lowers substantially with respect to that of the isolated molecules leading to molecular recognition without significant changes of molecular shape. (b) In the template-like interaction, one of the molecules is rigid (large energy gap), while the other one offers a quasi-continuum of conformational states. Among the latter, there is one that (despite of being a conformational excited state), due to the perfect matching of synthons results in considerable energy lowering, much below the energy of isolated molecules. Thus, one of the molecules has to distort in order to get perfect matching. (c) In the hand-glove type of interaction, the two interacting molecules offer quasi-continua of their conformational states. Two of the excited conformational states correspond to such molecular shapes as match each other perfectly (molecular recognition) and lower the total energy considerably. This lowering is so large that it is able to overcome the conformational excitation energy (an energy cost of molecular recognition).

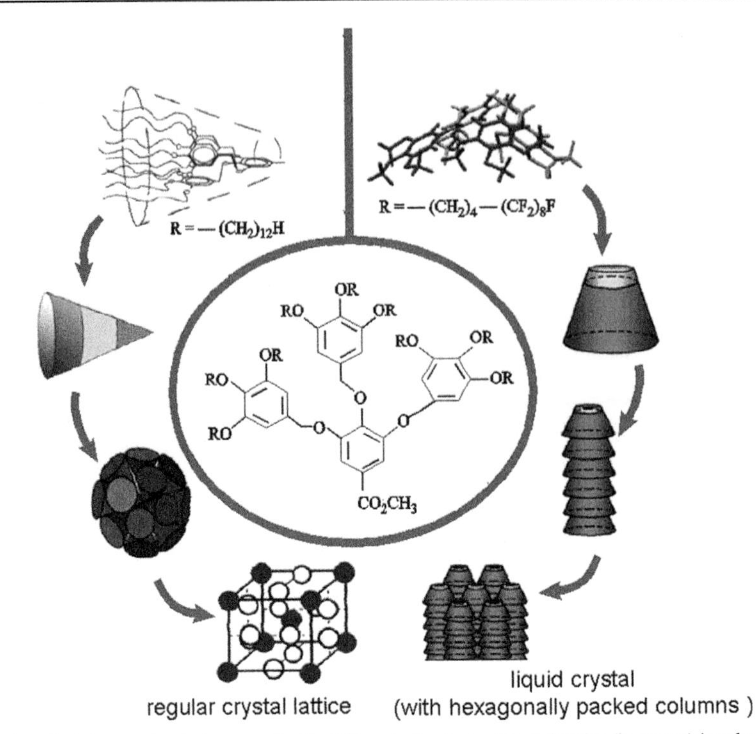

Fig. 13.20. The key-lock interaction in two marginally different situations: the molecules (in center) involved differ by replacing some of the hydrogens by the fluorine atoms in the substituent R. This picture shows how profound the consequences of this seemingly small detail are. In one case (left), we obtain a conelike molecule and then a crystal of cubic symmetry; in the other (right), the molecule has a platelike shape, and because of that, we get finally a liquid crystal with the hexagonal packing of columns of these molecular plates [after Donald A.Tomalia, *Nature Materials,* 2, 711(2003)].

Hermann Emil Fischer (1852–1919), German chemist and founding father of the domain of molecular recognition. Known mainly for his excellent work on the structure of sugar compounds. His correct determination of the absolute conformation of sugars (recognized decades later) was based solely on the analysis of their chemical properties. Even today, this would require advanced physicochemical investigations. In 1902, Fischer received the Nobel Prize *"for his work on sugar and purine syntheses."*

Another masterpiece of nature is self-organization of the tobacco virus, shown in Fig. 13.21. Such a complex system self-assembles because its parts not only fit one another (synthons), but also found themselves in solution and made perfect matching accompanied by an energy gain. Even more spectacular is the structure and functioning of bacteriophage T (Fig. 13.22).

13.14.3 Convex and Concave–The Basics of Strategy in the Nanoscale

Chemistry can be seen from various points of view. In particular, one may think of it (liberating oneself for a while from the overwhelming influence of the Mendeleev table) as of a general strategy to tailor matter in the nanoscale,[94] in order to achieve a suitable physicochemical goal. Then, we see an interesting feature, that: until recently, the concepts of chemistry have been based on the intermolecular interaction of *essentially convex molecules*.[95]

Practical use of molecular fitting in chemistry, taking place when *a convex molecule interacts with a concave one*, turned out to be a real breakthrough. To get to this stage, the chemists had to learn how to synthesize concave molecules, necessarily large ones to host the convex ones. The convex-concave interaction has some peculiar features, which make possible precise operation in the nanoscale through the following:

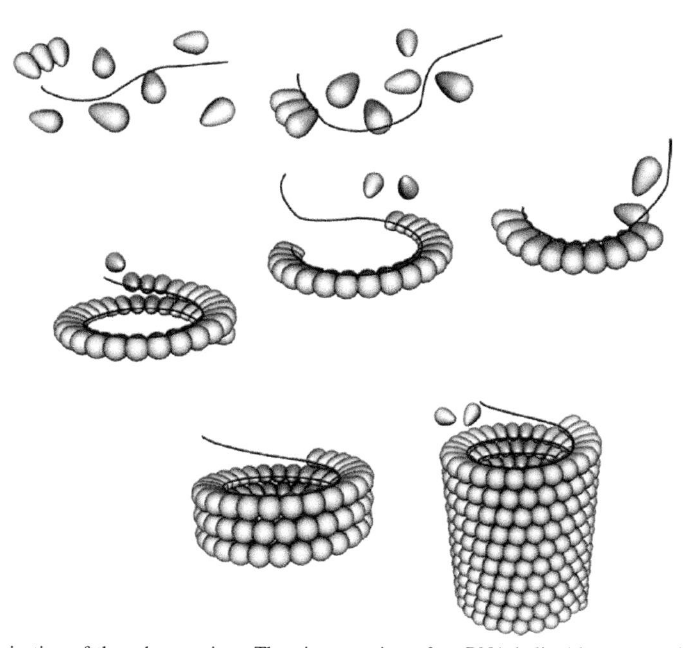

Fig. 13.21. Self-organization of the tobacco virus. The virus consists of an RNA helix (shown as a single "file") containing about 7000 nucleotides, which is sufficient genetic material to code the production of 5 – 10 proteins (first level of supramolecular self-organization). The RNA strand interacts very effectively with a certain protein (shown as a "drop," which is the second level). The protein molecules associate with the RNA strand forming a kind of necklace, and then the system folds (third level) into a rodlike shape typical for this virus. The rods are able to form a crystal (level four, not shown here), which melts after heating but is restored when cooled down.

---

[94] Let us stress that chemists always had to do with nanostructures, which are quite fashionable nowadays. The molecules have dimensions of the order of tens to hundreds of Å or more; i.e., 1–10 nm.

[95] We do not count here some small concave details of molecular surfaces, which were usually the size of a single atom and did not play any particular role in theoretical considerations.

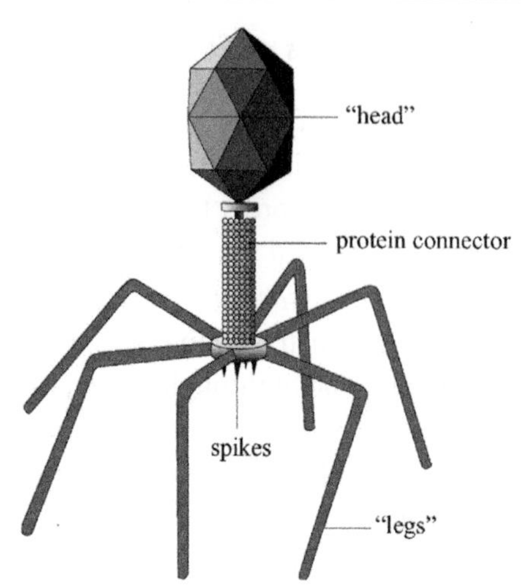

Fig. 13.22. Bacteriophage T represents a supramolecular construction that terrorizes bacteria. The hexagonal "head" contains a tightly packed double helix of DNA (the virus genetic material) wrapped in a coat made of protein subunits. The head is attached to a tubelike molecular connector built of 144 contractible protein molecules. On the other side of the connector, there is a plate with six spikes protruding from it, as well as six long, kinked "legs" made of several different protein molecules. The legs represent a "landing apparatus" which, using intermolecular interactions, attaches to a particular receptor on the bacterium cell wall. This reaction is reversible, but what happens next is highly irreversible. First, an enzyme belonging to the "monster" makes a hole in the cell wall of the bacterium. Then the 144 protein molecules contract at the expense of energy from hydrolysis of the ATP molecule (adenosine triphosphate - a universal energy source in biology), which the monster has at its disposal. This makes the head collapse, and the whole monster serves as a syringe. The bacteriophage's genetic material enters the bacterium body almost immediately. That is the end of the bacterium.

- A convex molecule of a particular shape may fit a specific concave one, much better than other molecules are able to do (*molecular recognition*).
- Elimination of some potential convex reactants ("*guest molecules*") to enter a reaction center inside a concave pocket of the "*host molecule,*" just because their shape makes the contact impossible (due to the steric hindrance; i.e., an excessive valence repulsion, as shown in Fig. 13.23a1).
- Isolating the reaction center from some unwanted neighborhood (Fig. 13.23a2), which in the extreme case leads to molecular reaction vessels, even such large ones as a biological cell.
- Positioning the reactants in space in some particular orientation (unlike that shown in Fig.13.23a3).
- A strong binding (many interatomic contacts distributed on *large contact surface*), in order to make a molecular complex. (Fig. 13.23a4).
- The same binding to be sufficiently weak intermolecular binding in order to allow for reorganization of the complex.
- Assuring the reaction centers (*predefined* by chemists) to be close in space and therefore forcing the reaction to proceed with a high yield (Fig. 13.23b1,b2).
- *Leaving the reaction pocket by the products,* because they do not have enough space in the pocket (Fig. 13.23b3).

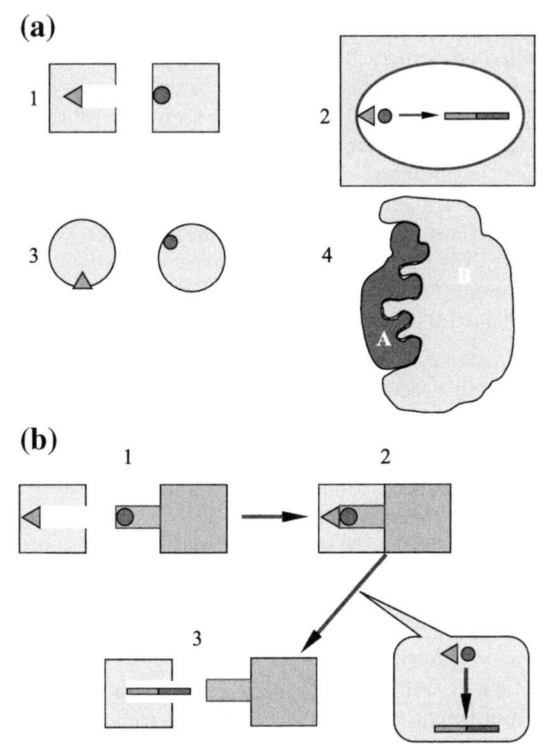

Fig. 13.23. How to make a particular chemical reaction of two molecules happen with high yield (scheme)? The figure shows several ways to get two reaction centers (one in each molecule, symbolized by triangle and circle) close in space. (a1) This architecture does not allow the reaction to proceed; (a2) the reaction will proceed with higher probability when in a reaction cavity (the cavity protects also the molecules from influence of other molecules of the neighborhood); (a3) the two centers to meet there must be a lot of unsuccessful attempts (what is known as entropic barrier); (a4) molecules A and B attract effectively in one configuration only, at which they have a large contact surface; (b1 and b2) a model of catalytic center (reaction cavity); the molecules fit best exactly at a configuration for which the reaction centers meet; (b3) the reaction products do not fit the cavity, which results in leaving the cavity.

Therefore,

> one can imagine an idea of a *molecular machine*, which offers a specific reaction space (pocket), selects the right objects from the neighborhood, fixes their positions as to have the key reaction centers close in space, makes the reaction proceed, then removes the products from the reaction space, leaving the pocket ready for the next reaction.

Summary

- Interaction energy of two molecules (at a given geometry) may be calculated within any reliable quantum mechanical method by subtracting from the total system energy the sum of the energies of the subsystems. This is called a *supermolecular method.*

- The supermolecular method has at least one important advantage: *it works independently of the interaction strength and of the intermolecular distance.* The method has the disadvantage that due to the subtraction, a loss of accuracy occurs and no information is obtained about the structure of the interaction energy.
- In the supermolecular method, there is a need to compensate for what is called the *basis set superposition error (BSSE).* The error appears because due to the incompleteness of the atomic basis set (Ω_A, Ω_B), the individual subsystem A with the interaction switched off profits from the Ω_A basis set only, while when interacting, the energy is lowered due to the total $\Omega_A \cup \Omega_B$ basis set (the same pertains to any of the subsystems). As a result, a part of the calculated interaction energy does not come from the interaction, but from the problem of the basis set used (BSSE) described above. A remedy is called the *counter-poise method,* in which all quantities (including the energies of the individual subsystems) are calculated within the $\Omega_A \cup \Omega_B$ basis set.
- Perturbational method has limited applicability:
  - At long intermolecular separations, what is called the *polarization approximation* may be used
  - At medium distances, a more general formalism called the *symmetry adapted perturbation theory* (SAPT) may be applied
  - At short distances (of the order of chemical bond lengths), perturbational approaches are inapplicable
- One of the advantages of a low-order perturbational approach is the possibility of dividing the interaction energy into well-defined, physically distinct energy contributions.
- In a polarization approximation approach, the unperturbed wave function is assumed as a *product* of the exact wave functions of the individual subsystems: $\psi_0^{(0)} = \psi_{A,0}\psi_{B,0}$. The corresponding zero-order energy is $E_0^{(0)} = E_{A,0} + E_{B,0}$.
- Then, the first-order correction to the energy represents what is called the *electrostatic interaction energy*: $E_0^{(1)} = E_{\text{elst}} = \langle \psi_{A,0}\psi_{B,0}|V\psi_{A,0}\psi_{B,0}\rangle$, which is the Coulombic interaction (at a given intermolecular distance) of the frozen charge density distributions of the individual, non-interacting molecules. After using the multipole expansion, E_{elst} can be divided into the sum of the multipole-multipole interactions plus a remainder, called the *penetration energy.* A multipole-multipole interaction corresponds to the permanent multipoles of the isolated molecules. An individual multipole-multipole interaction term (2^k–pole with 2^l–pole) vanishes asymptotically as $R^{-(k+l+1)}$; e.g., the dipole-dipole term decreases as $R^{-(1+1+1)} = R^{-3}$.
- In the second order, we obtain the sum of the induction and dispersion terms: $E^{(2)} = E_{\text{ind}} + E_{\text{disp}}$.
- The induction energy splits into $E_{\text{ind}}(A \to B) = \sum_{n_B}' \frac{|\langle \psi_{A,0}\psi_{B,n_B}|V\psi_{A,0}\psi_{B,0}\rangle|^2}{E_{B,0}-E_{B,n_B}}$, which pertains to polarization of molecule B by the unperturbed molecule A, and $E_{\text{ind}}(B \to A) = \sum_{n_A}' \frac{|\langle \psi_{A,n_A}\psi_{B,0}|V|\psi_{A,0}\psi_{B,0}\rangle|^2}{E_{A,0}-E_{A,n_A}}$, with the roles of the molecules reversed. The induction energy can be represented as the permanent multipole–induced dipole interaction, with asymptotic vanishing as $R^{-2(k+2)}$.
- The dispersion energy is defined as $E_{\text{disp}} = \sum_{n_A}' \sum_{n_B}' \frac{|\langle \psi_{A,n_A}\psi_{B,n_B}|V\psi_{A,0}\psi_{B,0}\rangle|^2}{(E_{A,0}-E_{A,n_A})+(E_{B,0}-E_{B,n_B})}$ and represents a result of the electronic correlation. After applying the multipole expansion, the effect can be described as a series of instantaneous multipole–instantaneous multipole interactions, with the individual terms decaying asymptotically as $R^{-2(k+l+1)}$. The most important contribution is the dipole–dipole ($k = l = 1$), which vanishes as R^{-6}.
- The polarization approximation fails for medium and short distances. For medium separations, we may use SAPT. The unperturbed wave function is symmetry-adapted; i.e., it has the same symmetry as the exact function. This is not true for the polarization approximation, where the productlike $\varphi^{(0)}$ does not exhibit the proper symmetry with respect to electron exchanges between the interacting molecules. The symmetry-adaptation is achieved by a projection of $\varphi^{(0)}$.
- SAPT *reproduces all the energy corrections that appear in the polarization approximation* ($E_{\text{elst}}, E_{\text{ind}}, E_{\text{disp}}, \ldots$) *plus provides some exchange-type terms* (in each order of the perturbation).
- The most important exchange term is the valence repulsion appearing in the first-order correction to the energy: $E_{\text{exch}}^{(1)} = \langle \psi_{A,0}\psi_{B,0}|V\hat{P}_{AB}\psi_{A,0}\psi_{B,0}\rangle - \langle \psi_{A,0}\psi_{B,0}|V\psi_{A,0}\psi_{B,0}\rangle\langle \psi_{A,0}\psi_{B,0}|\hat{P}_{AB}\psi_{A,0}\psi_{B,0}\rangle + O(S^4)$, where

$\hat{P}_{AB}$ stands for the single exchanges' permutation operator and $O(S^4)$ represents all the terms decaying as the fourth power of the overlap integral or faster.

- The interaction energy of N molecules *is not pairwise additive*; i.e., it is not the sum of the interactions of all possible pairs of molecules. Among the energy corrections up to the second order, the exchange and, first of all, the induction terms contribute to the non-additivity. The electrostatic and dispersion (in the second order) contributions are pairwise additive.
- The non-additivity is highlighted in what is called the *many-body expansion of the interaction energy*, where the interaction energy is expressed as the sum of two-body, three–body, etc. energy contributions. The $m-$body interaction, in a system of N molecules, is defined as that part of the interaction energy that is non-explicable by any interactions of $m' < m$ molecules, but explicable by the interactions among m molecules.
- The dispersion interaction in the third-order perturbation theory contributes to the three-body non-additivity and is called the *Axilrod-Teller energy*. The term represents a correlation effect. Note that the effect is negative for three bodies in a linear configuration.
- The most important contributions: electrostatic, valence repulsion, induction, and dispersion lead to a richness of supramolecular structures.
- Molecular surface (although not having an unambiguous definition) is one of the most important features of the molecules involved in the molecular recognition.
- The electrostatic interaction plays a particularly important role because it is of a long-range character as well as very sensible to relative orientation of the subsystems. The hydrogen bond X-H...Y represents an example of the domination of the electrostatic interaction. This results in its directionality, linearity and a small (as compared to typical chemical bonds) interaction energy of the order of −5 kcal/mol.
- Also, valence repulsion is one of the most important energy contributions because it controls how the interacting molecules fit together in space.
- The induction and dispersion interactions for polar systems, although contributing significantly to the binding energy, in most cases do not have a decisive role in forming structure and only slightly modify the geometry of the resulting structures.
- In aqueous solutions, the solvent structure contributes very strongly to the intermolecular interaction, thus leading to what is called the *hydrophobic effect.* The effect expels the non-polar subsystems from the solvent, thus causing them to approach, which *looks* like an attraction.
- A molecule may have such a shape that it fits that of another molecule (synthons, small valence repulsion, and a large number of attractive atom-atom interactions).
- In this way, molecular recognition may be achieved by the key-lock-type fit (the molecules non-distorted), template fit (one molecule distorted), or hand-glove-type fit (both molecules distorted).
- Molecular recognition may be planned by chemists and used to build complex molecular architectures, in a way similar to that in which living matter operates.

Main Concepts, New Terms

multipole moments (p. 813)
nanostructures (p. 867)
natural division (p. 798)
non-additivity (p. 847)
Padé approximants (p. 842)
Pauli blockade (p. 842)
penetration energy (p. 814)
permanent multipoles (p. 826)
polarization catastrophe (p. 858)
polarization collapse (p. 836)
polarization perturbation theory (p. 840)
rotaxans (p. 801)
SAPT (p. 827)
SE mechanism (p. 853)
supermolecular method (p. 804)
symmetrized polarization approximation (p. 842)
symmetry forcing (p. 835)
synthon (p. 870)
TE mechanism (p. 854)
template interaction (p. 871)
valence repulsion (p. 862)
van der Waals radius (p. 860)
van der Waals interaction energy (p. 796)

From the Research Front

Intermolecular interaction influences any liquid and solid-state measurements. Physicochemical measurement techniques give only some indications of the shape of a molecule, except NMR, X-ray, and neutron analyses, which provide the atomic positions in space, but are very expensive. This is why there is a need for theoretical tools that may offer such information in a less expensive way. For very large molecules, such an analysis uses the force fields described in Chapter 7. This is currently the most powerful theoretical tool for determining the approximate shape of molecules with a number of atoms even of the order of thousands. To obtain more reliable information about intermolecular interactions, we may perform calculations within a supermolecular approach, necessarily of an *ab initio* type, because other methods give rather low-quality results. The DFT method popular nowadays fails at its present stage of development, because the intermolecular interactions area, especially the dispersion interaction, is a particularly weak point of the method. If the particular method chosen is the Hartree-Fock approach (currently limited to about 300 atoms), we have to remember that it cannot take into account any dispersion contribution to the interaction energy by *definition*.[96] *Ab initio* calculations of the correlation energy still represent a challenge. High-quality calculations for a molecule with 100 atoms may be carried out using the MP2 method. Still more time consuming are the CCSD(T) or SAPT calculations, which are feasible only for systems with a dozen of atoms, but offer an accuracy of 1 kcal/mol required for chemical applications.

Ad Futurum

No doubt the computational techniques will continue to push the limits mentioned above. The more coarse the method used, the more spectacular this pushing will be. The most difficult to imagine would be a great progress in methods using explicitly correlated wave functions. It seems that pushing the experimental demands and calculation time required will cause experimentalists (they will perform the calculations[97]) to prefer a rough estimation using primitive methods rather than wait too long for a precise result (which still is not very appropriate because it was obtained without taking the influence of solvent, etc. into account). It seems that in the near future, we may expect theoretical methods exploiting the synthon concept. It is evident that a theoretician has to treat the synthons on an equal footing with other atoms, but a practice-oriented theoretician cannot do that. Otherwise, he would wait virtually forever for something to happen in the computer, while in reality, the reaction takes only a picosecond or so. Still further in the future, we will see the planning of hierarchic multilevel supramolecular systems, taking into account the kinetics and competitiveness among such structures. In the still more distant future, functions performed by such supramolecular structures, as well as their sensitivity to changing external conditions, will be investigated.

[96] The dispersion energy represents an electronic correlation effect, which is absent in the Hartree-Fock energy.
[97] So, what will theoreticians do? My answer is given in Chapter 15.

Additional Literature

J. O. Hirschfelder, C. F. Curtiss, and R. B. Bird, *Molecular Theory of Gases and Liquids,* Wiley, New York (1964).
A thick "bible" (1249 pages long) of intermolecular interactions, with most important facts that came before the advent of computers and SAPT.

H. Margenau and N. R. Kestner, *Theory of Intermolecular Forces,* Pergamon, Oxford (1969).
This book contain a lot of detailed derivations. There is also a chapter devoted to the non-additivity of the interaction energy–a rarity in textbooks.

H. Ratajczak and W. J. Orville, eds., *Molecular Interactions*, Wiley, Chichester (1980).
A three-volume edition containing a selection of articles by experts in the field.

A.J. Stone, *"The Theory of Intermolecular Forces,"* Oxford Univ.Press, Oxford, 1996.
The book contains basic facts about the field of intermolecular interactions given in the language of perturbation theory, as well as the multipole expansion (including a lot of useful formulas for the electrostatic, induction, and dispersion contributions). This very well written book presents many important problems in a clear and comprehensive way.

S. Scheiner, ed., *Molecular Interactions*, Wiley, Chichester (1997).
A selection of articles written by experts.

P. Hobza and R. Zahradnik, *Intermolecular Complexes*, Elsevier, Amsterdam (1988).
This book contains many useful details. This monograph establishes a strong link between the theoretical methods and the experimental investigations of intermolecular interactions, including those important in biology.

Questions

1. The rigid interaction energy of the subsystems A and B in the AB system
   a. does not represent a measurable quantity
   b. requires calculation of the electronic energy of AB, A, and B, where the configuration of the nuclei in A and B are the same as those in AB
   c. requires calculation of the electronic energy of AB, A and B, where the configurations of the nuclei in A, and B correspond to the minimum energy of these subsystems
   d. the interaction energy of two water molecules in the H_4O_2 system is a unambiguously defined quantity.
2. The Boys-Bernardi method of removing the basis set superposition error (BSSE) means
   a. a high-quality calculation: all calculations within the most extended basis set available, feasible separately for A, B, and AB.
   b. the energy of the total system should be calculated within the joint sets of the atomic orbitals centered on the individual subsystems, while the energy of the subsystems should be calculated within their individual basis sets.
   c. the atomic basis set for AB represents the sum of the basis sets for A and B
   d. all quantities are calculated within a common basis set being the sum of the sets for the individual molecules.
3. The zeroth order wave function in the polarization perturbation theory (for a finite intermolecular distance)
   a. does not satisfy the Pauli exclusion principle
   b. represents a product of the wave functions for the polarized molecules
   c. admits that two electrons with the same spin coordinates occupy the same point in space
   d. represents a product of the wave functions for the isolated molecules

4. Induction energy (R denotes the intermolecular distance)
   a. decays as R^{-7} for the interaction of two hydrogen molecules
   b. decays as R^{-6} for the immobilized water and ammonia molecules
   c. represents an attraction
   d. is an electronic correlation effect.
5. Dispersion energy (R denotes the intermolecular distance)
   a. is not equal to zero for two polar molecules
   b. the Hartree-Fock method overestimates its value
   c. represents an electronic correlation effect
   d. for R sufficiently large decays as R^{-6}
6. The multipole moments of a point particle of charge q and the coordinates x, y, z.
   a. the only nonzero multipole moment of a pointlike particle represents its charge (i.e., its monopole)
   b. the z-component of the dipole moment is equal to qz
   c. a pointlike particle cannot have a dipole moment, so we always have $\mu_z = 0$
   d. the values of the multipole moments of a particle depend on the coordinate system chosen.
7. In the symmetry adapted perturbation theory (SAPT) for $H_2O \ldots H_2O$
   a. the dispersion energy appears in the second-order
   b. one obtains a minimum of the electronic energy when the molecules approach each other
   c. the valence repulsion appears in the first order
   d. the zeroth-order wave function is asymmetric with respect to exchange of the coordinates of any two electrons.
8. In the symmetry adapted perturbation theory (SAPT) for Ne...Ne
   a. the wave function of the zeroth order represents an antisymmetrized product of the wave function for the isolated neon atoms
   b. the electrostatic energy appears in the first order and is equal zero within the multipole approximation
   c. the exchange corrections appear in every order
   d. the penetration part of the electrostatic energy is equal zero
9. Additivity and non-additivity of the interaction
   a. the electrostatic energy is always additive, while the induction energy is always non-additive
   b. the three-body contribution represents this part of the interaction energy of N subsystems, which cannot be explained by the pairwise interactions of the subsystems
   c. the dispersion energy (in the second order) contains only the pairwise interactions of the subsystems
   d. additivity means that the interaction energy is a sum of the pairwise interaction energies.
10. Supramolecular chemistry
    a. molecular recognition means a strong attraction of the molecules at their unique mutual configuration only
    b. the angle O-H...O in the hydrogen bond HO-H...OH_2 is equal to 180.0^0
    c. the term hydrogen bond pertains to the hydrogen-hydrogen interaction, where the hydrogens belong to different molecules
    d. the hand-glove interaction is also known as the key-lock interaction.

Answers

1a,b, 2d, 3a,c,d, 4b,c, 5a,c,d, 6b,d, 7a,b,c, 8a,b,c, 9a,c,d, 10a

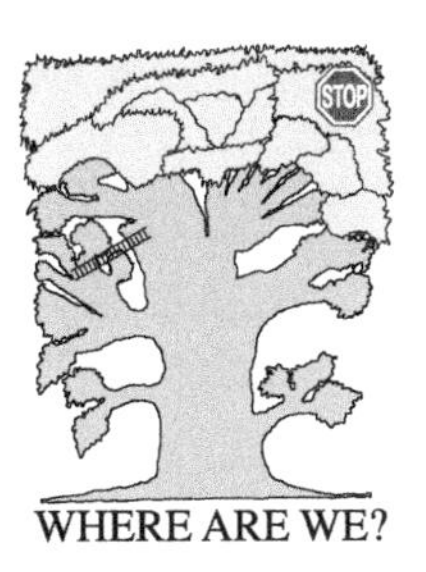

WHERE ARE WE?

CHAPTER 14

Chemical Reactions

"Enter through the narrow gate,
for the gate is wide and the way
is broad that leads to destruction..."
St. Matthew (7.13)

Where Are We?

We are already picking fruit in the crown of the TREE.

An Example

Why do two substances react and another two do not? Why does increasing temperature start a reaction? Why does a reaction mixture change color? As we know from Chapter 6, this tells us about some important electronic structure changes. On the other hand, the products (when compared to the reactants) tell us about profound changes in the positions of the nuclei that take place simultaneously. Something dramatic is going on. But what, and why?

What Is It All About?

How does atom A eliminate atom C from a diatomic molecule BC? How can a chemical reaction be described as a molecular event? Where does the reaction barrier come from? Such questions will be discussed in this chapter.

Ideas of Quantum Chemistry, Second Edition. http://dx.doi.org/10.1016/B978-0-444-59436-5.00014-3

We are already acquainted with the toolbox for describing the electronic structure at *any* position of the nuclei. It is time to look at possible *large* changes of the electronic structure at *large* changes of nuclear positions. The two motions of the electrons and nuclei will be coupled together (especially in a small region of the configurational space).

Our discussion consists of four parts:

- In the first part (after using the Born-Oppenheimer approximation, fundamental to this chapter), we assume that we have calculated the ground-state electronic energy (i.e., the potential energy for the nuclear motion). It will turn out that the hypersurface has a *characteristic drainpipe shape, and the bottom in the central section, in many cases, exhibits a barrier.* Taking a three-atom example, we will show how the problem *could* be solved, if we *were capable* of calculating the quantum dynamics of the system accurately.
- In the second part, we will concentrate on a specific representation of the system's energy that takes *explicitly* into account the abovementioned reaction drain-pipe ("*reaction path Hamiltonian*"). Then we will focus on how to describe the proceeding of a chemical reaction. Just to be more specific, an example will be shown in detail.
- In the third part (acceptor-donor theory of chemical reactions), we will find the answer to the question of *where the reaction barrier comes from and what happens to the electronic structure when the reaction proceeds.*
- The fourth part will pertain to the *reaction barrier height* in electron transfer (a subject closely related to the second and the third parts).

Why Is This Important?

Chemical reactions are at the heart of chemistry, making possible the achievement of its ultimate goals, which include synthesizing materials with desired properties. What happens in the chemist's flask is a complex phenomenon that consists of an astronomical number of elementary reactions of individual molecules. In order to control the reactions in the flask, it would be good to *understand first the rules which govern these elementary reaction acts*. This is the subject of this chapter.

What Is Needed?

- Hartree-Fock method (Chapter 8)
- Conical intersection (Chapter 6)
- Normal modes (Chapter 7)
- Appendices M, E, Z, I (recommended), G (just mentioned)
- Elementary statistical thermodynamics or even phenomenological thermodynamics: entropy, free energy

Classical Works

Everything in quantum chemistry began in the 1920s.

The first publications that considered conical intersection - a key concept for chemical reactions – were two articles from two Budapest schoolmates: Janos (John) von Neumann and Jenő Pál (Eugene) Wigner, "*Über merkwürdige diskrete Eigenwerte*" published in *Physikalische Zeitschrift, 30,* 465 (1929); and "*Über das Verhalten von Eigenwerten bei adiabatischen Prozessen*," which also appeared in *Physikalische Zeitschrift, 30,* 467 (1929). ★ Then a paper called "*The crossing of potential surfaces*," by their younger schoolmate Edward Teller, was published in the *Journal of Chemical Physics, 41,* 109 (1937). ★ A classical theory of the "*reaction drainpipe*" with entrance and exit channels was first proposed by Henry Eyring, Harold Gershinowitz, and Cheng E. Sun in "*Potential energy surface for linear* H_3," published in the *Journal of Chemical Physics, 3,* 786 (1935), and then by Joseph O. Hirschfelder, Henry Eyring and Brian Topley in an article called "*Reactions involving hydrogen molecules and atoms*," in *Journal of Chemical Physics, 4,* 170 (1936); and by Meredith G. Evans and Michael Polanyi in "*Inertia and driving force of chemical reactions*," which appeared in *Transactions of the Faraday Society, 34,* 11 (1938). ★ Hugh Christopher Longuet-Higgins, Uno Öpik, Maurice H.L. Pryce, and Robert A. Sack in a splendid paper called "*Studies of the Jahn-Teller effect*," published in *Proceedings of the Royal Society of London, A244,* 1 (1958), noted for the first time that the wave function changes its phase close to a conical intersection, which later on became known as the *Berry phase*. ★ The acceptor-donor description of chemical reactions was first proposed by Robert S.J. Mulliken in "*Molecular compounds and their spectra*," *Journal of the American Chemical Society, 74,* 811 (1952). ★ The idea of the intrinsic reaction coordinate (IRC) was first explained by Isaiah Shavitt in "*The tunnel effect corrections in the rates of reactions with parabolic and Eckart barriers*," Report WIS-AEC-23, Theoretical Chemistry Lab., University of Wisconsin, (1959), as well as by Morton A. Eliason and Joseph O. Hirschfelder in the *Journal of the Chemical Physics, 30,* 1426 (1959), in an article called "*General collision theory treatment for the rate of bimolecular, gas phase reactions*." ★ The symmetry rules allowing some reactions and forbidding others were first proposed by Robert B. Woodward and Roald Hoffmann in two letters to the editor: "*Stereochemistry of electrocyclic reactions*" and "*Selection rules for sigmatropic reactions*," *Journal of American Chemical Society, 87,* 395, 2511 (1965), as well as by Kenichi Fukui and Hiroshi Fujimoto in an article published in the *Bulletin of the Chemical Society of Japan, 41,* 1989 (1968). ★ The concept of the steepest descent method was formulated by Kenichi Fukui in an article called "*A formulation of the reaction coordinate*," which appeared in the *Journal of Physical Chemistry, 74,* 4161 (1970), although the idea seems to have a longer history. ★ Other classical papers include a seminal article by Sason S. Shaik called "*What happens to molecules as they react? Valence bond approach to reactivity*," published

John Charles Polanyi (b. 1929), Canadian chemist of Hungarian origin, son of Michael Polanyi (one of the pioneers in the field of chemical reaction dynamics), and professor at the University of Toronto. John was attracted to chemistry by Meredith G. Evans, who was a student of his father. Three scholars, John Polanyi, Yuan Lee, and Dudley Herschbach shared the 1986 Nobel Prize *"for their contributions concerning the dynamics of chemical elementary processes."*

Yuan Tseh Lee (b. 1936) is a native of Taiwan. He has been called *"a Mozart of physical chemistry"* by his collegues. His first scientific research was conducted at the National Taiwan University, in very poor financial conditions. He understood that he has to rely on his own hard work. Lacking the proper courses at the University, he decided to study by his own (quantum mechanics, statistical mechanics, but also German and Russian). He wrote that he was deeply impressed by a biography of Marie Curie and that her idealism decided his own path.

Dudley Herschbach (b. 1932) writes that he spent his childhood in a village close to San Jose, picking fruit, milking cows, etc. Thanks to his wonderful teacher, he became interested in chemistry. He graduated from Harvard University (majoring in physical chemistry), where as he says, he has found *"an exhilerating academic environment."*

In 1959, he became professor at the University of California at Berkeley. In 1967, his laboratory was joined by Yuan Lee and constructed a "supermachine" for studying crossing molecular beams and the reactions in them.

One of the topics was the alkali metal atom - iodine collisions.

in *Journal of the American Chemical Society, 103,* 3692 (1981). ★ The Hamiltonian path method was formulated by William H. Miller, Nicolas C. Handy, and John E. Adams, in an article called *"Reaction path Hamiltonian for polyatomic molecules,"* published in the *Journal of Chemical Physics, 72,* 99 (1980). ★ The first quantum dynamics simulation was performed by a Ph.D. student named George C. Schatz (under the supervision of Aron Kupperman) for the reaction $H_2 + H \rightarrow H + H_2$, reported in *"Role of direct and resonant processes and of their interferences in the quantum dynamics of the collinear $H + H_2$ exchange reaction,"* in *Journal of Chemical Physics, 59,* 964 (1973). ★ John Polanyi, Dudley Herschbach, and Yuan Lee proved that the lion's share of the reaction energy is delivered through the rotational degrees of freedom of the products; e.g., J.D. Barnwell, J.G. Loeser, and D.R. Herschbach, *"Angular correlations in chemical reactions. Statistical theory for four vector correlations,"* published in the *Journal of Physical Chemistry, 87,* 2781 (1983). ★ Ahmed Zewail (Egypt/USA) developed an amazing experimental technique known as *femtosecond spectroscopy*, which for the first time allowed the study of the reacting molecules at different stages of an ongoing reaction [*Femtochemistry – Ultrafast Dynamics of The Chemical Bond*, vols. I and II, A.H. Zewail, World Scientific, New Jersey, Singapore (1994)]. ★ Among others, Josef Michl, Lionel Salem, Donald G. Truhlar, Robert E. Wyatt, and W. Ronald Gentry also contributed to the theory of chemical reactions.

14.1 Hypersurface of the Potential Energy for Nuclear Motion

Theoretical chemistry is currently in a stage that experts in the field characterize as "*the primitive beginnings of chemical ab initio dynamics*".[1] The majority of the systems studied so far are *three-atomic*.[2]

The Born-Oppenheimer approximation works wonders, as it is possible to consider the (classical or quantum) dynamics of the *nuclei*, while the electrons disappear from the scene (their role became determining the potential energy for the motion of the nuclei, described in the electronic energy, the quantity corresponding to $E_0^0(R)$ from Eq. (6.8) on p. 266).

Even with this approximation, our job is not simple, for the following reasons:

- The reactants, as well as the products, may be quite large systems and the many-dimensional ground-state potential energy hypersurface $E_0^0(\boldsymbol{R})$ may have a very complex shape, whereas we are most often interested in the small fragment of the hypersurface that pertains to a particular one of many possible chemical reactions.
- We have many such hypersurfaces $E_k^0(\boldsymbol{R})$, $k = 0, 1, 2, \ldots$, each corresponding to an electronic state: $k = 0$ means the ground state, $k = 1, 2, \ldots$ corresponds to the excited states. There are processes that take place on a single hypersurface without changing the chemical bond pattern[3], but the very essence of chemical reaction is to change the bond pattern, and therefore excited states come into play.

It is quite easy to see where the fundamental difficulty is. Each of the hypersurfaces $E_k^0(\boldsymbol{R})$ for the motion of $N > 2$ nuclei depends on $3N - 6$ atomic coordinates (the number of translational and rotational degrees of freedom was subtracted).

Determining the hypersurface is not an easy matter, for the following reasons:

- A high accuracy of 1 kcal/mol is required, which is (for a fixed configuration) very difficult to achieve for *ab initio* methods,[4] and even more difficult for the semi-empirical or empirical methods.

---

[1] R.D. Levine and R.B. Bernstein, *Molecular Reaction Dynamics and Chemical Reactivity*, Oxford University Press (1987).

[2] John Polanyi recalls that the reaction dynamics specialists used to write as the first equation on the blackboard A + BC → AB + C, which made any audience burst out laughing. However, one of the outstanding specialists (Zare) said about the simplest of such reactions (H_3) (*Chem. Engin. News*, June 4 (1990)32): "*I am smiling, when somebody calls this reaction the simplest one. Experiments are extremely difficult, because one does not have atomic hydrogen in the stockroom, especially the high speed hydrogen atoms (only these react). Then, we have to detect the product i.e., the hydrogen, which is a transparent gas. On top of that it is not sufficient to detect the product in a definite spot, but we have to know which quantum state it is in.*"

[3] Strictly speaking, a change of conformation or formation of an intermolecular complex represents a chemical reaction. Chemists, however, reserve this notion for more profound changes of electronic structure.

[4] We have seen in Chapter 10 that the correlation energy is very difficult to calculate.

- The *number of points* on the hypersurface that have to be calculated is extremely large and increases exponentially with the system size.[5]
- There is no general methodology telling us what to do with the calculated points. There is a consensus that we should approximate the hypersurface by a smooth analytical function, but no general solution has yet been offered.[6]

14.1.1 Potential Energy Minima and Saddle Points

Let us denote $E_0^0(\boldsymbol{R}) \equiv V$. The most interesting points of the hypersurface V are its *critical points*; i.e., the points for which the gradient ∇V is equal to zero:

$$G_i = \frac{\partial V}{\partial X_i} = 0 \quad \text{for} \quad i = 1, 2, \ldots, 3N, \tag{14.1}$$

where X_i denote the Cartesian coordinates that describe the configurations of N nuclei. Since $-G_i$ represents the force acting along the axis X_i, no forces act on the atoms in the configuration of a critical point.

There are several types of critical points. Each type can be identified after considering the *Hessian*; i.e., the matrix with elements

$$V_{ij} = \frac{\partial^2 V}{\partial X_i \partial X_j} \tag{14.2}$$

calculated for the critical point. There are three types of critical points: maxima, minima and saddle points (cf., Chapter 7 and Fig. 7.12, as well as the Bader analysis, p. 667). The saddle points, as will be shown shortly, are of several classes depending on the signs of the Hessian eigenvalues. Six of the eigenvalues are equal to zero (rotations and translations of the total system, see, p. 359) because this type of motion proceeds without any change of the potential energy V.

We will concentrate on the remaining $3N - 6$ eigenvalues:

- In the minimum, the $3N - 6$ Hessian eigenvalues $\lambda_k \equiv \omega_k^2$ (ω is the angular frequency of the corresponding normal modes) are all positive,
- In the maximum, all are negative.

---

5 Indeed, if we assume that 10 values for each coordinate axis is sufficient (and this looks like a rather poor representation), then for N atoms, we have 10^{3N-6} quantum mechanical calculations of good quality to perform. This means that for $N = 3$, we may still pull it off, but for larger N, everybody has to give up. For example, for the reaction $HCl + NH_3 \rightarrow NH_4Cl$, we would have to perform 10^{12} quantum mechanical calculations.

6 Such an approximation is attractive for two reasons: first, we dispose the (approximate) values of the potential energy for *all* points in the configuration space (not only those for which the calculations were performed), and second, the analytical formula may be differentiated and the derivatives give the forces acting on the atoms.

It is advisable to construct the abovementioned analytical functions following some theoretical arguments. These are supplied by intermolecular interaction theory (see Chapter 13).

- For a saddle point of the nth order, $n = 1, 2, \ldots, 3N - 7$, the n eigenvalues are negative, while the rest are positive. Thus, a first-order saddle point corresponds to all but one of the Hessian positive eigenvalues; i.e., one of the angular frequencies ω is therefore imaginary.

The eigenvalues were obtained by diagonalization of the Hessian. Such diagonalization corresponds to a rotation of the local coordinate system (cf., p. 359). Imagine a two-dimensional surface that at the minimum could be locally approximated by an ellipsoidal valley. The diagonalization means such a rotation of the coordinate system x, y that both axes of the ellipse coincide with the new axes x', y' (as discussed in Chapter 7). On the other hand, if our surface locally resembled a cavalry saddle, diagonalization would lead to such a rotation of the coordinate system that one axis would be directed along the horse, and the other across.[7]

IR and Raman spectroscopies providing the vibration frequencies and force constants tell us a lot about how the energy hypersurface close to minima looks, both for the reactants and the products. On the other hand, theory, and recently also the *femtosecond spectroscopy*,[a] are the only source of information about the first-order saddle points. However, the latter are extremely important for determining reaction rates since any saddle point is a kind of pivot point – it is as important for the reaction as the Rubicon was for Caesar.[b]

[a] In this spectroscopy, we hit a molecule with a laser pulse of a few femtoseconds. The pulse perturbs the system, and when relaxing, it is probed by a series of new pulses, each giving a spectroscopic fingerprint of the system. A femtosecond is an incredibly short time, light is able to run only about $3 \cdot 10^{-5}$ cm. Ahmed Zewail, the discoverer of this type of spectroscopy, received the Nobel Prize in 1999.

[b] In 49 BC, Julius Caesar heading his Roman legions crossed the Rubicon River (the border of his province of Gaul), which initiated civil war with the central power in Rome. His words, "*alea iacta est*" (the die is cast) became a symbol of a final and irreversible decision.

The simplest chemical reactions are those that do not require crossing any reaction barrier. For example, the reaction $Na^+ + Cl^- \rightarrow NaCl$ or other similar reactions that are not accompanied by bond breaking or bond formation take place *without any barrier.*[8]

After the barrierless reactions, there is a group of reactions in which the reactants and the products are separated by a single saddle point (no intermediate products). *How do we describe such a reaction in a continuous way?*

A PES (Potential Energy Surface and therefore all possible configurations of the nuclei) can be divided into separate valleys with the saddle points at their borders. Let us focus on a particular energy valley of the ground-state PES. One may ask about the stability of the system occupying this valley. A suitable kinetic energy may be sufficient to allow the system pass the saddle points at the border of the valley (Fig. 14.1). The distances from the bottom of the valley to these saddle points may be different (depending on how large deformation of nuclear configuration is needed

[7] A cavalry saddle represents a good illustration of the first-order saddle of a 2-D surface.

[8] As a matter of fact, the formation of van der Waals complexes may also belong to this group. However, in large systems, when precise docking of molecules take place, the final docking may occur with a steric barrier.

Fig. 14.1. An analogy between the electronic ground-state PES and a mountain landscape, as well as between the stability of a molecular system and the stability of a ball in the landscape. On the right side is the lake, and its bottom's lowest point represents the most stable position of the ball. If the ball had a small kinetic energy, it would only circulate around the bottom of the valley, which is analogous to molecular vibrations (preserving the pattern of the chemical bonds). If the kinetic energy were larger, this would mean a larger amplitude of such oscillations, in particular, one that corresponds to passing over the barrier (the arrow toward the lowest saddle point) and getting to the next valley, which means a change of the chemical bond pattern, one reaction channel is open. One sees, that when the kinetic energy is getting larger, the number of possible reaction channels increases quickly (other arrows).

to reach a given saddle point) and, more importantly, may correspond to different energies ("*barriers for a given reaction*"). The destabilization is the easiest when moving toward the *lowest-energy saddle point,* which is often associated with the motion described by the lowest-frequency vibrational mode. The larger the kinetic energy of the system, the more numerous possibilities of crossing over the saddle points, each characterized by some particular products.[9]

A possibility to pass to next and next valleys (through the saddle points) means a chain of chemical reactions.

14.1.2 Distinguished Reaction Coordinate (DRC)

We often define a reaction path in the following way.

- First, we choose a particular distance (s) between the reacting molecules (e.g., an interatomic distance, one of the atoms belongs to molecule A, the other to B).
- Then we minimize the potential energy by optimization of all atomic positions, while keeping the s distance fixed.[10]

[9] In other words, we are opening new reaction channels.
[10] This is similar to calculating the relaxed interaction energy of Eq. (13.2).

- Change s by small increments from its reactant value until the product value is obtained (for each s optimizing all other distances).
- This defines a path (DRC) in the configurational space, and the progress along the path is measured by s.

A deficiency of the DRC is an arbitrary choice of the distance. The energy profile obtained (the potential energy versus s) depends on the choice. Often the DRC is reasonable close to the reactant geometry and becomes misleading when close to the product value (or *vice versa*). There is no guarantee that such a reaction path passes through the saddle point. Not only that, but other coordinates may undergo discontinuities, which feels a little catastrophic.

14.1.3 Steepest Descent Path (SDP)

Because of the Boltzmann distribution, the potential energy *minima* are most important, mainly low-energy ones.[11]

The saddle points of the first order are also important because it is believed that any two minima may be connected by a chain of the first-order saddle points. Several first-order saddle points to pass mean a multistage reaction that consists of several steps, each one representing a pass through a single first-order saddle point (elementary reaction). Thus, *the least energy-demanding path from the reactants to products goes via a saddle point of the first order.* This steepest descent path (SDP) is determined by the direction $-\nabla V$. First, we choose a first-order saddle point $\boldsymbol{R}_0$, then diagonalize the Hessian matrix calculated at this point and the eigenvector $\boldsymbol{L}$ corresponding to the single negative eigenvalue of the Hessian. Now, let us move all atoms *a little* from position $\boldsymbol{R}_0$ in the direction indicated in the configurational space by $\boldsymbol{L}$, and then let us follow vector $-\nabla V$ until it reduces to zero (then we are at the minimum). In this way, we have traced half the SDP. The other half will be determined starting from the other side of the saddle point and following the $-\boldsymbol{L}$ vector first.

Shortly, we will note a certain disadvantage of the SDP, which causes us to prefer another definition of the reaction path (see, p. 902).

14.1.4 Higher-Order Saddles

The first-order saddle points play the prominent role in the theory of chemical reactions. How can a chemist imagine a higher-order saddle point? Are they feasible at all in chemistry? The first-order saddle point may be modeled by a bond A-X with the X atom in a position with a repulsion with the atom B, from which a departure of X means an energy relief. A multiple-order saddle point may correspond to a geometry with several such atoms stuck in a high-energy position.

---

[11] Putting aside some subtleties (e.g., does the minimum support a vibrational level), the minima correspond to stable structures, since a small deviation from the minimum position causes a gradient of the potential to become nonzero, and this means a force pushing the system back toward the minimum position.

14.1.5 Our Goal

We would like to present a theory of elementary chemical reactions within the Born-Oppenheimer approximation, which describes nuclear motion on the potential energy hypersurface.

We have the following alternatives:

1. To perform *molecular dynamics*[12] on the hypersurface V (a point on the hypersurface represents the system under consideration).
2. To solve the *time-independent Schrödinger equation* $\hat{H}\psi = E\psi$ for the motion of the nuclei with potential energy V.
3. To solve the time-dependent *Schrödinger equation* with the boundary condition for $\psi(x, t = 0)$ in the form of a wave packet.[13] The wave packet may be directed into the entrance channel toward the reaction barrier (from various starting conditions). In the barrier range, the wave packet splits into a wave packet crossing the barrier and a wave packet reflected from the barrier (cf., p. 180).
4. To perform a semi-classical analysis that highlights the existence of the SDP, or a similar path, leading from the reactant to the product configuration.

Before going to more advanced approaches, let us consider possibility 1.

14.2 Chemical Reaction Dynamics (A Pioneers' Approach)

The SDP does not represent the only possible reaction path. It is only the *least-energy expensive path* from reactants to products. In real systems, the point representing the system will attempt to get through the pass in many different ways. Many such attempts are unsuccessful (non-reactive trajectories). If the system leaves the entrance channel (*reactive trajectories*), it will not necessarily pass through the saddle point, because it may have some extra kinetic energy, which may allow it to go with a higher energy than that of the barrier. Everything depends on the starting position and velocity of the point running through the entrance channel.

In the simplest case of a three-atom reaction,

$$A + BC \rightarrow AB + C,$$

the potential energy hypersurface represents a function of $3N - 6 = 3$ coordinates (the translations and rotations of the total system were separated). Therefore, even in such a simple case, it is difficult to draw this dependence. We may simplify the problem by considering only a limited set of geometries (e.g., the three atoms in a linear configuration). In such a case, we have only

[12] This is a classical approach. We have to ensure that the bonds may break; this is a very non-typical molecular dynamics.

[13] One example of this is a Gaussian function (in the nuclear coordinate space) moving from a position in this space with a starting velocity.

two independent variables[14] R_{AB} and R_{BC} and the function $V(R_{AB}, R_{BC})$ may be visualized by a map quite similar to those used in geography. The map has a characteristic shape, shown in Fig. 14.2.

- *"Reaction map"*. First of all we can see the characteristic *"drainpipe"* shape of the potential energy V for the motion of the nuclei [i.e., the function $V(R_{AB}, R_{BC}) \to \infty$ for $R_{AB} \to 0$ or for $R_{BC} \to 0$]; therefore, we have a high energy wall along the axes. When R_{AB} and R_{BC} are both large, we have a kind of plateau that goes gently downhill toward the bottom of the curved drain-pipe, that extends nearly parallel to the axes. The chemical reaction A + BC → AB + C means a motion, close to the bottom of the drainpipe, from a point corresponding to a large R_{AB}, while R_{BC} has a value corresponding to the equilibrium BC length (Fig. 14.2a, see the entrance arrow), until a large R_{BC} and R_{AB} with a value corresponding to the length of the isolated molecule AB (exit arrow).
- *Barrier.* A projection of the drainpipe bottom on the R_{AB} R_{BC} plane gives the SDP. Therefore, the SDP represents one of the important features of the *"landscape topography."*

---

[14] This is the case after separating the center-of-mass motion. The separation may be done in the following way. The kinetic energy operator has the form
$\hat{T} = -\frac{\hbar^2}{2M_A}\frac{\partial^2}{\partial X_A^2} - \frac{\hbar^2}{2M_B}\frac{\partial^2}{\partial X_B^2} - \frac{\hbar^2}{2M_C}\frac{\partial^2}{\partial X_C^2}$. We introduce some new coordinates:

- The center-of-mass coordinate $X_{CM} = \frac{M_A X_A + M_B X_B + M_C X_C}{M}$ with the total mass $M = M_A + M_B + M_C$
- $R_{AB} = X_B - X_A$
- $R_{BC} = X_C - X_B$

To write the kinetic energy operator in the new coordinates, we start with relations

$$\frac{\partial}{\partial X_A} = \frac{\partial R_{AB}}{\partial X_A}\frac{\partial}{\partial R_{AB}} + \frac{\partial X_{CM}}{\partial X_A}\frac{\partial}{\partial X_{CM}} = -\frac{\partial}{\partial R_{AB}} + \frac{M_A}{M}\frac{\partial}{\partial X_{CM}},$$

$$\frac{\partial}{\partial X_B} = \frac{\partial R_{AB}}{\partial X_B}\frac{\partial}{\partial R_{AB}} + \frac{\partial R_{BC}}{\partial X_B}\frac{\partial}{\partial R_{BC}} + \frac{\partial X_{CM}}{\partial X_B}\frac{\partial}{\partial X_{CM}} = \frac{\partial}{\partial R_{AB}} - \frac{\partial}{\partial R_{BC}} + \frac{M_B}{M}\frac{\partial}{\partial X_{CM}},$$

$$\frac{\partial}{\partial X_C} = \frac{\partial R_{BC}}{\partial X_C}\frac{\partial}{\partial R_{BC}} + \frac{\partial X_{CM}}{\partial X_C}\frac{\partial}{\partial X_{CM}} = \frac{\partial}{\partial R_{BC}} + \frac{M_C}{M}\frac{\partial}{\partial X_{CM}}.$$

After squaring these operators and substituting them into $\hat{T}$, we obtain, after a brief derivation, $\hat{T} = -\frac{\hbar^2}{2M}\frac{\partial^2}{\partial X_{CM}^2} - \frac{\hbar^2}{2\mu_{AB}}\frac{\partial^2}{\partial R_{AB}^2} - \frac{\hbar^2}{2\mu_{BC}}\frac{\partial^2}{\partial R_{BC}^2} + \hat{T}_{ABC}$, where the reduced masses $\frac{1}{\mu_{AB}} = \frac{1}{M_A} + \frac{1}{M_B}$, $\frac{1}{\mu_{BC}} = \frac{1}{M_B} + \frac{1}{M_C}$, whereas $\hat{T}_{ABC}$ stands for the mixed term

$$\hat{T}_{ABC} = \frac{\hbar^2}{M_B}\frac{\partial^2}{\partial R_{AB}\partial R_{BC}}.$$

In this way, we obtain the center-of-mass motion separation (the first term). The next two terms represent the kinetic energy operators for the independent pairs AB and BC, while the last one is the mixed term $\hat{T}_{ABC}$, whose presence is understandable: atom B participates in two motions, those associated with R_{AB} and R_{BC}. We may eventually get rid of $\hat{T}_{ABC}$ after introducing a skew coordinate system (the coordinates are determined by projections parallel to the axes). After rewriting $\hat{T}$, we obtain the following condition for the angle θ between the two axes, which ensures that the mixed terms vanish: $\cos\theta_{opt} = \frac{2}{M_B}\frac{\mu_{AB}\mu_{BC}}{\mu_{AB}+\mu_{BC}}$. If all the atoms have their masses equal, we obtain $\theta_{opt} = 60^0$.

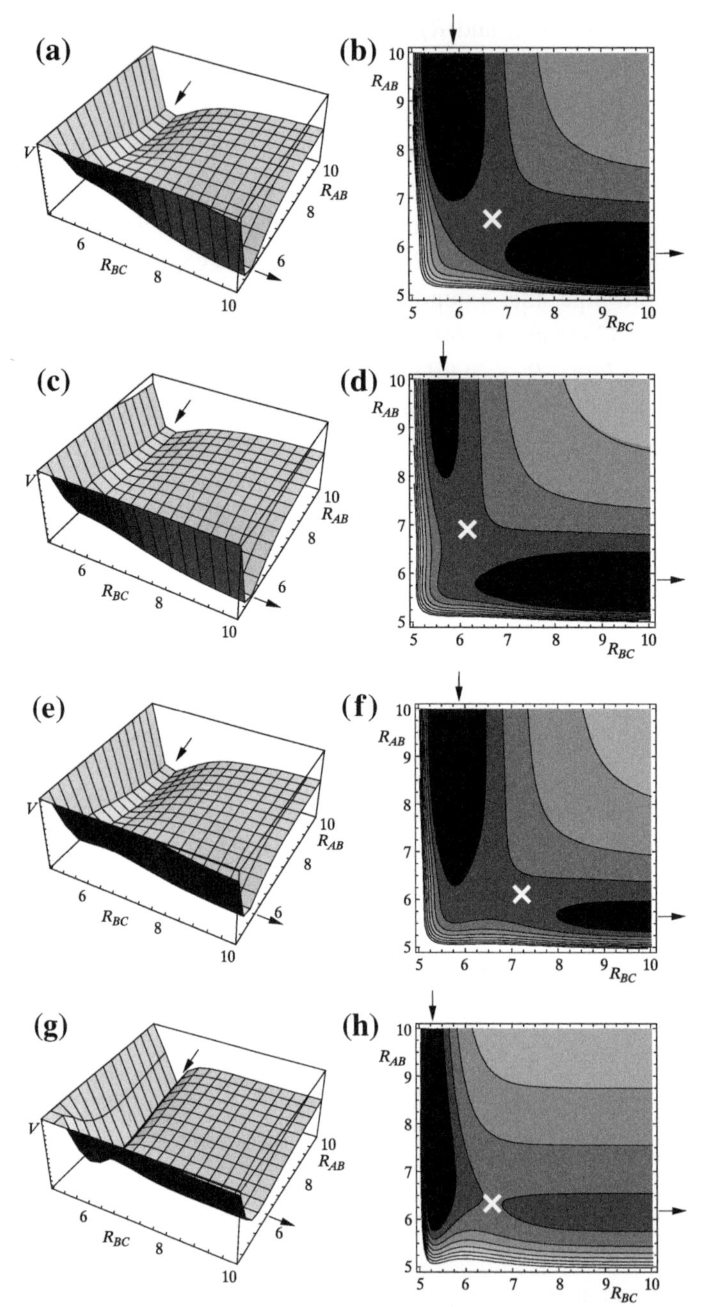

Fig. 14.2. The "drainpipe" A+BC → AB+C (for a fictitious collinear system). The surface of the potential energy for the motion of the nuclei is a function of distances R_{AB} and R_{BC}. On the left side, there is the view of the surface, while on the right side, the corresponding maps are shown. The barrier positions are given by the crosses on the right figures. Panels (a) and (b) show the symmetric entrance and exit channels with the separating barrier. Panels (c) and (d) correspond to an exothermic reaction with the barrier in the entrance channel ("*an early barrier*"). Panels (e) and (f) correspond to an endothermic reaction with the barrier in the exit channel ("*a late barrier*"). This endothermic reaction will not proceed spontaneously, because due to the equal width of the two channels, the reactant's free energy is lower than the product's free energy. Panels (g) and (h) correspond to a spontaneous endothermic reaction, because due to the much wider exit channel (as compared to the entrance channel), the free energy is lower for the products. Note that there is a van der Waals complex well in the entrance channel just before the barrier. There is no such well in the exit channel.

Travel on the potential energy surface along the SDP is not a flat trip, because the drainpipe consists of two valleys: the reactant valley (*entrance channel*) and the product valley (*exit channel*) separated by a pass (*saddle point*), which causes the *reaction barrier*. The saddle point corresponds to the situation, in which the old chemical bond is already weakened (but still exists), while the new bond is just emerging. This explains (as has been shown by Henry Eyring, Michael Polanyi, and Meredith Evans) why the energy required to go from the entrance to the exit barrier is much smaller than the dissociation energy of BC; e.g., for the reaction $H + H_2 \rightarrow H_2 + H$, the activation energy (to overcome the reaction barrier) amounts only to about 10% of the hydrogen molecule binding energy. Simply, when the BC bond breaks, a new bond AB forms *at the same time* compensating for the energy cost needed to break the BC bond.

The barrier may have different positions in the reaction drainpipe, e.g., it may be in the entrance channel (*early barrier*), Fig. 14.2c and d, or in the exit channel (*late barrier*), Fig. 14.2e and f, or, it may be in between (symmetric case, Figs. 14.2a,b). The barrier position influences the course of the reaction.

When determining the SDP, kinetic energy was neglected; i.e., the motion of the point representing the system resembles a "crawling." A chemical reaction does not, however, represent any crawling over the energy hypersurface, but rather a dynamic that begins in the entrance channel and ends in the exit channel, including motion "uphill" the potential energy V. Overcoming the barrier thus is possible only, when the system has an excess of kinetic energy.

What will happen, if we have an early barrier? A possible reactive trajectory for such a case is shown in Fig. 14.3a.

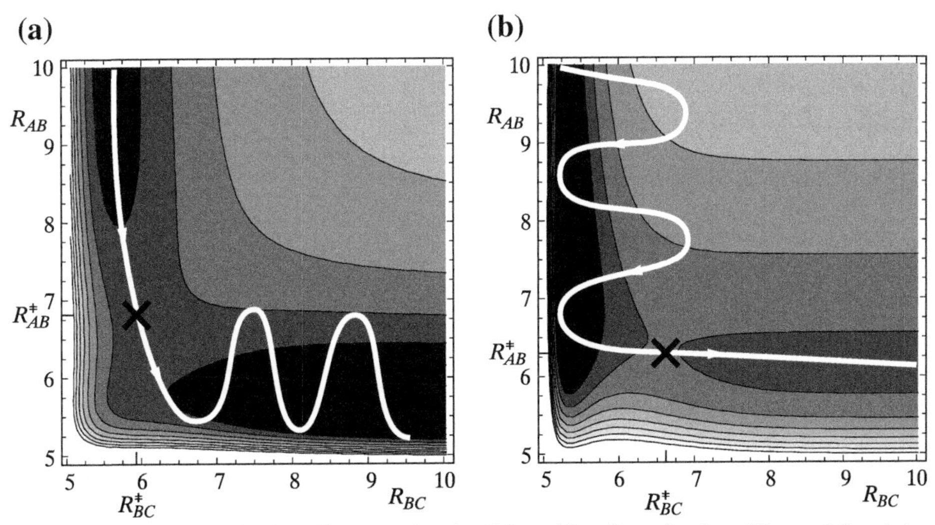

Fig. 14.3. A potential energy map for the collinear reaction $A + BC \rightarrow AB + C$ as a function of R_{AB} and R_{BC} (scheme). The distances $R^{\#}_{AB}$ and $R^{\#}_{BC}$ determine the saddle point position. (a) shows a reactive trajectory. If the point that represents the system runs sufficiently fast along the entrance channel toward the barrier, it will overcome the barrier by a "charge ahead." Then, in the exit channel, the point has to oscillate, which means product vibrations. (b) shows a reaction with a late barrier. In the entrance channel, a promising reactive trajectory is shown as a wavy line. This means that the system oscillates in the entrance channel in order to be able to attack the barrier directly after passing the corner area (the bobsleigh effect).

It is seen that the most effective way to pass the barrier is to set the point (representing the system) in fast motion along the entrance channel. This means that atom A has to have high kinetic energy when attacking the molecule BC. After passing the barrier the point slides downhill, entering the exit channel. Since, after sliding down, it has high kinetic energy, a *bobsleigh effect* is taking place; i.e., the point climbs up the potential wall (as a result of the repulsion of atoms A and B) and then moves by making zigzags similar to a bobsleigh team. This zigzag means, of course, that strong oscillations of AB take place (and the C atom leaves the rest of the system). Thus,

early location of a reaction barrier may result in a vibrational excited product.

A different thing happens when the barrier is *late*. A possible reactive (i.e., successful) trajectory is shown in Fig. 14.3b. For the point to overcome the barrier, it has to have a high momentum along the BC axis, because otherwise it would climb up the potential energy wall in vain, as the energy cost was too large. This may happen if the point moves along a zigzag way *in the entrance channel* (as shown in Fig. 14.4b). This means that

to overcome a late barrier, the *vibrational excitation* of the reactant BC is effective

because an increase in the kinetic energy of A will not produce much. Of course, the conditions for the reaction to occur matter less for high-collision energies of the reactants. On the other hand, a collision that is too fast may lead unwanted reactions to occur (e.g., dissociation of the system into $A + B + C$). Thus, there is an energy window for any given reaction.

14.3 Accurate Solutions (Three Atoms[15])

14.3.1 Coordinate System and Hamiltonian

This approach to the chemical reaction problem corresponds to point 2 on p. 892.

Jacobi Coordinates

For three atoms of masses M_1, M_2, M_3, with total mass $M = M_1 + M_2 + M_3$, we may introduce the Jacobi coordinates (see p. 341) in three different ways (Fig. 14.4).

Each of the coordinate systems (let us label them $k = 1, 2, 3$) highlights two atoms (i, j) "close" to each other and a third (k) that is "distant", Fig. 14.4a." Now, let us choose a pair of vectors $\boldsymbol{r}_k,\ \boldsymbol{R}_k$ for each of the choices of the Jacobi coordinates by the following procedure

[15] The method was generalized for an arbitrary number of atoms [D. Blume, and C.H. Greene, "*Monte Carlo Hyperspherical Description of Helium Cluster Excited States*", 2000].

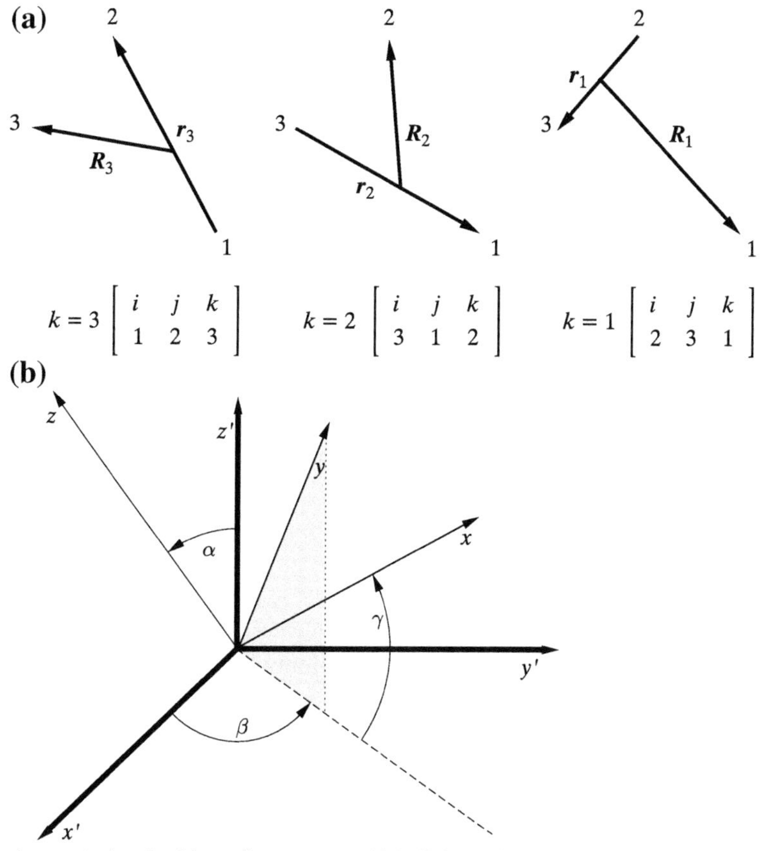

Fig. 14.4. (a) The three equivalent Jacobi coordinate systems; (b) the Euler angles show the mutual orientation of the two Cartesian coordinate systems. First, we project the y-axis on the x', y' plane (the result is the dashed line). The first angle α is the angle between axes z' and z, and the two other ones (β and γ) use the projection line described above. The relations among the coordinates are given by H.Eyring, J.Walter, and G.E.Kimball, *Quantum Chemistry* John Wiley, New York (1967).

($\boldsymbol{X}_i$ represents the vector identifying nucleus i in a space-fixed coordinate system, SFCS, cf., Appendix I available at booksite.elsevier.com/978-0-444-59436-5). First, let us define $\boldsymbol{r}_k$:

$$\boldsymbol{r}_k = \frac{1}{d_k}(\boldsymbol{X}_j - \boldsymbol{X}_i), \tag{14.3}$$

where the square of the mass scaling parameter equals

$$d_k^2 = \left(1 - \frac{M_k}{M}\right)\frac{M_k}{\mu}, \tag{14.4}$$

while μ represents the reduced mass (for three masses):

$$\mu = \sqrt{\frac{M_1 M_2 M_3}{M}}. \tag{14.5}$$

Now the second vector needed for the Jacobi coordinates is chosen as

$$\boldsymbol{R}_k = d_k \left[\boldsymbol{X}_k - \frac{M_i \boldsymbol{X}_i + M_j \boldsymbol{X}_j}{M_i + M_j} \right]. \tag{14.6}$$

The three Jacobi coordinate systems are related by the following formulas (cf., Fig. 14.4):

$$\begin{pmatrix} \boldsymbol{r}_i \\ \boldsymbol{R}_i \end{pmatrix} = \begin{pmatrix} \cos \beta_{ij} & \sin \beta_{ij} \\ -\sin \beta_{ij} & \cos \beta_{ij} \end{pmatrix} \begin{pmatrix} \boldsymbol{r}_j \\ \boldsymbol{R}_j \end{pmatrix}, \tag{14.7}$$

$$\tan \beta_{ij} = -\frac{M_k}{\mu}, \tag{14.8}$$

$$\beta_{ij} = -\beta_{ji}.$$

The Jacobi coordinates will now be used to define what is called the (more convenient) hyperspherical democratic coordinates.

Hyperspherical Democratic Coordinates

When a chemical reaction proceeds, the role of the atoms changes and using the same Jacobi coordinate system all the time leads to technical problems. In order not to favor any of the three atoms despite possible differences in their masses, we introduce *hyperspherical democratic coordinates*.

First, let us define the z-axis of a Cartesian coordinate system, which is perpendicular to the molecular plane at the center of mass i.e., parallel to $\boldsymbol{A} = \frac{1}{2}\boldsymbol{r} \times \boldsymbol{R}$, where $\boldsymbol{r}$ and $\boldsymbol{R}$ are *any* (just democracy, the result is the same) of the vectors $\boldsymbol{r}_k, \boldsymbol{R}_k$. Note that by definition, $|\boldsymbol{A}|$ represents the area of the triangle built of the atoms. Now, let us construct the x- and y-axes of the rotating with molecule coordinate system (RMCS; cf., p. 293) in the plane of the molecule, taking care of the following:

- The Cartesian coordinate system is right-handed
- The axes are oriented along the main axes of the moments of inertia,[16] with $I_{yy} \geq I_{xx}$.

Finally, we introduce hyperspherical democratic coordinates equivalent to the RMCS:

- The first coordinate measures the *size of the system*, or its "*radius*":

$$\rho = \sqrt{R_k^2 + r_k^2}, \tag{14.9}$$

where ρ has no subscript, because the result is independent of k (to check this, use Eq. (14.7)).

[16] These directions are determined by diagonalization of the inertia moment matrix (cf., Appendix K available at booksite.elsevier.com/978-0-444-59436-5).

- The second coordinate describes the system's *shape*:

$$\cos\theta = \frac{2\,|\boldsymbol{A}|}{\rho^2} \equiv u. \tag{14.10}$$

Since $|\boldsymbol{A}|$ is the area of the triangle, $2\,|\boldsymbol{A}|$ means, therefore, the area of the corresponding parallelogram. The last surface (in the numerator) is compared to the surface of a square with side ρ (in the denominator; if u is small, the system is elongated like an ellipse with three atoms on its circumference):

- As the third coordinate we choose the angle ϕ_k for *any* of the atoms (in this way, we determine, where the kth atom is on the ellipse):

$$\cos\phi_k = \frac{2(\boldsymbol{R}_k \cdot \boldsymbol{r}_k)}{\rho^2 \sin\theta} \equiv \cos\phi. \tag{14.11}$$

As chosen, the hyperspherical democratic coordinates which cover all possible atomic positions within the plane $z = 0$ have the following ranges: $0 \leq \rho < \infty, 0 \leq \theta \leq \frac{\pi}{2}, 0 \leq \phi \leq 4\pi$.

Hamiltonian in these Coordinates

The hyperspherical democratic coordinates represent a useful alternative for RMCS from Appendix I available at booksite.elsevier.com/978-0-444-59436-5 (they themselves form another RMCS), and therefore do not depend on the orientation with respect to the body-fixed coordinate system (BFCS). However, the molecule has somehow to "be informed" that it rotates (preserving the length and the direction of the total angular momentum), because a centrifugal force acts on its parts and the Hamiltonian expressed in BFCS (cf., Appendix I available at booksite.elsevier.com/978-0-444-59436-5) has to contain information about this rotation.

The exact kinetic energy expression for a polyatomic molecule in a space-fixed coordinate system (SFCS; cf. Appendix I available at booksite.elsevier.com/978-0-444-59436-5) has been given in Chapter 6 (Eq. 6.39). After separation of the center-of-mass motion, the Hamiltonian is equal to $\hat{H} = \hat{T} + V$, where V represents the electronic energy playing the role of the potential energy for the motion of the nuclei (an analog of $E_0^0\left(R\right)$ from Eq. (6.8), and we assume the Born-Oppenheimer approximation). In the hyperspherical democratic coordinates, we obtain[17]

$$\hat{H} = -\frac{\hbar^2}{2\mu\rho^5}\frac{\partial}{\partial\rho}\rho^5\frac{\partial}{\partial\rho} + \hat{\mathcal{H}} + \hat{\mathcal{C}} + V\left(\rho,\theta,\phi\right), \tag{14.12}$$

with

$$\hat{\mathcal{H}} = \frac{\hbar^2}{2\mu\rho^2}\left[-\frac{4}{u}\frac{\partial}{\partial u}u\left(1-u^2\right)\frac{\partial}{\partial u} - \frac{1}{1-u^2}\left(4\frac{\partial^2}{\partial\phi^2} - \hat{J}_z^2\right)\right], \tag{14.13}$$

$$\hat{\mathcal{C}} = \frac{\hbar^2}{2\mu\rho^2}\left[\frac{1}{1-u^2}4i\hat{J}_z u\frac{\partial}{\partial\phi} + \frac{2}{u^2}\left[\hat{J}_x^2 + \hat{J}_y^2 + \sqrt{1-u^2}\left(\hat{J}_x^2 - \hat{J}_y^2\right)\right]\right], \tag{14.14}$$

[17] J.G. Frey and B.J. Howard, *Chem.Phys.*, *99*, 415 (1985).

where the first part and the term with $\frac{\partial^2}{\partial \phi^2}$ in $\hat{\mathcal{H}}$ represent what are called *deformation terms*; next, there is a term with $\hat{J}_z^2$ describing the rotations about the z-axis, the terms in $\hat{C}$ contain the Coriolis term (with $i\hat{J}_z u \frac{\partial}{\partial \phi}$).

14.3.2 Solution to the Schrödinger Equation

Shortly, we will need some basis functions that depend on the angles θ and ϕ, preferentially each of them somehow adapted to the problem we are solving. These basis functions will be generated as the eigenfunctions of $\hat{\mathcal{H}}$ obtained at a fixed value $\rho = \rho_p$:

$$\hat{\mathcal{H}}\left(\rho_p\right) \Phi_{k\Omega}\left(\theta, \phi; \rho_p\right) = \varepsilon_{k\Omega}\left(\rho_p\right) \Phi_{k\Omega}\left(\theta, \phi; \rho_p\right), \tag{14.15}$$

where, because of two variables θ, ϕ we have two quantum numbers k and Ω (numbering the solutions of the equations).

The total wave function that also takes into account rotational degrees of freedom (θ, ϕ) is constructed as (the quantum number $J = 0, 1, 2, \ldots$ determines the length of the angular momentum of the system, while the quantum number $M = -J, -J+1, \ldots, 0, \ldots J$ gives the z-component of the angular momentum) a linear combination of the basis functions $U_{k\Omega} = D_{\Omega}^{JM}\left(\alpha, \beta, \gamma\right) \Phi_{k\Omega}\left(\theta, \phi; \rho_p\right)$:

$$\psi^{JM} = \rho^{-\frac{5}{2}} \sum_{k\Omega} F_{k\Omega}^{J}\left(\rho; \rho_p\right) U_{k\Omega}\left(\alpha, \beta, \gamma, \theta, \phi; \rho_p\right), \tag{14.16}$$

where α, β, γ are the three Euler angles (Fig. 14.4b) that define the orientation of the molecule with respect to the distant stars, $D_{\Omega}^{JM}\left(\alpha, \beta, \gamma\right)$ represents the eigenfunctions of the symmetric top,[18] and $\Phi_{k\Omega}$ is the solution to Eq. (14.15), while $F_{k\Omega}^{J}\left(\rho; \rho_p\right)$ stands for the ρ-dependent expansion coefficients [i.e., functions of ρ (centered at point ρ_p)]. Thanks to $D_{\Omega}^{JM}\left(\alpha, \beta, \gamma\right)$, the function ψ^{JM} is the eigenfunction of the operators $\hat{J}^2$ and $\hat{J}_z$.

In what is known as the *close coupling method*, the function from Eq. (14.16) is inserted into the Schrödinger equation $\hat{H}\psi^{JM} = E_J \psi^{JM}$. Then, the resulting equation is multiplied by a function $U_{k'\Omega'} = D_{\Omega'}^{JM}\left(\alpha, \beta, \gamma\right) \Phi_{k'\Omega'}\left(\theta, \phi; \rho_p\right)$ and integrated over angles $\alpha, \beta, \gamma, \theta, \phi$, which means taking into account all possible orientations of the molecule in space (α, β, γ) and all possible shapes of the molecule (θ, ϕ) which are allowed for a given size ρ. We obtain a set of linear equations for the unknowns $F_{k\Omega}^{J}\left(\rho_p; \rho\right)$:

$$\rho^{-\frac{5}{2}} \sum_{k\Omega} F_{k\Omega}^{J}(\rho; \rho_p) \langle U_{k'\Omega'} | (\hat{H} - E_J) U_{k\Omega} \rangle_{\omega} = 0. \tag{14.17}$$

The summation extends over some assumed set of k, Ω (where the number of k, Ω pairs is equal to the number of equations). The symbol $\omega \equiv (\alpha, \beta, \gamma, \theta, \phi)$ means integration over the angles. The system of equations is solved numerically.

[18] D.M. Brink, and G.R. Satchler, *Angular Momentum* Clarendon Press, Oxford (1975).

If, when solving the equations, we apply the boundary conditions suitable for a discrete spectrum (vanishing for $\rho = \infty$), we obtain the stationary states of the three-atomic molecule. We are interested in chemical reactions, in which one of the atoms comes to a diatomic molecule, and after a while, another atom flies out, leaving (after reaction) the remaining diatomic molecule. Therefore, we have to apply suitable boundary conditions. As a matter of fact, we are not interested in the details of the collision; we are positively interested in what comes to our detector from the spot where the reaction takes place. What may happen at a certain energy E to a given reactant state (i.e., what the product state is; such a reaction is called *state-to-state*) is determined by the corresponding *cross section*[19] $\sigma(E)$. The cross section can be calculated from what is called the ***S** matrix*, whose elements are constructed from the coefficients $F^J_{k\Omega}(\rho; \rho_p)$ found from Eq. (14.17). The $\boldsymbol{S}$ matrix plays a role of an energy dependent dispatcher: such a reactant state changes to such a product state with such and such probability.

We calculate the *reaction rate constant* k assuming all possible energies E of the system (satisfying the Boltzmann distribution) and taking into account that fast products arrive more often at the detector when counting per unit time:

$$k = \text{const} \int dE\ E\sigma(E) \exp\left(-\frac{E}{k_B T}\right), \tag{14.18}$$

where k_B is the Boltzmann constant.

The calculated reaction rate constant k may be compared with the result of the corresponding state-to-state experiment.

14.3.3 Berry Phase

When considering accurate quantum dynamics calculations (point 3 on p. 892), people met the problem of what is called the *Berry phase*.

In Chapter 6, wave function 6.9 corresponding to the adiabatic approximation, was assumed. In this approximation, the electronic wave function depends parametrically on the positions of the nuclei. Let us imagine we take one (or more) of the nuclei on an excursion. We set off, go slowly (in order to allow the electrons to adjust), the wave function deforms, and then, we go back home and put the nucleus exactly in place. Did the wave function come back exactly, too? Not necessarily. By definition (cf. Chapter 2), a class Q function has to be a unique function of coordinates. This, however, does not pertain to a parameter. What certainly came back is the probability density $\psi_k(\boldsymbol{r}; \boldsymbol{R})^* \psi_k(\boldsymbol{r}; \boldsymbol{R})$, because it decides that we cannot distinguish the starting and the final situations. *The wave function itself might undergo a phase change; i.e., the starting*

[19] After summing up the experimental results over all the angles, this is to be compared with the result of the abovementioned integration over angles.

function is equal to $\psi_k(\boldsymbol{r}; \boldsymbol{R}_0)$, *while the final function is* $\psi_k(\boldsymbol{r}; \boldsymbol{R}_0)\exp(i\phi)$ *and* $\phi \neq 0$. This phase shift is the Berry phase.[20] Did it happen or not? Sometimes we can tell.

Let us consider a quantum dynamics description of a chemical reaction according to point 3 cited earlier. For example, let us imagine a molecule BC fixed in space, with atom B directed to us. Now, atom A, represented by a wave packet, rushes toward atom B. We may imagine that the atom A approaches the molecule and makes a bond with the atom B (atom C leaves the diatomic molecule) or atom A may first approach atom C, then turn back and make a bond with atom B (as before). The two possibilities correspond to two waves, which finally meet and interfere. If the phases of the two waves differed, we would see this in the results of the interference. The scientific community was surprised that some details of the reaction $H + H_2 \rightarrow H_2 + H$ at higher energies are impossible to explain without taking the Berry phase[21] into account. One of the waves described above made a turn around the conical intersection point (because it had to by-pass the equilateral triangle configuration; cf. Chapter 6). As it was shown in the work of Longuet-Higgins *et al.* mentioned above, this is precisely the reason why the function acquires a phase shift. We have shown in Chapter 6 (p. 314) that such a trip around a conical intersection point results in changing the phase of the function by π.

The phase appears, when the system makes a "*trip*" in configurational space. We may make the problem of the Berry phase more familiar by taking an example from everyday life. Let us take a 3-D space. Put your arm down against your body with the thumb directed forward. During the operations described below, do not move the thumb with respect to your arm. Now stretch your arm horizontally sideways, rotate it to your front, and then put down along your body. Note that now your thumb is not directed toward your front anymore, but toward your body. When your arm went back, the thumb made a rotation of 90^0.

Your thumb corresponds to $\psi_k(\boldsymbol{r}; \boldsymbol{R})$ (i.e., a vector in the Hilbert space), which is coupled with a slowly varying neighborhood ($\boldsymbol{R}$ corresponds to the hand positions). When the neighborhood returns, the vector may have been rotated in the Hilbert space [i.e., multiplied by a phase $\exp(i\phi)$].

14.4 Intrinsic Reaction Coordinate (IRC) or Statics

This section addresses point 4 of our plan on p. 892.

Two reaction coordinates were proposed till now: DRC and SDP. Use of the first of these may lead to some serious difficulties (like energy discontinuities). The second reaction coordinate will undergo in a moment a useful modification and will be replaced by what is known as the *intrinsic reaction coordinate (IRC)*.

[20] The discoverers of this effect were H.C. Longuet-Higgins, U. Öpik, M.H.L. Pryce, and R.A. Sack, *Proc.Roy.Soc.London, A244,* 1 (1958). The problem of this geometric phase diffused into the consciousness of physicists much later, after an article by M.V. Berry, *Proc.Roy.Soc.London, A392*, 45 (1984).

[21] Y.-S.M. Wu, and A. Kupperman, *Chem.Phys.Letters*, *201,* 178 (1993).

What Is the IRC?

Let us use the Cartesian coordinate system once more, with $3N$ coordinates for the N nuclei: $X_i, i = 1, \ldots 3N$, where X_1, X_2, X_3 denote the x-, y-, and z-coordinates of atom 1 of mass M_1, etc. The ith coordinate is therefore associated with mass M_i of the *corresponding* atom. The classical Newtonian equation of motion for an atom of mass M_i and coordinate X_i is[22]:

$$M_i \ddot{X}_i = -\frac{\partial V}{\partial X_i} \quad \text{for} \quad i = 1, \ldots, 3N. \tag{14.19}$$

Let us introduce what are called *mass-weighted coordinates* (or, more precisely, weighted by the square root of mass):

$$x_i = \sqrt{M_i} X_i. \tag{14.20}$$

In such a case, we have

$$\sqrt{M_i}\sqrt{M_i}\ddot{X}_i = -\frac{\partial V}{\partial x_i}\frac{\partial x_i}{\partial X_i} = \sqrt{M_i}\left(-\frac{\partial V}{\partial x_i}\right) \tag{14.21}$$

or

$$\ddot{x}_i = -\frac{\partial V}{\partial x_i} \equiv -g_i, \tag{14.22}$$

where g_i stands for the ith component of the gradient calculated in the mass-weighted coordinates. This equation can be integrated for a short time t, and we obtain

$$\dot{x}_i = -g_i t + v_{0,i}, \tag{14.23}$$

or, for a small time increment $\mathrm{d}t$ and initial speed $v_{0,i} = 0$ (for the definition of the IRC as a path characteristic for potential energy V we want to neglect the influence of the kinetic energy), we obtain

$$\frac{\mathrm{d}x_i}{-g_i} = t\mathrm{d}t = \textit{independent} \text{ of } i. \tag{14.24}$$

Thus,

in the coordinates weighted by the square roots of the masses, a displacement of atom number i is proportional to the potential gradient (and does not depend on the atom mass).

If mass-weighted coordinates were not introduced, a displacement of the point representing the system on the potential energy map *would not follow the direction of the negative gradient or the steepest descent* (on a geographic map such a motion would look natural, because slow rivers flow this way). Indeed, the formula analogous to Eq. (14.24) would have the form

[22] Mass × acceleration equals force; a dot over the symbol means a time derivative.

$\frac{dX_i}{-G_i} = \frac{t}{M_i}dt$, where $G_i = \frac{\partial V}{\partial X_i}$ and therefore, during a single watch tick dt, light atoms would travel long distances while heavy atoms short distances. Thus, the trajectory would depend on mass, contrary to the discovery of Galileo Galilei. This means a counterintuitive sliding across the gradient direction.

Thus, after introducing mass-weighted coordinates, we may forget about masses, in particular about the atomic and the total mass, or equivalently, we may treat these as unit masses. The atomic displacements in this space will be measured in units of $\sqrt{\text{mass}} \times$ length, usually in: $\sqrt{u}a_0$, where $12u = {}^{12}$C atomic mass, $u = 1822.887\ m$ (m is the electron mass), and sometimes also in units of $\sqrt{u}$Å.

Eq. (14.24) takes into account our assumption about the zero initial speed of the atom in any of the integration steps (also called *trajectory-in-molasses*), because otherwise, we would have an additional term in dx_i: the initial velocity times time. Shortly speaking, when the watch ticks,

the system, represented by a point in $3N$-dimensional space, crawls over the potential energy hypersurface along the negative gradient of the hypersurface (in mass weighted coordinates). When the system starts from a saddle point of the first order, a small deviation of the position makes the system slide down on one or the other side of the saddle. The trajectory of the nuclei during such a motion is called the *intrinsic reaction coordinate (IRC)*.

The point that represents the system slides down with infinitesimal speed along the IRC.

Measuring the Travel Along the IRC

In the space of the mass-weighted coordinates, trajectory IRC represents a certain curve $\boldsymbol{x}_{\mathrm{IRC}}$ that depends on a parameter $s : \boldsymbol{x}_{\mathrm{IRC}}(s)$.

The parameter s measures the position along the reaction path IRC;

e.g., in $\sqrt{u}a_0$ or $\sqrt{u}$ Å. Let us take two close points on the IRC and construct the vector: $\boldsymbol{\xi}(s) = \boldsymbol{x}_{\mathrm{IRC}}(s + ds) - \boldsymbol{x}_{\mathrm{IRC}}(s)$, then

$$(ds)^2 = \sum_i [\xi_i(s)]^2. \tag{14.25}$$

We assume that $s = 0$ corresponds to the saddle point, $s = -\infty$ to the reactants, and $s = \infty$ to the products, (Fig. 14.5).

For each point on the IRC [i.e., on the curve $\boldsymbol{x}_{\mathrm{IRC}}(s)$], we may read the mass-weighted coordinates and use them to calculate the coordinates of each atom. Therefore, each point on the IRC corresponds to a certain spatial and electronic structure of the system.

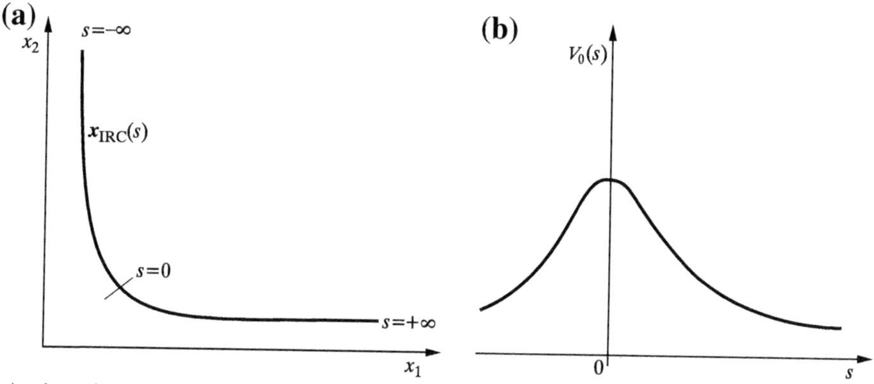

Fig. 14.5. A schematic representation of the IRC. (a) Curve $\boldsymbol{x}_{IRC}(s)$ and (b) energy profile when moving along the IRC [i.e., curve $V_0(\boldsymbol{x}_{IRC}(s))$] in the case of two mass-weighted coordinates x_1, x_2.

14.5 Reaction Path Hamiltonian Method

14.5.1 Energy Close to IRC

A hypersurface of the potential energy represents an expensive product. We have first to calculate the potential energy for a grid of points. If we assume that 10 points per coordinate is a sufficient number, then we have to perform 10^{3N-6} advanced quantum mechanical calculations, and for $N = 10$ atoms, this gives 10^{24} calculations, which is an impossible task. Now you see why specialists like three-atomic systems so much.

Are all the points necessary? For example, if we assume low energies, the system will, in practice, stay close to the IRC. If that is so, then why worry about other points? This idea was examined by Miller, Handy and Adams.[23] They decided to introduce the coordinates that are natural for the problem of motion in the reaction drainpipe. The approach corresponds to point 4, cited on p. 892.

The authors derived the

Reaction Path Hamiltonian:
an approximate expression for the energy of the reacting system in the form, that stresses the existence of the IRC and system's possible departures from it.

This formula (*Hamilton function of the reaction path*) takes the following form:

$$H(s, p_s, \{Q_k, P_k\}) = T(s, p_s, \{Q_k, P_k\}) + V(s, \{Q_k\}), \tag{14.26}$$

where T is the kinetic energy, V stands for the potential energy, s denotes the reaction coordinate along the IRC, $p_s = \frac{ds}{dt}$ represents the momentum coupled with s (mass = 1), $\{Q_k\}$, $k = 1, 2, \ldots 3N - 7$ stand for other coordinates orthogonal to the reaction path $\boldsymbol{x}_{IRC}(s)$ (this is why Q_k will depend on s) and the momenta $\{P_k\}$ conjugated with them.

[23] W.H. Miller, N.C. Handy, and J.E. Adams, *J.Chem.Phys.*, *72*, 99 (1980).

We obtain the coordinates Q_k in the following way. At point s of the IRC, we diagonalize the Hessian (i.e., the matrix of the second derivatives of the potential energy) and consider all the resulting normal modes [$\omega_k(s)$ are the corresponding frequencies; cf. Chapter 7] *other* than that which corresponds to the reaction coordinate s (the latter corresponds to the "imaginary"[24] frequency ω_k). The diagonalization also gives the orthonormal vectors $\boldsymbol{L}_k(s)$, each having a direction in the $(3N-6)$-dimensional configurational space (the mass-weighted coordinate system). The coordinate $Q_k \in (-\infty, +\infty)$ measures the displacement along the direction of $\boldsymbol{L}_k(s)$. The coordinates s and $\{Q_k\}$ are called the *natural coordinates*. To stress that Q_k is related to $\boldsymbol{L}_k(s)$, we will write it as $Q_k(s)$.

The potential energy close to the IRC can be approximated (*harmonic approximation*) by

$$V(s, \{Q_k\}) \cong V_0(s) + \frac{1}{2}\sum_{k=1}^{3N-7} \omega_k(s)^2 Q_k(s)^2, \tag{14.27}$$

where the term $V_0(s)$ represents the potential energy that corresponds to the bottom of the reaction drainpipe at a point s along the IRC, while the second term tells us what will happen to the potential energy if we displace the point (i.e., the system) perpendicular to $\boldsymbol{x}_{IRC}(s)$ along all the normal oscillator coordinates. In the *harmonic approximation* for the oscillator k, the energy goes up by half the force constant $\times$ the square of the normal coordinate Q_k^2. The force constant is equal to ω_k^2 because the mass is equal to 1.

The kinetic energy turns out to be more complicated:

$$T(s, p_s, \{Q_k, P_k\}) = \frac{1}{2}\frac{[p_s - \sum_{k=1}^{3N-7}\sum_{k'=1}^{3N-7} B_{kk'} Q_{k'} P_k]^2}{[1 + \sum_{k=1}^{3N-7} B_{ks} Q_k]^2} + \sum_{k=1}^{3N-7} \frac{P_k^2}{2} \tag{14.28}$$

The last term is recognized as the vibrational kinetic energy for the independent oscillations perpendicular to the reaction path (recall that the mass is treated as equal to 1). If in the first term we insert $B_{kk'} = 0$ and $B_{ks} = 0$, the term would be equal to $\frac{1}{2}p_s^2$ and, therefore, would describe the kinetic energy of a point moving as if the reaction coordinate were a straight line.

Coriolis and Curvature Couplings:

$B_{kk'}$ are called the *Coriolis coupling constants*. They couple the normal modes perpendicular to the IRC.

The B_{ks} are called the *curvature coupling constants*, because they would equal zero if the IRC were a straight line. They couple the translational motion along the reaction coordinate with the vibrational modes orthogonal to it. All the above coupling constants B depend on s.

[24] For large $|s|$, the corresponding ω^2 is close to zero. When $|s|$ decreases (we approach the saddle point), ω^2 becomes negative (i.e., ω is imaginary). For simplicity, we will call this the "imaginary frequency" for any s.

Therefore, in the reaction path Hamiltonian, we have the following quantities that characterize the reaction drainpipe:

- The reaction coordinate s that measures the progress of the reaction along the drainpipe.
- The value $V_0(s) \equiv V_0(\boldsymbol{x}_{IRC}(s))$ represents the energy that corresponds to the bottom of the drainpipe[25] at the reaction coordinate s.
- The width of the drainpipe is characterized[26] by a set $\{\omega_k(s)\}$.
- The curvature of the drainpipe is hidden in constants B; their definition will be given later in this chapter. Coefficient $B_{kk'}(s)$ tells us how normal modes k and k' are coupled together, while $B_{ks}(s)$ is responsible for a similar coupling between reaction path $\boldsymbol{x}_{\mathrm{IRC}}(s)$ and vibration k perpendicular to it.

14.5.2 Vibrational Adiabatic Approximation

Most often when moving along the bottom of the drainpipe, potential energy $V_0(s)$ changes only moderately when compared to the potential energy changes that the molecule undergoes when oscillating perpendicularly[27] to $\boldsymbol{x}_{IRC}(s)$. Simply, the valley bottom profile results from the fact that the molecule hardly holds together when moving along the reaction coordinate s, *a chemical bond breaks*, while *other bonds remain strong*, and it is not so easy to stretch their springs. This suggests that there is *slow* motion along s and *fast* oscillatory motion along the coordinates Q_k.

Since we are mostly interested in the slow motion along s, we may average over the fast motion.

The philosophy behind the idea is that while the system moves slowly along s, it undergoes a large number of oscillations along Q_k. After such vibrational averaging, the only information that remains about the oscillations are the vibrational quantum levels for each of the oscillators (the levels will depend on s).

Vibrational Adiabatic Approximation:
The fast vibrational motions will be treated quantum-mechanically and their total energy will enter the potential energy for the classical motion along s.

This approximation parallels the adiabatic approximation made in Chapter 6, where fast motion of electrons was separated from the slow motion of nuclei. There the total electronic

[25] That is, the classical potential energy corresponding to the point of the IRC given by s (this gives an idea of how the potential energy changes when walking along the IRC).

[26] A small ω corresponds to a wide valley; when measured along a given normal mode coordinate ("*soft*" vibration), a large ω means a narrow valley ("*hard*" vibration).

[27] That is, when moving along the coordinates Q_k, $k = 1, 2, \ldots 3N - 7$.

energy became the potential energy for motion of nuclei, here the total vibrational energy (the energy of the corresponding harmonic oscillators in their quantum states) becomes the potential energy for the slow motion along s. This concept is called the *vibrational adiabatic approximation.*

In this approximation, to determine the stage of the reaction we give two *classical* quantities: *where* the system is on the reaction path (s), *how fast* the system moves along the reaction path (p_s), as well as what are the *quantum states* of the oscillators vibrating perpendicularly to the reaction path (vibrational quantum number $v_k = 0, 1, 2, \ldots$ for each of the oscillators). Therefore, the potential energy for the (slow) motion along the reaction coordinate s is[28]

$$V_{adiab}(s; v_1, v_2, \ldots, v_{3N-7}) = V_0(s) + \sum_{k=1}^{3N-7} \left(v_k + \frac{1}{2}\right) \hbar[\omega_k(s) - \omega_k(-\infty)], \quad (14.29)$$

where we have chosen an arbitrary additive constant in the potential as being equal to the vibrational energy of the reactants (with minus sign): $-\sum_{k=1}^{3N-7} (v_k + \frac{1}{2})\hbar\omega_k(-\infty)$. Note that even if one would have $v_k = 0$ for each of the oscillators, there would be a nonzero vibrational correction to $V_0(s)$, because the zero-vibrational energy changes if s changes.

The vibrational adiabatic potential V_{adiab} was created for a given set of the vibrational quantum numbers v_k, fixed during the reaction process. Therefore, it is impossible to exchange energy between the vibrational modes (we assume, therefore, that the Coriolis coupling constants $B_{kk'} = 0$), as well as between the vibrational modes and the reaction path (we assume that the curvature coupling constants $B_{ks} = 0$). This would mean a change of v_k's.

From Eq. (14.29), we may draw the following conclusion:

When the frequency of a normal mode decreases dramatically (which corresponds to breaking of a chemical bond) during the reaction, the square bracket becomes negative. This means that an excitation of the bond, before the reaction, decreases the (vibrational adiabatic) reaction barrier and the reaction rate will increase.

As a matter of fact, this is a quite obvious: a vibrational excitation that engages the chemical bond to be broken already weakens the bond before the reaction.

[28] Even if (according to the vibrational adiabatic approximation) the vibrational quantum numbers were kept constant during the reaction, their energies, depending on s through ω, would change.

Why Do Chemical Reactions Proceed?

Exothermic Reactions

When the reactants (products) have kinetic energy higher than the barrier and the corresponding momentum p_s is large enough, there is a high probability that the barrier will be overcome (cf. p. 180). Even if the energy is lower than the barrier, there is still a nonzero probability of passing to the other side because of the tunneling effect. In both cases (passing over and under the barrier), it is easier when the kinetic energy is large.

The barrier *height* is usually different for the reaction reactants→products and for the products→reactants (Fig. 14.2). If the barrier height is smaller for the reactants, this *may* result in an excess of the product concentration over the reactant concentration.[29] Since the reactants have higher energy than the products, the potential energy excess will change into the kinetic energy[30] of the products (which is observed as a temperature increase–the reaction is exothermic). This may happen if the system has the possibility to pump the potential (i.e., electronic) energy excess into the translational and rotational degrees of freedom or to a "*third body or bodies*" (through collisions with the solvent molecules, for example) or has the possibility to emit light quanta. If the system has no such possibilities, the reaction will not take place.

Endothermic Reactions

The barrier height does not always decide.

Besides the barrier height, the *widths* of the entrance and exit channels also count.

For the time being, let us take an example with $V_0\left(-\infty\right) = V_0(\infty)$ i.e., the barrier calculated from IRC is *the same* in both directions. Imagine a narrow entrance channel, (i.e., large force constants for the oscillators,) and a wide exit channel [i.e., low force constants (Fig. 14.2g and h and 14.6)].

The vibrational levels in the entrance channel are high, while in the exit channel, they are low. This results in $V_{adiab}(-\infty; v_1, v_2, \ldots, v_{3N-7}) > V_{adiab}(\infty; v_1, v_2, \ldots, v_{3N-7})$; i.e., the barrier for the reaction reactants→products is lower, while for the reaction products→reactants, it is higher. The products will form more often than the reactants.

On top of that, if the entrance channel is narrow while the exit channel is wide, the density of the vibrational states will be small in the entrance channel and large in the exit channel. Therefore, for $T > 0$, there will be a lot of possibilities to occupy the low-energy vibrational

[29] This occurs because a lower barrier is easier to overcome.

[30] In most cases, this is rotational energy.

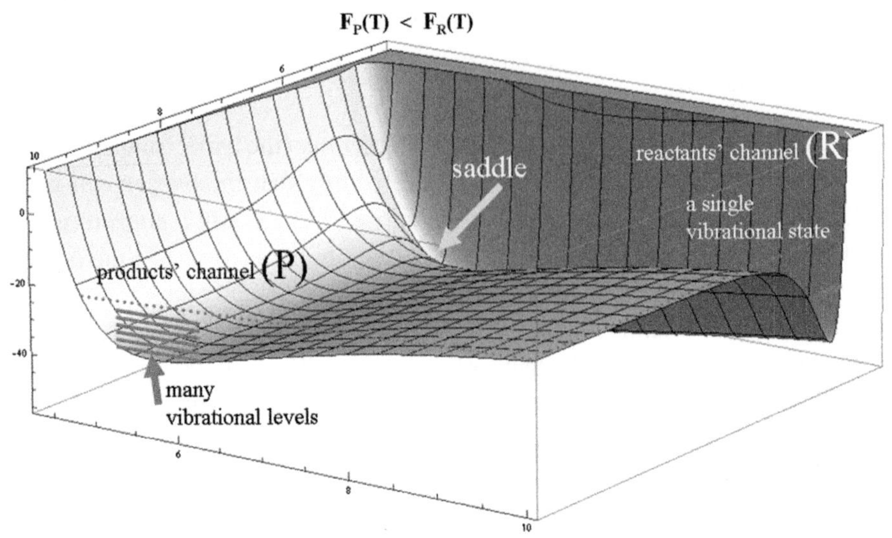

Fig. 14.6. Why do some endothermic reactions proceed spontaneously? As we can see, the reactants have lower energy than the products. Yet it is not V that decides the reaction to proceed, but the free energy $F = E - TS$, where T is the temperature and S the entropy. The free energy depends on the density of the vibrational states of the reactants and products. The more numerous the low-energy vibrational levels, the larger the entropy and the lower the free energy, if $T > 0$. As we can see, the reactant vibrational levels are scarce, while on the product side, they are densely distributed. When the energy gain related to the entropy overcomes the potential energy loss, then the (endothermic) reaction will proceed spontaneously.

levels for the products, while only a few possibilities for the reactants. This means a high entropy of the products and a small entropy of the reactants; i.e., the products will be more stabilized by the entropy than the reactants.[31] Once again we can see that

> while the energy in a spontaneous endothermic reaction increases, the decisive factor is the free energy, which *decreases*. The reactants→products reaction occurs "uphill" for the potential energy, but "downhill" for the free energy.

Kinetic and Thermodynamic Pictures

- In a macroscopic reaction carried out in a chemist's flask, we have a statistical ensemble of the systems that are in different microscopic stages of the reaction.
- The ensemble may be modeled by a reaction drainpipe (we assume here that the barrier exists) with a lot of points, each representing one of the reacting systems.

[31] It pertains to the term $-TS$ in the free energy. Biology is a masterpiece of chemistry, the endothermic reactions, which are not spontaneous, are often coupled with exothermic ones. In such a way, the whole process may go spontaneously.

- When the macroscopic reaction begins (e.g., we mix two reactants), a huge number of points appear in the entrance channel; i.e., we have the reactants only. As the reactant molecules assemble or dissociate, the points appear or disappear.
- If the barrier were high (no tunneling) and temperature $T = 0$, all the reactants would be in their zero-vibrational states[32] and in their ground rotational states. This state would be stable even when the product valley corresponded to a lower energy (this would be a metastable state).
- Raising the temperature causes some of the ensemble elements (points) in the entrance channel to acquire energy comparable to the barrier height. Those points might have a chance to pass the barrier either by tunneling (for energy smaller than the barrier) or by going over the barrier. Not all of the elements with sufficient energies would pass the barrier–only those with reactive trajectories.
- After passing the barrier, the energy is conserved, but in the case of exothermic reactions, the excess of energy changes into the translational, vibrational, and rotational energy of products or may be transferred to a "third body" (e.g., the solvent molecules). This increases the temperature.
- The probability of the reactive trajectories might be calculated in a way similar to that described in Chapter 4 (tunneling[33]), with additional taking into account the initial vibrational and rotational states of the reactants, as well as averaging over the energy-level populations.
- The products also would have a chance to pass the barrier back to the reactant side, but in the beginning, the number of the elements passing the barrier in the reactant-to-product direction would be larger (a non-equilibrium state).
- However, the higher the product concentration, the more often the products transform into the reactants. As an outcome, we arrive at the *thermodynamic equilibrium state*, in which the average numbers of the elements passing the barrier per unit time in either direction are equal.
- If the barrier is high and the energies are considered low, then the stationary states of the system could be divided into those (of energy $E_{i,R}$), with high amplitudes in the entrance channel (the reactant states) and those (of energy $E_{i,P}$) with high amplitudes in the exit channel (product states). In such a case, we may calculate the partition function for the reactants (R):

$$Z_R(T) = \sum_i g_i \exp\left(-\frac{E_{i,R} - E_{0,R}}{k_B T}\right),$$

and for the products (P),

$$Z_P(T) = \sum_i g_i \exp\left(-\frac{E_{i,P} - E_{0,R}}{k_B T}\right) = \sum_i g_i \exp\left(-\frac{E_{i,P} - E_{0,P} + \Delta E}{k_B T}\right),$$

[32] Note, that even in this case ($T = 0$), the energy of these states would not only depend on the bottom of the valley, $V_0(s)$, but also on the valley's width through $\omega_k(s)$, according to Eq. (14.29).

[33] See also H. Eyring, J. Walter, and G.E. Kimball, *Quantum Chemistry* John Wiley, New York (1967).

where g_i stands for the degeneracy of the ith energy level and the difference of the ground-state levels is $\Delta E = E_{0,P} - E_{0,R}$.

- Having the partition functions, we may calculate (at a given temperature, volume and a fixed number of particles[34]) the free or Helmholtz energy (F) corresponding to the entrance and to the exit channels (in thermodynamic equilibrium):

$$F_R(T) = -k_B T \frac{\partial}{\partial T} \ln Z_R(T), \tag{14.30}$$

$$F_P(T) = -k_B T \frac{\partial}{\partial T} \ln Z_P(T). \tag{14.31}$$

- The reaction goes in such a direction as to attain the minimum of free energy F.
- The higher the density of states in a given channel (this corresponds to higher entropy), the lower the value of F. The density of the vibrational states is higher for wider channels (see Fig. 14.6).

14.5.3 Vibrational Non-adiabatic Model

Coriolis Coupling

The vibrational adiabatic approximation is hardly justified because the reaction channel is curved. This means that motion along s couples with some vibrational modes, and also the vibrational modes couple among themselves. Therefore, we have to use the non-adiabatic theory, and this means that we need coupling coefficients B (p. 906). The Miller-Handy-Adams reaction path Hamiltonian theory gives the following expression for the $B_{kk'}$ (Fig. 14.7):

$$B_{kk'}(s) = \frac{\partial \boldsymbol{L}_k(s)}{\partial s} \cdot \boldsymbol{L}_{k'}(s), \tag{14.32}$$

where $\boldsymbol{L}_k, k = 1, 2, \ldots 3N - 7$ represent the orthonormal eigenvectors ($3N$-dimensional; cf. Chapter 7, p. 359) of the normal modes Q_k of frequency ω_k.

If the derivative in Eq. (14.32) is multiplied by an increment of the reaction path Δs, we obtain $\frac{\partial \boldsymbol{L}_k(s)}{\partial s} \Delta s$ which represents a change of normal mode vector $\boldsymbol{L}_k$ when the system moved along the IRC by the increment Δs. This change might be similar to normal mode eigenvector $\boldsymbol{L}_{k'}$. This means that $B_{kk'}$ measures how much eigenvector $\boldsymbol{L}_{k'}(s)$ resembles the *change* of eigenvector

[34] Similar considerations may be performed for a constant pressure (instead of volume). The quantity that then attains the minimum at the equilibrium state is the Gibbs potential G.

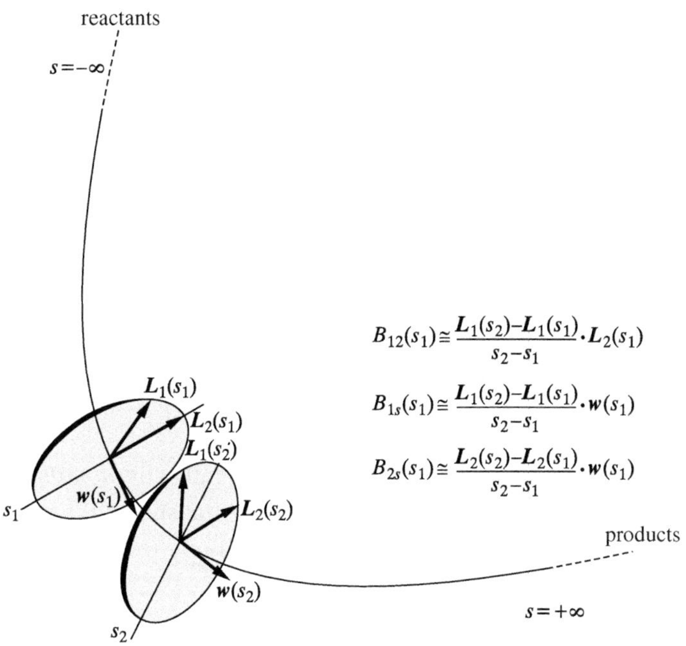

Fig. 14.7. A scheme: the Coriolis coupling coefficient (B_{12}) and the curvature coefficients (B_{1s} and B_{2s}) related to the normal modes 1 and 2 and reaction coordinate s. Diagonalization of the two Hessians calculated at points $s = s_1$ and $s = s_2$ gives two corresponding normal mode eigenvectors $\boldsymbol{L}_1(s_1)$ and $\boldsymbol{L}_2(s_1)$ as well as $\boldsymbol{L}_1(s_2)$ and $\boldsymbol{L}_2(s_2)$. At points s_1 and s_2, we also calculate the versors $\boldsymbol{w}(s_1)$ and $\boldsymbol{w}(s_1)$ that are tangent to the IRC (curved line). The calculated vectors inserted into the formulas give the approximations to B_{1s}, B_{2s} and B_{12}.

$\boldsymbol{L}_k(s)$ (when the system moves along the reaction path).[35] Coupling coefficient $B_{kk'}$ is usually especially large close to those values of s, for which $\omega_k \cong \omega_{k'}$ i.e., for the crossing points of the vibrational frequency (or energy) curves $\omega_k(s)$. These are the points where we may expect an important energy flow from one normal mode to another, because the energy quanta match ($\hbar\omega_k(s) \cong \hbar\omega_{k'}(s)$). Coriolis coupling means that the directions of $\boldsymbol{L}_k$ and $\boldsymbol{L}_{k'}$ change, when the reaction proceeds and this resembles a rotation in the configurational space about the IRC.

[35] From differentiating the orthonormality condition $\boldsymbol{L}_k(s) \cdot \boldsymbol{L}_{k'}(s) = \delta_{kk'}$, we obtain

$$\begin{aligned}
&\frac{\partial}{\partial s}\left[\boldsymbol{L}_k(s) \cdot \boldsymbol{L}_{k'}(s)\right] \\
&= \left[\frac{\partial \boldsymbol{L}_k(s)}{\partial s} \cdot \boldsymbol{L}_{k'}(s) + \boldsymbol{L}_k(s) \cdot \frac{\partial \boldsymbol{L}_{k'}(s)}{\partial s}\right] \\
&= \left[\frac{\partial \boldsymbol{L}_k(s)}{\partial s} \cdot \boldsymbol{L}_{k'}(s) + \frac{\partial \boldsymbol{L}_{k'}(s)}{\partial s} \cdot \boldsymbol{L}_k(s)\right] \\
&= B_{kk'} + B_{k'k} = 0.
\end{aligned}$$

Hence, $B_{kk'} = -B_{k'k}$.

Curvature Couplings

Curvature coupling constant B_{ks} links the motion along the reaction valley with the normal modes orthogonal to the IRC (Fig. 14.7):

$$B_{ks}(s) = \frac{\partial \boldsymbol{L}_k(s)}{\partial s} \cdot \boldsymbol{w}(s), \tag{14.33}$$

where $\boldsymbol{w}(s)$ represents the unit vector tangent to the intrinsic reaction path $\boldsymbol{x}_{IRC}(s)$ at point s. Coefficient $B_{ks}(s)$ therefore represents a measure of how the *change* in the normal mode eigenvector $\boldsymbol{L}_k(s)$ resembles a motion along the IRC. Large $B_{ks}(s)$ makes an energy flow from the normal mode to the reaction path (or *vice versa*) much easier.

Donating Modes:
The modes with large $B_{ks}(s)$ in the *entrance* channel are called the *donating modes*, because an excitation of such modes makes possible an energy transfer to the reaction coordinate degree of freedom (an increase of the kinetic energy along the reaction path). This will make the reaction rate increase.

In the vibrational adiabatic approximation, coefficients B_{ks} equal zero. This means that in such an approximation, an exothermic reaction would transform the net reaction energy (defined as the difference between the energy of the reactants and the products) into the kinetic energy of translational motion of products, because the energy of the system in the entrance channel could not be changed into the vibrational energy of the products (including the "*vibrations*" of a rotational character). However, as was shown by John Polanyi and Dudley Herschbach, the reactions do not go this way–a majority of the reaction energy goes into the rotational degrees of freedom (excited states of some modes). The rotations are hidden in these of the vibrations at $s = 0$, which are similar to the internal rotations and in the limit of $s \rightarrow +\infty$ transform into product rotations. Next, the excited products emit infrared quanta (heat) in the process of infrared fluorescence. This means that in order to have a realistic description of reaction, we have to abandon the vibrational adiabatic approximation.

14.5.4 *Application of the Reaction Path Hamiltonian Method to the Reaction $H_2 + OH \rightarrow H_2O + H$*

The reaction represents one of a few polyatomic systems for which precise calculations were performed.[36] It may be instructive to see how a practical implementation of the reaction path Hamiltonian method looks.

[36] G.C.J. Schatz, *J.Chem.Phys. 74,* 113 (1981); D.G. Truhlar, and A.D. Isaacson, *J.Chem.Phys. 77*, 3516 (1982); A.D. Isaacson, and D.G. Truhlar, *J.Chem.Phys. 76,* 380 (1982) and above all, the paper by Thom Dunning, Jr. and Elfi Kraka in *Advances in Molecular Electronic Structure Theory: The Calculation and Characterization of Molecular Potential Energy Surfaces*, ed. T.H. Dunning, Jr., JAI Press, Greenwich, CT, (1989), p. 1.

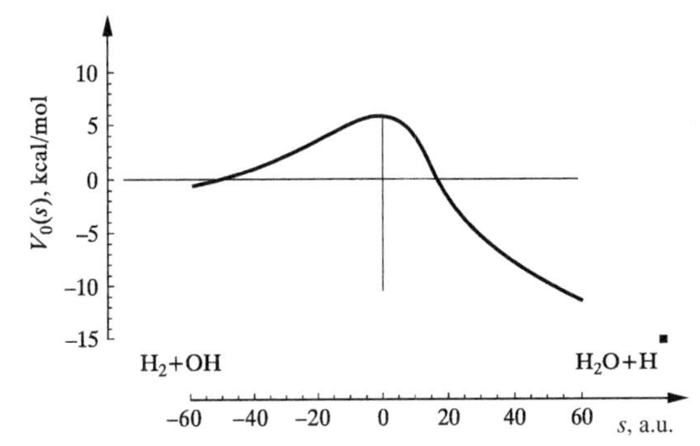

Fig. 14.8. The reaction $H_2 + OH \rightarrow H_2O + H$ energy profile $V_0(s)$ for $-\infty \leq s \leq \infty$. The value of the reaction coordinate $s = -\infty$ corresponds to the reactants, while $s = \infty$ corresponds to the products. It turns out that the product energy (shown by a small square on the right) is lower than the energy of the reactants (i.e., the reaction is exothermic). The barrier height in the entrance channel calculated as the difference of the top of the barrier and the lowest point of the entrance channel amounts to 6.2 kcal/mol. Source: T. Dunning, Jr. and E. Kraka, from *Advances in Molecular Electronic Structure Theory*, ed. T. Dunning, Jr., JAI Press, Greenwich, CT (1989), courtesy of the authors.

Potential Energy Hypersurface

The *ab initio* configuration interaction calculations (p. 615) of the electronic energy for the system under study were performed by Walsh and Dunning[37] within the Born-Oppenheimer ("*clamped nuclei*") approximation; see p. 269. The electronic energy obtained as a function of the nuclear configuration plays the role of the potential energy for the motion of the nuclei. The calculation gave the electronic energy for a relatively scarce set of the configurations of the nuclei, but then the numerical results were fitted by an analytical function.[38] The IRC energy profile obtained is shown in Fig. 14.8.

It is seen from this figure that the barrier height for the reactants is equal to about 6.2 kcal/mol, while the reaction energy calculated as the difference of the products minus the energy of the reactants is equal to about -15.2 kcal/mol (an exothermic reaction). What happens to the atoms when the system moves along the reaction path? This is shown in Fig. 14.9.

The saddle point configuration of the nuclei when compared to those corresponding to the reactants and to products tells us whether the barrier is early or late. The difference of the OH distances for the saddle point and for the product (H_2O) amounts to 0.26 Å, which represents $\frac{0.26}{0.97} = 27\%$, while the HH distance difference for the saddle point and of the reactant (H_2) is equal to 0.11 Å, which corresponds to $\frac{0.11}{0.74} = 15\%$. In conclusion, the saddle point resembles the reactants more than the products; i.e., the barrier is early.

Normal Mode Analysis

Let us see what the normal mode analysis gives when performed for some selected points along the IRC. The calculated frequencies are shown in Fig. 14.10 as wave numbers $\bar{\nu} = \omega/(2\pi c)$.

---

[37] S.P. Walsh, and T.H. Dunning, Jr., *J.Chem.Phys. 72,* 1303 (1980).

[38] G.C. Schatz, and H. Elgersma, *Chem.Phys.Letters, 73,* 21 (1980).

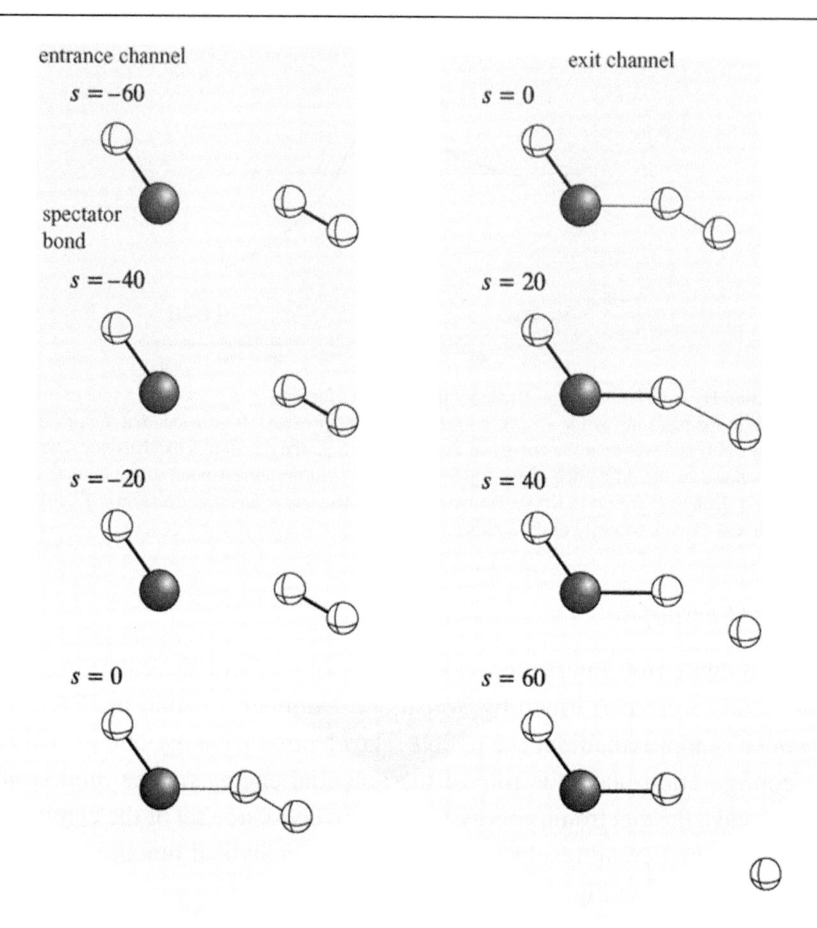

Fig. 14.9. The optimum atomic positions (read please as two columns) in the reacting system $H_2 + OH \rightarrow H_2O + H$ as functions of the reaction coordinate s. Source: T. Dunning, Jr. and E. Kraka, from *Advances in Molecular Electronic Structure Theory*, ed. T. Dunning, Jr., JAI Press, Greenwich, CT (1989), courtesy of the authors.

As we can see, before the reaction takes place, we have two normal mode frequencies ω_{HH} and ω_{OH}. When the two reacting subsystems approach one another, we have to treat them as a single entity. The corresponding number of vibrations is $3N - 6 = 3 \times 4 - 6 = 6$ normal modes. Fig. 14.10 shows five (real) frequencies. Two of them have frequencies close to those of HH and OH, and three others have frequencies close to zero and correspond to the vibrational and rotational motions of the loosely bound reactants.[39] The last "vibrational mode" (not shown) is connected with a motion along the reaction path and has imaginary frequency. Such a frequency means that the corresponding curvature of the potential energy is negative.[40] For example, at

[39] By the van der Waals interactions; see Chapter 13.

[40] Note that $\omega = \sqrt{k/m}$, where the force constant k stands for the second derivative of the potential energy (i.e., its curvature).

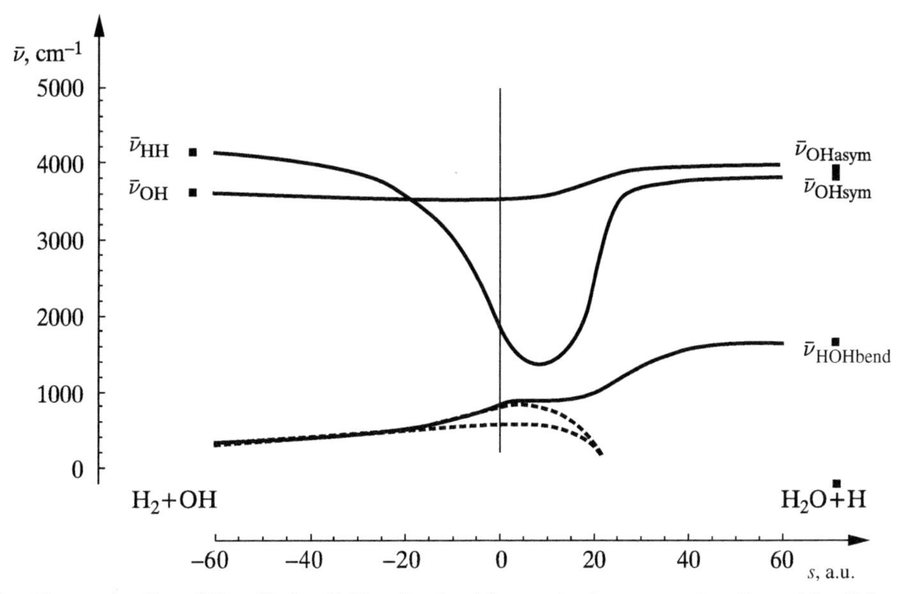

Fig. 14.10. The reaction $H_2 + OH \rightarrow H_2O + H$. The vibrational frequencies (in wave numbers $\bar{\nu} = \omega/(2\pi c)$) for the normal modes along the coordinate s. The little squares correspond to the asymptotic values. Only the real wave numbers are given (the "vibration" along s is imaginary and not given). Source: T. Dunning, Jr. and E. Kraka, from *Advances in Molecular Electronic Structure Theory*, ed. T. Dunning, Jr., JAI Press, Greenwich, CT (1989), courtesy of the authors.

the saddle point, when moving along the reaction path, we have a potential energy maximum instead of a minimum as would be for a regular oscillator.

The frequency ω_{HH} drops down close to the saddle point. This is precisely the bond to be broken. Interestingly, the frequency minimum is attained at 9 a.u. beyond the saddle point. Afterward the frequency increases fast, and when the reaction is completed, it turns out to be the OH symmetric stretching frequency of the water molecule. Only from this, we can tell what has happened: the HH bond was broken and a new OH bond was formed. At the end of the reaction path, we also have the antisymmetric stretching mode of the H_2O, which evolved from the starting value of ω_{OH} while changing only a little (this reflects that one OH bond exists all the time and in fact represents a *reaction spectator*) as well as the HOH bending mode, which appeared as a result of the strengthening of an intermolecular interaction in $H_2 + OH$ when the reaction proceeded. The calculations have shown that this vibration corresponds to the *symmetric* stretching mode[41] of the H_2O. The two other modes at the beginning of the reaction

[41] At first sight, this looks like contradicting chemical intuition since the antisymmetric mode is apparently compatible to the reaction path (one hydrogen atom being far away while the other is close to the oxygen atom). However, everything is all right. The SCF LCAO MO calculations for the water molecule within a medium size basis set give the OH bond length equal to 0.95 Å, whereas the OH radical bond length is equal to 1.01 Å. This means that when the hydrogen atom approaches the OH radical (making the water molecule), the hydrogen atom of the radical has to get *closer* to the oxygen atom. The resulting motion of both hydrogen atoms is similar to the symmetric (not antisymmetric) mode.

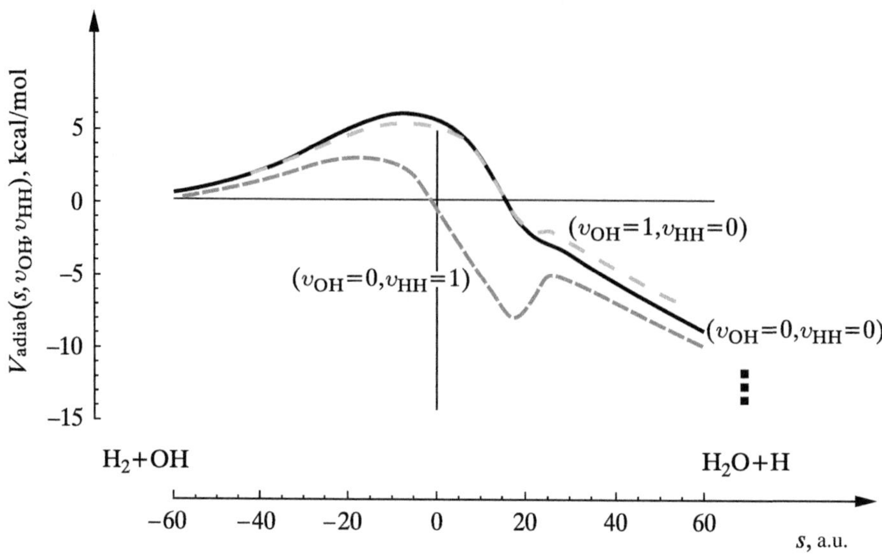

Fig. 14.11. The reaction $H_2 + OH \rightarrow H_2O + H$ (within the vibrational adiabatic approximation). Three sets of the vibrational numbers $(v_{OH}, v_{HH}) = (0,0), (1,0), (0,1)$ were chosen. Note that the height and position of the barrier depend on the vibrational quantum numbers assumed. An excitation of H_2 decreases considerably the barrier height. Source: T. Dunning, Jr. and E. Kraka, from *Advances in Molecular Electronic Structure Theory*, ed. T. Dunning, Jr., JAI Press, Greenwich, CT (1989), courtesy of the authors.

have almost negligible frequencies, and after an occasional increasing of their frequencies near the saddle point end up with zero frequencies for large s. Of course, at the end, we have to have $3 \times 3 - 6 = 3$ vibrational modes of the H_2O and so we do.

Example. Vibrational Adiabatic Approximation

Let us consider several versions of the reaction that differ by assuming various quantum vibrational states of the reactants.[42] Using Eq. (14.29), for each set of the vibrational quantum numbers we obtain the vibrational adiabatic potential V_{adiab} as a function of s (Fig. 14.11).

The adiabatic potentials obtained are instructive. It turns out that the following is true:

- The adiabatic potential corresponding to the vibrational ground state $(v_{OH}, v_{HH}) = (0,0)$ gives lower barrier height than the classical potential $V_0(s)$ (5.9 kcal/mol vs 6.1). The reason for this is the lower zero-vibration energy for the saddle point configuration than for the reactants.[43]
- The adiabatic potential for the vibrational ground state has its maximum at $s = -5$ a.u., not at the saddle point $s = 0$.

[42] We need the described frequencies of the modes that are orthogonal to the reaction path.

[43] This stands to reason because when the Rubicon is crossed, all the bonds are weakened with respect to the reactants.

- Excitation of the OH stretching vibration does not significantly change the energy profile, in particular the barrier is lowered by only about 0.3 kcal/mol. Thus, the OH is definitely a spectator bond.
- This contrasts with what happens when the H_2 molecule is excited. In such a case, the barrier is lowered by as much as about 3 kcal/mol. This conforms that the HH stretching vibration is a *donating mode*.

Example. Non-adiabatic Theory

Now let us consider the vibrational non-adiabatic procedure. To do this, we have to include the coupling constants B. This is done in the following way. Moving along the reaction coordinate s, we perform the normal mode analysis, resulting in the vibrational eigenvectors $\boldsymbol{L}_k(s)$. This enables us to calculate how these vectors change and to determine the derivatives $\partial \boldsymbol{L}_k/\partial s$. Now we may calculate the corresponding dot products [see Eqs. (14.32) and (14.33)] and obtain the coupling constants $B_{kk'}(s)$ and $B_{ks}(s)$ at each selected point s. A role of the coupling constants B in the reaction rate can be determined after dynamic studies assuming various starting conditions (the theory behind this approach will not be presented in this book). Yet some important information may be extracted just by inspecting functions $B(s)$. The functions $B_{ks}(s)$ are shown in Fig. 14.12.

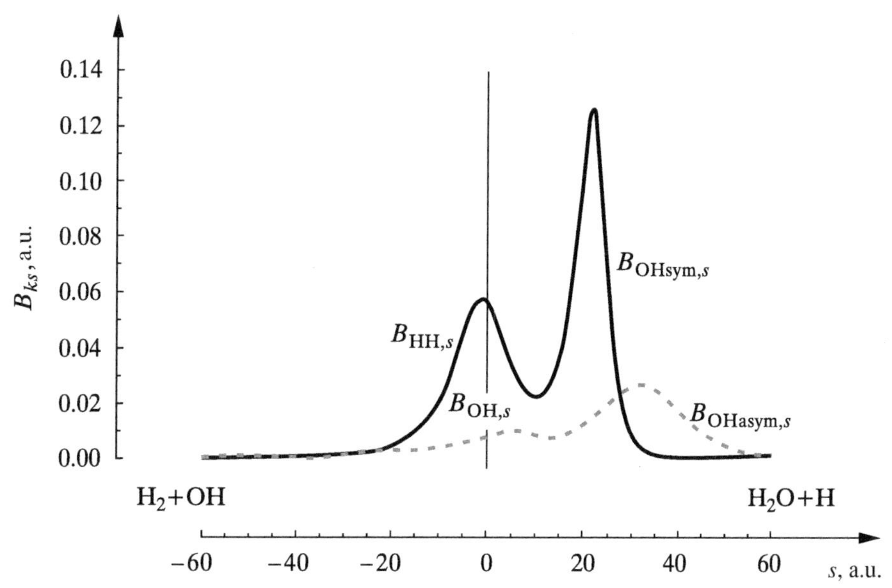

Fig. 14.12. The reaction $H_2 + OH \rightarrow H_2O + H$. The curvature coupling costants $B_{ks}(s)$ as functions of s. The $B_{ks}(s)$ characterize the coupling of the kth normal mode with the reaction coordinate s. Source: T. Dunning, Jr. and E. Kraka, from *Advances in Molecular Electronic Structure Theory*, ed. T. Dunning, Jr., JAI Press, Greenwich, CT (1989), courtesy of the authors.

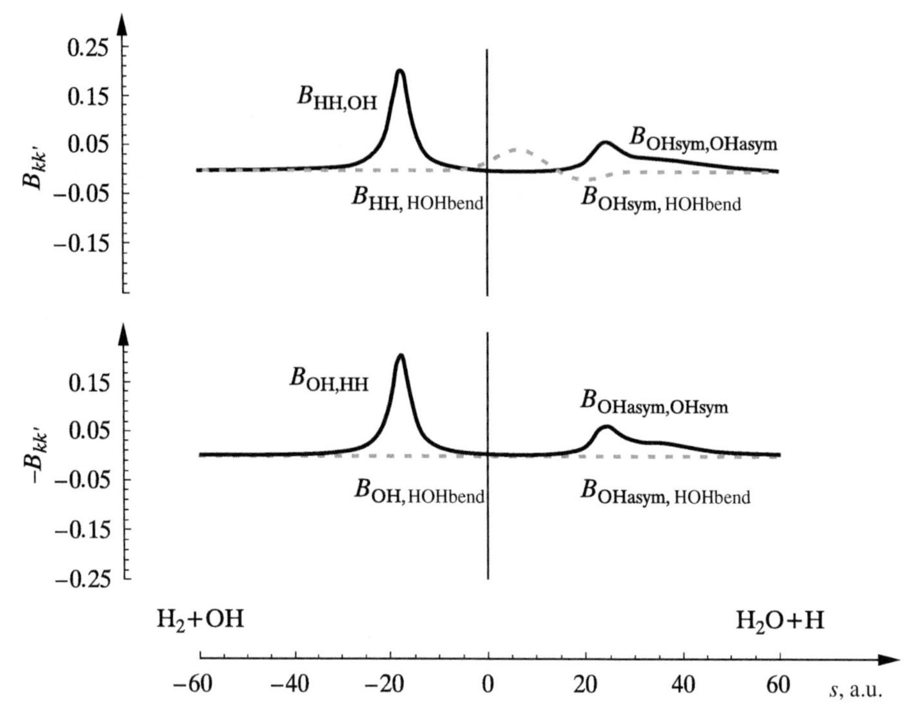

Fig. 14.13. The reaction $H_2 + OH \to H_2O + H$. The Coriolis coupling constants $B_{kk'}(s)$ as functions of s. A high value of $|B_{kk'}(s)|$ means that close to reaction coordinate s the changes of the k-th normal mode eigenvector resemble eigenvector k'. Source: T. Dunning, Jr. and E. Kraka, from *Advances in Molecular Electronic Structure Theory*, ed. T. Dunning, Jr., JAI Press, Greenwich, CT (1989), courtesy of the authors.

As we can see:

- In the entrance channel, the value of $B_{OH,s}$ is close to zero, so there is practically no coupling between the OH stretching vibrations and the reaction path. As a result, there will be practically no energy flow between those degrees of freedom. This might be expected from a weak dependence of ω_{OH} as a function of s. Once more, we see that the OH bond plays only the role of a reaction spectator.
- This is not the case for $B_{HH,s}$. This quantity attains its maximum just before the saddle point (let us recall that the barrier is early). Therefore, the energy may flow from the vibrational mode of H_2 to the reaction path (and *vice versa*) and a vibrational excitation of H_2 may have an important impact on the reaction rate (recall that the adiabatic barrier lowers when this mode is excited).

The Coriolis coupling constants $B_{kk'}$ as functions of s are plotted in Fig. 14.13 (only for the OH and HH stretching and HOH bending modes).

As we can see from Fig. 14.13:

- The maximum coupling for the HH and OH modes occurs long before the saddle point (close to $s = -18$ a.u.), enabling the system to exchange energy between the two vibrational modes.

- In the exit channel, we have quite significant couplings between the symmetric and anti-symmetric OH modes and the HOH bending mode.

This means that formation of H-O...H obviously influences both the OH stretching and HOH bending.

14.6 Acceptor-Donor (AD) Theory of Chemical Reactions

14.6.1 An Electrostatic Preludium–The Maps of the Molecular Potential

Chemical reaction dynamics is possible only for very simple systems. Chemists, however, have most often to do with medium-size or large molecules. Would it be possible to tell anything about the barriers for chemical reactions in such systems? Most of chemical reactions start from a situation when the molecules are far away but already interact. The main contribution is the electrostatic interaction energy, which is of the most long-range character (Chapter 13). Electrostatic interaction depends strongly on the mutual orientation of the two molecules (the *steric effect*). Therefore, the orientations are biased toward the privileged ones (energetically favorable). There is quite a lot of experimental data suggesting that privileged orientations lead, at smaller distances, to low reaction barriers. There is no guarantee of this, but it often happens for electrophilic and nucleophilic reactions, because the attacking molecule prefers those parts of the partner that correspond to high electron density (for *electrophilic attack*) or to low electron density (for *nucleophilic attack*).

We may use an electrostatic probe (e.g., a unit positive charge) to detect, which parts of the molecule "like" the approaching charge (energy lowering), and which do not (energy increasing).

The electrostatic interaction energy of the point-like unit charge (probe) in position $\boldsymbol{r}$ with molecule A is described by the following formula (the definition of the electrostatic potential produced by molecule A, Fig. 14.14a):

$$V_A(\boldsymbol{r}) = +\sum_a \frac{Z_a}{|\boldsymbol{r}_a - \boldsymbol{r}|} - \int \frac{\rho_A(\boldsymbol{r}')}{|\boldsymbol{r}' - \boldsymbol{r}|} d\boldsymbol{r}', \tag{14.34}$$

where the first term describes the interaction of the probe with the nuclei[a] denoted by index a, and the second means the interaction of the probe with the electron density distribution of the molecule A denoted by ρ_A (according to Chapter 11).

[a] By the way, to calculate the electrostatic interaction energy of the molecules A and B, we have to take (instead of a probe) the nuclei of the molecule B and sum the corresponding contributions, and then do the same with the electronic cloud of B. This corresponds to the following formula: $E_{elst} = \sum_b Z_b V_A(\boldsymbol{r}_b) - \int d\boldsymbol{r} \rho_B(\boldsymbol{r}) V_A(\boldsymbol{r})$, where b goes over the nuclei of B, and ρ_B represents its electronic cloud.

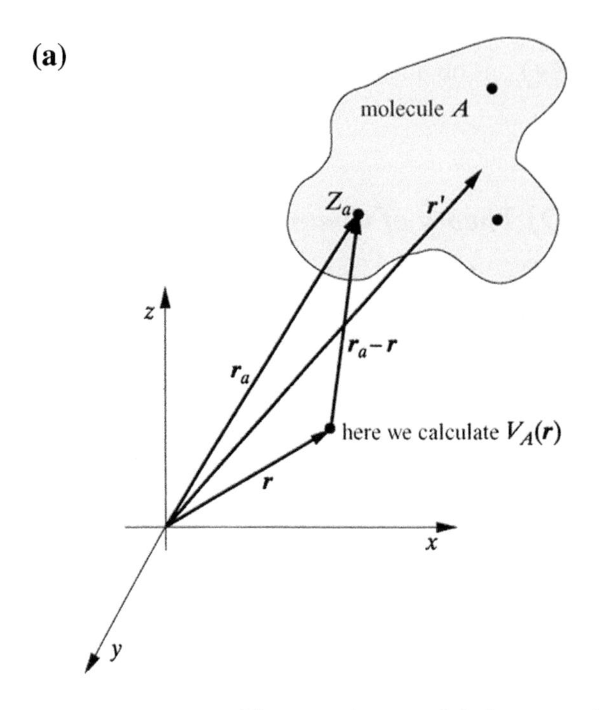

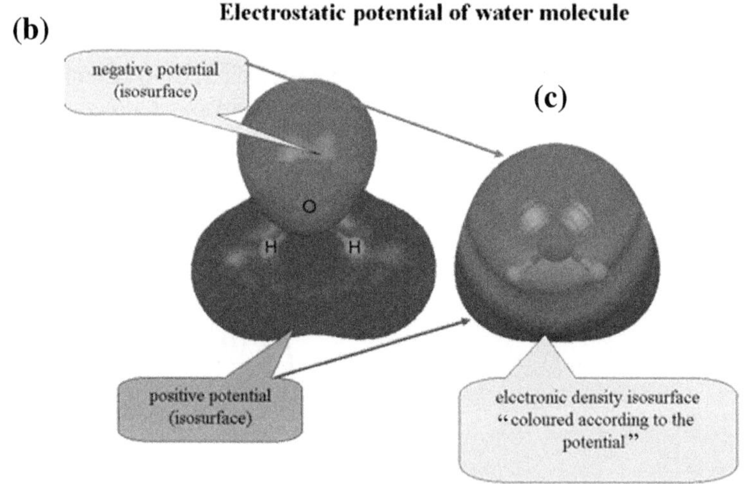

Fig. 14.14. Molecular electrostatic potential (MEP) represents the electrostatic interaction of the positive unit charge (probe) with molecule A. (a) the coordinate system and the vectors used in Eq. (14.34); (b) the equipotential surfaces $|V_A(\boldsymbol{r})|$ for the water molecule; (c) another way of showing the MEP: one computes $V_A(\boldsymbol{r})$ on the molecular surface (defined somehow). In more expensive books this is shown by coloring the surface using a certain convention: color $\leftrightarrow$ MEP. Such information is useful, because the role of the MEP is to predict the site of attack of another molecule, which is able to approach the surface. MEP in a.u. means the interaction of a proton with the molecule. It is seen that the proton will attack the region of the oxygen atom, while an attack of Cl^- would happen from the side of the hydrogens.

In the Hartree-Fock or Kohn-Sham approximation (as discussed in Chapter 11, p. 667, we assume the n_i-tuple occupation of the molecular orbital $\varphi_{A,i}$, $n_i = 0, 1, 2$),

$$\rho_A(\boldsymbol{r}) = \sum_i n_i |\varphi_{A,i}(\boldsymbol{r})|^2. \tag{14.35}$$

These molecular orbitals $\varphi_{A,i}$ are calculated in the literature for the isolated molecule, but one can imagine them to be computed *in the presence of the probe charge*. The first possibility has one advantage: we characterize the isolated molecule under study; however, we do not see how the charge flows in our molecule under the influence of the attacking object, and this should count during the attack. Taking the second possibility removes this deficiency, but only in part because the attacking object usually does not have a unit charge on it.

Therefore, in order to obtain $V_A(\boldsymbol{r})$ at point $\boldsymbol{r}$, it is sufficient to calculate the distances of the point from any of the nuclei (trivial) as well as the one-electron integrals, which appear after inserting into Eq. (14.34) $\rho_A(\boldsymbol{r}') = 2\sum_i |\varphi_{A,i}(\boldsymbol{r}')|^2$. Within the LCAO MO approximation, the electron density distribution ρ_A represents the sum of products of two atomic orbitals (in general centered at two different points). As a result, the task reduces to calculating typical one-electron three-center integrals of the nuclear attraction type (cf., Chapter 8 and Appendix P available at booksite.elsevier.com/978-0-444-59436-5), because the third center corresponds to the point $\boldsymbol{r}$ (Fig. 14.14). There is no computational problem with this for contemporary quantum chemistry.

In order to represent $V_A(\boldsymbol{r})$ graphically, we usually choose to show an equipotential surface corresponding to a given absolute value of the potential, while also indicating its sign (Fig. 14.14b). The sign tells us which parts of the molecule are preferred for the probelike object to attack and which are not. In this way, we obtain basic information about the reactivity of different parts of the molecule.[44]

Who Attacks Whom?

In chemistry, a probe will not be a point charge, but rather a neutral molecule or an ion. Nevertheless our new tool (electrostatic potential) will still be useful for the following reasons:

- If the probe represents a cation, it will attack those parts of the molecule A which are electron-rich (electrophilic reaction).
- If the probe represents an anion, it will attack the electron-deficient parts (nucleophilic reaction).
- If the probe represents a neutral molecule (B) with partial charges on its atoms, *its* electrostatic potential V_B is the most interesting. Those AB configurations that correspond to

[44] Having the potential calculated according to Eq. (14.34), we may ask about the set of atomic charges that reproduce it. Such charges are known as ESP (Electrostatic Potential).

the contacts of the associated sections of V_A and V_B with the opposite signs are the most electrostatically stable.

The site of the molecular probe (B) which attacks an electron-rich site of A itself has to be of electron-deficient character (and *vice versa*). Therefore, from the point of view of the attacked molecule (A), everything looks "upside down": an electrophilic reaction becomes nucleophilic and *vice versa*. When two objects exhibit an affinity to each other, who attacks whom represents a delicate and ambiguous problem and let it be that way. Therefore where does such nomenclature in chemistry come from? Well, it comes from the concept of ...didactics.

14.6.2 A Simple Model of Nucleophilic Substitution–MO, AD, and VB Formalisms

Let us take an example of a substitution reaction (we try to be as simple as possible):

$$H^- + H_2 \rightarrow H_2 + H^-, \tag{14.36}$$

and consider what is known as the *acceptor-donor formalism (AD)*. The formalism may be treated as intermediate between the configuration interaction (CI) and the valence bond (VB) formalisms (see Chapter 10). Any of the three formalisms is equivalent to the two others, provided they differ only by a linear transformation of many-electron basis functions.

In the CI formalism, the Slater determinants are built from the molecular spinorbitals.
In the VB formalism, the Slater determinants are built from the atomic spinorbitals.
In the AD formalism, the Slater determinants are built from the acceptor and donor spinorbitals.

14.6.3 MO Picture→AD Picture

As usual molecular orbitals for the total system $\varphi_1, \varphi_2, \varphi_3$ in a minimal basis set may be expressed (Fig. 14.15) using the molecular orbital of the donor (in our case, it is n, the $1s$ atomic orbital of H^-) and the acceptor molecular orbitals (bonding χ and antibonding χ^*):

$$\begin{aligned} \varphi_1 &= a_1 n + b_1 \chi - c_1 \chi^* \\ \varphi_2 &= a_2 n - b_2 \chi - c_2 \chi^* \\ \varphi_3 &= -a_3 n + b_3 \chi - c_3 \chi^*, \end{aligned} \tag{14.37}$$

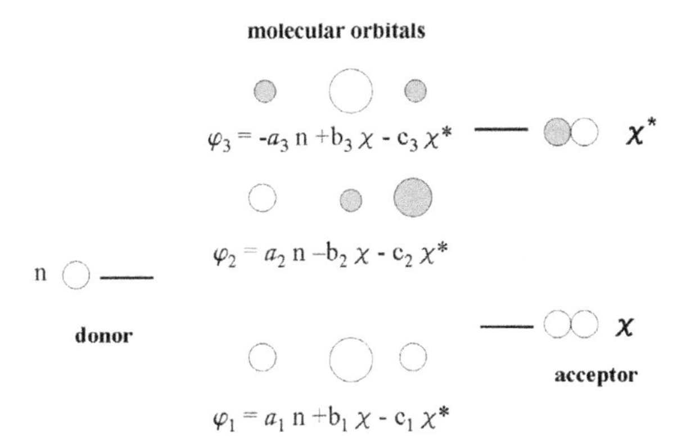

Fig. 14.15. A schematic representation of the molecular orbitals and their energies: of the donor (H^-, n representing the hydrogen atom $1s$ orbital), of the acceptor (H_2, bonding χ and antibonding χ^* of the hydrogen molecule) as well as of the total system H_3 in a linear configuration (center). The lowest-energy molecular orbital of H_3 does not have any node, the higher has one, while the highest has two nodes. In all cases, we use the approximation that the molecular orbitals are built from the three 1s hydrogen atomic orbitals only.

where $a_i, b_i, c_i > 0$, for $i = 1, 2, 3$. This convention comes from the fact that φ_1 is of the lowest energy and therefore exhibits no node, φ_2 has to play the role of the orbital second in energy scale and therefore has a single node, and φ_3 is the highest in energy and therefore has two nodes.[45]

Any N-electron Slater determinant Ψ^{MO} composed of the molecular spinorbitals $\{\phi_i\}, i = 1, 2, \ldots$ [cf., Eq. (M.1) on p. e109] may be written as a linear combination of the Slater determinants Ψ_i^{AD} composed of the spinorbitals $u_i, i = 1, 2, \ldots$ of the *acceptor and donor*[46] (AD picture)[47]:

$$\Psi_k^{MO} = \sum_i C_k(i)\, \Psi_i^{AD}, \tag{14.38}$$

Soon we will be interested in some of the coefficients $C_k(i)$. For example, the expansion for the ground-state Slater determinant (in the MO picture),

$$\Psi_0 = N_0\, |\varphi_1\bar{\varphi}_1\varphi_2\bar{\varphi}_2|, \tag{14.39}$$

[45] Positive values of a, b, c make possible the node structure described above.

[46] We start from the Slater determinant built of N molecular spinorbitals. Any of these is a linear combination of the spinorbitals of the donor and acceptor. We insert these combinations into the Slater determinant and expand the determinant according to the first row (Laplace expansion, see Appendix A available at booksite.elsevier.com/978-0-444-59436-5 on p. e1). As a result, we obtain a linear combination of the Slater determinants, all having the *donor or acceptor* spinorbitals in the first row. For each of the resulting Slater determinants, we repeat the procedure, but focusing on the second row, then the third row, etc. We end up with a linear combination of the Slater determinants that contain *only* the donor or acceptor spinorbitals. We concentrate on one of them, which contains some particular donor and acceptor orbitals. We are interested in the coefficient $C_k(i)$ that multiplies this Slater determinant number i.

[47] A similar expansion can also be written for the *atomic* spinorbitals (VB picture) instead of the donors and acceptors (AD picture).

gives

$$\Psi_0 = C_0(\mathrm{DA})\Psi_{DA} + C_0(\mathrm{D^+A^-})\Psi_{D^+A^-} + \cdots, \tag{14.40}$$

where $\bar{\varphi}_i$ denotes the spinorbital with spin function β, and φ_i, the spinorbital with spin function α, N_0 stands for the normalization coefficient, and Ψ_{DA}, $\Psi_{D^+A^-}$ represent the normalized Slater determinants with the following electronic configurations, in Ψ_{DA} : $n^2\chi^2$, in $\Psi_{D^+A^-}$: $n^1\chi^2(\chi^*)^1$, etc.

Roald Hoffmann (b.1937), Polish-born American chemist and professor at Cornell University in Ithaca, New York. Hoffmann discovered the symmetry rules that pertain to some reactions of organic compounds. In 1981, he shared the Nobel Prize with Kenichi Fukui "*for their theories, developed independently, concerning the course of chemical reactions.*" Hoffmann is also a poet and playwright. His poetry is influenced by chemistry, which, as he wrote, was inspired by Marie Curie.

His CV reads as a film script. When in 1941 the Germans entered Złoczów, Hoffmann's hometown, the four-year-old Roald was taken with his mother to a labor camp. One of the Jewish detainees betrayed a camp network of conspirators to Germans. They massacred the camp, but Roald and his mother had earlier been smuggled out of the camp by his father and hidden in a Ukrainian teacher's house. Soon after, his father was killed. The Red Army pushed the Germans out in 1944, and Roald and his mother went via Przemyśl to Cracow. In 1949 they finally reached America. Roald Hoffmann graduated from Stuyvesant High School, Columbia University, and Harvard University. At Harvard, Roald met the excellent chemist Robert Burns Woodward (syntheses of chlorophyl, quinine, strychnine, cholesterol, penicilline structure, vitamins), a Nobel Prize winner in 1965. Woodward alerted Hoffmann to the mysterious behavior of polyens in substitution reactions. Hoffmann clarified the problem using the symmetry properties of the molecular orbitals (now called the Woodward-Hoffmann symmetry rules; cf. p. 941).

We are mainly interested in the coefficient $C_0(\mathrm{DA})$. As shown by Fukui, Fujimoto and Hoffmann (see Appendix Z available at booksite.elsevier.com/978-0-444-59436-5, p. e191)[a]

$$C_0(\mathrm{DA}) \approx \langle \Psi_{DA} | \Psi_0 \rangle = \begin{vmatrix} a_1 & b_1 \\ a_2 & -b_2 \end{vmatrix}^2 = (a_1 b_2 + a_2 b_1)^2, \tag{14.41}$$

where in the determinant, the coefficients of the donor and acceptor orbitals appear in those molecular orbitals φ_i of the total system that are occupied in the ground-state Slater determinant Ψ_0 [the coefficients of n and χ in φ_1 are a_1 and b_1, respectively, while those in φ_2 are a_2 and $-b_2$, respectively, see Eq. (14.37)].

[a] We assume that the orbitals n, χ and χ^* are orthogonal (approximation).

Kenichi Fukui (1918–1998), Japanese chemist and professor at Kyoto University. One of the first scholars who stressed the importance of the IRC, and introduced what is called the frontier orbitals (mainly HOMO and LUMO), which govern practically all chemical processes. The HOMO-LUMO separation is crucial in chemistry. Fukui received the Nobel Prize in chemistry in 1981.

Function Ψ_0 will play a prominent role in our story, but now let us take two excited states, the doubly excited configurations of the total system[48]:

$$\Psi_{2d} = N_2|\varphi_1\bar{\varphi}_1\varphi_3\bar{\varphi}_3| \tag{14.42}$$

and

$$\Psi_{3d} = N_3|\varphi_2\bar{\varphi}_2\varphi_3\bar{\varphi}_3|, \tag{14.43}$$

where N_i stand for the normalization coefficients. Let us ask about the coefficients that *they* produce for the DA configuration [let us call these coefficients $C_2(\mathrm{DA})$ for Ψ_{2d} and $C_3(\mathrm{DA})$ for Ψ_{3d}]; i.e.,

$$\Psi_{2d} = C_2(\mathrm{DA})\Psi_{DA} + C_2(\mathrm{D^+A^-})\Psi_{D^+A^-} + \cdots, \tag{14.44}$$

$$\Psi_{3d} = C_3(\mathrm{DA})\Psi_{DA} + C_3(\mathrm{D^+A^-})\Psi_{D^+A^-} + \cdots \tag{14.45}$$

According to the result described above (see p. e191), we obtain:

$$C_2(\mathrm{DA}) = \begin{vmatrix} a_1 & b_1 \\ -a_3 & b_3 \end{vmatrix}^2 = (a_1b_3 + a_3b_1)^2, \tag{14.46}$$

$$C_3(\mathrm{DA}) = \begin{vmatrix} a_2 & -b_2 \\ -a_3 & b_3 \end{vmatrix}^2 = (a_2b_3 - a_3b_2)^2. \tag{14.47}$$

Such formulas enable us to calculate the contributions of the particular donor-acceptor resonance structures (e.g., DA, $\mathrm{D^+A^-}$, etc.; cf. p. 610) in the Slater determinants built of the molecular orbitals [Eq. (14.37)] of the total system. If one of these structures prevailed at a given stage of the reaction, this would represent important information about what has happened in the course of the reaction.

At every reaction stage, the main object of interest will be the ground-state of the system. The ground state will be dominated[a] by various resonance structures. As usual, the resonance structures are associated with the corresponding chemical structural formulas with the proper chemical bond pattern.

[a] That is, these structures will correspond to the largest expansion coefficients.

14.6.4 Reaction Stages

We would like to know the a, b, c values at *various reaction stages* because we could then calculate the coefficients C_0, C_2 and C_3 [Eqs. (14.41), (14.46), and (14.47)] for the DA, as well

[48] We will need this information later to estimate the configuration interaction role in calculating the CI ground state.

as for other donor-acceptor structures (e.g., D^+A^-, see below), and deduce what really happens during the reaction.

Reactant Stage (R)

The simplest situation is at the starting point. When H^- is far away from H-H, then of course (Fig. 14.15) $\varphi_1 = \chi, \varphi_2 = n, \varphi_3 = -\chi^*$. Hence, we have $b_1 = a_2 = c_3 = 1$, while the other $a, b, c = 0$, therefore:

i	a_i	b_i	c_i
1	0	1	0
2	1	0	0
3	0	0	1

Using Eqs. (14.41), (14.46), and (14.47), we get (the superscript R recalls that the results correspond to reactants):

$$C_0^R(\mathrm{DA}) = (0 \cdot 1 + 1 \cdot 1)^2 = 1, \tag{14.48}$$

$$C_2^R(\mathrm{DA}) = 0, \tag{14.49}$$

$$C_3^R(\mathrm{DA}) = (1 \cdot 0 - 0 \cdot 0)^2 = 0. \tag{14.50}$$

When the reaction begins, the reactants are correctly described as a Slater determinant with doubly occupied n and χ orbitals, which corresponds to the DA structure.

This is, of course, what we expected to obtain for the electronic configuration of the non-interacting reactants.

Intermediate Stage (I)

What happens at the intermediate stage (I)?

It will be useful to express first the atomic orbitals $1s_a, 1s_b, 1s_c$ through orbitals n, χ, χ^* (they span the same space). From Chapter 8, p. 439, we obtain

$$1s_a = n, \tag{14.51}$$

$$1s_b = \frac{1}{\sqrt{2}}\left(\chi - \chi^*\right), \tag{14.52}$$

$$1s_c = \frac{1}{\sqrt{2}}\left(\chi + \chi^*\right), \tag{14.53}$$

where we have assumed that the overlap integrals between different atomic orbitals are equal to zero.[49]

[49] In reality, they are small, but nonzero. Making them zero makes it possible to write down the molecular orbitals without any calculations.

The I stage corresponds to the situation in which the hydrogen atom in the middle (b) is at the same distance from a as from c, and therefore, the two atoms are equivalent. This implies that the nodeless, one-node, and two-node orbitals have the following form (where $\bigcirc$ stands for the 1s orbital and $\bigoplus$ for the $-1s$ orbital):

$$\begin{aligned}\varphi_1 &= \bigcirc\ \bigcirc\ \bigcirc = \frac{1}{\sqrt{3}}(1s_a + 1s_b + 1s_c)\\ \varphi_2 &= \bigcirc\ \cdot\ \bigoplus = \frac{1}{\sqrt{2}}(1s_a - 1s_c)\\ \varphi_3 &= \bigoplus\ \bigcirc\ \bigoplus = \frac{1}{\sqrt{3}}(-1s_a + 1s_b - 1s_c)\end{aligned} \quad (14.54)$$

Inserting Eq. (14.52), we obtain

$$\begin{aligned}\varphi_1 &= \frac{1}{\sqrt{3}}(n + \sqrt{2}\chi + 0 \cdot \chi^*)\\ \varphi_2 &= \frac{1}{\sqrt{2}}\left(n - \frac{1}{\sqrt{2}}(\chi + \chi^*)\right)\\ \varphi_3 &= \frac{1}{\sqrt{3}}\left(-n + 0 \cdot \chi - \sqrt{2}\chi^*\right)\end{aligned} \quad (14.55)$$

or

	a_i	b_i	c_i
$i=1$	$\frac{1}{\sqrt{3}}$	$\sqrt{\frac{2}{3}}$	0
$i=2$	$\frac{1}{\sqrt{2}}$	$\frac{1}{2}$	$\frac{1}{2}$
$i=3$	$\frac{1}{\sqrt{3}}$	0	$\sqrt{\frac{2}{3}}$

(14.56)

From Eqs. (14.41), (14.46), and (14.47), we have

$$C_0^I(\mathrm{DA}) = \left(\frac{1}{\sqrt{3}}\frac{1}{2} + \frac{1}{\sqrt{2}}\sqrt{\frac{2}{3}}\right)^2 = \frac{3}{4} = 0.75, \quad (14.57)$$

$$C_2^I(\mathrm{DA}) = \left(\frac{1}{\sqrt{3}} \cdot 0 + \sqrt{\frac{2}{3}}\frac{1}{\sqrt{3}}\right)^2 = \frac{2}{9} = 0.22, \quad (14.58)$$

$$C_3^I(\mathrm{DA}) = \left(\frac{1}{\sqrt{2}} \cdot 0 - \frac{1}{2}\frac{1}{\sqrt{3}}\right)^2 = \frac{1}{12} = 0.08, \quad (14.59)$$

The first is most important of these three numbers. Something happens to the electronic ground state of the system. At the starting point, the ground-state wave function had a DA contribution equal to $C_0^R(\mathrm{DA}) = 1$, while now this contribution has decreased to $C_0^I(\mathrm{DA}) = 0.75$. Let us see what will happen next.

Product Stage (P)

How does the reaction end up?

Let us see how molecular orbitals φ corresponding to the products are expressed by n, χ, and χ^*. At the end, we have the molecule H-H (made of the middle and left hydrogen atoms) and the outgoing ion H^- (made of the right hydrogen atom).

Therefore, the lowest-energy orbital at the end of the reaction has the form

$$\varphi_1 = \frac{1}{\sqrt{2}}(1s_a + 1s_b) = \frac{1}{\sqrt{2}}n + \frac{1}{2}\chi - \frac{1}{2}\chi^*, \tag{14.60}$$

which corresponds to $a_1 = \frac{1}{\sqrt{2}}, b_1 = \frac{1}{2}, c_1 = \frac{1}{2}$.

Since the φ_2 orbital is identified with $1s_c$, we obtain from Eqs. (14.41), (14.46), and (14.47): $a_2 = 0, b_2 = c_2 = \frac{1}{\sqrt{2}}$ (never mind that all the coefficients are multiplied by -1), and finally as φ_3, we obtain the antibonding orbital

$$\varphi_3 = \frac{1}{\sqrt{2}}(1s_a - 1s_b) = \frac{1}{\sqrt{2}}n - \frac{1}{2}\chi + \frac{1}{2}\chi^*; \tag{14.61}$$

i.e., $a_3 = \frac{1}{\sqrt{2}}, b_3 = \frac{1}{2}, c_3 = \frac{1}{2}$ (the sign is reversed as well). This leads to

i	a_i	b_i	c_i
1	$\frac{1}{\sqrt{2}}$	$\frac{1}{2}$	$\frac{1}{2}$
2	0	$\frac{1}{\sqrt{2}}$	$\frac{1}{\sqrt{2}}$
3	$\frac{1}{\sqrt{2}}$	$\frac{1}{2}$	$\frac{1}{2}$

(14.62)

Having a_i, b_i, c_i for the end of the reaction, we may easily calculate $C_0^P(\mathrm{DA})$ of Eq. (14.41), as well as $C_2^P(\mathrm{DA})$ and $C_3^P(\mathrm{DA})$ from Eqs. (14.46) and (14.47), respectively, for the reaction products

$$C_0^P(\mathrm{DA}) = \left(\frac{1}{\sqrt{2}} \cdot \frac{1}{\sqrt{2}} + 0 \cdot \frac{1}{2}\right)^2 = \frac{1}{4} \tag{14.63}$$

$$C_2^P(\mathrm{DA}) = \left(\frac{1}{\sqrt{2}} \cdot \frac{1}{2} + \frac{1}{\sqrt{2}} \cdot \frac{1}{2}\right)^2 = \frac{1}{2} \tag{14.64}$$

$$C_3^P(\mathrm{DA}) = \left(0 \cdot \frac{1}{2} - \frac{1}{\sqrt{2}} \cdot \frac{1}{\sqrt{2}}\right)^2 = \frac{1}{4}. \tag{14.65}$$

Now we can reflect on this for a while. It is seen that during the reaction, some important changes occur, namely

> when the reaction begins, the system is 100% described by the structure DA, while after the reaction it resembles this structure only by 25%.

Role of the Configuration Interaction

We may object that our conclusions seem quite naive. Indeed, there is something to worry about. We have assumed that, independent of the reaction stage, the ground-state wave function represents a single Slater determinant Ψ_0, whereas we should use a configuration interaction expansion. In such an expansion, besides the dominant contribution of Ψ_0, double excitations would be the most important (p. 653), which in our simple approximation of the three φ orbitals means a leading role for Ψ_{2d} and Ψ_{3d}:

$$\Psi_{CI} = \Psi_0 + \kappa_1 \Psi_{2d} + \kappa_2 \Psi_{3d} + \cdots$$

The two configurations are multiplied by some *small* coefficients κ (because we always deal with the electronic ground state dominated by Ψ_0). It will be shown that the κ coefficients in the CI expansion $\Psi = \Psi_0 + \kappa_1 \Psi_{2d} + \kappa_2 \Psi_{3d}$ are *negative*. This will serve us to make a more detailed analysis (than that performed so far) of the role of the DA structure at the beginning and end of the reaction.

The coefficients κ_1 and κ_2 may be estimated using perturbation theory, with Ψ_0 as unperturbed wave function. The first-order correction to the wave function is given by formula 5.24 on p. 245, where we may safely insert the total Hamiltonian $\hat{H}$ instead of the operator[50] $\hat{H}^{(1)}$ (this frees us from saying what $\hat{H}^{(1)}$ looks like). Then we obtain

$$\kappa_1 \cong \frac{\langle \varphi_2 \bar{\varphi}_2 | \varphi_3 \bar{\varphi}_3 \rangle}{E_0 - E_{2d}} < 0, \tag{14.66}$$

$$\kappa_2 \cong \frac{\langle \varphi_1 \bar{\varphi}_1 | \varphi_3 \bar{\varphi}_3 \rangle}{E_0 - E_{3d}} < 0, \tag{14.67}$$

because from the Slater-Condon rules (see Appendix M available at booksite.elsevier.com/978-0-444-59436-5) we have $\left\langle \Psi_0 | \hat{H} \Psi_{2d} \right\rangle = \langle \varphi_2 \bar{\varphi}_2 | \varphi_3 \bar{\varphi}_3 \rangle - \langle \varphi_2 \bar{\varphi}_2 | \bar{\varphi}_3 \varphi_3 \rangle = \langle \varphi_2 \bar{\varphi}_2 | \varphi_3 \bar{\varphi}_3 \rangle - 0 = \langle \varphi_2 \bar{\varphi}_2 | \varphi_3 \bar{\varphi}_3 \rangle$ and, similarly, $\left\langle \Psi_0 | \hat{H} \Psi_{3d} \right\rangle = \langle \varphi_1 \bar{\varphi}_1 | \varphi_3 \bar{\varphi}_3 \rangle$, where E_0, E_{2d}, E_{3d} represent the energies of the corresponding states. The integrals $\langle \varphi_2 \bar{\varphi}_2 | \varphi_3 \bar{\varphi}_3 \rangle = \left(\varphi_2 \varphi_2 | \varphi_3 \varphi_3 \right)$and $\langle \varphi_1 \bar{\varphi}_1 | \varphi_3 \bar{\varphi}_3 \rangle = \left(\varphi_1 \varphi_1 | \varphi_3 \varphi_3 \right)$ are Coulombic *repulsions* of a certain electron density distribution with *the same* charge distribution[51]; *therefore* $\langle \varphi_2 \bar{\varphi}_2 | \varphi_3 \bar{\varphi}_3 \rangle > 0$ *and* $\langle \varphi_1 \bar{\varphi}_1 | \varphi_3 \bar{\varphi}_3 \rangle > 0$, and indeed $\kappa_1 < 0$ and $\kappa_2 < 0$.

Thus, the contribution of the DA structure to the ground-state CI function results mainly from its contribution to the single Slater determinant Ψ_0 [coefficient C_0(DA)], but is modified by a small correction $\kappa_1 C_2(\mathrm{DA}) + \kappa_2 C_3(\mathrm{DA})$, where $\kappa < 0$.

What are the values of C_2(DA) and C_3(DA) at the beginning and at the end of the reaction? In the beginning, our calculations gave: $C_2^R(\mathrm{DA}) = 0$ and $C_3^R(\mathrm{DA}) = 0$. Note that $C_0^R(\mathrm{DA}) = 1$. Thus, the electronic ground state at the start of the reaction mainly represents the DA structure.

[50] This is because the unperturbed wave function Ψ_0 is an eigenfunction of the $\hat{H}^{(0)}$ Hamiltonian and is orthogonal to any of the expansion functions.

[51] These are $\varphi_2^* \varphi_3$ in the first case and $\varphi_1^* \varphi_3$ in the second one.

And what about the end of the reaction? We have calculated that $C_2^P(\text{DA}) = \frac{1}{2} > 0$ and $C_3^P(\text{DA}) = \frac{1}{4} > 0$. This means that at the end of the reaction, the coefficient corresponding to the DA structure will be *certainly smaller* than $C_0^P(\text{DA}) = 0.25$, the value obtained for the single determinant approximation for the ground-state wave function.

Thus,

> taking the CI expansion into account *makes our conclusion* based on the single Slater determinant even *sharper*. When the reaction starts, the wave function means the DA structure, while when it ends, this contribution is very strongly reduced.

14.6.5 Contributions of the Structures as Reaction Proceeds

What, therefore, represents the ground-state wave function at the end of the reaction?

To answer this question, let us consider first all possible occupations of the three energy levels (corresponding to n, χ, χ^*) by four electrons. As before we assume for the orbital energy levels: $\varepsilon_\chi < \varepsilon_n < \varepsilon_{\chi^*}$. In our simple model the number of such singlet-type occupations is equal only to six, Table 14.1 and Fig. 14.16.

Now, let us ask what is the contribution of each of these structures[52] in Ψ_0, Ψ_{2d}, and Ψ_{3d} in the three stages of the reaction. This question is especially important for Ψ_0, because this Slater determinant is dominant for the ground-state wave function. The corresponding contributions in Ψ_{2d} and Ψ_{3d} are less important because these configurations enter the ground-state CI wave function multiplied by the tiny coefficients κ. We have already calculated these contributions for the DA structure. The contributions of all the structures are given[53] in Table 14.2 (with the largest contributions in bold).

First, let us focus on which structures contribute mainly to Ψ_0 at the three stages of the reaction. As has been determined,

Table 14.1. All possible singlet-type occupations of the orbitals: n, χ and χ^* by four electrons. These occupations correspond to the resonance structures: DA, D^+A^-, DA*, D^+A^{-*}, $D^{+2}A^{-2}$ and DA**.

Ground state	DA	$(n)^2(\chi)^2$
Singly excited state	D^+A^-	$(n)^1(\chi)^2(\chi^*)^1$
Singly excited state	DA*	$(n)^2(\chi)^1(\chi^*)^1$
Doubly excited state	D^+A^{-*}	$(n)^1(\chi)^1(\chi^*)^2$
Doubly excited state	$D^{+2}A^{-2}$	$(\chi)^2(\chi^*)^2$
Doubly excited state	DA**	$(n)^2(\chi^*)^2$

[52] We have already calculated some of these contributions.

[53] Our calculations gave $C_0^I(\text{DA}) = 0.75$, $C_2^I(\text{DA}) = 0.22$, $C_3^I(\text{DA}) = 0.08$. In Table 14.2, these quantities are equal: 0.729, 0.250, and 0.020. The only reason for the discrepancy may be the nonzero overlap integrals, which were neglected in our calculations and were taken into account in those given in Table 14.2.

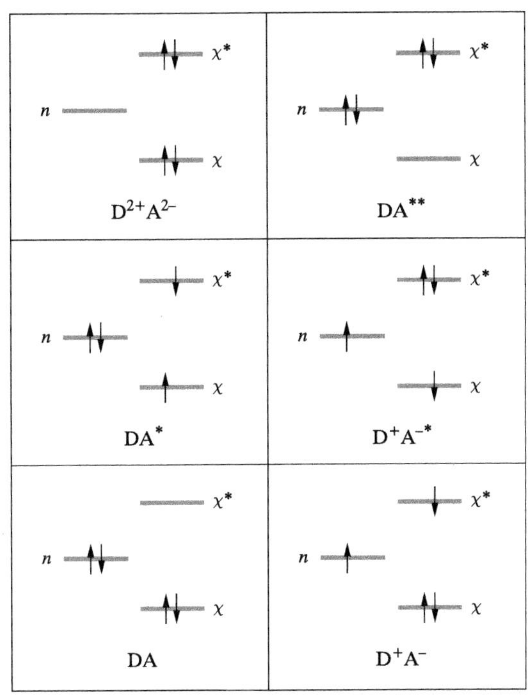

Fig. 14.16. The complete set of the six singlet wave functions ("*structures*"), that arise from occupation of the donor orbital n and of the two acceptor orbitals (χ and χ^*).

Table 14.2. Contributions $C_0(i)$, $C_2(i)$, $C_3(i)$ of the six donor-acceptor structures in the three Slater determinants Ψ_0, Ψ_{2d} i Ψ_{3d} built of molecular orbitals at the three reaction stages: reactant (R), intermediate (I), and product (P)[a]. The $|C_0(i)| > 0.5$ are in bold.

Structure	MO Determinant		R	I	P
DA	Ψ_0	$C_0(\text{DA})$	**1**	**0.729**	0.250
	Ψ_{2d}	$C_2(\text{DA})$	0	0.250	0.500
	Ψ_{3d}	$C_3(\text{DA})$	0	0.020	0.250
D^+A^-	Ψ_0	$c_0(\text{D}^+\text{A}^-)$	0	**−0.604**	**−0.500**
	Ψ_{2d}	$C_2(\text{D}^+\text{A}^-)$	0	0.500	0.000
	Ψ_{3d}	$C_3(\text{D}^+\text{A}^-)$	0	0.103	0.500
DA*	Ψ_0	$C_0(\text{DA}^*)$	0	0.177	0.354
	Ψ_{2d}	$C_2(\text{DA}^*)$	0	0.354	−0.707
	Ψ_{3d}	$C_3(\text{DA}^*)$	0	0.177	0.354
D^+A^{-*}	Ψ_0	$C_0(\text{D}^+\text{A}^{-*})$	0	0.103	**0.500**
	Ψ_{2d}	$C_2(\text{D}^+\text{A}^{-*})$	0	0.500	0.000
	Ψ_{3d}	$C_3(\text{D}^+\text{A}^{-*})$	0	−0.604	−0.500
DA**	Ψ_0	$C_0(\text{DA}^{**})$	0	0.021	0.250
	Ψ_{2d}	$C_2(\text{DA}^{**})$	0	0.250	0.500
	Ψ_{3d}	$C_3(\text{DA}^{**})$	1	0.729	0.250
$\text{D}^{+2}\text{A}^{-2}$	Ψ_0	$C_0(\text{D}^{+2}\text{A}^{-2})$	0	0.250	**0.500**
	Ψ_{2d}	$C_2(\text{D}^{+2}\text{A}^{-2})$	1	0.500	0.000
	Ψ_{3d}	$C_3(\text{D}^{+2}\text{A}^{-2})$	0	0.250	0.500

[a] S. Shaik, *J.Am.Chem.Soc.*, *103*, 3692 (1981)

- At point R, we have only the contribution of the DA structure.
- At point I, the contribution of DA decreases to 0.729, other structures come into play with the dominant D^+A^- structure (the coefficient equal to −0.604).
- At point P, there are three dominant structures: D^+A^-, D^+A^{-*} and $D^{+2}A^{-2}$.

Now we may think of going beyond the single determinant approximation by performing the CI. At the R stage, the DA structure dominates as before, but has some small admixtures of DA** (because of Ψ_{3d}) and $D^{+2}A^{-2}$ (because of Ψ_{2d}), while at the product stage, the contribution of the DA structure almost vanishes. Instead, some important contributions of the excited states appear, mainly of the D^+A^-, D^+A^{-*} and $D^{+2}A^{-2}$ structures, but also other structures of less importance.

The value of such qualitative conclusions comes from the fact that they do not depend on the approximation used (e.g., on the atomic basis set, neglecting the overlap integrals, etc).

For example, the contributions of the six structures in Ψ_0 calculated using the Gaussian atomic basis set STO-3G and within the extended Hückel method are given in Table 14.3. Despite the fact that even the geometries used for the R, I, and P stages are slightly different, the qualitative results are the same. It is rewarding to learn things that do not depend on detail.

Where Do the Final Structures D^+A^-, D^+A^{-*} and $D^{+2}A^{-2}$ Come From?

As seen from Table 14.2, the main contributions at the end of the reaction come from the D^+A^-, D^+A^{-*} and $D^{+2}A^{-2}$ structures. What do they correspond to *when the reaction starts*? From Table 14.2, it follows that the $D^{+2}A^{-2}$ structure simply represents Slater determinant Ψ_{2d} (Fig. 14.17). But where do the D^+A^- and D^+A^{-*} structures come from? There are no such contributions either in Ψ_0, or in Ψ_{2d} or in Ψ_{3d}. It turns out however that a similar analysis

Table 14.3. More advanced methods see the same! Contributions $C_0(i)$ of the six donor-acceptor structures (i) in the Ψ_0 Slater determinant at three different stages (R, I, and P) of the reaction[a]. Bold shows the most important contributions.

Structure	STO-3G			Extended Hückel		
	R	I	P	R	I	P
DA	**1.000**	**0.620**	0.122	**1.000**	**0.669**	0.130
D^+A^-	0.000	**−0.410**	**−0.304**	−0.012	**−0.492**	**−0.316**
DA*	0.000	0.203	0.177	0.000	0.137	0.179
D^+A^{-*}	0.000	0.125	**0.300**	0.000	0.072	**0.298**
DA**	0.000	0.117	**0.302**	0.000	0.176	**0.301**
$D^{+2}A^{-2}$	0.000	0.035	0.120	0.000	0.014	0.166

[a] S. Shaik, *J.Am.Chem.Soc., 103*, 3692 (1981).

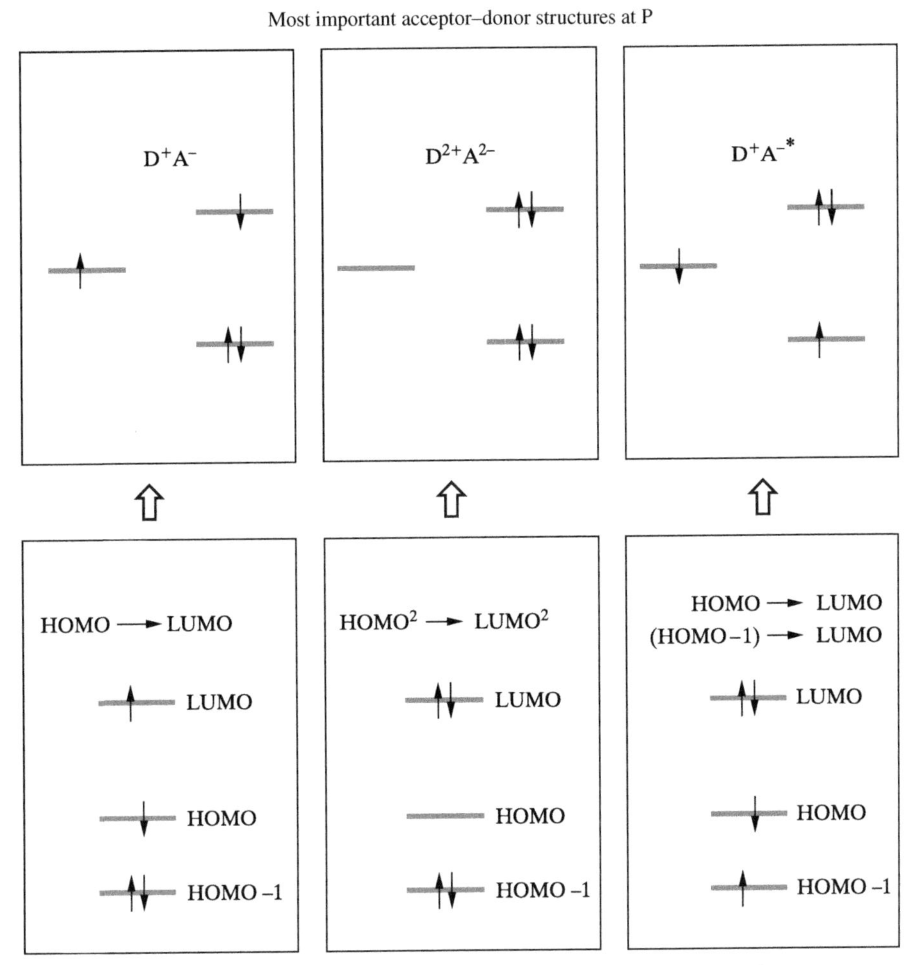

Fig. 14.17. What final structures are represented at the starting point?

applied to the normalized configuration[54] $N|\varphi_1\bar{\varphi}_1\varphi_2\bar{\varphi}_3|$ at stage R gives *exclusively* the D^+A^- structure, while applied to the $N|\varphi_1\bar{\varphi}_2\varphi_3\bar{\varphi}_3|$ determinant, it gives *exclusively* the D^+A^{-*} structure (Fig. 14.17). So we have traced them back. The first of these configurations corresponds to a single-electron excitation from HOMO to LUMO–this is, therefore, the lowest excited state of the reactants. Our picture is clarified:

the reaction starts from DA; at the intermediate stage (transition state), we have a large contribution of the first excited state that at the starting point was the D^+A^- structure related to the excitation of an electron from HOMO to LUMO.

[54] N stands for the normalization coefficient.

14.6.6 Nucleophilic Attack–The Model is More General: $H^- +$ Ethylene $\rightarrow$ Ethylene $+ H^-$

Maybe the acceptor-donor theory described above pertains only to the $H^- +$ H-H reaction? Fortunately, its applicability goes far beyond that. Let us consider a nucleophilic attack S_N2 of the H^- ion on the ethylene molecule (Fig. 14.18), perpendicular to the ethylene plane toward the position of one of the carbon atoms. The arriving ion binds to the carbon atom forming the CH bond, while another proton with two electrons (i.e., H^- ion) leaves the system. Such a reaction looks like it has only academic interest (except some isotopic molecules are involved, e.g.,

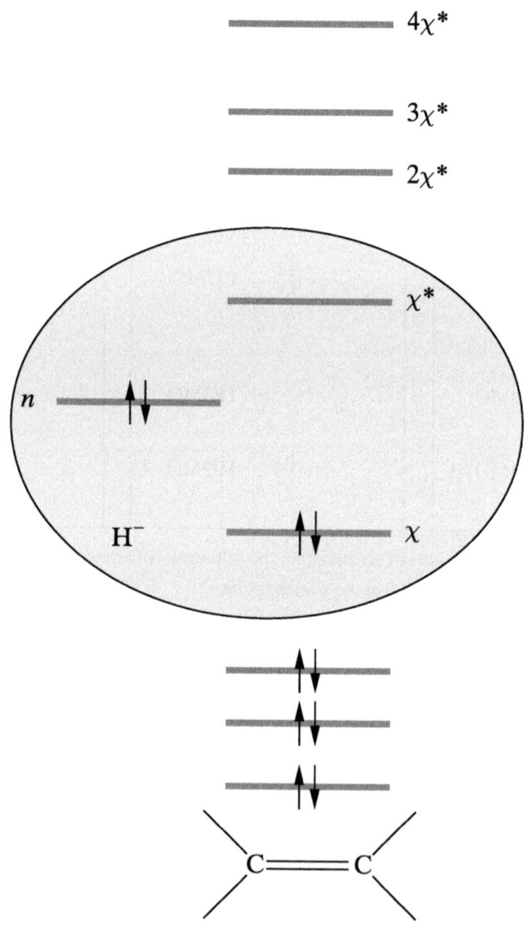

Fig. 14.18. The AD approach turns out to be more general. Nucleophilic substitution of ethylene by H^-. This aims to demonstrate that, despite considering a more complex system than the $H^- + H_2 \rightarrow H_2 + H^-$ reaction discussed so far, the machinery behind the scene works in the same way. The attack of H^- goes perpendicularly to the ethylene plane, onto one of the carbon atoms. The image shows the (orbital) energy levels of the donor (H^-, left side) and of the acceptor (ethylene, right side). Similarly as for $H^- + H_2$, the orbital energy of the donor orbital n is between the acceptor orbital energies χ and χ^* corresponding to the bonding π and antibonding π^* orbitals. Other molecular orbitals of the ethylene (occupied and virtual: $2\chi^*, 3\chi^*, \ldots$) play a marginal role due to high energetic separation from the energy level of n.

when one of the protons is replaced by a deuteron), but comprehension comes from the simplest examples possible, when the fewest things change.

The LCAO MO calculations for the ethylene molecule give the following result. The HOMO orbital is of the π bonding character, while the LUMO represents the antibonding π^* orbital (both are linear combinations of mainly carbon $2p_z$ atomic orbitals, z being the axis perpendicular to the ethylene plane). On the energy scale, the H^- $1s$ orbital goes between the π and π^* energies, similar to what happened with the χ and χ^* orbitals in the H^- + H-H reaction. The virtual orbitals (let us call them $2\chi^*$, $3\chi^*$, and $4\chi^*$) are far up in the energetic scale, while the doubly occupied σ-type orbitals are far down in the energetic scale. Thus, the H^- $n = 1s$ orbital energy is close to that of χ and χ^*, while other orbitals are well separated from them.

This energy level scheme allows for many possible excitations, far more numerous than considered before. *Despite this*, because of the effective mixing of only those donor and acceptor orbitals that are of comparable energies, *the key partners are, as before, n, χ and χ^**. The role of the other orbitals is only marginal: their admixtures will only slightly deform the shape of the main actors of the drama n, χ and χ^* known as the *frontier orbitals*. The coefficients at various acceptor-donor structures in the expansion of Ψ_0 are shown in Table 14.4. The calculations were performed using the extended Hückel method[55] at three stages of the reaction (R, in which the H^- ion is at a distance of 3 Å from the attacked carbon atom; I, with a distance 1.5 Å and P, with a distance equal to 1.08 Å; in all cases, the planar geometry of the ethylene was preserved). It is seen, (see Table 14.4), that despite the fact that a more complex method was used, the emerging picture is basically the same (see bold numbers in the Table): at the beginning the DA structure prevails; at the intermediate stage, we have a "hybrid" of the DA and D^+A^- structures; and at the end, we have a major role for the D^+A^- and D^+A^{-*} structures. We can see also that even if some higher excitations were taken into account (to the orbitals $2\sigma^*$, $3\sigma^*$), they play only a marginal role. The corresponding population analysis (not reported here) indicates a basically identical mechanism. This resemblance extends also to the S_N2 nucleophilic substitutions in aromatic compounds.

Table 14.4. Expansion coefficients $C_0(i)$ at the acceptor-donor structures in the ground-state wave function at various stages of the S_N2 substitution reaction of ethylene: reactant (R), intermediate (I), and product (P). The most important contributions are in bold.[a]

Structure	Coefficients		
	R	I	P
DA	**1.000**	**0.432**	0.140
$D^+A^-(n \to \pi^*)$	0.080	**0.454**	**0.380**
$DA^*(\pi \to \pi^*)$	−0.110	−0.272	−0.191
$D^+A^{-*}(n \to \pi^*, \pi \to \pi^*)$	−0.006	−0.126	**−0.278**
$D^+A^-(n \to 2\sigma^*)$	$<10^{-4}$	0.006	0.004
$D^+A^-(n \to 3\sigma^*)$	$<10^{-4}$	−0.070	−0.040

[a] S. Shaik, *J.Am.Chem.Soc., 103*, 3692 (1981).

[55] This model was introduced to chemistry by Roald Hoffmann. He used to say that he cultivates chemistry with an old, primitive tool, which because of this, ensures access to the wealth of the complete Mendeleyev periodic table.

Table 14.5. Expansion coefficients at the acceptor-donor structures for the reaction of proton with the hydrogen molecule at three different stages of the reaction: reactant (R), intermediate (I), and product (P).[a]

Structure	Coefficients		
	R	I	P
DA	**1.000**	**0.729**	0.250
D^+A^-	0	**0.604**	**0.500**
$D^{+*}A^-$	0	−0.104	**−0.500**
D^*A	0	−0.177	−0.354
$D^{**}A$	0	0.021	0.250
$D^{+2}A^{-2}$	0	0.250	0.500

[a] S. Shaik, *J.Am.Chem.Soc., 103,* 3692 (1981).

14.6.7 The Model Looks Even More General: The Electrophilic Attack $H^+ + H_2 \rightarrow H_2 + H^+$

Let us see whether this mechanism is even more general and consider the *electrophilic* substitution in the model reaction $H^+ + H\text{-}H \rightarrow H\text{-}H + H^+$. This time, the role of the donor is played by the hydrogen molecule, while that of the acceptor is taken by the proton. The total number of electrons is only two. The DA structure corresponds to $(\chi)^2(n)^0(\chi^*)^0$. Other structures are defined by a full analogy with the previous case of the H_3^- system: structure D^+A^- means $(\chi)^1(n)^1(\chi^*)^0$, structure $D^{+*}A^-$ obviously corresponds to $(\chi)^0(n)^1(\chi^*)^1$, structure D^*A to $(\chi)^1(n)^0(\chi^*)^1$, $D^{**}A$ to $(\chi)^0(n)^0(\chi^*)^2$ and $D^{+2}A^{-2}$ to $(\chi)^0(n)^2(\chi^*)^0$. As before, the ground-state Slater determinant may be expanded into the contributions of these structures. The results (the overlap neglected) are collected in Table 14.5.

It is worth stressing that we obtain essentially the same reaction machinery as before. First, at stage R, the DA structure prevails; next, at intermediate stage I, we have a mixture of the DA and D^+A^- structures, and we end up (stage P) with D^+A^- and $D^{+*}A^-$ (the energy levels for the donor are the same as the energy levels were previously for the acceptor; hence, we have $D^{+*}A^-$, and not D^+A^{-*} as before). This picture would not change qualitatively if we considered electrophilic substitution of the ethylene or benzene.

14.6.8 The Model Also Works for the Nucleophilic Attack on the Polarized Bond

Let us consider a nucleophilic attack of the species X on the polarized double bond $>C = Y$, where Y represents an atom more electronegative than carbon (say, oxygen):

$$X^- + {>C} = Y \rightarrow {>C} = X + Y^-.$$

The problem is similar to the ethylene case except the double bond is polarized. The arguments of the kind already used for ethylene make it possible to limit ourselves exclusively to the frontier orbitals n, $\pi \equiv \chi$ and $\pi^* \equiv \chi^*$.

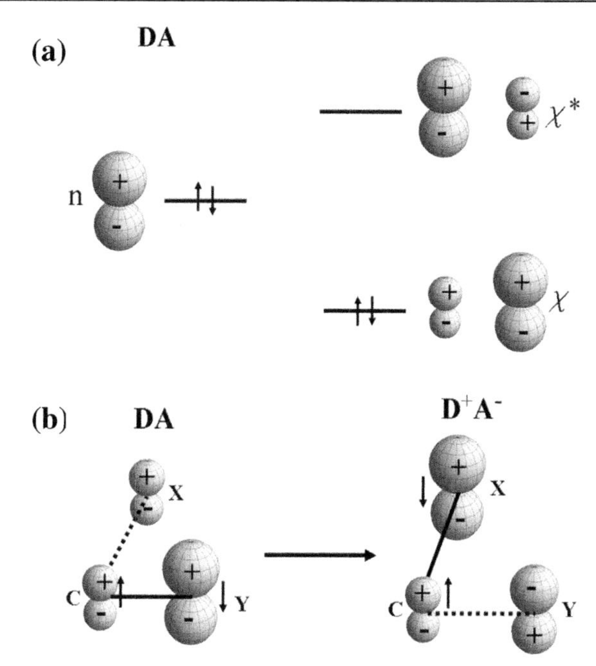

Fig. 14.19. Nucleophilic attack $X^- + >C = Y \rightarrow >C = X + Y^-$. The orbitals π and π^* are polarized (their polarizations are *opposite*, the + and − signs correspond to the positive and negative values of the orbitals). (a) The orbital energy levels with the starting electronic configuration (DA structure); (b) with the nucleophilic substitution reaction S_N2; within the plane of the three atoms XC Y, there are also two C-H bonds (not displayed).

This time the bonding π orbital is also a linear combination of the $2p_z$ atomic orbitals (the z axis is perpendicular to the $>C = Y$ plane, Fig. 14.19a), but is polarized what is shown as $(a \neq b)$

$$\pi = a \cdot (2p_z)_C + b \cdot (2p_z)_Y. \tag{14.68}$$

The role of the donor orbital n will be played by $(2p_z)_X$. We assume (by convention) that the coefficients are both positive with the normalization condition $a^2 + b^2 = 1$ (the overlap integrals between the atomic orbitals are neglected). Due to a higher electronegativity of Y we have $b > a$. In this situation the antibonding orbital π^* may be obtained directly from the orthogonality condition of the orbital π as:

$$\pi^* = b \cdot (2p_z)_C - a \cdot (2p_z)_Y. \tag{14.69}$$

Note that π^* has the opposite polarization to that of π, i.e. the electron described by π^* prefers to be close to the less electronegative carbon atom.

The DA structure corresponds to double occupation of n and π. From Eq. (10.28) we know that in the VB language the corresponding Slater determinant contains three structures: one of the Heitler-London type and the two ionic structures. In our case of a polarized bond, the *Heitler-London function would continue to treat the C and Y nuclei on the equal footing*, the polarity of the CY bond would be correctly restored by different weights of the ionic structures, $b^2 > a^2$. In conclusion: this will be the CY double bond but polarized.

Now, consider the all important D^+A^- structure in the new situation. According to Eq. (14.69):

$$D^+A^- \equiv \begin{vmatrix} \pi^*(1) & \pi^*(2) & \ldots & \pi^*(4) \\ \bar{\pi}(1) & \bar{\pi}(2) & \ldots & \bar{\pi}(4) \\ n(1) & n(2) & \ldots & n(4) \\ \bar{n}(1) & \bar{n}(2) & \ldots & \bar{n}(4) \end{vmatrix}$$

$$= b \begin{vmatrix} (2p_z)_C(1) & (2p_z)_C(2) & \ldots & (2p_z)_C(4) \\ \bar{\pi}(1) & \bar{\pi}(2) & \ldots & \bar{\pi}(4) \\ n(1) & n(2) & \ldots & n(4) \\ \bar{n}(1) & \bar{n}(2) & \ldots & \bar{n}(4) \end{vmatrix} - a \begin{vmatrix} (2p_z)_Y(1) & (2p_z)_Y(2) & \ldots & (2p_z)_Y(4) \\ \bar{\pi}(1) & \bar{\pi}(2) & \ldots & \bar{\pi}(4) \\ n(1) & n(2) & \ldots & n(4) \\ \bar{n}(1) & \bar{n}(2) & \ldots & \bar{n}(4) \end{vmatrix}$$

$$= -a \left\{ \begin{matrix} \dot{\mathrm{X}}\downarrow \\ | \\ \mathrm{C} - \dot{\mathrm{Y}}\uparrow \end{matrix} \quad \text{structure} \right\} + b \left\{ \begin{matrix} \dot{\mathrm{X}}\downarrow \\ | \\ \dot{\mathrm{C}}\uparrow - \mathrm{Y} \end{matrix} \quad \text{structure} \right\} \approx b \left\{ \begin{matrix} \dot{\mathrm{X}}\downarrow \\ | \\ \dot{\mathrm{C}}\uparrow - \mathrm{Y} \end{matrix} \quad \text{structure} \right\}.$$

We see that the spin pairing takes place between C and X atoms. Thus, $D^+ A^-$ corresponds to breaking the old CY bond and creating the new CX bond. This is what we have displayed in Fig. 14.19b and Fig. 14.20 shows the same in a pictorial representation.

The avoided crossing is needed to cause such a change of the electronic structure as to break the old bond and form the new one. Taking the leading VB structures only, we may say the following:

- The avoided crossing appears between two hypersurfaces, from which one corresponds to the old bond pattern (the first diabatic hypersurface) and the other to the new bond pattern (the second diabatic hypersurface).
- We see from the VB results why the variational method chose the D^+A^- structure among the six possible ones. This configuration was chosen because it corresponds exactly to the formation of the new bond: the two unpaired electrons with opposite spins localized on those atoms that are going to bind.
- The mechanism given is general and applies wherever at least one of the reactants has a closed shell. When both the reacting molecules are of the open-shell type, there will be no avoided crossing and no reaction barrier: the reactants are already prepared for the reaction.

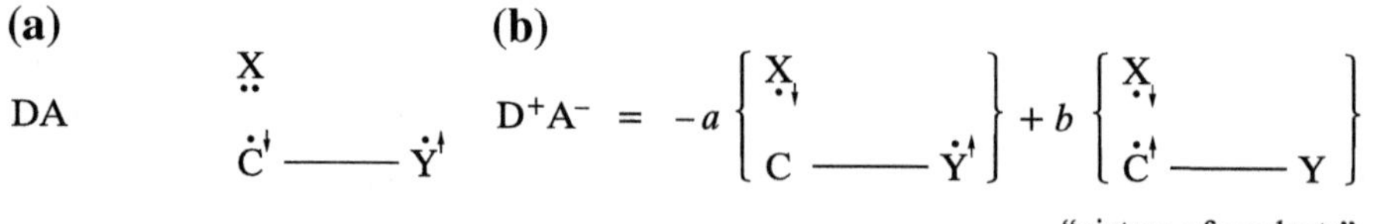

(a) DA

$$\begin{matrix} \mathrm{X} \\ \\ \dot{\mathrm{C}}^{\uparrow} \text{———} \dot{\mathrm{Y}}^{\uparrow} \end{matrix}$$

(b) $D^+A^- = -a \left\{ \begin{matrix} \mathrm{X} \\ \mathrm{C} \text{———} \dot{\mathrm{Y}}^{\uparrow} \end{matrix} \right\} + b \left\{ \begin{matrix} \mathrm{X} \\ \dot{\mathrm{C}}^{\uparrow} \text{———} \mathrm{Y} \end{matrix} \right\}$

"picture of products"

Fig. 14.20. Pictorial description of the DA and D^+A^- structures. For a large donor-acceptor distance, the electronic ground state is described by the DA structure (a). Structure D^+A^- already becomes very important for the intermediate stage (I). This structure, belonging to the acceptor-donor picture, is shown (b) in the VB representation, where the opposite spins of the electrons remind us that we have the corresponding covalent structure (note that $b > a$).

What Is Going on in the Chemist's Flask?

Let us imagine the molecular dynamics on energy hypersurface calculated using a quantum-mechanical method (classical force fields are not appropriate since they offer non-breakable chemical bonds). The system is represented by a point that slides downhill (with an increasing velocity) and climbs uphill (with deceasing velocity). The system has a certain kinetic energy because chemists usually heat their flasks.

Let us assume that first, the system wanders through those regions of the ground-state hypersurface that are far in the energy scale from other electronic states. In such a situation, the adiabatic approximation (Chapter 6) is justified and the electronic energy (corresponding to the hypersurface) represents potential energy for the motion of the nuclei. The system corresponds to a given chemical bond pattern (we may work out a structural formula). The point representing the system in the configurational space "orbits" at the bottom of an energy well, which means that the bond lengths vibrate as do bond angles and torsional angles, but a single bond remains single, double remains double, etc.

Due to heating the flask accidentally (to cite on example), the system *climbs up* the wall of the potential energy well. This may mean, however, that it approaches a region of the configurational space in which another diabatic hypersurface (corresponding to another electronic state) *lowers* its energy to such an extent that the two hypersurfaces tend to intersect. In this region, the adiabatic approximation fails, since we have *two* electronic states of comparable energies (both have to be taken into account), and the wave function cannot be taken as the product of an electronic function and a function describing the nuclear motion (as is required by the adiabatic approximation). As a result of mixing, the crossing is avoided and two *adiabatic* hypersurfaces (upper and lower) appear. Each is composed of two parts. One part corresponds to a molecule looking as if it had one bond pattern, while the other pertains to a different bond pattern. The bond pattern on each of the adiabatic hypersurfaces changes and the Rubicon for this change is represented by the boundary i.e., the region of the quasi-avoided crossing that separates the two diabatic parts of the adiabatic hypersurface. Therefore, when the system in its dynamics goes uphill and enters the boundary region, the corresponding bond pattern becomes fuzzy and changes to another pattern after crossing the boundary. The reaction is completed.

What will happen next? The point representing the system in the configurational space continues to move, and it may happen to arrive at another avoided-crossing region[56] and its energy is sufficient to overcome the corresponding barrier. This is the way multistep chemical reactions happen. It is important to realize that in experiments, we have to do with an ensemble of such points rather than with a single one. The points differ by their positions (configurations of the nuclei) and nuclear momenta. Only a fraction of them has sufficiently high kinetic energy to cross the reaction barrier. The rest wander through a superbasin (composed of numerous basins) of the initial region, thus undergoing vibrations, rotations including internal rotations, etc. Of

[56] This new avoided crossing may turn out to be the old one. In such a case, the system will cross the barrier in the opposite direction. Any chemical reaction is reversible (to different extents).

those which cross a barrier, only a fraction crosses the same barrier (i.e., the barrier of the same reaction). Others, depending on the initial conditions (nuclear positions and momenta) may cross other barriers. The art of chemistry means that in such a complicated situation, it is still possible to perform reactions with nearly 100% yield and obtain a pure chemical compound–which is the chemist's goal.

14.7 Symmetry-Allowed and Symmetry-Forbidden Reactions

14.7.1 Woodward-Hoffmann Symmetry Rules

The rules pertain to such an approach of two molecules that all the while, some symmetry elements of the nuclear framework are preserved (there is a symmetry group associated with the reaction, see Appendix C available at booksite.elsevier.com/978-0-444-59436-5). Then we have:

- The molecular orbitals belong to the irreducible representations of the group.
- During the approach the orbital energies change, *but their electron occupancies do not.*
- The reaction is symmetry-allowed when the total energy (often taken as the sum of the orbital energies) lowers; otherwise, it is symmetry-forbidden.

14.7.2 AD Formalism

A VB point of view is simple and beautiful, but sometimes the machinery gets stuck. For instance, this may happen when the described mechanism has to be rejected because it does not meet some symmetry requirements. Imagine that instead of a linear approach of H^- to H_2, we consider a T-shape configuration. In such a case, the all-important D^+A^- structure becomes *useless* for us because the resonance integral, which is proportional to the overlap integral between the $1s$ orbital of H^- (HOMO of the donor) and χ^* (LUMO of the acceptor), is equal to zero for symmetry reasons. If the reaction were to proceed, we would have had to form molecular orbitals from the above orbitals, which is impossible.

Yet there is an emergency exit from this situation. Let us turn our attention to the D^+A^{-*} structure, which corresponds to a doubly occupied χ^*, but a singly occupied χ. *This* structure would lead to the reaction because the overlap integral of $1s$ H^- and χ H-H has a nonzero value. In this way,

> a forbidden symmetry will simply cause the system to choose as the lead another structure, such that it allows the formation of new bonds in *this situation.*

The above example shows that symmetry can play an important role in chemical reactions. In what follows, the role of symmetry will be highlighted.

14.7.3 Electrocyclic Reactions

Robert Burns Woodward had an intriguing puzzle for his student, young Roald Hoffmann. The puzzle had to do with cyclization of alkens (molecules with alternant double and single carbon-carbon bonds). All these molecules had at the ends the planar $H_2C =$ group and in the cyclization reaction, these terminal groups had to *rotate* somehow in order to form a CC single bond: $-H_2C - CH_2-$ closing the cycle. In the case of $H_2C =$, the direction of the rotation (conrotation and disrotation; see Fig. 14.21a) is irrelevant for the final product. However, when a similar cyclization were carried out with the HRC= groups ($R \neq H$)), the two products possible *would differ*, as these are two distinct isomers.

One might, however, presume that both products will appear, since which rotation is carried out is a matter of a stochastic process.

The crux of the puzzle was, however, that only one of the two possible transformations took place, as if one of the rotations was forbidden, but another one allowed. It was even pire, since the allowed rotation was sometimes conrotation and sometimes disrotation, depending which alkene was considered.

Hoffmann solved the problem introducing the notion of symmetry-allowed and symmetry-forbidden chemical reactions. What really mattered was the symmetry of the π molecular orbitals involved. These are the orbitals, which certainly have to change their character when the reaction proceeds: one of them changes from π to σ. In the case of the cis-butadiene, we have two π orbitals (Fig. 14.21b): φ_1 and φ_2 (which are HOMO-1 and HOMO, respectively). The two electrons occupying φ_1 do not change their π character; they just go on a two-center CC π orbital in the cyclobutene.

The two electrons occupying φ_2 change their character from π to σ. Note that they transform differently under the conrotation or under the disrotation (see Fig. 14.21c). The conrotation preserves the C_2 axis in the carbon atoms' plane, while during the disrotation, the mirror plane perpendicular to the carbon atoms' plane is preserved. The conrotation results in the in-phase (i.e., *bonding*, or low-energy) overlapping of the terminal $2p$ atomic orbitals, while the disrotation would lead to the out-of-phase, high-energy σ^* orbital. Thus, the conrotation is symmetry-allowed (because there is an important energy gain), while the disrotation is forbidden (because it would correspond to an increase of the energy). For other alkenes, the same reasoning leads to the conclusion that $4n$ π-electrons give the conrotation allowed (the butadiene means $n = 1$) and the disrotation is forbidden, while $4n + 2$ π-electrons give the disrotation allowed, while the conrotation is forbidden. This is closely related to the type of isomer - the product of the cyclization reaction. The very reason for such a selection rule is that for the case of $4n + 2$, the HOMO π orbital has the two terminal $2p$ orbitals in-phase, while for the case of $4n$, the HOMO π orbital has the two terminal $2p$ orbitals out-of-phase (as in Fig. 14.21b).

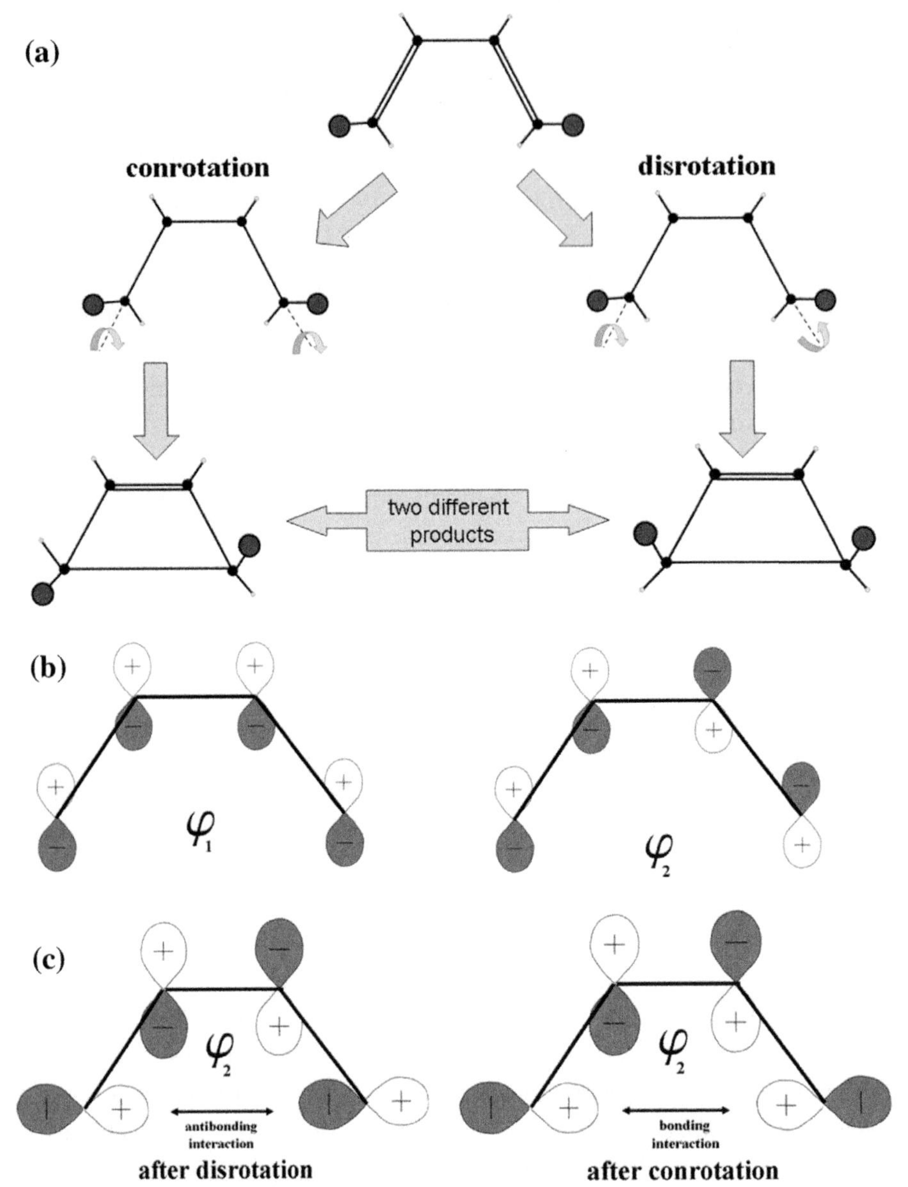

Fig. 14.21. Cyclization of the cis-butadiene. (a) Start from the top: conrotation (leftward) and disrotation (rightward) lead in general to different products, but only one of these transformation is symmetry allowed. (b) The doubly occupied π orbitals of the butadiene: φ_1 (HOMO-1) and φ_2 (HOMO). (c) The transformation of φ_2 under the conrotation and disrotation. The conrotation is symmetry-allowed, and the disrotation is symmetry-forbidden.

14.7.4 Cycloaddition Reaction

Let us take the example of the cycloaddition of two ethylene molecules when they bind together, forming the cyclobutane. The frontier orbitals of the ground-state ethylene molecule are the doubly occupied π (HOMO) and the empty π^* (LUMO) molecular orbitals.

The right side of Fig. 14.22a shows that the reaction would go toward the products if we prepared the reacting ethylene molecules in their triplet states. Such a triplet state has to be stabilized during the reaction, while the state corresponding to the old bond pattern should lose its importance. Is it reasonable to expect the triplet state to be of low energy in order to have the chance to be pulled sufficiently down the energy scale? Yes, it is, because the triplet state arises by exciting an electron from the HOMO (i.e., π) to the LUMO (i.e., π^*), and this energy cost is the lowest possible (in the orbital picture). Within the π-electron approximation, the Slater determinant corresponding to the triplet state (and representing the corresponding molecular orbitals as linear combination of the carbon $2p_z$ with atomic orbitals denoted simply as a and b) takes the form

$$\begin{aligned}
&N \det\left(\pi(1)\alpha(1)\pi^*(2)\alpha(2)\right) && (14.70)\\
&= N\left(\pi(1)\alpha(1)\pi^*(2)\alpha(2) - \pi(2)\alpha(2)\pi^*(1)\alpha(1)\right) && (14.71)\\
&= N\alpha(1)\alpha(2)\left(\left(a(1)+b(1)\right)\left(a(2)-b(2)\right) - \left(a(2)+b(2)\right)\left(a(1)-b(1)\right)\right) && (14.72)\\
&= -2N\alpha(1)\alpha(2)\left(a(1)b(2)-a(2)b(1)\right). && (14.73)
\end{aligned}$$

Such a function means that when one electron is on the first carbon atom, the other is on the second carbon atom (no ionic structures). The "*parallel*" electron spins of one molecule may be in the opposite direction to the similar electron spins of the second molecule. Everything is prepared for the cycloaddition (i.e., formation of the new chemical bonds).

Similar conclusions can be drawn from the Woodward-Hoffmann symmetry rules.

Example. The Diels-Alder Reaction (Woodward-Hoffmann approach)

The two ethylene molecules are oriented as shown in Fig. 14.22c. Let us focus on the frontier orbitals at long intermolecular distances. All are built of the symmetry orbitals composed of the four 2p carbon atomic orbitals (perpendicular to the planes corresponding to the individual molecules) and can be classified as symmetric (S) or antisymmetric (A) with respect to the symmetry planes P_1 and P_2. Moreover, by recognizing the bonding or antibonding interactions, without performing any calculations, we can tell that the SS-symmetry orbital, shown in Fig. 14.22c, is of the lowest energy (because of the bonding character of the intramolecular as well as intermolecular interactions), then the SA-symmetry (the bonding intramolecular - the π orbitals and the antibonding intermolecular, the intramolecular being more important) follows. Next, the AS-symmetry (the antibonding intra- and bonding intermolecular - orbitals π^*), and the highest-energy orbital AA (the antibonding intra- and intermolecular) follow. Since four electrons are involved, they occupy the SS and SA orbitals.[57] This is what we have at the beginning of the reaction, Fig. 14.22b, center.

What do we have at the end of the reaction? At the end, there are no π-electrons whatsoever; instead, we have two new σ chemical bonds, each built from the two sp^3 hybrids (Fig. 14.22d)

[57] What a nasty historical association.

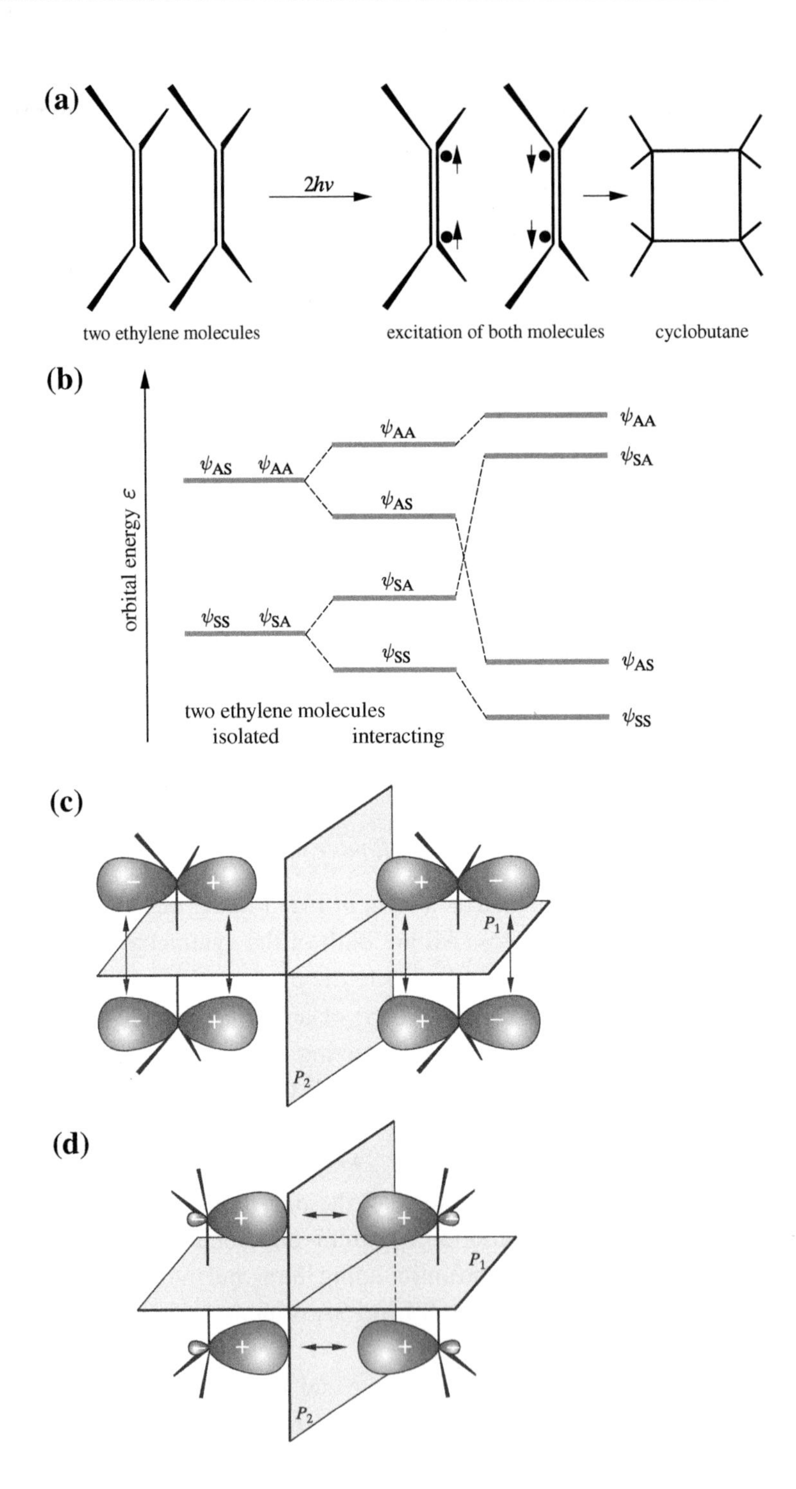
(a)
2hν
two ethylene molecules
excitation of both molecules
cyclobutane
(b)
orbital energy ε
ψ_{AS} ψ_{AA}
ψ_{SS} ψ_{SA}
ψ_{AA}
ψ_{AS}
ψ_{SA}
ψ_{SS}
ψ_{AA}
ψ_{SA}
ψ_{AS}
ψ_{SS}
two ethylene molecules
isolated
interacting
(c)
P_1
P_2
(d)
P_1
P_2

oriented from the first former ethylene molecule to the other.[58] Therefore, we may form the symmetry orbitals once again, recognize their bonding and antibonding character and hence the energetic order of their orbital energies without any calculations, just by inspection (Fig. 14.22b, right). The lowest energy corresponds, of course, to SS (because the newly formed σ chemical bonds correspond to the bonding combination and the lateral bonding overlap of the hybrids is also of the bonding character). The *next in energy, however, is the* AS (because of the bonding interactions in the newly formed σ bonds, while the lateral orbital interaction is weakly antibonding), and then the SA-symmetry orbital (antibonding interaction along the bonds that is only slightly compensated by the lateral in-phase overlap of the hybrids) follows. Finally, the highest-energy corresponds to the totally antibonding orbital of the AA-symmetry.

According to the Woodward-Hoffmann rules, the four π electrons, on which we are focusing, occupy the SS and SA orbitals from the beginning to the end of the reaction. This corresponds to low energy at the beginning of the reaction (R) but is very unfavorable at its end (P), because the unoccupied AS orbital is lower in the energy scale.

And what if we were smart and excited the reactants using a laser? This would allow double occupation of the AS orbital right at the beginning of the reaction and end up with a low energy configuration. To excite an electron per molecule means to put one on orbital π^*, while the second electron stays on orbital π. Of two possible spin states (singlet and triplet), the triplet state is lower in energy (see Chapter 8, p. 460). This situation was described by Eq. (14.73), and the result is that *when one electron sits on nucleus a, the other sits on b. These electrons have parallel spins–everything is prepared for the reaction.*

Therefore, the two ethylene molecules, when excited to the triplet state, open their closed shells in such a way that favors cycloaddition.

The cycloaddition is, therefore, forbidden in the ground state (no thermally activated reaction) and allowed *via* an excited state (photochemistry).

Fig. 14.22. Two equivalent schemes for the cycloaddition reaction of ethylene. Two ethylene molecules, after excitation to the triplet state, dimerize forming cyclobutane (a), because everything is prepared for electron pairing and formation of the new bonds. We obtain the same from the Woodward-Hoffmann rules (panels b, c, and d). According to these rules, we assume that the ethylene molecules preserve two planes of symmetry: P_1 and P_2 during all stages of the reaction. We concentrate on four π electrons–the main actors in the drama. In the beginning, the lowest-energy molecular orbital of the total system (b,c) is of the SS type (i.e., symmetric with respect to P_1 and P_2). The other three orbitals (not shown in Panel c) are of higher energies that increases in the following order: SA, AS, AA. Hence, the four electrons occupy SS and SA (b). Panel d shows the situation after the reaction. The four electrons are no longer of the π type—we now call them the σ type, and they occupy the hybrid orbitals shown here. Once more, the lowest energy (b) corresponds to the SS symmetry orbital (d). The three others (not shown in Panel d) have higher energy, but their order *is different from before* (b): AS, SA, AA. The four electrons should occupy, therefore, the SS and AS type orbitals, whereas (according to the Woodward-Hoffmann rule) they still occupy SS and SA. This is energetically unfavorable, and such a thermic reaction does not proceed. Yet, if before the reaction, the ethylene molecules were excited to the triplet state $(\pi)^1(\pi^*)^1$, then at the end of the reaction, they would correspond to the configuration: $(SS)^2(AS)^2$, of very low energy, and the photochemical reaction proceeds.

[58] We are dealing with a four-membered ring, so the sp^3 hybrids match the bond pattern only roughly.

14.7.5 Barrier Means a Cost of Opening the Closed Shells

Now we can answer more precisely the question of what happens when two molecules react. When the molecules are of the closed-shell character, a change of their electronic structure has to take place first. For that to happen, the kinetic energy of molecular collisions (the temperature plays an important role) has to be sufficiently high in order to push and distort[59] the nuclear framework, together with the electron cloud of each of the partners (i.e., the kinetic energy contra valence repulsion described in Chapter 13), to such an extent that the new configuration already corresponds to that behind the reaction barrier. For example, in the case of an electrophilic or nucleophilic attack, these changes correspond to the transformation $D \rightarrow D^+$ and $A \rightarrow A^-$, while in the case of the cycloaddition to the excitation of the reacting molecules to their triplet states. *These changes make the unpaired electrons move to the proper reaction centers. So long as this state is not achieved, the changes within the molecules are small and, at most, a molecular complex forms*, in which the partners preserve their integrity and their main properties. The profound changes follow from a quasi-avoided crossing of the DA diabatic hypersurface with an excited-state diabatic hypersurface, the excited state being to a large extent a "*picture of the product*."[60]

> Reaction barriers appear because the reactants have to open their valence shells and prepare the electronic structure to be able to form new bonds. This means that their energy goes up until the right excited structure lowers its energy so much that the system slides down the new diabatic hypersurface to the product configuration.

The right structure means the electronic distribution in which, for each to-be-formed chemical bond, there is a set of two properly localized unpaired electrons. *The barrier height depends on the energetic gap between the starting structure and the excited state which is the picture of the products.* By proper distortion of the geometry (due to the valence repulsion with neighbors) we achieve a "pulling down" of the excited state mentioned, but the same distortion causes the ground state to go up. The larger the initial energy gap, the harder to make the two states interchange their order. The reasoning is supported by the observation that the barrier height

---

[59] Two molecules cannot occupy the same volume due to the Pauli exclusion principle; cf. p. 861.

[60] Even the noble gases open their electronic shells when subject to extreme conditions. For example, xenon atoms under pressure of about 150 GPa change their electronic structure so much, that their famous closed-shell electronic structure ceases to be the ground state. Recall the Pauli Blockade, discussed in Chapter 13. Space restrictions for an atom or molecule by the excluded volume of other atoms; i.e., mechanical pushing leads to changes in its electronic structure. These changes may be very large under high pressure. The energy of some excited states lowers so much that the xenon atoms begin to exist in the *metallic state* [see, e.g., M.I. Eremetz, E.A. Gregoryantz, V.V. Struzhkin, H. Mao, R.J. Hemley, N. Muldero, and N.M. Zimmerman, *Phys.Rev.Letters, 85,* 2797 (2000)]. The xenon was metallic in the temperature range 300 K–25 m K. The pioneers of these investigations were R. Reichlin, K.E. Brister, A.K. McMahan, M. Ross, S. Martin, Y.K.Vohra, and A.L. Ruoff, *Phys.Rev.Letters, 62,* 669 (1989).

for electrophilic or nucleophilic attacks is roughly proportional to the *difference between the donor ionization energy and the acceptor electronic affinity, while the barrier for cycloaddition increases with the excitation energies of the donor and acceptor to their lowest triplet states.* Such relations show the great interpretative power of the acceptor-donor formalism. We would not see this in the VB picture because it would be difficult to correlate the VB structures based on the atomic orbitals with the ionization potentials or the electron affinities of the molecules involved. The best choice is to look at all three pictures (MO, AD, and VB) simultaneously. This is what we have done.

14.8 Barrier for the Electron-Transfer Reaction

In the AD theory, a chemical reaction of two closed-shell entities means opening their electronic shells (accompanied by an energy cost), and then forming of the new bonds (accompanied by an energy gain). The electronic shell opening might have been achieved in two ways: either an electron transfer from the donor to the acceptor, or an excitation of each molecule to the triplet state and subsequent electron pairing between the molecules.

Now we will be interested in the barrier *height* when the first of these possibilities occurs.

14.8.1 Diabatic and Adiabatic Potential

Example. Electron Transfer in $H_2^+ + H_2$

Let us imagine two molecules, H_2^+ and H_2, in a parallel configuration[61] at distance R from one another and having identical length 1.75 a.u. (see Fig. 14.23a). The value chosen is the arithmetic mean of the two equilibrium separations (2.1 a.u. for H_2^+, 1.4 a.u. for H_2).

There are two geometry parameters to change (Fig. 14.23): the length q_L of the left (or first) molecule and the length q_R of the right (or second) molecule. Instead of these two variables, we may consider the other two: their sum and their difference. Since our goal is to be as simple as possible, we will assume,[62] that $q_L + q_R = \text{const}$, and therefore the geometry of the total nuclear framework, may be described by a single variable: $q = q_L - q_R$, with $q \in (-\infty, \infty)$.

Now we assume that the extra electron is on the left molecule, so we have the $H_2 \ldots H_2^+$ system. It is quite easy to imagine what happens when q changes from 0 (i.e., from both bonds of equal length) to a somewhat larger value. Variable $q = q_L - q_R > 0$ means that $q_L > q_R$, so when q increases a bit, the energy of the system will decrease because the H_2^+ molecule elongates, while the H_2 shortens; i.e., both approach their equilibrium geometries. If q increases further, it will soon reach the value $q = q_0 = 2.1 - 1.4 = 0.7$ a.u., the optimum value for both molecules. A further increase of q will mean, however, a kind of discomfort for each of the molecules and

[61] We freeze all the translations and rotations.

[62] This assumption stands to reason because a shortening of one molecule will be accompanied by an almost identical lengthening of the other when they exchange an electron.

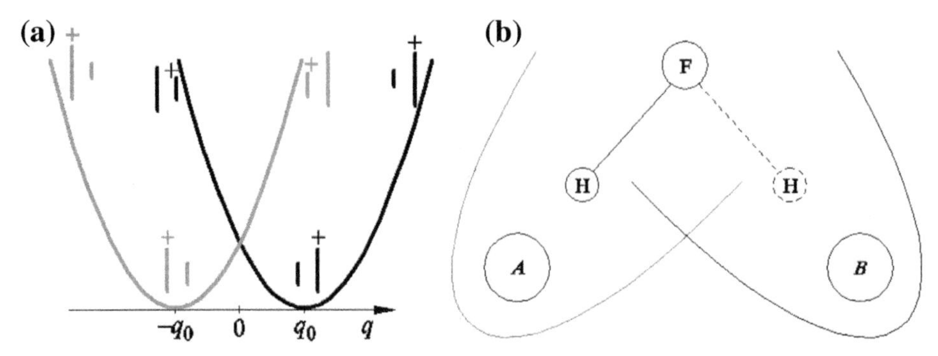

Fig. 14.23. An electron transfer is accompanied by a geometry change. (a) The black curve corresponds to the system $H_2 \ldots H_2^+$ (i.e., the extra electron is on the left molecule), and the gray curve pertains to the system $H_2^+ \ldots H_2$ (the electron is on the right molecule). Variable q equals the difference of the bond lengths of the r.h.s. molecule and the l.h.s. molecule. At $q = \pm q_0$, both molecules have their optimum bond lengths. (b) The HF pendulum oscillates between two sites, A and B, which accommodate an extra electron becoming either A^-B or AB^-. The curves that are similar to parabolas (the black one for AB^- and the gray one for A^-B) denote the energies of the diabatic states as functions of the pendulum angle.

the energy will go up–and for large q, it goes up very much. And what will happen for $q < 0$? $q < 0$ means an elongation of an already-too-long H_2 and a shortening of an already-too-short H_2^+. The potential energy goes up, and the total plot is similar to a parabola with the minimum at $q = q_0 > 0$ (see the black curve in Fig. 14.23a).

If, however, we assume that the extra electron resides all the time on the right molecule, so we have to do with $H_2^+ \ldots H_2$, then we will obtain the identical parabola-like curve as before, but with the minimum position at $q = -q_0 < 0$ (see the gray curve in Fig. 14.23a).

Diabatic and Adiabatic Potentials:

Each of these curves with a single minimum and with the extra electron residing all the time on a given molecule represents the *diabatic* potential energy curve for the motion of the nuclei. If, when the donor-acceptor distance changes, the electron *keeps pace* with it and jumps on the acceptor, then increasing or decreasing q from 0 gives a similar result: we obtain a single electronic ground-state potential energy curve with *two* minima in positions $\pm q_0$. This is the *adiabatic* curve.

Whether the adiabatic or diabatic potential has to be applied is equivalent to asking *whether the nuclei are slow enough that the electron keeps pace (adiabatic) or not (diabatic) with their motion.*[63] This is within the spirit of the adiabatic approximation; cf. Chapter 6, p. 302. Also, a

[63] In the electron transfer reaction $H_2^+ + H_2 \rightarrow H_2 + H_2^+$ the energy of the reactants is equal to the energy of the products because the reactants and the products represent the same system. Is it, therefore, a kind of fiction? Is there any reaction at all taking place? From the point of view of a bookkeeper (thermodynamics), no reaction took place, but from the point of view of a molecular observer (kinetics), such a reaction may take place. It is especially visible when instead of one of the hydrogen atoms, we use deuterium. Then the reaction $HD^+ + H_2 \rightarrow HD + H_2^+$ becomes real even for the bookkeeper (mainly because of the difference in the zero-vibration energies of the reactants and products).

diabatic curve corresponding to the same electronic structure (the extra electron sitting on one of the molecules all the time) is an analog of the diabatic hypersurface that preserved the same chemical bond pattern encountered before.

Example. The "HF Pendulum"

Similar conclusions come from another ideal system (Fig. 14.23b); namely, the hydrogen fluoride molecule is treated as the pendulum of a grandfather clock (the hydrogen atom down, the clock axis going through the fluorine atom) moving over two molecules: A and B, each of them may accommodate an extra electron.[64]

The electron is negatively charged, the hydrogen atom in the HF molecule carries a partial positive charge, and both objects attract each other. If the electron sits on the left molecule and during the pendulum motion does not keep pace and does not jump over to the right molecule, the potential energy has a single minimum for the angle $-\theta_0$ (the diabatic potential might be approximated by a parabola-like curve with the minimum at $-\theta_0$). A diabatic analogous curve with the minimum at θ_0 arises when the electron resides on B all the time. When the electron keeps pace with any position of the pendulum, we have a single adiabatic potential energy curve with two minima: at $-\theta_0$ and θ_0.

14.8.2 Marcus Theory

Rudolph Arthur Marcus (b. 1923), American chemist and professor at the University of Illinois in Urbana and at California Institute of Technology in Pasadena. In 1992, Marcus received the Nobel Prize *"for his contribution to the theory of electron transfer reactions in chemical systems."*

The contemporary theory of the electron transfer reaction was proposed by Rudolph Marcus.[65] The theory is based to a large extent on the harmonic approximation for the diabatic potentials involved (i.e., the diabatic curves represent parabolas). One of the parabolas corresponds to the reactants $V_R(q)$, the other to the products $V_P(q)$ of the electron transfer reaction (see Fig. 14.24).

Now let us assume that both parabolas have the same curvature (force constant f).[66] The reactants correspond to the parabola with the minimum at q_R (without losing generality, we adopt a convention that at $q = q_R$, the energy equals zero):

$$V_R(q) = \frac{1}{2} f (q - q_R)^2,$$

[64] Let the mysterious q be a single variable for a while, whose deeper meaning will be given later. In order to make the story more concrete, let us think about two reactant molecules (R) that transform into the product molecules (P): $A^- + B \rightarrow A + B^-$.

[65] The reader may find a good description of the theory in a review article by P.F. Barbara, T.J. Meyer, and M.A. Ratner, *J.Phys.Chem., 100,* 13148 (1996).

[66] This widely used assumption is better fulfilled for large molecules, when one electron more or less does not change much.

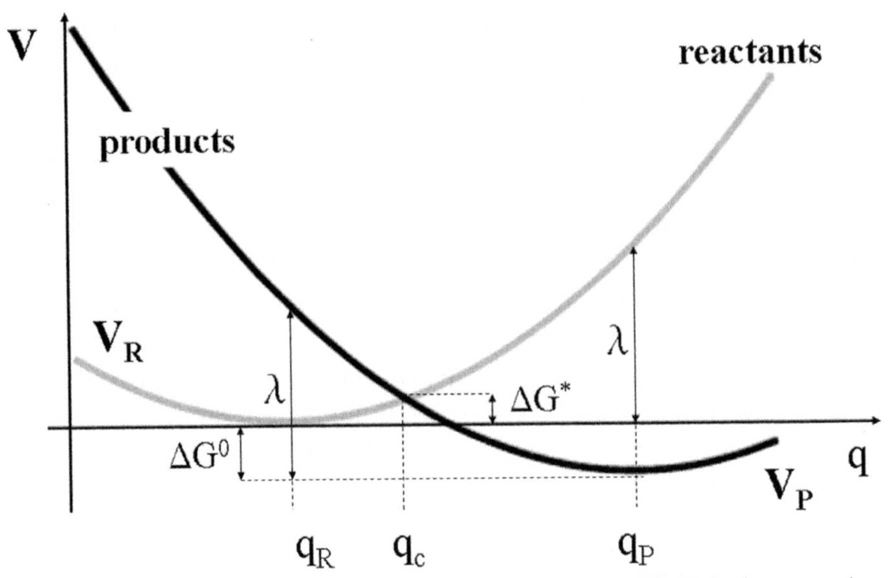

Fig. 14.24. The Marcus theory is based on two parabolic diabatic potentials $V_R(q)$ and $V_P(q)$ for the reactants (gray curve) and products (black curve), having minima at q_R and q_P, respectively. The quantity $\Delta G^0 \equiv V_P(q_P) - V_R(q_R)$ represents the energy difference between the products and the reactants at their equilibrium geometries, the reaction barrier $\Delta G^* \equiv V_R(q_c) - V_R(q_R) = V_R(q_c)$, where q_c corresponds to the intersection of the parabolas. The reorganization energy $\lambda \equiv V_R(q_P) - V_R(q_R) = V_R(q_P)$ represents the energy expense for making the geometry of the reactants identical with that of the products (and *vice versa*).

while the parabola with the minimum at q_P is shifted in the energy scale by ΔG^0 ($\Delta G^0 < 0$ corresponds to an exothermic reaction[67]):

$$V_P(q) = \frac{1}{2} f (q - q_P)^2 + \Delta G^0.$$

So far, we just have treated the quantity ΔG^0 as a potential energy difference $V_P\left(q_P\right) - V_R\left(q_R\right)$ of the model system under consideration ($H_2^+ + H_2$ or the "*pendulum*" HF)*, although the symbol suggests that this interpretation will be generalized in the future.*

Such parabolas represent a simple situation. The parabolas' intersection point q_c satisfies by definition $V_R(q_c) = V_P(q_c)$. This gives[68]

$$q_c = \frac{\Delta G^0}{f} \frac{1}{q_P - q_R} + \frac{q_P + q_R}{2}.$$

Of course, on the parabola diagram, the most important are the two minima, the intersection point q_c and the corresponding energy, which represents the reaction barrier reactants $\rightarrow$ products.

---

[67] That is, the energy of the reactants is higher than the energy of the products (as in Fig. 14.24).

[68] If the curves did not represent parabolas, we might have serious difficulties. This is why we need harmonicity.

Marcus Formula:
The electron-transfer reaction barrier is calculated as

$$\Delta G^* = V_R\left(q_c\right) = \frac{1}{4\lambda}\left(\lambda + \Delta G^0\right)^2, \tag{14.74}$$

where the *reorganization energy* λ represents the energy expense for distorting the products to the equilibrium configuration of the reactants, or distorting the reactants to get the equilibrium configuration of the products:

$$\lambda = V_P(q_R) - V_P(q_P) = \frac{1}{2}f(q_R - q_P)^2 + \Delta G^0 - \Delta G^0 = \frac{1}{2}f(q_R - q_P)^2.$$

The reorganization energy is, therefore, always positive.

Reorganization Energy:
Reorganization energy is the energy cost needed for making products in the nuclear configuration of the reactants.

If we ask about the energy needed to transform the optimal geometry of the reactants into the optimal geometry of the products, we obtain the same number. Indeed, we immediately obtain $V_R(q_P) - V_R(q_R) = \frac{1}{2}f(q_R - q_P)^2$, which is the same as before. This result is a consequence of the harmonic approximation and the same force constant assumed for V_R and V_P. It is seen that the barrier for the thermal electron transfer reaction is higher if the geometry change is larger upon the electron transfer [large $(q_R - q_P)^2$] and if the system is stiffer (large f).

Svante August Arrhenius (1859-1927), Swedish physical chemist, astrophysicist, professor at the Stockholm University, and originator of the electrolytic theory of ionic dissociation, measurements of the temperature of planets and of the solar corona, and also of the theory deriving life on Earth from outer space. In 1903, he received the Nobel Prize in chemistry *"for the services he has rendered to the advancement of chemistry by his electrolytic theory of dissociation."*

From the Arrhenius theory, the electron transfer reaction rate constant reads as

$$k_{ET} = Ae^{-\frac{\left(\lambda+\Delta G^0\right)^2}{4\lambda k_B T}}. \tag{14.75}$$

How would the reaction rate change if parabola $V_R(q)$ stays in place, while parabola $V_P(q)$ moves down in energy scale? In experimental chemistry, this corresponds to a *class* of the chemical reactions $A^- + B \rightarrow A + B^-$, with A (or B) from a homological series of compounds. The homology suggests that the parabolas are similar because the mechanism is the same (the reactions proceed similarly), and the situations considered differ only by a lowering the second parabola with respect to the

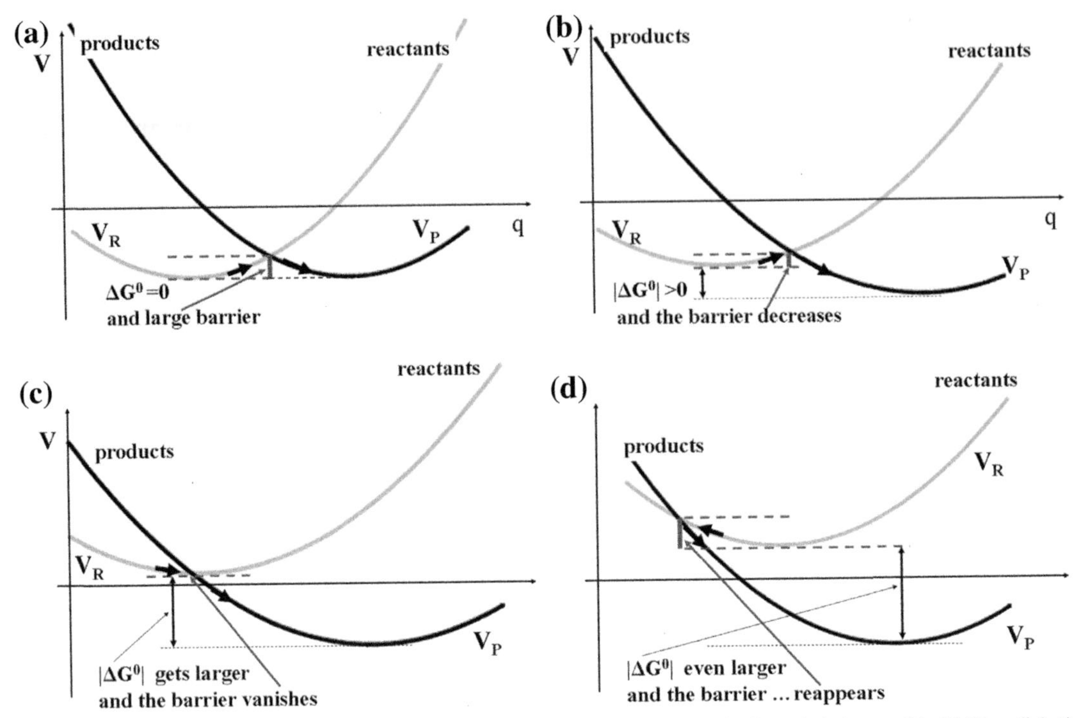

Fig. 14.25. Four qualitatively different cases in the Marcus theory. The reactant (product) parabola is gray (black). The adiabatic ground state energy is denoted as thick lower gray-and-black curve. We assume that $\Delta G^0 \leq 0$. (a) $\Delta G^0 = 0$, hence $\Delta G^* = \frac{\lambda}{4}$. (b) $\left|\Delta G^0\right| < \lambda$. (c) $\left|\Delta G^0\right| = \lambda$. (d) inverse Marcus region $\left|\Delta G^0\right| > \lambda$.

first; i.e., the reactions in the series become more and more exothermic. We may have four cases, Fig. 14.25:

Case 1: If the lowering is zero (i.e., $\Delta G^0 = 0$), the reaction barrier is equal to $\lambda/4$ (Fig. 14.25a).

Case 2: If $\left|\Delta G^0\right| < \lambda$ the reaction barrier is lower, because of the subtraction in the exponent, and the reaction rate *increases* (Fig. 14.25b). Hence, the ΔG^0 is the "driving force" in such reactions.

Case 3: When the $\left|\Delta G^0\right|$ keeps increasing, at $\left|\Delta G^0\right| = \lambda$, the reorganization energy cancels the driving force, and the barrier vanishes to zero. Note that this represents the highest reaction rate possible (Fig. 14.25c).

Case 4: Let us imagine now that we keep increasing the driving force. We have a reaction for which $\Delta G^0 < 0$ and $\left|\Delta G^0\right| > \lambda$. Compared to the previous case, the *driving force has increased, whereas the reaction rate decreases.* This might look like a possible surprise for experimentalists. A case like this is called the *inverse Marcus region* (Fig. 14.25d), foreseen by Marcus in the 1960s, using the two-parabola model. People could not believe this prediction until experimental proof[69] was established in 1984.

[69] J.R. Miller, L.T. Calcaterra, and G.L. Closs, *J.Am.Chem.Soc., 97,* 3047 (1984).

New Meaning of the Variable q

Let us make a subtraction:

$$V_R(q) - V_P(q) = f(q - q_R)^2/2 - f(q - q_P)^2/2 - \Delta G^0$$
$$= \frac{f}{2}[2q - q_R - q_P][q_P - q_R] - \Delta G^0 = Aq + B, \qquad (14.76)$$

where A and B represent constants. This means that in parabolic approximation,

> the diabatic potential energy difference depends *linearly* on coordinate q. In other words, measuring the stage of a given electron transfer reaction, we may use either q or $V_R(q) - V_P(q)$.

14.8.3 Solvent-Controlled Electron Transfer

The above examples and derivations pertain to a 1-D model of electron transfer (a single variable q), while in reality (imagine a solution), the problem pertains to a huge number of variables. What happens here? Let us take the example of electron transfer between Fe^{2+} and Fe^{3+} ions in an aqueous solution $Fe^{2+} + Fe^{3+} \rightarrow Fe^{3+} + Fe^{2+}$ (Fig. 14.26)[70]

It turns out that

> the solvent behavior is of key importance for the electron-transfer process.

For the reaction to proceed, the solvent has to *reorganize* itself next to both ions. The hydration shell of Fe^{2+} ion is of larger radius than the hydration shell of Fe^{3+} ion, because Fe^{3+} is smaller than Fe^{2+} and, in addition, creates a stronger electric field due to its higher charge. Both factors add to a stronger association of the water molecules with the Fe^{3+} ion than with Fe^{2+}. In a crude approximation, the state of the solvent may be characterized by two variable cavities (shown as circles), say: left and right (or, numbers 1 and 2) that could accommodate the rigid ions. Let us assume that the cavities have radii r_1 and r_2, whereas the fixed ionic radii are $r_{\text{Fe2+}}$ and $r_{\text{Fe3+}}$ (shown as vertical sections) with $r_{\text{Fe2+}} > r_{\text{Fe3+}}$ and that $r_1 + r_2 = const = r_{\text{Fe2+}} + r_{\text{Fe3+}}$ and introduce a single variable $q = r_2 - r_1$ that in this situation characterizes the state of the solvent. Let us see what happens when q changes.

We first consider that the extra electron sits on the left ion all the time (the gray reactant curve V_R in Fig. 14.26) and the variable q is a negative number (with a high absolute value; i.e., $r_1 \gg r_2$). As seen from Fig. 14.26, the energy is very high because the solvent squeezes the Fe^{3+} ion out (the second cavity is too small). It does not help that the Fe^{2+} ion has a lot of space in its cavity. Now we begin to move toward higher values of q. The first cavity begins to shrink, for a

[70] In this example, $\Delta G^0 = 0$; i.e., Case 1 discussed above.

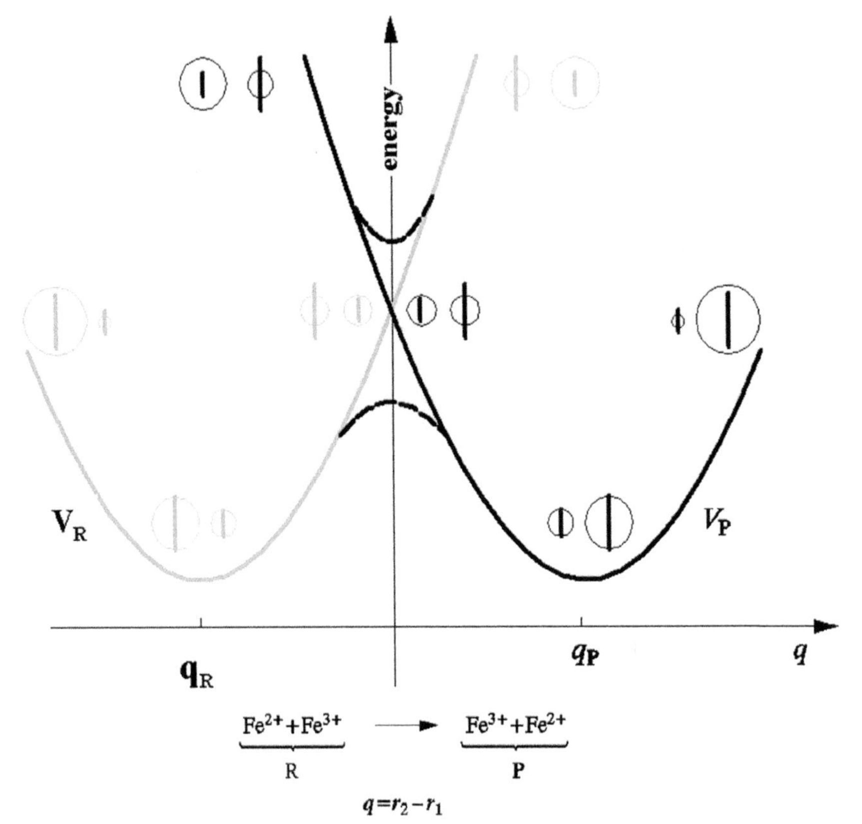

Fig. 14.26. The diabatic potential energy curves: V_R for the reactants (gray) and V_P for the products (black) pertaining to the electron transfer reaction $Fe^{2+} + Fe^{3+} \rightarrow Fe^{3+} + Fe^{2+}$ in an aqueous solution. The curves depend on the variable $q = r_2 - r_1$ that describes the *solvent,* which is characterized by the radius r_1 of the cavity for the first (say, left) ion and by the radius r_2 of the cavity for the second ion. For the sake of simplicity, we assume that $r_1 + r_2 = const$ and is equal to the sum of the ionic radii of Fe^{2+} and Fe^{3+}. For several points q, the cavities were drawn (shown as circles) as well as the vertical sections that symbolize the diameters of the left and right ions. In this situation, the plots V_R and V_P have to differ widely. The dashed lines represent the adiabatic curves (in the peripheral sections, they coincide with the diabatic curves).

while without any resistance from the Fe^{2+} ion, the second cavity begins to lose its pressure, thus making the Fe^{3+} ion happier. The energy decreases. Finally we reach the minimum of V_R, at $q = q_R$ and the radii of the cavities match the ions perfectly. Meanwhile, variable q continues to increase. Now the solvent squeezes the Fe^{2+} ion out, while the cavity for Fe^{3+} becomes too large. The energy increases again, mainly because of the first effect. We arrive at $q = 0$. The cavities are of equal size, but do not match either ion. This time, the Fe^{2+} ion experiences some discomfort, and after passing the point $q = 0$, the pain increases more and more, and the energy continues to increase. The whole story pertains to an extra electron sitting on the left ion all the time (no jump; i.e., the reactant situation). A similar dramatic story can be told when the electron is sitting all the time on the right ion (products situation). In this case, we obtain the V_P plot (black).

The V_R and V_P plots just described represent the diabatic potential energy curves for the motion of the nuclei, valid for the extra electron residing on the same ion all the time. Fig. 14.25

also shows the adiabatic curve when the extra electron has enough time to adjust to the motion of the approaching nuclei and the solvent, and jumps at the partner ion.

Taking a single parameter q to describe the electron transfer process in a solvent is certainly a crude simplification. Actually there are billions of variables in the game describing the degrees of freedom of the water molecules in the first and further hydration shells. One of the important steps toward successful description of the electron transfer reaction was the Marcus postulate,[71] that

> despite the multidimensionality of the problem, Eq. (14.76) is still valid; i.e., $V_R - V_P$ is a single variable describing the position of the system on the electron-transfer reaction path (it is, therefore, a collective coordinate that describes the positions of the solvent molecules).

No doubt low potential energy value is important, but it is also important how often this value can be reached by the system. This is connected to the *width* of the low-energy basin associated with the *entropy*,[72] as well as to the *free energy*. In statistical thermodynamics, we introduce the idea of the mean force, related to the free energy. Imagine a system in which we have two motions on different time scales: fast (e.g., of small solvent molecules) and slow (e.g., which changes the shape of a macromolecule). To focus on the slow motion, we average the energy over the fast motion (the Boltzmann factor needed will introduce a temperature dependence in the resulting energy). In this way, from the potential energy, we obtain the *mean force potential* depending only on the slow variables,[73] sometimes called the *free energy* (which is a function of geometry of the macromolecule); cf. p. 353.

> The second Marcus assumption is that the ordinate axis should be treated as the mean force potential, or the free energy G rather than just potential energy V.

It is very rare in theoretical chemistry that a many-dimensional problem can be transformed to a single variable problem. This is why the Marcus idea described above of a collective coordinate,

---

[71] Such collective variables are used very often in everyday life. Who cares about all the atomic positions when studying a ball rolling down an inclined plane? Instead, we use a single variable (the position of the center of the ball), which gives us a perfect description of the system in a certain energy range.

[72] A wide potential energy well can accommodate a lot of closely lying vibrational levels, and therefore the number of possible states of the system in a given energy range may be huge (large entropy). Recall the particle-in-a-box problem: the longer the box, the more dense the energy levels.

[73] The free energy is defined as $F(T) = -kT\frac{\partial}{\partial T}\ln Z$, where $Z = \sum_i \exp\left(-\frac{E_i}{kT}\right)$ represents the partition function (also known as the *sum of states*), E_i stands for the ith energy level. In the classical approach, this energy level corresponds to the potential energy $V(x)$, where x represents a point in configurational space, and the sum corresponds to an integral over the total configurational space $Z = \int dx \exp\left(-\frac{V}{kT}\right)$. Note that the free energy is a function of the temperature only, not of the spatial coordinates x. If however, the integration were only carried out over part of the variables (say, only the fast variables), then Z, and therefore also F, would become a function of the slow variables and of the temperature (mean force potential). Despite the incomplete integration, we sometimes use the name *free energy* for this mean force potential by saying that "*the free energy is a function of coordinates...*"

provokes the reaction: "*No way.*" However, as it turned out later, this simple postulate lead to a solution that grasps the essential features of electron transfer.

What Do the Marcus Parabolas Mean?

The example just considered of the electron transfer reaction : $Fe^{2+} + Fe^{3+} \rightarrow Fe^{3+} + Fe^{2+}$ reveals that in this case, the reaction barrier is controlled by the solvent (i.e., by billions of coordinates). As shown by Marcus, this plethora can be effectively replaced by a single collective variable. Only after this approximation may we draw the diabatic parabola-like curves. The intersection point of the two diabatic curves can be found easily only after assuming their parabolic character. And yet any collective variable means motion along a line in an extremely complex configurational space (solvent molecules plus reactants). Moving along this line means that, according to Marcus, we encounter the intersection of the ground and excited electronic states. As shown in Chapter 6, such a crossing occurs at the conical intersection. Is it, therefore, that during the electron transfer reaction, the system goes through the conical intersection point? How do we put together such notions as reaction barrier, reaction path, entrance and exit channels, not to mention the acceptor-donor theory? Looking at Fig. 14.27, pertaining to the reaction DA $\rightarrow$ D^+A^-, will be of some help to us in answering this question.

- The *diabatic hypersurfaces*, one corresponds to DA (i.e., the extra electron is on the donor all the time) and the second to D^+A^- (i.e., the extra electron resides on the acceptor), undergo the conical intersection (see Fig. 14.27a). For conical intersection to happen, at least three atoms are required. Imagine a simple model, with a diatomic acceptor A and an atom D as donor. Atom D has a dilemma: transfer the electron to either the first or the second atom of A. This dilemma means conical intersection. The variables ξ_1 and ξ_2 described in Chapter 6 were chosen (they lead to splitting of the adiabatic hypersurfaces; see Fig. 14.27b,c), which measure the deviation of the donor D with respect to the corner of the equilateral triangle of side equal to the length of the diatomic molecule A. The conical intersection point [i.e., (0, 0)] corresponds to the equilateral triangle. The figure also shows the upper and lower cones touching at (0, 0).
- The conical intersection led to two *adiabatic* hypersurfaces: lower (electronic ground state) and upper (electronic excited state). Each of the adiabatic hypersurfaces shown in Fig. 14.27

Fig. 14.27. Electron transfer in the reaction DA $\rightarrow$ D^+A^-, as well as the relation of the Marcus parabolas to the concepts of the conical intersection, diabatic and adiabatic states, entrance and exit channels and the reaction barrier. Panel (a) shows two diabatic surfaces as functions of the ξ_1 and ξ_2 variables that describe the deviation from the comical intersection point (within the bifurcation plane; cf., p. 312). Both surfaces are shown schematically in the form of the two intersecting paraboloids: one for the reactants (DA), and the second for products (D^+A^-). (b) The same as (a), but the hypersurfaces are presented more realistically. The upper and lower adiabatic surfaces touch at the conical intersection point. (c) A more detailed view of the same surfaces. On the ground-state adiabatic surface (the lower one), we can see two reaction channels I and II, each with its reaction barrier. On the upper adiabatic surface, an energy valley is visible that symbolizes a bound state that is separated from the conical intersection by a reaction barrier. (d) The Marcus parabolas represent the sections of the diabatic surfaces along the corresponding reaction channel, at a certain distance from the conical intersection. Hence, the parabolas in reality cannot intersect (undergo an avoided crossing).

(a)

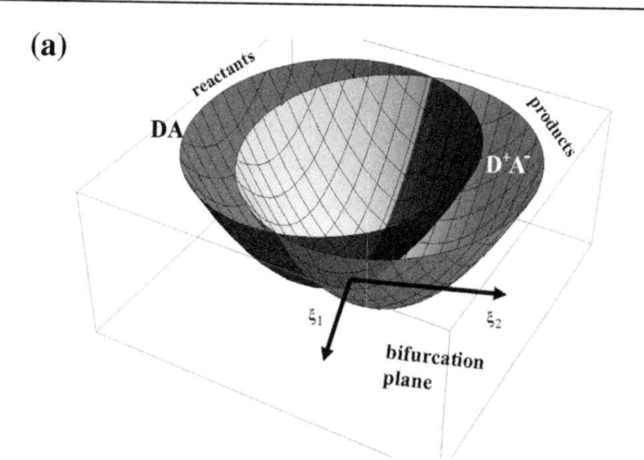
reactants
products
DA
D^+A^-
ξ_1
ξ_2
bifurcation plane

(b)

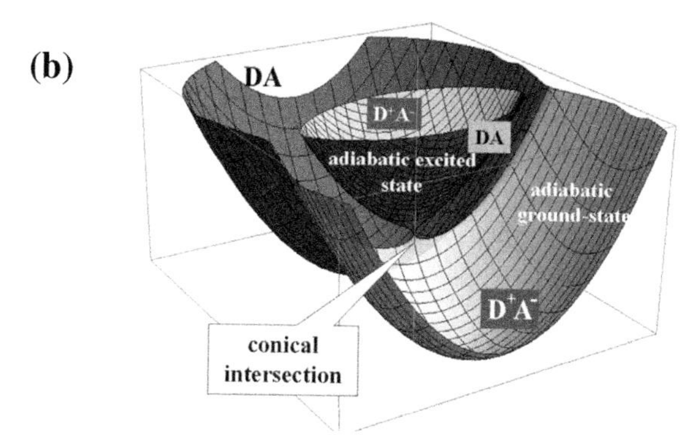
DA
D^+A^-
DA
adiabatic excited state
adiabatic ground-state
D^+A^-
conical intersection

(c)

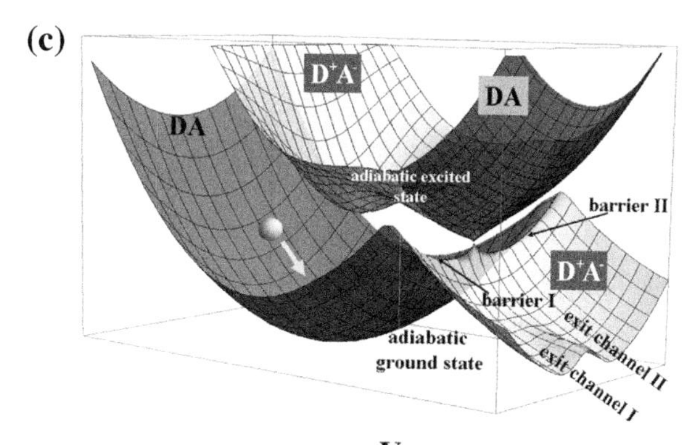
D^+A^-
DA
DA
adiabatic excited state
barrier II
D^+A^-
barrier I
adiabatic ground state
exit channel II
exit channel I

(d)

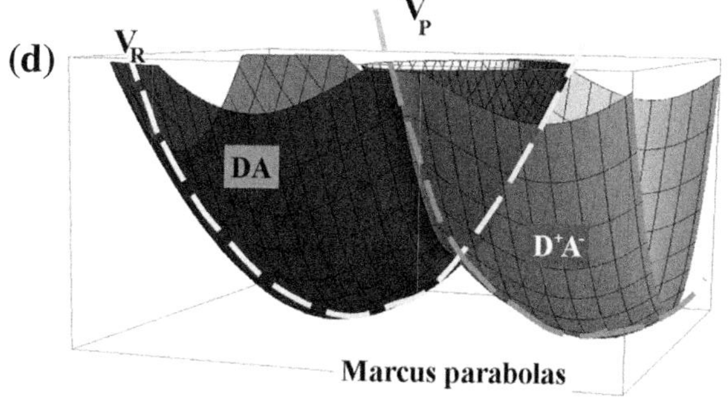
V_R
V_P
DA
D^+A^-
Marcus parabolas

consists of the dark gray half (the diabatic states of the reactants, DA) and the light gray half (the diabatic state of the products, D^+A^-). The border between them reveals the intersection of the two diabatic states and represents the line of change of the electronic structure reactants/products. Crossing the line means the chemical reaction happens.

- The "*avoided crossing*" occurs everywhere along the border except at the conical intersection. It is improbable that the reactive trajectory passes through the conical intersection because it usually corresponds to higher energy. It will usually pass by (this resembles an avoided crossing), and the electronic state DA changes to electronic state D^+A^- or *vice versa*. This is why we speak of the avoided crossing in a polyatomic molecule, whereas the concept pertains to diatomics only.
- Passing the border is easiest at *two* points (see Fig. 14.27c). These are the two saddle points (barriers I and II). A thermally induced electron transfer reaction goes through one of them. In each case, we obtain different products. Both saddle points differ in that D, when attacking A has the choice if joining either of the two ends of A, usually forming *two different* products. We therefore usually have *two* barriers. In the example given (H_3), they are identical, but in general they may differ. When the barrier heights are equal because of symmetry, it does not matter which is overcome. When they are different, one of them dominates (usually the lower barrier[74]).
- The Marcus parabolas (Fig. 14.27d) represent a special section (along the collective variable) of the hypersurfaces passing through the conical intersection (parabolas V_R and V_P). Each parabola represents a diabatic state, so a part of each reactant parabola is on the lower hypersurface, while the other one is on the upper hypersurface. We see that the parabolas are only an approximation to the hypersurface profile. The reaction is of a thermal character, and as a consequence, the parabolas should not pass through the conical intersection, because it corresponds to high energy, instead it passes through one of the saddle points.
- The "light gray" part of the excited state hypersurface (Fig. 14.27c) runs up to the "dark gray" part of the ground state hypersurface or *vice versa.* This means that photoexcitation (following the Franck-Condon rule this corresponds to a vertical excitation) means a profound change: the system looks as if it has already reacted (photoreaction).

Quantum Mechanical Modification

In Marcus equation (14.74), we assume that in order to make the electron transfer effective, we have to supply at least the energy equal to the barrier height. The formula does not obviously take into account the quantum nature of the transfer. The system may overcome the barrier not only by having its energy higher than the barrier, but also by tunneling,[75] when its energy is lower than the barrier height (cf. p. 174). Besides, the reactant and product energies are

---

[74] There may be some surprises. Barrier height is not all that matters. Sometimes what decides is access to the barrier region, in the sense of its width (this is where the entropy and free energy matter).

[75] We will also consider shortly an electron transfer due to optical excitation.

quantized (vibrational-rotational levels[76]). The reactants may be excited to one of such levels. The reactant vibrational levels will have different abilities to tunnel.

According to Chapter 2, only a time-dependent perturbation is able to change the system energy. As such a perturbation may serve the electric field of the electromagnetic wave. When the perturbation is periodic, with the angular frequency ω matching the energy difference of initial state k and one of the states of higher energy (n), then the transition probability between these states is equal to : $P_k^n(t) = \frac{2\pi t}{\hbar}|v_{kn}|^2 \delta\left(E_n^{(0)} - E_k^{(0)} - \hbar\omega\right)$ (the Fermi golden rule, Eq. (2.28), p. 92 is valid for relatively short times t), where $v_{kn} = \langle k|v|n\rangle$, with $v(\boldsymbol{r})$ representing the perturbation amplitude,[77] $V(\boldsymbol{r}, t) = v(\boldsymbol{r})e^{i\omega t}$. The Dirac delta function δ is a quantum-mechanical way of saying that the total energy has to be conserved. In phototransfer of the electron state, k represents the quantum mechanical state of the reactants, and n a product state, each of diabatic character.[78] In practice, the adiabatic approximation is used, in which the reactant and product wave functions are products of the electronic wave functions (which depend on the electronic coordinates $\boldsymbol{r}$ and, parametrically, on the nuclear configuration $\boldsymbol{R}$) and the vibrational functions $f(\boldsymbol{R})$ describing the motion of the nuclei: $\psi_{k,R}\left(\boldsymbol{r};\boldsymbol{R}\right) f_{v_1,R}(\boldsymbol{R})$ and $\psi_{n,P}\left(\boldsymbol{r};\boldsymbol{R}\right) f_{v_2,P}(\boldsymbol{R})$. The indices v_1 and v_2 in functions f denote the vibrational quantum numbers.

Then, the transition probability depends on the integral (Chapter 2)

$$|v_{kn}|^2 = \left|\langle\psi_{k,R}\left(\boldsymbol{r};\boldsymbol{R}\right) f_{v_1,R}(\boldsymbol{R})|v(\boldsymbol{r})|\psi_{n,P}\left(\boldsymbol{r};\boldsymbol{R}\right) f_{v_2,P}(\boldsymbol{R})\rangle\right|^2.$$

Let us rewrite it, making the integration over the nuclear and electronic coordinates explicit (where $\mathrm{d}V_{nucl}$ and $\mathrm{d}V_e$ mean that the integrations is over the nuclear and electronic coordinates, respectively):

$$v_{kn} = \int \mathrm{d}V_{nucl}\; f^*_{v_1,R}\left(\boldsymbol{R}\right)\; f_{v_2,P}\left(\boldsymbol{R}\right)\int \mathrm{d}V_e\; \psi^*_{k,R}\left(\boldsymbol{r};\boldsymbol{R}\right) v\left(\boldsymbol{r}\right) \psi_{n,P}\left(\boldsymbol{r};\boldsymbol{R}\right).$$

Now, let us use the Franck-Condon approximation that the optical perturbation makes the electrons move instantaneously while the nuclei do not keep pace with the electrons and stay in the same positions (we assume, therefore, equilibrium positions of the nuclei $\boldsymbol{R}_0$ in the reactants):

$$v_{kn} \approx \int \mathrm{d}V_{nucl}\; f^*_{v_1,R}\left(\boldsymbol{R}\right)\; f_{v_2,P}\left(\boldsymbol{R}\right)\int \mathrm{d}V_e\; \psi^*_{kR}\left(\boldsymbol{r};\boldsymbol{R}_0\right) v\left(\boldsymbol{r}\right) \psi_{n,P}\left(\boldsymbol{r};\boldsymbol{R}_0\right).$$

The last integral, therefore, represents a constant, and so

$$|v_{kn}|^2 = |V_{RP}|^2 \,|S_{osc}(v_1, v_2)|^2,$$

[76] For large molecules, we may forget the rotational spectrum since, because of the large inertia momentum, the rotational states form a quasi-continuum ("*no quantization*").

[77] $\boldsymbol{r}$ stands for those variables on which the wave functions depend.

[78] They will be denoted by the subscripts R and P.

where

$$V_{RP} = \int dV_e \, \psi^*_{k,R}\left(\boldsymbol{r}; \boldsymbol{R}_0\right) v\left(\boldsymbol{r}\right) \psi_{n,P}\left(\boldsymbol{r}; \boldsymbol{R}_0\right)$$

$$S_{osc}\left(v_1, v_2\right) = \int dV_{nucl} \; f^*_{v_1,R}\left(\boldsymbol{R}\right) \; f_{v_2,P}\left(\boldsymbol{R}\right). \tag{14.77}$$

The $|S_{osc}(v_1, v_2)|^2$ is called the *Franck-Condon factor*.

Franck-Condon Factor:
A Franck-Condon factor is the square of the absolute value of the overlap integral of the vibrational wave functions: one pertaining to the reactants with vibrational quantum number v_1 and the second pertaining to the products with vibrational quantum number v_2.

The calculation of V_{RP} is not an easy matter; therefore, we often prefer an empirical approach by modeling the integral as[79]

$$V_{RP} = V_0 exp[-\beta(R - R_0)],$$

where R_0 stands for the van der Waals distance of the donor and acceptor, R represents their distance, $\beta > 0$ is a constant, and V_0 means V_{RP} for the van der Waals distance.[80]

A large Franck-Condon factor means that by exciting the reactants to the vibrational state v_1 there is a particularly high probability for the electron transfer (by tunneling) with the products in vibrational state v_2.

Reorganization Energy

In the Marcus formula, reorganization energy plays an important role. This energy is the main reason for the electron-transfer reaction barrier.

The reorganization pertains to the neighborhood of the transferred electron[81] (i.e., to the solvent molecules, but also to the donors and acceptors themselves).[82] This is why the reorganization energy, in the first approximation consists of the internal reorganization energy (λ_i) that

[79] Sometimes the dependence is different. For example, in *Twisted Intramolecular Charge Transfer* (TICT), after the electron is transferred between the donor and acceptor moieties (a large V_{RP}) the molecule undergoes an internal rotaton of the moieties, which causes an important decreasing of the V_{RP} [K. Rotkiewicz, K.H. Grellmann, and Z.R. Grabowski, *Chem.Phys.Letters, 19,* 315 (1973)].

[80] As a matter of fact, such formulas only contain a simple message: V_{RP} decreases *very* fast when the donor and acceptor distance increases.

[81] The neighborhood is adjusted perfectly to the extra electron (to be transferred) in the reactant situation, and very unfavorable for its future position in the products. Thus, the neighborhood has to be reorganized to be adjusted for the electron transfer products.

[82] It does not matter for an electron what in particular prevents it from jumping.

pertains to the donor and acceptor molecules, and of the solvent reorganization energy (λ_0):

$$\lambda = \lambda_i + \lambda_0.$$

Internal Reorganization Energy

For the electron to have a chance of jumping from molecule A^- to molecule[83] B, it has to have the neighborhood reorganized in a special way. The changes should make the extra electron's life hard on A^- (together with solvation shells) and seduce it by the alluring shape of molecule B and its solvation shells. To do this, work has to be done. First, this is an energy cost for the proper deformation of A^- to the geometry of molecule A [i.e., already without the extra electron (the electron obviously does not like this–this is how it is forced out)]. Next, molecule B is deformed to the geometry of B^- (this is what makes B attractive to the extra electron–everything is prepared for it in B). These two energy effects correspond to λ_i.

Calculation of λ_i is simple:

$$\lambda_i = E(A^-B;\text{ geom } AB^-) - E(A^-B;\text{ geom } A^-B),$$

where $E(A^-B;\text{ geom } AB^-)$ denotes the energy of A^-B calculated for the equilibrium geometry of another species (namely AB^-), while $E(A^-B;\text{ geom } A^-B)$ stands for the energy of A^-B at its optimum geometry.

Usually the geometry changes in AB^- and A^-B attain several percentage points of the bond lengths or the bond angles. The change is therefore relatively small, and we may represent it by a superposition of the normal mode vectors[84] $\boldsymbol{L}_k$, $k = 1, 2, \ldots, 3N$ described in Chapter 7. We may use the normal modes of the molecule A^-B (when we are interested in electron transfer from A^- to B) or of the molecule AB^- (back transfer). Some normal modes are more effective than others in facilitating electron transfer. The normal mode analysis would show[85] that

> the *most effective normal mode of the reactants deforms them in such a way as to resemble the products*. This vibration reorganizes the neighborhood in the desired direction (for electron transfer to occur), and therefore effectively lowers the reaction barrier.

---

[83] "–" denotes the site of the extra electron. It does not necessarily mean that A^- represents an anion.

[84] Yet the normal modes are linear combinations of the Cartesian displacements.

[85] It usually turns out that there are several such vibrations. They will help electron transfer from A^- to B. The reason is obvious; e.g., the empirical formula for V_{RP} says that a vibration that makes the AB distance smaller will increase the transfer probability. This could be visible in what is known as *resonance Raman spectroscopy* close to a charge transfer optical transition. In such spectroscopy, we have the opportunity to observe particular vibronic transitions. The intensity of the vibrational transitions (usually from $v = 0$ to $v = 1$) of those normal modes that facilitate electron transfer will be highest.

Solvent Reorganization Energy

Spectroscopic investigations are unable to distinguish between the internal or solvent reorganization because nature does not distinguish between the solvent and the rest of the neighborhood. An approximation to the solvent reorganization energy may be calculated by assuming a continuous solvent model. Assuming that the mutual configuration of the donor and acceptor (separated by distance R) allows for enclosing them in non-overlapping spheres of radii a_1 and a_2, the following formula was derived by Marcus:

$$\lambda_0 = (\Delta e)^2 \left\{ \frac{1}{2a_1} + \frac{1}{2a_2} - \frac{1}{R} \right\} \left\{ \frac{1}{\epsilon_\infty} - \frac{1}{\epsilon_0} \right\},$$

where ϵ_∞ and ϵ_0 denote the dielectric screening constants measured at infinite and zero electromagnetic field frequency, respectively, and Δe is equal to the effective electric charge transferred between the donor and acceptor. The dielectric screening constant is related to the polarization of the medium. The value ϵ_0 is larger than ϵ_∞, because, at a constant electric field, the electrons as well as the nuclei (mainly an effect of the reorientation of the molecules) keep pace to adjust to the electric field. At high frequency, only the electrons keep pace; hence $\epsilon_\infty < \epsilon_0$. The last parenthesis takes care of the difference; i.e., of the reorientation of the molecules in space (cf. Chapter 12).

Summary

- A chemical reaction represents a molecular catastrophe, in which the electronic structure, as well as the nuclear framework of the system, changes qualitatively. Most often, a chemical reaction corresponds to the breaking of an old and the creation of a new bond.
- The simplest chemical reactions correspond to overcoming a single reaction barrier on the way from reactants to products through a saddle point along the intrinsic reaction coordinate (IRC). The IRC corresponds to the steepest descent trajectory (in the mass-weighted coordinates) from the saddle point to configurations of reactants and products.
- Such a process may be described as the system passing from the entrance channel (reactants) to the exit channel (products) on the electronic energy map as a function of the nuclear coordinates. For reaction $A + BC \rightarrow AB + C$, the map shows a characteristic reaction drainpipe. Passing along the drainpipe bottom usually requires overcoming a reaction barrier, its height being a fraction of the energy of breaking the old chemical bond.
- The reaction barrier reactants→products, is, as a rule, of different height to the corresponding barrier for the reverse reaction.
- We have shown how to obtain an accurate solution for three-atomic reactions. After introducing the hyperspherical democratic coordinates, it is possible to solve the Schrödinger equation (within the Ritz approach). We obtain the rate constant for the state-to-state elementary chemical reaction.

A chemical reaction may be described by the reaction path Hamiltonian in order to focus on the IRC measuring the motion along the drainpipe bottom (reaction path) and the motion orthogonal to the IRC.

- During the reaction, energy may be exchanged between the vibrational normal modes, as well as between the vibrational modes and the motion along the IRC.
- Two atoms or molecules may react in many different ways (reaction channels). Even if in some conditions they do not react (e.g., the noble gases), the reason for this is that their kinetic energy is too low with respect to the corresponding reaction barrier, and the opening of their electronic closed shells is prohibitively expensive in the

energy scale. If the kinetic energy increases, more and more reaction channels open because it is possible for higher and higher energy barriers to be overcome.

- A reaction barrier is a consequence of the "*quasi-avoided crossing*" of the corresponding diabatic hypersurfaces. As a result, we obtain two adiabatic hypersurfaces ("lower," or electronic ground state, and "upper," or electronic excited state). Each of the adiabatic hypersurfaces consists of two diabatic parts stitched along the border passing through the conical intersection point. On both sides of the conical intersection, there are usually two saddle points along the border line leading in general to two different reaction products (Fig. 14.27).
- The two intersecting diabatic hypersurfaces (at the reactant configuration) represent the electronic ground state DA and *that electronic excited state that resembles the electronic charge distribution of the products*–usually D^+A^-.
- The barrier appears, therefore, as the cost of opening the closed shell in such a way as to prepare the reactants for the formation of new bond(s).
- In Marcus electron transfer theory, the barrier also arises as a consequence of the intersection of the two diabatic potential energy curves. The barrier height depends mainly on the (solvent and reactant) reorganization energy.

Main Concepts, New Terms

From the Research Front

Chemical reactions represent a very difficult problem for quantum chemistry for the following reasons:

- There are a lot of possible reaction channels. Imagine the number of all combinations of atoms in a monomolecular dissociation reaction, also in their various electronic states. We have to select first which reaction to choose and a good clue may be the lowest possible reaction barrier.
- A huge change in the electronic structure is usually quite demanding for standard quantum mechanical methods.
- Given a chosen single reaction channel, we confront the problem of calculating the potential energy hypersurface. Let us recall (as detailed in Chapters 6 and 7) the number of quantum mechanical calculations to perform,

this is of the order of 10^{3N-6}. For as small number of nuclei as $N = 4$, we already have a million computation tasks to perform.
- Despite unprecedented progress in the computational technique, the cutting edge possibilities are limited in *ab initio* calculations to two diatomic molecules.

On the other hand, a chemist always has some additional information on which chemical reactions are expected to occur. Very often the most important changes happen in a limited set of atoms e.g., in functional groups, their reactivity being quite well understood. Freezing the positions of those atoms which are reaction spectators only, allows us to limit the number of degrees of freedom to consider.

Ad Futurum

Chemical reactions with the reactants precisely oriented in space will be more and more important in chemical experiments of the future. Here it will be helpful to favor some reactions by supramolecular recognition, docking in reaction cavities or reactions on prepared surfaces. For theoreticians, such control of orientation will mean the reduction of certain degrees of freedom. This, together with eliminating or simulating the spectator bonds, may reduce the task to manageable size. State-to-state calculations and experiments that will describe an effective chemical reaction that starts from a given quantum mechanical state of the reactants and ends up with another well-defined quantum mechanical state of the products will become more and more important. Even now, we may design with great precision practically any sequence of laser pulses (a superposition of the electromagnetic waves, each of a given duration, amplitude, frequency, and phase). For a chemist, this means that we are able to modulate the shape of the hypersurfaces (ground and excited states) in a controllable way because every nuclear configuration corresponds to a dipole moment that interacts with the electric field (cf., Chapter 12). The hypersurfaces may shake and undulate in such a way as to make the point representing the system move to the product region. In addition, there are possible excitations, and the products may be obtained *via* excited hypersurfaces. As a result, we may have selected bonds broken, and others created in a selective and highly efficient way. This technique demands important developments in the field of chemical reaction theory and experiment, because currently we are far from such a goal.

Note that the most important achievements in the chemical reaction theory pertained to ideas (von Neumann, Wigner, Teller, Woodward, Hoffmann, Fukui, Evans, Polanyi, Shaik) rather than computations. The potential energy hypersurfaces are so complicated that it took the scientists 50 years to elucidate their main machinery. Chemistry means first of all chemical reactions, and most chemical reactions still represent unbroken ground. This will change considerably in the years to come. In the longer term this will be the main area of quantum chemistry.

Additional Literature

R. D. Levine and R. B. Bernstein, *Molecular Reaction Dynamics and Chemical Reactivity,* Oxford University Press, Oxford (1987).

An accurate approach to the reactions of small molecules.

H. Eyring, J. Walter, and G. F. Kimball, *Quantum Chemistry,* John Wiley, New York (1967).

A good old textbook written by the outstanding specialists in the field. To my knowledge, no later textbook has covered this subject of quantum chemistry in more detail.

R. B. Woodward and R. Hoffmann, *The Conservation of Orbital Symmetry,* Acadamic Press, New York (1970).

A summary of the important discoveries made by these authors (Woodward-Hoffmann symmetry rules).

S. S. Shaik, "*What happens to molecules as they react? Valence bond approach to reactivity,*" *J. Amer. Chem. Soc., 103,* 3692 (1981).

An excellent paper that introduces many important concepts in a simple way.

Questions

1. The intrinsic reaction coordinate (IRC) represents (with N being the number of the nuclei):
   a. an arbitrary atom-atom distance, that changes from the value for the reactants' configuration to that of the products.
   b. a steepest descent path from the saddle point to the corresponding minima (using a Cartesian coordinate system in the nuclear configuration space).
   c. a steepest descent path from the saddle point to the corresponding minima (within a Cartesian coordinate system of the nuclei scaled by the square roots of the atomic masses).
   d. a curve in the $(3N-6)$ -dimensional space that connects two minima corresponding to two stable structures separated by a barrier.
2. In the vibrational adiabatic approximation within the reaction path Hamiltonian theory,
   a. the potential energy for the motion of the nuclei depends on the frequencies and excitations of the normal modes
   b. the vibrational contribution to the potential energy for the nuclei depends on the value of the coordinate s along the IRC
   c. excitations of some particular vibrational modes may lower the reaction barrier.
   d. no normal mode can change its vibrational quantum number at any s.
3. The donation mode
   a. is the one that offers the largest value of the zero-point energy in the entrance channel.
   b. when excited lowers the reaction barrier
   c. has a large value of the Coriolis coupling with at least one mode in the exit channel
   d. means a mode that has a large value of the curvature coupling in the entrance channel.
4. A spontaneous endothermic reaction proceeds (at $T > 0$), because
   a. the density of the vibrational energy levels is larger in the exit channel than in the entrance channel
   b. the energy of the bottom of the entrance channel is higher than that of the exit channel
   c. what decides about the direction is not the energy, but the free energy
   d. the exit channel is much wider than the entrance channel
5. In the acceptor-donor theory,
   a. the reactants correspond to the DA structure
   b. in the D^+A^- structure, the acceptor's bond becomes weaker, due to the electron occupation of its anti-bonding molecular orbital
   c. the D^+A^- structure represents a low-energy excited state of the products
   d. the reaction barrier comes from an intersection of the diabatic potential energy hypersurfaces for the DA and D^+A^- structures.
6. In the acceptor-donor theory at the intermediate stage of the reaction (I),
   a. an electron of the donor molecule jumps on the acceptor and occupies its antibonding orbital
   b. the structure $D^{2+}A^{2-}$ becomes the dominating one
   c. the dominating structures are D^+A^- and D^+A^{-*}
   d. the dominating structures are DA and D^+A^-
7. At the conical intersection for H_3,
   a. the excited state adiabatic PES increases linearly with the distance from the conical intersection point
   b. the ground state adiabatic PES increases linearly with the distance from the conical intersection point.

  c. the excited state adiabatic PES is composed of two diabatic surfaces corresponding to different electronic distributions
  d. the ground and the excited state adiabatic PESs coincide for any equilateral triangle configuration of H_3.

8. The conical intersection
  a. pertains to relative position (in energy scale) of two electronic states as functions of the configuration of the nuclei
  b. moving on the ground-state PES from its one diabatic part to the other corresponds to a thermally induced chemical reaction
  c. thanks to the conical intersection a UV excitation from the ground state may end up in a non-radiative transition back to the ground state.
  d. an electronic excitation (according to the Franck-Condon rule) of the reactants leads to a product-looking compound having the geometry of the reactants.

9. In the electron transfer theory of Marcus,
  a. the reorganization energy can be calculated as the energy needed to change the geometry of the reactants (from the equilibrium one) to the optimum nuclear configuration of the products
  b. the larger the absolute value of the energy difference between the products and the reactants the faster the reaction
  c. the activation energy is equal to the reorganization energy
  d. when the reactants and the products are the same, the barrier equals to $\frac{1}{4}$ of the reorganization energy.

10. In the electron transfer theory of Marcus,
  a. one assumes the same force constants for the reactants and for the products
  b. the reorganization energy can be calculated as the energy needed to change the geometry of the products (from the equilibrium one) to the optimum nuclear configuration of the reactants
  c. one assumes the harmonic approximation for either of the diabatic states
  d. the reason why there is an energy barrier for the electron transfer in the reaction $Fe^{2+} + Fe^{3+} \longrightarrow Fe^{3+} + Fe^{2+}$ is the reorganization energy of the solvent.

Answers

1c,d, 2a,b,c,d, 3b,d, 4a,c,d, 5a,b,d, 6a,d, 7a,c,d, 8a,b,c,d, 9a,d, 10a,b,c,d

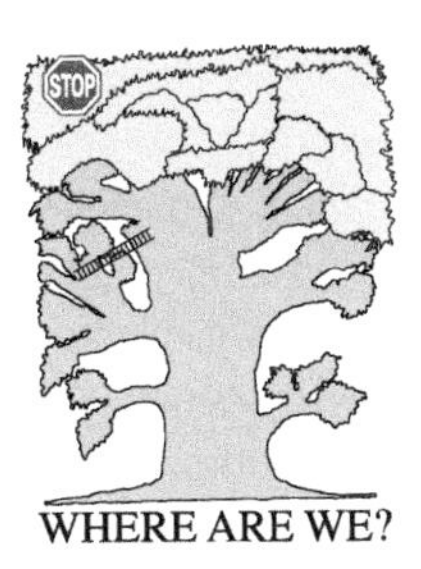

WHERE ARE WE?

Information Processing–The Mission of Chemistry[1]

"Concern for man and his fate
it must always form the chief interest
of all technical endeavors. Never forget
this in the midst of your diagrams and equations"
Albert Einstein

Where Are We?

We are on the top of the TREE crown.

An Example

Information is of key importance, not only for human civilization, but also for the functioning of any biological system. In those systems the hardware and software of information processing is based on chemistry. Thus, molecules, so interesting by themselves for chemists, theoreticians and experimentalists, and which are difficult to describe correctly, participate in their "second life," of a completely different character - they *process information*. In this special life, they cooperate with thousands of other molecules in a precise spatiotemporal algorithm that has its own goal. Do you recall the first words of the Introduction of this book, which talked about thrushes and finches and the program that has been written inside these animals? This book has made a loop. Imagine that molecules are devoid of this "second function," this would mean a disaster for our world.[2] *Despite of its striking importance this* "second life" *is virtually absent in chemistry nowadays,* which does not even have an appropriate language for describing information flows.

[1] This chapter is based in part on a lecture by the author delivered in Warsaw Academic Laser Center on December 7, 1999, as well as on his article in the *Reports of the Advanced Study Institute* of the Warsaw Technical University, 2(2013).

[2] In that case, we would be unable even to notice it.

Ideas of Quantum Chemistry, Second Edition. http://dx.doi.org/10.1016/B978-0-444-59436-5.00015-5

What Is It All About?

Chemistry has played, and continues to play, a prominent role in human civilization. If you doubt it, just touch *any* surface around you – no matter what it is, it probably represents a product of the chemical industry.[3] Pharmaceutical chemistry may be seen as a real benefactor, for it makes our lives longer and more comfortable. Is it the case, therefore, that the mission of chemistry is the production of better dyes, polymers, semiconductors, and drugs? *No, its true mission is much more exciting.*

Why Is This Important?

In this book, we have dealt with many problems of quantum chemistry. If this book were only about quantum chemistry, I would not write it. My goal was to focus on perspectives and images, rather than on pixel-like, separate problems. Before being quantum chemists, we are scientists, happy eyewitnesses of miracles going on around us. We are also human beings and have the right to ask ourselves where we are aiming. *Why* does the Schrödinger equation need to be solved? *Why* does we want to understand the chemical foundations of the world? Just for curiosity? Well, should curiosity legitimize *any* investigation?[4] What will the future role of chemistry be?

Chemistry is on the threshold of a big leap forward. Students of today will participate in this revolution. The limits will be set by our imagination, and maybe by our responsibility as well. The way that future progress in chemistry and biochemistry will be chosen, will determine the fate of human civilization. This *is* important.

[3] Here are just a few random examples in the room I'm sitting in: laptop (polymers), marble table (holes filled with a polymer), pencil (wood, but coated by a polymer), and box of paper tissue (dyes and polymer coat).

[4] Do not answer "*yes*" too quickly because then it gives other people the right to conduct any experiments they want on you and me.

What Is Needed?

- Elements of chemical kinetics
- Elements of differential equations
- Your natural curiosity

Classical Works

The classic papers pertain to three topics that appear at first sight to be unrelated: molecular recognition, differential equations, and information flow. These topics evolved virtually separately within chemistry, mathematics, and telecommunication, and only now[5] have begun to converge. ★ It seems that the first experiment with an oscillatory chemical reaction was reported by Robert Boyle in the 17th century (oxidation of phosphorus). Then several new reports on chemical oscillations were published (including books). *These results did not attract any significant interest in the scientific community because they contradicted the widely known, all-important, and successful equilibrium thermodynamics.* ★ Emil Hermann Fischer was the first to stress the importance of molecular recognition. In the article "*Einfluss der Configuration auf die Wirkung der Enzyme*," published in *Berichte* [*27*, 2,985 (1894)], Fischer used the self-explanatory words "*key-lock*" for the perfect fit of an enzyme and its ligand. ★ In 1903, Jules Henri Poincaré published in *Journal de Mathematiques Pures et Appliques* [*7*, 251 (1881)] an article called "*Mémoire sur les courbes définies par une équation différentielle*," in which he showed that a wide class of two coupled nonlinear differential equations leads to oscillating solutions that tend to a particular behavior *independent of the initial conditions* (called the *limit cycle*). ★ In 1910, Alfred J. Lotka in an article "*Contributions to the theory of chemical reactions,*" published in the *Journal of Physical Chemistry*, *14*, 271, (1910), proposed some differential equations that corresponded to the kinetics of an autocatalytic chemical reaction, and then, with Vito Volterra, gave a differential equation that describes a prey-predator *feedback* (oscillation) known as the Lotka-Volterra model. Chemistry of that time turned out to be non-prepared for such an idea. ★ In another domain, Harry Nyquist published an article called "*Certain factors affecting telegraph speed,*" in *The Bell Systems Technical Journal*, *3*, 324 (1924); and four years later, in the same journal [*7*, 535 (1928)], Ralph V.L. Hartley published a paper called "*Transmission of information,*" in which for the first time, the quantitative notion of information and the effectiveness of information transmission have been considered. ★ Twenty years later, the same topic was developed by Claude E. Shannon in "*A mathematical theory of communication,*" which was also published in *The Bell Systems Technical Journal*, *27*, 379 and 623 (1948), in which he related the notion of information and that of entropy. ★ The Soviet general Boris Belousov finally agreed to publish his only unclassified paper, "*Periodichesky deystvouyoushchaya rieakcya i yeyo miekhanism*," in an obscure Soviet medical journal called *Sbornik Riefieratow Radiacjonnoj Miediciny*, *Medgiz, Moskwa*, *1*, 145, (1959), reporting spectacular color oscillations in his test tube: yellow Ce^{4+} and then colorless Ce^{3+}, and again yellow, etc. (today called the Belousov-Zhabotinsky reaction). Information about this oscillatory reaction diffused to Western science in the 1960s and made a real breakthrough. ★ Belgian scientists Ilya Prigogine and Gregoire Nicolis in a paper called "*On symmetry breaking instabilities in dissipative systems,*" published in *Journal of Chemical Physics*, *46*, 3542, (1967), introduced the notion of the dissipative structures. ★ Charles John Pedersen reopened (after the pioneering work of Emil Fischer) the field of supramolecular chemistry, publishing an article called "*Cyclic polyethers and their complexes with metal salts,*" which appeared in the *Journal of the American Chemical Society*, *89*, 7017, (1967), and dealt with molecular recognition (cf. Chapter 13). This, together with later works of Jean-Marie Lehn and Donald J. Cram, introduced the new paradigm of chemistry known as *supramolecular chemistry* ★ Manfred Eigen and Peter Schuster, in three articles "*The hypercycle. A principle of natural self-organization,*" in *Naturwissenschaften* [11 (1977), and 1 and 7 (1978)] introduced to chemistry the idea of the hypercycles and of the natural selection of molecules. ★ Mathematician Leonard Adleman published in *Science*, [266, 1021 (1994)] an article called "*Molecular computation of solutions to combinatorial problems,*" in which he described *his own* chemical experiments that shed new light on the role that molecules can play in processing information. ★ Ivan Huc and Jean-Marie Lehn in a paper called "*Virtual combinatorial libraries: Dynamic generation of*

[5] The aim of this chapter is to highlight these connections.

molecular and supramolecular diversity by self-assembly," published in *Proceedings of the National Academy of Sciences (USA)*, *94*, 2106 (1997), stressed importance in chemistry of the idea of a molecular library as an easy-to-shift equilibrium *mixture* of molecular complexes, which was in contradiction with usual chemical practice of purifying substances.

What are the most important problems in chemistry? Usually we have no time to compose such a list, not to speak of presenting it to our students. The choice made reflects the author's personal point of view. The author tried to keep in mind that he is writing for mainly young (undergraduate and graduate) students, who are seeking not only for detailed research reports, but also for new *guidelines* in chemistry, for some general *trends* in it, and who want to establish strong and general *links* among mathematics, physics, chemistry, and biology. An effort was made to expose the ideas, not only to students' minds but also to their hearts.

It is good to recall from time to time that all of us: physicists, chemists, and biologists share the same electrons and nuclei as the objects of our investigation. It sounds trivial, but sometimes there is an impression that these disciplines investigate three different worlds. In the triad physics-chemistry-biology, chemistry plays a bridging role. By the middle of the twentieth century, chemistry had closed the period of the exploration of its basic building blocks: elements, chemical bonds and their typical lengths, and typical values of angles between chemical bonds. Future discoveries in this field are not expected to change our ideas fundamentally. Now we are in a period of using this knowledge for the construction of what we only could dream of. In this chapter, I will refer now and then to mathematicians and mathematics, who deal with ideal worlds. For some strange reason, at the foundation of (almost[6]) everything, there are logic and mathematics. Physics, while describing the real rather than the ideal world, more than other natural sciences is symbiotic with mathematics.

15.1 Multilevel Supramolecular Structures (Statics)

15.1.1 Complex Systems

Even a relatively simple system (e.g., an atom) often exhibits strange properties. Understanding simple objects seemed to represent a key for describing complex systems (e.g., molecules). Complexity *can* be explained using the first principles.[7] However, the complexity itself may add some important features. In a complex system, some phenomena may occur, which would be extremely difficult to foresee from knowledge of their component parts. Most important, sometimes the behavior of a complex system is universal; i.e., it is independent of the properties of the parts of which it is composed and related to the very fact that the system consists of many small parts interacting in a simple way.

[6] Yes, we mean almost: e.g., generosity is not included here.

[7] In the 1920s after presenting his equation (see Chapter 3), Paul Dirac said that now chemistry is explained. Yet, the journey from the equation to foreseeing the function of the ribosome in the human body is a long, long way.

The behavior of a large number of argon atoms represents a difficult task for theoretical description, but is still quite predictable. When the number of atoms increases, they pack together in compact clusters similar to those that we would have with the densest packing of tennis balls (the maximum number of contacts). We may be dealing with complicated phenomena (similar to chemical reactions) that is connected to the different stability of the clusters (e.g., "magic numbers" related to particularly robust closed shells[8]). Yet, the interaction of the argon atoms, however difficult for quantum mechanical description, comes from the quite primitive two-body, three-body etc. interactions (as discussed in Chapter 13).

15.1.2 Self-Organizing Complex Systems

Chemistry offers a wealth of intermolecular interactions.

Some intermolecular interactions are specific; i.e., a substrate A interacts with a particular molecule B_i from a set $B_1, B_2, \ldots B_N$ (N is large) much more strongly than with others. The reasons for this are their shape, the electric field[9] fitness, a favorable hydrophobic interaction, etc., resulting either in the "key-lock", template or "hand-glove" types of interaction (cf. Chapter 13). A molecule may provide a set of potential contacts localized in space (called a *synthon*), which may fit to another synthon of another molecule.

This idea is used in supramolecular chemistry.[10] Suppose that a particular reaction does not proceed with sufficient yield. Usually the reason is that, to run just this reaction, the molecules have to find themselves in a very specific position in space (a huge entropy barrier to overcome), but before this happens, they undergo some unwanted reactions. We may however "instruct" the reactants by substituting them with synthons such that the latter lock the reactants in the right position in space. The reaction that we want to happen becomes inevitable. The driving force for all this is the particularly high interaction energy of the reactants. Very often, however, the interaction energy has to be high, but not too high, in order to enable the reaction products to separate. This reversibility is one of the critically important features for "intelligent" molecules, which could adapt to external conditions in a flexible way.

If the system under consideration is relatively simple, even if the matching of corresponding synthons is completed, we would still have a relatively primitive spatial structure. However, we may imagine a far more interesting situation, when the following happens:

- The molecules were prepared in such a way as to assure *some* intermolecular interaction is particularly attractive. A specific matching is called *molecular recognition.*

---

[8] Similar closed shells are observed in nuclear matter, where the "tennis balls" correspond to nucleons.

[9] Both molecules carry their charge distributions, and their interaction at a certain geometry may considerably lower the Coulombic energy.

[10] C.J. Pedersen, *J.Am.Chem.Soc. 89*, 2495 and 7017 (1967); B. Dietrich, J.-M. Lehn, and J.-P. Sauvage, *Tetrahedron Lett.* 2885 and 2889 (1969), D.J. Cram, J.M. Cram, *Science (Washington), 183*, 803 (1974).

- The molecular complexes formed this way may recognize themselves again by using synthons previously existing or created in situ. In this way, a multilevel structure can be formed, each level characterized by its own stability.
- The multilevel molecular structure may depend very strongly on its environment. When this changes, the structure may decompose, and eventually another structure may emerge.

Therefore,

a *hierarchical* multilevel structure is formed, where the levels may exhibit different stabilities with regard to external perturbations.

An example is shown in Fig. 15.1.

There is nothing accidental in this system. The helices are composed of amino acids that ensure that the external surface is hydrophobic, and therefore they easily enter the hydrophobic lipid bilayer of the cell walls. The peptide links serve to *recognize* and dock some particular signaling molecules. Such seven-helix systems serve in biology as a universal sensor, with variations to make it specific for some particular molecular recognition and the processes that occur afterward. After docking with a ligand or by undergoing photochemical isomerization of the retinal, some conformational changes take place, which after involving several intermediates, finally result in a signal arriving at a nerve cell. We see what this structure is able to do in dynamics, not statics.

15.1.3 Cooperative Interactions

Some events may cooperate. Suppose that we have an extended object, which may undergo a set of events: A,B,C,..., each taking place separately and locally with a small probability. However, it may happen that for a less extended object, the events cooperate; i.e., event A makes it easier for event B to occur, and when A and then B happens, this makes it easier for event C to happen, etc.

Self-organization is possible without cooperativity, but cooperativity may greatly increase its effectiveness. The hemoglobin molecule may serve as an example of cooperativity in intermolecular interactions, where its interaction with the first oxygen molecule makes its interaction with the second easier, despite a considerable separation of the two binding events in space.

An example is shown in Fig. 15.2.

A schematic Fig. 15.2a shows a perfect molecular recognition of an endogenic (i.e., functioning in living organism, native) ligand by the receptor, followed by releasing a signaling molecule. Fig. 15.2b shows an agonist type of a ligand: its functioning is the same as that of the native ligand, but it differs from the latter. Finally, Fig. 15.2c shows an imperfect recognition: the ligand (antagonist) binds, but it fails to release the signaling molecule.

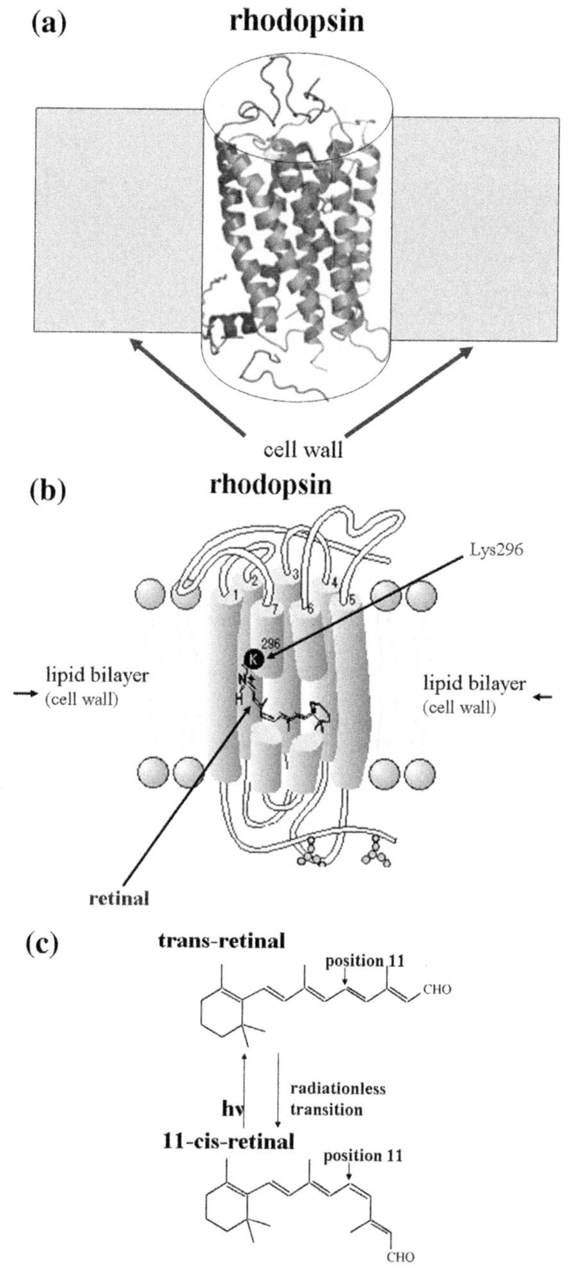

Fig. 15.1. A universal biological sensor based on rhodopsin (a protein), a schematic view. (a) The sensor consists of seven α-helices (shown here as ribbons) connected in a sequential way by some oligopeptide links. The molecule is anchored in the cell wall (lipid bilayer), due to the hydrophobic effect: the rhodopsin's lipophilic amino acid residues are distributed on the rhodopsin surface. (b) The α–helices (this time shown for simplicity as cylinders) form a cavity. Some of the cylinders have been cut out to display a cis-retinal molecule bound (in one of the versions of the sensor) to the amino acid 296 (lysine denoted as K, in helix 7). (c) The cis-retinal (a chain of alternating single and double bonds) is able to absorb a photon and change its conformation to trans (at position 11). This triggers the cascade of processes responsible for our vision. The protruding protein loops exhibit specific interactions with some drugs. Such a system is at the basis of the interactions with about 70% of drugs.

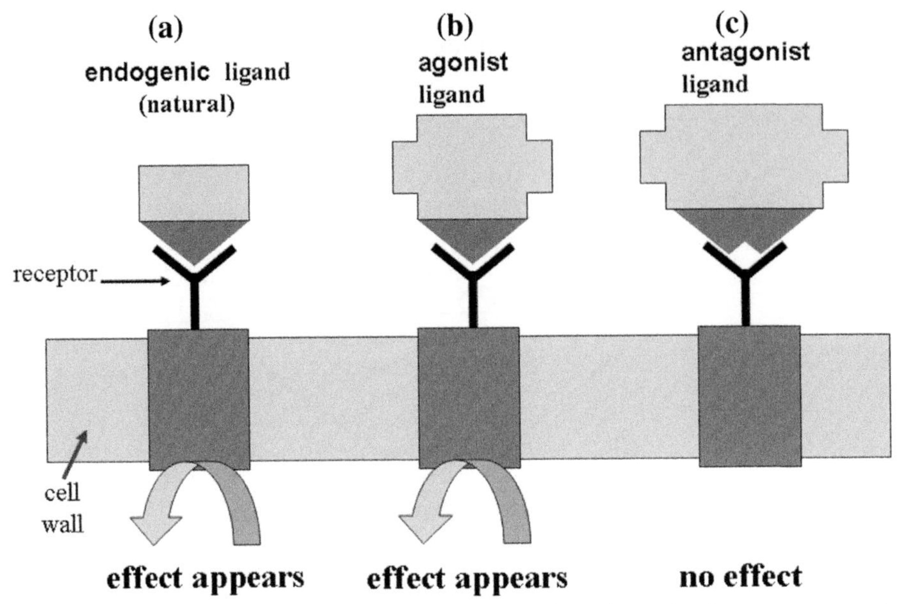

Fig. 15.2. Some examples of cooperativity (appearance of a phenomenon makes easier another phenomenon to appear). (a) Molecular recognition of synthons (upper side, endogenic ligand-receptor) causes an effect (e.g., releasing of a signaling molecule). (b) A similar perfect molecular recognition, where the effect appears, despite the fact that the ligand is not endogenic (this is known as agonist–its shape differs from the natural, endogenic one). (c) An example of antagonist: the recognition is so imperfect that the effect does not appear, although the ligand is blocking the receptor.

15.1.4 Combinatorial Chemistry–Molecular Libraries

Jean-Marie Lehn (b.1939), French chemist and professor of the University of Strasbourg. Lehn, together with Pedersen and Cram (all three received the Nobel Prize in 1987), changed the paradigm of chemistry by stressing the importance of molecular recognition, which developed into the field he termed supramolecular chemistry. By proposing dynamic molecular libraries, Lehn broke with another longstanding idea that of pure substances as the only desirable products of chemical reaction, by stressing the potential of diversity and instructed mixtures. *Courtesy of Professor Jean-Marie Lehn.*

Chemistry is often regarded as dealing with pure, uniquely defined substances, which is obviously a very demanding area. There are cases, however, when a chemist is interested in a mixture of all possible isomers instead of a single isomer or in a mixture of components that can form different complexes.

A complex system in a labile equilibrium may adjust itself to an external stimulus by changing its molecular composition. This is known as the *dynamic combinatorial library*,[11] but in fact it corresponds to quasi-dynamics, for we are interested mainly in shifting equilibrium

[11] I.Huc., and J.-M.Lehn, *Proc.Natl Acad.Sci.(USA)*, *94*, 2106 (1997).

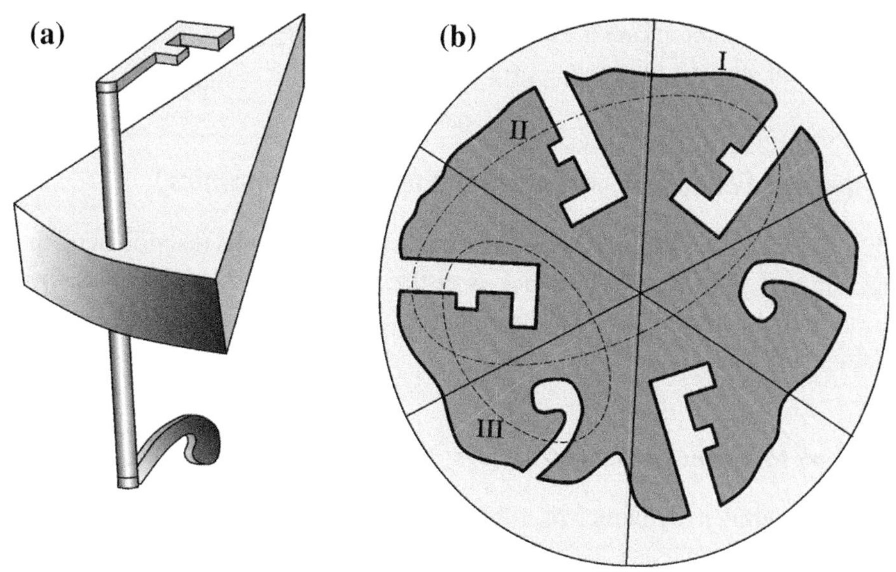

Fig. 15.3. A model of the immune system. (a) This image shows schematically some monomers in a solvent. They have the shape of a slice of pie with two synthons: protruding up and protruding down and differing in shape. The monomers form some side-by-side aggregates containing from two to six monomers, each aggregate resulting in some pattern of synthons on one face and the complementary pattern on the other face. Therefore we have a library of all possible associates in the thermodynamical equilibrium. There are plenty of monomers, a smaller number of dimers, even fewer trimers, etc., up to a tiny concentration of hexamers. (b) The attacking factor I (the irregular body shown) is best recognized and bound by one of the hexamers. If the concentration of I is sufficiently high, the *equilibrium among the aggregates shifts towards the hexamer mentioned above, which therefore binds all the molecules of I*, making them harmless. If the attacking factor were II and III, binding could be accomplished with some trimers or dimers (as well as some higher aggregates). The defense is highly specific and at the same time highly flexible (adjustable).

due to an external stimulus. Liquid water may be regarded as a molecular combinatorial library of various clusters, all of them being in an easy-to-shift equilibrium. This is why water is able to hydrate a nearly infinite variety of molecular shapes shifting the equilibrium toward the clusters that are just needed to wrap the solute by a water coat.

The immune system in our body is able to fight a lot of enemies and win, regardless of their shape and molecular properties (charge distribution). How is it possible? Would the organism be prepared for everything? Well, yes and no. Let us imagine a system of molecules (building blocks) having some synthons and able to create some van der Waals complexes (see Fig. 15.3). Since the van der Waals forces are quite weak, the complexes are in dynamic equilibrium. All complexes are present in the solution, but possibly none of the complexes dominates.

Now, let us introduce some "enemy-molecules." The building blocks use part of their synthons for binding the enemies (that have complementary synthons) and at the same time bind among themselves in order to strengthen the interaction. Some of the complexes are especially effective in this binding. Now, the Le Chatelier rule comes into the picture, and the equilibrium shifts to produce as many of the most effective binders as possible. On top of this, the most effective binder may undergo a chemical reaction that replaces the weak van der Waals forces by strong chemical forces (the reaction rate is enhanced by the supramolecular interaction). The enemy was tightly secured, and the invasion is over.[12]

15.2 Chemical Feedback–A Steering Element (Dynamics)

Steering is important to maintain the stability of the system and to its reaction upon the external conditions imposed. The idea of feedback is at the heart of any steering, since it is used to control the output of a device to correct its input data. In a sense, the feedback means also a use of information about itself.

15.2.1 A Link to Mathematics–Attractors

Systems often exhibit a dynamic, or time-dependent, behavior (chemical reactions, non-equilibrium thermodynamics).

Dynamic systems have been analyzed first in mathematics. When applying an iterative method of finding a solution to an equation, one first decides which operation is supposed to bring us closer to the solution, as well as what represents a reasonable zero-order guess (a starting point being a number, a function, or a sequence of functions). Then one forces an evolution ("dynamics") of the approximate solutions by applying the operation first to the starting point, then to the result obtained by the operation on the starting point, and then again and again until the convergence is achieved.

Mitchell Feigenbaum (b.1944), American physicist, employee of the Los Alamos National Laboratory, and then professor at the Cornell University and at the Rockefeller University. Feigenbaum discovered attractors after making some observations just playing with a pocket calculator.

He is also known for discovering some strange regularity in successive period-doubling bifurcation processes.

As an example, look at the equation $\sin(x^2 + 1) - x = 0$. There is an iterative way to solve this equation numerically: $x_{n+1} = \sin(x_n^2 + 1)$, where n stands for the iteration number. The iterative scheme means choosing any x_0, and then applying many times a sequence of four keys on the calculator keyboard (square, +, 1, sin). The *result* (0.0174577) *is independent of the starting point chosen*. The number 0.0174577 represents an attractor or a *fixed point* for the operation. As a chemical analog of the fixed point may serve as the thermodynamic equilibrium of a system (e.g., dissolving a substance in water), the same can be attained from any starting point (e.g., various versions of making solutions).

[12] A simple model of immunological defense, similar to that described here, was proposed by F. Cardull, M. Crego Calama, B.H.M. Snelling-Ruël, J.-L. Weidmann, A. Bielejewska, R. Fokkens, N.M.M. Nibbering, P. Timmerman, and D.N. Reinhoudt, *J.Chem.Soc.Chem.Commun.* 367 (2000).

There are other types of attractors. In 1881, Jules Henri Poincaré has shown that a wide class of two coupled nonlinear differential equations leads to the solutions that tend to a particular oscillatory behavior *independently of the initial conditions* (*limit cycle*).

Jules Henri Poincaré (1854–1912), French mathematician and physicist and professor at the Sorbonne, made important contributions to the theory of differential equations, topology, celestial mechanics, probability theory, and the theory of functions. Known as the "last universalist," Poincaré was excellent in virtually all domains of his time. Maybe the exceptions were music and physical education, where he was scored as, average at best.

Let us take an example of a set of two coupled differential equations:

$$\dot{r} = \mu r - r^2,$$
$$\dot{\theta} = \omega + br^2$$

(a dot over a symbol means a derivative with respect to time t), where r and θ are polar coordinates on a plane, (Fig. 15.4) and $\omega > 0, \mu > 0, b > 0$ are constants. When r is small, one may neglect the r^2 term with respect to μr. The resulting equation $\dot{r} = \mu r > 0$ means

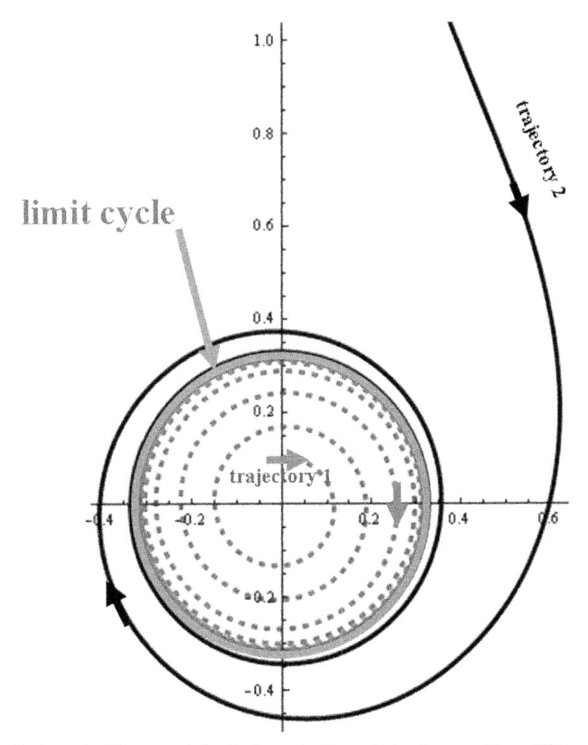

Fig. 15.4. An example of a limit cycle. Two special solutions (trajectories) of a set of two differential equations. Trajectories 1 (dashed line) and 2 (solid line) start from a point close to the origin and from a point very distant from the origin, respectively. As $t \to \infty$, the two trajectories merge into a single one (limit cycle).

that r starts to *increase* (while bringing the angular velocity $\dot{\theta} = \omega$). However, when r gets large enough, the $-r^2$ term starts to come into play, and this means an increasing tendency to *diminish* r. As a result, a compromise is achieved and one gets a stable trajectory.

Whatever the starting r and θ are, the trajectory tends to be at the circle of radius $r = \sqrt{\mu}$, and is rotating at a constant velocity $\dot{\theta} = \omega + b\mu$. This circle represents a limit cycle. Thus, *a limit cycle may be viewed as a feedback, a prototype of any steering, also a chemical one.*

The steering of chemical concentrations is at the heart of how the biological systems control the concentrations of thousands of substances.

This feedback-type behavior has been first considered in the domain of mathematics, with explicit targeting chemistry. In 1910, Alfred Lotka proposed some differential equations that corresponded to the kinetics of an autocatalytic chemical reaction,[13] and then with Vito Volterra derived a differential equation that describes a general feedback mechanism (oscillations) known as the *Lotka-Volterra model*. However, chemistry has not been ready yet for this link.

15.2.2 Bifurcations[14] and Chaos

Nonlinear dynamics turned out to be extremely sensitive to coupling with some external parameters (representing the "*neighborhood*").

Let us take what is called the *logistic equation*:

$$x = Kx(1 - x),$$

where $K > 0$ is a constant. This is a quadratic equation, and there is no problem with solving it by the traditional method. However, here we will focus *on an iterative scheme*:

$$x_{n+1} = Kx_n(1 - x_n),$$

which is obviously related to the iterative solution of the logistic equation. The biologist Robert May gave a numerical exercise to his Australian graduate students. They had to calculate how a rabbit population evolves when we let it grow according to the rule $x_{n+1} = Kx_n(1 - x_n)$, where the natural number n denotes the current year, while x_n stands for the (relative) population of, say, rabbits in a field, such that $0 \le x_n \le 1$. The number of the rabbits in year $(n + 1)$ is proportional to their population in the preceding year (x_n), because they reproduce very fast, but the rabbits eat grass, and the field has a finite size. The larger x_n is, the less the amount of

[13] A.J. Lotka, *J. Phys. Chem.*, 14, 271 (1910).

[14] A bifurcation (corresponding to a parameter p) denotes in mathematics a doubling of an object when the parameter exceeds a value p_0. For instance, when the object corresponds to the number of solutions of equation $x^2 + px + 1 = 0$, then the bifurcation point $p_0 = 2$. Another example of bifurcation is branching of roads, valleys, etc.

grass there is to eat, which makes the rabbits a bit weaker and less able to reproduce (this effect corresponds to $1 - x_n$).

The logistic equation contains a feedback mechanism.

The constant K measures the population-grass coupling strength (low-quality grass means a small K). What interests us is the fixed point of this operation (i.e., the final population the rabbits develop after many years at a given coupling constant K). For example, for $K = 1$, the evolution leads to a steadily self-reproducing population x_0, and x_0 depends on K (the larger K is, the larger x_0 is). The graduate students took various values of K. *Nobody* imagined that this quadratic equation could be hiding a mystery.

If K were small ($0 \leq K < 1$, extremely poor grass), the rabbit population would simply vanish (as shown in the first part of Fig. 15.5). If K increased (the second part of the plot, $1 \leq K < 3$), the population would flourish. When K exceeded 3, this flourishing would give a unexpected twist: instead of reaching a fixed point, the system would oscillate between two

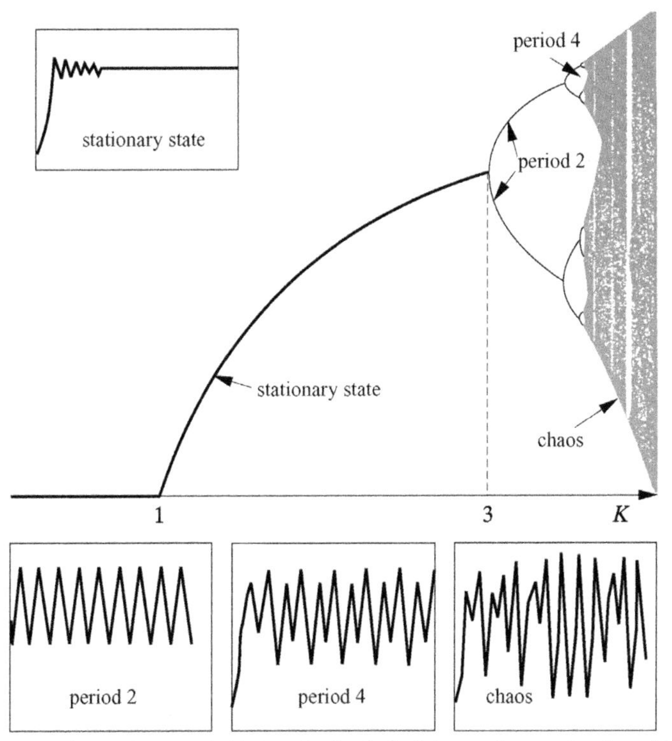

Fig. 15.5. The diagram of the fixed points and the limit cycles for the logistic equation as a function of the coupling constant K. From J. Gleick, *Chaos*, Viking, New York (1988). *Reproduced with permission of the author.*

sizes of the population (every second year, the population was the same, but two consecutive years have different populations). This resembles the limit cycle described above–the system just repeats the same cycle all the time.

This mathematical phenomenon was carefully investigated and the results were really amazing (Fig. 15.5). Further increase in K introduces further qualitative changes. First, for $3 \leq K < 3.44948$, the oscillations have period two (*bifurcation*), then at $3.44948 \leq K < 3.5441$, the oscillations have period four (next bifurcation, the four-member limit cycle), and then for $3.5441 \leq K < 3.5644$, the period is eight (next bifurcation).[15]

Then, the next surprise: exceeding $K = 3.56994$, we obtain populations that do not exhibit any regularity (no limit cycle, or just *chaos*). A further surprise is that this is not the end of the surprises. Some sections of K began to exhibit *odd*-period behavior, separated by some sections of chaotic behavior.

15.2.3 Brusselator Without Diffusion

Could we construct chemical feedback? Why would we want to do that? Those who have ever seen feedback working know the answer[16]–this is the very basis of control. Such control of chemical concentrations is at the heart of how biological systems operate.

Ilya Prigogine (1917–2003) Belgian physicist and professor at the Université Libre de Bruxelles. In 1977, he received the Nobel prize *"for his contributions to non-equilibrium thermodynamics, particularly the theory of dissipative structures."*

The dissipative structures appear in a medium as a result of matter and energy fluxes in open systems.

The first idea is to prepare such a system in which an increase in the concentration of species X triggers the process of its decreasing. The decreasing occurs by replacing X by a very special substance Y, each molecule of which, when disintegrating, produces several X molecules. Thus, we would have a scheme (X denotes a large concentration of X, and x is a small concentration of X; and this situation is similar for the species Y): (X, y) $\rightarrow$ (x, Y) $\rightarrow$ (X, y) or *oscillations of the concentration of X and Y in time.*

---

[15] Mitchell Feigenbaum was interested to see at which value $K(n)$ the next bifurcation into 2^n branches occurs. It turned out that there is a certain regularity–namely, $\lim_{n\to\infty} \frac{K_{n+1}-K_n}{K_{n+2}-K_{n+1}} = 4.669201609\ldots \equiv \delta$. To the astonishment of scientists, the value of δ turned out to be "*universal*"; i.e., characteristic for many *very different* mathematical problems and, therefore, reached a status similar to that of the numbers π and e. The numbers π and e satisfy the exact relation $e^{i\pi} = -1$, but so far, no similar relation was found for the Feigenbaum constant. There is an *approximate* relation (used by physicists in phase transition theory) which is satisfied: $\pi + \tan^{-1} e^{\pi} = 4.669201932 \approx \delta$.

[16] For example, an oven heats until the temperature exceeds an upper bound, then it switches off. When the temperature reaches a lower bound, the oven switches *itself* on (therefore, we have temperature *oscillations*).

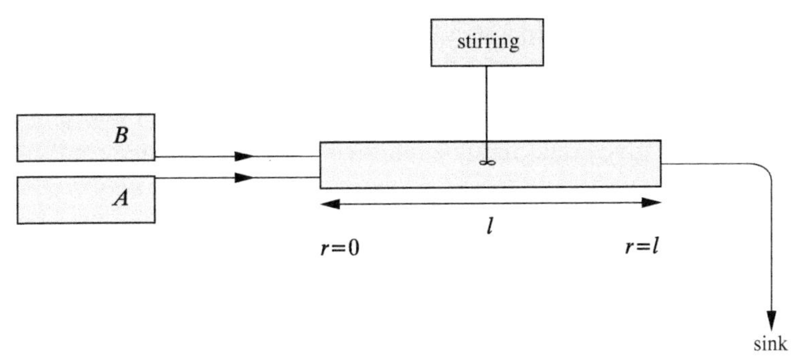

Fig. 15.6. A flow reactor (a narrow tube used to make a 1-D description possible) with stirring. The concentrations of A and B are kept constant at all times (the corresponding fluxes are constant).

Imagine that we carry out a complex chemical reaction in flow conditions[17]; i.e., the reactants A and B are pumped with a constant speed into a long, narrow tube reactor, there is intensive stirring in the reactor, then the products flow out to the sink (Fig. 15.6). After a while, a steady state is established.[18]

After A and B are supplied, the substances[19] X and Y appear, which play the role of catalysts; i.e., they participate in the reaction, but in total, their amounts do not change. To model such a situation, let us assume the following chain of chemical reactions[20]:

$$\begin{aligned} \mathrm{A} &\rightarrow \mathrm{X} \\ \mathrm{B}+\mathrm{X} &\rightarrow \mathrm{Y}+\mathrm{D} \\ 2\mathrm{X}+\mathrm{Y} &\rightarrow 3\mathrm{X} \\ \mathrm{X} &\rightarrow \mathrm{E} \\ \text{in total} &: \\ \mathrm{A}+\mathrm{B}+4\mathrm{X}+\mathrm{Y} &\rightarrow \mathrm{D}+\mathrm{E}+4\mathrm{X}+\mathrm{Y} \end{aligned}$$

This chain of reactions satisfies our feedback postulates. In step 1, the concentration of X increases; in step 2, Y is produced at the expense of X; in step 3, substance Y enhances the production of X (at the expense of itself–this is an *autocatalytic step*); then again X transforms to Y (step 2), etc.

If we shut down the fluxes in and out, a thermodynamic equilibrium is attained after a while with all the concentrations of the six substances (A,B,D,E,X,Y; their concentrations will be denoted as A, B, D, E, X, Y, respectively) being constant in space (along the reactor) and time. On the other hand, when we fix the in and out fluxes to be constant (but nonzero) for a

[17] Such reaction conditions are typical in industry.

[18] This is distinguished from the thermodynamic equilibrium state, where the system is isolated (no energy or matter flows).

[19] These substances appear due to the chemical reactions running.

[20] See, e.g., A.Babloyantz, *Molecules, Dynamics and Life*, Wiley, New York (1987).

long time, we force the system to be in a steady state and as far from thermodynamic equilibrium as we wish. In order to simplify the kinetic equations, let us assume the irreversibility of all the reactions considered (as shown in the reaction equations above) and put all the velocity constants equal to 1. This gives the kinetic equations for what is called the *brusselator model* (of the reactor):

$$\frac{dX}{dt} = A - (B+1)X + X^2Y$$
$$\frac{dY}{dt} = BX - X^2Y \tag{15.1}$$

These two equations, plus the initial concentrations of X and Y, totally determine the concentrations of all the species as functions of time (due to the stirring, there will be no dependence on the position in the reaction tube).

Steady State

A steady state (at constant fluxes of A and B) means $\frac{dX}{dt} = \frac{dY}{dt} = 0$, and therefore, we easily obtain the corresponding steady-state concentrations X_s, Y_s by solving Eq. (15.1)

$$0 = A - (B+1)X_s + X_s^2 Y_s$$
$$0 = BX_s - X_s^2 Y_s.$$

You can see that these equations are satisfied by

$$X_s = A,$$
$$Y_s = \frac{B}{A}.$$

Evolution of Fluctuations from the Steady State

Any system undergoes some spontaneous concentration fluctuations, or we may perturb the system by injecting a small amount of X and/or Y. What will happen to the stationary state found a while before, if such a fluctuation happens?

We have fluctuations x and y from the steady state:

$$X(t) = X_s + x(t),$$
$$Y(t) = Y_s + y(t). \tag{15.2}$$

What will happen next?

After inserting Eq. (15.2) into Eq. (15.1) we obtain the equations describing how the fluctuations evolve in time:

$$\frac{dx}{dt} = -(B+1)x + Y_s(2X_s x + x^2) + y(X_s^2 + 2xX_s + x^2),$$
$$\frac{dy}{dt} = Bx - Y_s(2X_s x + x^2) - y(X_s^2 + 2xX_s + x^2). \tag{15.3}$$

Since a mathematical theory for arbitrarily large fluctuations does not exist, we will limit ourselves to small x and y. Then, all the quadratic terms of these fluctuations can be neglected [the *linearization* of Eq. (15.3)]. We obtain

$$\frac{dx}{dt} = -(B+1)x + Y_s(2X_s x) + yX_s^2,$$
$$\frac{dy}{dt} = Bx - Y_s(2X_s x) - yX_s^2. \tag{15.4}$$

Let us assume fluctuations of the form[21]

$$x = x_0 \exp(\omega t),$$
$$y = y_0 \exp(\omega t). \tag{15.5}$$

and represent particular solutions of Eqs. (15.4) provided the proper values of ω, x_0 and y_0 are chosen. After inserting Eq. (15.5) into Eq. (15.4), we obtain the following set of equations for the unknowns ω, x_0, and y_0:

$$\omega x_0 = (B-1)x_0 + A^2 y_0$$
$$\omega y_0 = -Bx_0 - A^2 y_0. \tag{15.6}$$

This represents a set of homogeneous linear equations with unknown to x_0 and y_0. This means that we have to ensure that the determinant, composed of the coefficients multiplying the unknowns x_0 and y_0, vanishes (cf., secular equation, p. 238):

$$\begin{vmatrix} \omega - B + 1 & -A^2 \\ B & \omega + A^2 \end{vmatrix} = 0.$$

This equation is satisfied by some special values[22] of ω:

$$\omega_{1,2} = \frac{T \pm \sqrt{T^2 - 4\Delta}}{2}, \tag{15.7}$$

where

$$T = -\left(A^2 - B + 1\right), \tag{15.8}$$
$$\Delta = A^2. \tag{15.9}$$

[21] Such a form allows for very versatile behavior: exponential growth ($\omega > 0$), decaying ($\omega < 0$) or staying constant ($\omega = 0$), as well as for periodic behavior ($\mathrm{Re}\omega = 0, \mathrm{Im}\omega \neq 0$), quasiperiodic growth ($\mathrm{Re}\omega > 0, \mathrm{Im}\omega \neq 0$), or decay ($\mathrm{Re}\omega < 0, \mathrm{Im}\omega \neq 0$).

[22] They represent an analog of the normal mode frequencies from Chapter 7.

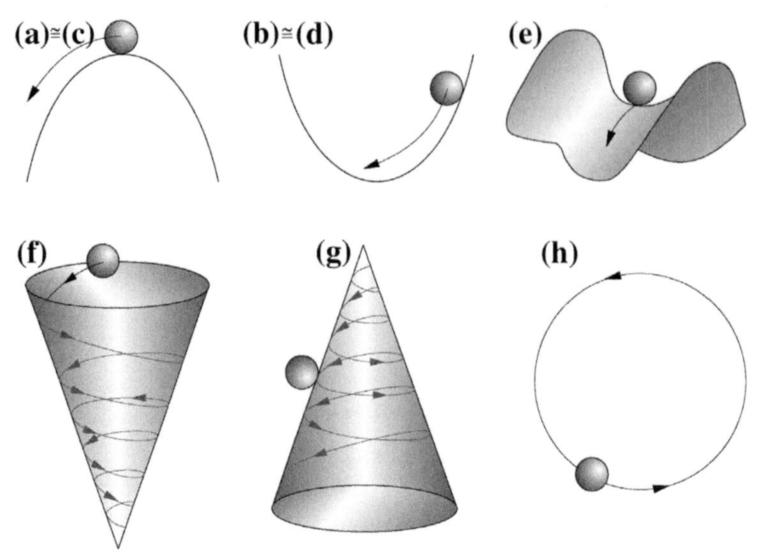

Fig. 15.7. Evolution types of fluctuations from the reaction steady state. The classification is based on the numbers ω_1 and ω_2 of Eq. (15.7). The individual figures correspond to the rows of Table 15.1. The behavior of the system (in the space of chemical concentrations) resembles the sliding of a point or rolling of a ball over certain surfaces in a gravitational field directed downward, as follows: (a) The unstable node resembles sliding from the top of a mountain. (b) The stable node resembles moving inside a bowl-like shape. (c) The unstable stellar node is similar to case (a), with a slightly different mathematical reason behind it. (d) A similar situation exists for the stable stellar node [resembles case (b)]. (e) Saddle–the corresponding motion is similar to a ball rolling over a cavalry saddle. (f) Stable focus–the motion resembles rolling a ball over the interior surface of a cone pointing downward. (g) Unstable focus–a similar rolling motion, but on the external surface of a cone that points up. (h) Center of marginal stability corresponds to a circular motion (oscillation).

Table 15.1. Fluctuation stability analysis (i.e., what happens if the concentrations undergo a fluctuation from the steady state values). The analysis is based on the values of ω_1 and ω_2 from Eq. (15.7); they may have real (subscript r) as well as imaginary (subscript i) parts, hence: $\omega_{r,1}, \omega_{i,1}, \omega_{r,2}, \omega_{i,2}$.

T	Δ	$T^2 - 4\Delta$	$\omega_{r,1}$	$\omega_{i,1}$	$\omega_{r,2}$	$\omega_{i,2}$	**Stability**
+	+	+	+	0	+	0	Unstable node
−	+	+	−	0	−	0	Stable node
−	+	0	−	0	−	0	Stable stellar node
+	+	0	+	0	+	0	Unstable stellar node
−	+	−	−	$i\omega$	−	$-i\omega$	Stable focus
+	+	−	+	$i\omega$	+	$-i\omega$	Unstable focus
0	+	−	0	$i\omega$	0	$-i\omega$	Center of marginal stability

Fluctuation Stability Analysis

Now it is time to pick the fruits of our hard work.

The way that the fluctuations depend on time is characterized by the roots $\omega_1(t)$ and $\omega_2(t)$ of Eq. (15.7), because x_0 and y_0 are nothing but some constant amplitudes of the changes. We have the following possibilities (see Fig. 15.7 and Table 15.1):

- Both roots are real, which happens only if $T^2 - 4\Delta \geq 0$. Since $\Delta > 0$, the two roots are of the same sign (sign of T). If $T > 0$, then both roots are positive, which means that the fluctuations $x = x_0 \exp(\omega t)$, $y = y_0 \exp(\omega t)$ *increase over time and the system will never return to the steady state (unstable node)*. Thus, the steady state represents a *repeller* of the concentrations X and Y
- If, as in the previous case at $T^2 - 4\Delta \geq 0$, but this time $T < 0$, then both roots are negative, and this means that the fluctuations from the steady state will vanish (*stable node*). It looks as if we had in the steady state, an *attractor* of the concentrations X and Y.
- Now let us take $T^2 - 4\Delta = 0$, which means that the two roots are equal ("degeneracy"). This case is similar to the two previous ones. If the two roots are positive, then the point is called the *stable stellar node* (attractor); if they are negative, it is called the *unstable stellar node* (repeller).
- If $T^2 - 4\Delta < 0$, we have an interesting situation: both roots are complex conjugates $\omega_1 = \omega_r + i\omega_i$, $\omega_2 = \omega_r - i\omega_i$, or $\exp \omega_{1,2} t = \exp \omega_r t \exp(\pm i\omega_i t) = \exp \omega_r t (\cos \omega_i t \pm i \sin \omega_i t)$. Note that $\omega_r = \frac{T}{2}$. We have, therefore, three special cases:
  - $T > 0$. Because of $\exp \omega_r t$ we have, therefore, a monotonic increase in the *fluctuations*, and at the same time because of $\cos \omega_i t \pm i \sin \omega_i t$ the two concentrations oscillate. Such a point is called the *unstable focus* (and represents a repeller).
  - $T < 0$. In a similar way, we obtain the *stable focus*, which means some damped vanishing concentration oscillations (attractor)
  - $T = 0$. In this case, $\exp \omega_{1,2} t = \exp(\pm i\omega_i t)$; i.e., we have the *undamped oscillations* of X *and* Y *in time (of Belousov-Zhabotinsky type) about the stationary point* X_s, Y_s, which is called, in this case, the *center of marginal stability*.

Qualitative Change

Can we qualitatively change the behavior of the reaction? Yes; it is sufficient just to change the concentrations of A or B (i.e., to rotate the reactor taps). For example, let us gradually change B. Then, from Eq. (15.8), it follows that the key parameter T begins to change, which leads to an *abrupt qualitative change* in the behavior. Such changes may be of great importance (also in the sense of information processing), as the control switch may serve to regulate the concentrations of some substances in the reaction mixture.

15.2.4 Brusselator with Diffusion–Dissipative Structures

If the stirrer were removed from the reactor, Eq. (15.1) has to be modified by adding diffusion terms:

$$\frac{dX}{dt} = A - (B+1)X + X^2 Y + D_X \frac{\partial^2 X}{\partial r^2}, \tag{15.10}$$

$$\frac{dY}{dt} = BX - X^2 Y + D_Y \frac{\partial^2 Y}{\partial r^2}. \tag{15.11}$$

Fig. 15.8. (a) Such an animal "*should not exist.*" Indeed, how did the *molecules* know that they have to make a beautiful pattern? I have looked at zebras many times, but only recently was I struck by the observation that what I see on the zebra's skin is described by the logistic equation. The skin on the zebra's neck exhibits quasiperiodic oscillations of black and white (period 2), and in the middle of the zebra's body, we have a *period doubling* (period 4), the zebra's back has period 8. Panel (b) shows the waves of the chemical information (concentration oscillations in space and time) in the Belousov-Zhabotinski reaction from several sources in space. A "freezing" (for any reason) of the chemical waves leads to a striking similarity with the zebra's skin. Panel (c) shows similar waves of an epidemic in a rather immobile society. The epidemic broke out in center A. Those who have contact with the sick person get sick, but after some time, they regain their health and become immune for some time. After the immune period is over, these people get sick again because there are a lot of microbes around. This is how epidemic waves propagate.

A stability analysis similar to that carried out before results *not only in oscillations in time, but also in space*; *i.e., in the reaction tube, there are waves of the concentrations* of X and Y *moving in space* (*dissipative structures*). Now, look at the photo of a zebra (Fig. 15.8) and at the bifurcation diagram in the logistic equation (Fig. 15.4).

15.2.5 Hypercycles

Let us imagine a system with a chain of consecutive chemical reactions. There are a lot of such reaction chains around, and it is difficult to single out an elementary reaction without such a chain being involved. They end up with a final product and everything stops. What would happen, however, if at a given point of the reaction chain, a substance X were created that was the same as one of the reactants at a previous stage of the reaction chain? The X would take control over its own fate, according to the Le Chatelier rule. In such a way, feedback would have been established, and instead of the chain, we would have a catalytic cycle. A system with feedback may adapt to changing external conditions, reaching a steady or oscillatory state. Moreover, in our system, a number of such independent cycles may be present. However, when two of them share a common reactant X, both cycles would begin to cooperate, usually exhibiting a very complicated stability/instability pattern or an oscillatory character. We may think of coupling many such cycles in a *hypercycle*, etc.[23]

Cooperating hypercycles based on multilevel supramolecular structures could behave in an extremely complex way when subject to variable fluxes of energy and matter. No wonder, then, that a few photons produced by the prey hidden in the dark and absorbed by the retinal in the lynx's eye may trigger an enormous variety of hunting behaviors. Or, maybe from another domain: a single glimpse of a girl may change the fates of many people,[24] and sometimes the fate of the world. This is because the retina in the eye, hit by the photon of a certain energy changes, its conformation from cis to trans. This triggers a cascade of further processes, which ends up as a nerve impulse traveling to the brain, and it is over.

15.2.6 From Self-Organization and Complexity to Information

Using the multilevel supramolecular architectures, one may tailor new materials exhibiting desired properties; e.g., adapting themselves to changes of the neighborhood ("smart materials"). The shape and the stability of such architectures may depend on a more or less subtle interplay of external stimuli. Such materials by the chemical synthesis of their building blocks are taught to have a function to perform (i.e., an action like ligand binding and/or releasing, transporting a ligand, an electron, or a photon). A molecule may perform several functions. Sometimes these functions may be *coupled* getting a functional cooperativity.

A vast majority of chemistry these days deals either with mastering the structure and/or studying how this structure behaves when allowing its dynamics in a closed system attaining equilibrium (a beakerlike approach). In the near future, chemistry will face the nonlinear, far from equilibrium dynamics of systems composed of multilevel supramolecular architectures. To my knowledge, no such endeavors have been undertaken yet. These will be very complex and very sensitive systems. It will probably be several years before we see such systems in action. It seems much more difficult for contemporary chemistry to control/foresee theoretically what kind of behavior to expect.

Biology teaches us that an unbelievable effect is possible: molecules may form spontaneously some large aggregates with very complex dynamics and the whole system is searching for energy-rich substances to keep itself running. The molecular functions of many molecules may be coupled in a complex space-temporal relationship at several time and space scales involving

[23] M.Eigen and P.Schuster, *Naturwissenschaften* 11 (1977) and 1 and 7 (1978).

[24] Well, think of a husband, children, grandchildren, etc.

enormous transport problems at huge distances of the size of our body. All this spans the time scale from femtoseconds to years, and on the spatial scale from angstroms to meters.

Future chemists will deal with molecular scenarios involving interplay of sophisticated, multilevel structures transforming in nonlinear dynamic processes into an object that has a purpose and plays a certain complex role. The achievements of today, such as molecular switches, molecular wires, etc. will be important, but they will represent just a few simple elements of a space-temporal molecular interplay that will come tomorrow.

15.3 Information and Informed Matter

The body acts as a medium in which a massive information processing is going on. What living organisms do is exchange information not only at the level of an individual, but also at molecular, cellular, and tissue levels. *The corresponding hardware and software are chemical systems.* It looks as if the chemical identity were much less important than the function the molecules perform. For example, the protein with the generic name *cytochrome C* is involved in electron transfer in all organisms, from yeast to humans. Each species, however, has a specific sequence of the amino acids in its *cytochrome C* that differs from the cytochromes of all other species. The differences range from a single amino acid to 50% of amino acids. Yet the function is preserved in all these molecules.

Thus, the most advanced chemistry we are dealing with is used for information processing. Chemistry is still in a stage in which one does not consider quantitatively the exchange of information. Information became, however, an object of quantitative research in telecommunication. As soon as 1924, Harry Nyquist was studying the efficiency of information channels when using a given number of electric potential entries in telegraphs. A few years later, Ralph Hartley published an article on measuring information. Twenty years after that, Claude E. Shannon introduced the notion of information entropy.

15.3.1 Abstract Theory of Information

As a natural measure of the amount of information contained in a binary sequence (a message, e.g., 00100010...) of length N, one may propose just the number N (i.e., the message length). To reconstruct the message, one has to ask N questions: *does the next position equal* 1?

The number of all possible messages of length N is equal to $M = 2^N$. Hence, the amount of information in a message (I, measured in bits) is

$$I \equiv N = \log_2 M.$$

Assumption of Equal Probabilities

If one assumes that the probability of picking out a particular message to be sent is the same for all the messages, $p = \frac{1}{M}$, one obtains the amount of information in a particular message as defined by Hartley:

$$I = -\log_2 p.$$

The receiver does not know M and therefore judges the amount of information (its importance in bits) by his own estimation of p *before* the particular information comes. Hence, I may be viewed as a measure of receiver's surprise that this particular information came true (a message about a marginally probable event contains a lot of information). *This means that the amount of information in a message does not represent a feature of this message, but in addition, it tells us about our knowledge about the message.* If one receives the same message twice, the amount of the first information coming is $I_1 = -\log_2 p_1$ (where p_1 is the probability of the event described by the message as judged by the receiver), while the second (identical) information carries $I_2 = -\log_2 1 = 0$, because there is no surprise. The situation becomes more ambiguous, when there are several receivers, each of them having his own estimation of p. Thus, the amount of information received by each of them may be different.[25]

The Probabilities May Differ–The Shannon Entropy of Information

Suppose that we have an alphabet of letters $a_1, a_2, \ldots a_m$. For a particular language, these are the letters of the language ($m = 26$ for the English language), for the genetic code $m = 4$ (adenine, thymine, guanine, cytosine), for proteins $m = 20$ (the number of the native amino acids), etc. A letter a_i appears N_i times in a large object (animal, plant, in biosphere, a given language etc.). Using the Laplace definition, one may calculate for each letter its probability to appear in the object $p_i = \frac{N_i}{\sum_i N_i}$ for $i = 1, 2, \ldots m$. Let a long message (e.g., a letter, a portion of DNA, a protein) contains N such letters. The number of different messages of length N that one is able to construct from a given set of letters can be computed if we knew how many times every letter appears in the message. One may estimate these numbers from the probabilities p_i, if one assumes that the message is not only long, but also typical for the language. Then the letter a_i

Claude Elwood Shannon (1916–2001), American mathematician and professor at the Massachusetts Institute of Technology (MIT), his professional life was associated with the Bell Laboratories. His idea, which is obvious today, that information may be transmitted as a sequence of 0s and 1s was shocking in 1948. It was said that Shannon used to understand problems *'in zero time'*.

He was equally successful in applying his information theory to the stock market.

[25] On top of that, it is not clear how to define p (i.e., a chance for some event to happen). Laplace's definition says that p can be calculated as frequencies if *one assumes that there is no reason to think* that some events are more probable than other events. This means subjectivity. Such a procedure is useless in case no repeating is possible, such as when estimating a chance to win a battle, to die in the next twenty minutes, etc. An alternative definition of p by Thomas Bayes is subjective as well ("*a degree of someone's conviction*"). Both definitions are used in practice.

will appear most probably Np_i times. Since the permutations of the same letters do not lead to different messages, the number of different messages of length N is

$$M = \frac{N!}{(Np_1)!(Np_2)!\ldots(Np_m)!}. \tag{15.12}$$

Analogous with the case of messages of the same probability, one defines the expected amount of information as the *entropy of information*, or the *Shannon information entropy*[26]:

$$\langle I \rangle = \log_2 M = \log_2 \frac{N!}{(Np_1)!(Np_2)!\ldots(Np_m)!} \approx N \sum_i^m \left[-p_i \log_2 p_i\right].$$

The expected amount of information per letter of the message, therefore, is

$$\langle I \rangle_N \equiv \frac{\langle I \rangle}{N} \approx \sum_{i=1}^m \left[-p_i \log_2 p_i\right]. \tag{15.13}$$

Properties of the Information Entropy

- $\langle I \rangle_N$ may be treated as a function of a discrete probability distribution[27]: $\langle I \rangle_N = \langle I \rangle_N (p_1, p_2, \ldots p_m)$, or as a functional of a continuous distribution $p\left(x\right)$, x being the random variable or variables.
- The information entropy attains a *minimum* (equal to 0) for any of the following situations: $\langle I \rangle_N (1, 0, \ldots 0) = \langle I \rangle_N (0, 1, \ldots 0) = \langle I \rangle_N (0, 0, \ldots 1) = 0$.
- The information entropy attains a *maximum* for uniform probability distribution $\langle I \rangle_N (\frac{1}{m}, \frac{1}{m}, \ldots \frac{1}{m})$, and in such a case, $\langle I \rangle_N = \sum_{i=1}^m \left[-\frac{1}{m} \log_2 \frac{1}{m}\right] = \log_2 m$.

Thus, the information entropy, $N \langle I \rangle_N$ or $\langle I \rangle_N$, represents a measure of our lack of knowledge about which particular message one may expect given the probability distribution of random variables. If the probability distribution is of the kind $(1, 0, \ldots 0)$, the message is fully determined. The worst case (i.e., the largest lack of knowledge what one may expect as a message) corresponds to a uniform distribution (maximum of the information entropy). Our lack of knowledge will decrease, if we are able to tell with higher probability what will happen.

15.3.2 Teaching Molecules

Molecular recognition represents one of many possible areas of application of the information theory. Suppose that we consider the formation of a molecular complex composed of two

---

[26] We use the Stirling formula, valid for large N : $\ln N! \approx N \ln N - N$. Indeed, $\log_2 \frac{N!}{(Np_1)!(Np_2)!\ldots(Np_m)!} = \log_2 N! - \sum_i \log_2 (Np_i)! = \frac{\ln N!}{\ln 2} - \sum_i \frac{\ln (Np_i)!}{\ln 2} \approx \frac{N \ln N - N}{\ln 2} - \sum_i \frac{Np_i \ln (Np_i) - Np_i}{\ln 2} = \frac{N \ln N}{\ln 2} - \sum_i \frac{Np_i \ln (Np_i)}{\ln 2} = \frac{1}{\ln 2}\left[N \ln N - \sum_i Np_i \ln N - \sum_i Np_i \ln p_i\right] = -N \sum_i p_i \log_2 p_i$.

[27] It is a symmetric function of variables $p_i > 0$; i.e., $\langle I \rangle_N (p_1, p_2, \ldots p_m) = \langle I \rangle_N (p_2, p_1, \ldots p_m)$, which is similar for all other permutations of variables.

molecules A and E. The molecule E is considered as a "teacher" and the A molecule as a "student." If all possible stable interaction energy minima on the potential energy surface (PES) are of the same depth, the lack of knowledge (information entropy) of which particular configuration one observes reaches its maximum. If the depth of one energy well is much larger than the depth of other energy valleys, the calculated Boltzmann probability is very close to 1 and the corresponding Shannon entropy reaches its minimum. In all other cases of calculating the Boltzmann probability distribution, one may calculate the information entropy $\langle I_{AE} \rangle$ for A recognizing E. Then the molecule A may be modified chemically (yielding molecule B) and the new molecule may fit better the molecule E. Also in this case, one can calculate the information entropy $\langle I_{BE} \rangle$. If $\langle I_{BE} \rangle < \langle I_{AE} \rangle$, the change from A to B meant acquiring a temperature-dependent amount of knowledge by the first molecule and we can calculate the information associated to this *molecular lesson* as $\Delta I = \langle I_{AE} \rangle - \langle I_{BE} \rangle$, also as a function of temperature. Such a lesson is meaningless if one considers the educated molecule alone, without any "teacher" to check the level of education. Information is related to the teaching-and-learning process through molecular interaction.

Fig. 15.9 shows an example of such a teaching[28] in a very simple model of the teacher molecule E and the taught molecule: $A \rightarrow B \rightarrow C \rightarrow D$ (in three lessons: $A \rightarrow B$, $B \rightarrow C$, $C \rightarrow D$). The model is simple, we intend only to bring the idea of molecular teaching, rather than to be interested in particular molecules, not speaking about currently prohibitively time-consuming study of several PES's for real molecules of necessarily considerable size (both in quantum-mechanical *ab initio* or semi-empirical type calculations). The following simplifying assumptions have been made in our model:

- The "model molecules" are planar objects (see Fig. 15.9) and approach each other within a common plane, with possible reversal of the plane of one of the molecules.
- For the sake of simplicity in counting the contact positions, each molecule on the contact side is built in a certain spatial resolution (the square structure unit shown at the bottom of Fig. 15.9).
- The interaction energy is zero unless the molecules form a contact along a line length larger than the structure unit side. The interaction energy is assumed to be equal to $\frac{1}{3}n$ (in kcals/mole), where the natural number n is the number of contact units.

All possible contacts have been taken into account, and the normalized Boltzmann probability distributions, as well as the Shannon information entropies, have been calculated for each pair of the molecules (i.e., AE, BE, CE, DE) as functions of temperature T within the range from 100 K to 400 K. The results (Fig. 15.9) show that the molecule A (with the simplest structure) does not recognize molecule E at any temperature T under study. The ignorance (Shannon's entropy) of A with respect to E is of the order of 5 bits. Transforming molecule A into B improves the

[28] L. Piela, *Rep.Advan.Study Instit.*, Warsaw Technical University, 2(2012).

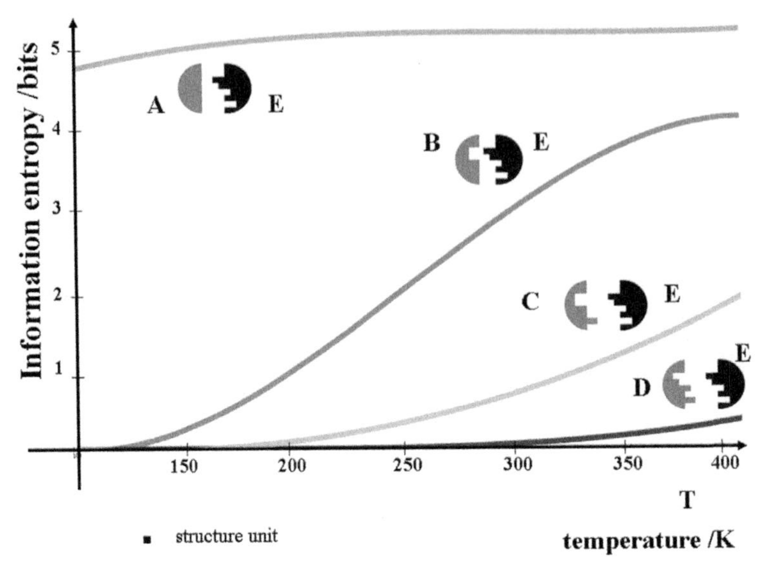

Fig. 15.9. A model of the temperature-dependent molecular teaching the left-side object ("student molecule") is recognize the shape of the right-side molecule ("teacher molecule") in three steps ("lessons"): A → B → C → D. Each of the complexes (AE, BE, CE, DE) is characterized by the state of ignorance (Shannon information entropy), while the knowledge learned in each lesson (A → B, B → C, C → D) is a measure of diminishing the ignorance of the student molecule (in bits).

situation, but only at low temperatures, resulting in molecules teaching and learning by 5 bits at $T = 100$ K, at room temperature by about 2 bits, and by only 1 bit at $T = 400$. The second lesson (teaching B to C) makes the recognition almost perfect below 200 K and very good matching below 300 K. Finally, the last step (C → D) gives perfect matching below 300 K and still very good matching even at 400 K.

15.3.3 Dynamic Information Processing of Chemical Waves

Mathematical Model

The Belousov-Zhabotinsky (BZ) reactive solution can be prepared in a special stable state that exhibits a remarkable sensitivity ("*excitable state*"). When a local small perturbation is applied in such a state, the system comes back to the initial state very quickly, but when the local perturbation exceeds a certain threshold, the system undergoes very large changes before returning a long way to the stable state. Another remarkable feature of such a system is that it becomes refractable to any consecutive excitation (at this locality) during a certain time after the first excitation (the refraction time). A local perturbation may propagate further in space, while the originally perturbed place relaxes to the stationary state. *This means that there are profile preserving traveling waves possible.*

Such a system can be described by a mathematical model known as the FitzHugh - Nagumo (FHN) scheme, originally applied to the propagation of electric excitation pulses in nerve tissues.[29] In the chemical context, these (mathematical) waves may correspond to time- and space-dependent concentration of some particular substance.
The BZ system may be viewed as a medium that can process information: for instance, an excess of a chemical species at a point of the reactor may be treated as 1 (or "*true*"), while its deficit as 0 (or "*false*"). Alternatively, the arrival or the absence of a wave in a predefined part of the reactor may be interpreted in a similar way.

Boris Pavlovich Belousov (1893–1970) looked for an inorganic analog of the biochemical Krebs cycle. The investigations began in 1950 in a Soviet secret military institute. Belousov studied mixtures of potassium bromate with citric acid, and a small admixture of a catalyst: a salt of cerium ions. He expected a monotonic transformation of the yellow Ce^{4+} ions into the colorless Ce^{3+}. Instead, he found oscillations of the color of the solvent (colorless-yellow-colorless-...etc, also called by Russians "vodka-cognac-vodka-..."). He wrote a paper and sent it to a Soviet journal, but the paper was rejected with a referee's remark that what the author had described was simply impossible. His involvement in classified research caused him to limit himself to bringing (by an intermediate) a piece of paper with reactants and his phone number written on it. He refused to meet anybody. Finally, Schnoll persuaded him to publish his results. Neither Schnoll nor his Ph.D. student Zhabotinsky ever met Belousov, though all they lived in Moscow.

Belousov's first chemistry experience was at the age of 12, while engaged in making bombs in the Marxist underground. Stalin thought of everything. When, formally underqualified, Belousov had problems as head of the lab, Stalin's handwritten instruction in ordinary blue pencil on a piece of paper, "*Has to be paid as a head of laboratory as long as he has this position,*" worked miracles.

After S.E.Schnoll, *Gheroy i zladieyi rossiyskoy nauki* Kron-Press, Moscow (1997).

This was the basis of chemical reactors using the BZ reaction, which can behave as logical gates (AND, OR, NOT).[30]

Following Sielewiesiuk and Górecki,[31] we consider here the FHN model of the BZ reaction in the case of a square planar reactor containing a thin layer of the BZ reactive solution [the position in the reactor is given by (x, y), the solution's depth is assumed being negligible]. Two quantities, $u(x, y)$ and $v(x, y)$, denote the concentration amplitudes of two chemical substances, one called the *activator* (which is present in the solution) and the other called the *inhibitor* (which is uniformly immobilized on the bottom of the reactor). The time evolution of their values is modeled by the following set of FHN equations[32]:

$$\tau \frac{\partial u}{\partial t} = -\gamma \left[ku(u-\alpha)(u-1) + v\right] + D_u \nabla^2 u, \tag{15.14}$$

[29] R. FitzHugh, *Biophysics J., 1*, 445 (1961); J. Nagumo, S. Arimoto, and S. Yoshizawa, *Proc. IRE*, 50, 2061 (1962).

[30] A. Toth, and K. Showalter, *J. Chem. Phys. 103*, 2058 (1995); O. Steinbock, P. Kettunen, and K. Showalter, *J. Phys. Chem. 100*, 18970 (1996).

[31] J. Sielewiesiuk and J. Górecki, *Acta Phys. Pol. B 32*, 1589 (2001); J. Sielewiesiuk and J. Górecki, *GAKUTO Intern. Series Math. Sci.Appl.* 17(2001).

[32] R. FitzHugh, *Biophysics J.*, 1, 445 (1961); J. Nagumo, S. Arimoto, and S. Yoshizawa, *Proc. IRE*, 50, 2061 (1962).

$$\frac{\partial v}{\partial t} = \gamma u, \tag{15.15}$$

where t stands for time and τ, k, α, D_u are constants, the last one being the diffusion constant for the activator (the diffusion of the inhibitor is neglected). The system of Eqs. (15.14) and (15.15) was found to correspond to an excitable state of the BZ solution for the following values of the parameters[33]: $k = 3$, $\tau = 0.03$, $\alpha = 0.02$, $D_u = 0.00045$. The quantity γ is a parameter that defines the architecture of the reactor: $\gamma = 1$ is set everywhere in the reactor, except the predefined regions called "passive" ones, where $\gamma = 0$. In the passive regions no production of the activator, only its diffusion, takes place according to Eq. (15.14). The passive regions are inhibitor-free, since $v = const$ follows from Eq. (15.15), and the constant is set to be zero.

Reactor's Geometry

After the excitation a traveling wave appears that propagates freely in the regions with $\gamma = 1$, while the passive regions ($\gamma = 0$) represent for the wave a kind of barrier to overcome. The wave penetrates the passive region. For a passive stripe of a certain width, the penetration depth depends on the stripe/incident wave impact angle[34], the most efficient being a perpendicular impact (with the parallel wave front and the stripe, as follows):

- If the stripe is too wide, the wave disappears.
- If the stripe is sufficiently narrow, the wave passes through it.

We will consider a chemical reactor called later on the double-cross reactor, in which the passive regions form a double cross (gray stripes) shown in Fig. 15.10a. The gray lines are the inhibitor-free stripes of a certain width (passive stripes). The role of the double cross is to provide a partitioning of the reactor into cells, within which the chemical wave can travel freely, while the stripes play the role of barriers for such a motion. In the model described here, their width is taken to be sufficiently small to allow the wave cross perpendicular stripes, but sufficiently large to prohibit leaking the wave sideways when the wave glides along the corridor between two parallel stripes (with the wave front at a right angle with the stripes).

Test Waves

This is seen in Figs. 15.10b and 15.10c, where a wave front 1 (2) moves with a constant velocity in the corridor eastward (northward). The wave is displayed at two instants of time:

[33] Given the values of parameters for the excitable solution, one can see that for small u, in Eq. (15.14), the terms *linear* and *quadratic* in u have the positive coefficient at the right side; i.e., those terms lead to the increase of u. Only at very large u does the negative $-\gamma k u^3$ term come into play, dumping finally the value of u. As one can see, v acts as an inhibitor, since it enters the right-side of Eq. (15.14) with the negative coefficient for $\frac{\partial u}{\partial t}$. As seen from Eq. (15.15), the production of the inhibitor v is enhanced by the presence of u, therefore diminishing u through Eq. (15.14). Thus, a stimulation of u up to very large amplitude before any significant damping by the $-\gamma k u^3$ term begins means that Eqs. (15.14 and (15.15) may indeed describe an excitable system.

[34] I.N. Motoike and K. Yoshikawa, *Phys. Rev. E*, *59*, 5354 (1999).

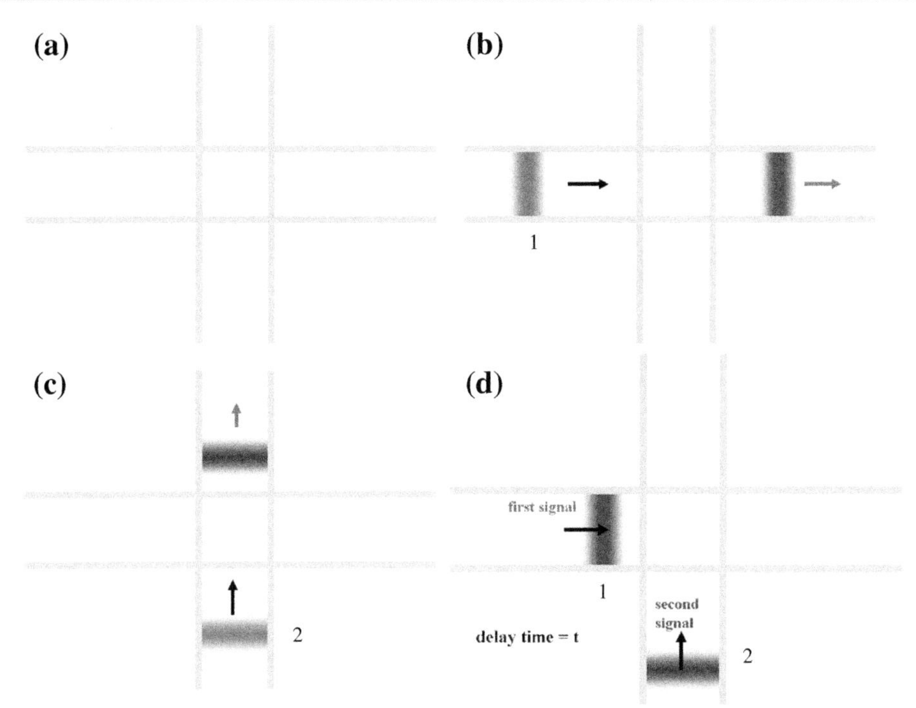

Fig. 15.10. Geometry of the double-cross reactor in the FHN mathematical model. (a) The gray stripes forming the double cross are passive regions; i.e., they correspond to the absence of the inhibitor ($v = 0, \gamma = 0$). This means that no autocatalytic reaction is taking place within the stripes, only diffusion is possible. (b,c) Two chemical waves of u are shown at two time values (earlier–lighter color; later–darker color), one heading east (b), the other one north (c). It was shown in calculations (Sielewiesiuk and Górecki) that both waves cross the perpendicular stripes, but are able to propagate along a single corridor without any leaking sideways. (d) Two such waves before the collision in the double-cross reactor. Wave 1 (heading East) comes to the reactor's center first, wave 2 (heading North) has a delay time t with respect to wave 1.

the lighter color for the earlier time snapshot, while the darker color for the later snapshot. From Figs. 15.10b and 15.10c, it is seen that the waves move straight along their corridors. The reason why the waves are numbered is possibly to distinguish them *after* two such waves collide. Fig. 15.10d shows preparation to such a collision: we see the two incoming wave fronts arriving at the center (the second one with delay t).

Waves' Collisions

After the collision, the outgoing waves (darker colors) are formed. Figs. 15.11–15.17 show how sensitive the output's dependence on the delay time between the two waves (1 and 2, lighter color going in and darker color going out) is. The output looks complicated, but in fact it can be understood by applying a simple analogy, related to the FHN equations. Imagine the white areas in Figs. 15.11–15.17 to be covered by grass, the double-cross's gray stripes being bare ground (no grass). The variable u may be thought to be a fire activator (like sparks). Since the stripes are sufficiently narrow, the sparks may diffuse through them (especially at the right-angle impact),

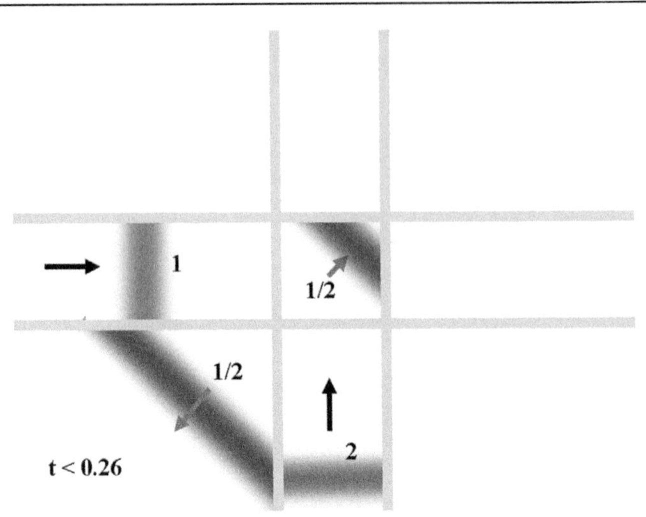

Fig. 15.11. Several solutions of the FHN equations simulating an excitable BZ reaction mixture in the double-cross reactor. The two incoming wave fronts (1 and 2) in two entrance channels (from the west and south directions–lighter color) differ in their time of arrival to the central square. The first wave (1) is heading to the center from the west direction, while the second one (2) goes north with a delay. The outcoming waves (darker color) depend on the delay time t. If an outcoming wave can be identified with the corresponding incoming one, they are denoted by the same label (number). For a delay of $t < 0.26$ time units with respect to wave 1, there are two outgoing chemical waves: one heading southwest, and the other one heading in the opposite direction. Both waves have been labeled "1/2" in order to stress that in this case, one cannot tell which of the incoming waves turns out to be the outcoming one.

despite the fact that there is no grass within the stripes that could support their production over there. If the fire front is perpendicular to the stripe, it is unable to set fire behind the stripe (unless it receives some help from another fire front). After the fire front passes, the grass behind begins to grow and then, after a certain refraction time, becomes again able to catch a fire. Indeed, note the following:

- When the delay ($t > 0$) is smaller than $t = 0.26$ time units (Fig. 15.11), there are two outgoing chemical waves: one heading southwest, the other one heading in the opposite direction. Both waves have been labeled 1/2 in order to stress that in this particular case, one cannot tell which of the incoming waves turn out to be the outcoming ones. Our analogy gives an explanation: both fire fronts when meeting about the lower-left corner of the central square help one another (across the stripes) to set fire in the corner, first in the outer one, then in the inner one (thus creating the two 1/2 fire fronts).
- When $0.28 < t < 3.45$ (Fig. 15.12), wave 1 goes through, while wave 2... disappears. Again, this stands to reason because fire front 1 comes first and leaves ashes in the center, thus causing the wave 2 to die.
- For $3.48 < t < 3.79$ (Fig. 15.13), wave 1 goes through, while wave 2 deflects, with a rather strange deflection angle of $\frac{3\pi}{4}$. In our analogy, this corresponds to the situation, when the fire front 1, coming first to the central square, just leaves the square and meets at the central

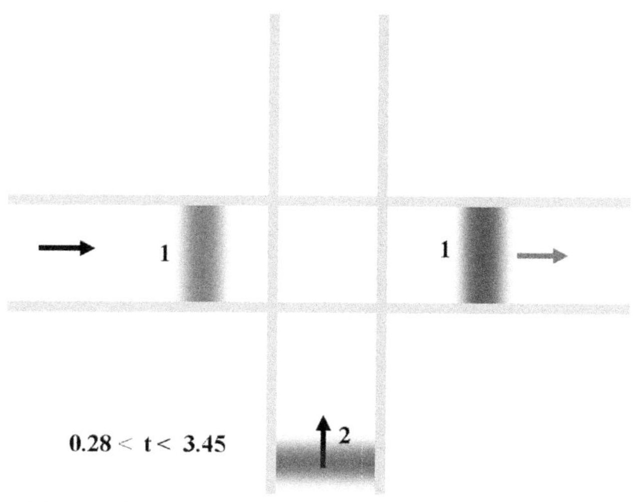

Fig. 15.12. The delay time $0.28 < t < 3.45$. See caption for Fig. 15.11.

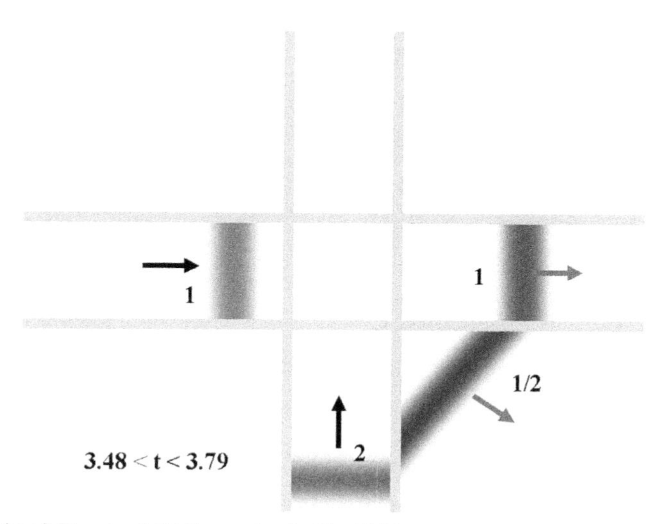

Fig. 15.13. The delay time $3.48 < t < 3.79$. See caption for Fig. 15.11.

lower-right corner the incoming fire front 2. Both are able to set fire front (denoted therefore 1/2) heading southeast.

- For $3.81 < t < 4.22$ (Fig. 15.14), one has a similar behavior as in the case of $0.28 < t < 3.45$ (Fig.15.12): wave 1 goes through, while wave 2 disappears. This also looks reasonable, since the fire front 1 is already in the exit corridor, while the fire front 2 does not find the grass in the center, because the refraction time is longer than the delay time.
- For $t = 4.25$ (Fig. 15.15), wave 1 continues its motion (eastward), while wave 2 turns right following the first one; thus, a double-front wave going eastward is formed. Comparing this to the previous case, now the refraction time is smaller than the delay time and the grass is

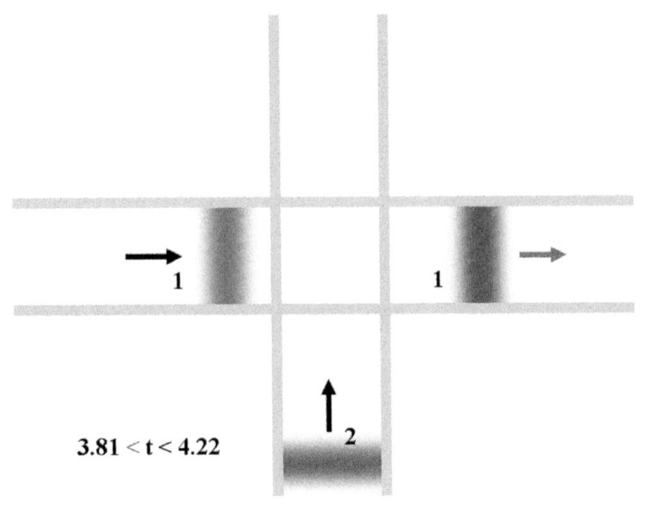

Fig. 15.14. The delay time $3.81 < t < 4.22$. See caption for Fig. 15.11.

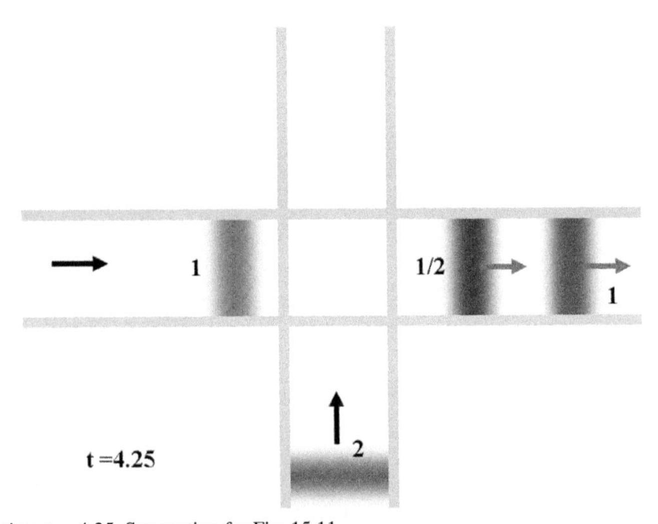

Fig. 15.15. The delay time $t = 4.25$. See caption for Fig. 15.11.

already grown, but there are also some sparks left from fire front 1. This causes a propensity for the fire front 2 to turn right.

- For $4.28 < t < 5.41$ (Fig. 15.16), there is a qualitative change once more: wave 2 splits, so in addition to the double wave described above, we get a second wave traveling north. In our analogy, since the refraction time is smaller that the delay time the two fire fronts move almost independent in their corridors, except that when fire front 2 passes the center it receives some additional sparks from the wave front 1, which already passed the center. This sets fire to the newly grown grass in the horizontal corridor.

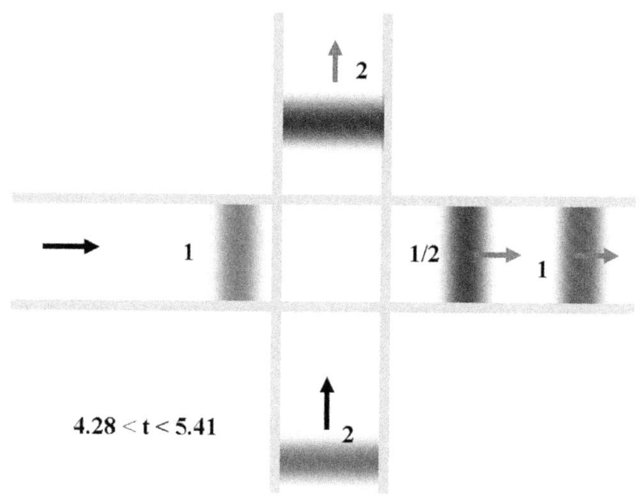

Fig. 15.16. The delay time $4.28 < t < 5.41$. See caption for Fig. 15.11.

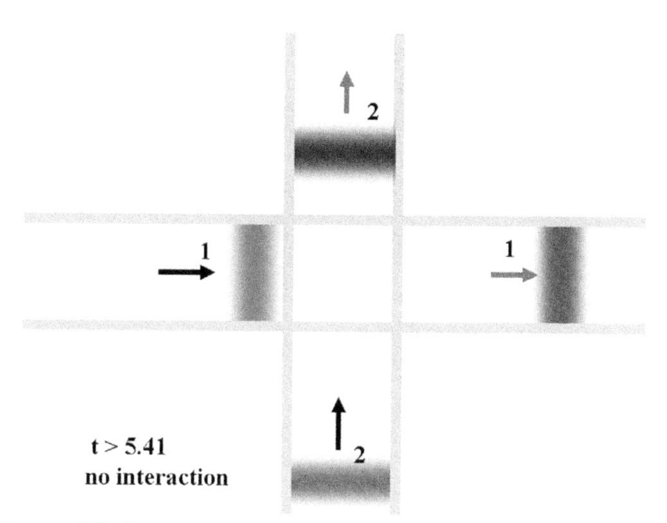

Fig. 15.17. The delay time $t > 5.41$. See caption for Fig. 15.11.

- Finally, for $t > 5.41$ (Fig. 15.17), the delay is so large that the two waves do not interact and pass through unchanged.

The versatility of this behavior witnesses about a large potential for such dynamic information processing. By using one parameter (time delay), one is able to ignite very different cells in the reactor. The chemical wave coming to a cell may trigger a cascade of other processes.

15.3.4 The Mission of Chemistry

There is an impression that in biology, chemistry is only a kind of substitute, a pretext, no more than a material carrier of the mission of the whole organism. Textbooks of biochemistry do not say much about chemistry, they talk about molecular functions to perform, so in a sense they are about metachemistry. A particular molecule seems not to be so important. What counts is its function. A good example is enzymes. One *type* of enzyme may perform the same or similar functions in many different organisms (from fungi to man). The function is the same, but the composition of the enzyme changes from species to species: two species may differ by as much as 70% of the amino acids. However, those amino acids that are crucial for the enzyme function are preserved in all species.

We may perceive chemistry as a potential medium for information processing. This unbelievable chemical task would be collecting, transporting, changing, dispatching, and transferring of information.

Chemistry, as we develop it, is far from such a masterpiece. What we are doing currently might be compared to chemical research by a Martian with a beautifully edited "*Auguries of Innocence*" by William Blake. The little green guy would perform a chemical analysis of the paper (he probably would even make a whole branch of science of that), examine the chemical composition of the printing dye; with other Martian professors, he would make some crazy hypotheses on the possible source of the leather cover, list the 26 different black signs, as well as their perpendicular and horizontal clusters, analyze their frequencies, etc. He would, however, be very far from the *information* the book contains, including the boring matrix of black marks:

To see a world in a grain of sand
And heaven in a wild flower
Hold infinity in the palm of your hand
And eternity in an hour

and most important, he could not even imagine his heart if any... beating any faster after reading this passage, because of thousands of associations *he could never have had*... We are close to what the Martian professor would do. We have wonderful matter in our hands from which we could make chemical poems, but so far we are able to do only very little.

Leonard M.Adleman (b.1945), American mathematician and professor of computer science and of molecular biology at the University of California, Los Angeles. As a young boy, he dreamed of becoming a chemist, then a medical doctor. These dreams led him to the discovery described in this chapter. He not only designed a new role for molecules, but also made the chemical experiments by himself.

Molecules could play much more demanding roles than those that we have foreseen for them: *they can process information*. The first achievement in this direction came from Leonard Adleman, a mathematician.

15.3.5 Molecules as Computer Processors

Computers have changed human civilization. Their speed doubles every year or so, but the expectations are even greater. A possible solution is parallel processing, or making lots of computations at the same time, another is miniaturization. As will be seen in a moment, both these possibilities could be offered by molecular computers, in which the elementary devices would be the individual molecules chemists work with all the time. This stage of technology is not yet achieved. The highly elaborated silicon lithographic technology makes it possible to create electronic devices of size of the order of 1000 Å. Chemists would be able to go down to the hundreds or even tens of angstroms. Besides, the new technology would be based on self-organization (supramolecular chemistry) and self-assembling. In 1 cm^3, we could store the information of a billion of today CD-ROMs. People thought a computer had to have the form of a metallic box. But then, things changed...

In 1994, mathematician Leonard M.Adleman,[35] began his experiments in a genetics lab, while learning biology in the evenings. Once, reading in bed Watson's textbook *The Molecular Biology of the Gene* he recognized that the features of the *polymerase* molecule interacting with the DNA strand described in the textbook perfectly match the features of what is called in mathematics a *Turing machine*, or, an *abstract representation of a computing device* elaborated just before World War II by Alan Turing.

Alan Mathison Turing (1912–1954), British mathematician, defined a device ("*Turing machine*") that consists of a read/write head that scans a 1-D tape divided into squares, each of which contains a 0 or a 1. The behavior of the machine is completely characterized by its current state, the content of the square it is just reading, and a table of instructions.

Such a theoretical concept was of importance in considering the feasibility of any program coded on the tape. Turing is known also for decoding further versions of the German Enigma code during the World War II. This was a continuation of the breaking of the Enigma code in 1933 by three young Polish mathematicians: Marian Rejewski, Jerzy Różycki and Henryk Zygalski. They constructed and at the outbreak of World War II in July, 1939, delivered to British and French the famous deciphering *bombe*. The British further massive development was especially successful, with thousands of top-secret German war documents deciphered. Alan Turing, a mathematical genius, was the key person in this effort.

Therefore, it was *certain* that the polymerase and the DNA (and certainly some other molecules) could be used as computers. If we think about it *now*, the computer in our head is more similar to water, than to a box with hard disks, etc. The achievement of Adleman was that he was able to translate a known and important mathematical problem into the language of

[35] L. Adleman, *Science*, *266*, 1021 (1994).

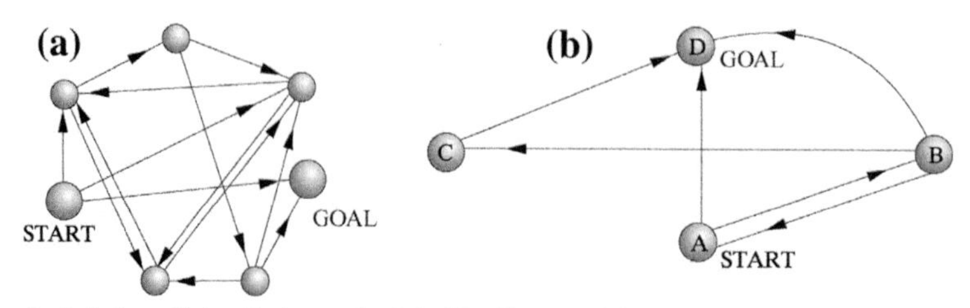

Fig. 15.18. A graph of airplane flights. Is the graph of the Hamilton type? This was a mathematical question for the molecular computer. (a) The graph from the Adleman's experiment; (b) a simplified graph described in this book.

laboratory recipes, and then using a chemical procedure he was able to solve the mathematical problem.

Fig. 15.18a shows the original problem that Adleman faced: a graph with 14 airplane flights involving seven cities.

The task is called the *traveling salesman problem*, notorious in mathematics as extremely difficult.[36] The salesman begins his journey from the city START and wants to go to the city GOAL, visiting every other city exactly once. This is feasible only for some flight patterns. Those graphs for which it is feasible are called *Hamilton graphs*. When the number of cities is small, such a problem may be quite effectively solved by the computer in our head. For seven cities, it takes on average 56 s, as stated by Adleman. For a little larger number, we need a desk computer, but *for 100* cities, *all the computers of the world would be unable to provide the answer*. But a molecular computer would have the answer within a second.

William Rowan Hamilton (1805–1865) was an Astronomer Royal in Ireland. At the age of 17, he found an error in the famous "*Celestial Mechanics*" by Laplace. This drew the attention of scientists and was the beginning of Hamilton's scientific career. In this book, his name is repeated many times (because of the word *Hamiltonian*). At the age of 13 he knew about the same number of languages, among others: Persian, Sanskrit, Arabic and Hindustani.

How Does a Molecular Computer Work?

Let us recall two important examples of complementary synthons: guanine and cytosine (GC) and adenine with thymine, which we discussed in Chapter 13, p. 869.

Let us repeat Adleman's algorithm for a much simpler graph (Fig. 15.18b). What Adleman did was the following.

[36] This problem belongs to what is called NP-hard (NP from *non-polynomial*), in which the difficulties increase faster than any polynomial with the size of the problem.

1. He assigned for every city a particular piece of DNA (sequence) composed of eight nucleic bases:

City A	A	C	T	T	G	C	A	G
City B	T	C	G	G	A	C	T	G
City C	G	G	C	T	A	T	G	T
City D	C	C	G	A	G	C	A	A

2. Then to each existing flight X → Y, another eight-base DNA sequence was assigned, which was composed of the second half of the sequence of X and the first half of the sequence of Y:

Flight A → B	G	C	A	G	T	C	G	G
Flight A → D	G	C	A	G	C	C	G	A
Flight B → C	A	C	T	G	G	G	C	T
Flight B → D	A	C	T	G	C	C	G	A
Flight B → A	A	C	T	G	A	C	T	T
Flight C → D	A	T	G	T	C	C	G	A

3. Then, Adleman ordered the synthesis of the DNA sequences[37] of the flights and the DNA sequences complementary to the cities, i.e.,

co-City A	T	G	A	A	C	G	T	C
co-City B	A	G	C	C	T	G	A	C
co-City C	C	C	G	A	T	A	C	A
co-City D	G	G	C	T	C	G	T	T

4. All these substances were mixed together and dissolved in water, and then you add a bit of salt and an enzyme called ligase.[38] Your computer just started.

How to Read the Solution

What happened in the test tube? First, matching and pairing of the corresponding synthons took place. For example, the DNA strand that codes the AB-flight (i.e., GCAGTCGG) found in the solution the complementary synthon of city B (i.e., the co-city AGCCTGAC), and because of the molecular recognition mechanism, it made a strong intermolecular complex:

```
flights     G  C  A  G  T  C  G  G
                        :  :  :  :                 ,
co-cities               A  G  C  C  T  G  A  C
```

[37] Nowadays is a matter of commercial activity.

[38] To be as effective as nature, we want to have conditions similar to those in living cells.

where the upper part is flights, and the lower part is co-cities. Note that the flights are the only feasible ones because only feasible flights' DNA sequences were synthesized. The role of a co-city's DNA is to provide the information that there is the possibility to land and take off in this particular city.

In the example just given, the complex will also find the synthon that corresponds to flight B → C i.e., ACTGGGCT, and we obtain a more extended strand

```
G C A G T C G G |A C T G G G C T
        ⋮ ⋮ ⋮ ⋮  ⋮ ⋮ ⋮ ⋮
        A G C C  T G A C.
```

In this way, from the upper part (from the lower part as well) of the intermolecular complexes, we can read a particular itinerary. The ligase was needed because this enzyme binds the loose ends of the DNA strands (thus removing the perpendicular separators above). Therefore,

> every possible itinerary is represented by a DNA oligomer. If the graph were Hamiltonian, then there would be in the solution the DNA molecule that encodes the right itinerary. In our example this molecule corresponds to itinerary A → B → C → D and is composed of 24 nucleotides: GCAGTCGGACTGGGCTATGTCCGA.

Eliminating Wrong Trajectories

Practically, independent of how large N is, after a second, the solution to the traveling salesman problem is ready. The only problem now is to be able to read the solution. This will currently take much more than a second, but in principle, it only depends linearly on the number of cities.

To get the solution, we use three techniques: polymerase chain reaction (PCR), electrophoresis, and separation through affinity. The machinery behind all this is supramolecular chemistry, with the recognition of synthons and co-synthons (called *hybridization* in biochemistry[39]).

The itineraries coded by the hybridization are mostly wrong. *One* of the reasons is that they do not start from the START CITY (A) and do not end up at the GOAL CITY (D). Using the PCR technique,[40] it is possible to increase the concentration of only those itineraries, which start from START and end at GOAL to such an extent that concentrations of all the other molecules may be treated as marginal.

[39] But this term is a bit misleading, if we think of the hybridization of atomic orbitals, p. 481.

[40] The PCR technique is able to copy a chosen DNA sequence and to grow its population, even from a single molecule to a high concentration, by using the repeated action of an enzyme, a polymerase.

The reaction was invented by Kary B.Mullis (b.1944), an American technical chemist in an industrial company. In 1983, Mullis was driving to his favorite California surfing area when the idea of a DNA copying molecular machine struck him suddenly. He stopped the car and made a note of the reaction. In 1993, Mullis received the Nobel Prize in chemistry *"for his invention of the polymerase chain reaction (PCR) method."*

Still there are a lot of wrong itineraries. First, there are a lot of itineraries that are too long or too short. This problem may be fixed by electrophoresis,[41] which allows the separation of DNA strands of a given length (in our case, the 24-city itineraries). In this way, we have itineraries starting from START and ending at GOAL and having 24 cities. They can be copied again by PCR to increase their concentration for further operations.

Now we have to eliminate wrong itineraries from these 24-city-long sequences: those which repeat some transit cities and leave others unvisited. This is done by the affinity separation method.[42] First, the co-synthon for the first transit city (in our case, C) on the list of transit cities (in our case, C and D) is attached to the surface of iron balls. The iron balls are then added to the solution and after allowing a second to bind to those itineraries that contain the city, they are picked out using a magnet. The balls are then placed in another test tube, the attached "itineraries" released from the surface of the iron balls and the empty iron balls are separated. Thus, we have in a test tube the "itineraries" that begin and end correctly, have the correct number of 24 nucleotides and certainly go through the first transit city (C) on our list of transit cities.

The process is repeated for the second transit city, and then the third, etc.

If, in the last test tube, any molecular "itinerary" swims, it has to be the Hamiltonian-like and is identified by the described procedure. The answer to the salesman problem is, therefore, positive. Otherwise, the answer is negative.

Thus, a mathematical problem was solved using a kind of molecular biocomputer. From the information processing point of view, this was possible because parallel processing was under way–a lot of DNA oligomers interacted with themselves at the same time. The number of such molecular processors is of the order of 10^{23}. This number is so huge, that such a biocomputer is able to check (virtually) all possibilities and to find the solution.

Summary

- Chemistry entered the second half of the twentieth century with detailed knowledge of the main building blocks of molecular structures: atoms, chemical bonds, bond angles and intermolecular interactions.
- The accumulated knowledge now serves to build more and more complex molecular architectures.

[41] Electrophoresis is able to physically separate DNA sequences according to their length. It is based on the electrolysis of a gel. Since DNA is an anion, it will travel through the gel to anode. The shorter the molecule, the longer distance it will reach. The DNA molecules of a given length can then be picked out by cutting the particular piece of gel and then they can be multiplied by PCR.

[42] The affinity separation method makes possible to separate particular sequences from a mixture of DNA sequences. This is achieved by providing its co-synthon attached to iron spheres. The particular sequence we are looking for binds to the surface of the iron ball, which may be separated from the solution afterward using a magnet.

- In these architectures, we may use *chemical bonds* (with energy of the order of 50–150 kcal/mol) to build the molecules, as well as *intermolecular interactions* (with energy of about 1–20 kcal/mol) to construct supramolecular structures from them.
- In supramolecular chemistry, we operate with synthons (i.e., some special systems of functional groups that fit together perfectly), giving rise to molecular recognition.
- The interaction leads to a molecular complex that facilitates further evolution of the system: either by a chemical reaction going on selectively at such a configuration of the molecules, or by further self-organization due to next-step molecular recognition of the newly formed synthons.
- This may result in forming complex systems of multilevel architecture, each level characterized by its own stability.
- The self-organization may take place with significant interaction non-additivity effects ("*nonlinearity*" in mathematical terms) that may lead to cooperation in forming the multilevel structure.
- The self-organized structures may interact with other such structures (chemical reactions or association).
- In particular, they may create the autocatalytic cycle, which represents chemical feedback.
- Such cycles may couple in a higher-order cycle, forming hypercycles.
- A dynamic system with hypercycles, when perturbed by an external stimulus, reacts in a complex and nonlinear way.
- One of the possibilities in non-equilibrium conditions are the limit cycles, which lead to dissipative structures, which may exhibit periodicity (in space and time) and chaotic behavior.
- Some dynamic systems may represent molecular libraries with the proportions of species strongly depending on external conditions (cf. the immune system).
- Molecules may act (e.g., transfer photon, electron, proton, ion, induce a conformational change, etc.) thus performing a function.
- Several functions may cooperate exhibiting a space/time organization of the individual functions.
- Some molecules may serve for effective information processing.
- Information processing seems to represent the ultimate goal of chemistry in the furture.
- Molecules can be "taught" by adjusting their shape to the requirements of a "teacher molecule," and the temperature-dependent amount of knowledge acquired can be measured in bits.
- Chemical waves can carry information very effectively changing their behavior that depends strongly on the spatial and time restrictions imposed.
- Molecules may serve as massively parallel computer processors in a way that is fundamentally distinct from performance of the contemporary computers.

Main Concepts, New Terms

activator (p. 995)
attractors (p. 978)
autocatalysis (p. 983)
Belousov-Zhabotinsky reaction (p. 994)
bifurcation (p. 980)
brusselator (p. 982)
center of marginal stability (p. 986)
chaos (p. 980)
chemical waves (p. 994)
combinatorial chemistry (p. 976)
complex systems (p. 976)
cooperativity (p. 974)
dissipative structures (p. 987)
DNA computing (p. 1003)
DNA hybridization (p. 1006)
excitable state (p. 994)
feedback (p. 978)
FitzHugh-Nagumo (FHN) model (p. 995)
fixed point (p. 978)
fluctuation (p. 984)
focus (stable and unstable, p. 986)
Hamilton graph (p. 1004)
hypercycles (p. 988)
information (p. 990)
limit cycle (p. 979)
logistic equation (p. 980)
molecular evolution (p. 1010)
molecular lesson (p. 993)

molecular libraries (p. 976)
nodes (stable and unstable, p. 987)
non-additivity (p. 1008)
NP-hard problem (p. 1004)
PCR (p. 1006)
polymerase (p. 1003)
reaction center (p. 997)
repellers (p. 987)
saddle point of reaction (p. 984)
self-organization (p. 989)
separation by affinity (p. 1004)
Shannon information entropy (p. 992)
stable focus (p. 986)
stable node (p. 986)
stable stellar node (p. 986)
stellar nodes (stable and unstable, p. 986)
student molecule (p. 984)
teacher molecule (p. 994)
teaching molecule (p. 992)
traveling salesman problem (p. 1004)
Turing machine (p. 1003)
unstable focus (p. 986)
unstable node (p. 986)
unstable stellar node (p. 986)

From the Research Front

One might think that to say that organic chemists are able to synthesize almost any molecule is certainly an exaggeration, but the statement seems sometimes to be very close to reality. Chemists were able to synthesize the five-Olympic-ring molecule, the three interlocked Borromean rings, the football made of carbon atoms, the "cuban," a hydrocarbon cube, "basketan," in the form of an apple basket (p. 802), a molecule in the form of Möbius band, etc. Now we may ask why the enormous synthetic effort was undertaken and what these molecules were synthesized for. Well, the answer seems to be that contemporary chemists are fascinated by their art of making complex and yet perfect and beautiful molecular objects. The main goal apparently was to demonstrate the mastership of modern chemistry. However, high symmetry does not necessarily means a particular usefulness. The synthetic targets should be identified by the careful planning of molecular *functions*, rather than molecular beauty.

Ad Futurum

We may expect that more and more often chemical research will focus on molecular function, and (later) on the space/time cooperation of the functions. Research projects will be formulated in a way that will highlight the role of the molecular function, and will consist of several (interrelated) steps:

1. First, the technical goal will be defined.
2. The molecular functions will be identified, which will make this goal achievable.
3. Theoreticians will design and test in computers ("*in silico*") the molecules that will exhibit the above functions.
4. Synthetic chemists will synthesize the molecules designed.
5. Physicochemists will check whether the molecular functions are there.
6. Finally, the material will be checked against the technical goal.

We will be able to produce "*intelligent*" materials that will respond to external conditions in a previously designed, complex, yet we hope, predictable way. The materials that will be created this way will not resemble the materials of today, which are mostly carrying out one often primitive function. The drugs of today are usually quite simple molecules, which enter the extremely complex system of our body. The drugs of tomorrow will involve much larger molecules (like proteins). Will we be clever enough to avoid unpredictable interactions with our body? What in principle do we want to achieve?

What will the motivation of our work be? Will we take into account the psychological needs of the human being, equilibrium of their minds?

What will the future of the human family be, which was able in the past to create such wonderful music, Chartres cathedral, breathtaking painting, moving poetry, abstract mathematics, proudly landed on other celestial bodies? In the past, nothing could stop their curiosity and ingeniousness–they were able to resist the harshest conditions on their planet. Humans have reached nowadays the technical level that probably will assure avoiding the next glaciation,[43] maybe allow a *small* asteroid to push by nuclear weapons off the target, if it were aimed dangerously at the Earth, also erasing in nuclear war most of its own population, together with the wonders of our civilization.

What is the goal of these beings and what will be the final limit of their existence? What they are aiming at? Do we want to know the smell of fresh bread, to be charmed by Chartres cathedral with all it is for, to use our knowledge to cherish the friendship of the human family, or will it be sufficient to pack a newborn into a personal container and make computers inject substances that will make his neural system as happy as in Seventh Heaven?

Which of the goals we want, as chemists, to participate in?

Additional Literature

M. Eigen and P. Schuster, *The Hypercycle. A Principle of Natural Organization*, Springer Verlag, Berlin (1979).
An excellent, comprehensible book, written by the leading specialists in the domain of molecular evolution.

I. Prigogine, *From Being to Becoming. Time and Complexity in Physical Sciences*, Freeman, San Francisco (1980).
A book written by the most prominent specialist in the field.

A. Bablyoyanz, *Molecules, Dynamics, and Life*, Wiley, New York (1987).
The author describes the scientific achievements of Prigogine and his group, which she participated in. An excellent, competent book, the most comprehensible of these recommended books.

J.-M. Lehn, *Supramolecular chemistry: Concept and Perspectives*, VCH, 1995.
A vision of supramolecular chemistry given by one of its founders.

Question

1. An oscillatory solution of differential equations
   a. represents an attractor
   b. has been discovered in the twentieth century
   c. when met in chemistry means concentration oscillations
   d. is a limit cycle
2. A dissipative structure
   a. may appear in thermodynamic equilibrium
   b. represents a space and/or time-dependent distribution of concentrations of chemical substances
   c. depends on the matter and energy fluxes in the system
   d. may appear, when the system is sufficiently far from a thermodynamic equilibrium
3. A molecular library
   a. cannot exist in thermodynamic equilibrium
   b. in case of mixture of A and B substances, contains molecular complexes A_nB_m with various m, n
   c. has the ability to shift the equilibrium under the influence of some other substances
   d. means a complete set of books on molecular physics.
4. A molecular self-organization
   a. means a spontaneous formation of molecular complexes resulting in the supramolecular structures exhibiting a short-range and/or a long-range order
   b. is possible only in non-equilibrium
   c. is a result of the molecular recognition through spatial and electrical matching
   d. is impossible without a planned chemist' action

[43] Well, it is expected within the next 500 years.

5. In the iterative solution $x_{n+1} = Kx_n(1 - x_n)$ of the logistic equation
   a. one obtains a fixed point at any K
   b. any attempt of increasing K leads to a bifurcation
   c. there is a range of K that corresponds to a chaotic behavior of the solution x
   d. at a sufficiently small K the population vanishes
6. In the brussellator without diffusion (x and y stand for fluctuations of the substances X and Y)
   a. a stable node corresponds to an exponential vanishing of x and y
   b. a stable focus means vanishing oscillations of x and y
   c. at least one of the reactions should have an autocatalytic character
   d. X is a catalyst but Y is not
7. In an isolated system
   a. the entropy does not change
   b. after a sufficiently long time the gradient of the temperature must attain zero
   c. the concentration gradients are zero
   d. one cannot observe dissipative structures
8. Information entropy
   a. is equal to $-\sum_i p_i \log_2 p_i$
   b. represents a measure of our ignorance about a coming message
   c. is equal to the mean number of questions necessary to define the probability distribution
   d. attains the minimum for all p_i being equal.
9. An event has only four possible outputs with *a priori* probabilities: $p_1 = p_2 = p_3 = p_4 = \frac{1}{4}$. Reliable information comes that in fact the probabilities are different: $p_1 = \frac{1}{2}, p_2 = \frac{1}{4}, p_3 = \frac{1}{8}, p_4 = \frac{1}{8}$. This information had I_1 bits, and I_1 equals to
   a. 1 bit
   b. 0.5 bit
   c. 2 bits
   d. 0.25 bit
10. The situation corresponds to Question 9, but a second piece of reliable information coming says that the situation changed once more and now: $p_1 = \frac{1}{2}, p_2 = 0, p_3 = 0, p_4 = \frac{1}{2}$. The second piece of information had I_2 bits. We pay for information in proportion to its quantity. Therefore, for the second piece of information we have to pay
    a. the same as for the first piece of information
    b. twice as much as for the first piece of information
    c. half of the price for the first piece of information
    d. three time more than for the first piece of information.

Answers

1a,c,d, 2b,c,d, 3b,c, 4a,c, 5c,d, 6a,b,c, 7b, 8a, b, 9d, 10d

Acronyms

AD acceptor-donor method: A theoretical description of a chemical reaction in terms of acceptor molecular orbitals (MOs) and donor MOs.

AIM atoms in molecules: An analysis of the critical points of the molecular electron density distribution that leads to its unique partition into atomic contributions.

AMO Alternant Molecular Orbitals: A version of the unrestricted Hartree-Fock (UHF) method, in which the occupied orbitals are modified by admixtures of virtual orbitals.

AO Atomic Orbital: A function of an electron's position in space, centered in a point and decaying exponentially, like Slater-type orbitals (STOs) or Gaussian type orbitals (GTOs), at large distances from the center.

BFCS Body-fixed coordinate system: The coordinate system fixed on the moving molecule.

BO Born-Oppenheimer approximation: An approximation assuming that the electrons move in the field of the clamped nuclei, while the nuclei move in the potential energy being the electronic energy.

BOAS Bond-Order Alternating Solution: The electronic density distribution that breaks the translational symmetry of the nuclear framework by doubling the period.

BSSE Basis Set Superposition Error: An error in the calculation of intermolecular interaction energy stemming from using an incomplete basis set of atomic orbitals (AOs) and calculation of the energies of isolated molecules by using only their own basis sets of AOs.

B3LYP Becke-Lee-Young-Parr Density Functional Theory: A semiempirical density functional theory (DFT) method of hybrid type; i.e., with the exchange-correlation potential composed of several empirical contributions.

CAS SCF Complete Active Space Self-Consistent Field: An iterative and variational method of solving the Schrödinger equation with the variational wave function in the form of a linear combination of all the Slater determinants (coefficients and spinorbitals are determined variationally) that can be built from a limited set of the spinorbitals (forming the active space).

Ideas of Quantum Chemistry, Second Edition. http://dx.doi.org/10.1016/B978-0-444-59436-5.00051-9

CC Coupled Cluster: A non-variational method of solving the Schrödinger equation with the wave function in the form of an exponential operator (to be determined) acting on the Hartree-Fock wave function.

CCSD Coupled Cluster Singles and Doubles: A non-variational method of solving the Schrödinger equation with the wave function in the form of an exponential operator (with the explicit presence of the single and double excitations, their contribution to be determined in the method) acting on the Hartree-Fock wave function.

CCSD(T) Coupled Cluster Singles and Doubles with Triples: A non-variational method of solving the Schrödinger equation with the wave function in the form of an exponential operator (with the explicit presence of the single and double excitations, their contribution to be determined in the method, and approximate contribution from the triple excitation) acting on the Hartree-Fock wave function.

CI Configuration Interaction: A variational method with the trial wave function in the form of a linear combination of the given set of the Slater determinants.

CIS Configuration Interaction Singles: A variational method with the trial wave function in the form of a linear combination of the given set of the singly excited Slater determinants.

CISD Configuration Interaction Singles and Doubles: A variational method with the trial wave function in the form of a linear combination of the given set of the singly and doubly excited Slater determinants.

CISDT Configuration Interaction Singles, Doubles, and Triples: A variational method with the trial wave function in the form of a linear combination of the given set of the singly, doubly, and triply excited Slater determinants.

CP Counter-poise: A method of elimination of the basis set superposition error (BSSE) in the intermolecular interaction energy by calculating all quantities using the basis set of atomic orbitals (AOs) of the whole system.

CSB Charge-Shift Bonding: Two maxima of the electron localization function (ELF) in a chemical bond, interpreted as the manifestation of resonance of two ionic structures.

CSF Configuration State Function: An expansion function in the configuration interaction (CI) method that has the same symmetry and spin state as those of the exact wave function.

CT Charge Transfer: The intermolecular transfer of an electric charge being a difference between the electric charge distribution in the isolated molecules and in their complex.

DC Dirac-Coulomb: An approximate and many-electron quasi-relativistic theory, in which the one-electron Hamiltonians are the Dirac relativistic Hamiltonians, whereas the electron-electron interaction operators are represented uniquely by the (non-relativistic) Coulomb interactions.

DFT Density Functional Theory: A theory in which the total energy of a molecule depends on its electron density distribution.

DODS Different Orbitals for Different Spins: Another name for the UHF method. *See* UHF.

DRC Distinguished Reaction Coordinate: A selected distance changing from the value of the reactants to the value of the products of an elementary chemical reaction.

ELF Electron Localization Function: A measure, defined in the density functional theory (DFT), of the tendency to occupy a point of space by an electron pair.

EOM-CC Equation-of-Motion Coupled Cluster: A non-variational method of solving the Schrödinger equation for excited states (related to the equation of motion, EOM), with the wave function calculated in the coupled cluster (CC) method.

FBZ First Brillouin Zone: The set of vectors of the inverse space in a periodic system, which correspond to all possible distinct Bloch functions.

FCI Full Configuration Interaction: A configuration interaction (CI) method with all possible excitations from a given finite set of molecular orbitals (MOs).

FEMO Free Electron Molecular Orbitals: π – electrons in a molecule treated as free electrons in a box.

FF Force Field: A simple mathematical expression mimicking the electronic energy as a function of the positions of the nuclei.

FF Finite Field: Method of solving the Schrödinger equation for a molecule in an external field with the molecule-field interaction term included in the Hamiltonian.

FVAO Field-Variant Atomic Orbital: An atomic orbital (AO) that depends on the external electric field intensity.

GEA Gradient Expansion Approximation: A class of the DFT functionals that take into account a nonlocal character of the exchange-correlation energy through a gradient correction.

GHF General Hartree-Fock: A Hartree-Fock method with spinorbitals of the most general form.

GIAO Gauge-Including Atomic Orbital: A method of calculating a molecule in the magnetic field that ensures the invariance of the results with respect to the choice of the origin of the vector potential describing the magnetic field. Formerly, the spellout was *gauge-invariant atomic orbital*.

GTO Gaussian Type Orbital: An atomic orbital (AO) with an exponential decaying $\exp(-\zeta r^2)$, where r stands for the distance from a given point in space ("*center*"), and $\zeta > 0$.

HF Hartree-Fock: A variational method with the trial wave function in the form of a single Slater determinant.

HOMO Highest Occupied Molecular Orbital: The highest (in terms of energy scale) occupied molecular orbital (MO).

HTS High-Temperature Superconductor: A crystalline substance exhibiting superconductivity with an unusually high critical temperature.

IRC Intrinsic Reaction Coordinate: The steepest descent curve in the space of the nuclear configurations (with the mass-weighted coordinates) that connects two electronic energy minima through the first-order saddle point (transition state).

KS Kohn-Sham: The density functional theory (DFT) method, in which the electronic density distribution results from a single Slater determinant (the Kohn-Sham determinant).

LCAO CO Linear Combination of Atomic Orbitals Crystal Orbitals: The expression of crystal orbitals (COs) as a linear combination of atomic orbitals (LCAOs).

LCAO MO Linear Combination of Atomic Orbitals Molecular Orbitals: The expression of molecular orbitals (MO) as a linear combination of atomic orbitals (LCAO).

LDA Local Density Approximation: A density functional theory (DFT) method that estimates the exchange-correlation energy from the energy in homogeneous electron gas.

LUMO Lowest Unoccupied Molecular Orbital: The lowest (in terms of energy scale) unoccupied molecular orbital (MO).

MBPT Many-Body Perturbation Theory: An iterative perturbation-based method of solving the Schrödinger equation.

MCD Monte Carlo Dynamics: A dynamic with a stochastic choice of configurations of the nuclei and a criterion for accepting or rejecting this choice.

MC SCF Multi-Configurational Self-Consistent Field: A variational iterative solution of the Schrödinger equation with the trial function in the form of linear combination of variable Slater determinants.

MD Molecular Dynamics: The solution of the Newton equation of motion for the nuclei.

MEP Molecular Electrostatic Potential: The electrostatic potential created by a molecule as a function of position in space.

MM Molecular Mechanics: Minimization of the molecular electronic energy, approximated by the force field (FF), as a function of positions of the nuclei.

MO Molecular Orbital: A one-electron function that is a solution of the Fock equation for a molecule.

MP, MP2, MP4 Møller-Plesset Perturbation Theory: A perturbational method [up to the second (MP2) or fourth (MP4) order] of the solution of the Schrödinger equation with the Hartree-Fock function as the zeroth approximation.

NLDA Non-Local Density Approximation: A density functional theory (DFT) method with the exchange-correlation energy correction containing the electron density gradient.

NMR Nuclear Magnetic Resonance: A spectroscopic method in which transitions between the energy levels of the nuclear magnetic moments result from their interaction with the local magnetic field and among themselves.

NO Natural Orbital: A molecular orbital corresponding to the diagonal form of the one-electron density matrix.

PES Potential Energy Surface: The electronic energy as a function of configuration of the nuclei.

PW Plane Wave: The function $A_{\mathbf{k}}(\mathbf{r}) = \exp(i\mathbf{k} \cdot \mathbf{r})$ used in descriptions of periodic systems with vector $\mathbf{k}$ belonging to the first Brillouin zone (FBZ).

PW91 Perdew-Wang 91: A semi-empirical method of finding the ground-state electronic density distribution within the density functional theory (DFT).

QED Quantum ElectroDynamics: A quantum theory of charged particles interacting with the electromagnetic field which goes beyond the Dirac theory.

RHF Restricted Hartree-Fock: A variational method with a single Slater determinant with doubly occupied molecular orbitals (MOs) as a trial function.

ROHF Restricted Open-Shell Hartree-Fock: A variational method with Slater determinant(s) with doubly occupied core molecular orbitals (MOs), but different valence MOs for different spins.

SAPT Symmetry-Adapted Perturbation Theory: A perturbational method of calculating intermolecular interaction energy while taking into account the Pauli exclusion principle.

SCF Self-Consistent Field: An iterative method of solving the Fock equation.

SCF LCAO CO Self-Consistent Field Linear Combination of Atomic Orbitals - Crystal Orbitals: An iterative method of solving the Fock equation for crystals (in the LCAO CO approximation).

SCF LCAO MO Self-Consistent Field Linear Combination of Atomic Orbitals - Molecular Orbitals: Iterative method of solving the Fock equation for molecule (in the LCAO MO approximation).

SDP Steepest Descent Path: The steepest descent trajectory (of lowering the electronic energy as a function of configuration of the nuclei) that connects a first-order saddle point with two adjacent energy minima corresponding to the stable configurations of the reactants and products.

SFCS Space-fixed coordinate system: The coordinate system of the laboratory in which the molecule is observed and measured.

SE Single-Exchange: A contribution to the exchange interaction (valence repulsion of molecules) non-additivity effect coming from the interaction of the Pauli deformation of the electron cloud due to two interacting molecules with the electric field created by the third molecule.

SHG Second Harmonic Generation: A frequency doubling of light in materials with nonlinear electric properties.

SOS Sum-Over-States: Perturbational corrections with summation over unperturbed states.

STO Slater Type Orbital: An atomic orbital (AO) with the asymptotic exponential decay $\exp(-\zeta r)$, where r means the distance from a certain point in space, and $\zeta > 0$.

SUSY SUperSYmmetry: A symmetry-like relation between two dissimilar systems that comes from a symmetry of mathematical expressions that describe them.

TE Triple-Exchange: A contribution to the exchange interaction (valence repulsion of molecules) non-additivity effect coming from a single electron-exchange between two molecules by mediation of a third one.

THG Third Harmonic Generation: A frequency tripling of light in materials with nonlinear electric properties.

UHF Unrestricted Hartree-Fock: The variational method with a single Slater determinant as a trial function [without a molecular orbital (MO) double occupancy restriction].

VB Valence Bond: A variational method with the wave function in the form of a linear combination of the Slater determinants built of atomic spinorbitals.

VSEPR Valence Shell Electron Pair Repulsion: An algorithm of predicting the spatial structure of a molecule by counting the electronic pairs in the valence shell of a central atom with substituents.

ZDO Zero-Differential Overlap: Neglecting any product of two atomic orbitals (AOs), that describe the same electron, but with different centers.

Tables

Table 1. Units of physical quantities.

Quantity	Unit	Symbol	Value
Light velocity		c	$299792.458\ \frac{\text{km}}{\text{s}}$
Planck constant		h	$6.6260755 \cdot 10^{-34}\ \text{J} \cdot \text{s}$
Mass	Electron rest mass	m_0	$9.1093897 \cdot 10^{-31}\ \text{kg}$
Charge	Element.charge = a.u. of charge	e	$1.60217733 \cdot 10^{-19}\ \text{C}$
Action	$\frac{h}{2\pi}$	$\hbar$	$1.05457266 \cdot 10^{-34}\ \text{J} \cdot \text{s}$
Length	bohr = j.at d“lugo”sci	a_0	$5.29177249 \cdot 10^{-11}\ \text{m}$
Energy	hartree = a.u. of energy	E_h	$4.3597482 \cdot 10^{-18}\ \text{J}$
Time	a.u. of time	$\frac{\hbar}{E_h}$	$2.418884 \cdot 10^{-17}\ \text{s}$
Velocity	a.u. of velocity	$\frac{a_0 E_h}{\hbar}$	$2.187691 \cdot 10^{6}\ \frac{\text{m}}{\text{s}}$
Momentum	a.u. of momentum	$\frac{\hbar}{a_0}$	$1.992853 \cdot 10^{-24}\ \frac{\text{kg m}}{\text{s}}$
Electr.dipole moment	a.u. of electr.dipole	ea_0	$8.478358 \cdot 10^{-30}\ \text{Cm}$
			(2.5415 D)
Magn.dipole	Bohr magneton	$\frac{e\hbar}{2m_0 c}$	$0.92731 \cdot 10^{-20}\ \frac{\text{erg}}{\text{gauss}}$
Polarizabil.		$\frac{e^2 a_0^2}{E_h}$	$1.648778 \cdot 10^{-41}\ \frac{\text{C}^2\ \text{m}^2}{\text{J}}$
Electric field		$\frac{E_h}{ea_0}$	$5.142208 \cdot 10^{11}\ \frac{\text{V}}{\text{m}}$
Boltzm.constant		k_B	$1.380658 \cdot 10^{-23}\ \frac{\text{J}}{\text{K}}$
Avogadro constant		N_A	$6.0221367 \cdot 10^{23}\ \text{mol}^{-1}$

Table 2. Conversion coefficients.

	a.u.	**erg**	**eV**	$\frac{\text{kcal}}{\text{mole}}$	**1 cm^{-1}**	**1 Hz**	**1 K**
1 a.u.	1	$4.35916 \cdot 10^{-11}$	27.2097	627.709	$2.194746 \cdot 10^{5}$	$6.579695 \cdot 10^{15}$	$3.15780 \cdot 10^{5}$
1 erg	$2.29402 \cdot 10^{10}$	1	$6.24197 \cdot 10^{11}$	$1.43998 \cdot 10^{13}$	$5.03480 \cdot 10^{15}$	$1.50940 \cdot 10^{26}$	$7.2441 \cdot 10^{15}$
1 eV	$3.67516 \cdot 10^{-2}$	$1.60206 \cdot 10^{-12}$	1	23.0693	$8.06604 \cdot 10^{3}$	$2.41814 \cdot 10^{14}$	$1.16054 \cdot 10^{4}$
1 $\frac{\text{kcal}}{\text{mol}}$	$1.59310 \cdot 10^{-3}$	$6.9446 \cdot 10^{-14}$	$4.33477 \cdot 10^{-2}$	1	$3.49644 \cdot 10^{2}$	$1.048209 \cdot 10^{13}$	$5.0307 \cdot 10^{2}$
1 cm^{-1}	$4.556336 \cdot 10^{-6}$	$1.98618 \cdot 10^{-16}$	$1.23977 \cdot 10^{-4}$	$2.86005 \cdot 10^{-3}$	1	$2.997930 \cdot 10^{10}$	1.43880
1 Hz	$1.519827 \cdot 10^{-16}$	$6.62517 \cdot 10^{-27}$	$4.13541 \cdot 10^{-15}$	$9.54009 \cdot 10^{-14}$	$3.335635 \cdot 10^{-11}$	1	$4.7993 \cdot 10^{-11}$
1 K	$3.16676 \cdot 10^{-6}$	$1.38044 \cdot 10^{-16}$	$8.6167 \cdot 10^{-5}$	$1.98780 \cdot 10^{-3}$	0.69502	$2.08363 \cdot 10^{10}$	1

Name Index

U

V

W

Y

Z

Subject Index

D

Printed and bound by CPI Group (UK) Ltd, Croydon, CR0 4YY

08/05/2025

01864919-0001